T0213215

CAMBRIDGE LIBRARY COLLECTION

Books of enduring scholarly value

Botany and Horticulture

Until the nineteenth century, the investigation of natural phenomena, plants and animals was considered either the preserve of elite scholars or a pastime for the leisured upper classes. As increasing academic rigour and systematisation was brought to the study of 'natural history', its subdisciplines were adopted into university curricula, and learned societies (such as the Royal Horticultural Society, founded in 1804) were established to support research in these areas. A related development was strong enthusiasm for exotic garden plants, which resulted in plant collecting expeditions to every corner of the globe, some-times with tragic consequences. This series includes accounts of some of those expeditions, detailed reference works on the flora of different regions, and practical advice for amateur and professional gardeners.

Flora Capensis

This seminal publication began life as a collaborative effort between the Irish botanist William Henry Harvey (1811–66) and his German counterpart Otto Wilhelm Sonder (1812–81). Relying on many contributors of specimens and descriptions from colonial South Africa – and building on the foundations laid by Carl Peter Thunberg, whose *Flora Capensis* (1823) is also reissued in this series – they published the first three volumes between 1860 and 1865. These were reprinted unchanged in 1894, and from 1896 the project was supervised by William Thiselton-Dyer (1843–1928), director of the Royal Botanic Gardens at Kew. A final supplement appeared in 1933. Reissued now in ten parts, this significant reference work catalogues more than 11,500 species of plant found in South Africa. Volume 5 appeared in three parts, the first comprising sections published between 1901 and 1912, covering Acanthaceae to Proteaceae.

CAMBRIDGE
UNIVERSITY PRESS

University Printing House, Cambridge, CB2 8BS, United Kingdom

Cambridge University Press is part of the University of Cambridge.

It furthers the University's mission by disseminating knowledge in the pursuit of education, learning and research at the highest international levels of excellence.

www.cambridge.org
Information on this title: www.cambridge.org/9781108068116

© in this compilation Cambridge University Press 2014

This edition first published 1912
This digitally printed version 2014

ISBN 978-1-108-06811-6 Paperback

This book reproduces the text of the original edition. The content and language reflect the beliefs, practices and terminology of their time, and have not been updated.

Cambridge University Press wishes to make clear that the book, unless originally published by Cambridge, is not being republished by, in association or collaboration with, or with the endorsement or approval of, the original publisher or its successors in title.

FLORA CAPENSIS.

VOL. V. SECT. 1.

DATES OF PUBLICATION OF THE SEVERAL PARTS
OF THIS VOLUME.

PART I., pp. 1–224, was published *June*, 1901.

PART II., pp. 225–448, was published *May*, 1910.

PART III., pp. 449–640, was published *January*, 1912.

PART IV., pp. 641–end, was published *June*, 1912.

FLORA CAPENSIS:

BEING A

Systematic Description of the Plants

OF THE

CAPE COLONY, CAFFRARIA, & PORT NATAL

(AND NEIGHBOURING TERRITORIES)

BY

VARIOUS BOTANISTS.

EDITED BY

SIR WILLIAM T. THISELTON-DYER, K.C.M.G.,
C.I.E., LL.D., D.Sc., F.R.S.

HONORARY STUDENT OF CHRIST CHURCH, OXFORD,
LATE DIRECTOR, ROYAL BOTANIC GARDENS, KEW.

*Published under the authority of the Governments of the
Cape of Good Hope, Natal and the Transvaal.*

VOLUME V. SECTION 1.
ACANTHACEÆ TO PROTEACEÆ.

LONDON:
LOVELL REEVE & CO., LTD.,
6, HENRIETTA STREET, COVENT GARDEN.
Publishers to the Home, Colonial and Indian Governments.
1912.

LONDON :
PRINTED BY WILLIAM CLOWES AND SONS, LIMITED,
DUKE STREET, STAMFORD STREET, S.E., AND GREAT WINDMILL STREET, W.

PREFACE.

ON the completion of Volume VII. the preparation of Volume IV. was taken up. Professor FRANCIS GUTHRIE and Dr. BOLUS undertook in South Africa the elaboration of *Ericaceæ*. Meanwhile, with the aid of contributors whom I was able to enlist at home, a commencement was made with the present volume, and the first part was published as long ago as 1901. Unhappily the death of Professor GUTHRIE in 1899 and the failing health of Dr. BOLUS left their contribution unfinished, and it had to be continued and completed at Kew. It became impracticable to make any further progress with Volume V. till the two sections of the preceding one had been disposed of.

In a vast undertaking like the present, the progress of which is necessarily protracted, it is the inevitable but melancholy task of the Editor to record the loss from time to time of those whose generous assistance and co-operation have made its ultimate accomplishment possible. That it should be so must be his justification for the somewhat erratic mode of publication which he has felt obliged to adopt. Particular orders have been as far as possible entrusted to those who had made them a special study. Had he waited to invoke their aid in following a continuous sequence, that aid would in at least two cases have been unavailable and with difficulty replaced.

CHARLES BARON CLARKE, M.A., F.R.S., died 25th August, 1906; he elaborated *Commelinaceæ* (1897), *Cyperaceæ* (1898) and *Acanthaceæ* (1901) in the present volume. (Obituary notice and bibliography, *Kew Bulletin*, 1906, pp. 271–281.) MAXWELL TYLDEN MASTERS, M.D., F.R.S., died 30th May, 1907; he worked out *Restiaceæ* for the *Flora* in 1897. (Obituary notice and bibliography, *Kew Bulletin*, 1907, pp. 325–334.) Dr. THEODORE

COOKE, C.I.E., F.L.S., on the completion in 1908 of his Bombay Flora on which he had been engaged for ten years, volunteered his aid for the *Flora Capensis*. He rendered great service by filling up many gaps in the present section undisposed of at the moment. I had hoped to receive from him more extended contributions, but while occupied on *Amarantaceæ* he was seized with illness which terminated fatally on 5th November, 1910. (Obituary notice, *Kew Bulletin*, 1910, pp. 350–352.) The death of HARRY BOLUS, D.Sc., F.L.S., in England· on 25th May, 1911 (obituary notice, *Kew Bulletin*, 1911, pp. 275–277) is something more than the loss of a contributor of specialized accomplishments. In his knowledge of the South African flora it may be said with confidence that Dr. BOLUS had no living rival. I cannot do better than quote a few words from Professor PEARSON'S penetrating appreciation of his work and character (*Kew Bulletin*, 1911, pp. 319–322):— "By common consent Dr. BOLUS occupied a unique and honoured place amongst botanical workers in South Africa. His death removes one of the most striking figures from the ranks of her scientific men, and leaves a vacancy which no man can fill. In the annals of South African Botany his name and his record will be written in large characters." Dr. BOLUS took more than a keen interest in the progress of this work. As has been acknowledged in previous prefaces, Kew has received from him a continuous stream of fresh and novel material. Nor can it be doubted that his position and reputation in South Africa weighed with the Legislature of Cape Colony in inducing it to make successive grants in aid of its preparation and publication. Having during his lifetime endowed the Chair of Botany in the South African College, he bequeathed to it his herbarium and library and a considerable portion of his fortune.

From Volume IV. onwards the area comprised in the *Flora* has been extended to the Tropic. In many of the regions so included material is scanty or wholly wanting. It is therefore with no small satisfaction that I am able to record that the PERCY SLADEN Memorial Expedition (assisted by a grant from the Royal Society of London) worked between Ceres Road in

Cape Colony and Lüderitzbucht in Great Namaqualand in the summer of 1908-9, and a second expedition under the same auspices, between Eendekuil in Cape Colony and Sendling's (or Bethany) Drift in the Orange River in the summer of 1910-11. The summer flora of the greater part of these regions was previously little known, and the material collected in Bushmanland and the eastern part of Great Namaqualand in the former journey and in the Richtersveld (between Ookiep and the Orange River) in 1910-11, furnishes many new records of distribution and contains a considerable number of new species.

The Natal Government has made no contribution to the work since 1907. On the other hand that of the Transvaal has given spontaneously a liberal grant, and this has been followed by a still more substantial one which it is hoped will provide sufficiently for the completion of the work.

In the present section I have been again fortunate in securing the aid of contributors who in many cases were able to bring to bear the advantage of previous study on the groups they undertook. Amongst these are Mr. C. B. CLARKE, F.R.S., who worked out *Acanthaceæ*, and Mr. R. A. ROLFE, A.L.S., *Selagineæ*. I have had the further advantage of the continued co-operation of South African botanists. Professor PEARSON has contributed *Verbenaceæ*, and I am indebted to the Trustees of the South African Museum for granting leave of absence to Mr. E. P. PHILLIPS, one of their staff, to come to Kew to work out *Proteaceæ*, an order with which he had obtained a first-hand acquaintance in the field. The expiration of Mr. PHILLIPS'S leave left his task in some respects incomplete, and I am indebted to Dr. STAPF, F.R.S., Keeper of the Herbarium, and to Mr. JOHN HUTCHINSON, Assistant for Tropical Africa, for supplementing what was needed to Mr. PHILLIPS'S work. The laborious task of elaborating *Labiatæ* was undertaken simultaneously by Messrs. BROWN and SKAN, and by Dr. COOKE.

I continue to be indebted for invaluable aid to Mr. C. H. WRIGHT, A.L.S., and to Mr. N. E. BROWN, A.L.S., Assistant Keepers of the Herbarium, the former in reading the proofs and in other ways, the latter for working out the localities and distribution.

For the limits of the regions under which the localities are cited in which the species have been found to occur, reference may be made to the Preface to Volume VI.

Besides the maps already cited in the Prefaces to Volumes VI. and VII., the following have also been used :—

Map of the Colony of the Cape of Good Hope and neighbouring territories. Compiled from the best available information. By JOHN TEMPLER HORNE, Surveyor-General, 1895.

Stanford's new Map of the Orange Free State and the southern part of the South African Republic, etc., 1899.

Carte du Théâtre de la Guerre Sud-Africaine. Par le Colonel CAMILLE FAVRE, 1902.

To many of the South African correspondents of Kew enumerated in previously published volumes I have again to tender my acknowledgments for the contribution of specimens in aid of the work to the Herbarium of the Royal Botanic Gardens.

I must further record my obligations to some new contributors, and to those whose kind assistance in various ways has been of the greatest value in the preparation of this section of Volume V.

Prof. G. BECK, Ritter von Mannagetta und Lerchenau, University of Prague. Loan of *Proteaceæ*.

HARRY BOLUS, Esq., D.Sc., F.L.S., contributed many specimens and lent portions of his Herbarium.

J. BURTT DAVY, Esq., F.L.S. Plants from the Transvaal.

Prof. H. H. DIXON, F.R.S., Trinity College, Dublin. Loan of the *Labiatæ* in Harvey's Herbarium.

R. A. DÜMMER, Esq. Plants from Namaqualand and Cape Division.

Prof. J. EICHLER, Curator of the Botanic Department, K Naturalienkabinet at Stuttgart. Loan of *Proteaceæ*.

Geheimrath Dr. A. ENGLER, Director of the Botanic Garden and Museum, Dahlem. Loan of *Selagineæ* and *Proteaceæ*.

Prof. C. FLAHAULT, Director of the Institute of Botany University of Montpellier. Collection of Basutoland Plants.

E. E. GALPIN, ESQ., F.L.S., Queenstown, Cape Colony. Large collections of South African plants and loan of portions of his private Herbarium.

Dr. H. O. JUEL, Director of the Botanic Garden, Upsala. Loan of portions of Thunberg's Herbarium.

Prof. C. A. M. LINDMAN, Curator of the Botanic Department of the Natural History Museum, Stockholm. Loan of *Proteaceæ*.

Dr. J. MUIR. Specimens from Riversdale District.

Dr. L. PÉRINGUEY, Director of the South African Museum, Cape Town. Various duplicates and loan of specimens from the South African Museum.

E. P. PHILLIPS, Esq., M.A. Collection of *Proteaceæ* and others.

Mrs. R. POTT (formerly Miss R. LEENDERTZ). Plants from Transvaal.

Prof. HANS SCHINZ, Director of the University Botanic Garden and Museum, Zürich. Large collections of South African plants and loan of specimens.

Dr. S. SCHONLAND, Curator of the Albany Museum, Grahamstown. Contribution and loan of specimens of *Proteaceæ* and others.

Prof. E. WARMING, late Director of the Botanic Garden, Copenhagen. Loan of *Selagineæ*.

J. MEDLEY WOOD, Esq., A.L.S., Director of the Botanic Garden, Durban. Collections of Natal plants and loan of specimens.

Dr. A. ZAHLBRUCKNER, Keeper of the Botanic Department of the Hofmuseum, Vienna. Loan of *Labiatæ* and *Proteaceæ*.

I must allow myself more personally to express my indebtedness to Lieut.-Colonel PRAIN, C.M.G., C.I.E., F.R.S. Director of the Royal Botanic Gardens, for kind and unfailing assistance in many ways, without which the task of editing a work of this kind at a distance from the resources of Kew could hardly be accomplished.

It has been the practice, at any rate in the more recent works that have emanated from Kew, to conform to the classification and sequence of orders adopted in BENTHAM and

HOOKER'S *Genera Plantarum*. This work was not available
to Professor HARVEY when he commenced the *Flora Capensis*,
and he appears to have based himself on the *Prodromus* of
AUGUSTIN PYRAMUS DE CANDOLLE, in the three volumes of
which he was the author with Dr. SONDER. From Volume IV.
onwards of the continuation the *Genera Plantarum* has been
followed. There is in consequence an inconsistency between
the earlier and later portions of the work in regard to the
delimitation of the sub-classes adopted, which, although of
little practical importance inasmuch as it scarcely affects the
sequence of the orders, it is desirable to clear up. How this
arises will be apparent from the statement with which Professor
HARVEY commences the preface to his third volume : "This
.... contains the orders of CALYCIFLORÆ with a monopetalous
corolla and an inferior ovary. The fourth volume .... will, it
is hoped, include the Heaths (*Ericeæ*) and all the *Monopetalæ*
with superior ovaries, *i.e.*, the COROLLIFLORÆ proper." BENTHAM
and HOOKER in the *Genera Plantarum* have adopted the sub-
class GAMOPETALÆ from Endlicher, who had established it in
1836 to include all orders with a monopetalous corolla. This
arrangement has been followed in the present work from
Volume IV. onwards. The removal of monopetalous orders
from *Calycifloræ* does not however affect the sequence in which
they are dealt with. But in order to make the classification
consistent throughout, it is necessary in the first place to
substitute the following new definitions of sub classes 1–3 for
those given by Professor HARVEY in Vol. I., p. xxxiii.

Sub-class I. THALAMIFLORÆ. Ord. I.–XLI. (Vol. I., pp. 1–
449). *Calyx and Corolla* (generally) present. *Petals*
separate, inserted, as are also the *stamens*, on the receptacle
(i.e., *hypogynous*). *Ovary* free.

Sub-class II. CALYCIFLORÆ. Ord. XLII.–LXX. (Vol. I.,
pp. 450–528; Vol. II., pp. 1–572). *Calyx and Corolla*
(generally) present. *Petals* separate. *Stamens* inserted
on the calyx (*perigynous*) or on the ovary (*epigynous*).
Ovary free or more or less adnate to the calyx-tube.
LXXI. *Balanophoreæ* (Vol. II., p. 572), and LXXII.

Loranthaceæ (Vol. II., p. 574) are transferred to MONO-CHLAMYDEÆ.

Sub-class III. GAMOPETALÆ. Ord. LXXIII.–CVI. (Vols. III., IV., and V., sect. 1, pp. 1–392). *Calyx and corolla* both present. *Petals* united in a gamopetalous corolla. *Stamens* inserted upon the corolla. *Ovary* free or more or less adnate to the calyx-tube.

In the second place the "Sequence of Orders" in Vol. III., p. ix., must be remodelled as below; the definitions of the orders themselves are unaffected.

Sub-class III. GAMOPETALÆ. Ord. LXXIII.–LXXVII.

Series I. INFERÆ. Ord. LXXIII.–LXXVII. *Ovary* inferior.

COHORT i. RUBIALES. *Stamens* epipetalous. *Ovary* 2–∞-celled, cells 1–∞-ovuled.

LXXIII. RUBIACEÆ (page 1).

COHORT ii. ASTERALES. *Stamen* epipetalous. *Ovary* 1-celled, 1-ovuled.

LXXIV. VALERIANEÆ (page 39).

LXXV. DIPSACEÆ (page 41).

LXXVI. COMPOSITÆ (page 44).

COHORT iii. CAMPANALES. *Stamens* usually epigynous and free from the corolla.

LXXVII. CAMPANULACEÆ (page 530).

W. T. T.-D.

WITCOMBE, 6th *May*, 1912.

SEQUENCE OF ORDERS CONTAINED IN VOL. V. SECT. 1, WITH BRIEF CHARACTERS.

Continuation of Series III. BICARPELLATÆ. Ord. CI.–CVI.

COHORT ix. PERSONALES (continued). *Corolla* usually irregular or oblique. *Stamens* 2 or 4. *Ovules* numerous or 2 superposed.

CI. ACANTHACEÆ (page 1). *Calyx* usually divided to the base. *Ovary* 2-celled; ovules 2 or few (rarely numerous) in each cell, superposed. *Capsule* loculicidally 2-valved; valves recurving elastically from the apex. *Seeds* exalbuminous, borne on processes of the placenta (*retinacula*). (*Herbs or shrubs. Leaves opposite, usually entire, exstipulate. Inflorescence various, sometimes of strobilate spikes.*)

COHORT x. LAMIALES. *Corolla* usually irregular or oblique. *Stamens* 4 or 2, a fifth sometimes represented by a staminode. *Ovary* 2- or 4-celled; ovules solitary or 2 collateral. *Fruit* usually included in the calyx, indehiscent, 1-seeded or dividing into 2 or 4 1-seeded nutlets.

CII. MYOPORINEÆ (page 92). *Leaves* alternate. *Flowers* axillary. *Anthers* 2-celled. *Radicle* superior. (*Erect or diffuse herbs, shrubs or rarely trees.*)

CIII. SELAGINEÆ (page 95). *Leaves* alternate, rarely the lower opposite. *Flowers* in terminal spikes or panicles or lateral towards the ends of the branches. *Anthers* 1-celled. *Radicle* superior. (*Small heath-like shrubs or undershrubs, tufted perennial herbs, or rarely small annuals.*)

CIV. VERBENACEÆ (page 180). *Leaves* usually opposite or whorled. *Ovary* entire, rarely shortly 4-lobed; style terminal. *Radicle* inferior. *Fruit* more or less drupaceous. (*Herbs, shrubs or trees.*)

CV. LABIATÆ (page 226). *Leaves* opposite or whorled. *Ovary* usually deeply 4-lobed; style gynobasic. *Radicle* inferior. *Fruit* of 4 nutlets. (*Herbs or shrubs, usually with square stems and branches.*)

ANOMALOUS ORDER.

CVI. PLANTAGINEÆ (page 387). *Corolla* regular, 4-lobed, scarious. *Stamens* 4, alternate with the corolla-lobes. *Ovary* entire, 2-celled. (*Perennial or annual herbs, either stemless with rosulate leaves or caulescent with alternate or opposite leaves. Inflorescence spicate.*)

Sub-class IV. MONOCHLAMYDEÆ. Ord. CVII.–CXVII.

SERIES i. CURVEMBRYEÆ. *Seeds* with farinaceous albumen; embryo curved, lateral or surrounding the albumen, rarely nearly straight, narrow and subcentral. *Ovule* solitary in each cell or carpel, or (in some *Amarantaceæ*) several. *Flowers* hermaphrodite, rarely 1-sexual or polygamous. *Stamens* as many as the perianth-segments or fewer, rarely more.

CVII. NYCTAGINEÆ (page 392). *Perianth* with a persistent base enclosing and sometimes adhering to the fruit. *Stamens* hypogynous, sometimes many. *Ovary* 1-celled; style simple. *Seeds* with inferior radicle. (*Herbs, shrubs or trees, with usually opposite entire exstipulate leaves. Flowers in cymes, panicles or corymbs, usually coloured.*)

CVIII. ILLECEBRACEÆ (page 398). *Perianth* herbaceous, or scarious at the margin, persistent. *Stamens* perigynous, as many as the perianth-segments and opposite to them. *Ovary* 1-celled; styles or style-arms 2–3. (*Annual or perennial herbs, rarely shrubs. Leaves usually opposite and stipulate. Flowers minute, usually green.*)

CIX. AMARANTACEÆ (page 402). *Perianth* dry, not herbaceous. *Stamens* hypogynous or perigynous, as many as the perianth-segments and opposite to them; filaments connate at the base. *Ovary* 1-celled; style simple or 2–3-fid; ovules 1 to many. *Utricle* indehiscent, or bursting irregularly or circumscissile. (*Herbs or undershrubs, rarely trees. Leaves opposite or alternate, exstipulate. Flowers small or minute, bracteate and bracteolate.*)

CX. CHENOPODIACEÆ (page 433). *Perianth* membranous or herbaceous. *Stamens* hypogynous or perigynous, as many as the perianth-segments and opposite to them or fewer; filaments free. *Ovary* 1-celled, 1-ovuled; style simple, 2–3-lobed, or styles 2–3 distinct. *Utricle* indehiscent. (*Annual or perennial herbs or shrubs, rarely small trees. Leaves alternate rarely opposite, exstipulate. Inflorescence various; flowers small.*)

CXI. PHYTOLACCACEÆ (page 454). *Perianth* herbaceous or coriaceous, rarely membranous, 5-lobed, usually persistent. *Stamens* hypogynous, 3–25; filaments sometimes connate at the base. *Carpels* 2 to many, free or united; styles as many as the carpels, free or united at the base. *Radicle* inferior or descending. (*Shrubs or herbs, rarely trees. Leaves alternate, exstipulate. Flowers usually racemose, small or medium-sized, green or whitish.*)

CXII. POLYGONACEÆ (page 459). *Perianth* herbaceous, membranous and often coloured, rarely adhering to the base of the ovary. *Stamens* perigynous, usually a few more than the perianth-segments; filaments free or connate at the base. *Ovary* 1-celled, 1-ovuled; styles or style-arms 2–3. *Fruit* a trigonous or lenticular nut. *Radicle* superior or ascending. (*Herbs or shrubs. Leaves alternate; petiole dilated into a membranous sheath below. Flowers small, racemose or axillary.*)

SERIES ii. MULTIOVULATÆ AQUATICÆ. Submerged herbs. *Ovary* syncarpous, 1–3-celled; ovules numerous.

CXIII. PODOSTEMACEÆ (page 482). *Perianth* small or absent. *Stamens* 1 to many; filaments free or united. *Ovary* superior, cells or placentas 2–3. *Seeds* exalbuminous. (*Submerged herbs of various habit, often resembling mosses, foliaceous or frondose hepaticæ or lichens.*)

SERIES iii. MULTIOVULATÆ TERRESTRES. Terrestrial parasitic herbs (in the South African genera). *Ovary* syncarpous; ovules numerous.

CXIV. CYTINACEÆ (page 485). *Ovary* inferior, 1-celled in the South African genera, with parietal or pendulous placentas, or ovuliferous all over. *Seeds* exalbuminous.

SERIES iv. MICREMBRYEÆ. *Ovary* syncarpous or apocarpous, or of a single carpel; ovules solitary in South African genera. *Seeds* with abundant fleshy or floury albumen; embryo very small.

CXV. PIPERACEÆ (page 487). *Ovary* superior, 1-celled, 1-ovuled. *Stamens* 2–4. *Flowers* hermaphrodite in the South African genera, very small. (*Herbs or shrubs, erect or climbing. Leaves alternate, opposite or whorled, stipulate or exstipulate. Inflorescence spicate or racemose.*)

CXVA. MONIMIACEÆ (page 492). *Ovary* superior, 1-celled, or carpels several, distinct. *Stamens* 10–15. *Flowers* diœcious in the only South African genus. (*Shrubs or small trees. Leaves subopposite, minutely pellucid-punctate. Inflorescence racemose.*)

SERIES V. DAPHNALES. *Ovary* of a single carpel, very rarely of several united; ovules solitary or 2 collateral very rarely few in superposed pairs. *Perianth* usually calycine; segments 1–2-seriate. *Stamens* perigynous, as many or twice as many as the perianth-lobes, occasionally fewer. *Flowers* usually hermaphrodite. (*Trees or shrubs, very rarely herbs.*)

CXVI. LAURINEÆ (page 493). *Perianth-segments* 6 or 4, 2-seriate, imbricate. *Stamens* typically in 4 whorls, some often reduced to staminodes or suppressed; anthers dehiscing by valves opening upwards. *Ovary* 1-celled; ovule 1, pendulous. *Radicle* superior.

CXVII. PROTEACEÆ (page 502). *Perianth-segments* 4, valvate. *Stamens* as many as the perianth-lobes and opposite to them; anthers dehiscing longitudinally. *Ovary* 1-celled; ovules solitary or 2 collateral, or rarely few in superposed pairs, pendulous or lateral. *Radicle* inferior.

FLORA CAPENSIS.

ORDER CI. **ACANTHACEÆ**.

(By C. B. CLARKE.)

Flowers hermaphrodite, irregular. *Calyx* inferior, free; segments 5 or 4, nearly separate or united. *Corolla* gamopetalous; tube campanulate or linear; limb 2-lipped or 5-lobed, more or less 1-sided. *Stamens* on the corolla, 4 didynamous, or 2 (with or without rudiments of others); anther-cells 2 or 1, rounded acute or tailed at the base, parallel at equal height, or one more or less below the other; pollen ellipsoid (then usually ribbed or banded longitudinally) or globose (then often honeycombed or echinulate); equatorial pores 2 or 3 (for the protusion of pollen-tubes) closed by stopples. *Ovary* superior, 2-celled; ovules 2 or several in each cell, superimposed, or sometimes the lower ovule in each cell rudimentary or wanting, or (in *Thunbergia*) 2 ovules, collateral in each cell; style long, simple, minutely 2-fid. *Capsule* loculicidal, often elastically dehiscent; in a few genera the placentæ, remaining attached to the top of the capsule, spring up elastically from the bottom, thus scattering the seeds. *Seeds* usually as many as the ovules (except in several genera where the lower ovule in each cell is imperfect), held up on the thickened upcurved outgrowth of the funicle (the *retinaculum*), except in *Thunbergia.*

Herbs or shrubs. Leaves opposite, nearly always simple, entire; stipules 0. Inflorescence very various (even in the same genus), in strobilate spikes, or in heads or clusters, or of remote solitary flowers. Bracts large, or small or 0; bracteoles 2 (*prophylla*) often present, large or small.

Species 2000, abundant in the Tropics, frequent in temperate climates, absent in cold regions, in Europe only represented by 3 or 4 species of the genus *Acanthus* which reach the Mediterranean.

The Order is marked (except *Thunbergia*) by the strong upcurved hooks which carry the seeds; cf. Dyer, Fl. Trop. Afr. v. 1, 2.

Tribe 1. *THUNBERGIEÆ.* *Corolla* not, or obscurely, 2-lipped; lobes contorted in the bud, i.e. no lobe wholly within or without the others. *Ovules* 2 in each cell, collateral. *Capsule* beaked. *Seeds* orbicular, without retinacula.

I. **Thunbergia.**—*Bracteoles* 2, large, enclosing the small calyx.

Tribe 2. *RUELLIEÆ.* *Corolla* lobes contorted in the bud. *Stamens* 4, or in *Chætacanthus* 2; anther-cells at nearly equal height. *Calyx* or *bracteoles* often conspicuous. *Seeds* discoid, covered, at least on the margins, with numerous fine white hairs which spring out on applying water.

Subtribe 1. HYGROPHILEÆ. *Ovules* 3 or more in each cell. *Capsule* (unless accidentally) with more than 4 seeds. *Corolla* distinctly 2-lipped.

II. **Hygrophila.**—*Stamens* 4; anther-cells muticous; pollen ellipsoid, ribbed longitudinally.

Subtribe 2. EU-RUELLIEÆ. *Ovules* 3 or more in each cell. *Capsule* usually with more than 4 seeds. *Corolla* not, or obscurely, 2-lipped.

III. **Ruellia.**—*Pollen* globose, honeycombed. *Capsule* with seeds in the upper part, cylindric and solid at the base.

IV. **Ruelliopsis.**—*Pollen* ellipsoid, longitudinally grooved. *Capsule* seed-bearing nearly from the base. *Anther-cells* tailed at the base.

Subtribe 3. STROBILANTHEÆ. *Ovules* 2 in each cell. *Capsule* with 4 (or fewer) seeds.

* *Placentæ not rising elastically from the base of the capsule.*

V. **Dyschoriste.**—*Stamens* 4, perfect.

VI. **Chætacanthus.**—*Stamens* 2, with or without minute rudiments of others.

** *Placentæ rising elastically from the base of the capsule with the ripe seeds.*

VII. **Phaylopsis.**—Floral leaf containing 3–1 ebracteolate flowers.

VIII. **Petalidium.**—Each flower with 2 large ovate or elliptic bracteoles.

Tribe 3. *ACANTHEÆ.* *Corolla* 1-lipped; tube short. *Stamens* 4; anthers 1-celled; pollen ellipsoid, longitudinally banded. *Ovules* 2 (or 1) in each cell.

* *Calyx 4-partite to the base (i.e. the 2 anticous segments connate to the tip or very nearly so). Corolla wanting the posticous lip.*

IX. **Blepharis.**—*Bract* large, acute, often ending in a spine. The two anticous *filaments* with a short process near the top. *Ovary* with 2 pits at the apex of the posticous face. *Seeds* with hygroscopic hairs.

X. **Acanthopsis.**—*Bract* large, obovate, ending in 5–3 compound spines. *Filaments* without processes. *Ovary* without pits at the apex of the posticous face. *Seeds* with hygroscopic hairs.

XI. **Acanthus.**—*Filaments* without processes. *Ovary* without pits at the apex of the posticous face. *Seeds* without hygroscopic hairs.

** *Calyx 5-partite to the base. Corolla-limb split down the posticous face, so that the lip has 5 lobes all on one side.*

XII. **Sclerochiton.**—*Sepals* all similar; posticous 1-nerved.

XIII. **Crossandra.**—Posticous *sepal* broader than the others, 2-nerved, often 2-toothed at the tip.

Tribe 4. *JUSTICIEÆ.* *Corolla-limb* subequally 5-lobed or 2-lipped, one lobe wholly within, one wholly without, in the bud. *Ovules* 2–1 in each cell (except in *Crabbea*).

Subtribe 1. TETRANDRÆ. *Stamens* 4, all fertile.

XIV. **Crabbea.**—*Flowers* in dense compound heads. *Ovules* 3 in each cell.

XV. **Glossochilus.**—*Flowers* solitary. *Leaves* narrowly cuneate-oblong, glabrate. *Ovules* 2 in each cell. *Capsule* hardly stalked.

XVI. **Asystasia.**—*Flowers* in racemes, spikes or heads. *Ovules* 2 in each cell. *Capsule* long-stalked.

Subtribe 2. ERANTHEMEÆ. *Stamens* 2 fertile. *Corolla* hardly 2-lipped.

XVII. **Mackaya.**—*Corolla* curved; tube much inflated in the upper half.

Subtribe 3. BARLERIEÆ. *Stamens* 2 fertile. *Calyx* large, 4-partite to the base, i.e. two anticous lobes connate nearly or quite to the tip.

XVIII. **Barleria.**—*Pollen* globose, reticulated. *Seeds* with hygroscopic hairs.

Subtribe 4. EU-JUSTICIEÆ. *Stamens* 2. *Corolla* distinctly 2-lipped. *Calyx* small, divided nearly to the base into 5 or 4 narrow segments. *Pollen* ellipsoid or globose, longitudinally banded, never honeycombed. *Seeds* without hygroscopic hairs.

* MONOTHECIEÆ. *Anthers* 1-celled. *Placentæ not rising elastically with the seeds from the base of the capsule.*

XIX. **Ruttya.**—*Corolla-tube* wide; segments 5, ovate, in two lips.

** Typicæ. *Anthers 2-celled. Placentæ not rising elastically from the base of the capsule. Flower not appearing as though enclosed between two opposite bracts.*

† One *anther-cell* below the other, distinctly tailed (but not tailed in two species here recorded under *Justicia*).

XX. **Justicia.**—*Corolla-tube* not much longer than the limb. *Seeds* usually 4 to the capsule, rough or tubercular.

XXI. **Monechma.**—*Corolla-tube* not much longer than the limb. *Seeds* 2 to the capsule, smooth, usually shining.

XXII. **Siphonoglossa.**—*Corolla-tube* slender, much longer than the limb.

†† One *anther-cell* slightly below the other, hardly tailed at the base (or very shortly tailed in *Adhatoda*). See also 20. *Justicia.*

XXIII. **Adhatoda.**—*Inflorescence* congested ; bracts conspicuous. *Corolla-tube* broad. *Capsule* 1-2-seeded.

XXIV. **Rhinacanthus.**—*Inflorescence* diffuse ; bracts inconspicuous. *Corolla-tube* long linear ; posticous lip small. *Capsule* 4-seeded.

XXV. **Ecbolium.**—*Bracts* conspicuous, strobilate. *Corolla-tube* cylindric ; lips not elongate. *Capsule* (where known) 2-seeded.

††† One *anther-cell* entirely below the other, not tailed.

XXVI. **Isoglossa.**—*Pollen* globose, flattened, with a stopple in the centre of each face.

*** Hypoesteæ. *Placentæ not rising elastically from the base of the capsule. Spikelet usually of 1 flower with an imperfect second flower, appearing included by the two bracts corresponding to the two flowers or overtopped by the 2 bracts when they are narrow.*

XXVII. **Peristrophe.**—*Anthers* 2-celled, one cell much above the other.

XXVIII. **Hypoestes.**—*Anthers* 1-celled.

**** Solutæ. *Placentæ rising elastically from the base of the capsule with the ripe seeds.*

XXIX. **Macrorungia.**—Shrubs. *Calyx* divided about half-way down. *Corolla* red ; lips 1 in. long.

XXX. **Dicliptera.**—Herbs. *Calyx* divided to the base. *Corolla* pink.

I. **THUNBERGIA,** Linn. f.

Bracteoles 2, large, elliptic or ovate. *Calyx* much shorter than the bracteoles, either subtruncate or of 10–14 small linear teeth. *Corolla :* tube oblique, more or less widened upwards ; lobes 5, rounded, spreading, contorted in the bud. *Stamens* 4, didynamous ; connective often produced at the top as a short horn ; anthers nearly similar, 2-celled, oblong, often spurred at the base and with beaded hairs. *Pollen* globose, obscurely banded, smooth or most minutely tubercled. *Ovary* ovoid, acute ; style funnel-shaped, or of two lobes one below the other ; ovules in each cell 2. *Capsule* globose, abruptly rostrate, loculicidally dehiscent. *Seeds* in each cell 2, near the base, subhemispheric or flattened ; hilum central on the ventral face, attached directly to the placenta without retinacula.

Rambling, twining or scandent, or small and suberect. Peduncles in the Cape species axillary, solitary, 1-flowered.

Species 96, viz. 15 in S.E. Asia, 60 in Tropical Africa, 10 in the Mascarene Islands, 17 in South Africa.

Section 1. THUNBERGIOPSIS, Lindau. Calyx-teeth about 5 (but irregular), short-ovate; stigma funnel-shaped ... ... (1) **natalensis.**

Section 2. EU-THUNBERGIA, Lindau. Calyx-teeth about 12, linear. One lobe of the stigma below the other.

     Anther-cells without linear spurs at the base:
        Leaves all sessile ... ... (2) **capensis.**
        Some leaves petioled, 1–2 in. long ... ... (3) **purpurata.**
     Some anther-cells (6 or 4 in each flower) with linear rigid spurs at the base:
        Leaves all sessile, i.e. petioles of the lower leaves 0–⅛ in. long:
           Glabrous ... ... (4) **Galpini.**
           Hairy:
           Leaves lanceolate to linear, 2 in. long:
              Bracteoles ¾ in. long ... ... (5) **venosa.**
              Bracteoles ⅓ in. long ... ... (6) **stenophylla.**
           Leaves elliptic or ovate:
              Bracteoles cordate at the base ... (7) **cordibracteata.**
              Bracteoles rounded or truncate at the base:
                 Style densely hispid in the upper part ... ... (8) **hirtistyla.**
                 Style glabrous:
                    Leaves ovate ... ... (9) **atriplicifolia.**
                    Leaves elliptic or oblong:
                      Leaves narrow-triangular at the top, pointed:
                         Some leaves toothed ... (10) **aspera.**
                         Leaves all entire ... (11) **xanthotricha.**
                      Leaves obtuse triangular at the tip (12) **Bachmanni.**
     Petioles of the middle and lower stem-leaves ¼–½ in. long:
        Hispid; bracteoles ⅓ in. long ... ... (13) **neglecta.**
        Softly villous; bracteoles ¾ in. long ... (14) **amœna.**
     Some petioles of the middle stem-leaves exceeding ¾ in. in length:
        Margin of leaves entire (base sometimes angular) ... ... (15) **pondoensis.**
        Leaves or many of them angular and toothed:
           Petioles not winged; corolla white ... (16) **dregeana.**
           Petioles winged; corolla (usually) yellow with dark eye ... ... (17) **alata.**

1. T. natalensis (Hook. Bot. Mag. t. 5082); thinly hairy or glabrate; stem shrubby, 2 ft. high, branched; leaves 2–4 in. long, oblong or elliptic, acute, cordate at the base, usually subentire but often sinuate-toothed and occasionally coarsely toothed, subhastate; petiole usually less than ⅛ in. long, sometimes ¼–½ in.; bracteoles ¾–1¼ in. long, lanceolate, acute, veined; calyx-tube ⅙ in. long; lobes 5, shallow, ovate, hardly 1/10 in. long; corolla-tube 1–1½ in. long, much inflated from $\frac{1}{4}$ in. above the base, curved, yellowish; lobes $\frac{1}{4}$–$\frac{1}{8}$ in. long, ovate, blue; anther-cells short, with few beaded hairs, one cell in each of the two larger spurred at the base; style funnel-shaped at the top, margin with short triangular lobes; capsule 1 in. long, densely and minutely hairy. *Harvey, Thes. Cap.* i. 25, *t.* 38; *T. Anders. in Journ. Linn. Soc.* vii. 18.

KALAHARI REGION: Transvaal; by the river at Lydenberg, *Wilms,* 1216! Crocodile River, 4800 ft., *Schlechter,* 3911!

EASTERN REGION: Pondoland; Port St. John, *Galpin,* 3399! Natal; Hills near Pinetown, 1800 ft., *Wood in MacOwan & Bolus Herb. Norm. Austr.-Afr.,* 1337! Inanda, *Wood,* 284! Umzimkulu River, *McNeil in Wood Herb.,* 1806! on the skirts of woody places at Attercliffe, Umhlali, and Maritzburg, 500–2500 ft., *Sanderson,* 169! Northdene, 400–500 ft., *Wood,* 4983! and without precise locality, *Sanderson,* 869! *Gerrard,* 75!

2. T. capensis (Retz. in Phys. Soellsk. Handl. i. [1776] 163);

more or less hispid; stems 4–20 in. long, decumbent from a woody root; leaves $\frac{3}{4}$–1 in. long, ovate, angular toothed or subentire, base truncate; petiole 0–$\frac{1}{2}$ in. long; bracteoles $\frac{2}{3}$ in. long, ovate or elliptic; calyx-teeth about 12, linear, $\frac{1}{8}$ in. long; corolla yellow; tube $\frac{2}{3}$ in. long; limb $\frac{3}{4}$ in. in diam.; anthers nearly or quite glabrous, cells not spurred (very rarely with a minute mucro) at the base; pollen globose, banded, with a few minute tubercles; style with one lobe far below the other; capsule $\frac{1}{2}$ in. long, glabrous; seeds $\frac{1}{8}$ in. long, not greatly flattened. *Linn. f. Suppl.* 292; *Gærtn. Fruct.* iii. 23, *t.* 183, *fig.* 4; *Thunb. Nov. Gen.* i. 21, 22. *Prod.* 106, *and Fl. Cap. ed. Schult.* 488; *Lam. Ill.* iii. 97, *t.* 549, *fig.* 1; *Drège, Zwei Pflanzengeogr. Documente,* 137, 226 (*capensis, letter c only*); *Krauss in Flora,* 1845, 72; *Lodd. Bot. Cab. t.* 1529; *Nees in Linnæa,* xv. 351, *and in DC. Prod.* xi. 55, *excl. var.* β; *T. Anders. in Journ. Linn. Soc.* vii. 20 *partly; Lindau in Engl. Jahrb.* xvii. *Beibl.* 41, 36, 39. *T. humilis, Eckl. et Zeyh. Cat. Sem. Pl. Cap. fide Nees in Linnæa,* xv. 351.

COAST REGION: Cape Div.; Cape Promontory, *Masson!* Mossel Bay, and between Gamtoos River and Swartbeck River, *Thunberg;* Knysna Div.; between Plettenberg Bay and Melville, *Burchell,* 5375! Humansdorp Div.; Kromme River, *Krauss,* 1651! Uitenhage Div.; between Vanstadens River and Galgebosch, *Burchell,* 4678! between Uitenhage and Drosdy Farm, *Burchell,* 4464! and without precise locality, *Ecklon & Zeyher,* 772! *Zeyher!* Port Elizabeth Div.; Algoa Bay, *Cooper,* 3025! between Krakakamma and the upper part of Leadmine River, *Burchell,* 4604! Alexandria Div.; on the Zuur Berg Range, 2500–3500 ft., *Drège!* Albany Div.; mountains near Grahamstown, 2000 ft., *MacOwan!* 800 ft., *Bolus,* 1676! between Assegai Bosch and Rautenbachs Drift, *Burchell,* 4199! and without precise locality, *Bowie!* Fort Beaufort Div.; without precise locality, *Cooper,* 452! 560 *partly!* British Kaffraria, *Gill!*

CENTRAL REGION: Albany Div.; on a rocky mountain east side of Zwartwater Poort, *Burchell,* 3431!

The type of var. *grandiflora,* Nees (in DC. Prod. xi. 55), has not been found; but from a scrap which Nees has marked " *T. capensis?* var. *grandiflora nana,*" it was probably some form of *T. atriplicifolia* or *T. aspera.*

3. T. purpurata (Harvey ms. in Herb. Hook.); sparingly hairy,

often glabrate except the innovations; stems twining, several feet long; leaves 2–3½ in. long, ovate or triangular, acuminate, purple beneath, base truncate or hastate often with acute angles, margin sinuate, sparingly toothed, often entire; petioles attaining 1–2 in., not winged; bracteoles ½–⅔ in. long, elliptic or oblong; calyx-teeth about 12, linear, ¼ in. long; corolla-tube ¾–1 in. long, pale yellow; limb 1¼ in. in diam., white; anthers glabrous, except for a dense tuft of long beaded hairs on the triangular acute base of each cell, without any linear curved spur; one lobe of the stigma much below the other; capsule ¾ in. long, glabrous; seeds ⅕ in. in diam., flattened. *T. angulata, T. Anders. in Journ. Linn. Soc.* vii. 19 *partly; Lindau in Engl. Jahrb.* xvii. *Beibl.* 41, 40 *partly; not of Hook. T. dregeana, Lindau in Engl. Jahrb.* xvii. *Beibl.* 41, 36, 38, *ex descript., not of Nees.*

EASTERN REGION: Natal; Inanda, *Wood*, 1218! and without precise locality, *Sanderson*, 442! *Gerrard*, 1965!

Lindau, in Engl. Jahrb. xvii. Beibl. 41, 37, places correctly *T. angulata*, Hook. (a Madagascar plant), in the section which has the anther-cells spurred at the base. The original picture of Hooker (Exot. Fl. t. 166) shows the Madagascar plant correctly, the separate enlarged anther incorrectly, but Hooker says his picture was taken partly from Cape material. T. Anderson, who united the two species, does not appear to have looked at the anthers.

4. **T. Galpini** (Lindau in Engl. Jahrb. xxiv. 310); glabrous, except for the beaded hairs on the anthers; stem repeatedly branched upwards; branchlets quadrangular; leaves 2 by ½ in., elongate-triangular, acute, from a subcordate base; petiole 0–$\frac{1}{16}$ in. long; bracteoles ¾ in. long, broadly oblong; corolla-tube shortly exceeding the bracteoles; limb 1¼ in. in diam.; anther-cells (some of them) spurred at the base; style glabrous; one lobe of the stigma below the other.

KALAHARI REGION: Transvaal; Saddleback Mountain, near Barberton, 3200–3500 ft., *Galpin*, 1277!

5. **T. venosa** (C. B. Clarke); fulvous hispid, becoming glabrate; stems 2 ft. long, straggling, little divided; leaves up to 3 by ½ in., lanceolate, obscurely-toothed or nearly entire, nerves primary and secondary much raised on the lower surface; petioles 0–⅙ in. long; bracteoles ¾ in. long, elliptic-lanceolate; calyx-teeth 12, linear, more than ¼ in. long; corolla yellow; tube 1 in. long; limb 1 in. in diam.; anther-cells with beaded hairs their whole length, basal spurs linear-conic, rather short, scarcely curved; style papillose and white hairy near the top; stigmas very large, one lobe below the other; capsule ¾ in. long, glabrous.

EASTERN REGION: Natal; Itafamasi, *Wood*, 643! Inanda, *Wood*, 696!

6. **T. stenophylla** (C. B. Clarke); viscous hairy; leaves 2½ by ¼ in., nearly entire, subsessile; bracteoles ⅓ in. long; calyx-teeth 12, scarcely ⅕ in. long, linear, viscous; corolla-tube nearly 1 in. long; anther-cells (some) spurred at the base.

KALAHARI REGION: Transvaal; Pilgrims Rest, *Greenstock* (in the British Museum).

7. T. cordibracteolata (C. B. Clarke); rather sparsely hispid; leaves 2 by $\frac{3}{4}$ in., elliptic, entire, base cordate or truncate; petioles 0–$\frac{1}{6}$ in. long; bracteoles $\frac{3}{4}$–1 by $\frac{1}{2}$ in., broad and cordate at the base; calyx-teeth 12, linear, $\frac{1}{5}$ in. long; anther-cells (some of them) spurred at the base, glabrous except for a basal tuft of hairs; style glabrous; two stigma-lobes but little separated in the young flower (alone seen); capsule $\frac{3}{4}$ in., glabrous.

KALAHARI REGION: Orange River Colony; without precise locality, *Cooper*, 895!

8. T. hirtistyla (C. B. Clarke); hairy; leaves 1$\frac{1}{4}$ by $\frac{1}{2}$ in., elliptic-oblong, entire, softly hairy; petioles 0–$\frac{1}{16}$ in. long; bracteoles $\frac{3}{4}$ in. long, elliptic; corolla-tube slightly exceeding the bracteoles; limb 1$\frac{1}{4}$ in. in diam.; anther-cells (some of them) spurred at the base; style densely brown-hispid towards the top; one stigma-lobe much below the other.

EASTERN REGION: Natal; without precise locality, *Gerrard*, 1274!

9. T. atriplicifolia (E. Mey. in Drège, Zwei Pflanzengeogr. Documente, 144, 226); patently hispid; stems decumbent from a woody root; ascending branches 6–12 in. long; leaves ovate, 1$\frac{1}{4}$ by 1 in. (in Drège's type), up to 2$\frac{3}{4}$ by 1$\frac{1}{2}$ in. (in form *megalantha*, C. B. Clarke), sparingly toothed and obscurely angular, or frequently subentire (occasionally much toothed, as the leaf of *Atriplex*), base rounded, 3–5 digitate basal nerves usually distinct beneath; petiole (even of the lower stem leaves) 0–$\frac{1}{6}$ in.; bracteoles $\frac{1}{2}$–$\frac{2}{3}$ in. long (in form *megalantha*, $\frac{3}{4}$ in. long and upwards), elliptic, not cordate at the base; calyx-teeth about 12, linear, $\frac{1}{5}$ in. long; corolla cream-coloured; tube $\frac{3}{4}$ in. long; limb 1 in. in diam.; or (in form *megalantha*) tube 1$\frac{1}{4}$ in. long; limb 2 in. (occasionally more) in diam.; anther-cells (usually 6 of the 8) with a strong curved spur at the base, also a tuft of beaded hairs at the base and usually with beaded hairs to the summit; pollen most minutely and sparsely tubercled; style with one lobe far below the other; capsule $\frac{1}{2}$–$\frac{2}{3}$ in. long, glabrous. *Nees in DC. Prod.* xi. 56 *partly*; *T. Anders. in Journ. Linn. Soc.* vii. 20 *partly*. *T. flavohirta, Lindau in Engl. Jahrb.* xxiv. 311.

VAR. β, **Kraussii** (C. B. Clarke), smaller, more densely hairy; leaves (largest) $\frac{3}{4}$ by $\frac{1}{4}$ in.; anthers glabrous except for a tuft of hairs at the base of each cell where is also a strong curved spur. *T. atriplicifolia, Krauss in Flora*, 1845, 72; *Nees in DC. Prod.* xi. 56.

COAST REGION: Komgha Div.; Kei River, 1000 ft., *Drège!*
KALAHARI REGION: Transvaal; Sabia River, *Mudd!* Apies River, near Pretoria, *Nelson*, 277! Houtbosch, *Rehmann*, 6183! Lydenberg, *Wilms*, 1209!
EASTERN REGION: Tembuland; Bazeia, 2000 ft., *Baur*, 58! Griqualand East; near Clydesdale, 2500 ft., *Tyson*, 1206! Mount Currie, 5000 ft., *Schlechter*, 6561! Natal; Inanda, 1800 ft., *Wood*, 42! 90! near Durban, *Wood*, 1! and without precise locality, *Grant! Gerrard*, 220! *Gerrard & McKen!* Zululand, Mrs. *McKenzie!* VAR. β: Natal; in grassy places throughout Natal, *Krauss*, 405!

Between the original type of this species collected by Drège and the extreme form *megalantha*, C. B. Clarke, there is a series, among which *T. flavohirta*, Lindau, is near Drège's type. *T. atriplicifolia*, Lindau (in Engl. Jahrb. xvii. Beibl. 41, 36, 39), with the anther-cells not spurred at the base is either a new species or a form of *T. capensis ;* in the type example of Drège's *atriplicifolia* (and in all the material above cited), several anther-cells (usually 6 or 4 in each flower) have at the base a linear curved strong smooth spur with a white hard point. This species is diagnosed here from *T. aspera* by the ovate (not elliptic-oblong) leaves ; the var. β differs from all the other material quite as much as *T. aspera* does, but having ovate leaves is arranged here.

10. T. aspera (Nees in DC. Prod. xi. 56) ; leaves $1\frac{1}{4}$ by $\frac{1}{2}$ in., elliptic-oblong, green on both faces, often with a tooth at the lower angle, otherwise subentire, glabrate when mature ; anther-cells with few hairs except the basal tuft ; otherwise as typical *T. atriplicifolia*, E. Mey. *T. atriplicifolia*, *T. Anders. in Journ. Linn. Soc.* vii. 20 *partly*. *T. Bachmanni, Lindau in Engl. Jahrb.* xvii. 94 *partly*.

VAR. β, **parvifolia** (Sonder in Linnæa, xxiii. 90); more hispid ; leaves very erect in all the dried material, subentire, upper surface brown, lower pale. *T. capensis*, near var. *grandiflora*, Nees ms.

COAST REGION : British Kaffraria (Caffer Land), *Gill !*
KALAHARI REGION : Var. β : Transvaal ; Magaliesberg, *Burke ! Zeyher*, 1418 ! Olifants Nek, *Burke !*
EASTERN REGION : Pondoland, *Bachmann*, 1266 ! Natal ; Coastland to 1000 ft., *Sutherland !*

The "type" is Gill's example, above described. In this (and in all the plants here placed) the corolla has the tube $\frac{3}{4}$ in. long, the limb 1 in. in diam., and is yellow (*Bolus*) ; the spur at the base of the anther-cell is strong ; so that *T. aspera*, Lindau (in Engl. Jahrb. xvii. Beibl. 41, 36, 39), is some remote species. Bachmann, 1266, matches exactly so far as it goes ; the example at Kew, however, has no flowers. Sonder says that Zeyher, 1418, was typical *T. aspera ;* and that the Olifants Nek plant was his var. *parvifolia ;* but Olifants Nek is in the Magaliesberg, and the two appear identical. The plants, arranged below as *T. Bachmanni*, var. *minor*, differ from *T. aspera*, var. *parvifolia* by their large flowers, hardly otherwise.

11. T. xanthotricha (Lindau in Engl. Jahrb. xxiv. 311) ; leaves up to 1 by $\frac{1}{4}-\frac{1}{3}$ in., elliptic-lanceolate, entire, hispid, base rounded ; petiole $0-\frac{1}{8}$ in. long ; corolla-tube nearly 1 in. long ; limb $1\frac{1}{3}$ in. in diam. ; otherwise nearly as *T. aspera*.

KALAHARI REGION : Transvaal : in grassy fields around Barberton, 2000–3000 ft., *Galpin*, 496 !

Flowers cream-coloured (*Galpin*). Leaves smaller, widest very near the base ; otherwise not separable from *T. Bachmanni*, var. *minor*. The hairs are tawny-yellow in nearly all the present group.

12. T. Bachmanni (Lindau in Engl. Jahrb. xvii. 94, and Beibl. 41, 38, 41 partly) ; robust, leaves $2\frac{3}{4}$ by $1-1\frac{1}{4}$ in., elliptic-oblong (scarcely ovate), entire ; otherwise as the large-flowered examples above called *T. atriplicifolia*.

VAR. β, **minor** (C. B. Clarke) ; leaves $1\frac{1}{2}$ by $\frac{2}{3}$ in., elliptic, widest near the middle, markedly 3–5-nerved on the under surface at the base, frequently with a tooth near the base.

KALAHARI REGION : Transvaal ; hill sides near Barberton, 3000 ft., *Galpin*,

933! Var. β: Orange River Colony; without precise locality, *Cooper*, 893!
3024! Transvaal; Pilgrims Rest, *Greenstock!* and without precise locality,
Sanderson!

EASTERN REGION: Pondoland; near Saugmeisters, *Bachmann*, 1267!

Leaves widest near the base in Bachmann, 1267. In Galpin, 933, the lower
leaves are ovate, and it would so far go better with *T. atriplicifolia*, form
megalantha, but the leaves are very entire. The var. *minor*, on the contrary,
differs from *T. xanthotricha*, Lindau, by having the leaves not rarely toothed. It
is quite an open question whether the series, from *T. atriplicifolia* to *T.
Bachmanni* with all their varieties, is not better united, as has been done by
T. Anderson.

13. T. neglecta (Sonder in Linnæa, xxiii. 89); scabrous hairy;
stems about a foot long, procumbent rambling, much branched;
leaves 1 in. long and broad, ovate, sinuate-toothed, triangular or
subcordate at the base; petioles of the middle and lower stem-leaves
$\frac{1}{4}$-$\frac{1}{2}$ in. long; bracteoles $\frac{1}{4}$ in. long, elliptic; calyx-teeth about 12,
linear, $\frac{1}{6}$ in. long; corolla-tube $\frac{1}{2}$-$\frac{3}{4}$ in. long; limb 1 in. in diam.;
anther-cells (6-4 of those in one flower) spurred and with beaded
hairs at the base; style glabrous; stigma with one lobe below the
other; capsule exceeding $\frac{1}{2}$ in. long, glabrous; seeds not much
compressed, surface wrinkled. *T. Anders. in Journ. Linn. Soc.* vii.
20. *T. hirta, Sonder in Linnæa*, xxiii. 88, *not of Lindau. T.
Bauri, Lindau in Engl. Jahrb.* xxiv. 312. *T. dregeana, partly
Nees ms.*

KALAHARI REGION: Transvaal; Magaliesberg, *Zeyher*, 1420! *Burke*, 252!
Waterfall River, near Lydenberg, *Wilms*, 1231! Orange River Colony; Great
Vet River, *Zeyher*, 1419! *Burke!*
EASTERN REGION: Tembuland; Bazeia, 2000 ft., *Baur*, 169!

Sonder says that *T. neglecta* differs from his *T. hirta* by having the corolla
only half the length. His example of *T. hirta* (Zeyher, 1419), is larger than his
type (Zeyher, 1420) *neglecta* in all its parts; the corolla-tube may be half as long
again in the dried example.

14. T. amœna (C. B. Clarke); a densely villous twiner; leaves up
to 1$\frac{1}{4}$ by 1 in., cordate-ovate, toothed; petioles up to $\frac{1}{3}$ in. long;
bracteoles $\frac{3}{4}$ in. long, cordate-triangular; calyx-teeth about 10, linear,
$\frac{1}{6}$ in. long; corolla-tube 1 in. long; limb 1$\frac{1}{4}$ in. in diam., yellow;
anther-cells (6-4 of those in each flower) spurred at the base; style
glabrous; stigma with 1 lobe much below the other.

KALAHARI REGION: Transvaal; Houtbosch Berg, *Nelson*, 498! Pilgrims Rest,
Greenstock!

15. T. pondoensis (Lindau in Engl. Jahrb. xvii. 93, and Beibl.
41, 37, 41); a hairy twiner; leaves up to 1$\frac{1}{2}$ by $\frac{1}{4}$ in., entire, acute
not acuminate, base deeply hastate or cordate; petioles $\frac{1}{4}$-$\frac{3}{4}$ in. long;
bracteoles $\frac{2}{5}$ in. long, elliptic; calyx-teeth 12, linear, $\frac{1}{6}$ in. long;
corolla-tube $\frac{3}{4}$ in. long; limb 1 in. in diam.; anther-cells short, with
a tuft of hairs at the base, several in each flower spurred at the base;
style glabrous; stigma with 1 lobe much below the other. *T.
angulata, Hook.*, var. *in Herb. Kew. T. angulata, Lindau, partly,
i.e. the Cape plant in Engl. Jahrb.* xvii. *Beibl.* 41, 37, 40.

EASTERN REGION: Pondoland; Backbeach, Durba, *Bachmann*, 1265; Zululand; Entumeni, *Wood*, 4015!

The differences of this from the Madagascar *T. angulata* are very small; in the latter the leaves are more acuminate, the basal lobes rather different. Lindau distinguishes *T. angulata* by the leaves being glabrous between the nerves.

16. T. dregeana (Nees in Linnæa, xv. 352); a hairy twiner; leaves 2 by 1¼ (or sometimes up to 4 by 2) in., cordate-ovate or hastate, toothed or angular; petioles ½–1½ in. long, not winged; bracteoles ⅔–¾ in. long, elliptic; calyx-teeth 12, linear, ⅕ in. long; corolla white; tube ¾ in. long; limb 1¼ in. in diam.; anther-cells (several of them) spurred at the base, with beaded hairs especially at the base; style glabrous; stigma with 1 lobe much below the other; capsule ¾ in. long, glabrous. *Presl, Bot. Bemerk.* 94; *Nees in DC. Prod.* xi. 58; *Drège in Linnæa*, xx. 200; *T. Anders. in Journ. Linn. Soc.* vii. 20, *not of Lindau*. *T. fragrans b.*, *E. Mey. in Drège, Zwei Pflanzengeogr. Documente*, 141, 226, *not of Roxburgh*.

COAST REGION: Uitenhage Div.; Zwartkops River, *Zeyher*, 3601! Sunday River, ex *Nees*. Alexandria Div.; Enon, 1000 ft., *Drège*. Albany Div.; Glenfilling, below 1000 ft., *Drège!* Fort Beaufort Div.; without precise locality, *Cooper*, 454! 560! Komgha Div.; near Komgha, 2000 ft., *Flanagan*, 1747!

CENTRAL REGION: Somerset Div.; on the Bosch Berg, *Burchell*, 3215! Philipstown and Kat River, ex *Nees*.

EASTERN REGION: Griqualand East; among shrubs by the sides of streams near Clydesdale, 2500 ft., *Tyson*, 2559! Natal; Coastland to 1000 ft., *Sutherland!* Inanda, *Wood*, 543! and without precise locality, *Sanderson*, 442!

17. T. alata (Boj. ex Sims in Bot. Mag. t. 2591); a hairy twiner; leaves 2¼ by 1½ in., broadly hastate, toothed or angular; petioles 1–2 in. long, winged; bracteoles ½–⅔ in. long, ovate-elliptic; calyx-teeth 12, linear, ⅕ in. long; corolla yellow with a purple eye (but paler or sometimes nearly white varieties occur); tube ¾ in. long; limb 1½ in. in diam.; anther-cells with beaded hairs, and (several in each flower) spurred at the base; style glabrous; stigma with 1 lobe much below the other; capsule ⅔ in. long, pubescent; seeds sub-hemispheric, reticulate. *Hook. Exot. Fl.* iii. *t.* 177; *Nees in DC. Prod.* xi. 58; *T. Anders. in Journ. Linn. Soc.* vii. 19; *C. B. Clarke in Hook. f. Fl. Brit. Ind.* iv. 391; *and Burkill in Dyer, Fl. Trop. Afr.* v. 16; *Lindau in Engl. Pfl. Ost-Afr. C.* 366, *and in Engl. Jahrb.* xvii. *Beibl.* 41, 37, 40.

EASTERN REGION: Natal; near Durban, *Wood*, 495! 3092! and without precise locality, *Cooper*, 2771!

In Tropical Africa and Natal indigenous; introduced into many warm parts of the World. Lindau records it (as though wild) from Namaqualand.

Lindau, in Engl. Jahrb. xvii. Beibl. 41, 37, diagnoses this plant from closely-allied species by its bracteoles being "keeled." The midrib of the bracteoles beneath is "slender"; at any rate, I cannot distinguish the present species by that character.

Imperfectly known Species.

18. T. hirta (Lindau in Engl. Jahrb. xvii. Beibl. 41, 36, 38, not of Sonder); hairy; leaves about ½ in. long, oval, pointed, cordate or somewhat truncate, toothed, petioled; flowers solitary, axillary; anther-cells none spurred at the base ; style 2-lobed.

KALAHARI REGION: Transvaal; Magaliesberg, ex *Lindau.*

19. **T. atriplicifolia** (Lindau in Engl. Jahrb. xvii. Beibl. 41, 36, 39, not of E. Mey) ; leaves sessile, toothed, rounded at the base, thickly and softly hairy; flowers solitary, axillary ; anther-cells none spurred at the base ; style 2-lobed.

From the Cape to Port Natal, ex *Lindau.*

20. **T. aspera** (Lindau in Engl. Jahrb. xvii. Beibl. 41, 36, 39, not of Nees) ; leaves sessile, toothed, truncate at the base, scabrous-hairy at least on the margins; flowers solitary, axillary; anther-cells none spurred at the base ; style 2-lobed.

EASTERN REGION: Kaffraria, ex *Lindau.*

II. HYGROPHILA, R. Br.

Bracts large, oblong ; bracteoles shorter than the calyx. *Calyx* deeply 5-fid, in the Cape species 4-fid. *Corolla* 2-lipped ; lobes 5, contorted in the bud. *Stamens* 4 ; anther-cells muticous; pollen ellipsoid, with longitudinal grooves. *Ovules* 3 or more in each cell. *Capsule* narrow-oblong, seed-bearing from the base; seeds usually very numerous, 2–8 only in the Cape species, with hygroscopic hairs.

Leaves entire.

Species 20, widely spread in the Tropics.

The single South African species belongs to the section *Asteracantha* (often considered generically distinct), which has the flowers packed in large dense axillary clusters surrounded by strong spines.

1. **H. spinosa** (T. Anders. in Thwaites, Enum. Pl. Zeyl. 225) ; stems 3–4 ft. long, stout, suberect, hispid; leaves 4–8 by ½–1 in., oblong, hispid, subsessile ; axillary inflorescences 1–2 in. in diam. ; spines ½–1½ in. long, stout ; bracts ½–1 in. long, lanceolate ; bracteoles ¼–⅓ in. long, narrow ; calyx ⅓–½ in. long; segments lanceolate ; corolla 1 in. long, pale blue-purple; capsule ⅓–½ in. long. *T. Anders. in Journ. Linn. Soc.* vii. 22 ; *C. B. Clarke in Hook. f. Fl. Brit. Ind.* iv. 408; *Burkill in Dyer, Fl. Trop. Afr.* v. 31. *Barleria longifolia, Linn. Amoen. Acad.* iv. 320. *Asteracantha longifolia, Nees in Wall. Pl. As. Rar.* iii. 90, *and in DC. Prod.* xi. 247 ; *Wight, Ic. t.* 449 ; *Lindau in Engl. Pfl. Ost-Afr. C.* 367.

KALAHARI REGION : Transvaal; between Spitz Kop and Komati River, *Wilms*, 1203 !

EASTERN REGION : Delagoa Bay, on the Umkomaas River, 50 ft., *Sanderson*, 575 !

Abundant in Tropical Africa and India.

III. RUELLIA, Linn.

Bracteoles 2, oblong or spathulate-elliptic. *Calyx* shorter than the bracteoles (or in *R. Zeyheri* scarcely longer), regular or 2-lipped in

Section *Fabria*, divided nearly to the base or only half-way. *Corolla* ¾–2 in. long, purplish or white, not 2-lipped; tube linear-oblong or dilated nearly from the base; lobes contorted in bud. *Stamens* 4, subsimilar; anther-cells 2, parallel, nearly at equal height, not spurred at the base; pollen globose, surface reticulate or honey-combed. *Style* linear, with one linear-oblong branch, the other suppressed; ovary with 6–16 ovules, glabrous or hairy. *Capsule* cylindric, narrowed and solid at the base, usually perfecting more than 4 seeds in the clavate upper part. *Seeds* on prominent retinacula, with many hygroscopic hairs on their margin.

Undershrubs; leaves nearly or quite entire, full of cystoliths, which are also conspicuous in the calyx; flowers axillary, not running into strobilate spikes.

Species 150 in the warm and temperate parts of both hemispheres, especially numerous in America.

R. spinescens and *R. depressa*, Thunb. Prod. 104, are referred to *Aptosimum*, among *Scrophulariaceæ*.

Section 1. DIPTERACANTHUS. Calyx equally 5-fid.

Calyx divided nearly to the base:
  Bracteoles spathulate-elliptic:
    Leaves ½–1½ in. long, tip triangular ... ... (1) **patula**.
    Leaves ¼–¾ in. long, tip rounded ... ... (2) **Zeyheri**.
  Bracteoles narrow-oblong; leaves 2 in. long ... (3) **Baurii**.
Calyx-tube ½ as long as the teeth... ... ... (4) **Woodii**.

Section 2. FABRIA. Calyx obscurely 2-lipped, viz. 2 teeth free nearly to the base, 3 teeth connate (sometimes nearly to the middle).

Thinly hairy; leaves ovate or cordate at the base ... (5) **ovata**.
Densely and softly hairy; leaves elliptic, narrowed into
  the petiole ... ... (6) **malacophylla**.
Softly hairy; leaves linear-oblong ... ... (7) **stenophylla**.

1. **R. patula** (Jacq. Misc. Bot. ii. 358); a small shrub, pubescent or nearly glabrous; branches 6–18 in. long; leaves ½–1½ in. long, ovate or elliptic, tip obtusely triangular, base suddenly narrowed; petiole ⅕–1 in. long; flowers axillary, solitary or a few clustered, grey-purple or more often white; bracteoles ⅓–⅔ in. long, spathulate-elliptic; calyx ⅕ in. long, divided to the base; teeth equal, linear; corolla ¾–1⅓ in. long; tube much inflated; stamens and pollen of the genus; ovary glabrous; style thinly hairy; capsule ½–⅔ in. long. *Jacq. Ic. Pl. Rar.* i. 12, *t.* 119; *T. Anders. in Journ. Linn. Soc.* vii. 24; *C. B. Clarke in Hook. f. Fl. Brit. Ind.* iv. 412, *and in Dyer, Fl. Trop. Afr.* v. 45; *Lindau in Engl. & Prantl, Pflanzenfam.* iv. 3B, 310, *fig.* 124 E, F, *in Engl. Jahrb.* xviii. 63, *t.* 1, *fig.* 24, *and in Engl. Pfl. Ost-Afr. C.* 368. *R. pilosa, T. Anders. in Journ. Linn. Soc.* vii. 25 *partly. R. Huttonii, T. Anders. in Journ. Linn. Soc.* vii. 25 *from descr. Dipteracanthus patulus, Nees in Wall. Pl. Asiat. Rar.* iii. 82, *and in DC. Prod.* xi. 126; *Wight, Ic. Pl. Ind. Or. t.* 1505; *Oersted in Vidensk. Meddel. Kjob.* 1854, 180, *t.* 4, *figs.* 19-21.

COAST REGION : Albany Div.; Grahamstown, *MacOwan!*
WESTERN REGION : Namaqualand, ex *Lindau.*
KALAHARI REGION : Transvaal; in natives gardens, north of Pretoria, *Nelson,*

279! Boschveld, at Klippan, *Rehmann*, 5248! Vaalbosch Fontein, 4300 ft., *Schlechter*, 4232! Bechuanaland; Bakwena Territory, 3500 ft., *Holub!*

EASTERN REGION : Natal; Inanda, *Wood*, 363! and without precise locality, *Sutherland! Gerrard*, 1682!

Also in Tropical Africa, Mascarene Islands, and India; a common weed among grass.

Ruellia pilosa of Linn. f., and of Thunb. Prod. p. 104, is an *Antirrhinum ;* the name is taken up by T. Anders. (in Journ. Linn. Soc. vii. 25) for Burke's plant of *Ruellia ovata* below. Whether *Ruellia pilosa*, Lindau in Engl. & Prantl, Pflanzenf. iv. 3B 309, refers to a plant or to T. Anderson's name is not known.

2. R. Zeyheri (T. Anders. in Journ. Linn. Soc. vii. 25); a small undershrub; branches with very short internodes; leaves $\frac{1}{4}$–$\frac{3}{4}$ in. long, elliptic, obtuse, some (especially of the lower) nearly orbicular; bracteoles $\frac{1}{3}$ in. long, about as long as the calyx; otherwise as *R. patula,* Jacq. *Lindau in Engl. & Prantl, Pflanzenfam.* iv. 3B, 310. *Dipteracanthus Zeyheri, Sonder in Linnæa,* xxiii. 90. *Dipteracanthus sp., Drège in Linnæa,* xx. 200.

SOUTH AFRICA : without precise locality, *Masson!*

COAST REGION : Swellendam Div.; Buffeljagts River, 1000–2000 ft., *Zeyher*, 3600 ; Riversdale Div.; between Great Vals River and Zoetemelks River, *Burchell*, 6589! near Zoetemelks River, *Burchell*, 6737! Mossel Bay Div.; between Little Brak River and Hartenbosch, *Burchell*, 6212!

EASTERN REGION : Natal; Mooi River Valley, 2000–3000 ft., *Sutherland !*

Very near *R. patula*, Jacq. Suborbicular leaves occur on each of the collections referred to *R. Zeyheri.* Sonder says the capsule was 4-seeded; the ovary is 8-ovuled.

3. R. Baurii (C. B. Clarke); innovations with many white long several-celled hairs; leaves up to 2–3 by 1$\frac{1}{4}$ in., obtusely triangular at the tip, narrowed into a petiole $\frac{1}{12}$–$\frac{1}{3}$ in. long; bracteoles $\frac{3}{4}$ by $\frac{1}{10}$ in., narrow-oblong; calyx $\frac{1}{4}$ in. long or rather more, divided nearly to the base into 5 equal linear teeth; corolla, capsule and seeds as of *R. patula.*

EASTERN REGION : Tembuland ; hilly spots near Bazeia, 2000 ft., *Baur*, 309 ! Griqualand East; Vaal Bank, near Kokstad, *Haygarth in Wood Herb.*, 4177! Natal; Umzinyati Valley, *Wood*, 1375 !

The branches collected are only 4–6 in. long, and the plant appears to resemble *R. suffruticosa*, Roxb. The flowers are noted as " white " both by Wood and Haygarth.

4. R. Woodii (C. B. Clarke); innovations with many white long several-celled hairs ; leaves up to 1$\frac{1}{2}$ by 1 in., elliptic or ovate, tip sub-obtuse, base narrowed or rounded ; petiole $\frac{1}{6}$ in. long ; bracteoles $\frac{2}{3}$ in. long, oblong ; calyx $\frac{1}{4}$–$\frac{1}{3}$ in. long, divided rather more than half-way down ; teeth linear-lanceolate unequal; corolla purple (*Wood*) otherwise as *R. Baurii.*

EASTERN REGION : Griqualand East; Clydesdale, 2500 ft., *Tyson*, 1308! Natal; Colenso, 3300 ft., *Wood*, 4053 ! Weenen Country, 3000–5000 ft., *Sutherland!*

The stems seen are less than 4 in. long, and this may be a variety only of *R. Baurii.* The difference in the colour of the flowers is of small account in this group. When the calyx-segments are connate near the base in this genus, the connection is excessively thin and often ruptures.

5. R. ovata (Thunb. Prodr. 104) ; innovations ciliate-pubescent, stems woody at the base; branches 6–20 in. long, often divided, procumbent or straggling; leaves ¼–1 in. long, ovate, tip (in the upper leaves) triangular often subacute, base (in the lower leaves) truncate or cordate; petiole 0—¹⁄₁₀ in. long; flowers pale mauve or white; bracteoles ⅔ in. long, spathulate-oblong ; calyx nearly ⅓ in. long; teeth linear or linear-lanceolate, 2 nearly free to the base, 3 connate less than ½ their length ; corolla, capsule and seeds nearly as of *R. patula,* Jacq. *Thunb. Fl. Cap. ed. Schult.* 480; *T. Anders. in Journ. Linn. Soc.* vii. 25 ; *Lindau in Engl. & Prantl, Pflanzenf.* iv. 3B, 309. *R. cordata, Thunb. Prod.* 104, *and Fl. Cap. ed. Schult.* 480; *Nees in Linnæa,* xv. 356. *R. ciliaris & R. pubescens. Pers. Syn.* ii. 176. *R. pilosa, Linn. f. Suppl.* 290(?); *Thunb. Prodr.* 104; *T. Anders. in Journ. Linn. Soc.* vii. 25, 114. *Dipteracanthus pilosus & D. cordifolius, Nees in Linnæa,* xv. 353, 354; *Drège in Linnæa,* xx. 200. *Fabria rigida, E. Mey. in Drège, Zwei Pflanzengeogr. Documente,* 134, 185. *F. cordifolia & F. pilosa, Nees in DC. Prod.* xi. 114.

COAST REGION: Alexandria Div.; Enon, 1000–2000 ft., *Drège!* Albany Div.; Glenfilling, *Drège;* between the source of Kasuga River and Sidbury, *Burchell,* 4162! Komgha Div.; near Komgha, 2000 ft., *Flanagan,* 1313!

KALAHARI REGION: Orange River Colony; *Cooper,* 3029! Transvaal; Pretoria. at Apies Poort, *Rehmann,* 4104! Macalisberg, 6000–7000 ft., *Zeyher,* 1414! *Burke!* near Lydenberg, *Wilms,* 1210! Kaap River Valley, near Barberton, 2000 ft., *Galpin,* 498! 1199!

EASTERN REGION: Tembuland, near streams at Bazeia, 2000 ft., *Baur,* 448! Griqualand East; near Clydesdale, 2500 ft., *Tyson,* 1292! Natal; Inanda, *Wood,* 397! Malvern, 5000–6000 ft., *Wood,* 4929! and without precise locality, *Gerrard,* 1267!

In Galpin, 1199, the upper leaves are ¾ by ½ in., narrowed into the petiole, the lower stem-leaves are 1 by 1½ in., depressed-orbicular with cordate base ; those between are intermediate. A similar tendency in the leaves is visible in much of the material.

6. R. malacophylla (C. B. Clarke) ; whole plant densely and softly hairy; lower leaves 1 by ⅓ in., elliptic, narrowed at the base ; petiole ¼ in. long; bracteoles, calyx, and corolla as of *R. ovata;* ovary densely hairy except at the base ; style densely hairy.

EASTERN REGION: Natal; without precise locality, *Gerrard,* 427!

7. R. stenophylla (C. B. Clarke); young parts soft with long white hairs; leaves 1¼–1¾ by ¼–⅓ in., linear-oblong; petioles 0–⅙ in. long; calyx ¼ in. long or rather more, obscurely 2-lipped ; segments narrow, acute, 3 connate at the base ; corolla 1¼–1½ in. long, pale mauve (*Galpin*); stamens 4 ; anthers short-oblong not spurred at the base ; pollen globose, reticulate ; ovary glabrous ; style with 1 linear-oblong branch the other suppressed ; capsule ½ in. long or rather more, clavate.

KALAHARI REGION: Transvaal; Barberton, 2800 ft., *Galpin,* 640!

This is close to *R. ovata*, Thunb., differing in the narrow leaves and the soft (scarcely scabrous) indumentum.

Imperfectly known Species.

8. R. aristata (Thunb. Prodr. 104, and Fl. Cap. ed. Schult. 479) ; a much branched shrub, scarcely 1 ft. high ; branches tetragonous, minutely pubescent ; leaves ¼ in. long, obovate, obtuse, entire, glabrous, petioled ; flowers axillary, verticillate ; bracts similar to the leaves, villous-scabrous ; calyx-segments subulate, aristate, scabrous.

CENTRAL REGION : Karoo, *Thunberg!*

This is reduced by Nees to his *Chætacanthus-Persoonii*, none of the specimens of which had more than 2 stamens. Schultes says of this that the " stamens were approximated in pairs."

IV. RUELLIOPSIS, C. B. Clarke.

Bracteoles 2, linear, shorter than the calyx. *Calyx* deeply divided ; segments 5, linear, unequal. *Corolla* 1–1½ in. long, not 2-lipped ; tube funnel-shaped for ⅔ its length ; segments 5, subequal, round, contorted in the bud. *Stamens* 4, subsimilar ; anthercells 2, oblong, at equal height, spurred at the base ; pollen nearly globose, many-ribbed. *Ovary* with 4 ovules in each cell ; style hairy, with 1 linear and 1 suppressed stigmatic branch. *Capsule* cylindric, 8-seeded from the base. *Seeds* hygroscopically hairy on the margin.

Small shrubs ; leaves linear, entire ; flowers solitary axillary.

Species 1 ; with another plant from the Tropical Kalahari imperfectly known and doubtfully referred here.

1. R. setosa (C. B. Clarke in Dyer, Fl. Trop. Afr. v. 59) ; root woody ; branches 15 in. long, trailing, hispid ; leaves 2 by ¼ in., obtuse, hispid with scattered long white hairs ; calyx ⅔–¾ in. long ; corolla 1¼ in. long, bright or pale blue ; tube more than ¾ in. long ; capsule ½–⅓ in. long, glabrate ; seeds 8, silky, white. *Calophanes setosus, Nees in DC. Prod.* xi. 112 ; *T. Anders. in Journ. Linn. Soc.* vii. 24.

COAST REGION : Grahamstown ; collector not indicated !
KALAHARI REGION : Transvaal ; Boschveld, at Klippan, *Rehmann*, 5250 ! Vaal River, *Nelson*, 177 ! Pienaars River, 4300 ft., *Schlechter*, 4222 ! Bechuanaland ; Batlapin Territory, *Holub!* Mafeking, *Bolus*, 6411 !

Also in the Tropical Kalahari.

V. DYSCHORISTE, Nees.

Bracteoles 2, linear or narrow-oblong, much shorter than the calyx. *Calyx* segments 5, acute, subequal, often aristate. *Corolla* ⅓–1½ in. long, more or less 2-lipped ; tube inflated upwards or rarely linear to the top ; segments 5, contorted in the bud. *Stamens* 4, subsimilar ; anther-cells 2, oblong, at equal height, usually spurred at the base ; pollen ellipsoid or subglobose, with several longitudinal ribs. *Ovary* with 2 ovules in each cell; style hairy, 1 stigmatic arm linear-oblong, the other suppressed. *Capsule* linear-cylindric, hardly clavate, solid at the base, usually perfecting 4 seeds. *Seeds* discoid, densely clothed with hygroscopic white hairs.

Small shrubs; leaves entire or obscurely cuneate; flowers subsessile, axillary, clustered or scattered and solitary.

Species 60, in the warmer parts of both hemispheres.

Corolla-tube less than $\frac{1}{2}$ in. long, inflated in the upper part:
  Flowers clustered:
    Anthers tailed; calyx tubular for $\frac{1}{3}$ its length ... ... (1) **depressa.**
    Anthers muticous; calyx divided nearly to the base ... ... (2) **mutica.**
  Flowers scattered, solitary:
    Leaves glabrous, linear-oblong, $\frac{1}{8}$–$\frac{1}{6}$ in. wide ... ... (3) **erecta.**
    Leaves hairy, elliptic-oblong, $\frac{1}{4}$–$\frac{1}{3}$ in. wide (4) **transvaalensis.**
Corolla-tube $\frac{3}{4}$ in. long, linear nearly to the top ... (5) **Fischeri.**

1. **D. depressa** (Nees in Wall. Pl. As. Rar. iii. 81); a small shrub, nearly glabrous, the innovations minutely scabrous hairy; stems trailing, the ends of the branches erect; leaves up to 1 by $\frac{1}{3}$ in. (usually smaller), elliptic, entire or obscurely crenate; petiole up to $\frac{1}{4}$–$\frac{1}{3}$ in. long; flowers in axillary clusters of 3–8; bracteoles scarcely $\frac{1}{6}$ in. long, linear-oblong; calyx $\frac{1}{3}$ in. long, divided $\frac{2}{3}$ the way down, with linear scabrous-hairy teeth; corolla about $\frac{1}{2}$ in. long, somewhat 2-lipped, upper part of the tube inflated; stamens 4, subsimilar; anthers tailed at the base; capsule $\frac{1}{4}$–$\frac{1}{2}$ in. long, suberect, glabrous. *Nees in DC. Prod.* xi. 106; *C. B. Clarke in Dyer, Fl. Trop. Afr.* v. 72. *Ruellia depressa, Linn. Syst. Veget. ed. Murr.* 576; *Thunb. Prodr.* 104, *and Fl. Cap. ed. Schult.* 479. *Calophanes Nagchana, Nees in DC. Prod.* xi. 109; *C. B. Clarke in Hook. f. Fl. Brit. Ind.* iv. 410. *C. natalensis, T. Anders. in Journ. Linn. Soc.* vii. 23. *C. crenatus, Schinz in Bull. Herb. Boiss.* iii. 415. *Linostylis ovata, Sonder in Linnæa,* xxiii. 94.

EASTERN REGION: Natal; amongst grass, near Phœnix Station, 400–500 ft., *Wood,* 4967! and without precise locality, *Gueinzius,* 46!

Also in Tropical Africa and in India.

Thunberg found this species in Oudshorn Div.; in the Karoo behind Attaquas Kloof, in Vanrhynsdorp Div., between Olifants River and the Bokke Veldt, in Calvinia Div., at Hantam, and in Sutherland Div., on the Rogge Veldt, but Thunberg's plant may have been *D. radicans,* Nees, or one of the other closely allied species. The example of Wood, 4967, at the British Museum is *Asystasia Schimperi,* T. Anders., but this is not the *Calophanes crenatus* of Schinz; cf. Lindau in Engl. Jahrb. xxii. 118.

2. **D. mutica** (C. B. Clarke in Dyer, Fl. Trop. Afr. v. 73); pubescent; branches up to 12 in. long, straight; calyx more than $\frac{1}{3}$ in. long, divided nearly to the base; segments linear, aristate, hispid; anthers not tailed at the base; otherwise as *D. depressa,*

Nees. *Calophanes radicans, var. mutica, S. Moore in Journ. Bot.* 1880, 198.

KALAHARI REGION : Transvaal; near Lydenberg, *Wilms,* 1190!

Also in Angola.

Wilms' plant is issued as *D. radicans,* Nees; the very straight branches are not found in any example of *D. radicans.*

3. **D. erecta** (C. B. Clarke); nearly glabrous, except the corolla; branches slender, woody, somewhat quadrangular and glaucous; leaves ¾ by ⅙ in., narrowly oblong, obtuse, subsessile; flowers scattered, solitary, subsessile; bracteoles small; calyx ¼ in. long, divided ⅓ the way down, full of cystoliths; teeth 5, subequal, lanceolate, acute; corolla ½ in. long, pubescent without; tube in its upper part inflated; stamens 4, subsimilar; anther-cells oblong, at equal height, tailed at the base; pollen subglobose with 16 meridional ribs reaching the pole.

KALAHARI REGION : Transvaal; near Lydenberg, *Wilms,* 1233!

4. **D. transvaalensis** (C. B. Clarke); pubescent, 15 in. high, with straight suberect branches; leaves 1 by ⅓ in., elliptic, narrowed to a triangular tip, entire or sinuate-dentate; petiole 0–1/10 in. long; flowers solitary, subsessile in the axils (in one case a loose cyme of 3 flowers); bracteoles linear-oblong, shorter than the calyx-tube; calyx ½ in. long and upwards, divided ¼ way down; teeth 5, lanceolate-linear, glandular hairy; corolla more than ¾ in. long, blue with yellow veins (*Schlechter*); tube funnel-shaped upwards; stamens 4, subsimilar; anther-cells 2, oblong, at equal height, tailed at the base; pollen short ellipsoid with 10 longitudinal ribs; capsule which is ½ in. long, and seeds as of the genus.

SOUTH AFRICA: without precise locality, *Zeyher,* 1391!
KALAHARI REGION : Transvaal; Bosch Veld, at Klippan, *Rehmann,* 5254! Upper Molopo River, *Holub,* 1977! 1978! 1979! 1980! Pietersburg, 4700 ft., *Schlechter,* 4354!

5. **D. Fischeri** (Lindau in Engl. Jahrb. xx. 11); innovations minutely grey pubescent, the plant otherwise nearly glabrous; branches somewhat robust; leaves ½–⅔ by ⅓ in., elliptic, obtuse, sometimes apiculate; petiole 0–1/12 in. long; flowers 3–1 in small axillary cymes; calyx ½ in. long, divided ½ way down, full of cystoliths, not hispid; corolla yellow; tube ¾ in. long, linear-cylindric nearly to the top, mouth 2-lipped; stamens 4, subsimilar; anther-cells oblong, at equal height, not tailed at the base; capsule nearly ½ in. long with the seeds as of the genus. *Lindau in Engl. Pfl. Ost-Afr. C.* 367, *and in Ann. Istit. Bot. Roma,* vi. 68; *C. B. Clarke in Dyer, Fl. Trop. Afr.* v. 77.

KALAHARI REGION : Transvaal; Waterfall River, near Lydenburg, *Wilms,* 1232!

Also in East Tropical Africa to Somaliland.

The above description is taken from Wilms, 1232, which was issued as *D.*

Fischeri, and which agrees well as to the corolla, stamens, capsule and all essentials. But the type examples of *D. Fischeri* are larger in all their parts, more hairy, and their calyx hispid.

VI. CHÆTACANTHUS, Nees.

Bracteoles 2, linear or oblong, much shorter than the calyx. *Calyx* divided ½–¾ the way down; teeth 5, linear. *Corolla* white wherever noted; tube linear nearly to the top; lobes 5, subequal, contorted in the bud. *Stamens* 2; rudiments of the other 2 filaments small; anther-cells oblong, at equal height, not tailed at the base; pollen ellipsoid, longitudinally ribbed. *Style* linear, thinly hairy; stigma with 1 linear-oblong branch, the other suppressed; ovules 2 in each cell. *Capsule* linear-cylindric, generally perfecting 4 seeds. *Seeds* discoid with much hygroscopic hair.

Rootstock woody, short, whence arise branches 8–24 in. long; leaves entire, with cystoliths which are conspicuous also in the calyx; flowers axillary, solitary or few together.

Species 4, endemic in South Africa, hardly more than varieties of one.

Leaves ¼–½ in. long, the lower and middle obtuse :
  Calyx-teeth sparingly hairy, usually longer than
    the corolla-tube ...   .   (1) **Persoonii.**
  Calyx-teeth glandular-hairy, usually shorter than
    the corolla-tube ...   .   (2) **glandulosus.**
Leaves ½–1¼ in. long, the lower subacutely triangular
    at the tip :
  Calyx-teeth patently hispid   ...   .   (3) **Burchellii.**
  Calyx-teeth nearly glabrous   ...   .   (4) **costatus.**

1. **C. Persoonii** (Nees in Linnæa, xv. 357, var. β only, excl. all syn.); stem very short, up to ⅕ in. in diam.; branches 6–15 in. long, nearly glabrous; leaves ¼–½ in. long, obovate, lower obtuse, nearly glabrous; petiole 0–1/12 in. long; calyx ½ in. long, divided more than ½ the way down; teeth linear, sparsely hairy; corolla-tube scarcely as long as the calyx-teeth; stamens, anthers and style as described for the genus; capsule ⅓ by 1/12–1/10 in., glabrous. *Nees in DC. Prod.* xi. 462, *var. β only*; *Hook. Journ. Bot.* ii. (1840) 126; *Hochst. in Flora*, 1845, 72 *partly*; *Drège in Linnæa*, xx. 199 (?). *C. setiger, Lindau in Engl. Jahrb.* xviii. 38, *and in Engl. & Prantl, Pflanzenfam.* iv. 3B, 281, *fig. J. Eranthemum obovatum, var. c, E. Meyer in Drège, Zwei Pflanzengeogr. Documente*, 131, 182. *Calophanes Persoonii, T. Anders. in Journ. Linn. Soc.* vii. 23 *partly*. *Ruellia thymifolia, Vahl MS. ex Nees.*

SOUTH AFRICA: without precise locality, *Masson! Oldenburg!*
COAST REGION: Mossel Bay Div.; Little Brak River, *Burchell*, 6173! Knysna Div.; on hills at Vlugt, *Bolus*, 1769! Uitenhage Div.; on hills between Coega River and Zwartkops River, *Ecklon & Zeyher*, 334! near the Zwartkops River, *Ecklon & Zeyher!* Slaay Kraal, *Burke!* between Van Stadens River and Galgebosch, *Burchell*, 4675! Alexandria Div.; on grassy hills at Addo, 1000–2000 ft., *Drège!* Albany Div.; near Grahamstown, *Burke! MacOwan!* between Kasuga River and Sidbury, *Burchell*, 4164! Blue Krantz, *Burchell*, 3635! Bathurst Div.; near Theopolis, between Riet Fontein and the sea, *Burchell*,

4087 ! Komgha Div. ; Komgha, 2000 ft., *Flanagan*, 811! British Kaffraria, *Cooper*, 235! *Gill!*

CENTRAL REGION : Somerset Div. ; Somerset East, 2600 ft., *Bolus!* Alexandria Div. ; on the rocks of Zwartwater Poort, *Burchell*, 3377 !

EASTERN REGION : Tembuland ; near Bazeia, 2000 ft., *Baur*, 273 ! Natal ; in muddy places around Durban Bay, *Krauss*, 306 !

The type-specimen of *Chætacanthus Persoonii*, *var. a*, Nees, inscribed by Nees' hand, is *Justicia protracta*, T. Anders. *Ruellia setigera*, Pers. Syn. ii. 176, another name for *Ruellia aristata*, Thunb. (Fl. Cap. ed. Schult. 479), was referred to the present plant by Nees in Linnæa, xv. 357, and in DC. Prod. xi. 462. *Ruellia setigera*, Pers., had, however, 4 stamens, a point on which the old authors are much safer than the modern, and was therefore neither of the two. It may have been *R. patula*, Jacq. (otherwise not included by Thunberg).

2. **C. glandulosus** (Nees in DC. Prod. xi. 462) ; stem and leaves pubescent ; calyx-teeth with many gland-tipped hairs, usually shorter than the corolla-tube ; otherwise as *C. Persoonii*, Nees. *Schinz in Mém. Herb. Boiss.* x. 63. *C. Persoonii*, *Hochst. in Flora*, 1845, 72 *partly*. *Eranthemum obovatum*, *var. b*, *E. Meyer in Drège*, *Zwei Pflanzengeogr. Documente*, 134, 182. *Calophanes Persoonii*, *T. Anders. in Journ. Linn. Soc.* vii. 23 *partly*.

COAST REGION : Port Elizabeth Div. ; Port Elizabeth, *Bolus!* Albany Div. ; Fish River Heights, *Hutton!*

KALAHARI REGION : Transvaal ; hills above Apies River, near Pretoria, *Rehmann*, 4249 ! Wonderboom Poort, *Rehmann*, 4512 !

EASTERN REGION : Griqualand East ; near Kokstad, 5000 ft., *Tyson!* Natal ; in muddy places around Durban Bay, *Krauss*, 380 ! near the coast, *Wood*, 1157 ! and without precise locality, *Gerrard*, 180 ! Delagoa Bay, *Junod*, 319.

In this species, imperfect anther-cells with little or no good pollen occur ; in Rehmann, 4512, which contains flowers of every age, the anthers appear never to open, so that the plant is female. Similar imperfect anther-cells occur occasionally in *C. Persoonii*.

3. **C. Burchellii** (Nees in DC. Prod. xi. 462) ; leaves up to ¾–1 in. long, sparingly hairy, elliptic, lower with a subacute triangular tip ; calyx-teeth with rather long several-celled white spreading hairs ; otherwise as *C. Persoonii*, Nees. *C. Burkei, Sonder in Linnæa*, xxiii. 94. *Calophanes Burkei, T. Anders. in Journ. Linn. Soc.* vii. 24.

SOUTH AFRICA : without precise locality, *Mudd!*

KALAHARI REGION : Orange River Colony ; Thaba Unchu, *Burke!* Transvaal ; Hooge Veld, at Trigards Fontein, *Rehmann*, 6707 ! near Lydenburg, *Wilms*, 1196! Mac Mac Hills, *Mudd!*

EASTERN REGION : Griqualand East ; near Clydesdale, 2500 ft., *Tyson*, 2078 ! near Kokstad, 5000 ft., *Tyson!* MacOwan Herb. Austr.-Afr., 1509! Natal ; near Gourton, 4000 ft., *Wood*, 3623 ! near Durban, 100 ft., *Wood*, 34! Inanda, *Wood*, 213 ! near the junction of the Tugela and Blaawkrantz Rivers, 2000–3000 ft., *Evans*, 669 ! Weenen County, 3000–5000 ft., *Sutherland!* and without precise locality, *Gerrard*, 186 !

Nees describes this species as having the corolla-tube scarcely longer than the calyx-teeth, as is the case in his specimen (collected by Burke, not by Burchell). But the length of the corolla-tube varies much in other examples, till in Gerrard, 186, it is very nearly 1 in. long.

4. **C. costatus** (Nees in DC. Prod. xi. 462) ; leaves ½–1 in. long, elliptic or oblong, lower subacutely triangular at the tip, nearly glabrous, nerves prominent beneath ; calyx-teeth glabrous or very

nearly so; otherwise as *C. Persoonii*, Nees. *Sonder in Linnæa*, xxiii.
94. *Calophanes costatus*, *T. Anders. in Journ. Linn. Soc.* vii. 23.

KALAHARI REGION : Transvaal; Apies Poort, near Pretoria, *Rehmann*, 4109 !
Rust Plaats, near Origstad, on the Bosch Veld, *Wilms*, 1193 ! near Lydenburg,
Wilms, 1194 ! 1195 ! Magalies Berg, *Zeyher*, 1402 ! *Burke !* Matebe Valley,
Holub ! Barberton, 2400 ft., *Galpin*, 644 ! and without precise locality, *McLea !*

Nees describes the leaves as prominently 3-nerved beneath. The leaves are
pinnate-nerved, as throughout the genus, and there are 7–9 nerves equally
prominent.

VII. PHAYLOPSIS, Willd.

Bracteoles 0. *Calyx* 5-lobed nearly to the base; 2 anticous
segments linear or linear-spathulate; posticous segment ovate;
2 inner segments shorter linear. *Corolla* small; tube $\frac{1}{6}$–$\frac{1}{3}$ in. long;
lobes 5, contorted in the bud. *Stamens* 4; anther-cells mucronate
at the base, hardly tailed; pollen short-ellipsoid, longitudinally
12-ribbed, with 3 stopples. *Style* thinly hairy; 1 stigmatic arm
linear-oblong, the other very short; ovary with 2 ovules in each cell.
Capsule ellipsoid, compressed, solid at the base, usually perfecting
4 seeds, dehiscing elastically : the placentæ (carrying the seeds)
separate from the capsule-wall and spring up from the bottom ;
margins of seeds with numerous hygroscopic hairs.

Small, shrubby ; leaves often oblique, those in 1 opposite pair unequal, elliptic,
entire or crenate ; inflorescence in cylindric or ovoid spikes, each broad floral leaf
enclosing a contracted cyme of usually 3 flowers ; bract to each flower 0.

Species 15, in Africa, Mascarene Isles and India.

Anticous 2 calyx-teeth linear-ligulate, acute ... ... (1) **parviflora.**
Anticous 2 calyx-teeth linear-spathulate ... ... (2) **longifolia.**

1. **P. parviflora** (Willd. Sp. Pl. iii. 342) ; pubescent ; stem 1–2 ft.
long, branched ; leaves up to 3 by $1\frac{1}{2}$ in. (some on the same stem
only $\frac{1}{2}$ by 1 in. long), acuminate at both ends, nearly entire ; petiole up
to $1\frac{1}{2}$ in. long; inflorescence 1–2 in. long, dense, strobilate; lower
floral leaves $\frac{1}{3}$ by $\frac{1}{2}$ in., rounded, subtruncate ; calyx $\frac{1}{4}$–$\frac{1}{3}$ in. long, hairy ;
2 anticous segments linear-ligulate, acute ; corolla $\frac{1}{3}$ in. long, white
or purplish ; tube funnel-shaped at the top ; ovary glabrous, sparingly
glandular at the top ; capsule $\frac{1}{6}$–$\frac{1}{4}$ in. long. *T. Anders. in Journ.
Linn. Soc.* vii. 26 ; *C. B. Clarke in Hook. f. Fl. Brit. Ind.* iv. 417, *in
Dyer, Fl. Trop. Afr.* v. 83. *P. longifolia, Sims in Bot. Mag. t.*
2433. *Micranthus oppositifolius, Wendl. Bot. Beobacht.* 39 ; *Lindau
in Engl. & Prantl, Pflanzenfam.* iv. 3B, 298, *and in Engl. Pfl. Ost-
Afr. C.* 367. *M. longifolius and M. imbricatus, O. Kuntze, Rev. Gen.*
493. *Ætheilema imbricatum, R. Br. Prod.* 478; *Nees in DC. Prod.*
xi. 262 *partly. Æ. reniforme, Nees in Wall. Pl. As. Rar.* iii. 94, *and
in DC. Prod.* xi. 261 ; *Wight, Ic. Pl. t.* 1533 ; *Hochst. in Flora,*
1845, 70. *Phaulopsis oppositifolius and P. longifolius, Lindau in
Engl. & Prantl, Pflanzenfam. Nachtr. zu* ii.–iv. 305.

EASTERN REGION : Natal ; in woods around Durban Bay, *Krauss*, 231 ! roadsides near Durban, *Wood*, 65 ! borders of woods near Durban, 150 ft., *Wood in*

MacOwan & Bolus, Herb. Norm. Austr.-Afr., 1338! Inanda, *Wood*, 152! and without precise locality, *Grant! Cooper*, 3040! *Gerrard*, 10 partly!

Also in India, Mascarene Isles and Tropical Africa.

Phaylopsis longifolia, Sims (Bot. Mag. t. 2433), is figured from a Sierra Leone plant, and is identical with the West African examples accepted by all authors as *P. parviflora*.

2. **P. longifolia** (T. Thoms. in Speke, Journ. Append. 643, not of Sims); 2 anticous calyx-segments linear-spathulate; otherwise as *T. parviflora*. *T. Anders. in Journ. Linn. Soc.* vii. 26. *Ætheilema anisophyllum*, *E. Meyer in Drège, Zwei Pflanzengeogr. Documente*, 160, 162. *Micranthus longifolius*, *Lindau in Engl. & Prantl, Pflanzenfam.* iv. 3B, 298, *fig.* 120, A—F, *not of O. Kuntze*.

EASTERN REGION: Natal; near Durban, below 500 ft., *Drège!* and without precise locality, *Gerrard*, 13!

Also in East Tropical Africa.

This species might be esteemed a variety of *P. parviflora*.

VIII. PETALIDIUM, Nees.

Bracteoles 2, very large, ovate or elliptic, ultimately more or less scarious, prominently veined. *Calyx* deeply divided into 5 unequal narrow-lanceolate segments. *Corolla-tube* dilated towards the top; lobes contorted in the bud. *Stamens* 4; anther-cells mucronate at the base; pollen ellipsoid, few-ribbed, stopples 3 with (nearly always) 1 tubercle above and 1 below each stopple. *Style* with 2 unequal branches; ovary with 2 ovules in each cell, glabrous. *Capsule* small (commonly $\frac{1}{4}$ in. long), ellipsoid, compressed, dehiscing with elasticity; the placentæ separate from the capsule-wall, and spring up from the bottom. *Seeds* 4 (or more often 2) to the capsule, with many hygroscopic hairs.

Small shrubs; leaves entire; inflorescence of contracted monopodial cymes, often reduced to single flowers, as nearly always in the South African species; bract to each flower small, narrow, or 0.

Species 18, in Africa and the Mascarene Islands, 1 in India.

Pseudobarleria hirsuta, T. Anders. (in Journ Linn. Soc. vii. 26), supposed by him to have been collected in Extra-tropical West Africa, is *Petalidium Currori*, S. Moore (cf. Dyer, Fl. Trop. Afr. v. 91); but was really collected in Angola.

Leaves linear, $\frac{1}{12}$–$\frac{1}{8}$ in. broad ... ... (1) **linifolium.**
Leaves oblong, $\frac{1}{4}$–$\frac{1}{3}$ in. broad ... ... (2) **oblongifolium.**

1. **P. linifolium** (Harv. Thes. Cap. ii. 27, t. 143); nearly glabrous; stem 2–3 ft., erect, branched; leaves 1 by $\frac{1}{8}$–$\frac{1}{6}$ in.; flowers scattered, mostly solitary; bracteoles $\frac{1}{2}$ by $\frac{1}{4}$ in., finally white with reticulating green veins; calyx $\frac{1}{4}$–$\frac{1}{3}$ in. long; teeth 5, unequal, much longer than the tube, minutely hairy; corolla-tube $\frac{3}{4}$ in. long; limb $1\frac{1}{4}$ in. in diam.; capsule $\frac{1}{4}$–$\frac{1}{3}$ by $\frac{1}{6}$ in. *T. Anders. in Journ. Linn. Soc.* vii. 25.

WESTERN REGION: Great Namaqualand; between Brashwater and Sunfarar, and on a table mountain near Bethany, *Schinz*, 21!

Also in Damaraland.

2. **P. oblongifolium** (C. B. Clarke); leaves 1 by $\frac{1}{4}$–$\frac{1}{3}$ in.; petioles $\frac{1}{8}$ in.; bracteoles $\frac{3}{4}$ by $\frac{1}{3}$ in., finally white-scarious with reticulating purple veins; calyx-teeth $\frac{1}{4}$–$\frac{1}{3}$ in. long, unequal, glandular-hairy; corolla, stamens, anthers, pollen and pistil, as of *P. linifolium*.

KALAHARI REGION: Transvaal; Piet Potgeiters Rust, 4300 ft., *Schlechter*, 4773!

IX. BLEPHARIS, Juss.

Bract large, ovate, acute or lanceolate, strongly nerved, often spine-toothed; bracteoles 2, narrow, or (in *B. boerhaaviæfolia*) 0. *Calyx* 4-partite nearly to the base; anticous segment of 2 sepals united nearly to the tip; posticous sepal longer, ovate at the base, oblong above, 3-nerved; 2 inner sepals much shorter and narrower. *Corolla* short-tubed; limb blue or white, throat often yellow, or according to Nees in *B. capensis* the corolla wholly yellow; posticous lip 0, represented by a horny rim; anticous lip nearly flat obovate with 3–5 rounded lobes. *Stamens* 4, didynamous: anthers 1-celled, oblong, with a fringe of long white hairs along the slit; 2 anticous filaments 2-fid near the top, one branch a short-oblong process or nearly obsolete, the other carrying the anther; pollen ellipsoid, smooth, with 3 longitudinal chinks (cf. Lindau in Engl. Jahrb. xviii. t. 1, fig. 36). *Ovary* with 2 (or 1) ovules in each cell; style-branches 2, subequal, linear-lanceolate; at the apex of the ovary on the posticous face are 2 hollows filled with glands. *Capsule* ellipsoid, flattened, woody, shining-brown, 2- (seldom 4-) seeded. *Seeds* much flattened, covered with rope-like hair-bundles, which on applying water unroll into very long 1-celled hairs each furnished with a spiral band within.

Weedy undershrubs, often spinous, or with spine-toothed leaves; leaves of the main stem approximated frequently in fours; the lower pair in each false whorl often smaller and sometimes reduced to compound spines; flowers in spikes, which are sometimes cylindric strobilate with imbricate bracts, sometimes ovoid, few-flowered and very loose; the spikes in spp. 1–4, 13–15, are reduced having all the bracts (except the uppermost) empty, and appear as single flowers closely enclosed by 4–8 bracts.

Species 50, mostly African; a few extending through Arabia and the Orient Region to India.

Section 1. Leaves entire, not spine-toothed on the margins (except *B. linariæfolia* in which the leaves have occasionally a few teeth).

Spikes small, reduced, with the terminal bract
  alone bearing a flower, axillary, mostly solitary;
  corolla $\frac{1}{2}$–$\frac{3}{8}$ in. long:
    Bracteoles 0 ...   . (1) **boerhaaviæfolia.**
    Bracteoles present:
      Bracts spine-toothed:
        Leaves slightly hairy or glabrescent (2) **molluginifolia.**
        Leaves scabrous with several-celled
          hairs ...   . (3) **setosa.**
      Bracts without spinous teeth   ...   ... (4) **innocua.**

Spikes with several flowers : corolla 1 in. long :
   Leaves linear :
      Spikes ovoid, loose, few-flowered ...   ... (5) **angusta.**
      Spikes cylindric ; bracts imbricated   ... (6) **linariæfolia.**
   Leaves oblong-elliptic or obovate   ...   ... (7) **Stainbankiæ.**

Section 2. Leaves more or less toothed or spinous on the margin; often
approximated in fours whereof the lower pair are reduced to pinnate spines.

   Bracts glabrous or puberulous :
      Bracts lanceolate ; corolla yellow or white   ... (8) **capensis.**
      Bracts obovate, shortly acuminate ; corolla
        blue ...   . (10) **mitrata.**
   Bracts villous at least on the nerves :
      Spikes few-flowered ; corolla-lips yellow or white (9) **Ecklonii.**
      Spikes many-flowered ; corolla-lips blue   ... (11) **hirtinervia.**

Section 3. Leaves more or less toothed or spinous on the margin ; often
approximated in fours, whereof the lower pair are not dissimilar to the upper
though frequently somewhat smaller.

   Spikes small, reduced, with the terminal bract alone
    supporting a flower :
      Flower-bract with 5 spinous marginal teeth ... (12) **uniflora.**
      Flower-bract mucronate ; lateral teeth obsolete (13) **inermis.**
   Spikes several-flowered :
      Branches with very short internodes ; leaves
        appearing clustered ; harsh small shrubs :
          Spike small, few-flowered ; bracts ending
            in a strong pinnate spine   ...   ... (14) **furcata.**
          Spike short cylindric ; bracts hairy ending
            in a simple spine ...   . (15) **villosa.**
          Spike loose, few-flowered ; bracts oblong,
            glabrous, like the leaves ...   . (16) **marginata.**
      Branches with many internodes 1 in. long or
        more ; many leaves approximated in fours :
          Spikes cylindric many-flowered ; subovoid
            spikes occur also :
            Bracts very hairy :
              Ovary with a large obversely
                mitriform cap ...   . (17) **obmitrata.**
              Ovary with no cap under the
                style-base   ...   . (19) **longispica.**
            Bracts minutely hairy, their large
              tips squarrose   ...   . (18) **squarrosa.**
          Spikes all ovoid, few-flowered, sometimes
            very shortly cylindric :
            Leaves elliptic or oblong :
              Outermost bracts needle-like :
                Bracts glabrous without ... (20) **pruinosa.**
                Bracts hairy without:
                  Leaves irregularly spine-
                    toothed   ...   ... (21) **diversispina.**
                  Leaves closely ciliate-
                    spinescent ...   ... (22) **serrulata.**
            Outermost bracts pinnatifid
              spinous ; leaves or spines pin-
              nate at base ; bracts minutely
              scabrous-hairy :
                Posticous sepal obtuse   ... (23) **subvolubilis.**
                Posticous sepal widened at
                  the top   ...   . (24) **dilatata.**

Leaves linear to linear-oblong :
      Leaves spine-toothed,   the
            lower pair subequal   ...   (25) **procumbens**.
      Leaves remotely toothed, the
            lower pair often smaller
            than the upper   ...   ...   (26) **inæqualis**.
      Leaves pinnatifid; elongated
            branches without flowers;
            spikes terminal on short
            basal branches   ...   ...   (27) **sinuata**.

1. **B. boerhaaviæfolia** (Pers. Syn. ii. 180); scabrous, and with some long white hairs; stems 6–24 in. long, much branched, procumbent; leaves up to 1¼ by ⅓–⅔ in., oblong or elliptic, obtuse, entire, scarcely petioled, without spine-teeth; reduced spikes (1-flowered) scattered, axillary, solitary or 2–3 together; empty lower bracts 5–8, upper gradually larger, uppermost supporting the flower ¼–⅓ in. long, spathulate, strongly veined, with long marginal retrorse scabrous bristles; bracteoles 0; posticous sepal ½ in. long; 2 inner sepals ¼ in. long; corolla ½ in. long; capsule ¼ in. long or rather more, ovoid, usually 2-seeded. *Wight, Ic. Pl. t.* 458; *Nees in DC. Prod.* xi. 266; *Sonder in Linnæa*, xxiii. 92; *T. Anders. in Journ. Linn. Soc.* vii. 34; *C. B. Clarke in Hook. f. Fl. Brit. Ind.* iv. 478, *and in Dyer Fl. Trop. Afr.* v. 96; *Lindau in Engl. & Prantl, Pflanzenfam.* iv. 3B, 316, *fig.* 126, B—H; *Schinz in Mém. Herb. Boiss.* x. 63. *B. Togodelia, Solms-Laub. in Schweinf. Beitr. Fl. Aethiop.* 108, 243; *Lindau in Engl. Pfl. Ost-Afr. C.* 369. *Acanthus maderaspatensis, Linn. Sp. Pl. ed.* 2, 892.

EASTERN REGION: Natal; Umzinyati Falls, *Wood*, 944! and without precise locality, *Gerrard*, 1677! Delagoa Bay, *Monteiro*, 13! *Junod*, 265, *Kuntze*.

Also plentiful in Tropical Africa and India, and to Java.

2. **B. molluginifolia** (Pers. Syn. ii. 180); leaves oblong or linear, usually less than ¼ in. broad; bracteoles ⅓–½ in. long, linear-lanceolate, mucronate, sometimes spathulate, boat-shaped at the top, more or less spinous on the margin; otherwise as *B. boerhaaviæfolia*, Pers. *Nees in DC. Prod.* xi. 266; *T. Anders. in Journ. Linn. Soc.* ix. 500; *C. B. Clarke in Hook. f. Fl. Brit. Ind.* iv. 479, *and in Dyer, Fl. Trop. Afr.* v. 98. *B. saturejæfolia, Pers. Syn.* ii. 180; *Nees in Linnæa*, xv. 360, *and in DC. Prod.* xi. 265; *Hook. Journ. Bot.* ii. (1840) 126; *Drège in Linnæa*, xx. 200; *T. Anders. in Journ. Linn. Soc.* vii. 34; *Lindau in Engl. & Prantl, Pflanzenfam.* iv. 3B, 317, *in Engl. Pfl. Ost-Afr. C.* 369. *B. integrifolia, E. Mey. in Drège, Zwei Pflanzengeogr. Documente*, 49, 136, 137, 138, 141, 142, 168; *Hochst. in Flora*, 1845, 70. *B. boerhaaviæfolia, var. micrantha, Sonder in Linnæa*, xxiii. 92, *ex descr.* *B. Gueinzii, T. Anders. in Journ. Linn. Soc.* vii. 34, *ex descr.* *Acanthus integrifolius, Linn. f. Suppl.* 294; *Thunb. Prod.* 97, *and Fl. Cap. ed. Schult.* 456.

SOUTH AFRICA: without precise locality, *Zeyher*, 1410!
COAST REGION: Riversdale Div.; Riversdale, 400 ft., *Schlechter*, 1799!

Mossel Bay Div.; hills near the landing-place at Mossel Bay, *Burchell*, 6293 ! between Little Brak River and Hartenbosch, *Burchell*, 6213 ! Humansdorp Div.; near Little Zekoe River, *MacOwan*, 304 ! Uitenhage Div.; near the Zwartkops River, *Ecklon & Zeyher*, 475 ! Sunday River, *Bowie !* Alexandria Div.; on the Zuur Berg Range, 2000–3000 ft., *Drège !* Enon and Fish River, *Baur*, 1081 ! Albany Div.; near Grahamstown, *Burke ! MacOwan*, 147 ! Lower Albany, at Glenfilling, 1000 ft., *Drège !* Queenstown Div.; near Queenstown, 3800 ft., *Galpin*, 1942 ! Komgha Div.; near the Kei River, 1800 ft., *Flanagan*, 1083 ! CENTRAL REGION : Willowmore Div.?; valley of the Gamtoos River, *Bolus*, 2421 ! Somerset Div.; without precise locality, *Bowie !* Albert Div.; between Stormberg Spruit and Braam Berg, 4500 ft., *Drège !*

KALAHARI REGION : Griqualand West, Hay Div.; at Griqua Town, *Burchell*, 1950 ! Orange River Colony; Vet River, *Burke !* Bechuanaland; near the source of the Kuruman River, *Burchell*, 2473 ! Transvaal; Magalies Berg, 4600 ft., *Schlechter*, 3682 ! Pienaars River, 4300 ft., *Schlechter*, 4208 !

EASTERN REGION : Griqualand East; mountain sides around Clydesdale, 2500 ft., *Tyson*, 2682 ! *MacOwan & Bolus*, Herb. Norm. Austr.-Afr., 829 ! Natal ; at the foot of Table Mountain, *Krauss*, 21 ! near Umkomaas River, *Wood*, 1413 ! in "Thorns" near Mooi River, *Wood*, 4429 ! and without precise locality, *Gerrard*, 1919 ! Delagoa Bay, *Kuntze*.

Also in Tropical Africa and India.

3. B. setosa (Nees in DC. Prod. xi. 265); leaves 1½ by ⅓ in., scabrous with several-celled white hairs; bracteoles ending in a white bristle; otherwise as *B. molluginifolia*, Pers. *C. B. Clarke in Dyer, Fl. Trop. Afr. v. 98.*

KALAHARI REGION : Transvaal; Apies River, near Pretoria, *Burke !*
EASTERN REGION : Natal; without precise locality, *Gerrard*, 1918 !
Also in Tropical Africa.

4. B. innocua (C. B. Clarke) ; minutely pubescent; stem 6–18 in. long, much divided, wiry, hispid-scabrous; leaves 1 by ¼–⅓ in., elliptic or oblong, entire, scarcely petioled ; spikes ¾ by ¼ in., axillary, subsessile, reduced to 5–7 closely imbricate bracts, the terminal supporting a flower, the rest empty;' bracts mucronate, nearly spineless, the uppermost (flower-bearing) ¼ in. long; bracteoles ⅔ in. long, boat-shaped, lanceolate ; anticous calyx-segment ⅔ in. long ; corolla ⅔ in. long; on the posticous face of the ovary at its top are 2 viscid hollows and 4 transverse glands.

KALAHARI REGION : Transvaal; Houtbosch, *Rehmann*, 6192 ! near Lydenberg, *Wilms*, 1215 ! Bronkhorst Spruit, *Wilms*, 1215a ! Pietersburg. 4700 ft., *Schlechter*, 4681 !

5. B. angusta (T. Anders. in Journ. Linn. Soc. vii. 35) ; sparsely and minute hairy ; stems 3–7 in. long, wiry ; leaves 1½–2½ by 1/10 in., linear, entire, without spines, often approximated in fours; spikes 1 by ¾ in., subsessile, ovoid, loose, of 2–4 flowers ; bracts ¾ in. long, ovate, acute, with 5–7 spinous teeth; bracteoles ½ by 1/16 in.; calyx hairy within and without; anticous segment ½ in. long ; corolla 1 in. long, blue. *Acanthodium angustum, Nees in DC. Prod.* xi. 273; *Sonder in Linnæa*, xxiii. 92.

KALAHARI REGION : Transvaal; near Schoen Spruit, *Burke !* and without precise locality, *MacLea in Bolus Herb.* 6475 !
EASTERN REGION : Natal, *Owen*, according to T. Anderson.

6. B. linariæfolia (Pers. Syn. ii. 180) ; pubescent ; stems 4–18 in. long ; leaves up to 3 by $\frac{1}{5}$ in., entire or occasionally with a few small spines ; spikes 3–1 by $\frac{3}{4}$ in., cylindric or ovoid, strobilate ; bracts imbricate, $\frac{3}{4}$–1 in. long, spinous ; bracteoles $\frac{1}{2}$–$\frac{2}{3}$ in. long, linear ; posticous calyx-segment $\frac{3}{4}$ in. long ; corolla 1 in. long. *T. Anders. in Journ. Linn. Soc.* vii. 36 ; *C. B. Clarke in Dyer, Fl. Trop. Afr.* v. 100. *B. sindica, T. Anders. in Journ. Linn. Soc.* ix. 500 ; *C. B. Clarke in Hook. f. Fl. Brit. Ind.* iv. 479. *Acanthodium grossum, Wight, Ic. Pl. t.* 1535, 1536, *not of Nees.*

KALAHARI REGION : Bechuanaland ; between Mafeking and Ramoutsa, *Lugard !*

Frequent in Tropical Africa, extending into South-west Asia.

7. B. Stainbankiæ (C. B. Clarke) ; hairy branches 3–8 in. long, slender, rambling ; leaves $\frac{1}{2}$–$1\frac{1}{2}$ by $\frac{1}{4}$–$\frac{1}{3}$ in., obovate or oblong, entire, without spines, often approximated in fours ; petiole 0–$\frac{1}{5}$ in. long ; spikes sessile, axillary, solitary, ovoid, up to $1\frac{1}{2}$ by 1 in., of 4–7 flowers ; bracts 1 by $\frac{1}{2}$ in., with spinous teeth up to $\frac{1}{5}$ in. long, softly hairy ; bracteoles $\frac{2}{3}$ in. long, linear ; calyx hairy within and without ; anticous segment $\frac{1}{2}$ in. long ; corolla 1 in. long, white (*Mrs. Stainbank*).

KALAHARI REGION : Transvaal ; *Mrs. Stainbank in Wood Herb.*, 3661 ! Johannesberg, 5000 ft., *Mrs. Galpin in Galpin Herb.*, 1399 !

8. B. capensis (Pers. Syn. ii. 180) ; nearly glabrous except the calyx and corolla ; small harsh shrub (but "sometimes attaining 6–8 ft.," *Niven*) ; stem-leaves often approximated in fours whereof the lower pair are reduced to 3–7-toothed pinnate whitened spines, the upper pair 1–$1\frac{1}{2}$ in. long, oblong or elliptic, more or less spine-toothed ; spikes near the end of the branches, 1–2 by $\frac{3}{4}$–1 in., loose, of 4–9 flowers ; bracts 1 by $\frac{1}{3}$ in., elliptic-lanceolate, glabrous, white-nerved, 3–9-toothed, terminal tooth long, narrow ; calyx pubescent without, hairy within ; anticous segment $\frac{1}{3}$–$\frac{1}{2}$ in. long ; corolla $\frac{3}{4}$ in. long, white (or yellow according to *Nees*) ; capsule frequently 4-seeded. *T. Anders. in Journ. Linn. Soc.* vii. 35 ; *Lindau in Engl. & Prantl, Pflanzenfam.* iv. 3B, 318 ; *Hochst. in Flora*, 1845, 70 ? *Acanthus capensis, Linn. f. Suppl.* 295 ; *Thunb. Prod.* 97, *and Fl. Cap. ed. Schult.* 455 ; *Drège, Zwei Pflanzengeogr. Documente*, 61, 161 *letter b. Acanthodium capense, Nees in Linnæa*, xv. 361, *and in DC. Prod.* xi. 276, *excl. var. villosa.*

SOUTH AFRICA : without precise locality, *Masson ! Oldenburg !*
COAST REGION : Div. ? Bickman's Kloof, *Niven*, 90 ! Riversdale Div. ; Riversdale, *Pappe !* Humansdorp Div. ; between Milk River and Gamtoos River, *Burchell*, 4793 ! and without precise locality, *Niven*, 23 ! Fort Beaufort Div. ; without precise locality, *Cooper*, 519 ! Queenstown Div. ; Griffithsville, near Queenstown, 3500 ft., *Galpin*, 1758 !
CENTRAL REGION : Jansenville Div. ; Zwart Ruggens, on the Karoo Flats, 2000–3000 ft., *Drège !* Somerset Div. ; between the Zuurberg Range and Klein Bruintjes Hoogte, 2000–2500 ft., *Drège !* near Somerset, *Bowker*, 18 ! 60 ! Graaff

Reinet Div.; hills near Graaff Reinet, 2900 ft., *Bolus*, 427! Murraysburg Div.; near Murraysburg, *Tyson!* Colesberg Div.; vley near Sea Cow (Zeekoe) River, *Shaw!* Middelburg Div.; at "Rock Station," near Middelburg, *Burchell*, 2797!

9. **B. Ecklonii** (C. B. Clarke); bracts $\frac{3}{4}$ by $\frac{1}{2}$ in., obovate, villous on undersurface of the nerves, tip short spinous; otherwise nearly as *B. capensis*, Pers. *B. hirtinervia, T. Anders. in Journ. Linn. Soc.* vii. 35, *partly*. *Acanthodium capense, Sonder in Linnæa*, xxiii. 93. *A. capense, var. β, Nees in DC. Prod.* xi. 276. *Blepharacanthus capensis, Drège in Linnæa*, xx. 200.

Coast Region: Riversdale Div.; 500 ft., *Schlechter*, 2012! Humansdorp Div.; by the Gamtoos River, *Bowie*, 23! Uitenhage Div.; on the Karoo-like hills by the Sunday River, *Ecklon & Zeyher*, 148! Grassrug, 500–1000 ft., *Zeyher*, 1406! Algoa Bay, *Bowie!*

Noted 8 ft. high; the upper branches $\frac{1}{5}$ in. in diam.

10. **B. mitrata** (C. B. Clarke); glabrescent, except the inflorescence; stems 2–6 in. long, harsh; leaves often approximated in fours, the lower pair reduced to pinnate spines, the upper pair $\frac{1}{2}$–$1\frac{1}{2}$ by $\frac{1}{6}$–$\frac{1}{3}$ in., oblong or elliptic, sinuate, remotely spine-toothed; petiole 0–$\frac{1}{5}$ in.; spikes 1 in. long, ovoid, subsessile, of 2–5 flowers; bracts 1 by $\frac{1}{3}$ in., obovate, puberulous glaucescent without, with 7–9 marginal spines; bracteoles $\frac{1}{2}$ in. long, linear-lanceolate; calyx pubescent without, hairy within; anticous segment $\frac{1}{3}$ in. long; posticous sepal $\frac{2}{3}$ in. long; corolla $\frac{3}{4}$ in. long, pale blue; ovary surmounted by a large ovoid cap, hollow with a ring of long white hairs inside near the top. *B. hirtinervia, T. Anders. in Journ. Linn. Soc.* vii. 35, *partly*. *Acanthus furcatus, var. α, E. Meyer in Drège, Zwei Pflanzengeogr. Documente*, 62, 161. *Blepharacanthus furcatus, Presl, Bot. Bemerk.* 98. *Acanthodium capense, Nees in DC. Prod.* xi. 276 *partly*.

Central Region: Carnarvon Div.; at the northern exit of Karree Bergen Poort, near Carnarvon, *Burchell*, 1563! Albert Div.; near Weltevrede, by the Gamka River, *Drège!*

11. **B. hirtinervia** (T. Anders. in Journ. Linn. Soc. vii. 35, excl. syn.); branches 10 in. long, stout, pubescent upwards; leaves often approximated in fours, the lower pair reduced to pinnate whitened spines, the upper $1\frac{1}{2}$ by $\frac{1}{3}$–$\frac{1}{2}$ in., subentire, spine-toothed, when mature nearly glabrous; spikes often terminal, up to $3\frac{1}{2}$ by $1\frac{1}{4}$ in., hairy; bracts $1\frac{1}{4}$ by $\frac{3}{4}$ in., villous; anticous calyx-segment $\frac{3}{4}$ in. long; posticous sepal 1 in. long; corolla $1\frac{1}{4}$–$1\frac{1}{2}$ in. long, blue in the dried state (*Nees* says yellow). *Acanthodium hirtinervium, Nees in DC. Prod.* xi. 277; *Sonder in Linnæa*, xxiii. 93.

Coast Region: Port Elizabeth Div.; on the sand-hills, near Port Elizabeth, *Drège*, 7932!

Central Region: Somerset Div.; near Somerset East, *Bowker!*

Kalahari Region: Transvaal; Magalies Berg, *Burke!*

12. **B. uniflora** (C. B. Clarke); pubescent; stem 12 in. long, rigid, branched; leaves mostly approximated in fours, 1 by $\frac{1}{4}$ in.,

oblong, nearly glabrous, with few irregular teeth, scarcely spinous; spikes (reduced to 1 terminal flower, with all the lower bracts empty), axillary, solitary; flower-bract $\frac{2}{3}$ by $\frac{1}{8}$ in., with 5 spinous teeth; bracteoles $\frac{3}{4}$ by $\frac{1}{10}$ in.; anticous calyx-segments $\frac{2}{3}$ in. long; posticous sepal $\frac{3}{4}$ in. long; corolla 1 in. long.

KALAHARI REGION: Transvaal; Makapans Berg, at Streyd Poort, *Rehmann,* 5440!

Extends to the Tropical Transvaal.

This differs from *B. pungens,* T. Anders., from Mozambique, by the presence of bracteoles.

13. B. inermis (C. B. Clarke); glabrescent, very spinous; stems 9 in. long, thick, harsh, branched; leaves 1 by $\frac{1}{6}$ in., lanceolate, spine-toothed, with numerous simple spines $\frac{1}{2}$–$\frac{2}{3}$ in. long in the axils (which form the lowest bracts of the spike developed the next year); spikes (with all the lower bracts empty, the uppermost containing a flower), axillary, 1–3 together; flower-bracts $\frac{2}{3}$–$\frac{1}{8}$ in., lanceolate, spine-tipped; lateral teeth 0 or minute and obscure; bracteoles $\frac{1}{2}$–$\frac{2}{3}$ in. long, linear; posticous sepal $\frac{2}{3}$ in. long; corolla $\frac{3}{4}$ in. long; capsule $\frac{1}{4}$ in. long, 2-seeded. *B. capensis, T. Anders. in Journ. Linn. Soc.* vii. 35, *partly. Acanthodium capense, var. inermis, Nees in DC. Prod.* xi. 277.

SOUTH AFRICA: without precise locality, *Thom,* 370! 375!

14. B. furcata (Pers. Syn. ii. 180); sparingly hairy, very spinous; stems 4–6 in. long, woody, branched, rigid, with short internodes; leaves 1–1$\frac{1}{2}$ by $\frac{1}{12}$–$\frac{1}{5}$ in., lanceolate or nearly linear, with few spinous teeth; petiole 0–$\frac{1}{6}$ in. long; spines in the leaf-axils (the basal bracts of the next year's spikes) $\frac{1}{4}$–$\frac{1}{3}$ in. long, mostly 3-fid or pinnate near the base; spikes axillary, $\frac{3}{4}$–1 by $\frac{1}{3}$ in., ovoid, few-flowered; bracts $\frac{2}{3}$ by $\frac{1}{3}$ in., nearly glabrous, terminated by a long rigid 3-fid or pinnate spine; bracteoles $\frac{1}{2}$ in. long, linear-lanceolate; calyx minutely pubescent without, hairy within; anticous segment scarcely $\frac{1}{3}$ in. long; posticous sepal exceeding $\frac{1}{2}$ in. in length, acute; corolla $\frac{3}{4}$ in. long, bluish; ovary with 3 glands at the top below the 2 hollows at the style-base; capsule $\frac{1}{4}$ in. long, 2-seeded. *T. Anders. in Journ. Linn. Soc.* vii. 35, *excl. the Port Natal example. Acanthus furcatus, Linn. f. Suppl.* 295; *Thunb. Prod.* 97, *and Fl. Cap. ed. Schult.* 455. *A. integrifolius, var. e, E. Meyer in Drège, Zwei Pflanzengeogr. Documente,* 93, 161. *A. macer, E. Meyer in Drège, Zwei Pflanzengeogr. Documente,* 67, 161. *Blepharacanthus integrifolius, Presl, Bot. Bemerk.* 98. *Acanthodium furcatum, Nees in DC. Prod.* xi. 276; *Sonder in Linnæa,* xxiii. 93, *partly. A. macrum, Nees in DC. Prod.* xi. 276; *Sonder in Linnæa,* xxiii. 93.

SOUTH AFRICA: without precise locality, *Masson*!

CENTRAL REGION; Calvinia Div.; Springbok Kuil River, 2000–3000 ft., *Zeyher,* 1408! Prince Albert Div.; Gamka River, *Burke!*

WESTERN REGION: Little Namaqualand; by the Orange River, near Verleptpram, *Drège!*

KALAHARI REGION: Prieska Div.; near the Orange River, between "Gariep Station and Shallow Ford," *Burchell,* 1650!

15. B. villosa (C. B. Clarke); glabrescent, except the inflorescence; stems 2–4 in. long, woody, with internodes less than $\frac{1}{4}$ in. long; leaves $1\frac{1}{4}$ by $\frac{1}{3}$ in., oblong, sinuate, remotely spine-toothed; petiole 0–$\frac{1}{5}$ in. long; spikes 1–$1\frac{1}{4}$ in. long, axillary, 4–8-flowered; bracts 1 by $\frac{1}{2}$ in., villous or pubescent without, obovate, suddenly acuminate into a tip with 3–5 simple spines; bracteoles $\frac{2}{3}$ in. long, linear; calyx pubescent without, hairy within; anticous segment $\frac{1}{3}$ in. long; posticous sepal $\frac{3}{4}$ in. long, acute; corolla nearly 1 in. long, blue; capsule $\frac{1}{4}$ in. long, 4-seeded. *B. hirtinervia, T. Anders. in Journ. Linn. Soc.* vii. 35 *partly. Acanthodium capense, var.* δ*, *villosum subacaule, Nees in DC. Prod.* xi. 277. *A. furcatum, Sonder in Linnæa*, xxiii. 93 *partly.*

Central Region: Middelburg Div.; "Rock Station," near Middelburg, *Burchell*, 2799/2! between Wolve Kop and Middelburg, *Burchell*, 2790! Cradock Div.; by the Fish River, near Cradock, *Burke! Zeyher*, 1405!

16. B. marginata (C. B. Clarke); nearly glabrous; stem 2–5 in. long, woody, branched, internodes mostly less than $\frac{1}{8}$ in. long; leaves in tufts, 1 by $\frac{1}{6}$ in., oblong or lanceolate, with 8–14 equidistant spines; spikes ovoid, of 2–4 flowers, very loose; bracts $\frac{3}{4}$ in. long, oblong, glabrous, closely resembling the smaller leaves; bracteoles $\frac{2}{3}$ by $\frac{1}{12}$ in., usually ciliate, subspinous; calyx closely pubescent without, densely hairy within; anticous segment $\frac{1}{2}$ in. long; posticous sepal $\frac{3}{4}$ in. long, acute; corolla $\frac{3}{4}$–1 in. long, blue; capsule $\frac{1}{3}$ in. long, ellipsoid, 4-seeded. *Acanthodium marginatum, Nees in DC. Prod.* xi. 275.

Kalahari Region: Griqualand West, Hay Div.; at Griqua Town, *Burchell*, 1859! 1902! Herbert Div.; in sandy soil near Backhouse (Douglas), *Shaw*, 59! Bechuanaland; between Hamapery and Kosi Fontein, *Burchell*, 2541!

17. B. obmitrata (C. B. Clarke in Dyer, Fl. Trop. Afr. v. 101); hairy; branches 10 in. long; internodes 1–4 in. long; leaves approximated in fours, up to $1\frac{1}{2}$ by $\frac{1}{3}$–$\frac{1}{2}$ in., densely and minutely hairy, spinous-toothed; spines in the axils (i.e. the basal spines of next year's spikes) up to $\frac{2}{3}$ in. long, simple or toothed near the base; spikes 2 by $\frac{3}{4}$ in., rather dense, 6–10-flowered, hairy; bracts $\frac{3}{4}$–1 by $\frac{1}{3}$ in., obovate, suddenly acuminate, with 7–9 spines, hairy within and without; bracteoles $\frac{3}{4}$–1 in. long, narrowly linear; anticous calyx-segment scarcely $\frac{1}{2}$ in. long; posticous sepal nearly $\frac{3}{4}$ in. long; corolla 1 in. long, blue; ovary crowned by a large pale brown cap, below the two hollows filled with glands at the style-base.

Kalahari Region: Transvaal: Boschveld, between Eland's River and Klippan, *Rehmann*, 5049!

Also in Angola.

18. B. squarrosa (T. Anders. in. Journ. Linn. Soc. vii. 35); nearly glabrous; branches 6–10 in. long, procumbent rambling, internodes 1–3 in. long; leaves mostly approximated in fours, up to

4 by $\frac{3}{4}$ in., lanceolate at either end, spinous-toothed ; petiole 0–$\frac{1}{5}$ in. long ; spikes up to 4 by 2 in., 8–14-flowered ; bracts 1$\frac{1}{2}$–2 in. long, upper half lanceolate, recurved, with 9–13 long spines ; bracteoles $\frac{3}{4}$ by $\frac{1}{8}$ in. ; calyx nearly glabrous without, densely hairy within ; anticous segment $\frac{1}{2}$ in. long; corolla more than 1 in. long, bluish. *Engl. in Engl. Jahrb.* x. 260. *Acanthodium squarrosum, Nees in DC. Prod.* xi. 275 ; *Sonder in Linnæa,* xxiii. 93.

KALAHARI REGION: Griqualand West; Boetsap, 4000 ft., *Marloth,* 962 ex *Engler.* Orange River Colony; Great Vet River, *Zeyher,* 1404! *Burke!* Transvaal; Klerksdorp, *Nelson,* 307! between Mamusa and Homans Vley, *Holub,* 2028! 2029!

19. B. longispica (C. B. Clarke) ; hairy ; stems branched, stout, with internodes 1–4 in. long ; leaves approximated in fours, upper pair up to 4$\frac{1}{2}$ by $\frac{3}{4}$ in., spinous-toothed, when mature glabrescent, lower pair similar but rather smaller ; spikes terminal (on quasi-peduncles 1–4 in. long), up to 5 by 1$\frac{1}{2}$ in., often 20-flowered, densely hairy ; on the peduncles are a few distant lanceolate bracts about $\frac{3}{4}$ in. long ; bracts (of the spikes) 1$\frac{1}{4}$ in. long, ovate, shortly acute, with 15 spines ; bracteoles $\frac{3}{4}$ by $\frac{1}{10}$ in. ; calyx hairy within and without ; anticous segment $\frac{2}{3}$ in. long ; corolla 1$\frac{1}{4}$ in. long, white.

EASTERN REGION : Natal; near Estcourt, 3600 ft., *Wood,* 671! 3509! Colenso, 3100 ft., *Wood,* 903! *MacOwan, Herb. Aust-Afr.,* 1510! 3200 ft., *Schlechter,* 6683!

This plant bears a general resemblance to *B. hirtinervia* in the long leaves and long hairy spikes, but, of the four approximated leaves the two outer are not here reduced to spines. As to the few scattered bracts on the quasi-peduncles, I do not recall anything like them in the genus.

20. B. pruinosa (Engl. in Engl. Jahrb. x. 260 ?) ; minutely scabrous-pubescent or glabrate ; stems 8 in. long, rigid, internodes up to 1$\frac{1}{2}$ in. long ; leaves approximated in fours, the upper pair 1–1$\frac{3}{4}$ by $\frac{1}{4}$–$\frac{1}{3}$ in., oblong or subelliptic, subentire (i.e. without lobes), with scattered spinous teeth, minutely scabrous ; petiole 0–$\frac{1}{6}$ in. long, axillary spines (i.e. the basal bracts of the next year's spikes) $\frac{1}{2}$ in. long, linear, simple ; spikes 1–1$\frac{1}{2}$ in. long, axillary, 2–5-flowered ; bracts $\frac{3}{4}$–1 by $\frac{1}{4}$–$\frac{1}{3}$ in., acuminate into a lanceolate tip, glabrous, pruinose, nerves conspicuously white, spinous teeth 9 and up to $\frac{1}{5}$ in. long; bracteoles $\frac{2}{3}$ in. long, linear; anticous calyx-segment $\frac{2}{3}$ in. long; posticous sepal $\frac{3}{4}$ in. long, hairy within; corolla $\frac{3}{4}$–1 in. long, blue. *Lindau in Engl. & Prantl, Pflanzenfam.* iv. 3B, 318. *B. furcata, T. Anders. in Journ. Linn. Soc.* vii. 35, *partly.*

COAST REGION : Uitenhage Div.; Uitenhage, *Bowie,* 2!
EASTERN REGION : Griqualand East; Shawbury, on the Tsitsa River, 1500–1800 ft., *Baur,* 201! Natal; near Durban, *Sutherland!* and without precise locality, *Gerrard,* 1273!

The plant above described is Sutherland's. Engler's *B. pruinosa* is not known to me by any authentic specimen. The description appears to agree very well with Sutherland's plant; but Engler's *B. pruinosa* was Marloth, 1444, from Hereroland, a remote locality, with a totally different climate.

21. B. diversispina (C. B. Clarke in Dyer, Fl. Trop. Afr. v. 104);
pubescent; branches 8–18 in. long, internodes 1–4 in. long; leaves
often 4 subsimilar and approximated, 1–1½ by ⅓–½ in., elliptic-oblong,
hardly lobed, spine-toothed; spines in the axils (i.e. the lowest
bracts to the next year's spikes) simple, ½ in. long, needle-like;
spikes 1–1¼ in. long, ovoid, 2–7-flowered; bracts ⅔ by ½ in., with
9 spinous teeth, hairy without; bracteoles more than ½ in. long,
linear; anticous calyx-segment ½ in. long; posticous sepal ¾ in. long;
corolla ¾–1 in. long, blue. *B. procumbens, T. Anders. in Journ.
Linn. Soc.* vii. 35 *partly. Acanthodium diversispinum, Sonder in
Linnæa,* xxiii. 92; *var. β, Nees in DC. Prod.* xi. 275.

KALAHARI REGION: Bechuanaland; Maadji Mountain, *Burchell,* 2377!
Transvaal; *Holub!*
Also in Tropical Bechuanaland and Angola.

22. B. serrulata (Ficalho et Hiern. in Trans. Linn. Soc. ser. 2, Bot.
ii. 24); pubescent, harsh, woody, branched; leaves often 4 together
subsimilar, 1 by ⅓ in., oblong, obtuse, scarcely toothed, with numerous
closely-placed forward-pointing spinous serratures; spikes axillary,
1½ by ⅔ in., short-cylindric or ovoid, 4–8-flowered; lowest bracts
pinnatifid near the base; bracts (to the flowers) ⅔ by ⅓ in., obovate,
suddenly and shortly acuminate, densely and shortly hairy, with
9–13 marginal spinous cilia; bracteoles ½ in. long, linear; calyx
hairy within and without; anticous segment ⅓ in. long; corolla
¾ in. long; capsule ¼ in. long, 2-seeded. *C. B. Clarke in Dyer, Fl.
Trop. Afr.* v. 102. *B. procumbens, T. Anders. in Journ. Linn. Soc.*
vii. 35 *partly. Acanthodium serrulatum, Nees in DC. Prod.* xi.
275; *Sonder in Linnæa,* xxiii. 93. *A. diversispinum, var. a, Nees
in DC. Prod.* xi. 275.

KALAHARI REGION: Transvaal; Apies River, *Burke!* Magalies Berg, *Burke!*
Also in Angola.

23. B. subvolubilis (C. B. Clarke); minutely pubescent or
glabrescent; stems 6–12 in. long, branched, slender, flexuose, internodes 1–3 in. long; leaves often approximated in fours, subsimilar,
1½ by ¼–⅓ in., narrowly elliptic, not lobed, spine-toothed; bracts
1 in. long, the lowest sterile foliaceous, pinnatifid (not simple needle-
like); corolla 1¼ in. long; otherwise as *B. diversispina.*

KALAHARI REGION: Orange River Colony; Thaba Unchu, *Burke!* Transvaal;
lower slopes of mountains near Barberton, 2800 ft., *Galpin,* 881!

24. B. dilatata (C. B. Clarke); minutely scabrous-pubescent;
stems 6–10 in. long, rigid, internodes 1–2 in. long; leaves often
approximated in fours, subsimilar, ¾–1¼ by ⅙–⅓ in., elliptic or
narrowly obovate, hardly lobed, with few irregular spine-tipped
teeth; spikes 1½ by 1 in., axillary, 2–5-flowered; bracts 1 by ½ in.,
ovate, shortly acuminate, puberulous with 9 spinous cilia, lowest
(sterile) foliaceous, pinnatifid; bracteoles ⅔ in. long, linear; anticous

calyx-segment $\frac{1}{2}$ in. long; posticous sepal $\frac{3}{4}$ in. long, ovate-oblong upwards, wider at the top, truncate, somewhat scarious and coloured; corolla 1 in. long, blue.

Var. *β*, **explicatior**, C. B. Clarke; larger in all parts; leaves up to $1\frac{3}{4}$ in. long; bracts up to $1\frac{1}{2}$ in. long; bracteoles nearly 1 in. long; calyx more hairy.

COAST REGION: Stutterheim Div.; Kabousie River, 2100 ft., *Flanagan*, 593!

CENTRAL REGION: Graaff Reinet Div.; near Graaff Reinet, 2500 ft., *Bolus*, 46!

KALAHARI REGION: Transvaal; Hooge Veld, at Trigards Fontein, *Rehmann*, 6711! Var. *β*: Transvaal; at Apies Poort, *Rehmann*, 4099! 4100! in a journey to Johannesburg, *Miss Saunders*, 2!

EASTERN REGION: Griqualand East; near Clydesdale and Ibisi, 2500 ft., *Tyson*, 1152! bank of a small stream near Ibisi River, *Wood*, 3060! Natal; Zuur Berg, *Wood*, 879!

25. B. procumbens (Pers. Syn. ii. 180); puberulous and sparsely hairy; stems 6–18 in. long, much branched, procumbent, internodes up to 1 in. long; leaves often approximated in fours, subsimilar, $\frac{1}{2}$–$1\frac{1}{2}$ by $\frac{1}{12}$–$\frac{1}{6}$ in., linear or linear-oblong, deeply toothed, spinous, scarcely petioled; spikes axillary, ovoid, up to $1\frac{1}{2}$ by 1 in., 1–8-flowered; bracts 1 in. long, ovate, shortly acute, with recurved points, and with 15–25 very narrow closely placed softly hairy marginal spines; bracteoles $\frac{3}{4}$ by $\frac{1}{10}$ in.; calyx hairy within and without; anticous segment $\frac{2}{3}$–$\frac{3}{4}$ in. long; corolla exceeding 1 in. in length, blue; capsule $\frac{2}{3}$ in. long, 4-seeded, or much shorter and with 2 larger seeds and the 2 lower ovules infertile. *T. Anders. in Journ. Linn. Soc.* vii. 35 *partly; Hochst. in Flora*, 1845, 71; *Lindau in Engl. & Prantl, Pflanzenfam.* iv. 3B, 318. *B. glomerata, Poir. Encycl. Suppl.* i. 89. *Acanthus procumbens, Linn. f. Suppl.* 294; *Thunb. Prod.* 97, *and Fl. Cap. ed. Schult.* 456; *E. Meyer in Drège, Zwei Pflanzengeogr. Documente*, 127, 131, 161. *A. glomeratus, Lam. Encycl.* i. 23. *Acanthodium procumbens, Nees in Linnæa*, xv. 362, *and in DC. Prod.* xi. 273; *Sonder in Linnæa*, xxiii. 92. *Blepharacanthus procumbens, Drège in Linnæa*, xx. 200. *Blepharanthus procumbens, Hook. Journ. Bot.* ii. (1840) 126.

SOUTH AFRICA: without precise locality, *Masson! Oldenburg!*
COAST REGION: Uitenhage Div.; by the Sunday River and Coega River, *Drège!* between Zwartkops River and Coega River, *Krauss*, 1118, near the Zwartkops River, *Ecklon & Zeyher*, 242! Grassrug, *Baur*, 1021! Port Elizabeth Div.; Algoa Bay, *Cooper*, 3039! Humansdorp Div., *Thunberg! Niven*, 24! 88! Alexandra Div.; Bushman's River, *Bowie*, 4!
CENTRAL REGION: Somerset Div.; Sunday River, *Bowie*, 5!

26. B. inæqualis (C. B. Clarke); hairy; branches 4–9 in. long, with internodes 1–$2\frac{1}{2}$ in. long; leaves often approximated in fours, whereof the upper pair is $1\frac{1}{2}$–$2\frac{1}{2}$ by $\frac{1}{4}$ in., linear, with few remote scarcely spinous teeth, the lower pair similar but smaller (or sometimes much smaller and subentire); spikes axillary, often in the forks of the branches, ovoid, 2–5-flowered; bracts $\frac{2}{3}$ by $\frac{1}{3}$ in., obovate, shortly acute, hairy, with 11 marginal spines $\frac{1}{8}$–$\frac{1}{5}$ in. long; bracteoles $\frac{2}{3}$ in. long, linear; anticous calyx-segment $\frac{1}{3}$ in. long; posticous sepal $\frac{3}{4}$ in. long, hairy within; corolla 1 in. long, blue.

KALAHARI REGION: Transvaal; Houtbosch Berg, *Nelson*, 419! Krans Kop, near Modimulle, on the Nyl River, *Nelson*, 126! Pietersburg, 4700 ft., *Schlechter*, 4352!

27. B. sinuata (C. B. Clarke); minutely pubescent; barren branches 7 in. long, with internodes 1–4 in. long; leaves often approximated in fours, subsimilar, 1–2$\frac{1}{2}$ by $\frac{1}{4}$ in., narrowly oblong, sinuate-pinnatifid, spinous-toothed; spikes 2 by 1$\frac{1}{2}$ in., on short peduncles near the crown of the root, ovoid, 3–5-flowered; bracts 1$\frac{1}{2}$ by $\frac{3}{4}$ in., ovate, shortly acuminate, puberulous, with 11 spines up to $\frac{1}{4}$ in. long; bracteoles nearly 1 in. long, linear; anticous calyx-segment $\frac{2}{3}$ in. long; posticous sepal 1 in. long; corolla 1$\frac{1}{4}$ in. long, blue. *Acanthodium sinuatum, Nees in Linnæa*, xv. 362, *and in DC. Prod.* xi. 274. *Acanthus humilis, Vahl ex Nees in DC. Prod. l.c.*

COAST REGION: Albany Div.; Grahamstown, *Ecklon;* Flats near Grahamstown, 2000 ft., *MacOwan*, 215! Bothas Berg, near Grahamstown, 800–1000 ft, *Baur*, 1095!

X. ACANTHOPSIS, Harv.

Bract obovate, truncate, terminated by 5–3 narrow-lanceolate toothed or compound spines. *Calyx-segments:* anticous of 2 sepals connate nearly to the tip, ovate-triangular, 10-nerved; posticous ovate-triangular, 11-nerved. *Stamens* 4; 2 anticous filaments without a process at the top. *Style* with 1 linear branch, the other very small or 0; no hollows at the style-base; otherwise as *Blepharis.*

Small, prickly; leaves linear-oblong, toothed, spinous; spikes dense; capsule generally 2-seeded.

Species, the 7 following, endemic in South Africa.

Branches short; internodes less than $\frac{1}{4}$ in. long:
　Bracts without secondary spines at the top
　　within:
　　　Bracts 5-fid at the top ...   .   (1) **carduifolia.**
　　　Bracts 3-fid at the top ...   .   (2) **glauca.**
　Bracts with secondary spines at the top
　　within:
　　　Bracts with white hairs, not viscid   ...   (3) **Disperma.**
　　　Bracts viscid with gland-tipped hairs ...   (4) **hoffmannseggiana.**
Branches 6–12 in. long; internodes up to 1 in.
　long:
　　Spikes cylindric:
　　　Bracts 5-fid at the top, softly hairy   ...   (5) **horrida.**
　　　Bracts 3-fid at the top, sparingly
　　　　hairy   ...   .   (6) **trispina.**
　　Spikes globose ...   .   (7) **spathularis.**

1. A. carduifolia (Schinz in Verh. Bot. Ver. Brandenb. xxxi. 200); villous (but see var. *β*); stem 1 in. long, woody, densely leafy; leaves 3 by $\frac{1}{3}$ in., pinnatifid, doubly spinous-toothed; spike 2–3 by $\frac{1}{2}$ in.; bract (without the spines) $\frac{1}{2}$ by $\frac{1}{3}$ in., obovate, truncate,

crowned by 5 narrow-lanceolate spinous teeth $\frac{1}{4}$–$\frac{1}{3}$ in. long, without
any secondary (apparently inner) spines round the top; bracteoles
$\frac{1}{2}$ in. long, linear; calyx hairy without and within; anticous segment
$\frac{1}{3}$ in. long; corolla $\frac{3}{4}$ in. long. *Lindau in Engl. & Prantl,
Pflanzenfam.* iv. 3B, 319. *Acanthus carduifolius, Linn. f. Suppl.*
294; *Thunb. Prod.* 97, *and Fl. Cap. ed. Schult.* 455; *E. Meyer in
Drège, Zwei Pflanzengeogr. Documente,* 69, 91, 161. *Acanthodium
carduifolium, Nees in DC. Prod.* xi. 278. *Blepharis carduifolia,
T. Anders. in Journ. Linn. Soc.* vii. 35.

Var. β, glabra (C. B. Clarke): leaves and bracts less hairy, in age glabrescent.
Acanthus glaber, E. Meyer in Drège. Zwei Pflanzengeogr. Documente, 67, 161.
Acanthodium glabrum, Nees in DC. Prod. xi. 278. *Blepharis carduifolia, var.
glabra, T. Anders. in Journ. Linn. Soc.* vii. 35; *Schinz in Verh. Bot. Ver.
Brandenb.* xxxi. 200.

WESTERN REGION: Little Namaqualand; Kamies Berg and Rood Berg,
3000–4000 ft., *Drège!* Kaus Mountains, *Drège!* Calvinia Div.; Hantam, ex
Nees. Var. β: Little Namaqualand; Kamies Berg, 3000–4000 ft., *Drège!*

2. **A. glauca** (Schinz in Verh. Bot. Ver. Brandenb. xxxi. 201);
nearly stemless and glabrous; leaves linear-lanceolate, remotely
spinous-toothed; spike terminal, cylindric, elongate; bracts obovate-
cuneate, bearing 3 terminal spines; 2 larger calyx-segments ovate-
oblong, silky. *Lindau in Engl & Prantl, Pflanzenfam.* iv. 3B, 319.
Acanthus glaucus, E. Meyer in Drège, Zwei Pflanzengeogr. Documente,
91, 161, *and in Linnæa,* xx. 200. *A. glaucescens, E. Meyer in Drège,
Zwei Pflanzengeogr. Documente,* 95. *Acanthodium glaucum, Nees in
DC. Prod.* xi. 277. *Blepharis glauca, T. Anders. in Journ. Linn.
Soc.* vii. 36.

SOUTH AFRICA: without precise locality. *Masson!*
WESTERN REGION: Little Namaqualand; Uitkomst, 2000–3000 ft., *Drège,*
Kook Fontein, 3000–4000 ft., *Drège.*

3. **A. Disperma** (Nees in DC. Prod. xi. 278); stemless; leaves
$3\frac{1}{2}$ by $\frac{3}{4}$ in., sinuate, spinous-toothed, minutely hairy; spikes 3–5,
sessile on the crown of the root, $2\frac{1}{4}$ by $1\frac{1}{4}$ in., very dense, resembling
teazel heads; bracts (without the teeth) $\frac{1}{4}$ by $\frac{1}{3}$ in., primary teeth 5,
lanceolate, $\frac{1}{4}$–$\frac{1}{3}$ in. long, pinnatifid-spinous, with 3–4-celled white
hairs, secondary teeth up to $\frac{1}{3}$ in. long apparently forming an inner
row, spinescent; bracteoles $\frac{1}{2}$ in. long, linear, hairy; calyx hairy
without and within; anticous segment $\frac{1}{3}$ in. long; corolla $\frac{3}{4}$ in. long,
bluish; lip narrow-obovate, obtusely 3-lobed; capsule $\frac{1}{5}$ in. long,
ellipsoid, flattened, shining-brown, 2-seeded; seeds covered with
bundles of long hairs, each spirally thickened within, elastically
separating when wetted. *T. Anders. in Journ. Linn. Soc.* vii. 37;
Schinz in Verh. Bot. Ver. Brandenb. xxxi. 201; *Lindau in Engl.
& Prantl, Pflanzenfam.* iv. 3B, 319, 281, *fig.* 110, *D. Acanthodium
dipsaceum, E. Meyer in Drège, Zwei Pflanzengeogr. Documente,* 92,
161. *Acanthopsis (genus), Harv. in Hook. Lond. Journ.* i.
(1842) 28.

WESTERN REGION: Namaqualand; between Holgat River and Orange River,
000–1500 ft., *Drège!* Groot (Gamtoos) River, *Niven,* 25!

The specific name "*dipsaceum*" was printed twice by E. Meyer, and refers to the resemblance of the spikes to the heads of teazel. In founding the genus *Acanthopsis*, Harvey created no species, but he mis-cited "*Acanthodium dispermum*"; succeeding authors have not gone further back than this.

4. A. hoffmannseggiana (C. B. Clarke); leaves subpinnatifid, doubly spinous-toothed ; bracts viscid with gland-tipped hairs besides short white hairs; otherwise as *A. Disperma*, Nees. *A. carduifolia, Schinz in Verh. Bot. Ver. Brandenb.* xxxi. 200 *partly. Acanthodium hoffmannseggianum, Nees in DC. Prod.* xi. 277 ; *Sonder in Linnæa,* xxiii. 93. *Blepharis carduifolia, T. Anders. in Journ. Linn. Soc.* vii. 35 *partly.*

KALAHARI REGION : Griqualand West, Hay Div. ; Kloof Village in the Asbestos Mountains, *Burchell,* 1654! Prieska Div. ; Zand Valley, *Burchell,* 1632/2 ! Griqua Town, *Mrs. Orpen in Bolus Herb.,* 6476 !

5. A. horrida (Nees in DC. Prod. xi. 278); branches 4–10 in. long, stout, divided, grey puberulous from minute white deflexed hairs ; internodes ¼–1 in. long ; leaves 2–2½ by ¼–⅓ in., coarsely sinuate-toothed, strongly spinous, most minutely pubescent ; spikes terminal, subsessile, up to 2½ by 1 in., dense, softly hairy ; bract (without the spines) ¼ by ⅓ in. ; primary terminal spines 5, ⅓–½ in. long, pinnatifid, with many long white hairs ; secondary spines much shorter ; calyx, corolla, capsule and seeds nearly as of *A. Disperma. T. Anders. in Journ. Linn. Soc.* vii. 36 ; *Schinz in Verh. Bot. Ver. Brandenb.* xxxi. 201 ; *Lindau in Engl. & Prantl, Pflanzenfam.* iv. 3B, 319. *Dilivaria horrida, Nees in Linnæa,* xv. 363 ; *Drège, in Linnæa,* xx. 200. *Acanthodium plumosum, E. Meyer in Drège, Zwei Pflanzengeogr. Documente,* 67, 161. *A. plumulosum, E. Meyer ex Nees in DC. Prod.* xi. 278.

WESTERN REGION : Little Namaqualand ; Kamies Berg, 3000–4000 ft., *Drège.*

6. A. trispina (C. B. Clarke); bract (without the teeth) ¼ by ⅙ in. ; terminal primary spines 3, ⅓ in. long, scarcely toothed, sparingly white hairy ; otherwise as *A. horrida.*

SOUTH AFRICA : without precise locality, *Mund !*

7. A. spathularis (Schinz in Verh. Bot. Ver. Brandenb. xxxi. 201) ; shrubby, erect ; stem silky-hoary ; leaves lanceolate, sinuate-toothed, spinous, velvety ; spikes lateral and terminal, subsessile, globose ; bracts broad wedge-shaped, truncate, with 5 terminal spines and strong intermediate spines, central spine of the upper bracts foliaceous spathulate-lanceolate mucronate. *Lindau in Engl. & Prantl, Pflanzenfam.* iv. 3B, 319. *Acanthus spathularis, E. Meyer in Drège, Zwei Pflanzengeogr. Documente,* 91, 161. *Acanthodium spathulare, Nees in DC. Prod.* xi. 277. *Blepharis spathularis, T. Anders. in Journ. Linn. Soc.* vii. 35.

WESTERN REGION : Little Namaqualand ; Silver Fontein, 2000–3000 ft. *Drège.*

No example seen ; the above description copied from Nees.

XI. ACANTHUS, Linn.

Anticous *filaments* not bifid, i.e. without a process at the top. *Style* without glandular hollows at the base; branches 2, most minute, subequal. *Seeds* without hairs, reticulated like brain-coral; otherwise as *Blepharis* or *Acanthopsis*.

Usually larger plants with larger flowers than the two preceding genera.

Species 10; from the Mediterranean to South Africa, South-east Asia and Polynesia.

1. **A. ilicifolius** (Linn. Sp. Pl. ed. ii. 892); glabrous, except the minutely ciliate calyx; stems 2–4 ft. long; leaves 4–6 by 2–2½ in., usually coarsely pinnatifid spinous-toothed like holly-leaves (but plants without spines occur); spikes 3 by 1–1½ in.; bract and bracteoles ¼–⅓ in. long, ovate; calyx ½–⅔ in. long; corolla 1–1⅔ in. long, blue; capsule up to 1¼ by ⅔ in.; seeds usually 4. *T. Anders. in Journ. Linn. Soc.* vii. 36; *C. B. Clarke in Hook. f. Fl. Brit. Ind.* iv. 481, *and in Dyer, Fl. Trop. Afr.* v. 108; *Lindau in Engl. & Prantl, Pflanzenfam.* iv. 3B, 319. *Dilivaria ilicifolia, Juss. Gen.* 103; *Nees in Linnæa,* xv. 363, *and in DC. Prod.* xi. 268; *Wight, Ic. Pl. t.* 459.

COAST REGION: Uitenhage, *Wendemann*. Introduced.

A native of the Seashores of India, Malay Archipelago, and Australia.

XII. SCLEROCHITON, Harv.

Bract elliptic or obovate, obtuse; bracteoles 2, nearly as ong as the bract, oblong. *Sepals* 5, free (or very nearly so), longer than the bracts, elliptic or lanceolate, inner narrower. *Corolla* short-tubed, 1-lipped; lip obovate with 3–5 rounded lobes. *Stamens* 4; anthers short, ovoid-triangular, 1-celled, a row of short hairs near the slit; filaments not branched at the top; pollen minute, short ellipsoid with 3 longitudinal chinks. *Style* without glandular hollows at the base; branches 2, subequal, short; ovules 2 or 1 in each cell. *Capsule* (where known) 2-seeded; seeds without hairs.

Without spines; flowers in spikes.

Species 9; in Tropical and South Africa.

1. **S. harveyanus** (Nees in DC. Prod. xi. 279); undershrub 2–6 ft. high; branches densely pubescent or hairy; leaves 1–1½ by 1 in. (or in some examples hardly half so large), elliptic or ovate, subobtuse, glabrescent, entire or wavy crenulate; petiole 0–⅙ in. long, hairy; flowers axillary, solitary, or running into spikes at the ends of branches; bract ¼ in. long, elliptic, obtuse; bracteoles ¼ in. long, narrower; sepals ⅓ in. long, rigid, minutely pubescent, obtuse, outer 3 obovate or elliptic, inner 2 oblong; corolla ¾ in. long, purplish-blue or white; capsule ⅓ by ⅕ in., ellipsoid, rigid, smooth, 2-seeded; seeds without hairs. *T. Anders. in Journ. Linn. Soc.* vii. 37; *Harv. Thes. Cap.* ii. 28, *t.* 145; *Lindau in Engl. & Prantl,*

Pflanzenfam. iv. 3B, 316. *Sclerochiton* (*genus*), *Harv. in Hook. Lond. Journ. Bot.* i. (1842) 27.

SOUTH AFRICA: without precise locality, *Drège*, 4037!
CENTRAL REGION: Somerset Div.; on the Bosch Berg, 3500 ft., *MacOwan*, 538!
KALAHARI REGION: Transvaal; in wooded ravines at Moodies, near Barberton, 4000 ft., *Galpin*, 964!
EASTERN REGION: Transkei, *Mrs. Barber*, 29! 36! Pondoland; Shawbury, in forests, 1800 ft., *Baur*, 200! Natal; Inanda, *Wood*, 1212! and without precise locality, *Gerrard*, 1678! *McKen*, 5!

XIII. CROSSANDRA, Salisb.

Bract large, ovate; bracteoles linear, nearly as long as the bract. *Calyx* 5-partite to the base; 2 anticous segments oblong, acute, 1-nerved; posticous segment broader, 2-nerved, often 2-toothed; 2 inmost segments lanceolate, rather shorter. *Corolla* orange or red; tube long, linear; limb of 5 segments imbricate in bud. *Stamens* 4, subsimilar; anthers 1-celled; pollen oblong, with 3 longitudinal (often very slender) chinks. *Ovary* oblong, with 2 ovules in each cell; stigma obscurely 2-fid. *Capsule* oblong-ellipsoid, usually 4-seeded. *Seeds* with fringed scales or tufts of hairs which are slightly hygroscopic.

Shrubs; leaves entire; flowers in strobilate spikes.

Species 14, African or Mascarene, 1 of these extending to India.

1. C. Greenstockii (S. Moore in Journ. Bot. 1880, 37); pubescent; stems 6–12 in. long; leaves usually 4–6 in. long, narrowly elliptic, tapering to the base, nearly glabrous beneath when mature; peduncles 0–6 in. long; spikes 1–6 by ¾–1 in., dense; bract ¾ in. long, herbaceous, ovate, hairy, margins with several innocuous teeth; bracteoles linear, nearly as long as the bract; posticous calyx-segment ¼ in. long, 2-nerved, ending in 2 spinous hairy teeth; corolla red (orange to bright scarlet, from notes of collectors); tube ¾ in. long; segments ½ in. long, turned to one side; anthers slightly hairy; pistil glabrous, except for a few hairs towards the base of the style; capsule ⅔ by ¼ in., 4-seeded. *Lindau in Engl. & Prantl, Pflanzenfam.* iv. 3B, 319, *and in Engl. Pfl. Ost-Afr. C.* 370; *C. B. Clarke in Dyer, Fl. Trop. Afr.* v. 113.

SOUTH AFRICA: without precise locality, *Mrs. Saunders*, 167 (*in Wood Herb.*, 3894)!
KALAHARI REGION: Transvaal; Pilgrims Rest, *Greenstock!* near Pretoria, *MacLea in Bolus Herb.*, 3388! near Lydenburg, *Atherstone! Wilms*, 1207! Bosch Veld, at Menaars Farm, *Rehmann*, 4378! near Barberton, 2500–3000 ft., *Galpin*, 491! Crocodile River, 5000 ft., *Schlechter*, 3978! Olifants River District, Hartebeest River, *Nelson*, 405! Nyl River near Makapans Kraals, *Nelson*, 92!
EASTERN REGION: Natal; Tugela River, *Gerrard*, 1679!

Also frequent in Tropical East Africa.

In Gerrard, 1679, the leaves are small (1½ in. long), hairy beneath; the calyx-segments are long-aristate at the tip, the posticous segment 2-aristate. This plant has been set aside as "distinct," but there are similar (and intermediate) forms in Tropical Africa.

XIV. CRABBEA, Harv. partly.

Bracts large, ovate or oblong, often spine-toothed; bracteoles 0. *Calyx* divided nearly to the base; segments 5, linear, aristate, with long unicellular hairs; posticous segment more lanceolate at the base. *Corolla* $\frac{1}{3}$–$\frac{3}{4}$ in. long, white or pale pink; tube cylindric, funnel-shaped upwards; anticous lip 3-lobed, folded down over the other in bud. *Stamens* 4, subsimilar; anther-cells 2, at nearly equal height, muticous at the base; pollen globose, reticulate. *Ovary* glabrous; ovules 4–2 in each cell; style glabrous, articulated on the ovary; 1 stigmatic lobe a short oblong plate, the other absent. *Capsule* small, narrow-oblong, 8–4-seeded; seeds discoid, covered with hygroscopic hairs.

Low shrubs; hairs simple, leaves entire, obtuse; inflorescence in dense compound axillary or peduncled heads; floral leaves (bracts of authors) ovate, outer very large empty, inner containing apparently a cluster of 3–5 flowers, which really form a condensed unilateral raceme; proper bract to each flower linear-lanceolate or 0.

Species 8, in South and East Tropical Africa, very closely allied.

The stems often grow on, above the highest axil that carries a head, the leaves above the head being much narrower than those below.

Floral leaves with innocuous marginal spines :
  Leaves ovate or oblong :
    Leaves 1–4 in. long :
      Outer floral leaves papery, minutely
       hairy ... ...   (1) **nana.**
      Outer floral leaves herbaceous, softly
       hairy ... ...   (2) **hirsuta.**
      Leaves 6 in. long ... ...   (3) **robusta.**
    Leaves, at least the upper, linear ... ...   (4) **angustifolia.**
  Floral leaves without marginal spines; heads peduncled :
    Leaves elliptic or somewhat obovate ... ...   (5) **pedunculata.**
    Leaves linear ... ...   (6) **Galpinii.**

1. **C. nana** (Nees in DC. Prod. xi. 162); branches 2–12 in. long, trailing, flexuose; leaves up to 4 by 1$\frac{1}{4}$ in., oblong-elliptic, when mature scabrid on the nerves beneath or nearly glabrous, base narrowed : petiole 0–$\frac{1}{6}$ in. long; heads 1–2 in. in diam., sessile or short-peduncled; outer floral leaves 1$\frac{1}{3}$ by $\frac{1}{2}$ in., acute, finally glabrescent, papery, marginal spinescent teeth $\frac{1}{6}$–$\frac{1}{4}$ in. long; calyx-segments $\frac{1}{2}$ in. long, usually thinly hairy at the top, sometimes densely hairy to the base; corolla $\frac{3}{4}$ in. long; capsule $\frac{1}{2}$ in. long. *T. Anders. in Journ. Linn. Soc.* vii. 32; *Lindau in Engl. & Prantl, Pflanzenfam.* iv. 3B, 313; *C. B. Clarke in Dyer, Fl. Trop. Afr.* v. 118. *C. cirsioides, Nees in DC. Prod.* xi. 163. *Ruellia nana, Nees in Linnæa,* xv. 355; *E. Meyer in Drège, Zwei Pflanzengeogr. Documente,* 135, 141. *R. cirsioides, Nees in Linnæa,* xv. 354; *Drège in Linnæa,* xx. 200.

COAST REGION : Humansdorp Div.; between Galgebosch and Melk River, *Burchell,* 4765! Alexandria Div.; Zuurberg Range, 2000–3000 ft., *Drège!* Lower Albany; Glenfilling, 1000 ft., *Drège!* Queenstown Div. (Tambukiland); *Zeyher!*

KALAHARI REGION : Orange River Colony; Bloemfontein, *Rehmann*, 3838 !
Seven Fountains, *Burke! Vet River, Burke! Vaal River, Burke!* Transvaal;
Black Kopies, 4700 ft., *Schlechter*, 4174 ! Lydenburg, *Wilms*, 1205 !

EASTERN REGION : Tembuland ; grassy places near Bazeia Mountains,
2000 ft., *Baur*, 65 ! Griqualand East; on the sides of mountains near Matatiele,
5000 ft., *Tyson*, 1633 ! Kokstad, 4500 ft., *Tyson*, 1158 ! Natal; bank of the Tugela
River, 4000 ft., *Wood*, 3635 ! near Colenso, 3000 ft., *Wood*, 4423 ! and
without precise locality, *Sutherland!*

Also in Tropical Africa.

In the type of *C. nana*, the branches have not grown on much above the sessile
heads of flowers ; in *C. cirsioides* the branches continued beyond the heads have
longer internodes and narrower leaves.

2. **C. hirsuta** (Harv. in Hook. Lond. Journ. Bot. i. (1842) 27) ;
mature leaves hairy on the lower surface ; heads sessile or shortly
peduncled ; outermost floral leaves herbaceous with long white hairs;
otherwise as *C. nana*, Nees. *Nees in DC. Prod.* xi. 163 ; *Lindau in
Engl. Jahrb.* xviii. 63, *t.* 1, *fig.* 28, *and in Engl. & Prantl, Pflanzenfam.* iv. 3B, 313 ; *T. Anders. in Journ. Linn. Soc.* vii. 32 ; *C. B.
Clarke in Dyer, Fl. Trop. Afr.* v. 119. *C. ovalifolia, Fic. et Hiern
in Trans. Linn. Soc. ser.* 2, *Bot.* ii. 24, *t.* 6, *fig.* A.

COAST REGION : Alexandria Div. ; Zuurberg Range, 2000 ft., *Bolus*, 9124 !
Komgha Div. ; Kei River mouth, *Flanagan*, 1163 !

KALAHARI REGION : Transvaal ; hill-sides near Barberton, 3000–3500 ft.,
Galpin, 856 ! 1303 ! hill-sides near Johannesberg, 5000 ft., *Galpin*, 1399a !
near Lydenberg, *Wilms*, 1206 ! Bosch Veld, between Elands River and Klippan,
Rehmann, 5050 !

EASTERN REGION : Transkei ; between Gekau (Geua or Geun) River and
Basche River, 1000–2000 ft., *Drège !* Griqualand East; near Clydesdale,
2500–3000 ft., *Tyson*, 2799 ! *MacOwan & Bolus, Herb. Norm. Aust.-Afr.*, 762 !
828 ! Natal; near Maritzburg, *Wood*, 3170 ! Clairmont, 100 ft., *Wood*, 646 !
Maxwell, Ixopo, 4000–5000 ft., *Evans*, 656 ! and without precise locality, *Cooper*,
3041 ! *Peddie! Williamson !*

The ordinary habit of this species is to have trailing flexuose branches 8–12 in.
long, exactly as in the form "*cirsioides*" of *C. nana*, without heads of flowers in
the upper axils; this is *C. ovalifolia*, as correctly figured l.c., the leaves in
which are not strictly ovate, i.e. they are not wider towards the base. Galpin's
1399a has narrowly oblong leaves, and is intermediate between *C. hirsuta* and
C. angustifolia.

3. **C. robusta** (N. E. Brown) ; branch ⅛ in. in diam. ; apparently
1–2 ft. long, thinly hirsute; leaves up to 6 by 2½ in., oblique
elliptic, obtuse, narrowed at the base, when mature nearly glabrate
beneath except the midrib; petioles ½ in. long ; heads in
appearance terminal on stout peduncles 1–3 in. long ; outer floral
leaves 1½ by ¼–⅓ in., elongate narrow triangular, slightly hairy,
marginal teeth weak, hardly ⅙ in. long; proper bracts and calyx
nearly as in *C. hirsuta*, with many hairs ; corolla 1 in. long.

EASTERN REGION : Swaziland; Horo Concession, 2000 ft., *Galpin*, 1265 !

4. **C. angustifolia** (Nees in DC. Prod. xi. 163) ; lower leaves
narrowly oblong, up to ⅔ in. broad, sterile shoots produced beyond
the heads with linear leaves 3 by ⅛–¼ in., often undulate subcrenulate

on the margins; otherwise as *C. hirsuta,* Nees. *Harv. Thes. Cap.* i. 40, *t.* 64; *T. Anders. in Journ. Linn. Soc.* vii. 32; *Engl. in Engl. Jahrb.* x. 263. *C. undulatifolia, Engl. in Engl. Jahrb.* x. 263; *Lindau in Engl. Prantl, Pflanzenfam.* iv. 3B, 313.

SOUTH AFRICA: without precise locality, but probably from the Magalies Berg, *Zeyher,* 1412!

KALAHARI REGION: Griqualand West; Groot Boetsap. 4000 ft., *Marloth,* 985! Bechuanaland; Mafeking, 4200 ft., *Bolus,* 6413! Groot Fontein, 4000 ft., *Marloth,* 1079; Transvaal; Magalies Berg, *Burke,* 405!

5. **C. pedunculata** (N. E. Brown); stems 2–6 in. long; leaves 2–5½ in. long, narrowly or broadly elliptic, obtuse, narrowed to the base, when mature nearly glabrous beneath; petioles 0–⅓ in. long; peduncles 1–4½ in. long; outer floral leaves 1 by ½–⅔ in., ovate, acute, more or less hairy, without marginal spines; flowers nearly as in *C. nana.*

EASTERN REGION: Natal; Inanda, *Wood,* 365! Krantz Kloof, 5000 ft., *Schlechter,* 3210! and without precise locality, *Sanderson,* 466!

6. **C. Galpinii** (C. B. Clarke); branches 3 in. long; leaves 2 by ⅕ in., linear; peduncle 1 in. long; heads small; outer floral leaves ¾ by ⅓ in., ovate, acute, sparsely hairy, without marginal spines; corolla white, scarcely ½ in. long.

KALAHARI REGION: Transvaal, Barberton, 2800 ft., *Galpin,* 1148!

This may possibly prove to be a variety of *C. pedunculata* from which it differs as *C. angustifolia* differs from *C. hirsuta.* The heads, floral leaves, and flowers are considerably smaller than in any other species of *Crabbea,* but the single branch perhaps is only a "depauperated" example.

XV. GLOSSOCHILUS, Nees.

Calyx 5-partite to the base; segments linear. *Corolla* medium-sized; tube gradually enlarged from the base much swollen at the top; limb of 5 segments (not contorted in the bud) evidently 2-lipped; anticous lip 3-lobed, the central lobe ovate, twice as broad as the lateral. *Stamens* 4; anther-cells 2, one much below the other, both minutely mucronate at the base; pollen nearly globose with 2 large pores and indistinct bands. *Ovary* with 2 ovules in each cell, glabrous except near the style-base; style-branches 2, subequal. *Capsule* narrowly-ellipsoid, shortly stalked, 4-seeded. *Seeds* much flattened, smooth, the thickened margins without hairs.

A small shrub; leaves narrow, glabrescent; flowers solitary, axillary, a few approximated towards the ends of the branches; the bract, or floral leaf, is a reduced upper leaf; bracteoles 0.

Species 1, endemic.

The material is small, and the æstivation of the corolla not certain (though, *fide* Bentham, not contorted); the anticous lip is probably wholly outside in the bud. The genus differs by its ovules entirely from *Ruelliopsis* and *Ruellia,* and by its seeds from *Dyschoriste.* If the corolla should really prove to be contorted it cannot be placed in any one of these genera.

1. **G. Burchellii** (Nees in DC. Prod. xi. 83); branches 4–6 in. long, woody, glabrate; leaves $\frac{3}{4}$ by $\frac{1}{8}-\frac{1}{6}$ in., narrowly cuneate-oblong, obtuse, glabrate; petiole $0-\frac{1}{8}$ in. long; bract $\frac{2}{3}$ by $\frac{1}{12}$ in.; calyx-segments exceeding $\frac{1}{4}$ in. in length, densely white-hairy on the margins; corolla $\frac{3}{4}$ in. long; tube nearly $\frac{1}{2}$ in. long, middle lobe of the anticous lip more than $\frac{1}{4}$ in. broad at the base; capsule $\frac{1}{3}-\frac{1}{2}$ in. long. *Lindau in Engl. & Prantl, Pflanzenfam.* iv. 3B, 312.

KALAHARI REGION: Griqualand West, Hay Div.; plains between Griqua Town and Witte Water, *Burchell*, 1976! Bechuanaland; near Kuruman, *Burchell*, 2434! near the source of the Kuruman River, *Burchell*, 2471! Hamapery, near Kuruman, *Burchell*, 2439!

XVI. **ASYSTASIA**, Blume.

Bracts small, or in *A. Schimperi* large; bracteoles minute or 0, or in *A. Schimperi* longer than the calyx. *Calyx* small, divided to the base; segments 5, equal, linear. *Corolla* curved, one-sided but not (or most obscurely) 2-lipped; tube long, at the base linear or narrow-cylindric, in the upper half much inflated (except in *A. stenosiphon*); segments 5, not very unequal; anticous wholly outside in the bud. *Stamens* 4, perfect, subsimilar; anther-cells oblong or linear, at equal height or one a little lower than the other, muticous or minutely tailed at the base; pollen oblong or subglobose, with longitudinal smooth bands or grooves which reach (or nearly reach) the poles, stopples 2 or 3. *Ovary* with 2 ovules in each cell; stigmas 2, very small, subequal, oblong. *Capsule* 4-seeded at the top on a long solid stalk, flattened laterally; seeds flattened, wrinkled or tubercled, without hygroscopic hairs.

Herbs or shrubs; leaves entire, often wavy or irregularly toothed on the margin; inflorescence various.

Species 35, in the warmer parts of the Old World.

The genus is here defined by the corolla not contorted nor 2-lipped, the 4 perfect stamens, the capsule 4-seeded at the top. *Mackaya* is kept separate, for convenience, as having only 2 fertile stamens. Lindau has put *Mackaya* in a separate tribe on pollen characters only. The structure of the pollen appears to me essentially the same in *Asystasia* and in *Mackaya*, the question whether the smooth longitudinal bands reach the pole or not is (for me) not essential, nor always easy to see. Moreover, in *A. natalensis*, *A. stenosiphon* and others, the pollen is that of *Mackaya* rather than of *A. coromandeliana*.

Racemes terminal, loose; bracts inconspicuous:
  Stems and capsules pubescent:
    Lower leaves petioled, ovate ... ... (1) **coromandeliana.**
    Lower leaves subsessile, oblong ... ... (2) **natalensis.**
  Stems and capsules glabrous or very nearly so:
    Corolla 1 in. long; tube very narrow ... (3) **stenosiphon.**
    Corolla 1$\frac{1}{2}$ in. long; tube inflated upwards ... ... (4) **varia.**
Spikes terminal, dense; bracts much exceeding the calyx ... ... (5) **Schimperi.**
Peduncles axillary, long, slender with about 2 flowers ... ... (6) **subbiflora.**

1. A. coromandeliana (Nees in Wall. Pl. As. Rar. iii. 89);
sparsely pubescent, perennial; branches 1–3 ft. long; leaves 3 by
1 in., ovate, suddenly narrowed at the base; petiole $0-\frac{1}{3}$ in. long;
racemes 2–6 in. long, one-sided, terminal or in a terminal panicle;
lower flowers distant; pedicels up to $\frac{1}{12}-\frac{1}{6}$ in. long; bracts minute;
sepals $\frac{1}{6}$ in. long, linear-lanceolate, pubescent; corolla from lurid
purple to pale yellow, $\frac{2}{3}-1\frac{1}{4}$ in. in total length; tube broadly
cylindric, inflated for $\frac{2}{3}$ its length; anthers muticous; ovary and
style-base pubescent; capsule 1 in. long, pubescent; seeds $\frac{1}{8}$ in.
in diam. *Nees in DC. Prod.* xi. 165; *C. B. Clarke in Hook. f.
Fl. Brit. Ind.* iv. 493, *and in Dyer, Fl. Trop. Afr.* v. 131; *Schinz
in Mém. Herb. Boiss.* x. 64. *A. gangetica, T. Anders. in Thwaites,
Enum. Pl. Zeyl.* 235, *and in Journ. Linn. Soc.* vii. 52; *Lindau in
Engl. Jahrb.* xviii. 63, *t. i. fig.* 49, *in Engl. & Prantl, Pflanzenfam.*
iv. 3B, 326, *fig.* 131, *and in Engl. Pfl. Ost-Afr. C.* 370; *Schinz in
Mém. Herb. Boiss.* x. 64. *A. capensis, Nees in Linnæa,* xv. 356,
and in DC. Prod. xi. 167. *Justicia gangetica, Linn. Amœn. Acad.*
iv. 299. *Ramusia nyctaginea, E. Meyer in Drège, Zwei Pflanzengeogr.
Documente,* 150, 160. 215. *Dyschoriste biloba and Ruellia biloba,
Hochst. in Flora,* 1845, 72.

COAST REGION: Ceres Div.; Mosterts Hoek, 3500 ft., *Schlechter*, 317!
EASTERN REGION: Pondoland; St. Johns River, below 1000 ft., *Drège!*
Port St. John, 50 ft., *Galpin,* 3402! Natal; Inanda, *Wood,* 147! Coastland,
Sutherland! near Durban, 150 ft., *Wood in MacOwan & Bolus Herb. Norm.
Aust.-Afr.,* 1006! *Krauss,* 262! *Drège!* and without precise locality, *Peddie!
Plant,* 55! *Grant!* Delagoa Bay, *Forbes! Junod, Kuntze.*

Widely spread throughout Africa and India.

This is a weed that has been divided into numerous species, the corolla varying
much in size and colour. The pollen is oblong-ellipsoid, banded, with 2 stopples,
and is correctly figured by Lindau in his picture cited. The description of the
pollen of the genus *Asystasia* by me in Dyer, Fl. Trop. Afr. v. 130, is taken
from other species there included in the genus, and is not correct for *A.
coromandeliana.*

2. A. natalensis (C. B. Clarke); branches quadrangular, pubescent;
upper internodes $3\frac{1}{2}$ in. long; stem-leaves 2 by $\frac{1}{2}$ in., oblong-
lanceolate, obtuse, subsessile; inflorescence as of *A. coromandeliana,*
but much more densely pubescent; corolla $\frac{3}{4}$ in. long, rather slender;
stamens 4, fertile, subsimilar; anthers oblong, at equal height,
muticous; pollen subglobose, bands not reaching the pole, stopples
large; capsule 4-seeded, on a solid stalk, densely pubescent.

EASTERN REGION: Natal; without precise locality, *Gerrard,* 1680!

The leaves of *A. coromandeliana* when subsessile are ovate; the present plant
has the pollen of *Mackaya* (not of *A. coromandeliana*), but the stopples (I believe)
are 2 only.

3. A. stenosiphon (C. B. Clarke); nearly glabrous, except the
corolla; leaves $2\frac{1}{4}$ by $\frac{3}{4}$ in., elongate-triangular, widest close to the
truncate (sometimes subcordate) base; petioles $0-\frac{1}{4}$ in. long; inflorescence, bracts, and calyx nearly as of *A. coromandeliana;* corolla 1 in.
in total length; tube linear, hardly $\frac{1}{16}$ in. in diam., funnel-shaped in

the upper $\frac{1}{3}$; ovary glabrous; pollen subglobose, bands not reaching the pole, stopples 3; capsule $\frac{1}{2}$–$\frac{2}{3}$ in. long, glabrous, with a cylindric stalk.

EASTERN REGION: Fort Beaufort Div.; Koonap Heights, *Baur*, 271! The pollen is as that of *Mackaya*.

4. A. varia (N. E. Brown in Gard. Chron. xii. (1892), 760); sparsely pubescent; leaves varying on one stem from ovate to linear, i.e. from $1\frac{3}{4}$ by $1\frac{1}{4}$ in. to $4\frac{1}{2}$ by $\frac{1}{8}$ in., margins undulate or obscurely toothed; petioles $\frac{1}{4}$–$\frac{3}{4}$ in. long; raceme terminal, $1\frac{1}{2}$–$3\frac{1}{2}$ in. long, loose, 4–8-flowered; pedicels 0–$\frac{1}{6}$ in. long; bracts hardly $\frac{1}{8}$ in long, narrow; calyx $\frac{1}{6}$ in. long, divided nearly to the base into 5 linear segments; corolla $1\frac{3}{4}$ in. long, slightly curved, hardly 2-lipped; tube much dilated in the upper half; lobes 5, ovate, nearly equal, pale mauve with dark brown veins; stamens 4, similar; anther-cells at nearly equal height; pollen ellipsoid, banded; capsule 1 in. long, 4-seeded, nearly glabrous; linear-cylindric stalk $\frac{1}{6}$ in. long.

EASTERN REGION: Zululand; in woods at Entumeni, *Wood*, 3976! Natal or Zululand; without precise locality, *Gerrard*, 1683!

5. A. Schimperi (T. Anders. in Journ. Linn. Soc. vii. 53); erect branched annual 6–18 in. high, young parts hispid-pubescent; leaves up to 3–2 by 1 in., ovate or oblong, narrowed at both ends; petioles 0–1 in. long; spikes terminal, 1–2 in. long, rather dense; bracts up to $\frac{1}{2}$–$\frac{2}{3}$ by $\frac{1}{8}$ in. narrowly elliptic; bracteoles $\frac{1}{8}$ in. long, falcate-lanceolate; calyx-segments $\frac{1}{6}$ in. long, linear; corolla small, pink; tube $\frac{1}{6}$ in. long, dilated upwards; stamens 4, subsimilar; anther-cells one a little below the other, minutely mucronate at the base; pollen broadly oblong, trigonous, with 3 stopples, the smooth bands hardly reaching the pole; capsule $\frac{3}{4}$ in. long, 4-seeded, minutely pubescent, linear-cylindric base hardly $\frac{1}{6}$ in. long; seeds flattened, coarsely tubercled. *C. B. Clarke in Dyer, Fl. Trop. Afr.* v. 135 *including var. Grantii, C. B. Clarke. Adhatoda rostrata, Solms-Laub. in Schweinf. Beitr. Fl. Aethiop.* 104; *Lindau in Engl. & Prantl, Pflanzenfam.* iv. 3B, 326.

EASTERN REGION: Natal; Phœnix Station, 400–500 ft., *Wood*, 4967 partly! and without precise locality, *Gerrard*, 181!

Frequent in East Tropical Africa.

These Natal examples must be at least 18 in. high, and are var. *Grantii* (C. B. Clarke in Dyer, Fl. Trop. Afr. v. 135). There have been distributed two very different plants under *Wood*, 4967. One is the present species, as see Lindau in Engl. Jahrb. xxii. 118; the other, on which Schinz founded his *Calophanes crenatus* is a true *Calophanes* (now termed *Dyschoriste*), for which, see above, p. 16.

6. A. subbiflora (C. B. Clarke); glabrous, except the corolla; branches 1 ft. long, trailing, flexuose, 4-angular and sub-4-winged, internodes 1–$2\frac{1}{2}$ in. long; leaves $1\frac{3}{4}$ by $\frac{1}{3}$ in., oblong, obtuse, much tapering at the base so that the petiole is obscure; peduncles axillary, 2 in. long, slender, 2-flowered; bracts $\frac{1}{4}$ in. long, linear; pedicels

$\frac{1}{6}$ in. long; calyx divided nearly to the base, glabrous, segments exceeding $\frac{1}{3}$ in. long, linear, scarcely lanceolate at the base; corolla $\frac{3}{4}$ in. long, pubescent without, pale blue, very nearly as that of *A. coromandeliana;* stamens 4, subsimilar; anthers linear-oblong, muticous, one a little below the other; pollen oblong, banded, with 2 stopples; ovary and style glabrous; stigmas 2, very small, oblong.

KALAHARI REGION: Transvaal; Upper Moodies, near Barberton, in stony ground, 4500 ft., *Galpin*, 1272!

The inflorescence is that of *A. ansellioides*, C. B. Clarke in Dyer, Fl. Trop. Afr. v. 136.

XVII. MACKAYA, Harv.

Corolla somewhat curved, hardly 2-lipped; tube much inflated in the upper half; lobes 5, ovate, not very unequal. *Stamens* 2 fertile, 2 represented by linear filaments without rudiments of an anther; anthers at equal height, muticous; pollen subglobose, with 3 pores, the bands not reaching the pole; seeds hardly rugose; otherwise as *Asystasia.*

Species, beside the one here described, 3 Indian.

This genus differs from *Asystasia* by having 2 fertile stamens only, from *Eranthemum* by the upper half of the corolla-tube being swollen. *Graptophyllum* differs little, but by the strongly 2-lipped corolla. The other 3 species of the genus are *Thyrsacanthus indicus*, Nees, *Eranthemum indicum*, Collett & Hemsl. (in Journ. Linn. Soc. xxviii. 105, not of C. B. Clarke), and *Eranthemum lateri-florum*, C. B. Clarke. The pollen does not differ from that of some *Asystasias.*

1. **M. bella** (Harv. Thes. Cap. i. 8, t. 13); nearly glabrous; leaves 5 by $1\frac{1}{2}$ in., narrowed at both ends, margins crenate or very obtusely toothed; petioles $\frac{1}{4}$–$\frac{1}{3}$ in. long; raceme terminal, 4–6 in. long, loose, of 4–12-flowers; pedicels 0–$\frac{1}{6}$ in. long, in opposite pairs; bracts hardly $\frac{1}{8}$ in. long, linear; calyx $\frac{1}{4}$ in. long, divided nearly to the base into 5 linear segments; corolla $1\frac{1}{2}$–2 in. long, curved but only obscurely 2-lipped, pale lilac; tube linear-cylindric in the lower half, campanulate in the upper; lobes ovate, not very unequal; anthers $\frac{1}{4}$ in. long; capsule $1\frac{1}{2}$ in. long, 4-seeded, nearly glabrous, cylindric stalk $\frac{3}{4}$ in. long; seeds nearly smooth. *T. Anders. in Journ. Linn. Soc.* vii. 53; *Lindau in Engl. & Prantl, Pflanzenfam.* iv. 3B, 336, *fig.* 135, A; *Bot. Mag. t.* 5797. *Asystasia sp., Benth. et Hook. f. Gen. Pl.* iii. 1095.

EASTERN REGION: Natal; Kruis Fontein, Tongat, *Sanderson*, 167! Umvoti River, *Adlam in MacOwan & Bolus Herb. Norm. Aust.-Afr.*, 240! Inanda, *Wood*, 1028!

XVIII. BARLERIA, Linn.

Bract a spine or a leaf, entire, toothed or pinnatifid, a second sterile bract often present; bracteoles 0. *Calyx* large, of 4 distinct segments; anticous segment of 2 sepals united nearly or frequently quite to the tip with 1 mid-nerve ; posticous segment of nearly equal length, rather narrower; 2 inner segments much smaller. *Corolla* with a cylindric tube, 2-lipped or subequally 5-fid; segments not contorted in bud. *Stamens* 2 perfect, 2 rudimentary often added ; anther-cells 2, parallel at equal height, not spurred at the base; pollen globose, honeycombed. *Ovary* with 2 (or 1) ovules in each cell; style long; stigmas 2, short or subconfluent. *Capsule* either ovoid 2-seeded, or ellipsoid 4-seeded ; flattened, sometimes (when 2-seeded) beaked, usually hard and shining ; seeds large, shaggy with hygroscopic hairs.

Inflorescence (when fully developed) of axillary often scorpioid cymes, sometimes forming dense 1-sided many-flowered strobilate spikes, sometimes reduced to 3–2-flowered clusters or (apparently) solitary flowers; the axillary quasi-solitary flowers again not rarely running into terminal heads or spikes.

Species 120, mostly in Africa and Tropical Asia, a few in America.

Series A. Interpetiolar spines (i.e. reduced leaves of sterile shoots or outermost bracts of axillary flowers or inflorescences) often present. Posticous sepal spine-tipped or with teeth on its margin.

Section 1. PRIONITIS. Lower ovule in each cell minute or 0; capsule never having more than 2 seeds, ovoid with conic beak. Posticous sepal spine-tipped or mucronate, without teeth on its margin. Corolla in the Cape species yellow.

> Corolla 2-lipped ; 4 posticous segments in 1 lip   ... (1) **Prionitis.**
> Corolla subequally 5-fid; tube long linear   ... (2) **Holubii.**

Section 2. ACANTHOIDEA. Ovary with 2 similar ovules in each cell; capsule often 4-seeded. Posticous sepal usually prominently (rarely obscurely) toothed on its margins. Corolla in the Cape species blue-purplish (or white).

> Leaves less than 1 in. long ; small shrubs ; racemes few-flowered :
>> Leaves ovate to lanceolate; prickly harsh plants :
>>> Posticous sepal spinous-ciliate on the margins :
>>>> Branches and leaves hispid beneath ... (3) **pungens.**
>>>> Branches and leaves pubescent beneath (4) **irritans.**
>>> Posticous sepal obscurely or obsoletely toothed   ...   . (5) **stimulans.**
>> Leaves linear to linear-oblong; twiggy, less harsh, plants :
>>> Corolla-tube ¼ in. long   ...   . (6) **Virgula.**
>>> Corolla-tube ¾ in. long   ...   . (7) **bechuanensis.**
> Leaves, many of them, 1 in. long ; racemes few-flowered :
>> Posticous sepal not (or rarely obscurely) spine-toothed ...   . (8) **acanthoides.**
>> Posticous sepal deeply and rigidly spine-toothed ...   . (9) **elegans.**
>> Posticous sepal with many short innocuous teeth :
>>> Innovations and young leaves without stellate hairs :
>>>> Bracts to raceme subspinous; leaves (some of them) petioled :
>>>>> Corolla 1¼ in. long ; petioles less than ⅙ in.   ...   . (10) **Gueinzii.**
>>>>> Corolla 1¾ in. long ; petioles up to ⅓ in. long   ...   ... (11) **barbata.**

Bracts to heads more spinous; leaves all subsessile :

Posticous sepal $\frac{3}{4}$ to $1\frac{1}{4}$ in. long ... (12) **Woodii.**

Posticous sepal less than $\frac{3}{4}$ in. long ... ... (13) **jasminiflora.**

Innovations and young leaves densely stellate-hairy ... ... (14) **affinis.**

Leaves usually less than 1 in. long; racemes normally many-flowered, in fruit very thick; trailing plants:

Thinly hispid; posticous sepal with short spine-tip ... ... (15) **macrostegia.**

Villous; posticous sepal with long spine-tip :

Posticous sepal strongly spine-toothed on the margins ... ... (16) **lichtensteiniana.**

Posticous sepal softly villous, not toothed on the margin ... ... (17) **media.**

Series B. Spineless. Posticous sepal without teeth (or very obscurely toothed) on its margin.

Flowers mainly in long terminal spikes :

Spikes long-cylindric ... ... (18) **crossandriformis.**

Spikes ovoid ... ... (19) **ovata.**

Flowers in loose cymes ... ... (20) **obtusa.**

Flowers solitary or in small clusters :

Plants, especially the inflorescence, hairy :

Capsule glabrous, or very nearly so :

Leaves narrowed into a very short petiole :

Stem thinly hispid; corolla blue (21) **meyeriana.**

Stem densely and minutely grey-hairy; corolla yellow ... ... (22) **cinereicaulis.**

Leaves subsessile, almost rounded at the base ... ... (23) **Rehmanni.**

Capsule densely hairy upwards ... ... (24) **pretoriensis.**

Plants very nearly glabrous :

Leaves subsessile, rounded at the base ... (25) **Wilmsii.**

Leaves narrowed at the base into a petiole 0–$\frac{1}{5}$ in. long ... ... (26) **Galpinii.**

1. **B. Prionitis** (Linn. Sp. Pl. ed. i. 636, ed. ii. 887); glabrous or nearly so; stems 1–3 ft. high; leaves 2–5 by $\frac{1}{3}$–1 in., elliptic, narrowed at both ends, mucronate; petiole 0–$\frac{1}{3}$ in. long; interpetiolar spines (i.e. basal bracts to the inflorescence of the succeeding year) simple, rigid, $\frac{1}{4}$–$\frac{3}{4}$ in. long; inflorescence reduced nearly to a simple spike; lower flowers more or less distant, sometimes 2 in each axil; flower-bract similar to the external bract, but rather longer, dilated lanceolate at the base; posticous sepal nearly $\frac{1}{2}$ in. long, elliptic, acuminate, mucronate, without teeth on its margins; anticous calyx-segment resembling the posticous sepal sometimes ending in 2 short spinous teeth; corolla yellow; tube hardly $\frac{1}{2}$ in. long; posticous lip of 4 connate segments $\frac{2}{3}$–1 in. long; anticous segment clawed, shorter; capsule $\frac{1}{2}$–$\frac{2}{3}$ in. long, ovoid-conic, beaked, 2-seeded. *Nees in DC. Prod.* xi. 237; *T. Anders. in Journ. Linn. Soc.* vii. 28 (*only as to non-African plants*); *C. B. Clarke in Hook. f. Fl. Brit. Ind.* iv. 482, *and in Dyer, Fl. Trop. Afr.* v. 145;

Lindau in Engl. Jahrb. xviii. 63, *t.* 1, *fig.* 29, *and in Engl. & Prantl, Pflanzenfam.* iv. 3B, 314, *fig.* 105, *C, fig.* 106, B, *and fig.* 125, E. *B. Hystrix, Linn. Mant.* 89. *B. prionitoides, Engl. in Engl. Jahrb.* x. 262.

KALAHARI REGION: Transvaal; Inkumpi River, *Nelson*, 370! Olifants River, *Nelson*, 400!

Also in Tropical Africa and Asia. This plant is often cultivated, and spreads as a weed.

2. **B. Holubii** (C. B. Clarke); branchlets retrorsely hairy; internodes ½ in. long, obovate, suddenly acuminate into a spine, rigid, glabrate; interpetiolar spines ½–⅔ in. long, simple, rigid; calyx-segments lanceolate, spine-tipped, without marginal teeth; corolla-tube ¾ in. long, linear-cylindric, not dilated in the upper part; limb of 5 subequal rounded segments ¼ in. long.

KALAHARI REGION: Transvaal; Marico District, *Holub*!

3. **B. pungens** (Linn. f. Suppl. 290); harsh, prickly; branches 4–18 in. long, hispid towards the tips; leaves ½–1 in. long, subsessile, ovate, acute, spine-tipped, mostly rounded at the base, hispid beneath at least on the nerve, margin not (or most obscurely) spine-toothed; racemes 8–1-flowered, axillary; outermost bracts (or axillary spines) mostly dilated, pinnatifid toothed near the base; posticous sepal ovate, up to ¾ by ⅓ in., spine-tipped, papery, veined, with spinous teeth on the margin; corolla blue; tube ⅔–1 in. long; capsule ½–⅔ by ¼ in., ellipsoid, flattened, glabrescent, 4-seeded. *Nees in Linnæa,* xv. 358, *and in DC. Prod.* xi. 236 (*excluding var. macrophylla*); *T. Anders. in Journ. Linn. Soc.* vii. 28; *Lindau in Engl. & Prantl, Pflanzenfam.* 3B, 314, *not of Thunb. B. pungens,* δ *only, E. Meyer in Drège, Zwei Pflanzengeogr. Documente,* 138, 168. *Crabbea pungens, Harv. Gen. South Afr. Pl.* 276. *Acanthus procumbens, Willd. ex Nees in DC. Prod.* xi. 236.

COAST REGION: Riversdale Div.; between Zoetemelks River and Little Vet River, *Burchell*, 6850! Mossel Bay Div.; Little Brak River, *Burchell*, 6192! Humansdorp Div.; Gamtoos River, *Niven*, 22! Uitenhage Div.; between Uitenhage and the Drostdy Farm, *Burchell*, 4453! and without precise locality, *Masson*, 45! *Bowie*, 7! *Tredgold*, 35! Port Elizabeth Div.; Algoa Bay, *Cooper*, 3038! near Port Elizabeth, *Forbes*! Alexandria Div.: Sam Tees Flats near Enon, *Baur*, 1093! Albany Div.; near Grahamstown, *MacOwan*! *Bowie*, 7! 8! Bothas Berg, *Baur*, 1093!

CENTRAL REGION: Somerset Div.; between Little Fish River and Great Fish River, 2000–3000 ft., *Drège*! near Somerset, *Bowker*, 212! Albany Div.; between Zwartwarter Poort and "Soutars Post" (Fish River Rand), *Burchell*, 3445! at Kurukuru River, *Burchell*, 3517!

4. **B. irritans** (Nees in Linnæa, xv. 359); branches nearly glabrous or pubescent towards the tips; leaves ovate, more or less lanceolate, acuminate, margins sinuate white-cartilaginous thickened, more or less spine-toothed; posticous sepal ½–⅔ by ⅙–⅕ in., elliptic-lanceolate; otherwise as *B. pungens. Nees in DC. Prod.* xi. 236, *incl. var.* β; *Sonder in Linnæa,* xxiii. 92; *T. Anders. in Journ.*

Linn. Soc. vii. 28; *Engl. in Engl. Jahrb.* x. 263; *Lindau in Engl. & Prantl, Pflanzenfam.* iv. 3B, 314. *B. pungens, Thunb. Fl. Cap. ed. Schult.* 458; *E. Meyer in Drège, Zwei Pflanzengeogr. Documente, only* 129, 168.

VAR. β, **rigida** (C. B. Clarke); leaves oblong, $\frac{3}{4}$ by $\frac{1}{6}$–$\frac{1}{5}$ in., nearly glabrous, narrowed at the base into a very short petiole, margins much sinuate, more spine-toothed. *B. rigida, Nees in DC. Prod.* xi. 242; *Sonder in Linnæa*, xxiii. 92.

SOUTH AFRICA: without precise locality, *Harvey!*

COAST REGION: Uitenhage Div.; by the Sunday River, 1000 ft., *Zeyher*, 1421! hills by the Zwartkops River, *Drège!*

CENTRAL REGION: Somerset Div.; Somerset East, *Bowker*, 161! 65! Graaff Reinet Div.; near Graaff Reinet, 2500 ft., *Bolus*, 563! Var. β: Calvinia Div., Springbok Kuil, 2000–3000 ft., *Zeyher*, 1417!

KALAHARI REGION: Griqualand West; Groot Boetsap, 4000 ft., *Marloth*, 966! Var. β: Griqualand West; Kimberley, *Mrs. Barber!* Hay Div.; plains between Griqua Town and Witte Water, *Burchell*, 1991!

WESTERN REGION: Great Namaqualand; Tiras, *Schinz*, 3!

5. **B. stimulans** (E. Meyer in Drège, Zwei Pflanzengeogr. Documente, 61, 62, 168); woody, densely branched and thorny with many nearly simple spines $\frac{1}{2}$–1 in. long; leaves $\frac{1}{2}$–$\frac{3}{4}$ in. long, ovate, mucronate, margins without spinous teeth; posticous sepal $\frac{1}{2}$–$\frac{2}{3}$ by $\frac{1}{3}$ in., ovate, margins not (or most obscurely) spine-toothed. *Nees in DC. Prod.* xi. 241; *Sonder in Linnæa*, xxiii. 92, *including var. macracantha; T. Anders. in Journ. Linn. Soc.* vii. 28; *Lindau in Engl. & Prantl, Pflanzenfam.* 3B, 315.

SOUTH AFRICA: without precise locality, *Zeyher*, 1416!

CENTRAL REGION: Prince Albert Div.; Gamka River, *Burke!* Graaff Reinet and Jansenville Divs.; by the Sunday River, 1500–2000 ft., *Drège!*

KALAHARI REGION: Prieska Div.; Keikams Poort (Modder Gat Poort), *Burchell*, 1621!

6. **B. Virgula** (C. B. Clarke); minutely and thinly strigose with simple white hairs; twigs 6 in. long, very slender, straight; leaves attaining $\frac{1}{2}$ by $\frac{1}{12}$ in., linear, nearly entire; axillary spines $\frac{1}{4}$ in. long, subpinnatifid at the base; flowers axillary, subsolitary; posticous sepal $\frac{1}{3}$ in. long, lanceolate, margins with a few rigid teeth; corolla-tube $\frac{1}{4}$ in. long; lobes $\frac{1}{4}$–$\frac{1}{3}$ in. long; capsule rather more than $\frac{1}{4}$ in. long, ellipsoid, flattened, glabrous, 4-seeded.

KALAHARI REGION: Transvaal, Marico District, *Holub!*

7. **B. bechuanensis** (C. B. Clarke); branchlets 1–2 in. long, from a woody stock, with minute lines of hairs; leaves attaining $\frac{1}{2}$ by $\frac{1}{16}$ in., linear, mucronate, glabrate, margins cartilaginous white spine-toothed; axillary spines $\frac{1}{4}$ in. long, linear-lanceolate, subfoliaceous; flowers solitary, axillary; posticous sepal $\frac{1}{3}$–$\frac{1}{2}$ in. long, elliptic-lanceolate, minutely hairy, margins spine-toothed; corolla deep blue; tube exceeding $\frac{3}{4}$ in. in length, linear-cylindric to the top; lobes $\frac{1}{3}$ in. long, orbicular.

Var. β, **espinulosa** (C. B. Clarke); leaves without marginal teeth, obtuse, not distinctly mucronate.

KALAHARI REGION: Bechuanaland; Batlapin Territory, *Holub!* Var. β: Griqualand West; by the Vaal River at the Diamond Fields, *Nelson*, 151!

8. B. acanthoides (Vahl, Symb. i. 47); undershrub; branches 6–30 in. long, canescent; leaves 1–2 in. long, oblong-elliptic, mucronate, minutely hairy; petiole 0–$\frac{1}{4}$ in. long; interpetiolar spines $\frac{1}{4}$–$\frac{3}{4}$ in. long, rigid, simple or subpinnatifid; racemes with 8–1 flowers, condensed, unilateral; bracts as the interpetiolar spines; posticous sepal $\frac{2}{3}$ by nearly $\frac{1}{2}$ in., ovate, veined, slightly ciliate; corolla white or pale blue (*Schinz*); tube 1$\frac{1}{2}$–3 in. long, linear-cylindric nearly to the top; capsule $\frac{1}{2}$ in. long, 4-seeded. *Nees in DC. Prod.* xi. 240; *T. Anders. in Journ. Linn. Soc.* vii. 27 *partly; C. B. Clarke in Hook. f. Fl. Brit. Ind.* iv. 484, *and in Dyer, Fl. Trop. Afr.* v. 152; *Lindau in Engl. & Prantl, Pflanzenfam.* iv. 3B, 314, *and in Engl. Pfl. Ost-Afr. C.* 369. *B. acanthoides, var. lanceolata, Schinz in Verh. bot. Ver. Brandenb.* xxxi. 199.

WESTERN REGION: Great Namaqualand; Gamochab, *Schinz*, 2!

Frequent in Tropical Africa and India.

9. B. elegans (S. Moore in Journ. Bot. 1880, 269); pubescent with simple hairs; branches 1–2 ft. long; leaves 2–3 by $\frac{1}{2}$–1$\frac{1}{4}$ in. entire; petiole 0–$\frac{1}{3}$ in. long; racemes 1–1$\frac{1}{2}$ in. long, 2–8-flowered, unilateral; bracts $\frac{1}{2}$–1 by $\frac{1}{8}$ in., linear-lanceolate, spine-toothed or subpinnatifid; posticous sepal $\frac{3}{4}$ in. long, elliptic, acute, marginal teeth many, strong, often $\frac{1}{6}$ in. long; corolla blue, mauve or white (*Galpin*); tube $\frac{1}{2}$–$\frac{2}{3}$ in. long, linear-cylindric, at the top narrowly funnel-shaped; capsule $\frac{1}{3}$ in. long, 4-seeded. *C. B. Clarke in Dyer, Fl. Trop. Afr.* v. 154. *B. pungens, var. macrophylla, Nees in DC. Prod.* xi. 237.

KALAHARI REGION: Transvaal; at Avoca, near Barberton, 1800 ft., *Galpin*, 886!

EASTERN REGION: Natal; without precise locality, *Gerrard*, 1681! Delagoa Bay, *Forbes!*

Also in Tropical Africa. It must not be inferred from Nees' synonym that this plant bears the slightest resemblance to *B. pungens.*

10. B. Gueinzii (Sonder in Linnæa, xxiii. 91); branches 16 in. long, hairy towards the ends; leaves 1$\frac{1}{2}$ by $\frac{3}{4}$ in., ovate, shortly acute, entire, with simple hairs on both faces, base obtuse or rounded; petioles 0–$\frac{1}{6}$ in. long; racemes 3–1-flowered; outermost bracts $\frac{1}{4}$–$\frac{1}{3}$ in. long, linear, recurved, innocuous, hardly spinescent; posticous sepal $\frac{1}{2}$–$\frac{2}{3}$ in. long, ovate, membranous, nervose, margins innocuously spine-toothed; corolla 1$\frac{1}{4}$ in. in total length, bluish-purple (*Galpin*). *T. Anders. in Journ. Linn. Soc.* vii. 30.

COAST REGION: East London Div.; East London, *Flanagan*, 1773!
KALAHARI REGION: Transvaal, Barberton, 3000 ft., *Galpin*, 842!
EASTERN REGION: Natal; Coastland, 0–1000 ft., *Sutherland!* without precise locality, *Cooper*, 3026! *Gueinzius*, 383.

Also in South Tropical Africa.

11. B. barbata (E. Meyer in Drège, Zwei Pflanzengeogr. Documente, 152, 168); petioles $\frac{1}{4}$ in. long and upwards; posticous sepal exceeding $\frac{3}{4}$ in. long, ovate-triangular, acute; corolla 1$\frac{3}{4}$ in. in total length; otherwise as *B. Gueinzii. B. obtusa, var., Nees in DC. Prod.* xi. 231.

EASTERN REGION: Pondoland; between St. Johns River and Umtsikaba River, 1000 ft., *Drège !* Natal; without precise locality, *Gerrard*, 1973!

This species may turn out to be only a large form of *B. Gueinzii*, but not a variety of *B. obtusa.*

12. **B. Woodii** (C. B. Clarke); branches 18–30 in. long, hairy; leaves 1¾ by ½–⅔ in., elliptic-oblong, entire, hairy on both faces; petioles 0–¹⁄₁₆ in.; racemes up to 6-flowered (mostly 3–1-flowered), axillary, condensed; outermost bracts foliaceous, lanceolate, not spinous; other bracts ¾–1¼ in. long, broadly elliptic, acute, with many innocuous marginal teeth; corolla blue or white, 1¼ in. in total length; tube cylindric, in the upper half funnel-shaped.

EASTERN REGION: Natal; Inanda, *Wood*, 803! Oakford, 1600 ft., *Wood*, 851!

13 **B. jasminiflora** (C. B. Clarke); branches 5 in. long, hairy, spreading from a woody root; leaves scarcely 1 by ¼ in., oblong, entire, mucronate, sparsely hispid with simple hairs, subsessile; flowers solitary; outer bracts ½ in. long, lanceolate, rather rigid, viscous, with a few bristles on the margins; posticous sepal ⅔ in. long, elliptic or ovate, spine-tipped, rather rigid, with bristles on the margin; corolla about 1 in. long (judging by the style), lake-coloured (*Burchell*).

COAST REGION: Uitenhage or Humansdorp Div.; between Galgebosch and Melk River, *Burchell*, 4759!

Burchell notes that the flowers were shaped as those of jasmine, with equal spreading corolla-lobes.

14. **B. affinis** (C. B. Clarke); whole plant stellate-hairy; branches 6–8 in. long, slender; leaves ¼–⅓ in. in diam., ovate or orbicular, often broader than long, even in age stellate-tomentose, base truncate; petiole up to ⅛–⅙ in. long; flowers axillary, solitary; outermost bracts ¼–½ in. long, linear, scarcely spinescent; posticous sepal ½ in. long, ovate, nervose, with many innocuous teeth on the margins; corolla (judging by the style) not less than ¾ in. long.

KALAHARI REGION: Transvaal; Marico District, *Holub !*

This plant is closely allied to *B. spinulosa*, Klotzsch, from Mozambique, and to its very close allies in Mozambique.

15. **B. macrostegia** (Nees in DC. Prod. xi. 235); thinly hispid; innovations neither silky nor woolly; trailing branches 1–2 ft. long, from the crown of a woody rootstock; leaves (in type examples) ¾–1 by ⅕ in., oblong, mucronate, minutely spine-ciliate on the margin, petiole 0–⅛ in. long (in other examples broader leaves occur); heads axillary, remote, in flower 1½ by 1 in., in fruit 2 in. and upwards in length, and very thick; calyx-segments (and bracts similar to them) very densely imbricated in 4 ranks, the 2 ranks on one side empty (bracts), the 2 ranks on the other side being sepals; anticous sepal (in fruit) ¾ by ⅔ in. broadly ovate, representing 2

sepals as throughout the genus *Barleria;* posticous sepal 1¼ by ⅔ in.,
ovate-lanceolate, mucronate, margins harshly ciliate, hardly toothed,
terminal mucro hardly excurrent as a spine; corolla pink (*Mrs.
Stainbank*); tube ¾ in. long, cylindric, very little narrowed at
the top; lobes ⅓ in. long; capsule exceeding ½ in. long, flattened,
shining-brown, 4-seeded. *B. burchelliana, Nees in DC. Prod.* xi.
235; *Engl. in Engl. Jahrb.* x. 262. *B. burkeana, Sonder in Linnæa,*
xxiii. 92; *T. Anders. in Journ. Linn. Soc.* vii. 31.

KALAHARI REGION: Orange River Colony; Vet River, *Burke,* 451! Riet
Fontein, *Rehmann,* 3689! Bechuanaland; between Kuruman and Matlareen
River, *Burchell,* 2191! near Mafeking, 4200 ft., *Bolus,* 6429! near Kachun,
3900 ft., *Marloth,* 1038; Batlapin territory, *Holub!* Transvaal; Bosch Veld,
between Elands River and Klippan, *Rehmann,* 5048! near Pretoria, *Roe in Bolus
Herb.,* 3045! and without precise locality, *Holub! Mrs. Stainbank in Wood
Herb.,* 3664! *Miss Saunders,* 10a!

This species, the two very closely allied which follow, and *B. capitata,* Klotzsch
(from Mozambique), form a striking group, recognized by the trailing stems with
small leaves and large thick densely strobilate heads of fruit. Nees supposed
" Burke " to be an abbreviation for " Burchell "; his two species, *B. macrostegia* and *B. burchelliana,* are founded on one collection. Sonder attempted to
correct the specific name. The difficulty is evaded by taking up *macrostegia.*

16. **B. lichtensteiniana** (Nees in DC. Prod. xi. 235); innovations
densely grey-silky-strigose; leaves up to 1¾ by ½ in.; posticous sepal
with a long spinous mucro and spinescent teeth ⅕ in. long on the
margins; otherwise as *B. macrostegia. Masters in Gard. Chron.*
1870, 73, *figs.* 12–13; *Lindau in Engl. & Prantl, Pflanzenfam.* iv.
3B, 314.

CENTRAL REGION: Hopetown Div., near Hopetown, *Muskett,* 42!
KALAHARI REGION: Griqualand West, Hay Div.; Asbestos Mountains at
Kloof Village, *Burchell,* 1652! on plains at the foot of the Asbestos Mountains,
between the Kloof Village and Witte Water, *Burchell,* 2068!

T. Anderson, either by accident or from supposing it not specifically distinct
from *B. macrostegia,* omits this plant altogether. The flower in Burchell, 2068,
is normal; arrested flowers, as those figured by Masters, also occur; such occur
in other species of the genus, and in other genera of the Order, and may be (*fide*
Masters) connected with self-fertilization.

17. **B. media** (C. B. Clarke); softly and somewhat thickly hairy;
innovations densely grey-strigose; leaves 1 by ½ in.; posticous sepal
with a long spine at the tip, softly ciliate, hardly toothed on the
margins; otherwise as *B. lichtensteiniana.*

KALAHARI REGION: Bechuanaland; on the rocks at Chue Vley, *Burchell,*
2386!

The young shoots are grey-strigose, often with deflexed hairs, exactly as in
B. lichtensteiniana. The spine-tip of the posticous sepal is as of *B. lichtensteiniana,* but the margins without teeth are as of *B. macrostegia.* I think
the South African plants of this group must be arranged either as three species
or as one.

18. **B. crossandriformis** (C. B. Clarke); nearly glabrous, except
the inflorescence; branches 12 in. long; leaves attaining 3½ by 1½ in.,

entire, narrowed to a petiole $\frac{1}{8}$–$\frac{1}{2}$ in. long; inflorescence a terminal strobilate spike 3 by $\frac{3}{4}$ in.; posticous sepal $\frac{1}{2}$–$\frac{2}{3}$ by $\frac{1}{3}$ in. ovate, acute, entire, softly hairy; corolla buff (*Galpin*); tube $\frac{3}{4}$ in. long; lobes $\frac{1}{3}$ in. long; pollen globose, honeycombed, capsule $\frac{2}{3}$ in. long, ovoid, compressed, beaked, pubescent, 2-seeded.

KALAHARI REGION: Transvaal; among scrub on a hillside at Avoca, near Barberton, 1800 ft., *Galpin*, 887!

This plant, in external appearance, is very like *Crossandra nilotica*, Oliv.

19. B. ovata (E. Meyer in Drège, Zwei Pflanzengeogr. Documente, 147, 149, 168); hirsute with tawny hair; branches 3–16 in long; leaves subsessile, ovate or elliptic, entire, in the type-specimen $2\frac{1}{2}$ in. long, in other examples only half as large; flowers all in the upper axils, running into a terminal spike; posticous sepal 1 by $\frac{1}{3}$ in., elliptic, acute, entire, hairy; corolla $1\frac{1}{2}$ in. in total length; capsule $\frac{1}{3}$ in. long, ellipsoid, compressed, shining, glabrous, 4-seeded. *Nees in DC. Prod.* xi. 230; *T. Anders. in Journ. Linn. Soc.* vii. 31. *B. natalensis, Lindau in Engl. Jahrb.* xx. 23.

KALAHARI REGION: Orange River Colony, *Cooper*, 841! Basutoland; Mont aux Sources, 9500 ft., *Guthrie*, 4889! in the mountains, *Thode!* Transvaal; on the Saddleback Range, near Barberton, 4000 ft.. *Galpin*, 832! near Lydenburg, *Wilms*, 1222! Spitzkop Goldmine, *Wilms*, 1223! Crocodile River, 4800 ft., *Schlechter*, 3899!

EASTERN REGION: Pondoland, between Umtata River and St. Johns River, 1000–2000 ft., *Drège!* Natal, Camperdown, 1800 ft., *Wood*, 470! 1936! on a hill near Oakford, *Wood*, 851.

There may be more than one species here. The variation in the size of the leaves is very great. Nees says that the (outer) calyx-segments are spinulose-serrate, and Lindau says that in his *B. natalensis* they are spinescent-toothed; in all the specimens I can find no teeth, far less any spines. Then both Nees and Lindau say the flowers are blue. All the dried examples here have flowers, and they are all deep blue in the dried state; but, on both his collections, Wilms has noted "flowers sulphur"; they must have turned from yellow to deep blue in drying. Wood has noted "flowers yellowish-white." In *B. ovata*, the hair is soft; in *B. natalensis*, the hair is stiffer; other difference I have found none.

20. B. obtusa (Nees in Linnæa, xv. 358); hairy; plants very variable in size, sometimes only 8 in. high, much branched, dense, with no internode so much as $\frac{1}{3}$ in. long, and no leaf so much as $\frac{3}{4}$ in. long, at other times with branches 20 in. long, internodes 3 in. long and leaves 2 in. long; various intermediate forms occur; leaves ovate or elliptic; petioles 0–$\frac{1}{4}$ in. long; flowers 4–1, $\frac{1}{8}$–$\frac{2}{3}$ in. apart, in loose axillary (usually monopodial) cymes; bracteoles linear $\frac{1}{6}$–$\frac{1}{3}$ in. long, often recurved; posticous sepal $\frac{1}{2}$–$\frac{1}{3}$ by $\frac{1}{8}$–$\frac{1}{4}$ in., narrowly oblong, obtuse, entire, often wider in its upper half, hairy; corolla 1–$1\frac{1}{4}$ in. in total length, blue; capsule $\frac{2}{3}$–$\frac{3}{4}$ in. long, ellipsoid, compressed, shining chestnut-coloured, 4-seeded. *Nees in DC. Prod.* xi. 231, *excl. var. β**; *Sonder in Linnæa*, xxiii. 92; *T. Anders. in Journ. Linn.* vii. 31 *excl. syn. B. barbata; Lindau in Engl. & Prantl, Pflanzenfam.* iv. 3B, 314. *B. obtusa var. cymulosa, Hochst. in Flora*, 1845, 72. *B. diandra, E. Meyer in Drège, Zwei Pflanzengeogr. Documente,* 157,

**168. B. barbata, E. Meyer in herb. Drège partly. B. uitenhagensis,
Hochst. and Ruellia ovata, Zeyher ex Nees in DC. Prod. xi. 231.**

SOUTH AFRICA : without precise locality, *Burke! Drège*, 3602a !
COAST REGION : Uitenhage Div., *Zeyher! Ecklon & Zeyher*, 930 ! Albany
Div., *Bowker*, 44 ! British Kaffraria, *Cooper*, 3048 !
CENTRAL REGION : Somerset Div. ; Somerset, *Bowker*, 23 ! 186 ! Graaff
Reinet Div. ; mountains on the south west side of Graaff Reinet, *Burchell*, 2936 !
near Graaff Reinet, 2500 ft., *Bolus*, 41 !
KALAHARI REGION : Orange River Colony ; Doorn Kop, *Burke!* Transvaal ;
near Lydenburg, *Wilms*, 1224 ! Apies Poort, near Pretoria, *Rehmann*, 4105 !
4107 ! Sheba Battery, Kaap Valley, near Barberton, 1900 ft., *Galpin*, 1330 !
EASTERN REGION : Pondoland ; between St. Johns River and Umtsikaba River,
under 1000 ft., *Drège!* Natal ; between Umzimkulu River and Umkomanzi
River, *Drège!* Ullahlane, 2000 ft., *Sutherland!* on a hill near Ladysmith, 3000–
4000 ft., *Wood*, 5627 ! Umkomaas and Buffels Draai at about 1000 ft., *Wood*,
880 ! without precise locality, *Grant! Plant*, 79 ! 80! *Sanderson! Cooper*, 1081 !
Mrs. Saunders, 183 ! *Gerrard*, 102 !

Also in South Tropical Africa.

This species is well-defined by the loose cymes, and narrowly oblong obtuse
posticous sepal. The variability in habit and size of leaves is very great, but
T. Anderson did not attempt to establish any varieties. As to Nees' varieties,
1-flowered and 3-flowered cymes occur pretty frequently on one plant. *B. barbata*,
Drège, is remote from *B. obtusa*, and is close to *B. Gueinzii*. But there is, in the
Kew Herbarium, a branch of the large state of *B. obtusa*, issued by Drège
as *B. barbata*, E. Meyer.

**21. B. meyeriana (Nees in DC. Prod. xi. 230); stems 1–2 ft. long,
branched, glabrous or thinly hispid, the tips and inflorescence thinly
hispid ; leaves 1–1¼ by ¼–⅘ in., elliptic-oblong, entire, when mature
glabrate ; petiole 0–⅛ in. long ; flowers 1 or 2–3 clustered in the axils ;
posticous sepal ⅓–½ by ⅛–⅙ in., oblong, entire, hairy ; corolla blue or
lavender, 1¼ in. in total length ; capsule ½ in. long, ovoid and flattened,
beaked, 2-seeded, glabrous or very nearly so. T. Anders. in Journ.
Linn. Soc. vii. 28. B. ciliata, E. Meyer in Drège, Zwei Pflanzengeogr.
Documente, 145, 154, 168.**

KALAHARI REGION : Transvaal ; near Barberton, 2800 ft., *Galpin*, 880 !
Thorncroft, 97 ! *Wood*, 4162 ! Pietersburg, 4700 ft., *Schlechter*, 4372 !
EASTERN REGION : Transkei ; near Gekau (Geua or Geuu) River, under
1000 ft., *Drège !* Pondoland or Natal ; between Umtentu River and Umzimkulu
River, under 500 ft., *Drège!* Natal ; near Durban, 200 ft., *Wood*, 7553 !
Camperdown,¯ *Rehmann*, 7715 ! Umhloti, *Wood*, 792 ! Umkomaas (Umkomanzi)
Valley, *Wood*, 915 ! and without precise locality, *Gueinzius! Gerrard*, 1268 !
Sutherland!

**22. B. cinereicaulis (N. E. Brown); stem and young leaves
densely minutely grey-hairy; posticous calyx-segment oblong to
elliptic, ⅕ in. broad; corolla yellow (Galpin); otherwise as B.
meyeriana.**

KALAHARI REGION : Transvaal ; on a hillside above Sheba water-race, at
Avoca near Barberton, 1900 ft., *Galpin*, 1331 !
EASTERN REGION : Natal ; without precise locality, *Gerrard*, 1266 !

Exceedingly like *B. meyeriana*; the indumentum differs in nature.

**23. B. Rehmanni (C. B. Clarke); branches 6–8 in. long, sparsely
hispid; leaves subsessile, 1 by ⅓ in., ovate-oblong, obtuse, entire**

almost rounded at the base, hispid-ciliate on the margins or nearly glabrate; flowers solitary, axillary, or pedicels hardly $\frac{1}{8}$ in. long; bracteoles (or rather floral leaves) $\frac{3}{4}$ by $\frac{1}{6}$ in., linear-lanceolate, ciliate with clustered hairs on the margins; anticous calyx-segment $\frac{1}{2}$ by nearly $\frac{1}{6}$ in., oblong, bifid at the tip into 2 teeth $\frac{1}{10}$ in. long; posticous sepal $\frac{1}{4}$ by $\frac{1}{8}$ in., narrow-lanceolate, ciliate on the margins; corolla $\frac{1}{3}$–1 in. in total length; pollen globose, honeycombed; ovary glabrous.

KALAHARI REGION: Transvaal; Bosch Veld, at Elands River, *Rehmann*, 4968!

24. **C. pretoriensis** (C. B. Clarke); innovations villous-sub-tomentose; branches 8 in. long, slender, woody, internodes $\frac{1}{2}$–1 in. long; leaves $1\frac{1}{2}$ by $\frac{1}{4}$–$\frac{1}{3}$ in., narrowly oblong, when mature nearly glabrate; flowers axillary, solitary, more than $1\frac{1}{2}$ in. long; posticous sepal $\frac{1}{2}$ by $\frac{1}{8}$–$\frac{1}{6}$ in., elliptic-lanceolate, entire, hairy, almost white-tomentose; capsule exceeding $\frac{1}{2}$ in. in length, ovoid with lanceolate beak, very hairy upwards, 2-seeded.

KALAHARI REGION: Transvaal; Apies Poort near Pretoria, *Rehmann*, 4103!

25. **B. Wilmsii** (Lindau ms.); very nearly glabrous; branches 6–11 in. long; leaves up to 2 by $\frac{3}{4}$ in., sessile, ovate, tip triangular and obtuse, base rounded or subcordate; flowers solitary or a few clustered at the tips of the branches; posticous sepal exceeding $\frac{3}{4}$ by $\frac{1}{4}$ in., narrow-elliptic, obtuse, entire, green, glabrous; corolla lilac-purple (*Galpin*), $1\frac{1}{4}$ in. long.

KALAHARI REGION: Transvaal; near Lydenburg, *Wilms*, 1217! grassy slopes near Barberton, 2800 ft., *Galpin*, 863!

26. **B. Galpinii** (C. B. Clarke); nearly glabrous; branch 1 ft. long, with 2 lines of minute white pubescence; leaves 3 by $1\frac{1}{4}$–$1\frac{1}{3}$ in., ovate, entire, glabrous, tip triangular-obtuse, base narrowed into a petiole $\frac{1}{6}$ in. long; inflorescence a few-flowered cluster in an upper axil; posticous sepal $\frac{1}{2}$–$\frac{2}{3}$ in. long, ovate, entire, green, glabrous or very nearly so; corolla yellow (*Galpin*).

KALAHARI REGION: Transvaal; at Sheba Battery, in Kaap Valley, near Barberton, 1900 ft., *Galpin*, 1331a!

Imperfectly known Species.

27. **B. repens** (Nees in DC. Prod. xi. 230); prostrate, 1–2 ft. long, young parts yellow-strigose; leaves $1\frac{1}{2}$ by $\frac{3}{4}$ in., shortly villous, afterwards glabrate, base attenuated sometimes to the base of the petiole; racemes scattered, 2–1-flowered; bract minute, linear; 2 outer calyx-segments subsimilar, up to $\frac{1}{4}$ by $\frac{1}{2}$–$\frac{2}{3}$ in., ovate, sparsely hairy, subentire, ultimately membranous, reticulate; corolla $1\frac{1}{2}$ in. long, pink, tube linear; pistil glabrous; capsule $\frac{1}{2}$–$\frac{3}{4}$ in. long, 2-seeded, or 4-seeded with the 2 lower seeds considerably smaller. *C. B. Clarke in Dyer, Fl. Trop. Afr.* v. 166; *Schinz in Mém. Herb. Boiss.* x. 63.

EASTERN REGION : Delagoa Bay, *Junod*, 114, 321.

Frequent in South-east Tropical Africa.

28. B. ilicina (E. Meyer ex T. Anders. in Journ. Linn. Soc. vii.
28) ; erect, glabrous ; stem terete ; leaves shortly petioled, ovate,
spinous at the tip, spinous-toothed on the margins ; bracts simple,
rigid, tip and margins spinous ; outer calyx-segments ovate, long
spinous-toothed on the margins, membranous, glabrous, reticulated ;
inner lanceolate, spinous, 1-nerved.

WESTERN REGION : Little Namaqualand ; between Holgat River and Orange
River, *Drège*.

Not seen. There is no *Barleria ilicina* in Drège, Zwei Pflanzengeogr.
Documente, but there is a *Blepharis? ilicina*, *E. Meyer*, at pp. 92 and 168.
T. Anderson's plant might, from the description, be *Barleria irritans*, var. *rigida.*

XIX. RUTTYA, Harv.

Bra ts and bracteoles small, linear. *Calyx* 5-partite nearly to the
base ; segments 5, equal, narrowly lanceolate. *Corolla-tube* wide ;
segments 5, ovate, in 2 lips. *Stamens* 2, fertile ; anthers 1-celled' ;
pollen subglobose, longitudinally ribbed. *Style* with 2 very small
equal branches. *Capsule* ellipsoid, stalked, 4-seeded in the upper
part ; seeds without hairs.

Panicles dense, terminal, appearing as short or long cylindric spikes ; leaves
entire ; the herbaceous petioles disarticulate at the base leaving small wooden
cup-like scars on the branches.

Species 5, African, whereof 1 extends to Arabia, 2 to Madagascar.

1. R. ovata (Harv. in Hook. Lond. Journ. Bot. i. (1842) 27) ; a
small glabrescent shrub, the young parts and inflorescence sparingly
pubescent ; branches rather thick ; leaves 3¼ by 1½ in., ovate,
narrowed suddenly into a petiole ⅓ in. long ; inflorescence 1½–3 by
1¼ in., dense, of numerous abbreviated cymes appearing almost
comose from the numerous caudate sepals ½ in. long ; corolla ¾ in.
long, white (*Wood*). *Nees in DC. Prod.* xi. 309 ; *Harv. Thes.
Cap.* ii. 27, *t.* 144 ; *T. Anders. in Journ. Linn. Soc.* vii. 51 ; *Lindau
in Engl. & Prantl, Pflanzenfam.* iv. 3B, 340. *Hypoestes fimbriata,
E. Meyer in Drège, Zwei Pflanzengeogr. Documente*, 156, 193 *fide
Nees.*

KALAHARI REGION : Transvaal ; Rimers Creek, near Barberton, *Wood*, 23 !
4016 ! Elaudspruit, 5000 ft., *Schlechter*, 3871 !
EASTERN REGION : Natal ; Umhloti, *Wood*, 1224 ! littoral, *Wood*, 462 !
near Durban, *Williamson !* near the mouth of Umzimkulu River, *Drège*, and
without precise locality, *Sutherland ! Peddie ! Gueinzius !*

XX. JUSTICIA, Linn.

Calyx small, divided nearly to the base into 5 or 4 narrow
segments. *Corolla* (in the Cape species) small or medium-sized,
2-lipped, hairy ; tube not (or scarcely) longer than the limb ;

posticous lip entire or shortly 2-lobed, anticous lip 3-lobed; palate often with spots or transverse wrinkles. *Stamens* 2; one anther-cell below the other, tailed at the base (but, in *J. mutica* and *J. campylostemon*, the anther-cells are nearly at equal height and the basal tail is absent or most minute); pollen ellipsoid, with 2 stopples, longitudinally banded, the rows of tubercles various. *Ovary* with 2 ovules in each cell; style thinly hairy or glabrous, branches minute. *Capsule* 4-seeded; stalk usually short; placenta not rising elastically from the base; seeds tubercular-rugose.

Herbs or shrubs; leaves entire or obscurely wavy; flowers white with purple or rose spots, or in a few species yellow; inflorescence very varied, on which variety the sections below are grounded; bracts in the first section large (the flowers in strobilate spikes), more commonly small, narrow; bracteoles generally small, narrow. *Justicia*, as here defined, differs from *Adhatoda* by the basal tail to the lower anther-cell (which character fails us in the 2 species of *Justicia* above mentioned) and in habit.

Species 200, throughout the warmer parts of the World; or at least 300 if (as proposed by Lindau) the genus *Dianthera* be sunk in *Justicia*. The genus *Isoglossa* (here admitted) differs by having a slender corolla-tube and globose (not ellipsoid) pollen. It is difficult to estimate the number of species in these closely-allied genera without a revision of the species of the World.

Section 1. BETONICA. Spikes strobilate; floral leaves ovate, often somewhat 4-ranked, often (always in the Cape species) reticulated with green veins.

Petioles often $\frac{1}{4}$–$\frac{1}{2}$ in. long ... ... (1) **Betonica.**
Petioles absent or nearly so:
  Stems and leaves hairy ... ... (2) **betonicoides.**
  Stems glabrous; leaves glabrate:
    Leaves $1\frac{1}{2}$ by $\frac{1}{3}$ in., ovate-oblong ... ... (3) **trinervia.**
    Leaves 1 by $\frac{1}{4}$ in., lanceolate, acute ... (4) **pallidior.**
    Leaves $2\frac{1}{4}$ by $\frac{1}{4}$–$\frac{1}{3}$ in., narrowly oblong ... (5) **cheiranthifolia.**

Section 2. ROSTELLULARIA. Flowers axillary, but running into terminal (not strobilate) spikes; floral leaves much narrower than in Sect. *Betonica*.

Leaves ovate or elliptic:
  Flowers yellow ... ... (6) **flava.**
  Flowers (not yellow) variously purple and white:
    Leaves up to 5 by 2 in., thin, green, glabrate ... ... (7) **petiolaris.**
    Leaves rarely exceeding 2 in. long, thick, hairy:
      Calyx $\frac{1}{3}$ in. long ... ... (8) **Bowiei.**
      Calyx $\frac{1}{8}$ in. long ... ... (9) **Burchellii.**
Leaves linear ... ... (10) **spergulæfolia.**

Section 3. CALOPHANOIDES. Flowers axillary, scattered, solitary or 2–3 clustered under 1 floral leaf, sometimes approximated towards the ends of the branches, but not running into a terminal spike. Bracts, bracteoles and calyx-segments narrow, never large.

Plants with woody branches and medium-sized leaves:
  Leaves usually 1–2 in. long; bracteoles 0 ... (11) **capensis.**
  Leaves usually $\frac{1}{3}$–$\frac{2}{3}$ in. long; bracteoles $\frac{1}{8}$ in. long ... ... (12) **cuneata.**
Plants with herbaceous branches or very small leaves:

Capsule glabrous or minutely hairy on the
sutures :
    Lower anther-cell not tailed   ...   ... (13) **mutica.**
    Lower anther-cell tailed :
       Calyx-segments hairy :
          Corolla yellow, ¾ in. long ...   ... (14) **odora.**
          Corolla white with pink or purple
          marks :
             Upper leaves very small, close
               together   ...   . (15) **pulegioides.**
             Upper leaves similar to the
             lower :
               Leaves narrowly elliptic ... (16) **Kraussii.**
               Leaves roundly ovate   ... (17) **rotundifolia.**
       Calyx-segments glabrous, or very nearly
       so :
          Flowers often shortly pedicelled   ... (18) **orchioides.**
          Flowers axillary; leaves very obtuse (19) **thymifolia.**
Capsule hairy all over ...   . (20) **Woodii.**

Section 4. GENDARUSSA. Peduncles towards the top of the branches, carrying simple or compound cymes; flowers clustered; bracts and bracteoles
inconspicuous.

    Petioles 1–2 in. long; cyme-branches very slender... (21) **campylostemon.**
    Petioles less than ½ in. long; cyme-branches
       rather rigid ...   . (22) **Bclusii.**

Section 5. ANSELLIA. Peduncles scattered, carrying 2 or few separate
flowers near the top; bracts and bracteoles narrow, small (or 0). (*Diantheræ*
spp., Benth.)

    Corolla ½ in. long ...   . (23) **anagalloides.**
    Corolla less than ¼ in. long :
       Hairy; peduncles with 3–5 flowers   ... (24) **matammensis.**
       Nearly glabrous; peduncles with 2–3 flowers... (25) **exigua.**

1. J. Betonica (Linn. Sp. Pl. 15); a nearly glabrous shrub;
leaves 2–3 by ⅔–1 in., ovate-lanceolate, base acuminate; petiole ¼ in.
or more; spikes 3–4 by ¾ in., terminal, strobilate; bracts somewhat
4-ranked ½ by ⅓ in., ovate, acute, white, green-veined; bracteoles
⅓ by ¼ in., similar to the bracts; calyx ¼ in. long, divided nearly to
the base; segments 5, lanceolate, subulate, densely viscous-pubescent;
corolla ½ in. long, white with rose spots; one anther-cell below the
other, long-tailed; ovary glabrous below, the top hispid; style thinly
hairy; capsule ½ by ¼ in., 4-seeded, clavate, pubescent nearly to the
base; seeds rugose. *T. Anders. in Journ. Linn. Soc.* vii. 38;
C. B. Clarke in. Dyer, Fl. Trop. Afr. v. 184. *J. lupulina,* a, *E.*
Meyer in Drège, Zwei Pflanzengeogr. Documente, 144, 195. *Adhatoda*
Betonica, Nees in Wall. Pl. As. Rar. iii. 103, *and in DC. Prod.* xi.
385. *Dicliptera lupulina, Presl, Bot. Bemerk.* 95. *Nicoteba*
Betonica, Lindau in Engl. Jahrb. xviii. 56, 63, *t.* 2, *fig.* 56, *in Engl.*
& Prantl, Pflanzenfam. iv. 3B, 329, *and in Engl. Pfl. Ost-Afr.*
C. 370.

COAST REGION: Albany Div.; Great Fish River, *Masson!* King
Williamstown Div.; between Buffalo River and Kei River, 1000–2000 ft., *Drège!*
Komgha Div.; Komgha, 2000 ft., *Flanagan,* 670 !

A frequent plant in India and Malaya; 1 example also from Mozambique seen.

2. **J. betonicoides** (C. B. Clarke in Dyer, Fl. Trop. Afr. v. 184); stems densely and shortly hairy; leaves 2–4 in. long, when mature hairy beneath, subsessile; otherwise as *J. Betonica*, Linn.

EASTERN REGION : Natal; Tongaat, *Wood,* 108! and without precise locality, *Gerrard,* 321!

Frequent from Mozambique to Abyssinia.

3. **J. trinervia** (Vahl, Enum. i. 156); leaves 1½ by ⅓ in., ovate-oblong, subsessile; flowers rather smaller than those of *J. Betonica,* Linn.; otherwise as *J. Betonica. C.B. Clarke in Dyer, Fl. Trop. Afr.* v. 185. *J. Betonica, T. Anders. in Journ. Linn. Soc.* ix. 510. *J. lupulina c, E. Meyer in Drège, Zwei Pflanzengeogr. Documente,* 158, 195. *Adhatoda trinervia, Nees in DC. Prod.* xi. 386. *Nicoteba trinervia, Lindau in Engl. & Prantl, Pflanzenfam.* iv. 3B, 329.

EASTERN REGION: Natal; Umlazi River, under 200 ft., *Drège!* near Durban, *Williamson!* and without precise locality, *Peddie!*

Also in East Tropical Africa and South India.

4. **J. pallidior** (C. B. Clarke); stems glabrous, except in var. β; leaves 1 by ¼ in., lanceolate, acute at both ends, subsessile; otherwise as *J. Betonica,* Linn. *J. Betonica, T. Anders. in Journ. Linn. Soc.* vii. 38 *partly. Adhatoda variegata, var. pallidior, Nees in DC. Prod.* xi. 385 ; *Sonder in Linnæa,* xxiii. 94.

VAR. β, **Cooperi** ; stems densely and shortly hairy in the upper part.

COAST REGION: Var. β : British Kaffraria, *Cooper,* 3114!

KALAHARI REGION: Transvaal; Apies River, *Burke,* 514! *Schlechter,* 3613! Magalies Berg, *Zeyher,* 1399! Pilgrims Rest, *Greenstock!*

This species only differs from *J. trinervia* by the smaller, narrower and more acute leaves.

5. **J. cheiranthifolia** (C. B. Clarke); leaves up to 2¼ by ¼–⅓ in., narrowly oblong, subsessile; otherwise as *J. Betonica,* Linn. *J. Betonica, T. Anders. in Journ. Linn. Soc.* vii. 38 *partly. Adhatoda cheiranthifolia, Nees in DC. Prod.* xi. 387 ; *Sonder in Linnæa,* xxiii. 94.

KALAHARI REGION: Transvaal; Magalies Berg, *Zeyher,* 1400! *Burke!* rocky places near Barberton, 3500 ft., *Thorncroft,* 127! *Wood,* 4164! Saddleback Range near Barberton, 4000–4500 ft., *Galpin,* 616!

EASTERN REGION: Natal; *Gerrard,* 1271!

6. **J. flava** (Vahl, Symb. ii. 15, not of Kurz) ; hairy, up to 2–4 ft. high; leaves 2 by ¾–1 in., ovate-lanceolate, decurrent on the petiole; petiole 0–⅔ in. long; spikes terminal, 4–8 by ¾ in., continuous or interrupted at the base, lower whorls sometimes distant, i.e. axillary clusters; floral leaves linear-oblong with 3–1 flowers; bracts exceeding ¼ in. long, linear, spathulate-tipped; bracteoles scarcely ¼ in. long, linear; calyx ⅙ in. long, 5-fid to the base, scarious brown, or hardly green; segments linear; corolla ⅓–½ in. long, yellow; one anther-cell much below the other, tailed; pollen globose with

2 stopples, banded, rows of tubercles obscure ; ovary pubescent
upwards ; style thinly hairy ; capsule $\frac{1}{2}$ in. long, pubescent, 4-seeded ;
seeds tubercular-rugose. *C. B. Clarke in Dyer, Fl. Trop. Afr.* v.
190 *with syn.* J. *fasciata,* E. *Meyer in Drège, Zwei Pflanzengeogr.
Documente,* 160, 195 ; *T. Anders. in Journ. Linn. Soc.* vii. 39 ;
Lindau in Engl. & Prantl, Pflanzenfam. iv. 3B, 349, *and in Engl.
Pfl. Ost-Afr.* C. 373. *Dianthera flava, Vahl, Symb.* i. 5. *Dicliptera
fasciata, Presl, Bot. Bemerk.* 95. *Adhatoda flava, Nees in DC.
Prod.* xi. 401. A. *fasciata, Nees in DC. Prod.* xi. 402. *Athlianthus,
Endl. Gen. Suppl.* ii. 63.

KALAHARI REGION : Transvaal ; Marico District, *Holub !*
EASTERN REGION : Natal ; Umgeni River, under 500 ft., *Drège !* Durban,
Wood, 831 ! Delagoa Bay, *Forbes !*

Abundant throughout Tropical Africa.

Nees says *Athlianthus,* Endl., is J. *petiolaris;* but, as it is said to be a
tropical species with yellow flowers, it must be J. *flava.*

7. J. petiolaris (E. Meyer in Drège, Zwei Pflanzengeogr. Documente,
150, 196) ; branches 12–20 in. long, more or less 4–6-angular
upwards, with white deflexed hairs, glabrate below ; leaves up to
5 by 2 in., acuminate at either end, decurrent on the petiole, thin,
green when mature, glabrous except for a few strigose hairs on the
nerves ; petiole often 1–1$\frac{1}{2}$ in. long ; spikes 5 by $\frac{3}{4}$ in., terminal,
usually interrupted at the base ; floral leaves oblong, with 3–1 flowers
under each ; bract $\frac{1}{4}$ in. long or rather more, lanceolate ; bracteoles
$\frac{1}{6}$ in. long, linear ; calyx $\frac{1}{3}$ in. long, 5-partite to the base ; segments
linear-lanceolate, acute ; corolla $\frac{3}{4}$ in. long, blue (*Wood*) ; one anther-
cell below the other, tailed ; pollen ellipsoid with 2 stopples, tubercles
obsolete ; pistil glabrous, except for a few scattered hairs on the
style ; capsule $\frac{1}{2}$ in. long, glabrous, 4-seeded ; seeds tubercular-
rugose. *T. Anders. in Journ. Linn. Soc.* vii. 39 ; *Lindau in Engl. &
Prantl, Pflanzenfam.* iv. 3B, 349. *Adhatoda petiolaris, Nees in DC.
Prod.* xi. 402.

EASTERN REGION : Pondoland ; St. Johns River, under 1000 ft., *Drège !*
Natal ; without precise locality, *Gerrard,* 1896 ! Zululand ; Indulindi, 1000–
1800 ft., *Wood,* 3953 !

8. J. Bowiei (C. B. Clarke) ; leaves up to 2$\frac{1}{2}$ by 1$\frac{1}{2}$ in., rhomboid-
ovate, rather thick, persistently hairy on both surfaces (in Bowie's
example); calyx green, $\frac{1}{3}$ in. long ; corolla red (*Bowie*) or blue
(*Flanagan*) ; capsule more than $\frac{1}{2}$ in. long ; otherwise as J. *petiolaris.*
Adhatoda petiolaris, var. ? β, *Nees in DC. Prod.* xi. 402.

SOUTH AFRICA : without precise locality, *Guthrie,* 4711 !
COAST REGION : moist situations in George, Uitenhage, and Albany Divs. ;
Bowie ! East London Div. ; near the mouth of the Kei River, *Flanagan,* 882 !

9. J. Burchellii (C. B. Clarke) ; leaves hardly attaining 1$\frac{1}{4}$ in.
long ; petiole less than $\frac{1}{4}$ in. long ; calyx hardly $\frac{1}{8}$ in. long ; capsule
$\frac{1}{3}$ in. long ; otherwise as J. *petiolaris,* E. Meyer.

COAST REGION: Bathurst Div. ; between Riet Fontein and the seashore, *Burchell,* 4107 !

This, like *J. Bowiei,* may be arranged merely as a variety of *J. petiolaris.*

10. J. spergulæfolia (T. Anders. in Journ. Linn. Soc. vii. 43) ; pubescent; branches 6–10 in. long, undivided; leaves 1 by $\frac{1}{16}$ in., linear; spike terminal, 2–3$\frac{1}{2}$ in. long, of 10–20 flowers, the lower $\frac{1}{4}$–$\frac{1}{6}$ in. apart, only 1 in each pair of bracts developed ; bracts $\frac{1}{6}$ in. long, narrowly lanceolate ; bracteoles $\frac{1}{12}$ in. long, narrowly lanceolate ; calyx $\frac{1}{8}$ in. long or rather more, 5-partite to the base ; segments linear-lanceolate ; corolla $\frac{1}{4}$ in. long ; one anther-cell much below the other, tailed ; pollen ellipsoid, with 2 stopples and 1 row of tubercles beside each stopple ; ovary glabrous below, with very long hair at the top ; capsule 4-seeded (*T. Anderson*).

KALAHARI REGION : Transvaal ; Maxalaquena River, 4250 ft., *Schlechter,* 4267 !

Also in Damaraland.

This is allied to *J. linearispica,* C. B. Clarke (in Dyer, Fl. Trop. Afr. v. 192), and the large group of *J. peploides* in India ; the constant absence of one flower in each pair gives it a marked aspect.

11. J. capensis (Thunb. Prod. 104) ; a rather stout shrub, attaining 3–5 ft. in height ; branches quadrangular, densely hairy when young, more or less glabrate in age, not rarely with the 2 opposite faces hairy, the alternate glabrous ; leaves 1–2$\frac{1}{2}$ by $\frac{2}{3}$–1 in., elliptic, hairy on both surfaces, becoming glabrate except on the nerves, tip rounded, base cuneate ; petiole hardly any ; flowers axillary, distant, but often 2–3 under 1 floral leaf, minutely pedicelled, without other bract ; bracteoles obsolete ; calyx $\frac{1}{6}$ in. long, 5-partite to the base ; segments linear ; corolla $\frac{2}{3}$ in. long, reddish (*Kensit*) ; one anther-cell below the other, tailed (in one example the anther-cells are nearly at equal height, both equally short-tailed) ; pollen ellipsoid with 2 stopples and 1 row of tubercles on either side of each stopple ; pistil glabrous ; capsule $\frac{3}{4}$ in. in length and upwards, 4-seeded ; seeds tubercular-rugose. *Thunb. Fl. Cap. ed. Schult.* 478 ; *T. Anders. in Journ. Linn. Soc.* vii. 41 ; *Engl. in Engl. Jahrb.* x. 264 ; *Lindau in Engl. & Prantl, Pflanzenfam.* iv. 3B, 349. *J. amygdalina, a, E. Meyer in Drège, Zwei Pflanzengeogr. Documente,* 129, 195. *J. oleæfolia, Schlechtend. ex Nees in DC. Prod.* xi. 391 ; *E. Meyer in Drège, l.c.* 151, 196. *Gendarussa capensis, Nees in Linnæa,* xv. 366, *including var. β* ; *Drège in Linnæa,* xx. 200 ; *Hook. Journ. Bot.* ii. (1840) 126. *Adhatoda capensis, Nees in DC. Prod.* xi. 391, *excl. var. arenosa.*

COAST REGION : Uitenhage Div. ; Cannon Hill, *Kensit,* 9 ! by the Zwartkops River, below 500 ft., *Drège ! Ecklon & Zeyher,* 82 ! in thickets bordering the plains, *Bowie !* near the mouths of the Coega and Zwartkops Rivers, *Zeyher,* 3593 ! East London Div. ; near the mouth of the Kei River, 200 ft., *Flanagan,* 2350 ! British Kaffraria, *Cooper,* 136, 394 !

CENTRAL REGION : Alexandria Div ; Zwartwater Poort, *Burchell,* 3399 ! Somerset Div. ; near Somerset East, *Bowker !*

KALAHARI REGION : Griqualand West ; Groot Boetsap, 4000 ft., *Marloth,* 1076 (fide Engler).

12. J. cuneata (Vahl, Symb. ii. 10); a rigid shrub 1–2 ft. high; branches thick, glabrous or at the top minutely pubescent; leaves $\frac{2}{3}$ by $\frac{1}{6}-\frac{1}{4}$ in., narrowly elliptic or obovate, obtuse, glabrous, thick, drying a dark reddish-brown, narrowed at the base; petiole $0-\frac{1}{12}$ in.; flowers scattered, axillary, nearly sessile; bracteoles lanceolate, nearly as long as the calyx; pedicel scarcely $\frac{1}{20}$ in. long, but distinct; calyx $\frac{1}{8}-\frac{1}{6}$ in. long, deeply 5-lobed; lobes lanceolate, obtuse, nearly glabrous except for scattered hairs on the margins; corolla $\frac{1}{2}$ in. long, white (*Bowie*); one anther-cell a little below the other, tailed; pollen ellipsoid, banded, with 2 stopples and 1 row of tubercles on each side of each stopple; pistil glabrous; very young capsule with 4 similar ovules apparently about to perfect seed. *T. Anders. in Journ. Linn. Soc.* vii. 41 *partly; Lindau in Engl. & Prantl, Pflanzenfam.* iv. 3B, 349. *Gendarussa cuneata, Nees in Linnæa,* xv. 367; *Hook. Journ. Bot.* ii. (1840) 126. *G. hyssopifolia, Nees in Linnæa,* xv. 368; *Drège in Linnæa,* xx. 200; *Adhatoda cuneata, Nees in DC. Prod.* xi. 392 *partly; Sonder in Linnæa,* xxiii. 94. *A. hyssopifolia, α, Nees in DC. Prod.* xi. 392 *partly; Sonder in Linnæa,* xxiii. 94.

SOUTH AFRICA: without precise locality, *Masson!*
COAST REGION: Uitenhage Div.; by the Zwartkops River, *Zeyher,* 260! on exposed heights by the Sunday River, *Bowie,* 119!

J. hyssopifolia, Linn., from the Canaries, is a shrub with very much larger leaves and flowers; the Cape plant which Nees has named *hyssopifolia,* is identical with that he has named *cuneata.* Nees has also named as *Tyloglossa cuneata* the remote *Monechma foliolosum* below, see p. 74, which does not match the present species in colour and has only 4 calyx-lobes (apart from the differences in capsule and seeds here treated as generic). T. Anderson does not mention *J. foliolosa,* Drège, but cites Nees in DC. Prod. "with syn."

13. J. mutica (C. B. Clarke); persistently hairy; branches 10–20 in. long; internodes 1–2 in. long; leaves $1\frac{1}{4}$ by $\frac{3}{4}$ in., ovate; petiole up to $\frac{1}{4}$ in. long; flowers axillary, scattered, solitary; calyx-segments exceeding $\frac{1}{4}$ in. in length, linear; corolla $\frac{1}{2}-\frac{2}{3}$ in. long; anther-cells at nearly equal height, muticous or most obscurely tailed at the base; pollen ellipsoid with 2 stopples and 1 row of tubercles on either side of each stopple; ovary glabrous; style thinly hairy; capsule $\frac{1}{2}$ by $\frac{1}{4}$ in., 4-seeded. *Adhatoda protracta, Nees in DC. Prod.* xi. 392 *partly.*

COAST REGION: in wooded situations in Uitenhage and Albany Divs., *Bowie!*

The anther-cells are not those of the genus *Justicia :* the larger calyx, capsule and leaves also do not agree with any example of *J. pulegioides.*

14. J. odora (Vahl, Enum. i. 164); an undershrub, 1–2 ft. high, glabrous, except the flowers; leaves 1 by $\frac{1}{4}-\frac{1}{3}$ in., narrowly elliptic, obtuse, base narrowed; petiole $0-\frac{1}{10}$ in. long; flowers axillary, solitary or more rarely 2–3 together; calyx-segments $\frac{1}{4}$ in. long, linear; corolla $\frac{2}{3}$ in. long, yellow; one anther-cell much below the other, tailed; capsule $\frac{1}{2}$ in. long, glabrous, 4-seeded; seeds tubercular-scabrous. *T. Anders. in Journ. Linn. Soc.* vii. 42 ; *C. B. Clarke in Dyer, Fl. Trop. Afr.* v. 201 (*excl. syn. J. leucodermis*). *J. polymorpha, Schinz in Verh. Bot. Brandenb.* xxxi. (1890) 203 ; *Lindau in Engl. & Prantl, Pflanzenfam.* iv. 3B, 349. *Adhatoda odora, Nees in DC. Prod.* xi. 399. *Gendarussa odora, Presl, Bot. Bernerk.* 95.

KALAHARI REGION : Transvaal; Avoca, near Barberton, 1900 ft., *Galpin,* 1238 !

Extends to Abyssinia.

The calyx, corolla, and capsule are all rather longer than in the type plant of Vahl.

15. **J. pulegioides** (E. Meyer in Drège, Zwei Pflanzengeogr. Documente, 151, 156, 196); sparingly pubescent, or frequently nearly glabrous except the flowers; branches 4–12 in. long, slender, but woody ; leaves on the ultimate branches about $\frac{1}{4}$ in. (rarely more than $\frac{1}{2}$ in.) long, elliptic, obtuse, very shortly petioled, internodes $\frac{1}{4}$ in. long and on the main branches of the same plant internodes $1\frac{1}{2}$ in. with leaves 1 in. long occur; flowers solitary, axillary, scattered ; calyx-segments $\frac{1}{6}$ in. long, narrowly linear ; corolla $\frac{1}{3}$–$\frac{1}{2}$ in. long, white marked with rose or purple ; one anther-cell clearly lower than the other, tailed ; capsule $\frac{1}{4}$ in. long, hardly $\frac{1}{8}$ in. broad, glabrous, 4-seeded. *J. convexa, E. Meyer in Drège, Zwei Pflanzengeogr. Documente,* 155, 195, *fide Nees. J. protracta, T. Anders. in Journ. Linn. Soc.* vii. 41 *partly ; Lindau in Engl. & Prantl, Pflanzenfam.* iv. 3B, 349. *Gendarussa protracta, Nees in Linnæa,* xv. 371 ; *Drège in Linnæa,* xx. 200. *G. prunellæfolia, Hochst. in Flora,* 1845, 71. *Chætacanthus Persoonii, a, Nees in Linnæa,* xv. 356, *and in DC. Prod.* xi. 462 *partly ; T. Anders. in Journ. Linn. Soc.* vii. 23 *partly* (?). *Adhatoda protracta, vars. β, γ, δ, Nees in DC. Prod.* xi. 390 ; *Sonder in Linnæa,* xxiii. 94.

VAR. β, late-ovata (C. B. Clarke); leaves broad-ovate (many $\frac{3}{4}$ in. wide), hispid and reticulately rugose beneath.

COAST REGION : Uitenhage Div., *Ecklon & Zeyher,* 436! Komgha Div ; Komgha, 2000 ft., *Flanagan,* 725 !

CENTRAL REGION : Somerset Div. ; at Commadagga, *Burchell,* 3300 ! Var. β : Alexandria Div. ; on the rocks of Zwartwater Poort, *Burchell,* 3364 ! 3405 !

KALAHARI REGION : Transvaal ; Houtbosch Rand, 4500 ft., *Schlechter,* 3321 !

EASTERN REGION : Pondoland ; between St. Johns River and Umtsikabi River, 1000–2000 ft., *Drège !* Natal ; borders of woods around Durban Bay, *Krauss,* 304 ! Inanda, *Wood,* 309 ! 718 ! Durban Flats, *Wood in MacOwan & Bolus, Herb. Norm. Aust.-Afr.,* 1019 ! and without precise locality, *Peddie ! Sanderson,* 433 ! *Grant !*

Krauss, 304, is a fine and typical example of this species, and is written up by Nees' hand as his *Chætacanthus Persoonii a.* The var. β, *late-ovata,* is probably specifically distinct, the leaves being very unlike in structure to those of any other plant referred to *J. pulegioides.*

16. **J. Kraussii** (C. B. Clarke); branches 12--16 in. long, pubescent; internodes 1–2$\frac{1}{2}$ in. long; leaves 1$\frac{1}{4}$ by $\frac{1}{2}$ in., elliptic-oblong, glabrate, petiole $\frac{1}{8}$–$\frac{1}{6}$ in. long; flowers white, nearly as of

J. pulegioides; capsule not seen.　*J. protracta, T. Anders. in Journ.
Linn. Soc.* vii. 41 *partly. Gendarussa mollis, Hochst. in Flora,*
1845, 71. (*Not Justicia mollis, E. Meyer in Drège, Zwei
Pflanzengeogr. Documente,* 92, 196).

VAR. β, florida (C. B. Clarke) ; branches long, repeatedly divided ; flowers in
the upper axils numerous ; the floral leaves (or bracts) much reduced in size.

EASTERN REGION: Natal ; in grassy flats between the Umlazi River and
Durban Bay, *Krauss,* 61! Inanda, *Wood,* 423! and without precise locality,
Gerrard, 1272! Var. β : Inanda, *Wood,* 566!

Justicia mollis, E. Meyer, is carried to *Monechma ;* to avoid confusion Krauss'
n. 61 is given here a new specific name. The series of plants included by T.
Anderson under *J. protracta* consists of closely-allied plants ; the present *J.
Krausii,* with its variety, is exceedingly near *J. filifolia,* Lindau, and its allies in
Mozambique.

17. **J. rotundifolia** (E. Meyer in Drège, Zwei Pflanzengeogr.
Documente, 154, 196) ; branches 9 in. long, weak, slightly pubescent ;
internodes up to $1\frac{1}{2}$ in. long ; leaves $\frac{2}{3}$ by $\frac{1}{3}$ in., ovate, obtuse, thin,
pubescent, ultimately glabrate ; petiole $0-\frac{1}{6}$ in. long ; flowers few,
remotely scattered, solitary or 2 together, axillary ; calyx-segments
scarcely $\frac{1}{6}$ in. long, linear ; corolla nearly $\frac{1}{2}$ in. long ; one anther-cell
below the other, tailed ; pistil glabrate, except for a few minute hairs
in the lower part of the style ; capsule unknown. *T. Anders. in
Journ. Linn. Soc.* vii. 41. *Adhatoda rotundifolia, Nees in DC.
Prod.* xi. 391.

EASTERN REGION : Pondoland or Natal ; between Umtentu River and Umzimkulu River, under 500 ft., *Drège!*

Only one fragment was seen by Nees and T. Anderson. It differs from
J. pulegioides, E. Meyer, in the weak stems and very thin leaves.

18. **J. orchioides** (Linn. f. Suppl. 85) ; a rugged undershrub,
8–24 in. high ; branches many, nearly all alternate, woody, often
$\frac{1}{4}$ in. in diam., glabrous or obscurely puberulous at the tips ; leaves
$\frac{1}{8}-\frac{1}{2}$ in. long, subsessile, oblong elliptic or ovate, puberulous or
glabrate, tip obtuse or acute, base narrowed-ovate or truncate ;
pedicels $0-\frac{1}{6}$ in. long, few, scattered ; bracteoles shorter than the
calyx and distant from it in the wild plant ; calyx scarcely $\frac{1}{6}$ in.
long ; lobes lanceolate with white margins, nearly glabrous not
glandular (in Bolus, 672, the calyx is minutely hairy, but scarcely
glandular) ; corolla $\frac{1}{3}-\frac{1}{2}$ in. long, white, hairy outside ; lower anther-
cell half-way below the other, with a long clavate tail ; pollen
ellipsoid, with 2 stopples, longitudinally 6–8-ribbed, hardly banded,
without tubercles ; ovary glabrous ; style thinly hairy in the lower
half ; capsule $\frac{1}{2}$ in. long, normally 4-seeded ; seeds tuberculate-
rugose. *Thunb. Prod.* 104, *and Fl. Cap. ed. Schult.* 479 ; *Lindau
in Engl. & Prantl, Pflanzenfam.* iv. 3B, 349. *J. diosmophylla,
Lindau in Engl. & Prantl, Pflanzenfam.* iv. 3B, 349. *J. patula,
T. Anders. in Journ. Linn. Soc.* vii. 42 ; *Lindau in Engl. &
Prantl, Pflanzenfam.* iv. 3B, 349 ; *Lichtenst. in Roem. et Schult.
Syst.* i. 164 ? *J. orchioides, a, E. Meyer in Drège Zwei Pflanzengeogr. Documente,* 138, 196. *Gendarussa patula, Nees in*

64   ACANTHACEÆ (Clarke).   [*Justicia.*

Linnœa, xv. 371. *G. orchioides*, *Nees in Linnœa*, xv. 369 ? *incl. var.*
β. *G. pygmœa*, *Nees in Linnœa*, xv. 369. *G. Linaria*, *Nees ex
Drège in Linnœa*, xx. 200 ? *G. diosmophylla*, *Nees in Linnœa*, xv.
370 ; *Drège in Linnœa*, xx. 200. *Adhatoda patula*, *Nees in DC.
Prod.* xi. 393. *A. pygmœa*, *Nees in DC. Prod.* xi. 394. *A.
diosmophyllu*, *Nees in DC. Prod.* xi. 394 ; *Sonder in Linnœa*,
xxiii. 94. *A. orchioides, var. latifolia*, *Nees in DC. Prod.* xi. 393.

SOUTH AFRICA: without precise locality, *Zeyher*, 1394!
COAST REGION: Clanwilliam Div.; Lange Kloof, 400 ft., *Schlechter*, 8052!
Uitenhage Div.; at Commando Kraal, *Burke!* by the Sunday River, near
Commando Kraal, *Bolus*, 2677! by the Sunday River, *Ecklon & Zeyher*, 860!
Zeyher, 1392! Port Elizabeth Div.; near Port Elizabeth, *Baur*, 109! Alexandria
Div.; near Enon, *Baur*, 1051! Albany Div.; in dry thickets, *Bowie*, 149!
CENTRAL REGION: Somerset Div.; between Great and Little Fish Rivers,
2000–3000 ft., *Drège!* Graaff Reinet Div.; plains near Graaff Reinet, 2500 ft.,
Bolus, 683!
WESTERN REGION: Little Namaqualand; between Annenous and Abbeolakte,
700 ft., *MacOwan & Bolus*, *Herb. Norm. Aust.-Afr.*, 672!

J. orchioides, Veut. Jard. Malm. t. 51, of which there are dried examples at
Kew, is probably a cultivated state of *J. patula*, T. Anders.; it is larger in all
its parts; some leaves exceed ¾ in. in length; some pedicels are nearly ¼ in. long,
and have 2 bracts near the calyx, which is hairy; the corolla is ⅔–¾ in. long;
Ventenat says the seeds were 4 or 5.—*J. patula*, Lichtenst. in Roem. et Schultes
Syst. i. 164, can only be guessed at from the description, which says that the
branches were all opposite and the bracts at the base of the calyx glandular;
these points strongly suggest that Lichtenstein's *patula* was Drège's *J. patula, b*
which was Nees' *J. orchioides, a.*

19. **J. thymifolia** (C. B. Clarke); nearly glabrous, except the
corolla; branches 12 in. long, divided, rigid, striate, internodes
mostly ¼–½ in. long; leaves ⅓–½ by ¼ in, very obtuse; petiole
0–1/10 in.; flowers towards the ends of the branches, axillary, solitary,
rarely 2 together, not running into a spike; calyx-lobes 5, ⅙–⅕ in.
long, narrowly lanceolate or oblong or somewhat obovate, whitish,
nearly glabrous; corolla ⅔ in. long or more, hairy outside; capsule
¼ in. long, and nearly ¼ in. broad upwards, glabrous, 4-seeded;
seeds tubercular-rugose. *Adhatoda thymifolia*, *Nees in DC. Prod.*
xi. 392.

KALAHARI REGION: Griqualand West, Hay Div.; between Griqua Town and
Spuigslang, *Burchell*, 1702!

This species appears to have been overlooked altogether by T. Anderson.

20. **J. Woodii** (C. B. Clarke); whole plant hairy; stems 3 in.
long; leaves ⅓ by ¼ in., ovate, obtuse; flowers scattered, axillary;
calyx-lobes ⅛ in. long, linear; corolla hardly ¼ in. long, hairy;
capsule hardly more than ¼ in. long, 4-seeded, sessile, hairy all
over.

EASTERN REGION: Natal; Noodsberg, 2000 ft., *Wood*, 112!

21. **J. campylostemon** (T. Anders. in Journ. Linn. Soc. vii. 44);
branches 1–2 ft. long, with pubescent lines, or glabrate; leaves
4–5 by 1½–2 in., elliptic, acuminate at each end, glabrate or pubescent on the nerves, margin wavy; petioles 1–2½ in. long; cymes

axillary, lax, compound, with slender branches, the flowers approximated at the tips ; bracts and bracteoles inconspicuous ; calyx-lobes $\frac{1}{6}$ in. long, linear, acute, scarcely pubescent ; corolla $\frac{1}{2}-\frac{2}{3}$ in. long, white with purple spots ; one anther-cell slightly below the other, with a minute white tail ; pollen ellipsoid, banded, stopples 2, tubercles minute ; ovary glabrous ; style slightly hairy in the lower part ; capsule $\frac{2}{3}$ in. long, glabrous, stalk very narrow, top clavate $\frac{1}{6}$ in. wide and 4-seeded ; seeds tubercular-rugose. *N. E. Brown in Gard. Chron.* xix. (1883) 44. *Campylostemon campanulatus, E. Meyer in Drège, Zwei Pflanzengeogr. Documente,* 153, 170. *Leptostachya campylostemon, Nees in DC. Prod.* xi. 378. *Rhaphidospora campylostemon, Lindau in Engl. & Prantl, Pflanzenfam.* iv. 3 B, 329.

Coast Region : Albany Div. ; Blue Krantz, *Burchell,* 3636! Stockenstrom Div. ; Kat Berg, 2000 ft., *Hutton!*
Central Region : Somerset Div. ; Somerset East, *Bowker!*
Kalahari Region : Transvaal ; Makwongwa Forest, near Barberton, 3200 ft., *Galpin,* 904!
Eastern Region : Pondoland ; between St. Johns River and Umtsikaba River, under 1000 ft., *Drège!* Griqualand East ; in a wood on Mount Malowe, 4000 ft., *Tyson,* 2083! and in *MacOwan & Bolus, Herb. Norm. Aust.-Afr.,* 892! Natal ; Inanda, *Wood,* 773! Tongaat, 3022! and without precise locality, *Gerrard,* 1898! 1899! 1900! *Cooper,* 1104! *Cordukes!*

22. J. Bolusii (C. B. Clarke) ; shrubby, nearly glabrous except the flowers ; branches 12 in. long, leafy ; leaves 2 by $\frac{1}{2}-\frac{2}{3}$ in., elliptic, obtuse, base narrowed ; petiole $\frac{1}{8}-\frac{1}{4}$ in. long ; peduncles many towards the top of the branches, about 1 in. long, rigid, mostly simple, with 2–5 flowers clustered at the top ; bracts and bracteoles inconspicuous ; calyx $\frac{1}{4}$ in. long, divided to the base ; segments 5, very narrowly lanceolate, minutely pubescent and glandular ; corolla $\frac{2}{3}$ in. long ; one anther-cell much below the other, with a large forked tail ; pollen ellipsoid, longitudinally striated, stopples 2, tubercles 0 ; ovary densely shaggy ; style slightly hairy.

Coast Region : Komgha Div. ; margins of woods near Komgha, 1800 ft., *Flanagan,* 608!

This species comes between *J. Gendarussa,* Linn. f., and *J. cordata,* T. Anders.

23. J. anagalloides (T. Anders. in Journ. Linn. Soc. vii. 42) ; branches 6–16 in. long, weak, pubescent or hispid, often in lines ; leaves $\frac{2}{3}$ by $\frac{1}{4}$ in., when mature nearly glabrous, tip obtuse, base cuneate ; petiole 0–$\frac{1}{8}$ in. long ; peduncles scattered, $\frac{1}{4}$–1 in. long, with 2 or 3 flowers near the top ; bracteoles linear, shorter than the calyx ; calyx $\frac{1}{4}$ in. long, 5-lobed to the base ; segments linear, nearly glabrous ; corolla $\frac{1}{8}$ in. long, white ; lower anther-cell with a clavate tail ; pollen ellipsoid with 2 stopples and 1 row of tubercles on either side of each stopple ; style-base and top of ovary hairy ; capsule more than $\frac{1}{4}$ in. long, ellipsoid, widened upwards, glabrate, 4-seeded ; seeds tubercular-rugose. *Adhatoda anagalloides, Nees in DC. Prod.* xi. 403.

Kalahari Region : Transvaal ; Apies River, *Burke!* Macmac Creek, *Mudd!* Saddleback Range, near Barberton, 4000–5000 ft., *Galpin,* 682! Elands Fontein, near Johannisburg, 5500 ft., *Gilfillan*, 1418! near Lydenburg, *Atherstone!* *Wilms*, 1201! Hoogeveld, near Standerton, *Wilms*, 1201b! Mooifontein, 5500 ft., *Schlechter*, 3565!

T. Anderson cites (Journ. Linn. Soc. l.c.) *Adhatoda patula*, Nees in part, as a synonym.

24. **J. matammensis** (Oliv. in Trans. Linn. Soc. xxix. 130); softly white hairy on the stems, at the nodes and the base of the leaves; branches 8–12 in. long, divided, weak; leaves $1\frac{1}{4}$ by $\frac{1}{4}$–$\frac{1}{3}$ in., elliptic or subovate; peduncles numerous, 1–$1\frac{3}{4}$ in. long, bearing in the upper half 3–5 separate (mostly alternate) flowers; bracts or bracteoles linear, smaller than the calyx; calyx-lobes $\frac{1}{8}$ in. long, linear, nearly glabrous; corolla $\frac{1}{4}$ in. long; capsule $\frac{1}{8}$ in. long, glabrate, with 4 rugose seeds. *C. B. Clarke in Dyer, Fl. Trop. Afr.* v. 209. *J. anselliana, Lindau in Engl. & Prantl, Pflanzenfam.* iv. 3B, 283, fig. 112, *D, E*.

KALAHARI REGION: Transvaal; Hammans Kraal, 4400 ft., *Schlechter*, 4203!

Common in Tropical Africa, south to Matabeleland. It represents the section *Ansellia*, T. Anders., which Bentham (Benth. et Hook. f. Gen. Pl. ii. 1113) placed under the genus *Dianthera*.

25. **J. exigua** (S. Moore in Journ. Bot. 1900, 204); nearly glabrous; stem 8 in. long, much branched; leaves $\frac{1}{2}$ by $\frac{1}{4}$ in., ovate or elliptic; petiole 0–$\frac{1}{10}$ in. long; peduncles scattered, $\frac{1}{4}$–$\frac{1}{3}$ in. long, carrying 2 (sometimes 3) flowers near the top; bracteoles much shorter than the calyx; calyx $\frac{1}{8}$ in. long, divided to the base into 4 linear nearly glabrous segments; corolla less than $\frac{1}{6}$ in. long; lower anther-cell tailed; pollen most minute, ellipsoid, banded, with 2 stopples and a row of strong tubercles on either side of each stopple; ovary nearly glabrous, except at the tip; capsule $\frac{1}{8}$ in. long, 4-seeded, nearly sessile, glabrate; seeds tubercular-rugose.

KALAHARI REGION: South African Goldfields, *Baines!*

Also in Tropical Africa near Bulawayo.

Imperfectly known Species.

26. **J. incerta** (C. B. Clarke); a small undershrub; branches 4–6 in. long, divided, hairy; stem-leaves $\frac{1}{2}$–$\frac{3}{4}$ by $\frac{1}{3}$–$\frac{1}{2}$ in., ovate, subobtuse, pubescent, base narrowed; petiole $\frac{1}{12}$–$\frac{1}{8}$ in. long; upper leaves narrower, obovate, passing into the spathulate floral leaves; flowers solitary, opposite, in 6–12 of the uppermost axils, almost running into a terminal spike; bracteoles more than $\frac{1}{4}$ in. long, linear-spathulate, similar to the bract (or floral leaf); calyx $\frac{1}{5}$ in. long, divided to the base; segments 5, linear, pubescent; corolla $\frac{1}{2}$ in. long; one anther-cell much the lower, long-tailed; pollen ellipsoid, banded, with 2 stopples, tubercles very obscure; pistil glabrous except for a few scattered hairs on the style; capsule exceeding $\frac{1}{4}$ by $\frac{1}{8}$ in., subsessile, 4-seeded; seeds tubercular-rugose.

KALAHARI REGION: Transvaal; Boschveld, between Elands River and Klippan, *Rehmann*, 5058!

This might be located artificially next *J. petiolaris*. The spathulate upper leaves and bracteoles make it look not a *Justicia*.

27. J. (?) Brycei (C. B. Clarke); hispid with white many-celled hairs; branches 1–1½ in. long; leaves ⅛ by ⅛–⅙ in., elliptic, obtuse, cuneate at the base; petiole hardly any; flowers solitary, approximate in the upper axils; bract ⅓–½ in. long, linear; bracteoles 0; calyx 5-partite to the base; segments ¼–⅓ in. long, linear, green, hispid with white hairs; corolla ½–⅔ in. long, pink, palate spotted with purple; stamens 2; lower anther-cell shortly and obtusely tailed; pollen ellipsoid, with 2 stopples and longitudinal bands, with no rows of tubercles on either side of the stopples; ovary glabrous; base of style sparsely and minutely hairy.

EASTERN REGION: Basutoland; near the summit of Machacha, 10,000 ft., *Bryce!*

28. J. (Adhatoda) hantamensis (Lindau in Engl. Jahrb. xx. 66); branches and leaves glabrous; leaves ½ in. long, sessile, lanceolate, distant; flowers solitary, short-pedicelled; bracteoles lanceolate; calyx-segments ⅓ by $\frac{1}{10}$ in., broadly lanceolate, thinly hairy; corolla ⅔ in. long; lower anthers separate, spurred; pollen ellipsoid, banded longitudinally, with 2 pores, and 2 rows of tubercles on either side of each pore; capsule glabrous. *Lindau in Engl. & Prantl, Pflanzenfam.* iv. 3B, 349.

CENTRAL REGION: Calvinia Div.; Hantam Mountain, near Calvinia, *Meyer.*

29. Rhytiglossa rubicunda (Hochst. in Flora, 1845, 71); a small shrub; branches roundish, clothed with deflexed hairs; leaves ½–1 in. long, ovate, pubescent, petioled; spikes short, terminal and axillary; bracts villous, lanceolate or narrowly spathulate; calyx-segments linear, hairy, glandular, shorter than the bracts; corolla red; capsules glabrous; seeds rugose.

COAST REGION: Knysna Div.; in Zitzikamma forest, *Krauss*, 1128!

XXI. MONECHMA, Hochst.

Bracts inconspicuous (except in *M. bracteatum*); bracteoles small, linear. *Calyx* divided nearly to the base; segments 5 or 4, narrow or linear, rarely more than ⅓ in. long. *Corolla* small, rarely attaining ¾ in. long, 2-lipped; posticous lip subentire. *Stamens* 2; one anther-cell distinctly below the other, tailed; pollen ellipsoid, longitudinally banded, with 2 stopples; one or more rows of tubercles on each side of either stopple, frequently very obscure. *Ovary* with 1 or 2 ovules in each cell; style shortly 2-lobed at the tip or subentire. *Capsule* small, rarely ⅔ in. long, usually pubescent at the top, 2-seeded; placentæ not rising elastically from the base of the valves; seeds discoid, quite smooth, often shining and blotched, without tubercles or corrugations.

Leaves entire, small; flowers axillary, often few scattered, less commonly approximated towards the tips of the branches or (in *M. bractealum*) strobilate. The genus, as here understood, differs from *Justicia* by having only 2 (very smooth) seeds to the capsule.

Species about 27, African, whereof one (*M. bracteatum*) extends through Arabia to Bombay.

Section 1. BRACTEATÆ. Flowers loosely strobilate ;
bracts round-ovate ...   .   (1) **bracteatum.**

Section 2. SOLITARIÆ. Flowers solitary ; bracts
inconspicuous :
   Whole plant hoary   ...   .   (2) **incanum.**
   Plant variously hairy or glabrate, not hoary :
     Calyx 5-fid :
      Leaves ovate or obovate :
       Whole plant softly hairy   ...   ...   (3) **molle.**
       Plant glabrate or with a little scat-
        tered hair   ...   .   (4) **leucoderme.**
      Leaves from linear to narrow-elliptic :
       Flowers few, distant, scattered   ...   (5) **pseudopatulum.**
       Flowers approximated towards the
       ends of the branches :
        Calyx-lobes more than $\frac{1}{3}$ in.
        long   ...   .   (6) **Linaria.**
        Calyx-lobes less than $\frac{1}{3}$ in.
        long :
         Leaves and calyx-lobes
         acute :
          Leaves glabrate   ...   (7) **acutum.**
          Leaves viscous-hairy   ...   (8) **arenicola.**
         Leaves and calyx-lobes ob-
         tuse :
          Leaves oblong ...   ...   (9) **Atherstonei.**
          Leaves very few, subu-
          late ...   .   (10) **spartioides.**
     Calyx 4-fid :
      Bracteoles densely margined by long
      white hairs   ...   .   (11) **fimbriatum.**
      Bracteoles not densely margined, usually
      hairy or pubescent :
       Leaves linear-spathulate, with a re-
       curved tip :
        Branches round ...   .   (12) **divaricatum.**
        Branches quadrangular ...   ...   (13) **nepetoides.**
       Leaves flat at the tip, not re-
       curved :
        Leaves narrowly linear-lanceo-
        late, petioled ...   .   (14) **namaense.**
        Leaves cuneate-elliptic, sub-
        sessile ...   .   (15) **foliosum.**

1. **M. bracteatum** (Hochst. in Flora, 1841, 375), pubescent ;
stems 1–3 ft. long ; leaves up to $4\frac{1}{2}$ by 1–$1\frac{1}{2}$ in. (usually much
smaller), narrowed at both ends ; petiole $\frac{1}{2}$–1 in. long ; spikes
subsessile (sometimes more than 2 at one node), 2 by $\frac{2}{3}$ in., strobilate ;
bracts $\frac{1}{4}$–$\frac{1}{3}$ in. in diam., ovate or orbicular ; calyx $\frac{1}{5}$ in. long ; corolla
$\frac{1}{3}$–$\frac{1}{2}$ in. long, white ; lower anther-cell long-tailed ; pollen oblong-
ellipsoid, stopples 2, longitudinal bands rather obscure ; capsule $\frac{1}{4}$ in.
long, hairy, 2-seeded ; seeds smooth, yellow-brown, often with black
blotches. *Nees in DC. Prod.* xi. 411 ; *C. B. Clarke in Dyer, Fl.
Trop. Afr.* v. 214 *with all syn. M. debile, Schinz in Mém. Herb.
Boiss.* x. 64. *M. angustifolium, Nees in DC. Prod.* xi. 412.

*Justicia blepharostegia, E. Meyer in Drège, Zwei Pflanzengeogr.
Documente,* 160, 195 ; *T. Anders. in Journ. Linn. Soc.* vii. 43;
Lindau in Engl. & Prantl, Pflanzenfam. iv. 3B, 349. *Dicliptera
blepharostegia, Presl, Bot. Bemerk.* 95.

KALAHARI REGION : Transvaal; Klipdam, 4500 ft., *Schlechter,* 4492 !
EASTERN REGION: Natal; near Durban, below 500 ft., *Drège!* Umhloti
Flat, *Wood,* 1256! and without precise locality, *Gerrard,* 392! Delagoa Bay,
Junod, 145 !

Common in Tropical Africa; also in Bombay.

2. M. incanum (C. B. Clarke) ; a persistently hoary rugged undershrub; branches 6–15 in. long, sometimes $\frac{1}{4}$ in. in diam., with very
short internodes ; leaves $\frac{1}{2}$–$\frac{3}{4}$ by $\frac{1}{10}$ in., linear-obovate, obtuse;
flowers scattered, axillary, mostly solitary; bracts small, narrow;
calyx $\frac{1}{6}$ in. long ; corolla $\frac{1}{2}$ in. long ; lower anther-cell long-tailed;
pollen ellipsoid with 2 stopples, distinctly banded; capsule nearly
$\frac{1}{2}$ in. long, 2-seeded, glabrous; seeds large, hard, shining, dark
brown with black blotches, subglobose, a little compressed. *Justicia
incana, T. Anders. in Journ. Linn. Soc.* vii. 42 ; *Engl. in Engl.
Jahrb.* x. 264; *Lindau in Engl. & Prantl, Pflanzenfam.* iv. 3B,
349. *Gendarussa incana, Nees in Linnæa,* xv. 367, *excl. var. β ;
Drège in Linnæa,* xx. 200 ; *G. capensis, Presl, Bot. Bemerk.* 95.
Adhatoda incana, Nees in DC. Prod. xi. 393. *Justicia capensis,
E. Meyer in Drège, Zwei Pflanzengeogr. Documente,* 62, 195, *not of
Thunb.*

SOUTH AFRICA : without precise locality, *Zeyher,* 1393!
CENTRAL REGION: Beaufort West Div.; between Beaufort West and
Rhinoster Kop, 2500–3500 ft., *Drège!* Karoo near Beaufort West, *Henderson!*
Murraysburg Div. ; on the Sneeuwberg Range, near Murraysburg, *Bolus,* 1783!
rough slopes near Murraysburg, 4100 ft., *Tyson,* 81 ! Cradock Div.; Brack
River, *Burke!* Albert Div.; hills near Burgersdorp, *Mrs. Barber!*
WESTERN REGION : Great Namaqualand ; Tiras, *Schinz,* 20 !
KALAHARI REGION : Prieska Div.; Zand Valley, *Burchell,* 1632! Orange
River Colony (?) Rance (? Rands) Bosch, *Burke,* 280 ! Griqualand West; Great
Boetsap, 4000 ft., *Marloth,* 967 ! British Bechuanaland ; Huss Hills, *Holub!*

3. M. molle (C. B. Clarke); densely and shortly pubescent;
stems 6–10 in. long, stout, branched; internodes mostly $\frac{1}{4}$–$\frac{1}{2}$ in. long,
some up to 1$\frac{1}{4}$ in. long ; leaves $\frac{1}{3}$ to $\frac{1}{2}$ in. in diam., round-ovate,
rather thick, nearly sessile ; flowers solitary, scattered, not clustered
towards the ends of the branches ; pedicels 0–$\frac{1}{4}$ in. long; bracteoles 2,
$\frac{1}{4}$ in. long, linear-oblong, obtuse ; calyx nearly $\frac{1}{4}$ in. long; lobes 5,
oblong to linear, hardly acute, pubescent and with some gland-headed
hairs ; corolla $\frac{2}{3}$–$\frac{3}{4}$ in. long, white (*Bolus*) ; one anther-cell much
below the other, with a long clavate tail; pollen ellipsoid, with
2 stopples and 2 longitudinal rows of tubercles on either side of each
stopple ; capsule $\frac{1}{2}$ in. long, stalk rather thick, slightly puberulous at
the tip, 2-seeded ; seeds discoid, margined, smooth. *Justicia mollis,
E. Meyer in Drège, Zwei Pflanzengeogr. Documente,* 92, 196 ; *T.
Anders. in Journ. Linn. Soc.* vii. 42 ; *Lindau in Engl. & Prantl,*

Pflanzenfam. iv. 3B, 349. *Gendarussa mollis, Presl, Bot. Bemerk.* 95. *Adhatoda mollissima, Nees in DC. Prod.* xi. 391.

WESTERN REGION: Little Namaqualand; between Holgat River and Orange River, 1000–1500 ft., *Drège!* in dry stony places near Spektakel, 800 ft., *Bolus & MacOwan, Herb. Norm. Aust.-Afr.*, 670!

4. M. leucoderme (C. B. Clarke); minutely puberulous; stem short, stout, branches many, 4–9 in. long; leaves $\frac{1}{3}$ by $\frac{1}{4}$ in., tip obtuse, triangular, base narrowed; petiole 0–$\frac{1}{12}$ in. long; flowers solitary, axillary; bracteoles 2, $\frac{1}{6}$ in. long, linear; calyx $\frac{1}{6}$ in. long, 5-partite to the base; segments linear, minutely pubescent, and in *Schinz*, 29, with gland-headed hairs; corolla scarcely $\frac{1}{2}$ in. long; one anther-cell below the other, tailed; pollen ellipsoid with 2 stopples, bands and tubercles obscure; capsule $\frac{1}{6}$–$\frac{1}{5}$ in. long, ovoid, on a short narrow stalk, 2-seeded, puberulous to the base; seeds discoid, hard, brown, smooth, shining, blotched. *Justicia leucodermis, Schinz in Verh. Bot. Ver. Brandenb.* xxxi. 202; *Lindau in Engl. & Prantl, Pflanzenfam.* iv. 3B, 349.

WESTERN REGION: Great Namaqualand; between Tiras and Rehoboth, *Schinz*, 1! Tschirub Mountain, *Schinz*, 29!

Wrongly placed under *J. odora*, Vahl, by me in Dyer, Fl Trop. Afr. v. 201.

5. M. pseudopatulum (C. B. Clarke); nearly glabrous, branches 1 ft. long (or more), terete, not striated, smooth, the internodes up to $1\frac{1}{2}$–2 in. long; leaves up to $1\frac{1}{2}$ by $\frac{1}{6}$ in., linear-oblong (no elliptic or subovate leaves except in the variety, *latifolia*); petioles 0–$\frac{1}{6}$ in. long; flowers few, scattered; pedicels $\frac{1}{8}$–$\frac{1}{6}$ in. long; bracteoles 2, $\frac{1}{8}$ in. long, oblong, obtuse: calyx hardly $\frac{1}{6}$ in. long, deeply 5-lobed; lobes linear, subacute, puberulous or glabrate; corolla $\frac{1}{3}$–$\frac{1}{2}$ in. long; capsule $\frac{1}{2}$ in. long, stalked, narrowly ellipsoid, subquadrangular, compressed, 2-seeded; seeds discoid, smooth, hard, blotched. *Gendarussa patula, Drège in Linnæa,* xx. 200.

VAR. β, **latifolium** (C. B. Clarke); some leaves on the main stems 1 by $\frac{1}{8}$ in. long (similar to those on the type plant); other leaves on the main stem 1 by nearly $\frac{1}{2}$ in., elliptic, obtuse, with a definite petiole $\frac{1}{4}$ in. long; other leaves on the upper branches are $\frac{1}{2}$ by $\frac{1}{4}$ in., obovate-elliptic, obtuse; calyx, corolla, capsule and seeds as in the type.

CENTRAL REGION: Calvinia Div.; Kamos (near Lospers Flats), 2000–3000 ft., *Zeyher*, 1395! Var. β: Somerset Div.; Somerset East, *Bowker*, 163! Graaff Reinet Div.; amongst bushes near Graaff Reinet, 2500 ft., *Bolus*, 61!

Of the two plants here treated, the material, though scanty, is complete. The capsule and seeds, the calyx, the scattered short-pedicelled flowers are similar. The short calyx, on a pedicel, resembles very much *Justicia patula* (which Drège determined it to be), from which the capsule and seeds show it to be remote.

6. M. Linaria (C. B. Clarke); minutely pubescent; branchlets 13 on the fragment seen, about 6 in. long, 6–8-ribbed, rising in a close cluster, but little divided; leaves $\frac{1}{2}$–$\frac{2}{3}$ by $\frac{1}{8}$–$\frac{1}{6}$ in.; flowers solitary, approximated in the upper axils; bracteoles 2, very small, linear; calyx deeply 5-partite, more than $\frac{1}{3}$ in. long; lobes linear; corolla

$\frac{1}{3}$–$\frac{1}{2}$ in. long; one anther-cell below the other, tailed; pollen ellipsoid, bands and tubercles obscure; ovary below glabrous, with a few gland-headed hairs in the upper part, base of style with scattered simple hairs. *Justicia Linaria, T. Anders. in Journ. Linn. Soc.* vii. 42. *J. patula, E. Meyer in Drège, Zwei Pflanzengeogr. Documente*, 108, 196 (letter *a* only).

COAST REGION: Vanrhynsdorp Div.; near Ebenezer, below 500 ft., *Drège!*

There is no capsule on the single fragmentary example, so that it is doubtful whether it is a *Monechma* or a *Justicia*; it differs from the similar species of *Monechma* by its longer calyx. It is really an undescribed species, for T. Anderson only cites the name as equivalent to *Gendarussa Linaria*, Nees, which again is unpublished except so far that Nees himself says it was his *Adhatoda orchioides*, var. *angustifolia*, to which again he says this fragment of Drège's belonged. I suspect, from the small piece of the branch on which the shoots stand, that *M. Linaria* was 2–3 ft. high or more; *Justicia orchioides*, E. Meyer, from locality b (Uitenhage Div.; between Coega River and Sunday River, 1000 ft.), has also been matched with *M. Linaria*; but it has neither flowers, fruit, nor indication of inflorescence.

7. **M. acutum** (C. B. Clarke); undershrub; branches 6–15 in. long, nearly round or subquadrangular, with minute white deflexed hairs, becoming glabrate; internodes attaining 1 in. in length; leaves $\frac{3}{4}$ by $\frac{1}{6}$ in., linear-lanceolate, acute at either end, subsessile, becoming nearly glabrous; flowers rather numerous in the upper axils, in one branch forming an interrupted spike 6 in. long; bracteoles $\frac{1}{8}$ in long, linear-lanceolate; calyx $\frac{1}{8}$ in. long, deeply 5-partite; lobes linear-lanceolate, acute, with white hairs on the lower margins; corolla $\frac{1}{3}$ in. long, white (*Burchell*); lower anther-cell much lower than the other, tailed; pollen ellipsoid, bands and tubercles obscure; ovary glabrous; style thinly hairy nearly to the top, young capsules 2-seeded.

COAST REGION: Humansdorp Div.; between Galgebosch and Melk River, *Burchell*, 5761! 4785! Uitenhage Div.; Grassrug, near Uitenhage, *Baur!*

Baur's plant has broader much less obtuse leaves than in the type collections of Burchell above described; but it agrees so closely in the inflorescence, the indumentum of the stem and calyx, the small flowers and the (imperfectly ripe) capsule, that it cannot conveniently be treated as a species from our scanty material.

8. **M. arenicola** (C. B. Clarke in Dyer, Fl. Trop. Afr. v. 218); a viscous pubescent undershrub; branches 6–10 in. long, internodes mostly $\frac{1}{6}$–$\frac{1}{3}$ in. long; leaves up to 1 by $\frac{1}{4}$–$\frac{1}{3}$ in., lanceolate, acute, subsessile, with long white hairs; flowers solitary, few, approximate towards the ends of the branchlets; bracteoles 2, scarcely $\frac{1}{4}$ in. long, linear; calyx $\frac{1}{4}$ in. long, 4-partite nearly to the base; lobes linear; corolla $\frac{1}{2}$ in. long; one anther-cell lower than the other, tailed, pollen ellipsoid, with 2 stopples, tubercles minute in 2 rows on each side of each stopple; capsule $\frac{1}{3}$ in. long, pubescent, 2-seeded; seeds smooth, blotched. *Justicia arenicola, Engl. in Engl. Jahrb.* x. 264; *Lindau in Engl. Jahrb.* xix. 151, *and in Engl. & Prantl, Pflanzenfam.* iv. 3B, 349.

KALAHARI REGION: Prieska Div.; Zand Valley, *Burchell*, 1631! Hopetown Div.; near Hopetown, 4500 ft., *Muskett in Bolus Herb.*, 2568!

Also frequent in Lower Guinea.

9. M. Atherstonei (C. B. Clarke); viscous-pubescent, branches 8 in. long, round, not striate; leaves up to $\frac{3}{4}$ by $\frac{1}{6}$ in., narrow, lanceolate-obovate, hardly acute, subsessile; flowers few, solitary, approximate in the upper axils; bracteoles 2, $\frac{1}{6}$ in. long, linear-oblong; calyx $\frac{1}{6}$ in. long, deeply 5-partite; lobes linear, hardly acute; corolla nearly $\frac{3}{4}$ in. long; one anther-cell below the other, with a very long tail; pollen ellipsoid with 2 stopples, banded, and with 1 row of tubercles beside each stopple; capsule $\frac{1}{3}$ in. long, stalked, 2-seeded, glabrate; seeds brown, smooth.

WESTERN REGION : Little Namaqualand; sandy flats near the Aur Aap River, a tributary of the Orange River, *Atherstone*, 11!

10. M. spartioides (C. B. Clarke); a straggling nearly glabrous and leafless shrub; branches up to 2 ft. long, round, with many internodes 2 in. long; leaves very few, $\frac{3}{4}$ by $\frac{1}{16}$ in., subulate; flowers solitary, few, approximated in the upper axils; bracteoles $\frac{1}{6}$ in. long, linear; calyx $\frac{1}{6}$ in. long, minutely pubescent and with gland-headed hairs, deeply 5-partite; lobes linear; corolla $\frac{1}{2}-\frac{2}{3}$ in. long, white with violet tips (*Schinz*) ; one anther-cell much below the other, tailed ; pollen ellipsoid with 2 stopples, slightly banded, scarcely tubercled ; pistil glabrous; capsule $\frac{1}{3}$ in. long, 2 seeded. *Justicia spartioides*, *T. Anders. in Journ. Linn. Soc.* vii. 43.

WESTERN REGION : Great Namaqualand; on the bed of a river at Cannas, *Schinz*, 22! and without precise locality, *Schinz*, 28! Little Namaqualand, Jus, 2800 ft., *Schlechter*, 11407!

11. M. fimbriatum (C. B. Clarke); a pubescent small shrub; branches 14 in. long, frequently divided, quadrangular, many inter-nodes $1\frac{1}{2}$ in. long; leaves $\frac{3}{4}$ by $\frac{1}{16}$ in., narrowly linear ; flowers solitary, approximate in the upper axils; bracteoles 2, $\frac{1}{5}$ in. long, linear, the margins most densely fringed by long white many-celled hairs; calyx $\frac{1}{5}$ in. long, deeply 5-fid, minutely pubescent and with some gland-headed hairs ; corolla $\frac{1}{2}$ in. long ; one anther-cell much below the other, tailed; pollen ellipsoid with 2 stopples and 1 obscure row of tubercles on either side of each stopple ; capsule $\frac{1}{3}$ in. long, sparsely and minutely hairy, 2-seeded ; seeds smooth.

EASTERN REGION : Natal ; without precise locality, *Gerrard*, 1269!

Also lately collected by Schlechter in Tropical Transvaal.

The dense white fringe to the bracteoles in *Gerrard*, 1269, catches the eye at once ; the fringe is less prominent in Schlechter's plant.

12. M. divaricatum (C. B. Clarke); an undershrub, attaining 2 ft. in height (at least) ; branches terete, not striate, minutely hairy, many internodes $1-2\frac{1}{2}$ in. long; leaves $\frac{1}{2}-\frac{2}{3}$ by $\frac{1}{8}-\frac{1}{6}$ in., linear-obovate, pubescent, long-tapering to the base, tip obtuse, recurved, with a minute mucro; flowers scattered, axillary, few, mostly towards the ends of the branches; bracteoles 2, $\frac{1}{4}$ in. long, linear ; calyx $\frac{1}{4}$ in. long, deeply 4-fid ; lobes linear, glandular-pubescent, very white on the margins ; corolla $\frac{1}{2}$ in. long, rose-coloured (*Bolus*) ; one anther-cell much lower than the other, tailed ; pollen ellipsoid with

2 stopples and 2 rows of small tubercles on either side of each stopple; ovary glabrous; style thinly hairy; capsule nearly ½ in. long, 2-seeded; seeds smooth, blotched. *Justicia divaricata, Willd. ex Nees in DC. Prod.* xi. 391. *T. Anders. in Journ. Linn. Soc.* vii. 42; *Lindau in Engl. & Prantl, Pflanzenfam.* iv. 3B, 349. *J. patula, E. Meyer in Drège, Zwei Pflanzengeogr. Documente,* 91, 196 (letter *b* only). *J. orchidioides, T. Anders. in Journ. Linn. Soc.* vii. 42. *Gendarussa incana, var. villosa, Nees in Linnæa,* xv. 367. *G. patula, Presl, Bot. Bemerk.* 95. *Adhatoda divaricata, Nees in DC. Prod.* xi. 391. *A. orchioides var.* a, *Nees l.c.* 393. *Ruellia setigera, Zeyher ex Nees, l.c.* 391.

CENTRAL REGION: Graaff Reinet Div.; banks of the Zwart River near Graaff Reinet, 2700 ft., *Bolus*, 748!

WESTERN REGION: Little Namaqualand; Kaus Mountains, 3000–4000 ft., *Drège!* between Spektakel and Komaggas, 1000 ft., *Bolus & MacOwan, Herb. Norm. Aust.-Afr.*, 671!

This plant is exceedingly unlike *Justicia orchioides*, Linn. f.; yet Nees has written on Drège's *J. patula* from locality b "*Tyloglossa orchioides*, var. *a*."

13. **M. nepetoides** (C. B. Clarke); branches quadrangular, glandular pubescent; leaves ¼–½ in. long; corolla ⅓ in. long; capsule hardly exceeding ¼ in. long; otherwise as *M. divaricatum. Adhatoda capensis, var. arenosa, Nees in DC. Prod.* xi. 391.

KALAHARI REGION: Prieska Div.; Keikams Poort (Modder Gat Poort), *Burchell*, 1616! Griqualand West, Hay Div.; Klipfontein, *Burchell*, 2149!

This might be a variety of *M. divaricatum*, with smaller leaves, flowers, and seeds; in *M. divaricatum* the branches are terete. It must not be inferred, from Nees having determined this plant as a variety of *Justicia capensis*, Thunb., that it bears any kind even of external resemblance to that species.

14. **M. namaense** (C. B. Clarke); a minutely pubescent undershrub; branches 16 in. long, nearly terete, with internodes up to 2 in. long; leaves 1–1½ by ⅛–⅕ in., narrowly linear-lanceolate, glabrate, flat, the tip straight (not recurved); flowers few, scattered, axillary; bracteoles ¼ in. long, linear, densely and shortly viscous pubescent; calyx ⅙–⅕ in. long, 4-partite to the base; segments linear, minutely pubescent; corolla ½ in. long; one anther-cell below the other, tailed; pollen ellipsoid with 2 stopples, banded, 1 row of tubercles on either side of each stopple; capsule ⅓ in. long, 2-seeded, pubescent at the top; seeds smooth. *Justicia namaensis, Schinz in Verh. Bot. Ver. Brandenb.* xxxi. 202; *Lindau in Engl. & Prantl, Pflanzenfam.* iv. 3B, 349.

WESTERN REGION: Great Namaqualand; Gamosab, *Schinz*, 25! Little Namaqualand; between Garies and Springbok, 2500 ft., *Schlechter*, 11175!

KALAHARI REGION: Griqualand West, Hay Div.; between Kloof Village and Witte Water, on the plains at the foot of the Asbestos Mountains, *Burchell*, 2072!

15. **M. foliosum** (C. B. Clarke); an undershrub, nearly glabrous except the innovations; branches 4–8 in. long, with internodes mostly very short, some up to 1 in. long; leaves ½ by ⅙ in., cuneate-

elliptic, subsessile, tip obtuse, base narrow; flowers few, axillary, scattered; bracteoles $\frac{1}{6}$ in. long, linear-oblong; calyx 4-partite to the base; segments exceeding $\frac{1}{4}$ in. long, linear, broader and minutely hairy in the lower half; corolla $\frac{1}{2}$–$\frac{2}{3}$ in. long; one anther-cell much below the other, tailed; pollen ellipsoid, with 2 stopples and several rows of small tubercles; capsule exceeding $\frac{1}{3}$ in. long, with 2 smooth seeds. *Justicia foliosa, E. Meyer in Drège, Zwei Pflanzengeogr. Documente,* 62, 195. *J. cuneata, T. Anders. in Journ. Linn. Soc.* vii. 41 *partly. Gendarussa foliolosa, Presl, Bot Bemerk.* 95. *G. cuneata, Drège in Linnæa,* xx. 200. *Adhatoda cuneata, Nees in DC. Prod.* xi. 392 *partly.*

COAST REGION: Kuysna Div.; without precise locality, *Bolus,* 2422! Uitenhage Div.; near Roodewal, *Bolus,* 1873!
CENTRAL REGION: Beaufort West Div.; between Beaufort West and Rhinoster Kop, 2500–3000 ft., *Drège!*

Nees has written " *Tyloglossa cuneata,* N. ab E.," on Drège's example marked " *Justicia foliosa,* E. M." ; the leaves bear a resemblance to those of *Justicia cuneata,* Vahl.

XXII. SIPHONOGLOSSA, Oerst.

Corolla with a long linear tube, much longer than the 2-lipped limb; otherwise as *Justicia.*

Species 8, viz. 5 in Tropical America and the 3 following :—

S. *tubulosa* was removed from *Justicia* to the American genus *Siphonoglossa* by Bentham (Benth. et Hook. f. Gen. Pl. ii. 1110). S. Moore added *S. Nummularia* which is beyond question congeneric with *S. tubulosa.* Baillon (Hist. des Plantes, x. 441) records *S. tubulosa* under *Siphonoglossa,* but does not appear to have examined or considered it. Lindau (in Engl. & Prantl, Pflanzenfam. iv. 3B, 338) says that these two species can scarcely be referred to *Siphonoglossa,* and (l.c. p. 349) records *S. tubulosa* (under a different name) as a true *Justicia.* The question is greatly complicated by the arrival of a third South African species which has the corolla of *Beleropone,* not of *Siphonoglossa.*

Corolla in total length under 1 in.; lips less than
  $\frac{1}{4}$ in. long :
    Leaves 1–1$\frac{1}{2}$ in. long, elliptic, long-petioled   ...   (1) **tubulosa.**
    Leaves $\frac{1}{4}$–$\frac{1}{3}$ in. long, round, subsessile   ...   ...   (2) **Nummularia.**
Corolla in total length 2 in. and upwards; lips $\frac{2}{3}$ in.
    long   ...   .   (3) **linifolia.**

1. S. tubulosa (Lindau in Engl. & Prantl, Pflanzenfam. iv. 3B, 338); minutely and thinly hairy; branches 6–12 in. long, weak; leaves 1$\frac{2}{4}$ by $\frac{3}{4}$ in., elliptic, narrowed at both ends, thin; petiole $\frac{1}{4}$–$\frac{3}{4}$ in. long; flowers few, scattered, in the upper axils; bracts and bracteoles inconspicuous; calyx-segments $\frac{1}{6}$ in. long, linear; corolla-tube $\frac{3}{4}$ by $\frac{1}{20}$ in., slightly widened upwards; lips $\frac{1}{4}$ in. long, with rounded lobes; one anther-cell much below the other, tailed; pollen ellipsoid, stopples 2, tubercles none or most obscure; capsule $\frac{1}{4}$–$\frac{1}{3}$ by $\frac{1}{10}$ in., glabrate, 4-seeded. *Justicia tubulosa, E. Meyer in Drège, Zwei Pflanzengeogr. Documente,* 150, 196; *T. Anders. in Journ. Linn. Soc.* vii. 41. *J. suffruticosa, E. Meyer in Drège, Zwei Pflanzengeogr. Documente,* 153, 196. *J. leptantha, Lindau in Engl.*

& *Prantl, Pflanzenfam.* iv. 3 B, 349. *J. prostrata, Schlechtend. ex Nees in DC. Prod.* xi. 390. *Gendarussa leptantha, Nees in Linnæa*, xv. 372. *Rhinacanthus tubulosus, Presl, Bot. Bemerk.* 95. *Adhatoda tubulosa, Nees in DC. Prod.* xi. 392. *A. leptantha, Nees in DC. Prod.* xi. 390.

COAST REGION: Alexandria Div.; Bushman River, *Ecklon.* EASTERN REGION: Pondoland; St. Johns River, 1000 ft., *Drège!* between St. Johns River and Umtsikaba River, under 1000 ft., *Drège!* Griqualand East; Zuur Berg Mountains, near Kokstad, 4000 ft., *Tyson*, 1700! 5000–6000 ft., *Tyson*, 1166!

The plant meant by Schinz (*Mém. Herb. Boiss.* x. 64) is doubtful, as he appears to have confused the *prostrata* of Nees with the *protracta* of T. Anderson.

2. **S. Nummularia** (S. Moore in Journ. Bot. 1880, 40); softly hairy, branches 6–10 in. long, rigid; leaves $\frac{1}{4}$–$\frac{1}{3}$ in. long, round, subsessile; flowers few, scattered in the upper axils; bracteoles minute; calyx-segments 5, hardly $\frac{1}{8}$ in. long, narrow-lanceolate; corolla-tube $\frac{3}{4}$ by $\frac{1}{20}$ in., hardly widened upwards; lips $\frac{1}{4}$ in. long with rounded lobes; one anther-cell much below the other, tailed; pollen ellipsoid with 2 stopples, and 2 or 3 rows of tubercles on either side of each stopple.

EASTERN REGION: King Williamstown Div.; Keiskamma Hoek, *Cooper*, 370!

3. **S. (?) linifolia** (C. B. Clarke); stem short, woody, whence arise annual simple flowering stems 6–9 in. long; leaves $1\frac{1}{4}$ by $\frac{1}{4}$ in., narrow-lanceolate, glabrous or obscurely hispidulous, sessile; flowers solitary in the 2 or 3 upper axils; bracts linear; calyx-lobes $\frac{1}{4}$–$\frac{1}{3}$ in. long, linear, thinly hispidulous; corolla hairy, "pale purple, lower lip blotched deep purple" (*Galpin*); tube $1\frac{1}{2}$–$1\frac{3}{4}$ in. long, the lower half about $\frac{1}{20}$ in. broad, the upper half tubular $\frac{1}{12}$–$\frac{1}{10}$ in. broad; posticous lip $\frac{2}{3}$ in. long, oblong-linear, subentire; anticous lip $\frac{3}{4}$ in. long, narrowly cuneate, cut halfway down into 3 narrow-lanceolate lobes; anthers exserted, rather large, purple-rose, one (nearly entirely) below the other with a short distinct white tail; pollen ellipsoid with 2 stopples and 3 rows of tubercles on either side of each stopple; ovary glabrous, oblong, with 4 ovules; style very long, glabrous, slightly swollen in the middle. *Aulojusticia linifolia, Lindau in Engl. Jahrb.* xxiv. 325 *and Engl. & Prantl, Pflanzenfam. Nachtr. zu* ii.-iv. 309.

KALAHARI REGION: Transvaal; mountain sides of the Saddleback Range, near Barberton, 4000–5000 ft., *Galpin*. 826! stony places on mountains near Barberton, 3500 ft., *Thorncroft*, 81 (in *Wood. Herb.* 4160)!

XXIII. **ADHATODA**, Nees.

Anther-cells at nearly equal height, the lower one obscurely mucronate or very shortly tailed; pollen without rows of tubercles; otherwise as *Justicia.*

Mostly large or stout leafy shrubs ; the inflorescence terminal, of many flowers collected in compound oblong or capitate heads ; corolla rather broad with a short tube, white, rose- or purple-spotted, the posticous subentire lip arched ; capsule large stout ; seeds scabrous, flattened.

A genus of 8 species, in the warm parts of the Old World, very indistinctly diagnosed from *Justicia ;* but forming a natural group, usually recognizable from *Justicia* by one or other of the marks above given. As to the distinction in the pollen, regarded as decisive by Lindau, there are very many species referred by him to *Justicia* in which the rows of tubercles are very obscure, and in several, as in *A. natalensis* below, I cannot see a trace of them.

Bentham, unwilling to multiply genera, where the distinctions between them are so unsatisfactory, has sunk *Duvernoia* in *Adhatoda.* The whole structure of the calyx in *Duvernoia* is so unlike that of any other *Adhatoda*, that I think the genus might be well maintained.

Subgenus 1. EU-ADHATODA. Calyx 5-partite to the base ; segments linear-lanceolate, subulate, imbricated in the bud.

Leaves broadly-elliptic, hairy   ...   .   (1) **natalensis.**
Leaves narrowly-elliptic, glabrate   ...   .   (2) **Andromeda.**

Subgenus 2. DUVERNOIA. Calyx 5-fid hardly half-way down, splitting into 5–2 narrow triangular lobes ; segments broad-triangular, rigid, glabrous   ...   ...   (3) **Duvernoia.**

1. A. natalensis (Nees in DC. Prod. xi. 391) ; a stout hairy shrub ; leaves $2\frac{1}{4}$ by 1 in., elliptic, narrowed at either end ; petiole $0-\frac{1}{6}$ in. long ; inflorescence terminal, a condensed panicle 2 by 1 in. ; flowers numerous ; bract $\frac{2}{3}$ by $\frac{1}{8}$ in. ; bracteoles $\frac{1}{2}$ by $\frac{1}{12}$ in. ; sepals 5, $\frac{1}{2}$ by $\frac{1}{12}$ in., linear-subulate, softly hairy ; corolla $\frac{2}{3}$ in. long ; tube short, broad ; posticous subentire lip arched ; one anther-cell slightly lower than the other, with a very short tail ; pollen shortly ellipsoid, longitudinally banded, without tubercles ; ovary densely hairy ; capsule 1 by $\frac{1}{5}$ in., stout, 4-seeded. *Gendarussa densiflora, Hochst. in Flora*, 1845, 71. *Justicia natalensis, T. Anders. in Journ. Linn. Soc.* vii. 38 ; *Lindau in Engl. & Prantl, Pflanzenfam.* iv. 3B, 349. *Duvernoia trichocalyx, Lindau in Engl. Jahrb.* xxii. 122.

KALAHARI REGION : Transvaal ; *Sanderson !*
EASTERN REGION : Natal ; hills near Pietermaritzburg, 1000–2000 ft., *Krauss*, 453 ! Inanda, *Wood*, 174 ! 304 ! and without precise locality, *Sanderson ! Sutherland !*

Duvernoia trichocalyx (collected by O. Kuntze at Durban) has not been seen. The description of Lindau agrees with that of *A. natalensis*, except that the leaves are said to be glabrous.

2. A. Andromeda (C. B. Clarke) ; branches 6–12 in. long, rather stout, minutely hairy ; leaves $1\frac{3}{4}$ by $\frac{1}{2}$ in., narrowly elliptic, glabrate, narrowed at the base, subsessile ; flowers many in a condensed terminal panicle 2 by 1 in. ; bract $\frac{1}{2}$ by $\frac{1}{8}$ in. ; bracteoles $\frac{1}{3}$ by $\frac{1}{12}$ in. ; sepals nearly $\frac{1}{2}$ by $\frac{1}{12}$ in., linear-subulate, hairy ; corolla $\frac{3}{4}$ in. long, white with rose spots ; tube very short, broad ; posticous subentire lip arched ; one anther-cell slightly below the other, obscurely mucronate at the base ; pollen ellipsoid, longitudinally banded, without tubercles ; ovary minutely densely pubescent. *Duvernoia Andromeda, Lindau in Engl. & Jahrb.* xx. 42, *and in Engl. & Prantl, Pflanzenfam.* iv. 3B, 339.

EASTERN REGION : Griqualand East; mountains around Clydesdale, 3500 ft., *Tyson*, 2063! Pondoland; without precise locality, *Bachmann*, 1273! Natal; stony fields near Durban, *Wood in MacOwan, Herb. Aust.-Afr.*, 1512!

This species is closely allied to *A. natalensis.* Lindau calls this a *Duvernoia*, while he keeps *A. natalensis* in *Justicia*, on account of differences in the pollen only; but I see no difference in the pollen.

3. **A Duvernoia** (C. B. Clarke); a stout minutely pubescent shrub, attaining 8 ft. high; leaves often 7 by 3 in., base attenuate; petioles ¼–1 in. long; inflorescence a compound terminal panicle; lower peduncles often 2–4 in. long with a compound spike 2–4 by 1¼ in. at their tops; floral leaves conspicuous, numerous, ⅔ by ⅙ in.; flower bract ⅓ by $\frac{1}{12}$ in.; bracteoles ¼ by $\frac{1}{16}$ in.; calyx ¼ in. long, when mature divided half-way down into 5 triangular lobes, nearly glabrous; corolla 1 in. long; tube short, broad; posticous subentire lip arched, throat within densely hairy; anther-cells at nearly equal height, 1 obscurely mucronate at the base; pollen ellipsoid, banded longitudinally, without tubercles; ovary densely shaggy; capsule 1¼ by ¼–⅓ in., oblong-clavate, very stout, 4-seeded, grey-pubescent; seeds ¼–⅓ in. in diam., very flat, tubercular-scabrous. *Duvernoia adhatodioides, E. Meyer in Drège, Zwei Pflanzengeogr. Documente*, 150, 180; *Nees in DC. Prod.* xi. 323; *T. Anders. in Journ. Linn. Soc.* vii. 37; *Lindau in Engl. & Prantl, Pflanzenfam.* iv. 3B, 336, 339, *fig.* 135, F.

EASTERN REGION : Transkei, *MacOwan*, 2001! Kaffraria; Kreilis Country, *Bowker!* mountains of Kaffraria, *Mrs. Barber!* Pondoland; St. Johns River, below 1000 ft., *Drège!* in open woods near the confluence of the Tsitza and Umzimvubu Rivers, 1000 ft., *Baur in MacOwan & Bolus, Herb. Norm. Aust.-Afr.*, 558! Natal; Tongaat, *Cooper*, 3031! Noodsberg, *Wood*, 931! and without precise locality, *Gerrard*, 264!

Imperfectly known Species.

4. **Duvernoia tenuis** (Lindau in Engl. Jahrb. xx. 44); a shrub 5–10 ft. high, with glabrous branches; leaves 5 by 2 in., narrowed at either end, hairy on the nerves; petioles ½ .in. long or more; inflorescence very loosely paniculate, of few flowers; peduncles slender, bifariously hairy; bracts and bracteoles small; calyx ⅙ in. long; corolla green-yellow, smooth; tube ⅛ in. long; lip ⅛ in. long; one anther-cell lower than the other, mucronate at the base; pollen of *Duvernoia.*

EASTERN REGION : Pondoland; in Egosa Bush, near Dorkin, *Bachmann*, 1275!

From the description this would appear to be something very remote from *Duvernoia*, with which it is placed on the character of the pollen. Might it be *Justicia campylostemon*, T. Anders. ?

XXIV. RHINACANTHUS, Nees.

Calyx small, divided nearly to the base; segments 5, linear. *Corolla* 2-lipped; tube linear, much longer than the lips; posticous lip oblong, subentire, erect or recurved. *Stamens* 2; anther-cells 2, one a very little below the other, not tailed but sometimes mucronate at the base; pollen ellipsoid, with 3 longitudinal smooth bands. *Ovary* hairy. *Capsule* oblong, 4-seeded at the top, with a linear cylindric stalk; placentæ not rising elastically from the base of the valves; seeds covered with tubercles.

Rambling. Flowers in sessile distant clusters on the branches of the panicle; bracts and bracteoles small linear.

Species 3 or 4, in Africa and India.

This genus differs from *Siphonoglossa* in the inflorescence and in the pubescent ovary.

1. **R. communis** (Nees in Wall. Pl. As. Rar. iii. 109); sparingly pubescent; leaves 2–5 in. long, ovate to lanceolate; petioles 0–$\frac{1}{3}$ in. long; flowers in small clusters on the branches of loose divaricating panicles; bracts $\frac{1}{8}$–$\frac{1}{4}$ in. long, linear; calyx $\frac{1}{8}$–$\frac{1}{4}$ in. long; corolla-tube $\frac{3}{4}$ by $\frac{1}{16}$–$\frac{1}{12}$ in., linear to the top; lip $\frac{1}{3}$ in. long, posticous linear-oblong; capsule 1 in. long, pubescent, finally glabrate, linear-cylindric stalk more than $\frac{1}{2}$ in. long; seeds 4, covered with tubercles. *Nees in DC. Prod.* xi. 442; *T. Anders. in Journ. Linn. Soc.* vii. 51; *Lindau in Engl. Jahrb.* xviii. 63, *t.* 2, *fig.* 68; *C. B. Clarke in Hook. f. Fl. Brit. Ind.* iv. 541, *and in Dyer, Fl. Trop. Afr.* vi. 224. *R. macilentus, Presl, Bot. Bemerk.* 95. *R. oblongus, Nees in DC. Prod.* xi. 444; *T. Anders. in Journ. Linn. Soc.* vii. 51. *R. nasutus, Lindau in Engl. & Prantl, Pflanzenfam.* iv. 3B, 339, *and in Engl. Pfl. Ost-Afr. C.* 371. *Justicia macilenta, E. Meyer in Drège, Zwei Pflanzengeogr. Documente,* 159, 196. *Peristrophe oblonga, Nees in Linnæa,* xv. 375. *Pseuderanthemum dichotomum, Lindau in Engl. Jahrb.* xx. 40, *and in Engl. Pfl. Ost-Afr. C.* 371.

EASTERN REGION: Pondoland, *Bachmann*, 1270! Natal; in thickets by the Upper Umlazi River, below 500 ft., *Drège!* Inanda, *Wood*, 146! 534! Coastland up to 1000 ft., *Sutherland!* and without precise locality, *Sanderson! Gerrard*, 391!

An abundant plant in Tropical Africa and India.

XXV. ECBOLIUM, Kurz.

Calyx small, deeply 5-fid; segments linear. *Corolla-tube* cylindric; lips 2, not elongated. *Stamens* 2; anther-cells 2, oblong, muticous, at equal height; pollen globose, with longitudinal bands. *Ovary* with 2 ovules in each cell.

Shrubs; leaves entire, rather thick; spikes strobilate, terminal; bracts large, enclosing the calyx; bracteoles small, linear; capsule (where known) with 2 flat rough seeds.

Species 8–12, natives of Africa, Madagascar, Arabia and India.

1. **E. Flanagani** (C. B. Clarke); a yellowish-green shrub, nearly glabrate, the young parts hairy; leaves $1\frac{1}{2}$ by $\frac{3}{4}$ in., ovate, subobtuse, base broad-triangular; petioles $\frac{1}{4}$–$\frac{3}{4}$ in. long; spikes terminal, $1\frac{1}{2}$ by $\frac{2}{3}$ in., strobilate; bracts $\frac{2}{3}$ by $\frac{1}{5}$ in.; calyx scarcely $\frac{1}{6}$ in. long; segments linear, pubescent; corolla-tube $\frac{1}{2}$–$\frac{2}{3}$ by $\frac{1}{10}$ in., subcylindric; lobes $\frac{1}{4}$ in. long; anther-cells oblong, muticous, at equal height;

pollen subglobose with 3 large stopples and 3 bands continued to the poles ; ovary glabrous, lower part of style thinly hairy.

COAST REGION : Komgha Div. ; near the mouth of the Kei River, 200 ft., *Flanagan,* 2351 !

This species is very close to *E. barlerioides,* Lindau ; it differs by the narrower bracts and long-petioled leaves.

Imperfectly known Species.

2. E. protractum (Schinz in Mém. Herb. Boiss. x. 64).

EASTERN REGION : Delagoa Bay, *Kuntze.*

This is defined as *Ecbolium protractum,*. O. Kuntze, Rev. Gen. Pl. iii. pt. 2, 248. This may not improbably mean *Justicia pulegeioides,* E. Meyer ; but it must be recollected that the plant *Justicia pulegioides,* E. Meyer, is the type of the genus *Chætacanthus,* Nees.

XXVI. ISOGLOSSA, Oerst.

Calyx 5-partite nearly to the base ; segments 5, linear, $\frac{1}{8}$–$\frac{1}{8}$ in. long. *Corolla* white, with some rose, yellow or purple spots, 2-lipped, $\frac{1}{3}$–$\frac{2}{3}$ in. long (much longer in the subgenus *Ramusia*) ; anticous lip not much shorter than the tube, with 3 oblong or ovate lobes. *Stamens* 2, without rudiments of others ; filaments glabrous ; one anther-cell completely above the other, elliptic ; pollen globose, much flattened, almost lenticular, with one stopple in the centre of each smooth circular face. *Pistil* glabrous ; style hardly 2-lobed ; ovary oblong-ellipsoid with 2 ovules in each cell. *Capsule* $\frac{1}{2}$–$\frac{3}{4}$ in. long, usually 4-seeded at the top, lower half much narrower (except in *I. origanoides*). *Seeds* tubercular-scabrous.

Leaves ovate or elliptic ; inflorescences panicled ; floral leaves and bracts small, or larger and lanceolate or spathulate-obovate.

Species about 30, in Africa and the Mascarene Islands, besides a few in India.

The Cape species (except subgenus *Ramusia*) agree most closely in habit, corolla, stamens, and pistil, and are consequently difficult to distinguish, both among themselves and from some Tropical African species.

Subgenus 1. EU-ISOGLOSSA. Corolla $\frac{1}{3}$–$\frac{2}{3}$ in. long ; tube cylindric.

Calyx with hairs none gland-headed :
    Inflorescences $\frac{1}{2}$–2 in. long :
        Bracteoles not white-margined   ...   ...   (1) **ciliata.**
        Bracteoles white-margined ...   ...   (2) **sylvatica.**
    Inflorescences 3–6 in. long   ...   ...   (3) **ovata.**
Calyx with numerous gland-headed hairs :
    Leaves large, some up to 5 by 2–3 in.   ...   (4) **Woodii.**
    Leaves medium-sized, some up to 3 by 1–1½ in. :
        Flowers very loosely paniculate, nearly all
          solitary   ...   .   (5) **prolixa.**
        Flowers approximated :
          Bracts linear-lanceolate   ...   ...   (6) **eckloniana.**
          Bracts obovate, obtuse with a short
            acumination   ...   .   (7) **stipitata.**
    Leaves attaining 1–1½ in. in length, mostly
        shorter :

Spikes slightly interrupted, manifestly viscid-hairy:
  Bracts $\frac{1}{2}$–$\frac{2}{3}$ in. long, broadly lanceolate (8) **origanoides.**
  Bracts $\frac{1}{4}$–$\frac{1}{3}$ in. long, spathulate-obovate (9) **Grantii.**
  Bracts $\frac{1}{8}$–$\frac{1}{6}$ in. long, oblong ... ... (10) **Bolusii.**
Spikes interrupted, loose, sparingly hairy (except the calyx):
  Corolla $\frac{3}{4}$ in. long, rather broad ... (11) **Macowanii.**
  Corolla $\frac{1}{3}$–$\frac{1}{2}$ in. long, slender ... ... (12) **delicatula.**

Subgenus 2. RAMUSIA.—Corolla $1\frac{1}{2}$ in. long; tube slender.
  Bracteoles and calyx thinly hispid with white non-glandular hairs ... ... (13) **hypoestiflora.**
  Bracteoles and calyx viscid with many gland-headed hairs ... ... (14) **Cooperi.**

1. **I. ciliata** (Lindau in Engl. & Prantl, Pflanzenfam. iv. 3B, 344); pubescent; branches 2–3 ft. long, slender, not virgate; leaves up to 2 by $\frac{3}{4}$ in., elliptic, narrowed at either end; petiole 0–$\frac{2}{3}$ in.; inflorescences terminal and axillary, short, dense, $\frac{1}{2}$–1 by $\frac{1}{2}$–$\frac{3}{4}$ in., without any solitary flowers, compound; floral leaves $\frac{1}{2}$–$\frac{2}{3}$ in. long, lanceolate or narrowly obovate, imbricate, prominent to the top of the inflorescence; bracteoles $\frac{1}{4}$–$\frac{1}{3}$ in. long, linear; calyx $\frac{1}{4}$ in. long, scarious, hardly at all green; segments linear, with many long many-celled hairs, none gland-headed; corolla $\frac{1}{2}$ in. long. *Rhytiglossa ciliata, Nees in Lindl. Nat. Syst. Bot. ed.* ii. 445 *in Linnæa,* xv. 364, *and in DC. Prod.* xi. 335; *Sonder in Linnæa,* xxiii. 93; *Hook. Journ. Bot.* ii. (1840) 126; *Oerst. in Kjob. Vidensk. Meddel.* 1854, 155 *in obs. t.* 5. *fig.* 27. *Justicia intercepta, E. Meyer in Drège, Zwei Pflanzengeogr. Documente,* 160, 195. *J. divaricata, Zeyher ex Nees in DC. Prod.* xi. 336. *J. capensis, Ecklon ex Nees l.c. Ecteinanthus divaricatus, T. Anders. in Journ. Linn. Soc.* vii. 45.

COAST REGION: Uitenhage Div.; amongst bushes by the Zwartkops River, *Ecklon & Zeyher,* 73! and without precise locality, *Ecklon & Zeyher,* 928! Alexandria Div.; Addo, *Zeyher,* 1401! East London Div.; near East London, 100 ft., *Galpin,* 3186!

EASTERN REGION: Natal; Umgeni River, 500 ft., *Drège!*

2. **I. sylvatica** (C. B. Clarke); branches and leaves nearly glabrate; inflorescences somewhat interrupted at the base; bracteoles $\frac{1}{3}$ in. long, linear from a lanceolate base, widened by scarious margins; corolla wholly white (*Burchell*); otherwise as *I. ciliata. Dianthera sylvatica, Burchell ms.*

COAST REGION: Knysna Div.; in the forest, near Melville, *Burchell,* 5438!

3. **I. ovata** (Lindau in Engl. & Prantl, Pflanzenfam. iv. 3B, 344); hairy; branches 2 ft. long at least, rather stout; leaves up to $2\frac{1}{2}$ by $1\frac{1}{4}$ in., ovate, base often rounded, nerves prominent on the lower surface; petioles hardly $\frac{1}{6}$ in. long; inflorescences up to 4–6 by $\frac{1}{2}$ in., terminal and lateral (forming terminal panicles 6–16 in. long), more or less interrupted below; floral leaves $\frac{1}{2}$ by $\frac{1}{6}$ in. (lower hardly or not imbricate), including 1–2 (rarely 3) flowers; bracteoles $\frac{1}{3}$ in. long, linear-lanceolate; calyx $\frac{1}{3}$ in. long, green, not at all scarious;

segments linear, with many short few-celled hairs none gland-
headed; corolla $\frac{2}{3}$–$\frac{3}{4}$ in. long. *I. Bachmanni, Lindau in Engl.
Jahrb.* xx. 57, *and in Engl. & Prantl, Pflanzenfam.* iv. 3 B, 344.
Justicia ovata, E. Meyer in Drège, Zwei Pflanzengeogr. Documente,
149, 196. *Dicliptera ovata, Presl, Bot. Bemerk.* 95. *Rhytiglossa
ovata, Nees in DC. Prod.* xi. 336. *Ecteinanthus ovatus, T. Anders.
in Journ. Linn. Soc.* vii. 45.

COAST REGION : Komgha Div. ; Komgha, 2000 ft., *Flanagan,* 680!

EASTERN REGION : Pondoland ; on stony heights between Umtata River and
St. Johns River, 1000–2000 ft., *Drège!* and without precise locality, *Bachmann,*
1272 ! Natal; Inanda, 1800 ft., *Wood,* 45! near Verulam, *Wood,* 799 !

4. I. Woodii (C. B. Clarke); stems and leaves sparingly hairy;
leaves up to 5 by 2–3 in., ovate, acuminate, tapering into a petiole
1–2 in. long ; inflorescences 2 by $\frac{1}{3}$–$\frac{1}{2}$ in., hardly interrupted, forming
terminal panicles 2–4 in. in diam., and also remote axillary inflo-
rescences ; floral leaves $\frac{1}{8}$–$\frac{1}{5}$ in. long, obovate, suddenly contracted
into a small acute tip ; bracteoles $\frac{1}{8}$ in. long, linear; calyx $\frac{1}{8}$ in. long ;
segments linear, with many gland-headed hairs ; corolla $\frac{1}{3}$ in. long ;
capsule $\frac{1}{3}$ in. long, stalked, with 4 tubercular-scabrous flattened seeds
in the upper half.

EASTERN REGION : Natal; Inanda, *Wood,* 786 ! edges of woods, Berea, near
Durban, 100–200 ft., *Wood,* 3945 ! and without precise locality, *Gerrard,*
1897 !

"This plant does not flower every year; it is commonly believed to flower
once in seven years only ; this year, 1888, it has produced flowers in great
abundance " (*Wood*).

5. I. prolixa (Lindau in Engl. & Prantl, Pflanzenfam. iv. 3 B,
344) ; stems hairy ; leaves up to 3 by 1$\frac{1}{2}$ in., ovate, pubescent on
the nerves, apex acuminate, base obtusely rhomboid; petioles up to
1$\frac{1}{3}$ in. long ; panicles terminal and axillary, loose, most of the lower
flowers solitary· but some pedicelled ; floral leaves (here mostly
1-flowered, i.e. bracts) lower ovate-lanceolate, upper linear-lanceo-
late ; bracteoles $\frac{1}{8}$ in. long, linear ; calyx $\frac{1}{3}$ in. long; segments linear,
densely viscous with gland-headed hairs ; corolla $\frac{1}{2}$–$\frac{1}{3}$ in. long;
capsule $\frac{1}{2}$ in. long, stalked, with 4 rough flattened seeds in the upper
half. *Justicia prolixa, E. Meyer in Drège Zwei Pflanzengeogr.
Documente,* 150, 196. *Rhytiglossa prolixa, Nees in DC. Prod.*
xi. 336 *partly. Ecteinanthus prolixus, T. Anders. in Journ. Linn.
Soc.* vii. 45.

EASTERN REGION : Pondoland ; St. Johns River, below 1000 ft., *Drège!*

Various neighbouring species have been sorted with this single collection of
Drège which differs from them by the loose inflorescence ; in the lower flowers,
the pedicel is sometimes $\frac{1}{4}$–$\frac{2}{3}$ in. long between the bract and the bracteoles.

6. I. eckloniana (Lindau in Engl. & Prantl, Pflanzenfam. iv. 3 B,
344) ; inflorescence much more dense, with no solitary flowers ; calyx-
segments broader, linear, but scarcely acute, otherwise as *I. prolixa,*
Lindau. *Rhytiglossa eckloniana, Nees in Lindl. Nat. Syst. Bot.*
ed. ii. 445, *in Linnœa,* xv. 365, *and in DC. Prod.* xi. 336. *Ecteinanthus ecklonianus, T. Anders. in Journ. Linn. Soc.* vii. 45.

COAST REGION: Uitenhage Div.; in shaded situations, *Bowie!* Komgha Div.; near Komgha, 2000 ft., *Flanagan*, 668!
EASTERN REGION: Natal; Groen Berg, *Wood*, 1619!

7. I. stipitata (C. B. Clarke); hairy; leaves up to 3 by 1¼ in., ovate, tip acuminate, base rounded or rhomboidal; petioles up to 1 in. long; inflorescences 2½ by ½ in., slightly interrupted at the base; bracts ⅕ in. long, obovate, obtuse with a short linear tip, obscure in the fruiting spikes; calyx exceeding ¼ in. in length; segments linear, with many gland-headed hairs; capsule ⅓ in. long, stalked, with 4 small seeds in the upper half. *Rhytiglossa glandulosa, Hochst. in Flora*, 1845, 71. *R. prolixa, Nees in DC. Prod.* xi. 336 *partly*. *Ecteinanthus origanoides, T. Anders. in Journ. Linn. Soc.* vii. 45.

COAST REGION: Komgha Div.; near the Kei River, 600 ft., *Flanagan*, 2322!
EASTERN REGION: Natal; in shady woods around Durban Bay, *Krauss*, 302 (distributed as 502) partly!

The species *Rhytiglossa glandulosa*, Hochst., is founded on *Krauss*, 302; Nees has written "*Rhytiglossa prolixa*" and T. Anderson "*Ecteinanthus origanoides*" on the Kew sheet. There are two good fruiting branches and two small flowering branchlets added. It resembles much *I. origanoides*, which has the bracts not obovate, and a much less stipitate capsule. The two small flowering branchlets I have named *Isoglossa Grantii*.

8. I. origanoides (Lindau in Engl. & Prantl, Pflanzenfam. iv. 3B, 344); branches rigid; leaves attaining 1 by ⅓–½ in., pubescent; petioles 0–⅛ in. long; inflorescences terminal, and on the main stem on short lateral branches, 2½ by ½ in., hardly interrupted, with brown viscid hairs; bracts or floral leaves ½–⅔ in. long, broadly lanceolate; calyx ½ in. long; segments linear, with many gland-headed hairs; capsule ½ by ⅛ in., 4-seeded, very little narrower in the lower half. *Rhytiglossa origanoides, Nees in Lindl. Nat. Syst. Bot. ed.* ii. 445, *in Linnæa*, xv. 365, *and in DC. Prod.* xi. 336. *R. glandulifera, Presl, Bot. Bemerk.* 95. *Justicia glandulifera, E. Meyer in Drège, Zwei Pflanzengeogr. Documente*, 134, 195 (letter *a* only). *Ecteinanthus origanoides, T. Anders. in Journ. Linn. Soc.* vii. 45.

COAST REGION: Alexandria Div.; in Johannes Kloof, between Enon and the Zuurberg Range, 1000–2000 ft., *Drège!*

9. I. Grantii (C. B. Clarke); branches pubescent; leaves up to 1⅓ by 1 in., ovate, acuminate, hairy, base broadly rhomboid; petiole up to ½ in. long; inflorescences terminal on the main stem, and on short lateral branches, 2 by ⅓ in., viscid-hairy, hardly interrupted; bracts or floral leaves conspicuous, green, ¼–⅓ in. long, spathulate, broadly obovate, suddenly acuminate; calyx ⅙ in. long; segments linear, with many gland-headed hairs. *Rhytiglossa prolixa, Nees in DC. Prodr.* xi. 336 *partly*.

EASTERN REGION: Natal; in shady woods around Durban Bay, *Krauss*, 302 (distributed as 502) partly! and without precise locality, *Grant!*

10. I. Bolusii (C. B. Clarke); stems 2 ft. long, much branched upwards, somewhat hairy; leaves attaining 1¼ by ¾ in., ovate, acuminate, slightly hairy, base decurrent ; petiole rarely attaining $\frac{1}{10}$ in. in length ; inflorescences terminal on the main stem, and on short lateral branches, attaining 2 by $\frac{1}{3}$ in., slender, interrupted viscid-hairy ; floral leaves (or bracts) in the upper half of the inflorescence oblong, very small, often less than $\frac{1}{6}$ in. long ; bracteoles similar but smaller ; calyx $\frac{1}{6}-\frac{1}{5}$ in. long ; segments linear, with many gland-headed hairs ; corolla $\frac{1}{4}$ in. long, rather slender ; capsule $\frac{1}{3}$ in. long, stalked, 4-seeded.

SOUTH AFRICA: without precise locality, *Masson!*

CENTRAL REGION : Stockenstrom Div. ; Kat Berg, *Shaw in Bolus Herb.*, 299!

KALAHARI REGION : Transvaal ; in wooded ravines near Barberton, 3000 ft., *Galpin*, 456 ! 958 !

11. I. Macowanii (C. B. Clarke) ; stems and leaves sparingly pubescent ; leaves 1–1$\frac{1}{3}$ by $\frac{2}{3}$ in., broadly elliptic, not acuminate ; petioles $\frac{1}{4}-\frac{3}{4}$ in. long ; inflorescences scattered, many on the lower branches, interrupted, few-flowered, only slightly viscous ; bracts (even the upper ones) ovate or broadly lanceolate, often exceeding the calyx ; calyx $\frac{1}{4}$ in. long ; segments oblong-linear, white-hispid and also with some gland-headed hairs ; corolla $\frac{3}{4}$ by $\frac{1}{4}$ in., considerably larger than in the 2 preceding species.

COAST REGION : Albany Div. ; without precise locality, *Cooper*, 1519!

CENTRAL REGION : Somerset Div. ; edges of woods on the Bosch Berg, 2000–2800 ft., *MacOwan*, 933 ! *Bolus*, 299 !

EASTERN REGION : Natal ; without precise locality, *Cooper*, 2876 !

12. I. delicatula (C. B. Clarke) ; stems and leaves sparingly pubescent ; leaves up to 1$\frac{1}{4}$ by 1 in., elliptic, acuminate at either end ; petioles up to $\frac{1}{2}$ in. long ; inflorescences up to 1$\frac{1}{3}$ by $\frac{1}{3}$ in., interrupted, loose, sparingly hairy ; bracts lanceolate, about $\frac{1}{8}$ in. long, bracteoles similar but rather shorter ; calyx $\frac{1}{6}$ in. long ; segments linear, with some gland-headed hairs ; corolla $\frac{1}{3}-\frac{1}{2}$ in. long.

COAST REGION : King Williamstown Div. ; Perie woods, 2500 ft., *Tyson*, 1046!

EASTERN REGION : Natal ; Umkomaas, 4000–5000 ft., *Wood*, 4607 !

13. I. hypoestiflora (Lindau in Engl. Jahrb. xx. 58) ; sparingly hairy, branches 8–20 in. long ; leaves up to 3 by 1 in., elliptic, acuminate, base narrowed ; petiole up to 1$\frac{1}{4}$ in. long ; flowers in small loose panicles, mostly towards the ends of the branches ; bracts up to $\frac{3}{4}$ by $\frac{1}{8}$ in., linear-oblong ; bracteoles much smaller, linear-oblong, thinly hispid with non-glandular white hairs ; calyx $\frac{1}{4}$ in. long ; segments linear, pubescent ; corolla-tube $\frac{3}{4}$ by $\frac{1}{20}$ in., hardly widened upwards ; anticous lip $\frac{2}{3}$ by $\frac{1}{3}$ in., ovate, nearly entire ; posticous lip $\frac{1}{2}-\frac{2}{3}$ in. long, linear-oblong ; stamens 2 ; anther-cells oblong, muticous, one completely above the other ; pollen globose, compressed, circular faces smooth with a few scattered prickles and a stopple in the centre of each ; pistil glabrous ; style very shortly 2-lobed ; capsule $\frac{1}{2}$ by $\frac{1}{6}$ in., 4-seeded, the placentæ not rising elastically from the base ; seeds roughly wrinkled. *Justicia*

G 2

tridentata, E. Meyer in Drège, Zwei Pflanzengeogr. Documente, 150, 196. *Gendarussa tridentata, Presl, Bot. Bemerk.* 95. *Ramusia tridentata, Nees in DC. Prod.* xi. 309 ; *T. Anders. in Journ. Linn. Soc.* vii. 50. *Peristrophe sp., Benth. in Benth. & Hook. f. Gen. Pl.* ii. 1122.

EASTERN REGION: Pondoland; St. Johns River, under 1000 ft., *Drège!* and without precise locality, *Bachmann,* 1283 ; Griqualand East ; Clydesdale, 4000 ft., *Tyson,* 2142! Natal ; near Durban, 200 ft., *Wood,* 7477! and without precise locality, *Gueinzius! Gerrard,* 1898!

A shrub 12–16 ft. high (*Lindau*).

14. I. Cooperi (C. B. Clarke) ; branchlets, bracts and bracteoles viscid-hairy with gland-headed hairs ; leaves up to 2 by ¾ in.; corolla 1¼ in. long ; otherwise as *I. hypoestiflora.*

EASTERN REGION: Griqualand East; Zuurberg Range, 4000 ft., *Tyson,* 1774! Natal; Umgeni Falls, *Cooper,* 1183!

This appears, from the stoutness of the branches in herbarium specimens, to be a large shrub like *I. hypoestiflora ;* the leaves are smaller, the bracts rather narrower, and the corolla rather smaller.

XXVII. PERISTROPHE, Nees.

Calyx small; segments 5, nearly separate, linear-lanceolate. *Corolla* medium-sized, pink ; tube linear, hardly widened upwards; lips 2, long, posticous narrow, subentire. *Stamens* 2 ; anther-cells 2, muticous, one much above the other; pollen oblong-ellipsoid, banded longitudinally. *Style* with 2 very short oblong branches. *Capsule* 4-seeded, on a cylindric stalk ; placentæ not rising elastically from the capsule-base. *Spikelets* panicled, pedicelled, about ¼–½ in. by ¹⁄₁₀ in., glabrous or minutely puberulous, containing usually 1 perfect and 1 imperfect flower; bracts 2, rather longer than the calyx, linear-lanceolate; bracteoles to each flower 2, similar, rather shorter.

Species 20 in the warmer regions of the Old World.

This genus differs from *Dicliptera* only in that the placentæ do not, in the ripe fruit, rise elastically from the base of the capsule valves. The 5 Cape species, here recorded, are easily separated from everything else, though perhaps they form only one large species. The Cape *Ramusia,* reduced to *Peristrophe* by Bentham, is here, as by Lindau, placed in *Isoglossa,* of which it has the watch-shaped pollen.

Whole plant glabrate :
    Panicles rather dense, interspersed with leaves   ...   (1) **caulopsila.**
    Panicles very straggling, almost leafless   ...   ...   (2) **bicalyculata.**
    Panicles stout, rather dense, almost leafless   ...   (3) **Hensii.**
Leaves persistently thinly hairy :
    Spikelets ¼ in. long; corolla ⅔ in. long   ...   ...   (4) **cernua.**
    Spikelets ⅓–½ in. long; corolla 1 in. long   ...   ...   (5) **natalensis.**

1. P. caulopsila (Presl, Bot. Bemerk. 95) ; glabrate ; leaves (lower) attaining 1½ by ⅔ in., elliptic, acuminate at either end ; petiole up to ⅓ in. long, but all the upper leaves much smaller; panicle 14 by 5 in., compound, dense with numerous pedicelled spikelets and small leaves ; spikelets about $\frac{1}{4}$ by $\frac{1}{12}$ in., 2 outer bracts $\frac{1}{4}$ by $\frac{1}{16}$ in. ; corolla $\frac{3}{8}$–$\frac{3}{4}$ in. long. *Nees in DC. Prod.* xi. 498 ; *T. Anders. in Journ. Linn. Soc.* vii. 48 *partly ; Lindau in Engl. & Prantl, Pflanzenfam.* iv. 3B, 331. *Justicia caulopsila, E. Meyer in Drège, Zwei Pflanzengeogr. Documente*, 137, 195. *J. acinoides, Hort. Kew. ex Nees in DC. Prod.* xi. 498.

COAST REGION : Uitenhage Div. ; in thickets by the Sunday River, *Bowie !* Queenstown Div. ; Gwatyn, 2900 ft., *Galpin*, 2044 !
CENTRAL REGION : Somerset Div. ; between the Zuurberg Range and Klein Bruintjeshoogte, 2000–2500 ft., *Drège !* "Otter Station " on Little Fish River, *Burchell*, 3263 ! Somerset East ; 1500 ft., *Bolus*, 1652 partly !

2. P. bicalyculata (Nees in Wall. Pl. As. Rar. iii. 113) ; panicle large, straggling, thin, with elongated branches and scantily leafy ; otherwise as *P. caulopsila*, Presl. *Nees in DC. Prod.* xi. 496 ; *T. Anders. in Journ. Linn. Soc.* vii. 47 ; *Lindau in Engl. & Prantl, Pflanzenfam.* iv. 3B, 331, *and in Engl. Pfl. Ost-Afr. C.* 371 ; *C. B. Clarke in Dyer, Fl. Trop. Afr.* v. 242. *Dianthera bicalyculata, Retz. in Vet. Acad. Handl.* [1775] 297, *t.* 9.

WESTERN REGION : Great Namaqualand ; Homeib River, *Schinz*, 19 !
An abundant plant in Tropical Africa and India.

Dr. Schinz's example cited is quite typical *P. bicalyculata*, but among the quantity of specimens, admitted as *P. bicalyculata*, some are very near *P. caulopsila*.

3. P. Hensii (C. B. Clarke in Dyer, Fl. Trop. Afr. v. 243) ; stems stout, strongly hexagonal even in the panicle ; panicle rigid ; ultimate peduncles short, stout ; leaves few ; corolla often exceeding 1 in. in length : otherwise as *P. bicalyculata*, Nees. *Dicliptera Hensii, Lindau in Engl. Jahrb.* xxii. 120.

COAST REGION : Uitenhage Div. ; *Tredgold!* Komgha Div. ; near Komgha, 2000 ft., *Flanagan*, 721 ! British Kaffraria, *Cooper*, 161 !
EASTERN REGION : Natal ; Umhlanga, *Wood*, 609 !
Also in Tropical Africa.

4. P. cernua (Nees in Linnæa, xv. 374) ; leafy upwards ; leaves persistently hairy ; inflorescences very small, from the lower part of the branches, of 1–4 spikelets ; otherwise as *P. caulopsila*, Presl. *Nees in DC. Prod.* xi. 498 ; *Hook. Journ. Bot.* ii. (1840) 126. *P. caulopsila, T. Anders. in Journ. Linn. Soc.* vii. 48 *partly. Justicia capensis, Ecklon ex Nees in DC. Prodr.* xi. 498.

COAST REGION : Uitenhage Div. ; near the Zwartkops River, *Ecklon & Zeyher*, 40 !

T. Anderson had good reason for calling this only a form of *P. caulopsila ;* but it appears to me quite as well separable from *P. caulopsila* as is *P. bicalyculata*, which T. Anderson admitted as specifically distinct.

5. P. natalensis (T. Anders. in Journ. Linn. Soc. vii. 48) ; branches stout, hexagonal, often scabrous or hispid on the angles (as is *P. bicalyculata* sometimes) ; leaves up to 2 by $\frac{3}{4}$ in., ovate-lanceolate, rhomboid at the base, persistently hairy ; inflorescence 16 by

6 in., leafy, dense with numerous spikelets; outer bracts up to
½ iu. long, linear, often rather widened upwards (i.e. narrowly
subspathulate); calyx nearly ¼ in. long; corolla exceeding 1 in. in
length.

EASTERN REGION : Natal; without precise locality, *Gueinzius! Grant!*

Imperfectly known Species.

6. P. Krebsii (Presl, Bot. Bemerk. 94) ; a glabrous shrub ; leaves
nearly 1 by ⅓ in., lanceolate, cuspidate, narrowed at the base,
petioled; peduncles axillary, solitary, as long as the petiole, bifid ;
pedicels shorter than the leaf, with setaceous bracts at their base;
bracteoles 2, unequal, enclosing the calyx, one similar to the calyx-
segments, the other longer, cuspidate ; corolla white, thrice the
length of the calyx; anther-cells divaricate ; style very long.

SOUTH AFRICA: without precise locality, *Krebs,* 251 !

XXVIII. HYPOESTES.

Spikelets containing 1 flower with a rudiment of a second;
bracts 2, free or united at the base into a tube. *Calyx* much
smaller than the bracts, divided nearly to the base; sepals 5, linear.
Corolla pink or white ; tube slender, dilated near the top; lips 2,
long. *Stamens* 2 ; anthers 1-celled, muticous; pollen ellipsoid,
longitudinally banded. *Ovary* with 2-1 ovules in each cell; style
shortly and equally 2-fid. *Capsule* small, stalked, 4- or 2-seeded ;
placentæ not rising elastically from the base of the capsule ; seeds
smooth or rough, not hairy.

Herbs or shrubs ; leaves entire, often wavy on the margins ; cymes of spikelets axillary and terminal, in heads or elongate into dense or loose spikes.

Species 60 ; extending from Africa to Australia.

Bracts of the spikelet free to the base :
  Bracts of the spikelet oblong at the base, linear
  in the upper half:
    Leaves usually 1–2½ in. long   ... ... **(1) aristata.**
    Leaves usually 2–4 iu. long ... ... **(2) antennifera.**
  Bracts of the spikelet narrowly obovate, very
  obtuse :
    Lower bract ⅓–½ in. long ... ... **(3) triflora.**
    Lower bract ⅔–¾ in. long ... ... **(4) phaylopsoides.**
  Bracts of the spikelet connate at the base into a
  distinct tube:
    Bracts of the spikelet hairy or densely pubes-
    cent ... ... **(5) verticillaris.**
    Bracts of the spikelet glabrate or scarcely
    puberulous ... ... **(6) Forskalei.**

1. H. aristata (R. Br. Prod. 474, in Obs.) ; more or less pubescent;
branches 1–2 ft. long; leaves 1–2 by ½–1 in., ovate, narrowed at
both ends ; petiole 0–⅔ in. long ; heads globose, axillary, of numerous
1-flowered spikelets ; bracts of the spikelet free, ⅖ in. long, lower
half linear-oblong, upper half bristle-like ; two bracteoles ⅙–⅕ in.

long, linear-oblong ; second flower with its bracteoles generally
wanting ; corolla ⅔ in. in total length ; capsule ¼ in. long, glabrous ;
seeds 4, smooth. *Nees in Linnæa*, xv. 375, *and in DC. Prod.* xi.
509 ; *E. Meyer in Drège, Zwei Pflanzengeogr. Documente*, 129, 193 ;
Hochst. in Flora, 1845, 70 ; *T. Anders. in Journ. Linn. Soc.* vii.
48 ; *Lindau in Engl. & Prantl, Pflanzenfam.* iv. 3B, 333, *and in
Engl. Pfl. Ost-Afr. C.* 371 ; *C. B. Clarke in Dyer, Fl. Trop. Afr.*
v. 245. *Justicia aristata, Vahl, Symb.* ii. 2.

COAST REGION : Knysna Div. ; near the mouth of the Knysna River, *Bowie !*
Humansdorp Div. ; in a wooded kloof, near Humansdorp, 450 ft., *Galpin*, 4394 !
amongst shrubs by the Kabeljouws River, *Bolus*, 2423 ! Uitenhage Div. ; without
precise locality, *Cooper*, 1495 ! *Tredgold*, 36 ! Port Elizabeth Div. ; on sand-hills
and rocky shores near Port Elizabeth, below 100 ft., *Drège !* near Port Elizabeth,
Burchell, 4319 ! 4338 ! Algoa Bay, *Forbes !* Albany Div. ; near Grahamstown,
Ecklon & Zeyher, 876 ! *Atherstone*, 11 ! *Read ! Burchell*, 3584/2 ! *Bunbury !*
Bathurst Div. ; near Theopolis, *Burchell*, 4110 ! King Williamstown Div. ;
Keiskamma, *Hutton !*
  KALAHARI REGION : Orange River Colony, *Cooper*, 1037 ! 3042 ! Transvaal ;
Houtbosch, *Rehmann*, 6185 ! Barberton, 2000 ft., *Galpin*, 644 !
  EASTERN REGION : Natal ; De Beers Pass, 5000–6000 ft., *Wood*, 6022 !

Also in South-east Tropical Africa.

2. H. antennifera (S. Moore in Journ. Bot. 1880, 41) ; leaves up
to 4–5 by 2–2½ in., pubescent on the underface or sometimes with
many long hairs along the nerves, often very thin in texture ;
petioles up to 1–1½ in. long ; bracts of the spikelets often ⅔ in.
long ; corolla often 1 in. long ; otherwise as *H. aristata*, R. Br.
Lindau in Engl. & Prantl, Pflanzenfam. iv. 3B, 333, *and in Engl.
Pfl. Ost-Afr. C.* 371 ; *C. B. Clarke in Dyer, Fl. Trop. Afr.* v. 245.
H. aristata, var. macrophylla, Nees in DC. Prod. xi. 510. *H. plumosa,
E. Meyer in Drège, Zwei Pflanzengeogr. Documente*, 153, 193.

SOUTH AFRICA : without precise locality, *Zeyher*, 1396 !
  COAST REGION : British Kuffraria ; near Breakfast Vley, *Cooper*, 3037 !
  KALAHARI REGION : Orange River Colony, *Cooper*, 3036 !
  EASTERN REGION : Pondoland ; between St. Johns River and Umtsikaba
River, 1000 ft., *Drège !* Natal ; without precise locality, *Plant*, 99 ! *Gerrard*, 10 !
Grant !

Also in East Tropical Africa.

3. H. triflora (Roem. et Schult. Syst. i. 141) ; more or less hairy,
stems 1–3 ft. long ; leaves 2½ by 1 in. (or often smaller), ovate,
narrowed at either end ; petioles ¼–⅓ in. long ; spikelets 5 or fewer
(often 3) together in scattered heads, sometimes solitary, mostly 1-
flowered ; 2 outer bracts free, ⅓–½ by ⅙–⅛ in., narrowly obovate-oblong,
subobtuse ; corolla ¾ in. in total length ; capsule nearly ½ in. long,
glabrous, 4-seeded. *Nees in DC. Prod.* xi. 506 ; *T. Anders. in Journ.
Linn. Soc.* vii. 50 ; *Lindau in Engl. & Prantl, Pflanzenfam.* iv. 3B,
333 ; *C. B. Clarke in Hook. f. Fl. Brit. Ind.* iv. 557, *and in Dyer,
Fl. Trop. Afr.* v. 247. *Justicia triflora, Forsk. Fl. Ægypt.-Arab.* 4.

EASTERN REGION : Tembuland ; Bazeia, 2500 ft., *Baur*, 109 ! Griqualand
East ; Umzimkulu district, in Enyembi woods, 5000 ft., *Tyson*, 2547 ! Natal ;

Mount West, 5200 ft., *Schlechter*, 6829! Polela District, *Fourcade in Wood Herb.* 4282!

Frequent in Tropical Africa and India.

4. H. phaylopsoides (S. Moore in Trans. Linn. Soc. ser. 2, Bot.

iv. 34); lower bract of the spikelet attaining ¾ by ¼ in., upper considerably smaller; otherwise as *H. triflora. Lindau in Engl. Pfl. Ost-Afr. C.* 371; *C. B. Clarke in Dyer, Fl. Trop. Afr.* v. 248.

KALAHARI REGION: Orange River Colony, *Cooper*, 3035! Transvaal; Barberton, 3000 ft., *Galpin*, 959!

EASTERN REGION: Natal; Noods Berg, *Wood*, 1058!

Also in South-east Tropical Africa.

This species resembles closely the large-leaved forms of *H. triflora*, Roem. & Schult.

5. H. verticillaris (R. Br. Prod. 474, in Obs.); more or less hairy;

branches 1–3 ft. long; leaves 2½ by 1 in., elliptic-lanceolate, sometimes larger, wavy-crenate on the margins, narrowed at either end; petioles ¼-¾ in. long; spikelets in axillary clusters or more often the short peduncle carrying an oblong or linear spike of clusters; 2 outer bracts to each spikelet connate at the base into a distinct tube, ⅛ in. in total length, pubescent or hairy, the free tips narrowly oblong; corolla ⅓-½ in. long; capsule ¼-⅓ in. long, smooth, often 2-seeded. *Nees in Linnæa*, xv. 376, *in DC. Prod.* xi. 507; *Krauss in Flora*, 1845, 70; *T. Anders. in Journ. Linn Soc.* vii. 48; *Lindau in Engl. Jahrb.* xviii. 63, *t.* 2, *fig.* 73, *in Engl. & Prantl, Pflanzenfam.* iv. 3B, 332, 333, *fig.* 134 A–C, *and in Engl. Pfl. Ost-Afr. C.* 371; *Rolfe in Oates, Matabeleland, ed.* ii. 406; *C. B. Clarke in Dyer, Fl. Trop. Afr.* v. 250; *Schinz in Mém. Herb. Boiss.* x. 64. *H. polymorpha, E. Meyer in Drège, Zwei Pflanzengeogr. Documente,* 123, 159, 193. *H. clinopodia, Nees in DC. Prod.* xi. 508, *fide T. Anders. l.c. Justicia verticillaris, Linn. f. Suppl.* 85; *Thunb. Prod.* 104, *and Fl. Cap. ed. Schult.* 479. *Dicliptera verticillaris, Juss. in Ann. Mus. Paris,* ix. (1807) 268.

SOUTH AFRICA: without precise locality, *Zeyher*, 1397!
COAST REGION: Caledon Div.; Hang Klip, *Mund & Maire!* Swellendam Div.! in moist situations, *Bowie!* George Div.; near George, *Drège!* Humansdorp Div.; by the Kabeljouws River, *Bolus*, 2424! Uitenhage Div.; Grassrug, near Uitenhage, *Baur!* and without precise locality, *Ecklon & Zeyher,* 78! *Pappe! Masson! Tredgold,* 28! *Rehmann!* Albany Div.; Grahamstown, *Bunbury!* and without precise locality, *Cooper,* 1540! *Atherstone,* 41! East London Div.; sea coast, *Galpin,* 1830! King Williams Town Div.; 1500 ft., *Tyson,* 1016! Cathcart Div.; Windvogel Mountain near Goshen, 3500 ft., *Baur,* 931!
CENTRAL REGION: Somerset Div.; Somerset East, 2000 ft., *Bolus,* 298! *Bowker!* Bruintjes Hoogte, on the lower part, *Burchell,* 2987!
KALAHARI REGION: Orange River Colony; Vaal River, *Burke!* Transvaal; Apies Poort, near Pretoria, *Rehmann,* 4106! Barberton, 3000 ft., *Galpin,* 957! between Spitz Kop and Komati River, *Wilms,* 1198!
EASTERN REGION: Tembuland; Bazeia, 2500 ft., *Baur,* 167! Natal; near Durban, *Drège!* near Byrne, *Wood,* 1818! Inanda, *Wood,* 874! Pietermaritzburg, *Oates!* and without precise locality, *Gerrard,* 1270! *Cooper,* 3033! Zululand; Entumeni, *Wood,* 3968! Delagoa Bay, *Junod,* 380.

Abundant in Tropical Africa.

Nees cites " *Hypoestes clinopodia,* E. Meyer in Cat. Pl. Drège, *a* 1837."
I find no such *Hypoestes* in Drège's book of 1843. Nees (as usual) omits to
cite the page. The *Hypoestes clinopodia* cited by Nees is not an error for
Justicia clinopodia, E. Meyer, which is a dissimilar plant, referred by Nees
himself to *Dicliptera.* Moreover, T. Anderson in reducing the *Hypoestes
clinopodia,* Nees, says that his was the plant collected by Drège.

6. H. **Forskalei** (R. Br. Prod. 474, in Obs.); stem and leaves
nearly glabrous; leaves up to $1\frac{1}{4}$ by $\frac{1}{8}$ in., lanceolate, obtuse ;
2 outer bracts of the spikelet glabrate or scarcely puberulous ;
otherwise as *H. verticillaris,* R. Br. *Nees in DC. Prod.* xi. 507 ; *T.
Anders. in Journ. Linn. Soc.* vii. 49; *Lindau in Engl. & Prantl,
Pflanzenfam.* iv. 3B, 333, *and in Engl. Pfl. Ost-Afr. C.* 371 ; *C. B.
Clarke in Dyer, Fl. Trop. Afr.* v. 249. *H. depauperata, Lindau in
Engl. Jahrb.* xx. 52. *Justicia paniculata, Forsk. Fl. Ægypt.-Arab.* 4.
J. Forskalei, Vahl, Symb. i. 2.

KALAHARI REGION : Bechuanaland ; near the pass in Kamhanin Mountains,
Burchell, 2182! Kuruman, *Marloth,* 1120!

Frequent in Tropical Africa.

Imperfectly known Species.

7. H. **menthæfolia** (E. Meyer in Drège, Zwei Pflanzengeogr.
Documente, 160, 193).

SOUTH AFRICA : Natal; hills near Durban, below 500 ft., *Drège.*

This name is not mentioned either by Nees or by T. Anderson.

8. H. **glabrata** (Presl, Bot. Bemerk. 96).

SOUTH AFRICA : without precise locality, *Krebs,* 252.

XXIX. MACRORUNGIA, C. B. Clarke.

Calyx divided about half-way down into 5 lanceolate segments.
Corolla red ; tube about $\frac{1}{4}$ in. long ; anticous lip 1 in. long or more,
outside in the bud ; posticous lip exceeding 1 by $\frac{1}{5}$ in., emarginate.
Stamens 2 ; filaments much exserted ; anther-cells 2, oblong, muticous, one a little below the other ; pollen ellipsoid, with several
rows of small tubercles and 2 stopples, longitudinal bands obscure.
Capsule ovoid or oblong-ovoid ; placentæ rising elastically from the
base of the valves.

Shrubs; leaves entire, ultimately glabrous ; flowers in strobilate spikes ;
bracts prominent, imbricate.

Species 3 in Tropical Africa, besides the one here described.

If the genus is sunk, the species must be placed in *Rungia,* as by T. Anderson.

1. M. **longistrobus** (C. B. Clarke) ; a shrub ; branches, young
leaves beneath, and innovations densely shortly white hairy ; leaves
5 by $1\frac{1}{2}$ in., narrowed at either end, soon glabrate, tip obtuse ;
petiole $\frac{1}{5}$ in. long ; spikes 3 by $\frac{1}{4}$ in., terminal on the main stem and
on short axillary branches ; bracts $\frac{2}{3}$ by $\frac{1}{3}$ in., imbricate, green with

scarious margins, nearly glabrous ; bracteoles minute ; calyx $\frac{1}{3}$ in. long, minutely but densely hairy within and without ; styles 1$\frac{3}{4}$ in. long.

KALAHARI REGION : Transvaal ; Avoca, near Barberton, 1800 ft., *Galpin*, 888 !

There is no fruit on Galpin's excellent specimen; but the corolla and stamens are so closely like those of *M. pubinervis* (of which the ripe fruit is known) that these two plants must be congeneric.

XXX. DICLIPTERA, Juss.

Bracts 2, much longer than the calyx, containing 1 (more rarely 2) flowers; bracteoles 2, linear. *Calyx* $\frac{1}{4}$ in. long or less, divided to the base ; segments 5, linear, hairy. *Corolla* $\frac{1}{2}$–1 in. long, pink, deeply 2-lipped; tube linear-funnel-shaped. *Stamens* 2 ; anther-cells 2, muticous, one much below the other ; pollen ellipsoid, longitudinally banded, without tubercles. *Ovules* 2 in each cell ; style with 2 subequal very short lobes. *Capsule* $\frac{1}{4}$–$\frac{1}{3}$ in. long, ovoid, very much flattened laterally, 4-seeded ; the placentæ rising elastically from the base of the capsule-valves; seeds rough or tubercled.

Herbs ; leaves entire ; spikelets solitary, clustered or capitate.

Species 60 ; in the tropical and subtropical regions of both hemispheres.

Bracts broad, ovate or somewhat obovate :
  Leaves 2–3 in. long ; spikelets from one axil
   numerous ... ...   (1) heterostegia.
  Leaves 2–3 in. long ; spikelets 1–5 (commonly
   3) from one axil ... ...   (2) zeylanica.
  Leaves $\frac{3}{4}$–1 in. long ; spikelets 1–3 from one
   axil ... ...   (3) capensis.
Bracts cuneate-oblong, acute or obtuse or lanceolate :
  Bracts $\frac{1}{2}$–$\frac{2}{3}$ by $\frac{1}{8}$ in., lanceolate, not widened
   upwards :
    Bracts $\frac{2}{3}$ in. long, acute, mucronate ...   (4) clinopodia.
    Bracts $\frac{1}{2}$ in. long, not mucronate...   (5) transvaalensis.
  Bracts $\frac{1}{3}$ by $\frac{1}{4}$ in. cuneate-oblong with rounded
   tip ... ...   (6) Quintasii.
  Bracts $\frac{1}{3}$ by $\frac{1}{10}$–$\frac{1}{8}$ in., broadly-lanceolate ...   (7) minor.

1. D. heterostegia (Presl, Bot. Bemerk. 95) ; stems 2 ft. long, hairy or nearly glabrous ; leaves up to 3$\frac{1}{2}$ by 2 in., much narrowed at either end ; petioles long, some attaining 2$\frac{1}{2}$ in. ; inflorescences 1$\frac{1}{2}$–2 in. in diam., of many spikelets, rather dense, in many of the axils ; bracts paired, shortly pedicelled, $\frac{1}{4}$–$\frac{1}{2}$ in. broad, sparsely hairy, enclosing 1 (or 1–3) flowers, lower ovate, acuminate, acute, or mucronate, upper rather shorter, rounded at the top, somewhat scarious ; bracteoles $\frac{1}{4}$ in. long, linear-lanceolate ; sepals $\frac{1}{6}$ in. long, linear ; corolla $\frac{2}{3}$ in. long ; capsule $\frac{1}{3}$ in. long, very hairy on its narrow margins. *Nees in DC. Prod.* xi. 478 ; *T. Anders. in Journ. Linn. Soc.* vii. 47. *Justicia heterostegia, E. Meyer in Drège, Zwei Pflanzengeogr. Documente,* 152, 195.

EASTERN REGION: Pondoland; between St. Johns River and Umsikaba River, 1000–2000 ft., *Drège!* Natal; Clairmont, below 400 ft., *Wood*, 1309! and without precise locality, *Gerrard*, 26!

2. D. zeylanica (Nees in DC. Prod. xi. 474); stems 2 ft. long, sparsely hairy; leaves up to $3\frac{1}{2}$ by $1\frac{3}{4}$ in., ovate-lanceolate, thinly hairy, secondary nerves raised conspicuously on the under face; petioles up to 1–$1\frac{1}{2}$ in.; inflorescences axillary, loose, of 3–1 (rarely 5) spikelets on pedicels $\frac{1}{6}$–$\frac{1}{3}$ in. long; bracts $\frac{1}{2}$–$\frac{2}{3}$ by $\frac{1}{3}$ in., obovate-elliptic, acute, mucronate; calyx, corolla and fruit as in *D. heterostegia*, Presl. *C. B. Clarke in Hook. f. Fl. Brit. Ind.* iv. 552. *D. bivalvis, Nees in DC. Prod.* xi. 475 *partly; Wight, Ill. Nat. Ord. Ind. Pl.* ii. 191, *t.* 164 *b, fig.* 10, *and Ic. Plant.* v. *t.* 1551; *T. Anders. in Journ. Linn. Soc.* ix. 519, *not of Jussieu.*

COAST REGION: Cape Div.; near the Cape of Good Hope Promontory, *Oldenburg!* Komgha Div.; in woods near the mouth of the Kei River, 100 ft., *Flanagan*, 801!

Frequent in the Indian Peninsula and Ceylon. Hardly differs specifically from *D. heterostegia.*

3. D. capensis (Nees in Linnæa, xv. 373); sparingly hairy; stems 4–18 in. long, slender; leaves 1 by $\frac{1}{2}$ in., tip triangular, base obtuse; petioles up to $\frac{1}{2}$ in. long; inflorescences axillary, thin, of 3–1 spikelets; bracts $\frac{1}{3}$ in. long and nearly as broad, round or ovate or somewhat obovate, both usually shortly acuminate, mucronate; corolla $\frac{1}{2}$ in. long; capsule $\frac{1}{3}$ in. long, very hairy on the margins. *Nees in DC. Prod.* xi. 481; *T. Anders. in Journ. Linn. Soc.* vii. 47; *Lindau in Engl. & Prantl, Pflanzenfam.* iv. 3B, 333. *D. propinqua, Nees in Linnæa*, xv. 373, *and in DC. Prod.* xi. 477. *Tyloglossa pubescens, Hochst. in Flora*, 1845, 71?

COAST REGION: Riversdale Div.; between the Gauritz River and Great Vals River, *Burchell*, 6523! Uitenhage Div.; near the Zwartkops River, *Ecklon!* King Williamstown Div.; near King Williamstown, 1600 ft., *Tyson*, 1013!

CENTRAL REGION: Somerset Div.; between Little Fish River and Commadagga, *Burchell*, 3280! Albert Div.; without precise locality, *Cooper*, 1767!

4. D. clinopodia (Nees in DC. Prod. xi. 483); pubescent or hairy, the branches and leaves often becoming nearly glabrate; branches 1–2 ft. long; leaves $2\frac{1}{4}$ by 1 in., narrowed at either end; petioles up to $\frac{1}{2}$–1 in. long; spikelets in dense heads, the terminal $1\frac{1}{2}$ by 1 in., the axillary often much shorter; bracts $\frac{2}{3}$ by $\frac{1}{8}$ in., oblong-lanceolate, acute, not widened upwards; bracteoles $\frac{1}{4}$ in. long, linear-lanceolate; sepals $\frac{1}{5}$ in. long, linear; corolla 1 in. long; capsule hardly exceeding $\frac{1}{4}$ in. in length, minutely hairy on the margins. *T. Anders. in Journ. Linn. Soc.* vii. 47. *Justicia clinopodia, E. Meyer in Drège, Zwei Pflanzengeogr. Documente*, 158, 195.

KALAHARI REGION: Orange River Colony; Vaal River, *Burke!* Transvaal; at Sterk Spruit near Lydenburg, *Wilms*, 1191!

EASTERN REGION: Natal; in the valley of the Umlazi River, *Drège!* Great Noods Berg, 3000 ft., *Wood*, 4280! Delagoa Bay; by the Crocodile River near Louws Creek, 1400 ft., *MacOwan & Bolus, Herb. Norm. Aust.-Afr.*, 1339!

5. D. transvaalensis (C. B. Clarke); branches patently hairy; leaves (upper only seen) $1\frac{1}{4}$ by $\frac{1}{3}$ in., elliptic-lanceolate, acute, hairy, base cuneate; petiole $\frac{1}{10}$ in. long; inflorescence 4 by 1 in., terminal, compound, of rather close panicles; bracts $\frac{1}{2}-\frac{2}{3}$ by $\frac{1}{8}$ in., lanceolate or linear-oblong, closely and shortly hairy, not mucronate; corolla $\frac{2}{3}$ in. long; stamens 2; anther-cells 2, muticous, one above the other.

KALAHARI REGION : Transvaal ; without precise locality, *Holub !*

6. D. Quintasii (Lindau in Engl. Jahrb. xxii. 121); branches 1 ft. long, 6-angular, hairy or almost hispid; leaves up to $1\frac{3}{4}$ by 1 in. long, rhomboid-elliptic, narrowed at the base, pubescent; petioles up to $\frac{2}{3}$ in. long; heads $\frac{2}{3}-1$ in. in diam., of few or many spikelets, terminal or axillary and on axillary peduncles 1–2 in. long; lower bract $\frac{1}{3}-\frac{1}{2}$ by $\frac{1}{6}$ in., cuneate-oblong, rounded at the tip with a minute bristle, slightly hairy; bracteoles $\frac{1}{4}$ in. long, linear-lanceolate; sepals $\frac{1}{6}$ in. long, linear; corolla $\frac{3}{4}$ in. long; capsule $\frac{1}{4}-\frac{1}{3}$ in. long, minutely hairy on the margins. *Schinz in Mém. Herb. Boiss.* x. 64.

EASTERN REGION : Tembuland; along river banks near Bazeia, 2000 ft., *Baur*, 150! Natal ; Biggars Berg, *Rehmann*, 7100! Lorenzo Marques; *Quintas*, 85!

This species is very close to *D. angolensis*, S. Moore, which has wider bracteoles.

7. D. minor (C. B. Clarke); nearly glabrate except the calyx; branches 5–7 in. long, slender ; leaves up to $\frac{3}{4}$ by $\frac{1}{4}$ in., narrowed at either end, tip obtuse ; petiole up to $\frac{1}{8}$ in. long; spikelets in clusters of 3–5, axillary and on axillary peduncles $\frac{1}{4}$ in. long; lower bract $\frac{1}{3}$ by $\frac{1}{10}-\frac{1}{8}$ in , broadly lanceolate, hardly acute ; bracteoles $\frac{1}{5}$ in. long, linear-lanceolate ; sepals less than $\frac{1}{6}$ in. long, linear, minutely hairy; corolla $\frac{1}{2}$ in. long ; capsule $\frac{1}{4}$ in. long.

KALAHARI REGION: Griqualand West, Hay Div. ; Klipfontein, *Burchell*, 2147! Bechuanaland; Bakwena Territory, 3500 ft., *Holub!* Transvaal, without precise locality, *Holub!*

ORDER CII. **MYOPORINEÆ.**

(By R. A. ROLFE.)

Flowers hermaphrodite, irregular or nearly regular. *Calyx* inferior, 5-partite or 5-lobed. *Corolla* gamopetalous; tube short and some-what campanulate, or elongate and infundibuliform ; limb subequal, oblique or bilabiate, 5- (or rarely 6-) lobed ; lobes imbricate, the two posticous often exterior, sometimes deeply connate. *Stamens* 4, didynamous or subequal, rarely as many as the corolla-lobes, inserted on the corolla and alternating with its lobes, included or exserted ; filaments filiform or thickened at the base ; anthers normally 2-celled ; cells at first parallel, afterwards diverging from the confluent apex, dehiscing longitudinally. *Disc* hypogynous, small or nearly obsolete. *Ovary* superior, 2-celled or more or less perfectly 3–10-celled by the intrusion of septa between the ovules; ovules when the ovary is 2-celled 2 in each cell and collateral or 4–8 in superposed pairs, when the ovary is many-celled solitary, anatropous, with a superior micropyle; style simple, terminal, short or somewhat elongated, filiform; stigma terminal, small, entire or obscurely emarginate, rarely oblique. *Fruit* drupaceous, indehiscent; exocarp fleshy, succulent or rarely dry; endocarp hard or thin, 2-celled or the cells as numerous as the seeds, rarely breaking up into pyrenes. *Seeds* 2–10, usually solitary, in cells arranged in one series round the axis, very rarely superposed (the upper ovules being generally abortive), pendulous, oblong; testa membranous or somewhat thickened; albumen fleshy, slender or nearly absent; embryo straight or slightly curved; radicle terete, superior; cotyledons semiterete, slightly broader and shorter (rarely longer) than the radicle.

Erect or diffuse herbs, shrubs or rarely trees, glabrous, tomentose, canescent, lepidote or pubescent. Leaves alternate or rarely opposite, entire or rarely dentate, exstipulate. Flowers axillary, solitary or fascicled, subsessile or pedicellate. Bracts small or absent.

DISTRIB.—Genera 6, and species about 80, mostly Australian, with a few Polynesian representatives; 1 in the Sandwich Islands, 1 in Mauritius, 2 others in China and Japan, 1 in the West Indies, a somewhat doubtful one in Tropical Africa, and the two following in South Africa.

Baillon (Hist. Pl. ix. 420) reduces *Myoporineæ* to the rank of a tribe of *Scrophulariaceæ*, but Wettstein (Engl. & Prantl, Pflanzenfam. iv. 3B, 354) follows Bentham in regarding it as a distinct order, which is more in accordance with its characters.

I. OFTIA, Adans.

Calyx 5-partite; segments narrow, acuminate, not enlarging in fruit. *Corolla*-tube cylindrical, equal or slightly dilated at the villous throat; limb spreading, 5-lobed; lobes obovate, subequal. *Stamens* 4, subequal, affixed to the middle of the tube, included; filaments short; anthers oblong, affixed about the middle of the back; cells parallel, ultimately somewhat confluent at the apex. *Ovary* 2-celled; style included within the corolla-tube; stigma oblong, oblique, somewhat thickened; ovules 4–6 in each cell, superposed in pairs. *Drupe* small, globose, succulent; putamen 1–2-celled. *Seeds* often solitary through abortion, oblong, somewhat curved; testa somewhat thickened; albumen fleshy, slender; embryo small, rather shorter than the albumen (*Bocquillon*).

Villous or sometimes viscid-pubescent much-branched shrubs; leaves alternate, or the lower opposite or somewhat verticillate, sessile, serrulate, often small; flowers white, sessile or shortly pedicelled in the upper axils, ebracteate.

DISTRIB. Species 2, endemic.

Leaves ovate or broad, ⅛–½ in. long, nearly flat ... ... (1) **africana.**
Leaves lanceolate or narrow, ¼–1 in. long, with more or less revolute margin ... ... (2) **revoluta.**

1. O. africana (Bocq. ex Baill. Adansonia, ii. 11); branches striate, strongly pubescent, the younger having narrow dentate wings formed by the decurrent leaf-bases; leaves numerous and crowded, sessile, ovate or ovate-oblong, acute, strongly dentate, pubescent or hispidulous, $\frac{1}{2}$–$1\frac{1}{4}$ in. long, 3–10 lin. broad; flowers axillary, solitary, shortly pedicelled; calyx narrowly campanulate, glandular-hispidulous, $2\frac{1}{2}$–3 lin. long; lobes subulate, about six times as long as the tube; corolla white, with a slender blue streak at the base of each lobe; tube narrowly oblong, 3–4 lin. long, hispidulous outside; lobes subequal, obovate-orbicular, crenulate, 2–3 lin. long; fruit subglobose, 2–$2\frac{1}{2}$ lin. diam. *Lantana ? africana, Linn. Sp. Pl. ed.* i. 628; *Thunb. Prod.* 98, *and Fl. Cap. ed. Schult.* 458; *Ait. Hort. Kew. ed.* i. ii. 352. *L. crispa, Thunb. Prod.* 98; *and Fl. Cap. ed. Schult.* 458. *L. capensis, Thunb. Prod.* 98; *and Fl. Cap. ed. Schult.* 459. *Spielmannia Jasminum, Medic. in Act. Acad. Theod. Palat.* iii., *Phys.* 196, *t.* 15; *Schauer in DC. Prod.* xi. 526. *S. africana, Willd. Sp. Pl.* iii. 321; *Ait. Hort. Kew. ed.* ii. iv. 45; *Pers. Syn.* i. 141; *Lam. Ill.* i. 337, *t.* 85; *Bot. Mag. t.* 1899; *E. Meyer, Comm.* 273, *and in Drège, Zwei Pflanzengeogr. Documente,* 71, 86, 223. *S. decurrens, Mœnch, Meth.* 479.

SOUTH AFRICA: without precise locality, *Ecklon,* 765! *Harvey,* 511! *Sieber,* 156! *Thom,* 268! 677!
COAST REGION: Vanrhynsdorp Div.; Gift Berg, 1500–2500 ft., *Drège!* Clanwilliam Div.; Clanwilliam, *Mader in MacOwan Herb.* 2174! Cape Div.; Table Mountain, 500 ft., *Bolus,* 2927! *Thunberg.* Lions Head, *Pappe!* Devils Peak, *Wolley Dod,* 635! Cape Flats, near Rondebosch, *Burchell,* 220! Simons Bay, *Mac Gillivray,* 668! *Milne,* 1127! Paarl Div.; Paarl Mountains, 800–1500 ft., *Drège!* Worcester Div., Baines Kloof, *Hutton!* Caledon Div.; near the River Zonder Einde, 500 ft., *Galpin,* 4396! Riversdale Div.; Garcias Pass, on the summit of a mountain ridge, 3000 ft., *Galpin,* 4395!

Spielmannia Jasminum var. nitida (Schauer in DC. Prod. xi. 526) based on a specimen collected by Drège and said to have leaves only half as large, scabrid, and shining on both surfaces, is a plant which I have not seen, unless Drège's Paarl specimens be identical.

2. O. revoluta (Bocq. ex Baill. Adansonia, ii. 12); branches villous-pubescent; leaves numerous, subadpressed to the stem and much imbricate, narrowly ovate-lanceolate, subacute, strongly dentate and revolute at the margin, densely villous, 2–4 lin. long; flowers axillary, solitary, subsessile; calyx narrowly campanulate, $1\frac{3}{4}$–2 lin. long, glandular-hispidulous; lobes subulate-linear, acute, about eight times as long as the tube; corolla white; tube narrowly-oblong, $2\frac{1}{2}$–3 lin. long; lobes obovate-oblong, 2 lin. long; fruit globose, $1\frac{1}{4}$ lin. diam. *Spielmannia revoluta, E. Meyer, Comm.* 274, *and in Drège, Zwei Pflanzengeogr. Documente,* 67, 223. *S. Desertorum, Ecklon & Zeyher ex Schauer in DC. Prod.* xi. 526 *in syn.*

SOUTH AFRICA: without precise locality, *Forsyth!*
WESTERN REGION: Little Namaqualand; Kamies Bergen, near Kaspars Kloof, Elleboog Fontein and Geelbeks Kraal, 3000–4000 ft., *Drège!* near Ookiep, in stony places, 3000 ft., *MacOwan & Bolus Herb. Norm. Aust.-Afr.,* 673! Modder Fontein, *Whitehead!* and without precise locality, *Ecklon!*

ORDER CIII. SELAGINEÆ.

(By R. A. ROLFE.)

Flowers hermaphrodite, irregular. *Calyx* inferior, 5-fid, 5-partite, or (through the segments being variously connate or deficient) 3- or 2-partite or spathaceous. *Corolla* gamopetalous; base shortly or slenderly tubular; throat usually broader; limb spreading, normally 5-fid, sometimes 4-fid through the two posticous lobes being united or the anticous absent; sometimes more or less bilabiate; lobes equal or the posticous pair shorter. *Stamens* 4, didynamous, or reduced to 2, inserted on the corolla-tube, exserted or included, alternating with the corolla-lobes; filaments filiform or very slightly thickened at the base; anthers 1-celled, obliquely basifixed or versatile, dehiscing longitudinally. *Disc* hypogynous, annular, unilateral and gland-like or inconspicuous. *Ovary* superior, 2-celled or rarely by abortion obliquely 1-celled; style terminal, filiform, simple, acute, obtuse or minutely bifid, stigmatiferous at the apex but not or only slightly thickened; ovules solitary, pendulous, anatropous. *Fruit* small, included within the calyx, 2-celled, or by abortion 1-celled, indehiscent, sometimes separating into two 1-seeded nutlets; pericarp slightly fleshy, crustaceous or somewhat woody, rarely membranous, sometimes corky and with a pair of spurious lateral cells in either carpel. *Seeds* pendulous, usually oblong and terete; testa membranous; albumen fleshy; embryo terete, straight; radicle superior, cotyledons narrow.

Small heath-like shrubs or undershrubs, tufted perennial herbs, or rarely small annuals. Leaves alternate, fascicled, or rarely the lower opposite, cauline or rarely radical, entire or toothed, often narrow and rigid or coriaceous. Flowers small, solitary in the axils of the bracts or rarely shortly pedicelled with the bract adnate to the pedicel, arranged in terminal elongated spikes, corymbose panicles, or sometimes lateral towards the ends of the branches. Corolla lilac, pink, various shades of purple, white, or rarely yellow.

DISTRIB. Genera 10, and species about 240, mostly concentrated in South Africa, with about 20 representatives in Tropical Africa, one in Madagascar, and an outlying genus (*Lagotis*) widely dispersed through the north temperate zone, another in the Mediterranean region (*Globularia*), and a single monotype in Socotra (*Cockburnia*). *Dischisma ciliatum* was collected by Drummond in the Swan River district, W. Australia, where it is reported to be abundant, though believed to be only an accidental introduction.

Much difference of opinion exists as to the relationship of these plants. Baillon (Hist. Pl. ix.) reduces them to *Scrophularineæ*, making three distinct tribes, *Selagineæ*, *Hebenstreitieæ* and *Globularieæ*, while he refers *Lagotis* to *Digitaleæ*; Wettstein (in Engl. & Prantl, Pflanzenfam. iv. 3B) considers *Globulariaceæ* as distinct, but refers the remainder to *Scrophularineæ*, making of the South African species a distinct tribe, called *Antirrhinoideæ-Selagineæ*, and referring *Lagotis* to *Rhinanthoideæ-Digitaleæ*; but this arrangement hardly gives full value to the marked peculiarities of the organs of fructification.

I. **Hebenstreitia.**—*Calyx* spathaceous, subhyaline. *Corolla* expanded behind into 4 lobes, divided in front down to the middle of the tube. *Stamens* 4

II. **Dischisma.**—*Calyx* bipartite; segments lateral, entire. *Corolla* expanded behind into 4 lobes, divided in front down to the middle of the tube. *Stamens* 4.

III. **Walafrida.**—*Calyx* tripartite; middle lobe usually smaller than the lateral, occasionally wanting. *Corolla* 5-lobed; lobes more or less unequal. *Stamens* 4.

IV. **Selago.**—*Calyx* equally or unequally 5-lobed or 5-partite, not adnate to the bract. *Corolla* 5-lobed; lobes more or less unequal. *Stamens* 4.

V. **Microdon.**—*Calyx* 5-dentate, adnate at the base to the bract; teeth subequal. *Corolla* 5-lobed; lobes more or less unequal. *Stamens* 4.

VI. **Gosela.**—*Calyx* subequally 5-lobed. *Corolla*-tube long and slender; limb with 5 spreading subequal lobes. *Stamens* 2 perfect, with 2 linear staminodes.

VII. **Agathelpis.**—*Calyx* tubular, subequally 5-dentate. *Corolla*-tube long and slender; limb with 5 spreading subequal lobes. *Stamens* 2, without staminodes.

I. HEBENSTREITIA, Linn.

Calyx membranous or hyaline, spathaceous, the apex posticous, entire or emarginate. *Corolla*-tube slender, divided in front down to or below the middle, expanded behind into a flat or concave 4-lobed limb; lobes subequal or the intermediate pair longer or deeply connate, occasionally with a fifth minute lobe in the fissure of the tube. *Stamens* 4, didynamous, affixed to the margins of the divided tube below the lobes; filaments short; anthers oblong or linear. *Ovary* 2-celled; style entire. *Fruit* oblong, ovate or broad, subterete or compressed, rarely separating spontaneously into distinct cocci, both cells perfect or one abortive; pericarp equally indurated round the cells or variously dilated and corky, with a pair of spurious cells at their adjacent margins. *Seeds* oblong, cylindrical.

Shrubs, undershrubs or annual herbs; leaves alternate or the lower opposite, often narrow, entire or often toothed, sometimes short and broad; spikes terminal, often dense, short or elongate; flowers sessile, white, yellow or rose; bracts broad or narrow, or the lower somewhat leaf-like, imbricate or lax, exceeding the calyx.

Species 30 South African, one of which extends into Tropical Africa as far as the mountains of Abyssinia, with one other only known from Tropical Africa.

Section 1. EU-HEBENSTREITIA. Fruit oblong, rarely ovoid or globose, without vacuoles or spurious cells.

Perennial with woody stem, or sometimes flowering
the first year and thus appearing annual, about
¾–4 ft. high :
Spikes usually numerous, aggregated into panicles
near the summit of the branches :
Spikes very dense; bracts ovate-lanceolate,
long acuminate ... ... (1) **polystachya**
Spikes more lax; bracts broadly ovate,
rather abruptly acuminate ... ... (2) **Oatesii.**
Spikes solitary or not aggregated into panicles at
the summit of the branches :
Leaves glabrous or nearly so :
Leaves lanceolate or linear, serrulate or
more or less dentate, or if entire, not
filiform :
Leaves rather lax, entire or closely
serrulate, or if dentate over 2½ lin.
broad :

Leaves entire ... ... (3) **elongata.**
Leaves more or less toothed :
　Spikes 1½–2 in. long, narrow (4) **Sutherlandi.**
　Spikes 3–6 in. long, brca l :
　　Leaves narrowly lanceo-
　　　late or cuneate-oblong;
　　　bracts long acuminate (5) **comosa.**
　　Leaves broadly lanceo-
　　　late ; bracts shortly
　　　acuminate ... ... (6) **Cooperi.**
Leaves dense, linear or rarely lanceo-
　late, closely and often strongly
　denticulate or dentate ... ... (7) **fruticosa.**
Leaves rather lax, linear, distinotly
　denticulate or dentate, rarely entire (8) **dentata.**
Leaves filiform or linear-filiform, entire :
　Bracts suberect :
　　Bracts broadly ovate-lanceolate,
　　　shortly acuminate ... ... (9) **integrifelia.**
　　Bracts narrowly ovate-lanceolate,
　　　long acuminate ... ... (10) **Watsoni.**
　　Bracts spreading or recurved ... ... (11) **Rehmanni.**
　Leaves tomentose ... ... (12) **robusta.**
Annual with herbaceous stems or rarely somewhat
　woody, from 2 or 3 up to about 9 in. high :
　Corolla under 3 lin. long :
　　Fruit nearly straight :
　　　Leaves strongly dentate ... ... (13) **ramosissima.**
　　　Leaves entire or rarely obscurely
　　　　denticulate :
　　　　Corolla-lobes short or oblong :
　　　　　Bracts strongly recurved
　　　　　　towards the apex :
　　　　　　Spikes lax and some-
　　　　　　　what elongated ... (14) **hamulosa.**
　　　　　　Spikes rather dense and
　　　　　　　short ... ... (15) **glaucescens.**
　　　　　Bracts not or scarcely re-
　　　　　　curved towards the apex :
　　　　　　Corolla 1½ lin. long ... (16) **minutiflora.**
　　　　　　Corolla 2–2½ lin. long :
　　　　　　　Leaves ¼–½ lin.
　　　　　　　　long ... ... (17) **parviflora.**
　　　　　　　Leaves ½–1½ lin.
　　　　　　　　long ... ... (18) **stenocarpa.**
　　　　　Inner pair of corolla lobes linear,
　　　　　　much longer than the outer ... (19) **macra.**
　　　Fruit much incurved ... ... (20) **discoidea.**
　Corolla 5–6 lin. long :
　　Bracts ovate, with short reflexed acumi-
　　　nate apex ... ... (21) **crassifolia.**
　　Bracts ovate-lanceolate, long acuminate :
　　　Bracts ciliate and somewhat pubes-
　　　　cent ... ... (22) **pubescens.**
　　　Bracts glabrous ... ... (23) **sarcocarpa.**

Section 2. **POLYCENIA.** Fruit ovoid or globose, with a pair of vacuoles or
spurious cells in each carpel.

　Annual, from 2 or 3 up to about 10 in. high :
　　Plant sparingly branched at the base, usually

2–5 in. high ; branches more or less straight ;
spikes not numerous ... ... (24) **fastigiosa.**
Plant much branched, usually 5–10 in. high ;
  branches more or less diffuse ; spikes numerous
  and often crowded :
    Corolla-lobes oblong, inner pair not much
      longer than the outer ... ... (25) **repens.**
    Inner pair of corolla-lobes linear, much longer
      than the broad outer pair ... ... (26) **fenestrata.**
Perennial, 1 ft. or more high :
  Leaves lanceolate or oblong-lanceolate :
    Branches glabrous or minutely puberulous ... (27) **Dregei.**
    Branches pubescent :
      Branches closely pubescent ; calyx minutely
        tridenticulate ... ... (28) **lanceolata.**
      Branches sparsely pubescent, calyx strongly
        tridenticulate ... ... (29) **leucostachys.**
  Leaves broadly ovate or cordate-ovate ... ... (30) **cordata.**

1. **H. polystachya** (Harv., MSS.) ; perennial, stout and much branched, up to 4 ft. high (*Galpin*) or 6 ft. (*Gerrard*) ; branches glabrous or puberulous in decurrent lines from the leaf-bases ; leaves lanceolate, acute, closely serrate except near the base, glabrous, ¾–3 in. long ; spikes very numerous, 1–4 in. long, dense, often congested into panicles ; bracts ovate-lanceolate, acuminate, 2½–3 lin. long ; calyx ovate-oblong, subobtuse, 1¼–1½ lin. long, 2-nerved, herbaceous between the nerves ; corolla white (*Galpin*), 4–5 lin. long ; tube rather slender ; lobes oblong, inner pair much narrower but scarcely longer than the outer ; fruit oblong, 1½ lin. long.

KALAHARI REGION : Orange River Colony ; Drakensberg, *Cooper*, 1014 ! and without precise locality, *Cooper*, 3015 ! Transvaal ; near Snitzkop, *Wilms*, 1167 ! between Middelburg and Crocodile River, *Wilms*, 1168 ! Houtbosch, *Rehmann*, 6203 ! in fields by the Elands River, 6500 ft., *Schlechter*, 3834 ! Umlomati Valley, near Barberton, in swampy ground, *Galpin*, 1271 !

EASTERN REGION : Natal ; Tugela River, *Gerrard*, 376 ! Umvoti district, in a swamp, *Gerrard*, 1248 ! Noods Berg, 2500 ft., *Wood*, 104 ! near Enon, *Wood*, 104 ! Murchison, 1800 ft., *Wood*, 301 ! Arnolds Farm, Newcastle, *Rehmann*, 7032 ! Oliviers Hoek, sources of the Tugela River, 5000 ft., *Allison* ! in marshes, near Richmond, 3000 ft., *Schlechter*, 6730 ! and without precise locality, *Cooper*, 1150 !

2. **H. Oatesii** (Rolfe in Oates' Matabele Land, ed. ii. 406, t. 12) ; perennial, much branched, over 1 ft. high ; branches pubescent in decurrent lines from the leaf-bases ; leaves narrowly lanceolate or linear, subacute, serrulate except near the base, glabrous, ½–1½ in. long ; spikes elongate, 1–3 in. long in fruit, dense ; bracts ovate, acuminate, glabrous, 2–2½ lin. long ; calyx ovate, obtuse, with 2 nerves near the centre, 1–1¼ lin. long ; corolla 4–5 lin. long ; tube slender ; lobes oblong, 1 lin. long ; fruit oblong, 2 lin. long. *Hebenstreitia near dentata, Thunb., Oliv. in Oates' Matabele Land, ed. i. 368.*

EASTERN REGION (?) : Between Pietermaritzburg and the Crocodile River, *Oates* !

The plant from Faku's Territory, doubtfully referred here when the species

was described, is now made a distinct species under the name of *H. Sutherlandi,*
Rolfe.

3. H. elongata (Bolus); perennial, branched at the base, 1–1½ ft.
or more high; branches puberulous, chiefly in decurrent lines
from the leaf-bases; leaves numerous, linear or linear-lanceolate,
subobtuse, entire or slightly denticulate above the middle, glabrous,
½–1¾ in. long; spikes elongate, 2–8 in. long in fruit, dense; bracts
ovate-lanceolate, acuminate, 2½–3½ lin. long; calyx ovate-oblong,
obtuse, 2 lin. long, with 2 prominent nerves, herbaceous between the
nerves; corolla deep orange and red (*Wood*), 5–7 lin. long; tube
slender; lobes oblong, inner pair narrower and rather longer than the
outer; fruit oblong, 2½ lin. long.

KALAHARI REGION : Transvaal; Pretoria, on stony hills at 4000 ft., *McLea
in Bolus Herb.,* 3083 ! grassy mountain slopes around Barberton, *Galpin,* 523 !
Jeppestown Ridge, near Johannesberg, 6000 ft., *Gilfillan in Galpin Herb.,*
6061 ! Rhenoster Poort. Botsabelo district, *Nelson,* 403 !

EASTERN REGION : Natal; near Currys Post, 4000 ft., *Wood,* 3620 ! Biggars
Berg, *Wood,* 4223 ! Riet Vlei, Greenwich Farm, *Fry in Galpin Herb.,* 2726 !
and a cultivated specimen, *Hort. Kew !*

A more luxuriant plant than *H. dentata,* L., and confined to the eastern side of
the continent.

4. H. Sutherlandi (Rolfe); perennial, about 1 ft. high, much
branched; branches minutely puberulous in decurrent lines from the
leaf-bases; leaves linear-lanceolate, subacute, regularly serrulate to
below the middle, 4–12 lin. long; spikes narrow, somewhat elongate,
1–2 in. long, many-flowered; bracts ovate-lanceolate, acuminate,
2–2½ lin. long; calyx broadly ovate-oblong, obtuse, 1¼ lin. long,
with 2 strongly keeled nerves and a second slender pair between
them; corolla 3½–4 lin. long; tube slender; lobes oblong, inner
pair not longer than the outer, and only half as broad.

KALAHARI REGION : Basutoland, 8000 ft., *Melleish in Sanderson's Herb.,*
634!

EASTERN REGION : Pondoland; Faku's Territory, *Sutherland !*

5. H. comosa (Hochst. in Flora, 1845, 70); perennial, branched,
1–4 ft. high; branches minutely puberulous, sometimes only in
decurrent lines from the leaf-bases; leaves numerous, lanceolate or
elliptic-lanceolate, subacute, serrate to near the base, glabrous, ½–2 in.
long; spikes elongate, 2–6 in. long, dense; bracts ovate-lanceolate,
acuminate, glabrous, 2½–4 lin. long; calyx ovate-oblong, obtuse,
2–2½ lin. long, 2-nerved, more or less herbaceous along the centre;
corolla yellow or white, with an orange-red blotch on the limb,
5–6 lin. long; tube slender; lobes oblong, inner pair much narrower
and scarcely longer than the outer. *Hochst. Beitr. Fl. Cap. und
Natal.* 134 ; *Walp. Rep.* iv. 147; *Choisy in DC. Prod.* xii. 5. *H.
comosa, var. serratifolia, Gard. Chron.* 1892, xii. 34, 188.

VAR. β (?) integrifolia (Rolfe); leaves narrowly linear, entire, ¾–1½ in. long,
otherwise much as in the type.

KALAHARI REGION : Transvaal; Pilgrims Rest, *Mudd !*

EASTERN REGION : Griqualand East; Vaal Bank, *Haygarth in Wood Herb.*,
4194! Natal; near Durban, among grasses, *Krauss*, 327! *Peddie! Wood*, 45!
Wood in MacOwan and Bolus, Herb. Norm. Aust.-Afr., 1340 ! Tugela;
common about Durban, *Mc Ken,* 375! *Williamson!* between Durban and
Pietermaritzburg, 0–2000 ft., plentiful, *Sanderson!* 62! Potgieters Hill,
Sanderson! Umgeni waterfall, *Rehmann*, 7474! Camperdown. *Rehmann*, 7797!
and without precise locality, *Gerrard*, 418 ! *Grant!* near Pietermaritzburg,
Wilms, 2204 ! Zululand, *Gerrard & Mc Ken*, 2045 !

VAR. β : Griqualand East; on the slopes of Mount Malowe, near the Umzimkulu River, 4000 ft., *Tyson in MacOwan Herb. Aust.-Afr.*, 1513 !

The variety *integrifolia* may be distinct, but except in leaf character is so
similar to the type that I do not see how to separate it.

6. H. Cooperi (Rolfe) ; perennial, branched, over $1\frac{1}{2}$ ft. high ;
puberulous, chiefly in decurrent lines from the leaf-bases; leaves
lanceolate, acute, strongly dentate to near the base, glabrous, 3–10 lin.
long ; spikes oblong or elongate, up to $3\frac{1}{2}$ in. long in fruit ; bracts
ovate-oblong, acute, $2\frac{1}{2}$ lin. long ; calyx oblong, obtuse, 2 lin. long;
corolla 5 lin. long ; tube slender; lobes oblong, inner pair rather
narrower than the outer ; fruit oblong, 2 lin. long.

KALAHARI REGION : Basutoland; without precise locality, *Cooper*, 737!

7. H. fruticosa (Sims in Bot. Mag. t. 1970, not of Linn. fil.);
perennial ; much branched, $\frac{3}{4}$–$1\frac{1}{2}$ ft. high; branches puberulous in
decurrent lines from the leaf-bases or sometimes glabrous ; leaves
usually very numerous and crowded, linear or lanceolate-linear, acute
or subobtuse, more or less closely and acutely denticulate, rarely subentire, glabrous, 2–10 lin. long ; spikes oblong or elongate, 1–5 in.
long, dense ; bracts ovate-lanceolate, acuminate, $2\frac{1}{2}$–4 lin. long;
calyx ovate, obtuse, 2–$2\frac{1}{2}$ lin. long, with 2 herbaceous nerves;
corolla white with or without a deep orange blotch on the limb,
4–5 lin. long ; tube slender; lobes oblong, inner pair much narrower
than the outer and scarcely exceeding them; fruit oblong, 2 lin.
long. *E. Meyer, Comm.* 247, *and in Drège, Zwei Pflanzengeogr.
Documente*, 69, 78, 82, 83, 98, 189 ; *Walp. Rep.* iv. 146 ; *Choisy
in DC. Prod.* xii. 4 ; *Rolfe in Journ. Linn. Soc.* xx. 351, *in note.
H. scabrida, Burm. fil. Prod. Cap.* 17.—*Burm. Rar. Afr. Pl.* 109,
t. 41, *fig.* 1.

VAR. α, **dura** (Rolfe); branches rather short; leaves crowded, lanceolate
linear, $2\frac{1}{2}$–7 lin. long, teeth small or minute. *H. dura, Choisy in DC. Prod.*
xii. 4.

VAR. β, **lanceolata** (Choisy in DC. Prod. xii. 4); rather loosely branched;
leaves rather lax, lanceolate, $2\frac{1}{2}$–6 lin. long, very strougly toothed.

VAR. γ, **robusta** (Rolfe): branches rather long and lax; leaves narrowly
lanceolate, $\frac{1}{2}$–$1\frac{1}{2}$ in. long, strongly and often very acutely toothed; corolla up to
7 lin. long.

SOUTH AFRICA : var. α, without precise locality, *Zeyher*, 1381 !
COAST REGION : Paarl Div.; in sandy places at Achter de Paarl, 6000 ft.,
Drège! Tulbagh Div. ; on rocks in New Kloof, 1000 ft., *Drège!* Worcester Div.;
Dutoits Kloof, 3000–4000 ft., *Drège!* Drakenstein Mountains, 3000–4000 ft.,
Drège! Queenstown Div.; Hangklip Mountain, 6300 ft., *Galpin*, 1618! East
London Div.; Panmure, *Hutton!* and without precise locality, *Cooper*, 119!
Stockenstrom Div.; summit of Elands Berg, *Scott-Elliot*, 390!

CENTRAL REGION: Somerset Div.; on the upper part of Bruintjes Hoogte, *Burchell*, 3030! Var. β: Graaff Reinet Div.; Wagenpads Berg, *Burchell*, 2823!

WESTERN REGION: Little Namaqualand; in rocky mountainous places near Ezels Fontein and Modder Fontein, 4000–5000 ft., *Drège!*

KALAHARI REGION: Orange River Colony; Wolfe Kop, near the Caledon River, *Burke*, 438! Var. γ: Orange River Colony; Bloemfontein, *Rehmann*, 3883! Witte Bergen, near the Caledon River, *Rehmann*, 3941! Transvaal; without precise locality, on mountains, *Nelson*, 3!

EASTERN REGION: Tembuland; Bazeia, 2500 ft., *Baur*, 157! Natal, Camperdown, *Rehmann*, 7796!

This species is extremely variable, if all the above forms are correctly referred here. They pass so gradually into each other that I do not see how to separate them. Some indeed have been referred to the following species, and a more complete series of specimens would perhaps necessitate some modification of the present arrangement. The species was originally described and figured from a specimen that flowered in the Fulham nursery of Messrs. Whitley Brame and Milne. *H. fruticosa*, Linn. fil., to which it was doubtfully referred, is a true *Dischisma* (*D. fruticosum*, Rolfe), and must be excluded from most of the above references. *H. scabrida*, Burm fil., is cited here with some doubt, for although the habit agrees well, especially with the variety *dura*, the structure of the flower, as represented in Burmann's rude figure (on which it was based) does not agree with the genus, or indeed with the Order.

8. H. dentata (Linn. Sp. Pl. ed. i. 629, ed. ii. 878); perennial, 1–2 ft. high, more or less copiously branched; branches puberulous in decurrent lines from the leaf-bases or nearly glabrous; leaves numerous, linear or lanceolate-linear, subacute, more or less toothed on the upper half, glabrous, ¼–1½ lin. long, or the radical sometimes longer; spikes elongate, up to 6 in. long in fruit, dense or somewhat lax; bracts ovate or ovate-lanceolate, acute or acuminate, 2–3 lin. long; calyx oblong-lanceolate, acute, 1½–2 lin. long, with 2 herbaceous nerves; corolla yellow or white, usually with a darker or orange blotch on the centre of the limb, 4–6 lin. long; tube slender; lobes oblong, middle pair often narrower than the outer; fruit oblong or ovoid-oblong, 1½–2 lin. long. *Berg. Pl. Cap.* 153; *Lam. Encycl.* iii. 77; *Ill. t.* 521; *Bot. Mag. t.* 483; *Gærtn. Fruct.* i. 238, *t.* 51, *fig.* 5; *Choisy, Mém. Selag.* 22, *partly, t.* 1, *fig.* 1, *t.* 2, *fig.* 2, *and in DC. Prod.* xii. 3, *excl. vars.*; *E. Meyer, Comm.* 247, *excl. syn., and in Drège, Zwei Pflanzengeogr. Documente*, 45, 153, 189; *Hochst in Flora*, 1845, 70; *Rolfe in Journ. Linn. Soc.* xx. 343, 347. *H. pulchella, Salisb. Prod.* 93.—*J. & C. Commelin, Hort. Amstel.* ii. 217, *t.* 109; *Burm. Rar. Afr. Pl.* 114, *t.* 42, *fig.* 2.

Var. β, **integrifolia** (E. Meyer, Comm. 247, partly, excl. syn.); leaves quite entire. *E. Meyer in Drège, Zwei Planzengeogr. Documente*, 44, 45, 50, 77, 112, 151, 189.

SOUTH AFRICA: without precise locality, *Alexander! Forster! Masson! Harvey*, 446! *Wright*, 516! and cultivated specimens!

COAST REGION: Malmesbury Div.; Darling, *Bachmann*. 504! Hopefield, *Bachmann*, 103! Clanwilliam Div.; Modder Fontein, *Whitehead!* Koude Berg, 2400 ft., *Schlechter*, 8728! Cape Div.; Devils Mountain, *Ecklon & Zeyher! Krauss*, 1093; Cape Flats, *Ecklon*, 738! in sandy places at Hout Bay, *Ecklon & Zeyher!* by the Blockhouse, Cape Peninsula, *Wolley Dod*, 502! Camps Bay, *Alexander!* on rocky and sandy soils of the Cape district, *Bowie!* Tulbagh Div.; Tulbagh, *Pappe!* Mitchells Pass, 1200 ft., *Schlechter*, 8953! Caledon Div.;

Caledon, *Pappe!* Genadendal, *Grey!* Albany Div.; margins of woods near Grahamstown, *MacOwan*, 984! Stockenstrom Div.; Kat Berg, in grassy fields, 4000–5000 ft., *Drège.* Queenstown Div.: Andries Berg, 6100–6400 ft., *Galpin*, 2018! Elandsberg, *Cooper*, 215! 243! Winter Berg, *Mrs. Barber*, 685! British Kaffraria; without precise locality, *Cooper*, 119! Var. β, Clanwilliam Div.; Wuppertl al, *Drège.* Cape Div.; Sand Flats between Tyger Berg and Blue Berg, below 5000 ft., *Drège.* Queenstown Div.; Storm Bergen, 5000–6000 ft., *Drège!*

CENTRAL REGION: Graaff Reinet Div.; Oude Berg, near Graaff Reinet, 4500 ft., *Bolus.* 154!

WESTERN REGION: Vanrhynsdorp Div.; Karee Bergen, 6800 ft., *Schlechter*, 8273!

KALAHARI REGION: Basutoland; without precise locality, *Cooper*, 3014! Transvaal; by the river near Lydenberg, *Wilms*, 1169! Pilgrims Rest, *Greenstock!*

EASTERN REGION: Transkei, *Mrs. Barber*, 265! Pondoland; Umtsikaba River, below 500 ft., *Drège.* Natal; between Pietermaritzburg and Greytown, *Wilms*, 2203a! Dargle Farm, *Mrs. Fannin!* Var. β, Pondoland; between St. Johns River and Umtsikaba River, in grassy fields, 1000–2000 ft., *Drège.*

Also widely diffused in Tropical Africa.

This has been much confused with the preceding and following species, which are remarkably polymorphic, and in some cases difficult to distinguish from each other. It is usually perennial, but seedlings which flower during the first year have all the appearance of being annual, as in the case of Commelin's coloured figure, on which the species was chiefly founded, and such specimens I am unable to distinguish in other respects from shrubby perennial ones. Many specimens are imperfect in this and other respects, and observations in the field might lead to some modification of the above arrangement. Thunberg confused the species with *H. repens*, Jarosz, for all his herbarium specimens belong to the latter.

9. **H. integrifolia** (Linn. Sp. Pl. ed. i. 629, ed. ii. 878); perennial or sometimes annual (?), more or less branched, chiefly near the base, ¾–2 ft. high; branches puberulous; leaves numerous, filiform or linear-filiform, subacute, entire, glabrous or somewhat scaberulous; ¼–1¼ in. long; spikes sometimes short, but usually much elongated, 1½ up to several inches, and occasionally 1 ft. long in fruit, dense or somewhat lax; bracts ovate, acuminate, 2–3 lin. long; calyx ovate-oblong, subobtuse, 1½ lin. long, with a pair of slender herbaceous nerves; corolla 4–5 lin. long; tube slender; lobes oblong, inner pair narrower and rather longer than the outer; fruit narrowly oblong, 2 lin. long. *Murr. Syst. Veg. ed.* 14, 570; *Rolfe in Journ. Linn. Soc.* xx. 343. *H. dentata, var. integrifolia, E. Meyer, Comm.* 247, *in part, and in Drège, Zwei Pflanzengeogr. Documente*, 45. *H. dentata, var. β, integrifolia, and var. γ parvifolia*(?), *Choisy in DC. Prod.* xii. 4. *H. scabra, Thunb. Prod.* 103, *and Fl. Cap. ed. Schult.* 477; *Choisy in DC. Prod.* xii. 4; *Rolfe in Journ. Linn. Soc.* xx. 354. *H. aurea, Andr. Bot. Rep. t.* 252; *Choisy in DC. Prod.* xii. 4. *H. tenuifolia, Schrad. Cat. Hort. Götting. ex Reichb. Ic. Bot. Exot.* ii. 13, *t.* 133; *E. Meyer, Comm.* 248, *partly, and in Drege, Zwei Pflanzengeogr. Documente*, 96, 189 (*letter a only*). *H. virgata, E. Meyer, Comm.* 249; *Choisy in DC. Prod.* xii. 5.

VAR. α, **laxiflora** (Rolfe); a luxuriant form: leaves up to 1½ in. long; spikes very lax; corollas 5–5½ lin. long.

SOUTH AFRICA: without precise locality, *Forster! Forsyth! Hutton! Masson!*

COAST REGION: Clanwilliam Div.; Modder Fontein, *Whitehead!* Riversdale Div.; Riversdale, *Rust*, 89! Mossel Bay Div.; on sandy hills near Fish Bay, below 200 ft., *Drège!* Klein Berg, 800 ft., *Galpin*, 4398! Knysna Div.; on sand-hills at the western end of Groene Valley, *Burchell*, 5659! Uniondale Div.; Long Kloof at Wagenbooms River, *Bolus*, 2427! Uitenhage Div.; in sandy places near the Zwartskop River, *Zeyher*, 958! 3583! Port Elizabeth Div.; Algoa Bay, *Cooper*, 3076! Port Elizabeth, *Cape Herb.! Albany* Div.; near Grahamstown, *Bolton! Burke!* and without precise locality, *Cooper*, 1558! *Atherstone*, 91! *Miss Bowker! Williamson!* Stockenstrom Div.; Kat Berg, 3000–5000 ft., *Drège!* Stutterheim Div.; Stutterheim, *Cape Herb.! Bathurst* Div.; at Riet Fontein, between Theopolis and Port Alfred, *Burchell*, 3937! British Kaffraria, *Cooper*, 113!

CENTRAL REGION: Aliwal North or Wodehouse Div., *Zeyher!*

WESTERN REGION: Vanrhynsdorp Div.; near Mieren Kasteel, below 100 ft., *Drège!*

KALAHARI REGION: Bechuanaland; Eastern Bamanguato Territory? between Shoshong and Molopolole, *Holub!* Transvaal; near Lydenburg, *Atherstone!* Var. β, Transvaal; without precise locality, *Mrs. Stainbank in Wood Herb.*, 3645!

EASTERN REGION: Natal; on a sandy flat near Durban, 50 ft., *Wood*, 1720! 100 ft., *Wood*, 330!

A very variable species, which seems to include all the different forms above cited, though Choisy keeps H. *aurea*, Andr., *H. scabra*, Thunb., and *H. virgata*, E. Meyer, distinct, while making *H. tenuifolia*, Schrad., synonymous with *H. integrifolia*, Linn., which he enumerates as *H. dentata*, Linn., var. β, *integrifolia*. He also refers the Abyssinian specimens to this, in which I cannot follow him. The species was based on *H. foliis integerrimis*, Linn. Hort. Cliff. 497, and the characters given as separating it from *H. dentata*, Linn., point distinctly to the present one. No specimen appears to have been preserved. It is easily separated by its nearly filiform leaves and long narrow flower-spikes. From this there seems to be a regular gradation down to the more stunted forms represented by *H. virgata*, E. Meyer and *H. scabra*, Thunb.

10. H. Watsoni (Rolfe); perennial (?), somewhat branched, over 1 ft. high; branches glabrous or very minutely puberulous; leaves numerous, filiform-linear, subacute, entire or minutely denticulate, minutely canescent, $\frac{1}{2}$–$2\frac{1}{2}$ in. long; spikes elongate (young in the specimens seen); bracts ovate-lanceolate, long acuminate, 2–4 lin. long; calyx ovate-oblong, obtuse, $1\frac{1}{4}$–$1\frac{1}{2}$ lin. long, with 2 very slender or obscure nerves; corolla white, with or without an orange blotch in the centre of the limb, 4–5 lin. long; tube slender; lobes oblong, inner pair longer and narrower than the outer.

COAST REGION: East London Div.; East London, on the sea coast, 50 ft., *Galpin*, 1869! and a cultivated specimen from the same locality, *Watson!*

Nearly allied to *H. integrifolia*, Linn., but differing in its more lanceolate and acuminate bracts. It is apparently a coast plant.

11. H. Rehmanni (Rolfe); annual (?), much branched, more or less diffuse, 6–10 in. or more high; branches rather slender, puberulous; leaves numerous, narrowly linear, subobtuse, entire, 3–12 lin. long; bracts spreading or somewhat reflexed, ovate or ovate-oblong, shortly acuminate, 2–$2\frac{1}{2}$ lin. long; calyx ovate-oblong, obtuse, $1\frac{1}{4}$ lin. long, with 2 herbaceous nerves; corolla $3\frac{1}{2}$–$4\frac{1}{2}$ lin. long; tube slender; lobes oblong, inner pair longer and narrower than the outer; fruit oblong, 2 lin. long.

KALAHARI REGION: Transvaal; Hooge Veld, between Trigards Fontein and Standerton, *Rehmann*, 6766! 6767!

A dwarf, slender species, readily separated from its allies by its small flowers and very spreading bracts.

12. H. robusta (E. Meyer, Comm. 246); perennial, 1½ ft. or more high; branches closely pubescent; leaves numerous, linear, subobtuse, entire, somewhat fleshy, puberulous, ½–1 in. long; spikes elongate, up to 9 in. long in fruit; bracts ovate, mucronate, 2½ lin. long; calyx ovate-oblong, 2-dentate, 2-nerved, pubescent on and between the nerves, 2 lin. long; corolla 5 lin. long; tube stoutish; lobes oblong, inner pair narrower than the outer but scarcely longer; fruit ovoid-oblong, 2 lin. long. *E. Meyer in Drège, Zwei Pflanzengeogr. Documente,* 109, 189; *Endl. Iconogr.* 12, *t.* 76; *Walp. Rep.* iv. 145; *Choisy in DC. Prod.* xii. 4. *H. augusta, Pritz. Ic. Bot. Index,* 526.

COAST REGION: Clanwilliam Div. ; between Jackals River and Oliphants River, at 500–1000 ft., *Drège*!

Easily distinguished by its robust habit, closely pubescent branches and puberulous leaves.

13. H. ramosissima (Jarosz, Pl. Nov. Cap. 14); annual, branched chiefly at the base, 3–6 in. high; branches erect, pubescent in decurrent lines from the leaf-bases; leaves linear-lanceolate, acute, with 2 or 3 pairs of prominent teeth above the middle, narrowed at the base, 4–8 in. long; spikes oblong or somewhat elongated, rather dense, ¼–2 in. long; bracts spreading or recurved, ovate, shortly acuminate, subobtuse, 2–2½ lin. long; calyx ovate-oblong, subobtuse, 1½ lin. long, pubescent; corolla 2½–3 lin. long; tube stout; lobes broadly oblong, inner pair very minute and shorter than the outer; fruit oblong, 1½ lin. long. *Choisy in DC. Prod.* xii. 5; *Rolfe in Journ. Linn. Soc.* xx. 357. *Selago squarrosa, Choisy in DC. Prod.* xii. 16.

SOUTH AFRICA : without precise locality, *Link! Masson!*
COAST REGION: Stellenbosch Div. ; Stellenbosch, *Alexander!* Swellendam Div. ; in caroo-like soil by the River Zonder Einde, at Hassaquas Kloof and by the Breede River, *Zeyher*, 3581!

Jarosz wrongly described this as perennial, as his own type specimen shows.

14. H. hamulosa (E. Meyer, Comm. 249); annual, much branched, 2–6 in. high; branches puberulous; leaves few, linear, subobtuse, entire, lower ½–1½ in. long, upper diminishing into the bracts; spikes elongate, lax, 1–3 in. long; bracts spreading or recurved, ovate-oblong, subobtuse, 1½–2 lin. long; calyx ovate-oblong, obtuse, 1 lin. long; corolla yellow (*Bolus*), 2¼ lin. long; tube stoutish; lobes very short and broad, subequal. *E. Meyer in Drège, Zwei Pflanzengeogr. Documente,* 91, 189; *Walp. Rep.* iv. 146; *Choisy in DC. Prod.* xii. 5.

WESTERN REGION: Little Namaqualand; between Silver Fontein, Koper Berg, and Kaus Mountain, *Drège.* Spektakel Mountain, *Morris in Bolus Herb.*, 5747! in fields formerly cultivated near Klip Fontein, 3000 ft., *MacOwan & Bolus, Herb. Norm. Aust.-Afr.*, 675!

15. **H. glaucescens** (Schlechter in Engl. Jahrb. xxvii. 185); annual, sparingly branched at the base, diffuse, 2–6 in. high; branches minutely puberulous; leaves linear, subobtuse, entire, glabrous, not numerous, $\frac{1}{2}$–1 in. long; spikes oblong or somewhat elongated, $\frac{1}{2}$–1$\frac{1}{2}$ in. long; bracts spreading and recurved, ovate-oblong, obtuse, 1$\frac{1}{2}$–2 lin. long; calyx oblong, obtuse, without prominent nerves, 1$\frac{1}{4}$ lin. long; corolla 2 lin. long; tube stoutish; lobes oblong, inner pair narrower and longer than the outer; fruit oblong, 1$\frac{1}{4}$ lin. long, each carpel having 3 obtuse longitudinal ridges.

CENTRAL REGION: Calvinia Div.; Hantam Mountains, *Meyer!* Brand Vlei, *Johanssen,* 16!
WESTERN REGION: Vanrhynsdorp Div.; Zout River, 750 ft., *Schlechter,* 8112!

16. **H. minutiflora** (Rolfe); annual, branched at the base, diffuse, 1$\frac{1}{2}$–3 in. high; branches minutely puberulous; leaves linear, subobtuse, somewhat denticulate above the middle or subentire, glabrous, 2–6 lin. long; spikes ovoid or slightly elongated in fruit and rather lax, $\frac{1}{4}$–$\frac{3}{4}$ in. long; bracts spreading, ovate-oblong, obtuse, 1 lin. long; calyx oblong, shortly tridentate, with 3 slender nerves, $\frac{3}{4}$ lin. long; corolla 1$\frac{1}{4}$ lin. long; tube stoutish; lobes broadly oblong, short, subequal; fruit oblong, $\frac{3}{4}$ lin. long. *H. parviflora, E. Meyer, Comm.* 249, *partly, and in Drège, Zwei Pflanzengeogr. Documente,* 69, 189 (*letter c only*). *H. parviflora, var. β, denticulata, Choisy in DC. Prod.* xii. 5.

WESTERN REGION: Little Namaqualand; Modder Fonteins Berg, Rood Berg, and Ezels Kop, 4000–5000 ft., *Drège!*

This species was confused with *H. parviflora* by E. Meyer, but is readily distinguished by its spreading habit and smaller flowers, with a tridentate trinerved calyx. The specimens turn blackish in drying.

17. **H. parviflora** (E. Meyer, Comm. 249, partly); annual, much branched, erect, 5–8 in. high; branches minutely puberulous; leaves linear, subobtuse, entire, glabrous, 3–6 lin. long; spikes narrow, elongated, 1–3 in. long, dense; bracts ovate-oblong, acuminate, 1$\frac{1}{2}$–2 lin. long; calyx ovate-oblong, subobtuse, without prominent nerves, 1 lin. long; corolla 2 lin. long; tube stoutish; lobes oblong, the inner pair exceeding the outer; fruit oblong, 1$\frac{1}{4}$ lin. long. *E. Meyer in Drège, Zwei Pflanzengeogr. Documente,* 62, 90, 91, 189 (*excl. letter c*); *Walp. Rep.* iv. 146; *Choisy in DC. Prod.* xii. 5, *excl. var. β.*

COAST REGION: Malmesbury Div.; Hopefield, *Bachmann,* 1149!
CENTRAL REGION: Beaufort West Div.; Nieuwveld Mountains, near Beaufort West, *Drège.*
WESTERN REGION: Little Namaqualand; between Kousies (Buffels) River

and Silver Fontein, *Drège!* between Silver Fontein, Koper Berg, and Kaus Mountain, *Drège!* at Zabies, *Schlechter*, 95! and without precise locality, *Bolus*, 6651!

This species is dimorphic, as Drège's and Schlechter's specimens include both short- and long-styled forms : in the latter case the short-styled have the corolla at least a third longer than the long-styled.

18. H. **stenocarpa** (Schlechter in Engl. Jahrb. xxvii. 186); annual, simple or branched at the base, 3–4 in. high; branches puberulous; leaves narrowly linear, subobtuse, entire, glabrous, $\frac{1}{2}$–1$\frac{1}{2}$ in. long ; spikes oblong, $\frac{1}{2}$–1$\frac{1}{2}$ in. long, many-flowered ; bracts ovate, acuminate, 1$\frac{1}{2}$–2 lin. long; calyx ovate-oblong, obtuse, without prominent nerves, $\frac{3}{4}$ lin. long; corolla 2$\frac{1}{2}$ lin. long; tube stoutish; lobes oblong, inner pair narrower and rather longer than the outer; fruit oblong, 1 lin. long.

WESTERN REGION : Vanrhynsdorp Div.; Karee Bergen, 1000 ft., *Schlechter*, 8168!

19. H. **macra** (E. Meyer, Comm. 248); annual(?); branches herbaceous, erect or ascending ; leaves lax, very narrowly linear ; bracts ovate-lanceolate ; corolla-tube filiform, shorter than the bracts, white ; inner pair of lobes linear, outer pair subovate, oblique and shorter than the inner pair ; fruit oblique, oblong ; carpels subequal, not sulcate, the posterior margin much expanded, the anterior contracted. *E. Meyer in Drège, Zwei Pflanzengeogr. Documente*, 138, 189 ; *Walp. Rep.* iv. 146 ; *Choisy in DC. Prod.* xii. 5.

CENTRAL REGION : Somerset Div. ; near Little Fish River and Great Fish River, 2000–3000 ft., *Drège.*

Only known to me from the description. Whether annual or perennial is not stated, but both Meyer and Choisy associate it with species belonging to the former group, where also the herbaceous stem would place it. I have not succeeded in identifying anything with the description, which, however, is very imperfect.

20. H. **discoidea** (E. Meyer, Comm. 249) ; annual (?) ; branches herbaceous, erect ; leaves narrow, "subtridactyloid" ; bracts ovate-lanceolate ; corolla-tube exceeding the bracts ; lobes oblique, ovate, subequal ; filaments longer than the anthers ; fruit orbicular ; carpels much curved, so that the base and apex are nearly contiguous, cohering by the thickened margins. *Walp. Rep.* iv. 147 ; *Choisy in DC. Prod.* xii. 5.

SOUTH AFRICA : without precise locality, *Drège.*

Only known to me from the description. The fruit is very remarkable in shape, and I cannot find anything at all approaching it. The herbaceous stem would indicate that it is an annual, and as the authors above named both associate it with annual species, it probably belongs to this group.

21. H. **crassifolia** (Choisy in DC. Prod. xii. 4) ; annual (?), much branched, 6–9 in. high, more or less diffuse ; branches stoutish, puberulous ; leaves linear or oblong, obtuse, somewhat fleshy, entire, glabrous, 3–9 lin. long, occasionally appearing fascicled owing to the arrest of axillary branchlets ; spikes oblong or elongated, 1–4 in.

long, many-flowered ; bracts ovate or ovate-oblong, shortly acuminate
with a recurved subacute apex, 2–2½ lin. long ; calyx ovate-oblong,
obtuse, nerves not prominent, 1¾ lin. long ; corolla 5 lin. long ; tube
narrow ; lobes narrowly oblong, ¾ lin. long, inner pair narrower than
the outer ; fruit oblong, 1¾ lin. long. *H. robusta, var. β, glabrata,
F. Meyer, Comm.* 247, *and in Drège, Zwei Pflanzengeogr. Documente,*
55, 62, 109, 189 ; *Walp. Rep.* iv. 145. *H. tenuifolia, E. Meyer,
Comm.* 248, *partly* (*not of Schrad.*), *and in Drège, Zwei Pflanzengeogr.
Documente,* 96, 189 (*letter a only*).

SOUTH AFRICA: without precise locality, *Masson !*

COAST REGION : Clanwilliam Div. ; Ezels Bank, in the Ceder Bergen,
4000 ft., *Schlechter*, 8807! between Lange Valley and Oliphants River,
1000–1500 ft., *Drège;* by the Oliphants River, *Ecklon & Zeyher !* Tulbagh Div. ;
without precise locality, *Ecklon & Zeyher !*

CENTRAL REGION :·Fraserburg Div. ; between Patrys Fontein and Great Brak
River, *Burchell*, 1516 ! Sutherland Div. ; between Knileuberg and Great Reed
River, *Burchell*, 1345 ! Beaufort West Div. ; Nieuwveld Mountains near Beaufort
West, 3000-5000 ft., *Drège.* Graaff Reinet Div. ; in stony places near Zuure
Plaats, in the Sneeuwberg Range, 5000 ft., *Drège!*

WESTERN REGION : Little Namaqualand ; between Kousie (Buffels) River and
Silver Fontein, 2000 ft., *Drège!* in sandy places near Port Nolloth, *Bolus*, 6650 !
and without precise locality, *Scully*, 73 !

KALAHARI REGION : Orange River Colony; Witte Bergen, on shady sides of
the mountains, *Mrs Barber and Mrs. Bowker !* near the Caledon River, *Burke !*

22. H. pubescens (Rolfe); annual (?), erect, 3–6 in. high, unbranched (?) ; branches puberulous ; leaves linear, subobtuse, with 2
or 3 pairs of small teeth above the middle, glabrous, 4–9 lin. long ;
spikes oblong, 1 in. or more long, dense ; bracts ovate-lanceolate,
acuminate, ciliate and somewhat pubescent on the lower half,
4–4½ lin. long; calyx lanceolate-linear, subobtuse, regularly ciliate,
nearly 2 lin. long ; corolla 7–8½ lin. long ; tube very slender ;
lobes linear, inner pair 2 lin. long and nearly twice as long as the
outer.

CENTRAL REGION : Calvinia Div. ; Hantam Mountains, *Meyer !*

This has the general appearance of a small seedling state of *H. dentata*, Linn.,
flowering the first year, but is readily distinguished by the pubescent lower part
of the bracts, the ciliate calyx, and by the long and very narrow corolla lobes.
It has all the appearance of being an annual.

23. H. sarcocarpa (Bolus); annual, much branched, somewhat
diffuse, 6–8 in. high ; branches minutely puberulous ; leaves lax,
linear, subobtuse, entire, glabrous, ½–1¼ in. long ; spikes oblong or
elongated, 1–3 in. long, rather dense ; bracts ovate-lanceolate, subobtuse, 3 lin. long ; calyx oblong, obtuse, nerves not prominent,
1½ lin. long ; corolla rose (*Bolus*), 5–7 lin. long; tube narrow ; lobes
linear-oblong, 1–1½ lin. long, the inner pair exceeding the outer ;
fruit globose, fleshy, 2 lin. broad.

WESTERN.REGION : Little Namaqualand ; in stony places near Klip Fontein,
3000 ft., *Bolus in MacOwan and Bolus, Herb. Norm. Aust.-Afr.*, 674! Great
Namaqualand ; Aro Ass, *Fleck*, 437 !

24. H. fastigiosa (Jarosz, Pl. Nov. Cap. 14) ; annual, branched chiefly at the base, suberect or somewhat diffuse, 2–5 in. high; branches more or less straight, puberulous; leaves linear, subobtuse, entire or slightly denticulate above the middle, glabrous, 3–12 lin. long; spikes oblong or somewhat elongated, ½–2 in. long, rather dense; bracts spreading, ovate-oblong, subobtuse, glabrous, 2 lin. long; calyx ovate-oblong, subobtuse, 1 lin. long; corolla yellow (*Jarosz*), 3 lin. long; tube slender; lobes oblong, minute; fruit ovoid, 1¼ lin. long. *Choisy in DC. Prod.* xii. 6. *H. fastigiata, Steud. Nom. ed.* 2, i. 724. *H. macrostylis, Schlechter in Journ. Bot.* 1898, 317.

SOUTH AFRICA: without precise locality, *Link!*
COAST REGION: Clanwilliam Div.; in sandy places near Clanwilliam, at 3500 ft., *Schlechter.* Cape Div.; Simons Bay, *Wright*, 512! in sandy places on the mountain behind Simons Town, *Bolus*, 4872! Cape Peninsula, between Red Hill and Slang Kop, *Wolley Dod*, 3023!

Jarosz wrongly defined this species as a perennial with a frutescent stem, which led Choisy, who had not seen it, to doubt the genus. The original specimen, however, in the Berlin Herbarium enables the mystery to be cleared up. It is very distinct from the two following species, being especially well marked in habit. *H. macrostylis*, Schlechter, is cited here with some doubt, for I have not seen the original specimen. But Major Wolley Dod's Cape Peninsula plant, which is said to be identical, appears to be n. 3023 of the collection (the author mentions having lost the ticket), which is unquestionably *H. fastigiosa*, Jarosz. The fruit, however, does not agree with the description, and the style is only shortly exserted.

25. **H. repens** (Jarosz, Pl. Nov. Cap. 15), annual, much branched and diffuse, 3–10 in. high; branches puberulous in decurrent lines from the leaf-bases; leaves linear or lanceolate, subobtuse, with 2 or 3 pairs of teeth above the middle or rarely subentire, puberulous, 3–12 lin. long; spikes oblong or somewhat elongated, ½ up to 2 in. long in fruit, usually dense; bracts spreading, ovate or ovate-oblong, subacute, 1½–2 in. long, glabrous; calyx ovate-oblong, subobtuse, 1 lin. long; corolla 2½ lin. long; lobes short, the inner pair narrower than the outer; fruit ovoid or subglobose, 1–1¼ lin. long. *Choisy in DC. Prod.* xii. 6. *H. dentata, Thunb. Prod.* 103 (*excl. syn., not of Linn.*); *Thunb. Fl. Cap. ed. Schult.* 477 (*excl. var.* γ); *Rolfe in Journ. Linn. Soc.* xx. 356. *H. tenera, Spreng. ex Walp. Rep.* iv. 147. *Polycenia hebenstreitioides, Choisy in Mém. Soc. Phys. Genèv.* ii. ii. 91, *t.* 2, *fig.* 1; *Choisy, Mém. Selag.* 21, *t.* 2, *fig.* 1; *E. Meyer, Comm.* 246, *and in Drège, Zwei Pflanzengeogr. Documente*, 98, 105, 112, 212; *Hochst in Flora*, 1845. 70; *and Beitr. Fl. Cap. und Natal.* 134; *Walp. Rep.* iv. 142; *Choisy in DC. Prod.* xii. 2. *P. tenera, Walp. Rep.* iv. 143; *Choisy in DC. Prod.* xii. 2.

SOUTH AFRICA: without precise locality, *Forbes! Grey! Auge! Harvey! Oldenburg! Link! Zeyher.* 314! and cultivated specimens!
COAST REGION: Clanwilliam Div.; Lange Kloof, 400 ft., *Schlechter*, 8387! Malmesbury Div.; Darling, *Bachmann*, 401! near Hopefield, *Bachmann*, 101! 1148! 2133! 2162! on the way to Moorrees Berg, *Bachmann*, 1143! Vogelstruis Fontein, near Hopefield, *Bachmann*, 1144! by the Berg River, *Zeyher*, 1398! Piquetberg Div.; Alexanders Hoek, 300 ft., *Schlechter*, 5151! Cape Div.; Signal Hill, near Cape Town, *Wilms*, 3513! Lion Mountain, *Ecklon*, 382! mountain flats near Cape Town, *Ecklon & Zeyher!* in sandy places between Cape Town and Duiker Valley, below 100 ft., *Drège!* Table Mountain and Devils

Mountain, *Drège*; Cape Peninsula, near Maitland, *Wolley Dod*, 637! Camps Bay, *Wolley Dod*, 1362! Upper North Battery at Simonstown, *Wolley Dod*, 1875! Tyger Berg, *Krauss*, 1005. Paarl Div.; Achter de Paarl, *Drège*; Paarl, *Alexander!* Stellenbosch Div.; Hottentots Holland, *Ecklon & Zeyher!* Div.? Kuil River, *Zeyher!*

This varies greatly in stature; luxuriant specimens, having a stouter, less diffuse habit, and more regularly toothed leaves, look at first rather different, but grade into the ordinary form. *H. tenera*, Spreng., is only a very slender form.

26. H. fenestrata (Rolfe); annual, erect or diffuse, 6–8 in. high, much branched; branches terete, sparsely pubescent; leaves linear, obtuse, remotely denticulate near the apex, glabrous, 3–6 lin. long; spikes ovoid, dense, 6 lin. long (in early stage of flowering); bracts ovate or ovate-lanceolate, subobtuse, 1½ lin. long; calyx ovate, subobtuse or shortly emarginate, 1 lin. long; corolla 3 lin. long; lobes oblong, the inner narrower and much exceeding the outer; fruit ovate, subacute, the spurious cells on the ventral side incomplete? *Polycenia fenestrata, E. Meyer, Comm.* i. 246, *and in Drège, Zwei Pflanzengeogr. Documente*, 98, 112, 212; *Walp. Rep.* iv. 143; *Choisy in DC. Prod.* xii. 2.

COAST REGION: Cape Div.; sand-flats between Tyger Berg and Blue Berg, under 500 ft., *Drège*. Paarl Div.; near Paarl, 400–800 ft., *Drège!*

An imperfectly known species; the only specimen seen is in a very young state.

27. H. Dregei (Rolfe); perennial, much branched, 1½ ft. or more high; branches terete, mostly glabrous; leaves numerous, lanceolate, acute, with 2 to 4 pairs of acute teeth above the middle, decurrent at the base, nearly glabrous, 3–6 lin. long, the lower larger, broader and subspathulate; spikes oblong or elongated, 1 up to 6 in. long in fruit, many-flowered; bracts broadly ovate, acuminate, 2½–3 lin. long; calyx oblong, with 3 acute teeth, 3-nerved, 1½ lin. long; corolla white and yellow (*Galpin*), 4–5 lin. long; tube slender; lobes oblong, short, the middle pair exceeding the outer; fruit globose, apiculate, 1 lin. long. *Polycenia fruticosa, E. Meyer, Comm.* i. 245, *and in Drège, Zwei Pflanzengeogr. Documente*, 115, 212; *Walp. Rep.* iv. 143 (*excl. syn.*); *Choisy in DC. Prod.* xii. 3.

COAST REGION: Caledon Div.: Baviaans Kloof, near Genadendal, 1000–2000 ft., *Drège!* Genadendal Mountains, 1500 ft., *Galpin*, 4397! *Schlechter*, 9800!

28. H. lanceolata (Rolfe); perennial, much branched, over 1¼ ft. high; branches terete, pubescent; leaves somewhat lax, lanceolate, acute, with 1 to 3 pairs of acute teeth above the middle or subentire, more or less pubescent, not decurrent, 4–14 lin. long; spikes oblong or elongated, 1 up to 5 in. long in fruit, many-flowered; bracts ovate-lanceolate, acuminate, 2½–3 lin. long; calyx oblong, with 3 acute teeth, 3-nerved, 2 lin. long; corolla 5 lin. long; tube slender; lobes oblong, the middle pair exceeding the outer; fruit ovoid-globose, 1½ lin. long. *Polycenia lanceolata, E. Meyer, Comm.* i. 245, *and in*

Drège, Zwei Pflanzengeogr. Documente, 76, 212; *Walp. Rep.* iv. 144;
Choisy in DC. Prod. xii. 3. *P. lanceolata, var. β pubescenti-*
hirsuta, Walp. Rep. iv. 144.

COAST REGION: Piquetberg Div.; in stony places on the Piquetberg Range,
1500–3000 ft., *Drège!*

Polycenia lanceolata var. *β. glabrata* (E. Meyer, Comm. 245, and in Drège,
Zwei Pflanzengeogr. Documente, 125, 160, 212; Choisy in DC. Prod. xii. 3.
P. lanceolata var. *glabra,* Walp. Rep. iv. 144), is based on material collected by
Drège in COAST REGION: Knysna Div., Groene Valley, 500 ft., and EASTERN
REGION: Natal, below 100 ft., which I have not seen. It is described as having
more or less glabrous branches and leaves, very glabrous bracts, and the fruit
as having only one spurious cell in each achene.

29. H. leucostachys (Schlechter in Engl. Jahrb. xxvii. 186);
perennial, somewhat branched, 1–1½ ft. high; branches erect or
ascending, pubescent; leaves oblong-lanceolate, acute, acutely
dentate, more or less pubescent, ½–1¼ in. long; spikes elongate,
3 in. or more long, many-flowered; bracts oblong-lanceolate, acute,
somewhat pubescent, 2½ lin. long; calyx oblong, with 3 acute teeth,
3-nerved, 2 lin. long; corolla 5 lin. long: tube slender; lobes
oblong, inner rather narrower and longer than the outer; fruit
oblong, obtuse, achenes equal, semiterete, glabrous (*Schlechter*).

COAST REGION: Worcester Div.; Bains Kloof, 1300 ft., *Schlechter,* 9158!

I have not seen the fruit, but the flowering specimen is so similar to the
preceding species that I think it must belong to this section. It chiefly differs
from *H. lanceolata* in its looser pubescence, less hairy bracts, and more strongly
toothed calyx. Schlechter compares it with *H. dentata,* Linn., which belongs to
Eu-Hebenstreitia, but as he has labelled a good fruiting example of *H. Dregei,*
Rolfe (*Polycenia fruticosa,* E. Meyer), "*H. crassifolia,* Choisy, var.?" (which
also belongs to the other section) it would appear that he did not examine the
structure of the fruit.

30. H. cordata (Linn. Syst. ed. 13, ii. 420); perennial, much
branched, 1 ft. or more high; branches hispidulous; leaves
numerous, spreading, broadly ovate or cordate-ovate, obtuse or sub-
mucronate, somewhat fleshy, 1–3 lin. long; spikes oblong or somewhat elongate, 1–2½ in. long, very dense; bracts ovate, mucronate,
2½–3 lin. long; calyx oblong or obovate-oblong, obscurely 3-denticulate,
2½–3 lin. long; corolla white with yellow throat (*Burchell*), 4–5 lin.
long; tube stoutish; lobes broadly oblong, subequal; fruit ovoid-
globose, 2–2½ lin. long. *Linn. Mant.* 420; *Murr. Syst. Veg. ed.* 14,
570; *Lam. Encycl.* iii. 78; *Ait. Hort. Kew.,* ed. 1, ii. 356, *ed.* 2,
iv. 48; *Thunb. Prodr.* 103, *and Fl. Cap. ed. Schult.* 478; *Willd.*
Sp. Pl. iii. 332; *Jarosz, Pl. Nov. Cap.* 13; *Choisy Mém. Selag.* 23;
Spreng. Syst. Veg. ii. 754; *Rolfe in Journ. Linn. Soc.* xx. 347, 356.
Polycenia cordata, E. Meyer, *Comm.* 245, *and in Drège, Zwei*
Pflanzengeogr. Documente, 94, 129, 212; *Walp. Rep.* iv. 143;
Choisy in DC. Prod. xii. 3.

SOUTH AFRICA: without precise locality, *Bowker! Wallich!*
COAST REGION: Malmesbury Div.; Hopefield, *Bachmann,* 1151! Cape Div.;
on the sea coast at False Bay, *Bolus,* 2890! about Camps Bay, *Burchell,* 305!
sea shore at Camps Bay, *Harvey,* 513! Cape Peninsula, Railway at St. James,

Wolley Dod, 2117! Doorn Hoogte, *Zeyher!* Worcester Div.; Elandsberg, *Wallich!* Uitenhage Div.; by the Zwartkops River, *Zeyher*, 3582! on sand hills near the mouth of the Zwartkops River, *Ecklon & Zeyher*, 413! on the sea coast near the mouth of the Van Stadens River, *MacOwan*, 732! Port Elizabeth Div.; on the sand flats and rocky shores at Port Elizabeth, below 100 ft., *Drège;* Port Elizabeth, near the Burying ground, *Burchell*, 4318! Bathurst Div.; sand dunes on the beach at Port Alfred, *Galpin*, 323!

CENTRAL REGION: Somerset Div.; at Somerset East, *Bowker!* probably an error as to locality.

WESTERN REGION: Little Namaqualand, near the mouth of the Orange River, *Drège!*

Remarkably different in habit from all the preceding, and apparently largely a coast plant.

II. DISCHISMA, Choisy.

Calyx 2-partite; segments lateral, linear or lanceolate-oblong. *Corolla* tube slender, divided in front to or below the middle, dilated behind and expanded into a flat or concave 4-lobed limb; lobes subequal or the middle pair longer. *Stamens* 4, didynamous, affixed to the margins of the fissure below the lobes; filaments short; anthers oblong or linear. *Ovary* 2-celled; style entire. *Fruit* oblong or ovate, subterete or compressed, rarely spontaneously breaking up into cocci, both cells usually perfect; pericarp indurated. *Seeds* oblong, cylindrical.

Small branched shrubs or annual herbs, resembling *Hebenstreitia* in habit; leaves alternate or the lower opposite, often narrow and toothed, rarely short and broad; spikes terminal, dense, short or elongated, rarely capitate; bracts imbricate, ovate or lanceolate, sometimes more or less leaf-like.

Species 11, all South African, one of them naturalized in Western Australia.

Leaves petiolate or narrowed at the base, linear, lanceolate, or rarely elliptic :
  Inflorescence capitate or ovoid :
    Bracts 2–3 lin. long, shortly acuminate ... (1) **arenarium.**
    Bracts 6–8 lin. long, with a long attenuate
      apex ... ... (2) **capitatum.**
  Inflorescence spicate :
    Annuals, 2–6 in. high :
      Corolla exserted, 5 lin. long ... ... (3) **spicatum.**
      Corolla not exserted, 2 lin. long ... (4) **clandestinum.**
    Perennials, ½–2 ft. high :
      Branches and bracts more or less pubescent :
        Spikes elongate, narrow, and rather
          lax ... ... (5) **leptostachyum.**
        Spikes short in the flowering
          stage, broader, rather dense :
          Leaves tomentose, entire or
            nearly so ... ... (6) **tomentosum.**
          Leaves glabrous, mostly
            strongly toothed :
            Leaves lanceolate to
              elliptical ... ... (7) **erinoides.**
            Leaves linear ... ... (8) **ciliatum.**
        Branches and bracts glabrous ... ... (9) **fruticosum.**
  Leaves sessile, ovate or nearly as broad as long :
    Leaves longer than broad, rather scattered ... (10) **squarrosum.**
    Leaves broader than long, much crowded ... (11) **crassum.**

1. **D. arenarium** (E. Meyer, Comm. 251); annual, decumbent,
usually much branched, 1–6 in. high; branches sparingly pubescent;
leaves sessile, linear or lanceolate-linear, subacute, entire or denticulate
near the apex, glabrous, 3–6 lin. long; spikes ovoid or oblong,
3–7 lin. long; bracts ovate, acuminate, ciliate near the base, 2–3 lin.
long; calyx lobes oblong or lanceolate, acute, somewhat villous,
1 lin. long: corolla yellowish-white, 1¼ lin. long; tube slender;
lobes oblong, minute; fruit oblong, 1 lin. long. *E. Meyer in Drège,*
Zwei Pflanzengeogr. Documente, 113, 179; *Hochst. in Flora,* 1845,
70, *and Beitr. Fl. Cap. und Natal.* 134; *Walp. Rep.* iv. 149;
Choisy in DC. Prod. xii. 7.

COAST REGION : Clanwilliam Div. ; Zuur Fontein, 150 ft., *Schlechter,* 8532!
Malmesbury Div. ; between Gr··ene Kloof and Saldanha Bay, *Drège !* Cape Div. ;
Simons Bay, *Wright,* 510! Oatlands Point, *Wolley Dod,* 2933! near Green
Point, *Krauss,* 1629. Caledon Div. ; Babylons Tower, *Cape Herb. !* Div. ?
Kuils River, *Cape Herb. ;* and *cultivated specimens !*

2. **D. capitatum** (Choisy in Mém. Soc. Phys. Genèv. ii. ii. 94,
t. 1, fig. 2); annual, decumbent, usually much branched, 2–8 in.
high; branches pubescent; leaves sessile, lax, narrowly linear,
subacute, entire or with 1–3 pairs of teeth near the apex, glabrous,
3–7 lin. long; spikes ovoid or oblong, ½–1¼ in. long: bracts 6–8 lin.
long, with a broad ovate ciliate and pubescent base, and a long
alternate linear glabrous apex, sometimes toothed and more or less
leaf-like; calyx-lobes linear-lanceolate, acute, somewhat villous, 1 lin.
long; corolla 2 lin. long; tube slender; lobes oblong, scarcely ¼ lin.
long; fruit oblong, minutely verrucose, 1 lin. long. *Choisy, Mém.*
Selag. 24, *t.* 1, *fig.* 2; *E. Meyer, Comm.* 251, *and in Drège, Zwei*
Pflanzengeogr. Documente, 98, 107, 179; *Hochst. in Flora,* 1845,
70, *and Beitr. Fl. Cap. und Natal.* 134; *Walp. Rep.* iv. 149;
Choisy in DC. Prod. xii. 7; *Benth. Fl. Austral.* v. 31; *Rolfe in*
Journ. Linn. Soc. xx. 355. *Hebenstreitia capitata, Thunb. Prodr.*
103, *and Fl. Cap. ed. Schult.* 477; *Jarosz, Pl. Nov. Cap.* 13; *Spreng.*
Syst. Veg. ii. 754. *Selago hispida, Sieb. ex Choisy in DC. Prod.*
xii. 7 (*not of Linn. fil.*) *Dischisma claudestinum, Schlechter in*
Engl. Jahrb. xxvii. 187, *in note* (*not of E. Meyer*).

SOUTH AFRICA: without precise locality, *Forster ! Forbes ! Masson !* and
cultivated specimens !
COAST REGION : Vanrhynsdorp Div. ; in sandy places near Ebenezer, below
500 ft., *Drège !* Clanwilliam Div. ; Pakhuis Berg, 2500 ft., *Schlechter,* 8617!
Malmesbury Div. ; near Hopefield, *Bachmann,* 1388 ! 2142 ! Cape Div. ; near
Cape Town, *Thunberg ! Harvey,* 517 ! Cape Flats, *Krauss,* 1640 ; Roudebosch,
Wolley Dod, 641 ! Simons Bay, *Wright !* Paarl Div. ; near Paarl, 400 ft., *Drège !*
Caledon Div. ; River Zonder Einde, *Zeyher,* 3579!

This species was also collected by *Drummond* in the Swan River district,
Western Australia, where it is believed to be naturalized.

3. **D. spicatum** (Choisy in Mém. Soc. Phys. Genèv. ii. ii. 94);
annual, somewhat branched at the base; 3–6 in. high; branches
densely pubescent; leaves linear or subspathulate, subobtuse, entire,
glabrous, ½–2 in. long; spikes dense, many-flowered, 1–4 in. long;
bracts 3½ lin. long, with a broad lanceolate more or less lanate base,

and an attenuate subobtuse and nearly glabrous apex ; corolla white, 5½ lin. long ; tube slender ; lobes oblong, ¾ lin. long ; fruit oblong, 1¼ lin. long. *Choisy, Mém. Selag.* 24 ; *E. Meyer, Comm.* 251, *and in Drège, Zwei Pflanzengeogr. Documente,* 71, 73, 94, 179 ; *Walp. Rep.* iv. 149 ; *Choisy in DC. Prod.* xii. 7 ; *Rolfe in Journ. Linn. Soc.* xx. 355. *Hebenstreitia spicata, Thunb. Prodr.* 103. *Dischisma affine, Schlechter in Engl. Jahrb.* xxvii. 187.

COAST REGION : Vanrhynsdorp Div.; Giftberg, 1500–2000 ft., *Drège!* Clanwilliam Div. ; near Honig Vallei and Koude Berg, 3000–4000 ft., *Drège.*

WESTERN REGION : Little Namaqualand ; in dry sandy plains at the mouth of the Orange River, *Drège!* on hills at 1500–2000 ft., *Cape Herb.!* Vanrhynsdorp Div.; Zout River, 450 ft., *Schlechter,* 8119 !

4. **D. clandestinum** (E. Meyer, Comm. 251) ; annual, erect, rarely slightly branched at the base, 2–5 in. high ; branches somewhat pubescent ; leaves sessile, linear, subobtuse, entire or with 1–2 pairs of minute teeth near the apex, glabrous, ½–1¼ in. long ; spikes dense, many-flowered, ½–2 in. long ; bracts 4–8 lin. long, with a broad lanceolate lanate base, and an attenuated linear obtuse glabrous apex ; calyx-lobes linear, acute, villous, 1½–2 lin. long ; corolla 2 lin. long ; tube slender ; lobes oblong, minute. *E. Meyer in Drege, Zwei Pflanzengeogr. Documente,* 94, 179 ; *Walp. Rep.* iv. 149 ; *Choisy in DC. Prod.* xii. 7. *Hebenstreitia capitata, Hort. Berol. ex Choisy in DC. Prod.* xii. 6, *not of Thunb. Dischisma occludens, Schlechter in Engl. Jahrb.* xxvii. 188.

COAST REGION : Clanwilliam Div. ; Koude Berg, 3400 ft., *Schlechter,* 8748 !

WESTERN REGION : Little Namaqualand ; in sandy places near the Haazenkraals River, 2000 ft., *Drège!*

5. **D. leptostachyum** (E. Meyer, Comm. 251) ; apparently a small shrub, much branched and over 1 ft. high ; branches more or less lanate or pubescent when young ; leaves linear, subobtuse, entire or nearly so, glabrous or slightly canescent, ½–1 in. long ; spikes elongated, many-flowered, 2–6 in. long ; bracts lanceolate, more or less lanate at the margins and base, apex attenuate or acuminate, 4–5 lin. long ; calyx-lobes linear-lanceolate, somewhat pubescent, with the margin membranous and ciliate, 2 lin. long ; corolla ½ in. long ; lobes linear, subequal, 1 lin. long ; fruit oblong. *E. Meyer in Drège, Zwei Pflanzengeogr. Documente,* 91, 109, 179 ; *Walp. Rep.* iv. 149 ; *Choisy in DC. Prod.* xii. 6.

COAST REGION : Clanwilliam Div. ; in sandy places between Piquiniers Kloof and Pretoris Kloof, 1000–1500 ft., *Drège!*

WESTERN REGION : Little Namaqualand ; in dry plains between Goedemans Kraal and Kaus Mountain, 2000–2500 ft., *Drège!*

6. **D. tomentosum** (Schlechter in Engl. Jahrb. xxvii. 189) ; a herb or perennial shrub, much branched, 6–9 in. high ; branches tomentose ; leaves rather dense, linear, subobtuse, 2–2½ lin. long ; spikes dense, rather broad, ½–1¼ in. long (the young flowering state only seen) ; bracts elliptic-oblong, subobtuse, tomentose, 2½–3 lin. long ; calyx-lobes linear, acute, ciliate, 1¼–1½ lin. long ; corolla

4–5 lin. long; tube slender; lobes elliptic-oblong, 1–1¼ lin. long, outer pair broader than the inner.

CENTRAL REGION : Ceres Div. ; Cold Bokkeveld, 3500 ft., *Schlechter*, 8879!

7. **D. erinoides** (Sweet, Hort. Brit. ed. 2, 414) ; a much-branched straggling shrub, ½–2 ft. high; branches pubescent; leaves elliptic or lanceolate, subacute, strongly serrate or dentate, glabrous, ¼–1 in. long, sometimes more or less fascicled ; spikes dense or ultimately somewhat lax, 1–6 in. long; bracts 3–5 lin. long, ovate and ciliate at the base, very acuminate or almost setaceous and nearly glabrous above ; calyx-lobes linear-lanceolate, acute, puberulous, 1½ lin. long; corolla white, 4–5 lin. long; tube slender; lobes oblong, subequal, ¾–1 lin. long; fruit oblong, 1½ lin. long. *Hebenstreitia erinoides, Linn. fil. Suppl.* 286 ; *Murr. Syst. Veg. ed.* 14, 570 ; *Thunb. Prodr.* 103, *and Fl. Cap. ed. Schult.* 478 ; *Jarosz, Pl. Nov. Cap.* 13 ; *Spreng. Syst. Veg.* ii. 754; *Rolfe in Journ. Linn. Soc.* xx. 351. *H. chamœdryfolia, Link ex Jarosz, Pl. Nov. Cap.* 14 ; *Link, Enum. Pl. Hort. Berol.* ii. 125. *D. flaccum, E. Meyer, Comm.* 250 ; *and in Drège, Zwei Pflanzengeogr. Documente,* 129, 179 ; *Walp. Rep.* iv. 147 ; *Choisy in DC. Prod.* xii. 7 ; *Rolfe in Journ. Linn. Soc.* xx. 351. *Dischisma chamœdryfolium, Walp. Rep.* iv. 148; *Choisy in DC. Prod.* xii. 7.

SOUTH AFRICA: without precise locality, *Bowie! Forbes! Masson!* and *cultivated specimens!*
COAST REGION: Cape Div. ; path towards Smitswinkel Bay, *Wolley Dod*, 3024! in shady situations on the Cape Downs, *Bowie!* Stellenbosch Div. ; mountains of Hottentots Holland, *Cape Herb.!* Riversdale Div. ; Tygerfontein, 800 ft., *Galpin*, 4400! Knysna Div.; on sand-hills near the landing-place at Plettenberg Bay, *Burchell*, 5313! near the Goukamma River, *Burchell*, 5594! Port Elizabeth Div. ; on sandy hills below 100 ft., at Port Elizabeth, *Drège!*

8. **D. ciliatum** (Choisy in Mém. Soc. Phys. Genèv. ii. ii. 94, excl. var. β) ; a much-branched straggling shrub, 1–2 ft. or more high; branches pubescent; leaves subsessile, linear, acute, with 1–3 pairs of teeth near the apex or subentire, glabrous, ¼–1 in. long, often more or less fascicled ; spikes usually dense, 1–6 in. long; bracts 3–5 lin. long; ovate and ciliate at the base, very acuminate and glabrous above ; calyx-lobes linear, concave, keeled, membranous and ciliate at the margin, 1¼ lin. long; corolla white, 4–6 lin. long ; tube slender; lobes oblong, subequal, 1 lin. long ; fruit oblong, 1–1½ lin. long. *Choisy, Mém. Selag.* 24, *excl. var. β ; E. Meyer, Comm.* 250, *and in Drège, Zwei Pflanzengeogr. Documente,* 78, 85, 106, 115, 120, 179 ; *Walp. Rep.* iv. 148 ; *Choisy in DC. Prod.* xii. 6 ; *Rolfe in Journ. Linn. Soc.* xx. 346, 356. *Hebenstreitia ciliata, Berg. Pl. Cap.* 154 ; *Linn. Mant.* 420 ; *Thunb. Prodr.* 103, *and Fl. Cap. ed. Schult.* 477 ; *Willd. Sp. Pl.* iii. 331; *Jarosz, Pl. Nov. Cap.* 14 ; *Spreng. Syst. Veg.* ii. 754. *H. hispida, Lam. Encycl.* iii. 78. *H. albiflora, Jarosz, Pl. Nov. Cap.* 13 ; *Link, Enum. Pl. Hort. Berol.* ii. 124. *H. alba, Jacq. Eclog.* ii. *t.* 151. *Dischisma hispidum, Sweet, Hort. Brit. ed.* 2, 415 ; *Choisy in DC. Prod.* xii. 7.

VAR. β, **crassifolium** (E. Meyer, Comm. 250) ; leaves quite entire, somewhat fleshy, broader at the base. *E. Meyer, in Drège, Zwei Pflanzengeogr. Documente,* 90, 179.

SOUTH AFRICA : without precise locality, *Alexander ! Miss Cole ! Forster ! Mund ! Nelson ! Oldenburg ! Sieber,* 147 ! *Thom ! Wallich ! Wright,* 509, 515 ! *Zeyher,* 3586 ! and cultivated specimens !

COAST REGION : Malmesbury Div. ; Darling, *Bachmann,* 429 ! Zwartland, *Zeyher,* 3585 ! Cape Div. ; Wynberg, 200–600 ft., *Drège !* in shady places on the Cape Downs, *Bowie !* Simons Bay, *MacGillivray,* 626 ! Camps Bay, *Burchell,* 303 ! hill at Muizenberg, *Wallich !* foot of Table Mountain, *Pappe ! Ecklon,* 739 ! *Wallich !* Table Mountain, above Camps Bay, 800 ft., *Galpin,* 4399 ! near Cape Town, *Harvey,* 514 ! in stony grassy places on Lion Mountain, near Cape Town, 800 ft., *MacOwan & Bolus, Herb. Norm. Aust. Afr.,* 934 ! Devils Peak, *Wilms,* 3522 ! *Wolley Dod,* 639 ! Flats near Rondebosch, *Zeyher,* 3584 ! *Wolley Dod,* 640 ! Flats near Doornhoogde, *Wolley Dod,* 638 ! Paarl Div. ; *Paarl, Elliott !* rivers on Paarl Berg, *Drège.* Tulbagh Div. ; at Steendal, *Pappe !* near Tulbagh Waterfall, 1000–2000 ft., *Ecklon & Zeyher !* mountains near New Kloof, 1000 ft., *Drège ;* Mitchells Pass, 1300 ft., *Schlechter,* 8943 ! Caledon Div. ; mountains near Genadendal, 1000–1500 ft., *Drège !* Zonder Einde River, 600 ft., *Drège !* at Caledons Institution, *Bowie !* Riversdale Div. ; Riversdale, *Rust,* 446 ! Mossel Bay Div. ; in a dry channel of an arm of the Gauritz River, *Burchell,* 6501 ! on the road between Hartenbosch and Mossel Bay, *Burchell,* 6225 ! Knysna Div, ; on the sands at Plettenberg Bay, *Bowie !* Port Elizabeth Div. ; at Port Elizabeth, *Hewitson,* 140 !

WESTERN REGION : Var. β : Little Namaqualand ; between Silver Fontein and Kousies (Buffels) River, 1000–2000 ft., *Drège.*

The variety *crassifolium* is only known to me from the description.

9. D. fruticosum (Rolfe in Journ. Linn. Soc. xx. 351) ; perennial, stout, erect, much branched, 1–2 ft. or more high ; branches glabrous ; leaves sessile, somewhat crowded near the base of the branches, linear or linear-lanceolate, subobtuse, with 3–5 pairs of small teeth above the middle, glabrous, 4–9 lin. long ; spikes elongated, dense, up to 3 in. long in fruit ; bracts lanceolate or ovate, acute, glabrous, $1\frac{1}{2}$–2 lin. long ; calyx-lobes lanceolate or elliptic-oblong, apiculate, 1–$1\frac{1}{4}$ lin. long ; corolla 3–4 lin long ; tube slender ; lobes oblong, nearly 1 lin. long ; fruit ovoid, $1\frac{1}{2}$ lin. long. *Hebenstreitia fruticosa, Linn. fil. Suppl.* 287 ; *Murr. Syst. Veg. ed.* 14, 570 ; *Poir. Encycl.* iii. 78 ; *Thunb. Prodr.* 103 ; *and Fl. Cap. ed. Schult.* 478, *excl. syn. Berg. ; Willd. Sp. Pl.* iii. 331 ; *Spreng. Syst. Veg.* ii. 754 ; *not of Sims, E. Meyer, Walpers, or Choisy. Selago ramulosa, Choisy in DC. Prod.* xii. 20, *not of Link or E. Meyer.*

SOUTH AFRICA : without precise locality, *Masson !*

COAST REGION : Clanwilliam Div. ; Lamberts or Alexanders Kloof, *Wallich !* Piquetberg Div. ; Piquetberg Mountain, *Thunberg !*

This species has been much confused by authors.

10. D. squarrosum (Schlechter in Engl. Jahrb. xxvii. 188) ; perennial, much branched, 1–2 ft. high ; branches minutely puberulous ; leaves sessile, squarrose, broadly ovate, subacute, entire or denticulate, $1\frac{1}{2}$–2 lin. long ; spikes oblong, very dense, $\frac{3}{4}$–$1\frac{1}{2}$ in. long : bracts elliptic-ovate, acute, ciliate at the base, apex recurved, $2\frac{1}{2}$ lin. long ; calyx-lobes linear-oblong, obtuse, keeled, ciliate, $1\frac{1}{4}$ lin. long ;

corolla 4–5 lin. long; tube slender; lobes linear-oblong, subequal, ¾–1 lin. long; fruit oblong, 1¼ lin. long.

SOUTH AFRICA: without precise locality, *Forsyth! Masson!*
COAST REGION: Clanwilliam Div.; Clanwilliam, 250 ft., *Schlechter*, 8426! sandy dunes near Clanwilliam, 260 ft., very common, *Leipoldt*, 272!

11. D. crassum (Rolfe); perennial, much branched, 1 ft. or more high; branches divaricate, glabrous; leaves sessile, dense, spreading, very broadly ovate, subobtuse or apiculate, crenulate, glabrous, 2 lin. long, 2¼ lin. broad; gradually passing into bracts at the summit of the branches; spikes dense, ½–1 in. long; bracts spreading, very broadly ovate, somewhat cuneate at the base, entire, 2½ lin. long and broad; calyx-lobes linear-oblong, subacute, concave, ciliate, 1 lin. long; corolla 3–4 lin. long; tube slender; lobes oblong, subequal, 1 lin. long; fruit oblong, 1 lin. long.

SOUTH AFRICA: without precise locality, *Masson!*
COAST REGION: Piquetberg Div.; sea shore near St. Helena Fontein, *Wallich!*

A very remarkable species, bearing some resemblance to *Hebenstreitia cordata*, Linn., with which Masson appears to have collected it. Owing to the peculiar leaf-arrangement the branches have a remarkable thickened appearance.

III. **WALAFRIDA**, E. Meyer.

(SELAGO, Sect. MACRIA, E. Meyer.)

Calyx 3-lobed or 3-partite, with the middle lobe usually smaller, sometimes minute or wanting. *Corolla*: tube short or somewhat elongate, often narrow at the base, broader in the throat; limb subequally or unequally 5-lobed, sometimes partially bilabiate, the two posterior lobes sometimes shorter, and the middle anterior lobe rather longer than the others. *Stamens* 4, didynamous, inserted in the corolla throat, more or less exserted; filaments filiform; anthers perfectly 1-celled. *Ovary* 2-celled; style slender, obtuse, slightly clavate or minutely bidentate at the apex. *Fruit* ovoid, globose or oblong, included within the calyx, often separating into distinct cocci when mature; pericarp slender, crustaceous or equally indurated round the cells, rarely with a pair of spurious cells at their adjacent margins. *Seeds* short or oblong.

Shrubs or undershrubs, often heath-like, dwarf and much branched, sometimes small annual herbs; leaves narrow or small, sometimes minute, alternate, often crowded in axillary fascicles, entire; flowers sessile in the axils of the bracts, arranged in short terminal spikes or heads, which are often aggregated into corymbs or narrow panicles at the summit of the branches; bracts ovate or narrow, more or less imbricate.

Species 31 in South Africa, one of which extends into Tropical Africa, with four additional representatives in the latter region, and a single outlying species in Central Madagascar.

This genus was originally founded by E. Meyer on a single species, characterized by the presence in the cocci of a pair of additional spurious cells (analagous to these found in *Hebenstreitia*, sect. *Polycenia*), but in Bentham and Hooker's

Genera Plantarum it was reduced to *Selago*, because the calyx is identical with that of *Selago*, sect. *Macria*, E. Meyer. The character of the calyx, however, in this Order has been so much relied upon for generic limitations that I have here restored *Walafrida*, enlarging it by the addition of the species of *Selago* having a three-lobed calyx (sometimes 2-lobed by abortion), thus restricting *Selago* to the species having a 5-lobed calyx. The extra South African species are *W. lacunosa,* Rolfe (*Selago lacunosa*, Klotzsch), *W. alopecuroides*, Rolfe (*S. alopecuroides,* Rolfe), *W. angolensis*, Rolfe (*S. angolensis*, Rolfe), and *W. Dinteri*, Rolfe (*S. Dinteri*, Rolfe), all from Tropical Africa, and *W. muralis*, Rolfe (*S. muralis,* Benth. & Hook. f.), from Madagascar.

Section 1. EU-WALAFRIDA. Fruit globose or ovoid-globose, with a pair of vacuoles or spurious cells in each carpel ...   .   (1) **nitida.**

Section 2. MACRIA. Fruit oblong or ovoid-oblong, without vacuoles or spurious cells :
　*Perennials, very rarely erect annuals :
　　Leaves not fascicled :
　　　Leaves entire :
　　　　Leaves strongly ciliate   ...   ...   (2) **ciliata.**
　　　　Leaves not ciliate, rarely hispid at the
　　　　　margin :
　　　　　　Leaves ovate or lanceolate   ...   (3) **myrtifolia.**
　　　　　　Leaves linear or lanceolate-
　　　　　　　linear :
　　　　　　　　Corolla-tube 2½–3 lin. long...   (4) **albanensis.**
　　　　　　　　Corolla-tube 1½–2 lin. long :
　　　　　　　　　Leaves hispid at the
　　　　　　　　　margin, not turning
　　　　　　　　　black in drying   ...   (5) **recurva.**
　　　　　　　　　Leaves not hispid at the
　　　　　　　　　margin, turning black
　　　　　　　　　in drying :
　　　　　　　　　　Spikes somewhat
　　　　　　　　　　elongated and
　　　　　　　　　　rather lax   ...   (6) **zuurbergensis.**
　　　　　　　　　　Spikes short and
　　　　　　　　　　rather dense   ...   (7) **Zeyheri.**
　　　Leaves crenulate or denticulate   ...   ...   (8) **apiculata.**
　　Leaves more or less fascicled :
　　　†Flowers in compact or lax corymbose
　　　　panicles ; corolla-tube 1½–2½ lin. long :
　　　　　Leaves orbicular or broadly-elliptic ...   (9) **rotundifolia.**
　　　　　Leaves linear or lanceolate-linear :
　　　　　　Leaves 3–6 lin. long ; calyx-lobes
　　　　　　　subacute   ...   .   (10) **cinerea.**
　　　　　　Leaves 2–4 lin. long ; calyx lobes
　　　　　　　acute :
　　　　　　　　Corymb usually small and
　　　　　　　　dense ; corolla-tube 2–2½
　　　　　　　　lin. long   ...   .   (11) **Macowani.**
　　　　　　　　Corymb usually lax ; corolla-
　　　　　　　　tube 1½–2 lin. long   ...   (12) **decipiens.**
　　　††Flowers aggregated in dense ovoid or
　　　　oblong terminal heads :
　　　　　Corolla-tube 1½–2½ lin. long :
　　　　　　Middle lobe of calyx rather smaller
　　　　　　than the side lobes :
　　　　　　　Calyx 1 lin. long ; corolla-
　　　　　　　tube 1½ lin. long ...   ...   (13) **witbergensis.**

Calyx 1¼–1½ lin. long; corolla-
    tube 2–2¼ lin. long   ... (14) **polycephala.**
Middle lobe of calyx minute or
    obsolete... ... ... (15) **congesta.**
Corolla-tube ¾–1¼ lin. (rarely 1½ lin.)
    long :
    Largest leaves 4–7 lin. long :
        Leaves linear, often rather
        fleshy :
            Bracts ovate-oblong, and
            as well as the leaves
            distinctly fleshy ... (16) **crassifolia.**
            Bracts oblong-lanceo-
            late, not or scarcely
            fleshy... ... ... (17) **Nachtigali.**
        Leaves filiform or nearly so .. (18) **tenuifolia.**
    Largest leaves 1–3 lin. long :
        Leaves with minute blackish
        dots ... ... (19) **distans.**
        Leaves concolorous :
            Branches puberulous,
            often minutely so :
                Bracts linear-ob-
                long :
                    Side-lobes of
                    calyx sub-
                    acute ... (20) **diffusa.**
                    Side-lobes of
                    calyx obtuse (21) **gracilis.**
                Bracts ovate-oblong :
                    Calyx ¾ lin.
                    long; leaves
                    spreading ... (22) **squarrosa.**
                    Calyx 1 lin.
                    long; leaves
                    not spread-
                    ing :.. ... (23) **articulata.**
            Branches strongly and
            closely pubescent ... (24) **pubescens.**
†††Flowers in roundish or oblong heads,
    which are numerous and mostly arranged
    on short lateral branchlets, forming elon-
    gated narrow panicles towards the
    summit of the branches; corolla-tube
    ¾–1¼ lin. long :
    Leaves oblong-linear and rather rigid ;
        panicles short :
            Bracts linear-oblong... ... ... (25) **micrantha.**
            Bracts ovate-oblong ... ... (26) **saxatilis.**
    Leaves not rigid ; panicles generally
        much elongated :
        Leaves narrow, often more or less
            filiform ... ... (27) **densiflora.**
        Leaves linear, sometimes lanceolate-
            linear, not filiform ... ... (28) **paniculata.**
††††Flowers usually in elongated narrow spikes :
    Spikes usually aggregated into lax
        panicles ... ... (29) **geniculata.**
    Spikes numerous, aggregated into com-
        pact panicles ... ... (30) **polystachya.**
Annual, decumbent, 2–3 in. high ... ... (31) **minuta.

1. W. nitida (E. Meyer, Comm. 272) ; perennial, much branched, 1–2 ft. high ; branches divaricate, pubescent ; leaves not fascicled, broadly ovate or ovate-oblong, subobtuse or apiculate, entire, glabrous, 3–6 lin. long ; spikes ovoid or oblong, dense, ½–1 in. long ; bracts ovate-lanceolate, acute, pubescent, 2–3 lin. long ; calyx 1½ lin. long ; lobes ovate-lanceolate, acute, puberulous, subequal ; corolla rose-purple ; tube slender, 4 lin. long ; lobes elliptic-oblong, subequal, 1 lin. long ; fruit ovoid, globose, 1–1¼ lin. long. *E. Meyer in Drège, Zwei Pflanzengeogr. Documente,* 118, 126, 229 ; *Walp. Rep.* iv. 144 ; *Choisy in DC. Prod.* xii. 21. *W. trimera, Hochst. in Flora,* 1845, 70 ; *Krauss, Beitr. Fl. Cap. und Natal.* 134 ; *Walp. Rep.* iv. 144. *Selago nitida, Schlechter in Engl. Jahrb.* xxvii. 187 (*in note*).

SOUTH AFRICA : on sands of the south-east coast, *Bowie!*
COAST REGION : Caledon Div. ; River Zonder Einde, *Zeyher,* 3561 ! Humansdorp Div. ; between Gamtoos River and Kabeljouw River, below 500 ft., *Drège!* Uitenhage Div. ; Algoa Bay, *Forbes! Cooper,* 2486! Uitenhage, *Zeyher! Pappe!* stony places near the Elands River, *Zeyher,* 936 ! 3558! Van Stadens Berg, on rocks at 1500–2000 ft., *Drège! Zeyher! Ecklon & Zeyher!* on the road between Galgebosch and Melk River, *Burchell,* 4776 ! on the sides of Winterhoek Mountains, *Krauss,* 1106 ! Port Elizabeth Div. ; slopes of Baakens Valley, *Tyson,* 2189 ! Port Elizabeth, *Hewitson,* 139 !

2. W. ciliata (Rolfe) ; perennial, much branched, 1–2 ft. high ; branches pubescent, often stout ; leaves ovate or ovate-lanceolate, acute, strongly ciliate, 3–5 lin. long, generally crowded ; spikes oblong, dense, ½–1¼ lin. long ; bracts ovate-lanceolate, acute or acuminate, ciliate, 2–3 lin. long ; calyx 1½ lin. long ; lobes subulate-lanceolate, acute, ciliate ; corolla purple or sometimes white ; tube slender, 2–2½ lin. long ; lobes broadly-oblong, 1–1½ lin. long ; fruit ovoid-oblong, 1 lin. long (immature). *Selago ciliata, Linn. fil. Suppl.* 285 ; *Murr. Syst. Veg. ed.* 14, 568 ; *Thunb. Prodr.* 100, *and Fl. Cap. ed. Schult.* 465 ; *Choisy, Mém. Selag.* 39, *t.* 5 ; *E. Meyer, Comm.* 271, *and in Drège, Zwei Pflanzengeogr. Documente,* 120, 219 ; *Hochst. in Flora,* 1845, 70 ; *Krauss, Beitr. Fl. Cap. und Natal.* 134 ; *Drège in Linnæa,* xx. 202 ; *Walp. Rep.* iv. 165 ; *Choisy in DC. Prod.* xii. 20 ; *Rolfe in Journ. Linn. Soc.* xx. 350.

SOUTH AFRICA : without precise locality, *Thunberg! Masson! Oldenburg, Forsyth!*
COAST REGION : Swellendam Div. ; on hills between Buffeljagts River and Karmelks River, 500–1000 ft., *Drège!* on hills near the Buffeljagts River, *Zeyher,* 3561 ! near Swellendam, *Pappe!* Riversdale Div. ; between Gauritz River and Great Vals River, *Burchell,* 6518! between Great Vals River and Zoetemelks River, *Burchell,* 6573 ! hills near Zoetemelks River, *Burchell,* 6749 ! George Div. ; margins of woods near George, *Krauss,* 1135 !

3. W. myrtifolia (Rolfe) ; perennial, loosely branched, 1–2 ft. high ; branches puberulous ; leaves lanceolate or oblong, subacute, 3–8 lin. long ; spikes ovoid or oblong, dense, ½–1¼ lin. long ; bracts lanceolate, acute, ciliate, 2–3 lin. long ; calyx 1–1½ lin. long ; lobes lanceolate, acute, ciliate ; corolla purple ; tube slender, 4 lin. long ; lobes oblong, 1 lin. long. *Selago myrtifolia, Reichb. Mittheil.* 1829,

68, *ex Reichb. Ic. Exot. Cent.* iii. 10, *t.* 223; *Walp. Rep.* iv. 166.
S. Gillii, Hook. Bot. Mag. t. 3028; *Lindl. Bot. Reg. t.* 1504; *Choisy in DC. Prod.* xii. 20. *S. ohlendorffiana, Lehm. Delectus Seminum Hort. Hamb.* 1831, 6; *and Linnæa,* vi. *Litt.-Ber.* 74; *Walp. Rep.* iv. 167; *Choisy in DC. Prod.* xii. 21.

SOUTH AFRICA; without precise locality, *cultivated specimens!*

4. **W. albanensis** (Rolfe); perennial, 1–2 ft. high; branches erect, virgate, puberulous; leaves linear-lanceolate, acute, hispidulous at the margins, imbricate, 2–4 lin. long; spikes broadly oblong, dense, ½–¾ in. long; bracts subulate-lanceolate, acute, hispidulous and ciliate at the margin, 2–3 lin. long; calyx 1½–2 lin. long; lobes subulate, acute, hispidulous-ciliate; corolla mauve (*Galpin*); tube slender, 2½–3 lin. long; lobes broadly oblong, 1 lin. long. *Selago albanensis, Schlechter in Journ. Bot.* 1896, 503.

COAST REGION: Western districts? without precise locality, *Cooper,* 3077! Albany Div.; Grahams Town, *Cape Herb.!* Bathurst Div.; Kasuga River, *MacOuan,* 732! Port Alfred, at 200 ft., *Galpin,* 3022!
CENTRAL REGION: Somerset Div.; Somerset East, *Bowker!*

5. **W. recurva** (Rolfe); perennial, much branched, 1–2 ft. high; branches divaricate, minutely puberulous; leaves lanceolate-linear, acute, rigid, hispidulous at the margin, carinate, somewhat recurved at the apex, 2–4 lin. long; not turning black in drying; spikes ovoid or oblong, dense, ½–1¼ lin. long; bracts ovate-lanceolate, acuminate, hispidulous and somewhat ciliate, 2 lin. long; calyx 1 lin. long; lobes subulate-lanceolate, acute, ciliate, unequal; corolla-tube slender, 1½ lin. long; lobes orbicular-oblong, ½ lin. long; fruit oblong, 1 lin. long. *Selago recurva, E. Meyer, Comm.* 271; *and in Drège, Zwei Pflanzengeogr. Documente,* 130, 219; *Walp. Rep.* iv. 166; *Choisy in DC. Prod.* xii. 19, *partly.*

COAST REGION: Humansdorp Div.; between Gamtoos River and Kabeljouw River, 600 ft., *Drège!*

Choisy (l.c.) has confused this species with *Selago fruticosa,* Linn., to which it bears some superficial resemblance.

6. **W. zuurbergensis** (Rolfe); perennial, much branched, 1–1½ ft. high; branches minutely puberulous in decurrent lines from the leaf-bases; leaves not fascicled, somewhat crowded, linear-oblong, sub-acute, hispidulous, 2–3 lin. long; spikes oblong, somewhat lax, 1–2 in. long; bracts linear-lanceolate, acute, hispidulous, 2½–3 lin. long; calyx 1¼–1½ lin. long, puberulous; lobes ovate-lanceolate, acute, ciliate; the middle lobe rather smaller than the lateral pair; corolla-tube linear-oblong, 1½–2 lin. long; lobes broadly oblong, unequal, about half as long as the tube.

COAST REGION: Alexandria Div.; Zuurberg Range, at 2000 ft., *Bolus,* 9123!

7. **W. Zeyheri** (Rolfe); perennial, much branched, 1–2 ft. high; branches more or less divaricate, puberulous or pubescent; leaves

linear or linear-oblong, subacute, often recurved, fleshy, glabrous,
2–5 lin. long, turning black in drying; spikes oblong or somewhat
elongated, dense, $\frac{1}{2}$–1 in. long; bracts lanceolate, subacute, ciliate,
1$\frac{1}{2}$–2$\frac{1}{2}$ lin. long; calyx $\frac{3}{4}$–1$\frac{1}{4}$ lin. long; lobes subulate-lanceolate,
acute, ciliate, unequal; corolla-tube slender, 1$\frac{1}{2}$–2 lin. long; lobes
orbicular-oblong, $\frac{3}{4}$–1 lin. long; fruit subglobose, $\frac{3}{4}$ lin. long. *Selago
Zeyheri, Choisy in DC. Prod.* xii. 19. *S. sp., Drège in Linnæa,*
xx. 202.

SOUTH AFRICA: without precise locality, *Masson!*
COAST REGION: Uitenhage Div.; fields near the Zwartkops River, *Zeyher*, 83!
hills of Addo, *Zeyher*, 769! calcareous places on the flats between the Coega and
Zwartkops Rivers, *Zeyher*, 961! 3559! between the Zwartkops and Sunday
Rivers, under 1000 ft., *Ecklon & Zeyher!*

8. W. apiculata (Rolfe); perennial, branched chiefly at the base,
about $\frac{3}{4}$ ft. high; branches puberulous; leaves solitary, somewhat
crowded, elliptic-lanceolate or oblong, apiculate, denticulate, puberulous, 3–4 lin. long; heads ovoid, dense, 5–8 lin. long; bracts broadly
obovate- or elliptic-oblong, apiculate, concave, slightly ciliate, 1$\frac{1}{2}$ lin.
long; calyx 1$\frac{1}{2}$ lin. long, unilateral; lobes broadly oblong, subobtuse or
apiculate, ciliate, nearly as long as the tube; corolla-tube broadly
oblong, 1$\frac{1}{4}$ lin. long; lobes broadly oblong or suborbicular, shorter
than the tube. *Selago apiculata, E. Meyer, Comm.* 256, *and in
Drège, Zwei Pflanzengeogr. Documente,* 53, 219; *Walp. Rep.* iv. 151;
Choisy in DC. Prod. xii. 16.

CENTRAL REGION: Aliwal North Div.; summit of the Witte Bergen, 7000–
7500 ft., *Drège!*

9. W. rotundifolia (Rolfe); perennial, much branched, 1–2 ft.
or more high; branches erect, virgate, puberulous; leaves numerous
and dense, orbicular or elliptic-oblong, obtuse, glaucous, 2–7 lin.
long; flowers arranged in dense corymbose cymes or panicles, $\frac{3}{4}$–2 in.
broad; bracts lanceolate-oblong, subacute, puberulous, 1$\frac{1}{2}$ lin. long;
calyx 1 lin. long; lobes subulate-lanceolate, puberulous; corolla-tube
slender, 2$\frac{1}{2}$ lin. long; lobes broadly-oblong, $\frac{1}{2}$–$\frac{3}{4}$ lin. long; fruit
oblong, 1 lin. long. *Selago rotundifolia, Linn. fil. Suppl.* 285;
Murr. Syst. Veg. ed. 14, 568; *Thunb. Prodr.* 100, *and Fl. Cap. ed.
Schult.* 465; *E. Meyer, Comm.* 271, *and in Drège, Zwei Pflanzengeogr.
Documente,* 128, 219; *Walp. Rep.* iv. 166; *Drège in Linnæa,* xx. 202;
Choisy in DC. Prod. xii. 20; *Rolfe in Journ. Linn. Soc.* xx. 350.

SOUTH AFRICA: without precise locality, *Masson!*
COAST REGION: Knysna Div.; near Knysna, among heaths, *Tyson in Mac-
Owan & Bolus, Herb. Norm.Aust. Afr.,* 974! Port Elizabeth Div.; on grassy hills
at Krakakamma, *Zeyher*, 586! 3562! in moist depressions between Klaasniemand
Fontein and Bethelsdorp, 500 ft., *Drège!* Uitenhage Div.; on the flats near Van
Stadens River Mountains, *Zeyher,* 586! between Krakakamma and Van Stadens
Berg, *Ecklon & Zeyher!*

10. W. cinerea (Rolfe); perennial, much branched, 1–2 ft. or more
high; branches erect, puberulous; leaves more or less fascicled,
linear or linear-oblong, subobtuse, glaucous or minutely canescent,

3–6 lin. long; flowers arranged in dense corymbose cymes
or panicles, 1–3 in. broad; bracts linear-lanceolate, subacute,
minutely ciliate, 1–1¼ lin. long; calyx ¾–1 lin. long, canescent;
lobes lanceolate, subacute, ciliate; corolla white to blue (*Flanaghan*);
tube slender, 1½–2 lin. long; lobes broadly oblong, 1 lin. long.
Selago cinerea, Linn. fil. Suppl. 285, *not of E. Meyer or Choisy;
Murr. Syst. Veg. ed.* 14, 568; *Thunb. Prodr.* 99, *and Fl. Cap. ed.
Schult.* 463; *Rolfe in Journ. Linn. Soc.* xx. 350. *S. cinerascens,
E. Meyer, Comm.* 270, *not of Choisy; Walp. Rep.* iv. 165.

SOUTH AFRICA : without precise locality, *Drège!*
COAST REGION : Riversdale Div.; near Milkwood Fontein, 600 ft., *Galpin,*
4403! Riversdale, *Rust,* 167! Knysna Div.; Knysna, *Bowie!* East London
Div.; banks of Kahoon River, East London, 50 ft., *Galpin,* 3332! Kahoon
Drift, near East London, 2000 ft., *Flanaghan,* 214!

"*S. cinerascens,* E. Meyer," quoted by Hochst. in Flora, 1845, 70, and Krauss,
Beitr. Fl. Cap. und Natal, 134, as collected on Devils Mountain, Cape Div., by
Krauss, is probably different.

11. W. Macowani (Rolfe); perennial, much branched, 1–2 ft. or
more high; branches puberulous; leaves in approximate fascicles,
linear or linear-oblong; flower-heads dense, more or less arranged
in corymbs, ¾–1¼ in. broad; bracts linear-lanceolate, acute,
minutely ciliate, 2 lin. long; calyx 1–1¼ lin. long; lobes subulate,
acute, puberulous; corolla-tube slender, 2–2½ lin. long; lobes oblong,
1 lin. long; fruit oblong, 1 lin. long.

COAST REGION : Albany Div.; Bothas Berg, *MacOwan,* 970! 981!

12. W. decipiens (Rolfe); perennial, much branched, 1–2 ft. or
more high; branches more or less erect, minutely puberulous; leaves
more or less fascicled, linear or oblong-linear, acute, minutely
canescent, 2–3 lin. long; spikes short, dense-flowered, often more or
less aggregated into compact or loose corymbs at the summit of the
branches; bracts ovate-lanceolate, acuminate, minutely ciliate,
1½–2 lin. long; calyx ¾–1 lin. long; lobes triangular-lanceolate,
acuminate, unequal; corolla-tube slender, 1½–2 lin. long; lobes
elliptic-oblong, subequal, ¾ lin. long; fruit ovoid-oblong, 1 lin. long.
Selago decipiens, E. Meyer, Comm. 270; *Walp. Rep.* iv. 165;
Choisy in DC. Prod. xii. 19. *S. canescens, Drège in Linnæa,* xx.
202, *not of Linn. fil.*

SOUTH AFRICA: without precise locality, *Drège!*
COAST REGION : Uitenhage Div.; stony plains on the hills above Elands River,
Zeyher, 929! between Coega River and Sunday River, *Zeyher,* 1377! Grasrug,
300 ft., *Baur,* 1013! Uitenhage, *Pappe!* Port Elizabeth Div.; near Port
Elizabeth, *Wilms,* 2456! Alexandria Div.; hills of Addo, *Zeyher,* 769! *Ecklon
& Zeyher!* Albany Div.; near Grahamstown, *Burke,* 419!

There seems to have been some confusion with Zeyher's 769, which in the
British Museum Herbarium and at Dublin belongs to this species, while at Kew
it is attached to *W. Zeyheri,* Rolfe.

13. W. witbergensis (Rolfe); perennial, much branched, 1 ft. or
more high; branches puberulous or canescent; leaves more or less

fascicled, linear, subobtuse, rather fleshy, canescent, 2-3 lin. long ; spikes ovoid, 3-6 lin. long, more or less aggregated into a corymb at the summit of the branches ; bracts linear-lanceolate, subobtuse, puberulous and ciliate, 1½ lin. long ; calyx pubescent, 1 lin. long ; side-lobes ovate, middle lobe subulate, all acute and ciliate ; corolla white (*Galpin*) ; tube 1½ lin. long ; lobes elliptic-oblong, subequal, ½ lin. long. *Selago witbergensis, E. Meyer, Comm.* 270, *and in Drège, Zwei Pflanzengeogr. Documente,* 52, 219 ; *Walp. Rep.* iv. 165 ; *Choisy in DC. Prod.* xii. 18.

SOUTH AFRICA : without precise locality, *Verreaux!*
CENTRAL REGION : Aliwal North Div.; Witte Bergen, among stones and grass, 5000-6000 ft., *Drège.*
COAST REGION : Uitenhage Div. ; Zwartkops River, *Ecklon & Zeyher,* 39! 51!

I have not seen Drège's original specimen.

14. W. polycephala (Rolfe); perennial, much branched, 1 ft. or more high ; branches puberulous ; leaves densely fascicled, usually spreading, linear, obtuse, nearly glabrous, 2-3 lin. long; spikes ovoid, very dense, seldom arranged in corymbs, 6-10 lin. long; bracts ovate-lanceolate, acuminate, nearly glabrous, 2-2½ lin. long; calyx 1¼-1½ lin. long; lateral lobes elliptic-oblong, acute, middle lobe subulate-lanceolate, acute, all ciliate ; corolla-tube linear-oblong, 2-2¼ lin. long; lobes elliptic-oblong, subequal, ¾ lin. long. *Selago polycephala, Otto ex Walp. Rep.* iv. 164 ; *Choisy in DC. Prod.* xii. 19.

SOUTH AFRICA: without precise locality ; cultivated specimens.
COAST REGION : Uitenhage Div.; on the sand-hills near the Zwartkops River, *Zeyher,* 3576! " Kakkerlak Valley," *Zeyher,* 3566!

The original description of this species was made from cultivated specimens, which I have not seen.

15. W. congesta (Rolfe) ; perennial, much branched, over 6 in. high ; branches puberulous ; leaves fascicled, linear, subobtuse, nearly glabrous, 1½-2½ lin. long; spikes terminal, solitary, ovoid, dense, 6 lin. long ; bracts linear-lanceolate, acute, hispidulous, 1½-2 lin. long ; calyx 1-1¼ lin. long, hispidulous ; lateral lobes linear-oblong, obtuse, ciliate ; middle lobe very minute or obsolete ; corolla-tube linear-oblong, 2 lin. long ; lobes obovate-oblong, subequal, about a third as long as the tube. *Selago congesta, Rolfe in Journ. Linn. Soc.* xx. 356. *S. fruticosa, Thunb. Prodr.* 98, *and Fl. Cap. ed. Schult.* 460, *partly, not of Linn. fil.*

SOUTH AFRICA : without precise locality, *Thunberg! Masson!*

16. W. crassifolia (Rolfe) ; perennial, over 6 in. high ; branches minutely canescent ; leaves mostly somewhat fascicled, linear, subobtuse, minutely canescent, fleshy, 3-6 lin. long; spikes capitate or short, dense, corymbosely arranged and congested at the summit of the branches ; bracts ovate-oblong, subobtuse, fleshy, slightly ciliate near the base, 1½ lin. long; calyx ¾ lin. long ; lateral lobes broadly elliptic-ovate, obtuse, ciliate; middle lobe similar but rather smaller; corolla-tube campanulate-oblong, $1\frac{1}{2}$ lin. long; lobes orbicular-oblong, scarcely half as long as the tube.

CENTRAL REGION: Murraysburg Div.; near Murraysburg, 4000 ft., *Tyson*, 177!

17. **W. Nachtigali** (Rolfe); perennial ?, branched chiefly near the base, about 6–9 in. high; branches minutely canescent; leaves somewhat fascicled or rarely solitary, linear or oblong-linear, subacute, minutely canescent, 1–7 lin. long; spikes capitate or oblong, up to $\frac{3}{4}$ in. long, often numerous and more or less corymbosely arranged at the summit of the branches; bracts oblong-lanceolate to linear-oblong, subacute, minutely canescent, 1–$1\frac{3}{4}$ lin. long; calyx $\frac{3}{4}$–1 lin. long; lateral lobes elliptic-oblong, obtuse, minutely ciliate; middle lobe usually smaller, sometimes nearly obsolete; corolla white (*Galpin*); tube oblong, $\frac{3}{4}$–1 lin. long; lobes elliptic-oblong, about half as long as the tube; fruit broadly ovoid or subglobose, slightly compressed, $\frac{3}{4}$ lin. long. *Selago Nachtigali, Rolfe in Verhandl. Bot. Ver. Brandenb.* xxxi. 205.

SOUTH AFRICA: without precise locality, *Nachtigal!*
COAST REGION: Queenstown Div.; Shiloh, 3500 ft., *Baur*, 773! plains near Queenstown, 3500 ft., *Galpin*, 1646!
KALAHARI REGION: Griqualand West; plains between Griqua Town and Witte Water, *Burchell*, 1895! near Griqua Town, *Orpen in Bolus Herb.*, 5752! and without precise locality, *Marloth*, 1030! Orange River Colony; Mud River Drift, *Rehmann*, 3603! Bechuanaland; Masupa River, in Banquaketse Territory, *Holub!*
EASTERN REGION: Natal! Port Natal, *Miss Owen!*

18. **W. tenuifolia** (Rolfe); perennial, branched chiefly at the base, about 1 ft. high; branches minutely puberulous; leaves in approximate fascicles, filiform, subobtuse, minutely puberulous, 3–7 lin. long; spikes short, subcapitate when young, arranged in dense corymbose panicles at the end of the branches, $\frac{3}{4}$–$2\frac{1}{4}$ in. broad; bracts oblong, subobtuse, concave, very minutely puberulous, 1 lin. long; calyx 1 lin. long; lobes linear-oblong, subobtuse, minutely ciliate; corolla-tube oblong, 1 lin. long; lobes subequal, nearly half as long as the tube; fruit subglobose, compressed, $\frac{3}{4}$ lin. long.

KALAHARI REGION: Transvaal; near Lydenburg, *Wilms*, 1161! 1162! Brankhorst Spruit, in the Middelburg District, *Wilms*, 1161a!

19. **W. distans** (Rolfe); perennial, much branched, 6 in. or more high; branches puberulous or minutely hispidulous when old; leaves in approximate fascicles, linear, obtuse, somewhat thickened at the margin, glaucous and bearing numerous minute brown dots, 2–3 lin. long; spikes solitary, short; bracts linear, obtuse, slightly thickened, glaucous and bearing numerous minute brown dots, $1\frac{3}{4}$ lin. long; calyx 1 lin. long; lateral lobes ovate-oblong, obtuse, sometimes bidentate at the apex, very minutely ciliate; middle lobe much smaller; corolla-tube campanulate, 1 lin. long; lobes elliptic-oblong, subequal, $\frac{1}{3}$ as long as the tube. *Selago distans, E. Meyer, Comm.* 266, *and in*

Drège, Zwei Pflanzengeogr. Documente, 65, 219, *not of Lindl.* ; *Walp. Rep.* iv. 160 ; *Choisy in DC. Prod.* xii. 11.

CENTRAL REGION : Prince Albert Div. ; Kendo, on dry hills at 3000–4000 ft., *Drège!*

20. W. diffusa (Rolfe) ; perennial, much branched, more or less diffuse, about 6 in. high ; branches minutely puberulous ; leaves somewhat fascicled, linear, subacute, nearly glabrous, $1\frac{1}{4}$–2 lin. long ; bracts linear, subacute, nearly glabrous, 1–$1\frac{1}{4}$ lin. long ; calyx $\frac{1}{3}$ lin. long ; lateral lobes elliptic-oblong, subobtuse, ciliate, sometimes minutely bidentate at the apex ; middle lobe acute, only half as long as the lateral lobes ; corolla-tube oblong, 1 lin. long ; lobes broadly oblong, subequal, scarcely half as long as the tube ; fruit ovoid-oblong, $\frac{3}{4}$ lin. long. *Selago diffusa, Hochst in Flora,* 1845, 68 ; *Krauss, Beitr. Fl. Cap. und Natal.* 132, *not of Thunb.; Choisy in DC. Prod.* xii. 12.

COAST REGION ! Humansdorp Div. ; in clay soil near the Zekoe River, *Krauss,* 1137 !

21. W. gracilis (Rolfe) ; perennial, much branched, about 6 in. high ; branches puberulous ; leaves in approximate fascicles, linear, obtuse, minutely puberulous, $\frac{3}{4}$–$1\frac{1}{4}$ lin. long ; spikes oblong, narrow, $\frac{1}{2}$–$\frac{3}{4}$ lin. long ; bracts lanceolate-linear, subacute, somewhat curved, slightly ciliate near the base, $\frac{3}{4}$–1 lin. long ; calyx $\frac{3}{4}$ lin. long ; lateral lobes oblong, obtuse, ciliate; middle lobe acute, minute ; corolla-tube oblong, 1–$1\frac{1}{4}$ lin. long ; lobes broadly oblong, unequal, about half as long as the tube. *Selago sp., Drège in Linnæa,* xx. 202.

COAST REGION : Swellendam Div. ; Hassaquas Kloof, *Zeyher,* 3577 !

22. W. squarrosa (Rolfe) ; perennial, much branched, about 6 in. high ; branches spreading, puberulous ; leaves more or less fascicled, spreading, linear, obtuse, minutely canescent, $\frac{3}{4}$–2 lin. long; spikes oblong, narrow, $\frac{1}{2}$–$\frac{3}{4}$ in. long ; bracts ovate, apiculate and subobtuse, ciliate near the base, 1 lin. long ; calyx $\frac{3}{4}$ lin. long ; lateral lobes ovate, obtuse, ciliate ; middle lobe much smaller ; corolla-tube oblong, 1 lin. long ; lobes elliptic-oblong, unequal, not half as long as the tube ; fruit ovoid, $\frac{3}{4}$ lin. long.

COAST REGION : Riversdale Div.; Riversdale, *Rust,* 259! Uitenhage Div.; in stony and sandy places, Koegas Kop, *Cape Herb.!*

23. W. articulata (Rolfe) ; perennial, much branched, 3–8 in. high ; branches puberulous or pubescent ; leaves densely fascicled, linear or oblong-linear, obtuse, minutely canescent, $\frac{1}{2}$–1 lin. long; spikes capitate or short, corymbosely arranged at the summit of the branches or solitary ; bracts oblong or ovate-oblong, obtuse, velvety, 1 lin. long ; calyx 1 lin. long, pubescent ; lateral lobes linear-oblong, subobtuse, ciliate ; middle lobe about half as long as the lateral lobes ; corolla-tube about $1\frac{1}{4}$ lin. long ; lobes elliptic-oblong, unequal, scarcely half as long as the tube ; fruit ovoid-oblong, $\frac{1}{2}$–3 lin. long. *Selago articulata, Thunb. Prodr.* 99, *and Fl. Cap. ed. Schult.* 460 ;

Choisy in DC. Prod. xii. 21 ; *Rolfe in Journ. Linn. Soc.* xx. 351.
S. geniculata, Choisy in DC. Prod. xii. 9, *not of Linn. fil. S.
geniculata, var. β, Choisy, Mém. Selag.* 33, *fide Choisy.*

SOUTH AFRICA: without precise locality, *Thunberg! Masson! Lichtenstein,*
386!

CENTRAL REGION: Calvinia Div. ; Hantam Mountains, *Meyer!*

24. W. pubescens (Rolfe) ; perennial, much branched, about 1 ft.
high ; branches densely and closely pubescent ; leaves fascicled,
linear, obtuse, minutely canescent, ½–1 lin. long ; spikes capitate or
oblong, rather narrow, 2–4 lin. long ; bracts linear-oblong, obtuse,
canescent, ciliate at the base, ¾–1 lin. long ; calyx ¾ lin. long ; lateral
lobes ovate-oblong, obtuse, pubescent and ciliate ; middle lobe much
smaller ; corolla-tube oblong, 1 lin. long ; lobes elliptic-oblong,
unequal, not half as long as the tube.

COAST REGION: Cape Div. ; Simons Town, *Schlechter,* 663!

25. W. micrantha (Rolfe) ; perennial much branched, ½–1¼ ft.
high ; branches densely puberulous ; leaves in approximate fascicles,
more or less spreading, linear, obtuse, minutely puberulous, fleshy,
½–2 lin. long ; spikes capitate or oblong in fruit, arranged in a more
or less compact somewhat elongated panicle towards the summit of
the branches ; bracts oblong or linear-oblong, obtuse, minutely
puberulous, ¾–1¼ lin. long ; calyx ½–¾ lin. long ; lateral lobes
elliptic-oblong, obtuse, usually more or less ciliate, united to near the
middle ; middle lobe generally much smaller and narrower ; corolla-
tube oblong, ¾–1 lin. long ; lobes elliptic-oblong, subequal, scarcely
half as long as the tube ; fruit ovoid, ¾ lin. long. *Selago micrantha,
Choisy in Mém. Soc. Phys. Genèv.* ii. ii. 98, *and in Mém. Selag.* 28 ;
E. Meyer, Comm. 269, *and in Drège, Zwei Pflanzengeogr. Documente,*
54, 219 ; *Walp. Rep.* iv. 164; *Choisy in DC. Prod.* xii. 19. *S.
appressa, Drège in Linnæa,* xx. 202, *not of Choisy. S. glabrata,
var., Choisy in DC. Prod.* xii. 9. *S. fruticosa, Choisy in DC. Prod.*
xii. 19, *not of Linn.*

SOUTH AFRICA: without precise locality, *Masson! Thom,* 146! 193! *Krebs!
Alexander!*

COAST REGION : Swellendam Div.; between Swellendam and Cogmans Kloof,
800 ft., *Bolus,* 8073! plains of Swellendam, *Bowie!* Uitenhage Div.; fields near
the Zwartkops River, *Ecklon,* 29! Commando Kraal, between Karroogebosch and
Sunday River, *Zeyher,* 3575! Zwartkops River Hoogte, *Cape Herb.!* Enon,
400–500 ft., *Baur,* 1050!

CENTRAL REGION : Somerset Div.; without locality or collector! Philipstown Div.; Bavers Pan, *Burchell,* 2713 ; Colesberg Div.; on a rocky mountain
at Naauw Poort, *Burchell,* 2767! Albert Div.; without precise locality,
Cooper, 1359! New Hantam or Zeekoe River? *Drège.*

KALAHARI REGION : Orange River Colony, without precise locality, *Cooper,*
827!

This species has been much confused if all the above are correctly referred here.
The original specimen I have not seen.

26. W. saxatilis (Rolfe) ; perennial, much branched, ½–1 ft. or
more high ; branches minutely puberulous ; leaves in approximate
fascicles, linear or oblong-linear, obtuse, nearly glabrous, ½–2 lin. long ;

spikes small, capitate or oblong, dense, laterally arranged on very
short lateral branchlets in a compact somewhat elongated panicle
towards the summit of the branches; bracts ovate-oblong, obtuse,
rather fleshy, ciliate at the base, $\frac{3}{4}$-1 lin. long; calyx $\frac{1}{2}$-$\frac{3}{4}$ lin. long;
lateral lobes elliptic-oblong, obtuse, strongly ciliate; middle lobe
very small or obsolete; corolla-tube oblong, $\frac{3}{4}$ lin. long; lobes
broadly-oblong, subequal, not half as long as the tube. *Selago
saxatilis, E. Meyer, Comm.* 269, *and in Drège, Zwei Pflanzengeogr.
Documente,* 54, 219; *Walp. Rep.* iv. 163; *Choisy in DC. Prod.*
xii. 18.

SOUTH AFRICA: without precise locality, *Scott-Elliot,* 537!
CENTRAL REGION: Colesberg Div.; among stones near Colesberg, 4500 ft.,
Drège! Shaw, 53! 56!
KALAHARI REGION: Griqualand West; Eitalers Fontein, *Rehmann,* 3347!
Orange River Colony; Bloemfontein, *Rehmann,* 3895! 3896!

27. W. densiflora (Rolfe); perennial, much branched, more
or less diffuse, $\frac{1}{2}$-1$\frac{1}{2}$ ft. high; branches puberulous, somewhat
flexuose; leaves more or less fascicled, linear, obtuse, glabrous or
nearly so, 1$\frac{1}{2}$-5 (rarely to 8) lin. long; spikes globose, at length
somewhat elongated, dense, laterally arranged on short branchlets
and forming dense somewhat elongated panicles at the summit of the
branches; bracts linear or oblong-linear, subobtuse, nearly glabrous,
curved, $\frac{1}{2}$-$\frac{3}{4}$ lin. long; calyx $\frac{1}{2}$ lin. long; lateral lobes oblong or
elliptic-oblong, middle lobe subulate, all sparingly ciliate or nearly
glabrous; corolla white (*Burchell*) or mauve (*Galpin*); tube
oblong, broader above, $\frac{3}{4}$ lin. long; lobes elliptic-oblong, subequal,
half as long as the tube; fruit broadly-ovoid, $\frac{1}{3}$-$\frac{1}{2}$ lin. long. *Selago
densiflora, Rolfe in Bull. Herb. Boiss.* ii. 222.

KALAHARI REGION: Griqualand West; between Griqua Town and Moses
Fontein, *Burchell,* 2134! Honeynest Kloof, *Rehmann,* 3396! and without precise
locality, *Marloth,* 910! Orange River Colony; Caledon River, *Burke,* 422!
Zeyher, 1380! Oliphants Fontein, *Rehmann,* 3513! Draai Fontein, *Rehmann,*
3631! 3634! hills near the Vaal River, *Mrs. Bowker,* 650! Bechuanaland; plains
near Mafeking, *Bolus,* 6434! Transvaal; Kudus Poort, near Pretoria, *Rehmann,*
4678! hill-sides near Johannesburg, 5000 ft., *Galpin,* 1380! Houtbosch, *Rehmann,* 6205! Hooge Veld, at Standerton, *Rehmann,* 6822! Kalk Spruit, between
the Vaal River and Heidelberg, *Schenck!* near Little Oliphants River, 5100 ft.,
Schlechter, 3807!
EASTERN REGION: Natal; Roadsides between Mooi River and Estcourt,
3500 ft., *Wood,* 3485! between Pietermaritzburg and Newcastle, *Wilms,* 2119!
near Howick, 3500 ft., *Schlechter,* 6788! Natal, *Miss Owen!*

28. W. paniculata (Rolfe); perennial, much branched, $\frac{1}{2}$-2 ft.
or more high; branches minutely puberulous, more or less flexuose;
leaves more or less fascicled, linear or linear-oblong, subobtuse,
glabrous or nearly so, 1$\frac{1}{2}$-8 lin. long; spikes ovoid or oblong,
usually dense, laterally arranged on short branchlets and forming
elongated narrow panicles towards the summit of the branches;
bracts oblong or ovate-oblong, obtuse, nearly glabrous, $\frac{3}{4}$-1 lin.
long; calyx $\frac{1}{3}$-$\frac{1}{2}$ lin. long; lateral lobes oblong, obtuse, ciliate;
the middle lobe very small or often absent; corolla white (*Bowker*);

tube oblong, ½ lin. long; lobes elliptic-oblong or rounded, shorter
than the tube; fruit ovoid, somewhat compressed, ⅓ lin. long.
Selago paniculata, Thunb. Prodr. 99, *and Fl. Cap. ed. Schult.*
462; *Spreng. Syst. Veg.* ii. 745; *Rolfe in Journ. Linn. Soc.* xx.
353. *S. choisiana, E. Meyer, Comm.* 268, *and in Drège, Zwei
Pflanzengeogr. Documente,* 62, 219; *Walp. Rep.* iv. 163; *Choisy in
DC. Prod.* xii. 18. *S. amboensis, Rolfe in Dyer, Fl. Trop. Afr.*
v. 272.

COAST REGION: Clanwwilliam Div.; at Oliphants River and near Villa
Brackfontein, *Ecklon & Zeyher,* 33! 69! Mossel Bay Div.; in the dry channel
of an arm of the Gauritz River, *Burchell,* 6460!
CENTRAL REGION: Sutherland Div.; at the Great Reed River, *Burchell,*
1365! Beaufort West Div.; near Rhenoster Kop, 3000 ft., *Drège!* Graaff Reinet
Div.; near Graaff Reinet, 2500 ft., *Bolus,* 293! *Bowker!* Colesberg Div.;
Colesberg, *Shaw,* 50! 51! 52! 55! Albert Div.; *Cooper,* 655!
WESTERN REGION: Namaqualand, without precise locality, *Wyley!*
KALAHARI REGION: Griqualand West; between Kuruman and the Vaal
River, *Cruikshank, in Bolus Herb.,* 293! Klipdrift, *Barber!* Kimberley, *Reh-
mann,* 3435! and without precise locality, *Rehmann,* 3430! Vaal River,
Nelson, 164! Orange River Colony; bed of Umdelu River, *Mrs. Bowker,* 474!
and without precise locality, *Hutton!* Bechuanaland; Batlapin Territory,
Holub!
EASTERN REGION: Tembuland; Imvane, *Baur,* 81!
Also in Tropical Africa.
A polymorphic and widely-diffused species, if all the above are correctly referred
here.

29. **W. geniculata** (Rolfe); perennial, much branched, ½–1¼ ft.
high; branches often more or less spreading, minutely canescent;
leaves in approximate fascicles, linear, subobtuse, minutely canescent,
1½–8 lin. long; spikes elongate, often somewhat lax, ½–4 in. long,
often numerous and arranged in lax panicles at the summit of the
branches; bracts lanceolate-oblong, acute, ciliate at the base, 1 lin.
long; calyx ¾ lin. long, pubescent; lateral lobes lanceolate-oblong,
acute, ciliate; middle lobe not half as large as the lateral lobes;
corolla purple or sometimes white; tube oblong, ¾–1 lin. long; lobes
broadly oblong, unequal, scarcely half as long as the tube; fruit
ovoid-oblong, ¾ lin. long. *Selago geniculata, Linn. fil. Suppl.* 284;
Thunb. Prodr. 98, *and Fl. Cap. ed. Schult.* 460; *Rolfe in Journ.
Linn. Soc.* xx. 350; *not of E. Meyer or Choisy. S. leptostachya,
E. Meyer, Comm.* 266, *and in Drège, Pflanzengeogr. Documente,*
54, 56, 59, 65, 129, 219; *Walp. Rep.* iv. 161; *Choisy in DC. Prod.*
xii. 11, *excl. var. eckloniana. S. polygaloides, Choisy in DC. Prod.*
xii. 11, *not of Mém. Selag.*

SOUTH AFRICA: without precise locality, *Thunberg! Masson! Thom,* 294!
COAST REGION: Riversdale Div.; near the Gauritz River, *Bowie!* Mossel
Bay Div.; on dry hills on the east side of the Gauritz River, *Burchell,* 6441!
Humansdorp Div.; between Melk River and Gamtoos River, *Burchell,* 4792!
Uitenhage Div.; Zwartkops River, among shrubs below 100 ft., *Drège! Ecklon!
Zeyher,* 497! 1382! Komando Kraal, east of the Sunday River, *Zeyher,* 862!
Albany Div.; Grahamstown, *Bolton! MacOwan,* 218! and *Miss Bowker!* Victoria
East Div.; Alice, *Pappe!* Queenstown Div.; Shiloh, 3500 ft., *Baur,* 860!
Zwartkei River, 3500 ft., *Baur,* 989! plains near Queenstown, 3500–3600 ft.,
Galpin, 1701!

CENTRAL REGION: Prince Albert Div.; near Kendo, 2500–3000 ft., *Drège!*
Somerset Div.; open ground near Somerset East, *Scott-Elliot*, 332! *Bowker!*
and without precise locality, *Mrs. Barber*, 411! Richmond Div.; near Styl Kloof,
4000–5000 ft., *Drege; Graaff* Reinet Div.; stony hills near Graaff Reinet, *Bolus*,
355! Aberdeen Div.; near Camdeboo Mountain, 2000–3000 ft., *Drège;* Hanover
Div.; inundated places near the Zeekoe River on the Sneeuwberg Range,
4500–5000 ft., *Drège!* south side of the Snowy Mountains, *Burke!* Colesberg
Div.; between Plettenbergs Beacon and "Flat Station," *Burchell*, 2749! Colesberg, *Shaw!*

KALAHARI REGION: Orange River Colony, hills and valleys, *Mrs. Barber!*

According to Mrs. Barber, this is a valuable plant to sheep farmers, and its
colonial name is "*Aasbasjes.*"

30. W. polystachya (Rolfe); perennial, much branched, 9 in.
high; branches puberulous; leaves in approximate fascicles, linear,
subobtuse, nearly glabrous, 1–4 lin. long; spikes oblong or somewhat
elongated, $\frac{1}{2}$–2 in. long, arranged in a dense corymbose or somewhat
elongated panicle; bracts linear, subobtuse, ciliate at the base,
1 lin. long; calyx $\frac{1}{2}$–$\frac{3}{4}$ lin. long; lateral lobes linear-oblong, obtuse,
ciliate; middle lobe acute, about half as long as the lateral lobes;
corolla white; tube oblong, $\frac{3}{4}$ lin. long; lobes elliptic-oblong, not
half as long as the tube; fruit ovoid-oblong, $\frac{1}{2}$ lin. long.

COAST REGION: Komgha Div.; grassy hills near Komgha, 2000 ft.,
Flanaghan, 369!
KALAHARI REGION: Orange River Colony; Liedenbergs Vley, *Rehmann!*

31. W. minuta (Rolfe); annual, much branched, decumbent,
about 2–3 in. high; branches puberulous; leaves solitary or somewhat fascicled, petiolate or spathulate with an ovate-oblong limb,
obtuse, glabrous, $\frac{3}{4}$–1$\frac{1}{2}$ lin. long; spikes capitate, solitary, 1$\frac{1}{2}$–2 lin.
long; bracts subspathulate-oblong, obtuse, $\frac{3}{4}$ lin. long; calyx $\frac{1}{2}$ lin.
long; lateral lobes broadly-oblong, obtuse, ciliate; middle lobe rather
smaller and more acute; corolla pale rose (*Burchell*); tube oblong,
$\frac{1}{2}$–$\frac{3}{4}$ lin. long; lobes broadly oblong, scarcely half as long as the
tube; fruit ovoid-globose, $\frac{1}{2}$ lin. long.

CENTRAL REGION: Calvinia Div.; Hantam Mountains, *Meyer!*
KALAHARI REGION: Prieska Div.; Ongars River, at "Rushy Station,"
Burchell, 2125!

IV. SELAGO, Linn.

Calyx shortly or deeply 5-lobed. *Corolla*-tube short and broad or
elongated and narrow, always more or less dilated in the throat;
limb subequally 5-lobed, or somewhat bilabiate; posticous lobes
shorter than the anticous; intermediate anticous lobe usually longer
than the outer pair. *Stamens* 4, didynamous, inserted at the base of
the corolla-throat, shortly or much exserted; filaments filiform;
anthers perfectly 1-celled; staminode usually absent, if present,
small. *Ovary* 2-celled; style exserted, slender, obtuse, slightly
thickened or minutely tridentate at the apex. *Fruit* oblong, ovoid
or subglobose, included within the calyx, often breaking up into

cocci ; pericarp crustaceous or indurated, without vacuoles or spurious
cells. *Seeds* oblong or rounded.

Shrubs or undershrubs, usually dwarf, much branched and heath-like, sometimes annual herbs, more or less decumbent at the base ; leaves solitary or
fascicled, alternate or the lower sometimes opposite or subopposite, often small
and narrow, sometimes broader, oblong, elliptic or spathulate, often entire,
sometimes more or less toothed ; flowers sessile or subsessile, or sometimes more
or less pedicelled, spicate, capitate or paniculate, or frequently with the spikes
or heads disposed in elongated panicles or broad corymbs ; bracts narrow or broad,
often more or less imbricate, sometimes adnate to the pedicels.

Species 112 South African, 2 of which extend into Tropical Africa, with
17 endemic species in the latter region, mostly on the hills.

The species with a trilobed calyx, sometimes bilobed by abortion of the middle
lobe, are now referred to *Walafrida*.

Section 1. EU-SELAGO. Leaves more or less fascicled ; calyx campanulate;
tube usually as long as or nearly as long as the lobes.

*Perennials, with more or less woody branches and
    sessile or subsessile flowers, rarely annuals
    with the flowers pedicelled :
†Flowers in lax or compact panicles, or in heads
    or spikes on short lateral branchlets form-
    ing elongated narrow panicles near the
    summit of the branches, rarely in simple
    spikes or heads :
Individual heads rounded :
Leaves puberulous or hispidulous :
Inflorescence thyrsoid, not 1-sided :
    Calyx 1¼ lin. long, villous ... ...   (1) **villicalyx.**
    Calyx 1 lin. long, pubescent ... ...   (2) **pachypoda.**
Inflorescence invariably 1-sided :
    Plant 1 ft. or more high :
    Calyx 1–1½ lin. long :
        Calyx-lobes about as long as the
        tube ... ...   (3) **Cooperi.**
        Calyx-lobes about ⅓ as long as
        the tube ... ...   (4) **Sandersoni.**
    Calyx ¾ lin. long :
        Corolla-tube 1½–1¾ lin. long ...   (5) **Barbula.**
        Corolla-tube 1 lin. long ...   (6) **capitellata.**
    Plant 6 in. high ... ...   (7) **Galpinii.**
Leaves closely pubescent or velvety :
    Leaves oblong ... ...   (8) **lithospermoides.**
    Leaves linear ... ...   (9) **Holubii.**
Individual heads oblong, somewhat elongated
    or lax :
‡Panicles ample, or if small, the leaves not
    small and rigid :
Branches pubescent or puberulous :
    Leaves oblong :
        Leaves 2–4 lin. long ... ...   (10) **Flanaganii.**
        Leaves 3–7 lin. long... ... ...   (11) **pubescens.**
    Leaves linear or lanceolate-linear :
        Pubescence close, short, and uni-
        formly dense ... ...   (12) **Schlechteri.**
        Pubescence soft, lax, or more or less
        in decurrent lines :

Leaves lanceolate-linear ... ... (13) **Burchellii.**
Leaves linear :
   Inflorescence more or less
     compact :
     Bracts lanceolate or oblong-
      lanceolate :
      Bracts and calyx-lobes
       slightly ciliate ... ... (14) **Forbesii.**
      Bracts and calyx-lobes
       strongly ciliate or
       villous :
       Branches pubescent or
        puberulous :
        Calyx-lobes very un-
         equal, broader at
         the base... ... (15) **canescens.**
        Calyx-lobes subequal,
         scarcely broader
         at the base ... (16) **ramulosa.**
        Branches villous ... (17) **villicaulis.**
      Bracts linear or subulate ... (18) **linearis.**
   Inflorescence ample, lax :
     Panicles somewhat dense ;
      individual heads globose
      or nearly so ... ... (19) **Thunbergii.**
     Panicles very lax ; individual
      spikes more or less elon-
      gated ... ... (20) **glabrata.**
Leaves filiform or nearly so :
   Flowers in ample lax panicles :
     Bracts and calyx villous ... (21) **laxiflora.**
     Bracts and calyx pubescent ... (22) **tephrodes.**
   Flowers in narrow, usually dense
     panicles ... ... (23) **adpressa.**
Branches closely and minutely canes-
   cent or rarely glabrous :
   Leaves oblong to ovate ; corolla-tube
     1½–2 lin. long :
   Leaves tomentose or velvety ... (24) **hermannioides.**
   Leaves puberulous, hispidulous or
     nearly glabrous :
     Leaves 1–4 lin. long :
      Leaves oblong ; calyx 1½ lin.
       long ... ... (25) **pinguicula.**
      Leaves ovate-oblong ; calyx
       2 lin. long ... ... (26) **ovata.**
     Leaves 3–8 lin. long :
      Leaves flat ... ... (27) **namaquensis.**
      Leaves more or less revolute at
       the margins ... ... (28) **robusta.**
Leaves linear, or if broader the corolla-
   tube stout and rarely 1½ lin.
   long :
   Leaves linear, 2–8 lin. long, rarely
     oblong :
     Flowers in compact thyrsoid
      panicles ... ... (29) **speciosa.**
     Flowers in dense spikes or heads,
      which are usually more or less
      arranged in panicles :
      Leaves 5–8 lin. long ... ... (30) **lineatifolia.**

Leaves 2–4, rarely 6 lin. long :
  Heads numerous, arranged
    in dense narrow panicles   (31) **Saundersiæ.**
  Heads usually few, arranged
    in lax panicles :
    Leaves crowded right up
      to the inflorescence ...   (32) **Bolusii.**
    Leaves small or lax near
      the inflorescence   ...   (33) **albida.**
Leaves oblong, 1–2 lin. long, rarely
  longer and slender :
  Flowers in   compact   narrow
    panicles   ...   .   (34) **minutissima.**
  Flowers in short or oblong spikes   (35) **divaricata.**
‡‡Panicles small, often reduced to simple
  spikes ; leaves small and rigid :
  Calyx ¾ lin. long ; corolla-tube ¾–1 lin.
    long :
    Branches canescent or minutely pu-
      berulous :
      Leaves not squarrose :
        Leaves 1–3 lin. long   ...   ...   (36) **Burkei.**
        Leaves ¾–1½ lin. long   ...   ...   (37) **Zeyheri.**
      Leaves squarrose or recurved   ...   (38) **tenuis.**
    Branches pubescent or puberulous with
      ferruginous hairs...   ...   ...   (39) **ferruginea.**
  Calyx 1 lin. long ; corolla-tube 1–1¼ lin.
    long ...   .   (40) **rigida.**
††Flowers in broad, often dense, more or less
  corymbose panicles, rarely reduced to
  simple spikes or heads :
  ‡ Flowers sessile or subsessile :
  Leaves generally oblong, rigid and more
    or less spreading :
    Calyx pubescent or villous :
      Leaves not half as broad as long :
        Corolla-tube 2½–3 lin. long   ...   (41) **polystachya.**
        Corolla-tube 1½–2 lin. long :
        Leaves strongly hispid or setu-
          lose   ...   .   (42) **scabrida.**
        Leaves minutely hispidulous or
          nearly glabrous :
          Calyx 1 lin. long ; corolla-tube
            1½ lin. long   ...   ...   (43) **luxurians.**
          Calyx 1½ lin. long ; corolla-
            tube 2 lin. long   ...   ...   (44) **Dregei.**
      Leaves half as broad as long :
        Leaves very villous...   ...   ...   (45) **setulosa.**
        Leaves glabrous   ...   .   (46) **brevifolia.**
    Calyx puberulous :
      Leaves uniformly green ; bracts sub-
        acute .   ...   (47) **glomerata.**
      Leaves minutely black punctate ;
        bracts obtuse   ...   .   (48) **punctata.**
Leaves broad or narrow, but not rigid
  (except in short-leaved forms of 57,
  S. corymbosa) :
  Inflorescence broadly and densely
    corymbose or the heads corym-
    bosely arranged and usually
    narrow :

§ Leaves dense, rarely lax and then
   broad :
   Leaves lanceolate or oblong, more
     or less toothed, rarely linear-
     lanceolate or entire and then
     more or less elongated :
     Corolla-tube 3–4 lin. long   ... (49) **longituba.**
     Corolla-tube 1½–2½ lin. long :
      Leaves dense right up to the
       inflorescence :
       Leaves lanceolate :
        Bracts lanceolate ; corolla-
         tube 2½ lin. long ... (50) **Wilmsii.**
        Bracts linear-lanceolate ;
         corolla-tube 2 lin.
         long :
       Leaves linear-lanceolate :
        Leaves ½–1½ in. long ;
         bracts linear-oblong . (51) **natalensis.**
        Leaves 2–8 lin. long ;
         bracts oblong :
         Bracts linear-oblong ;
          calyx 1–1¼ lin. long (52) **aggregata.**
         Bracts broadly oblong ;
          calyx ¾–1 lin. long (53) **Nelsoni.**
      Leaves becoming lax near the
       inflorescence   ...   ... (54) **foliosa.**
     Corolla-tube about 1 lin. long :
      Bracts lanceolate-linear, not
       concave ... ... (55) **transvaalensis.**
      Bracts elliptic-oblong, obtuse,
       concave ... ... (56) **hyssopifolia.**
   Leaves filiform or slender, entire,
     rarely linear and very short :
     Calyx ¼ lin. long ; corolla-tube
      ½–¾ lin. long   ...   ... (57) **corymbosa.**
     Calyx 1 lin. long; corolla-tube
      1½–2 lin. long   ...   ... (58) **stricta.**
§§ Leaves lax, linear or oblong :
   Leaves linear   ...   . (59) **Woodii.**
   Leaves oblong or linear-oblong ... (60) **monticola.**
Inflorescence more or less elongated, or
  the heads solitary or broad and long
  peduncled :
  Heads long peduncled :
   Branches puberulous ; plant about
    6 in. high   ...   . (61) **compacta.**
   Branches pubescent or villous ;
    plant about 1–2 ft. high   ... (62) **villosa.**
  Heads sessile or subsessile :
   Leaves ½–1¼ in. long, oblong-
    lanceolate or elliptical   ... (63) **elata.**
   Leaves 2–6 lin. long, lanceolate or
    oblong-lanceolate :
    Stems puberulous ; heads often
     rather lax   ...   . (64) **lydenbergensis.**
    Stems pubescent ; heads usually
     dense :
     Heads not solitary   ...   ... (65) **Atherstonei.**
     Heads solitary ...   . (66) **Muddii.**
    Leaves 3–6 lin. long, linear   ... (67) **Rehmanni.**

‡‡Flowers distinctly (sometimes long) pedicelled; bracts adnate to the pedicels :
  Corolla-tube 2 lin. long or under :
    Bracts broad or subobtuse :
      Corolla-tube 1 lin. long :
        Bracts suborbicular-ovate ...   ...   (68) **peduncularis.**
        Bracts elliptic-oblong   ...   ...   (69) **lepidioides.**
      Corolla-tube 1½ lin. long :
        Leaves more or less dentate   ...   (70) **Rustii.**
        Leaves entire   ...   .   (71) **Tysoni.**
    Bracts narrow or subacute :
      Leaves entire :
        Calyx-lobes broadly linear, often
          nearly as long as the bracts ...   (72) **trinervia.**
        Calyx-lobes narrowly linear, often
          not half as long as the bracts .   (73) **racemosa.**
      Leaves strongly dentate   ...   ...   (74) **longipedicellata.**
  Corolla-tube 3–3½ lin. long   ...   ...   (75) **longiflora.**
†††Flowers in solitary nodding heads   ...   ...   (76) **nutans.**
**Annuals with sessile flowers :
  Branches erect or only decumbent at the
    base :
    Flowers in broad or oblong heads :
      Leaves entire, linear   ...   .   (77) **elegans.**
      Leaves more or less toothed, lanceolate to
        spathulate:
        Leaves coriaceous   ...   .   (78) **heterophylla.**
        Leaves herbaceous :
          Leaves all more or less spathulate   ...   (79) **herbacea.**
          Upper leaves all lanceolate or narrow :
            Calyx 1½ lin. long ...   .   (80) **cephalophora.**
            Calyx ¾ lin. long ...   .   (81) **phyllopodioides.**
    Flowers in narrow spikes :
      Plant strongly pubescent ...   .   (82) **hirta.**
      Plant puberulous ...   .   (83) **hamulosa.**
  Branches invariably weak and decumbent :
    Leaves broadly spathulate, toothed ...   ...   (84) **decumbens.**
    Leaves narrowly spathulate, subentire   ...   (85) **corrigioloides.**

Section 2. SPURIÆ. Leaves not fascicled, generally more or less toothed,
never heath-like ; calyx-lobes invariably elongated and narrow, often more than
twice as long as the tube.

Branches erect or suberect; spikes or heads
  numerous :
  Inflorescence ample and loosely corymbose   ...   (86) **verbenacea.**
  Inflorescence compactly or densely corymbose :
    Leaves obovate or elliptic-oblong :
      Stems and leaves glabrous ...   .   (87) **serrata.**
      Stems and leaves pubescent   ...   ...   (88) **quadrangularis.**
    Leaves lanceolate or linear :
      Corolla-tube 2–4 lin. long :
        Leaves lanceolate, never turning black
          in drying   ...   .   (89) **Burmanni.**
        Leaves linear or lanceolate-linear, in-
          variably turning black in drying ...   (90) **spuria.**
      Corolla-tube 1½ lin. long ...   .   (91) **guttata.**
  Branches more or less decumbent at the base;
    spikes or heads usually few or solitary :
    Calyx 1 lin. long, glabrous or nearly so   ...   (92) **incisa.**
    Calyx 1¼–1½ lin. long :

1. S. villicalyx (Rolfe) ; perennial, branched chiefly at the base,
about 6–12 in. high ; branches puberulous ; leaves in approximate
fascicles, crowded, lanceolate or linear-lanceolate, subobtuse, puberulous, 4–9 lin. long ; heads roundish, dense, usually numerous and

aggregated into small compact panicles at the summit of the branches; bracts lanceolate-oblong, subacute, villous, strongly ciliate, 1¾–2 lin. long; calyx 1¼ lin. long, villous; lobes oblong, subobtuse, strongly ciliate, unequal, about as long as the tube; corolla pink (*Wood*); tube linear-oblong, 2¼–2½ lin. long; lobes broadly-oblong, scarcely half as long as the tube.

EASTERN REGION: Natal; near Kar Kloof, 3000–4000 ft., *Wood*, 4453! Kar Kloof, *Rehmann*, 7394! between Kar Kloof and the Umgeni River, *Rehmann*, 7427! and without precise locality, *Sutherland!*

2. **S. pachypoda** (Rolfe) : perennial, branched chiefly at the woody sometimes thickened base, ½–1 ft. high; branches minutely puberulous; leaves in approximate fascicles, linear or lanceolate-linear, subacute, minutely puberulous, 3–9 lin. long; heads roundish, dense, often numerous and arranged in small compact thyrsoid panicles at the summit of the branches; bracts lanceolate-oblong, subobtuse, ciliate, 1½–1¾ lin. long; calyx 1 lin. long, pubescent; lobes oblong, subobtuse, strongly ciliate, unequal, nearly as long as the tube; corolla lilac (*Wood*) or dark blue (*Tyson*); tube linear-oblong, 1½–2 lin. long; lobes broadly oblong, unequal, nearly half as long as the tube; fruit ovoid-oblong, 1 lin. long.

EASTERN REGION: Tembuland; Bazeia Mountain, 4000 ft., *Baur*, 610! Pondoland; Faku's Territory, *Sutherland!* Griqualand East; summit of Ingeli Mountains, 6000 ft., *Tyson*, 1237! 1324! *Tyson in MacOwan and Bolus, Herb Norm. Aust.-Afr.*, 1342! sides of the Zuurberg Range, 3500 ft., *Tyson*, 1713! Natal; near the summit of Amawahqua Mountain, 6800 ft., *Wood*, 4575! among stones near Curries Post, 5000 ft., *Schlechter*, 6807!

3. **S. Cooperi** (Rolfe); perennial, much branched, 1½–2 ft. high; branches minutely puberulous; leaves more or less fascicled, usually lax, narrowly lanceolate or linear-lanceolate, acute, minutely puberulous, 4–9 lin. long; heads roundish, small, numerous, aggregated into a compact narrow one-sided panicle; bracts lanceolate or lanceolate-oblong, subobtuse, puberulous, ciliate, 1¼–2½ lin. long; calyx 1–1¼ lin. long, villous; lobes oblong, obtuse, ciliate, about as long as the tube; corolla-tube linear-oblong, 1½–2 lin. long; lobes broadly oblong, unequal, about a third as long as the tube.

CENTRAL REGION: Albert Div.; without precise locality, *Cooper*, 602! 1378!
EASTERN REGION: Pondoland; Faku's Territory, *Sutherland!* Natal; on the Rovelo hills, at 70.0 ft., *Sutherland!*

4. **S. Sandersoni** (Rolfe); perennial, branched chiefly towards the base, 1–1½ ft. or more high; branches puberulous; leaves in approximate fascicles, linear, subobtuse, puberulous, 3–7 lin. long; heads roundish, 3–4 lin. long, numerous, and arranged in a narrow or somewhat thyrsoid panicle towards the summit of the branches; bracts oblong or lanceolate-oblong, subobtuse, ciliate, 1¼–1½ lin. long; calyx 1 lin. long, villous; lobes oblong, subobtuse, about a third as

long as the tube : corolla mauve (*Evans*); tube oblong, 1–1½ lin. long; lobes broadly oblong, unequal, about half as long as the tube.

KALAHARI REGION : Basutoland ; without precise locality, *Cooper*, 3011 ! Transvaal, *Sanderson !*
EASTERN REGION : Natal ; below Mont-aux-Sources, 7000–8000 ft., *Evans*, 755 ! and without precise locality, *Sanderson*, 185 ! in Dublin Herbarium, but is probably from the Transvaal.

5. **S. Barbula** (Harv.); perennial, more or less branched towards the base, 1–2 ft. high (*Gerrard*); branches puberulous ; leaves more or less fascicled, linear-lanceolate, subobtuse, puberulous, 3–5 lin. long; heads roundish, small, numerous, arranged on short lateral branchlets and forming long and narrow one-sided panicles towards the summit of the branches; bracts oblong, subobtuse, concave, somewhat ciliate, 1 lin. long; calyx ¾ lin. long, puberulous ; lobes oblong, obtuse, strongly ciliate or barbate, nearly as long as the tube ; corolla deep blue (*Gerrard*); tube linear-oblong, 1½–1¾ lin. long ; lobes broadly oblong, about a third as long as the tube.

KALAHARI REGION : Transvaal; Ingoma Hill, in the Vryheid district, *Gerrard*, 1241 !

6. **S. capitellata** (Schlechter in Journ. Bot. 1897, 345) ; perennial, much branched, 1–1½ ft. high ; branches minutely puberulous ; leaves more or less fascicled, sometimes crowded, linear, subacute, minutely puberulous, 3–8 lin. long ; heads roundish, small, numerous, and arranged in a narrow more or less compact panicle towards the summit of the branches ; bracts lanceolate or oblong, subobtuse, 1 lin. long; calyx ¾ lin. long ; lobes oblong, obtuse, strongly ciliate ; corolla blue (*Galpin*); tube oblong, 1 lin. long; lobes broadly oblong, unequal, more than half as long as the tube.

KALAHARI REGION : Transvaal : hill-sides at Johannesburg, 7000 ft., *Galpin*, 1398 ! Lydenburg district, near Paarde Plaats, *Wilms*, 1171 ! 1172 !
EASTERN REGION : Natal ; Imbazami River, *Nelson*, 15 ! and without precise locality, *Wood*, 3905 ! *Mrs. Saunders*, 159 !

7. **S. Galpinii** (Schlechter in Journ. Bot. 1897, 281) ; perennial, branched chiefly at the woody base, about 6 in. high ; branches, puberulous; leaves numerous, somewhat fascicled, linear, subacute, minutely puberulous, 4–6 lin. long ; heads globose, dense, numerous and arranged in compact narrow panicles at the summit of the branches ; bracts oblong-lanceolate, subobtuse, concave, ciliate, 1½ lin. long ; calyx 1 lin. long, villous ; lobes oblong, subobtuse, about as long as the tube ; corolla purple (*Galpin*); tube linear-oblong, 1½ lin. long ; lobes broadly oblong, unequal, about half as long as the tube.

COAST REGION : Queenstown Div. ; summit of Hangklip Mountain, 6600 ft., *Galpin*, 1508 !
KALAHARI REGION : Orange River Colony, without precise locality, *Thomas !*

8. S. lithospermoides (Rolfe); perennial, more or less branched, ½ ft. or more high ; branches pubescent; leaves more or less fascicled, oblong, subobtuse, closely pubescent or velvety, 2½–4 lin. long ; bracts oblong, subobtuse, pubescent, 1½–2 lin. long ; calyx 1 lin. long, pubescent; lobes oblong, subobtuse, strongly ciliate, shorter than the tube ; corolla-tube oblong, 1 lin. long ; lobes broadly oblong, nearly as long as the tube.

EASTERN REGION : Natal; in the Rovelo hills at 7000 ft., *Sutherland!*

9. S. Holubii (Rolfe in Dyer, Fl. Trop. Afr. v. 271); perennial, much branched, ½–1½ ft. high ; branches closely puberulous or nearly tomentose ; leaves more or less fascicled, sometimes crowded, linear or oblong-linear, subobtuse, closely pubescent or velvety, 2–5 lin. long ; heads roundish or short, numerous, racemosely disposed on short lateral branchlets near the summit of the branches; bracts oblong, obtuse, hispidulous, 1¼–1½ lin. long; calyx 1 lin. long, puberulous ; lobes linear, subobtuse, ciliate, longer than the tube ; corolla-tube oblong, 1½ lin. long ; lobes broadly oblong, a quarter as long as the tube ; fruit broadly ovoid-globose, somewhat compressed, ½–¾ lin. long.

COAST REGION : Albany Div.; Grahams Town, without collector! (the correctness of this record seems open to question. It is not improbable that it may have been collected by *Burke* in the Transvaal).

KALAHARI REGION : Bechuanaland ; Barolong Territory, *Holub!* Batlapiu Territory, *Holub!*

Also in Tropical Africa.

10. S. Flanaganii (Rolfe); perennial, branched chiefly towards the base, 1 ft. or more high ; branches pubescent; leaves in approximate fascicles, oblong or linear-oblong, subobtuse, pubescent, 2–4 lin. long ; spikes oblong, numerous, aggregated into a thyrsoid panicle at the summit of the branches ; bracts oblong-lanceolate, subobtuse, pubescent, 2 lin. long; calyx 1 lin. long, pubescent ; lobes oblong, subacute, ciliate, about as long as the tube ; corolla blue (*Flanagan*) ; tube linear-oblong, 2½–3 lin. long ; lobes broadly oblong, unequal, about a third as long as the tube.

KALAHARI REGION : Orange River Colony ; summit of Mont-aux-Sources, at 9500 ft., *Flanagan*, 2108 !

11. S. pubescens (Rolfe) ; perennial, much branched, 1 ft. or more high ; branches softly pubescent ; leaves in approximate fascicles, oblong-lanceolate, subobtuse, pubescent, 3–7 lin. long ; heads short or somewhat elongated, forming narrow, somewhat lax panicles near the apex of the branches; bracts elliptic-lanceolate or oblong-lanceolate, subacute, strongly ciliate; calyx ¾–1 lin. long ; lobes unequal, strongly ciliate, longer than the tube ; corolla-tube oblong, 1¼ lin. long ; lobes unequal, nearly as long as the tube.

COAST REGION : George Div. ; Woodville, 800 ft., *Galpin*, 4401 !

12. S. Schlechteri (Rolfe); perennial, much branched, 1 ft. or more high; branches densely puberulous; leaves more or less fascicled, linear or lanceolate-linear, subobtuse, minutely puberulous, 3–8 lin. long; flowers mostly aggregated in small compact thyrsoid panicles at the ends of the branches; bracts linear, subacute, incurved, villous, 1½ lin. long; calyx ½ lin. long; lobes oblong, subobtuse, strongly ciliate, longer than the tube; corolla-tube oblong, 1 lin. long; lobes broadly oblong, unequal, about half as long as the tube; fruit ovoid, ¾ lin. long.

EASTERN REGION: Natal; among stones near Curries Post, 5000 ft., *Schlechter*, 6810!

13. S. Burchellii (Rolfe); perennial, much branched, 2–3 ft. high (*Burchell*); branches pubescent, chiefly in decurrent lines from the leaf-bases; leaves more or less fascicled, sometimes crowded, lanceolate-linear, subacute, minutely puberulous, 3–5 lin. long; heads roundish, small, numerous and aggregated into compact narrow panicles at the summit of the branches; bracts lanceolate-oblong, subobtuse, slightly ciliate, 1–1¼ lin. long; calyx ¾ lin. long; lobes triangular-oblong, subacute, ciliate, longer than the tube; corolla purple (*Burchell*); tube oblong, ¾ lin. long; lobes broadly oblong, unequal, nearly as long as the tube.

COAST REGION: George Div.; between Touw River and Kaymans River, *Burchell*, 5775! near George, *Burchell*, 5997!

14. S. Forbesii (Rolfe); perennial, much branched, 1½ ft. or more high; branches puberulous, or sometimes pubescent in decurrent lines from the leaf-bases; leaves in approximate fascicles, crowded, linear, subobtuse, hispidulous, 2–4 lin. long; flowers arranged in compact or dense thyrsoid panicles at the summit of the branches; bracts lanceolate-oblong, subobtuse, slightly ciliate at the base, 1–1¼ lin. long; calyx 1 lin. long; lobes oblong, subobtuse, slightly ciliate, nearly as long as the tube, corolla-tube oblong, 1–1½ lin. long; lobes broadly oblong, more than half as long as the tube.

COAST REGION: Uitenhage Div.; Algoa Bay, *Forbes!* Port Elizabeth Div.; Cape Recife, *Burchell*, 4383!

15. S. canescens (Linn. fil. Suppl. 284); perennial, much branched, 1–1¼ ft. high; branches puberulous or pubescent; leaves in approximate fascicles, linear, subobtuse, puberulous, 2–3 lin. long; flowers arranged in compact or somewhat lax thyrsoid panicles at the summit of the branches; bracts lanceolate or lanceolate-oblong, subobtuse, strongly ciliate at the base, 1–1¼ lin. long; calyx ¾ lin. long; lobes broadly oblong, subobtuse, strongly ciliate, unequal, nearly as long as the tube; corolla-tube oblong, 1–1¼ lin. long; lobes broadly oblong, unequal, nearly as long as the tube. *Thunb. Prodr. 98, and Fl. Cap. ed. Schultes,* 460; *Rolfe in Journ. Linn. Soc.* xx. 349 (*not of E. Meyer and Choisy*).

SOUTH AFRICA : without precise locality, *Thunberg!*
COAST REGION : Swellendam Div. ; Grootvaders Bosch Mountains, *Bowie !*
Port Elizabeth Div. : near Port Elizabeth, *Holub!* at 1100 ft., *West in Mac Owan, Herb. Aust.-Afr* , 1939 !

This species has been much confused, and most of the specimens referred to *S. canescens,* Linn. fil., by authors belong to n. 33, *S. albida,* Choisy.

16. S. ramulosa (E. Meyer, Comm. 265, not of Link) ; perennial, much branched, 1–2 ft. or (ex *Burchell*) up to 5 ft. high ; branches puberulous or pubescent ; leaves in approximate fascicles, linear, subobtuse, minutely puberulous or hispidulous, 3–5 lin. long ; flowers arranged in compact or somewhat lax panicles at the end of the branches, bracts oblong-lanceolate, obtuse, strongly ciliate at the base, 1–1½ lin. long ; calyx nearly ¾ lin. long ; lobes very unequal, oblong, broader at the base, subobtuse, strongly ciliate, shorter than the tube ; corolla lilac (*Galpin*) ; tube oblong, 1 lin. long ; lobes broadly oblong, unequal, about as long as the tube. *E. Meyer in Drège, Zwei Pflanzengeogr. Documente,* 120, 219 ; *Walp. Rep.* iv. 160. *S. Meyeri, Choisy in DC. Prod.* xii. 8.

COAST REGION : Swellendam Div. ; near Karmelks River, below 1000 ft., *Drège!* George Div. ; near Woodville, 700 ft., *Tyson in MacOwan and Bolus, Herb. Norm. Aust.-Afr.,* 984! Humansdorp Div. ; hill-sides at Humansdorp, 300 ft., *Galpin,* 4402!

S. ramulosa, Link, remains altogether doubtful. See note at the end of the genus.

17. S. villicaulis (Rolfe) ; perennial, much branched, 1½ ft. or more high ; branches villous ; leaves in approximate fascicles, linear, subobtuse, puberulous, 3–5 lin. long ; flowers in a compact thyrsoid or narrow panicle at the summit of the branches ; bracts oblong-lanceolate, subacute, villous, 1¼–1¾ lin. long ; calyx ¾ lin. long, villous ; lobes oblong, subacute, nearly as long as the tube ; corolla-tube oblong, 1½ lin. long ; lobes broadly oblong, unequal, nearly as long as the tube.

COAST REGION : Knysna Div. ; near the Gowkamma River, *Burchell,* 5598! Knysna, *Pappe!*

18. S. linearis (Rolfe) ; perennial, much branched, 1 ft. or more high ; branches puberulous ; leaves in approximate fascicles, linear, subobtuse, puberulous or hispidulous, 2–3 lin. long ; spikes short, terminal ; bracts linear or subulate, subobtuse, hispidulous or nearly glabrous, 1½ lin. long ; calyx ¾ lin. long ; lobes oblong, subobtuse, ciliate, about as long as the tube ; corolla-tube oblong, 1 lin. long ; lobes broadly oblong, unequal, rather shorter than the tube.

COAST REGION : Uniondale Div. ; Long Kloof, near Ongelegen, *Bolus,* 2426!

This specimen is in poor condition.

19. S. Thunbergii (Choisy in DC. Prod. xii. 9) ; perennial, much branched, 1–2 ft. or more high ; branches puberulous ; leaves in approximate fascicles, linear, subobtuse, puberulous, 3–4 lin. long ; flowers in small heads arranged in ample rather dense elongated panicles towards the summit of the branches; bracts lanceolate-oblong, subobtuse, ciliate, 1–1¼ lin. long ; calyx ¾ lin. long; lobes oblong, subobtuse, ciliate, about half as long as the tube ; corolla-tube ¾–1 lin. long ; lobes broadly oblong, unequal, nearly as long as the tube. *S. glabrata, var.* β, *Choisy in Mém. Soc. Phys. Genèv.* ii. ii. 105, *excl. syn., and in Mém. Selag.* 35. *S. tephrodes, Drège, in Linnæa*, xx. 202 (*not of E. Meyer*).

SOUTH AFRICA : without precise locality, *Thom!* and cultivated specimens ! COAST REGION : Caledon Div. ; Zwarteberg, near the hot springs, 1000–2000 ft., *Zeyher*, 3574 ! between Bot River and the Zwart Berg, *Ecklon & Zeyher!* Swellendam Div. ; Grootvaders Bosch, *Bowie!* George Div. ; Outeniqua Mountains, *Mund & Maire*, 155 ! Knysna Div. ; Plettenberg Bay, *Bowie!*

20. S. glabrata (Choisy in Mém. Soc. Phys. Genèv. ii. ii. 104 ? excl. syn. and var. β) ; perennial, much branched, 1–1½ ft. or more high ; branches puberulous ; leaves in approximate fascicles, linear, subobtuse, nearly glabrous, revolute at the margin, 2–4 lin. long ; flowers in elongated lax spikes arranged in a narrow lax panicle towards the summit of the branches ; bracts linear, subacute, nearly glabrous, 1 lin. long ; calyx ½ lin. long, nearly glabrous ; lobes oblong, subobtuse, glabrous or slightly ciliate, about as long as the tube ; corolla-tube oblong, 1–1¼ lin. long ; lobes broadly oblong, unequal, about half as long as the tube ; fruit ovoid-globose, ½–¾ lin. long. *Choisy in Mém. Selag.* 34, *excl. syn. and var.* β ; *Walp. Ann.* iv. 166, *partly ; Choisy in DC. Prod.* xii. 9 *partly ; Hochst. in Flora*, 1845, 69 ; *Krauss, Fl. Cap. und Natal.* 133.

SOUTH AFRICA : without precise locality, *Bowie!* COAST REGION : Caledon Div. ; at Caledon, *Alexander!* Swellendam Div. ; in Grootvaders Bosch, *Cape Herb.!* Riversdale Div. ; near Zoetemelks River, *Burchell*, 6636 ! Aasvogel Berg, *Zeyher* (*ex Choisy*) ; foot of Aasvogel Berg, *Krauss*, 1099. Uitenhage Div. ; near " Kreg," *Zeyher* (*ex Choisy*).

This species has been much confused, and it is not absolutely certain what the original was. The variety β was afterwards referred to the preceding species by the author himself.

21. S. laxiflora (Choisy in DC. Prod. xii. 8) ; perennial, much branched, 1 ft. or more high ; branches puberulous or pubescent ; leaves in approximate fascicles, filiform or linear, subobtuse, puberulous, 2–3 lin. long; flowers arranged in lax panicles at the summit of the branches ; bracts ovate-elliptic, acute or apiculate, villous, 1¼ lin. long ; calyx ¾ lin. long, villous ; teeth oblong, obtuse, about half as long as the tube ; corolla-tube oblong, 1 lin. long ; lobes oblong, unequal, about as long as the tube. *S. geniculata, Choisy in Mém. Soc. Phys. Genèv.* ii. ii. 102, *excl. var.* β (*ex Choisy*).

SOUTH AFRICA : without precise locality, *Lambert* (*ex Choisy*), *Zeyher* (*ex Choisy*), *Nelson!* COAST REGION : Clanwilliam Div. ; Brak Fontein, near Olifants River, *Zeyher*, 26 !

22. S. tephrodes (E. Meyer, Comm. 264, partly); perennial, much branched, $\frac{1}{2}$–$1\frac{1}{2}$ ft. high; branches canescent or puberulous; leaves in approximate fascicles, filiform or linear, subobtuse, puberulous, 2–5 lin. long; flowers numerous, arranged in ample lax, or rarely smaller and more compact panicles towards the end of the branches; bracts ovate-oblong, subobtuse, pubescent, $1\frac{1}{4}$–$1\frac{1}{2}$ lin. long, adnate at the base to the short pedicel; calyx $\frac{3}{4}$ lin. long, pubescent; teeth oblong, subobtuse, a third as long as the tube; corolla-tube oblong, $1\frac{1}{4}$–$1\frac{1}{2}$ lin. long; lobes broadly oblong, more than half as long as the tube. *E. Meyer in Drège, Zwei Pflanzengeogr. Documente*, 97, 219, *letter a only; Walp. Rep.* iv. 159, *partly; Rolfe in Journ. Linn. Soc.* xx. 349, *partly. S. stricta, Choisy in DC. Prod.* xii. 8, *partly* (*not of Berg.*).

SOUTH AFRICA: without precise locality, *Masson! Alexander!*
COAST REGION: Malmesbury Div.; on stony hills near Malmesbury, 600 ft., *Bolus*, 4318! Paarl Div.; near Paarl, 500–1000 ft., *Drège!* among stones on Paarl Mountains, 700 ft., *Bolus*, 2889! *MacOwan & Bolus, Herb. Norm. Aust.-Afr.*, 847! sandy plains at Paarl, *Niven*, 79! Caledon Div.; between Bot River, and Caledon, *Cape Herb.!*

S. stricta, Hochst in Flora, 1845, 68, and Krauss, Fl. Cap. und Natal. 132, from COAST REGION: Cape Div.; sides of Tyger Berg, *Krauss*, 1089, may belong here, but 1 have not seen it. The species has been much confused.

23. S. adpressa (Choisy in Mém. Soc. Phys. Genèv. ii. ii. 103, t. 4); perennial, much branched, 1–$1\frac{1}{2}$ ft. high; branches minutely puberulous; leaves in approximate fascicles, spreading, linear, subobtuse, minutely puberulous, $1\frac{1}{2}$–2 lin. long; heads roundish, subsessile, or on short lateral branchlets, forming a compact narrow panicle at the summit of the branches; bracts oblong-lanceolate, acute, puberulous and ciliate, $1\frac{1}{4}$–$1\frac{1}{2}$ lin. long; calyx 1 lin. long, villous; lobes oblong, subobtuse, ciliate, nearly as long as the tube; corolla-tube oblong, 1–$1\frac{1}{4}$ lin. long; lobes broadly oblong, unequal, half as long as the tube. *Choisy, Mém. Selag.* 33, *t.* 4; *Walp. Rep.* iv. 161; *Choisy in DC. Prod.* xii. 8; *Hochst. in Flora*, 1845, 69; *Krauss in Fl. Cap. und Natal.* 133; *Rolfe in Journ. Linn. Soc.* xx. 349, *in note. S. tephrodes, E. Meyer, Comm.* 264, *partly, and in Drège, Zwei Pflanzengeogr. Documente*, 87, 219, *letter b only.*

SOUTH AFRICA: without precise locality, *Thunberg! Niven! Nelson! Mund! Verreaux! Miss Cole!*
COAST REGION: Malmesbury Div.; Mooresbury, *Bachmann*, 725! Cape Div.; Signal Hill, behind Sea Point, Cape Peninsula, *Wolley Dod*, 3529! sides of Tyger Berg, *Krauss*, 1108, Paarl Div.; Paarl Mountains, 1500–2000 ft., *Drège!* Paarl, *Alexander!* Julbagh Div.; Winterhoeks Berg, 900 ft., *Bolus*, 5216! *Pappe!* Worcester Div.; hills near Worcester, *Cape Herb.!* Stellenbosch Div.; near Somerset West, *Ecklon & Zeyher!* Stellenbosch, *Harvey!*

24. S. hermannioides (E. Meyer, Comm. 267); perennial, much branched, 1 ft. or more high; branches canescent or minutely puberulous; leaves solitary or somewhat fascicled, linear-oblong, subobtuse, tomentose or velvety, 3–6 lin. long; flowers arranged in small compact thyrsoid panicles at the summit of the branches; bracts elliptic-oblong, subobtuse, concave, villous, $1\frac{1}{2}$–2 lin. long; calyx 1 lin. long, villous; lobes triangular-oblong, obtuse, unequal, about as long as the tube; corolla-tube oblong, 2 lin. long; lobes broadly oblong, unequal, about half as long as the tube. *E. Meyer in Drège, Zwei Pflanzengeogr. Documente,* 95, 219; *Walp. Rep.* iv. 162; *Choisy in DC. Prod.* xii. 16.

SOUTH AFRICA: without precise locality, *Masson!*
WESTERN REGION: Little Namaqualand; near Uitkomst, 2000–2500 ft., *Drège!*

25. S. pinguicula (E. Meyer, Comm. 255); perennial, much branched, up to 6 in. or more high; branches sparsely puberulous; leaves fascicled, oblong-linear, obtuse, somewhat thickened, minutely puberulous or hispidulous, 3–4 lin. long; spikes oblong, short; bracts ovate-oblong, subobtuse, somewhat thickened, hispidulous, 2 lin. long; calyx $1\frac{1}{2}$ lin. long, somewhat villous; lobes oblong, obtuse, strongly ciliate, longer than the tube; fruit ovoid-globose, 1 lin. long. *Walp. Rep.* iv. 151; *Choisy in DC. Prod.* xii. 12.

SOUTH AFRICA: without precise locality, *Drège!*

I have only seen fruiting specimens of this species.

26. S. ovata (Rolfe); perennial, branched chiefly at the base, 1 ft. or more high; branches minutely canescent or puberulous; leaves solitary or fascicled, ovate-oblong, minutely canescent or hispidulous, 1–2 lin. long; spikes oblong, $\frac{1}{2}$–$1\frac{1}{2}$ lin. long, mostly terminal; bracts ovate-oblong, subobtuse, concave, somewhat thickened, ciliate near the base, 2–3 lin. long; calyx 2 lin. long; lobes oblong, obtuse, ciliate, about a third as long as the tube; corolla-tube oblong, 2 lin. long; lobes broadly oblong, nearly half as long as the tube.

CENTRAL REGION: Calvinia Div.; Hantam Mountains, *Meyer!*

27. S. namaquensis (Schlechter in Engl. Jahrb. xxvii. 189); perennial, much branched, 6–8 in. or more high; branches minutely puberulous; leaves generally more or less fascicled, lanceolate-oblong, subobtuse, minutely puberulous, 3–7 lin. long; spikes oblong, numerous, forming compact narrow panicles at the summit of the branches; bracts elliptic-oblong, subobtuse, minutely puberulous, ciliate at the base, $1\frac{1}{4}$–$1\frac{3}{4}$ lin. long; calyx 1 lin. long, puberulous; lobes oblong, subobtuse, strongly ciliate, unequal, about as long as the tube; corolla-tube oblong, $1\frac{1}{2}$ lin. long; lobes broadly oblong, unequal, about half as long as the tube.

WESTERN REGION: Vanrhynsdorp Div.; Karee Bergen, 1200 ft., *Schlechter,* 8179!

28. S. robusta (Rolfe); perennial, much branched, 3 ft. high (*Burchell*); branches stout, closely and minutely puberulous; leaves more or less fascicled, linear, obtuse, revolute at the margin, glabrous, 2–8 lin. long; spikes roundish or oblong, dense, ½–1 in. long, solitary or rarely subpaniculate by the addition of a few spikelets on short lateral branches; bracts oblong or ovate-oblong, obtuse, glabrous, slightly ciliate near the base, 2 lin. long; calyx 1¼ lin. long, slightly pubescent; lobes triangular-oblong, subacute, ciliate, unequal, rather longer than the tube; corolla white (*Burchell*); tube oblong, 2 lin. long; lobes broadly oblong, subequal, scarcely a third as long as the tube; fruit ovoid-oblong, 1¼ lin. long.

CENTRAL REGION: Fraserburg Div.; at the Zak River, *Burchell*, 1513!

29. S. speciosa (Rolfe); perennial, stout, much branched, 1–1½ ft. high (*Bowker*); branches minutely canescent; leaves in approximate fascicles, linear, subacute, revolute at the margin. minutely canescent, 2–6 lin. long; spikes short, usually arranged in a small compact thyrsoid panicle at the summit of the branches; bracts oblong or ovate-oblong, obtuse, minutely ciliate at the base, 1–1¼ lin. long; calyx campanulate, 1 lin. long; lobes oblong, obtuse, ciliate, nearly as long as the tube; corolla white; tube oblong, 1½–1¾ lin. long; lobes broadly oblong, subequal, scarcely half as long as the tube.

COAST REGION: Queenstown Div.: on hill-sides on the flats of the Zwart Kei River, *Mrs. Bowker*, 312! Fiuchams Nek, 3900 ft., *Galpin*, 1936!

30. S. linearifolia (Rolfe); perennial, much branched, 1 ft. or more high; branches canescent or puberulous; leaves in approximate fascicles, linear, subobtuse, puberulous, 5–8 lin. long; flowers arranged in small compact thyrsoid panicles at the summit of the branches; bracts oblong-lanceolate, subobtuse, puberulous, 2 lin. long; calyx 1 lin. long, puberulous; lobes oblong, subobtuse, unequal, ciliate, twice as long as the tube; corolla-tube oblong, 2 lin. long; lobes broadly oblong, unequal, scarcely half as long as the tube.

COAST REGION: Clanwilliam Div.; Bull Hoek, 600 ft., *Schlechter*, 8379!

31. S. Saundersiæ (Rolfe); perennial, much branched, 1 ft. or more high; branches canescent or minutely puberulous; leaves in approximate fascicles, linear or oblong-linear, subobtuse, canescent or minutely puberulous, 2–5 lin. long; heads rounded. numerous, small, aggregated into compact narrow panicles towards the summit of the branches; bracts oblong-lanceolate, subacute, 1–1½ lin. long; calyx ¾ lin. long, puberulous; lobes broadly triangular-oblong, subacute, ciliate, unequal, about as long as the tube; corolla-tube oblong, 1 lin. long; lobes broadly oblong, unequal, rather shorter than the tube; fruit ovoid, ¾ lin. long.

EASTERN REGION: without precise locality, *Mrs. Saunders*, 68! (*in Wood Herb.* 3882)!

32. S. Bolusii (Rolfe); perennial, much branched, 1–1½ ft. high; branches minutely canescent; leaves in approximate fascicles, linear, subobtuse, nearly glabrous, 2–3 lin. long; spikes roundish or oblong, 3–4 lin. long, solitary or arranged in small panicles at the end of the branches; bracts linear-oblong, subobtuse, nearly glabrous, 1–1¼ lin. long; calyx ¾ lin. long, pubescent; lobes broadly-oblong, obtuse, strongly ciliate, subequal, much longer than the tube; corolla white; tube oblong, 1¼ lin. long; lobes elliptic-oblong, unequal, shorter than the tube; fruit ovoid-oblong, ⅜ lin. long. *S. distans, Lindl. in Bot. Reg.* 1845, *t.* 46 (*not of E. Meyer*).

CENTRAL REGION: Graaff Reinet Div.; at the summit of mountains near Graaff Reinet, 4300 ft., *Bolus,* 695!

Mr. Bolus notes that in other localities the plant occurs up to an altitude of 8000 ft.

33. S. albida (Choisy in DC. Prod. xii. 18); perennial, much branched, 1–2 ft. or up to 4 or 5 ft. (*Cooper*) high; branches minutely canescent; leaves in approximate fascicles, linear or oblong-linear, subobtuse, more or less canescent or minutely scaberulous, 2–5 lin. long; spikes roundish or oblong, often numerous and arranged in a more or less elongated panicle at the summit of the branches; bracts linear, subobtuse, pubescent, 1–1¼ lin. long; calyx ¾–1 lin. long, pubescent or villous; lobes oblong, subobtuse, rather longer than the tube; corolla blue or white; tube oblong, ¾–1¼ lin. long; lobes broadly oblong, unequal, about ½ as long as the tube; fruit ovoid-oblong, ½–1¼ lin. long. *S. fruticosa, Thunb. Prodr.* 98, *and Fl. Cap. ed. Schult.* 460, *not of Linn.; Rolfe in Journ. Linn. Soc.* xx. 355. *S. canescens, E. Meyer, Comm.* 265, *and in Drège, Zwei Pflanzengeogr. Documente,* 50, 55, 60, 90, 219; *Walp. Rep.* iv. 160; *Choisy in DC. Prod.* xii. 10 (*not of Linn. fil.*). *S. cinerascens, Choisy in DC. Prod.* xii. 9 (*not of E. Meyer*). *S. capituliflora, Rolfe in Journ. Linn Soc.* xx. 355.

SOUTH AFRICA: without precise locality, *Thunberg! Masson! Forster! Zeyher,* 1374! 1379!
COAST REGION: Queenstown Div.; Queenstown, *Cooper,* 3013!
CENTRAL REGION: Calvinia Div.; Brandvlei, *Johanssen,* 1! Prince Albert Div.; on the Karroo near Constable, 3200 ft., *MacOwan and Bolus Herb. Norm. Aust.-Afr.* 676! Karroo Veld, *Hutton!* Div.? Elephant River, *Mund!* Karroo, *Ecklon!* Somerset Div.; upper part of Bruintjes Hoogte, *Burchell,* 3042! Somerset East, *Bowker,* 145! Fraserburg Div.; at Dwaal River, *Burchell,* 1464! Sutherland Div.; Roggeveld Mountains, *Burchell,* 1310/2! Murraysburg Div.; on stony cliffs near Murraysburg, 4000 ft., *Tyson,* 268! Graaff Reinet Div.; on stony slopes near Graaff Reinet, 4000 ft., *Bolus,* 32! Graaff Reinet, *Bowker!* Compass Berg, 6000–7000 ft., *Drège!* Aberdeen Div.; Camdeboo, on hills near Hamerkuil, 3000 ft., *Drège.* Colesberg Div.; Colesberg, *Shaw,* 54! 57! Middelburg Div.; Sneeuwberg Range, between Compass Berg and Rhinoster Berg, 5000 ft., *Drège.* Conway Farm, *Gillfillan in Herb. Galpin,* 2991! Sneeuw Berg, *Wyley!* Rhinoster Kop and Buffel Fontein, *Wyley!* Craddock Div.; without precise locality, *Cooper,* 503! Albert Div.; without precise locality, *Cooper,* 576! 1755! on dry rocky hills near Gaatje, 4500–5000 ft., *Drège!*
WESTERN REGION: Great Namaqualand; Keetmanshoop, *Fenshel,* 105! Aus, *Schinz,* Gamokab, *Schinz!* Daberas, *Fleck,* 434! Little Namaqualand; between Koussies (Buffels) River and Silver Fontein, 1500–2000 ft., *Drège!*

KALAHARI REGION: Griqualand West Div.; plains between Kloof Village and Wittewater, *Burchell*, 2097 ! at Griqua Town, *Burchell*, 2114! near Griqua Town, *Orpen in Herb. Bolus*, 5749! between Griqua Town and Witte Water, *Burchell*, 1975 ! Orange River Colony ; Caledon River, *Burke* ! Vaal River, *Mrs. Bowker*, 476 ! Bloemfontein, *Rehmann*, 3714 ! 3894 ! Basutoland ; without precise locality, *Cooper*, 3012 !

Also in Tropical Africa.

34. S. minutissima (Choisy in Mém. Soc. Phys. Genèv. ii. ii. 100, t. 3) ; perennial, much branched, 1–2 ft. high ; branches minutely canescent or puberulous ; leaves in approximate fascicles, oblong or linear, obtuse, minutely puberulous or nearly glabrous, $\frac{3}{4}$–4 lin. long ; spikes capitate or short, on short lateral branchlets, arranged in a narrow more or less compact and elongated panicle towards the summit of the branches ; bracts linear-oblong, subobtuse, 1–1$\frac{1}{2}$ lin. long, nearly glabrous ; calyx $\frac{3}{4}$–1 lin. long ; lobes broadly oblong, obtuse, slightly ciliate, nearly as long as the tube ; corolla white (*Burchell*) ; tube oblong, 1–1$\frac{1}{4}$ lin. long ; lobes broadly oblong, subequal, scarcely half as long as the tube. *Choisy, Mém. Selag.* 30, t. 3 ; *E. Meyer, Comm.* 268, *and in Drège, Zwei Pflanzengeogr. Documente,* 54, 66, 91, 219 ; *Walp. Rep.* iv. 163 ; *Choisy in DC. Prod.* xii. 13. *S. appressa, E. Meyer, Comm.* 267, *not S. adpressa, Choisy.*

COAST REGION : Clanwilliam Div. ; Zuur Fontein, 150 ft., *Schlechter*, 8554 ! Malmesbury Div. ; Hopefield, *Bachmann*, 110 ! Leliefontein, near Hopefield, *Bachmann*, 1600 ! between Leliefontein and Rondekuil, near Hopefield, *Bachmann*, 2194 ! Worcester Div. ; mountain sides in Hex River Valley, 1600 ft., *Tyson*, 671 !

CENTRAL REGION: Prince Albert Div. ; between Hex River Mountains and Driekoppen, 2500–3000 ft., *Drège*. Fraserburg Div. ; between Klein Quaggas Fontein and Dwaal River, *Burchell*, 1452! Carnarvon Div. ; between Carnarvon and Elands Valley, *Burchell*, 1581 ! Hanover Div. ; on rocks between Riviertje and Nieuwkerks Hoogte, 4000–5000 ft., *Drège* ! Colesberg Div. ; near the Orange River, *Knobel* ! Middelburg Div.; Sneeuwberg Range, between Compass Berg and Rhenoster Berg, 5000 ft., *Drège* !

WESTERN REGION : Little Namaqualand ; on rocky hills near Uknip, 2000 ft., *Drège* ! among stones near Klip Fontein, 3000 ft., *MacOwan and Bolus Herb. Norm. Aust.-Afr.*, 679 ! and without precise locality, *Morris in Herb. Bolus.* 5753 ! *Scully*, 77 !

35. S. divaricata (Linn. f. Suppl. 284) ; perennial, much branched, 1 ft. or more high ; branches minutely canescent or puberulous ; leaves fascicled, linear or oblong-linear, subobtuse, minutely pubescent, 1–2 lin. long ; spikes subcapitate or short ; bracts oblong, subobtuse, minutely puberulous, 1 lin. long ; calyx $\frac{3}{4}$ lin. long ; lobes oblong, obtuse, ciliate, rather longer than the tube ; corolla-tube oblong, $\frac{3}{4}$ lin. long ; lobes broadly oblong, half as long as the tube. *Thunb. Prodr.* 99, *and Fl. Cap. ed. Schult.* 460 ; *Rolfe in Journ. Linn. Soc.* xx. 349.

SOUTH AFRICA : without precise locality, *Thunberg* !

36. S. Burkei (Rolfe) ; perennial, much branched, 1–1$\frac{1}{2}$ ft. high ; branches canescent or minutely puberulous ; leaves in approximate

fascicles, linear, subobtuse, minutely puberulous, 1–3 lin. long;
spikes narrow, 3–6 lin. long; bracts elliptic-oblong, subobtuse,
concave, minutely puberulous, 1 lin. long; calyx $\frac{3}{4}$ lin. long,
minutely puberulous; lobes oblong, subobtuse, ciliate, rather longer
than the tube; corolla-tube oblong, 1 lin. long; lobes broadly oblong,
unequal, more than half as long as the tube; fruit ovoid-oblong,
1 lin. long.

SOUTH AFRICA: without precise locality, *Zeyher*, 1379!
KALAHARI REGION: Transvaal; Mooi River, *Burke*, 500!

37. S. Zeyheri (Rolfe); perennial, much branched, $\frac{1}{2}$–1 ft. high;
branches minutely puberulous, leaves in approximate fascicles,
oblong, subobtuse, minutely puberulous, $\frac{3}{4}$–1$\frac{1}{2}$ lin. long; spikes
oblong, $\frac{1}{4}$–$\frac{3}{4}$ in. long; bracts oblong or linear-oblong, subobtuse,
minutely puberulous, 1 lin. long; calyx $\frac{3}{4}$ lin. long, pubescent;
lobes oblong, subobtuse, rather shorter than the tube; corolla-tube
oblong, 1 lin. long; lobes broadly oblong or rounded, unequal, rather
shorter than the tube.

SOUTH AFRICA: without precise locality, *Zeyher*, 1375! 1378!
CENTRAL REGION: Calvinia Div.; Hantam Mountains, *Meyer!* Prince
Albert Div.; Gamka River, *Burke!* Beaufort West Div.; Beaufort West,
Schenck!

38. S. tenuis (E. Meyer, Comm. 266); perennial, much branched,
$\frac{1}{2}$–1 ft. high; branches puberulous; leaves in approximate fascicles,
spreading, oblong-linear, subobtuse, puberulous, 1–1$\frac{1}{2}$ lin. long;
spikes narrow, mostly solitary, $\frac{1}{4}$–$\frac{1}{2}$ in. long; bracts oblong,
subobtuse, fleshy, minutely puberulous, 1 lin. long; calyx $\frac{3}{4}$ lin.
long; lobes oblong, subobtuse, ciliate, unequal, about twice as long
as the tube; corolla-tube oblong, 1 lin. long; lobes broadly oblong,
shorter than the tube. *E. Meyer, in Drège, Zwei Pflanzengeogr.
Documente*, 95, 219; *Walp. Rep.* iv. 161; *Choisy in DC. Prod.*
xii. 11, *partly.*

WESTERN REGION: Little Namaqualand; in dry places at the foot of the
mountains near Geelbekskraal, 2500 ft., *Drège!*

39. S. ferruginea (Rolfe); perennial, much branched, $\frac{3}{4}$–1 ft.
high; branches pubescent or hispidulous with ferruginous hairs;
leaves in approximate fascicles, oblong, subobtuse, hispidulous,
1–2 lin. long; spikes oblong or somewhat elongate, $\frac{1}{4}$–1 in. long;
bracts triangular-oblong, subacute, hispidulous, $\frac{3}{4}$ lin. long; calyx
$\frac{3}{4}$ lin. long, pubescent; lobes narrowly triangular-oblong, subacute,
longer than the tube; corolla-tube oblong, $\frac{3}{4}$ lin. long; lobes broadly
oblong, subequal, not half as long as the tube; fruit ovoid-oblong,
$\frac{3}{4}$ lin. long. *S. leptostachya, var. eckloniana, Choisy in DC. Prod.*
xii. 11.

COAST REGION: Div.? between Gauritz River and Lange Kloof, *Ecklon &
Zeyher*, 66!

40. S. rigida (Rolfe); perennial, much branched, $\frac{1}{2}$–1$\frac{1}{2}$ ft. high;
branches puberulous or hispidulous; leaves in approximate fascicles,

spreading, hispidulous, 1–2 lin. long ; spikes ovoid or oblong, dense, 4–12 lin. long; bracts ovate-lanceolate or lanceolate-oblong, sub-obtuse, puberulous, ciliate at the base, about 1¼ lin. long; calyx 1 lin. long; lobes oblong, subobtuse, ciliate, the two upper about as long as the tube, the rest smaller; corolla white (*Burchell*) ; tube oblong, 1–1¼ lin. long ; lobes broadly oblong, unequal, scarcely half as long as the tube.

CENTRAL REGION : Sutherland Div. ; Roggeveld Mountains, *Burchell*, 1310/1 ! on the Wind Heuvel, Koedoes Mountains, *Burchell*, 1287 ! between Kuilenberg and Great Riet River, *Burchell*, 1348 !

41. **S. polystachya** (Linn. Mant. 250, not of E. Meyer) ; perennial, much branched, 1–2 ft. high ; branches canescent or minutely puberulous ; leaves in approximate fascicles, linear or oblong-linear, subobtuse, canescent or sometimes hispidulous, 3–5 lin. long ; flowers in short oblong spikes, which are sometimes numerous and aggregated into compact corymbose panicles at the summit of the branches ; bracts lanceolate-linear, subobtuse, ciliate near the base, 1½–2 lin. long ; calyx 1¼ lin. long, villous ; lobes oblong-linear, subobtuse, longer than the tube ; corolla-tube linear-oblong, 2½–3 lin. long ; lobes broadly oblong, about a quarter as long as the tube. *Choisy, Mém. Selag.* 40 ; *Rolfe in Journ. Linn. Soc.* xx. 348. *S. cinerea, E. Meyer, Comm.* 263, *and in Drège, Zwei Pflanzengeogr. Documente,* 78, 109, 219 (*not of Linn. fil.*) ; *Walp. Rep.* iv. 158 ; *Choisy in DC. Prod.* xii. 10 *partly ; Rolfe in Journ. Linn. Soc.* xx. 348.

SOUTH AFRICA : without precise locality, *Masson ! Nelson ! Mund ! Thom ! Harvey,* 451 !

COAST REGION : Clanwilliam Div. ; in sandy places between Berg Vallei and Zwartbast Kraal, 600–1000 ft., *Drège !* between Piquiniers Kloof and Pretoris Kloof, 1500 ft., *Drège !* mountains near Oliphants River, *Bolus,* 5748 ! Piquet-berg Div. ; Piquet Berg, *Schlechter,* 7916 ! Cape Div. ; Cape Flats, *Rehmann,* 1955 ! 1959 ! Tulbagh Div. ; New Kloof, 1000 ft., *Drège.* Stellenbosch Div. ; near Somerset West, *Cape Herb. !* George Div. ; Outeniqua Mountains, in Montagu Pass, *Rehmann,* 290 !

S. cinerea, Hochst. in Flora, 1845, 69, and Krauss, Fl. Cap. und Natal. 133, from among shrubs near the Koega River in Uitenhage Div., *Krauss,* 1090, may belong here. The species has been much confused by authors.

42. **S. scabrida** (Thunb. Prodr. 99) ; perennial, much branched, ½–1½ ft. high ; branches puberulous or hispidulous ; leaves more or less fascicled (the fascicles more or less approximate) ; oblong or linear-oblong, subobtuse, scaberulous or hispid, 1½–5 lin. long ; heads roundish or oblong, mostly numerous and arranged in compact or somewhat lax corymbs at the summit of the branches ; bracts linear-lanceolate, subobtuse, hispidulous or pubescent, 2–2½ lin. long ; calyx 1½–2 lin. long, villous ; lobes subulate-linear, subacute, twice as long as the tube ; corolla-tube linear-oblong, 1½–2 lin. long ; lobes broadly oblong, subequal, about a third as long as the tube ; fruit ovoid-oblong, 1 lin. long. *Thunb. Fl. Cap. ed. Schult.* 461 ; *Choisy, Mém. Selag.* 41 ; *E. Meyer, Comm.* 265, *and in Drège, Zwei Pflanzengeogr. Documente,* 74, 115, 219 ; *Walp. Rep.* iv. 160, 167 ;

Choisy in DC. Prod. xii. 13 ; *Hochst. in Flora,* 1845, 70 ; *Krauss, Fl. Cap. und Natal.* 134 ; *Rolfe in Journ. Linn. Soc.* xx. 352. *S. cinerea, Drège in Linnæa,* xx. 201 (*not of Linn. fil.*). *S. glandulosa, Choisy in DC. Prod.* xii. 20 ; *Rolfe in Journ. Linn. Soc.* xx. 352.

SOUTH AFRICA : without precise locality, *Thunberg ! Thom ! Harvey ! Wallich !*

COAST REGION : Clanwilliam Div. ; on rocks near Wupperthal, 2500 ft., *Drège !* Sneeuw Kop, *Wallich !* Pakhuis Berg, 2600 ft., *Schlechter,* 8658 ! Piquetberg Div. ; Kardouw, *Zeyher,* 3553 ! Malmesbury Div. ; near Hopefield, *Bachmann,* 1598 ! 1599 ! Cape Div. ; Muizenberg Mountain, *Wallich,* 379 ! *Wolley Dod,* 749 ! *Bolus,* 2892 ! 4533 ! flats near Rondebosch, *Wolley Dod,* 320 ! sandy ground near Simonstown, *Wolley Dod,* 428 ! top of Steen Berg slopes, *Wolley Dod,* 2736 ! *Ecklon,* 31 ! Blauw Berg, *Cape Herb. !* sides of Tyger Berg, *Krauss,* 1092 ! Stellenbosh Div. ; Lowrys Pass, *Pappe !* Caledon Div. ; on the great mountain at Genadendal, *Burchell,* 8628 ! mountains near Genadendal, 2000–4800 ft., *Drège ! Bolus,* 7411 ! 7412 ! Grabouw, near the Palmiet River, *Bolus,* 4186 ! Houw Hoek, *Zeyher,* 3560 ! *Schlechter,* 7559 ! Caledon, *Pappe !* Swellendam Div. ; hills of Swellendam, *Bowie !* near Swellendam, *Bolus,* 8072 ! *Eklon & Zeyher,* 31 ! Riversdale Div. : at Riversdale, *Schlechter,* 1871 !

KALAHARI REGION : Basutoland ; Kornet Spruit, between the Orange River and Caledon River, at the foot of the Witte Bergen, 5000–6000 ft., *Ecklon & Zeyher !*

43. S. luxurians (Choisy in DC. Prod. xii. 9) ; perennial, much branched, about 1 ft. high ; branches closely pubescent ; leaves in approximate fascicles, spreading, oblong-linear, subobtuse, nearly glabrous, 1½–2 lin. long ; spikes oblong, about ½ in. long, terminal and lateral near the apex of the branches ; bracts oblong, subobtuse, ciliate, 1–1¼ lin. long ; calyx campanulate, pubescent, 1 lin. long ; lobes oblong, subobtuse, ciliate, twice as long as the tube ; corolla-tube oblong, 1½ lin. long ; lobes broadly oblong, subequal, about half as long as the tube ; fruit ovoid, 1 lin. long.

COAST REGION : Mossel Bay Div. ; Attaquas Kloof Mountains, *Mund & Maire* ; Attaquas Kloof, *Gill !* Uitenhage Div. ; Van Stadens Berg, *Ecklon & Zeyher,* 54 (*ex Choisy*).

44. S. Dregei (Rolfe in Journ. Linn. Soc. xx. 353) ; perennial, much branched, 1–1½ ft. high ; branches pubescent ; leaves in approximate fascicles, spreading, oblong-linear, subobtuse, hispidulous, 1–4 lin. long ; flowers arranged in dense corymbose panicles 1–3 in. broad at the summit of the branches ; bracts linear-lanceolate, subacute, villous, 2 lin. long ; calyx 1¼ lin. long, villous ; lobes subulate-linear, acute, three times as long as the tube ; corolla white (*Burchell*) ; tube linear-oblong, about 2 lin. long ; lobes broadly oblong, unequal, scarcely half as long as the tube ; fruit ovoid, 1 lin. long. *S. glomerata, E. Meyer, Comm.* 264, *partly, and in Drège, Zwei Pflanzengeogr. Documente,* 128, 219, *letter b only ; Drège in Linnæa,* xx. 202 ; *Walp. Rep.* iv. 159 ; *Choisy in DC. Prod.* xii. 9, *partly* (*not of Thunb.*).

SOUTH AFRICA : without precise locality, *Thom,* 293 ! 323 !

COAST REGION : Cape Div. ; near Cape Town, *Harvey !* Riversdale Div. ; hills near Zoetemelks River, *Burchell,* 6748 ! Uitenhage Div. ; between Klasniemand Fontein and Bethelsdorp, 500–800 ft., *Drège !* Kakkerlak Valley and summit of

Van Stadens Berg, *Zeyher*, 3564! and without precise locality, *Zeyher*, 807 in Kew Herb.!
This species has been confused with *S. glomerata*, Thunb.

45. S. setulosa (Rolfe); perennial, much branched, 1–1¼ ft. high; branches setulose-pubescent; leaves in approximate fascicles, spreading, ovate-oblong or suborbicular, obtuse, puberulous, often setulose at the margin, 1½ lin. long: bracts oblong or lanceolate-oblong, subobtuse, very villous, 1½–2 lin. long; calyx 1½ lin. long, villous; lobes linear-oblong, subobtuse, nearly as long as the tube; corolla-tube linear-oblong, 2½–3 lin. long; lobes broadly oblong, about 1 lin. long.

SOUTH AFRICA: without precise locality, *Thom*, 46!

46. S. brevifolia (Rolfe); perennial, much branched, 1 ft. or more high; branches pubescent; leaves in approximate fascicles, oblong, subobtuse, somewhat fleshy, nearly glabrous, ¾–1 lin. long; spikes oblong, short, solitary or somewhat aggregated at the summit of the branches; bracts linear-lanceolate, subacute, hispid, ciliate, 1½–2 lin. long; calyx 1½ lin. long, villous; lobes linear, subobtuse, twice as long as the tube; corolla white (*Galpin*); tube linear-oblong, 2 lin. long; lobes broadly oblong, unequal, about half as long as the tube; fruit oblong, ¾ lin. long.

COAST REGION: Riversdale Div.; Muiskraal, near Garcias Pass, 1200 ft., *Galpin*, 4405!

47. S. glomerata (Thunb. Prodr. 99); perennial, much branched, 1–2 ft. high; branches puberulous; leaves in approximate fascicles, oblong or linear-oblong, subobtuse or apiculate, minutely puberulous or glabrous, 1½–5 lin. long; flowers arranged in globose or sub-corymbose, usually dense panicles ¾–1¼ in. (or occasionally more) broad; bracts lanceolate-oblong, subacute, concave, ciliate, 1–1½ lin. long; calyx 1–1½ lin. long, puberulous; lobes linear-oblong, subobtuse, ciliate, rather longer than the tube; corolla blue (*Burchell*) or white (*Galpin*); tube oblong or linear-oblong, 1–2 lin. long; lobes broadly oblong or obovate-oblong, unequal, nearly half as long as the tube. *Thunb. Fl. Cap. ed. Schult.* 461; *E. Meyer, Comm.* 264, *and in Drège, Zwei Pflanzengeogr. Documente*, 121, 137, 219 (*excl. letter b*); *Walp. Rep.* iv. 159; *Choisy in DC. Prod.* xii. 9 (*partly*); *Rolfe in Journ. Linn. Soc.* xx. 353. *S. sp., Drège in Linnæa*, xx. 202. *S. cærulea, Burch. ex Hochst. in Flora*, 1845, 70; *Krauss, Fl. Cap. und Natal.* 134; *Choisy in DC. Prod.* xii. 9 (*in note*).

SOUTH AFRICA: without precise locality, *Thunberg! Thom*, 503! 512!
COAST REGION: Ceres Div.; plains near Ceres, 1500 ft., *MacOwan and Bolus, Herb. Norm. Aust.-Afr.*, 1341! Prince Alfred, 3000 ft., *Schlechter*, 9984! Worcester Div.; Hex River Mountains, at Axells Farm, *Rehmann*, 2699! Hex River Valley, *Tyson in Herb. Bolus*, 5964! Caledon Div.; mountains of Baviaans Kloof, near Genadendal, *Burchell*, 7856! Swellendam Div.; plains at Swellendam, *Bowie!* Riversdale Div.; Riversdale, *Rust*, 445! Mossel Bay Div.; Gauritz River, *Pappe!* George Div.; west side of Kaymans River, *Burchell*,

5805! Outeniqua Mountains, Montagu Pass, *Rehmann*, 291! Knysna Div.;
Knysna, *Pappe!* Uniondale Div.; Lange Kloof. on rocks near the waterfall,
2000 ft., *Drège!* Lange Kloof, *Cape Herb.! Humansdorp Div.;* hill-side at
Humansdorp, 400 ft., *Galpin*, 4404! Uitenhage Div.; fields by the Zwartkops
River, *Zeyher*, 3563! and without precise locality, *Zeyher*, 807 *in Dublin Herb.!*
foot of Winterhoek Mountain, *Krauss*, 1091! Alexandria Div.; rocks on the
north side of the Zuur Bergen, *Drège!* Albany Div. ; Fish River Rand, *Burchell*,
3449!
CENTRAL REGION : Ceres Div. ; Cold Bokkeveld, at Sand Fontein, 4000 ft.,
Schlechter, 10138 !

48. S. punctata (Rolfe) ; perennial, much branched, $\frac{3}{4}$–$1\frac{1}{4}$ ft. high ;
branches puberulous ; leaves in approximate fascicles, linear or
oblong-linear, subobtuse, minutely glaucous, and bearing numerous
black dots, 2–3 lin. long ; flowers aggregated in corymbs $\frac{1}{2}$–$1\frac{1}{2}$ in.
broad at the summit of the branches ; bracts oblong, subobtuse,
concave, minutely canescent, $1\frac{1}{4}$ lin. long ; calyx 1–$1\frac{1}{4}$ lin. long,
minutely puberulous ; lobes oblong, subobtuse, rather shorter than the
tube ; corolla white (*Galpin*) ; tube oblong, 1 lin. long ; lobes broadly
oblong, unequal, about as long as the tube.

CENTRAL REGION : Queenstown Div.; Hangklip Mountain, 6600 ft., *Galpin*,
1809! Andries Berg, near Bailey, 6500 ft., *Galpin*, 1924!

49. S. longituba (Rolfe) ; perennial, over 1 ft. high ; branches
pubescent ; leaves solitary or somewhat fascicled, usually crowded,
elliptic-lanceolate, subacute, serrate, 3-nerved, hispidulous, 5–8 lin.
long ; flowers aggregated in a dense corymb 2 in. broad ; bracts
lanceolate-oblong, subobtuse, hispidulous, somewhat ciliate, $1\frac{1}{4}$ lin.
long ; calyx $1\frac{1}{4}$ lin. long, nearly glabrous ; lobes oblong, subobtuse,
ciliate, about a third as long as the tube ; corolla purple (*Galpin*) ;
tube linear, 3–4 lin. long ; lobes oblong, about a quarter as long as
the tube.

KALAHARI REGION : Transvaal, on hill-sides near Barberton, 2800–3000 ft.,
Galpin, 398!

50. S. Wilmsii (Rolfe) ; perennial, more or less branched, $1\frac{1}{2}$ ft. or
more high ; branches puberulous ; leaves in approximate fascicles,
lanceolate or oblong-lanceolate, acute, serrate or denticulate, hispidulous, 3–8 lin. long ; flowers aggregated in dense corymbs 1–$1\frac{3}{4}$ in.
broad at the summit of the branches ; bracts lanceolate, subacute,
2 lin. long ; calyx $1\frac{1}{2}$ lin. long, narrow, glabrous ; lobes oblong,
subacute, scarcely half as long as the tube ; corolla-tube linear,
$2\frac{1}{2}$ lin. long ; lobes broadly oblong, subequal, about a quarter as long
as the tube.

KALAHARI REGION : Transvaal, near Paarde Plants, in the Lydenberg
district, *Wilms*, 1163 !

51. S. natalensis (Rolfe) ; perennial, branched chiefly at the base,
$1\frac{1}{2}$–4 ft. high ; branches minutely puberulous ; leaves in approximate
fascicles, linear or lanceolate-linear, acute, denticulate above the
middle, with acute teeth, nearly glabrous, $\frac{1}{2}$–$1\frac{1}{2}$ lin. long ; flowers
arranged in dense corymbose panicles 1–4 in. broad at the summit of

the branches ; bracts linear-oblong, subobtuse, 1 lin. long, adnate at
the base to the short pedicel; calyx 1 lin. long, nearly glabrous;
lobes oblong, subacute, about a third as long as the tube ; corolla
violet-blue (*Sanderson*) ; tube linear-oblong, 2 lin. long ; lobes broadly
oblong, subequal, about a quarter as long as the tube.

EASTERN REGION: Natal; on stony slopes near Inanda, 1800 ft., *Wood*, 53!
521! Bothas Railway Station, *Wood*, 4863! Bothas Hill, overlooking Potgieters
Farm, 1200–1500 ft., *Sanderson*, 88! Transvaal; Ingoma in Vryheid District,
Gerrard & McKen, 1240!

52. S. aggregata (Rolfe); perennial, branched chiefly near the
base, 1½ ft. or more high; branches pubescent; leaves more or less
fascicled, crowded, lanceolate or oblong-lanceolate, subacute, serrate
or denticulate, puberulous, 4–8 lin. long ; flowers aggregated in
corymbose or rarely somewhat thyrsoid heads 2–3½ in. broad at the
summit of the branches; bracts linear-oblong, subobtuse, nearly
glabrous, 2 lin. long; calyx 1¼ lin. long, nearly glabrous; lobes
oblong, subobtuse, about a quarter as long as the tube ; corolla blue
(*Wilms*); tube linear-oblong, 2 lin. long ; lobes broadly oblong,
about a quarter as long as the tube.

KALAHARI REGION : Transvaal; near Lydenburg, *Wilms*, 1165! near Paarde
Plaats, in the Lydenburg District, *Wilms*, 1165a!
EASTERN REGION : Natal; near Greytown, *Wilms*, 2193!

53. S. Nelsoni (Rolfe); perennial, chiefly branched at the woody
base, ½–1½ ft. high ; branches closely pubescent; leaves in approximate fascicles, lanceolate or lanceolate-linear, subobtuse, entire or
denticulate near the apex, nearly glabrous, 2–6 lin. long; spikes
capitate or short, corymbosely arranged in compact heads ½–1½ in.
broad at the summit of the branches; bracts oblong, subobtuse,
puberulous, 1–1¼ lin. long ; calyx 1 lin. long; lobes broadly oblong,
obtuse, slightly ciliate, about a third as long as the tube ; corolla-
tube oblong, 1½–1¾ lin. long ; lobes broadly oblong, subequal, about
a third as long as the tube.

KALAHARI REGION : Transvaal; Houtbosch Berg, *Nelson*, 439! *Rehmann*,
6208! 6209! 6210!

54. S. foliosa (Rolfe); perennial, branched chiefly towards the
base, 1½–2½ ft. high ; branches pubescent ; leaves more or less
fascicled, crowded, oblong-lanceolate or elliptic-oblong, subobtuse,
pubescent, ¼–1½ in. long ; flowers aggregated in compact (rarely
somewhat lax) corymbs at the ends of the branches; bracts oblong,
obtuse, concave, puberulous and ciliate, 1¼–1½ lin. long ; calyx
1½–1¾ lin. long, minutely puberulous ; lobes oblong, obtuse, ciliate,
about a quarter as long as the tube ; corolla purple (*Tyson*) ; tube
oblong, 2 lin. long ; lobes broadly oblong, subequal, about a third as
long as the tube.

EASTERN REGION : Griqualand East; mountain slopes at Clydesdale, near the
Umzimkulu River, 3000 ft., *Tyson*, 2523! *Tyson in MacOwan, Herb. Aust.-Afr.*,
1514!

55. S. transvaalensis (Rolfe); perennial, branched chiefly near the base, over 1½ ft. high; branches pubescent; leaves in approximate fascicles, lanceolate-linear, subacute, dentate above the middle with two to four pairs of teeth, hispidulous, ½–1 in. long; flowers aggregated in a dense corymbose panicle 2–2½ lin. broad at the summit of the branches; bracts lanceolate-linear, subobtuse, puberulous, 1 lin. long; calyx nearly glabrous, 1 lin. long; lobes linear, subobtuse, ciliate, rather shorter than the tube; corolla-tube linear-oblong, 1 lin. long; lobes oblong, unequal, not half as long as the tube.

KALAHARI REGION: Transvaal; Houtbosch, *Rehmann*, 6211!

56. S. hyssopifolia (E. Meyer, Comm. 262); perennial, much branched, 1½–5 ft. high; branches pubescent or puberulous; leaves more or less fascicled, sometimes crowded, lanceolate or linear, or sometimes oblong, subobtuse, entire or more or less denticulate or dentate, softly pubescent or puberulous, ¼–1½ in. long, 1–5 lin. broad; flowers very numerous, corymbosely arranged at the summit of the branches in a broad lax panicle or more compact head; bracts elliptic-oblong, obtuse, concave, usually more or less ciliate, adnate at the base to the pedicel, 1 lin. long; pedicels ½–1 lin. long; calyx 1 lin. long; lobes broadly oblong, obtuse, minutely ciliate, shorter than the tube; corolla white or light blue; tube campanulate-oblong, 1 lin. long; lobes broadly oblong or rounded, shorter than the tube; fruit ovoid-oblong, subcompressed, 1 lin. long. *E. Meyer in Drège, Zwei Pflanzengeogr. Documente*, 150, 151, 219; *Walp. Rep.* iv. 158; *Choisy in DC. Prod.* xii. 15; *Hochst. in Flora*, 1845, 69 (*var. lanceolata*); *Krauss, Fl. Cap. und Natal*. 133.

COAST REGION: Komgha Div.; grassy hills near Komgha, 1800 ft., *Flanagan*, 581!

KALAHARI REGION: Transvaal; mountain slopes at Macamac, 5000 ft., *McLea in Herb. Bolus*, 3135! near Lydenburg, *Wilms*, 1164!

EASTERN REGION: Tembuland; Bazeia, 2500 ft., *Baur*, 559! Pondoland; among grasses, between the great waterfall and Umtsikaba River, 1000–1500 ft., *Drège !* between St. Johns River and Umtsikaba River, *Drège*. Fakus Territory, *Sutherland!* Port St. John, summit of West Gate, 1100 ft., *Galpin*, 3477! and without precise locality, *Bachmann*, 1219! 1220! Griqualand East; banks of Umzimkulu River, near Clydesdale, 2500 ft., *Tyson in MacOwan & Bolus Herb. Norm. Aust.-Afr.*, 817! and *Tyson*, 2797! summit of the Zuurberg Range between Kokstad and Clydesdale, 6000 ft., *Tyson*, 1197! Natal; Inanda, *Wood*, 48, 575! Umzumbi, *Wood*, 3018! stony places near Highlands, 5000 ft., *Schlechter*, 6850! Attercliffe, *Sanderson*, 250! at the foot of Table Mountain, *Krauss*, 379! near Durban, *Wood*, 231! *Sanderson*, 591! *Plant*, 47! *Peddie! Grant! Gerrard and McKen*, 294! Umkomaas, *McKen*, 1242! Tugela River, *Gerrard and McKen*, 1670! and without precise locality, *Cooper*, 3501! *Sutherland !* Transvaal; Ingoma in Vryheid District, *Gerrard*, 1242! Swaziland; Piggs Peak, 4000 ft., *Galpin*, 1336!

57. S. corymbosa (Linn. Sp. Pl. ed. i., 629); perennial, much branched, 1–2 ft. or more high; branches pubescent or puberulous; leaves in approximate fascicles, linear or filiform, subobtuse, puberulous or nearly glabrous, 1–4 lin. long; flowers very numerous, corymbosely arranged at the summit of the branches in a compact or

somewhat lax head $\frac{1}{2}$–4 in. broad; bracts oblong, obtuse, more or less
ciliate, slightly curved, often adnate at the base to the pedicel, $\frac{1}{2}$–$\frac{3}{4}$ lin.
long; calyx $\frac{1}{2}$ lin. long; lobes broadly oblong, obtuse, ciliate, generally
longer than the tube; corolla white; tube oblong, $\frac{1}{2}$–$\frac{3}{4}$ lin. long;
lobes broadly oblong, unequal, as long as or shorter than the tube;
fruit ovoid-oblong, $\frac{1}{2}$ lin. long. *Linn. Sp. Pl. ed.* 2, 876; *Berg. Pl.
Cap.* 156; *Ait. Hort. Kew. ed.* 2, iii. 431; *Thunb. Fl. Cap. ed.
Schult.* 459; *Choisy, Mém. Selag.* 30; *E. Meyer, Comm.* 263, *and
in Drège, Zwei Pflanzengeogr. Documente,* 55, 73, 88, 99, 100, 128,
138; *Walp. Rep.* iv. 158; *Choisy in DC. Prod.* xii. 10; *Hochst. in
Flora,* 1845, 69; *Krauss, Fl. Cap. und Natal.* 133; *Rolfe in Journ.
Linn. Soc.* xx. 342. *S. corymbosa,* β *polystachya, E. Meyer,
Comm.* 263 (*excl. syn.*); *Choisy in DC. Prod.* xii. 10.—*J. and C.
Commelin, Hort. Amstel.* ii. 79, *t.* 40.

SOUTH AFRICA: without precise locality, *Thunberg! Banks & Solander!
Forsyth! Grey! Harvey,* 401! 467! *Mund! Thom,* 536! *Villette! Walton!*
and cultivated specimens!

COAST REGION: Clanwilliam Div.; rocky places on Blue Berg, between
Bosch Kloof and Honig Vallei, *Drège!* Cape Div.; near Cape Town, *Pappe!*
between Cape Town and the foot of Table Mountain, *Burchell,* 59! Table
Mountain, *Ecklon,* 740! *Fleck,* 436! *MacGillivray,* 615! Lion Mountain and
Table Mountain, 500–1500 ft., *Drège!* Wynberg, *Wallich!* Devils Peak, *Wilms,*
3512! *Wolley Dod,* 2405! Cape Flats, *Krauss,* 1100, *Rehmann,* 1957! Paarl
Div.; among shrubs at Berg River, near Paarl, 400 ft., *Drège!* Paarl, *Elliott!*
Riversdale Div.; between Great Vals River and Zoetemelks River, *Burchell,*
6585! Mossel Bay Div.; in a dry channel of an arm of the Gauritz River,
Burchell, 6494! Knysna Div.; Melville, *Burchell,* 5440! Uniondale Div.;
Lange Kloof, at Apies River, *Burchell,* 4948! Uitenhage Div.; Van Stadens
Berg, below 1000 ft., *Drège!* between Van Stadens Berg and Bethelsdorp, below
1000 ft., *Drège!* among shrubs on the Van Stadens Mountains, *Zeyher,* 268!
Zwartkops River, *Zeyher,* 3568! 3569! Grasrug, *Baur!* Algoa Bay, *Cooper,*
1487! Albany Div.; Howisons Poort, near Grahamstown, *Baur! Hutton!*
Bothas Hill, 2000 ft., *Drège!* near Grahamstown, *Bolton!* and without precise
locality, *Williamson!* King Williamstown Div.; fields near King Williamstown,
Cooper, 40! *Tyson in MacOwan & Bolus, Herb. Norm. Aust.-Afr.,* 845! Durban,
101! British Kaffraria; without precise locality, *Cooper,* 3010!

CENTRAL REGION: Graaff Reinet Div.; Sneeuw Berg Range, 4500 ft.,
Bolus, 1977! between Compas Berg and Rhinoster Berg, 4500–5000 ft., *Drège.*
Graaff Reinet, *Bolus,* 611! Colesberg Div.; at Naauw Poort, *Burchell,* 2760!
Colesberg, *Shaw!* Albert Div.; *Cooper,* 1771!

EASTERN REGION: Pondoland; Fakus Territory, *Sutherland!* Natal, without
precise locality, *Cooper,* 3009!

S. polystachya, Hochst. in Flora, 1845, 69, and Krauss, Fl. Cap. und Natal.
133, collected by Krauss (1101) in clayey soil throughout the Swellendam
district, may belong here. *S. polystachya* of Linnæus was erroneously referred
here by E. Meyer.

58. S. stricta (Berg. Pl. Cap. 155); perennial, much branched,
$\frac{1}{2}$–$1\frac{1}{2}$ ft. high; branches puberulous or pubescent; leaves in approximate fascicles, more or less spreading, linear, subobtuse, puberulous,
2–6 lin. long; heads short, often numerous and arranged in compact
corymbose panicles at the summit of the branches, sometimes solitary;
bracts oblong-lanceolate, subacute, concave, villous, strongly ciliate,
$1\frac{1}{2}$–2 lin. long; calyx 1 lin. long, very villous, lobes triangular-
oblong, subobtuse, strongly ciliate, about as long as the tube;

corolla-tube linear-oblong, 1½–2 lin. long; lobes broadly-oblong, more than half as long as the tube. *Thunb. Fl. Cap. ed. Schult.* 461 ; *Choisy, Mém. Selag.* 33 *partly, excl. syn.; Rolfe in Journ. Linn. Soc.* xx. 346. *S. tephrodes, E. Meyer, Comm.* 264 *partly, and in Drège, Zwei Pflanzengeogr. Documente,* 78, 219, *letter c only. S. hispida, Choisy in DC. Prod.* xii. 10 *partly (not of Linn. fil.). S. hispida, var. nana, Choisy in DC. Prod.* xii. 10. *S. sp., Drège in Linnæa,* xx. 202.

SOUTH AFRICA : without precise locality, *Bergius! Thunberg! Masson!*
COAST REGION : Cape Div.; Table Mountain, 3000 ft., *Schlechter,* 126 ! Tulbagh Div. ; in rocky places, New Kloof, 800–1200 ft., *Drège! Ecklon & Zeyher,* 28! summit of Witsen Berg, *Zeyher,* 3817 ! *Pappe!* Saron, 2000 ft., *Schlechter,* 10655 !

This species has been much confused by authors.

59. S. Woodii (Rolfe) ; perennial, branched chiefly near the base, 1¾–2 ft. high; branches pubescent; leaves somewhat fascicled, linear, subacute, hispidulous, 1–6 lin. long ; flowers aggregated into a compact corymbose panicle 1¼–1½ in. broad at the summit of the branches ; bracts elliptic-oblong, subobtuse or apiculate, concave, hispidulous, 1–1¼ lin. long, adnate at the base to the pedicel ; pedicel slender, puberulous, 1–1¼ lin. long ; calyx campanulate, puberulous, 1¼–1½ lin. long ; lobes broadly oblong, obtuse, rather shorter than the tube ; corolla white (*Wood*) ; tube oblong, 1–1¼ lin. long ; lobes oblong, subequal, scarcely half as long as the tube ; fruit oblong, 1 lin. long.

EASTERN REGION : Natal; near Murchison, *Wood,* 3006 !

60. S. monticola (Wood and Evans in Journ. Bot. 1897, 489) ; perennial, branched chiefly at the woody base, 3–4 ft. high ; branches pubescent ; leaves in approximate fascicles, linear-oblong, subobtuse, pubescent or hispidulous, 3–6 lin. long ; heads roundish, 3–4 lin. diam., numerous and arranged in broad lax corymbs at the summit of the branches ; bracts oblong, subobtuse, concave, hispidulous, 1–1¼ lin. long, adnate at the base to the short pedicel ; calyx 1–1¼ lin. long, hispidulous ; lobes broadly oblong, obtuse, slightly ciliate, about half as long as the tube ; corolla white (*Evans*) ; tube oblong, 1 lin. long ; lobes broadly oblong, unequal, nearly as long as the tube ; fruit oblong, 1–1¼ lin. long.

EASTERN REGION : Natal; on the Drakensberg Range at the sources of the Inyasuti River, 6000–7000 ft., *Evans,* 655 !

61. S. compacta (Rolfe) ; perennial, branched chiefly near the base, 6 in. or more high; branches minutely puberulous ; leaves more or less fascicled, crowded, linear, subobtuse, minutely puberulous, 2–5 lin. long ; heads roundish, 4–5 lin. diam., distinctly peduncled, arranged in compact corymbs at the summit of the branches ; bracts linear-oblong, obtuse, minutely puberulous, 2 lin. long ; calyx 2 lin. long, minutely puberulous ; lobes broadly oblong or suborbicular, obtuse, a third as long as the tube ; corolla-tube linear-oblong, $2\frac{1}{2}$ lin. long ; lobes broadly oblong, subequal, about a quarter as long as the tube.

KALAHARI REGION : Transvaal; near Paarde Plaats, in the Lydenberg district, *Wilms*, 1160 !

62. S. villosa (Rolfe); perennial, much branched, 1–2 ft. or more high ; branches villous ; leaves solitary or somewhat fascicled, crowded, oblong or linear-oblong, subobtuse, villous or almost velvety, 3–9 lin. long; heads roundish, peduncled, 5–6 lin. diam., numerous towards the summit of the branches ; bracts linear-oblong, subobtuse, villous, 2 lin. long, adnate at the base to the short pedicel ; calyx 2 lin. long, villous ; lobes broadly oblong, subobtuse, a third as long as the tube ; corolla white or cream-white (*Galpin*) ; tube linear-oblong, $2\frac{1}{2}$ lin. long ; lobes broadly oblong, subequal, a third as long as the tube.

KALAHARI REGION : Transvaal; among rocks on the summit of Saddleback Mountain, near Barberton, 5000 ft., *Galpin*, 948 ! Umlomati (Lomati) Valley, Barberton, 4000 ft., *Galpin*, 1307 !

63. S. elata (Rolfe); perennial, much branched, 2–3 ft. high ; branches pubescent or puberulous ; leaves solitary or with a few small leaflets in the axils, lanceolate-oblong or elliptical, subobtuse, puberulous, $\frac{1}{2}$–$1\frac{1}{2}$ in. long ; racemes short, subcapitate when young, numerous, arranged in a lax or narrow panicle towards the summit of the branches ; bracts lanceolate-oblong, subobtuse, puberulous, 1–$1\frac{1}{2}$ lin. long ; adnate at the base to the pedicel ; pedicels up to $\frac{1}{2}$ lin. long ; calyx $1\frac{1}{2}$–2 lin. long, puberulous ; lobes linear, subobtuse, ciliate, about a third as long as the tube ; corolla white (*Thorncroft*) ; tube linear, $2\frac{1}{2}$–3 lin. long; lobes broadly oblong, subequal, about a third as long as the tube.

KALAHARI REGION : Transvaal; in scrub on a hill-side at Barberton, 3500 ft., *Galpin*, 862 ! Houtbosch, *Rehmann*, 6206 ! and without precise locality, at 5000 ft., *Thorncroft*, 287 (in *Herb. Wood*, 4347) !

64. S. lydenbergensis (Rolfe); perennial, much branched, 1 ft. or more high ; branches closely pubescent ; leaves in approximate fascicles, oblong or lanceolate-oblong, subobtuse, puberulous, 2–4 lin. long ; heads roundish, numerous, arranged in a more or less compact panicle at the summit of the branches ; bracts lanceolate-oblong, subobtuse, puberulous, adnate at the base to the very short pedicel, 1–$1\frac{1}{2}$ lin. long ; calyx 1 lin. long, puberulous ; lobes oblong, subobtuse, about a quarter as long as the tube ; corolla-tube oblong, $1\frac{1}{2}$–$1\frac{3}{4}$ lin. long; lobes broadly oblong, a quarter as long as the tube.

KALAHARI REGION : Transvaal; on mountains at Lynsklip Spruit, in the Lydenberg district, *Nelson*, 388 ! 533 !

65. S. Atherstonei (Rolfe) ; perennial, much branched, 1 ft. or more high ; branches pubescent; leaves more or less fascicled, crowded, oblong, subobtuse, pubescent, 2–6 lin. long; heads 5–6 lin. diam., dense, terminal and lateral at the summit of the branches; bracts lanceolate-oblong, pubescent, adnate at the base to the short pedicel, about 1½ lin. long; calyx 1¼ lin. long, pubescent; lobes oblong, obtuse, ciliate, about a quarter as long as the tube; corolla-tube linear-oblong, about 1¾ lin. long; lobes broadly oblong, obtuse, subequal, about a third as long as the tube.

KALAHARI REGION: Transvaal; near Lydenberg, *Atherstone!*

66. **S. Muddii** (Rolfe); perennial, much branched, 6–12 in. high; branches pubescent; leaves somewhat fascicled, rather crowded; ovate-oblong, subacute, puberulous, 2½–4 lin. long; heads mostly solitary, dense-flowered, 5–7 lin. diam.; bracts oblong-lanceolate, subobtuse, puberulous, 2 lin. long; calyx 1½ lin. long, minutely puberulous; lobes oblong, subobtuse, ciliate, about a quarter as long as the tube; corolla lilac (*Mudd*); tube linear-oblong, 1¾ lin. long; lobes broadly oblong, subequal, about a third as long as the tube.

KALAHARI REGION: Transvaal; Berg Plateau, *Mudd!*

67. **S. Rehmanni** (Rolfe); perennial, much branched, ½–¾ ft. high; branches puberulous; leaves in approximate fascicles, linear, subacute, puberulous, 3–5 lin. long; heads mostly solitary, dense-flowered, ¾ lin. diam.; bracts lanceolate-oblong, subobtuse, very minutely puberulous, 2 lin. long; calyx 1¼–1½ lin. long, nearly glabrous; lobes oblong, subobtuse, about a third as long as the tube; corolla-tube linear-oblong, 1¾ lin. long; lobes broadly oblong, subequal, about a quarter as long as the tube.

KALAHARI REGION: Transvaal; Houtbosch, *Rehmann*, 6212!

Some of the flower-heads are completely abnormal, owing to the ovaries being transformed into globose galls, as the result of puncture by insects.

68. **S. peduncularis** (E. Meyer, Comm. 262); erect, herbaceous, resembling *S. trinervia*, E. Meyer, but more slender, and with fewer branches; leaves alternate, linear, acute at both ends, with some small leaflets in the axils; flowers 1 lin. long, 20–30 arranged in subcorymbose cymes, slightly larger than those of *S. trinervia*, E. Meyer; bracts suborbicular-ovate, subacute, concave, scarcely exceeding the calyx-tube; lower pedicels 2 lin. long; calyx-lobes 5, very obtuse, as long as the corolla-tube; fruit subglobose. *Walp. Rep.* iv. 157; *Choisy in DC. Prod.* xii. 13.

EASTERN REGION: Natal; between Umzimkulu River and Umkomanzi River, 200 ft., *Drège.*

Only known to me from the description, in which it is compared with *S. trinervia*, E. Meyer.

69. **S. lepidioides** (Rolfe); annual (?), branched chiefly at the base, 1 ft. or more high; branches pubescent; leaves somewhat fascicled, linear or oblong-linear, subacute, hispidulous, 2–4 lin.

long ; flowers in racemes, which are often congested when young, but elongated in fruit up to 2 in. or more long; bracts elliptic-oblong, obtuse, concave, glabrous, adnate at the base to the pedicel, 1 lin. long; pedicels slender, 1 lin. long; calyx campanulate, $\frac{3}{4}$ lin. long, nearly glabrous ; lobes triangular-oblong, subobtuse, shorter than the tube ; corolla-tube oblong, 1 lin. long; lobes broadly-oblong, unequal, shorter than the tube ; fruit ovoid, subcompressed, 1 lin. long.

EASTERN REGION : Pondoland, *Bachmann*, 1217 ! 1221 !

70. **S. Rustii** (Rolfe) ; annual (?), branched chiefly at the base, $\frac{1}{2}$ ft. or more high ; branches puberulous ; leaves somewhat fascicled, lanceolate or lanceolate-linear, subobtuse, denticulate above the middle, puberulous, 3–5 lin. long; flowers arranged in dense heads or arrested racemes 3–4 lin. long; bracts subspathulate-oblong, obtuse, ciliate, 1$\frac{1}{2}$ lin. long, adnate at the base to the pedicel; pedicel $\frac{1}{3}$–$\frac{1}{2}$ lin. long ; calyx 1$\frac{1}{2}$ lin. long; lobes oblong, obtuse, strongly ciliate, shorter than the tube ; corolla-tube linear-oblong, 1$\frac{1}{2}$ lin. long ; lobes broadly oblong, subequal, about a third as long as the tube.

COAST REGION : Riversdale Div.; Riversdale, *Rust*, 100 !

71. **S. Tysoni** (Rolfe) ; perennial, much branched, about 1 ft. high; branches puberulous ; leaves usually solitary, sometimes with a few small leaflets in the axils, lanceolate or oblong-lanceolate, sub-obtuse, minutely puberulous, 3–6 lin. long; racemes subcapitate when young, afterwards elongated up to 2 in. ; bracts lanceolate-oblong, subobtuse, puberulous, adnate at the base to the pedicel, 2–2$\frac{1}{2}$ lin. long; pedicels $\frac{1}{2}$ lin. long; calyx 1$\frac{1}{4}$ lin. long, puberulous; lobes lanceolate-oblong, acute, ciliate, nearly twice as long as the tube; corolla-tube linear-oblong, 1$\frac{1}{2}$ lin. long; lobes broadly oblong, scarcely half as long as the tube ; fruit ovoid-oblong, 1–1$\frac{1}{4}$ lin. long.

EASTERN REGION : Griqualand East; at the summit of Mount Currie, near Kokstad, 7000 ft., *Tyson*, 1238 ! *Tyson in MacOwan and Bolus, Herb. Norm. Aust.-Afr.*, 969 !

72. **S. trinervia** (E. Meyer, Comm. 261); perennial, branched, 1 ft. or more high; branches minutely puberulous ; leaves mostly solitary, lanceolate-oblong to linear-lanceolate, subobtuse, nearly glabrous, 4–12 lin. long; racemes subcorymbose at first, elongated in fruit up to 1 in.; bracts lanceolate, subacute, adnate at the base to the pedicel, 1$\frac{1}{2}$–2 lin. long; pedicels slender, 1 lin. long; calyx 1$\frac{1}{4}$ lin. long ; lobes linear, subobtuse, ciliate, about three times as long as the tube; corolla-tube linear-oblong, 1$\frac{3}{4}$ lin. long; lobes broadly oblong, subequal, about a third as long as the tube; fruit ovoid-oblong, subcompressed, 1 lin. long. *E. Meyer in Drège, Zwei Pflan-zengeogr. Documente*, 157, 219; *Walp. Rep.* iv. 157 ; *Choisy in DC. Prod.* xii. 13.

EASTERN REGION : Natal ; between Umzimkulu River and Umkomanzi River, 200 ft., *Drège!*

73. **S. racemosa** (Bernh. in Flora, 1845, 69) ; annual or perennial (?), branched chiefly at the base, 1–1½ ft. high ; branches minutely puberulous ; leaves solitary, lanceolate or lanceolate-linear, subobtuse, puberulous, ¼–2 in. long; racemes short when young, afterwards elongated up to 3 in. ; pedicels ½–1 lin. long; bracts linear or oblong-linear, subobtuse, puberulous, 2–3 lin. long ; calyx 1–1¼ lin. long ; lobes linear, acute, ciliate, more or less spreading, two to four times as long as the tube ; corolla-tube linear-oblong, 2 lin. long; lobes broadly oblong, unequal, not half as long as the tube ; fruit ovoid-oblong, subcompressed, ¾–1 lin. long. *Walp. Rep.* iv. 157 ; *Choisy in DC. Prod.* xii. 13.

EASTERN REGION : Griqualand East ; among stones on the Zuurberg Range, 5500 ft., *Schlechter,* 6582 ! Natal; near Hermans Berg, *Gerrard,* 1244 ! in grassy places near Durban Bay, *Krauss,* 225 ! Inanda, 1800 ft., *Wood,* 64 ! 331 ! 408 ! 412 ! near Durban, *Gerrard & McKen,* 575 ! 576 ! and without precise locality, *Sanderson,* 139 ! *Sutherland ! Gerrard,* 320 !

74. **S. longipedicellata** (Rolfe) ; perennial, much branched, 1 ft. or more high ; branches pubescent ; leaves in approximate fascicles, spreading, narrowly subspathulate-lanceolate, acute, with about two pairs of acute teeth near the apex, somewhat hispidulous, 2–4 lin. long ; racemes short or elongated up to 2 in. or more in fruit ; bracts linear, acute, adnate at the base to the pedicel, 1½ lin. long ; pedicels slender, 2 lin. long ; calyx 1 lin. long ; lobes subulate-linear, acute, slightly ciliate, somewhat spreading at the apex, about twice as long as the tube ; corolla pink (*Wood*) ; tube oblong, 2 lin. long ; lobes broadly oblong, subequal, about a quarter as long as the tube ; fruit oblong. 1 lin. long.

EASTERN REGION : Zululand ; Entumeni, 2000–3000 ft., *Wood,* 3967 !

75. **S. longiflora** (Rolfe); perennial, much branched, 1½–2 ft. or more high ; branches pubescent ; leaves in approximate fascicles, spreading, linear-lanceolate, subacute, puberulous or hispidulous, 2–3 lin. long ; racemes oblong or elongated in fruit up to 4 or 5 in. ; bracts linear-lanceolate, acute, somewhat ciliate, 2–2½ lin. long, adnate at the base to the pedicel ; pedicels about ½ lin. long ; calyx 1–1¼ lin. long ; lobes linear, acute, ciliate, about twice as long as the tube ; corolla blue (*Wood*); tube linear, somewhat pubescent, 3–3½ lin. long ; lobes broadly elliptic-oblong, unequal, about 1 lin. long; fruit ovoid-oblong, 1¼ lin. long.

EASTERN REGION : Natal; near Enon, Upper Illovo, 3000 ft., *Wood,* 1858 ! near Byrne, 3000 ft., *Wood,* 326 ! *Wylie,* 5217 ! and without precise locality, *Mrs. K. Saunders!*

76. **S. nutans** (Rolfe in Journ. Linn. Soc. xx. 354, 358); perennial, branched chiefly at the base, 1–1½ ft. high ; branches puberulous or nearly tomentose ; leaves solitary, the lower opposite or sub-

opposite, linear or oblanceolate-linear, obtuse, sometimes denticulate towards the apex, narrowed at the base, pubescent, $\frac{1}{2}$–1 in. long; heads solitary, nodding, short or oblong, $\frac{1}{2}$–1 in. long; bracts lanceolate or elliptic-lanceolate, subobtuse or apiculate, concave, 4–5 lin. long; calyx tubular, puberulous, 3 lin. long; lobes oblong, obtuse, ciliate, a quarter as long as the tube; corolla-tube linear, 6–7 lin. long; lobes oblong or obovate-oblong, subequal, 1 lin. long. *Selago cephalophora, E. Meyer, Comm.* 256, *and in Drège, Zwei Pflanzengeogr. Documente,* 51, 53, 219 (*not of Thunb.*); *Walp. Rep.* iv. 152; *Choisy in DC. Prod.* xii. 17 *partly.*

CENTRAL REGION: Aliwal North Div.; Witte Bergen, in valleys at 4500–5000 ft., *Drège,* and in rugged grassy places at 7000–7500 ft., *Drège!*

This species has been confused with *S. cephalophora,* Thunb.

77. S. elegans (Choisy in DC. Prod. xii. 14); annual (?), somewhat branched at the base, 1–1$\frac{1}{4}$ ft. high; branches minutely puberulous, hairs more or less in lines from the leaf-bases on the lower part; leaves mostly solitary, somewhat crowded near the base, linear, subacute, minutely puberulous, 3–8 lin. long; heads ovoid or oblong, dense, $\frac{1}{2}$–1$\frac{1}{2}$ in. long; bracts lanceolate-linear, subacute, ciliate, 2–2$\frac{1}{2}$ lin. long; calyx 2 lin. long; lobes subulate-linear, subacute, ciliate, about twice as long as the tube; corolla-tube linear-oblong, about 2$\frac{1}{2}$ lin. long; lobes broadly oblong, subequal, a third as long as the tube.

COAST REGION: Clanwilliam Div.; Brak Fontein, *Cape Herb.!* Swellendam Div.; Rietkuil, near the Buffeljagts River, *Zeyher,* 3571! Oudshorn Div.; Cango, *Mund!*

78. S. heterophylla (E. Meyer, Comm. 256, not of Thunb.); annual (?), branched chiefly at the base, $\frac{3}{4}$–1$\frac{1}{4}$ ft. high; branches nearly glabrous; leaves lanceolate or linear-lanceolate, acute, often denticulate or dentate above the middle, glabrous, 2–6 lin. long; heads ovoid or oblong, dense, 5–8 lin. long; bracts lanceolate-oblong, ciliate, 2 lin. long; calyx 2 lin. long; lobes linear, subobtuse, ciliate, twice as long as the tube; corolla-tube linear-oblong, 2 lin. long; lobes oblong, unequal, shorter than the tube. *E. Meyer in Drège, Zwei Pflanzengeogr. Documente,* 74, 219; *Walp. Rep.* iv. 151; *Choisy in DC. Prod.* xii. 17; *Rolfe in Journ. Linn. Soc.* xx. 354.

COAST REGION: Clanwilliam Div.; Ezels Bank, on the Cederbergen, 3000–4000 ft., *Drège!* Tulbagh Div.; on the Skurfde Berg, *Cape Herb.!* Caledon Div.; by the River Zonder Einde, *Zeyher,* 3572!

This species was wrongly identified by E. Meyer with *S. heterophylla,* Thunb., but as the latter is a young seedling of *S. spuria,* Linn., the name *heterophylla* may be retained for the present one.

79. S. herbacea (Choisy in Mém. Soc. Phys. Genèv. ii. ii. 109); annual, erect, more or less branched, 1–1$\frac{1}{2}$ ft. high; branches puberulous; leaves mostly opposite, solitary or with some small leaves in the axils, spathulate, subobtuse, crenulate, nearly glabrous, $\frac{1}{2}$–1 in. long; heads roundish or oblong, 3–6 lin. long, more elongated in fruit, solitary, or several corymbosely arranged at the summit of the branches; bracts linear-oblong, subobtuse, ciliate at the base, 2 lin. long; calyx 1¼ lin. long; lobes linear-oblong, subobtuse, ciliate, about three times as long as the tube; corolla-tube linear-oblong, 2 lin. long; lobes broadly oblong, subequal, about a quarter as long as the tube; fruit oblong, 1½ lin. long. *Choisy, Mém. Selag.* 39; *Walp. Rep.* iv. 153; *Choisy in DC. Prod.* xii. 17; *Hochst. in Flora,* 1845, 69; *Krauss, Fl. Cap. und Natal.* 133. *S. lobeliacea, Hochst. in Flora,* 1845, 69.

COAST REGION : Knysna Div.; on dunes at Zitzikamma, *Krauss,* 1104, 1105. Uitenhage Div.; Olifants Hoek, between Bushman and Sunday Rivers, below 300 ft., *Ecklon & Zeyher,* 16! 41! Algoa Bay, *Berlin Herb.!*

80. **S. cephalophora** (Thunb. Prodr. 100); annual, erect or decumbent at the base, sometimes not much branched, ½–1¼ ft. high; branches pubescent; leaves usually solitary, the lower opposite or subopposite, oblanceolate-linear or subspathulate, subobtuse, dentate or denticulate, narrowed at the base, pubescent, ¼–1 in. long; heads dense, 3–6 lin. long, mostly numerous and corymbosely arranged at the summit of the branches; bracts oblong, subobtuse, ciliate, 2 lin. long; calyx 1½ lin. long; lobes oblong, subobtuse, ciliate, about a quarter as long as the tube; corolla-tube oblong, 1½ lin. long; lobes broadly oblong, about half as long as the tube; fruit oblong, 1¼ lin. long. *Thunb. Fl. Cap. ed. Schult.* 464; *Choisy, Mém. Selag.* 34 ; *Walp. Rep.* iv. 152 *partly; Choisy in DC. Prod.* xii. 17 *partly; Rolfe in Journ. Linn. Soc.* xx. 354 (*not of E. Meyer*).

SOUTH AFRICA : without precise locality, *Masson! Auge! Mund & Maire!* COAST REGION : Malmesbury Div.; near Hopefield, *Bachmann,* 177! 1491! 2164! scrubby flats north of Houtjes Bay, in Saldanha Bay, *Yorke!* Cape Div.; sand hills near Duine Fontein, Cape Peninsula, *Wolley Dod,* 1860! Riversdale Div.; sandy plains at Kuils (Kafferkuils?) River, *Cape Herb.!* Uniondale Div.; Lange Kloof, 300 ft., *Schlechter,* 8385!

S. cephalophora, E. Meyer, is now referred to *S. nutans,* Rolfe.

81. **S. phyllopodioides** (Schlechter in Engl. Jahrb. xxvii. 190); annual, erect, simple or branched in the upper part, 4–6 in. high; branches strigillose-puberulous with retrorse hairs; leaves narrowly oblanceolate-linear, obtuse, sparsely puberulous, lower obscurely dentate, attenuated at the base into a short petiole, 1 in. long, upper sessile, entire, smaller; flowers in terminal heads, sometimes disposed in subcorymbose panicles; bracts oblong, obtuse, hispidulous, slightly longer than the calyx; calyx somewhat bilabiate, ¾ lin. long, puberulous; upper lip shortly bifid; lower shortly trilobed; lobes obtuse, subequal; corolla 2 lin. long; tube swollen above the cylindrical base; lobes rounded, very obtuse.

COAST REGION : Vanrhynsdorp Div.; hills near Drooge River, 1200 ft., *Schlechter,* 8322.

Only known to me from the description, in which it is compared with *S. cephalophora,* Thunb.

82. S. hirta (Linn. fil. Suppl. 285); annual, decumbent, much branched, 4–9 in. high; branches pubescent; leaves opposite or sub-opposite, solitary, spathulate with a broadly elliptic-oblong blade, obtuse, crenulate or dentate, somewhat pubescent, ½–1¼ in. long; spikes narrow, elongate, dense, 1–5 in. long in fruit; bracts lanceolate-oblong, subobtuse, more or less recurved at the apex, pubescent, 1½–2 lin. long; calyx 1 lin. long, villous; lobes subulate-oblong, scarcely a quarter as long as the tube; corolla-tube oblong, ¾ lin. long; lobes oblong, minute, about a third as long as the tube; fruit ovoid, ¾–1 lin. long. *Thunb. Prodr.* 100, *and Fl. Cap. ed. Schult.* 464; *E. Meyer, Comm.* 258, *in Drège, Zwei Pflanzengeogr. Documente,* 98, 219, *and in Linnæa,* xx. 201; *Walp. Rep.* iv. 152; *Choisy in DC. Prod.* xii. 16, *excl. syn.* (*not of Mém. Selag.*); *Rolfe in Journ. Linn. Soc.* xx. 351.

COAST REGION: Malmesbury Div.; Moorrees Berg, near Hopefield, *Bachmann,* 1152! Paarl Div.; near Paarl, 400–600 ft., *Drège!* Tulbagh Div.; Ceres Road, 900 ft., *Schlechter,* 9080! Tulbagh Waterfall, *Cape Herb.!* Worcester Div.; in fields at Roode Zand, *Thunberg!* Swellendam Div.; Hassaquas Kloof, *Zeyher,* 3578!

S. hirta, Choisy, in Mém. Soc. Phys. Genèv. ii. ii. 107, is now referred to *S. quadrangularis,* Choisy.

83. S. hamulosa (E. Meyer, Comm. 257); annual, much branched, erect or somewhat decumbent, 2½–3½ in. high; branches puberulous; leaves opposite or subopposite, solitary, spathulate-lanceolate, sub-obtuse, nearly glabrous, 3–8 lin. long; spikes oblong, rather lax, 6–12 lin. long; bracts lanceolate-oblong, subobtuse, more or less recurved towards the apex, puberulous, 1½ lin. long; calyx ¾ lin. long, pubescent; lobes triangular-oblong, subacute, ciliate, shorter than the tube; corolla-tube oblong, ½ lin. long; lobes oblong, minute, subequal, scarcely a quarter as long as the tube; fruit ovoid, 1 lin. long. *E. Meyer in Drège, Zwei Pflanzengeogr. Documente,* 94, 219; *Walp. Rep.* iv. 152; *Choisy in DC. Prod.* xii. 16.

WESTERN REGION: Little Namaqualand; among grasses at Haazenkraals River, 1500–2000 ft., *Drège!*

84. S. decumbens (Thunb. Prodr. 100); annual, often much branched, diffuse or decumbent, 4–8 in. high; branches pubescent; leaves solitary or somewhat fascicled, spathulate, or petiolate with an ovate limb, subobtuse, denticulate or dentate, pubescent, 2–6 lin. long; flowers arranged in short dense heads 3–5 lin. long, mostly terminal; bracts lanceolate-oblong, subobtuse, pubescent, 1½–2 lin. long, the lower usually foliaceous; calyx 1 lin. long, villous; lobes linear-oblong, subobtuse, longer than the tube; corolla-tube oblong, 1–1¼ lin. long; lobes broadly oblong, unequal, scarcely half as long as the tube; fruit ovoid-oblong, ¾ lin. long. *Thunb. Fl. Cap. ed. Schult.* 465; *Rolfe in Journ. Linn. Soc.* xx. 354. *S. cordata, E. Meyer, Comm.* 257, *and in Drège, Zwei Pflanzengeogr. Documente,* 55,

62, 69, 219; *Walp. Rep.* iv. 152; *Choisy in DC. Prod.* xii. 16 (*not of Thunb.*).

CENTRAL REGION : Calvinia Div.; Hantam Mountains, *Meyer!* Beaufort West Div.; Nieuweld Mountains, near Beaufort West, 3000–4000 ft., *Drège!* Graaff Reinet Div.; in dry river channels near Zuure Plaats, 4000 ft., *Drège!*
WESTERN REGION : Little Namaqualand ; in rocky places on Roode Berg, 4500–5000 ft., *Drège!*

S. cordata, Thunb. Prodr. 100, is now referred to *Phyllopodium heterophyllum*, Benth.

85. S. corrigioloides (Rolfe); annual, diffuse or procumbent, much branched, 2–4 in. high ; branches minutely puberulous ; leaves solitary or slightly fascicled, subspathulate-oblong, obtuse, nearly glabrous, 2–5 lin. long ; flowers arranged in short dense heads 2–4 lin. long ; heads terminal and lateral, numerous ; bracts ovate or ovate-oblong, obtuse, puberulous, 1½ lin. long, the lower sometimes foliaceous ; calyx campanulate, puberulous, 1 lin. long ; lobes oblong, subobtuse, ciliate, shorter than the tube ; corolla-tube oblong, 1 lin. long ; lobes broadly oblong, unequal, nearly half as long as the tube.

CENTRAL REGION : Calvinia Div.; Hantam Mountains, *Meyer!* Fraserburg Div. ; at Stink Fontein, *Burchell*, 1396! Sutherland Div.; near Sutherland, between Jackals Fontein and Kuilenberg, *Burchell*, 1335 !

86. S. verbenacea (Linn. fil. Suppl. 285); annual, loosely branched, erect, 1–3 ft. or more high ; branches quadrangular, glabrous or pubescent ; leaves solitary or rarely with a few small axillary leaves, usually lax, the lower opposite or nearly so, lanceolate, oblong or obovate, acute or apiculate, strongly and acutely dentate, glabrous or pubescent, ½–4 in. long ; flowers arranged in ample lax (rarely small compact) corymbs at the summit of the branches, individual heads growing out into short spikes in fruit ; bracts linear or subulate, acute, minutely puberulous, 1½–2 lin. long ; calyx 1–1¼ lin. long, minutely puberulous ; lobes subulate or linear, acute, unequal, about three to four times as long as the tube ; corolla blue or white (*Burchell*); tube linear, 2–2½ lin. long ; lobes broadly oblong, unequal, a third to half as long as the tube. *Thunb. Prodr.* 100, *and Fl. Cap. ed. Schult.* 464; *Choisy, Mém. Selag.* 38 ; *E. Meyer, Comm.* 258, *and in Drège, Zwei Pflanzengeogr. Documente,* 79, 81, 82, 84, 87, 99, 100, 101, 219; *Walp. Rep.* iv. 153 ; *Choisy in DC. Prod.* xii. 17 ; *Hochst. in Flora,* 1845, 69 ; *Krauss, Fl. Cap. und Natal.* 133 ; *Rolfe in Journ. Linn. Soc.* xx. 350.

SOUTH AFRICA : without precise locality, *Thunberg! Masson! Nelson!* and cultivated specimens !
COAST REGION : Paarl Div. ; in moist places, among shrubs, by the Berg River, 400 ft., *Drège!* between Paarl and French Hoek, 500 ft., *Drège!* in shady places on Paarl Mountains, 900–1500 ft., *Drège! Bolus,* 4612 B ! Drakenstein Mountains, near the waterfall, at 2000 ft., *Drège! Paarl, Alexander! Elliott!* Worcester Div.; Bains Kloof, *Rehmann,* 2295! Dutoits Kloof, 1500–3000 ft., *Drège! Bolus,* 5217 B ! mountains above Worcester, *Rehmann,* 2482! Caledon Div. ; Caledon Baths, *Pappe!* mountains near Klein River, *Cape Herb.!* between Donker Hoek and Houw Hoek Mountains, *Burchell,* 8023 !

in stony places near Hangklip, 1000 ft., *Krauss*, ⁻1107, Vogelgat, 300 ft., *Schlechter*, 9527! Swellendam Div.; between Grootvaders Bosch and Zuur-braak, *Burchell*, 7262! and without precise locality, *Bowie!* George Div.; without precise locality, *Bowie!* Riversdale Div.; moist places on the lower part of the Lange Bergen, near Kampsche Berg, *Burchell*, 7014!

87. S. serrata (Berg. Pl. Cap. 159); perennial, much branched, 1–2 ft. or more high; branches stout, somewhat angular, glabrous; leaves solitary but usually crowded, cuneate-obovate, oblong or lanceolate, obtuse or subacute, serrate, crenulate or dentate, glabrous, $\frac{1}{2}$–1 in. long; flowers aggregated into dense corymbs 1–5 in. broad at the summit of the branches; bracts linear or linear-lanceolate, acuminate, glabrous or nearly so, $1\frac{1}{2}$–3 lin. long; calyx about $1\frac{1}{2}$ lin. long, minutely puberulous; lobes subulate, slightly broader at the base, unequal, about twice to four times as long as the tube; corolla crimson, purple or lilac; tube linear, 3–4 lin. long; lobes broadly oblong or rounded, unequal, about a third as long as the tube; fruit oblong, 1 lin. long. *Rolfe in Journ. Linn. Soc.* xx. 347. *S. fasciculata. Linn. Mant.* 256; *Jacq. Ic.* iii. 7, *t.* 496; *Gærtn. De Fruct.* i. 239, *t.* 51, *fig.* 6; *Lam. Ill. t.* 521, *fig.* 2; *Thunb. Prodr.* 100, *and Fl. Cap. ed. Schult.* 464; *Choisy, Mém. Selag.* 37; *Bot. Reg. t.* 184; *Lodd. Bot. Cab. t.* 1423; *E. Meyer, Comm.* 259, *excl. var. and syn., and in Drège, Zwei Pflanzengeogr. Documente,* 73, 82, 88, 109, 219; *Walp. Rep.* iv. 154, *excl. var. hirta; Choisy in DC. Prod.* xii. 15; *Hochst. in Flora,* 1845, 69; *Krauss, Fl. Cap. und Natal.* 133; *Rolfe in Journ. Linn. Soc.* xx. 348. *S. lanceolata, Choisy in DC. Prod.* xii. 15.

SOUTH AFRICA: without precise locality, *Banks and Solander! Forsyth! Grey! Hooker! Masson! Oldenburgh!* and *cultivated specimens!*

COAST REGION: Clanwilliam Div.; between Berg Vallei and Lange Vallei, near Zwartbast Kraal, 800 ft., *Drège!* near Honig Vallei, at 3000 ft., *Drège!* Cape Div.; Table Mountain, 1000–3500 ft., *Thunberg! Ecklon,* 61! *Bowie! Burchell,* 520! *MacOwan and Bolus, Herb. Norm. Afr. Aust.,* 239! *Drège!* Boers Mountain, *Wallich,* 347! Cape Town, at Stinkwater, *Rehmann,* 1236! near Cape Town, *Harvey!* Muizen Berg, 1500 ft., *Bolus,* 4535! Klaasjagers Berg, *Wolley Dod,* 293! False Bay, *Robertson!* Tulbagh Div.; Witsen Berg, *Cape Herb.!* Worcester Div.; Dutoits Kloof, near Uitkyk, 3000–4000 ft., *Drège.* Bredasdorp Div.? Elands Kloof, 1000 ft., *Schlechter,* 9752! Caledon Div.; Lowrys Pass, 2000 ft., *Schlechter,* 7235! Baviaans Kloof, near Genadendal, *Burchell,* 7878! *Krauss,* 1102, Genadendal, *Roser!* Knysna Div.; Plettenberg Bay, *Bowie!*

S. fasciculata, var. *hirta,* E. Meyer, is now referred to the following species.

88. S. quadrangularis (Choisy in DC. Prod. xii. 15); perennial, much branched, $\frac{1}{2}$–2 ft. or more high; branches somewhat quad-rangular, villous; leaves solitary or somewhat fascicled, usually crowded, cuneate-obovate, often much narrowed at the base, crenate or dentate, pubescent, $\frac{1}{2}$–$1\frac{1}{2}$ in. long; flowers aggregated in dense corymbs 1–5 in. broad at the summit of the branches; bracts linear or subulate, acute, glabrous or nearly so, $1\frac{1}{2}$–$2\frac{1}{2}$ lin. long; calyx $1\frac{1}{2}$–2 lin. long, minutely puberulous; lobes subulate, slightly broader at the base, unequal, about twice to three times as long as the tube; corolla pale pink (*Galpin*); tube linear, 2 lin. long; lobes broadly oblong, unequal, about a quarter as long as the tube; fruit oblong, 1 lin. long. *S. hirta, Choisy in Mém. Soc. Phys. Genèv.* ii. ii. 107, *and in Mém. Selag.* 37, *not of Linn. fil. S. fasciculata, var. β hirta, E. Meyer, Comm.* 259, *and in Drège, Zwei Pflanzengeogr. Documente,* 77, 88, 219; *Rolfe in Journ. Linn. Soc.* xx. 351. *S. decumbens, Choisy in DC. Prod.* xii. 18, *not of Thunb.*

SOUTH AFRICA: without precise locality, *Forster! Roxburgh! Bergius!*
COAST REGION: Cape Div.; Table Mountain. 1500–3500 ft., *Drège! Mac-Gillivray,* 654! 655! *Ecklon,* 734! *Galpin,* 4416! *Bolus,* 4503! 4612! *Harvey,* 36! Table and Devils Mountains, *Zeyher,* 3572! Newlands, near Devils Peak, *Wilms,* 3516! Constantia Berg, *Wolley Dod,* 1916! "around the Cape," *Milne,* 124! Tulbagh Div.; rugged places, New Kloof, 15J0–2000 ft., *Drège!*

89. S. Burmanni (Choisy in Mém. Soc. Phys. Genèv. ii. ii. 108); annual, branched chiefly at the base, erect, $\frac{1}{2}$–$\frac{3}{4}$ ft. high; branches minutely puberulous; leaves mostly solitary, somewhat crowded, lanceolate, acute, dentate or denticulate, glabrous or nearly so, 3–9 lin. long; flowers aggregated in dense corymbs 1–3 in. broad at the summit of the branches; bracts linear or lanceolate-linear, acute, glabrous or minutely puberulous, $1\frac{1}{2}$–2 lin. long; calyx 1–$1\frac{1}{4}$ lin. long, minutely puberulous; lobes linear or subulate, four to five times as long as the tube; corolla-tube linear, 2–$2\frac{1}{2}$ lin. long; lobes broadly oblong, very unequal, from half to nearly as long as the tube. *Choisy, Mém. Selag.* 38, *and in DC. Prod.* xii. 15. *S. arguta, E. Meyer, Comm.* 260, *and in Drège, Zwei Pflanzengeogr. Documente,* 71 ("*argentea*" *by error*), 219; *Walp. Rep.* iv. 155.

SOUTH AFRICA: without precise locality, *Burmann (ex Choisy).*
COAST REGION: Vanrhynsdorp Div.; in a valley on Gift Berg, at 1500 ft., *Drège!*

90. S. spuria (Linn. Sp. Pl. ed. i. 629); perennial or annual, branched chiefly at the base, erect, $\frac{1}{2}$–2 ft. high; branches puberulous or glabrous; leaves solitary or slightly fascicled, crowded near the base, linear or lanceolate-linear, subacute, acutely dentate in the upper part, glabrous or rarely pubescent, $\frac{1}{4}$–$1\frac{1}{4}$ in. long; heads roundish when young, elongating into spikes in fruit, numerous, and aggregated into dense corymbs $\frac{1}{2}$–5 in. broad at the summit of the branches; bracts linear or lanceolate-linear, subacute, glabrous, $1\frac{1}{2}$–3 lin. long; calyx $1\frac{1}{2}$–2 lin. long, glabrous; lobes linear or subulate, subacute, very unequal, three to four times as long as the tube; corolla purple, lilac or white; tube linear or filiform, 2–4 lin. long; lobes broadly oblong, very unequal, two to three times shorter than the tube; fruit oblong, 1 lin. long. *Linn. Sp. Pl. ed.* 2, 877; *Thunb. Prodr.* 99, *and Fl. Cap. ed. Schult.* 463; *Choisy, Mém. Selag.* 36; *E. Meyer, Comm.* 260, *and in Drege, Zwei Pflanzengeogr. Documente,* 71, 74, 78, 81, 87, 101, 106, 118, 219; *Walp. Rep.* iv. 156; *Choisy in DC. Prod.* xii. 14; *Hochst. in Flora,* 1845, 69; *Krauss, Fl. Cap. und Natal.* 133; *Rolfe in Journ. Linn. Soc.* xx. 342. *S. rapunculoides, Linn. Amœn. Acad.* iv. 319; *Thunb. Prodr.* 99 *and Fl. Cap. ed. Schult.* 463; *Choisy, Mém. Selag.* 35; *E.*

Meyer, Comm. 260. *and in Drège, Zwei Pflanzengeogr. Documente,*
76, 77, 119, 219; *Walp. Rep.* vi. 156; *Choisy in DC. Prod.* xii.
14; *Hochst. in Flora,* 1845, 69; *Krauss, Fl. Cap. und Natal.* 133;
Rolfe in Journ. Linn. Soc. xx. 344. *S. coccinea, Linn. Amœn.
Acad.* vi. 89; *Choisy, Mém. Selag.* 36; *E. Meyer, Comm.* 261,
and in Drège, Zwei Pflanzengeogr. Documente, 81, 83, 98, 101, 219;
Walp. Rep. iv. 157; *Choisy in DC. Prod.* xii. 14; *Hochst. in Flora,*
1845, 69; *Krauss, Fl. Cap. und Natal.* 133; *Rolfe in Journ. Linn.
Soc.* xx. 344. *S. heterophylla, Thunb. Prodr.* 99, *and Fl. Cap. ed.
Schult.* 463 (*not of E. Meyer*); *Rolfe in Journ. Linn. Soc.* xx. 353.
S. pallida, Salisb. Prodr. 93. *S. pulchella, Salisb. Prodr.* 93. *S.
dentata, Poir. Encycl.* vii. 57. *S. fulvomaculata, Link, Enum. Hort.
Berol.* ii. 123; *Spreng. Syst. Veg.* ii. 746. *S. teretifolia, Link,
Enum. Hort. Berol.* ii. 124; *Spreng. Syst. Veg.* ii. 746 (*not of
Walp.*). *S. densifolia, Hochst. in Flora,* 1845, 69. *S. spicata,
Hochst. ex Choisy in DC. Prod.* xii. 14.—*Burm. Rar. Afr. Pl.*
115, *t.* 42, *fig.* 1, 3.

SOUTH AFRICA: without precise locality, *Thunberg! Masson! Roxburgh!
Nelson! Oldenburgh! Bunbury! Forster! Grey! Sieber,* 135! *Thom,* 761!
Villet! Zeyher, 1386! 1387! *Harvey,* 249! 394! *Miss Cole! Niven!*
COAST REGION: Vanrhynsdorp Div.; Gift Berg, 1500–2000 ft., *Drège!* Clan-
william Div.; Ceder Bergen, 2000–3000 ft., *Drège.* Piquetberg Div; on the
top of Piquet Berg, 2000 ft., *Drège.* Malmesbury Div.; Hopefield, *Bachmann,*
1602! Cape Div.; near Cape Town, *Burchell,* 462! Cape Flats, *Harvey!
Krauss,* 1096! Table Mountain, *Burchell,* 587! *Ecklon,* 736! *Brown,* 198!
MacGillivray, 652! *Milne,* 123! Simons Bay, *MacGillivray,* 653! *Wright,* 513!
Van Camps Bay, *Burchell,* 321! False Bay, *Robertson!* road to Constantia,
Wallich! Wynberg, Wallich! Banks and Solander! Drège! Vyges Kraal,
Wolley Dod, 427! Klaver Vley, *Wolley Dod,* 1981! near Durban Road Station,
100 ft., *Bolus,* 3865! Paarl Div.; Paarl Mountains, 1500 ft., *Drège!* in sandy
places at French Hoek, 500 ft., *Drège!* plain by the Berg River, near Paarl,
400 ft., *Drège!* between Paarl and Lady Grey Railway Bridge, *Drège.* Tulbagh
Div.; in moist mountainous places between New Kloof and Tulbagh Waterfall,
1000–1500 ft., *Drège!* New Kloof Mountains, 1000 ft., *Drège! Ecklon & Zeyher!*
Great Winter Hoek, 1000–1500 ft., *Drège! Pappe!* Tulbagh Waterfall, 1200 ft.,
Pappe! Schlechter, 9054! Worcester Div.; Dutoits Kloof, 2000–3000 ft.,
Drège! Drakenstein Mountains, at 2000–3000 ft., *Drège!* Breede River, 800 ft.,
Drège! Stellenbosch Div.; near Somerset, *Ecklon & Zeyher,* 37! mountain sides,
Hottentots Holland, *Krauss,* 1098, and between Hottentots Holland and Houw
Hoek, *Krauss,* 1097; Caledon Div.; stony hill, near Ganze Kraal, *Burchell,*
7561! Baviaans Kloof, near Genadendal, *Burchell,* 7837! Donker Hoek and
Ezelsjagt Mountains, 1500–2000 ft., *Drège,* Zwarte Berg, *Thom,* 635! Genadendal,
Roser! on hills at Grietjes Gat, near Palmiet River, *Bolus,* 4184! Swellendam
Div.; on mountain ridges along the lower part of the Zonder Einde River,
Ecklon & Zeyher! plains of Swellendam, *Bowie!* Riversdale Div.; between
Little Vet River and Kampsche Berg, *Burchell,* 6888! George Div.; Wolf Drift,
Malgat River, *Burchell,* 6120! near the Great Brak River, 350 ft., *Young in
Herb. Bolus,* 5529! Cradock Berg, 800 ft., *Galpin,* 4410! plains of George,
Bowie!

A very common and polymorphic plant, which is usually divided into three
species by authors, though I cannot find a single character by which to sub-
divide it.

91. S. guttata (E. Meyer, Comm. 259); annual, branched chiefly
at the base, about 1 ft. high; branches glabrous or nearly so; leaves

lanceolate or oblong-lanceolate, acute, strongly and acutely dentate, glabrous or nearly so, 3–6 lin. long; flowers aggregated in dense corymbose heads ½–1½ in. broad at the summit of the branches; bracts linear-lanceolate, acute, glabrous, 1½ lin. long; calyx 1 lin. long, glabrous ; lobes linear or subulate, acute, three to four times as long as the tube ; corolla-tube linear, 1½ lin. long; lobes broadly oblong, three to four times as long as the tube; fruit oblong, ½ lin. long. *E. Meyer in Drège, Zwei Pflanzengeogr. Documente,* 75, 219 ; *Walp. Rep.* iv. 154 ; *Choisy in DC. Prod.* xii. 16.

COAST REGION : Clanwilliam Div. ; rocky places on the Cederberg Range near Ezelsbank, 4000–5000 ft., *Drège !* South Kloof, on descent from Sneeuw Kop, *Wallich !*

92. S. incisa (Hochst. in Flora, 1845, 69); annual, chiefly branched at the somewhat decumbent base, about ¾ ft. high ; branches somewhat angular, glabrous or somewhat pubescent; leaves oblong-lanceolate or cuneate-oblong, acute, strongly and acutely dentate, narrowed at the base, glabrous or slightly pubescent, the lower subspathulate, 2–8 lin. long; spikes oblong, up to 1 in. long, sometimes three to six arranged in a lax corymb at the summit of the branches ; bracts linear or lanceolate-linear, acute, glabrous or minutely puberulous, 1–1½ lin. long; calyx 1 lin. long, glabrous or minutely puberulous ; lobes linear, acute, three to four times as long as the tube ; corolla white (*Bolus*) ; tube linear, 1½ lin. long; lobes broadly oblong, unequal, about a quarter as long as the tube. *Krauss, Beitr. Fl. Cap. und Natal.* 133 ; *Choisy in DC. Prod.* xii. 17.

COAST REGION: Caledon Div. ; mountain slopes behind Genadendal, 2750 ft., *MacOwan & Bolus, Herb. Norm. Aust.-Afr.,* 678! sides of Baviaans Kloof, *Krauss,* 1103!

93. S. ascendens (E. Meyer, Comm. 259); annual, more or less decumbent at the base, much branched, ½–1 ft. high ; branches pubescent or puberulous; leaves cuneate-obovate or elliptic-lanceolate, acute or apiculate, strongly and acutely dentate, narrowed at the base, puberulous or pubescent, 2–9 lin. long ; spikes oblong, up to 1 in. long in fruit, sometimes aggregated in small corymbs ; bracts linear or lanceolate-linear, acute, sometimes with a pair of acute teeth near the apex, villous or nearly glabrous, 1½–2 lin. long ; calyx 1¼–1½ lin. long, villous or nearly glabrous ; teeth linear or subulate, acute, three to four times as long as the tube ; corolla purple (*Bolus*) ; tube linear, 2½–3 lin. long; lobes broadly oblong, unequal, about a third as long as the tube ; fruit narrowly oblong, 1 lin. long. *E. Meyer in Drège, Zwei Pflanzengeogr. Documente,* 82, 219 ; *Walp. Rep.* iv. 154 ; *Choisy in DC. Prod.* xii. 17.

COAST REGION : Worcester Div. ; rocky mountainous places in Dutoits Kloof, between Uitkyk and Slang Hoek, 4000 ft., *Drège !*
CENTRAL REGION : Ceres Div. ; in stony shady places on the Skurfde Berg, near Gydouw, 4800 ft., *Bolus,* 7555 ! Cold Bokkeveld, near Gydouw, 5000 ft., *Schlechter,* 10007 !

94. S. humilis (Rolfe) ; annual (?), decumbent, branched chiefly at the base, 2–6 in. high ; branches with pubescence chiefly in decurrent lines from the leaf-bases ; leaves usually solitary, crowded, oblong-lanceolate, elliptical, or the lower subspathulate, acute or subobtuse, dentate or sometimes nearly entire, puberulous or nearly glabrous, 2–8 lin. long ; spikes roundish or oblong, dense, 4–9 lin. long ; bracts lanceolate, subacute, slightly ciliate, 2–2¼ lin. long ; calyx 1½ lin. long, minutely puberulous ; lobes subulate or linear, subobtuse, about twice as long as the tube ; corolla-tube linear-oblong, 1½–2 lin. long ; lobes broadly oblong, subequal, about three or four times shorter than the tube ; fruit oblong, 1 lin. long.

CENTRAL REGION : Ceres Div.; Cold Bokkeveld, K yn Vley, 6000 ft., *Schlechter*, 10205 !

95. S. Mundii (Rolfe) ; perennial, much branched, 1–1½ ft. or more high ; branches canescent or minutely puberulous ; leaves not fascicled, lax or somewhat crowded, spreading, linear, subacute, glaucous or very minutely canescent, 2½–4 lin. long ; spikes broadly oblong, subcapitate when young, dense, up to 2 in. long ; bracts linear, subobtuse, minutely puberulous, 2–2½ lin. long ; calyx 1½ lin. long, minutely puberulous ; lobes subulate-linear, acute, ciliate, nearly twice as long as the tube ; corolla white (*MacOwan*) ; tube linear, 3–4 lin. long ; lobes broadly oblong, unequal, a third to half as long as the tube ; fruit 1–1¼ lin. long.

SOUTH AFRICA : without precise locality, *Mund & Maire !*
COAST REGION : Tulbagh Div.; in stony places on the sides of New Kloof, 900–1000 ft., *MacOwan*, 2913 ! *MacOwan, Herb. Aust.-Afr.*, 1938 ! Tulbagh Waterfall, 1000–2000 ft., *Ecklon & Zeyher*, 31 ! slopes of Winterhoek Berg, near Tulbagh, 1200 ft., *Bolus*, 5215 !

96. S. diffusa (Thunb. Prodr. 99) ; perennial, much branched, about 1 ft. high ; branches more or less diffuse, very minutely puberulous ; leaves not fascicled, numerous, somewhat lax, spreading or recurved, linear, subobtuse, canescent or minutely puberulous ; spikes oblong or nearly capitate when young, rather lax, up to ½ in. long ; bracts oblong-lanceolate, subacute, 1½–2 lin. long ; calyx 1½ lin. long, puberulous ; lobes oblong, subobtuse, ciliate, about as long as the tube ; corolla-tube oblong, 2 lin. long ; lobes broadly oblong, subequal, about a third as long as the tube. *Thunb. Fl. Cap. ed. Schult.* 461, *not of Choisy ; Rolfe in Journ. Linn. Soc.* xx. 352.

SOUTH AFRICA : without precise locality, *Masson !*
COAST REGION : Malmesbury Div.; Saldanha Bay, *Thunberg !*
S. diffusa, Choisy, is now referred to *Walafrida Kraussii*, Rolfe.

97. S. fruticosa (Linn. Mant. 87) ; perennial, much branched, 1–1½ ft. high ; branches minutely puberulous, more or less divaricate ; leaves not fascicled, imbricate or somewhat spreading, linear or oblong, obtuse, canescent or puberulous, 1–3 lin. long ; spikes broad, oblong, ½–1¼ lin. long ; bracts lanceolate-linear, subacute, puberulous or pubescent, 2–3 lin. long ; calyx 1–1½ lin. long, pubescent ; lobes

linear-oblong, subacute, strongly ciliate, rather longer than the tube ;
corolla white (*Burchell, Galpin*), or purple (*Bolus*) ; tube linear or
oblong-linear, 2–3 lin. long ; lobes broadly oblong, about a third to a
quarter as long as the tube ; fruit oblong, 1 lin. long. *Rolfe in
Journ. Linn. Soc.* xx. 347, *not of other authors. S. sp., Drège in
Linnæa,* xx. 201. *S. recurva, Choisy in DC. Prod.* xii. 19 *partly, not
of E. Meyer.*

SOUTH AFRICA : without precise locality, *Masson ! Mund ! Thom,* 29 ! 662 !
COAST REGION : Malmesbury Div. ; near Groene Kloof (Mamre), 300 ft.,
Bolus, 4319 ! Caledon Div. ; in clay soil near Caledon, 800 ft., *MacOwan and
Bolus, Herb. Norm. Aust.-Afr.,* 677 ! Zwarte Berg, 1000 ft., *Schlechter,* 10362 !
Swellendam Div. ; on a carroo-like plain below Voormans Bosch, *Zeyher,* 3570 !
Mossel Bay Div. ; between Mossel Bay and Zout River, *Burchell,* 6335 ! Klein
Berg, 800 ft., *Galpin,* 4414 ! Knysna Div. ; Plettenberg Bay, *Bowie !* Albany
Div. ; without precise locality, *Bowie !* Fort Beaufort Div. ; on hills by the Kat
River, *Cape Herb. !*

This species has been much confused by authors.

98. S. diosmoides (Rolfe) ; perennial, much branched, ½ ft. high ;
branches minutely puberulous ; leaves not fascicled, somewhat
adpressed to the stem and imbricate, linear, obtuse, hispidulous, 1 lin.
long ; heads roundish or ovoid, about 5 lin. long ; bracts linear,
subobtuse, hispidulous, 1½–2 lin. long ; calyx 1½ lin. long, very
minutely puberulous ; lobes triangular, subobtuse, ciliate, about half
as long as the tube ; corolla-tube linear-oblong, 2 lin. long ; lobes
broadly oblong, a third to half as long as the tube ; fruit ovoid,
1 lin. long.

COAST REGION : Swellendam Div. ; Swellendam, *Ecklon & Zeyher !* George
Div. ; Gauritz River, *Ecklon & Zeyher !*

99. S. fruticulosa (Rolfe) ; perennial, much branched, ½–1¼ ft.
high ; branches more or less divaricate, minutely puberulous ; leaves
not fascicled, mostly spreading or reflexed at the apex, somewhat
crowded, oblong-linear, obtuse, canescent or minutely puberulous,
¾–1¼ lin. long ; spikes broad, short or oblong, ½–¾ in. long, dense ;
bracts lanceolate-linear, obtuse and often somewhat recurved at the
apex, hispidulous or puberulous, about 1½–2 lin. long ; calyx 1 lin.
long, pubescent ; lobes oblong, subobtuse, ciliate, about as long as the
tube ; corolla-tube linear, about 2 lin. long ; lobes broadly oblong,
unequal, about a third as long as the tube. *S. triquetra, E. Meyer,
Comm.* 254, *not of Thunb. ; Walp. Rep.* iv. 150 ; *Choisy in DC. Prod.*
xii. 11 *partly ; Rolfe in Journ. Linn. Soc.* xx. 350 *partly.*

SOUTH AFRICA : without precise locality, *Masson ! Drège ! Thom,* 437 ! 584 !
Miss Cole ! Ecklon & Zeyher !
COAST REGION : Malmesbury Div. ; Hopefield, *Bachmann,* 104 ! 142 ! 1603 !
between Blauw Berg and Groene Kloof, *Cape Herb. !* Cape Div. ; sides of Devils
Mountain, *Pappe !* hills near Cape Town, 500 ft., *Bolus,* 4530 ! *Harvey,* 600 !
slopes of Lion Mountain, 250 ft., *Bolus,* 2893 ! *Wolley Dod,* 3097 ! 3097a !
Schlechter, 971 ! Paarl Div., *Elliott !* Worcester Div. ; Drakenstein Mountains
Rehmann, 2254 ! mountains above Worcester, *Tyson in MacOwan Herb.,* 2923 !
Caledon Div. ; Caledon, 1000 ft., *Schlechter,* 7596 !

This species has hitherto been confused with n. 103, *S. triquetra,* Linn. fil.

100. S. ramosissima (Rolfe); perennial, profusely branched, 4–9 in. high·; branches minutely puberulous; leaves not fascicled, crowded, imbricate, more or less adpressed to the branches, oblong, obtuse, hispidulous, ½–1 lin. long; heads roundish or shortly oblong, dense, 3–6 lin. long; bracts oblong-lanceolate, obtuse, hispidulous and ciliate, 1–1¼ lin. long; calyx 1–1¼ lin. long, closely villous; lobes oblong, obtuse, ciliate, nearly as long as the tube; corolla white (*Galpin*); tube linear-oblong, 1½ lin. long ; lobes broadly oblong, unequal, about half as long as the tube. *S. ericina, Drège in Linnæa,* xx. 201 ; *Choisy in DC. Prod.* xii. 12, *not of E. Meyer.*

COAST REGION: Cape Div.; Table Mountain, 500 ft., *Schlechter,* 160! Bredasdorp Div.; Zeekoe Vley, 100 ft., *Schlechter,* 10546! Swellendam Div. ; on mountain ridges along the lower part of the River Zonder Einde, *Zeyher,* 3573! on dry hills near the eastern bank of the Breede River, *Burchell,* 7479! Mossel Bay Div. ; near Great Brak River, 300 ft., *Galpin,* 4415!

S. ericina, E. Meyer, is now referred to *S. triquetra,* Linn. fil.

101. S. Morrisii (Rolfe); perennial, much branched, ¾–1 ft. high; branches minutely puberulous, more or less divaricate ; leaves not fascicled, more or less adpressed to the branches and imbricate, oblong-linear, obtuse, minutely puberulous, ½–¾ lin. long; spikes broad, ovoid-oblong, dense, 4–7 lin. long; bracts lanceolate-linear, subacute, not recurved at the apex, hispidulous, strongly ciliate below the middle, 1¼–1½ lin. long; calyx 1 lin. long, pubescent; lobes oblong, obtuse, strongly ciliate, rather shorter than the tube ; corolla-tube linear-oblong, 1–1¼ lin. long ; lobes broadly oblong, subequal, about a third as long as the tube.

WESTERN REGION: Namaqualand ; without precise locality, *Morris in Herb. Bolus,* 5750!

102. S. lamprocarpa (Schlechter) ; perennial, much branched, about 1 ft. high; branches pubescent or hispidulous ; leaves not fascicled, crowded, somewhat spreading, lanceolate-linear, subobtuse, hispidulous or minutely puberulous ; spikes short or oblong, dense, ¼–¾ in. long; bracts ovate-oblong, subobtuse, hispidulous, slightly ciliate at the base, 1½ lin. long; calyx 1 lin. long, villous ; lobes triangular-oblong, subacute, rather shorter than the tube ; corolla-tube oblong, 1¼ lin. long ; lobes broadly oblong, unequal, rather more than half as long as the tube.

VAR. β, **major** (Schlechter) ; leaves rather larger ; spikes bearing fewer and much larger flowers ; bracts 2½ lin. long ; calyx 1½ lin. long ; corolla-tube 2¼ lin. long; fruit ovoid-oblong, 1¼–1½ lin. long.

CENTRAL REGION: Ceres Div.; Cold Bokkeveld, on Gydouw Berg, 6000 ft., *Schlechter,* 10047! Var. *major:* Ceres Div. ; Cold Bokkeveld, on Tafel Berg, 6200 ft., *Schlechter,* 10093 !

The difference in the size of the flowers in the two forms is remarkable, but they agree well in other respects.

103. S. triquetra (Linn. fil. Suppl. 284); perennial, much branched, ½–1 ft. high ; branches puberulous; leaves not fascicled, crowded and somewhat spreading, oblong-linear, obtuse, canescent or minutely puberulous, $\frac{3}{4}$–$1\frac{1}{4}$ lin. long ; spikes narrow, oblong or somewhat elongated, dense or somewhat lax, $\frac{1}{2}$–$1\frac{1}{2}$ in. long; bracts lanceolate-linear, subobtuse, puberulous or hispidulous, ciliate near the base, $1\frac{1}{2}$ lin. long; calyx 1 lin. long, pubescent; lobes oblong, obtuse, ciliate, rather longer than the tube; corolla-tube oblong, $1\frac{1}{4}$–$1\frac{1}{2}$ lin. long; lobes broadly oblong, unequal, about a third as long as the tube. *Thunb. Prodr.* 99, *and Fl. Cap. ed. Schult.* 461 ; *Rolfe in Journ. Linn. Soc.* xx. 350 *partly, not of E. Meyer nor Choisy. S. ericina, E. Meyer, Comm.* 254, *not of Choisy.*

SOUTH AFRICA: without precise locality, *Thunberg! Masson! Roxburgh! Thom,* 341!

COAST REGION : Worcester Div.; mountains near Worcester, 800 ft., *Tyson in MacOwan and Bolus, Herb. Norm. Aust-Afr.,* 970! *and in Herb. MacOwan,* 2922! Albany Div.; road-sides near Zondagh (Sunday) River, *Bowie!*

WESTERN REGION : Vanrhynsdorp Div. ; Bokkeveld Berg, *Schenck!*

A much confused species, which I had previously failed to distinguish from the plant called *S. triquetra* by E. Meyer and Choisy. The latter is now called *S. fruticulosa,* Rolfe.

104. **S. nigrescens** (Rolfe in Journ. Linn. Soc. xx 352) ; perennial, much branched, 1–$1\frac{1}{2}$ ft. high ; branches erect, minutely puberulous ; leaves not fascicled, somewhat lax, not spreading, linear, subobtuse, somewhat keeled, hispidulous ; spikes oblong, rather dense, 1–2 in. long; bracts linear-lanceolate, subacute, ciliate, $1\frac{1}{2}$ lin. long; calyx $\frac{3}{4}$–1 lin. long, pubescent ; lobes linear, subacute, ciliate, unequal, about as long as the tube; corolla-tube linear-oblong, $1\frac{1}{2}$–2 lin. long ; lobes broadly oblong, subequal, about a third as long as the tube.

SOUTH AFRICA: without precise locality, *Thunberg!*

Only known from the original specimen in Thunberg's Herbarium which is labelled "*Selago diffusa,*" (sheet γ), though it apparently has nothing to do with the description of that species. The leaves turn black in drying.

105. **S. aspera** (Choisy in DC. Prod. xii. 12) ; perennial, much branched, $\frac{1}{2}$–1 ft. or more high; branches canescent or minutely puberulous ; leaves not fascicled, numerous and crowded, linear, subobtuse, canescent, 1–3 lin. long; spikes oblong, dense, $\frac{1}{2}$–2 in. long ; bracts linear, subacute, puberulous, ciliate, $1\frac{1}{2}$–2 lin. long; calyx 1–$1\frac{1}{4}$ lin. long, villous; lobes triangular-oblong, subacute, ciliate, nearly as long as the tube ; corolla-tube linear-oblong, 2 lin. long ; lobes broadly oblong, subequal, about a third as long as the tube.

COAST REGION : Caledon Div.; Caledon, *Cape Herb.!* Swellendam Div. ; between Kochmans Kloof and Gauritz River, *Ecklon & Zeyher !* Bathurst Div.; Port Alfred, *Souta! Div. ?*; "calcareous hills behind Kaos River," *Cape Herb. !*

106. **S. Thomii** (Rolfe) ; perennial, much branched, $\frac{3}{4}$–$1\frac{1}{4}$ ft. high ; branches very minutely puberulous ; leaves not fascicled, somewhat crowded, linear, subobtuse, keeled, hispidulous, 1–2 lin. long; spikes oblong, dense, $\frac{1}{2}$–$\frac{3}{4}$ in. long; bracts linear, subacute, hispidulous,

2 lin. long; calyx 1–1¼ lin. long, puberulous; lobes oblong, sub-obtuse, slightly ciliate, shorter than the tube; corolla white (*Burchell*); tube linear-oblong, 2 lin. long; lobes broadly oblong, subequal, about a third as long as the tube.

SOUTH AFRICA: without precise locality, *Thom*, 66! 307!

COAST REGION: Riversdale Div.; between Gauritz River and Great Vals River, *Burchell*, 6522! Riversdale, *Rust*, 161!

The leaves turn black in drying.

107. S. elata (Choisy in DC. Prod. xii. 12); perennial, much branched, ½–1 ft. or more high; branches puberulous; leaves not fascicled, numerous and crowded, linear, subacute, hispidulous, 2–5 lin. long; spikes oblong, dense, ½–1¼ in. long; bracts oblong-lanceolate, subacute, somewhat villous and strongly ciliate, 2 lin. long; calyx 1–1¼ lin. long, villous; lobes oblong, subobtuse, strongly ciliate, about as long as the tube; corolla-tube linear-oblong, 2–2¼ lin. long; lobes broadly oblong, subequal, about a third as long as the tube.

SOUTH AFRICA: without precise locality, *Mund and Maire!*

COAST REGION: Swellendam Div.; at Hassaquas Kloof, near the Breede River, *Zeyher*, 3550!

108. S. spinea (Link, Enum. Pl. Hort. Berol. ii. 123); perennial, much branched, ½–¾ ft. or more high; branches puberulous; leaves not fascicled, crowded, linear, subobtuse, hispidulous, 3–5 lin. long; spikes oblong, dense, ½–¾ in. long; bracts lanceolate, acute, puberulous and strongly ciliate, 1½–2 lin. long; calyx 1 lin. long, villous; lobes oblong, obtuse, strongly ciliate, rather shorter than the tube; corolla-tube linear-oblong, 2 lin. long; lobes broadly oblong, subequal, about a third as long as the tube. *Choisy in DC. Prod.* xii. 12. *S. pinea, E. Meyer, Comm.* 255, *and in Drège, Zwei Pflanzengeogr. Documente,* 119, 219; *Walp. Rep.* iv. 151.

COAST REGION: Tulbagh Div.; hills near Roode Zand, 500–1000 ft., *Drège!* on Witsen Berg, near Tulbagh, *Burchell*, 8706!

109. S. eckloniana (Choisy in DC. Prod. xii. 13); perennial, much branched, 1 ft. or more high; branches minutely puberulous; leaves not fascicled, numerous and crowded, spreading and somewhat recurved near the apex, linear, acute, minutely puberulous or hispidulous, 3–4 lin. long; spikes oblong, dense, 1–2½ lin. long; bracts lanceolate, acute, villous and strongly ciliate, 2–2¼ lin. long; calyx 1 lin. long, villous; lobes oblong, obtuse, ciliate, shorter than the tube; corolla-tube linear-oblong, 2–2½ lin. long; lobes broadly oblong, about a third as long as the tube. *S. pinea, Drège in Linnæa,* xx. 201, *not of Link.*

COAST REGION: Swellendam Div.; on hills at Hassaquas Kloof, near the Brede River, *Zeyher*, 3551!

110. S. glutinosa (E. Meyer, Comm. 255); perennial, much branched, ½–1½ ft. high; branches puberulous; leaves not fascicled,

.densely crowded, more or less spreading, linear, subacute, hispidulous or puberulous, 3–7 lin. long; spikes oblong, dense, 1–2 lin. long; bracts linear, subacute, pubescent or villous, $1\frac{1}{2}$–$2\frac{1}{2}$ lin. long; calyx $1\frac{1}{2}$–2 lin. long, densely villous; lobes narrowly triangular, acute, rather shorter than the tube; corolla-tube oblong, $2\frac{1}{2}$ lin. long; lobes broadly oblong, about half as long as the tube. *E. Meyer in Drège, Zwei Pflanzengeogr. Documente,* 68, 69, 73, 219; *Walp. Rep.* iv. 151; *Choisy in DC. Prod.* xii. 13.

COAST REGION: Clanwilliam Div.; Ceder Bergen, between Blue Berg and Honig Vallei, 2000–3000 ft., *Drège!* Elephants River and near Brak Fontein, *Ecklon & Zeyher!* mountain slopes at Wupperthal, 1800 ft., *MacOwan, Herb. Aust.-Afr.,* 1940! 3216! Kers Kop, near Wupperthal, 3000 ft., *Schlechter,* 8790! Worcester Div.; Hex River Valley, on Groote Tafel Berg, *Rehmann,* 2763!
WESTERN REGION: Little Namaqualand; on rocks at Roode Berg and Modderfonteins Berg, 3500–4000 ft., *Drège!* and without precise locality, *Morris in Herb. Bolus,* 5751! *Zeyher,* 3553!

111. S. hispida (Linn. fil. Suppl. 284); perennial, much branched, $\frac{1}{2}$–1 ft. or more high; branches pubescent; leaves not fascicled, numerous and crowded, spreading or recurved, linear, subacute, hispidulous or pubescent, $1\frac{1}{2}$–2 lin. long; spikes oblong, dense, $\frac{1}{2}$–$1\frac{1}{4}$ lin. long; bracts linear, subacute, pubescent or hispid, $1\frac{1}{2}$–2 lin. long; calyx 1–$1\frac{1}{4}$ lin. long, villous; lobes triangular-oblong, subacute, rather shorter than the tube; corolla-tube linear-oblong, $2\frac{1}{4}$ lin. long; lobes broadly oblong, subequal, about a third as long as the tube. *Thunb. Prodr.* 99. *and Fl. Cap. ed. Schult.* 461; *Choisy, Mém. Selag.* 32, *and in DC. Prod.* xii. 10 *partly; Rolfe in Journ. Linn. Soc.* xx. 350.

SOUTH AFRICA: without precise locality, *Thunberg!* *Masson!*
Choisy in his later monograph confounded this species with *S. stricta,* Berg., a very different plant, which again has been confused with *S. paniculata,* Thunb., (now *Walafrida paniculata,* Rolfe) and *S. tephrodes,* E. Meyer.

112. S. curvifolia (Rolfe); perennial, much branched, 1 ft. or more high; branches minutely puberulous; leaves not fascicled, numerous and crowded, spreading and more or less recurved, linear, subacute, hispidulous or minutely puberulous, $1\frac{1}{2}$–2 lin. long; spikes oblong, dense, $\frac{1}{2}$–1 in. long; bracts linear, subobtuse, hispidulous, $1\frac{1}{4}$ lin. long; calyx 1 lin. long, shortly pubescent; lobes oblong, obtuse, somewhat ciliate, about as long as the tube; corolla-tube oblong, $1\frac{1}{4}$ lin. long; lobes broadly oblong, subequal, about half as long as the tube; fruit ovoid, 1 lin. long.

SOUTH AFRICA: without precise locality, *Thom,* 327!
COAST REGION: Worcester Div.; Hex River Valley on Groote Tafel Berg, *Rehmann,* 2759!

Imperfectly known species.

113. S. abietina (Burm. fil. Fl. Cap. Prod. 17); stems prostrate, naked; branches simple, leafy; leaves imbricate, very minute; spikes capitate, sessile.

SOUTH AFRICA : without precise locality, *Burmann.*

This species probably belongs to the section *Ericoideæ*, but the description is totally inadequate to determine its affinity.

114. S. teucriifolia (Burm. fil. Fl. Cap. Prodr. 17); branched; leaves ovate, serrate; flowers terminal, sessile.

SOUTH AFRICA : without precise locality, *Burmann.*

This may be either *S. serrata*, Berg.. or an allied species, or may not belong to the genus at all, for some of Burmann's species are now transferred elsewhere.

115. S. Walpersii (Choisy in DC. Prod. xii. 20); stems suberect; branches terete, glabrous; leaves fascicled, sessile, 2 lin. or slightly more long; corolla white, becoming rose-coloured when dried; tube 3 lin. long, somewhat curved; fruit as in *S. ascendens*, E. Meyer. *S. teretifolia, Walp. Rep.* iv. 154, *not of Link.*

SOUTH AFRICA : without precise locality; described from cultivated specimens.

This should belong to the section *Spuriæ* according to the position assigned to it by Walpers, but the fascicled leaves would exclude it, and its affinity remains quite doubtful.

116. S. spuria (Lodd. Bot. Cab. t. 391, not of Linn.); a small erect shrub; branches slender; leaves apparently seldom (if at all) fascicled, linear, entire, 2–3½ lin. long; spikes oblong, about 1 in. long; corolla lilac-purple; tube slender. *Choisy in DC. Prod.* xii. 10, *in syn.*

SOUTH AFRICA : without precise locality, described from a cultivated specimen.

This is not *S. spuria*, Linn. Choisy cites it doubtfully under *S. cinerea*, but his plant of that name is *S. polystachya*, Linn., and the figure is wanting in the details which would enable the point to be settled.

117. S. comosa (E. Meyer, Comm. 255); branches elongate, flexuose, pubescent; leaves oblong, obtuse, imbricate, glabrous, with the midrib thickened below; spikes elongate, flexuose, bracts oblong-lanceolate, hirsute; calyx hispidulous; lobes acuminate, the posterior somewhat exceeding the others. *E. Meyer in Drège, Zwei Pflanzen-geogr. Documente,* 141, 219; *Walp. Rep.* iv. 150; *Choisy in DC. Prod.* xii. 12.

COAST REGION : Bathurst Div.; among grasses near Kowie River, 400–800 ft., *Drège.*

I fail to identify this species, though from the position assigned to it by E. Meyer, it ought to belong to the section *Ericoideæ.*

118. S. pterophylla (Otto ex Sweet, Hort. Brit. ed. ii. 415). Known by name only.

119. S. purpurea (Cels, Cat. Arb. et Pl. 1817, 34). Known by name only.

V. MICRODON, Choisy.

Calyx campanulate or tubular, shortly and subequally 5-toothed, adnate at the base to the bract. *Corolla-tube* cylindrical or slightly enlarged at the throat; limb spreading; lobes 5, subequal, the upper united to about the middle. *Stamens* 4, didynamous, affixed to the throat of the corolla, the outer pair slightly exserted, the other shorter than the lobes; anthers short, perfectly 1-celled. *Ovary* 2-celled, one cell smaller and with the ovule abortive; style exserted, obtuse at the apex. *Fruit* included within the calyx, semi-ovoid or oblong, one cell perfect and with crustaceous endocarp, the other smaller, membranous, empty, and usually adherent to the perfect cell.

Small, much branched heath-like shrubs. Leaves linear, oblong or suborbicular, sometimes fascicled. Spikes ovate, oblong or somewhat elongated. Bracts ovate or suborbicular, often large and spreading, adnate at the base to the calyx.

Species 5, endemic.

Inflorescence more or less elongated :
　　Leaves oblong to suborbicular　...　...　...　(1) **lucidus.**
　　Leaves linear　...　...　...　...　...　(2) **cylindricus.**
Inflorescence ovoid or suborbicular, sometimes oblong
　　in fruit :
　　　Corolla-tube 4½–6 lin. long　...　...　...　(3) **orbicularis.**
　　　Corolla-tube 2–3 lin. long :
　　　　Leaves 6–9 lin. long　...　...　...　(4) **ovatus.**
　　　　Leaves 4–5 lin. long　...　...　...　(5) **linearis.**

1. **M. lucidus** (Choisy in Mém. Soc. Phys. Genèv. ii. ii. 97, excl. syn. Linn. fil.); perennial, much branched, erect, 1–2 ft. or more high; branches more or less puberulous; leaves rarely fascicled, crowded, ovate-elliptic, oblong or sometimes suborbicular, subobtuse or apiculate, coriaceous, somewhat glaucous, 3–5 lin. long; spikes elongated or oblong, dense or somewhat lax, 1–4 in. long; bracts ovate or suborbicular-ovate, mucronate or acute, sometimes denticulate, usually shining, minutely puncticulate, 2–3 lin. long; calyx narrowly campanulate, 2 lin. long, minutely puncticulate, sometimes hispidulous; lobes triangular, acute, about a fifth as long as the tube; corolla white; tube linear-oblong, 3–4 lin. long; lobes broadly oblong, subequal, about a quarter as long as the tube. *Choisy, Mém. Selag.* 27, *excl. syn. Linn. fil.*; *E. Meyer, Comm.* 253, *and in Drège, Zwei Pflanzengeogr. Documente,* 76, 77, 202; *Walp. Rep.* iv. 168, *excl. syn. Linn. fil.*; *Choisy in DC. Prod.* xii. 22. *Selago lucida, Vent. Jard. Malmais. t.* 26. *S. bracteata, Thunb. Prodr.* 100, *and Fl. Cap. ed. Schult.* 465 ; *Rolfe in Journ. Linn. Soc.* xx. 354.

SOUTH AFRICA : without precise locality, *Thunberg! Masson! Niven! Verreaux! Wallich! Ecklon & Zeyher!*

COAST REGION : Piquetberg Div.; Piquetberg, 1500–2000 ft., *Drège!* Tulbagh Div.; rocky places in New Kloof, 1500–2000 ft., *Drège!* at Witsen Berg and Skurfde Berg, *Zeyher,* 1384! Winterhoek Berg, 1300 ft., *Bolus,* 5218! *Pappe!*

176 SELAGINEÆ (Rolfe). [*Microdon.*

CENTRAL REGION: Ceres Div.; Cold Bokkeveld, near Gydouw, 5000 ft., *Schlechter*, 10002! *Bodkin in Herb. Bolus*, 7556!

2. **M. cylindricus** (E. Meyer, Comm. 253); perennial, much branched, 1–2 ft. high; branches puberulous; leaves not or rarely fascicled, crowded, linear, subacute, glabrous or nearly so, 5–9 lin. long; spikes elongated or rarely oblong, dense, 1–4 in. long; bracts ovate, acute or mucronate, minutely puncticulate, 2–2½ lin. long; calyx 1–1¼ lin. long, glabrous or hispidulous on the angles, minutely puncticulate; lobes broadly triangular, obtuse, minute; corolla-tube slender, 3–4 lin. long; lobes broadly oblong, about a fifth as long as the tube; fruit oblong, included within the calyx. *E. Meyer in Drège, Zwei Pflanzengeogr. Documente*, 80, 82, 202; *Walp. Rep.* iv. 169; *Choisy in DC. Prod.* xii. 23. *M. Linkii, Walp. Rep.* iv. 169; *Selago polygaloides, Linn. fil. Suppl.* 284, *not of Choisy; Thunb. Prodr.* 99, *and Fl. Cap. ed. Schult.* 462; *Rolfe in Journ. Linn. Soc.* xx. 350. *S. spicata, Link, Enum. Pl. Hort. Berol.* ii. 124.

SOUTH AFRICA: without precise locality, *Thunberg! Masson! Oldenburg!*
COAST REGION: Worcester Div.; in stony places in Dutoits Kloof, 1500–3500 ft., *Drège!* 2200 ft., *Bolus*, 5219! mountains near the Hex River, *Bolus*, 6042! Hex River Mountains, at Axells Farm, *Rehmann*, 2700!

3. **M. orbicularis** (Choisy in DC. Prod. xii. 22); perennial, much branched, 1 ft. or more high; branches puberulous; leaves not or rarely slightly fascicled, generally lax, linear-oblong, subobtuse, glabrous or nearly so, 3–6 lin. long; heads rounded or oblong in fruit, 1–2 in. long; bracts orbicular or ovate-orbicular, obtuse or abruptly apiculate, membranous, more or less shining and minutely puncticulate, 3–5 lin. long; calyx oblong, somewhat swollen near the base, shining, 1¾–2 lin. long; lobes triangular-oblong, subacute, short; corolla-tube slender, 4½–6 lin. long; lobes broadly oblong, about a quarter as long as the tube; fruit oblong, 1 lin. long, included within the calyx.

SOUTH AFRICA: without precise locality, *Masson!*
COAST REGION: Clanwilliam Div.; Oliphants River, 500 ft., *Schlechter*, 8478! Tulbagh Div.; without precise locality, *Ecklon & Zeyher!*

4. **M. ovatus** (Choisy in Mém. Soc. Phys. Genèv. ii. ii. 97, t. 1, fig. 4); perennial, much branched, 1–2 ft. high; branches puberulous or nearly pubescent; leaves in approximate fascicles, linear, subacute, hispidulous or nearly glabrous, 6–9 lin. long; spikes ovoid, very dense, somewhat elongated in fruit, ¾–2¼ in. long; bracts closely imbricate, spreading or reflexed at the apex, broadly ovate or reniform-ovate, mucronate, glabrous, minutely puncticulate, 2–3 lin. long; calyx oblong 2 lin. long; lobes broadly triangular, acute, unequal, the two upper about a quarter as long as the tube, the three lower smaller; corolla-tube oblong, 2½–3 lin. long; lobes broadly oblong or rounded, unequal, about half as long as the tube. *Choisy, Mém. Selag.* 27, *t.* 1, *fig.* 4.; *Walp. Rep.* iv. 168; *Choisy in DC. Prod.* ix. 22. *M. sp., Drège in Linnæa*, xx. 201. *Selago*

capitata, Berg. Pl. Cap. 157. *Lippia ovata, Linn. Mant.* 89. *Selago ovata, Thunb. Prodr.* 99, *and Fl. Cap. ed. Schult.* 462; *Ait. Hort. Kew. ed.* 2, iii. 432; *Lam. Ill.* iii. 77, *t.* 521, *fig.* 1; *Curt. Bot. Mag. t.* 186. *Dalea lippiastrum, Gærtn. Fruct.* i. 235, *t.* 51, *fig.* 2.

SOUTH AFRICA : without precise locality, *Thunberg ! Masson ! Forsyth !* and *cultivated specimens !*

COAST REGION : Clanwilliam Div.; sandy places at Driefontein, *Zeyher*, 3552! *Wallich !* Cape Div.; Cape Flats, *Harvey!* mountains at Muizenberg, *Wallich !*

5. M. linearis (Choisy in DC. Prod. xii. 23); perennial, much branched, 1–1½ ft. high; branches puberulous; leaves in approximate fascicles, linear or slender, subobtuse, hispidulous or nearly glabrous, 4–5 lin. long; spikes ovoid or oblong, very dense, ½–1½ in. long; bracts spreading, densely crowded, broadly ovate, often broader than long, apiculate or mucronate, glabrous and minutely puncticulate, sometimes crenulate, 2–2½ lin. long; calyx oblong, somewhat curved, 2 lin. long, densely puncticulate; lobes triangular, subacute, denticulate, short; corolla-tube linear-oblong, 2–2½ lin. long; lobes broadly oblong, about a quarter as long as the tube; fruit oblong, 1 lin. long, included within the calyx. *M. sp., Drège in Linnæa*, xx. 201.

COAST REGION : Malmesbury Div.; Groene Kloof and " Predikstael," *Zeyher*, 1385 ! on hills at Katzenberg, near Groene Kloof, 300 ft., *Bolus*, 4317 ! Zwartland, in sandy soil, *Cape Herb. !*

VI. GOSELA, Choisy.

Calyx campanulate, 5-lobed nearly to the middle, with nearly equal lobes. *Corolla-tube* slender, elongated, very little enlarged at the throat, mouth slightly constricted; lobes 5, obovate-oblong, subequal, spreading. *Stamens* 2 perfect, inserted above the middle of the tube, included; filaments short; anthers linear; staminodes 2, affixed to the apex of the tube; filaments very short; anthers small and empty. *Ovary* 2-celled; style entire. *Fruit* included within the calyx, by abortion 1-celled and 1-seeded.

A small much-branched heath-like shrub. Leaves linear, small, more or less fascicled. Spikes short or somewhat elongated, many-flowered, with the rhachis, bracts and calyx densely hirsute-villose. Flowers sessile.

DISTRIB. Species 1, endemic.

1. G. eckloniana (Choisy in DC. Prod. xii. 22); branches slender, puberulous; leaves more or less fascicled, linear, obtuse, revolute at the margin, 2–5 lin. long; spikes dense, oblong, becoming elongated in fruit, 2–5 in. long; bracts ovate-lanceolate, subacute, densely villous, especially at the margin, 2½–3¼ lin. long; calyx 1½–2 lin. long, very villous; lobes subulate-linear, rather longer than the tube; corolla-tube slender, 7–9 lin. long; lobes spreading, obovate-oblong, subequal, 1–1¼ lin. long.

SOUTH AFRICA : without precise locality, *Ecklon,* 123 ! *Ecklon & Zeyher,* 36 !
COAST REGION : Piquetberg Div. ; Piquiniers Kloof, *Ecklon & Zeyher !*
Kardouw, *Zeyher !*

VII. AGATHELPIS, Choisy.

Calyx tubular, shortly 5-toothed, adnate at the base to the bract.
Corolla-tube slender, elongated, very little enlarged at the throat,
mouth slightly constricted ; lobes 5, obovate-oblong, subequal,
spreading. *Stamens* 2, inserted above the middle of the tube,
included ; filaments short ; anthers oblong or linear ; staminodes 0.
Ovary 2-celled ; style entire. *Fruit* included within the calyx,
oblong, by abortion 1-celled and 1-seeded.

Small, much branched, heath-like shrubs. Leaves small or linear, often
fascicled. Spikes usually elongated and narrow, many-flowered. Bracts ovate or
oblong, more or less adnate to the calyx and including it. Flowers sessile.

Species 3, endemic.

Bracts not shining :
     Leaves linear ...   . (1) **angustifolia.**
     Leaves oblong ...   . (2) **parvifolia.**
Bracts shining   ...   . (3) **nitida.**

1. A. **angustifolia** (Choisy in Mém. Soc. Phys. Genèv. ii. ii. 95,
(excl. syn. *Selago polygaloides*) t. 1, fig. 3) ; branches terete, puberulous ; leaves numerous, sessile, linear, subacute, glabrous, 3–9 lin.long ;
spikes more or less elongated, 1–8 in. long, dense, many-flowered ;
bracts ovate-oblong, acute or subacuminate, keeled, glabrous, 2–3 lin.
long ; calyx oblong, 1–1¼ lin. long, 5-ribbed, with the ribs more or
less scaberulous, 5-dentate ; teeth very short, oblong, obtuse ;
corolla-tube slender, generally more or less curved, 4–5 lin. long,
lobes obovate-oblong, obtuse, ¾ lin. long. *Choisy, Mém. Selag.* 25
(excl. *syn. Selago polygaloides*) t. 1, fig. 3 ; *E. Meyer, Comm.* 252,
and in *Drège, Zwei Pflanzengeogr. Documente,* 87, 88, 89, 162 ;
Drège in Linnæa, xx. 201 ; *Walp. Rep.* iv. 170 ; *Choisy in DC.
Prod.* xii. 23. *Selago dubia, Linn. Sp. Pl. ed.* i. 629. *Eranthemum
angustatum, Linn. Mant.* 171. *Eranthemum angustifolium, Murr.
Syst. ed.* 13, 55. *Selago angustifolia, Thunb. Prodr.* 99, *and Fl.
Cap. ed. Schult.* 462. *Agathelpis adunca, E. Meyer, Comm.* 252,
and in Drège, Zwei Pflanzengeogr. Documente, 71, 74, 109, 113,
162 ; *Walp. Rep.* iv. 171 ; *Choisy in DC. Prod.* xii. 23. *A.
mucronata, E. Meyer, Comm.* 252, *and in Drège, Zwei Pflanzengeogr.
Documente,* 76, 80, 115, 162 ; *Walp. Rep.* iv. 170 ; *Choisy in DC.
Prod.* xii. 24. *A. sp., Drège in Linnæa,* xx. 201.

SOUTH AFRICA: without precise locality, *Auge ! Bunbury,* 169 ! *Forsyth !
Forster ! Niven ! Roxburgh ! Thom ! Sieber,* 66 ! *Ecklon,* 78 ! *Forbes ! Harvey,*
416 ! *Bowie !*
COAST REGION: Vanrhynsdorp Div. ; Gift Berg, 1500–2000 ft., *Drège.*
Clanwilliam Div. ; between Wupperthal and Ezels Bank, 3000 ft., *Drège !*
between Piquiniers Kloof and Pretoris Kloof, 1500 ft., *Drège !* Clanwilliam,
Mader in Herb. MacOwan, 2181 ! Piquetberg Div. ; Piquet Berg, 1000–2000 ft.,
Drège ! Malmesbury Div. ; between Groene Kloof and Klip Berg, below 400 ft.,

Drège. Groene Kloof, *Cape Herb.!* Hopefield, *Bachmann,* 879! Cape Div.;
False Bay, *Thunberg!* Camps Bay, *Burchell,* 310! Devils Mountain, *Burchell,*
8462! *Pappe! Bolus,* 2914! near the Blockhouse, *Wolley Dod,* 631! Muizen
Berg, *Bolus,* 2914! Table Mountain, up to 3000 ft., *Drège! Ecklon,* 273! *Fleck,*
433! *Milne,* 159! *Galpin,* 4418! Simons Bay, *MacGillivray,* 560! sandy places
near the Cape Flats, *Zeyher,* 3557! hills near Cape Town, *Harvey!* Wynberg,
Harvey! Hout Bay, Harvey, 201! coast at Riet Valei, *Zeyher,* 3556! Liou
Mouutniu, *Cape Herb.!* between Cape Town and Table Mountain, *Burchell,* 918!
Kommetjes, 100 ft., *Galpin,* 4419! Constantia Berg, *Schlechter,* 1! Paarl Div.;
Paarl Mountains, 800–1500 ft., *Drège!* Tulbagh Div.; Tulbagh, *Pappe!* Tulbagh
Waterfall, 1500 ft., *Schlechter,* 9062! *Ecklon & Zeyher!* Worcester Div.; rocky
places between Wagenmakers Valei and Dutoits Kloof, 1500–2500 ft., *Drège!*
Bains Kloof, *Rehmann,* 2290! Stellenbosch Div.; Hottentots Holland, *Cape
Herb.!* Stellenbosch, 1000–3000 ft., *Zeyher!* Caledon Div.; by the Zonder Einde
River, *Burchell,* 7536! on Donker Hoek Mountain, *Burchell,* 7973! Baviaans
Kloof, *Burchell,* 7651! among shrubs near Genadendal, 2000–3000 ft., *Drège.*
Caledon, *Pappe!* Klein Rivers Berg, 1000–3000 ft., *Ecklon & Zeyher!* Houw
Hoek Berg, 1000–3000 ft., *Ecklon & Zeyher!* Great Houw Hoek, *Zeyher,*
1883! Swellendam Div.; on the right bank of the Zonder Einde River, *Burchell,*
7501!

CENTRAL REGION: Ceres Div.; Cold Bokkeveld, at Klyn Vley, 4500 ft.,
Schlechter, 10196!

2. **A. parvifolia** (Choisy in Mém. Soc. Phys. Genèv. ii. ii. 95
partly); branches terete, puberulous; leaves numerous, sessile, oblong
or elliptic-oblong, subobtuse, rigid, 2–3 lin. long; spikes 1–3 in. long,
dense, many-flowered; bracts ovate, acute or mucronate, $1\frac{3}{4}$–$2\frac{1}{2}$ lin.
long; calyx oblong, $1\frac{1}{3}$ lin. long, 5-ribbed, with the ribs more or less
muricate, 5-dentate; teeth minute, oblong, obtuse; corolla-tube
slender, more or less curved or sometimes nearly straight, 4–$4\frac{1}{2}$ lin.
long; lobes oblong, obtuse, subequal, 1 lin. long. *Choisy, Mém.
Selag.* 26 *partly; Walp. Rep.* iv. 170; *Choisy in DC. Prod.* xii. 23.
A. brevifolia, E. Meyer, Comm. 253, *and in Drège, Zwei Pflanzengeogr. Documente,* 73, 75, 162; *Walp. Rep.* iv. 171; *Choisy in DC.
Prod.* xii. 24. *Eranthemum parviflorum, Berg. Pl. Cap.* 2. *E.
parvifolium, Linn. Mant.* 171; *Lam. Ill.* i. 60, *t.* 17, *fig.* 2.—*J. and
C. Commelin, Hort. Amstel.* ii. 119, *t.* 60.

SOUTH AFRICA: without precise locality, *Bergius?*
COAST REGION: Clanwilliam Div.; in rocky places near Honig Valei, 3000 ft.,
Drège! near Ezels Bank, 4000–5000 ft., *Drège!* Paarl Div.; French Hoek,
3500 ft., *Schlechter,* 9250!

A. parvifolia, Choisy, remains somewhat doubtful, but, if it belongs to the genus
at all, the character, "foliis ovato-linearibus brevibus," given by Bergius should
place it here. I have cited the figures on which the description seems to have
been chiefly based, though I am in doubt whether they belong to the genus
at all.

3. **A. nitida** (E. Meyer, Comm. 252); branches terete, puberulous
or pubescent; leaves numerous, sessile, linear, subobtuse or acute,
glabrous, 6–11 lin. long; spikes oblong or somewhat elongated,
1–3 in. long; bracts ovate-lanceolate, acute, keeled, rather thin and
shining, 4–5 lin. long; calyx oblong, $2\frac{1}{2}$–$3\frac{1}{2}$ lin. long, 5-ribbed,
5-dentate; teeth oblong, obtuse, about one-sixth as long as the tube;
corolla white; tube slender, rarely straight, 5–6 lin. long; lobes
oblong, obtuse, 1–$1\frac{1}{4}$ lin. long. *E. Meyer in Drège, Zwei Pflanzen-*

geogr. Documente, 89, 162; *Walp. Rep.* iv. 169; *Choisy in DC. Prod.* xii. 23.

SOUTH AFRICA: without precise locality, *Pappe*, 52!
COAST REGION: Cape Div.; near the waterfall on Devil's Mountain, 1000–1500 ft., *Drège! Wolley Dod,* 632! *Harvey!* Stellenbosch Div.; Stellenbosch, *Harvey!*

ORDER CIV. **VERBENACEÆ.**

(By H. H. W. PEARSON.)

Flowers hermaphrodite, rarely polygamous, irregular or, in a few genera, regular. *Calyx* inferior, persistent, gamosepalous, campanulate, tubular or cup-shaped with 4, 5, or rarely 6–8 lobes or teeth, rarely subtruncate. *Corolla* gamopetalous; tube usually cylindric or dilated above, often curved, rarely very short and broadly campanulate; limb 5-, 4- or rarely many-lobed, regular or more or less 2-lipped; lobes spreading and flat or the posterior 1–2 suberect, the anterior frequently larger, sometimes concave, imbricate in bud, the posterior rarely the lateral being outermost, the anterior innermost. *Stamens* 4, perfect, didynamous, or 2, inserted on the corolla-tube and alternating with the lobes, the posterior (or posterior 3) usually small, anantherous, reduced to a staminode or altogether absent; filaments free, filiform or slightly thickened or broader at the base, inappendiculate; anthers dorsifixed, introrse, with 2 distinct parallel or divergent cells, opening by longitudinal slits. *Disc* usually inconspicuous, sometimes thickened and fleshy beneath the ovary, very rarely annular. *Ovary* superior, sessile, formed of 2 (or, by abortion, 1) carpels, syncarpous, acute, obtuse, or retuse, 4-furrowed or rarely shortly 4-lobed, in the young condition normally 1-celled becoming later 2-celled by the intrusion of the ovuliferous margins of the carpels; cells 2- or (by abortion) 1-ovuled, becoming later 2-chambered by the formation of a spurious septum between the ovules; style terminal, entire; stigma terminal, usually oblique or bifid at the apex. *Ovules* erect, pendulous or laterally attached to the infolded edges of the carpels or, rarely, to a central column. *Fruit* usually more or less drupaceous; mesocarp juicy, fleshy or dry; endocarp hard, bony (rarely thin). *Seeds* always separate in distinct cells or chambers; albumen in section *Stilbeæ* fleshy, otherwise usually 0 or scanty; embryo straight; cotyledons flat or a little thickened at the base, free or rarely much thickened and fused together; radicle inferior, short, sometimes minute.

Herbs, shrubs or trees; leaves, except in a few genera, opposite or whorled, entire, dentate or incised, in *Vitex* usually digitately compound; inflorescence spicate or racemose or with the ultimate branching cymose, the cymes being centripetally developed, opposite or trichotomously paniculate; bracts usually small; flowers often brightly coloured.

About 73 genera, including 700 species in the tropical regions of both hemispheres; very few in subtropical and temperate areas.

It is stated in Harvey's "Genera of South African Plants" (2nd edition) that the genus *Stachytarpheta* is represented by "1 or 2 Cape species, probably naturalized" (l. c. 290). No specimens, however, are preserved in the Kew, British Museum, or Harvey's Herbarium, and I have failed to find any authority for the statement.

A cultivated specimen of *Holmskioldia sanguinea*, Retz., from Griqualand East is in the Kew Herbarium.

A.—Inflorescence centripetal (spicate or racemose); ovule basal, erect, anatropous.

Tribe 1. *STILBEÆ. Seed* albuminous. Low shrubs of ericoid habit.

* *Corolla 4- or 5-lobed, regular.*

† Calyx subequally 5-lobed or 5-toothed ; anther-cells parallel.

I. **Campylostachys.**—*Calyx* deeply 5-lobed or polysepalous. *Corolla* 4-lobed. *Fruit* dehiscing by 4 valves.

II. **Stilbe.**—*Calyx* 5-toothed or 5-partite. *Corolla* 5-lobed. *Fruit* indehiscent.

†† Calyx 2-lipped ; anther-cells diverging.

III. **Euthystachys.**—Only South African genus.

** *Corolla 2-lipped.*

IV. **Eurylobium.**—*Calyx* symmetric.
V. **Xeroplana.**—*Calyx* 2-lipped.

Tribe 2. *VERBENEÆ. Seed* exalbuminous. Herbs or shrubs.

* *Inflorescence a spike or unbranched spicate raceme.*

† Ovary 2-celled with 1 ovule in each cell; fruit of 2 pyrenes (or, by abortion, 1).

§ Calyx small (about 1 lin. long).

VI. **Lantana.**—*Calyx* truncate or obscurely toothed. *Fruit* drupaceous.
VII. **Lippia.**—*Calyx* 2-4-lobed or -toothed. *Fruit* dry, hard.

§§ Calyx not less than 3 lin. long.

VIII. **Bouchea.**—Only South African genus.

†† Ovary 4-chambered with 1 ovule in each chamber.

IX. **Priva.**—*Calyx* accrescent in fruit. *Pyrenes* 2, each 2-seeded.
X.—**Verbena.** *Calyx* unchanged or very slightly accrescent in fruit. *Pyrenes* 4, each 1-seeded.

** *Inflorescence a branched raceme.*

XI. **Duranta.**—*Calyx* accrescent. *Pyrenes* 4, each 2-chambered and 2-seeded.

B.—Inflorescence cymose.

Tribe 3. *VITICEÆ. Cymes* panicled, corymbose, umbellate or, sometimes, reduced to a single flower. *Ovule* inserted laterally.

XII. **Vitex.** *Drupe* entire containing a single 4-chambered pyrene. *Leaves* usually digitately compound.
XIII. **Clerodendron.**—*Drupe* 4-lobed or -furrowed. *Pyrenes* 4. *Leaves* simple.

Tribe 4. *AVICENNIEÆ. Cymes* capitate. *Ovule* pendulous.

XIV. **Avicennia.**—Only South African genus.

I. CAMPYLOSTACHYS, Kunth.

Calyx deeply 5-lobed or polysepalous with subequal narrow slightly imbricating lobes (or sepals). *Corolla* subequally 4- (sometimes 5-) lobed; tube short, broad, hairy within at the throat. *Stamens* 4, equal, inserted between the corolla-lobes in the upper part of the tube; anthers ovate, with parallel cells distinctly separated below. *Ovary* 2-lobed, 2-celled; ovules 2, erect, basal; style glabrous, minutely 2-lobed at the apex. *Fruit* shorter than the calyx, oblong-ovate, at first 2-grooved, later dehiscing by 4 valves. *Seed* albuminous, solitary, erect, large, with a reticulately wrinkled testa.

Erect ericoid shrubs; leaves crowded in whorls, subulate or linear-subulate, thick, hard, with revolute margins; spike short, capitate, terminal, sessile among the leaves; flowers solitary, sessile in the axils of the bracts; bracts subulate above the middle, much broadened and sheathing at the base; bracteoles 2, subulate, oblique, keeled.

1 species, endemic.

1. **C. cernua** (Kunth in Abh. Akad. Berlin, 1831, 207); a low shrub, 1–2 ft. high; branches erect, terete, puberulous or shortly pubescent; leaves crowded, in whorls of 4–6, erect, spreading, or somewhat recurved, linear-subulate, narrowed at the base, minutely apiculate, glabrous or minutely puberulous, 3–6 lin. long, $\frac{1}{3}$–1 lin. broad; spike subglobose, more or less nodding, about 6 lin. in diam.; flowering bract broadly cuneate, membranous and sheathing below, in the distal half abruptly linear-subulate, minutely apiculate, puberulous or villous on the back and on the distal margin of the sheath, $2\frac{1}{2}$–$3\frac{1}{2}$ lin. long, $1\frac{1}{2}$–$2\frac{1}{2}$ lin. broad near the base; bracteoles boat-shaped, falcate, acuminate, acute, strongly keeled, glabrous within, villous on the keel and margin, 2–$2\frac{3}{4}$ lin. long, $\frac{1}{2}$ lin. broad; calyx membranous, 5-partite or polysepalous, 2–$2\frac{1}{4}$ lin. long; lobes (or sepals) linear, obliquely acuminate, acute, keeled, puberulous or glabrous; corolla-tube funnel-shaped, glabrous without, $1\frac{1}{2}$–2 lin. long; lobes linear, obtuse or subacute, glabrous, 3-nerved, 2 lin. long, $\frac{1}{2}$–$\frac{3}{4}$ lin. broad; stamens 2–3 lin. long; style $2\frac{1}{2}$–$3\frac{1}{2}$ lin. long. *Walp. Rep.* iv. 173; *A.DC. in DC. Prod.* xii. 605. *Stilbe cernua, Linn. fil. Suppl.* 441; *Thunb. Prodr.* 29, *and Fl. Cap. ed. Schult.* 146; *Willd. Sp. Pl.* iv. 1116; *Lam. Encycl. Suppl.* v. 251.

SOUTH AFRICA: without precise locality, *Masson! Thom,* 614! 628! *Pappe! Harvey,* 199! *Nelson! Drège;*
COAST REGION: Cape Div.; on Table Mountain, *Burchell,* 609! Camps Bay, *Burchell,* 335! on Devils Mountain, *Wolley Dod,* 895! *Krauss.* near Cape Town, up to 2000 ft., *Zeyher,* 3588! *Burchell!* False Bay, *Robertson! Caledon Div.:* Houw Hoek, *Schlechter,* 9434! Stellenbosch Div.; on the western slopes of Hottentots Holland, near Lowrys Pass, 800 ft., *Bolus,* 4188! on Donker Hoek, *Burchell,* 7940! Swellendam Div.; on the Tradouw Mountains, *Bowie!*

II. STILBE, Berg.

Calyx leathery or membranous, more or less campanulate, 5-toothed or 5-partite, symmetric or with the three posterior teeth or lobes slightly larger. *Corolla* with a narrow tube, somewhat widening upwards, bearing within at the throat a wide ring of erect white hairs; lobes 5, equal. *Stamens* 4, subequal, inserted at the top of the corolla-tube between the lobes; anthers oval, with parallel cells continuous at the summit, separated below. *Ovary* of 2 carpels, 2-celled, 2-ovuled (or, by abortion, 1-celled, 1-ovuled), glabrous; style entire or minutely bifid, glabrous. *Fruit* oblong, enclosed in the calyx, 2-lobed, 2-celled (or, by abortion, 1-celled, 1-seeded), indehiscent; pericarp thin, membranous. *Seed* albuminous, erect; testa pitted or reticulately wrinkled.

Erect, glabrous, or occasionally hairy, ericoid shrubs; leaves in whorls of 3–7, erect, spreading, or reflexed, narrow, linear-subulate, hard, with revolute margins, passing upwards into bracts; spike dense, terminal, sessile among the upper leaves; flowers white or rose-coloured, solitary, sessile in the axils of the bracts; bracteoles 2, narrow, usually obliquely acuminate.

5 species, endemic.

Section 1. AMPHISTILBE. Calyx 5-partite.
　Adult leaves erect or spreading　...　...　...　(1) phylicoides.
　Adult leaves strongly reflexed　...　...　...　(2) mucronata.

Section 2. EUSTILBE. Calyx 5-toothed.
　Calyx membranous　...　...　...　...　...　(3) ericoides.
　Calyx cartilaginous:
　　Corolla-lobes 3-nerved, glabrous　...　...　(4) albiflora.
　　Corolla-lobes 1-nerved, more or less villous　...　(5) vestita.

1. **S. phylicoides** (A.DC. in DC. Prod. xii. 606); a low shrub, 1–2 ft. high; branches erect, glabrous or minutely pubescent, with prominent leaf-scars; leaves crowded, in whorls of 6 or 7, erect or spreading, shortly apiculate, with incurved apex, abruptly narrowed at the base, glabrous, $2\frac{1}{2}$–$3\frac{1}{4}$ lin. long, $\frac{1}{2}$–$\frac{3}{4}$ lin. broad; spike sub-globose, white, about 6 lin. in diam.; flowering bract coriaceous, boat-shaped, acuminate, acute, villous, 2–$2\frac{1}{4}$ lin. long, $\frac{1}{2}$ lin. broad; bracteoles narrowly linear, falcate, acuminate, acute, villous along the margins, 2–$2\frac{3}{4}$ lin. long, $\frac{1}{4}$–$\frac{1}{3}$ lin. broad; calyx with a tuft of white hairs at the base externally on the upper side, 5-partite, $2\frac{1}{2}$–3 lin. long; lobes lanceolate, acute, mucronate, in the upper half densely villous; corolla-tube glabrous, 2 lin. long; lobes linear, densely silky villous, 1–$1\frac{1}{2}$ lin. long; stamens 2 lin. long; style about 2 lin. long. *Campylostachys phylicoides, Sonder in Linnæa,* xx. 202 (*name only*).

COAST REGION: Swellendam Div.; near Swellendam, *Zeyher*, 3589! *Shand in Herb. Bolus,* 6256; Riversdale Div.; Lower part of the Lange Bergen near Kampsche Berg, *Burchell,* 6937! on Kampsche Berg, *Burchell,* 7083! summit of Kampsche Berg, *Burchell,* 7127! mountains of Garcias Pass, 1200 ft., *Galpin,* 4420!

2. S. mucronata (N. E. Br. in Hook. fil. Ic. Pl. t. 2526) : branches densely villous-tomentose; leaves crowded, in whorls of 4 or 5, reflexed or spreading, very rarely erect, with incurved mucronate apex, silky-tomentose when young, later glabrous above, tomentose beneath, 2–4 lin. long, $\frac{1}{2}$–1 lin. broad; spike subglobose, about 6 lin. in diam.; flowering bract linear, acuminate, concave, sheathing below, densely silky-villous on the back and margins, 2–2$\frac{1}{2}$ lin. long, 1 lin. broad; bracteoles narrowly lanceolate or oblanceolate, acute, more or less mucronate, densely silky-villous in the upper half and along the margins, 1$\frac{1}{2}$–2$\frac{1}{2}$ lin. long, $\frac{1}{3}$–$\frac{2}{4}$ lin. broad; calyx 1$\frac{1}{2}$–2 lin. long, deeply 5- or 6-partite; lobes lanceolate, acute, densely silky-villous in the upper half and along the margins; corolla-tube glabrous, 1$\frac{1}{4}$–1$\frac{1}{2}$ lin. long; lobes narrowly linear, acute, densely villous on the inner face, 1 lin. long, $\frac{1}{4}$ lin. broad at the base; stamens 1$\frac{1}{2}$ lin. long; style about 2 lin. long. *Phylica mucronata, E. Meyer in Drège, Zwei Pflanzengeogr. Documente,* 84 (*name only*).

VAR. β, cuspidata (H. H. W. Pearson); branches densely silky-villous when young; leaves linear-subulate, cuspidate at the apex, silky-villous when young, 5–6 lin. long; flowering bract linear-subulate with a brown mucro at the apex, narrowly sheathing at the base, densely silky-villous about the middle, 2–3 lin. long; bracteoles and calyx-lobes with brown acuminate tips.

SOUTH AFRICA: without precise locality, *Niven! Roxburgh,* "*Brunia,* No. 7"!

COAST REGION: Stellenbosch Div.; Lowrys Pass, 1000–2000 ft., *Drège! Burchell,* 8221! Caledon Div.; between Palmiet River and Lowrys Pass, *Burchell,* 8172! Houw Hoek, 1400–3600 ft., *Schlechter,* 7574! *Bolus,* 8409! Bredasdorp Div.; Elim, 800 ft., *Schlechter,* 7636! Var. β, Caledon Div.; on the Zwarte Berg, near Caledon, 3000 ft., *Bolus!*

3. S. ericoides (Linn. Mant. 305); a small erect or somewhat straggling (*Niven*) shrub branching chiefly at the base, $\frac{1}{2}$–2 ft. high; adult branches glabrous; leaves crowded in whorls of 3–5, erect or spreading, abruptly narrowed at the base, truncate or minutely apiculate, glabrous, 1$\frac{1}{2}$–3 lin. long, $\frac{1}{4}$–$\frac{1}{2}$ lin. broad; spike erect, cylindric, $\frac{1}{2}$–$\frac{3}{4}$ in. long, $\frac{1}{4}$–$\frac{1}{2}$ in. in diam.; flowering bract lanceolate with revolute margins, pubescent or puberulous, 1$\frac{1}{2}$–2$\frac{1}{2}$ lin. long; bracteoles linear-subulate, with revolute margins, sheathing towards the base, falcate, acuminate, acute, puberulous, 1$\frac{1}{2}$ lin. long; flowers pink; calyx tubular, membranous, 5-toothed, glabrous, silky-puberulous or -pubescent without, 1–1$\frac{1}{2}$ lin. long; teeth narrowly deltoid, acute, with thickened villous margins, shorter than the tube; corolla-tube glabrous, 1$\frac{1}{2}$–2 lin. long; lobes linear, obtuse, with a single median nerve branching near the apex, glabrous, 1–1$\frac{1}{2}$ lin. long; stamens 1$\frac{1}{4}$–1$\frac{1}{2}$ lin. long; style 1$\frac{1}{2}$–3$\frac{1}{2}$ lin. long. *Murr. Syst. Veg. ed.* 13, 772 ; *Thunb. Prodr.* 29, *and Fl. Cap. ed. Schult.* 146; *Willd. Sp. Pl.* iv. 1117 ; *Poir. in Lam. Encycl. Suppl.* v. 251 *excl. fig.*; *E. Meyer, Comm.* 280; *Walp. Rep.* iv. 171 ; *A.DC. in DC. Prod.* xii. 607. *S. virgata, Poir. l. c.* 252 ; *Lam. Ill. t.* 856, *fig.* 3. *Selago ericoides, Linn. Mant.* 87. *Luehea ericoides, F. W. Schmidt in Usteri, Ann.* vi. (1793) 118.

SOUTH AFRICA: without precise locality, *Wright*, 428! *Pappe! Villet! Masson! Zeyher! Roxburgh! Lehmann!*

COAST REGION: Malmesbury Div.; neighbourhood of the Berg River and in Zwartland, *Ecklon & Zeyher!* Cape Div.; sand-flats at Wynberg, *Drège!* Cape Flats, *Zeyher*, 1390! *Harvey! Burchell*, 8526! 8552! flats near Newlands, *Wolley Dod*, 633! near Cape Town, *Burchell!* in fields near Smitswinkel Bay, 500 ft., *MacOwan and Bolus*, *Herb. Norm. Aust.-Afr.*, 243! in open fields at Rondebosch, near Cape Town, below 100 ft., *Bolus*, 3969! near Cape Town, *Harvey!* flats near Wynberg, *Drège! Niven!* Bredasdorp Div.; Elim, *Schlechter*, 7728! Swellendam Div.; on the mountains of Swellendam, *Bowie!* Uitenhage Div.; Algoa Bay, *Forbes!*

The type specimen in the Linnæan Herbarium has a glabrous calyx.

4. S. albiflora (E. Meyer, Comm. 279); a low shrub, 1–2 ft. high, with stems erect, villous in the young parts, frequently unbranched; leaves crowded, in whorls of 4–6, erect or spreading, with incurved mucronate apex, narrowed at the base, glabrous, 3–5 lin. long, $\frac{1}{2}$ lin. broad; spike ovoid or ovoid-cylindric, 7–12 lin. long, 5–8 lin. in diam., with a silky-pubescent axis; flowering bract lanceolate, sheathing below, puberulous, $2\frac{1}{2}$ lin. long, $\frac{1}{2}$–$\frac{3}{4}$ lin. broad; bracteoles narrowly lanceolate, falcate, slightly keeled, villous along the margins above the middle, $2\frac{1}{4}$ lin. long, $\frac{1}{4}$–$\frac{1}{2}$ lin. broad; calyx tubular, cartilaginous, brown, glabrous, 5-toothed; tube distinctly 5-nerved, $1\frac{1}{2}$ lin. long; teeth narrowly deltoid, acute, with thickened villous margins and a single median nerve, $\frac{1}{2}$ lin. long, $\frac{1}{4}$–$\frac{1}{3}$ lin. broad at the base; corolla-tube glabrous without, puberulous within above the middle, 2 lin. long; lobes linear, 3-nerved, glabrous, 2 lin. long, $\frac{1}{3}$ lin. broad; stamens $2\frac{1}{2}$–3 lin. long; style $1\frac{1}{2}$–5 lin. long. *Walp. Rep.* iv. 172; *A.DC. in DC. Prod.* xii. 607. *S. albiflora, var. pilosa A.DC. l.c. S. albiflora, var., Sonder in Linnæa*, xx. 202 (*name only*). *S. ericoides, Lam. Ill. t.* 856, *fig.* 2.

SOUTH AFRICA: without precise locality, *Zeyher*, 199! *Masson! Nelson! Niven*, 45!

COAST REGION: Clanwilliam Div.; between Biedow and Honig Vallei, 2500–3000 ft., *Drège.* Piquetberg Div.; on the Piquet Berg, 1500–2000 ft., *Drège*, and without precise locality, *Zeyher*, 1389! Tulbagh Div.; Tulbagh Waterfall, 1000–1500 ft., *Drège!* Roode Zand Cascade, *Niven! Roxburgh*, "Stilbe, No. 2"! mountains above Tulbagh, *Pappe!* Caledon Div.; near Caledon, *Pappe!* on the Zwarte Berg and region of the hot springs, *Ecklon & Zeyher!* mountains of Baviaans Kloof, near Genadendal, *Burchell*, 7633! Genadendal, 2800 ft., *Schlechter*, 9855! between Palmiet River and Lowrys Pass, *Burchell*, 8177! Lowrys Pass, 3000 ft., *Schlechter*, 7225! Swellendam Div.; on Grootvaders Bosch Mountains, *Bowie!*

5. S. vestita (Berg. Pl. Cap. 30, t. 4, fig. 6); a low, branched shrub, 1–2 ft. high; branches erect, glabrous when adult; leaves densely crowded, in whorls of 4–6, erect, spreading, or somewhat reflexed, with incurved mucronate apex, glabrous, $3\frac{1}{2}$–7 lin. long, $\frac{1}{2}$–1 lin. broad; spike ovoid-cylindric, 7–12 lin. long, 5–7 lin. in diam., with a densely silky-pubescent axis persisting after the fall of the flowers; flowers white; flowering bract linear-acuminate, mucronate, sheathing towards the base, villous along the margins of the sheath and below the acumen, 4 lin. long, 1 lin. broad;

bracteoles falcate, acuminate, sheathing towards the base, keeled, villous in the upper two-thirds, $2\frac{1}{2}$–3 lin. long, 1 lin. broad; calyx tubular, cartilaginous, brown, glabrous, 5-toothed; tube indistinctly 5-nerved, $1\frac{1}{4}$–$1\frac{1}{2}$ lin. long; teeth narrowly deltoid, acute, villous along the margins, with a single distinct median nerve, $\frac{1}{2}$ lin. long, $\frac{1}{4}$ lin. broad at the base ; corolla-tube glabrous, $1\frac{1}{2}$–2 lin. long; lobes linear with a single median nerve, silky-villous along the margins and on the inner face or with only a few long white hairs, $2\frac{1}{4}$–$2\frac{1}{2}$ lin. long, $\frac{1}{4}$ lin. broad ; stamens 2 lin. long; style 2–3 lin. long. *S. pinastra, Linn. Mant.* 305 ; *Murr. Syst. Veg. ed.* 13, 772 ; *Thunb. Prodr.* 29, *and Fl. Cap. ed. Schult.* 146 ; *Willd. Sp. Pl.* iv. 1116 ; *Poir. in Lam. Encycl. Suppl.* v. 251 ; *Lam. Ill. t.* 856, *fig.* 1 ; *Walp. Rep.* iv. 172 ; *A.DC. in DC. Prod.* xii. 607 ; *Lindl. Veg. Kingd.* 607, *fig.* 411. *Selago Prunastri, Linn. Syst. ed.* 10, 1117. *S. pinastra, Linn. Sp. Pl. ed.* ii. 876, *excl. syn. S. Pinastri, Linn. MSS. in Herb. propr.*

SOUTH AFRICA : without precise locality, *Boivin,* 645 ! *Masson ! Roxburgh ! Bunbury,* 167 ! *Zeyher !*
COAST REGION : Cape Div.; mountain sides between Constantia and Hout Bay, 1000 ft., *Bolus,* 3959 ! Muizen Berg, 1000 ft., *Bolus,* 3348 ! in stony places on Table Mountain above Slang Kuil, *MacOwan and Bolus, Herb. Norm. Aust.-Afr.,* 926 ! Simons Bay, *Wright,* 429 ! 430 ! Table Mountain, 600–2500 ft., *Harvey,* 198 ! *Galpin,* 4421 ! *Krauss,* 1060. Table Mountain and near Witteboom, 600–1000 ft., *Drège !* Swellendam Div. ; on the Zwarte Berg and Tradouw Mountains, *Bowie !*

III. EUTHYSTACHYS, A. DC.

Calyx of 5 subequal sepals; the 3 posterior connate into a broad 3-toothed lip, the 2 anterior free. *Corolla-tube* funnel-shaped, equally 5-fid, bearded in the throat. *Stamens* 4, subequal or didynamous (? sometimes 5, one being sterile), inserted in the throat of the corolla between the lobes, with introrse anthers; anther-cells continuous at the apex, divergent below, widely so when mature, dehiscing by a continuous longitudinal slit. *Ovary* 2-celled, with 2 basal erect ovules ; style undivided. *Fruit* not seen.

Erect ericoid shrubs about a foot high ; leaves numerous, crowded, verticillate, hard, linear-subulate or -triquetrous, with revolute margins ; spike terminal or lateral, few-flowered, compact, sessile among the leaves ; flowers sessile, solitary in the axils of leaf-like bracts which overtop them, 2-bracteolate.

1 species, endemic.

1. E. abbreviata (A. DC. in DC. Prod. xii. 606) ; branches with prominent leaf-scars, pubescent or puberulous ; leaves in whorls of 4, erect, linear-subulate or -triquetrous, acute, glabrous, $4\frac{1}{2}$–6 lin. long, $\frac{1}{2}$–$\frac{2}{3}$ lin. broad at the base; spike terminal or lateral, 6–10-flowered, subglobose, with a densely silky-pubescent axis, about 4 lin. in diam. ; flowering bract linear and sheathing towards the base, triquetrous towards the apex, acute, with a median ridge on the upper surface extending from the middle to the apex, puberulous on

the margins of the sheath and at the base, 6–7 lin. long; 1 lin. broad near the base; bracteoles narrowly boat-shaped, acuminate, acute, strongly keeled, puberulous on the margins, 3–3½ lin. long, $\frac{1}{3}$–$\frac{1}{2}$ lin. broad; flowers concealed by the bracts and leaves; calyx membranous, puberulous within and along the free margins; posterior lip ovate, with 3 (or 2) short deltoid-acuminate teeth, 2½ lin. long, 1¼ lin. broad; anterior lip 2–lobed almost to the base; lobes ovate-elliptic, acuminate, acute, 1–nerved, 2½ lin. long, $\frac{1}{2}$–$\frac{3}{4}$ lin. broad; corolla narrowly campanulate, equally 5-lobed, glabrous without, marked with scattered groups of dark cells; lobes elliptic, obtuse, obscurely nerved, with thickened and revolute margins, keeled, villous on the inner face; tube 1½–2 lin. long; lobes 1 lin. long, ½ lin. broad at the base; stamens included; ovary obscurely 4-angled, glabrous; style 2 lin. long. *Campylostachys abbreviata, E. Meyer, Comm.* 279; *Walp. Rep.* iv. 173.

COAST REGION: Worcester Div.; Drakenstein Mountains, 2000–3000 ft., *Drège!*

IV. EURYLOBIUM, Hochst.

Calyx membranous, subcampanulate, prominently 5-angled, equally 5-toothed. *Corolla* 2-lipped; tube narrowly funnel-shaped, bearded within in the throat; upper lip 2-lobed to about the middle, lower 3-partite, smaller than the upper; lobes broader than in *Stilbe*, ovate, obtuse, 3-nerved. *Stamens* as in *Euthystachys*. *Ovary* 2-celled with 1 erect ovule in each cell; style filiform, not conspicuously divided at the apex. Mature *fruit* unknown.

Low, ericoid, glabrous shrubs in habit closely resembling *Stilbe;* leaves crowded in whorls, linear-subulate, hard, entire or minutely serrate, with revolute margins; spike terminal, compact, sessile among the upper leaves; flowers solitary, sessile in the axils of leaf-like bracts, 2-bracteolate.

1 species, endemic.

1. E. serrulatum (Hochst. in Flora, 1842, 229); branches pubescent, with prominent leaf-scars; leaves erect, in whorls of 4, minutely serrate, acute, glabrous, impressed-punctate, narrowed at the base, 4–4½ lin. long, $\frac{1}{2}$–$\frac{3}{4}$ lin. broad; spike ovoid-cylindric, ½–1½ in. long, 3–4 lin. in diam.; flowers white; flowering bract leaf-like, 2½–5 lin. long, $\frac{3}{4}$–1 lin. broad; bracteoles narrowly boat-shaped, falcate, acuminate, minutely serrate, glabrous, 2½–3 lin. long, ⅓ lin. broad; calyx membranous, 5-toothed; tube prominently 5-ribbed, glabrous, 1–1½ lin. long; teeth deltoid, acute, 1-nerved, minutely ciliate along the margins, ¾ lin. long, ½ lin. broad at the base; corolla-tube glabrous without, 1½–2 lin. long; posterior lip 1¼–1¾ lin. broad; anterior lip deeply lobed into 3 ovate obtuse segments, each with 3 parallel nerves anastomosing at the apex; filaments 1–2 lin. long, slightly broader at the insertion; ovary glabrous, conic; style 2½ lin. long. *Hochst. in Flora,* 1845, 70;

Walp. Rep. iv. 173; *A.DC. in DC. Prod.* xii. 608. *Stilbe serrulata, Hochst. in Flora,* 1842, 229.

COAST REGION: Caledon Div.; Genadendal, 4700 ft., *Schlechter,* 9820! on the southern slopes of the mountains near Genadendal, *MacOwan and Bolus, Herb. Norm. Aust.-Afr.,* 375! tops of the mountains of Baviaans Kloof near Genadendal, *Burchell,* 7740! among rocks (Bunter Sandstone) on the summit of the mountain near Genadendal, 3000 ft., *Krauss,* 1110.

V. XEROPLANA, Briquet.

Calyx tubular, narrow, 5-toothed, 2-lipped; upper lip 3-toothed, the lower consisting of 2 free lobes. *Corolla-tube* slender, cylindric, slightly dilated at the mouth, more or less pilose within the throat; limb bilabiate, spreading, with oblong-linear narrow flat lobes; upper lip consisting of 2 free segments, much longer than the 3 equal segments of the lower. *Stamens* 4, sub-equal, inserted in the corolla-throat between the segments, exserted; filaments glabrous, erect; anthers ovate, with distinct parallel cells. *Ovary* 2-celled, 2-ovuled; style exserted, with an entire slightly capitellate stigma; ovule erect from the base of the cell, anatropous. Mature *fruit* unknown.

A low branched shrub, glabrous, ericoid, resembling *Stilbe* or *Eurylobium* in habit; leaves in whorls of 3, crowded, linear, thick, with revolute margins; spike terminal, sessile among the upper leaves: flowers solitary, sessile in the axils of the leaf-like bracts, with 2 lateral narrow bracteoles shorter than the flowers.

1 species, endemic.

1. **X. Zeyheri** (Briquet in Bull. Herb. Boiss. iv. 336); a low much-branched shrub, 4–6 in. high; primary root (apparently) persistent, erect, woody; branches villous at the apex, erect or ascending, slender, with internodes 1–1½ lin. long, with grey bark; leaves sessile, in whorls of 3, hard, revolute, ericoid, villous above, beneath green, shining and quite glabrous, crowded in somewhat villous fascicles at the apices of the sterile branches, 2 lin. long; spike about 3 lin. long, 2 lin. broad; calyx membranous, glabrous or subglabrous, with nerves obscure without, about 1½ lin. long; tube about 1 lin. long; upper lip of 3 lobes connate for half their length; lower lip of 2 free lobes, less than ½ lin. long; lobes ovate-oblong, obtuse, villous within and at the margins, violet; corolla-tube pilose within in the upper part, glabrous or glabrescent without, 2 lin. long; limb spreading; lobes of the upper lip oblong-linear, obtuse, 1-nerved, about ¾ lin. long; lobes of the lower lip much shorter than (but otherwise similar to) those of the upper; stamens erect, exserted, glabrous, about 2½ lin. long; style very little shorter than the ovary. *Stilbe Zeyheri, Briquet, l.c.* 338.

SOUTH AFRICA: without precise locality, *Ecklon and Zeyher,* 8 *in Herb. Delessert.*

VI. **LANTANA**, Linn.

Calyx small, membranous, tubular, truncate or lobed. *Corolla-tube* narrow, cylindric, usually somewhat wider above the middle; limb spreading, regular or obscurely 2-lipped, with 4–5 broad obtuse or emarginate lobes. *Stamens* 4, didynamous, inserted about the middle of the tube, included; anthers ovate, with parallel cells. *Ovary* of 1 carpel, 2-celled; cells 1-ovuled; ovule erect, inserted at or near the base of the cell; style usually short; stigma thickened, oblique or sublateral. *Fruit* drupaceous with more or less fleshy mesocarp; endocarp hard, 2-celled or spontaneously separating when ripe into 2 1-celled 1-seeded portions (pyrenes).

Shrubs, seldom herbs, erect, scabrid, pubescent or tomentose with simple hairs; leaves opposite, toothed, serrate or crenate, usually rugose; spikes (in the South African species) axillary, pedunculate, many-flowered, contracted into small subglobose or cylindric heads; flowers red, orange, white or variegated, small or medium-sized, sessile, solitary in the axils of the bracts; bracts ovate or oblong, broad at the base; bracteoles minute, or 0.

Species about 50, mostly Tropical American; several extensively introduced into the Tropical and sub-Tropical regions of the Old World; 3 in South Africa, including 1 introduced.

Branches unarmed:
    Calyx prominently 4-ribbed; bracts not exceeding
       1½ lin. long ... ... (1) **galpiniana.**
    Calyx not ribbed; bracts (at least the lower ones)
       exceeding 2 lin. long   ...   . (2) **salvifolia.**
Branches armed ...   . (3) **Camara.**

1. **L. galpiniana** (H. H. W. Pearson); an erect, branched leafy shrub; branches unarmed, prominently tetragonal, scabrid, with striated reddish-brown bark, pubescent in the young parts; internodes 1–2 in. long; leaves opposite, bearing in their axils short leafy branches, shortly petiolate or subsessile, oblong-ovate, somewhat rounded or narrowed at the base, obtuse, with crenate-serrate revolute and thickened margins, scabrid and rugose above, scabrid-pubescent on the midrib and primary and secondary nerves beneath, profusely punctate-glandular, with 4–5 curved ascending primary nerves on each side impressed above, very prominent beneath, ¾–1¼ in. long, ¼–½ in. broad; petiole not exceeding 2 lin. long; spikes 1–3 in the leaf-axils, subglobose or cylindric, up to 2½ lin. in diam.; peduncle pubescent, shorter than the leaves; bracts crowded, imbricate, obovate or subcuneate, with a truncate base, abruptly acuminate, pubescent towards the apex and along the margins, 1–1¼ lin. long, about ¾ lin. broad; calyx shortly 2-lobed, with 4 prominent ribs, densely pubescent without, glabrous within, ½–¾ lin. long; lobes rounded; corolla pubescent without in the upper part, glabrous within, slightly exceeding the bract, about 1½ lin. long; young fruit oblong, apiculate owing to the persistent base of the style, drupaceous, glabrous.

KALAHARI REGION : Transvaal; Johannesburg, Jeppes Town Ridges, 6000 ft., *Gilfillan in Herb. Galpin*, 6165 !

2. L. salvifolia (Jacq. Hort. Schoenbr. iii. 18, t. 285); an erect,
much-branched, aromatic shrub, 3–6 ft. high; branches unarmed,
tetragonal, scabrid, densely pubescent or villous in the younger parts;
leaves opposite, or in whorls of 3 or 4, shortly petiolate, oblong-
ovate, rounded or subcordate at the base, more or less decurrent on
the petiole, obtuse or subacute, coarsely crenate or crenate-serrate,
scabrid-pubescent or pubescent and more or less rugose above,
villous-tomentose, pubescent, scabrid-pubescent or -puberulous on the
veins beneath, profusely punctate-glandular, with 4–5 ascending
primary nerves on each side impressed above, prominent beneath,
$\frac{3}{4}$–1$\frac{1}{2}$ in. long, $\frac{1}{3}$–1 in. broad; petiole 1–4 lin. long; spike pedunculate, axillary, subglobose, becoming cylindric, $\frac{1}{4}$–$\frac{1}{3}$ in. in diam.,
up to 1 in. long; peduncle acutely tetragonal, scabrid-pubescent,
shorter or longer than the leaves; bracts herbaceous, imbricate,
becoming separated in fruit by the elongation of the internodes of
the axis, deciduous, the lower ones frequently barren, sessile, broadly
oblong-ovate, acuminate, obtuse, with entire revolute or flat margins,
5–7-nerved at the base, profusely punctate-glandular, puberulous or
pubescent, 2–6 lin. long, 1–3$\frac{1}{2}$ lin. broad; calyx tubular, loosely
investing the base of the corolla-tube, obscurely 2-lobed or 4-toothed,
glandular, glabrous within, more or less distinctly 4-nerved,
pubescent without and on the margin, $\frac{3}{4}$–1 lin. long; corolla-tube
straight, slightly dilated about the middle, pubescent and glandular
without above the middle, glabrous within, $\frac{1}{4}$–$\frac{1}{2}$ in. long; ovary and
style glabrous, about 1 lin. long; drupe subglobose, purple, with a
sweet edible pulp, glabrescent, about 2 lin. long, 1$\frac{1}{4}$ lin. broad;
endocarp hard, bony, deeply furrowed; pyrene ovoid, acute, flattened
at the commissure, 1$\frac{1}{2}$–1$\frac{3}{4}$ lin. long, $\frac{3}{4}$–1 lin. broad. *Linn. Sp. Pl.*
ed. 2, 875; *Thunb. Prodr.* 98, *and Fl. Cap. ed. Schult.* 459; *Hiern*
in Cat. Afr. Pl. Welw. i. 827; *Baker in Dyer, Fl. Trop. Afr.*
v. 276 *partly. L. alba, Mill. ex Link, Enum. Pl. Hort. Berol.*
ii. 126; *Walp. Rep.* iv. 63; *Schauer in DC. Prod.* xi. 606.
L. salviæfolia, E. Meyer, Comm. 273; *Drège in Linnæa* xx. 202;
Walp. Rep. iv. 64; *Schauer in DC. Prod.* xi. 605. *L. rugosa,*
Thunb. Prodr. 98, *and Fl. Cap. ed. Schult.* 459. *L. indica, Roxb.*
Hort. Beng. 46; *Wight, Ic. Pl. t.* 1464; *Clarke in Hook. f. Fl.*
Brit. Ind. iv. 562. *Lippia caffra, Sonder in Linnæa,* xxiii. 88.
Camara salviæfolia, O. Kuntze, Rev. Gen. Pl. iii. 250. *C.*
salviæfolia, var. transvalensis, O. Kuntze, l. c.

SOUTH AFRICA: without precise locality, *Harvey,* 572! *Zeyher,* 1370! 1371!
Masson! Alexander!

COAST REGION : Mossel Bay Div.; on dry hills on the eastern side of Gauritz
River, *Burchell,* 6413! Uitenhage Div.; near the Zwartkops River, between
Villa Paul Maria and Uitenhage, 50–500 ft., *Zeyher,* 1373! *Ecklon & Zeyher!*
in thickets near Uitenhage, *MacOwan, Herb. Aust.-Afr.,* 1385! amongst other
shrubs near the Zwartkops River, *Zeyher,* 454! near brooks at the foot of Winter-
hoek Mountains, *Krauss,* 1134, and without precise locality, *Pappe!* Alexandria
Div.; Oliphants Hoek, *Pappe!* Albany Div.; Fish River Heights, *Hutton!* near
Grahamstown, *Bolton!* and without precise locality, *Williamson!* Fort Beaufort
Div.; on grassy hills near the Kat River, *Drège,* and without precise locality,
Cooper, 416! 453! British Kaffraria, *Cooper,* 143! Eastern Frontier, *Hutton!*

CENTRAL REGION : Beaufort West Div.; near Rhenoster Kop, 3000 ft., *Drège.* Graaff Reinet Div.; near Graaff Reinet, on rugged hills, 2600 ft., *Bolus,* 52! *Drège.* Cradock Div.; near Cradock, *Kuntze.* Somerset Div.; Modder Fontein, near Brak River, 2500 ft., *Drège !*

KALAHARI REGION : Hay Div.; on the Asbestos Mountains, near the Kloof Village, *Burchell,* 2055 ! Bechuanaland; plains between "Olive Tree Station and Last Water Station, *Burchell,* 2325 ! on Maadji Mountain, *Burchell,* 2368 ! Orange River Colony, Vaal River, *Burke !* Transvaal; hills above Aapjes River, *Rehmann,* 4260 ! *Zeyher (October) !* Houtbosch, *Rehmann,* 6183 ! near Pretoria, *Wilms,* 1177 ! Johannesburg, *Kuntze.*

EASTERN REGION : Transkei; near the Bashee River, 500 ft., *Drège.* Krelis Country, *Bowker,* 9 ! near Butterworth, *Bowker,* 387 ! Tembuland ; Qumancu River, *Baur,* 473 ! Natal; near Pietermaritzburg, *Wilms,* 2207 ! near Greytown, *Wilms,* 2208 ! near Durban, *Sanderson,* 147 ! *Gerrard & McKen,* 42 ! 599 ! 634 ! 637 ! Inanda, *Wood,* 246 ! on the hills between Umzimculu River and Umko-manzi River, among tall grasses, below 500 ft., *Drège,* and without precise locality, *Harvey !* *Cooper,* 1237 ! 3017 ! Delagoa Bay ; Rikatla, *Junod,* 65.

Also in Tropical Africa and India.

The berries are used for food in Zululand in times of scarcity; native names "Uguguvama" and (?) "Umpema" (Kew Bulletin, 1898, 53).

This species, as defined above, is very variable; in thus treating it I have followed Hiern and Baker. Jacquin's type (which I have seen) is finely silky-tomentose (or pubescent) on the under surfaces of the leaves which are hardly rugose. A Durban specimen (*Sanderson,* 147) exactly represents this typical form which, however, is not confined to Natal. Kuntze distinguishes his var. *transvalensis* by its narrow leaves which are not rugose ; although the material examined includes specimens possessing these characters, I cannot regard them as constituting a definite variety ; the relative length and breadth and the rugosity of the leaves often vary considerably in the same specimen. The plant upon which Sonder founded his *Lippia caffra* (Aapjes River, *Zeyher*) is very small and imperfect ; it is, however, exactly matched by a fruit-bearing specimen from Bechuanaland (*Burchell,* 2325 *in Herb. Kew.*) which is merely a villous form of *Lantana salvifolia.*

3. **L. Camara** (Linn. Sp. Pl. 627) ; an erect shrub, 4–8 ft. high; branches tetragonal, furrowed, armed with few or many irregular, recurved prickles, hispid at the nodes, with a few scattered stiff hairs on the internodes ; leaves opposite, petiolate, ovate, cordate or sub-cordate at the base, obtuse or subacute, with crenate-serrate margins, slightly rugose, scabrid-pubescent above and on the veins beneath, punctate-glandular, with 4–6 ascending primary nerves on each side slightly impressed above, prominent beneath ; petiole $\frac{1}{4}$–$\frac{3}{4}$ in. long ; blade $1\frac{3}{4}$–$2\frac{1}{2}$ in. long, 1–$1\frac{1}{2}$ in. broad ; spike pedunculate, axillary, solitary, subglobose, $\frac{1}{2}$–1 in. in diam. ; peduncle subtetragonal, scabrid-pubescent, 1–$2\frac{1}{2}$ in. long ; bracts herbaceous, oblong-lanceo-late, acute, 3-nerved, puberulous or pubescent, $2\frac{1}{2}$–4 lin. long, $\frac{3}{4}$ lin. broad ; outer flowers red, inner yellowish-white ; calyx tubular, loosely investing the base of the corolla-tube, obscurely 2-lobed, minutely 4-toothed, distinctly 4-nerved, glabrous within, pubescent without and on the margin, $\frac{3}{4}$–1 lin. long ; corolla-tube straight, dilated above the middle, pubescent without, glabrous within, $\frac{1}{6}$–$\frac{1}{4}$ in. long; posterior lip more or less emarginate ; anterior sinuate, obscurely crenate ; drupe about the size of a small pea, black, shining, glabrous. *Lam. Encycl.* i. 565, *and Ill: t.* 540, *fig.* 1 ; *Murr. Syst.*

Veg. ed. 14, 566 ; *Walp. Rep.* iv. 61 ; *Schauer in DC. Prod.* xi. 598 ; *Clarke in Hook. f. Fl. Brit. Ind.* iv. 562 ; *Baker in Dyer, Fl. Trop. Afr.* v. 275. *L. aculeata, Linn. Sp. Pl.* 627, *and Mant.* 419 ; *Lam. Encycl.* i. 566, *and Ill. t.* 540, *fig.* 2 ; *Murr. Syst. Veg. ed.* 14, 566 ; *Gærtn. Fruct.* i. 267, *t.* 56, *fig.* 4 ; *Walp. l. c.* 59 ; *Bot. Mag. t.* 96. *L. scabrida, Ait. Hort. Kew. ed.* 1, ii. 352 ; *Walp. l. c.* 60.

COAST REGION : Cape Div. ; on the Devils Mountain near Rondebosch, *Wilms,* 3530!

EASTERN REGION : Natal ; near Durban, *Wilms,* 2205! and without precise locality, *Cooper,* 3018!

A Tropical American species, widely introduced in the Old World.

VII. LIPPIA, Linn.

Calyx small, membranous, ovoid-campanulate or compressed, 2–4-lobed, 4-toothed, more or less truncate, 2-keeled, slightly accrescent, ultimately 2-valved enclosing (sometimes adhering to) the fruit. *Corolla* with cylindric, straight or curved tube, somewhat widened at the throat, rarely shorter than the bract ; limb spreading, oblique, more or less 2-lipped, 4-lobed ; lobes broad, frequently emarginate, the anterior (lower) being somewhat larger than the posterior (upper) and the 2 lateral equal and smaller than the posterior. *Stamens* 4, didynamous, inserted near the middle of the corolla-tube, included or somewhat exserted ; anthers ovate, with parallel cells. *Ovary* of 1 carpel, 2-celled ; cells 1-ovuled ; ovule erect, inserted at or near the base of the cell ; style usually short ; stigma terminal, oblique or recurved, thickened. *Fruit* small, with a hard dry epicarp, enclosed in the slightly accrescent, closely adpressed calyx ; endocarp hard and bony, easily separated (or falling asunder spontaneously) into 2 1-seeded portions (pyrenes). *Seeds* exalbuminous.

Shrubs or undershrubs, rarely herbs, with variously hairy, rarely glabrous epidermis ; leaves opposite or in whorls of 3 (occasionally 4), rarely alternate ; spike slender, elongate and lax, cylindric and dense, or short, and subglobose, becoming more or less cylindric as the fruit matures ; flowers small, sessile, solitary in the axils of broadly imbricate (in the denser spikes) or small bracts.

About 110 species, chiefly in Tropical America ; 10 in Tropical Africa.

The South African species belong to the subgenus *Zapania,* Benth., Section 4. *Euzapania,* Briq. § A, *Axillifloræ,* Briq., of which the characters are :— Spikes short, contracted, usually capituliform becoming more or less elongated during and after flowering, pedunculate, axillary. Bracts broad, persistent, imbricate, concave or flat, concealing the calyx. Calyx short, tubular, sometimes compressed, not winged.

Prostrate herbs, rooting at the nodes :
   Bracts rounded or shortly apiculate ; adult leaves
     more than 1 in. long ... ... (1) **nodiflora.**
   Bracts caudate-acuminate ; adult leaves less than
     1 in. long ... ... (2) **reptans.**

Erect undershrubs :
  Calyx distinctly 2-lobed :
    Bracts more than 2 lin. long and 1½ lin.
      broad, exceeding the flowers   ...   ...   (3) **scaberrima.**
    Bracts less than 2 lin. long and 1½ lin. broad,
      not exceeding the flowers ...   .   (4) **asperifolia.**
  Calyx truncate, subtruncate or obscurely 2-
    lobed :
    Leaves serrate or crenate-serrate :
      Bracts more than 2 lin. long; calyx
        truncate or obscurely 4-toothed   ...   (5) **Wilmsii.**
      Bracts less than 2 lin. long; calyx
        obscurely 2-lobed :
        Adult leaves elliptic, not more than
          1 in. long ...   .   (6) **Rehmanni.**
        Adult leaves ovate, more than
          1½ in. long   ...   .   (7) **bazeiana.**
    Leaves crenate...   ...   .   (8) **pretoriensis.**

1. **L. nodiflora** (Michx. Flor. Bor. Am. ii. 15) ; a creeping perennial herb rooting at the nodes, obscurely pubescent, with closely
adpressed silvery-white canoe-shaped unicellular hairs attached
by the middle, acuminate and serrulate at both ends ; stem
ridged, with internodes 1½–3 in. long ; leaves opposite, sessile
or petiolate, obovate, cuneate and entire in the basal half, sharply
and coarsely dentate towards the rounded or subacute apex, with the
midrib and ascending primary lateral nerves obscure or slightly
prominent, 1¼–1½ in. long, ⅓–⅔ in. broad ; spike subglobose, solitary,
up to ¾ in. long, 2–2½ lin. in diam. ; peduncle 1–2¼ in. long ; bracts
not exceeding the flowers, obovate or subrhomboid, cuneate at the
base, truncate, more or less rounded, apiculate or shortly mucronate
at the apex, with a narrow membranous sinuate entire or obscurely
serrate margin, glabrous above, 1¼–1¼ lin. long, 1⅛–1¼ lin. broad ;
calyx deeply 2-lobed, compressed, mitre-shaped, puberulous on the
prominent keels with simple adpressed hairs attached by their bases,
elsewhere thinly membranous and glabrous, 1–1¼ lin. long ; lobes
acuminate, ciliate ; corolla white, minutely and obscurely puberulous
without beneath the lower (anterior) lobe, otherwise glabrous,
1½–2 lin. long, later raised as a calyptra by the ripening fruit ;
upper lip erect, bifid ; lower larger than the upper, obscurely
3-lobed with the middle lobe oblong, about ½ lin. long ; pyrene
very shortly oblong, plano-convex, flattened at the commissure,
acute, obtuse or rounded at the apex, glabrous, minutely rugose,
½ lin. long. *Walp. Rep.* iv. 49 ; *Schauer in DC. Prod.* xi. 585 ;
C. B. Clarke in Hook. f. Fl. Brit. Ind. iv. 563 ; *Baker in Dyer,
Fl. Trop. Afr.* v. 279. *Verbena nodiflora, Linn. Sp. Pl.* 20 ;
Burm. Fl. Ind. t. 6, *fig.* 1. *V. capitata, Forsk. Fl. Ægypt.-Arab.*
10. *Blairea nodiflora, Gaertn. Fruct.* i. 266, *t.* 56. *Zapania nodiflora, Lam. Ill.* 59, *t.* 17, *fig.* 3 ; *R. Br. Prodr.* i. 514 ; *Hochst. in
Flora,* 1845, 68.

EASTERN REGION : Natal; Mount Edgecumbe, *Wood,* 1127! Durban, *Rehmann,* 8814 ! *Gerrard and McKen,* 817 ! in sandy places near the mouth of the
Umlaas (Umlazi) River, *Krauss,* 182 ! in sandy places near the seashore, between

Umtentu River and Umzimkulu River, *Drège !* and without precise locality, *Sanderson*, 384! Zululand; without precise locality, *Gerrard*, 511! Delagoa Bay; without precise locality, *Forbes! Junod*, 257.

Common in waste places in the warmer regions of both hemispheres.

The South African specimens include Schauer's two varieties, *sarmentosa* and *repens* with intermediate forms.

2. **L. reptans** (H. B. & K. Nov. Gen. et Sp. ii. 263); a creeping perennial herb, obscurely pubescent with closely-adpressed hairs similar in form to those of *S. nodiflora*; stem prostrate and rooting at the nodes or ascending, terete or subterete; internodes about 1 in. long; leaves opposite, petiolate, obovate, cuneate and entire towards the base, coarsely and acutely serrate-dentate towards the rounded or acute apex with obscure ascending primary lateral nerves, punctate-glandular, about ¾ in. long, ⅙–⅓ in. broad; spike ovoid, solitary, about 4 lin. long, 2–3 lin. in diam.; peduncle slender, terete, glabrescent, 2–2¾ in. long; bracts herbaceous, shorter than the flowers, obovate-cuneate or suborbicular, shortly caudate-acuminate, keeled, with a ciliate membranous margin, glabrous above, pubescent beneath, 1½–2 lin. long, ½–1 lin. broad; calyx with 2 short acuminate ciliate lobes, compressed, mitre-shaped, pubescent on the keels with simple spreading hairs attached by their bases, elsewhere thinly membranous, minutely pubescent or glabrous, 1–1½ lin. long; corolla white becoming red, glabrous, 2½–3 lin. long; posterior lobe erect, broadly oblong, shortly bifid; anterior obscurely 3-lobed; ovary and style ½–¾ lin. long; pyrene (immature) ovoid, flattened at the commissure, ¾ lin. long, ¼–½ lin. broad. *Walp. Rep.* iv. 48; *Schauer in DC. Prod.* xi. 584. *L. strigulosa, Mart. et Gal. in Bull. Acad. Roy. Brux.* xi. (1844) 319.

COAST REGION: Cape Div.; damp ground in Raapenberg Vley, *Wolley Dod*, 3522!

A native of the West Indies and Tropical America; introduced into South Africa.

3. **L. scaberrima** (Sonder in Linnæa, xxiii. 87); an erect aromatic scabrid shrub, 1–2 ft. high; stem and branches tetragonal, striate, scabrid, glandular; leaves opposite, narrowly lanceolate, or ellipticlanceolate, narrowed into the very short petiole, 3-nerved at the base, with margins scabrid, entire towards the base, crenate-serrulate from about the middle to the apex, subacute, scabrid on both surfaces, with the primary nerves impressed above, prominent beneath, profusely punctate-glandular, 1–1½ in. long, 2–3 lin. broad; spike ovoid, solitary, up to 6 lin. long, 2–5 lin. in diam.; peduncle striate, scabrid-puberulous, 1–2 in. long; bracts exceeding the flowers, broadly ovate, shortly cuspidate, sometimes rounded and obtuse at the apex, with numerous parallel nerves, glandular, puberulous or pubescent, with an entire ciliate margin, 2–5 lin. long, 1½–3 lin. broad; calyx 2-lobed, compressed, mitre-shaped, pubescent without, glabrous within, ⅓–½ as long as the corolla; lobes acute, obtuse or rounded at the apex, ¼–½ lin. long; corolla

glandular, pubescent without and within, $1\frac{1}{2}$–$2\frac{1}{4}$ lin. long; anterior lip broad, with a sinuate margin, rounded or subemarginate, larger than the broadly oblong posterior; ovary and style 1–$1\frac{1}{2}$ lin. long; pyrene smooth, glabrous, semiglobose, $\frac{3}{4}$–1 lin. long.

KALAHARI REGION : Orange River Colony ; Sand River, *Zeyher*, 1372! *Burke!* between the Vaal River and Rhenoster Berg, *Mrs. Bowker*, 665! Griqualand West; banks of the Vaal and Harts Rivers, *Holub!* Bechuanaland; Barolong Territory, *Holub!* Transvaal; Jeppes Town Ridges, Johannesburg, about 6000 ft., *Gilfillan in Herb., Galpin*, 6246!

4. **L. asperifolia** (Rich. Cat. Hort. Med. Par. 67, ex H. B. & K. Nov. Gen. et Sp. ii. 265); an erect much-branched aromatic shrub, 4–5 ft. high (*Gerrard*); adult stems terete or subangular, striate, scabrid-pubescent; leaves opposite or in whorls of 3 or 4, shortly petioled or sessile, oblong or oblong-lanceolate, cuneate at the base, obtuse or subacute, serrate or crenate-serrate, rugose, with 4–7 ascending primary nerves on each side impressed above, prominent beneath, scabrid-pubescent above, pubescent on the veins beneath, $\frac{3}{4}$–$1\frac{1}{4}$ in. long, $\frac{1}{3}$–$\frac{1}{2}$ in. broad; spikes small, globose, with a pubescent axis, solitary or 2–4 together, up to 5 lin. long, 2–3 lin. in diam.; peduncle $\frac{1}{4}$–$1\frac{3}{4}$ in. long; bracts not exceeding the flower, broadly ovate, obovate or slightly obcordate, shortly and abruptly acuminate or caudate-acuminate, silky-pubescent and glandular beneath, glabrous above, 1–$1\frac{1}{2}$ lin. long, $\frac{3}{4}$–1 lin. broad; calyx 2-lobed, compressed, densely pubescent without, glabrous within, $\frac{3}{4}$–1 lin. long; lobes shorter than the tube, subacute, obtuse or rounded ; corolla white, glandular and pubescent without in the upper part (more densely so in the lateral regions) glabrous or minutely pubescent within, $1\frac{1}{4}$–2 lin. long; posterior lobe broadly triangular, somewhat cordate at the base, smaller than the anterior; pyrene oblong, plano-convex, flattened at the commissure. *Kunth, Syn.* ii. 54 ; *Walp. Rep.* iv. 47 ; *Schauer in DC. Prod.* xi. 583 ; *Baker in Dyer, Fl. Trop. Afr.* v. 280. *L. capensis, Spreng. Syst. Veg.* ii. 751. *L. scabra, Hochst. in Flora*, 1845, 68 ; *Walp. l. c.* 134. *Verbena globiflora, L'Herit. Stirp. Nov.* 23, *t.* 12 (*excl. syn.*); *Willd. Sp. Pl.* i. 116. *V. capensis, Thunb. Fl. Cap. ed. Schult.* 447. *Zapania odoratissima, Scop. Delic.* i. 34, *t.* 15. *Z. lantanoides, Lam. Ill.* i. 58. *Z. odorata, Pers. Syn.* ii. 140. *Z. globiflora, Poir. in Lam. Encycl.* viii. 840. *Lantana lavandulacea, Willd. Sp. Pl.* iii. 319 ; *Jacq. Hort. Schoenbr.* iii. 59, *t.* 361 ; *Walp. l. c.* 64.

SOUTH AFRICA : without precise locality, *Drège !*

COAST REGION : Albany Div. ; in thickets near Grahamstown, *MacOwan!* King Williamstown Div. ; on the banks of the Buffalo River, near King Williams-town, *Pappe! Drège!* in rough places on the mountains near King Williamstown, 1500 ft., *Tyson in MacOwan and Bolus, Herb. Norm. Aust.-Afr.*, 848! Kruntz Kloof, *Kuntze*. Keiskamma, *Mrs. Hutton !* Eastern frontier, *MacOwan*, 503! *Hutton !* British Kaffraria, *Cooper*, 156!

KALAHARI REGION : Transvaal; near Lydenburg, *Wilms*, 1182!

EASTERN REGION : Transkei; Kreilis Country, *Bowker*, 276! Natal ; Inanda, 1800 ft., *Wood*, 32! coast land, *Sutherland !* at the edges of the woods around Durban Bay, *Krauss*, 247! between Durban and Maritzburg, *Sanderson*, 97 !

near Durban, *Gerrard and McKen,* 638! near the Tugela River, *Gerrard,* 635! and without precise locality, *Cooper,* 1006! *Gerrard,* 61! *Grant!*

Native name "Um-Suswane"; used medicinally (J. Medley Wood, Ann. Rep. Col. Herb. 1892). In British Kaffraria known as "Fever-tea."

Also in Tropical Africa and Tropical America.

5. **L. Wilmsii** (H. H. W. Pearson); an erect shrub, 1–2 ft. high; stem terete or subangular, scabrid-pubescent, glandular; leaves opposite, narrowly elliptic or elliptic-oblong, obtuse or subacute, cuneately narrowed into the short petiole, serrate, rugose, with 5–7 ascending primary nerves on each side impressed above, prominent beneath, scabrid-pubescent above and on the primary and secondary nerves beneath, elsewhere glabrous, punctate-glandular, 1–3 in. long, $\frac{1}{3}$–$\frac{2}{3}$ in. broad; spikes subglobose, solitary or in pairs, up to 7 lin. long, 3–4 lin. in diam.; peduncle terete towards the base, tetragonal above, striate, glandular, pubescent or scabrid-pubescent, $\frac{3}{4}$–1$\frac{3}{4}$ in. long; bracts narrowly ovate with a truncate base, acuminate, pubescent, glandular; outer ones 4 lin. long, 1$\frac{1}{4}$ lin. broad, inner 2$\frac{1}{2}$–3 lin. long, $\frac{3}{4}$–1 lin. broad; calyx truncate or obscurely 4-toothed, with 4 distinct nerves, pubescent without, glabrous within, $\frac{3}{4}$–1$\frac{3}{4}$ lin. long; corolla pubescent without in the upper part, more densely so in the lateral regions, glandular, glabrous within, about twice as long as the calyx; lobes entire, the anterior larger; ovary and style 1$\frac{1}{2}$–1$\frac{3}{4}$ lin. long; pyrene shortly oblong, plano-convex, flattened at the commissure, smooth or very delicately sculptured, glabrous, 1 lin. long, $\frac{1}{2}$–$\frac{3}{4}$ lin. broad.

KALAHARI REGION: Transvaal; near Lydenburg, *Wilms,* 1180!

The following Tropical African specimens also belong to this species, which should be added to the Flora of Tropical Africa:—British East Africa: Ukamba; *Scott-Elliot,* 6484! British Central Africa: Nyasaland; Mount Zomba, 4000–6000 ft., *Whyte!* and without precise locality, *Buchanan,* 1381! Rhodesia: Inyanga Mountains, 6000–7000 ft., *Evelyn Cecil,* 219!

6. **L. Rehmanni** (H. H. W. Pearson); an erect shrub, much branched above, exceeding 1 ft. in height; adult stems tetragonal, striate, scabrid-pubescent, with rounded angles; internodes 1–2 in. long; leaves opposite, sessile or shortly petioled, elliptic, cuneate at the base, acute or obtuse, with crenate-serrate slightly recurved margins, with 5 ascending primary nerves on each side impressed above, prominent beneath, rugose, scabrid-pubescent above and on the nerves and veins beneath, punctate-glandular, 1–1$\frac{1}{4}$ in. long, 5–6 lin. broad; spike (in fruit) cylindric or ovoid-cylindric, with a pubescent axis, solitary, up to 7 lin. long, 3 lin. in diam.; peduncle terete in the lower $\frac{3}{4}$, dilated and tetragonal above, pubescent, $\frac{1}{2}$–1 in. long; bracts broadly ovate or obovate, with a truncate base, shortly acuminate, densely pubescent, 1$\frac{1}{2}$–1$\frac{3}{4}$ lin. long, 1 lin. broad; calyx subtruncate or with 2 very short rounded lobes, distinctly nerved, densely pubescent without, glabrous within, $\frac{1}{2}$ lin. long; corolla pubescent without and within above the middle, 1$\frac{1}{2}$ lin. long; pyrene

semiglobose or shortly obovoid, plano-convex, flattened at the commissure, glabrous, $\frac{1}{2}-\frac{3}{4}$ lin. long.

KALAHARI REGION: Transvaal; hills above Aapies River, *Rehmann,* 4259!

7. **L. bazeiana** (H. H. W. Pearson); a low erect shrub; adult stems terete, striate, glandular, scabrid-puberulous or -pubescent; leaves opposite, shortly petioled, ovate, acute or obtuse, coarsely serrate, slightly rugose, with 4–5 ascending primary nerves on each side impressed above, prominent beneath, scabrid above and on the principal nerves beneath, punctate-glandular, $1\frac{1}{2}-2\frac{1}{4}$ in. long, $\frac{1}{2}-1$ in. broad; spike globose, solitary, up to 8 lin. long, about 4 lin. in diam.; peduncle angular, 4-grooved, scabrid-puberulous or -pubescent, $1\frac{1}{2}-2$ in. long; bracts ovate, acuminate or caudate-acuminate, densely pubescent and profusely glandular, $1\frac{3}{4}-3$ lin. long; calyx compressed, very shortly 2-lobed, adpressed-pubescent without, glabrous within, $\frac{1}{2}-\frac{3}{4}$ lin. long; lobes rounded; corolla pubescent without in the upper part, puberulous within, about 2 lin. long; anterior lobe broad, entire or subemarginate; posterior sinuate, smaller than the anterior; ovary and style about 1 lin. long; pyrene shortly oblong, plano-convex, flattened at the commissure, smooth, glabrous, $\frac{3}{4}$ lin. long, $\frac{1}{4}-\frac{1}{2}$ lin. broad.

EASTERN REGION: Tembuland; Bazeia, 2000 ft., *Baur,* 462!

8. **L. pretoriensis** (H. H. W. Pearson); an erect shrub exceeding $1\frac{1}{2}$ ft. in height; adult stems tetragonal, striate, scabrid-pubescent, with rounded glabrescent angles; internodes about $2\frac{1}{2}$ in. long; leaves opposite, shortly petioled, elliptic-oblong, more or less narrowed at both ends, obtuse, with crenate thickened and recurved margins, with 4–6 ascending primary nerves on each side impressed above, prominent beneath, scabrid-puberulous above, scabrid-pubescent on the nerves and veins beneath, punctate-glandular, $2-2\frac{1}{2}$ in. long, $\frac{3}{4}$ in. broad; spike small, globose, with a glabrous axis, solitary, $2-2\frac{1}{4}$ lin. in diam.; peduncle slender, terete, pubescent, $1\frac{1}{2}-1\frac{3}{4}$ in. long; bracts broadly ovate, acuminate, with a truncate or sub-cordate base, 3–7-nerved at the base, with a prominent midrib, pubescent; outer ones $2\frac{1}{2}-3$ lin. long, $1\frac{1}{2}$ lin. broad, inner $1\frac{1}{2}-2$ lin. long, $\frac{3}{4}-1$ lin. broad; calyx very shortly 2-lobed with rounded lobes, glabrous within, pubescent without and on the margin, $\frac{1}{2}-\frac{3}{4}$ lin. long; corolla densely pubescent on the outside above the middle, glabrous within, 1–2 lin. long; fruit not seen.

KALAHARI REGION: Transvaal; Wonderboom Poort, near Pretoria, *Rehmann,* 4523!

VIII. **BOUCHEA**, Chamisso.

Calyx narrowly tubular, prominently 5-ribbed, obliquely truncate or 5-toothed, usually becoming more or less dilated below as the fruit ripens, at length splitting longitudinally in front. *Corolla*: tube long, slender, cylindric, slightly widening at the throat, straight or curved ; limb spreading, oblique, with 5 subequal short broad obtuse or emarginate lobes. *Stamens* 4, perfect, didynamous, inserted in the upper part of the corolla-tube, included; filaments very short ; anthers ovate, with parallel cells. *Ovary* of 1 carpel, 2-celled, 2-ovuled, glabrous ; gynophore short, fleshy, and basal or flattened and forming an anterior scale ; ovule basal, erect; style terminal, filiform, somewhat thickened above, at the apex unequally divided into an anterior club-shaped stigmatic lobe and a posterior small or minute tooth ; stigma terminal, oblique, laminar, sub-bilobed. *Fruit* usually shorter than the calyx-tube and enclosed in it, oblong, hard, with a smooth, striated or sculptured surface, attached to the gynophore by a basal or anterior more or less concave area (areole), when ripe remaining intact or spontaneously separating into 2 1-seeded cocci.

Perennial herbs or low undershrubs ; leaves opposite or subopposite, toothed or incised, seldom entire ; inflorescence a terminal spike or simple spicate raceme ; flowers sessile or shortly pedicellate, solitary in the axils of persistent bracts ; bracts small, usually narrow ; bracteoles minute or 0.

About 25 species in Tropical and South Africa, Tropical America and India.

The South African species are here grouped in Schauer's sections *Rhagocarpium* and *Chascanum* (Schauer in DC. Prod. xi. 557), the characters of which are, however, somewhat modified.

In the second group of the section *Chascanum*, the aborted carpel is probably represented by the anterior scale-like gynophore.

Section 1. RHAGOCARPIUM. Cocci separating spontaneously when ripe; each with an obliquely basal pit-like areole ; bracteoles 0.

Leaves oblong or elliptic, acute or subacute at the apex :
    Bracts obtuse or subacute ; flowers sessile ;
      corolla-tube not exceeding 1 in. long   ...   (1) **adenostachya**.
    Bracts acuminate; flowers shortly stalked ;
      corolla-tube $1\frac{1}{4}$–$1\frac{1}{2}$ in. long ... ... (2) **longipetala**.
Leaves cuneate, cuneate-orbicular or -ovate, rounded or obtuse at the apex :
    Leaves acutely toothed, glandular-pubescent
      beneath ... ... (3) **hederacea**.
    Leaves obtusely toothed or crenate, hirsute on
      the nerves beneath ... ... (4) **Wilmsii**.
Leaves lanceolate ... ... (5) **Schlechteri**.

Section 2. CHASCANUM. Ripe fruit not separating spontaneously into cocci; areole basal or on the anterior face ; flowers bracteolate.

Areole basal :
    Leaves toothed, not more than $\frac{1}{2}$ in.
      broad ... ... (6) **cuneifolia**.
    Leaves entire, not less than 1 in. broad ... (7) **latifolia**.
Areole on the anterior face of the fruit :
    Spike 4–12 in. long. (See also 13, *pinnati-*
      fida.) ... ... (8) **garepensis**.
    Spike not more than 3 in. long :
      Leaves whorled (rarely opposite),
        crowded ... ... (9) **cernua**.

Leaves opposite, distant :
  Leaves dentate or serrate :
    Whole plant externally glab-
      rous   ...   .   (10) **glandulifera.**
    Plant minutely pubescent or
      puberulous :
        Leaves cuneate or ob-
          long-cuneate,   livid ;
          entire except at the
          rounded apex   ...   (11) **namaquana.**
        Leaves   ovate - oblong
          with distantly toothed
          margins   ...   ...   (12) **pumila.**
  Leaves deeply pinnatipartite   ...   (13) **pinnatifida.**

1. **B. adenostachya** (Schauer in DC. Prod. xi. 560) ; a low undershrub, branched at the base, about 1 ft. high ; stems terete, hirsute
with spreading and reflexed hairs ; leaves opposite, subopposite or
scattered, oblong or elliptic, narrowed from the middle into the very
short petiole, obtuse or subacute, acutely 5–9-toothed in the apical
half, 3-nerved at the base, glandular-pubescent, especially on the
nerves beneath, $\frac{2}{3}$–$1\frac{1}{4}$ in. long, $\frac{1}{4}$–$\frac{1}{2}$ in. broad ; spike elongate,
bearing numerous closely adpressed white flowers, pubescent with
spreading and reflexed glandular hairs, 3–4 in. long ; bracts sessile,
linear-subulate, obtuse, glandular-pubescent, $2\frac{1}{2}$–4 lin. long ; bracteoles 0 ; calyx obliquely and subequally 5-toothed, glandular-
pubescent without, glabrous within, 4–5 lin. long ; corolla-tube
narrow, curved, glabrous without, sparsely villous in the throat,
10–12 lin. long ; lobes broadly obovate, entire, glabrous, 2–$2\frac{1}{2}$ lin.
long, $1\frac{1}{2}$–2 lin. broad ; filaments glandular-puberulous ; fruit separating
spontaneously into 2 cocci when ripe ; coccus cylindric, flattened at
the commissure, finely striate below, reticulately sculptured above,
$2\frac{1}{2}$ lin. long.

KALAHARI REGION : Griqualand West ; Hebron, *Nelson*, 189 ! Bechuanaland ;
near the ruins at Kuruman, *Burchell*, 2426 ! near Hamapery, *Burchell*, 2495/1 !
at Kosi Fontein, *Burchell*, 2561 ! 2582 ! Orange River Colony ; low situations on
the Witte Bergen, *Bowker*, 705 !

Schauer, not having seen the fruit, placed this species in the section
Chascanum (Schauer, l. c.). More complete material shows that its affinities are
with species of the group *Rhagocarpium.*

2. **B. longipetala** (H. H. W. Pearson) ; a low undershrub,
1–2 ft. high ; stem unbranched, erect, terete, clothed with a white
reflexed pubescence among which are scattered long white hairs ;
leaves alternate, petiolate, oblong, cuneately narrowed from the
middle to the base, decurrent on the petiole, obtuse or rounded at the
apex, coarsely and acutely dentate-serrate above the middle,
glandular-puberulous above, pubescent with white hairs beneath
especially on the nerves ; nerves very distinct, impressed above,
prominent beneath ; petiole pubescent, 1–3 lin. long ; lamina
8–10 lin. long, 5–7 lin. broad ; raceme elongate, rather lax, bearing
many closely-adpressed white flowers, shortly pubescent with white

glandular hairs, about 8 in. long; bracts sessile, linear-lanceolate, acute, with entire membranous ciliate margins, glandular-pubescent, 3 lin. long; calyx obliquely and unequally 5-toothed, glandular-pubescent without, finely pubescent within; tube 4–5 lin. long; teeth not exceeding ¾ lin.; corolla-tube curved, glabrous without, villous in the throat, 1⅓–1½ in. long; filaments glandular-puberulous; fruit spontaneously separating when ripe into 2 cocci; coccus cylindric, flattened at the commissure, reticulately sculptured at the apex, longitudinally ridged below, 2¼–3 lin. long.

KALAHARI REGION: Transvaal; eastern slopes of the Saddleback Mountain, Barberton, 4500 ft., *Galpin*, 1171!

3. **B. hederacea** (Sonder in Linnæa, xxiii. 86); a low undershrub, branching at the base, 1–1½ ft. high; stems erect, terete, hirsute and glandular-pubescent; leaves opposite, subopposite or scattered, shortly petioled or subsessile, cuneate or cuneate-oblong, with margins entire in the basal half, broadly and acutely dentate towards the obtuse or rounded apex, glandular-pubescent, hirsute on the nerves beneath, with nerves impressed above, prominent beneath, ¾–1 in. long, 6–10 lin. broad; raceme elongate, bearing many closely adpressed flowers, glandular-pubescent with spreading or reflexed hairs, 4–6 in. long; bracts sessile, linear-acuminate, with membranous margins extending from the base to above the middle, glandular-pubescent beneath, puberulous above, ciliate on the margins, 2½–3 lin. long, ¾–1 lin. broad near the base; bracteoles 0; calyx shortly and subequally 5-toothed, glandular-pubescent without, minutely puberulous within, 5–5½ lin. long; corolla-tube curved, glabrous without, villous in the throat, 10–15 lin. long; lobes oblong or obovate, obtuse or somewhat emarginate, about 2 lin. long; filaments glandular-puberulous; fruit separating spontaneously when ripe into 2 cocci; coccus cylindric, somewhat flattened at the commissure, longitudinally ridged or striate below, reticulately sculptured above, 2–2½ lin. long.

VAR. β, **natalensis** (H. H. W. Pearson); leaves cuneate or cuneate-ovate, obtuse or rounded at the apex, hirsute on the very prominent nerves beneath, 1–2¼ in. long, ¾–1¼ in. broad; bracts spreading or recurved at the tips.

KALAHARI REGION: Transvaal; Magalies Berg, *Zeyher*, 1367! *Burke!* between Bronkhorst Spruit and Middelburg, *Wilms*, 1184! Pretoria, on the hills above Apies River, *Rehmann*, 4261! Matebe Valley, *Holub*, 1945! EASTERN REGION: Var. β. Natal; near the Tugela River, *Gerrard & McKen*, 1246! on a rocky hill above Ladysmith, *Wood*, 4246!

4. **B. Wilmsii** (Gürke in Notizblatt Königl. bot. Gart. Berlin, iii. (1900) 74); a low undershrub, about 1 ft. high; stem unbranched, erect, terete, glandular-pubescent and sparsely villous with long soft white hairs; leaves opposite, shortly petioled or subsessile, cuneate-orbicular or -ovate, with margins entire in the basal third, coarsely crenate or dentate-crenate towards the rounded or obtuse apex, punctate-glandular, adpressed puberulous above, glandular-pubescent and hirsute on the nerves beneath, with the nerves depressed above,

prominent beneath, 1-1½ in. broad; petioles not exceeding ¼ in.
long; spike bearing many closely-adpressed flowers, densely
glandular-pubescent, 7-9 lin. long; bracts sessile, linear, obtuse,
glandular-pubescent, ¼-⅓ in. long, ½-1 lin. broad; bracteoles 0;
calyx narrowly tubular, obliquely and unequally 5-toothed, glandular-
pubescent without, finely puberulous within, 4-5 lin. long; teeth less
than 1 lin. long; corolla-tube thin, slightly curved in the upper part,
glabrous, about 1 in. long; lobes subequal, glabrous, 1½ lin. broad;
stamens inserted in the throat of the corolla-tube; filaments
glandular-puberulous, ½-¾ lin. long; fruit separating spontaneously
when ripe into 2 cocci; coccus cylindric, flattened at the commissure,
longitudinally striate below the middle, finely reticulately sculptured
near the apex, 2¼-3 lin. long.

KALAHARI REGION: Transvaal; near Lydenburg, *Wilms*, 1183 (*December,*
1895)! *Wilms*, 1183 (*October,* 1895). Komati Poort, *Schlechter,* 11771!

Gürke (l.c.) places this species in section *Chascanum,* on the ground that the cocci
are still coherent in the ripe fruit. This is not the case in the specimens
at Kew and the British Museum (*Wilms,* 1183, collected in December, 1895),
which have the fruit characters of section *Rhagocarpium.* Gürke further states
that the leaves are "obovate" and the bracts "lanceolate," with which also
our specimens do not agree. At the same time there is little doubt that our
specimens and those quoted by Gürke belong to the same species.

5. B. Schlechteri (Gürke in Notizblatt Königl. bot. Gart. Berlin,
iii. (1900) 75);, a low undershrub exceeding 1 ft. high; stems erect,
little-branched, glabrous at the base, pubescent above with very
short reflexed rather stiff hairs; leaves opposite, sessile, lanceolate,
acuminate, narrowed at the base, entire, pubescent on both sides
with very short reflexed rather coarse hairs, up to 1½ in. long,
2-2½ lin. broad; spike short, lax-flowered, 2¼-3½ in. long; bracts
sessile, lanceolate, acuminate, rough with reflexed hairs, 1-2 lin.
long; bracteoles 0; calyx narrowly tubular, shortly 5-toothed,
shortly pubescent, 4-5 lin. long; teeth about ½ lin. long; corolla-
tube very thin, slightly curved, 6-8 lin. long; lobes 5, subequal;
fruit unknown.

KALAHARI REGION: Transvaal; on the hills, near Komati Poort, *Schlechter,*
11764!

Of this species I have not seen a specimen. The characters of the fruit being
unknown its position is somewhat doubtful: it is, however, provisionally placed
here on account of the absence of bracteoles.

6. B. cuneifolia (Schauer in DC. Prod. xi. 559); a much branched
low undershrub, 1-1½ ft. high; branches tetragonal, with 2 opposite
lines of pubescence alternating in successive internodes; leaves
broadly cuneate or oblong-cuneate, narrowed into the short petiole,
with a rounded apex, coarsely and acutely dentate, thick, obscurely
veined, sparsely puberulous on the margins and midrib near the base,
otherwise glabrous, ½-1 in. long, ¼-½ in. broad; raceme lax, minutely
puberulous, 3-6 in. long; flowers white; bracts linear-subulate,
minutely puberulous, 1-2 lin. long; bracteoles triangular-subulate,

¼-½ lin. long; calyx equally 5-toothed, obscurely pubescent, 4-5 lin. long; teeth ⅙-¼ lin. long; corolla-tube curved, glabrous, 9-10 lin. long; lobes subequal, oblong or obovate, rounded or slightly emarginate at the apex, glabrous, 2-3 lin. long; filaments glandular-puberulous; fruit oblong, sculptured, 2½-3 lin. long, 1 lin. broad; areole basal. *Harvey, Thes. Cap.* i. 18, *t.* 28. *Buchnera cuneifolia, Thunb. Fl. Cap. ed. Schult.* 466. *B. cernua, Houtt. Handl.* ix. 542, *t.* 58, *fig.* 2. *Phryma dehiscens, Linn. fil. Suppl.* 277. *Chascanum cuneifolium, E. Meyer, Comm.* 276 ; *Maund and Hensl. Botanist,* iv. 196 ; *Hochst. in Flora,* 1845, 68. *Deniseia dehiscens, O. Kuntze, Rev. Gen. Pl.* iii. 250.

SOUTH AFRICA : without precise locality, *Oldenburg! Forster! Holland,* 60 !

COAST REGION : Mossel Bay Div.; on dry hills on the Eastern side of Gauritz River, *Burchell,* 6427 ! Uitenhage Div.; on the slopes of Winterhoek Mountains, 1000 ft., *Krauss,* 1129 ! Addo, 1000–2000 ft., *Ecklon & Zeyher,* 3547 ! *Zeyher,* 842 ! and without precise locality, *Alexander!* Alexandria Div.; on the Zuurberg Range, 2000 ft., *Drège!* Albany Div.; Fish River Heights, *Hutton!* near Grahamstown, *MacOwan,* 125 ! near Glenfilling, 1000 ft., *Drège!* in calcareous places on the hills of the eastern side of the Bushmans River, *Zeyher,* 842 ! Fort Beaufort Div. : without precise locality, *Ecklon & Zeyher!* between Konap River and Enon, *Baur,* 1053 ! Peddie Div. ; Fredricksburg, *Gill!* King Williamstown Div.; on stony hills around King Williamstown, 1500 ft., *Tyson in MacOwan & Bolus, Herb. Norm. Aust.-Afr.,* 841 ! *Kuntze.* Queenstown Div. ; lower slopes of mountains near Queenstown, 3900 ft., *Galpin,* 1665 ! British Kaffraria ; without precise locality, *Cooper,* 209 ! 377 ! 378 ! Eastern Frontier, *Hallack!*

CENTRAL REGION : Prince Albert Div. ; on rocky hills near Weltevrede, 2500–3000 ft., *Drège!* Somerset Div. ; near Somerset East, *Bowker!* Graaff Reinet Div. ; in stony places near Graaff Reinet, 2500 ft., *Bolus!* in rocky soils of the Karoo country, *Bowie!*

EASTERN REGION : Natal ; Umkomaas (Umkomanzi) Cutting, *Wood,* 877 ! near Durban, below 400 ft., *Drège!* near the Mooi River, *Gerrard and McKen,* 1245 !

7. **B. latifolia** (Harvey, Thes. Cap. ii. 57, t. 190) ; a low undershrub, 2-3 ft. high ; stems little-branched, angular, softly pubescent with short spreading white hairs ; leaves erect, subsessile, broadly obovate or ovate, acute or obtuse, cuneate at the base, entire, gland-dotted, softly pubescent, with 2-4 ascending primary lateral nerves on each side, obscure above, prominent beneath, 1¼-2¾ in. long, 1-1½ in. broad ; raceme densely pubescent, bearing many crowded spreading white flowers, 2½-5 in. long ; bracts linear-subulate, acute, puberulous above, pubescent beneath, 3-6 lin. long ; bracteoles 1 lin. long ; calyx-tube 5-angled, punctate-glandular, densely pubescent, 5-6 lin. long ; teeth subulate, 1 lin. long ; corolla-tube curved, glabrous without, villous in the throat, 7 lin. long ; lobes oblong, glabrous or ciliate on the margin, 3-4 lin. long, 2-3 lin. broad ; filaments minutely glandular-puberulous ; fruit oblong or slightly broader below the middle, finely sculptured in the upper ¾, 3 lin. long, 1½ lin. broad ; areole basal.

VAR. β, **glabrescens** (H. H. W. Pearson) ; leaves adpressed-puberulous along the margins and occasionally on the nerves beneath, otherwise glabrous, 1¼–3¼ in. long ; raceme puberulous ; corolla pale-pink.

EASTERN REGION : Natal ; Noods Berg, *Wood*, 106 ! on grassy slopes of Bothas Hill, 2100 ft., *Wood in Herb. Natal Bot. Gard.*, 553 ! *Wood in Mac-Owan, Herb. Aust.-Afr.*, 1511 ! and without precise locality, *Mrs. K. Saunders !* Zululand ; on dry plains, *Gerrard*, 1247 !

KALAHARI REGION : Var. β, Transvaal ; near Lydenburg, *Atherstone !* Kaap Valley plains, at the foot of Devils Kantoor, near Barberton, 3000 ft., *Galpin,* 666 ! on grassy hills near the Crocodile River, *Mitford Barber,* 9 !

8. B. garepensis (Schauer in DC. Prod. xi. 560) ; an erect glabrous undershrub, 2–3 ft. high ; stem subangular, striate, with internodes 2–3 in. long ; leaves opposite or subopposite, petiolate, oblong-ovate or obovate, shortly cuneate at the base, thick, coarsely obtusely and deeply dentate-serrate, ⅓–1 in. long, 2–5 lin. broad ; raceme narrow, elongate, bearing numerous crowded erect adpressed flowers, ⅓–1 ft. long ; bracts linear-subulate, puberulous on the margin, otherwise glabrous, 3 lin. long ; bracteoles setaceous, ½ lin. long ; calyx shortly and unequally toothed, glabrous without, puberulous within, 3–4 lin. long ; corolla-tube straight or slightly curved, puberulous within at the throat, otherwise glabrous, 8 lin. long ; lobes broad, 1–1½ lin. long, 1 lin. broad ; filaments glandular-puberulous ; fruit oblong, slightly curved, reticulately sculptured, 1½ lin. long, ½–¾ lin. broad, not separating into 2 cocci when mature ; areole bounded by a dentate margin, covering nearly half the anterior face of the fruit. *Chascanum garipense, E. Meyer, Comm.* i. 277.

WESTERN REGION : Namaqualand ; sandy flats near the Auraap River, *Atherstone*, 13 ! rugged hills by the Orange River near Verleptpram, *Drège !* and without precise locality, *Wylie*, 98 !

A small-leaved variety occurs in Lower Guinea, see Baker in Dyer, Fl. Trop. Afr. v. 282.

9. B. cernua (Schauer in DC. Prod. xi. 559) ; a low much branched undershrub, 1–2 ft. high ; branches angular, pubescent along 2 opposite lines alternating in successive internodes ; internodes ¼–½ in. long ; leaves whorled, rarely opposite, crowded, imbricate, sessile, oblong, cuneate and entire towards the base, acutely 3–7-toothed at the rounded apex, thick, coriaceous, puberulous on the midrib beneath near the base, otherwise glabrous, punctate-glandular, ½–¾ in. long, 3–5 lin. broad ; raceme puberulous, 2–3 in. long, bearing numerous white flowers ; bracts linear-lanceolate, acute, glabrous or sparsely puberulous, with ciliate margins, 3 lin. long, about 1 lin. broad ; bracteoles subulate, ciliate, 1 lin. long ; calyx-tube 5-toothed, glabrous without, minutely puberulous within, 3½–4 lin. long ; teeth subequal, triangular-subulate, ciliate, ½ lin. long ; corolla-tube curved, glabrous without, softly hairy in the throat, about ¾ in. long ; lobes oblong, obtuse, glabrous, 2 lin. long ; filaments glandular-puberulous ; fruit oblong, black when ripe, 1½–2 lin. long, 1–1¼ lin. broad, with the posterior face strongly curved, longitudinally grooved and finely reticulate ; areole on the anterior face, oblong, surrounded by an entire bevelled margin, about 1½ lin. long. *Buchnera cernua, Linn. Mant.* 251; *Thunb. Prodr.* 100, *and Fl. Cap. ed. Schult.* 466. *Chascanum cernuum, E. Meyer, Comm.* 276; *Hochst. in Flora,* 1845, 68.

SOUTH AFRICA : without precise locality, *Masson! Grey! Forbes! Boivin,* 647 *! Roxburgh!*

COAST REGION : Cape Div. ; Simonstown, near Oatlands, *Wolley Dod,* 634! rocky places on the hills behind Fish Hoek, near False Bay, *MacOwan & Bolus, Herb. Norm. Aust.-Afr.,* 242! by a stream at Smitswinkel Bay, *Wolley Dod,* 1308! Simons Bay, *MacWilliam!* Caledon Div. ; mountains near Hemel en Aarde, 500–2000 ft., *Zeyher,* 3548! Riversdale Div. ; near Karmelks River, *Drège!* Milkwood Fontein, 600 ft., *Galpin,* 4422! Uitenhage Div. ; in sandy places near Winterhoek Mountains, *Krauss,* 1086!

10. B. glandulifera (H. H. W. Pearson); a low glaucous undershrub, 1–2 ft. high ; stem much branched, subangular, glabrous, with internodes 1½–3 in. long; leaves opposite, petiolate, oblong or subrhomboid, obtuse, with margins obtusely incised, glabrous, thick, with 2–3 indistinct ascending primary nerves on each side, ½–1 in. long, ⅓–¾ in. broad ; petiole slender, 4–7 lin. long; raceme dense, bracteate, glabrous, 1–2 in. long (immature), bearing numerous flowers ; bracts subulate, acute, with flat membranous margins bearing stalked capitate glands on the upper surface and margins, about 2½ lin. long ; bracteoles subulate, minute, glabrous ; calyx-tube erect, prominently ridged, glandular, nearly black when dry, 3½–4 lin. long; teeth triangular, glandular, about ¼ lin. long ; corolla-tube curved, glabrous without, glandular within in the upper two-thirds, about ¾ in. long ; glands capitate with long minutely tuberculate stalk-cells which are shorter near the mouth than in the lower part of the tube ; lobes unequal, oblong or ovate-oblong, emarginate, glabrous, 1–1½ lin. long; filaments short, glandular; ovary oblong, glabrous; fruit not seen.

KALAHARI REGION : Little Bushman Land! Stickhand, *Schlechter,* 76!

In the absence of fruit, this species is provisionally placed in section *Chascanum* on account of the presence of bracteoles.

11. B. namaquana (Bolus); a low, branched, livid or slate-coloured undershrub, ½–1 ft. high ; stem terete, finely adpressed-pubescent ; leaves shortly petiolate, cuneate or oblong-cuneate, obtuse, coarsely 3–5-toothed or -crenate towards the apex, rather thick, with obsolete nerves, finely adpressed-puberulous, 5–7 lin. long, 2½–3½ lin. broad ; raceme lax, with a finely pubescent axis, bearing few reddish-yellow very shortly pedicelled flowers, about 1 in. long; bracts linear, obtuse or subacute, thick, finely pubescent, 2½ lin. long ; bracteoles subulate, pubescent, ¾ lin. long ; calyx unequally toothed, finely adpressed-pubescent without, glabrous within, 5–6 lin. long; corolla-tube straight, glabrous, about 1 in. long; lobes oblong, emarginate, glabrous, 2 lin. long, 1–1½ lin. broad ; filaments glandular-puberulous; fruit oblong, slightly narrowed towards the apex, finely sculptured, 3–3¼ lin. long, 1½ lin. broad, not separating into 2 cocci when ripe; areole oblong, on the anterior face, surrounded by a dentate margin, about 2 lin. long.

WESTERN REGION : Little Namaqualand ; on the stony slopes of Spektakel Mountain, 2500 ft., *MacOwan and Bolus, Herb. Norm. Aust.-Afr.*, 680 !

12. **B. pumila** (Schauer in DC. Prod. xi. 560); a low erect undershrub; stem terete, clothed with fine reflexed pubescence; leaves ovate-oblong, petiolate, obtuse, cuneate at the base, coarsely remotely and deeply dentate or dentate-serrate, finely reflexed-pubescent, rather thick, with obscure nerves, $\frac{1}{2}$–1 in. long, $\frac{1}{4}$–$\frac{1}{3}$ in. broad ; petiole $\frac{1}{4}$–$\frac{2}{3}$ in. long; spike short, finely pubescent, bearing few subsessile flowers, $\frac{1}{2}$–$\frac{3}{4}$ in. long ; bracts ovate, acuminate, acute, finely puberulous above, pubescent beneath, $1\frac{1}{2}$ lin. long, $\frac{1}{2}$ lin. broad; bracteoles subulate, finely pubescent, $\frac{3}{4}$ lin. long; calyx subequally 5-toothed, obscurely 5-ridged, clothed with a fine spreading or reflexed pubescence without, minutely puberulous within, 6–7 lin. long; corolla-tube narrow, straight, glabrous without, puberulous within in the upper part, 9–13 lin. long; lobes oblong, finely puberulous, $1\frac{3}{4}$ lin. long, 1 lin. broad; filaments glandular-puberulous; fruit oblong, finely and reticulately sculptured, $2\frac{1}{2}$–$3\frac{1}{2}$ lin. long, 1–$1\frac{1}{4}$ lin. broad, not separating spontaneously into 2 cocci when ripe; areole on the anterior face, bounded by a dentate margin. *B. pubescens, Schauer in DC. Prod.* xi. 560. *Chascanum pumilum, E. Meyer, Comm.* 277.

SOUTH AFRICA : without precise locality, *Bowker!*
CENTRAL REGION : Calvinia Div. ; between Lospers Plaats and Springbok Kuil River, *Zeyher*, 1366! Prince Albert Div.; hills between Blauw Krans and Wilgebosch Fontein, 2500–3000 ft., *Drège*, 4856 !
WESTERN REGION : Great Namaqualand ; in the dry bed of Scap River, *Schinz*, 48! and without precise locality, *Schinz*, 46! Little Namaqualand, *Scully*, 240!
KALAHARI REGION : Hay Div.; on the Asbestos Mountains, between Reit Fontein and Kloof Village, *Burchell*, 2017 ! Transvaal; on the Boshveld at Elands River, *Rehmann*, 4995!

13. **B. pinnatifida** (Schauer in DC. Prod. xi. 560) ; a low undershrub, much branched at the base, about 1 ft. high; branches terete, densely and finely pubescent; leaves opposite or subopposite, petiolate, deeply pinnatipartite, rather thick, finely pubescent, 1–$1\frac{1}{2}$ in. long, $\frac{3}{4}$ in. broad; segments linear, entire, obtuse or subacute, 1-nerved, $\frac{1}{4}$–$\frac{1}{3}$ in. long; raceme lax, few-flowered, finely pubescent, about 1 in. (very rarely up to 6 in.) long ; flowers white ; bracts linear, obtuse or acute, spreading or recurved, finely pubescent, about 2 lin. long ; bracteoles subulate, finely pubescent, $\frac{1}{2}$ lin. long; calyx-tube unequally and shortly 5-toothed, finely pubescent, 5 lin. long; corolla-tube erect, glabrous without, puberulous in the throat, 10 lin. long ; filaments glabrous ; fruit oblong, reticulately sculptured, 2–3 lin. long, $\frac{3}{4}$–1 lin. broad; areole on the anterior face, oblong, surrounded by a toothed margin, 1–$1\frac{1}{2}$ lin. long. *Buchnera pinnatifida, Linn. fil. Suppl.* 288 ; *Murr. Syst. Veg. ed.* 14, 572 ; *Thunb.*

Prodr. 100, *and Fl. Cap. ed. Schult.* 466.   *Chascanum pinnatifidum,*
E. Meyer, *Comm.* 277.

SOUTH AFRICA: without precise locality, *Wallich ! Masson!*
COAST REGION : Queenstown Div. ; Gwatyn, 2900 ft., *Galpin,* 2018 !
CENTRAL REGION : Carnarvon Div. ; at the northern exit of the Karree Bergen
Poort, *Burchell,* 1558 ! Richmond Div. ; near Limoen Fontein, 3000–4000 ft.,
Drège! Graaff Reinet Div. ; stony hills near Graaff Reinet, 2500 ft., *Bolus,* 193 !
Albert Div. ; between Leeuwen Fontein and Slenger Fontein, 4500 ft., *Drège !*
rocky hills near Burghersdorp, *Mrs. Barber !* Colesberg Div. ; Colesberg, *Shaw !*
Somerset Div. ; near Somerset East, *Bowker,* 82 !
KALAHARI REGION : Griqualand West, Hay Div. ; Asbestos Mountains, near
the Kloof Village, *Burchell,* 1664 ! 2045/2 ! Hunerneat Kloof, *Rehmann,* 3392 !
Kimberley, *Rehmann,* 3432 ! Barkly Div.; Hebron, *Nelson,* 188 ! near the
Orange River, *Burke !* and without precise locality, *Mrs. Barber !* Transvaal ;
northern slopes of the Mugalies Berg, 6000–7000 ft., *Zeyher,* 1368 ! *Burke !*
Boshveld at Klippan, *Rehmann,* 5311 !

Also in German South-west Africa : Amboland ? Omatope and Oshando,
Schinz ! Damaraland, *Een !* Buluwayo, *Rand !* but not included in the Flora of
Tropical Africa.

IX. PRIVA, Adans.

Calyx of the flower tubular, prominently 5-ribbed, the ribs
terminating in short teeth, in fruit dilated below, contracted at the
throat and closely applied to the pericarp. *Corolla* : tube cylindric,
slightly widening upwards, straight or curved ; limb spreading,
oblique, 5-lobed, sub-2-lipped, the 2 posterior lobes being usually
shorter than the anterior. *Stamens* 4, didynamous, inserted in the
middle of the tube, included ; anther-cells parallel or slightly
divergent below ; the posterior staminode minute or 0. *Ovary* of
2 carpels, 4-chambered, with 1 ovule in each chamber ; style filiform,
divided at the apex into an anterior stigmatic lobe and a posterior
minute or obsolete tooth ; stigma small, oblique, lamellar. *Fruit*
dry, enclosed in the dilated calyx-tube, separating when ripe into 2
2-chambered, 2-seeded (or by abortion, 1-seeded) cocci ; coccus
hard, muricate, rugose or smooth without, flat, concave or excavated
on the commissural face. *Seed* subterete, completely filling the
chamber.

Erect, glabrous, pubescent or villous herbs with opposite membranous toothed
leaves ; inflorescence a simple spike or spicate raceme, terminal or axillary,
pedunculate, elongate, slender ; flowers solitary in the axils of the small narrow
bracts ; bracteoles minute or absent.

About 10 species in the warmer regions of both hemispheres.

1. **P. leptostachya** (Juss. in Ann. Mus. Par. vii. 70) ; an erect
branched perennial herb, 1–2 ft. high ; branches slender, prominently 4-angled, striate, puberulous or pubescent, especially at the
nodes, with fine spreading or recurved hairs ; internodes 1½–4 in.
long; leaves ovate or ovate-triangular, obtuse or subacute at the
apex, cuneate, rounded or subcordate at the base, petiolate, coarsely
crenate-serrate, puberulous with delicately-hooked hairs on both
surfaces and with a few rigid adpressed hairs above, pale beneath,

1–3 in. long, $\frac{3}{4}$–2 in. broad; petiole $\frac{1}{4}$–$\frac{2}{3}$ in. long; raceme terminal or axillary, elongate, bearing many distant shortly pedicelled white flowers on a slender 4-angled striate pubescent axis, $\frac{1}{2}$–1 ft. long; bracts linear-subulate or linear-lanceolate, glabrous above, pubescent beneath, $\frac{1}{2}$–1 lin. long; bracteoles 0; calyx of the flower cylindric, globose in fruit, densely pubescent with fine spreading hooked hairs without, minutely puberulous within, $\frac{1}{6}$–$\frac{1}{4}$ in. long; corolla-tube glabrous without, puberulous within about the middle, about $\frac{1}{3}$ in. long; upper lip deeply 2-lobed; lower larger than the upper, deeply 3-lobed; anther-cells divergent below; staminode 0; fruit ovate-emarginate or obcordate, composed of 2 slightly coherent 1- or 2-seeded cocci separated by an excavated commissure, muricate on the back, $1\frac{1}{2}$–2 lin. long, $1\frac{1}{2}$–3 lin. broad. *Pers. Syn.* ii. 139; *Wall. Cat.* 2657 C; *Walp. Rep.* iv. 35; *Schauer in DC. Prod.* xi. 533; *Clarke in Hook. f. Fl. Brit. Ind.* iv. 565; *Baker in Dyer, Fl. Trop. Afr.* v. 285. *P. dentata, Juss. l. c.; Pers. l. c.; Walp. Rep.* iv. 35; *Schauer l. c.; Hochst. in Flora*, 1845, 68; *O. Kuntze, Rev. Gen. Pl.* iii. 254. *P. Forskaolii, E. Meyer, Comm.* 275; *Jaub. & Spach, Ill. Plant. Or.* v. 59, *t.* 455; *Walp. l. c. P. abyssinica, Jaub. & Spach, l. c.* 58, *tt.* 453, 454. *P. Meyeri, Jaub. & Spach. l. c.* 57. *Verbena Forskâlii, Vahl, Symb.* iii. 6. *Streptium asperum, Roxb. Pl. Corom.* ii. 25, *t.* 146; *Spreng. Syst. Veg.* ii. 754; *Wight in Hook. Journ. Bot.* i. (1834) 230, *t.* 130. *Tortula aspera, Roxb. in Willd. Sp. Pl.* iii. 359. *Zapania arabica, Poir. in Lam. Encycl.* viii. 844.

SOUTH AFRICA: without precise locality, *Alexander!*

COAST REGION: Uitenhage Div.; grassy mountainous places near Enon, 1500 ft., *Drège!* Albany Div.; Blue Krantz, *Burchell*, 3625! in thickets near Grahamstown, *MacOwan*, 1433! King Williamstown Div.; Zand Plaat, 1500 ft., *Drège;* British Kaffraria, in thickets, *Mrs. Barber*, 21!

CENTRAL REGION: Somerset Div.; on the Bosch Berg, *Burchell*, 3224! Albert Div.; without precise locality, *Cooper*, 1770!

KALAHARI REGION: Transvaal; river banks on the plains around Barberton, 2800 ft., *Galpin*, 746!

EASTERN REGION: Transkei Div.; banks of the River Bashee, 500 ft., *Drège.* Natal; Berea, *Wood*, 4092! around Durban Bay, *Krauss*, 420! near Durban, *Wilms*, 2191! 2202! *Kuntze*, and without precise locality, *Grant! Peddie! Sanderson*, 319! Delagoa Bay! without precise locality, *Junod*, 301, *Forbes!*

Also in Tropical Africa, Socotra and India.

X. VERBENA, Linn.

Calyx tubular, 5-ribbed, 5-toothed, unchanged or only slightly dilated at the base in the fruiting stage. *Corolla*: tube straight or curved, cylindric or slightly dilated upwards; limb spreading, sub-2-lipped, with 5 obtuse rounded or emarginate lobes, the 2 posterior being outside and the anterior innermost in the bud. *Stamens* 4, didynamous, included, inserted about or above the middle of the corolla-tube; anthers ovate with parallel or somewhat diverging cells, all inappendiculate or with the connective of the anterior pair produced above into a clavate or glanduliform appendage. *Ovary* of 2 carpels, entire at the apex or very shortly 4-lobed, 4-chambered at the time of flowering, each chamber containing 1 ovule attached laterally near the base ; style short, divided at the apex into a short anterior stigmatic lobe and an acute posterior tooth. *Fruit* with a dry hard pericarp enclosed in the calyx, separating when ripe into 4 narrow cocci.

Herbs or low shrubs with prostrate or erect stems, glabrous or hairy ; leaves opposite, seldom whorled or alternate, toothed, often incised or partite, seldom entire ; spikes terminal, seldom axillary, densely crowded or elongate with distant flowers, often corymbose or panicled ; flowers small, sessile, usually solitary in the axils of narrow bracts.

About 80 species in the tropical and extra-tropical regions of the New World ; a few also in the Old World. A few American species are widely introduced in the Eastern Hemisphere ; 2 in South Africa.

Burmann enumerates *V. hastata*, Linn., a North American plant, in the Floræ Capensis Prodromus, 1. I have seen no specimens. [See also Epistolæ ined. Caroli Linnæi (Van Hall, 1830), 95 ; Linn. Amœn. Acad. vi. 81.]

The 3 species known from South Africa belong to the sub-group *Verbenaca*, characterized by the inappendiculate anthers.

Section 1. PACHYSTACHYA. Flowers crowded in heads or spikes.
    Corolla-tube more than twice as long as the calyx ; primary lateral nerves excurrent in the teeth ... · ... ...   (1) **venosa**.
    Corolla-tube less than twice as long as the calyx ; primary lateral nerves not excurrent in the teeth ... ...   (2) **bonariensis**.

Section 2. LEPTOSTACHYA. Flowers small, loosely arranged in long narrow spikes ... ...   (3) **officinalis**.

1. **V. venosa** (Gill. et Hook. in Hook. Bot. Misc. i. 167) ; a perennial herb, with a creeping rhizome ; stem erect, simple or branched, acutely 4-angled, furrowed, hispid, about 1 ft. high ; internodes 1½–2 in. long ; leaves opposite, sessile, semi-amplexicaul, oblong or oblong-lanceolate, acute, rounded or subauriculate at the base, stiff, scabrid, hispid on the nerves beneath, with margins entire or coarsely and acutely dentate-serrate, with 4–7 curved ascending primary lateral nerves on each side impressed above, prominent beneath, excurrent in the teeth, 1½–3½ in. long, ½–¾ in. broad ; spike terminal, simple or dichotomously branched, cylindric, dense, bearing many bracteate lilac or blue flowers, ½–2 in. long ; bracts lanceolate, long-acuminate, with a strongly marked midrib, glabrous above, hirsute beneath and on the margins, 2–5 lin. long, ¼–1¼ lin. broad ; calyx of the flower cylindric, dilated below in fruit, coloured, obliquely and acutely toothed, pubescent without, hirsute on the ribs, minutely pubescent within, 2–2½ lin. long ; corolla-tube cylindric, pubescent without in the upper part and within, 4½–6 lin. long ; stamens inserted below the middle of the corolla-tube ; ovary and style about 2 lin. long, glabrous ; fruit enclosed in the dilated calyx ; coccus shortly oblong, striate, about 1 lin. long.

Hook. Bot. Mag. t. 3127; *Walp. Rep.* iv. 27 ; *Schauer in DC. Prod.*
xi. 541.

KALAHARI REGION: Transvaal; near Lydenburg, *Wilms*, 1176 !
A native of the Pampas of Buenos Ayres; introduced also into Texas, Madeira,
and St. Helena.

2. **V. bonariensis** (Linn. Sp. Pl. 20) ; a tall perennial herb ; stem
erect, unbranched below, acutely 4-angled, striate, scabrid-pubescent;
internodes $2\frac{1}{2}$–5 in. long; leaves opposite, amplexicaul, auriculate,
oblong-lanceolate, acute, stiff, scabrid, rugose above, hispid on the
nerves beneath, with margins strongly revolute, coarsely and acutely
dentate-serrate, with 4–6 sharply ascending primary lateral nerves
on each side, impressed above, prominent beneath, not excurrent
in the teeth, 2–$4\frac{1}{4}$ in. long, 3–5 lin. broad ; panicle terminal,
very lax, dichotomously branched, with fastigiate branches,
bracteate, 4–12 in. long; spike cylindric, dense, bearing numerous
bracteate lilac flowers, $\frac{1}{3}$–$\frac{2}{3}$ in. long, $\frac{1}{6}$ in. in diam.; bracts
lanceolate, acuminate, with a strongly marked midrib, hispid,
2–7 lin. long, $\frac{3}{4}$–$1\frac{1}{4}$ lin. broad ; calyx of the flower cylindric,
slightly dilated below in fruit, coloured, obliquely and acutely
toothed, pubescent without, hirsute on the ribs, minutely pubescent within, $1\frac{1}{2}$–2 lin. long; corolla-tube cylindric, pubescent
without and within in the upper part, 2–3 lin. long; stamens inserted
below the middle of the corolla-tube ; ovary and style about $1\frac{1}{2}$ lin.
long, glabrous; fruit enclosed in the dilated calyx; coccus shortly
oblong, striate, $\frac{1}{2}$–$\frac{3}{4}$ lin. long.   *Kniph. Orig. Cent.* 2, *n.* 98 ; *Gærtn.*
Fruct. i. 315, *t.* 66, *fig.* 1 ; *E. Meyer, Comm.* 274 ; *Walp. Rep.* iv.
19 ; *Schauer in DC. Prod.* xi. 541 ; *Clarke in Hook. f. Fl. Brit.*
Ind. iv. 565. *V. capensis, Thunb. Fl. Cap. ed. Schult.* 447 *partly.*
V. quadrangularis, Vellozo, Fl. Flum. t. 39.

COAST REGION : Cape Div. ; near Rondebosch, in damp places, below 400 ft.,
Drège! Rehmann, 1704! Tygerberg, near Pampoenkraal, at 500 ft., *Drège,*
Newlands Avenue, *Wolley Dod,* 481! Stellenbosch Div. ; near Somerset West,
Ecklon & Zeyher!

A native of Brazil, introduced into various parts of the Old World.

3. **V. officinalis** (Linn. Sp. Pl. 20) ; a tall perennial herb ; stems
erect, 4-angled, striate, scabrid on the angles, otherwise glabrous ;
leaves opposite, sessile or subsessile, sheathing at the base, oblong,
oblong-lanceolate or rhomboid-ovate, narrowed towards the base,
more or less deeply trifid, pinnatifid or bipinnatifid, with the lobes
acute or obtuse, coarsely inciso-dentate, adpressed puberulous or
glabrescent, thin, 1–4 in. long; panicle terminal, much branched,
wide, more or less leafy below, $\frac{1}{2}$–$1\frac{1}{2}$ ft. long; branches slender ;
spikes bracteate, very lax, slender, bearing numerous lilac flowers,
distant below, crowded above, 6–9 in. long ; bracts ovate, acute,
with the midrib very prominent beneath, pubescent beneath, glabrous
above, not exceeding 1 lin. long; calyx cylindric, minutely toothed,
glabrous within, pubescent without, hispid on the ribs, about 1 lin.

long ; corolla-tube delicate, cylindric, about twice as long as the calyx;
stamens inserted about the middle of the corolla-tube ; ovary and style
about 1 lin. long; cocci shortly oblong, striate, $\frac{3}{4}$–1 lin. long. *Burm. Fl.*
Cap. Prodr. 1 ; *Flor. Dan. t.* 628; *E. Meyer, Comm.* 274; *Walp.*
Rep. iv. 25 ; *Schauer in DC. Prod.* xi. 547 ; *Clarke in Hook. f.*
Fl. Brit. Ind. iv. 565; *Baker in Dyer, Fl. Trop. Afr.* v. 286; *var.*
natalensis, Hochst. in Flora, 1845, 68. *V. spuria, Linn. Sp. Pl.* 20;
Walp. l. c. V. sororia, D. Don, Prodr. Fl. Nepal. 104; *Sweet,*
Brit. Fl. Gard. iii. *t.* 202. *V. setosa, Mart. & Gal. Bull. Acad.*
Brux. xi. ii. 321 ; *Walp. l. c.* vi. 687.

SOUTH AFRICA : without precise locality, *Zeyher,* 1364! 1365! *Miller! Harvey,*
405 !

COAST REGION : Cape Div. ; near Cape Town, *Burchell,* 503! Newlands
Avenue, *Wolley Dod,* 492 ! roadside near Rondebosch. *Pappe!*

KALAHARI REGION : Transvaal ; near Lydenburg, *Wilms,* 1175! near
Pretoria, *Wilms,* 1175a! on the Magalies Berg, *Burke,* 59 ! Linokana, in the
Marico District, *Holub !*

EASTERN REGION : Transkei Div. ; on the banks of the Bashee River, 500 ft ,
Drège; Griqualand East; by streams near Clydesdale, 2500 ft., *Tyson,* 2105 !
Natal ; near the Umlaas (Umlazi) River, *Krauss,* 151! near Durban, *Sanderson,*
92! Camperdown, *Haygarth,* 473 (*in Herb. Wood,* 1964) ! near the Mooi River,
Gerrard, 1249 !

XI. DURANTA, Linn.

Calyx tubular or subcampanulate, truncate or minutely 5-toothed,
in fruit accrescent, closely adpressed to the enclosed drupe but free
from it, and usually constricted at the mouth. *Corolla:* tube
cylindric, straight or curved ; limb spreading, oblique or regular,
5-lobed. *Stamens* 4, didynamous, inserted at or above the middle of
the corolla-tube, included ; anthers ovate, inappendiculate, with
distinct, parallel cells. *Ovary* of 4 carpels more or less perfectly
8-chambered, containing 1 ovule in each chamber ; style short ;
stigma terminal, obliquely dilated, very short, unequally 4-lobed.
Drupe quite (rarely almost) enclosed in the accrescent calyx, with
juicy epicarp and bony endocarp ; pyrenes 4, each 2-celled and
2-seeded.

Glabrous or tomentose shrubs, unarmed or with axillary or supra-axillary
spines ; leaves opposite or whorled, entire or toothed ; racemes terminal, rarely
axillary, usually panicled, long or short ; flowers small, shortly pedicelled in the
axils of small bracts.

About 8 species, ranging from Bolivia and Brazil to the West Indies and
Mexico. One species is introduced in Tropical and South Africa.

1. **D. Plumieri** (Jacq. Select. Stirp. Amer. Hist. 186, t. 176, fig. 76);
an unarmed or spinous shrub, 5–10 ft. high ; branches angular or
terete, glabrous or finely pubescent, with tawny bark and prominent
lenticels ; spines (when present) in or above the leaf-axils, spreading,
straight or slightly curved, $\frac{1}{3}$–$\frac{2}{3}$ in. long ; leaves opposite, shortly
petioled, oblong, elliptic or ovate, acute or obtuse, cuneate or
rounded at the base, entire or serrate above the middle, glabrous ;

petiole finely pubescent, 2–4 lin. long; lamina 2–2¼ in. long,
¾–1¼ in. broad; racemes terminal and axillary, simple or panicled,
many-flowered, lax, erect or drooping, 2–6 in. long; flowers blue,
bracteate, on short pubescent pedicels; bracts very small, subulate,
pubescent, the lower sometimes leafy; calyx of the flower tubular,
with 5 very short subulate teeth, minutely puberulous without,
1½–2½ lin. long; corolla-tube at least twice as long as the calyx,
curved, pubescent without in the upper half, puberulous within;
limb pubescent, unequally lobed; drupe (when mature) globose,
about the size of a pea, deeply 4-furrowed, completely enclosed in
the accrescent calyx; pyrene 2–2½ lin. long. *Bot. Reg. t.* 244;
Walp. Rep. iv. 79; *Schauer in DC. Prod.* xi. 615; *Baker in Dyer,
Fl. Trop. Afr.* v. 287.

KALAHARI REGION: Transvaal; Magalies Berg, *Burke*, 32! by the Nyl
River, north of the Mission Station, *Nelson*, 109! Kaap River Valley, near
Barberton, *Galpin*, 1248!

EASTERN REGION: Natal; without precise locality, *Sanderson*, 294 !

A native of Tropical America, widely introduced (frequently cultivated) in
the Old World.

XII. **VITEX**, Linn.

Calyx campanulate, rarely more or less funnel-shaped, 5-toothed
or 5-lobed, very rarely 3-lobed, usually enlarged in the fruit.
Corolla: tube cylindric, slightly dilated at the throat, usually short,
erect or curved; limb spreading, oblique, sub-2-lipped, 5-lobed;
2 posterior lobes shorter than the other 3 and outside them in bud,
helmet-shaped, erect or reflexed, the anterior lobe the largest, entire
or emarginate. *Stamens* 4, didynamous, exserted or included;
anther-cells distinct, parallel, diverging or curved, affixed to the
filament by their apices. *Ovary* of 2 carpels, during flowering
4-celled with 1 ovule in each cell affixed laterally at or above the
middle of the septum; style slender, shortly and acutely bifid at the
apex. *Drupe* sessile, rarely enclosed in the usually accrescent calyx,
with more or less fleshy epicarp and a hard or bony 4-celled endocarp. *Seeds* obovate or oblong, exalbuminous.

Trees or shrubs, glabrous, tomentose or villous, usually with depressed sessile
glands on the leaves and flowers; leaves opposite, rarely in whorls of 3,
frequently digitately compound, with 3–7 petiolulate or sessile, entire or dentate,
coriaceous or membranous leaflets, sometimes 1-foliolate or simple; cymes axillary,
sessile or pedunculate, dense or loosely divaricate or arranged in a terminal
racemose panicle or, rarely, contracted and capitate; flowers white, blue, violet
or yellow; bracts small, seldom exceeding the calyx.

About 120 species in the warm regions of both hemispheres, a few extending
to the temperate regions in South Europe and Asia.

The following South African species belong to the subgenus *Agnus-Castus*
(Endl.) the characters of which are:—Calyx cup-shaped or campanulate with a
short truncate or 5-lobed or -toothed limb. Corolla with an erect or reflexed
upper lip.

Section 1. TERMINALES. Cymes arranged in a terminal
panicle ... ... (1) **mooiensis.**

Section 2.   AXILLARES.   Cymes axillary, simple or panicled.

Fruiting calyx cup-shaped, minutely toothed :
  Leaves opposite ; leaflets usually serrate ... (2) **harveyana.**
  Leaves in whorls of 3 ; leaflets entire ... ... (3) **geminata.**
Fruiting calyx campanulate, 5-lobed or -toothed :
  Adult leaflets glabrous (or glabrescent) above :
    Petiole less than 1 in. long ; leaflets obo-
    vate or elliptic-obovate, less than 1 in.
    broad ... ... (4) **obovata.**
    Petiole about 1 in. long ; leaflets oblong-
    elliptic, less than 1 in. broad ... ... (5) **Rehmanni.**
    Petiole about 1½ in. long ; leaflets oblong-
    elliptic or obovate, more than 1 in. broad (6) **reflexa.**
  Adult leaflets tomentose or pubescent above :
    Cymes exceeding the petioles :
      Leaflets canescent or finely tomentose... (7) **Zeyheri.**
      Leaflets lanate-tomentose on the
      principal nerves beneath ... ... (8) **Wilmsii.**
    Cymes shorter than the petioles ... ... (9) **gürkeana.**

1. **V. mooiensis** (H. H. W. Pearson in Hook. Ic. Pl. t. 2705 ined.) ; a low tree ; branches subangular, glabrous, with prominent leaf-scars ; leaves opposite (rarely alternate), simple, petiolate, thin, membranous, ovate or elliptic, subacute or obtuse, cuneate at the base, entire or sometimes coarsely serrate towards the apex, entirely glabrous or finely scabrid-pubescent on the nerves, eglandular, with 3–5 distinct primary lateral nerves on each side, ¾–1 in. long, 5–7 lin. broad ; petiole slender, glabrous, somewhat thickened at the base, 2–3 lin. long ; cymes 2-flowered, shortly pedunculate, loosely panicled ; panicle simple, racemose, terminal, pedunculate, slender, bracteate, with 3–4 nodes, 1–2½ in. long ; peduncle and axis minutely pubescent along 2 opposite lines ; bracts linear-subulate, about 1½ lin. long ; flowers shortly pedicelled, white ; calyx of the flower 5-lobed to the middle, glandular, minutely puberulous, promi-nently nerved, 2–2¼ lin. long ; lobes oblong, subacute, ½–¾ lin. broad ; corolla-tube short, cylindric, curved, obscurely puberulous without, finely villous within above the middle, about 2 lin. long ; posterior lip erect, shortly 2-lobed ; anterior lip 3-lobed ; stamens inserted about the middle of the tube, shortly exserted ; anther-cells horizontal, spherical, divaricate, dehiscing by longitudinal slits ; filaments glabrous ; ovary globose, glandular towards the apex ; drupe pear-shaped, glabrous, exserted from the slightly accrescent calyx, 3 lin. long, 1½–2 lin. broad.

EASTERN REGION : Natal ; near the Mooi River, *Gerrard & McKen,* 1238 !

2. **V. harveyana** (H. H. W. Pearson) ; a shrub, 6–8 ft. high ; branches with distant nodes, tetragonal, striate, adpressed-puberulous or -pubescent when young, especially at the nodes ; leaves opposite, petiolate, 3- (rarely 5-) foliolate ; petiole slender, terete, adpressed-puberulous, ¼–1 in. long ; leaflets subcoriaceous, obovate, shortly

acuminate, acute, obtuse or rounded at the apex, cuneate at the base, subsessile or on grooved petiolules not exceeding 3½ lin., entire or 1–7 serrate on each side above the middle, with 5–8 primary nerves on each side obscure above, prominent beneath, usually with tufts of hairs in their axils beneath, otherwise glabrous, minutely glandular, 1¼–2 in. long, ½–⅜ in. broad, lateral leaflets often much smaller; cymes pedunculate, axillary, loosely divaricate, few-flowered, usually exceeding the leaves, bracteate, pubescent at the nodes, otherwise adpressed-puberulous or glabrous; peduncle flattened, puberulous, 1–1¾ in. long; bracts subulate or narrowly spathulate, puberulous or glabrous, 1½–4½ lin. long; flowers shortly pedicelled, 2-bracteolate, pale-blue; calyx of the flower cup-shaped, subtruncate or very shortly 5-toothed, 10-nerved, glabrous within, pubescent and minutely glandular without, 1¼–1¾ lin. long; corolla-tube slightly curved, glabrous without below the middle, pubescent above, glabrous within, 3–4 lin. long; limb ultimately reflexed; the anterior lip pubescent within at the base, minutely so without; stamens inserted about the middle of the corolla-tube, included; filaments broad at the base and villous; ovary subglobose, glabrous; drupe spherical or suboblong, far exserted from the slightly accrescent calyx, glabrous, with a thick woody endocarp, about 4 lin. in diam.

EASTERN REGION : Natal; banks of the Upper Tugela River, *Gerrard and McKen,* 1250! Zululand; without precise locality, 100 ft., *Haygarth in Herb. Wood,* 7462!

3. **V. geminata** (H. H. W. Pearson) ; a large shrub ; branches stout, 6-angled, with long internodes ; striate, glabrescent, with a smooth purplish bark ; leaves in whorls of 3, petiolate, 5-foliolate ; petiole stout, subterete, broader at the base, puberulous or pubescent at base and apex, otherwise glabrous, 1–1¼ in. long ; leaflets coriaceous, oblanceolate, very shortly acuminate, obtuse, on grooved puberulous or glabrous petiolules, 1–3 lin. long, with entire, somewhat thickened and revolute margins, with 8–12 spreading primary nerves on each side, depressed above, prominent beneath, glabrous and dark-brown above, lighter beneath, puberulous along the nerves, otherwise glabrous, eglandular, the terminal leaflet 2½–2¾ in. long, 10–11 lin. broad, the lateral smaller ; cymes lax, divaricate, pedunculate, 2 from the axil of each leaf, equalling or somewhat exceeding the leaves, bracteate ; peduncle angular, prominently ribbed, adpressed puberulous, 1¾–2½ in. long ; bracts linear-elliptic, acute, puberulous or pubescent, attenuate at the base into a short petiole about one-half as long as the lamina, 5–8 lin. long ; flowers purplish (*Gerrard and McKen*), shortly pedicelled, ebracteolate, about ½ in. long ; calyx of the flower cup-shaped, unequally and minutely 5-toothed, glabrous within, finely pubescent without, obscurely nerved, about 1½ lin. long ; corolla-tube slightly curved, glabrous below, adpressed-pubescent above on the outside, puberulous within above the insertion of the stamens, pubescent in the throat, 4–5 lin. long ; limb reflexed, densely pubescent without,

minutely so within ; stamens inserted below the middle of the tube, exserted ; filaments broadened and villous towards the base ; ovary subglobose, with a glabrous wrinkled surface ; drupe unknown.

EASTERN REGION : Zululand ; by the Umlatusi River, *Gerrard and McKen*, 2027 !

4. V. obovata (E. Meyer, Comm. 273) ; a tree ; young branches angular, with short internodes, finely tawny-tomentose, later glabrous, with prominent leaf-scars; leaves opposite, petiolate, 5-(rarely 3-) foliolate ; petiole tomentose, 5–9 lin. long; leaflets coriaceous, obovate or elliptic-obovate, very shortly apiculate, cuneate at the base, subsessile or shortly petiolulate, entire, pubescent or tomentose when young especially along the midrib and margins, later finely puberulous beneath or entirely glabrous, profusely glandular, with 6–10 obscure primary nerves on each side, 1–1¾ in. long, ½–1 in. broad ; cymes axillary, pedunculate, divaricate, about equal to the leaves, bracteate, tomentose ; bracts elliptic or oblong-linear, narrowed at the base into a short petiole, pubescent along the margins below the middle, glandular, lower ones 5–6 lin. long, 1½–2 lin. broad ; flowers shortly pedicelled, 2-bracteolate ; calyx of the flower campanulate, subequally 5-lobed; tube 10-nerved, profusely glandular and minutely pubescent without, glabrous within, about 2 lin. long ; lobes rounded or subacute, strongly 1-nerved, glandular and finely pubescent, ¾–1 lin. long, 1–1¼ lin. broad at the base ; corolla-tube glabrous below the middle, finely puberulous and profusely glandular without, pubescent within above the insertion of the stamens along 2 anterior parallel lines extending to the base of the anterior lobe, about 4 lin. long; lobes glandular and finely pubescent on the back and along the margins, glabrous within; posterior lobes reflexed ; stamens inserted about the middle of the corolla-tube, included; filaments villous, dilated towards the base ; ovary globose-ovoid, glandular and pubescent in the upper two-thirds ; drupe obconic, shortly apiculate owing to the persistent base of the style, finely pubescent and glandular, shorter than the accrescent calyx, 2½ lin. long, 2 lin. in diam. at the apex. *Walp. Rep.* iv. 87 ; *Schauer in DC. Prod.* xi. 693.

COAST REGION: Komgha Div. ; in a valley near the Kei River, below 1000 ft., *Drège !* among rocks near Kei Hill, Komgha, 1500 ft., *Flanagan*, 578 ! among rocks on a hill between Komgha and the Kei River, 1500 ft., *Flanagan in MacOwan, Herb. Aust.-Afr.*, 1515 !

5. V. Rehmanni (Gürke in Bull. Herb. Boiss. iv. 818); a shrub ; branches with short internodes and prominent leaf-scars, finely tomentose in the younger parts ; leaves opposite, petiolate, 5- or 3-foliolate; petiole subangular, finely pubescent, ⅓–1½ in. long ; leaflets subcoriaceous, oblong-elliptic or elliptic-lanceolate, acuminate, acute, subacute, obtuse at the apex, cuneate at the base, subsessile or on finely pubescent petiolules 1–3½ lin. long, with entire sinuate margins, profusely glandular, glabrous above, finely pubescent on the nerves beneath, elsewhere puberulous or glabrous, with 8–16 ascending primary nerves on each side, obscure above, distinct beneath, $\frac{3}{4}$–$2\frac{3}{4}$ in. long, $2\frac{1}{2}$–11 lin. broad, lateral leaflets often much smaller; cymes pedunculate, axillary, divaricate, bracteate, finely tomentose, equalling or slightly exceeding the leaves; bracts linear-spathulate, pubescent and glandular, lower ones $2\frac{1}{2}$–5 lin. long; flowers shortly pedicelled, bracteolate, about $\frac{1}{3}$ in. long; calyx of the flower campanulate, with a spreading shortly 5-lobed limb; tube 10-nerved, prominently 10-ribbed, glabrous within, finely pubescent and profusely glandular without, 1–2 lin. long; lobes ovate-triangular, acute or shortly apiculate, keeled, finely pubescent and glandular, $\frac{1}{2}$–$\frac{3}{4}$ lin. long, $\frac{1}{2}$–$\frac{3}{4}$ lin. broad at the base; corolla-tube straight, profusely glandular and pubescent without in the upper three-quarters, puberulous in the anterior portion, otherwise glabrous within, about 3 lin. long; lobes pubescent and glandular; stamens inserted below the middle of the corolla-tube, included; filaments villous at the base; ovary densely pubescent and glandular in the upper half; drupe obconic, shorter than the accrescent calyx, profusely glandular, $2\frac{1}{2}$–3 lin. long, $1\frac{1}{2}$–2 lin. in diam.

KALAHARI REGION: Transvaal; by the Nyl River, *Nelson,* 101! Makapans Berg, at Stryd Poort, *Rehmann,* 5422!

EASTERN REGION: Natal; on the hillside in "Thorns" near the Mooi River, *Wood,* 4463! and without precise locality, *Gerrard,* 1510! *Sutherland!*

6. **V. reflexa** (H. H. W. Pearson); a tree about 15 ft. high; branches subterete, finely tawny-tomentose, when adult with light striate lenticellate bark and very prominent leaf-scars; leaves opposite or subopposite, petiolate, 5-foliolate; petiole finely tomentose, $1\frac{1}{2}$–2 in. long; leaflets membranous, oblong-elliptic or obovate, subacute or obtuse, cuneate at the base, on grooved tomentose petiolules 1–5 lin. long, with entire slightly thickened pubescent margins, with 8–12 primary nerves on each side, depressed above, prominent beneath, profusely glandular, glabrescent and dark brown above, puberulous especially along the midrib and principal nerves and lighter in colour beneath, $2\frac{1}{4}$–$3\frac{1}{4}$ in. long, 1–$1\frac{1}{2}$ in. broad; cymes axillary, pedunculate, divaricate, about as long as the leaves, finely tomentose, bracteate; peduncle flattened, 2–$2\frac{1}{2}$ in. long; bracts linear-oblong or -elliptic, acute, puberulous, attenuate at the base, about 5 lin. long; flowers shortly pedicelled, bracteolate, 4–5 lin. long; calyx of the flower campanulate, with a spreading subequally 5-lobed limb; tube 10-nerved, glabrous within, profusely glandular and minutely pubescent without, about 2 lin. long; lobes broadly and shortly ovate, acute, with prominent reticulate venation, profusely glandular, pubescent, 1 lin. long, $1\frac{1}{4}$–$1\frac{1}{2}$ lin. broad; corolla-tube glabrous below, minutely pubescent and profusely glandular above the middle on the outside, and with a ring of hairs about the

middle and pubescent above in the anterior portion within the tube ; lobes strongly reflexed, finely pubescent and profusely glandular ; stamens inserted about the middle of the corolla-tube, shortly exserted ; filaments broader and villous towards the base ; ovary subglobose, pubescent and glandular above the middle ; drupe obconic, shorter than the accrescent calyx, light-coloured, glandular, pubescent in the upper part, 2 lin. long, 1½ lin. in diam. near the apex.

KALAHARI REGION : Transvaal ; in dongas around Barberton, 2800 ft., *Galpin,* 602 !

7. **V. Zeyheri** (Sonder ex Schauer in DC. Prod. xi. 693) ; branches terete, densely clothed with a light tawny tomentum ; leaves opposite or subopposite, 3–5-foliolate, petiolate, canescent, densely glandular ; petiole tomentose, ¾–1½ in. long ; leaflets coriaceous, oblong or oblong-lanceolate, acuminate, shortly acute or rounded at the apex, cuneate at the base, sessile or subsessile, entire, with 8–13 obscure primary nerves on each side, 1¾–3 in. long, 7–16 lin. broad ; cymes pedunculate, axillary, divaricate, equalling or slightly exceeding the leaves, bracteate, densely and finely tomentose ; peduncle flattened, tomentose, 1½–2 in. long ; bracts linear-subulate or narrowly spathulate, canescent, lower ones ¼–½ in. long ; calyx of the flower campanulate, with a spreading 5-lobed limb ; tube 10-nerved, glabrous within, finely tomentose and glandular without, about 2 lin. long ; lobes broadly ovate, acute, spreading, tomentose, glandular, ½–¾ lin. long, about 1 lin. broad ; corolla-tube straight, about twice as long as the calyx, pubescent and densely glandular without, villous within above the insertion of the stamens ; lobes pubescent ; stamens inserted at or below the middle of the corolla-tube; shortly exserted ; filaments villous towards the base ; ovary globose, pubescent and glandular ; drupe obconic, minutely pubescent and glandular, shorter than the accrescent spreading calyx, 1½ lin. long, 1–1½ lin. broad. *O. Kuntze, Rev. Gen. Pl.* iii. 258.

Var. β **brevipes** (H. H. W. Pearson) ; leaflets 1½–2 in. long, ⅓–⅔ in. broad, on canescent petiolules 1–3 lin. long.

KALAHARI REGION : Bechuanaland ; Banquaketse Territory, on the Malau Hills, *Holub !* and Naprstek Hills, *Holub!* Transvaal ; Magalies Berg, *Burke,* 73 ! on the banks of the Crocodile River, *Zeyher,* 73. Var. β : Transvaal ; on the northern slopes of the Magalies Berg, near the Crocodile River, *Zeyher,* 1369 ! near Aapies River, *Burke !*

The type of this species (*Zeyher,* 73) is probably the plant which is in the Kew and British Museum Herbaria as *Burke,* 73, and which agrees very closely with Schauer's description. The fact that Burke and Zeyher visited the Magalies Berg in company supports this view.

The Kew specimens of var. *brevipes* are in an immature state. It is possible that more advanced material may justify its separation as a species.

8. **V. Wilmsii** (Gürke in Notizbl. Königl. bot. Gart. Berlin, iii. 76) ; a large shrub ; branches terete, densely lanate-tomentose ; leaves opposite, 3–5-foliolate, petiolate ; petiole stout, more or less

densely lanate-tomentose, ¾–2 in. long; leaflets subcoriaceous, ovate
or elliptic, shortly acuminate, acute or obtuse, cuneate or somewhat rounded at the base, sessile or shortly petiolulate, with
entire sinuate and ciliate margins, with 7–10 primary nerves on
each side depressed above, prominent beneath, pubescent or
puberulous above, more or less lanate-tomentose along the midrib
and primary nerves beneath, otherwise glabrous or puberulous,
profusely glandular, 1¾–4¼ in. long, ¾–2¼ in. broad; cymes axillary,
divaricate, not exceeding the leaves, bracteate, with lanate-tomentose
peduncle and branches; peduncle 1½–3 in. long; bracts linear or
linear-oblong, narrowed at the base, acute, more or less falcate,
puberulous or pubescent, lower ones 3–6 lin. long; flowers shortly
pedicelled, 2-bracteolate, white; calyx of the flower campanulate,
with a spreading 5-lobed limb; tube prominently 10-nerved,
pubescent and profusely glandular without, glabrous within,
1½–1¾ lin. long; lobes broad, rounded, apiculate or subdeltoid,
acute, with ciliate margins, pubescent and glandular, ¾ lin.
long, 1 lin. broad; corolla-tube glabrous without in the lower
part, pubescent and glandular above, pubescent within in the throat,
3 lin. long; lobes reflexed, pubescent and profusely glandular without, with ciliate margins, glabrous within; stamens inserted in the
middle of the corolla-tube, shortly exserted; filaments dilated and
villous towards the base; drupe obconic, shorter than the accrescent
calyx, glabrous below, pubescent and glandular above, 3 lin. long,
1½–2 lin. in diam. at the apex.

KALAHARI REGION: Transvaal; near Lydenburg, *Wilms*, 158! 159! Rimers
Creek, near Barberton, 2900 ft., *Thorncroft*, 13 (*in Herb. Wood*, 4156)!

9. **V. gürkeana** (H. H. W. Pearson); a small bush; branches
with short internodes and prominent leaf-scars, densely tawny-
pubescent when young; leaves opposite, petiolate, 5-foliolate; petiole
tawny-pubescent, 2–2½ in. long; leaflets membranous, ovate,
acuminate or caudate-acuminate, obtuse or minutely apiculate,
rounded or subcuneate at the base, on tawny-pubescent petiolules
2–6 lin. long, entire, with 9–12 obscure primary nerves on each
side slightly prominent beneath, sparsely puberulous above, tawny-
pubescent beneath, profusely glandular, 2–3 in. long, ¾–1¼ in. broad;
cymes pedunculate, axillary, divaricate, about one-half as long as
the petioles, with tawny-pubescent branches and bracts; flowers
shortly pedicelled, bracteolate, lavender-coloured (*Monteiro*); calyx
of the flower shortly campanulate, tawny-pubescent and glandular
without, glabrous within; tube 1–1½ lin. long; limb spreading
broadly, 5-lobed; lobes deltoid, subacute or rounded at the apex,
¼–½ lin. long; corolla-tube curved, glabrous below, puberulous and
glandular in the upper half without; pubescent within above the
insertion of the stamens, about 3 lin. long; lobes ultimately reflexed,
minutely pubescent and villous, profusely glandular without, with many
long multicellular hairs at the base of the anterior lobe within;

stamens inserted above the middle of the corolla-tube, far exserted; filaments densely villous at the base; ovary conic, densely villous and glandular in the upper half; drupe not seen.

EASTERN REGION : Delagoa Bay; *Mrs. Monteiro,* 20!

XIII. CLERODENDRON, Linn.

Calyx campanulate, rarely tubular, truncate, 5-toothed or 5-lobed, unchanged or accrescent in fruit. *Corolla :* tube narrow cylindric, straight or curved, equal or somewhat wider at the throat; limb spreading or reflexed, 5-lobed; lobes subequal or the 4 upper shorter and the anterior produced, sometimes concave. *Stamens* 4, inserted at the base of the corolla-throat, exserted, incurved in bud; anthers ovate or oblong with parallel cells. *Ovary* of 2 carpels imperfectly 4-chambered, with 1 ovule in each chamber inserted laterally above the middle of the septum; style long; stigma apical, shortly and acutely 2-lobed. *Drupe* globose or obovate, usually 4-furrowed or 4-lobed, with more or less fleshy epicarp and bony or crustaceous endocarp, separating into 4 (or, by abortion, fewer) pyrenes, distinct or cohering in pairs. *Seeds* oblong, exalbuminous.

Glabrous, hairy or rarely felted, sometimes climbing, usually unarmed trees or shrubs; leaves opposite or whorled, entire, rarely toothed or lobed; cymes usually lax, pedunculate in the axils of the upper leaves or paniculate at the apices of the branches or crowded in a terminal corymb or head; flowers usually large and beautiful, white, blue, violet or red; calyx frequently coloured like the corolla, white or green.

About 100 species, chiefly in the warm parts of the Old World; a few in Tropical America.

Section 1. EU-CLERODENDRON. Corolla funnel-shaped with a broad sub-equally 5-lobed limb; tube straight, less than 1 in. long.

Leaves profusely gland-dotted beneath; corolla-tube
    not exceeding ⅓ in. long   ...   .   **(1) glabrum.**
Leaves not gland-dotted; corolla-tube not less than
    ⅓ in. long   ...   .   **(2) Rehmanni.**

Section 2. CYCLONEMA. Corolla-limb obliquely 5-lobed; anterior lobe exceeding the posterior 4 and more or less concave; tube bent, less than 1 in. long.

*Cymes axillary :
    Leaves entire :
        Unarmed shrubs :
            Whole plant glabrous (very rarely more
                or less hirsute); leaves profusely and
                minutely gland-dotted beneath   ...   **(3) triphyllum.**
            Leaves and branches hirsute; leaves
                with a few relatively large sessile
                black glands beneath ...   .   **(4) hirsutum.**
        Shrub armed with spines   ...   .   **(5) spinescens.**
    Leaves toothed   ...   .   **(6) cæruleum.**
**Cymes long-peduncled, terminal or clustered at the
    ends of short leafy axillary branches   ...   ...   **(7) myricoides.**

***Cymes forming a terminal panicle :
 Leaves and branches puberulous or pubescent;
 leaves covered with minute scales above :
  Leaves opposite ... ... (8) **Wilmsii.**
  Leaves in whorls of 3 ... ... (9) **simile.**
 Leaves and branches glabrous ... ... (10) **Schlechteri.**

1. C. glabrum (E. Meyer, Comm. 273); a shrub or small tree, 4–15 ft. high; branches finely pubescent when young, later becoming glabrous, with light-grey bark and prominent lenticels; leaves opposite or in whorls of 3 or 4, petiolate, subcoriaceous, ovate, elliptic or lanceolate, acute or obtuse at the apex, cuneate at the base, entire, glabrous (or puberulous on the nerves), shining, profusely glandular-punctate beneath, with 4–7 primary lateral nerves on each side, prominent beneath, $\frac{1}{2}$–4 in. long, $\frac{1}{2}$–3 in. broad; petiole slender, $\frac{1}{4}$–$\frac{3}{4}$ in. long; cymes many-flowered, bracteate, contracted into a dense pyramidal or corymbose terminal panicle, lower ones in the axils of the upper leaves; bracts and bracteoles linear or linear-subulate, glabrous or pubescent, 2–3 lin. long; flowers pedicelled, white; calyx campanulate, 5-lobed, glabrous, puberulous or pubescent, with numerous sessile spherical glands without, $1\frac{1}{2}$–3 lin. long; lobes about half as long as the tube, subulate, acute; corolla-tube straight, finely puberulous and glandular, 3–4 lin. long; lobes subequal; ovary oblong, glabrous, 2-celled; ovules 2 in each cell; style far exserted; drupe about the size of a pea, glabrous, slightly exserted from the spreading accrescent calyx, containing 1 or 2 1-seeded pyrenes. *Walp. Rep.* iv. 110; *Schauer in DC. Prod.* xi. 661; *Baker in Dyer, Fl. Trop. Afr.* v. 297; *Wood, Natal Plants,* t. 45. *C. glabrum, var. angustifolia, E. Meyer, l.c.; Walp. l.c. C. capense, Ecklon & Zeyher ex Schauer l. c. C. ovale, Baker in Dyer, Fl. Trop. Afr.* v. 298 *partly. Ehretia triphylla, Hochst. in Flora,* 1844, 830. *Amerina triphylla, A.DC. in DC. Prod.* ix. 513. *Siphonanthus glabra, Hiern in Cat. Afr. Pl. Welw.* i. 842. *S. glabra, var. vaga, Hiern l.c.*

VAR. β, **ovale** (H. H. W. Pearson); young branches pubescent; leaves pubescent beneath and on the nerves above; panicle, branches, bracts and calyx pubescent or villous. *C. ovale, Klotsch in Peters, Reise Mossamb. Bot.* 257; *Vatke in Linnæa,* xliii. 537; *O. Kuntze, Rev. Gen. Pl.* iii. 250; *Baker in Dyer, Fl. Trop. Afr.,* v. 298 *partly.*

SOUTH AFRICA : among the stunted bush of the sand-hills near the coast, *MacOwan,* 748!

COAST REGION : Albany Div.; without precise locality, *Cooper,* 3496! Bathurst Div.; Port Alfred, 50 ft., *Galpin,* 2937! Komgha Div.; on the banks of the River Kei, 500 ft., *Drège.*

KALAHARI REGION : Transvaal; near Barberton, *Thorncroft,* 48 (*in Herb. Wood,* 4170)! Highland Creek near Barberton, 2900 ft., *Galpin,* 774!

EASTERN REGION: Transkei; banks of the Bashee River, *Drège!* Fort Bowker, on the Xnabara River, *Bowker,* 549; Natal; near Durban, *Gerrard and McKen,* 661! *Krauss,* 100! *Hewitson!* Inanda, 1000–2000 ft., *Wood,* 7551! and without precise locality, *Sanderson! Gerrard,* 726! *Cooper,* 1220! Zululand; without precise locality, *Gerrard,* 638! Var. β: Natal; Inanda, *Wood,* 1204! Clairmont, *Kuntze,* and without precise locality, *Cooper,* 1214!

Zululand ; near streams, *Gerrard and McKen,* 2026 ! and without precise locality, *Gerrard and McKen,* 2156!

Also in Tropical Africa.

I have not seen the type of var. *angustifolia,* but have reduced it, as there is a series of forms connecting a Natal plant (*Sanderson*), which corresponds closely with Meyer's description, with the typical form. For a similar reason I have not kept up Hiern's var. *vaga* which is also represented by a Natal specimen (*Hewitson*).

Native names (Zulu) "um-Quaquane," "um-Quaqongo"; some parts of the tree are used as a purgative for calves (Wood, l.c.).

2. **C. Rehmanni** (Gürke in Engl. Jahrb. xxviii. 294); a shrub; branches terete, when young clothed with a dense light-grey velvety tomentum, later becoming glabrous, with light-grey bark and prominent lenticels; leaves opposite, shortly petiolate, coriaceous, ovate or ovate-lanceolate, obtuse or acute at the apex, cuneate at the base, entire, pubescent or glabrescent above, pubescent beneath, with 4–5 primary lateral nerves on each side, obscure above, prominent beneath, 1–2½ in. long, 5–9 lin. broad; petiole densely pubescent, 2–3 lin. long; panicle as in *C. glabrum,* with villous-tomentose branches, 1½–2 in. long, 1–1½ in. broad; bracts and bracteoles linear or linear-subulate, tomentose, 2–4 lin. long; flowers pedicelled; calyx tubular, 5-toothed, villous without, 1–2 lin. long; corolla-tube straight, cylindric, pubescent and glandular in the upper part, ⅓–½ in. long; lobes subequal, oblong, subacute, pubescent and glandular beneath, glabrous above, about ⅙ in. long; ovary oblong, glabrous, 2-celled; ovules 2 in each cell.

KALAHARI REGION: Transvaal; Boshveld, between Elands River and Klippan, *Rehmann,* 5066! Houtbosch, *Rehmann,* 6199! 6200! Macapans Berg, at Stryd Poort, *Rehmann,* 5468! Waterval River, near Lydenburg, *Wilms,* 601!

EASTERN REGION: Delagoa Bay, *Junod,* 161 !

3. **C. triphyllum** (H. H. W. Pearson); a low undershrub, ½–2 ft. high; stems erect from a woody rootstock, unbranched, angular, striate, usually puberulous at the nodes, otherwise glabrous when adult; leaves in whorls of 3 or 4 or opposite, sessile, coriaceous, linear-oblong, acute or subacute, narrowed at the base, entire, with a distinct midrib and obscure ascending lateral nerves, glabrous, profusely gland-dotted beneath (rarely hirsute on the nerves beneath and margins), pale green, ½–2½ in. long, 1–6 lin. broad; cymes 1- to few-flowered, pedunculate, axillary; peduncle solitary, slender, with 2 opposite lanceolate bracts near the summit, up to 1 in. long; flowers pedicelled, blue or deep purple; calyx campanulate, 5-lobed, 5-ribbed, glabrous, 1½–3 lin. long, the tube equalling or slightly exceeding the ovate acute segments; corolla about three times as long as the calyx; tube bent, villous within at the throat or entirely glabrous, 1½–3 lin. long; 4 upper lobes subequal, obliquely obovate or elliptic, obtuse; lower obovate or oblong, exceeding the upper; filaments glabrous; drupe 1–2-seeded, ovoid, smooth, 5–8 lin. long,

4–6 lin. broad. *C. natalense, Gürke in Engl. Jahrb.* xviii. 183. *Cyclonema triphyllum, Harvey, Thes. Cap.* i. 17, *t.* 27.

SOUTH AFRICA : without precise locality, *Zeyher,* 1362! 1363!
KALAHARI REGION : Orange River Colony ; high situations near the Sand and Vals (Valsch) Rivers, *Barber,* 744! and without precise locality, *Cooper,* 897! Transvaal ; Aapies Poort, near Pretoria, *Rehmann,* 4236! Wonderboom Poort, near Pretoria, *Rehmann,* 4560! in fields near Pretoria, 4000 ft., *McLea in Herb. Bolus,* 3137! Magalies Berg, *Burke,* 117! 365! near Lydenburg, *Wilms,* 1153! Jeppes Ridges, near Johannesburg, 6000 ft., *Gilfillan in Herb. Galpin,* 6166! grassy plains around Barberton, 2500–4000 ft., *Galpin,* 506! Pilgrims Rest, *Greenstock!* and without precise locality, *Tuck,* 7!
EASTERN REGION : Natal ; Klip River, 3500–4500 ft., *Sutherland!* near Newcastle, *Wilms,* 2196! near Pieter Maritzburg, *Wilms,* 2125! near Gourton, amongst grass, 3000–4000 ft., *Wood!* Zululand ; without precise locality, *Gerrard,* 1251!

4. C. hirsutum (H. H. W. Pearson); a low undershrub up to 1½ ft. high ; stems erect from a woody rootstock, simple or branched at the base, 4-angled, hirsute ; leaves opposite or in whorls of 3, sessile or very shortly petioled, membranous, elliptic or oblong, acute, obtuse or rounded at the apex, narrowed at the base, with an entire ciliate margin, more or less hirsute, profusely glandular, with 1–3 primary lateral nerves on each side, ⅓–1¼ in. long, 2–5 lin. broad ; cymes 1–2-flowered, pedunculate, axillary ; peduncle solitary, slender, with 2 opposite linear or linear-lanceolate bracts near the summit, hirsute, ¾–1¼ in. long ; flowers pedicelled, sky-blue (*Gerrard*) ; calyx campanulate, 5-lobed, obscurely 5-ribbed, more or less hirsute, about 2 lin. long, the tube equalling the lanceolate hirsute segments ; corolla about ½ in. long ; tube bent, villous within at the throat, otherwise glabrous, 1¼–2¼ lin. long ; 4 upper lobes subequal, cuneate or subrotund ; lower oblong-cuneate, obtuse, exceeding the upper ; stamens scarcely exserted ; filaments glabrous ; ovary densely hirsute, later glabrescent ; drupe not seen. *Cyclonema? hirsutum, Hochst. in Flora,* 1842, 228, *and* 1845, 68 ; *Walp. Rep.* iv. 101 ; *Schauer in DC. Prod.* xi. 676.

VAR. β, ciliatum (H. H. W. Pearson); stems glabrous or glabrescent ; adult leaves ciliate, otherwise glabrous, or with a few scattered hairs near the margins ; peduncles glabrous ; bracts and calyx-lobes ciliate, otherwise glabrous. *Cyclonema ciliatum, Harv. MSS.*

EASTERN REGION : Griqualand East ; near the River Ibisi, 2500 ft., *Tyson in MacOwan, Herb. Norm. Aust.-Afr.,* 1516 ! Natal ; in grassy plains between the Umlaas (Umlazi) River and Durban Bay, *Krauss,* 106 ! near Greytown, *Wilms,* 2197 ! Pieter Maritzburg, 2000–3000 ft., *Sutherland!* near Durban, 3000–4000 ft., *Sutherland!* Gerrard and McKen, 800 ! Inanda. *Wood,* 41 ! and without precise locality, *Gerrard,* 324 ! Swaziland ; Havelock Concession, 4000 ft., *Saltmarshe in Herb. Galpin,* 985 ! var β : Natal ; Umgeni Falls, *Sanderson!* Attercliffe, *Sanderson,* 243 ! and without precise locality, *Sanderson,* 237 !

5. C. spinescens (Gürke in Engl. Jahrb. xviii. 180); a shrub ; branches terete, densely pubescent, frequently armed with axillary or supra-axillary straight or recurved pubescent spines shorter than the leaves; leaves opposite, shortly petioled, elliptic or suborbicular, minute apiculate, entire, subcoriaceous, pubescent, with 3–5 spreading primary lateral nerves on each side, obscure above, distinct beneath, about 1 in. long, $\frac{1}{2}$–$\frac{3}{4}$ in. broad; petiole up to 3 lin. long; cymes pedunculate in the axils of the upper leaves, usually 1-flowered; peduncle 2-bracteate above the middle, pubescent, shorter than or slightly exceeding the leaf; bracts linear, pubescent, 2–3 lin. long; calyx campanulate, densely hirsute, glandular, $\frac{1}{4}$–$\frac{1}{2}$ in. long; lobes 5, unequal, ovate, acute, not exceeding the tube; corolla glandular-pubescent; tube curved, puberulous within, twice as long as the calyx; 4 upper lobes obovate, obtuse or subacute, $3\frac{1}{2}$–5 lin. long, $2\frac{1}{2}$–$3\frac{1}{2}$ lin. broad; lower 1 cuneate-obovate, concave, rounded or subtruncate at the apex, exceeding the upper; stamens exserted; filaments puberulous, $\frac{1}{2}$–1 in. long; ovary glabrous; style slightly dilated at the base, $\frac{3}{4}$–1 in. long; drupe not seen. *Baker in Dyer, Fl. Trop. Afr.* v. 313. *C. uncinatum, Schinz in Verhandl. Bot. Vereins Brandenb.* xxxi. (1890) 206. *Cyclonema spinescens, Oliv. in Journ. Linn. Soc.* xv. 96; *Hook. Ic. Pl. t.* 1221. *Kalaharia spinipes, Baill. Hist. Plant.* xi. 111. *K. spinescens, Gürke in Engl. Pfl. Ost-Afr. C.* 340; *Henriques in Bolet. Soc. Brot.* xvi. 69.

KALAHARI REGION: Bechuanaland; Bakwena Territory, in Sirorume Valley, 3500 ft., *Holub!*

Also in Tropical Africa.

The plant upon which Schinz founded his *C. uncinatum* was collected in the North West Kalahari at Gorekas, about 90 miles North of the Tropic. The range of the species in South Africa will therefore probably be extended to the Western Kalahari and Namaqualand.

6. C. cæruleum (N. E. Br. in Kew Bulletin, 1895, 115); a low much-branched shrub, 2–3 ft. high; young branches dark brown, 4-angled, puberulous along 2 opposite lines when young, later with glabrous, cinereous, more or less tuberculate, wrinkled bark; leaves opposite, petioled, membranous, ovate, ovate-lanceolate or suboblong, acute, cuneate at the base, coarsely and acutely 3–4-toothed or serrate, rarely entire, sparsely puberulous above and on the nerves beneath, ciliolate along the margins, with 3–5 primary lateral nerves on each side, prominent beneath, $\frac{1}{3}$–$2\frac{1}{4}$ in. long, $\frac{1}{6}$–1 in. broad; petiole puberulous, 1–6 lin. long; cyme 1–3-flowered, pedunculate, solitary, axillary; peduncle slender, 2-bracteate near the summit, puberulous along the posterior line, otherwise glabrous, $\frac{3}{4}$–$1\frac{1}{4}$ in. long; flowers pedicelled, deep blue (*Gerrard*); bracts subulate, entirely glabrous or ciliolate on the margin, 1–$1\frac{1}{4}$ lin. long; calyx campanulate, 5-toothed, strongly 5-nerved, glabrous or minutely puberulous, 1–$2\frac{1}{2}$ lin. long; teeth 5, distant, narrowly deltoid, long-acuminate, acute, obscurely ciliolate on the margins, equalling the tube; corolla-tube bent, villous within at the throat, otherwise glabrous, about twice as long as the calyx; 4 upper lobes subequal, elliptic, obtuse; lower 1 cuneate-obovate, subtruncate; stamens and style far exserted, incurved; drupe 4-lobed, 2-seeded, glabrous.

EASTERN REGION : Natal; Mooi River Valley, 2000-3000 ft., *Gerrard and McKen*, 1252 ! *Sutherland !* and without precise locality, *Gerrard and McKen*, 2024! Swaziland ; near Jackson's Mission, *Mrs. K. Saunders !*

The foliage does not possess a disagreeable odour (*Gerrard*). See *C. myricoides.*

7. C. myricoides (R. Br. in Salt, Abyss. Append. lxv.) ; a low erect or scandent shrub, 3–6 ft. high, with leaves and young parts pubescent with short multicellular hairs, or glabrescent ; old branches angled, glabrous, striate, with light-brown bark and prominent leaf-scars and lenticels ; leaves opposite or whorled, petiolate or sub-sessile, membranous, oblong, acuminate, acute or rounded at the apex, cuneate at the base, coarsely, irregularly and acutely or obtusely serrate or inciso-serrate towards the apex or subentire, with 4–6 primary lateral nerves on each side, conspicuous beneath, $1\frac{1}{2}$–$3\frac{1}{2}$ in. long, $\frac{1}{2}$–2 in. broad ; petiole puberulous or pubescent, $\frac{1}{4}$–$\frac{1}{2}$ in. long ; cymes bracteate, lax, 1–3-flowered, forming short, loose, glabrous or puberulous panicles, terminal from short leafy axillary branches ; bracts and bracteoles linear or linear-lanceolate, pubescent, $1\frac{1}{2}$–$3\frac{1}{2}$ lin. long, lower ones frequently larger and leafy ; calyx broadly cam-panulate, 5-lobed, glabrous or glabrescent, $\frac{1}{6}$–$\frac{1}{3}$ in. in diam. ; lobes spreading, broadly ovate, obtuse or rounded at the apex, profusely glandular, leafy, slightly exceeding the tube ; corolla-tube short, bent, villous in the throat, otherwise glabrous, up to $\frac{1}{3}$ in. long ; 4 upper lobes subequal, oblong, obtuse or rounded at the apex, greenish-white, about $\frac{1}{3}$ in. long ; the lower 1 obovate-spathulate, about twice as long as the upper, concave, usually pale blue ; stamens and style far exserted ; filaments thickened and densely villous, with shaggy hairs in the lower half ; ovary globose, black, glabrous, glandular, 2-celled ; ovules 2 in each cell ; drupe deeply lobed, 2–3-seeded, 5–6 lin. in diam. near the apex. *Vatke in Linnæa,* xliii. 535 ; *Baker in Dyer, Fl. Trop. Afr.* v. 310. *Cyclonema myricoides, Hochst., in Flora,* 1842, 226 ; *Walp. Rep.* iv. 101 ; *Schauer in DC. Prod.* xi. 675 ; *Hook. f. Bot. Mag. t.* 5838. *C. myricoides, Hochst., var. sylvaticum, Schauer l.c.* 676. *C. sylvaticum,* and *C. serratum, Hochst. in Flora,* 1842, 227, and 1845, 68 ; *Walp. Rep. l.c.*

VAR. β. **cuneatum** (H. H. W. Pearson); leaves cuneate or cuneate-oblong, pubescent above, densely so beneath, $1\frac{1}{2}$–$2\frac{1}{2}$ in. long, $1\frac{1}{4}$–2 in. broad; cyme branches pubescent. *C. cuneatum, Gürke in Engl. Jahrb.* xxviii. 303.

EASTERN REGION: Natal; in woods near the Umlaas (Umlazi) River, *Krauss,* 333 ! 335! near Durban, *Gueinzius ! Gerrard and McKen,* 798 ! Inanda, *Wood,* 657 ! Congella, *Sanderson,* 718 ! and without precise locality, *Gerrard,* 21 ! 382 ! Zululand ; Eshowe, *Mrs. K. Saunders !* and without precise locality, *Mrs. K. Saunders !*

KALAHARI REGION : Var. β : Transvaal ; in a kloof near Schoemanns farm, in the Lydenburg District, *Wilms,* 160 ! amongst scrub near the water at Umvoti Creek, near Barberton, 3000 ft., *Galpin,* 601 ! Houtbosch, *Rehmann,* 6188.

Also in Tropical Africa.

Gerrard states (under *C. cæruleum*) that the foliage possesses a disagreeable odour.

8. C. Wilmsii. (Gürke in Engl. Jahrb. xxviii. 304); a low
undershrub; young branches subangular, dark brown; finely pubescent, later with glabrous cinereous more or less tuberculate wrinkled
bark; leaves opposite, sessile or shortly petioled, subcoriaceous,
oblong or oblong-lanceolate, acute, obtuse or rounded at the apex,
cuneate at the base, puberulous on the nerves beneath, otherwise
glabrous, with very numerous minute scales on the upper surface,
9–19 lin. long, 2½–5 lin. broad; primary lateral nerves 4–5 on each side,
ascending, obscure above, prominent beneath, with margins entire or
with 1–3 serrations towards the apex; cyme terminal, few-flowered,
bracteate, 1¼–1¾ in. long; bracts leafy, sessile, lanceolate, entire or
serrate, puberulous, 3–8 lin. long; calyx tubular, narrowed at the
base into the short pedicel, with a sinuate shortly 5-toothed limb,
puberulous or glabrous without, 2–4 lin. long; corolla-tube narrow,
straight, glandular-puberulous without, 3–4 times as long as the
calyx; 4 upper lobes subequal, subrotund; lower 1 oblong, exceeding the upper; ovary 4-lobed, glabrous; drupe not seen.

KALAHARI REGION: Transvaal; near Lydenburg, *Wilms,* 1082! Waterval
River, near Lydenburg, *Wilms,* 1159! Bosch Veld, at Kameel Poort, *Rehmann,*
4825! Komati Poort, 1000 ft., *Schlechter,* 11861!

9. C. simile (H. H. W. Pearson); a low undershrub about 1 ft.
high; stems erect from a woody rootstock, subangular, dark coloured,
finely pubescent, later with glabrous cinereous wrinkled bark;
leaves in whorls of 3, sometimes opposite, sessile, coriaceous, oblong
or elliptic, subacute, obtuse or rounded at the apex, cuneate at the
base, distantly and acutely serrate towards the apex, puberulous on
the nerves beneath, otherwise glabrous, with very numerous minute
scales on the upper surface, with 4–5 ascending primary lateral
nerves on each side, obscure above, prominent beneath, 1–1½ in.
long, 3–5 lin. broad; cymes 1–3-flowered, shortly pedunculate, forming a loose terminal few-flowered panicle about 4 in. long; bracts
leafy below, above linear-subulate, ciliate on the margin, otherwise
glabrous, ½–1 lin. long; calyx campanulate with a 5-toothed spreading sinuate limb, 5-ribbed, puberulous without, 1½–2 lin. long;
corolla-tube narrow, straight, glandular-puberulous without, ¾–1 in.
long; 4 upper lobes subequal, obovate, glandular-puberulous
without, about 2 lin. long; lower 1 oblong, concave, exceeding the
upper; ovary black, glabrous, 4-lobed; drupe not seen.

KALAHARI REGION: Transvaal; Boshveld, at Klippan, *Rehmann,* 5210!
"Gold-fields," *Baines!*

In habit this species resembles *C. triphyllum* and *C. hirsutum.* In the
characters of the inflorescence and flower it is allied to *C. Wilmsii,* chiefly
differing in the longer terminal panicle, longer corolla-tube and the whorled
leaves.

10. C. Schlechteri (Gürke in Engl. Jahrb. xxviii. 302); a low
shrub, with glabrous branches; leaves coriaceous, opposite, shortly
petiolate, broadly ovate, acute, cuneate at the base, coarsely serrate,
glabrous, 2¾–4 in. long, 1½–2½ in. broad; petiole 5–8 lin. long; panicle

terminal, few-flowered ; bracts leafy below, diminishing upwards, the uppermost being scarcely 5 lin. long ; bracteoles sessile, lanceolate, acute, glabrous, $2\frac{1}{2}$–$3\frac{1}{2}$ lin. long, $\frac{1}{2}$–1 lin. broad ; flowers supported on pedicels about $2\frac{1}{2}$ lin. long ; calyx broadly campanulate, 5-lobed to about the middle, 2–3 lin. long ; lobes suborbicular, obtuse, ciliate, otherwise glabrous ; stamens exserted, villous at the base of the filaments.

KALAHARI REGION : Transvaal ; Lions Creek, in shady places at 1000 ft., *Schlechter*, 12197 !

Imperfectly known Species.

11. **C. capense** (Don ex Steudel, Nomencl. ed. 2, i. 382). This may be a synonym of *C. glabrum*, E. Meyer.

XIV. AVICENNIA, Linn.

Calyx short, 5-partite, unchanged in the fruit, with broadly ovate imbricate lobes. *Corolla* with a short wide cylindric straight tube ; limb spreading, 4-lobed ; lobes subequal or the posterior a little broader. *Stamens* 4, inserted in the throat of the corolla-tube ; filaments very short ; anthers scarcely exserted, with parallel cells. *Ovary* with a central 4-winged conical column, imperfectly 4-celled ; ovules 4 (1 in each cell), pendulous from the apex of the axile placental column. *Fruit* dry, compressed, dehiscing by 2 thickened valves, 1-seeded. *Embryo* naked on account of the arrested development of the integuments of the ovule ; cotyledons large, longitudinally folded ; radicle inferior, villous ; plumule commencing to grow before the fruit falls (cf. *Rhizophora*).

Glabrous or canescent shrubs ; leaves opposite, entire, coriaceous ; cymes contracted, capituliform, pedunculate, usually paired in the axils of the upper leaves or arranged in a short thyrsus or trichotomous corymb at the apex of the branch ; flowers small, sessile, each with 2 bracteoles in the axil of a bract ; bracts and bracteoles shorter than the calyx.

Three species, on the Tropical shores of both hemispheres.

1. **A. officinalis** (Linn. Sp. Pl. ed. i. 110) ; a shrub or small tree ; young branches terete, furrowed, clothed with a dense minute white or yellowish-white indumentum, later glabrescent ; leaves oblong, oblong-lanceolate or elliptic, acuminate, acute or obtuse, narrowed at the base into the short petiole, entire, coriaceous, thick, green, glabrous and prominently reticulately veined above, densely clothed by a minute white or yellowish-white indumentum beneath, 2–3 in. long, 9–13 lin. broad ; spikes dense, globose, terminal and axillary, supported on short acutely 4-angled peduncles ; bracts and bracteoles broadly ovate, densely silvery tomentose along the margin and on the back or glabrescent, glabrous above, about $1\frac{1}{4}$ lin. long ; calyx-tube very short ; lobes broadly oblong or elliptic, rounded at the apex, glabrous within, pubescent on the back and margins, about 2 lin. long ; corolla-tube $\frac{1}{2}$–1 lin. long ; lobes 4, oblong, yellow, glabrous

within, densely and minutely tomentose without except at the tip, 1½–2 lin long ; ovary deeply grooved, tomentose.   *Schauer in DC. Prodr.* xi. 700 ; *Clarke in Hook. f. Fl. Brit. Ind.* iv. 604 ; *Baker in Dyer, Fl. Trop. Afr.* v. 332.   *A. resinifera, Forst. Prodr.* 45. *A. tomentosa, Jacq. Select. Stirp. Amer. Hist.* 178, *t.* 112, *fig.* 2 ; *Wight, Ic. Pl. t.* 1481 ; *R. Br. Prodr.* 518 ; *Roxb. Fl. Ind.* iii. 88 ; *Wall. Pl. As. Rar.* iii. 44, *t.* 271 ; *E. Meyer, Comm.* 277 ; *Walp. Rep.* iv. 131 ; *Schauer, l.c.* 699, 700 ; *Hochst. in Flora,* 1845, 68 ; *var. arabica, Walp. l.c.* 133.   *A. africana, Beauv. Fl. Owar.* i. 80, *t.* 47. *A. Meyeri, Miq. in Linnæa,* xviii. 262.

EASTERN REGION : Bomvanaland ; at the mouth of the River Xara, *Soga in MacOwan, Herb. Austr.-Afr.,* 1941 ! *MacOwan,* 3203 ! Natal ; on the muddy shore of Durban Bay, *Drège* ! *Krauss,* 241 ! *Sanderson,* 886 ! *Plant,* 21 ! *Wood,* 395 ! 1360 ! *Wilms,* 2229 ! *Rehmann,* 9004 ! *Peddie* ! *Cooper,* 1233 ! Delagoa Bay, *Junod,* 500 !

Found growing with *Rhizophora mucronata,* Lam., and *Bruguiera gymnorrhiza,* Lam., Drège, Zwei Pflanzengeogr. Documente, 159, no. 32 ; Harvey & Sonder, Flora Capensis, ii. 513, 514.   Also on the Tropical shores of both hemispheres.

ORDER CV.   **LABIATÆ.**

(By N. E. BROWN, T. COOKE and S. A. SKAN.)

Flowers irregular or more rarely regular or subregular, hermaphrodite or rarely unisexual.   *Calyx* tubular, campanulate or funnel-shaped, regularly or irregularly 3–10- (usually 5- rarely many-) toothed or with 2 entire or toothed lips, very rarely truncate or 5-partite, persistent and often enlarged in fertile flowers, very rarely deciduous above the base at the ripening of the fruit. *Corolla* gamopetalous, 2- (rarely 1-) lipped, or oblique or subregular and 4–5-lobed, deciduous.   *Stamens* usually 4, in 2 pairs of unequal length or subequal, all fertile or the upper pair sterile, occasionally 2 only, inserted in or at the mouth of the corolla-tube ; anthers 1- or 2-celled, opening longitudinally.   *Ovary* superior, seated on an entire or lobed disk, deeply or rarely shortly 4-lobed, 4-celled ; style central, arising from the base between the lobes, filiform ; stigma bifid or entire ; ovules solitary in each lobe or cell, erect, anatropous.   *Fruits* of 4 or by abortion fewer dry 1-seeded nutlets (the ripened lobes of the ovary).   *Seed* erect ; testa thin ; albumen little or usually none ; embryo straight or rarely curved, with fleshy cotyledons ; radicle next to the hilum.

Herbs or shrubs, usually with square stems and branches ; leaves opposite, whorled or rarely alternate, entire, toothed or lobed, usually gland-dotted ; flowers solitary and opposite or more usually 3 to many in a whorl and the pairs or whorls spaced out in terminal racemes or along the branches of a panicle, or crowded into a spike or head or corymb, or seated in the axils of foliage leaves, bracteate or with bracteoles mingled with the flowers.

DISTRIB. Genera about 170 ; species about 3400, in all warm and temperate regions, rare in arctic or alpine areas.

Burmann (*Fl. Cap. Prodr.* 16) enumerates the following species, which are mostly natives of Europe, the Mediterranean Region and the Orient, and of which he may have seen specimens of plants cultivated or introduced into South Africa. If introduced they have now apparently disappeared, as no South African material of any has been found in the Kew Herbarium : *Origanum sipyleum*, Linn., *O. syriacum*, Linn., *O. majoranum* [= *O. Majorana*, Linn.], *Thymus vulgaris*, Linn., *T. serpillum* [*T. Serpyllum*, Linn.], *Satureia hortensis*, Linn., *Hyssopus officinalis*, Linn., *H. nepethoides* [= *Lophanthus nepetoides*, Benth.], *Thymbra spicata*, Linn., *Nepeta Cataria*, Linn., *Prunella hyssopifolia*, Linn., *Marrubium peregrinum*, Linn., *Lamium Orvala*, Linn., *Phlomis zeylanica*, Linn. [= *Leucas zeylanica*, R. Br.], and *Prasium majus*, Linn.

Rosmarinus officinalis, Linn., was collected by Cooper (no. 3110) at Mequatling in Basutoland, where it had been introduced by a missionary some time before 1861.

Tribe 1. OCIMOIDEÆ.—*Calyx* equally or unequally 3–5-toothed, with the upper tooth often much larger than the others and sometimes decurrent on the tube. *Corolla* 2-lipped, oblique or nearly regular ; upper lip flattish or not hooded. *Stamens* 4, absent or rudimentary in female flowers, in pairs and all directed upon the lower side or lip of the corolla or about equally spreading, never all ascending ; anthers perfectly or imperfectly 1-celled by the confluence of the cells at the apex.

* *Flowers hermaphrodite, all with a fertile ovary.*

† Calyx enlarged, but not fleshy in fruit ; stamens all with fertile anthers.
‡ Corolla either distinctly 2-lipped or 5-lobed ; stamens exserted from the corolla-tube.
§ Calyx persistent in fruit, distinctly 3–5-toothed.
‖ Filaments all free, those of the upper pair bent like a knee or toothed or crested near the base ; upper calyx-tooth broadly ovate or sub-orbicular, decurrent on the tube.

I. **Becium.**—*Calyx* 3–5-toothed, its tube with a broad oblique or truncate space at its mouth, ciliate and sometimes with small teeth along its margin, separating the upper tooth from the others. *Corolla-tube* equalling or exserted beyond the calyx-teeth.

II. **Ocimum.**—*Calyx* 5-toothed, with no broad space at the mouth of the tube separating the upper tooth from the lateral teeth. *Corolla-tube* not or scarcely longer than the calyx-tube.

‖ ‖ Lower pair of filaments free or united, those of the upper pair some-times bearded or ciliate, but without a knee (except in *Orthosiphon bracteosus*) tooth or crest near the base.

III. **Orthosiphon.**—*Calyx* unequally 5-toothed ; upper tooth larger than the others, suborbicular to elliptic-oblong, sometimes decurrent on the tube. *Corolla-tube* usually much exserted, but sometimes not exceeding the calyx-teeth, straight or nearly so. *Stamens* all free or the filaments of the lower pair variably united, exserted.

IV. **Synclostemon.**—*Calyx* equally or subequally 5-toothed ; upper tooth not or scarcely larger than the others, not decurrent on the tube. *Corolla-tube* exserted, straight. *Filaments* of the lower pair of stamens united, all exserted.

V. **Plectranthus.**—*Calyx* equally or unequally 5-toothed ; upper tooth sometimes decurrent on the tube. *Corolla-tube* exserted, straight or decurved, often with a gibbosity or spur-like projection near the base ; lower lip compressed-boat-shaped. *Filaments* free.

VI. **Coleus.**—*Calyx* unequally 5-toothed ; upper tooth much broader than the others, sometimes decurrent on the tube. *Corolla-tube* exserted, usually very abruptly bent at about the middle ; lower lip compressed-boat-shaped. *Filaments* all shortly united above their insertion.

VII. **Pycnostachys.**—*Flowers* in a very dense spike. *Calyx* with 5 equal rigid spine-like teeth. *Corolla-tube* exserted, deflexed ; lower lip compressed-boat-shaped. *Filaments* all free.

VIII. **Geniosporum.**—*Calyx* 4–5-toothed, with the 3 upper teeth subequal or the middle one much larger than the others, not or scarcely decurrent on the tube, more rarely with 5 subequal teeth ; tube with prominent transverse veins when in fruit. *Corolla* very small ; tube exserted.

XI. **Hyptis.**—*Calyx* subequally 5-toothed ; upper tooth not decurrent on the tube. *Corolla* small, 5-lobed ; 4 of the lobes flat, variously directed, the fifth compressed-concave or saccate.

§§ Calyx falling away by a clean cut just above the base in fruit.

IX. **Æolanthus.**—*Calyx* very small, truncate or with a very short truncate upper lobe and 2–3 minute lower teeth. *Corolla* with a much exserted tube, 2-lipped. *Stamens* all free.

‡‡ Corolla with 4 subequal or slightly unequal lobes, not distinctly 2-lipped ; stamens included in the corolla-tube.

X. **Endostemon.**—*Calyx* unequally 5-toothed ; upper tooth suborbicular, very slightly decurrent on the tube. *Corolla* very small, straight.

†† Calyx enlarged and fleshy in fruit ; upper pair of stamens reduced to staminodes with rudimentary anthers.

XII. **Hoslundia.**—*Flowers* small. *Calyx* subequally 5-toothed, with the upper tooth slightly larger than the others, not decurrent on the tube.

** *Flowers unisexual, but with an ovary in the male flowers, which is never fertile, the sexes on different plants.*

XIII. **Iboza.**—*Calyx* minute. *Corolla* very small, subequally 5- (rarely 4-) lobed, or the lower lobe rather larger than the others. *Stamens* equally spreading.

Tribe 2. SATURINEÆ.—*Calyx* 5–10- (sometimes 13- rarely 15-) nerved, subequally 5-toothed or 2-lipped. *Corolla* subequally or unequally 4–5-lobed or 2-lipped ; lobes or lips usually very small and flat. *Stamens* (in the South African genera) 4, equal or the lower pair longer, distant, divergent or ascending under the upper lip ; anthers 2-celled or 1-celled by confluence of the cells at the apex.

XIV. **Mentha.**—*Calyx* 10-nerved. *Corolla* subequally 4-lobed. *Stamens* equal, erect, distant.

XV. **Micromeria.**—*Calyx* 13- (rarely 15-) nerved. *Corolla* 2-lipped. *Stamens* didynamous, the lower pair longer, all ascending under the upper lip.

Tribe 3. MONARDEÆ.—*Calyx* usually 2-lipped, more rarely subequally lobed. *Corolla* 2-lipped ; lips often large, the upper concave. *Stamens* 2, ascending under the upper lip ; anthers often with a long connective bearing 1 or 2 (sometimes 1 perfect and the other imperfect) more or less widely separated linear or oblong cells.

XVI. **Salvia.**—The only South African genus. *Connective* of the anthers jointed to a short filament, the upper part ascending and bearing a perfect cell, the lower deflexed or horizontal and bearing a smaller polliniferous or empty cell, or sometimes quite naked.

Tribe 4.   NEPETEÆ.—*Calyx* usually 15-nerved, equal or often oblique, 5-toothed or somewhat 2-lipped ; upper teeth usually largest.   *Corolla* 2-lipped ; lips rather large, the upper concave.   *Stamens* 4 (rarely 2), the upper pair longer, all ascending under the upper lip or sometimes divergent ; anthers 2-celled ; cells more or less divaricate, sometimes parallel.

XVII.   **Cedronella.**—The only genus in South Africa, where it is naturalized. *Leaves* 3-foliolate.   *Anther-cells* parallel.

Tribe 5.   STACHYDEÆ.—*Calyx* 5–10-nerved, equally or unequally 5–10-toothed (rarely many-toothed), rarely sometimes 2-lipped.   *Corolla* 2-lipped ; lips usually large, the upper erect and usually concave.   *Stamens* 4, the lower pair longer, all ascending under the upper lip or sometimes included in the tube ; anthers 2-celled or rarely 1-celled ; cells divergent or divaricate or confluent, sometimes parallel.

XVIII.   **Acrotome.**—*Calyx* tubular or campanulate, prominently 10-nerved, 5–10-toothed.   *Corolla-tube* often exserted ; upper lip somewhat arched.   *Stamens* included in the tube ; anther-cells confluent. *Nutlets* truncate at the apex.

XIX.   **Stachys.**—*Calyx* usually campanulate, 5–10-nerved, equally or subequally 5-toothed, rarely somewhat 2-lipped.   *Corolla-tube* included or exserted.   *Stamens* exserted ; anther-cells finally divaricate.   *Nutlets* rounded at the apex.

XX.   **Ballota.**—*Calyx* somewhat funnel-shaped, tubular at the base, 10-nerved, 5–10- or sometimes many-toothed ; teeth dilated at the base or connate in a spreading orbicular or oblique limb.   *Corolla-tube* subincluded.   *Stamens* exserted ; anther-cells finally divaricate. *Style* subequally 2-lobed.   *Nutlets* rounded at the apex.

XXI.   **Leucas.**—*Calyx* tubular, campanulate or sometimes inflated, equal or oblique at the mouth, equally or subequally 6–10-toothed.   *Corolla* usually white ; tube included ; lower lip usually about as large as the upper, not soon withering ; median lobe largest.   *Stamens* exserted ; anther-cells divaricate, finally confluent.   *Style* very unequally 2-lobed.   *Nutlets* scarcely truncate at the apex.

XXII.   **Lasiocorys.**—*Calyx* 5-toothed.   Otherwise as in *Leucas.*

XXIII.   **Leonotis.**—*Calyx* tubular, oblique at the mouth, 10-nerved, 8–10-toothed ; upper tooth usually much the largest.   *Corolla* often orange ; tube often exserted ; upper lip elongated ; lower lip much smaller, subequally 3-lobed, soon withering.   *Stamens* exserted ; anther-cells divaricate, subconfluent.   *Nutlets* ovoid-triquetrous, obtuse or truncate at the apex.

Tribe 6.   AJUGOIDEÆ.—*Calyx* 10- or irregularly many-nerved, equally 5-toothed or with the uppermost tooth largest, sometimes 2-lipped, with entire lips. *Corolla* in the South African genera 1-lipped or with a very short upper lip and a relatively large lower lip.   *Stamens* 4, more rarely 2, the lower pair longer, all as ending ; anthers 2-celled ; cells parallel, divergent or divaricate, sometimes confluent.   *Ovary* shortly 4-lobed or lobed to the middle.   *Nutlets* with an oblique or lateral usually large areole.   (In the other tribes the ovary is deeply 4-lobed and the nutlets have a small basal or slightly oblique areole.)

XXIV.   **Tinnea.**—*Calyx* 2-lipped, inflated in fruit.   *Corolla* 2-lipped.   *Stamens* rarely exserted.

XXV.   **Teucrium.**—*Calyx* equally 5-toothed or the uppermost tooth broadest and sometimes very large.   *Corolla* 1-lipped.   *Stamens* long-exserted.

XXVI.   **Ajuga.**—*Calyx* subequally 5-toothed.   *Corolla* 2-lipped.   *Stamens* exserted.

I. BECIUM, Lindley.

Calyx 3- (rarely 5-) toothed, two-lipped; upper lip or tooth broadly ovate, decurrent on the tube; lower lip formed by the oblique or truncate, ciliate or denticulate mouth of the tube and 2 subulate or bristle-like teeth, lateral teeth none or rarely developed. *Corolla* two-lipped; tube exserted beyond the calyx-teeth or about equalling them; upper lip 4-lobed, erect; lower lip concave or boat-shaped. *Stamens* 4, directed towards the lower lip, exserted; filaments all free, the upper pair abruptly bent or kneed and toothed or crested near the base; anthers dorsifixed, 1-celled. *Ovary* 4-lobed; style filiform; stigma bifid, of 2 linear lobes. *Nutlets* ellipsoid or oblong, slightly compressed dorsally, glabrous. *Lindl. Bot. Reg.* 1842, *Misc.* 42, and 1843, t. 15.

Herbs or small shrubs; leaves opposite or fascicled, simple, gland-dotted; inflorescence terminal, spike-like, with distant or crowded flower-whorls; flowers variable in size.

DISTRIB. A genus of several species, 2 in India, 1 in Arabia, the rest African.

As remarked by Lindley when he established the genus, *Becium* is well distinguished from *Ocimum* by the peculiar form of the calyx; the distinct deltoid acute lateral teeth of *Ocimum* being replaced in *Becium* by the obliquely truncate entire or denticulate, densely ciliate sides of the mouth of the tube, and the 2 lower teeth are small and bristle-like, and often divergent. In the general appearance of the inflorescence and flowers *Becium* is also somewhat different from *Ocimum* and easily recognised. *N. E. Br.*

Flower-whorls densely crowded into a head- or short
spike-like raceme or only 1–3 of the lower separate;
leaves 3–15 lin. broad ... ... (1) **obovatum.**

Flower-whorls all separate, $\frac{1}{8}$–$\frac{1}{2}$ in. apart; leaves $\frac{1}{2}$–$3\frac{1}{2}$
lin. broad:

Leaves $\frac{1}{2}$–$1\frac{1}{2}$ in. long, linear or linear-lanceolate ... (2) **angustifolium.**

Leaves $\frac{1}{3}$–$\frac{1}{2}$ in. long, subspathulate-oblanceolate ... (3) **burchellianum.**

1. B. obovatum (N. E. Br.); stems several from a perennial woody rootstock, herbaceous, 5–10 in. high, puberulous; leaves with petioles $\frac{1}{2}$–2 lin. long; blade $\frac{3}{4}$–$2\frac{1}{4}$ in. long, $\frac{1}{4}$–$1\frac{1}{4}$ in. broad, varying from lanceolate to suborbicular, acute or obtuse, usually tapering into the petiole, entire or slightly toothed, usually glabrous on both sides, occasionally puberulous on the veins beneath, secondary veins usually distinctly visible beneath; flower-whorls all crowded into a dense head-like or short spike-like raceme or the lower 1–3 distant, all with a pair of rather large crater-like glands at their base; racemes often crowned with a tuft of small dark (purple?) leafy bracts; pedicels $\frac{1}{2}$–$\frac{3}{4}$ (in fruit up to $1\frac{1}{2}$) lin. long; calyx campanulate, whitish-pubescent outside, glabrous within, much reticulated, dark purple on the upper side; tube $1\frac{1}{2}$ (in fruit 3) lin. long, ciliate and entire or finely denticulate at the mouth between the dorsal and lower teeth ; dorsal tooth about 1 lin. long, elliptic-ovate, obtuse, decurrent on the tube ; lower teeth $\frac{1}{4}$–$\frac{3}{4}$ lin. long, bristle-like ; corolla white or pink (*Wood*), pubescent on the lips outside ; tube 3 lin. long ; upper lip $3\frac{1}{2}$–5 lin. long from the base of gape, with 4 subquadrate somewhat toothed or crisped lobes, 2 lateral shorter and broader than the others ; lower lip $\frac{1}{5}$–$\frac{1}{4}$ in. long, concave, elliptic, obtuse ; stamens all free, much exserted ; upper pair 7–9 lin. long, with a retrorse ciliate crest at the bend near the base of the filaments ; lower pair $\frac{3}{4}$ in. long ; nutlets slightly compressed, orbicular, $\frac{1}{18}$ in. in diam., pale brown. *Ocimum obovatum, E. Meyer, Comm.* 226 ; *Drège, Zwei Pfl. Documente,* 205 ; *Benth. in DC. Prodr.* xii. 35 ; *Wood, Natal Pl.* iii. *t.* 257. *O. serpyllifolium, var. glabrior, Benth. in E. Meyer, Comm.* 226 (*excl. syn.*). *O. striatum, Hochst. in Flora,* 1845, 66 ; *Krauss, Beitr. Fl. Cap und Natal.* 130.

VAR. β, **hians** (N. E. Br.) ; leaves $\frac{1}{2}$–1 in. long, $\frac{1}{4}$–$\frac{1}{2}$ in. broad, lanceolate or ovate, acute, cuneate or rounded into the not more than $\frac{1}{4}$ lin.-long petiole, varying from glabrous to pubescent on both sides ; secondary veins rarely evident ; corolla white, tingęd with 5 longitudinal violet stripes and violet towards the margin on the upper lip (*Miss Leendertz*). blue (*Cooper*), otherwise as in the type. *Ocimum hians, Benth. in DC. Prodr.* xii. 36 ; *S. Moore in Journ. Bot.* 1903, 405.

VAR. γ, **Galpinii** (N. E. Br.) ; stem and both sides of the leaves pubescent or villous with jointed hairs, otherwise as in the type. *Ocimum Galpinii, Guerke in Engl. Jahrb.* xxvi. 78.

COAST REGION : Bedford Div. ; near Bedford, *Mrs. Hutton* ! East London Div. ; near Panmure, *Mrs. Hutton* !

KALAHARI REGION : Var. β : Orange River Colony ; on the Drakensberg Range, *Cooper,* 824 ! Transvaal ; many localities, common, *Burke,* 515 ! *Zeyher,* 1353 ! *McLea in Herb. Bolus,* 3106 ! *Wilms,* 1112 ! 1119 ! *Rand,* 710 ! 1277 ! *Gilfillan in Herb. Galpin,* 6062 ! *Bolus,* 9740 ! *Gough* ! *Schlechter,* 3775 ! 4684 ! *Burtt Davy,* 2094 ! 5040 ! 7292 ! *Miss Leendertz,* 210 ! 1516 (a pubescent form) ! 1730 ! Var. γ : Transvaal ; Saddle-back Range, near Barberton, *Galpin,* 413 ex *Guerke, Thorncroft,* 4334 ! Devils Kantoor, *Bolus,* 9739 ! Woodbush Mountains, *Barber,* 17 !

EASTERN REGION : Transkei ; between Gekau (Geua) River and Bashee River, *Drège,* 4770 ! Fort Bowker, *Bowker,* 559 ! Griqualand East ; near Clydesdale, *Tyson,* 1061 ! and in *MacOwan & Bolus, Herb. Norm. Austr.-Afr.* 471 ! Natal ; various localities, *Drège,* 4769 ! *Krauss,* 390 ! *Sutherland* ! *Sanderson* ! *Gerrard,* 325 ! *Wood,* 75 ! 86 ! 1340 ! 1397 ! *Wilms,* 3188 ! between Delagoa Bay and Pretoria, a pubescent form, *Bolus,* 9738 ! Var. γ : Natal ; *Gerrard,* 326 ! *Sanderson,* 269 ! Zululand, Eshowe, *Mrs. K. Saunders,* 7 !

Also in Tropical Africa.

An exceedingly variable species in the form and indumentum of the leaves. The variety *hians* can usually be readily distinguished by its appearance, but I believe this is merely due to the drier and higher region in which it occurs, some of the narrow-leaved specimens of the lower level typical form being almost indistinguishable from it, I find no difference in the flowers.

2. **B. angustifolium** (N. E. Br.) ; perennial, apparently 1–1$\frac{1}{2}$ ft. high, much branched above ; stems rather slender, obtusely 4-angled, 4-grooved, puberulous ; leaves $\frac{1}{2}$–1$\frac{1}{2}$ in. long, $\frac{1}{2}$–3$\frac{1}{2}$ lin. broad, linear or linear-lanceolate, subacute, tapering below into the petiole, entire or obscurely and minutely toothed, glabrous or

thinly puberulous beneath; flower-whorls 3–5 lin. apart, 4–6-flowered; bracts falling away when the flowers are in very young bud, $\frac{1}{2}$–$\frac{3}{4}$ lin. long, lanceolate, acute, often forming a small tuft at the apex of the spike-like raceme; pedicels $\frac{1}{2}$–$\frac{2}{3}$ lin. long, puberulous; calyx-tube 1–1$\frac{1}{4}$ (in fruit 2) lin. long, campanulate, white-puberulous and glandular outside, glabrous within; dorsal tooth $\frac{1}{3}$ (in fruit 1) lin. long, very broadly rounded or subtruncate, minutely apiculate; lower lip truncate and densely white-ciliate at the sides, with 2 bristle-like teeth $\frac{1}{3}$–$\frac{1}{2}$ lin. long, contiguous at the base, then diverging; corolla 2–2$\frac{1}{4}$ lin. long, thinly puberulous on the lips, glandular; tube 1$\frac{1}{4}$ lin. long and nearly as broad at the mouth, compressed-funnel-shaped, straight; upper lip with 4 short obtuse lobes, with the lateral much broader and more rounded than the others; lower lip elliptic, concave, obtuse; stamens very much exserted, spirally coiled, subequal, all free; upper pair inserted near the base of the corolla-tube, 4 lin. long, glabrous, with a rather long linear hairy tooth at the bend of the filaments; lower pair inserted at the base of the lower lip, glabrous. *Ocimum angustifolium*, *Benth. in DC. Prodr.* xii. 37. *O. filiforme*, *Guerke in Bull. Herb. Boiss.* vi. 556. *O. polycladum*, *Briq. in Bull. Herb. Boiss.* 2me *sér.* iii. 982.

KALAHARI REGION: Transvaal; Magaliesberg Range, *Burke*! *Zeyher*, 1356! hills near Pretoria, *Rehmann*, 4272! *Kirk*, 45! *Bolus*, 10843! hills near Rustenberg and near Woodstock, *Miss Pegler*, 1023! *Collins*, 57! Marico District, *Holub*! Daspoort, *Miss Leendertz*, 155! Hartebeeste Nek, near Pretoria, *Burtt Davy*, 774! near Warm Bath, *Bolus*, 12252! Kudus Poort, *Rehmann*, 4614 ex *Guerke*. British Bechuanaland; Baralong Territory, *Holub*!

The calyx is covered with glands, which, when boiled in water, swell, turn white and have the appearance of very minute pearl-like papillæ.

3. **B. burchellianum** (N. E. Br.); a woody much-branched shrub, growing to 3 ft. in height; branches whitish-puberulous; leaves subfasciculate at the nodes, $\frac{1}{3}$–$\frac{1}{2}$ in. long, $\frac{2}{3}$–1$\frac{1}{4}$ lin. broad, subspathulate-oblanceolate, obtuse, tapering into a short petiole at the base, longitudinally folded, rather thick, densely white-puberulous, sometimes glabrous above; whorls 6-flowered, $\frac{1}{3}$–$\frac{3}{4}$ in. apart; pedicels 1–1$\frac{1}{2}$ lin. long, puberulous; calyx 1$\frac{1}{2}$–2 lin. long and campanulate in flower, 3–3$\frac{1}{2}$ lin. long and somewhat tubular in fruit, whitish-puberulous outside, glabrous within; upper tooth orbicular-ovate, apiculate; lower lip with the obliquely truncate sides projecting beyond the base of the lower teeth into short minutely denticulate ciliate lobes; lower teeth $\frac{1}{3}$ lin. long, subulate; corolla rose-pink; tube about 2 lin. long, glabrous; upper lip 1–1$\frac{1}{4}$ lin. long, with 4 subequal obtusely rounded lobes, pubescent outside; lower lip 1$\frac{1}{2}$ lin. long and about as broad, suborbicular, obtuse, concave, glabrous; stamens 4–4$\frac{1}{2}$ lin. long; upper pair dilated near the base into a broad flattened retrorse densely ciliate tooth, otherwise glabrous; lower pair adnate to the corolla-tube; nutlets rather more than 1 lin. long, oblong, very pale brown, glabrous, smooth.

Ocimum burchellianum, Benth. Lab. 8, 707, *and in DC. Prodr.* xii. 36.
O. serpyllifolium, Benth. Lab. 707 ; *E. Meyer, Comm.* 226, *not of Forsk. O. helianthemifolium, Hochst. in Flora,* 1845, 67, *and Krauss, Beitr. Fl. Cap- und Natal.* 131 ; *Benth. in DC. Prodr.* xii. 36.

COAST REGION : Uitenhage Div. ; Sand Fontein, *Burke!* near Uitenhage, *Prior!* *Krauss,* 1121 ! Albany Div. ; near Grahamstown, *Burke! Galpin,* 201 ! between Konap River and Enon, 500–600 ft., *Baur,* 1082 ! and without precise locality, *Bowker* ! British Kaffraria, *Cooper,* 541 !
CENTRAL REGION : Somerset Div. ; near Little Fish River, 2000–3000 ft., *Drège,* 2320b! and without precise locality, *Bowker,* 22 ! Graaff Reinet Div. ; mountain sides near Graaff Reinet, 3000–4000 ft., *Drège,* 2320a ! *Bolus,* 56 ! Middelburg Div. ; on dry mountains between Seven Fonteins and Wagenpads Berg, *Burchell,* 2812 ! on the Sneeuwberg Range between Compass Berg and Rhenoster Berg, 5000–6000 ft., *Drège,* 3587 ! Tarka Div. ; near Tarkastad, *Shaw* ! Cradock Div. ; Rietfontein Plantation, near Cradock, *Sim* !

II. OCIMUM, Linn.

Calyx 5-toothed, deflexed in fruit, campanulate or tubular-campanulate ; dorsal tooth much larger than the rest and more or less decurrent on the tube ; 4 lower teeth usually unequal. *Corolla* 2-lipped ; tube about as long as the calyx-tube or shortly exserted ; upper lip erect, 4-lobed ; lower lip concave or nearly flat. *Stamens* 4, declinate, exserted ; filaments free, the upper pair abruptly bent near the base, with a crest, tuft of hairs or retrorse process at the knee ; anthers 1-celled. *Disk* 1-4-lobed. *Style* filiform ; stigma shortly bifid. *Nutlets* oblong, ellipsoid or sub-globose.

Herbs or small shrubs ; leaves opposite, simple, gland-dotted ; inflorescence terminal, spike-like ; flowers small, in 6-flowered whorls.

DISTRIB. A genus of many species, distributed throughout the warmer regions of the earth.

Leaves ¾–2½ in. broad, ovate or ovate-lanceolate, very
　distinctly and rather coarsely serrate ; calyx-tube
　glabrous inside *...*　...　...　...　...　...　(1) **suave.**

Leaves ⅙–⅞ in. broad, linear-lanceolate, lanceolate or
　ovate, entire or obscurely toothed :
　Calyx-tube hairy inside ; leaves ½–2½ in. long :
　　Fruiting racemes about ⅔ in. in diam. ; upper tooth
　　　of fruiting calyx 2¼–2½ lin. broad　...　...　(2) **simile.**

　Fruiting racemes ½ in. or less in diam. ; upper tooth
　　of fruiting calyx 1½ lin. broad :
　　Annual, with 1 main stem branching above　...　(3) **americanum.**

　　Perennial, with several or many stems from a
　　　woody branching base ...　...　...　...　(4) **fruticulosum.**

"Calyx puberulous, especially towards the base,
　becoming glabrous on other parts ; leaves up to
　¾ in. long "　...　...　...　...　...　...　(5) **Dinteri.**

1. O. suave (Willd. Enum. Pl. Hort. Bot. Berol. 629); a stout branching herb; stems square, with rather sharp angles, shortly and densely pilose to subglabrous; leaves spreading; petiole $\frac{1}{2}$–2$\frac{1}{4}$ in. long, pubescent or pilose; blade 1$\frac{1}{2}$–3$\frac{1}{2}$ in. long, $\frac{2}{3}$–2$\frac{1}{2}$ in. broad, flat, ovate or ovate-lanceolate, acute, cuneate at the base, obtusely and somewhat coarsely serrate, pubescent on both sides or occasionally glabrous; inflorescence of 3 or more terminal spike-like racemes 3–6 in. long, bearing numerous closely placed 6-flowered whorls of small flowers; bracts 1$\frac{1}{2}$–2 lin. long, ovate, very acuminate, reflexed, persistent; pedicels $\frac{1}{2}$–1 lin. long, pubescent; flowering-calyx deflexed, thinly pubescent to densely white pilose-pubescent outside, glabrous within; tube 1 lin. long, campanulate; upper tooth about 1 lin. long and broad, very broadly ovate with recurved margins; lateral teeth minute, filiform-subulate, with the mouth of the tube above them produced into a lobe or auricle under the upper tooth; lower teeth united into a deltoid body $\frac{1}{2}$ lin. long, minutely 2-toothed at the apex; fruiting-calyx enlarged, about 2$\frac{1}{2}$ lin. long, with the lobe formed by the lower teeth pressed against the auricles under the upper tooth and closing the mouth of the tube; corolla scarcely exserted from the calyx, white; tube 1 lin. long, glabrous; upper lip $\frac{3}{4}$ lin. long, subequally 4-lobed; lobes oblong, obtuse, pubescent on the back; lower lip $\frac{3}{4}$ lin. long, elliptic, obtuse, very concave, pubescent on the back; stamens exserted, unequal; filaments free; upper pair arising near the base of the corolla-tube, 2 lin. long, with a stout obtuse hairy reflexed process at the knee; lower pair inserted just below the base of the lower lip, 1$\frac{3}{4}$ lin. long; nutlets subglobose, $\frac{2}{3}$ lin. in diam., slightly rugulose, dark brown. *Benth. Lab.* 7, and in *DC. Prodr.* xii. 35; *Baker in Dyer, Fl. Trop. Afr.* v. 338; *Wood, Natal Pl.* iv. *t.* 325.

KALAHARI REGION: Transvaal; hills near Shilovane, *Junod*, 1144! near Rustenburg, *Nation*, 146! near Aapies River, *Schlechter*, 4169!

EASTERN REGION: Natal; Inanda, *Wood*, 1242! in the Botanic Garden, but indigenous, *Wood*, 1812! Nonoti River, *Gerrard*, 1228! Zululand; near Eshowe, 1500 ft., *Wood*, 3975!

Also in Tropical Africa and Tropical Asia.

2. O. simile (N. E. Br.); stem erect, branching above, more than 1 ft. high, subterete, scarcely 4-angled except at the inflorescence, glabrous and brown below, puberulous and purple at the inflorescence; leaves recurving, $\frac{3}{4}$–2 in. long including the $\frac{1}{4}$–$\frac{2}{3}$ in.-long petiole, $\frac{1}{4}$–$\frac{2}{3}$ in. broad, lanceolate, acute, tapering into the petiole at the base, entire or obscurely toothed, flat or longitudinally folded, glabrous on both sides with the exception of a few minute hairs on the midrib and veins beneath; racemes erect, laxly panicled, 2–5 in. long, with the whorls $\frac{1}{2}$–$\frac{5}{8}$ in. apart, 6-flowered; lower bracts leaf-like, upper about $\frac{1}{4}$ in. long, petiolate, lanceolate, acute, ciliate with long jointed hairs, deflexed, dark purple, persistent; pedicels $\frac{1}{8}$–$\frac{1}{4}$ in. long, puberulous; calyx deflexed, campanulate, puberulous outside, densely bearded with hairs at the middle of the tube inside; tube

$\frac{1}{8}$ (in fruit $\frac{1}{6}$–$\frac{1}{5}$) in. long; upper tooth 1 (in fruit $1\frac{1}{4}$) lin. long, sub-orbicular, apiculate; lateral teeth $\frac{3}{4}$ lin. long, deltoid, subulate-acute; lower teeth 1–$1\frac{1}{4}$ lin. long, subulate from a deltoid base; corolla puberulous on the back of the lips, otherwise glabrous, white; tube not exserted from the calyx-tube, about $\frac{1}{8}$ in. long; upper lip $\frac{1}{8}$ in. long, with 4 short rounded subequal lobes; lower lip $\frac{1}{8}$ in. long, $\frac{1}{10}$ in. broad, oblong, obtusely rounded at the apex; stamens exserted, free; upper pair inserted at the middle of the tube, nearly $\frac{1}{4}$ in. long, with a reflexed obtuse pubescent tooth near the base; lower pair rather shorter than the upper, inserted at the base of the lower lip, not toothed at the base, glabrous; nutlets nearly $1\frac{1}{4}$ lin. long, ellipsoid-oblong, with a narrow wing-like angle on each side of the lower half, glabrous, blackish.

KALAHARI REGION : Transvaal; on Madjadjes Mountains in Zoutpansberg district, *Burtt Davy*, 2714 ! 5288 !

Very similar in structural characters to *O. fruticulosum*, Burch., but is readily distinguished by its different appearance. It is evidently a larger plant, with larger leaves and flowers, the racemes being much stouter and when in fruit $\frac{5}{8}$–$\frac{3}{4}$ in. in diam., those of *O. fruticulosum* being less than or scarcely $\frac{1}{2}$ in. in diam., the narrow wing on each side of the nutlets seems wanting in the latter species, in which the nutlets are also smaller.

3. **O. americanum** (Linn. Amoen. Acad. ed. 1, iv. 276); a much-branched herb; stems obtusely or obscurely 4-angled, pubescent with short recurved hairs, usually bearded with long hairs at the nodes; leaves spreading, thinly pubescent to glabrous; petiole 2–4 lin. long; blade $\frac{1}{2}$–$2\frac{1}{4}$ in. long, $\frac{1}{6}$–$\frac{7}{8}$ in. broad, flat, lanceolate or ovate-lanceolate, about equally acute at each end, entire or obscurely toothed; racemes numerous, 4–8 in. long, of many equidistant 6-flowered whorls; bracts $1\frac{1}{2}$–2 lin. long, ovate, acute, persistent or deciduous; pedicels $1\frac{1}{2}$ lin. long, pubescent; calyx deflexed, $1\frac{3}{4}$–2 (in fruit $2\frac{1}{2}$) lin. long, campanulate, 5-toothed to about the middle, pilose outside and inside and ciliate with long white hairs; upper tooth orbicular, slightly apiculate, with a tuft of long white hairs at its base; lateral teeth broad at the base, mucronate, much shorter than the subulate lower teeth; corolla more or less pubescent outside; tube $1\frac{1}{4}$–$1\frac{1}{2}$ lin. long; upper lip about $1\frac{1}{4}$ lin. long, shortly and subequally 4-lobed; lobes rounded; lower lip $1\frac{1}{4}$ lin. long, oblong-obovate, obtuse, concave; stamens much exserted; upper pair inserted near the base of the corolla-tube, $3\frac{1}{2}$ lin. long, with a rather long linear hairy tooth at the knee; lower pair inserted at the base of the lower lip, $2\frac{1}{2}$ lin. long; nutlets $\frac{3}{4}$ lin. long, oblong, slightly shouldered at the base, glabrous, black. *Linn. Sp. Pl. ed.* 2, ii. 833. *O. canum, Sims in Bot. Mag. t.* 2452 ; *Benth. Lab.* 3, *in E. Meyer, Comm.* 226, *and in DC. Prodr.* xii. 32 ; *Drège, Zwei Pfl. Documente*, 205 ; *D. Dietr. Syn. Pl.* iii. 374 ; *Baker in Dyer, Fl. Trop. Afr.* v. 337 ; *Schinz in Mém. Herb. Boiss.* x. 61. *O. stamineum, Sims in Bot. Mag. under t.* 2452. *O. hispidulum, Schum. and Thonn. Beskr. Guin. Pl.* 266.

COAST REGION : Komgha Div. ; on the banks of the Kei River, *Drège,* 3583 !
KALAHARI REGION : Transvaal ; Warmbath, *Miss Leendertz,* 1541 !
EASTERN REGION : Delagoa Bay, *Forbes* ! *Bolus,* 9737 ! *Kuntze.*

Also in Tropical Africa, India and Brazil.

O. americanum is very similar to *O. Basilicum,* Linn., differing in its much
smaller flowers and fruiting-calyx and in the usually bearded nodes of the stem.

4. O. fruticulosum (Burchell, Trav. S. Afr. ii. 264); plant
8–12 in. high, branching from the woody base ; branches obtusely
4-angled, pubescent to thinly and minutely puberulous with reflexed
curved hairs ; leaves shortly petiolate, $\frac{1}{2}$–1$\frac{1}{4}$ in. long, 2–3$\frac{1}{2}$ lin.
broad, linear-lanceolate, acute at both ends, entire or obscurely
toothed, longitudinally folded (in dried specimens), glabrous on
both sides with the exception of a few hairs along the midrib;
racemes 2–4$\frac{1}{2}$ in. long, sometimes arranged in a corymb-like panicle,
with the whorls $\frac{1}{3}$–$\frac{1}{2}$ in. apart, about 6-flowered ; lowermost bracts
leaf-like, the remainder reflexed, petiolate, 1–2 lin. long, $\frac{1}{2}$–$\frac{3}{4}$ lin.
broad, ovate or ovate-lanceolate, acute, cuneate at the base, glabrous,
obscurely and minutely ciliate, persistent ; pedicels 1–1$\frac{1}{2}$ lin. long,
thinly pubescent ; calyx campanulate, pubescent and glandular outside, ciliate on the teeth, densely hairy within the tube, especially
when in fruit ; tube 1 (in fruit 1$\frac{1}{2}$) lin. long ; upper tooth $\frac{3}{4}$ lin.
long, orbicular, apiculate ; lateral teeth $\frac{1}{2}$ lin. long, deltoid, with a
short subulate point ; lower teeth $\frac{3}{4}$ lin. long, subulate ; corolla
thinly pubescent outside ; tube 1$\frac{1}{4}$ lin. long ; upper lip 1$\frac{1}{3}$ lin. long,
shortly 4-lobed ; lobes subequal, obtusely rounded at the apex;
lower lip 1$\frac{2}{3}$ lin. long, $\frac{2}{3}$ lin. broad, oblong, obtuse ; stamens exserted
much beyond the lower lip of the corolla, all free ; upper pair
inserted near the base of the corolla-tube, $\frac{1}{4}$–$\frac{1}{3}$ in. long, with a
reflexed pubescent tooth near the base ; lower pair inserted at the
base of the lower lip, about $\frac{1}{4}$ in. long, glabrous ; nutlets $\frac{3}{4}$ lin. long,
ellipsoid, sometimes slightly angular at the sides, glabrous, black.
Benth. in DC. Prodr. xii. 34. *O. canum, var. integrifolium, Engl.
Jahrb.* x. 267.

KALAHARI REGION : Griqualand West ; on mountains near Klip Fontein,
Burchell, 2160 ! 2634 ! Bechuanaland ; Batlapin Territory, *Holub* ! South African
Gold Fields, *Baines* ! Transvaal ; near the Aapies River, *Burke,* 346 !

Also in Tropical German South-west Africa.

5. O. Dinteri (Briq. in Bull. Herb. Boiss. 2me sér. iii. 980) ; stem
with internodes $\frac{2}{3}$–$\frac{5}{6}$ in. long, puberulous with very minute reflexed
hairs, becoming glabrous ; leaves up to $\frac{3}{4}$ in. long and 3 lin. broad,
lanceolate or oblong-lanceolate, acute or subacuminate, cuneately
tapering at the base into a petiole 2 lin. long, entire, green on both
sides, glabrous ; lateral veins not evident ; inflorescence up to 3$\frac{1}{4}$ in.
long ; whorls 6-flowered, distant, the lowest seated in the axils of
the uppermost leaves, the others with obovate, oblong or lanceolate
shortly petiolate bracts as long as or longer than the flowers ;

pedicels 1 lin. long, minutely puberulous; calyx 1¼ (in fruit 2½) lin.
long, campanulate, gland-dotted, puberulous, especially towards the
base, becoming glabrous on the other parts; upper tooth rounded,
ciliate; lateral teeth 1 lin. long, lanceolate; lower teeth 1½ lin. long,
setaceous, all ciliate; corolla exserted 1½ lin. beyond the mouth of
the calyx; puberulous on the outside of the lips; tube included in
the calyx; upper lip 1 lin. long, rounded-ovate; lower lip 1¼ lin.
long, spreading; stamens 1½–2 lin. longer than the corolla; filaments of the upper pair toothed at the base.

WESTERN REGION: Great Namaqualand, without precise locality, *Dinter*, 1549.

This may possibly have been collected north of the Tropic.

III. ORTHOSIPHON, Benth.

(HEMIZYGIA, Briq. in Engl. & Prantl, Pflanzenfam. iv. 3A, 368.)

Calyx campanulate or tubular-campanulate, unequally 5-toothed;
upper tooth much larger than the others, suborbicular, very broadly
ovate or oblong, sometimes decurrent upon the tube. *Corolla*
distinctly 2-lipped; tube usually exserted much beyond the calyx-
teeth, rarely about equalling them, straight or nearly so; upper lip
often partly formed by the truncate mouth of the tube, very
shortly 3–4-lobed; lower lip very concave or boat-shaped. *Stamens* 4,
exserted, directed towards the lower lip; upper pair with free
filaments, not toothed and very rarely bent like a knee near the
base, inserted at various heights in the corolla-tube; lower pair with
their filaments free to the base or variably united, inserted at or
just below the base of the lower lip; anthers 1-celled. *Disk* usually
unequally lobed. *Ovary* deeply 4-lobed; style filiform, exserted from
the corolla-tube; stigma slightly thickened and minutely bifid or
subentire or divided into 2 short filiform or subulate lobes. *Nutlets*
oblong, ellipsoid or suborbicular, usually slightly compressed dorsally.

Herbs, perennial or annual, with erect simple or branching stems; leaves
opposite or rarely whorled, gland-dotted; inflorescence terminal, simple or
branched, sometimes with 2 or more pairs of flowerless coloured bracts at the
apex; flower-whorls separate, 2–6-flowered; flowers variable in size.

DISTRIB. Species over 100, extending into Tropical Africa, Madagascar, Socotra,
Arabia, India, the Malay Archipelago and Australia.

The genus *Hemizygia* was established by Briquet for the reception of those
species of *Orthosiphon* having the filaments of the lower pair of stamens united.
But I find that the amount of union of these filaments varies considerably in some
species and in *O. subvelutinus* and *O. montanus*, although more usually united,
they are sometimes quite free; whilst *O. persimilis*, in which the filaments are
usually but not always free, is so like *O. Thorncroftii* in which they are united (see
note under *O. persimilis*) and grows with it at the same locality, that I cannot
consider these species and others similarly related should be placed in different
genera, as there is no other character than that of the union of the lower filaments
to separate generically *Hemizygia* and *Orthosiphon*. But besides this I have also
seen specimens in which the filaments were united in some flowers and free in

others on the same raceme of *O. persimilis* or in different racemes of the same plant in *O. montanus,* that is with both genera represented in the same inflorescence or by the same plant ! It appears to me probable that the union or non-union of the filaments has some connection with fertilisation or is of a semisexual nature, and as in at least three cases they may be either free or united in the same specimen or different specimens of the same species, it is evident that it cannot be utilised as a generic character. On this ground I have not upheld *Hemizygia,* which, if retained, should include all the species in the second division of the following key, where it has be-n utilised as affording a distinguishing character for most of the species. All measurements of the corolla in the key and descriptions are from flowers that have been soaked in boiling water and will not apply to them when dried. *N. E. Br.*

*Filaments of the lower pair of stamens free to their
    base :
  Leaves minutely petiolate, ⅛-⅓ in. long, mostly linear
      with revolute margins, or, if expanded, with a
      broadly rounded base   ...   . (18) **subvelutinus.**

  Leaves distinctly petiolate or acutely tapering to a
      subsessile base, more than ½ in. long including
      the petiole, not linear :
    Leaves thinly or not very densely pubescent with
        simple hairs on both sides :
      Leaves 3 in a whorl, serrate ; corolla-tube far
          exceeding the calyx-teeth ...   . (29) **serratus.**

      Leaves opposite :
        Leaves entire : corolla-tube not or scarcely
            exceeding the calyx-teeth   ...   ... (9) **persimilis.**

        Leaves more or less distinctly toothed ;
            corolla-tube exceeding the calyx-teeth :

          Petioles 4-9 lin. long ; upper lip of corolla
              distinct from and forming an angle
              with the mouth of the tube ...   ... (6) **labiatus.**

          Petioles 1-2 lin. long ; upper lip of corolla
              not distinct from the mouth of the
              tube, and except the very small terminal
              lobe continuous with it   ...   ... (26) **Bolusii.**

    Leaves glabrous on both sides or with a minute
        pubescence on the midrib and veins beneath ;
        corolla-tube ⅛-¼ in. long :
      Calyx with evident nerves, especially in fruit ... (21) **Wilmsii.**

      Calyx without evident nerves ...   . (22) **inconcinnus.**

**Filaments of the lower pair of stamens variably united,
    sometimes at the very base only :
  †Lateral and lower calyx-teeth linear-lanceolate and
      acute or narrowly deltoid and gradually tapering
      to a subulate point, flat for all or most of their
      length, not at all bristle-like :
    Leaves ovate or broadly lanceolate 1-2¼ in. long :
      Panicle 1 ft. or more long, very lax, with some-
          what spreading branches ; leaves densely
          white-tomentose on both sides   ...   ... (1) **macrophyllus.**

      Panicle 7-10 in. long, compact, narrow, with
          nearly erect branches :
        Leaves with thinly scattered hairs above, thickly
            pubescent on the veins beneath ; corolla-
            tube ½-⅝ in. long ...   . (2) **latidens.**

Leaves almost glabrous to the eye, slightly
   rough to the touch on both sides ; corolla-
   tube $\frac{3}{4}$–1 in. long   ...   .   (3) **macranthus.**

Leaves mostly linear with very revolute margins, a
   few sometimes ovate, less than $\frac{1}{2}$ in. long   ...   (18) **subvelutinus.**

††Lateral and lower (rarely only the lower) calyx-teeth
   bristle-like, flat only at the base :
  ‡Upper and terminal bracts rather large and very
    conspicuous, coloured, somewhat leaf-like,
    usually lax, spreading, the uppermost often
    without flowers in their axils, persistent (see
    also *O. humilis*) :
   Flowering calyx-tube $\frac{1}{8}$ in. long (larger in fruit) ;
    leaves 1$\frac{1}{2}$–3 in long, $\frac{1}{4}$–$\frac{1}{2}$ in. broad, sessile,
    narrowly lanceolate   ...   .   (11) **bracteosus.**

   Flowering calyx-tube $\frac{1}{6}$–$\frac{1}{3}$ in. long :
   Leaves of main stems 2–3 in. long, 1–1$\frac{1}{4}$ in.
    broad, elliptic-lanceolate ; corolla-tube about
    5 lin. long   ...   .   (4) **foliosus.**

   Leaves of main stems $\frac{2}{3}$–2 in. long, $\frac{1}{8}$–$\frac{3}{4}$ in.
    broad :
   Corolla-tube 9–10$\frac{1}{2}$ lin. (or more ?) long, very
    slender ; leaves densely tomentose with
    stellately branched hairs on both sides ...   (13) **Gerrardi.**

   Corolla-tube 6–8 lin. long :
   Leaves flat, not white-woolly beneath :
    Stems 1–3 ft. high, 1–2 lin. thick ;
     leaves ovate or lanceolate, with very
     prominent subparallel rib-like veins
     beneath   ...   .   (5) **transvaalensis.**

    Stems $\frac{3}{4}$–1 ft. high, about $\frac{1}{2}$ lin. thick ;
     leaves cuneately obovate or oblanceo-
     late, veins scarcely or but slightly
     prominent beneath ...   .   (7) **Muddii.**

   Leaves with revolute margins, linear or
    linear-lanceolate, white-woolly with
    simple hairs beneath   ...   .   (17) **decipiens.**

   Corolla-tube 3–5 lin. long :
   Leaves lanceolate, white- or greyish-
    tomentose on both sides   ...   ...   (12) **Elliottii.**

   Leaves lanceolate, green above, greyish-
    white with minute dense tomentum
    beneath ...   .   (14) **stenophyllus.**

   Leaves lanceolate or elliptic, green on both
    sides, no dense tomentum :
    Bracts with long tapering acute points ;
     lower stamens much longer than the
     lower lip of the corolla   ...   ...   (8) **Thorncroftii.**

    Bracts acute, not long-pointed ; lower
     stamens not exceeding the lower lip
     of the corolla ...   .   (10) **Rogersii.**

‡‡Upper and terminal bracts either small and incon-
　　spicuous or if larger neither much coloured
　　(violet in *O. humilis*), leaf-like, lax nor spreading,
　　deciduous or persistent :
　　Leaves green and pubescent (usually thinly) to
　　　　nearly glabrous beneath ; corolla-tube $\frac{1}{3}$–$\frac{1}{2}$ in.
　　　　long :
　　　　Lower stamens not exceeding the lower corolla-
　　　　　　lip ; leaves $\frac{1}{2}$–$\frac{3}{4}$ in. long, oblanceolate or
　　　　　　obovate　 ...　 ...　 ...　 ...　 ... (20) **Pretoriæ.**
　　　　Lower stamens very much longer than the
　　　　　　lower corolla-lip :
　　　　　　Leaves serrate with small teeth :
　　　　　　　Leaves lanceolate or ovate, acute ; veins
　　　　　　　　impressed above ; corolla-tube 2$\frac{3}{4}$–5 lin.
　　　　　　　　long :
　　　　　　　　Leaves distinctly petiolate :
　　　　　　　　　Leaves with moderately long and
　　　　　　　　　　apparently outstanding hairs on
　　　　　　　　　　both sides, subciliate ; petioles
　　　　　　　　　　2–3$\frac{1}{2}$ lin. long　 ...　 ...　 ... (23) **varians.**
　　　　　　　　　Leaves with very minute adpressed
　　　　　　　　　　hairs on both sides, not ciliate ;
　　　　　　　　　　petioles 1–2 lin. long　 ...　 ... (24) **affinis.**
　　　　　　　　Leaves narrowed to the base, scarcely
　　　　　　　　　petiolate, nearly or quite glabrous
　　　　　　　　　above, distinctly pubescent on the
　　　　　　　　　veins beneath, ciliate　 ...　 ... (25) **Holubii.**
　　　　　　　Leaves elliptic, obtuse or rounded at the
　　　　　　　　apex ; veins not impressed above ;
　　　　　　　　corolla-tube 5–6 lin. long　 ...　 ... (26) **Bolusii.**
　　　　　　Leaves entire, lanceolate or elliptic, finely
　　　　　　　puberulous on both sides ; veins not
　　　　　　　impressed and indistinct above ...　 ... (27) **humilis.**
Leaves densely whitish-pubescent or white- (rarely
　　yellowish-) tomentose beneath, excluding the
　　sometimes revolute margins :
　　Tomentum or pubescence on underside of leaves
　　　composed of unbranched hairs :
　　　Leaves flat or longitudinally folded, not
　　　　revolute at the margins, $\frac{3}{4}$–1$\frac{3}{4}$ in. long ;
　　　　inflorescence branched　 ...　 ...　 ... (28) **canescens.**
　　　Leaves revolute at the margins, $\frac{1}{4}$–$\frac{3}{4}$ in. long ;
　　　　inflorescence not branched :
　　　　Entire flowering-calyx less than $\frac{1}{4}$ in. long ;
　　　　　corolla-tube 3$\frac{1}{2}$–4$\frac{1}{2}$ lin. long　 ...　 ... (16) **albiflorus.**
　　　　Entire flowering-calyx $\frac{1}{4}$ in. or more long ;
　　　　　corolla-tube 7–7$\frac{1}{2}$ lin. long　 ...　 ... (17) **decipiens.**
　　Tomentum on underside of leaves composed of
　　　very minute branched hairs ; corolla-tube
　　　3$\frac{1}{2}$–9 lin. long :
　　　Pubescence on the calyx chiefly of long spread-
　　　　ing simple or slightly branched hairs, with
　　　　or without an admixture of (but not
　　　　tomentose with) minute branched hairs :
　　　　Stems 1–1$\frac{1}{2}$ ft. high, $\frac{3}{4}$–1 lin. thick ; lower
　　　　　stamens twice as long as the lower
　　　　　corolla-lip　 ...　 ...　 ...　 ... (15) **Rehmanii.**

Stems 6–10 in. high, $\frac{1}{3}$–$\frac{2}{3}$ lin. thick; lower stamens about as long as the lower corolla-lip ... ... (19) **teucriifolius,** var. β.

Pubescence on the calyx chiefly a dense tomentum of minute branched hairs, with or without some long simple hairs; lower stamens not or scarcely longer than the lower corolla-lip:

Stems 6–10 in. high, $\frac{1}{2}$–1 in. thick; filaments of lower stamens united almost to the apex ... ... (19) **teucriifolius.**

Stems 7–27 in. high, $\frac{3}{4}$–$1\frac{1}{2}$ lin. thick; filaments of lower stamens united for $\frac{1}{2}$–$\frac{2}{3}$ of their length ... ... (18) **subvelutinus.**

1. **O. macrophyllus** (N. E. Br.); stems 2 ft. or more high, stout, $1\frac{3}{4}$–2 lin. thick in the lower part, obtusely 4-angled, rather thinly subtomentose; leaves of the main stems $1\frac{3}{4}$–$2\frac{1}{2}$ in. long, $\frac{3}{4}$–1 in. broad (bearing short leafy shoots in their axils), lanceolate, acute or subacute, cuneately tapering into a short petiole, shortly serrate in the upper $\frac{2}{3}$, densely and very shortly white-tomentose on both sides, with the veins impressed above, prominent beneath, much reticulate; leaves of the axillary shoots similar, but smaller and usually very obtuse at the apex; panicle lax, 1–$1\frac{1}{4}$ ft. long, with simple or branched lateral branches 3–10 in. long; whorls $\frac{1}{2}$–1 in. apart, 4–6-flowered; bracts very caducous (only one pair at the summit of a branch seen), small, ovate, acuminate, deeply gibbous-concave, pubescent, white-woolly on the margins; pedicels $\frac{1}{2}$ lin. long, pubescent; calyx 5-toothed, pubescent outside, glabrous within, densely ciliate with white-woolly hairs on the teeth; tube 2–$2\frac{1}{2}$ lin. long, ovoid in fruit; upper tooth 1–$1\frac{1}{4}$ lin. long, 1 lin. broad, broadly subquadrate-ovate, somewhat angular at the sides, abruptly and shortly mucronate; lateral and lower teeth subequal, about $1\frac{1}{4}$ lin. long, rather narrowly deltoid-subulate; corolla about 3 times as long as the calyx, apparently dark red; tube $4\frac{1}{2}$–5 lin. long, very slightly curved just above the base, whence it becomes gradually dilated, compressed, glabrous outside, pubescent inside at the base and with a transverse band of hairs near the mouth on the upper side; upper lip 3–4 lin. long, 3-lobed at the apex; middle lobe 1 lin. long and nearly as broad, elliptic, obtuse, slightly recurved at the apex, minutely pubescent on the back; lateral lobes small, rounded; lower lip $3\frac{3}{4}$–$4\frac{1}{2}$ lin. long, subcordate-ovate, obtuse, longitudinally folded, concave, ciliate with woolly hairs, minutely pubescent on the back; stamens much exserted; upper pair arising at the middle of the corolla-tube, 8–9 lin. long, their filaments free, flattened, ciliate on both margins at the base only; lower pair arising at the base of the lower lip, $5\frac{1}{2}$–$6\frac{1}{2}$ lin. long, their filaments united to the apex, glabrous; style filiform, much exserted, glabrous; stigma very slightly thickened, unequally bifid; nutlets nearly $1\frac{1}{2}$ lin. long, oblong, compressed, light brown, shining. *Syncolostemon macrophyllus, Guerke in Bull. Herb. Boiss.* vi. 555.

EASTERN REGION : Natal ; on the Drakensberg at Ingagane, *Rehmann*, 7016!
and between Ingogo and Charlestown, 4000–5000 ft., *Wood*, 6398 !

2. O. latidens (N. E. Br.) ; herbaceous, 4–5 ft. high (*Gerrard*) ;
stems obtusely 4-angled, pubescent ; leaves with very short leafy
shoots in their axils scarcely longer than the 3–4 lin.-long petioles ;
blade 1¼–1¾ in. long, 10–14 lin. broad, broadly ovate, acute, rounded
or subcordate at the base, serrate, sparsely and minutely pubescent
above, more densely and finely greyish-pubescent beneath, with
prominent reticulate veins ; panicle elongated, narrow, 8–9 in. long,
1¾–2 in. broad, with suberect branches 1½–2 in. long ; whorls
¼–½ in. apart, 3–6-flowered ? ; bracts 2–6 lin. long, ½–1 lin. broad,
lanceolate, acute, about two of the terminal pairs without flowers in
their axils, purple, persistent, the others very deciduous, pubescent ;
pedicels 1–1½ lin. long, pubescent ; calyx tubular, 5-toothed, pubescent outside, puberulous within, apparently reddish or purplish-
tinted ; tube about 4 lin. long ; upper tooth 2 lin. long, 1¼ lin. broad,
oblong, obtuse ; lateral and lower teeth subequal, 1½–1¾ lin. long,
½ lin. broad at the middle, flat, linear-lanceolate, acute, ciliate ;
corolla nearly twice as long as the calyx, thinly pubescent on the
outside of the lips, glabrous within, rose-coloured ; tube 6–7½ lin.
long, dilated and compressed in the upper part, forming a vault to
the upper lip, which is 3½–4 lin. long, oblong, unequally 3-lobed at
the apex ; middle lobe about 1 lin. long and broad, orbicular,
obtuse ; lateral lobes very small, rounded ; lower lip 3–3½ lin. long,
oblong, obtuse, boat-shaped, about 1 lin. deep ; stamens much
exserted, unequal ; upper pair inserted slightly above the middle of
the tube, free, about ¾ in. long ; lower pair inserted at the base of
the lower lip, ¾ in. long, exserted 3–4 lin. beyond the upper pair,
their filaments united to the apex ; style filiform, exserted ; stigma
minutely bifid.

EASTERN REGION : Natal ; Umvoti district, *Gerrard*, 1233 !

3. O. macranthus (Guerke in Engl. Jahrb. xxvi. 84) ; plant
3–5 ft. high (*Cooper, Wood*), woody below, much branched (*Wood*) ;
branches obtusely 4-angled above, terete below, minutely puberulous ;
leaves spreading, with very short leafy shoots or fascicles of small
leaves in their axils ; petiole 1–4 lin. long ; blade (of main-stem
leaves) rather thick, 1–1¾ in. long, ½–1 in. broad, elliptic-ovate,
ovate-lanceolate or elliptic, obtuse to acute, cuneate at the base,
usually with 3–9 small teeth along the middle or upper two-thirds
on each side, sometimes nearly or quite entire, minutely rough to
the touch on both surfaces, almost glabrous to the eye, densely
gland-dotted ; panicle narrow, about 7–10 in. long and 2 in. broad,
with 6 or fewer pairs of suberect branches 1¾–3 in. long, puberulous ;
whorls ¼–⅝ in. apart, 2–6-flowered ; bracts ¼–⅓ in. long, ⅛–½ in.
broad, elliptic or elliptic-ovate, subacute, with short recurved points,
puberulous and with a minute cottony ciliation, often purplish-
tinted, very deciduous, but at first closely imbricate over the buds

and forming ovoid tips to the flower-branches; pedicels $\frac{1}{3}-\frac{2}{3}$ lin.
long, pubescent; calyx obconic-tubular, somewhat acute at the base,
harshly puberulous outside, puberulous within; tube $3\frac{1}{2}-4$ lin. long;
upper tooth slightly spreading, $1\frac{1}{2}-2\frac{1}{2}$ lin. long, $1-1\frac{1}{2}$ lin. broad,
elliptic-oblong or slightly obovate-oblong, obtuse or subacute, cilio-
late; lateral and lower teeth subequal, $1\frac{1}{2}-2$ lin. long, flat (not
bristle-like), gradually tapering from the $\frac{1}{2}$-lin. broad base to an
acute point, ciliate; corolla "pink and white" (*Wood*), "purple"
(*Cooper*), pubescent nearly to the base outside, puberulous within;
tube $\frac{3}{4}-1$ in. long, nearly straight, slender at the basal half,
widening to about $\frac{1}{4}$ in. in vertical diameter at the compressed
subtruncate mouth; upper lip with 3 small lobules; middle
lobule $\frac{3}{4}$ lin. long and broad, suborbicular or subquadrate?; lateral
lobules much smaller, rounded; lower lip deflexed, $2\frac{1}{2}-3$ lin. long,
1 lin. broad, boat-shaped, obtuse; stamens exserted, unequal;
upper pair inserted $\frac{1}{4}-\frac{1}{3}$ in. below the mouth of the tube, 5–6 lin.
long, free, flat, ciliate on the basal part; lower pair inserted at the
base of the lower lip, $3\frac{1}{2}-4$ lin. long, exserted beyond the upper
pair, with their filaments united to the apex, glabrous; stigma
minutely bifid. *Hemizygia Cooperi, Briq. in Bull. Herb. Boiss.*
2me *sér.* iii. 992.

KALAHARI REGION : Orange River Colony, *Cooper*, 1015 !
EASTERN REGION: Natal ; Mohlamba Range, 5000–6000 ft., *Sutherland* ! edge
of a ravine near Van Reenens Pass, 5500 ft., and banks of the Tugela, 4000 ft.,
Wood, 3573 (*Natal Herb.* 949) ! under shade on a rocky hill near Berlin Mission
Station, *Wood*, 3554 !

4. O. foliosus (N. E. Br.) ; stems woody at the base, pubescent
or pilose above, very leafy ; leaves ascending-spreading, 2–3 in.
long, $1-1\frac{1}{2}$ in. broad at the middle, broadly lanceolate or elliptic-
lanceolate, acute to obtuse, cuneately narrowed into a petiole
$1-1\frac{1}{2}$ lin. long, shortly serrate on the margin, slightly shining,
varying from rather thinly puberulous on both sides, with rather
longer pubescence on the prominent nerves beneath, to softly pilose
with long hairs on both sides ; panicle 6 in. or more long, with 2–3
pairs of ascending-spreading branches 3–6 in. long, sometimes
reduced to a simple raceme, somewhat harshly pubescent ; whorls
$\frac{1}{3}-\frac{3}{4}$ in. apart, 2-flowered ; upper and terminal bracts $\frac{1}{3}-\frac{2}{3}$ in. long,
$\frac{1}{5}-\frac{1}{4}$ in. broad, very spreading, lanceolate, acute, cuneate at the
base, rather thin, pubescent along the midrib beneath and ciliate,
otherwise glabrous, apparently carmine or rosy-purple, deciduous or
perhaps the upper persistent ; pedicels 1–2 lin. long, pubescent ;
calyx-tube $\frac{1}{4}$ (in fruit about $\frac{1}{3}$) in. long, campanulate, somewhat
harshly puberulous on the nerves ; upper tooth $\frac{1}{8}-\frac{1}{4}$ in. long, $\frac{1}{8}-\frac{1}{4}$ in.
broad, orbicular-ovate or transverse, very obtuse ; lateral and lower
teeth unequal, bristle-like, the lower pair $\frac{1}{8}$ in. long; corolla
apparently white, puberulous on the back of the upper and lower
lips, elsewhere glabrous ; tube $\frac{1}{2}$ in. long, compressed and dilated at
the upper part, truncate at the mouth or upper lip, which has a

small erect orbicular-ovate middle lobule recurved at the margins ;
lower lip directed forwards, $\frac{1}{4}$ in. long, boat-shaped, obtuse ;
stamens exserted much beyond the lower lip ; upper pair inserted
at about the middle of the corolla-tube, with free filaments,
puberulous on their basal part ; lower pair inserted at the base
of the lower lip, with the filaments united to the apex and
exceeding the upper pair ; nutlets compressed-orbicular, 1 lin. in
diam., smooth, brown. *Hemizygia foliosa*, S. Moore in Journ. Bot.
1905, 172.

KALAHARI REGION : Swaziland ; near Mbabane (Embabaan), 4600 ft., *Burtt
Davy*, 2833 ! *Bolus*, 12250 ! 12254 !

5. O. transvaalensis (Schlechter in Journ. Bot. 1897, 281,
transvaalense) ; perennial ; stems 2–3 ft. high, probably woody at the
base, branching at the upper part, obtusely 4-angled above, terete and
striate below, very thinly to rather densely pubescent with spreading
hairs ; leaves of the main stems (including the $\frac{1}{2}$–2 lin.-long petioles)
$\frac{1}{2}$–1$\frac{1}{3}$ in. long, $\frac{1}{4}$–$\frac{3}{4}$ in. broad, those on the branches smaller, ovate or
lanceolate, rarely elliptic, acute or obtuse, rounded at the base,
serrate, denticulate or rarely some entire, glabrous or with thin
minute adpressed pubescence or with longer scattered hairs above,
greyish beneath, sometimes with exceedingly minute pubescence,
sometimes thinly pilose on the conspicuously prominent veins ;
panicle with 1–2 pair of simple branches and a terminal one,
3$\frac{1}{2}$–6 in. long, each with 4–9 whorls $\frac{1}{2}$–1$\frac{1}{2}$ in. apart, 6-flowered ;
upper 2–4 pairs of bracts flowerless, persistent, $\frac{1}{4}$–1 in. long, 1–3 lin.
broad, lanceolate or linear-lanceolate, acute, cuneately tapering
into a short petiole, thin, glabrous above, minutely puberulous beneath, ciliate, coloured (pink ?) ; flowering bracts very
deciduous, leaf-like, thin ?, lanceolate, acute, perhaps more or less
coloured ; pedicels 1–2 lin. long, pubescent ; calyx-tube 3–4 lin.
long, tubular-campanulate, pubescent outside, minutely puberulous
within ; upper tooth 1$\frac{1}{2}$–2 lin. long, orbicular, obtuse ; lateral and
lower teeth bristle-like, the lower 1–1$\frac{1}{2}$ (in fruit up to 2$\frac{3}{4}$) lin. long ;
corolla moderately large, with widely gaping lips, minutely pube-
rulous at the tips of the lips outside and more minutely within the
tube, otherwise glabrous, pink (*Thorncroft*), lilac (*Bolus*) ; tube
much exserted, $\frac{1}{2}$–$\frac{2}{3}$ in. long, very slightly curved, slender at the
basal $\frac{2}{3}$, dilated and compressed at the truncate mouth which forms
the 3–5 lin.-long upper lip, which is abruptly constricted at the top
of the very small auricle-like lateral lobes into a small subquadrate
obtuse middle lobe ; lower lip 3–5 lin. long, boat-shaped, obtuse ;
stamens much exserted, very unequal, upcurved at the tips ; upper
pair inserted at the middle of the corolla-tube, 7 lin. long, with free
flattened filaments, minutely ciliate on the lower part ; lower pair
5–8 lin. long, extending about 3 lin. beyond the upper, with the
filaments united nearly or quite to the apex, glabrous ; stigma not
thickened, minutely bifid. *Ocimum Wilmsii, Guerke in Engl. Jahrb.*
xxvi. 79.

KALAHARI REGION: Transvaal; hills near Barberton, 3000–4500 ft., *Galpin,* 468 ex *Guerke, Bolus,* 7604! Concession Creek, near Barberton, *Thorncroft,* 175! 3125! *Wood,* 4289! near Lydenburg, *Wilms,* 1107! and 1108 ex *Guerke.* Rietfontein, Lydenburg, *Burtt Davy,* 7256! near the Crocodile River, *Schlechter,* 3916!

6. O. labiatus (N. E. Br.); apparently a branching herb (or shrub?) 2 ft. or more high; branches 4-angled, pubescent; leaves spreading; petiole very slender, $\frac{1}{3}$–$\frac{3}{4}$ in. long, pubescent; blade $\frac{3}{4}$–1$\frac{1}{2}$ in. long, $\frac{1}{2}$–1 in. broad, ovate or elliptic-ovate, obtuse or acute, very shortly and broadly cuneate or rounded at the base, obtusely serrate, very thinly and rather minutely pubescent on both sides, paler on the under side; racemes with 4–6 whorls $\frac{1}{4}$–$\frac{1}{2}$ in. apart, 4–6-flowered; lower bracts deciduous, not seen; upper more or less persistent, purple, 3$\frac{1}{2}$–4 lin. long, 2$\frac{1}{2}$ lin. broad, orbicular, abruptly cuspidate-acuminate; pedicels 2$\frac{1}{2}$–3 lin. long, slender, puberulous; calyx thin, purple; tube 2$\frac{1}{2}$ lin. long, campanulate, thinly puberulous; upper tooth 1$\frac{1}{2}$ lin. long, 1$\frac{3}{4}$ lin. broad, orbicular-ovate, decurrent on the tube; lateral teeth $\frac{3}{4}$ lin. long and the lower pair 1$\frac{1}{2}$ lin. long, all subulate from a broader base, flat for the greater part of their length; corolla-tube 4–4$\frac{1}{2}$ lin. long, with the upper part bent downwards from the middle and enlarging to about 2 lin. in vertical diameter at the mouth, glabrous; upper and lower lips subequal, 4–5 lin. long, measured from the base of the gape, about 1$\frac{3}{4}$ lin. broad, oblong; upper lip standing nearly at a right angle to the tube, with parallel sides, 3-toothed at the apex, with the middle tooth larger and rounder than the lateral pair; lower lip directed forwards or perhaps ultimately reflexed, obtuse, concave; stamens ultimately exserted beyond the lower lip; filaments all free, those of the upper pair inserted close to the base of the corolla-tube, ciliate at the basal part.

KALAHARI REGION: Transvaal; Woodbush Mountains, 6400 ft., *Schlechter,* 4434!

The long slender petioles and long upper lip of the corolla readily distinguish this from all the other South African species.

7. O. Muddii (N. E. Br.); stems probably several from a perennial rootstock, $\frac{3}{4}$–1 ft. high, slender, about $\frac{1}{2}$ lin. thick, erect, simple or slightly branched, 4-angled, pubescent; leaves in 3–4 distant pairs, $\frac{3}{4}$–1$\frac{1}{2}$ in. long, $\frac{1}{4}$–$\frac{3}{4}$ in. broad, those on the axillary shoots smaller, obovate or oblanceolate, acute or obtuse, very acutely tapering into the 1–1$\frac{1}{2}$ lin.-long petiole, acutely toothed in the upper half, very sparsely pubescent on both sides or nearly glabrous beneath and the veins but slightly prominent; raceme simple, with 2–4 whorls $\frac{1}{2}$–1 in. apart, 6-flowered; flowering-bracts $\frac{1}{2}$ in. long, 1$\frac{1}{2}$–2$\frac{1}{2}$ lin. broad, lanceolate, acuminate, ciliate, deciduous; terminal 2–3 pairs of bracts flowerless, persistent, spreading, pinkish-red, 6–10 lin. long, 1$\frac{1}{2}$–3 lin. broad, lanceolate, very acuminate, tapering into a short petiole, slightly pubescent, ciliate;

pedicels 1–2 lin. long, pubescent ; calyx-tube 3–3½ lin. long, tubular-campanulate, pubescent outside, minutely puberulous within, darkly coloured ; upper tooth 1–1½ lin. long, 1¾ lin. broad, obtusely rounded ; lateral and lower teeth 1–1⅔ lin. long, bristle-like from a deltoid base ; corolla slightly and minutely pubescent outside ; tube 7–7½ lin. long, dilated and compressed in the upper part, glabrous inside ; upper lip 3 lin. long, oblong, complicate-concave or some-what hooded, shortly and obtusely 3-lobed at the apex, the middle lobe ½ lin. long, nearly 1 lin. broad, twice as long as the lateral lobes ; lower lip 3–3½ lin. long, oblong, obtuse, concave ; stamens unequal, the united pair (or all when fully exserted) straight, except at the very apex, which is upcurved ; upper pair inserted about ¼ in. above the base of the corolla-tube, 7 lin. long, at length exserted, their filaments free, flat, ciliate ; lower pair inserted at the base of the lower lip, 4 lin. long, their filaments united nearly to the apex, glabrous ; stigma shortly bifid, with linear lobes.

KALAHARI REGION: Transvaal ; on the Drakensberg Plateau, *Mudd*! near Spitzkop, Lydenburg district, *Burtt Davy*, 1570 !

8. **O. Thorncroftii** (N. E. Br.) ; stems apparently several from the same rootstock, 5–8 in. high, about ⅔ lin. thick at the base, scabe-rulous ; leaves in 3–4 pairs, with or without short (usually shorter than the leaves) shoots in their axils, ascending-spreading, ½–⅞ in. long, ⅛–¼ in. broad, lanceolate, acute, acutely narrowed into a short petiole, with thinly scattered pubescence on both sides ; whorls 2–4 in a simple raceme, ⅜–1 in. apart, 3–6-flowered ; lower bracts ¼–⅓ in. long, lanceolate, acute, like reduced leaves, deciduous ; upper bracts persistent, thin and coloured, purple ?, spreading, ⅔–⅞ in. long, ¼–⅓ in. broad, lanceolate, tapering into a long acuminate point at the apex and into a distinct petiole at the base, often without flowers in their axils ; pedicels 1–1½ lin. long, pubescent ; calyx-tube ⅛–⅕ in. long, pubescent outside ; upper tooth ⅛–⅙ in. long, broadly ovate, acute ; lateral and lower teeth bristle-like, the lower ⅛ in. long ; corolla-tube 4½ lin. long, exceeding the calyx-teeth, slightly curved, compressed, truncate at the mouth, minutely puberulous ; upper lip 3-lobed, middle lobe erect, about ¾ lin. long and ⅔ lin. broad, oblong, obtuse, lateral lobes much smaller ; lower lip ⅛ in. long, elliptic-boat-shaped, obtuse ; upper pair of stamens inserted just below the middle of the tube, free, ciliate on the basal part, exserted as far as the lower pair or only to the tip of the lower lip ; lower pair inserted at the base of the lower lip and exserted far beyond it, 4–4½ lin. long, with their filaments united to the apex.

KALAHARI REGION : Transvaal ; near Barberton, *Thorncroft*, 3123 ! *Bolus*, 9743 ! Saddleback Range, near Barberton, *Galpin*, 465 (mingled with *O. persimilis*) !

9. **O. persimilis** (N. E. Br.) ; stems ½–1 ft. (or more ?) high, simple or branched, about 1 lin. thick at the base, 4-angled, pilose-pubescent ; leaves ½–1 in. long, ⅛–⅓ in. broad, lanceolate, acute,

cuneately tapering to an acute subsessile or scarcely petiolate base, thinly pilose on both sides; raceme with 5–7 whorls $\frac{1}{2}$–$\frac{3}{4}$ in. apart, 6-flowered; bracts on the lower part of the raceme very deciduous, not seen, upper persistent, thin and coloured, leaf-like, spreading, $\frac{1}{2}$–$\frac{7}{8}$ in. long, $\frac{1}{6}$–$\frac{1}{3}$ in. broad, lanceolate, acute, cuneate at the subsessile base, minutely puberulous and ciliate, often without flowers in their axils; pedicels 1$\frac{1}{2}$–2 lin. long, pubescent; calyx $\frac{1}{6}$–$\frac{1}{4}$ (in fruit $\frac{1}{3}$) in. long, tubular-campanulate, pubescent; upper tooth $\frac{1}{6}$ in. long, broadly ovate, acute; lateral and lower teeth bristle-like, the lower $\frac{1}{4}$ in. long; corolla-tube not or scarcely exceeding the lower calyx-teeth, $\frac{3}{8}$ in. long, compressed, truncate at the mouth, glabrous; upper lip 3-lobed; middle lobe $\frac{1}{16}$ in. long, 1 lin. broad, subquadrate, subtruncate, slightly puberulous on the back; lateral lobes much smaller; lower lip $\frac{1}{6}$ in. long, orbicular-boat-shaped, shortly stalked, very obtuse; stamens exserted and extending nearly or quite to the tip of the lower lip, both pairs with free filaments, or the lower pair united for $\frac{1}{3}$–$\frac{1}{2}$ of their length, sometimes both forms on the same stem; upper pair inserted at the middle of the corolla-tube, rather more than $\frac{1}{4}$ in. long, their filaments rather densely ciliate nearly to the apex; lower pair inserted at the base of the lower lip, $\frac{1}{6}$ in. long, their filaments ciliate for half-way along one margin.

KALAHARI REGION : Transvaal; near Barberton, *Thorncroft*, 3132! Saddleback Range, near 'Barberton, *Galpin*, 465 (mingled with *O. Thorncroftii*)! Nel Spruit, *Rogers*, 308 !

Dried specimens of this plant so closely resemble those of *O. Thorncroftii* and both have been collected by Galpin (465) as being one species, that in spite of the very obvious difference in the shorter corolla-tube and shorter stamens I am very doubtful if it be more than a sexual condition of that species, but this can only be decided by growing the plants from seed. As in the case of *O. subvelutinus* and *O. montanus* (which see) it is clear that the mere union or freedom of the lower pair of filaments is useless as a generic character since in a specimen of Galpin 465 I find flowers with united and others with free filaments in the same raceme !

10. **O. Rogersii** (N. E. Br.); stems probably several from the same rootstock, 5–8 in. high, sharply 4-angled, pilose, with 3–4 pairs of leaves, having short flowerless branches in the axils of the upper 2–3 pairs; leaves subsessile, 5–8 lin. long, 1$\frac{1}{2}$–2$\frac{1}{2}$ lin. broad, lanceolate or elliptic, acute or obtuse, green and thinly pilose on both sides or the upper surface with very few hairs; glands very conspicuous; inflorescence a simple raceme of 5–7 whorls $\frac{1}{4}$–1 in. apart and 3–6-flowered; bracts 4–5 lin. long, 2–2$\frac{1}{2}$ lin. broad, lanceolate to elliptic, subsessile, with the lower pair more or less leaf-like and persistent and the upper thinner, rosy-purple, persistent, those at the middle of the raceme more or less deciduous, both sides minutely puberulous and sometimes with a few longer hairs on the back; pedicels 1$\frac{1}{4}$–2 lin. long, pilose; calyx-tube 2$\frac{1}{2}$–2$\frac{3}{4}$ lin. long, tubular-campanulate, pilose outside; upper tooth 1$\frac{1}{2}$ lin. long, very broadly ovate, obtuse, sometimes apiculate; lateral and lower teeth

subequal, 1 lin. long, the lateral flat and deltoid-subulate, the lower more or less bristle-like, ciliate ; corolla minutely puberulous on the upper side of the tube outside and within, and on the back of the upper lip, otherwise glabrous, drying dark brownish-orange ; tube sometimes scarcely exceeding the calyx-teeth, sometimes exserted 1½ lin. beyond them, 3–4½ lin. long; upper lip formed by the very oblique mouth of the tube, 2–2½ lin. long, very obtusely 3-lobed at the apex, with the middle lobe larger than the lateral, ½–⅔ lin. long, 1–1½ lin. broad; lower lip 2–2¼ lin. long, boat-shaped, obtuse; stamens not exceeding the lower lip, all exserted from the tube and subequal or the upper pair included and scarcely reaching to the base of the lower pair ; upper pair with free filaments inserted near the base of the corolla-tube, flat, linear or linear-lanceolate, densely ciliate on both margins ; lower pair inserted at the base of the lower lip, 1–2¼ lin. long, with their filaments free or united only at the very base, glabrous.

KALAHARI REGION : Transvaal ; Nel Spruit, *Rogers,* 4740 ! Devils Kantoor, *Bolus,* 9742 !

11. **O. bracteosus** (Baker in Dyer, Fl. Trop. Afr. v. 375) ; plant ¾–2 ft. high, simple or branched ; branches thinly pubescent ; leaves sessile, 1½–3 in. long, ¼–½ in. broad, narrowly lanceolate, acute, cuneately narrowed at the base, slightly and shortly serrate, thinly and minutely puberulous on both sides ; raceme 1½–5 in. long, of 5–12 distant 6-flowered whorls ; upper and terminal bracts ¼–⅔ in. long, 2–5 lin. broad, often without flowers in their axils, persistent, purple-red, ovate-lanceolate, ovate or orbicular-ovate, acute to very obtuse ; lower bracts very deciduous, not seen ; pedicels about 1 lin. long, puberulous ; calyx-tube 1½ (in fruit nearly 2) lin. long, campanulate, puberulous ; upper tooth ¾–1 lin. long, broadly ovate, obtuse ; lateral and lower teeth bristle-like, the lower 1¼–1½ (in fruit up to 2) lin. long ; corolla white, minutely glandular outside, slightly and minutely puberulous above the insertion of the upper stamens within ; tube 3–4 lin. long, compressed, gradually enlarging from the base to the truncate mouth, which forms the upper lip and is very shortly and obtusely 3-lobed ; lower lip 1–1½ lin. long, deeply concave, obtuse ; stamens exserted ; upper pair 3½–4 lin. long, inserted below the middle of the tube, free, abruptly curved or kneed, but not toothed at the base, slightly pubescent ; lower pair inserted at the base of the lower lip, united nearly or quite to their apex, 2¼ lin. long ; stigma clavate, entire ; nutlets 1–1⅛ lin. long and broad, compressed-subglobose, with a keel on each side and one down the back on the apical part, slightly and coarsely reticulate, light brown. *Ocimum bracteosum, Benth. Lab.* 14, *and in DC. Prodr.* xii. 41 ; *Schinz in Mém. Herb. Boiss.* x. 61. *Hemizygia bracteosa, Briq., and H. Junodi, Briq., and var. Quintasii, Briq. in Ann. Conserv. Jard. Bot. Genèv.* ii. 248, 249 ; *Schinz in Bull. Herb. Boiss.* 2ᵐᵉ *sér.* iii. 661.

EASTERN REGION : Delagoa Bay, *Monteiro,* 20 ! *Junod,* 61, 235, *Bolus,* 9741 !

Briquet has separated this plant into two species, chiefly distinguishable by the size of the flowers ; but in different gatherings and often in different specimens of the same gathering (both Tropical and South African) I find the flowers vary much in size and exsertion of the corolla. The leaves and bracts are also very variable.

12. O. Elliottii (Baker in Dyer, Fl. Trop. Afr. v. 376); a branching shrub or shrublet ; young branches obtusely 4-angled, tomentose with a felt of minute branched hairs, becoming woody with a glabrous brown bark ; leaves ½–1¼ in. long, ⅓–½ in. broad, lanceolate, ovate or elliptic, acute or obtuse, cuneately narrowed into a petiole ½–1 lin. long, densely white- or greyish-tomentose on both sides with minute branched hairs, reticulated with rather thick prominent veins beneath ; racemes simple, 1½–3 in. long, with 4–8 2-flowered whorls ¼–⅓ in. apart ; bracts ¼–½ in. long, ⅙–⅓ in. broad, ovate or obovate, obtuse or broadly rounded at the apex, thin, glabrous, minutely ciliate, bright rosy-purple, all at first imbricating, forming a subglobose or oblong bud-like termination to the raceme, deciduous as the raceme develops, with some of the terminal pairs persistent and without flowers in their axils ; pedicels ⅔–1 lin. long, tomentose ; calyx-tube 2 lin. long, tubular-campanulate, thinly covered with minute branched hairs outside, glabrous within ; upper tooth ⅔–1 lin. long, 1½ lin. broad, semi-circular, decurrent on the tube ; lateral and lower teeth bristle-like, ⅔–1 lin. long ; corolla-tube 3½ lin. long, glabrous outside, puberulous on the lower part within, 1½–1¾ lin. in vertical diam. at the obliquely truncate mouth, which forms the upper lip and has a small transverse emarginate lobe ½ lin. long at the apex, and a small tooth on each side of it ; lower lip 2–2½ lin. long, concave or somewhat boat-shaped, obtuse ; stamens not exserted beyond the lower lip ; upper pair inserted at about ⅓ of the way up the corolla-tube, with free filaments, densely ciliate on their lower half ; lower pair exceeding the upper, inserted at the base of the lower lip, 2 lin. long, with their filaments united for a very short distance at the base.

KALAHARI REGION : Bechuanaland ; Bakwena Territory, 3500 ft., *Holub* !

Also in Tropical Africa.

13. O. Gerrardi (N. E. Br.) ; a shrub 3–4 ft. high, woody below, the branchlets, both sides of the leaves and the outside of the calyx densely clothed with white, somewhat flocculent tomentum of stellate hairs ; leaves rather thick ; petiole about 2 lin. long ; blade 5–8 lin. long, 2½–5 lin. broad, ovate or elliptic-oblong, obtuse or subacute, somewhat cuneate at the base ; inflorescence a raceme or few-branched panicle 1½–2 in. long ; flowers solitary in the axils of the bracts, opposite ; lower bracts 3–4 lin. long, 1–1½ lin. broad, narrowly elliptic or oblong-oblanceolate, obtuse, narrowed into a short petiole, thick and white-tomentose like the leaves, spreading ; upper bracts similar, but rather larger, thinner, rose-coloured (*Gerrard*), less tomentose ; pedicels 1–1¼ lin. long ; calyx-tube 2¾–

3¼ lin. long, subcylindric, enlarged in fruit; upper tooth ¾ lin. long, very broadly ovate, obtuse; lateral and lower teeth bristle-like, the latter 1 lin. long, twice as long as the former; corolla nearly 4 times as long as the calyx, rose-coloured (*Gerrard*), pubescent with short spreading hairs outside; tube 8½–10½ lin. long, slender, cylindric, straight; upper lip 3–3½ lin. long, 1¼ lin. broad, narrowly oblong, very shortly 3-lobed at the obtuse apex; lower lip 3–3½ lin. long, ½–⅔ lin. broad, linear-oblong, obtuse, boat-shaped; stamens all inserted close together at the base of the lower lip, glabrous, the upper pair free, 1¼ lin. long, lower pair 2 lin. long, their filaments connate to half-way up; stigma entire, scarcely enlarged; nutlets 1¼ lin. long, oblong, obtuse, smooth, light brown.

EASTERN REGION: Natal; in rocky ground near Ingoma, *Gerrard*, 1239!

14. **O. stenophyllus** (Guerke in Engl. Jahrb. xxvi. 84); stems woody below, 1½ ft. (or more?) high, obtusely and rather obscurely 4-angled, minutely tomentose with branched hairs, very leafy and beset with short leafy branchlets; leaves on the main stems ¾–1½ in. long, 1½–4½ lin. broad, those on the axillary branchlets smaller, all subsessile or very shortly petiolate, narrowly linear-lanceolate to oblong-lanceolate, acute or obtuse, entire or rarely a few of the largest toothed, glabrous above, with a very minutely papillate surface, minutely white-tomentose beneath; veins subparallel, impressed above, prominent beneath; inflorescence simple or branched at the base; whorls 5–9 lin. apart, 4–6-flowered; flowering bracts very caducous, 6–7 lin. long, 3–3½ lin. broad, ovate or ovate-lanceolate, acuminate, very glandular, minutely tomentose on the back; terminal 2–3 pairs of bracts flowerless, persistent, 5–11 lin. long, 1½–3 lin. broad, lanceolate, acute, rosy; pedicels 1–2 lin. long, pubescent with very short stout jointed hairs; calyx pubescent with short jointed hairs mingled with stalked glands; tube 2¼–2½ lin. long, campanulate, enlarged in fruit; upper tooth scarcely 1 lin. long, 1¼ lin. broad, broadly rounded, very obtuse, sometimes minutely crenulate; lateral and lower teeth unequal, ⅔–1 lin. long, bristle-like from a broader base, ciliate; corolla about twice as long as the calyx, rosy (*Tyson*), lilac (*Wood*); tube 5 lin. long, slightly curved near the base, dilated and compressed in the upper part, glabrous outside, very minutely pubescent within; upper lip 2–2½ lin. long, unequally 3-lobed; terminal lobe ½–¾ lin. long, 1 lin. broad, suborbicular or transverse, entire or emarginate, reflexed at the sides; lateral lobes very small, rounded, reflexed; lower lip 1⅔–2 lin. long, boat-shaped, ¾–1 lin. deep, obtuse, pubescent outside; stamens very much exserted, unequal; upper pair inserted at the middle of the corolla-tube, 4½–7 lin. long, their filaments free, flattened, ciliate on both margins at the base; lower pair inserted at the base of the lower lip, 3½–5½ lin. long, exceeding the upper pair, their filaments variously united, sometimes for ⅓–⅔ of their length, at others quite to the apex, glabrous; stigma slightly

thickened, shortly and obtusely 2-lobed ; nutlets 1 lin. long, oblong, slightly compressed, light brown, shining.

EASTERN REGION : Pondoland ; Fakus Territory, *Sutherland* ! Griqualand East ; in grassy places around Fort Donald, 5000 ft., *Tyson*, 1666 ! and at Emyembi, 5000 ft., *Tyson*, 2137 ! on both sides of a mountain near Emyembe Forest, 5000 ft., *Tyson in MacOwan and Bolus, Herb. Norm. Austr.-Afr.*, 1293 ! eastern side of the Zuurberg Range, 4500 ft., *Tyson*, 1720 ! Natal ; on a grassy hill near Umtamouma River, *Wood*, 3107 ! near Boston, 3000–4000 ft., *Wood*, 4624 (*Natal Herb.* 966) ! Mawaqa Mountain, 6000–7000 ft., *Wood*, 8126 !

15. **O. Rehmannii** (Guerke in Bull. Herb. Boiss. vi. 557) ; stems several from a woody rootstock, 1–1½ ft. high, ¾–1 in. thick, erect, very leafy and beset with short leafy axillary shoots, tomentose with minute branched hairs intermingled with long hairs ; leaves of the main stems 5–9 lin. long, 2–3 lin. broad, subsessile, rather thick, some flat, cuneately oblanceolate, subacute, toothed above the middle, revolute along the margins, others linear or linear-lanceolate from the margins being very revolute, minutely and somewhat thinly tomentose above, white-tomentose with minute branched hairs beneath ; veins subparallel, very stout and prominent beneath, impressed above ; leaves of the axillary shoots much smaller, linear-lanceolate, entire, with very revolute margins ; inflorescence 3–4 in. long, simple or with 1–2 pairs of branches at the base ; whorls 4–8 lin. apart, 4–6-flowered ; bracts caducous, not seen ; pedicels 1–1½ lin. long, densely villous with long simple hairs ; calyx villous outside and ciliate on the teeth with long simple hairs intermingled with stalked glands, puberulous inside ; tube 2½–3 lin. long, campanulate ; upper tooth 1–1¼ lin. long, 1⅓ lin. broad, orbicular-ovate, subacute ; lateral and lower teeth unequal, ⅔–1 lin. long, bristle-like from a broad base ; corolla about twice as long as the calyx, slightly curved below the middle ; tube 6–9 lin. long, dilated and compressed in the upper part, glabrous outside, minutely pubescent in the lower part inside ; upper lip or truncate mouth of the tube 2–2½ lin. long, 3-lobed ; middle lobe about ⅔ lin. long, 1 lin. broad, oblate or sub-orbicular, tomentose on the back ; lateral lobes small, rounded ; lower lip 2–2½ lin. long, very deeply boat-shaped, obtuse, tomentose on the back ; stamens unequal, exserted much beyond the lower lip of the corolla ; upper pair inserted at about ⅓ the way up the corolla-tube, their filaments free, flattened, tomentose on the lower part ; lower pair inserted at the base of the lower lip, united almost to the apex, glabrous ; stigma slightly thickened, with 2 short acute lobes.

KALAHARI REGION : Transvaal ; Houtbosch, *Rehmann*, 6172 ! *Schlechter*, 4442 !

I have only seen rather imperfect or unopened flowers of this species, and the measurements given are perhaps inaccurate.

16. **O. albiflorus** (N. E. Br.) ; a branching shrub 2–3 ft. high ; branches with internodes ½–1 in. long, pubescent, beset with numerous short axillary densely leafy branchlets ; leaves subsessile,

3–9 lin. long, $\frac{2}{3}$–1$\frac{1}{2}$ lin. broad, linear (probably sometimes also ovate), obtuse, strongly revolute along the margins, entire, pubescent above with rather long silky hairs, densely white-tomentose beneath with similar hairs ; whorls 3–5, about 3–6 lin. apart, 4–6-flowered, in simple racemes $\frac{3}{4}$–1$\frac{3}{4}$ in. long ; bracts falling away before the corolla is exserted from the calyx, 3 lin. (or more?) long, very broadly ovate, cuspidate-acuminate, deeply concave, pubescent ; pedicels $\frac{1}{2}$–$\frac{3}{4}$ lin. long, densely pilose with simple and gland-tipped hairs ; calyx-tube 1$\frac{1}{3}$–1$\frac{3}{4}$ lin. long, slightly enlarging in fruit, pilose like the pedicels outside ; upper tooth $\frac{3}{4}$–1 lin. long and broad, broadly ovate, acute, 4 lower teeth $\frac{1}{2}$–$\frac{3}{4}$ lin. long, bristle-like, ciliate chiefly with glandular hairs ; corolla twice as long as the calyx, white ; tube 3$\frac{1}{2}$–4$\frac{1}{2}$ lin. long, curved, rather abruptly dilated in the upper half, glabrous outside, pubescent within ; upper lip 1$\frac{1}{4}$ lin. long, unequally 3-lobed ; terminal lobe $\frac{1}{3}$ lin. long, $\frac{1}{2}$ lin. broad, transversely oblong, obtuse, with a few minute hairs on the back ; lateral lobes very small, rounded ; lower lip $\frac{3}{4}$ lin. long, deeply concave, with incurved margins, obtuse, pubescent outside ; stamens exserted, slightly unequal ; upper pair inserted 1$\frac{1}{2}$–2 lin. above the base of the corolla-tube, 3$\frac{1}{4}$ lin. long, their filaments free, contiguous, flattened, ciliate along one margin for about $\frac{2}{3}$ of their length ; lower pair inserted at the base of the lower lip, 1$\frac{1}{2}$ lin. long, their filaments united nearly to the apex, glabrous ; stigma clavate, emarginate or minutely bifid ; nutlets 1 lin. long, oblong, slightly compressed, light brown, shining.

KALAHARI REGION : Transvaal ; Mac Mac, *Mudd*!

17. **O. decipiens** (N. E. Br.) ; branches and leaves exactly as in *O. albiflorus* ; whorls $\frac{1}{2}$–$\frac{3}{4}$ in. apart, 4–6-flowered, in a raceme 1$\frac{1}{2}$–2$\frac{1}{2}$ in. long ; bracts 4–5 lin. long, 2–3 lin. broad, ovate, acuminate, thinly adpressed-pubescent, somewhat tomentose-ciliate, apparently rosy-purple, at first closely imbricate, covering the buds and forming an ovoid tip to the inflorescence, deciduous ; pedicels 1–1$\frac{1}{2}$ lin. long, pubescent ; calyx-tube 3–3$\frac{1}{4}$ in. long, tubular, pubescent and glandular outside ; upper tooth 1$\frac{1}{3}$–1$\frac{1}{2}$ lin. long, 1$\frac{1}{2}$ lin. broad, orbicular-ovate, very shortly and obtusely pointed ; 4 lower teeth subequal, 1 lin. long, bristle-like from a broader base ; corolla twice as long as the calyx, glabrous, curved at the middle ; tube 7–7$\frac{1}{2}$ lin. long, slender in the lower half, dilated and compressed at the upper part ; upper lip 2 lin. long, unequally 3-lobed, with the terminal lobe $\frac{2}{3}$ lin. long, $\frac{1}{2}$ lin. broad, ovate, obtuse ; lateral lobes very small, rounded ; lower lip 1–1$\frac{1}{4}$ lin. long, deeply boat-shaped, obtuse ; stamens slightly unequal, much exserted ; upper pair inserted at the middle of the corolla-tube, 6–6$\frac{1}{2}$ lin. long, with free flattened filaments, ciliate on one margin at the basal part ; lower pair inserted shortly below the base of the lower lip, 3 lin. long, slightly exceeding the upper pair, with their filaments united nearly to the apex, glabrous.

KALAHARI REGION : Transvaal ; Mac Mac, *Mudd* !

This species only differs from *O. albiflorus*, N. E. Br., in its more conspicuously bracteate inflorescence and larger flowers. It may possibly be a sexual state of that species, but this can only be determined from observation of the living plants.

18. **O. subvelutinus** (Guerke in Engl. Jahrb. xxvi. 80) ; stems several from a woody rootstock, 7–27 in. high, $\frac{3}{4}$–$1\frac{1}{2}$ lin. thick, minutely tomentose with branched hairs ; internodes 3–5 lin. long ; leaves with very short densely leafy branchlets in their axils, subsessile, ascending or spreading, 2–5 lin. long, $\frac{1}{3}$–$2\frac{1}{2}$ lin. broad, rather thick, linear or some of the primary ovate, acute, very revolute at the margins, minutely tomentose with branched hairs on both sides ; flower-whorls $\frac{1}{3}$–$\frac{3}{4}$ in. apart, 2–6-flowered, in racemes $1\frac{1}{4}$–$3\frac{1}{2}$ in. long ; bracts $1\frac{1}{2}$–3 lin. long, 1–$1\frac{1}{2}$ lin. broad, ovate, acute, tomentose like the leaves, persistent ; pedicels $\frac{3}{4}$–1 lin. long ; calyx-tube $2\frac{1}{4}$–$2\frac{1}{2}$ (in fruit 3) lin. long, campanulate, at first more or less tomentose outside with very minute branched hairs, mixed with long simple and gland-tipped hairs, minutely puberulous within ; upper tooth 1–$1\frac{1}{3}$ lin. long, $1\frac{1}{4}$–$1\frac{1}{2}$ lin. broad, very broadly ovate, obtuse ; lateral and lower teeth somewhat variable, bristle-like from a deltoid base or narrowly deltoid-attenuate, the lower $\frac{3}{4}$–$1\frac{1}{4}$ lin. long, all ciliate ; corolla much exserted, more or less tomentose (or sometimes nearly glabrous on the tube) outside with minute branched white hairs ; tube $4\frac{1}{2}$–6 lin. long, rather variable in outline curvature, slightly enlarged and compressed at the upper part and $1\frac{1}{4}$–$1\frac{3}{4}$ lin. broad vertically ; upper lip or truncate mouth of the tube $1\frac{3}{4}$–2 lin. long, its broadly rounded lateral lobes much shorter than the small rounded or subquadrate obtusely bifid terminal lobe ; lower lip 1–$1\frac{3}{4}$ lin. long, elliptic or suborbicular, obtuse or subacute, very concave ; stamens all exserted or the upper pair included and inserted $1\frac{1}{2}$–2 lin. above the base of the corolla-tube, 3–5 lin. long, with free flat filaments, ciliate along one margin on the basal part ; lower pair 1–$1\frac{3}{4}$ lin. long, free to the base or their filaments united for $\frac{1}{3}$–$\frac{1}{2}$ of their length, glabrous ; stigma very slightly thickened, very shortly bifid ; nutlets $\frac{1}{8}$ in. long, ellipsoid-oblong, dark brown, shining. *O. heterophyllus, Guerke in Engl. Jahrb.* xxvi. 82.

KALAHARI REGION : Transvaal, Lydenburg district ; near Lydenburg, *Atherstone* ! *Burtt Davy*, 1643 ! near Paarde Plaats, *Wilms*, 1152 ! near Spitz Kop, *Wilms*, 1155 ! and 1148 ex *Guerke*, Mac Mac Falls, *Burtt Davy*, 2536 !

In the authentic specimen of *O. heterophyllus* (Wilms 1155) at Kew the filaments of the lower pair of stamens are free to the base, whilst in the corresponding specimen at the British Museum they are united. This character taken alone would, upon the acceptance of Briquet's genus *Hemizygia*, place the different specimens of this plant under two genera, for I can find no other distinction between these two specimens, nor between them and *O. subvelutinus* (Wilms 1152), in which the filaments of the lower pair of stamens are also united in the specimen at Kew. I believe the union or non-union of the filaments in this plant to be either some sexual condition or to have some connection with different modes of fertilisation. The flowers on different specimens and sometimes even on the same specimen are distinctly variable, so that it may be of hybrid origin.

19. O. teucriifolius (N. E. Br.) ; perennial ; stems numerous from a woody rootstock, "in dense tufts" (*Wood*), 6–10 in. high, rather slender, terete, tomentose with minute branching hairs ; leaves very shortly petiolate, ¼–¾ in. long, ⅜–4 lin. broad, lanceolate, linear-lanceolate, linear or more rarely ovate, acute or subobtuse at each end, revolute along the margins, pubescent above, white-tomentose with minute branching hairs beneath ; racemes 1–4 in. long, with 3–6 distant 4–6-flowered whorls ; lower bracts 2–5 lin. long, like the leaves, persistent, upper deciduous ; pedicels 1–2 lin. long, tomentose ; calyx-tube 1½–1¾ (in fruit 2½) lin. long, campanulate, tomentose with minute branching (often mingled with long simple) hairs outside, puberulous within ; upper tooth suborbicular, minutely apiculate ; lateral and lower teeth subulate, the latter longer than the rest ; corolla twice as long as the calyx, tomentose on the lower lip outside, otherwise glabrous, pink or purple ; tube 3–3½ lin. long, 1½ lin. broad, curved below the middle, laterally flattened, truncate at the mouth, with a recurved 4-lobed margin or upper lip ; lobes unequal ½ lin. long ; lower lip 1–1¼ lin. long, boat-shaped, straight ; stamens exserted, equal, about equalling the lower lip of the corolla ; upper pair inserted below the middle of the tube, 2½–3½ lin. long, with the filaments free, ciliate on one margin ; lower pair inserted at the base of the lip, with the filaments united almost to the apex, 1–1¼ lin. long, glabrous ; stigma slightly thickened, shortly bifid ; nutlets rather more than 1 lin. long, oblong, slightly compressed, smooth, shining, brown. *Ocimum teucriifolium, Hochst. in Flora,* 1845, 66 ; *Krauss, Beitr. Fl. Cap- und Natal.* 130 ; *Benth. in DC. Prodr.* xii. 41. *Hemizygia teucriifolia, Briq. in Engl. & Prantl, Pflanzenfam.* iv. 3A, 369, *and in Ann. Conserv. Jard. Bot. Genèv.* ii. 247.

VAR. β, **galpiniana** (N. E. Br.) ; calyx-tube 2–2½ lin. long, with the pubescence on it chiefly consisting of long simple hairs, the minute branched hairs mostly confined to the base of the tube and nerves ; otherwise as in the type. *Hemizygia galpiniana, Briq. in Bull. Herb. Boiss.* 2ᵐᵉ sér. iii. 993.

COAST REGION : King Williamstown Div. ; near Pirie, 4000 ft., *Sim,* 107 !
KALAHARI REGION : Transvaal ; without precise locality, *Sanderson* ! Var. β : Transvaal ; eastern slopes of Saddleback Mountain, 4500 ft., *Galpin,* 1217 !
EASTERN REGION : Tembuland ; Bazeia, 2500 ft., *Baur,* 558 ! Griqualand East ; eastern slopes of the Zuurberg Range, 4500 ft.. *Tyson,* 1561 ! Natal ; at the foot of Table Mountain, *Krauss,* 448 ! on a grassy hill at Illovo, *Wood,* 1877 ! near Curry's Post, near Howick, 3000–4000 ft., *Wood,* 3567 ! and without precise locality, *Sutherland* ! Zululand ; on a grassy hill at Entumeni, *Wood,* 3964 (*Natal Herb.* 783) !

20. O. Pretoriæ (Guerke in Engl. Jahrb. xxvi. 81) ; perennial, producing from a woody rootstock numerous slender branching stems 4–10 in. high, with internodes ½–1 in. long, pubescent ; leaves ½–¾ in. long, 2–6 lin. broad, obovate, oblanceolate or elliptic, obtuse or subacute, cuneately tapering into a very short petiole, entire, often folded longitudinally, varying from nearly glabrous to thinly puberulous on one or both sides, usually more or less ciliate ;

whorls 2–8 in a simple raceme, ½–1 in. apart, 4–6-flowered ; bracts
persistent, ⅛–½ in. long, like reduced leaves, usually acute ; pedicels
1–2 (in fruit up to 3) lin. long, pubescent ; calyx-tube 2½ (in fruit
up to 3½) lin. long, tubular, much enlarged in fruit, pubescent with
spreading hairs ; upper tooth about 1 lin. long and broad, suborbicular, obtuse, lateral and lower teeth bristle-like, the lower
about 1–1½ lin. long ; corolla twice as long as the calyx, puberulous,
pinkish-white (*Wood*) ; tube 5–6 lin. long, rather slender, straight,
slightly enlarged at the upper half ; upper lip 2–2¼ lin. long, erect,
oblong, shortly 3-lobed at the apex, middle lobe twice as long as
the lateral, obovate or somewhat obcordate, slightly emarginate ;
lower lip ⅛–⅙ in. long, concave, obtuse ; upper stamens included,
inserted ⅔ up the tube, with free filaments 1–1¼ lin. long,
glabrous ; lower stamens inserted at the base of the lower lip and
nearly equalling it in length, 1½–1¾ lin. long, with their filaments
united for ¾ of their length, glabrous ; nutlets ⅛ in. long, oblong,
glabrous, brown. *S. Moore in Journ. Bot.* 1902, 385, *and* 1903, 405.
O. natalensis, Guerke in Engl. Jahrb. xxvi. 82.

KALAHARI REGION : Transvaal ; without precise locality, *McLea in Herb. Bolus,*
5776 ! Rietfontein, *Conrath,* 1060 ! near Pretoria, *Wilms,* 1151 ! near Johannesburg, *Rand,* 877 ! *Gilfillan in Herb. Galpin,* 6063 ! *Ommanney,* 13 ! *Miss Leendertz,*
1721 ! Bronkhorst Spruit, *Janse,* 59 !
EASTERN REGION : Natal ; on the Drakensberg Range near Coldstream, *Rehmann,*
6918 ! near Newcastle, 4000 ft., *Schlechter,* 3420 ! on the Biggars Berg, near
Glencoe, 4000–5000 ft., *Wood,* 4756 ! *Kuntze ex Guerke.* Zululand ; without
precise locality, *Gerrard,* 1219 !

21. **O. Wilmsii** (Guerke in Engl. Jahrb. xxvi. 81) ; perennial,
½–1 ft. high ; stems several branching from a woody rootstock,
simple or with erect branches, square, puberulous with minute
recurved hairs along the angles only ; leaves spreading ; petiole
1–3 lin. long ; blade ½–1⅓ in. long, 3–7 lin. broad, ovate or lanceolate, acute or obtuse, broadly or narrowly cuneate at the base,
subentire to acutely serrate, minutely puberulous on the veins
beneath, otherwise glabrous on both sides ; racemes simple, 1¼–5 in.
long ; whorls ¼–½ in. apart, 6-flowered ; bracts reflexed, ¾–1½ lin.
long, ovate, acute or apiculate, subpetiolate or sessile, glabrous,
ciliate, often bordered with purple ; pedicels 1–2 lin. long, pubescent ; calyx pubescent or subtomentose outside, glabrous within, pur-
plish-brown ; tube ⅛–⅙ (in fruit ⅕–¼) in. long, tubular-campanulate,
with distinct nerves, especially in fruit ; upper tooth ½–¾ lin. long,
orbicular-ovate, obtuse ; lateral and lower teeth bristle-like, the
lower ⅔–1 (in fruit up to 1¾) lin. long ; corolla-tube 2–3 lin. long,
slightly exceeding the lower calyx-teeth, straight, puberulous outside ; upper lip 1¼–2 lin. long, with 4 small rounded lobes ; lower
lip 1½–2 lin. long, boat-shaped, obtuse ; stamens all free, curved,
glabrous ; upper pair 1 lin. long, inserted in the throat of the tube
and just exserted from it or included ; lower pair 1¼ lin. long,
inserted at the base of the lower lip and shorter than it ; style
very minutely bifid or emarginate at the clavate apex ; nutlets

¾ lin. long, ellipsoid, obtuse, very minutely and faintly tuberculate, brown. *Orthosiphon glabratus*, Benth., var. *africanus*, Benth. *in DC. Prodr.* xii. 51. *O. neglectus*, *Briq. in Bull. Herb. Boiss.* 2me sér. iii. 988. *Plectranthus Bolusii*, *T. Cooke in Kew Bulletin*, 1909, 377, as to *Bolus* 11011 only.

VAR. β, **komghensis** (N. E. Br.); lateral and lower calyx-teeth deltoid-attenuate, not bristle-like, otherwise in appearance, structure and dimensions like the type.

COAST REGION: Var. β : Komgha Div. ; among stones near the Kei River, *Flanagan*, 477 !

KALAHARI REGION : Transvaal; Crocodile River, *Burke*, 162 ! *Zeyher*, 1357 ! Magalies Berg, *Burke*! Matebe Valley, *Holub*, 1952 ! 1953 ! Linokana, *Holub*! near Lydenburg, *Wilms*, 1114, 1115 ! Krugers Post, *Burtt Davy*, 7276 ! Springbok Flats, *Burtt Davy*, 1744 ! Wonderboom Poort, *Rehmann*, 4510 ! Koude River, *Schlechter*, 3728 ! Potgieters Rust, *Miss Leendertz*, 1439 ! *Bolus*, 11011 !

EASTERN REGION: Natal ; Sydenham, near Durban, 500 ft., *Wood*, 8538 ! near Pietermaritzburg, *Wilms*, 2189 !

The variety *komghensis* may prove to be distinct, as the locality is so widely different from that of the type, but the single dried specimen I have seen is so similar in all characters except the calyx-teeth, that I am inclined to think that the plant may have been an accidental introduction from the Transvaal, and the moister maritime climate may have influenced the development of the calyx-teeth. *O. Wilmsii* is quite distinct from the Indian *O. glabratus*, Benth., of which Bentham considered it to be a variety.

22. **O. inconcinnus** (Briq. in Bull. Herb. Boiss. 2me sér. iii. 991); branches with internodes ¾–1¼ in. long, minutely puberulous or glabrous ; leaves spreading ; petiole up to 5 lin. long ; blade up to 1¼ in. long and ¾ in. broad, ovate, obtuse to acute, roundedly tapering at the base, crenately toothed, nearly or quite glabrous above, pubescent on the veins beneath ; inflorescence up to 4 in. long ; whorls at length distant, 6-flowered ; bracts very minute, ovate, apiculate, deciduous, purplish ; pedicels 1½–2 lin. long, shortly pubescent ; calyx-tube nearly 1½ lin. long, campanulate-tubular, minutely and thinly pilose or nearly glabrous, without evident nerves, purplish-violet ; upper lobe or tooth rounded, ½–¾ lin. long ; lateral and lower teeth ⅓–½ lin. long, lanceolate-subulate ; corolla exserted about 2½ lin. beyond the mouth of the calyx ; tube cylindric ; lips about 1½ lin. long ; stamens reaching to the mouth of the corolla, but shorter than the lips.

EASTERN REGION : Natal ; hills near Camperdown, *Wood*, 4963.

I doubt if this is distinct from *O. Wilmsii*, Guerke. I have not seen it, and Mr. Wood informs me that no specimen of 4963 has been retained in the Natal Herbarium.

23. **O. varians** (N. E. Br.); stems probably several from the same rootstock, about 1 ft. high or less and ¾ lin. thick, 4-angled, pilose-pubescent between the angles, with 3–4 pairs of short flower-less ascending branchlets shorter than to twice as long as the leaves from whose axils they arise ; leaves spreading ; petiole ⅛–¼ in. long, slender ; blade ½–1 in. long, ¼–½ in. broad, lanceolate or ovate-lanceolate, acute, cuneately acute at the base, rather finely serrate,

thinly pilose-pubescent above and more thickly so beneath, green on
both sides, but apparently darker above; veins impressed above,
prominent beneath; whorls 5–8, about $\frac{1}{2}$–$\frac{3}{4}$ in. apart, 4–6-flowered,
in simple racemes; bracts not seen, very deciduous; pedicels
1–1$\frac{1}{2}$ lin. long, puberulous; calyx-tube 1$\frac{1}{2}$ (in fruit 2) lin. long,
campanulate, puberulous outside, glabrous within; upper tooth $\frac{2}{3}$ lin.
long and broad, suborbicular, decurrent on the tube; lateral and
lower teeth bristle-like, with the lower pair $\frac{2}{3}$ lin. long, lateral
shorter; corolla very variable; tube 2$\frac{2}{3}$–4$\frac{1}{2}$ lin. long, slender,
enlarging to 1–1$\frac{1}{4}$ lin. in vertical diameter at the compressed truncate or oblique mouth, which forms the upper lip and has a small
apical lobe $\frac{1}{4}$ lin. long, $\frac{1}{3}$ lin. broad and transverse in the short-
tubed and $\frac{1}{2}$ lin. long and broad and subquadrate in the long-tubed
corollas; lower lip 1$\frac{1}{4}$–1$\frac{1}{2}$ lin. long, elliptic-ovate, rather abruptly
acute; stamens all exserted far beyond the lower lip; filaments of
the lower pair united for the greater part of their length, exceeding
the upper pair, 2$\frac{1}{2}$ lin. long in the short-tubed and 4 lin. long in
the long-tubed flowers; nutlets $\frac{3}{4}$ lin. long, compressed, elliptic in
outline, dorsally keeled, smooth, pale brown.

KALAHARI REGION: Transvaal; Komati Poort, *Schlechter*, 11746!

The range in the variation of the length of the corolla-tube of this plant, even
on the same raceme, is remarkable; many species vary in this character, but I
have seen no other in which it is so great.

24. **O. affinis** (N. E. Br.); stems much branched at the upper
part, lower not seen, obtusely 4-angled, rather thinly pubescent
with small upcurved hairs; leaves $\frac{3}{4}$–1$\frac{1}{4}$ in. long, $\frac{1}{8}$–$\frac{1}{4}$ in. broad,
rather narrowly lanceolate, acute, cuneately tapering below into
the 1–2 lin.-long petiole, finely toothed along the margins, green
and minutely pubescent on both sides, not ciliate; veins impressed
on the upper surface; racemes all attaining about the same level,
2–3$\frac{1}{2}$ in. long; whorls 4–9, about $\frac{1}{2}$–$\frac{3}{4}$ in. apart, 4–6-flowered;
bracts very deciduous, about $\frac{1}{8}$ in. long, ovate, shortly acuminate,
concave, with a few minute hairs on the back, white-ciliate;
pedicels 1–1$\frac{1}{2}$ lin. long, puberulous; calyx-tube 1$\frac{3}{4}$ lin. long, puberulous outside; upper tooth 1 lin. long and rather more in breadth,
suborbicular, shortly decurrent on the tube, dark purple; lateral
and lower teeth subequal, $\frac{2}{3}$–$\frac{3}{4}$ lin. long, bristle-like; corolla-tube
about 3$\frac{1}{2}$ lin. long, gradually enlarging to the $\frac{1}{8}$ in.-high obliquely
truncate mouth which forms the upper 3-lobed lip, thinly puberu-
lous outside; middle lobe of upper lip $\frac{1}{2}$ lin. long, $\frac{3}{4}$ lin. broad,
transversely oblong, subtruncate; lateral lobes very small; lower
lip 1$\frac{1}{3}$–1$\frac{1}{2}$ lin. long, elliptic-ovate, obtuse, concave; stamens exserted
far beyond the lower lip; upper pair inserted at about $\frac{1}{3}$ of the
way up the corolla-tube; lower pair $\frac{1}{4}$ in. long, united nearly or
quite to their apex.

KALAHARI REGION: Transvaal; Woodbush Mountains, 5500 ft., *Schlechter*,
4737! near Potgieters, 3700 ft., *Bolus*, 11146!

The specimens seen appear to be portions broken off the upper part of the plant and are 8–9 in. in length, with 2–4 pairs of flowering branches, which are simple or again branched.   Allied to *O. Holubii*, but readily distinguished by the much more minute pubescence and absence of ciliation on the leaves.

25. **O. Holubii** (N. E. Br.); stems (only terminal pieces 7 in. long seen) 4-angled, with internodes 1–1½ in. long, sprinkled with short spreading hairs; leaves (upper only seen) spreading, subsessile, 1–1½ in. long, 2–4 lin. broad, lanceolate, acute or acuminate, tapering to an acute base, but with scarcely any petiole, bearing tufts of small leaves in their axils, glabrous above, with spreading pubescence on the nerves beneath, ciliate; inflorescence 4–5 in. long, terminal, with about 1 pair of branches at the base; whorls ½–1 in. apart, 2–6-flowered; bracts small, caducous, about 1 lin. long, ½–⅔ lin. broad, ovate, acute; pedicels 1–1½ lin. long, pubescent; calyx-tube 1¾ lin. long, larger in fruit, campanulate, 5-toothed, pubescent with simple and gland-tipped hairs outside, puberulous with gland-tipped hairs within; upper tooth 1 lin. long, 1⅓ lin. broad, broadly orbi-cular-ovate, obtuse; lateral and lower teeth unequal, bristle-like, pubescent with simple and gland-tipped hairs, the lower pair about 1 lin. long; corolla pubescent outside and on the lower part within; tube 4–5 lin. long, dilated and compressed in the upper part; upper lip 2½ lin. long, 3-lobed at the apex; middle lobe ¾ lin. long and broad, suborbicular, emarginate at the apex, reflexed at the sides; lateral lobes very small, rounded; lower lip about 2 lin. long, broadly elliptic, obtuse, concave; stamens much exserted, sub-equal; upper pair inserted ⅓ of the way up the tube, 6–7 lin. long, their filaments free, flat, retrorsely ciliate along the lower part of one margin; lower pair inserted at the base of the lower lip, 4½ lin. long, their filaments united to the apex, glabrous; stigma abruptly thickened, oblong, obtuse, slightly emarginate.

KALAHARI REGION : Eastern Bechuanaland ; Molopo River, *Holub* !

26. **O. Bolusii** (N. E. Br.) ; stems probably several from a perennial rootstock, 10–13 in. high, obtusely 4-angled, more or less pilose, with a pair of short leafy barren branchlets at 2–4 of the nodes; leaves of main stems in 4–5 pairs, 1–2½ in. apart, spreading; petiole 1–2 lin. long, sometimes almost wanting in the uppermost pair ; blade ½–1 in. long, ⅓–¾ in. broad, elliptic or elliptic-ovate, obtuse to broadly rounded at the apex, rounded or broadly cuneate at the base, with 3–7 small teeth on each side, mostly towards the apex, green and pubescent (but not densely) on both sides ; whorls 4–6 in a simple raceme, ½–1¼ in. apart, 4–10-flowered; bracts very deciduous, not seen; pedicels 1½–2 lin. long, pilose ; calyx-tube 2½–3¼ lin. long, campanulate, pilose-pubescent outside, glabrous within; corolla-tube about ½ in. long, its oblique mouth forming the 3-lobed upper lip, glabrous ; upper lobe 1¼ lin. long, 1½ lin. broad, broadly ovate, obtuse ; lateral lobes smaller ; lower lip ¼ in. long, concave-elliptic, obtuse; stamens all exserted much

beyond the lower lip ; upper pair inserted at about $\frac{1}{3}$ of the way up
the corolla-tube, with free filaments, flat and minutely ciliate on
the basal part; lower pair exceeding the upper, about $\frac{1}{2}$ in. long
when fully grown, inserted at the base of the lower lip of the
corolla, with their filaments variably free to the base or united
quite to the apex.

EASTERN REGION : Natal ; Giants Castle, Drakensberg Range, 9000 ft., *Bolus in
Herb. Guthrie*, 4894 !

This species is interesting as affording a good instance that the union or freedom
of the filaments of the lower pair of stamens is quite untenable as a generic
character, both *Orthosiphon* and *Hemizygia* being represented on the same plant.
The specimen is now in the Herbarium of Dr. Bolus.

27. **O. humilis** (N. E. Br.) ; stems arising from a woody rootstock, erect, 7–10 in. high, thinly villous below, more densely so
above, with 1–2 pairs of short flowerless ascending branches below
and a paniculate inflorescence of 2–3 pairs of lateral racemes and
a terminal one; leaves in 4 pairs, $\frac{1}{2}$–$\frac{3}{4}$ in. or rather more in
length, $\frac{1}{4}$–$\frac{1}{3}$ in. broad, lanceolate to elliptic, acute or obtuse,
cuneately narrowed to a subsessile base in the leaves on the main
stem and into a distinct petiole $\frac{1}{2}$–$\frac{2}{3}$ lin. long on the lateral
branches, entire, green and with very thinly scattered pubescence
of minute adpressed hairs on both sides, with longer and more
conspicuous hairs on the veins beneath, shortly ciliate ; glands
inconspicuous ; bracts $1\frac{1}{2}$–3 lin. long, 1–$1\frac{3}{4}$ lin. broad, elliptic-ovate,
acute or acuminate, with a few minute scattered adpressed hairs
on the back, violet or dark bluish-purple, very deciduous or perhaps
the uppermost persistent, but not very conspicuous ; pedicels
1–$1\frac{1}{2}$ lin. long, densely villous ; calyx-tube $\frac{1}{4}$ in. long, tubular-
campanulate, villous outside, microscopically puberulous within ;
upper tooth 2 lin. long and broad, suborbicular, not decurrent on
the tube, violet ; lateral and lower teeth subequal, $1\frac{1}{4}$–$1\frac{1}{2}$ lin. long,
bristle-like from a deltoid base, ciliate ; corolla partly destroyed in
the only two flowers seen ; tube not or scarcely exceeding the
calyx-teeth, $4\frac{1}{2}$–5 lin. long, dilated to about 2 lin. in vertical diam.
at the subtruncate mouth, which forms the upper lip with a small
transverse erect or reflexed lobe at the apex having a small
rounded tooth on each side of it, puberulous outside on the upper
part ; lower lip about $2\frac{1}{2}$ lin. long, concave, apparently directed
forwards, puberulous on the back ; stamens broken off in the flowers
seen ; upper pair inserted at the middle of the corolla-tube, with
free filaments ciliate at the lower part ; lower pair evidently longer
than the lower lip, with united filaments.

KALAHARI REGION : Transvaal ; Waterval Onder, *Rogers*, 4375 !

28. **O. canescens** (Guerke in Bull. Herb. Boiss. vi. 557) ; stem
1 ft. or more high, shortly branched, whitish-pubescent to subtomentose ; leaves of main stems, including the 1–2 lin.-long
petiole $\frac{3}{4}$–$1\frac{3}{4}$ in. long, $2\frac{1}{2}$–7 lin. broad, spreading or deflexed, lanceo-

late, acute, serrulate, cuneate into the petiole at the base, whitish-pubescent to subtomentose, especially beneath; panicle 6–8 in. long, with 2–3 pairs of branches, pubescent; whorls several, about 6-flowered, 4–6 lin. apart; bracts $\frac{1}{2}$–1 lin. long, ovate, acute, caducous; pedicels $\frac{3}{4}$–1$\frac{1}{2}$ lin. long; calyx-tube 1$\frac{3}{4}$ lin. long, campanulate, enlarged in fruit; upper tooth about 1 lin. long and broad, suborbicular, obtuse; lateral and lower teeth bristle-like, the latter $\frac{3}{4}$ lin. long; corolla more than twice as long as the calyx, slightly pubescent; tube 3–6$\frac{1}{2}$ lin. long, straight, compressed and dilated in the upper part; upper lip (truncate mouth of the tube) erect, 1–1$\frac{1}{2}$ lin. long, shortly 3-lobed; side-lobes minute, rounded, recurved; terminal lobe broadly rounded, very obtuse; lower lip $\frac{2}{3}$–1$\frac{1}{4}$ lin. long, saccate, obtuse; upper stamens inserted about 1$\frac{1}{3}$ lin. above the base of the corolla-tube, much exserted; about 5 lin. long, flattened, ciliate along one margin on the basal half; lower stamens inserted at the base of the lower lip, equalling or exceeding the upper stamens, 2$\frac{1}{2}$–3 lin. long, with their filaments united nearly to the apex, glabrous.

KALAHARI REGION: Transvaal; Wonderboom Poort, near Pretoria, *Rehmann*, 4507! *Miss Leendertz*, 553! Heidelberg, *Miss Leendertz*, 1027! near Botsabelo, *Schlechter*, 4070! Aapies Poort, *Rehmann*, 4114, and Elands River, *Rehmann*, 4891, ex *Guerke*; Rustenburg, *Collins*, 30!

29. **O. serratus** (Schlechter in Journ. Bot. 1897, 431, *serratum*); stem herbaceous, thickly covered with spreading white hairs; leaves in whorls of 3, recurved, $\frac{2}{3}$–1$\frac{1}{2}$ in. long, 5–7 lin. broad, oblanceolate or obovate, acute, tapering into a short petiole at the base, serrate, somewhat folded longitudinally, pubescent on both sides, pilose on the midrib and petiole; whorls several in a simple raceme, $\frac{1}{2}$–1 in. apart, about 9-flowered; bracts 3–4 lin. long, 1$\frac{1}{2}$–2 lin. broad, ovate, cuspidate-acuminate, nearly glabrous, purplish (?), persistent; pedicels 1$\frac{1}{2}$–2 lin. long, pubescent; calyx-tube 2–3 lin. long, tubular-campanulate, pubescent outside, reddish or purplish, much enlarged when in fruit; upper tooth 2 lin. long, 1$\frac{1}{2}$ lin. broad, elliptic, obtuse, apiculate, recurved at the sides; 4 lower teeth $\frac{2}{3}$–1 lin. long, bristle-like from a broad base; corolla purple, pubescent outside; tube $\frac{1}{2}$ in. long, much exserted, subcylindric; upper lip erect, 2$\frac{1}{2}$–3 lin. long, 1$\frac{1}{2}$–1$\frac{2}{3}$ lin. broad, oblong, shortly and unequally 4-lobed; lateral lobes somewhat triangular, obtuse; terminal lobes suborbicular; lower lip about 2 lin. long, 1$\frac{1}{2}$ lin. broad, oblong or elliptic-oblong, obtuse, deeply concave, deflexed; stamens much exserted, subequal; filaments all free to the base, flattened; upper pair $\frac{1}{2}$ in. long, inserted towards the base of the corolla-tube, rather densely villous-pubescent on the lower half; lower pair 2$\frac{1}{2}$ lin. long, inserted at the base of the lower lip, glabrous; stigma bifid, lobes $\frac{1}{4}$ lin. long, slender; nutlets about 1$\frac{1}{2}$ lin. long, oblong, glabrous.

KALAHARI REGION: Transvaal; in stony places on the lower hill-slopes near Barberton, 2800 ft., *Galpin*, 499! near Mafutane, 1500 ft., *Bolus*, 12249! Elandspruit Mountains, *Schlechter*, 3866! Potgieters Rust, *Miss Leendertz*, 1494!

Imperfectly known species.

30. O. ambiguus (Bolus in Journ. Linn. Soc. xviii. 394); stems ascending, scarcely 1 ft. high, slender, simple or three-branched at the base, pubescent; leaves with petioles about 1 in. long; blades 1–1¼ in. long, ⅞ in. broad, ovate, subobtuse, coarsely toothed, base cuneate, thinly pubescent above, with jointed hairs on the petiole and veins beneath; whorls 4–6-flowered, crowded in racemes scarcely 1 in. long; lower bracts leaf-like, subsessile, upper minute, entire; pedicels 2 lin. long; calyx 2½ lin. long; tube short, pubescent, glabrous within; teeth coloured, upper one scarcely decurrent, ovate, obtusely pointed; lower 4 subequal, lanceolate-subulate; corolla pale blue; tube ¾ in. long, slender, slightly incurved, thinly pubescent; upper lip obreniform, with a triangular lobe on each side at the base; lower lip boat-shaped; stamens exserted; stigma bifid, with subequal subulate lobes.

COAST REGION: Albany Div.; in woods on a mountain near Grahamstown, 2000 ft., *MacOwan,* 987.

IV. SYNCOLOSTEMON, E. Meyer.

Calyx subequally 5-toothed, with the dorsal tooth scarcely broader than the others, not decurrent on the tube. *Corolla* exserted from the calyx, 2-lipped; upper lip unequally 3-lobed; lower lip concave or boat-shaped, entire, often reflexed. *Stamens* 4, all perfect; upper pair with free filaments, neither bent nor toothed near the base; lower pair with the filaments united. *Disk* unequally lobed. *Ovary* 4-lobed; style filiform; stigma minutely bifid. *Nutlets* erect, oblong, obtuse, glabrous.

Herbs or in one species shrubby; leaves opposite, simple, gland-dotted; inflorescence terminal, paniculate, lax or crowded into a dense mass; flowers in pairs or 6-flowered whorls.

DISTRIB. Species 7, all endemic.

This genus only differs from *Orthosiphon* in its subequally toothed calyx. *N. E. Br.*

Corolla-tube 3–5 lin. long; lower or united pair of
    stamens 2–6 lin. long; whorls 2-flowered, distant:
    Leaves ½–1¾ in. long, narrowly lanceolate, 4–8 times
      as long as broad:
      Leaves with thin minute adpressed pubescence ... **(1) lanceolatus.**
      Leaves silvery-white with dense silky adpressed
        pubescence ... ... **(2) argenteus.**
    Leaves ¼–⅝ in. long, lanceolate to elliptic, varying
      from as long as broad to 3 times as long as broad:
      Pubescence on leaves of simple hairs... ... ... **(3) parviflorus.**
      Pubescence on leaves of very minute stellate hairs... **(4) concinnus.**
Corolla-tube 7–10 lin. long; lower or united pair of
    stamens 4–9 lin. long; whorls 2–6-flowered, crowded:
    Calyx-tube 2½–3½ lin. long; lower lip of corolla 1½–3½
      lin. long:
      Panicle narrow, its short branches ¼–1 in. apart;
        corolla-tube 7–7½ lin. long ... ... **(5) ramulosus.**

Panicle with its short branches and flowers all crowded
into a thick dense spike-like mass ;. corolla-tube
8-10 lin. long   ...   .   (7) **densiflorus.**

Calyx-tube 4 lin. long; lower lip of corolla 4 5½ lin.
long   ...   .   (6) **rotundifolius.**

1. **S. lanceolatus** (Guerke in Engl. Jahrb. xxvi. 77); stems 2–3
ft. high, erect, square, minutely and somewhat harshly puberulous ;
leaves with very short leafy shoots in their axils ; petiole ½–5 lin.
long ; blade ½–1¾ in. long, ¾–3 (rarely 4–5) lin. broad, narrowly or
linear-lanceolate, or the broader oblanceolate, obtuse or subacute,
cuneately tapering at the base, rather thick, minutely adpressed-
puberulous on both sides ; inflorescence 9–12 in. long, 4–8 in. broad,
paniculate or corymbosely paniculate, with several pairs of simple
or branched branches ; whorls 3–9 lin. apart, never more than
2-flowered ; bracts caducous, 1¾–2 lin. long, ¾–1 lin. broad, ovate
or ovate-lanceolate, acuminate, puberulous ; pedicels ¾–1½ lin. long,
puberulous ; calyx-tube 2 lin. long, campanulate, becoming ovoid
and somewhat constricted at the mouth in fruit, minutely pubescent outside, glabrous within and without woolly hairs on the teeth ;
teeth all alike, ¾–1¼ lin. long, narrowly deltoid and acuminate or
subulate from a deltoid base ; corolla white ; tube exserted, straight,
3–3¼ lin. long, dilated above, glabrous outside, with a transverse
pubescent band within, just above the insertion of the upper stamens ;
upper lip 2½–3 lin. long, unequally 3-lobed, pubescent on the back,
ciliate ; middle lobe suborbicular, subquadrate or transverse, notched
or obtuse ; lateral lobes broadly rounded ; lower lip 1½–2 lin. long,
very deeply concave, very obtuse, puberulous outside, ciliate ;
stamens exserted much beyond the lower lip, slightly unequal ;
upper pair inserted just above the middle of the corolla-tube, with
free flattened filaments, slightly ciliate at the base ; lower pair
inserted at the base of the lower lip, 3–5 lin. long, with the filaments
united nearly or quite to the apex, glabrous ; nutlets 1½–1¾ lin. long,
oblong, obtuse, with a deep rim or puckered frill or crest at the
basal end, light brown or grey green, often with a slender line down
the back and one on each side, shining.

VAR. *β*, **grandiflorus** (N. E. Br.); calyx-tube 3–3½ lin. long, tubular-
campanulate ; teeth 1¼–1¾ lin. long, subulate or narrowly deltoid-subulate, somewhat white-woolly at their margins ; corolla white ; tube 4½–5 lin. long, glabrous
outside ; upper lip 3½ lin. long ; stamens scarcely or not at all exceeding the lower
lip of the corolla, with the lower pair 2 lin. long; otherwise as in the type.

VAR. *γ*, **Cooperi** (N. E. Br.); calyx with white woolly hairs at its mouth and on
the margins and inner side of the teeth ; otherwise as in the type.   *S. Cooperi,*
Briq. in Herb. Boiss. 2ᵐᵉ *sér.* iii. 979.

KALAHARI REGION : Var. *γ* : Orange River Colony, *Cooper,* 2895 !
EASTERN REGION : Griqualand East ; near Clydesdale, *Schlechter,* 6616 ! Mount
Malowe, 4000 ft., *Tyson,* 2770 ! and in *MacOwan & Bolus, Herb. Norm. Austr-
Afr.* 1294 ! Var. *β* : Natal ; grassy hill near Enon, *Wood,* 1882 ! Var. *γ* : Natal ;
between Fark Kop and Camperdown, *Rehmann,* 7686 ! grassy hill near Umkomaas,
Wood, 1994 ! and without precise locality, *Sanderson* ! *Cooper,* 1151 !

I can find no specific difference between the three above forms, which may possibly be only sexual conditions. Guerke did not even distinguish var. *Cooperi* from the type.

2. **S. argenteus** (N. E. Br.); in height, habit, inflorescence, size and shape of flowers exactly as in *S. lanceolatus*, differing as follows:—stem much concealed by the numerous short axillary densely leafy shoots; leaves $\frac{1}{2}$–$\frac{3}{4}$ lin. long, 1–5 lin. broad, linear-lanceolate to broadly elliptic, acute to very obtuse, silvery-white on both sides from dense adpressed pubescence of silky hairs; branches of the panicle rather densely white-pubescent, with tufts of longer hairs at the nodes; pedicels $\frac{1}{2}$–$\frac{3}{4}$ lin. long; calyx-tube $1\frac{1}{2}$–2 lin. long, rather densely pubescent; teeth $\frac{1}{2}$ lin. long, narrowly deltoid, densely white-woolly on the margins and inner surface; corolla white.

EASTERN REGION : Zululand ; near Inyezaan, *Wood*, 3875 (*Natal Herb.* 726) !

This may be only a variety of *S. lanceolatus*, Guerke, as there seems to be no evident distinction in its flowers, but the more densely leafy stem and silvery leaves give it such a different appearance that I hesitate to unite them.

3. **S. parviflorus** (E. Meyer, Comm. 231); stems several from a woody rootstock, about 15 in. high, obtusely 4-angled, softly pubescent with slightly spreading hairs ; leaves of the main stems with very short leafy shoots in their axils; petiole $\frac{1}{3}$–1 lin. long; blades 3–6 lin. long, $1\frac{2}{3}$–3 lin. broad, those on the axillary shoots smaller, lanceolate-oblong or somewhat obovate-oblong or the lower elliptic, obtuse, somewhat cuneate at the base, slightly toothed in the upper half or entire, softly pubescent and densely gland-dotted on both sides ; inflorescence terminal, 1–3 in. long, simple or slightly branched ; whorls 4–6 lin. apart, 2-flowered ; bracts very caducous, $1\frac{1}{2}$ lin. long, ovate, acute, concave, pubescent ; pedicels $\frac{1}{2}$–$\frac{2}{3}$ lin. long, pubescent ; calyx pubescent and densely gland-dotted outside, glabrous inside the tube, ciliate and pubescent on the inside of the teeth ; tube 2 lin. long, campanulate ; teeth subequal, erect, almost 1 lin. long, subulate from a broad base or the upper tooth shorter than the rest and oblong, obtuse ; corolla exserted, about twice as long as the calyx ; tube straight, $3\frac{1}{4}$–$3\frac{1}{3}$ lin. long, gradually enlarging upwards, compressed, pubescent outside on the upper $\frac{1}{3}$, and inside at about the middle ; upper lip $2\frac{1}{2}$ lin. long, erect, 3-lobed, pubescent and gland-dotted on the back of the lobes ; terminal lobe 1 lin. long and almost as broad, oblong, notched at the obtuse apex, ciliate with rather long hairs ; lateral lobes $\frac{1}{2}$ lin. long, very broadly rounded, ciliate ; lower lip about $1\frac{2}{3}$ lin. long, deeply concave, obtuse, slightly pubescent and gland-dotted outside ; stamens much exserted, subequal ; upper pair inserted about 2 lin. above the base of the corolla, 4 lin. long, their filaments free, flattened, with a few minute hairs on the face of the basal part ; lower pair inserted at the base of the lower lip, 3 lin. long, their filaments united to the apex, glabrous. *Dietr. Syn. Pl.* iii. 385 ; *Benth. in DC. Prodr.* xii. 54.

VAR. β, dissitiflorus (N. E. Br.); stems 16–30 in. high; leaves entire, elliptic-lanceolate or sometimes suborbicular, more numerous or more crowded on the short axillary shoots than in the type; panicle lax, 4–7 in. long, 2–3½ in. broad; calyx-teeth usually somewhat spreading or recurved; terminal lobe of the upper lip of the corolla 1–1½ lin. long, about 1 lin. broad, obcordate-oblong or sub-quadrate, notched at the apex, less glandular on the back than in the type; nutlets 1⅓ lin. long, oblong, obtuse, with a thickened puckered rim at the base, white on the basal ¼, light brown above, shining. *S. dissitiflorus, Benth. in DC. Prodr.* xii. 54.

EASTERN REGION: Pondoland; between St. Johns River and Umsikaba River, *Drège*, 4749! Var. β : Natal; near Durban, *Drège*! plain between Umlaas River and Durban Bay, *Krauss*, 145! coast land, *Sutherland*! Inanda, 1800 ft., *Wood*, 52!

4. S. concinnus (N. E. Br.); stem rather thickly beset with numerous slender suberect leafy branches, with indumentum of scattered very minute branched tuft-like hairs, brownish; leaves ⅖–⅔ in. long, obovate-oblong or elliptic, rounded at the apex, shortly cuneate into a slender petiole ½–1 lin. long, entire or some of those on the main stem with a few small teeth towards the apex, with very minute stellately branched hairs on both sides and densely glandular, of a brownish-olive when dried; racemes in a corymb-like panicle 4–5 in. long and broad, with 2-flowered whorls 4–5 lin. apart; bracts (except the lowest leaf-like pair) caducous, about 1 lin. long, ovate, acute, apiculate, concave; calyx-tube 2 lin. long, cam-panulate, shortly pubescent on the nerves outside, glabrous within, ciliate at the mouth; teeth subulate from a broad base, 4 equal and ¾ lin. long, the dorsal one rather more deeply separated from them and 1 lin. long; corolla-tube 4–4½ lin. long, glabrous, obliquely truncate at the mouth, forming the shortly 3-lobed upper lip with the middle lobe 1–1⅓ lin. long, oblong, slightly notched at the truncate apex; lateral lobes smaller, rounded; lower lip 1½ lin. long, 3 lin. broad, transverse,. truncate, ciliate along the front margin; stamens far exserted, unequal, with the lower (united) pair ½ in. long, much exceeding the upper pair.

KALAHARI REGION: Transvaal; Elandspruit Mountains, 6000 ft., *Schlechter*, 3891!

5. S. ramulosus (E. Meyer, Comm. 231); plant 2 ft. or more high; stems obtusely 4-angled and grooved, minutely and somewhat velvety pubescent, bearing numerous short leafy shoots on the lower part; leaves (including the very short petioles) 2–4 lin. long, 1½–2 lin. broad, elliptic, obtuse, cuneately narrowed into the petiole, glabrous above, thinly and very minutely pubescent beneath, densely gland-dotted; panicle narrow, 6–8 in. long, 1½–2 in. broad, with ascending branches ¾–1¼ in. apart; whorls 6-flowered; bracts caducous, about 3½ lin. long, 2–2½ lin. broad, broadly ovate, cuspidate-acuminate, concave, thinly and minutely puberulous; pedicels ¼–½ lin. long, pubescent; calyx-tube 2½–3 lin. long, cylindric, scaberulous-pubescent outside, glabrous within; upper tooth 1–1¼ lin. long, deltoid, acute; lateral and lower teeth equal,

1⅓ lin. long, subulate, acute ; corolla much exserted, glabrous outside, minutely papillate at the base inside ; tube curved at the
base, 7–7½ lin. long, gradually enlarging upwards ; upper lip erect,
2–2¼ lin. long, 3-lobed ; middle lobe ¾ lin. long and about as broad,
oblong or obovate, obtuse or emarginate ; lateral lobes smaller,
rounded ; lower lip 1½–1¾ lin. long, concave, obtuse, reflexed ;
stamens much exserted, but spirally coiled, slightly unequal ; upper
pair inserted about 2 lin. above the base of the corolla, 10 lin. long,
their filaments free, flat, ciliate on one margin at the base ; lower
pair inserted at the base of the lower lip, 4½ lin. long, their filaments united to the apex, glabrous ; stigma unequally bifid. *Dietr.
Syn. Pl.* iii. 385 ; *Benth. in DC. Prodr.* xii. 54. *S. ramulosum,
Hochstetter in Flora,* 1845, 68 ; *Krauss, Beitr. Fl. Cap- und Natal.*
132.

EASTERN REGION : Tembuland; near Morley, *Drège,* 4744b!

6. **S. rotundifolius** (E. Meyer, Comm. 231) ; apparently shrubby,
branching ; branches obtusely 4-angled, minutely velvety-pubescent ;
leaves with very short leafy shoots in their axils ; petiole 1–3 lin.
long ; blade 3–9 lin. long, 2½–9 lin. broad, elliptic or orbicular,
obtuse at both ends or broadly cuneate at the base, with 2–4 small
teeth on each side, minutely pubescent or nearly glabrous on both
sides, densely gland-dotted ; panicles compact, spike-like, 1–6 in.
long ; whorls rather crowded, 2-flowered ; bracts caducous,
2½–3 lin. long, elliptic, acute, deeply concave, puberulous ; pedicels
1 lin. long, minutely pubescent ; calyx-tube 4–4¼ lin. long, subcylindric, becoming somewhat ovoid in fruit, red, puberulous or
pubescent outside and within ; teeth ¾–1¼ lin. long, all subequal in
length and deltoid, or the two lower rather longer and more subulate, always narrower than the rest ; corolla much exserted, straight,
glabrous, pink ; tube 8 lin. long, gradually enlarging upwards ;
upper lip 5–5½ lin. long, very obliquely directed forwards, 3-lobed
at the apex ; terminal lobe 1¼ lin. long, ¾ lin. broad, elliptic or
elliptic-oblong, very obtuse ; lateral lobes small, rounded ; lower lip
4–5½ lin. long, boat-shaped, obtuse ; stamens much exserted ; upper
pair inserted about 5 lin. above the base of the tube, about 1 in.
long, their filaments free, flattened, ciliate on both margins in the
basal part ; lower pair inserted at the base of the lower lip, 7–9 lin.
long, their filaments united to the apex, glabrous ; stigma not
enlarged, very minutely bifid. *Dietr. Syn. Pl.* iii. 385 ; *Benth. in
DC. Prodr.* xii. 53.

EASTERN REGION : Pondoland ; between the great waterfall and Umsikaba
River, *Drège,* 4743! between Umtentu River and Umzimkulu River, *Drège!*
Natal ; near Murchison, *Wood,* 3116!

In technical characters this species is very like *S. densiflorus,* but the leaves are
much less crowded, larger, more orbicular and more obtuse, the inflorescence is
less dense and the flowers larger, seen side by side they are easily distinguished.

7. **S. densiflorus** (E. Meyer, Comm. 231) ; apparently a branching
shrub ; stems or branches rather stout, 1¼–3 lin. thick, obtusely

4-angled, grooved down the sides, very shortly tomentose; leaves small, petiolate, lanceolate, ovate, elliptic or orbicular, acute or obtuse and apiculate, entire or slightly toothed, densely gland-dotted, thinly and minutely pubescent or glabrous on both sides; petiole $\frac{1}{2}$–3 lin. long; blade 2–6 lin. long, $1\frac{1}{3}$–5 lin. broad; panicle a dense oblong spike-like mass $1\frac{1}{4}$–5 in. long, $1\frac{1}{2}$–$2\frac{1}{2}$ in. in diam., with short erect branches; bracts caducous, about 3 lin. long, 2–$2\frac{2}{3}$ lin. broad, orbicular, abruptly cuspidate, thinly and minutely pubescent or glabrous, with a few glands on the back, often woolly-ciliate; pedicels $\frac{1}{2}$–$\frac{3}{4}$ lin. long, pubescent; calyx unequally 5-toothed; tube $3\frac{1}{2}$ lin. long, tubular, thinly pubescent outside, glabrous within; upper tooth 1–$1\frac{3}{4}$ lin. long, $\frac{1}{2}$–$\frac{2}{3}$ lin. broad at the base, whence it gradually tapers to an acute point, not decurrent on the tube; lateral teeth $1\frac{1}{2}$–$2\frac{1}{2}$ lin. long, connate in pairs for about $\frac{1}{3}$ of their length, with the connate part deltoid and the free tips bristle-like or subulate, thinly ciliate or glabrous; corolla very much exserted, glabrous, pink, crimson or white; tube 8–10 lin. long, gradually enlarging upwards, straight, compressed; slightly arched behind the apex; upper lip $2\frac{1}{2}$–3 lin. long, unequally 3-lobed; terminal lobe about $\frac{3}{4}$ lin. long, $\frac{2}{3}$ lin. broad, oblong, obtuse; lateral lobes much smaller, rounded; lower lip reflexed, $1\frac{1}{2}$–$3\frac{1}{4}$ lin. long, boat-shaped, obtuse, slightly and minutely ciliate; stamens much exserted, slightly unequal, at length (at least in some specimens) recurved or spirally coiled under the lower side of the corolla; upper pair inserted about $3\frac{1}{2}$ lin. above the base of the corolla, 8–10 lin. long, their filaments free, glabrous; lower pair inserted at the base of the lip, 4–5 lin. long, their filaments united to the apex, glabrous; stigma minutely bifid; nutlets $1\frac{1}{4}$–$1\frac{1}{2}$ lin. long, oblong, obtuse, compressed, light brown, shining. *Dietr. Syn. Pl.* iii. 385; *Benth. in DC. Prodr.* xii. 54. *S. densiflorum, Hochstetter in Flora*, 1845, 67; *Krauss, Beitr. Fl. Cap- und Natal.* 131.

COAST REGION: King Williamstown Div.; Buffalo Mountain, near King Williamstown, *Tyson*, 606! East London Div.; Panmure, *Mrs. Hutton*! Kaffraria, *Mrs. Barber*! *Dugmore*!

EASTERN REGION: Transkei; near Kentani, *Miss Pegler*, 386! Tembuland; Bazeia Mountain, *Baur*, 121! Griqualand East; mountains around Clydesdale, 3000 ft., *Tyson*, 2545! and in *MacOwan & Bolus, Herb. Norm. Austr.-Afr.* 861! Zuurberg, *Wood*, 1984! Pondoland; between St. Johns River and Umtsikaba River, *Drège*, 4744c! Natal; near the Umlaas River, *Krauss*, 96! Inanda, *Wood*, 6! 17! Inyangwine, *Wood*, 3017; Maritzburg, *Rehmann*, 7530! Coastland, *Sutherland*! Dumisa, *Rudatis*, 232!

The recurving and coiling of the stamens is very evident in some specimens, and not at all so in others, and probably has some connection with the fertilisation of the plant or is a sexual condition.

V. PLECTRANTHUS, L'Hérit.

Calyx campanulate when flowering, 5-toothed, enlarged in fruit, declinate or erect, usually 2-lipped; teeth subequal or the upper widest. *Corolla* 2-lipped; tube exserted, long or short, straight or decurved; throat equal or obliquely swollen; upper lip usually short, broad, 3–4-fid; lower lip usually much longer than the upper, entire, boat-shaped. *Stamens* 4, didynamous, declinate; filaments simple, free; anther-cells usually confluent. *Disk* usually produced into a gland in front which is sometimes longer than the 4-partite ovary. *Style* subequally 2-fid. *Fruit* of 4 orbicular ovoid or oblong smooth granulate or punctate nutlets.

Annual or perennial herbs or undershrubs; flowers usually small in lax (rarely close) paniculate or racemose cymes; bracts usually small, foliaceous.

DISTRIB. Species about 250, widely distributed through Africa, India, China, Japan, the Malay Archipelago, Australia and Polynesia. *T. C.*

A. ISODON. Calyx-teeth subequal or united into 2 lips; upper lip 3-toothed; cymes laxly many-flowered with a common peduncle (rarely contracted into dense verticils).

Leaves less than 2 in. long, sub-fleshy, cuneate at
the base ... ... (1) **spicatus.**
Leaves exceeding 2 in. long:
Leaves subsessile, opposite or ternately verticillate,
ovate-lanceolate ... ... (2) **calycinus.**
Leaves with petioles reaching 2 in. long, broadly
ovate with truncate base ... ... (3) **myrianthus.**

B. GERMANEA. Calyx 2-lipped for ¼ of its length; upper lip ovate, seldom resembling the teeth of the lower lip, but if so, broader and more distant; lower lips with narrow acuminate or subulate teeth.

Section 1. EU-GERMANEA. Ripe calyx declinate; upper lip ovate; lower lip with lanceolate-subulate teeth, 2 lowermost connate at the base; corolla-tube saccate or spurred at the base behind.

Corolla-tube spurred at the base behind:
A shrub ... ... (4) **fruticosus.**
Herbs:
Upper tooth of calyx broadly ovate, subobtuse, as broad as long; pedicels
exceeding 2 lin. long ... ... (5) **petiolaris.**
Upper tooth of calyx ovate-lanceolate;
pedicels 1 lin. long ... ... (6) **arthropodus.**
Corolla-tube not spurred at the base behind:
Suffruticose:
Leafless at flowering time; leaves 2–3 in.
long, rounded or subcordate at the base;
inflorescence a leafless panicle 12–18 in.
long ... ... (7) **floribundus.**
Leafy at flowering time; leaves less than
2 in. long, truncate or cuneate at the
base; inflorescence a simple lax raceme (8) **saccatus.**
Herbaceous:
Stem puberulous; verticils of 2 shortly
pedunculate 3–5-flowered, opposite,
branched cymes ... ... (9) **Rehmannii.**
Stem pubescent with articulate glanduliferous hairs; verticils 6–8-flowered,
not pedunculate ... ... (10) **Krookii.**
Stem villous; verticils 6-flowered, not or
scarcely pedunculate ... ... (11) **ciliatus.**

Section 2. STACHYANTHI. Cymes almost sessile, with very short
pedicellate flowers in elongated thick spike-like racemes ; upper lip of
calyx ovate, only moderately different from the teeth of the lower lip ;
corolla-tube deflexed about the middle.

Leaves 1 in. long, obovate-oblong, sessile ;
   bracts as broad as long ; corolla lilac ... (12) **villosus.**

Leaves 2 in. long, orbicular, shortly petiolate ;
   bracts narrower than long ; corolla
   yellow ... ... (13) **densiflorus.**

Section 3. COLEOIDES. Ripe calyx declinate ; upper tooth usually ovate,
more rarely triangular or lanceolate ; teeth of the lower lip acute and
narrower ; corolla-tube gibbously enlarged above the base, but not
spurred ; plants of various habit.

* Verticils pedunculate.
   Verticils distant, developing into many-
      flowered opposite racemes ; pedicels 2 lin.
      long ... ... (14) **Tysoni.**
   Verticils in opposite sets of 3 on long
      peduncles :
   Verticils 6-flowered :
      Stem villous ; leaves hairy ; bracts 2–4
         lin. long ; corolla-tube 4 lin. long,
         deflexed about the middle ... ... (15) **laxiflorus.**
      Stem and leaves sparsely hairy ; bracts
         1½ lin. long ; corolla-tube 2 lin. long,
         nearly straight ... ... (16) **hylophilus.**
   Verticils 4-flowered ; stem puberulous ;
      leaves glabrous ... ... (17) **Kuntzei.**
   Peduncles of verticils very short (sometimes
      present and absent on the same plant) :
   Pubescent or hairy herbs :
      Villous all over ; leaves suborbicular,
         deeply cut all round into oblong or
         deltoid segments ; verticils in 2
         sets of 3–7 flowers ... ... (18) **grandidentatus.**
      Pubescent ; leaves crenate, not deeply
         cut :
         Verticils 6 – 12 - flowered ; corolla
            ½ in. long ; pedicels reaching
            4 lin. long ; nutlets ¾ lin. in
            diam. smooth ... ... (19) **Cooperi.**
         Verticils 6-flowered ; corolla 1 in.
            or more long ; pedicels 2½ lin.
            long ; nutlets ¾ lin. in diam.
            rugulose ... ... (20) **coloratus.**
      A glabrous herb ; verticils 5-flowered ;
         corolla ¾ in. long ; pedicels reaching
         4 lin. long ; nutlets 1¼ lin. long,
         smooth ... ... (21) **Eckloni.**
** Verticils not pedunculate :
   Corolla-tube straight or nearly so :
   Stems procumbent, often rooting at the
      lower nodes :
   Leaves glabrous :
      Leaves less than 1 in. long ; verticils
         4–6-flowered ; corolla white with
         crimson lines ... ... (22) **Thunbergii.**

Leaves densely hispid :
   Verticils 4–6-flowered ; fruiting calyx
     4 lin. long ... ... (23) **strigosus.**

   Verticils 6-flowered ; fruiting calyx
     2½–3 lin. long ... ... (24) **parviflorus.**

Stems erect :
   Leaves less than 2 in. long, glabrous :
     Leaves 1½ in. long ; verticils 4–6-
     flowered ; pedicels 1½ lin. long ... (25) **zuluensis.**

     Leaves 1 in. long ; verticils 4–6-
     flowered ; pedicels 2 lin. long ;
     corolla 4¼ lin. long ... ... (26) **Bolusi.**

     Leaves ¾ in. long ; verticils 6-flowered ;
     pedicels 2 lin. long ; corolla 3½ lin.
     long ... ... (27) **purpuratus.**

   Leaves reaching 4 in. or more long :
     Verticils 6-flowered :
       Leaves truncate or subcordate at
         the base, puberulous ; pedicels
         2 lin. long ; nutlets almost
         black ; a tall undershrub 2–3
         ft. high ... ... (28) **Galpinii.**

       Leaves cuneate at the base, pubescent ;
         pedicels 4 lin. long ; nutlets
         dark yellow ; a tall herb ... (29) **Pegleræ.**

     Verticils 4–6-flowered ; leaves cuneate
       at the base, glabrous ; pedicels
       3 lin. long ; nutlets smooth, brown (30) **natalensis.**

Corolla-tube deflexed (usually sharply so) :
   Stems procumbent, sometimes rooting at
     the lower nodes :
     Leaves densely hispid ; lower teeth of
       calyx longer than the lateral ;
       verticils 6–16-flowered ... ... (31) **hirtus.**

     Leaves with short hairs ; lower teeth of
       calyx nearly equal to the lateral ;
       verticils 6-flowered ... ... (32) **nummularius.**

   Stems erect, suberect or ascending :
     Leaves sessile or subsessile :
       Shrubby ; root fibrous ; inflorescence
         of elongate spike-like racemes ;
         bracts 4 lin. long and as broad as
         long ... ... (33) **neochilus.**

       Herbaceous ; root tuberous, edible ;
         racemes short ; bracts 1 lin. long,
         elliptic ... ... (34) **esculentus.**

     Leaves petiolate :
       Low plants not exceeding 1 ft. high :
         Stems 6 in. high ; leaves fleshy,
         ½–¾ in. long ; verticils 6-flowered (35) **pachyphyllus.**

         Stems 8–12 in. high, slender ; leaves
           membranous, scarcely 1 in. long ;
           verticils 6–10-flowered ... (36) **elegantulus.**

Plants exceeding 1 ft. high :
  Suffruticose, 16–20 in. high, tomen-
  tose, pubescent or villous :
    Verticils 6–12-flowered ; nutlets
      dark brown   ... ... (37) **tomentosus.**
    Verticils 12–16-flowered ; nut-
      lets yellow   ... ... (38) **Woodii.**
  Herbaceous :
    Petioles 2 in. or more long :
      Stems 16 in. high ; petioles
      reaching 2½ in. long ; ver-
      ticils 6–8-flowered ; pedi-
      cels 2 lin. long ... ... (39) **dolichopodus.**
      A branched herb with ascend-
      ing branches; petioles reach-
      ing 2 in. long ; verticils
      6-flowered ; pedicels 4 lin.
      long ... ... (40) **grallatus.**
    Petioles less than 2 in. long :
      Stems 16 in. high, erect, rufous-
      pubescent ; leaves reaching
      3½ in. long, rhomboid-ovate,
      reticulately veined and ruf-
      ous-pubescent beneath ; in-
      florescence a large panicle ;
      verticils 4–6-flowered ; co-
      rolla 5 lin. long ... ... (41) **transvaalensis.**
    Stems suberect, densely pubes-
    cent ; leaves 1½ in. long,
    obovate, scabrous-pubescent
    on both sides ; verticils
    6–10-flowered; corolla 3 lin.
    long ... ... (42) **Draconis.**

1. **P. spicatus** (E. Meyer, Comm. 230); stems erect, pubescent ;
leaves small, broadly ovate, subfleshy, glabrous or nearly so, cuneate
at the base ; racemes elongate, simple or branched below ; verticils
contracted, the lower ½ in., the upper ¼ in. apart, pubescent ; bracts
reaching 2½ lin. long, lanceolate, acute ; fruiting calyx not seen ;
flowering calyx 1 lin. long, oblique, hairy and dotted with bright
red glands ; teeth equalling the tube, subequal, deltoid, acute,
corolla reaching 4 lin. long ; tube 2 lin. long, sharply deflexed about
the middle ; upper lip 1 lin. long, 4-lobed, 2 terminal lobes small,
rounded, lateral lobes minute ; lower lip 2 lin. long, ¾ lin. deep,
boat-shaped. *Benth. in DC. Prodr.* xii. 60.

COAST REGION : Uitenage Div. ; hill flats near Bethelsdorp, *Zeyher,* 3542 !
Bathurst Div. ; Glenfilling, *Drège,* 4731*b* !

2. **P. calycinus** (Benth. in E. Meyer, Comm. 230); an erect
coarse herb ; stems stout, fulvous-hairy or tomentose ; leaves subsessile,
opposite or often ternately verticillate, 2½–4 in. long, ½–1½ in.
broad, ovate-lanceolate or oblong-lanceolate, acute, serrate, sparsely
hairy above with scattered hairs, fulvously woolly-tomentose below ;

inflorescence in many-flowered dense pyramidal panicles 4–10 in.
long, more or less branched ; verticils formed of shortly pedunculate
cymes, each cyme in the axil of a lanceolate-oblong acute foliaceous
bract, which is ½ in. or more long, 2 lin. broad ; fruiting calyx
tubular, 4 lin. long, hairy ; flowering calyx 1¼ lin. long, hairy ;
teeth as long as the tube, equal, deltoid, acute, ciliolate ; corolla
6¾ lin. long, whitish with purple lip (*Galpin*) ; tube 3¼ lin. long,
contracted and suddenly bent close to the base, swollen and cylindric
above ; upper lip 2 lin. long, 4-lobed, 2 upper lobes orbicular, 1 lin.
broad, lateral lobes ¾ lin. long, orbicular ; lower lip 2¾ lin. long,
1 lin. deep, boat-shaped ; upper pair of stamens 2 lin. long ; lower
2½ lin. long ; style as long as the lower stamens ; stigma minutely
2-fid ; nutlets 1 lin. long, ¾ lin. broad, oblong, obtuse, brown. *Benth.
in DC. Prodr.* xii. 61 ; *Schinz in Mém. Herb. Boiss.* x. 60.
P. pyramidatus, Guerke in Bull. Herb. Boiss. vi. 552.

KALAHARI REGION : Orange River Colony, *Cooper*, 1016 ! Transvaal ; Barberton,
mountain-tops Saddleback range, *Galpin*, 1820 ! Hooge Veld, *Rehmann*, 6870 !
Houtbosch, *Rehmann*, 6179 !

EASTERN REGION : Transkei ; Krielis country, *Bowker* ! Kentani, *Miss Pegler*,
162 ! Tembuland ; Bazeia, 2500 ft., *Baur*, 97 ! between Morley and Umtata River,
Drège ! Pondoland ; between St. Johns River and Umsikaba River, *Drège*, 3584 !
Natal ; near Ladysmith, *Wilms*, 2201 ! *Gerrard*, 183 ! Inanda, *Wood*, 489 ! Drakensberg, Laingsnek, *Rehmann*, 6961 !

VAR. β, **pachystachyus** (T. Cooke) ; leaves narrower, hardly exceeding ½ in.
broad ; flowering calyx smaller, only 1 lin. long. *P. pachystachyus, Briq. in Bull.
Herb. Boiss.* 2me *sér.* iii. 1003.

EASTERN REGION : Griqualand East ; mountains about Clydesdale, *Tyson*, 1145 !
2749 ! and in *MacOwan & Bolus, Herb. Norm. Austr.-Afr.* 862 ! Natal ; near
Unkomaas, 5000 ft., *Wood*, 4621 !

3. **P. myrianthus** (Briq. in Bull. Herb. Boiss. 2me sér. iii. 1001) ;
an erect herb ; branches ascending, tomentosely pubescent, often
canescent ; leaves 3¼ in. long, 2¼ in. broad, broadly ovate, obtuse or
subacute, coarsely inciso-serrate, green, membranous, sparsely hairy
above, more densely adpressedly pubescent beneath, base broadly
and obliquely truncate ; petiole reaching 2 in. long ; inflorescence
densely many-flowered, 4–6 in. long, 1½–2 in. broad ; cymes ½–2 in.
long ; bracts small, ovate, deciduous ; pedicels very unequal ; flowering calyx ½–¾ lin. long, pubescent ; teeth 5, lanceolate, subequal,
very short ; fruiting calyx 2 lin. long, tubular-urceolate, pubescent
with short forward pointing hairs ; corolla blue, shortly pubescent ;
tube slender, cylindric for 1½–2 lin., then suddenly deflexed and
enlarged in a throat 1½ lin. long ; upper lip ¾–1 lin. long ; lower
lip separated by a broad sinus 1½ lin. long, ¾–1 lin. deep, obtusely
boat-shaped ; nutlets scarcely ½ lin. long, ovoid, smooth, yellow.
Germanea myriantha, Briq. l.c.

KALAHARI REGION : Transvaal ; Witwatersrand, *Hutton*, 877 !

4. **P. fruticosus** (L'Herit. Stirp. i. 85, t. 41) ; a shrub 3–4 ft.
high ; stems, petioles and nerves of the leaves pubescent or villous
with adpressed rufous hairs ; leaves opposite, 3–6 in. long, 2–3¼ in.

broad, ovate, acute, coarsely doubly crenate-serrate, the younger
hispid above with scattered hairs, at length glabrate, punctate,
base subcordate, truncate or cuneate; petioles $\frac{3}{4}$–1 in. long; inflorescence a lax sparingly-branched panicle 6–9 in. long, often with
2–4 opposite spreading racemes near the base, sometimes an unbranched raceme; verticils about 6-flowered, distant; bracts
foliaceous, ovate, acuminate, the lower rather large, reaching $\frac{1}{2}$ in.
long, becoming smaller upwards; pedicels slender, up to $\frac{3}{8}$ in. long;
fruiting calyx reaching $4\frac{1}{2}$ lin. long; tube nearly glabrous, with a
reflexed upper tooth; flowering calyx 2 lin. long, hairy; upper tooth
1 lin. long, $\frac{3}{4}$ lin. broad, ovate, acute; 4 other teeth lanceolate,
subulate, 2 lower 1 lin. long, connate below, 2 lateral $\frac{1}{2}$ lin. long;
corolla blue, nearly $\frac{1}{2}$ in. long; tube 3 lin. long, with a spur $\frac{1}{2}$ lin.
long at the base on the upper side; upper lip 2–$2\frac{1}{2}$ lin. long, 4-lobed,
2 upper lobes large, suborbicular, $1\frac{1}{2}$ lin. broad, 2 lateral lobes
small, $\frac{1}{2}$ lin. long and broad, rounded; lower lip 2 lin. long, $\frac{1}{2}$ lin.
deep, boat-shaped; upper pair of stamens 2 lin. long; lower pair
3 lin. long; style a little shorter than the lower stamens; stigma
equally shortly 2-fid; nutlets ovoid, 1 lin. long, $\frac{1}{2}$ lin. broad, brown,
dull. *Thunb. Fl. Cap. ed. Schult.* 448; *Willd. Sp. Pl.* iii. 168; *Ait. Hort.
Kew. ed.* 1, ii. 322; *Benth. Lab.* 32; *S. Moore in Journ. Bot.* 1903, 406.

SOUTH AFRICA : cultivated specimen, *Gouan*!
COAST REGION : Caledon Div. ; Genadendal, *Prior*! Swellendam Div. ; in
woods near Swellendam, *Drège*! Knysna Div. ; in the forest near Yzer Nek,
Burchell, 5207! and at Doukamma, *Drège*, 7947! George Div. ; near George in
the forest, *Burchell*, 6051! *Prior*!
KALAHARI REGION : Transvaal; Lydenburg, *Wilms*, 1128!

5. **P. petiolaris** (E. Meyer, Comm. 228); stem herbaceous,
puberulous; leaves $1\frac{1}{4}$–4 in. long, 1–$3\frac{3}{4}$ in. broad, broadly ovate,
subobtuse, coarsely and obtusely toothed, thinly puberulous above
and on the veins beneath, base cordate; petioles $\frac{3}{4}$–3 in. long; inflorescence terminal, simple or with 1–2 ascending branches at the
base; verticils $\frac{1}{2}$–1 in. apart, 4–6-flowered; bracts 1–2 lin. long, $\frac{3}{4}$–2
lin. broad, somewhat rhomboid-ovate, acute, glabrous, ciliolate;
pedicels 2–$3\frac{1}{2}$ lin. long; fruiting calyx 3–4 lin. long; flowering calyx
$1\frac{1}{4}$–$1\frac{1}{2}$ lin. long, puberulous or nearly glabrous; tube $\frac{3}{4}$–1 lin. long,
campanulate; upper tooth $\frac{1}{2}$–$\frac{3}{4}$ lin. long and about as broad, broadly
ovate, subobtuse; 4 other teeth subequal, subulate from a trian-
gular base; corolla dark red (*Wood*); tube about 5 lin. long,
abruptly deflexed at the middle, enlarged above, with a minute
triangular spur; upper lip about 4 lin. long, erect, pubescent,
unequally 4-lobed, terminal lobes about $1\frac{1}{2}$ lin. long and broad,
oblong, obtuse, lateral lobes rounded; lower lip 4 lin. long, $1\frac{1}{2}$ lin.
deep, boat-shaped, obtuse; stamens unequal; filaments all free,
glabrous; style about as long as the stamens; stigma shortly 2-fid;
nutlets $\frac{3}{4}$ lin. long, subquadrate-ellipsoid, light brown, not shining.

EASTERN REGION : Pondoland ; between Umtata River and St. Johns River,
Drège, 4773*b*! Natal; Coast land, *Sutherland*! Berea, near Durban, *Wood*,
3390! Dumisa, *Rudatis*, 339!

6. **P. arthropodus** (Briq. in Bull. Herb. Boiss. 2me sér. iii. 1073); an erect herb; branches green, glabrous or minutely puberulous; leaves 4 in. long, $2\frac{3}{4}$ in. broad, broadly ovate, acute or subacuminate, attenuated into a rather long shortly puberulous petiole, membranous, thin, puberulous on the nerves beneath, otherwise glabrous, regularly crenate; petiole $1\frac{1}{2}$ in. long, articulated with the blade, the articulation leaving (after the fall of the leaf) a prominent circular scar; racemes moderately slender, about $2\frac{1}{2}$ in. long; verticils 2–6-flowered; bracts $1\frac{1}{4}$–$1\frac{5}{8}$ in. long and broad, ovate, subpersistent; pedicels $\frac{1}{2}$–1 lin. long; flowering calyx scarcely $1\frac{1}{2}$ lin. long, campanulate, shortly puberulous; upper tooth the largest, 1 lin. long, ovate-lanceolate; other 4 teeth about $\frac{1}{2}$ lin. long; corolla-tube 1–$1\frac{1}{2}$ lin. long, with a small spur $\frac{1}{2}$ lin. long on the back; upper lip $\frac{3}{4}$ lin. long; lower lip $1\frac{1}{4}$ lin. long.

KALAHARI REGION: Transvaal; Houtbosch, *Rehmann*, 6151!

7. **P. floribundus** (N. E. Br. in Kew Bulletin, 1894, 12); stems erect, 2–4 ft. high, stout, striate, pubescent, leafless at flowering time; leaves 2–3 in. long, $\frac{1}{2}$–1 in. broad, oblong, obtuse, scabridly pubescent on both sides, base rounded or subcordate; inflorescence a leafless panicle 12–18 in. long, of numerous simple or branched pubescent racemes 1–$2\frac{1}{2}$ in. long, floriferous nearly to the base; verticils about 2-flowered; bracts $1\frac{1}{2}$–2 lin. long, about 1 lin. broad, elliptic-oblong, obtuse, pubescent; pedicels as long as the bracts or slightly longer; fruiting calyx $3\frac{1}{2}$–4 lin. long, campanulate, scabrid-pubescent and glandular; flowering calyx $2\frac{1}{2}$ lin. long; upper tooth $1\frac{1}{4}$–$1\frac{1}{2}$ lin. long and broad, suborbicular, apiculate, ciliate; 4 other teeth subequal, $1\frac{1}{2}$–$1\frac{3}{4}$ lin. long, $\frac{1}{2}$ lin. broad at the base, tapering to an acute apex, ciliate; corolla nearly $\frac{3}{4}$ in. long, bright golden yellow, pubescent; tube $4\frac{1}{2}$ lin. long, abruptly deflexed at $\frac{1}{3}$ its length from the base, enlarged towards the mouth; upper lip $2\frac{1}{2}$ lin. long, erect, 4-lobed, terminal lobes about $\frac{3}{4}$ lin. long, $1\frac{1}{4}$ lin. broad, suborbicular, lateral lobes erect, 2 lin. long, oblong, obtuse; lower lip $4\frac{1}{2}$ lin. long, 2 lin. deep, subacute, boat-shaped; upper pair of stamens 2 lin. long; lower pair 3–$3\frac{1}{2}$ lin. long; style filiform, glabrous, as long as the stamens; stigma minutely 2-fid. *Hook. Ic. Pl. t.* 2489.

KALAHARI REGION: Transvaal; margin of a wood at Upper Moodies and on the bank of a river on the De Kaap Valley Flats, near Barberton, 3000–4000 ft., rare, only 2 plants seen, *Galpin*, 591! Shiluvane, *Junod*, 574!
EASTERN REGION: Natal; Inanda, *Wood*, 646! 3843!
The variety *longipes* in Tropical Africa.

8. **P. saccatus** (Benth. in E. Meyer, Comm. 227); suffruticose, sparsely hairy on the stem and leaves; branches spreading horizontally, reaching 1 ft. long; leaves thick, rather succulent, laxly hairy or nearly glabrous, $\frac{7}{8}$–$1\frac{1}{2}$ in. long, $\frac{3}{4}$–$1\frac{1}{4}$ in. broad, deltoid or rhomboid, coarsely toothed, truncate or cuneate at the base; petioles

½–1½ in. long slender; inflorescence a simple lax raceme; verticils distant, 2–6-flowered, not pedunculate; bracts small, lanceolate, deciduous; pedicels 2–3½ lin. long; fruiting calyx 3 lin. or more long, glabrous; flowering calyx 1 lin. long, campanulate, hairy; upper tooth ovate, acute; 4 other teeth lanceolate, 2 lower slightly the longer; corolla blue; tube ½ in. long, ¼ in. broad, straight, obtusely saccate above a very short narrow base; upper lip ½ in. long and as broad as long, orbicular-oblong, divided at the apex by a triangular notch about 1 lin. deep into 2 rounded lobes; lateral lobes deltoid; lower lip of corolla ½ in. long; upper pair of stamens 3½ lin. long; lower pair 4½ lin. long; style shorter than the stamens, equally 2-fid; nutlets not seen. *Benth. in DC. Prodr.* xii. 62; *Bot. Mag. t.* 7841; *Wood & Evans, Natal Pl.* i. *t.* 85.

EASTERN REGION: Pondoland; near St. Johns River, *Drège,* 4771! forest at Ismuka, near Port St. John, 100 ft., *Galpin,* 2840! Transkei; Kentani, *Miss Pegler,* 338! Tsomo River, *Mrs. Barber,* 6! Natal; Inanda, *Wood,* 323! Izingolweni, *Wood,* 3037! and without precise locality, *Gerrard,* 1676! *Sanderson!*

9. **P. Rehmannii** (Guerke in Bull. Herb. Boiss. vi. 553); stem erect, puberulous; leaves 2–2½ in. long, 1–1½ in. broad, ovate or ovate-lanceolate, acute or acuminate, irregularly serrate, puberulous on both sides, reticulately nerved beneath, rounded at the base; petioles ½–¾ in. long; inflorescence a branched panicle reaching 10 in. long; verticils of 2 shortly pedunculate 3–5-flowered opposite cymes reaching in flower ⅜ in. long; bracts ovate-lanceolate, acute, 1¼ lin. long; fruiting calyx 4½ lin. long, tubular, externally puberulous, rigid; flowering calyx 1¼ lin. long; upper tooth ovate, acute; other 4 teeth subulate from a deltoid base, 2 lower longer than the lateral; corolla white (*Wood*); tube 1½ lin. long, deflexed at the very base, enlarged upwards; upper lip about 1 lin. long; nutlets ¾ lin. long, ½ lin. broad, broadly ellipsoid-oblong, rounded at both ends; dull brown.

KALAHARI REGION: Transvaal; Barberton, *Thorncroft,* 3259!
EASTERN REGION: Natal; Karkloof, *Rehmann,* 7359! summit of Peak of Byrne, *Wood,* 3167!

10. **P. Krookii** (Guerke ex Zahlbr. in Ann. Naturhist. Hofmus. Wien, xx. 1905, 48); a herb, pubescent with articulate glanduliferous hairs; stem branched, 16 in. high, glabrescent at the base; leaves ¾–1½ in. long, ⅜–1¼ in. broad, broadly ovate, acute, membranous, the lower attenuated into a longish petiole, pubescent on both sides and with irregularly toothed margins; petioles ⅜–1¼ in. long; racemes lax, 4–5 in. long; verticils about 5–7 lin. apart, 6–8-flowered; bracts ovate-lanceolate, acute; pedicels slender, 4–5 lin. long; fruiting calyx 4 lin. long; flowering calyx 2 lin. long; upper tooth broadly ovate, acute; other 4 teeth deltoid-lanceolate, acuminate, 2 lowest the longest; corolla saccately enlarged at the base.

COAST REGION : Alexandria Div. ? ; Zour Flats forest, *Tyson*, 1765 !
EASTERN REGION : Griqualand East ; between the Insizwa Range and the River
Umzimhlava, *Krook*, 1698 !

VAR. β, **grandifolia** (T. Cooke); less hairy ; leaves larger, 3½–4½ in. long,
2¼–3 in. broad irregularly and deeply coarsely crenate-serrate ; upper leaves
beneath the panicle with very short petioles ; lower leaves with petioles 2 in. long.

EASTERN REGION : Griqualand East ; woods near Kokstad, 5100 ft., *Tyson*,
1793 ! and in *MacOwan & Bolus, Herb. Norm.* 1344 !

11. **P. ciliatus** (E. Meyer, Comm. 227); stem and branches
usually villous ; leaves petiolate, 2–3 in. long, 1¼–1¾ in. broad,
ovate, acute, crenate-serrate, cuneate at the base, sparsely hairy
above and on the nerves beneath with short rather stiff hairs ;
petioles ½–1¼ in. long, villous ; inflorescence a simple or slightly
branched panicle ; verticils about 6-flowered, rather distant ;
peduncles very short or 0 ; pedicels reaching 3 lin. long ; fruiting
calyx 4 lin. long, campanulate ; upper tooth 1¼ lin. long, ¾ lin.
broad, ovate, acute, ciliate ; other 4 teeth lanceolate, acute, ciliate,
2 lateral 1 lin. long, 2 lower 1¾ lin. long, connate at the base ;
corolla ½ in. long or more ; tube 3 lin. long, nearly straight or
slightly deflexed at the very base, obtusely saccate above the base ;
upper lip 4 lin. long and as broad as long, obovate-cuneate ; terminal
lobes about 1 lin. deep, rounded ; lateral lobes small, obtuse ; lower
lip 4 lin. long ; nutlets obtusely ovoid, ¾ lin. long, ½ lin. broad,
brown.

COAST REGION : Albany Div. ; near Grahamstown, *Burchell*, 3580 ! *Kitching !
Atherstone*, 17 !
KALAHARI REGION : Swaziland ; Embabane, 4700 ft., *Bolus*, 12247 !
EASTERN REGION : Transkei ; Kentani, edge of woods near water, *Miss Pegler*,
352 ! Tembuland ; Bazeia, 2500 ft., *Baur*, 37 ! Pondoland ; near St. Johns
River, *Drège*, 4777 ! Natal ; Inanda, *Wood*, 63 !

12. **P. villosus** (T. Cooke in Kew Bulletin, 1909, 378, not of
Sieber) ; suffruticose ; stems erect, stout, pubescent ; leaves
1 in. long, sessile, obovate-oblong, obtuse, obscurely crenate,
densely finely villous on both sides, base cuneate ; inflorescence
of several villous dense spike-like racemes forming a branched
panicle ; verticils many-flowered, villous, closely packed ; bracts
3 lin. long and as broad as long, broadly ovate, acute,
villous on both sides ; pedicels about 1 lin. long ; flowering
calyx 1 lin. long, densely villous ; upper tooth larger than the
others, ¼ lin. broad, ovate-oblong, subacute, densely ciliate ; other
4 teeth subequal or the 2 lower smaller, oblong, much shorter
than the upper tooth, densely ciliate ; corolla 2½ lin. long, lilac
(*Wood*) ; tube 1½ lin. long, deflexed about the middle, enlarged
above the deflexion ; upper lip ¾ lin. long ; lower lip 1 lin. long,
½ lin. deep, boat-shaped ; upper pair of stamens 1¼ lin. long ; lower
pair 1¾ lin. long ; style as long as the lower stamens ; nutlets
scarcely ½ lin. long and nearly as broad as long, angular, pale brown,
quite smooth and shining.

EASTERN REGION : Zululand ; Entumeni, 2000–3000 ft., *Wood*, 3955!

This has a general similarity in appearance to *P. marrubioides*, but the calyx, the upper tooth of which is twice as long as the lower ones, is entirely different.

13. P. densiflorus (T. Cooke in Kew Bulletin, 1909, 378); suffruticose; stems stout, pubescent; leaves 2 in. long and as broad as long, orbicular, somewhat distinctly and obscurely crenate, softly villous on both sides; base cuneate running down into a short stout petiole, the junction of which with the leaf-blade is obscure; inflorescence paniculate, the racemes composing the panicle spike-like, reaching 10 in. long, densely villous; verticils many-flowered, close; bracts ovate-lanceolate, acute, deciduous, 1½ lin. long, 3 lin. broad, pubescent on the back and with ciliate margins; pedicels 1 lin. long; fruiting calyx 2¼ lin. long, campanulate, straight, villous; upper tooth 1¼ lin. long, ½ lin. broad, ovate-oblong, subacute, ciliate; 2 lateral teeth ¾ lin. long, ovate-oblong, acute, ciliate; 2 lower teeth similar but slightly smaller; flowering calyx 1¼ lin. long, densely villous; upper tooth rather more than ¼ lin. broad; other 4 teeth subequal or the lower smaller, oblong, ciliate; corolla yellow (*Wood*); tube 1½ lin. long, nearly straight, enlarged above; nutlets nearly ½ lin. long, ¼ lin. broad, almost flat on one face, angular on the other, smooth, shining, yellowish-brown.

EASTERN REGION : Natal ; near the Mooi River, 3000–4000 ft., *Wood*, 4475!

14. P. Tysoni (Guerke in Engl. Jahrb. xxvi. 77); stem erect, 16–20 in. high, pubescent; leaves petiolate, 1½ in. long, 1 in. broad, ovate, acute, coarsely crenate-serrate, slightly pubescent above, more densely so and copiously gland-dotted beneath, base shortly cuneate into the petiole; petioles ⅙–½ in. long; inflorescence a long simple raceme reaching 1 ft. or more long; verticils pedunculate, rather distant, developing into many-flowered opposite racemes reaching 6 lin. long; bracts small, deciduous; pedicels 2 lin. long; fruiting calyx 3 lin. long, pubescent and glandular; upper tooth 1¼ lin. long, ½–¾ lin. broad, ovate-oblong, subacute; lateral teeth 1 lin. long, ½ lin. broad, oblong, rounded; 2 lower teeth 2 in. long, lanceolate-subulate, united nearly to the top; corolla purple, ½ in. long, sparsely gland-dotted; tube ¼ in. long, sharply deflexed near the base, enlarged and funnel-shaped above; upper lip 1 lin. long; lower lip ¼ in. long, 1⁄10 in. deep, boat-shaped; nutlets globosely ellipsoid, ½ lin. long and nearly as broad as long, smooth, brown.

EASTERN REGION : Griqualand East ; in rough places on the banks of the Umzimkulu River, near Clydesdale, *Tyson*, 2769! and in *MacOwan & Bolus, Herb. Austr.-Afr.*, 1295! Natal ; Dumisa, *Rudatis*, 262!

15. P. laxiflorus (Benth. in E. Meyer, Comm. 228); stem erect, villous; leaves 2–3½ lin. long, ¼–1¼ in. broad, broadly ovate, acute or acuminate, sparsely hairy on both sides, crenate-serrate, base

cordate; petioles variable in length, $\frac{1}{2}$–3 in. long, villous; inflorescence a laxly branched (rarely simple) panicle 6–10 in. long;
verticils usually in 2 opposite sets of 3 on long peduncles (peduncles
rarely absent); pedicels often $\frac{1}{4}$ in. long, slender; bracts 2–4 lin.
long, ovate, acute; fruiting calyx 4 lin. long, tubular-campanulate;
flowering calyx 2 lin. long, densely villous, campanulate; teeth
deltoid, subequal in length, as long as the tube; uppermost tooth
broader than the 4 subequal lower ones; corolla pubescent outside;
tube 4 lin. long, much deflexed about the middle; upper lip 3 lin.
long, orbicular; lower lip 3 lin. long, 1 lin. deep, boat-shaped;
upper pair of stamens 2 lin. long; lower pair $2\frac{1}{2}$ lin. long; style
longer than the stamens; stigma equally 2-fid; nutlets globosely
ovoid, 1 lin. long, 3 lin. broad, yellowish brown. *Benth. in DC.
Prodr.* xii. 63.

COAST REGION : Uitenhage Div., *Zeyher,* 196 ! Albany Div. ; Howisons Poort,
near Grahamstown, *Zeyher,* 876 ! 3544 ! Stockenstrom Div. ; Kat Berg, *Shaw* !
KALAHARI REGION : Transvaal ; Houtbosch Berg, *Schlechter,* 4762 !
EASTERN REGION : Natal ; between Umzimkulu River and Umkomanzi River,
Drège, 3586 ! near Murchison, *Wood,* 3117 ! Inanda, *Wood,* 1047 ! and without
precise locality, *Gerrard,* 1222 ! *Sanderson,* 392 ! *Wood,* 1864 ! 4237 ! Transkei ;
Kentani, *Miss Pegler,* 161 ! Pondoland ; Port St. John, above Tiger Flat, *Galpin,*
2844 !

16. **P. hylophilus** (Guerke in Engl. Jahrb. xix. 203) ; an undershrub $2\frac{1}{2}$–4 ft. high; stems sparsely hairy; leaves 2–4 in. long,
$1\frac{1}{2}$–3 in. broad, opposite, broadly ovate, acute, crenate, sparsely
short-hairy on both sides, base deeply cordate; petioles $\frac{1}{2}$–3 in. long,
hairy; inflorescence a sparingly branched panicle or a simple
raceme 3–8 in. long; verticils distant, 6-flowered, 3 flowers on each
side usually on a peduncle $1\frac{1}{2}$–3 lin. long (the peduncle rarely
absent); flowers white (*Junod*); pedicels 2–3 lin. long, puberulous;
bracts $1\frac{1}{2}$ lin. long, ovate, acute; fruiting calyx 4 lin. long, tubular-
campanulate; flowering calyx $1\frac{1}{4}$ lin. long, campanulate, hairy;
upper tooth the largest, ovate, acute; other 4 teeth deltoid, acute,
2 lateral teeth a little shorter than the 2 lower; corolla about 4 lin.
long; tube 2 lin. long, nearly straight; upper lip 2 lin. long;
lower lip 2 lin. long, 1 lin. deep, boat-shaped; nutlets not seen.

KALAHARI REGION : Transvaal; Shiluvane, *Junod,* 777 !

I have seen only 2 rather meagre specimens of this species.

17. **P. Kuntzei** (Guerke in Kuntze, Rev. Gen. iii. ii. 260) ; plant
10–15 in. high; stem ascending, simple or more or less branched,
puberulous; lower leaves 2 in. long, $1\frac{5}{8}$ in. broad, broadly ovate or
suborbicular, obtuse or subacute, obtuse at the base or attenuated
into the petiole, coarsely crenate-serrate, thin, glabrous or nearly
so; petiole $1\frac{5}{8}$ in. long; inflorescence lax; verticils very distant,
$\frac{1}{2}$–$1\frac{1}{4}$ in. apart, 4-flowered; bracts lanceolate, 1–$1\frac{1}{2}$ lin. long;
pedicels $2\frac{1}{2}$–$3\frac{1}{2}$ lin. long; calyx slightly glandular; fruiting calyx
declinate, ovate-cylindric, 4–5 lin. long; upper tooth orbicular,

very shortly apiculate; lateral teeth broadly deltoid, acuminate; 2 lower teeth longer and narrower, acuminate; corolla bluish-purple, much longer than the calyx; nutlets yellowish brown.

EASTERN REGION : Natal; Clairmont, *Kuntze.*

The plant is less upright than *P. laxiflorus.* It has glabrous leaves and 4-flowered verticils, while *P. laxiflorus* has more or less hairy leaves and 6-flowered verticils. The teeth of the calyx are also different. I have not seen any specimens and Guerke has not described the corolla-tube.

18. **P. grandidentatus** (Guerke in Bull. Herb. Boiss. vi. 554); villous all over; leaves 1½–2½ in. long and as broad as long, orbicular, deeply cut all round, except at the base, into oblong or deltoid acute or subobtuse teeth, sometimes ½ in. deep, base sub-cordate, truncate or cuneate; petioles ¼–1¾ in. long; inflorescence a panicle near the base, the racemes forming it reaching as much as 9 in. long; verticils somewhat distant, of 3–7 sets of white flowers at each side; peduncles 0, or rarely present and very short; pedicels 1–1½ lin. long; bracts small, deciduous; fruiting calyx 2 lin. long, oblique, hairy and covered with red shining glandular dots; flower-ing calyx 1 lin. long, hairy; upper tooth ½ lin. long, and about as broad, suborbicular-oblong; other 4 teeth subequal, deltoid, acute; corolla 4 lin. long; tube 2 lin. long, deflexed at the very base, funnel-shaped above; upper lip 1½ lin. long, ciliate at the apex, and with red glandular dots; lower lip 2 lin. long, 1 lin. deep, boat-shaped; nutlets ½ lin. long, scarcely ½ lin. broad, ovoid, obtuse, shining, polished, brown.

COAST REGION: Queenstown Div.; Shiloh, 3500 ft., *Baur,* 797! Uitenhage Div. ; Zuurberg Range, *Tyson,* 1177!
EASTERN REGION : Griqualand East ; on moist rocks at Emyembe, in Umzimkulu district, 5500 ft., *Tyson,* 2163! and in *MacOwan, Herb. Austr.-Afr.,* 1517! Natal; in woods at Umkomaas, 4000–5000 ft., *Wood,* 4606!

19. **P. Cooperi** (T. Cooke in Kew Bulletin, 1909, 377); erect; stem obtusely quadrangular, grooved, pubescent; leaves up to 3½ in. long, 3 in. broad, deltoid-ovate, acute or acuminate, coarsely and somewhat irregularly crenate, sparsely pubescent above and on the nerves beneath; petioles of lower leaves reaching 2¼ in. long, those of the upper leaves shorter, more or less pubescent ; inflorescence in long racemes reaching 10 in. long, simple or forming a panicle branched at the base, the rhachis scaberulous with short gland-tipped hairs; verticils 6–12-flowered, often nearly 1 in. apart, not or scarcely pe-dunculate; bracts 1–1½ lin. long, glandular-pubescent; fruiting calyx 3 lin. long, curved, nearly glabrous, often purple-coloured; upper tooth nearly 1¼ lin. long, ovate, acuminate, erect; other 4 teeth subulate from a deltoid base, the lower teeth the longest; flowering calyx 1¼ lin. long, coloured, tube pubescent ; narrow basal portion of the corolla-tube ¾ lin. long, about ½ lin. broad, sharply deflexed; upper portion of tube above the narrow part

enlarged, almost cylindric, 2½ lin. long, 1 lin. broad, saccate at the base ; upper lip 1½ lin. long ; lower lip 2 lin. long, nearly 1 lin. deep, boat-shaped, acute ; stamens exserted ; nutlets ¾ lin. in diam., subglobose, compressed, smooth, dark yellow.

KALAHARI REGION : Orange River Colony, *Cooper*, 2982 !
EASTERN REGION : Natal ; in bush at Byrne, 3000 ft., *Wood*, 1843 ! and without precise locality, *Gerrard*, 1673 !

20. **P. coloratus** (E. Meyer, Comm. 228) ; erect, 1½–3 ft. high ; stems pubescent ; leaves 2–5 in. long, 1–3 in. broad, ovate, acute, irregularly crenate, sparsely hairy above, more densely so on the nerves beneath, base cuneate ; petioles 1–3 in. long, hairy ; inflorescence racemose or paniculate ; verticils laxly 6-flowered, not or scarcely pedunculate ; bracts 2–2½ lin. long, 1 lin. broad, ovate-lanceolate, acuminate ; pedicels 2–2½ lin. long, slender, pubescent ; fruiting calyx 4 lin. long ; upper tooth erect ; lower teeth ciliate ; flowering calyx campanulate, 1½–2 lin. long, usually coloured ; upper tooth broadly ovate, subacute ; other 4 teeth subulate, ciliate, 2 lower the longer ; corolla purple, 1 in. or more long ; tube ¾ in. long, straight, very narrow, scarcely enlarged above ; upper lip ¼ in. long, 2 terminal lobes shallow, rounded ; lateral lobes small, rounded ; stamens long, the longer pair reaching 6 lin. long, the shorter pair up to 4 lin. long ; nutlets ¾ lin. long, ½ lin. broad, ellipsoid-oblong, rounded at the ends, rugulose, dark brown. *Benth. in DC. Prodr.* xii. 64.

COAST REGION : King Williamstown Div. ; mountains near King Williamstown, *Mrs. Barber* ! East London Div. ; forest at Fort Grey, *Galpin*, 7826 !
EASTERN REGION : Transkei ; Tsomo River, *Mrs. Bowker* ! Kentani, *Miss Pegler*, 907 ! Pondoland ; between Umtata River and St. Johns River, *Drège*, 4778 ! Port St. John, above Tiger Flat, *Galpin*, 2843 ! Natal ; Inanda, *Wood*, 480 ! near Murchison, *Wood*, 3036 ! near Durban, *Guenzius* ! Dumisa, *Rudatis*, 314 ! and without precise locality, *Gerrard*, 1671 ! Zululand ; near Eshowe, *Wood*, 7591 !

21. **P. Eckloni** (Benth. in DC. Prodr. xii. 64) ; an erect herb, glabrous or nearly so ; stems striate, glabrous ; leaves 3½–6 in. long, 1½–3 in. broad, ovate, acute or acuminate, crenate-serrate, glabrous, cuneate (usually unequally) at the base ; petioles ¼–3 in. long ; inflorescence a branched panicle 2–9 in. long, formed of several opposite lax racemes of various lengths ; verticils laxly 6-flowered ; peduncles scarcely any ; pedicels reaching ⅓ in. long, slender ; bracts small, oblong, deciduous ; fruiting calyx reaching 5 lin. long, tubular-campanulate, ribbed, glabrous ; upper tooth erect ; flowering calyx campanulate, 2 lin. long ; upper tooth 1¼ lin. long, ¾ lin. broad, ovate, acute ; 4 other teeth linear-subulate, ciliate, 2 lateral ¾ lin. long, 2 lower 1¼ lin. long ; corolla ¾ in. long ; tube straight, funnel-shaped at the mouth, 5½ lin. long ; upper lip 2–2½ lin. long and as broad, 2 terminal lobes rounded ; lower lip 2¼ lin. long, 1 lin. deep, boat-shaped ; stamens very long, much exserted, the longer pair reaching ¾ in. long ; style shorter than the stamens ; nutlets 1¼ liu. long, ¾ lin. broad, ellipsoid, smooth, brown.

COAST REGION : Stockenstrom Div. ; near the Kat River and slopes of the Kat
Berg, *Ecklon* ! Eastern Frontier, *MacOwan,* 500 !
CENTRAL REGION : Somerset Div. ; Bosch Berg, *Burchell,* 3139 ! and without
precise locality, *Miss Bowker* !
EASTERN REGION : Transkei ; Kentani, *Miss Pegler,* 376 ! Tembuland; Bazeia,
Baur, 187 !

22. **P. Thunbergii** (Benth. Lab. 37) ; a procumbent plant ; stems
slender, more or less puberulous ; leaves opposite, $\frac{1}{2}$–$\frac{3}{4}$ in. long,
almost as broad as long, subfleshy, orbicular-obovate, cuneate at the
base, decurrent into a long slender petiole, with deeply crenate
margins, glabrous, gland-dotted ; inflorescence a simple raceme ;
verticils laxly 4–6-flowered, not pedunculate ;. bracts $\frac{3}{4}$–1 lin. long,
ovate-lanceolate, acute ; pedicels reaching 3 lin. long, filiform ;
fruiting calyx 3$\frac{1}{2}$ lin. long, becoming glabrous, gland-dotted ;
flowering calyx 1$\frac{1}{2}$ lin. long ; upper tooth $\frac{3}{4}$ lin. long, ovate, acute ;
other 4 teeth deltoid, acute, the lateral $\frac{3}{4}$ lin. long, 2 lower 1 lin.
long ; corolla white with crimson lines (*Mrs. Monteiro*) ; tube
reaching 4$\frac{1}{2}$ lin. long, straight, scarcely enlarged above ; upper lip
up to 3$\frac{1}{2}$–4 lin. long, obovate-oblong, rounded at the apex ; lower
lip reaching 3$\frac{1}{2}$ lin. long ; upper pair of stamens 2 lin. long ; lower
pair 3 lin. long. *Benth. in DC. Prodr.* xii. 67 ; *Schinz in Mém. Herb.
Boiss.* x. 60. *Ocymum racemosum, Thunb. Prodr.* 96.

COAST REGION : Knysna Div. ; in the forest at Doukamma, *Drège* ! Uniondale
Div. ; near the Keurbooms River, *Burchell,* 5150 ! Uitenhage Div. ; near
Uitenhage, *Schlechter,* 2501 ! *Prior* ! woods of Adow and by the Zwartkops River,
Zeyher, 121 ! Albany Div. ; in a wooded kloof west of Grahamstown, *Burchell,*
3579 ! Victoria East Div., *Cooper,* 399 !
EASTERN REGION : Tembuland ; between the Bashee River and Morley, *Drège,*
4779 ! Griqualand East ; in rocky places near streams round Clydesdale, 2500 ft.,
Tyson, 2802 ! and in *MacOwan & Bolus, Herb. Austr.-Afr.,* 1296 ! Delagoa Bay,
Mrs. Monteiro !

23. **P. strigosus** (Benth. in E. Meyer, Comm. 229) ; stem procumbent, rufous-pubescent, often rooting at the lower nodes ; leaves
opposite, variable in size, $\frac{1}{2}$–1$\frac{3}{4}$ in. long, $\frac{1}{2}$–1$\frac{1}{2}$ in. broad, orbicular,
obscurely crenate, fleshy, hispid above, rufous-pubescent on the
nerves beneath ; petioles variable in length from 1 lin. to 1$\frac{1}{4}$ in.
long ; inflorescence a simple (rarely branched) raceme ; verticils
4–6-flowered, not pedunculate ; bracts 1$\frac{1}{2}$–2 lin. long, ovate-lanceolate ; pedicels 2–3 lin. long, slender, pubescent ; fruiting calyx 4 lin.
long, campanulate ; flowering calyx 1$\frac{1}{2}$ lin. long, campanulate ; upper
tooth $\frac{3}{4}$ lin. long, ovate, acute ; other 4 teeth lanceolate, acute,
lateral $\frac{1}{2}$–$\frac{3}{4}$ lin. long, lower twice as long ; corolla 5 lin. long ; tube
2 lin. long, nearly straight ; upper lip 3 lin. long, oblong-obovate,
rounded at the apex ; lower lip 2 lin. long. *Benth. in DC. Prodr.*
xii. 68.

VAR. β, **lucidus** (Benth. in DC. Prodr. xii. 68) ; leaves more glabrous, distinctly
crenate ; stem and veins of the leaves rufous-pubescent. *P. lucidus, Burchell ex
Benth. l.c.* 68.

COAST REGION : Alexandria Div. ; Oliphants Hoek Forest, *Ecklon* ! Var. β :
Bathurst Div. ; Riet Fontein, *Burchell,* 3924 !

EASTERN REGION : Pondoland ; between St. Johns River and Umsikaba River, *Drège,* 4779c !

Closely allied to *P. Thunbergii,* Benth., from which it differs by the hispid, obscurely crenate leaves and a smaller corolla. There are however only 2 sheets of the plant at Kew.

24. **P. parviflorus** (Guerke in Kuntze, Rev. Gen. Pl. iii. ii. 261, not of others) ; stem 12 in. long, creeping, villous-pubescent, rooting at the lower nodes, thickly covered with short reddish brown hairs ; leaves 7 lin. long and almost as broad, suborbicular, subacute at the apex, thick, margins coarsely serrate near the top, villous-pubescent on both sides, obtuse at the base or slightly attenuated into the petiole ; petioles hardly exceeding 2½ lin. long ; racemes short ; verticils 6-flowered, ½–¾ in. apart ; bracts oblong-lanceolate, acute, sessile ; pedicels 2–3 lin. long ; fruiting calyx 2½–3 lin. long, glabrous ; upper tooth broadly ovate, subacute ; other 4 teeth deltoid-subulate, 2 lower the longest ; corolla white with a tinge of red.

COAST REGION : East London Div. ; near East London, *Kuntze.*

I have not seen any specimens. The above description is Guerke's. From the description there would seem little to separate the plant from *P. strigosus,* Benth. Guerke gives no description of the corolla.

25. **P. zuluensis** (T. Cooke in Kew Bulletin, 1909, 379) ; a slender herb ; stems obtusely quadrangular, puberulous ; leaves 1–1½ in. long, 1 in. broad, ovate, acute, glabrous or nearly so, with regularly serrate margins, truncate or shortly cuneate at the base ; petioles ¼–¾ in. long, puberulous ; inflorescence a simple terminal raceme 2½–4½ in. long ; verticils 4–6-flowered, 3–5 lin. apart, not pedunculate ; bracts 1½–2 lin. long, obovate, acuminate ; pedicels 1½ lin. long ; flowering calyx 2 lin. long, campanulate, pubescent ; upper tooth ½ lin. long, broadly ovate, subacute ; lateral teeth ½ lin. long, deltoid, acute ; 2 lower teeth twice as long as the lateral, lanceolate, acute ; corolla-tube nearly 3 lin. long, narrow cylindric basal portion ½–¾ lin. long, tube nearly straight and subcylindric above, saccate above the narrow part ; upper lip 2 lin. long, 1½ lin. broad, obovate, rounded at the apex, with 2 lateral rounded lobes near the base ; stamens exserted ; nutlets not seen.

EASTERN REGION : Natal or Zululand ; *Gerrard,* 1675 !

[26. **P. Bolusi** (T. Cooke in Kew Bulletin, 1909, 377) ; stems erect, obtusely quadrangular, simple or branched, leafy, hairy ; leaves reaching 1 in. long, ¾ in. broad, broadly ovate, acute or obtuse, green and sparsely hairy or almost glabrous above, pale and hairy on the nerves beneath, serrate (more or less irregularly) ; petioles 2–4 lin. long, pubescent ; inflorescence of simple or paniculate terminal racemes 4–6 in. long ; verticils 4–6-flowered, nearly ½ in. apart ; bracts ovate, acuminate ; pedicels scarcely 2 lin. long, pubescent ; fruiting calyx reaching

nearly ½ in. long, usually coloured, tubular-campanulate, glabrous
or nearly so ; flowering calyx 2 lin. long, coloured, tubular-campanulate, pubescent ; upper tooth nearly ¾ lin. long, ½ lin. broad, ovate,
acute ; lateral teeth short, oblong, obtuse, cuspidate ; 2 lower teeth
much longer than the lateral, lanceolate-subulate ; corolla 4¼ lin.
long ; tube 2½ lin. long, nearly straight, cylindric ; upper lip 1½ lin.
long and as broad as long, with rounded crenulate apex ; lower lip
1¾ lin. long, ¾ lin. deep, boat-shaped, acute ; stamens included.

KALAHARI REGION : Transvaal ; Houtbosch, *Rehmann*, 6167 ! A more robust
fruiting specimen (flowers said to be purple) from near Weenen, Natal, 4000 ft.
(*Wood*, 4488 !)

The description is retained for convenience, though *Bolus* 11011 is transferred
to *Orthosiphon Wilmsii* and the remaining material scarcely sustains the species.]

27. **P. purpuratus** (Harv. Thes. Capens. i. 53, t. 83) ; stems
12–14 in. high, branching, succulent and brittle, thinly puberulous
or nearly glabrous ; leaves ¾ in. long, nearly as broad as long, in
spreading subdistant decussate pairs, ovate or suborbicular, obtusely
or obsoletely crenate, glabrous or nearly so, purple beneath ; petioles
3–4 lin. long ; inflorescence of paniculately arranged racemes ; verticils laxly 6-flowered, not pedunculate ; bracts 1¼ lin. long, ½ lin. broad,
ovate-lanceolate, acute ; pedicels 2 lin. long ; fruiting calyx 3 lin.
long ; flowering calyx 1¼ lin. long, campanulate ; upper tooth
broadly ovate, acute ; other 4 teeth lanceolate, 2 lower the longest ;
corolla white (*Wood*) ; tube 2½ lin. long, nearly straight ; upper lip
1 lin. long, 4-lobed and with crenate margins, 2 terminal lobes
obovate, lateral lobes oblong rounded ; lower lip as long as the
upper ; nutlets ¾ lin. long and broad, subglobose, dark brown,
almost black.

EASTERN REGION : Natal ; Umzinyati Falls, *Wood*, 1223 ! and without precise
locality, *Cooper*, 3106 ! Originally described from a plant raised at Kew from seeds
sent from Natal by Mr. Vanse.

28. **P. Galpinii** (Schlechter in Journ. Bot. 1896, 393) ; a tall
branched undershrub, 2–3 ft. high (*Schlechter*), 9 ft. high (*Nelson*) ;
branches obtusely quadrangular, puberulous ; leaves 3½–7 in. long,
2½–6 in. broad, broadly ovate, acuminate, with irregularly crenate-
dentate margins, more or less puberulous, truncate or subcordate
at the base ; petioles 1–1½ in. long, subtomentose with rufous hairs ;
inflorescence a branched panicle consisting of several long lax
racemes, terminal one sometimes reaching 10 in. long ; verticils
about 6-flowered, not pedunculate ; bracts 1–1½ lin. long, ovate,
cuspidately acuminate ; pedicels 2 lin. long, filiform ; fruiting calyx
3½ lin. long, curved, glabrous ; flowering calyx 2 lin. long, campanulate, pubescent ; upper tooth ¾ lin. long, broadly ovate, acute ;
4 other teeth lanceolate, 2 lower twice as long as the lateral ones ;
corolla purple (*Schlechter*) ; tube 2¼ lin. long, not deflexed, gibbously
enlarged above the base on the upper, straight on the lower side ;
upper lip 2½ lin. long, 2-lobed at the apex by a triangular notch ;

lower lip 2 lin. long; stamens much exserted; nutlets ¾ lin. long, ½ lin. broad, globosely ovoid, dark brown (almost black), smooth.

KALAHARI REGION : Transvaal; near Barberton in wooded ravines, 3000–4000 ft., *Galpin*, 939 ! Houtbosch Berg, *Nelson*, 431 !

29. **P. Pegleræ** (T. Cooke in Kew Bulletin, 1909, 378); a tall plant; stems erect, nearly glabrous, purplish, grooved, finely striate; leaves 3–8 in. long, reaching 4 in. broad, broadly ovate, acuminate, coarsely and irregularly serrate, glabrous or nearly so above, pubescent on the nerves and closely gland-dotted beneath, base cuneate; petioles 1–3 in. long, purplish, glabrous or pubescent; inflorescence reaching 1 ft. or more long, the racemes simple or laxly paniculate; flowers purple; verticils 6-flowered, ½–⅔ in. apart, not pedunculate; bracts 1½ lin. long, ovate-lanceolate; pedicels reaching 4 lin. long, filiform; fruiting calyx 4¼ lin. long, curved, glabrous or nearly so; upper tooth ¾ lin. long, ovate, obtuse; other 4 teeth subulate, 2 lower nearly twice as long as the lateral; corolla about 5 lin. long; tube short, enlarged and funnel-shaped above; lower lip about 2½ lin. long; stamens exserted; nutlets 1 lin. long, ¾ lin. broad at the base, ovoid, dark yellow.

EASTERN REGION : Transkei; Kentani, *Miss Pegler*, 377 ! Natal or Zululand, *Gerrard*, 1235 !

30. **P. natalensis** (Guerke in Bull. Herb. Boiss. vi. 552); erect; stems glabrous or nearly so, obtusely quadrangular, grooved, often purple when young; leaves 1½–4 in. long, ¾–2 in. broad, elliptic-lanceolate or oblong-lanceolate, acuminate, sharply serrate, pubescent or nearly glabrous and with purple nerves and veins below when young, cuneate at the base; petioles 1–1¼ in. long, pubescent or nearly glabrous, articulated at the base and leaving after falling a large conspicuous scar on the stem; inflorescence a simple raceme (rarely a branched panicle); verticils 4–6-flowered, 3–5 lin. apart, not pedunculate; bracts 1½–2 lin. long, lanceolate, acuminate; pedicels reaching 3 lin. long; fruiting calyx 4 lin. long; flowering calyx reaching 2½ lin. long, very oblique, coloured, thinly membranous; upper tooth 1 lin. long, ovate, subobtuse, purple; lateral teeth 1 lin. long, deltoid-ovate, acuminate; 2 lower teeth 1¾ lin. long, lanceolate-subulate; corolla rather more than ½ in. long, glabrous; tube 2½ lin. long over all, lower narrow portion very short (about ½ lin. long), upper slightly enlarged part of the tube 3 lin. long, nearly straight, rather broader below than at the mouth, obtusely saccate, the saccate base appearing like a rounded spur ¼ lin. long on the upper side of the tube; upper lip 3 lin. long, 2¼ lin. broad, obovate-oblong, rounded and slightly lobed at the apex; lower lip 3 lin. long, 1½ lin. deep, boat-shaped, obtuse; stamens exserted, the longer pair 5½, the shorter 4 lin. long; style reaching to the level of the shorter stamens; nutlets ½ lin. long and nearly as broad at the base, ovoid, smooth, brown.

EASTERN REGION: Griqualand East; Zuurberg Range, *Tyson*, 1764! near Kokstad, 5200 ft., *Tyson*, 1793! Natal; Camperdown, *Rehmann*, 7701! Inanda, *Wood*, 558! Zululand; Entumeni, *Wood*, 3997!

31. P. hirtus (Benth. Lab. 38); procumbent, hispid; leaves $\frac{1}{2}$–1 in. long, $\frac{3}{8}$–$\frac{3}{4}$ in. broad, ovate, obtuse, crenate, coarsely hispid, cuneate at the base, thick; petioles 4–8 lin. long, densely hispid; inflorescence an elongate subsimple raceme reaching 6 in. long; verticils 6–16-flowered, not pedunculate; pedicels reaching 2 lin. long, hairy; bracts minute, deciduous; fruiting calyx 3 lin. long, incurved, striate; flowering calyx 1$\frac{1}{2}$ lin. long, hairy, campanulate; upper tooth ovate, acute; other 4 teeth lanceolate, acute, 2 lower longer and narrower than the lateral; corolla $\frac{1}{2}$ in. long; tube broad, 3 lin. long, deflexed at about the middle; upper lip 1$\frac{1}{2}$ lin. long, oblong, shallowly 2-lobed at the apex, lobes rounded; lower lip 3 lin. long; stamens exserted; nutlets globosely ovoid, $\frac{1}{2}$ in. long and broad, smooth, yellowish brown. *P. madagascarensis, Benth. Lab.* 37 (*excl. syn.*). *Ocymum hirtum, Herb. Banks. ex Benth. l.c.* 38. *O. tomentosum, Thunb. Prodr.* 96 *et Fl. Cap. ed. Schult.* 448.

COAST REGION: Uitenhage Div.; various localities, *Drège! Prior! Schlechter*, 2599! Port Elizabeth Div.; Earn Cliff, 150 ft., *Galpin*, 6466! Bathurst Div.; by the Bushmans River, *Zeyher*, 898! 1359! Albany Div.; on the rocks of Zwartwater Poort, *Burchell*, 3397!
KALAHARI REGION: Orange River Colony and Transvaal, *Burke!* Transvaal; near Lydenburg,, *Atherstone!*
EASTERN REGION: Transkei; around Kentani, *Miss Pegler*, 1516! Pondoland; between St. Johns River and Umsikaba River, *Drège!* Natal; near Durban, *Drège! Peddie! Krauss*, 75! near Mooi River, *Wood*, 4444!

32. P. nummularius (Briq. in Bull. Herb. Boiss. 2me sér. iii. 1072); a procumbent herb; stems hairy with long slender hairs; leaves 1–1$\frac{1}{4}$ in. long, as broad or sometimes broader than long, with short stout hairs on both sides, obtuse, upper portion crenate, truncate or shortly cuneate at the base; petiole 1 in. long, hairy; racemes reaching 4$\frac{1}{2}$ in. long, rhachis pubescent; verticils about 6-flowered, not pedunculate; bracts small, ovate, deciduous; pedicels 2 lin. long, pubescent; flowering calyx 1 lin. long, hairy; upper tooth ovate, slightly more than $\frac{1}{2}$ lin. long; 4 other teeth lanceolate; lateral teeth scarcely $\frac{1}{2}$ lin. long; lower teeth nearly equal to or slightly longer than the lateral; corolla pale purple (*Wood*), 4 lin. long; tube 1$\frac{1}{2}$ lin. long, deflexed below the middle and slightly enlarged above; upper lip 1 lin. long and as broad as long, pubescent and glandular, with 2 rounded strongly ciliate lobes at the apex and 2 short rounded lateral lobes; lower lip 2 lin. long, $\frac{3}{4}$ lin. deep, boat-shaped, separated from the upper lip by a broad sinus; stamens included or scarcely exserted.

EASTERN REGION: Natal; Camperdown, *Rehmann*, 7702! Zululand, Indulindi, 2000 ft., *Wood*, 3980! 3981!

33. P. neochilus (Schlechter in Journ. Bot. 1896, 394); an erect branched undershrub ; stems obtusely quadrangular, hirsute with fine whitish hairs ; leaves subsessile or shortly petiolate, $1\frac{1}{4}-1\frac{3}{4}$ in. long, $\frac{3}{4}-1$ in. broad, fleshy, elliptic or obovate, usually rounded at the apex, hairy on both surfaces, cuneate at the base ; inflorescence of elongate many-flowered spike-like racemes ; verticils about 6-flowered, close ; bracts submembranous, large, 4 lin. long and broad, broadly ovate or suborbicular, with a long acumen, ciliate, deciduous ; pedicels short, 1–2 lin. long ; fruiting calyx 3 lin. long ; upper tooth $1\frac{3}{4}$ lin. long, broader than long, bluntly and shortly acuminate, strongly nerved ; other 4 teeth lanceolate, acute, ciliate, 2 lower the longest ; flowering calyx $1\frac{1}{4}$ lin. long, campanulate, hairy ; corolla deep purple (*Galpin*) ; tube subcylindric, 3 lin. long, deflexed about the middle ; upper lip erect, subquadrate, $1\frac{1}{4}$ lin. long, 4-lobed ; lower lip 4 lin. long, boat-shaped ; nutlets $\frac{3}{4}$ lin. long, $\frac{1}{2}$ lin. broad, globosely ovoid, smooth, dark brown.

KALAHARI REGION : Transvaal ; Rimers Creek, near Barberton, 3000 ft., *Galpin*, 968 !
EASTERN REGION : Natal ; *Gerrard*, 1237 !

34. P. esculentus (N. E. Br. in Kew Bulletin, 1894, 12); root tuberous, edible ; stems at first erect, then bending towards the ground and branching, pubescent ; leaves sessile or nearly so, $1\frac{1}{2}-3$ in. long, $\frac{1}{2}-\frac{7}{8}$ in. broad, oblong, obtuse, minutely pubescent on both sides, cuneately narrowed at the base ; racemes solitary or fasci-culate, simple, $\frac{3}{4}-1$ in. long, hairy ; bracts 1 lin. long and broad, opposite, elliptic, obtuse, scabrous ; pedicels $1-1\frac{1}{4}$ lin. long ; flowering calyx $1\frac{1}{2}-2$ in. long, campanulate, scabrous with short stout hairs ; upper tooth ovate, subotuse ; other 4 teeth lanceolate, acute, subequal ; corolla yellow, 7 lin. long ; tube 3 lin. long, deflexed 1 lin. from the base, portion above the deflexion funnel-shaped ; upper lip $2-2\frac{1}{2}$ lin. long, obovate, 4-lobed, 2 terminal lobes made by a triangular notch $\frac{1}{2}$ lin. deep at the apex, rounded ; lateral lobes small, oblong, obtuse ; lower lip $3\frac{1}{2}$ lin. long, 1 lin. deep, boat-shaped, acute ; upper pair of stamens 2 lin. long ; lower pair 3 lin. long ; style as long as the lower stamens ; nutlets not seen.   *Hook. Ic. Pl. t.* 2488.

EASTERN REGION : Natal ; *Wood*, 3633 ! 5620 !

The plant is known as "*Umbondwe*" or "*Kafir Potato.*" The Kafirs eat the tubers. Cultivated specimens grown in the Botanic Gardens at Durban, at Skinners Court, Pretoria, and at Kew, are the only examples that have been seen, no wild specimens.

35. P. pachyphyllus (Guerke MS. in Bot. Mus. Univers. Zürich) ; a low plant about 6 in. high ; stem stout, branched from the base hairy ; leaves thick and fleshy, $\frac{1}{2}-\frac{3}{4}$ in. long, nearly as broad as long, shortly petiolate, obovate-oblong, cuneate at the base, shallowly crenate, very densely hairy on both sides ; flowers in simple racemes ; rhachis hairy ; verticils about 6-flowered, rather distant in the lower part of the raceme ; pedicels $1\frac{1}{2}$ lin. long, hairy ; flowering calyx $1\frac{1}{4}$ lin. long, densely hairy ; tube very short ; upper tooth rather less than 1 lin. long, and as broad as long, broadly ovate, ciliate ; lateral teeth shorter than the upper tooth, triangular ; lower teeth as long as the upper, lanceolate, acute ; corolla 5 lin. long ; tube narrow, deflexed about $\frac{1}{2}$ lin. from the base, slightly funnel-shaped upwards ; upper lip $\frac{1}{2}$ lin. long, central lobe rounded, 2-fid, lateral lobes less than $\frac{1}{2}$ lin. long, oblong ; lower lip 2 lin. long, $\frac{1}{2}$ lin. deep, boat-shaped, separated from the upper lip by a wide sinus, pubescent near the apex outside ; stamens included.

EASTERN REGION : Natal ; *Rehmann,* 7878 !

36. **P. elegantulus** (Briq. in Bull. Herb. Boiss. 2me sér. iii. 1005) ; a slender herb, 8–12 in. high ; stems at the base ascending, more or less branched, shortly crisply pilose or glandular-pilose ; leaves small, scarcely 1 in. long, $\frac{3}{4}$ in. broad, ovate or deltoid-ovate, acute or subacute, membranous, green, sparingly hairy or nearly glabrous, crenate-serrate ; petioles $\frac{1}{2}$–$\frac{3}{4}$ in. long ; racemes 4–7 in. long ; verticils 6–10-flowered, not pedunculate ; bracts small, deciduous ; pedicels shortly hairy, unequal, 4 lin. long ; fruiting calyx $\frac{1}{4}$ in. long, tubular-ovoid ; flowering calyx $\frac{1}{2}$–$\frac{3}{4}$ lin. long, broadly campanulate ; upper tooth ovate, small, shortly acuminate, recurved ; 4 other teeth lanceolate, acuminate, 2 lower the longest, teeth separated each by a distinct sinus ; corolla-tube shortly cylindric at the base for $\frac{3}{4}$ lin., then suddenly enlarged for 2 lin. ; upper lip small, erect ; lower lip 1 lin. long ; stamens included ; nutlets $\frac{1}{2}$ lin. long and about as broad, ovoid, black, shining.

EASTERN REGION : Natal ; Karkloof, *Rehmann,* 7368 !

37. **P. tomentosus** (Benth. in E. Meyer, Comm. 229) ; suffruti-cose ; stems stout, erect, tomentose or villous ; leaves 1$\frac{1}{4}$–3$\frac{1}{2}$ in. long, 1$\frac{1}{4}$–3 in. broad, broadly ovate, obtuse, coarsely crenate, tomentosely villous on both sides, rounded or cuneate at the base ; petioles $\frac{1}{2}$–1$\frac{1}{2}$ in. long, tomentose ; inflorescence rarely a simple raceme, the racemes usually forming a large panicle ; verticils 6–12-flowered, not pedun-culate ; bracts rhomboid, deciduous ; pedicels 2–3 lin. long, tomentose ; fruiting calyx 3–4 lin. long, hairy ; flowering calyx 1$\frac{3}{4}$ lin. long, campanulate, hairy ; upper tooth 1 lin. long, $\frac{3}{4}$ lin. broad, ovate-oblong, obtuse, ciliate ; other 4 teeth lanceolate, ciliate, 2 lower slightly longer than the lateral ; corolla $\frac{1}{2}$ in. long or more, purple ; tube 3 lin. long, deflexed about the middle, enlarged above ; upper lip 1$\frac{1}{4}$ lin. long, obovate, rounded at the apex, hairy ; lower lip 3 lin. long, scarcely 1 lin. deep, boat-shaped, acute, hairy ; upper pair of stamens 1$\frac{1}{4}$ lin. long ; lower 2$\frac{1}{4}$ lin. long ; style as long as the upper stamens, equally 2-fid ; nutlets $\frac{1}{2}$ lin. long and as broad as long, compressed, smooth and shining, dark brown. *Benth. in DC. Prodr.* xii. 67 ; *Wood, Natal Pl.* iv. *t.* 316.

KALAHARI REGION : Transvaal ; Lydenburg, *Wilms,* 1125 ! Barberton, *Thorncroft in Herb. Wood,* 4295 !
EASTERN REGION : Natal ; various localities, *Gerrard,* 394 ! 1674 ! *Wood,* 488 ! 3199 ! 4340 ! 4775 ! *Sanderson,* 126 ! 550 ! *Krauss,* 75 ! *Drège,* 4783 ! *Grant* !

38. **P. Woodii** (Guerke in Engl. Jahrb. xxvi. 76, by error *Wodii*) ;
stems 16–20 in. high, erect, branched, pubescent ; leaves 2–2½ in.
in diam., suborbicular, irregularly coarsely crenate, pubescent on
both sides ; petioles ¾–1 in. long ; inflorescence laxly branched ;
verticils ⅜–1 in. apart, 12–16-flowered ; fruiting calyx 2–2¼ lin.
long, ovoid, puberulous ; upper and lower lips equal ; upper tooth
ovate, acute ; 4 other teeth subequal, lanceolate-deltoid, acute,
lateral teeth slightly broader than the lower ; corolla ½ in. or more
long, purple ; tube 3 lin. long, deflexed about the middle, enlarged
above ; nutlets small, shining, smooth, yellowish brown.

EASTERN REGION : Natal ; Ipolweni, in open ground, 3000–4000 ft., *Wood* ;
Pinetown, *Rehmann,* 8032 !

Very similar to *P. tomentosus,* Benth., the chief differences being that *P. tomentosus* has a much more compact inflorescence and darker seeds than *P. Woodii.*

39. **P. dolichopodus** (Briq. in Bull. Herb. Boiss. 2ᵐᵉ sér. iii.
1069) ; an erect more or less branched herb, 16 in. high or more ;
stems and branches shortly crisply hairy ; leaves 2 in. long, 1¾ in.
broad, membranous, glabrous or sparsely hairy, ovate, subacute,
base truncate or shortly cuneate ; petioles of the lower leaves reaching
2¼ in. long, pubescent ; racemes simple ; verticils 6–8-flowered, the
lower 5 lin. apart, not pedunculate ; bracts small, deciduous ; pedicels
2 lin. long, pubescent ; flowering calyx 1½ lin. long, campanulate,
hairy ; upper tooth broadly ovate, acuminate, ciliate ; other 4 teeth
lanceolate, lateral ½ lin. long, 2 lower ¾ lin. long ; corolla externally
minutely puberulous ; tube cylindric at the base, straight portion
1 lin. long, then deflexed and enlarged into a funnel-shaped throat ;
upper lip ¾–1¼ lin. long ; lower lip 2 lin. long, 1 lin. deep.

COAST REGION : Komgha Div. ; in woods near Komgha, *Flanagan,* 740 !
EASTERN REGION : Natal ; Karkloof, *Rehmann,* 7383 !

The unusually long petioles are remarkable. The type specimen (*Rehmann,*
7383) is very incomplete, having neither flowers nor fruit, while its leaves are
rather dilapidated.

40. **P. grallatus** (Briq. in Bull. Herb. Boiss. 2ᵐᵉ sér. iii. 1004) ; a
branched herb ; branches ascending, crisply pilose, more or less
fistular ; leaves 2–3 in. long and nearly as broad as long, broadly
ovate, shortly acuminate, laxly crisply hairy on both sides, coarsely
crenate, truncate or subcordate at the base ; petioles 1¼–2 in. long,
crisply hairy ; racemes 2–5 in. long ; verticils 6-flowered, ⅜–⅞ in.
apart, not pedunculate ; bracts minute, deciduous ; pedicels 4 lin.
long, much exceeding the calyx and as well as the rhachis minutely
shortly puberulous ; flowering calyx 1–1½ lin. long, campanulate,
more or less hairy ; upper tooth less than ¾ lin. long, ovate ; 4 other
teeth lanceolate, lateral ¾ lin., lower 1 lin. long ; corolla shortly
puberulous externally ; tube ⅕ in. long, enlarged above ; upper lip 2 lin.
long, erect ; lower lip ⅕ in. long, 1 lin. deep, boat-shaped, separated
from the upper lip by a wide sinus ; stamens scarcely exserted.

EASTERN REGION : Natal ; stony places near Mount Frere, 4500 ft., *Schlechter,*
6415 !

41. P. transvaalensis (Briq. in Bull. Herb. Boiss. 2me sér. iii. 1005); herb about 16 in. high; stems erect, grooved, rufous-pubescent; branches ascending; leaves 1½–3½ in. long, 1–3 in. broad, rhomboid-ovate, acute, irregularly coarsely crenate, shortly hairy above, rufous-pubescent on the nerves and gland-dotted beneath, reticulately veined, base cuneate; petioles ½–1½ in. long, pubescent; inflorescence a panicle reaching 7 in. long, formed of several racemes; rhachis pubescent; verticils 4–6-flowered, not pedunculate, ⅜–⅗ lin. apart; bracts reaching 2½ lin. long, lanceolate, acute; pedicels up to 3 lin. long, slender, pubescent; flowering calyx 1½ lin. long, campanulate, hairy; upper tooth ovate, obtuse, shorter than the others; other 4 teeth lanceolate-subulate, ciliate, 2 lower much longer than the lateral, connate below; corolla pubescent and glandular, 5 lin. long, whitish (*Wood*); tube 2–2¾ in. long, shortly deflexed at ⅓ the way from the base; upper lip 1½ lin. long; lower lip 2–2½ lin. long, 1 lin. deep, obtusely boat-shaped, separated from the upper lip by a broad sinus; stamens slightly exserted; nutlets not seen.

KALAHARI REGION : Orange River Colony; Harrismith, 6500 ft., *Sankey*, 231! and without precise locality, *Cooper*, 994! Transvaal; Houtbosch, *Rehmann*, 6154!

EASTERN REGION : Natal; South Downs, Weenen County, *Wood*, 4378!

42. P. Draconis (Briq. in Bull. Herb. Boiss. 2me sér. iii. 1071); a herb indurated at the base; stem suberect, densely pubescent; leaves (including petioles) reaching 1½ in. long, obovate, cuneate, running down into a pubescent petiole of which the junction with the blade is obscure, numerous, deciduous, subobtuse at the apex, thick, rugose, crenate, scabrous-pubescent on both sides with short rigid hairs; inflorescence in simple or rarely branching racemes short or rather long; verticils 6–10-flowered, distant, not pedunculate; bracts small, deciduous; pedicels 2½ lin. long, densely pubescent, unequal; fruiting calyx 2 lin. long, incurved, membranous, strongly nerved, nerves scabrous with rigid hairs, ovoid-campanulate; flowering calyx 1 lin. long, pubescent; upper tooth broadly ovate, nearly as broad as long; other 4 teeth lanceolate, acute, ciliolate, 2 lower longer and narrower than the lateral; corolla 3 lin. long, purple (*Wood*), externally pubescent; tube straight for about ½ lin., then deflexed into a throat ¾ lin. long; upper lip less than 1 lin. long, pubescent, cleft at the apex into 2 deltoid ciliate lobes, lateral lobes about ¼ lin. long, oblong, obtuse; lower lip 1½ lin. long, ½ lin. deep, boat-shaped; stamens included; nutlets ½ lin. long and broad, subglobose, pale brown, smooth.

EASTERN REGION : Natal ; Biggarsberg, *Rehmann*, 7092 ex *Briquet*; in crevices of flat rocks near Botha's Hill Railway Station, *Wood*, 4574 !

Closely allied to *P. Thunbergii*, from which it differs by the densely hairy leaves and the calyces scabrous on the nerves.

VI. COLEUS, Lour.

Calyx ovoid-campanulate, usually declinate in fruit ; upper tooth ovate, broader than the other 4 ; lateral teeth ovate-truncate or narrow and acute ; 2 lower teeth acute, often connate beyond the middle. *Corolla* usually purple or lilac ; tube much longer than the calyx, usually deflexed about the middle ; throat funnel-shaped ; limb bilabiate ; upper lip obtusely 3–4-lobed, the lower oblong, deeply concave. *Stamens* 4, didynamous ; filaments shortly connate into a tube above their insertion in the corolla-throat ; anther-cells confluent. *Disk* produced into a gland on the lower side of the ovary ; style equally 2-fid at the apex. Nutlets ovoid or subglobose, smooth.

Annual or perennial herbs, shrubs or undershrubs ; leaves various in shape, sessile or petiolate ; inflorescence spicate, racemose or subpaniculate ; verticils simple or in more or less developed opposite cymes ; bracts deflexed or deciduous, or the uppermost sometimes coloured and comose.

Distrib. Species about 150, extending through Tropical Africa, the Mascarene Isles, India, China, the Malay Archipelago, Australia and Polynesia.

Leaves ovate, subcordate ; bracts minute ; flowering
　　calyx ¾ lin. long　　…　　…　　…　　…　　…　　(1) **Rehmanii.**

Leaves obovate, cuneate at the base ; bracts large, 5 lin.
　　long, 4 lin. broad ; flowering calyx 1½–2 lin. long　　(2) **Pentheri.**

1. C. Rehmannii (Briq. in Bull. Herb. Boiss. 2me sér. iii. 1075) ; an erect herb 1 ft. high with ascending crisply pilose dull green or purple branches ; leaves scarcely reaching 1¼ in. long 1$\frac{3}{16}$ in. broad, ovate, shortly acuminate, with slightly cordate or subcordate base, firm, more or less crisply pilose, dull green more or less reticulately nerved, regularly crenate ; petioles crisply hairy, shorter in the upper than in the lower leaves ; inflorescence a spike-like raceme reaching 6 in. long ; verticils 6-flowered ; bracts minute, deciduous ; pedicels ½–1½ lin. long ; flowering calyx minute, ¾ lin. long, broadly campanulate, shortly hairy ; fruiting calyx 2 lin. long, membranous, nerved ; upper tooth rounded, 1 lin. long ; lateral teeth rotund-truncate, ½ lin. deep ; 2 lower 1¼ lin. long, separated from the lateral by a deep sinus, connate below ; corolla small, externally puberulous ; tube scarcely exserted ; upper lip erect, shortly 3-lobed, ½ lin. long ; lower lip 1¼ lin. long ; staminal tube ½ lin. long ; nutlets less than ⅓ lin. long, smooth, pale yellow.

Kalahari Region : Transvaal ; Houtbosch, *Rehmann,* 6156 ex *Briquet.*

2. C. Pentheri (Guerke in Ann. Naturhist. Hofmus. Wien, xx. 48) ; stem herbaceous, 8 in. high, softly pubescent with long white hairs ; leaves ½–1½ in. long, ⅛–¾ in. broad, obovate, obtuse or subacute, moderately thick, hairy on both sides, margins crenate, base cuneate ; petioles very short, hairy ; inflorescence a simple spike-like raceme up to 5½ in. long ; rhachis densely pubescent, quadrangular ; verticils 4–6-flowered, not pedunculate, separated in fruit 4–5 lin. in the lower, closer in the upper part of the raceme ; bracts large, reaching as much as 5 lin. long (including the mucro), 4 lin. broad, broadly ovate, acute, with ciliate margins and a purplish mucro reaching 1 lin. long ; pedicels 2–2½ lin. long, erect, pubescent, drooping at the tip; flowering calyx 1½–2 lin. long, campanulate, hairy ; upper tooth 1 lin long and as broad or broader than long, orbicular-ovate, very shortly acuminate ; 2 lateral teeth deltoid, acute, ciliate, ¾ lin. long; 2 lower teeth linear-lanceolate, acute, slightly longer than the lateral ; fruiting calyx 2½–3 lin. long, curved, densely villous in the throat within ; upper tooth shortly acuminate, strongly nerved ; corolla purple (*Wood*), ½ in. long ; tube 2¾ lin. long, deflexed about the middle, slightly enlarged upwards ; upper lip 1–2 lin. long, obovate, cuneate, 4-lobed ; lower lip 3½ lin. long, boat-shaped ; filaments connate below for 1½ lin. ; upper pair of filaments 3 lin. long ; lower pair 4 lin. long ; style longer than the stamens ; nutlets nearly ¾ lin. long and broad, subglobosely ovoid, smooth, brown-black, minutely punctate.

CoAST REGION : Peddie Div. ; Breakfast Vley, *Krook*, 1716 ! East London Div. ; near Kintza River mouth, 50 ft., *Galpin*, 6554 !
KALAHARI REGION : Transvaal ; near Barberton, 3000 ft., *Thorncroft*, 109 ! *Wood*, 4295 ; near Lydenburg, *Atherstone* !
EASTERN REGION : Natal ; Byrne, *Wood*, 3199 ! Mooi River, 2000–3000 ft., *Wood*, 4340 !

VII. PYCNOSTACHYS, Hook.

Calyx ovoid-campanulate, slightly accrescent in fruit ; teeth 5 subulate, rigid, at length spreading, subspinescent. *Corolla* blue, pink or violet ; tube exserted, deflexed, enlarged in the throat; limb 2-lipped ; upper lip 4-toothed, shorter than the lower ; lower entire, concave. *Stamens* 4, didynamous, declinate ; filaments free, without teeth ; anthers confluent, 1-celled. *Disk* subequal. *Style* very shortly 2-fid at the tip; lobes subulate, equal. *Nutlets* subrotund, smooth.

Perennial erect herbs ; leaves opposite or whorled, broad or narrow, sessile or petiolate ; flowers densely crowded into simple terminal spikes, sessile ; bracts shorter or longer than the calyx.

DISTRIB. Species about 40, in Tropical and South Africa.

Leaves on long petioles, ovate ... ... (1) **urticifolia.**
Leaves sessile or subsessile, lanceolate or linear-lanceolate :
  Leaves usually broadly lanceolate, densely puberulous
    on all the veins beneath ; secondary veins
    horizontal ... ... (2) **reticulata.**

  Leaves linear-lanceolate or narrowly lanceolate, very
    minutely puberulous or subglabrous on the
    primary veins beneath ; secondary veins obliquely
    ascending ... ... (3) **purpurascens.**

1. P. urticifolia (Hook. in Bot. Mag. t. 5365); a much-branched perennial herb 5–8 ft. high; stems and branches obtusely quadrangular, pubescent; leaves long-petioled, ovate, acute, crenate, densely pubescent on both sides, rounded, truncate or shortly cuneate at the base; petioles reaching 2 in. long, pubescent; flowers in dense spikes growing out to 2–3 in. long, 1½ in. in diam. (excl. the corollas); bracts 2–3 lin. long, narrowly linear, slightly wider at the top than the base, pubescent; calyx reaching ½ in. long; tube 1 lin. long in flower, narrowly campanulate, pubescent, reaching in fruit 2 lin. long, strongly nerved, curved, with 5 submembranous processes which project above the mouth of the calyx alternating with its teeth and close over its mouth; teeth reaching nearly 5 lin. long, becoming very rigid and spine-like in fruit, pubescent; corolla bright blue; tube sharply deflexed at about 2¼ lin. from its base and narrowly cylindric (scarcely ½ lin. in diam.) above the deflexion, much dilated to about 1¾ lin. broad for 2 lin. below it; upper lip 1¾ lin. long, cuneate-oblong, with 4 rounded apical lobes; lower lip 3–3½ lin. long, nearly 2 lin. deep, boat-shaped, with intruded apex, pubescent outside; stamens not exserted beyond the lower lip, upper pair 2 lin. long, lower 3 lin. long; style longer than the stamens, shortly 2-fid. *Guerke in Engl. Jahrb.* xxii. 146; *Baker in Dyer, Fl. Trop. Afr.* v. 386.

SOUTH AFRICA : without locality, *Mrs. Saunders*, 3890 !
KALAHARI REGION : near Barberton; Louws Spruit, 1200 ft., *Bolus*, 9745 ! wooded ravines near Barberton, 3000 ft., *Galpin*, 943 ! Woodbush mountains, *Mrs. Barber*, 4 !

Also in Tropical Africa.

2. P. reticulata (Benth. in DC. Prodr. xii. 83); a perennial herb, 2–3 ft. high; stems obtusely quadrangular, more or less densely pubescent, erect, simple or branched; leaves sessile or nearly so, oblong or oblong-lanceolate or lanceolate, acute or subobtuse, irregularly crenate-serrate, strongly nerved and reticulately veined, pubescent above and densely so on the veins beneath, lower leaves 3–4 in. long, ½–1½ in. broad, becoming shorter upwards; flowers in dense cylindric spikes 1–2 in. long, ½–¾ in. in diam. (excl. the corollas), forming a lax panicle; bracts lanceolate, acute, 2½ lin. long, pubescent; calyx-tube in flower shortly pubescent; teeth reaching 2 lin. long, linear-subulate, pubescent; corolla pubescent outside; tube sharply deflexed about the middle, very narrow and straight for 2 lin. below the deflexion, then much dilated for 2 lin. above it; upper lip 1¼ lin. long, 1 lin. broad, quadrate, 4-lobed; lower lip nearly 3 lin. long, 1¼ lin. deep, boat-shaped; upper pair of stamens 2¼ lin. long; lower pair 3 lin. long; style longer than the stamens, shortly 2-fid.

KALAHARI REGION: Transvaal; Woodbush, *Rehmann*, 6174 !
EASTERN REGION: Natal; near Durban, *Drège*! *Krauss*, 329 ! *Wood in MacOwan & Bolus, Herb. Norm.*, 1016 ! Inanda, *Wood*, 59 ! and without precise locality, *Gerrard*, 101 !

3. P. purpurascens (Briq. in Bull. Herb. Boiss. 2me sér. iii. 998);
a herb 2–3 ft. (or more ?) high ; stem simple and with only 1 flower-
spike or paniculately branched with 3 to several spikes, obtusely
4-angled or terete, with 6–8 ribs, thinly to densely adpressed-
puberulous with very minute reflexed hairs ; leaves opposite or 3 in
a whorl, ascending or spreading, 2–5 in. long, ⅙–1⅙ in. broad, linear-
lanceolate or narrowly lanceolate, acute, tapering from below the
middle to the sessile base, serrate, some of the upper leaves nearly
entire, glabrous or thinly sprinkled with microscopic hairs on both
sides and more densely on the veins beneath ; secondary veins
obliquely ascending (not horizontal) ; flower-spike 1–3¾ in. long and
including the corollas 1–1¼ in. in diam. when dried ; bracts
deflexed, 3–5 lin. long, linear-lanceolate and acuminate or linear-
subulate, ciliolate ; calyx-tube ½ lin. long and cup-like when in
flower, elongating to 2 lin. long and gibbous on the lower side at the
base in fruit ; teeth 2–3 lin. long, spine-like, puberulous, with small
membranous lobules between them, closing the mouth of the tube
whilst the seeds are maturing, spreading outwards when they ripen ;
corolla abruptly bent at about the middle of the tube, thinly or
thickly puberulous outside on the upper part, pale pink or rosy-
purple (*Tyson*) ; tube 4–5 lin. long when measured along the bend,
very slender at the basal half, abruptly dilating to 1¼ lin. in vertical
diam. at the upper part ; upper lip 1 lin. long, 3-toothed ; lower lip
2–2½ lin. long, boat-shaped, curved ; nutlets ½–⅔ lin. long, oblong or
elliptic-oblong, flattened on the back, keeled down the inner face,
smooth, at first pale brown, finally dark brown. *P. Schlechteri,
Briq. l.c.* 999. *P. holophylla, Briq. l.c.* 1000. *P. reticulata, var.
angustifolia, Benth. in DC. Prodr.* xii. 83, *as to description and
Burke's specimen.*

KALAHARI REGION : Orange River Colony, *Cooper*, 1070 ! Transvaal ; Magaliesberg Range, *Burke*, 111 ; Pilgrims Rest, *Roe in Herb. Bolus*, 2647 ! Aapies Poort
and River, near Pretoria, *Rehmann*, 4111 ! *Miss Leendertz*, 1108 ! Woodbush
mountains, *Nelson*, 438 ! near Johannesburg, *E.S.C.A. Herb.* 347 (or 847 ?) ! near
Barberton, *Galpin*, 1318 ! *Thorncroft*, 4344 ! Elandspruit Mountains, *Schlechter*,
3884 ! Witwatersrand, *Mrs. Hutton*, 878 ! near Lydenburg, *Wilms*, 1122 ! Spitzkop
Goldmine, *Wilms*, 1122b !
EASTERN REGION : Griqualand East ; by mountain streams near Clydesdale,
Tyson, 2753 ! and in *MacOwan & Bolus, Herb. Norm.*, 859 ! near Mount Frere,
Schlechter, 6406 !

The three specimens which Briquet has described as three distinct species have
been examined and dissected and found to be identical. Of the five characters
mentioned as distinguishing them—the purple colour is the result of sun-exposure ;
the closed or open mouth of the calyx and colour of the nutlets are conditions
varying with the maturing of the seeds, both forms of calyx and colours of nutlets
may be found on the same spike ; the size of the corolla, &c., is variable according
to the vigour of the specimen and other causes, as in most *Labiatæ*, and the
toothing and pubescence of the leaves is nearly the same in all. *P. purpurascens* is
very closely allied to *P. reticulata*, but is readily distinguished by its narrower and
more glabrous leaves, whilst the flowers are stated to be different in colour.

VIII. **GENIOSPORUM**, Wallich.

Calyx campanulate, 4–5-toothed ; tube elongating and cylindric
in fruit, with prominent transverse veins on the upper part ; teeth
variable ; in the 4-toothed species, with the 3 upper teeth subequal and smaller than the lowest tooth, which is emarginate or
minutely bifid at the apex and inflexed and closes the tube in fruit ;
in the 5-toothed species, with the 3 upper larger than the other
2 or the teeth all subequal ; dorsal tooth scarcely decurrent on
the tube. *Corolla* small, exserted, 2-lipped ; tube short ; upper lip
shortly 4-lobed ; lower lip boat-shaped. *Stamens* 4, directed
towards the lower lip ; filaments all free, bearded, but neither toothed
nor crested near the base ; anthers versatile, reniform, 1-celled.
Style filiform ; stigma bifid. *Nutlets* erect, oblong or ellipsoid.

Perennial herbs ; leaves opposite ; inflorescence terminal, spike-like, with the
whorls many-flowered, usually crowded ; flowers very small.

DISTRIB. A small genus of 16 or 17 species, some undescribed, natives of India,
Madagascar, and Tropical Africa, one species extending into South Africa.

1. **G. angolense** (Briq. in Engl. Jahrb. xix. 164) ; a perennial
herb ; stems 2–3 ft. high, simple or branched, obtusely 4-(rarely 6-)
angled, obscurely puberulous ; leaves spreading ; petiole 2–5 lin.
long ; blade 1–2¼ in. long, ⅓–1 in. broad, lanceolate, acute, tapering
into the petiole at the base, serrate, subglabrous or with minute
scattered pubescence on both sides ; inflorescence dense, spike-like,
1¼–5 in. long, of numerous many-flowered crowded whorls, with
1 or 2 pairs of ovate acuminate parti-coloured leafy bracts at its
base ; pedicels about 1 lin. long, pubescent ; calyx 1 lin. long,
becoming 2 lin. long in fruit, campanulate, 4-toothed, pubescent
outside, glabrous within ; teeth about ⅔ as long as the tube, 3 upper
ovate, acute, lower one ovate-oblong, obtuse, emarginate, or shortly
bidentate, inflexed and closing the mouth of the tube in the fruiting
stage ; corolla small, unequally 5-lobed, pubescent outside and in
the upper part within ; tube 1¼–1½ lin. long, exceeding the calyx ;
4 upper lobes subequal, ½ lin. long, flat, ovate, subacute ; lower
lobe about 1¼ lin. long, ½ lin. broad, boat-shaped, subacute and
recurved at the apex ; stamens much exserted ; filaments free,
bearded at the base, 2–2½ lin. long ; upper pair inserted just below
the middle of the tube, lower pair inserted just below the base of
the lower corolla-lobe ; nutlets ½ lin. long, ellipsoidal, slightly compressed, obtuse, glabrous. *Baker in Dyer, Fl. Trop. Afr.* v. 351.

KALAHARI REGION : Transvaal ; in the district of Lydenburg, *Roe in Herb. Bolus,*
2648 ! by the river near Lydenburg, *Wilms,* 1139 ! Houtbosch, *Rehmann,* 6173 !
marshy places in Umlomati Valley, near Barberton, 4000 ft., *Galpin,* 1317 !
Thorncroft, 7214 ! near Mbabane, 4600 ft., *Bolus,* 12248 !

Also in Angola.

IX. ÆOLANTHUS, Mart.

Calyx small, ovoid, campanulate or tubular, truncate or obscurely 2-labiate, finally circumscissile near the base, in fruit usually accrescent, often contracted near the apex. *Corolla-tube* exserted, straight or decurved, slightly dilated or not above; limb 2-labiate; upper lip obtusely 4-toothed; lower lip larger, concave, entire. *Stamens* 4, didynamous, declinate; filaments free, without teeth; anthers confluent, 1-celled, at length flattened. *Disk* glandular. *Style* shortly 2-fid, with subulate arms. *Nutlets* rotundate or oblong compressed, smooth.

Annual or perennial herbs or small shrubs; leaves usually fleshy; cymes forming a laxly paniculate raceme, with sessile or very shortly pedicellate secund flowers; bracts small, caducous; corolla lilac.

DISTRIB. Species about 50, the others in Tropical Africa.

Calyx ¼ lin. long; bracts minute ... ... (1) **parvifolius.**
Calyx ½-¾ lin. long; bracts large:
  Leaves suborbicular; bracts broadly ovate, green ... (2) **canescens.**
  Leaves obovate-lanceolate; bracts elliptic-lanceolate,
    red ... ... (3) **Rehmannii.**

1. **Æ. parvifolius** (Benth. in DC. Prodr. xii. 80); an erect plant reaching 1½-2 ft. in height; stems puberulous, more or less woody below; leaves petiolate, ⅜-⅝ in. long, nearly as broad as long, subfleshy, broadly ovate or suborbicular, obtuse, irregularly and shortly toothed, glabrous or puberulous; petiole nearly as long as the blade; flowers sessile or subsessile, in unilateral puberulous cymes racemosely arranged; bracteoles minute; calyx puberulous, ¼ lin. long, tubular-campanulate, 2-lipped; upper lip 3-toothed, the lower subtruncate; corolla externally puberulous; tube 3 lin. long, curved, slightly enlarged towards the mouth; upper lip 1½ lin. long, with 2 tolerably deep rounded apical lobes and 2 obtuse lateral ones; lower lip 2 lin. long; longer pair of stamens 2 lin. long; shorter pair 1½ lin. long; anthers orbicular, flattened; style as long as the longer pair of stamens, shortly 2-fid.

EASTERN REGION: Natal; Inanda, *Wood*, 518! 828! *Gerrard*, 1236! Pondoland; between St. Johns River and Umsikaba River, *Drège*, 4765!

2. **Æ. canescens** (Guerke in Engl. Jahrb. xxii. 147); branched undershrub 9-18 in. high; stems and branches softly pubescent or puberulous; leaves ½-1 in. long and about as broad as long, suborbicular, obtuse or subacute, narrowed at the base into the petiole, coarsely serrate, hoary-puberulous on both sides; cymes dense; flowers sessile; bracts foliaceous, green, reaching 1¼ lin. long, ovate, obtuse or subacute, pubescent on both sides, ciliolate; calyx tubular, truncate or obscurely toothed, pubescent, circumscissile near the base, leaving the lower portion as a persistent scutelliform disk, in

flower ½ lin. long, in fruit reaching 1¼ lin. long ; corolla white or
lilac with purple spots (*Wood*); tube 1¼ lin. long, nearly straight,
not enlarged at the mouth ; upper lip ¾ lin. long, with 4 deep
rounded lobes ; lower lip ¾ lin. long ; style as long as or slightly
longer than the stamens, slender, bifid at the tip, with reflexed
lobes ; seeds ½ lin. long, oblong, slightly angular on the face,
rounded on the back, smooth, brown.

SOUTH AFRICA : without locality, *Burke* ! *Zeyher*, 1331 !
CENTRAL REGION : on the mountains near Graaff Reinet, 4400 ft. in clefts of
rocks, *Bolus*, 383 ! 1345 !
KALAHARI REGION : Transvaal ; Hooge Veld, *Rehmann*, 6856 ! Petersburg
District ; Houtbosch mountains, 5200 ft., *Bolus*, 10983 ! Dasport, *Leendertz*, 586 !
Johannesburg, Jeppes Town, *Gilfillan* !
EASTERN REGION : Natal ; Drakensburg mountains near Van Reenens Pass, in
rocky places, *Wood*, 7187 !

3. **Æ. Rehmannii** (Guerke in Bull. Herb. Boiss. iv. 819) ; suffruticose, 15–16 in. high, trailing (*Mudd*) ; stems and branches puberulous ; leaves petiolate, ovate-lanceolate, 7–12 lin. long, 6 lin.
broad, obtuse or subacute, subfleshy, irregularly crenate on the
margins, pubescent on both sides ; inflorescence lax, branches dark
red ; bracts 1 lin. long, elliptic-lanceolate, subacute, usually dark
red and with ciliolate margins ; calyx (flowering) ¾ lin. long,
accrescent in fruit, tubular, puberulous, usually red, circumscissile
near the base, leaving the lower portion as a scutelliform persistent circular disk ; corolla-tube 2½ lin. long, scarcely enlarged
towards the mouth ; upper lip 1¼ lin. long ; lower lip 1½ lin.
long, oblong ; longer stamens 2½ lin. long, shorter ones 1¾ lin.
long ; style as long as the longer stamens, 2-fid at the tip with
reflexed lobes.

KALAHARI REGION : Transvaal ; mountains around Houtbosch, 4700 ft., *Bolus*,
10938 ! Mac Mac, *Mudd*.

This differs from *Æ. canescens* in the scantier pubescence, the shape of the
leaves which are almost orbicular in *Æ. canescens*, in the lanceolate bracts and
the laxer inflorescence.

X. **ENDOSTEMON**, N. E. Br.

Calyx tubular-campanulate, unequally 5-toothed ; upper tooth
suborbicular, slightly decurrent on the tube ; lateral and lower
teeth narrowly deltoid-attenuate or deltoid-subulate. *Corolla* small,
exserted from the calyx, straight or nearly so, slightly oblique at
the mouth, shortly 4-lobed, with the upper and lower lobes larger
than the lateral and subequal or the lower slightly the larger, all
flat or the lower one slightly concave, not 2-lipped. *Stamens* 4, in
2 pairs, inserted on the lower side of the corolla-tube and not
exserted from it ; filaments all free, straight and without a tooth at
the base, upcurved at the apex ; anthers imperfectly 1-celled, from

the cells being almost confluent at their apex.   *Disk* indistinctly or
but slightly lobed.   *Ovary* 4-lobed to the base; style filiform,
abruptly enlarged at the base between the lobes of the ovary;
stigma large, at first oblong-clavate and entire, bifid at maturity,
included in the corolla-tube.   *Nutlets* oblong, 3-angled, truncate at
the apex.

A branching herb; leaves opposite; racemes paniculately arranged, elongated,
with numerous flower-whorls; flowers small.

DISTRIB.   Species 1, also in Tropical Africa.

This plant differs so much in appearance and structure from *Ocimum* that I have
generically separated it, and would place the genus near *Hyptis*, to which it seems
much more nearly allied than to *Ocimum*.

1. E. obtusifolius (N. E. Br.); a branching herb 1–3½ ft. high;
stem and branches 4-angled, pilose-pubescent; leaves spreading;
petiole 1½–5 lin. long; blade ½–1⅔ in. long, ⅓–1 in. broad, elliptic or
ovate, obtuse or acute, serrate with small teeth, more or less
pubescent on both sides; racemes 3–6 in. long, with numerous
distant irregularly 3–8-flowered whorls; bracts persistent, 1–2 lin.
long, ovate or lanceolate, acute, spreading, pubescent; pedicels
unequal, 1½–2½ lin. long, spreading-pubescent; calyx-tube 1–1¼ lin.
long, larger in fruit, pubescent outside, glabrous within; upper
tooth 1–1¼ lin. long and broad, suborbicular, apiculate; lateral
and lower teeth ¾–1 lin. long, deltoid-attenuate or deltoid-subulate;
corolla puberulous outside, white, spotted with rosy (*Wood*); tube
about twice as long as the calyx-tube, 2–2½ lin. long, straight or
nearly so, tubular-funnel-shaped, enlarging to 1 lin. in diam. at the
slightly oblique mouth; upper and lower lobes subequal, ¾–1 lin.
long, 1¼–2½ lin. broad, suborbicular or transversely oblong and
subtruncately rounded, entire or the upper slightly notched, lateral
lobes smaller and more rounded, all flat or the lower one slightly
concave; stamens inserted in the throat of the tube; upper pair
quite included; lower pair just reaching to the mouth of the tube;
filaments all pubescent, very short, those of the lower ½ lin. long;
ovary glabrous or shortly pubescent at the apex of its lobes; stigma
very thick and large in proportion to the size of the flower, included;
nutlets oblong, 3-angled, truncate and glabrous or pubescent at the
apex, dark brown.   *Ocimum obtusifolium, E. Meyer, Comm.* 227;
Drège, Zwei Pfl. Documente, 205; *Dietr. Syn. Pl.* iii. 376; *Benth. in
DC. Prodr.* xii. 38; *Schinz in Mém. Herb. Boiss.* 10, 61.   *O. rari-
florum, Hochst. in Flora,* 1845, 67; *Krauss, Beitr. Fl. Cap- und Natal.*
131.   *O. laxiflorum, Baker in Dyer, Fl. Trop. Afr.* v. 348.

KALAHARI REGION : Transvaal; common around Shilovane, *Junod,* 956 !
EASTERN REGION : Pondoland; near the Gwenyana River, 1000 ft., *Bolus,*
10259! Natal; between Umtentu River and Umzimkulu River, *Drège*! near
Durban, *Drège*; among reeds by the Umlaas River, *Krauss,* 8! Illovo, *Wood,*
6418! Delagoa Bay, *Junod,* 225 ex *Schinz.*

Also in Tropical Africa.

XI. HYPTIS, Jacq.

Calyx ovoid-campanulate or tubular, more or less accrescent in fruit; teeth 5, subequal, subulate. *Corolla* small; tube cylindric or slightly ventricose, equal or slightly enlarged in the throat; limb sub-bilabiate, 5-lobed; upper lobes flat, erect or spreading; lateral lobes usually similar; lower lobe (lip) saccate, abruptly deflexed in flower, entire or emarginate. *Stamens* 4, didynamous, declinate; filaments free, without teeth; anthers confluent, 1-celled. *Disk* entire or swollen in front into a short gland. *Style* shortly 2-fid or entire at the tip. *Nutlets* ovoid or oblong, smooth or punctate-rugulose, in a few species surrounded by a membranous wing.

Annual or perennial herbs or shrubs; inflorescence and habit polymorphous.

DISTRIB. Species 250, all American, a few naturalised in the Old World.

1. **H. pectinata** (Poit. in Ann. Mus. Par. vii. 1806, 474, t. 30); a tall variable annual 4–8 ft. high; stems erect, branched, pubescent, subwoody at the base, obtusely quadrangular; leaves (in the South African specimens) 1–2 in. long, $\frac{1}{2}$–$1\frac{1}{4}$ in. broad, ovate, obtuse or acute, shortly and sparsely hairy and green above, pallid and tomentose below, margins irregularly crenate, base rounded or subcordate; petioles reaching 1 in. long, pubescent; cymes racemosely arranged, often forming a panicle, after flowering elongate, secund, pectinate; bracts $1\frac{1}{2}$ lin. long, setaceous, ciliate; calyx (flowering) 1 lin. long, (fruiting) 2 lin. long, tubular; tube pubescent, truncate and hairy within at the mouth; teeth $\frac{1}{2}$ lin. long, setaceous, pilose; corolla pale purple or yellowish; tube scarcely reaching 1 lin. long, straight, slightly enlarged towards the mouth; stamens not exserted; style exserted, shortly 2-fid at the tip; nutlets $\frac{1}{2}$ lin. long, ovoid-oblong, smooth, brown. *Benth. in DC. Prodr.* xii. 127; *A. Rich. Tent. Fl. Abyss.* ii. 186; *Baker in Dyer, Fl. Trop. Afr.* v. 448; *Briq. in Bull. Soc. Bot. Belg.* xxxvii. 61. *Nepeta pectinata, Linn. Syst.* ed. x. 1097.

KALAHARI REGION: Transvaal; Barberton, Berea, 2900 ft., *Thorncroft*, 413!
EASTERN REGION: Natal, *Gerrard*, 405! and 1669! *Cooper*, 2985! *Krauss*, 330! Inanda, *Wood*, 524!

Also in Tropical Africa.

XII. HOSLUNDIA, Vahl.

Calyx small, subequally 5-toothed, becoming much enlarged and fleshy when in fruit; dorsal tooth scarcely larger than the rest, not decurrent on the tube. *Corolla* very small, exceeding the calyx, 2-lipped; upper lip 3-lobed; lower lip concave, entire. *Stamens* 4; upper pair reduced to staminodes, with very rudimentary anthers; lower pair fertile, with rather large reniform 1-celled anthers. *Disk* small, crenate. *Ovary* 4-partite; style filiform, incurved at the apex; stigma minutely 2-lobed. *Nutlets* enclosed in the fleshy calyx, dorsally compressed; testa thick and slightly mucilaginous.

Herbs or small shrubs; leaves opposite or whorled; inflorescence terminal, paniculate, with the primary branches simple or branched; flowers small; fruit berry-like.

DISTRIB. Species 3 or more, the others in Tropical Africa.

1. **H. decumbens** (Benth. in DC. Prodr. xii. 54); apparently a tall herb; stems angular, velvety; leaves in whorls of 3; petiole 1–2½ lin. long; blade 1–2½ in. long, 5–13 lin. broad, lanceolate or oblong-lanceolate, acute or acuminate, cuneate at the base, serrulate or crenate-dentate, minutely adpressed-pubescent above, minutely subtomentose beneath; panicle somewhat corymbose, with simple raceme-like branches; whorls 2–3 lin. apart, 2–3-flowered; bracts ½–2 lin. long, about ¼ lin. broad, linear, acute, minutely pubescent, persistent; pedicels ¾–2 lin. long, pubescent; flowering-calyx somewhat ovoid-campanulate, shortly 5-toothed, pubescent and densely covered with minute glands outside, glandular inside at the upper part and pubescent on the teeth; teeth subequal, ⅓–½ lin. long, ovate or deltoid-ovate, subobtuse; fruiting-calyx much enlarged, becoming fleshy, subglobose, nearly closed at the mouth, ¼ in. in diam., orange-yellow; corolla about 2½ lin. long, exserted, glandular outside on the upper part; tube 2 lin. long, straight, slightly enlarging upwards; upper lip erect, about 1 lin. long, with 4 small unequal lobes, pubescent on the back; lower lip ¾ lin. long, orbicular, concave, obtuse; stamens all free, inserted at the upper part of the corolla-tube; upper pair reduced to staminodes, with very rudimentary anthers, included; lower pair fertile, 1¼ lin. long, exserted; style stout in proportion to the flower, glabrous, exserted, incurved at the apex; stigma minutely 2-lobed, obtuse; nutlets dorsally compressed, orbicular in outline, 1–1¼ lin. in diam., brown. *Briq. in Bull. Herb. Boiss. 2me sér. iii. 661. H. verticillata, Briq. in Mém. Herb. Boiss.* x. 60, *not of Vahl. H. opposita, Vahl, var. decumbens, Baker in Dyer, Fl. Trop. Afr.* v. 377.

KALAHARI REGION: Transvaal; Avoca, near Barberton, 1900 ft., *Galpin*, 1246! Swazieland; near Mafutane, 1500 ft., *Bolus*, 12251!
EASTERN REGION: Delagoa Bay, *Forbes*! Matolla, *Schlechter*, 11695!

Also in Tropical Africa.

This is easily distinguished from the other species by the pubescence on the leaves and simple raceme-like branches of the panicle.

XIII. IBOZA, N. E. Br.

(MOSCHOSMA, Auct., not of Reichb.)

Flowers very small, diœcious; male larger than the female and having an abortive ovary or style. *Calyx* minute, similar in both sexes, campanulate, 3-lobed to the middle, with the lateral lobes

minutely bifid or emarginate at the apex, or unequally or subequally
5-lobed, 5-nerved ; dorsal lobe ovate to suborbicular, not decurrent
on the tube, usually not or but slightly larger than the other lobes.
Corolla very small, similar in both sexes, but larger in the male ;
tube funnel-shaped ; limb subequally or unequally 5- (rarely 4-) lobed ;
lobes more or less spreading, flat or slightly concave, the lower slightly
larger than the others. *Stamens* in the male flower 4, free, separate,
not contiguous in pairs and apparently not all directed towards the
lower lobe, exserted ; anthers reniform, dorsifixed, one-celled,
opening longitudinally and forming a peltate flattish disk or with
the margins recurved nearly or quite to the filament ; in the female
flower 0 or 4 and abortive. *Disk* unequally 4-lobed or minute.
Ovary 4-lobed, apparently perfect in the male flowers, but never
producing seed ; style not or but slightly exserted and subentire or
bifid at the apex in the male, much exserted, with linear or oblong-
linear spreading stigmas in the female. *Nutlets* erect, oblong or
ovoid and dorsally compressed, obtuse or acute.

Stout perennial herbs, sometimes (at least as to dried specimens) nearly leafless
at the time of flowering ; leaves opposite, petiolate, ovate or cordate, toothed ;
flowers very small, in large terminal much-branched panicles, with the ultimate
branches laxly or densely spike-like (termed spikes in the descriptions) and the
separate whorls 6-flowered.

DISTRIB. Species more than 12, the others in Tropical Africa.

The plants belonging to this genus have hitherto been referred to *Moschosma,*
but they differ entirely from that genus in habit, calyx and corolla, and in having
unisexual flowers, with the sexes on different plants. The small size and form of
the corolla and the arrangement and spread of the stamens is somewhat like that
of *Mentha,* next which I consider this genus should be placed. The generic name
Iboza is that by which *I. riparia* is known to the Kaffirs.

The differential characters of the species, although in most cases readily
distinguishable to the eye, are not easily expressed in words, as many of them,
which are very evident upon dissection, are such as cannot be used for purposes of
a key.

Leaves bullate-rugose above, densely tomentose beneath ;
   male spikes long, dense ; calyx 5-lobed, tomentose   (5) **Barberæ.**

Leaves not rugose above; calyx 3-lobed, lateral lobes
  acutely bifid or emarginate at the apex :
Male or falsely hermaphrodite flowers in dense spikes
  ½–⅔ in. long ; hairs on the underside of the leaves
  shorter than the thickness of the very prominent
  veins on which they stand ...   . (4) **brevispicata.**

Male or falsely hermaphrodite flowers in somewhat lax
  spikes 1–3 in. long, with the whorls distinctly
  separated ; female spikes dense, ¼–1 in. long ; hairs
  on the underside of the leaves longer than the
  thickness of the veins on which they stand :
Underside of leaves thinly to thickly pubescent, but
  the hairs not hiding the surface between the
  veins :
Leaves (except sometimes on the panicle) notched
  at the base ; tertiary veins not prominent
  beneath ; male flower-whorls 1½–3, female
  ½–1 lin. apart ...   . (1) **riparia.**

Leaves broadly rounded (not notched) at the base; tertiary veins prominent beneath; male flower-whorls 1–1½ lin. apart, female crowded (2) **Galpini.**

Underside of leaves densely tomentose, the hairs quite hiding the whole surface; male flower-whorls ⅜–1 lin. apart, female densely crowded (3) **Bainesii.**

1. **I. riparia** (N. E. Br.); a stout herb 2½–5 ft. high; stems obtusely 4-angled, branching above, puberulous, at least on the upper part; leaves with a pubescent petiole ½–3½ in. long and blade 1¼–6 in. long, 1–5 in. broad, broadly ovate, notched (or some on the panicle entire and rounded) at the base, acute or subacute at the apex, coarsely and obtusely toothed, with the lower margin of the teeth denticulate, not rugose and the tertiary veins not at all prominent beneath, pubescent on both sides, with the hairs longer than the thickness of the veins on which they stand and not hiding the surface between them; panicle 6–15 in. long, 4–8 in. in diam.; male spikes 1–3 in. long, ¼–⅓ in. in diam., lax, with the flower-whorls ⅛–¼ in. apart; bracts about 1 lin. long, 1¼–1⅓ lin. broad, very broadly triangular-ovate, subacute, concave, minutely pubescent, ciliate; pedicels ¼ lin. long; calyx ½ lin. long, cup-like, 3-lobed to the middle, puberulous; dorsal lobe elliptic or suborbicular, obtuse; lateral lobes elliptic-oblong, acutely bifid at the apex; corolla unequally 5- (or occasionally 4-) lobed, minutely and thinly pubescent outside, white or pale-lilac; tube about 1 lin. long, ¾ lin. in diam. at the oblique mouth, with a tuft of hairs near the base inside; lower lobe larger than the others, ¾–1 lin. long, oblong, obtuse; stamens 4, about ¾ lin. long, subequal, exserted; ovary 4-lobed, abortive; style shortly bifid at the apex; female spikes ½–1 in. long, ⅙ in. in diam., with flower-whorls somewhat crowded or ½–1 lin. apart; bracts ½–⅔ lin. long, otherwise as in the male; pedicels as in the male; corolla very like that of the male but smaller, ⅔–¾ lin. long; stamens none; ovary perfecting seeds; styles much exserted; stigmas linear, spreading. *Moschosma riparium, Hochst. in Flora*, 1845, 67; *Krauss, Beitr. Fl. Cap- und Natal.* 131; *Benth. in DC. Prodr.* xii. 49; *Wood & Evans, Natal Pl.* i. *tt.* 1 *and* 2; *Gard. Chron.* 1902, xxxi. 122, *fig.* 35, *and* 1904, xxxv. 30, *fig.* 13.

KALAHARI REGION: Transvaal; near Shilovane, *Junod*, 538! Marovunye forest, *Junod*, 1275!

EASTERN REGION: Natal; banks of streams, *Krauss*, 331! Inanda, *Wood*, 141! near Durban, *Wood in MacOwan & Bolus, Herb. Norm.* 1001! Tugela Valley, *Sutherland*! common throughout the colony, *Gerrard*, 1889!

2. **I. Galpini** (N. E. Br.); a stout herb 3 ft. high; stems obtusely 4-angled, densely pubescent, slightly harsh to the touch; leaves absent from the flowering panicle; petiole ¾–1 in. long; blade 1½–3½ in. long, 1¼–2½ in. broad, ovate, obtuse or subacute, broadly rounded at the base, coarsely toothed, not rugose, with the tertiary veins distinctly prominent beneath, densely covered with minute adpressed hairs above, shortly and thickly pubescent,

but the hairs not hiding the surface between the veins beneath; panicle 8–10 in. (or more?) long, about 6 in. broad in the male and 4 in. broad in the female plant, compact or the spikes somewhat crowded; male spikes 1–1$\frac{3}{4}$ in. long, $\frac{1}{4}$ in. in diam., with the flower-whorls about 1 lin. apart; bracts very caducous, only a few of the uppermost seen, about $\frac{3}{4}$ lin. long, $\frac{3}{4}$ lin. broad, broadly triangular-ovate, acute, broadly cuneate at the base, puberulous on the back, ciliate with rather long hairs; pedicels minute, scarcely $\frac{1}{8}$ lin. long; calyx $\frac{1}{3}$ lin. long, 3-lobed to the middle or below, minutely puberulous, ciliate; dorsal lobe ovate, obtuse; lateral lobes deltoid-oblong, minutely and acutely 2-toothed at the apex; corolla unequally 5-lobed, minutely and thinly pubescent outside, white; tube $\frac{3}{4}$–$\frac{4}{5}$ lin. long, funnel-shaped, about $\frac{2}{3}$ lin. in diam. at the mouth; lower lobe larger than the others, rather more than $\frac{1}{2}$ lin. long, oblong, very obtuse; stamens 4, subequal, exserted, $\frac{3}{4}$–1 lin. long; ovary aborting; style slightly exserted, minutely bifid at the apex; female spikes $\frac{1}{3}$–$\frac{2}{3}$ in. long, about $\frac{1}{6}$ in. in diam., with crowded flower-whorls; bracts more persistent than in the male, 3 lin. long, 1 lin. broad, broadly cordate, otherwise and as well as the pedicels and calyx as in the male; corolla subequally 5-lobed; tube $\frac{1}{4}$–$\frac{1}{3}$ lin. long, cylindric; lobes $\frac{1}{5}$–$\frac{1}{4}$ lin. long; style much exserted, about twice as long as the corolla-tube; stigmas widely spreading.

KALAHARI REGION: Transvaal; on hill-sides near Barberton, 3000 ft., *Galpin*, 972 ♂ and ♀!

3. **I. Bainesii** (N. E. Br.); a stout herb, probably 2–3 or more ft. high, the panicle only seen; stem subterete, softly pubescent or somewhat tomentose; lower leaves not seen, those on the panicle with a petiole about 1 lin. long and blade 5–10 lin. long, 3$\frac{1}{2}$–5 lin. broad, ovate, acute, cuneate or rounded at the base, coarsely toothed, rather thick, not rugose, densely and somewhat harshly pubescent above, densely tomentose with the hairs quite hiding the surface between the veins beneath; panicle 6–15 in. long, 3$\frac{1}{2}$–6 in. in diam., much branched, rather compact; male spikes 1–2 in. long, 3–3$\frac{1}{2}$ lin. in diam., somewhat lax, with whorls $\frac{2}{3}$–1 lin. apart; female spikes $\frac{1}{4}$–$\frac{1}{2}$ in. long, $\frac{1}{6}$–$\frac{1}{5}$ in. in diam., with densely crowded whorls; bracts about 1 lin. long and nearly as broad in the male, 1$\frac{1}{4}$–1$\frac{1}{3}$ lin. long and 1 lin. broad in the female, triangular-ovate, acute, slightly and broadly cordate at the base in the female, nearly truncate in the male, concave, puberulous, ciliate; pedicels scarcely $\frac{1}{4}$ lin. long; calyx in both sexes about $\frac{1}{3}$ lin. long, 3-lobed to the middle or below, rather densely pubescent; dorsal lobe ovate, obtuse; lateral lobes subquadrate-oblong, minutely 2-toothed or emarginate at the apex; male flowers with the corolla unequally 5-lobed, puberulous outside and with a tuft of woolly hairs near the base of the tube inside; tube 1$\frac{1}{4}$–1$\frac{1}{3}$ lin. long, narrowly funnel-shaped, about $\frac{1}{2}$ lin. in diam. at the mouth; lower lobe about twice as long as the rest, $\frac{3}{4}$ lin. long, oblong, obtuse; stamens 4, slightly

exserted, nearly equal, about $\frac{1}{2}$ lin. long; style slightly exserted, very much thickened and curved or distorted at the apical part, bifid at the apex, with the stigmas applied to each other; female flowers with the corolla very small, scarcely equalling the bracts; tube $\frac{1}{2}$ lin. long, about $\frac{1}{3}$ lin. in diam. at the mouth; lower lobe about $\frac{1}{4}$ lin. long, broadly ovate, otherwise as in the male; stamens none; style much exserted, comparatively rather stout; stigmas about $\frac{1}{2}$ lin. long, rather stout, widely spreading.

KALAHARI REGION : South African Gold Fields, *Baines* !

4. **I. brevispicata** (N. E. Br.); panicle of the male plant (the only part seen) 9–13 in. long, 4–5 in. in diam., somewhat laxly branched, very shortly and densely pubescent, nearly leafless when in flower, the only leaves on the specimen seen being two of the uppermost, $\frac{1}{2}$ in. long, $\frac{1}{3}$ in. broad, ovate, obtuse, rounded into a short petiole, obtusely toothed, not or scarcely rugose above, reticulated with very prominent veins beneath, puberulous on both sides with hairs shorter than the thickness of the veins beneath; flowers crowded into dense spikes $\frac{1}{2}$–$\frac{2}{3}$ in. long and rather more than $\frac{1}{4}$ in. thick ; bracts caducous, $\frac{2}{3}$–$\frac{3}{4}$ lin. long, $\frac{3}{4}$–1 lin. broad, sessile, broadly subcordate-ovate, acute, densely puberulous; pedicels very minute, not more than $\frac{1}{8}$ lin. long ; calyx rather less than $\frac{1}{2}$ lin. long, cup-like, 3-lobed, puberulous, ciliate, covered outside with yellow glands; dorsal lobe broader than long, rounded, obtuse, rather smaller than the shortly and acutely bifid lateral lobes ; corolla unequally 5-lobed, puberulous outside, apparently rosy-purple ; tube $\frac{2}{3}$ lin. long, funnel-shaped, slightly curved ; lower lobe larger than the rest, $\frac{2}{3}$ lin. long, $\frac{1}{2}$ lin. broad, oblong, obtusely rounded at the apex ; stamens 4, exserted ; anthers dark violet ; ovary becoming abortive ; style bifid, with linear stigmas.

KALAHARI REGION : Transvaal ; among rocks, summit of nek on Wonderboom Farm, near Pretoria, *Burtt Davy,* 1844 !

5. **I. Barberæ** (N. E. Br.) ; stems erect, probably tall, terete, very minutely puberulous, soft to the touch ; leaves very imperfectly represented and not more than $\frac{2}{3}$ in. long on the specimens seen shortly petiolate, suborbicular or broadly ovate, coarsely toothed, bullate-rugose and slightly scabrous above, densely tomentose beneath ; male panicle composed of several spike-like branches $1\frac{1}{2}$–4 in. long, $3\frac{1}{2}$–4 lin. in diam., with densely crowded 6-flowered whorls ; female plant not seen ; bracts $1\frac{1}{2}$–$1\frac{2}{3}$ lin. long, $1\frac{1}{3}$ lin. broad, triangular-ovate, acute, very broadly and shortly cuneate at the base, densely pubescent or subtomentose on the back, ciliate with rather long hairs ; pedicels about $\frac{1}{2}$ lin. long ; calyx nearly $\frac{2}{3}$ lin. long, 5-lobed to half-way down, tomentose, long-ciliate on the lobes ; upper lobe broadly deltoid-ovate, obtuse ; 4 lower lobes deltoid, acute ; corolla unequally 5-lobed, tomentose outside ; tube $1\frac{1}{4}$ lin. long, $\frac{3}{4}$ lin. in diam. at the mouth, narrowly funnel-shaped ; lobes

spreading; 4 upper $\frac{1}{3}$–$\frac{1}{2}$ lin. long, ovate, obtuse; lower lobe $\frac{3}{4}$ lin. long, $\frac{1}{2}$ lin. broad, oblong, obtuse; stamens very much exserted, all inserted at about the same height at the mouth of the corolla-tube, subequal; filaments free, glabrous, $1\frac{3}{4}$ lin. long; style filiform, glabrous, exserted; stigma bifid, with erect lobes $\frac{1}{8}$ lin. long; female flowers not seen.

KALAHARI REGION: Orange River Colony; without precise locality, *Mrs. Barber*, 7 !

This differs from all the other South African species in its long dense-flowered male spikes, and 5-lobed calyx.

XIV. MENTHA, Linn.

Calyx tubular or campanulate, 10–13-nerved, 5-toothed, equal or sub-bilabiate. *Corolla-tube* funnel-shaped; limb 4-lobed; lobes subequal or the upper rather broader, entire or emarginate. *Stamens* 4, subequal, erect, distant; filaments free; anthers 2-celled. *Disk* equal, subentire. *Style* shortly 2-fid at the apex. *Nutlets* ovoid, smooth.

Herbs with opposite, usually toothed leaves; flowers small; verticils usually many-flowered, aggregated in spike-like racemes or dispersed and axillary; floral leaves reduced to small bracts; bracteoles usually small, minute or obsolete; stamens dimorphic.

DISTRIB. Species variably estimated by various authors, probably between 20 and 30, distributed throughout the world, but more prevalent in temperate regions.

Leaves sessile (except in *longifolia*, sub-sp. *capensis*, var.
　　salicina); verticils in spike-like racemes :
　　Base of calyx and pedicels hairy　　...　　...　　... (1) **longifolia.**
　　Base of calyx and pedicels glabrous　...　...　... (2) **viridis.**
Leaves petiolate; verticils approximate in a head　... (3) **aquatica.**

1. **M. longifolia** (Huds. Fl. Angl. ed. i. 1762, 221 ; leaves sessile (except in var. *salicina* of subsp. *capensis*) ; and as well as the base of the calyx and pedicels hairy. *M. sylvestris, Linn. Sp. Pl. ed.* ii. (1763) 804.

A very polymorphic plant of which Briquet makes no less than 160 varieties in 21 sub-species.

The name *longifolia*, Huds., is prior to that of *sylvestris*, Linn., and has been adopted by Briquet.

SUBSP. **polyadena** (Briq. in Engl. & Prantl, Pflanzenfam. iv. 3A, 321) ; a herb reaching 20 in. high ; stem densely adpressedly retrorsely pubescent on the angles ; internodes $1\frac{1}{4}$–$1\frac{3}{4}$ in. long ; leaves lanceolate or narrowly ovate-lanceolate, acute, $1\frac{1}{2}$–$2\frac{1}{4}$ in. long, $\frac{1}{4}$–$\frac{1}{2}$ in. broad, green, quite glabrous and very glandular on both sides, margins more or less revolute and with distant triangular teeth, cordiform at the base ; nerves simple, conspicuous below, deeply impressed above ; inflorescence in spike-like racemes 1–$2\frac{1}{2}$ in. long ;

verticils close, very shortly pedunculate ; bracts linear-lanceolate, acute, 1 lin. long, more or less hairy ; pedicels 1 lin. long ; calyx 1¼ lin. long, tubular-campanulate, 10-nerved, densely pubescent ; teeth ⅓–½ lin. long, subulate from a triangular base and separated by a rather wide sinus ; corolla 2 lin. long ; lobes about ¾ lin. long, upper lobe suborbicular, emarginate, the others narrowly elliptic, obtuse. *M. silvestris, subsp. polyadena, Briq. Fragm. Monogr. Lab. in Bull. Soc. Bot. Genèv.* v. 84.

KALAHARI REGION : Transvaal ; near Pretoria, *Wilms,* 1142 ! *Miss Leendertz,* 420 ! along the Aapies River, *Burtt Davy,* 1077 !

SUBSP. capensis (Briq. in Engl. & Prantl, Pflanzenfam. iv. 3A, 321) ; whole plant 16–24 in. high ; stem pubescent, branched above ; branches erect ; leaves subtriangular, acute, broadest and more or less cordate at the base, sessile or subsessile, entire or rarely toothed, reaching 3 in. long, ½ in. broad, densely adpressedly pubescent above, white-tomentose beneath ; nerves impressed on the upper surface rendering it rugose, prominent on the lower ; flowers hermaphrodite, mauve (*Galpin*) ; racemes spike-like, 1–3 in. long ; verticils close above, often more or less distant in the lower part of the raceme ; bracts linear-subulate ; pedicels very short, retrorsely pilose ; calyx ₁₀⁻¹ in. long, hairy with forward-pointing hairs ; teeth ½ lin. long, triangular, hairy ; upper tooth broader than the others, emarginate ; lower teeth subequal, oblong, rounded at the apex ; corolla externally pubescent, about 1 lin. longer than the calyx. *M. capensis, Thunb. Prodr.* 95 ; *subsp. capensis, Briq. Fragm. Monogr. Lab. in Bull. Soc. Bot. Genèv.* v. 75.

COAST REGION : Cape Div. ; kloof on the west slope of Lions Head, *Wolley Dod,* 2337 ! Swellendam Div. ; near Swellendam, *Pappe* ! Uitenhage Div. ; by the Witte River, near Enon, *Drège,* 4766 c ! Albany Div., *Bowker* ! Queenstown Div. ; near Shiloh, *Baur,* 53 ! 946 ! watercourses on the Andriesberg, *Galpin,* 2016 !
CENTRAL REGION : Ceres Div. ; Prince Alfred, *Schlechter,* 9985 ! Somerset Div. ; Somerset East, *Bowker,* 97 ! by streams at the foot of the Bosch Berg, *MacOwan,* 225 ! Graaff Reinet Div. ; Compass Berg, *Shaw* ! Aliwal North Div. ; Witteberg Range, *Drège,* 4766a ! Colesberg, *Shaw* !
WESTERN REGION : Namaqualand, *Wyley* !
KALAHARI REGION : Griqualand West ; Ongeluk, *Burchell,* 2645 !
EASTERN REGION : Griqualand East ; Vaal Bank, *Wood,* 4198 !

VAR. α, salicina (Briq. in Engl. & Prantl, Pflanzenfam. iv. 3A, 321) ; leaves distinctly petiolate, narrow. *Mentha salicina, Benth. Lab.* 170, and in *DC. Prodr.* xii. 168.
CENTRAL REGION : Sutherland Div. ; by the great Riet River, *Burchell,* 1372 !

VAR. β, Cooperi (Briq. mss. in Herb. Kew.) ; leaves sessile ; densely tomentose beneath ; nervation not very prominent beneath ; lower verticils as much as ¾ in. apart ; spike-like racemes reaching 6 in. long.

COAST REGION : Fort Beaufort Div., *Cooper,* 555 !

VAR. γ, obscuriceps (Briq. Fragm. Monog. Lab. in Bull. Herb. Boiss. ii. 695) ; leaves sessile, lanceolate, with straight margins ; nervation simply pinnate, prominent beneath ; marginal teeth small, distant ; verticils close ; calyx-teeth ¼ in. long.

SOUTH AFRICA : without locality, *Drège* !

VAR. δ, **doratophylla** (Briq. Fragm. Monog. Lab. in Bull. Herb. Boiss. ii. 695); leaves sessile, heteromorphous, the upper and middle narrowly long-lanceolate, more or less sword-shaped, with marginal teeth ½ lin. long, the lower leaves sub-triangular, broad at the base, with subentire margins ; nervation somewhat obscure ; verticils congested ; calyx-teeth ½ lin. long.

SOUTH AFRICA : without locality, *Mund & Maire in Berlin Herb. ex Briquet.*

2. **M. viridis** (Linn. Sp. Pl. ed. ii. 804) ; stems erect, quadrangular, glabrous or nearly so ; leaves 1–2¾ in. long, ⅜–1 in. broad, lanceolate or ovate-lanceolate, sessile or subsessile, with serrate margins, glabrous on both sides, strongly nerved beneath, green on both surfaces ; floral leaves bract-like ; inflorescence in spike-like paniculately arranged racemes 2–3 in. long ; verticils many-flowered, scarcely pedunculate, usually closely packed upwards, more or less distant towards the base of the racemes ; pedicels ½ lin. long, glabrous ; bracts reaching 1¼–1½ lin. long, linear-lanceolate, acute, glabrous or nearly so ; calyx 1–1¼ lin. long, tubular, glabrous ; teeth nearly ½ lin. long, subulate from a triangular base ; corolla nearly 2 lin. long ; tube cylindric, about 1 lin. long ; upper lobe suborbicular, scarcely broader than the lower ; other 3 lobes broadly ovate-oblong, obtuse ; stamens much exserted ; nutlets rather more than ¼ lin. long, oblong, rounded at the apex, smooth, dark brown. *M. spicata, var. viridis, Linn. Sp. Pl. ed. i. 576. M. spicata, Huds. Fl. Angl. ed. i. 221.*

COAST REGION: Mossel Bay Div. ; dry channel of an arm of the Gauritz River, *Burchell*, 6457 ! Humansdorp Div. ; between Melk River and Gamtoos River, *Burchell*, 4798 !

Differs from *M. longifolia*, Huds., by its glabrous calyx and pedicels.

Widely distributed through cultivation.

3. **M. aquatica** (Linn. Sp. Pl. ed. ii. 805) ; stems usually erect, branched, more or less villous or hairy, at least on the angles ; leaves very variable in shape, ¾–1½ in. long, broadly ovate or more rarely ovate-lanceolate, sometimes almost orbicular, always petiolate, serrate, hairy on both sides (rarely subglabrous), with rounded or subcordate base ; petioles variable in length, hairy ; inflorescence capitate, subglobose or oblong ; verticils crowded ; pedicels reaching 1¼ lin. long, pubescent or hairy ; bracts up to 1¾ lin. long, linear-lanceolate, acute ; calyx about 2 lin. long, tubular, 13-nerved, hairy ; tube about 1¼ lin. long ; teeth subulate from a deltoid base ; corolla 2½ lin. long, pink or purple ; lobes about as long as the tube, upper lobe oblong, emarginate, other lobes narrower, oblong, obtuse ; stamens much exserted. *Benth. Lab.* 176.

COAST REGION : Cape Div. ; Muizenberg Vley, *Wolley Dod*, 999 ! Paarl Div. ; near the Berg River, *Drège* ! Knysna Div. ; hills near Knysna, *Burchell*, 5450 ! Port Elizabeth Div. ; around Krakakamma, *Burchell*, 4545 ! Queenstown Div. ; upper Zwartkei River, *Galpin*, 2680 !

KALAHARI REGION: Griqualand West; at Griqua Town, *Burchell*, 1930! Upper Campbell, *Burchell*, 1830! Orange River Colony, *Cooper*, 2885! Transvaal, *Mrs. Stainbank*, 3637!
EASTERN REGION: Transkei; Kentani, *Miss Pegler*, 394! Griqualand East; by streams about Clydesdale, *Tyson*, 2881! on the banks of streams, Zuurberg Range, *Tyson*, 1721! Natal; Mohlamba Range, 5000–6000 ft., *Sutherland*! Umzinati River, above the falls, 1000 ft., *Wood*, 1301! and without precise locality, *Gerrard*, 1231! *Cooper*, 2884!

Widely distributed through cultivation.

XV. MICROMERIA, Benth.

Calyx tubular or tubular-campanulate, 13–15-nerved; teeth sub-equal. *Corolla-tube* straight, shorter than the calyx or exserted; limb short, 2-lipped; upper lip erect, nearly flat, entire or emarginate; lower lip spreading, 3-lobed. *Stamens* 4, didynamous (lower pair the longer), ascending, arcuate-connivent at the apex, shorter than the corolla or less frequently exserted and divaricate at the apex; anthers 2-celled; cells distinct, parallel, divergent or divaricate. *Disk* equal or reduced to an anticous gland. *Style* bifid at the apex; lobes equal or the upper smaller than the lower. *Nutlets* ovoid or oblong, smooth.

Herbs or undershrubs; leaves usually small, entire or toothed; whorls of flowers axillary or crowded in a terminal spicate panicle; flowers usually small.

DISTRIB. Species about 60, widely spread in both hemispheres.

Leaves entire, 2–4 lin. long, 1–1½ lin. broad; flowers
   usually in few- to many-flowered cymes; calyx
   tubular   ...   .   (1) **biflora.**

Leaves toothed, 6–8 lin. long, 5–6 lin. broad; flowers
   solitary; calyx tubular-campanulate   ...   ...   (2) **pilosa.**

1. **M. biflora** (Benth. Lab. 378); a small tufted undershrub, pubescent, much-branched; stems ascending, slender, 3–12 in. long, densely leafy; leaves very shortly petiolate, elliptic to ovate-lanceolate, 2–4 lin. long, 1–2½ lin. broad, acute or somewhat obtuse, entire, flat or revolute, covered on the underside with yellowish sessile glands, fragrant; flowers solitary or usually in few- or several-flowered cymes; cymes subsessile or shortly pedunculate; bracts linear-subulate, ¾ lin. long; calyx tubular, 1¾ lin. long, prominently 15-ribbed, covered with fine spreading hairs; teeth triangular-subulate, scarcely ½ lin. long; corolla white, 2½–3 lin. long, pubescent outside; upper lip ¾ lin. long, ¾–1 lin. broad, emarginate; lower lip ¾–1¼ lin. long, 1¼–2 lin. broad; lobes rounded, emarginate. *DC. Prodr.* xii. 220; *Baker in Dyer, Fl. Trop. Afr.* v. 452; *S. Moore in Journ. Bot.* 1903, 406; *Hook. f. Fl. Brit. Ind.* iv. 650. *M. ovata, Benth. Lab.* 377, *and in DC. Prodr.* xii. 219. *Satureia Biflora Briq. in Engl. & Prantl, Pflanzenfam.* iv. 3A, 299.

COAST REGION : Queenstown Div. ; foot of the Winter Berg Mountains, near the source of the Konap River, *Mrs. Barber*, 121 !
KALAHARI REGION : Transvaal ; around Johannesburg, *Rand*, 881.
EASTERN REGION: Tembuland ; St. Augustine, Bazeia and Shawbury, 1800–2000 ft., *Baur*, 220 ! Griqualand East ; Mt. Malowe, near Clydesdale, 4000 ft., *Tyson*, 2129 !

Also in Tropical Africa, Arabia, Afghanistan and India.

2. **M. pilosa** (*Benth. in Benth. et Hook. f. Gen. Plant.* ii. 1188) ; a pilose herb ; stems weak, prostrate, sparingly branched, 1–1½ ft. long ; leaves shortly petiolate, ovate, 6–8 lin. long, 5–6 lin. broad, obtuse or somewhat acute, few-toothed, long-pilose on both sides especially on the nerves, gland-dotted on the underside ; petiole about 1 lin. long ; flowers solitary in the axils of the upper leaves ; peduncles up to 3½ lin. long, slender, bibracteate about the middle ; calyx tubular-campanulate, 1¾ lin. long, 15-nerved (5 of the nerves more distinct than the others), gland-dotted, pilose ; teeth deltoid to triangular-lanceolate, ⅔ lin. long, acute or subacute ; corolla 4–5 lin. long, pilose inside on the lower side ; upper lip 1½ lin. long and broad, emarginate ; lower lip 2¾ lin. long and broad ; lobes rounded, the median larger than the lateral. *Oliv. in Hook. Ic. Pl. t.* 1522.

KALAHARI REGION : Orange River Colony, *Cooper*, 2903 !
EASTERN REGION : Pondoland ; Fakus Territory, *Sutherland* ! Natal ; hills above Byrne, *Wood*, 3172 !

XVI. SALVIA, Linn.

Calyx ovoid, tubular or campanulate, 2-lipped ; upper lip entire or 3-toothed ; lower lip bifid. *Corolla-tube* included or exserted, equal, ventricose or enlarged above, naked or annular-pilose inside ; limb 2-lipped ; upper lip erect or falcate, usually concave, entire or emarginate ; lower lip spreading, 3-lobed. *Stamens* 2, anticous, arcuate ; filaments short ; connective jointed to the filament, linear, elongated, the upper part ascending and bearing an oblong or linear perfect anther-cell, the lower deflexed or horizontal, bearing a smaller polliniferous or empty anther-cell, or quite naked ; staminodes 2, posticous, very small, or wanting. *Disk* usually more prominent on the lower side. *Style* shortly 2-fid at the apex. *Nutlets* ovoid-triquetrous or somewhat compressed, smooth.

Herbs, undershrubs or shrubs ; leaves entire, toothed or more or less deeply lobed ; bracts small or large, rarely similar to the upper leaves ; whorls 2- to many-flowered, in spikes, racemes or panicles ; flowers variously coloured, large and showy or sometimes small and inconspicuous.

DISTRIB. Species about 700, widely dispersed in the temperate and tropical regions of both hemispheres.

S. acetabulosa, Linn., *S. Æthiopis*, Linn., *S. pratensis*, Linn., *S. Sclarea*, Linn., and *S. verticillata*, Linn., are enumerated in *Burm. f. Fl. Cap. Prodr.* 1, but we have seen no South African material of them. They are natives of Europe, some of them extending into North Africa and the Orient. *S. Æthiopis* is described as growing on sand dunes around Cape Town by Thunberg (*Fl. Cap. ed. Schult.* 452) ;

x 2

it is a stout herb covered with a white wool, having cordate-amplexicaul stem-leaves, large spinescent bracts and yellow flowers.

S. coccinea, Juss. ex *Murr. Comm. Gotting.* i. (1778) 86, t. 1, and *Linn. f. Suppl.* 88, is a native of Tropical America, and has been introduced into many warm countries. Specimens collected by Mund in South Africa and by Burtt-Davy (no. 2724) from Swaziland, "an escape," are in the Kew Herbarium ; it has a herbaceous canescent-puberulous stem, ovate petiolate acute crenate leaves, a striate calyx with short broad acute teeth, and a bright red long-exserted corolla.

S. pseudococcinea, Jacq. (*S. coccinea*, var. *pseudococcinea*, Gray), also a Tropical American plant, and differing from *S. coccinea* chiefly by having conspicuously hirsute stems, is recorded from Natal (*Guerke in Ann. Hofmus. Wien*, xx. 47), and recently Burtt-Davy has sent a specimen (no. 3005) from Bremmersdorp, Swaziland, collected near an old house.

S. patens, Cav., a Mexican species with tuberous roots, ovate-deltoid leaves often hastate at the base, and very large bright blue flowers, has been collected in Natal and sent to Kew by Mr. A. Hislop. He believed it to be wild in Natal.

Horminum foliosum, Burm. f. Fl. Cap. Prodr. 16 (without description), is unknown.

```
*Shrubs ; calyx usually much accrescent ; corolla-tube
    annular-pilose :
  †Bracts persistent :
    ‡Branches and leaves pubescent but not canescent :
      Leaves ovate or ovate-deltoid, rounded to
        cordate at the base ; calyx distinctly
        2-lipped ; upper lip minutely 3-toothed ;
        lower lip with 2 ovate acuminate lobes :
      Leaf-blade 6–12 lin. long ; upper lip of calyx
        truncate, recurved :
      Leaf-blade 6–8 lin. long ; branches glandular-
        pubescent ; upper lip of corolla 7–8 lin.
        long    ...    ...    ...    ...    ...   (1) garipensis.
      Leaf-blade 12 lin. long ; branches curled-
        pilose ; upper lip of corolla 4–4½ lin.
        long    ...    ...    ...    ...    ...   (2) Dinteri.
      Leaf-blade 2½–5 lin. long ; upper lip of calyx
        not truncate nor recurved    ...    ...   (3) Steingroeveri.
      Leaves obovate-oblong, narrowed to the base ;
        calyx less distinctly 2-lipped, with 4 short
        broad apiculate lobes    ...    ...    ...  (10) undulata.
    ‡‡Branches and leaves canescent :
      Corolla golden-yellow or ferruginous, 1¼–2 in.
        long ; upper lip often 8–15 lin. long   ...  (5) aurea.
      Corolla purple, violet or blue, 8–12 lin. long ;
        upper lip 4–6 lin. long :
      Calyx densely covered with long hairs ; lobes
        apiculate or acute ...   ...    ...    ...  (8) africana.
      Calyx shortly hairy ; lobes rounded ...   ...  (9) dentata.
  ††Bracts soon deciduous :
    Calyx distinctly accrescent ; lobes broad and
      rounded :
    Leaves orbicular, 1½–1¾ in. broad, thin   ...  (4) eckloniana.
    Leaves oblong-lanceolate to ovate, up to 7 lin.
      broad, rather thick :
    Corolla about 1½ in. long ; upper lip 8–9 lin.
      long...    ...    ...    ...    ...    ...  (6) nivea.
```

Corolla about 1 in. long ; upper lip about
   6 lin. long ...   . **(7) hastæfolia.**
Calyx scarcely accrescent ; lobes of the lower lip
   acuminate :
   Plant viscid ; calyx moderately hairy ; hairs
      rather short ...   . **(11) paniculata.**
   Plant not viscid ; calyx densely hairy ; hairs
      long, whitish ...   . **(12) albicaulis.**
**Herbs ; calyx not or only slightly accrescent ; upper
   lip with 3 very short connivent teeth ; corolla-tube
   not annular-pilose :
   Corolla 8–10 lin. long :
   Branches and leaves softly pubescent ...   ... **(13) rugosa.**
   Branches and underside of the leaves shortly
      whitish-tomentose   ...   . **(14) Radula.**
   Corolla 2–6 lin. long :
   Leaves toothed or sometimes shortly lobed ; calyx
      not densely hairy at the mouth   ...   ... **(15) disermas.**
   Leaves often deeply lobed ; calyx densely white-
      hairy at the mouth ...   . **(16) clandestina,** var.
***Herbs or rarely shrubs ; calyx not or only slightly   **angustifolia.**
   accrescent ; upper lip finally more or less truncate,
   with 3 usually small acute teeth having broad
   sinuses between them; corolla annular-pilose or not :
   †Leaves with an ovate-triangular, ovate-suborbicular,
      elliptic or ovate blade or terminal lobe, some-
      times auriculate at the base, sometimes lyrate ;
      lateral lobes when present few, usually 1 or 2
      pairs, very much smaller than the terminal
      lobe ; corolla-tube usually narrow in proportion
      to its length, often much exserted, not annular-
      pilose (rather broad and imperfectly annular-
      pilose in *S. Pegleræ*) :
   Branches and leaves glabrous or with a few rather
      long hairs chiefly on the petioles   ...   ... **(23) obtusata.**
   Branches (and usually the leaves) more or less
      covered with hairs, or in *S. lasiostachys* the
      branches are scabrous with the callous bases
      of hairs :
   Whorls very densely flowered (up to 24-flowered
      or more) ; terminal lobe of the leaves
      usually distinctly triangular and acute   ... **(17) Tysonii.**
   Whorls few- to about 10-flowered ; leaf-blade or
      terminal lobe often rounded, sometimes
      ovate, rarely acute :
      Corolla 12–14 lin. long, more than twice
         as long as the calyx   ...   . **(18) scabra.**
      Corolla 5–9 lin. long, usually not more than
         twice as long as the calyx :
      Branches scabrous with the callous bases of
         scattered hairs ...   . **(24) lasiostachys.**
      Branches scarcely scabrous, often more or
         less covered with short soft hairs :
         Corolla-tube very slender, usually only
            about $\frac{1}{2}$ lin. broad at the middle,
            not ventricose nor annular-pilose :
            Leaves with an undivided blade or
              terminal lobe up to $2\frac{1}{2}$ in. long
              and 2 in. broad ...   . **(20) aurita.**

Leaves with an undivided blade or
terminal lobe up to 1¼ in. long
and broad :
Leaves (excluding petiole) usually
more than 1½ in. long, all
divided, usually with 2 pairs of
rather large lateral lobes and a
larger terminal lobe ... ... (19) **Galpinii.**

Leaves (excluding petiole) usually
less than 1¼ in. long, undivided
or with 1–3 small lobes at the
base of a much larger terminal
lobe :
Plant puberulous ; leaves usually
with a terminal lobe and 1–3
smaller lateral lobes ; petiole
1–4 lin. long ... ... (21) **pallidifolia.**

Plant rather densely covered with
soft fine relatively long hairs ;
leaves often undivided ; petiole
up to 1 in. long ... ... (22) **triangularis.**

Corolla-tube rather broad, 1 lin. broad or
more at the middle, distinctly
ventricose, imperfectly annular-
pilose ... ... (33) **Pegleræ.**

††Leaves more or less oblong or obovate-oblong, rarely
elliptic, undivided or often deeply and many-
lobed, sometimes only few-lobed at the base ;
lateral lobes often narrow but usually large in
proportion to the terminal lobe ; corolla-tube
usually broad in proportion to its length (except
S. namaensis, which has a rather long very
narrow tube) included or exserted, annular-
pilose :
Pedicels when mature often 2½ lin. long ; corolla-
tube very narrow, much exserted ; stamens
much exserted ... ... (25) **namaensis.**

Pedicels when mature usually much less than
2 lin. long ; corolla-tube broad, or if narrow the corolla is very small ; stamens not or
only slightly exserted :
Stem and leaves densely covered with a short
somewhat velvety pubescence ; upper leaves
not lobed or only at the base :
Leaves usually oblong, up to ¾ in. broad, the
lower cauline with several lobes ; petiole
short or none ... ... (32) **raphanifolia.**

Leaves usually elliptic up to 1⅓ in. broad,
undivided or sometimes slightly lobed at
the base ; petiole up to 11 lin. long ... (22) **Pegleræ.**

Stem and leaves variously hairy, but the hairs
are usually much stiffer, sometimes shortly
whitish-tomentose (when the leaves are
much dissected), rarely nearly glabrous :
‡Corolla 2¾–4½ lin. long :
Stem woody at the base ; leaves up to
1 in. long ... ... (26) **Burchelli.**

Stem herbaceous ; mature leaves often 2-3
　in. long or more :
Branches sprinkled with minute stiff
　hooked hairs or sometimes almost
　glabrous ; leaves usually somewhat
　regularly toothed or lobed, narrow
　in proportion to their length, often
　only 2-4 lin. broad ... 　... 　... (27) **stenophylla.**

Branches rather densely covered with
　short fine hairs or with rather long
　coarse spreading or curled hairs ;
　leaves usually broadly and irregularly
　lobed, rather broad in proportion to
　their length, often more than 1 in.
　broad :
Leaves often deeply lobed, but usually
　not so far as the midrib ; lobes
　often triangular and acute ; corolla
　up to 4½ lin. long... 　... 　... (28) **runcinata.**

Leaves often lobed as far as the midrib ;
　lobes usually oblong and rounded
　or obtuse ; corolla up to 3 lin.
　long... 　... 　... 　... 　... (29) **sisymbrifolia.**

‡‡Corolla 5-7 lin. long :
Stem usually very much elongated, up to
　2 ft. high or more, usually unbranched (30) **repens.**

Stem often only about 1 ft. high or less,
　more or less branched :
Most of the leaves much lobed, often with
　long narrow lobes 　... 　... 　... (31) **monticola.**

Most of the leaves undivided, or with
　few rather short broad lobes :
Upper leaves less than twice as long
　as broad 　... 　... 　... 　... (34) **rudis.**

Upper leaves more than twice as long
　as broad :
Corolla about 5 lin. long :
　Middle stem-leaves 1¾-2½ in. long,
　　6-9 lin. broad, mostly with
　　very small teeth, rarely lobed (35) **Woodii.**

Middle stem-leaves usually 1-1½ in.
　long, 3-7 lin. broad, often
　coarsely toothed or lobed, if
　undivided they are only 3-4
　lin. broad 　... 　... 　... (28) **runcinata,** var.
　　　　　　　　　　　　　　　　　　　　　　grandiflora.

Corolla 6-7 lin. long 　... 　... (36) **Cooperi.**

1. **S. garipensis** (E. Meyer, Comm. 232) ; a softly glandular-
pubescent branched undershrub ; leaves rather thick, ovate, 6-8 lin.
long, 5-7 lin. broad, obtuse, rounded, truncate or cordate at the
base, irregularly crenate, rugose ; petiole 2-4 lin. long ; whorls
2-flowered, forming loose axillary or terminal unilateral spikes ;
flowers subsessile ; bracts ovate, 2-3 lin. long, acute ; calyx campanulate, deeply bilabiate, 5 lin. long, 4-5 lin. broad at the apex,

6 lin. long and 7–9 lin. broad when mature, villous, glandular, somewhat transparent, distinctly 13-nerved ; upper lip suborbicular, recurved, truncate, very minutely 3-toothed ; lower lip bifid ; lobes obliquely and narrowly ovate, 2 lin. long, shortly acuminate ; corolla 1 in. long, shortly glandular-pubescent outside ; tube funnel-shaped, 4 lin. long ; upper lip falcate, 7–8 lin. long, compressed, shortly and broadly 2-lobed ; lower lip deeply 3-lobed, about 4½ lin. long ; median lobe about 5 lin. broad, deeply concave ; lateral lobes falcate-oblong, about 1⅓ lin. broad. *Drège, Zwei Pflanzengeogr. Documente,* 93. *S. gariepensis, Benth. in DC. Prodr.* xii. 273.

WESTERN REGION : Little Namaqualand ; between Verleptpram and the mouth of the Orange River, under 1000 ft., *Drège*, 3112 !

2. **S. Dinteri** (Briq. in Bull. Herb.ʼ Boiss. 2ᵐᵉ sér. iii. 1075) ; an undershrub with erect curled-pilose somewhat twiggy branches ; leaves broadly ovate-deltoid, obtuse, broadly truncate-cordate at the base, somewhat lobed, erose-crenulate, bullate-rugose, dark green on both sides, sparingly pilose ; blade 1 in. long and broad ; petiole curled pilose, 5–7½ lin. long ; sinus between the obtuse auricles very large ; whorls 2-flowered, approximate, in a dorsiventral spike about 3½ in. long ; rhachis spreading-pilose, glandular ; bracts broadly ovate, shorter than the calyx ; calyx sessile or subsessile, campanulate, 3½–4 lin. long, 8 lin. long when mature, clothed with spreading somewhat curled hairs mixed with glands ; lips membranous-dilated, veined, scarcely coloured ; upper lip broadly ovate, recurved, subtruncate, 3-toothed, up to 4 lin. long ; teeth ¼ lin. long, acute, approximate ; lower lip 2-lobed ; lobes ovate, 4 lin. long, 3 lin. broad, subacute ; corolla-tube included ; upper lip narrowly compressed-falcate, oblong, 4–4½ lin. long ; lower lip 4–4½ lin. long ; median lobe obcordate ; lateral lobes smaller, rounded.

WESTERN REGION : Great Namaqualand ; Gubub, *Dinter*, 1111.

3. **S. Steingroeveri** (Briq. in Engl. Jahrb. xix. 191) ; a shrub ; older branches naked, white, the younger curled-pubescent ; internodes short ; leaves petiolate, cordate-ovate, 2½–5 lin. long and broad, obtuse at the apex, cordate at the base, irregularly incised-crenate, thick, pubescent above, tomentose beneath, net-veined ; whorls 2-flowered, secund, in short spikes ; bracts oblong or ovate, 2–2½ lin. long, 1–1½ lin. broad, membranous, entire ; calyx sessile, broadly campanulate, after flowering accrescent and 7½ lin. long ; tube 4 lin. long, enlarged towards the mouth ; lips ovate, membranous, veined, sparingly pilose, green ; upper lip subentire, not recurved ; lower lip with 2 entire lobes 2½ lin. long ; corolla exceeding the calyx by 7½ lin. ; tube included, annular-pilose inside, broad ; upper lip falcate, 5 lin. long ; lower lip with an oblong median lobe 4 lin. long ; lateral lobes rounded, spreading, 1–1½ lin. long ; style long-exserted.

WESTERN REGION : Great Namaqualand ; near Aus, *Steingroever*, 55.

4. S. eckloniana (Benth. in DC. Prodr. xii. 273); a shrub, shortly pubescent and glandular on the branches and leaves; leaves papery, orbicular, 1½–1¾ in. in diam., erose-dentate or shortly lobed, subcordate at the base; petiole ¼–½ in. long; whorls 2-flowered, approximate, forming a short terminal raceme; calyx broadly campanulate, much dilated at the apex, 7–8 lin. long, 7–8 lin. broad at the apex, villous and gland-dotted outside, shortly pubescent inside, irregularly veined, shortly 2-lipped; upper lip very broad, slightly retuse or undulate; lower lip shortly and broadly 2-lobed; corolla 1¼ in. long, minutely glandular-puberulous above outside, with a dense ring of long hairs near the base of the tube inside; tube 7 lin. long, about 5 lin. broad at the apex; upper lip subfalcate, 11 lin. long, compressed; lower lip 6 lin. long, 8 lin. broad; median lobe broadly obovate, 4½ lin. long, 5 lin. broad; lateral lobes rounded, slightly smaller. *S. rotundifolia, Benth. l.c.*

COAST REGION: Clanwilliam, *Ecklon*!

5. S. aurea (Linn. Sp. Pl. ed. ii. 38); a much-branched shrub, 2–4 ft. high or more, densely covered with a short white tomentum, minutely gland-dotted especially on the leaves; branches short, spreading; leaves suborbicular, elliptic-lanceolate or narrowly obovate, 5–18 lin. long, 2½–10 lin. broad, rounded, obtuse or rarely acute, more or less narrowed at the base, entire or rarely crenulate; petiole 1–5 (usually 2–3) lin. long; whorls 2-flowered, rather crowded, forming terminal racemes 2–3 in. long; calyx broadly campanulate, 6–8 lin. long and broad, 8–10 lin. long and broad and membranous after flowering, veined, purplish, villous; upper lip broad, undulate or emarginate; lower lip shortly and broadly 2-lobed; corolla 1¼–2 in. long, ferruginous or golden-yellow, glandular-pubescent outside, with a broad band of woolly hairs at the base of the tube inside; tube 7 lin. long, 4–5 lin. broad in the upper part; upper lip subfalcate, 8–15 lin. long, emarginate; lower lip 7 lin. long, 8 lin. broad, pilose at the base inside; median lobe broadly obovate, 4 lin. long, 6½ lin. broad; lateral lobes about 2 lin. long and 3 lin. broad, rounded. *Linn. Mant.* ii. 319; *Curt. Bot. Mag. t.* 182; *Thunb. Prodr.* 96, *and Fl. Cap. ed. Schult.* 448; *Pers. Syn.* i. 28; *Vahl, Enum.* i. 231; *Roem. et Schultes, Syst.* i. 217, *and Mant.* i. 181; *A. Dietr. Sp. Pl.* i. 264; *Benth. Lab.* 216, *in E. Meyer, Comm.* 233, *and in DC. Prodr.* xii. 273. *S. colorata, Linn. Syst. Nat. ed.* 12, ii. 66? *not of Vahl. S. africana lutea, Linn. Sp. Pl. ed.* i. 26. *S. africana fruticans, etc., Commel. Hort. Amst.* ii. 183, *t.* 92.

SOUTH AFRICA: without locality, *Grey*! *Villett*! *Pappe*! & cultivated specimens!
COAST REGION: Clanwilliam Div.? Bull Hoek, 800 ft., *Schlechter*, 8376! Malmesbury Div.; Groene Kloof, *Drège*, 1341a! Laauws Kloof, *Drège*, 1341b! Tulbagh Div.; Tulbagh, about 650 ft., *Schlechter*, 1396! Cape Div.; Table Mountain, *Cooper*, 2890! near Blockhouse, *Wolley Dod*, 491! Camps Bay, *MacOwan & Bolus, Herb. Norm. Austr.-Afr.* 820! *Prior*! Lion Mountain, *Drège*! Bredasdorp Div.; Strand Veld, between Cape Agulhas and Pottsberg, *Drège*,

314 LABIATÆ (Skan). [*Salvia.*

1341c! Mossel Bay Div.; Mossel Bay, under 500 ft., *Drège*, 1341d! Knysna Div.; between Knysna River Ford and Goukamma River, *Burchell*, 5556! Uitenhage Div., *Zeyher*, 144! Port Elizabeth Div.; along the Baakens River near Port Elizabeth, *Burchell*, 4344! Algoa Bay, *Forbes*! Bathurst Div.; sea coast, *Mrs. Hutton*! mouth of the Fish River, *MacOwan*, 419! Port Alfred, *Miss Sole*, 460!

6. S. nivea (Thunb. Prodr. 96); a branched erect shrub, 3 ft. high, more or less covered with short grey tomentum; leaves petiolate or the upper subsessile, thick, subcoriaceous, sometimes slightly rugose, oblong-lanceolate, lanceolate or sometimes ovate, $\frac{1}{2}$–$1\frac{1}{2}$ in. long, 2–7 lin. broad, acute, entire or the lower auriculate or dentate; petiole 1–7 lin. long; whorls 2-flowered, in short usually branched terminal racemes; calyx broadly campanulate, 7–8 lin. long, accrescent, membranous after flowering, veined, purplish, villous; upper lip broad, undulate; lower lip shortly and broadly 2-lobed; corolla about $1\frac{1}{2}$ in. long, purple, sparingly pubescent outside; tube 10–11 lin. long, $2\frac{3}{4}$–4 lin. broad above; upper lip 8–9 lin. long, slightly falcate, compressed, emarginate; lower lip about 5 lin. long and broad, flat; median lobe 3 lin. long and broad; lateral lobes much smaller, rounded. *Thunb. Fl. Cap. ed. Schult.* 450; *Vahl, Enum.* i. 231; *Roem. et Schultes, Syst.* i. 238, *and Mant.* i. 200; *A. Dietr. Sp. Pl.* i. 306; *Benth. Lab.* 218, *in E. Meyer, Comm.* 233, *and in DC. Prodr.* xii. 273. *S. lanceolata, Lam. Ill.* i. 72; *Poir. Encycl.* vi. 591. *S. diversifolia, Benth. in DC. Prodr.* xii. 274, *in syn. S. nitida, Drège, Zwei Pflanzengeogr. Documente,* 103.

SOUTH AFRICA: without locality, *Ecklon*!
COAST REGION: Vanrhynsdorp Div.; Ebenezer, *Drège*, 1340d! Clanwilliam Div.; between Berg Vallei and Lange Vallei, *Drège*, 1340c! Malmesbury Div.; Zwartland and Groenekloof, *Thunberg*! Riebecks Castle, *Drège*, 1340b. Tulbagh Div.; Vogel Vallei, under 1000 ft., *Drège*, 1340a! Cape Div.; near Cape Town, *Bolus*, 4765! Cape Flats, *Zeyher, Burke*! Vygeskraal Farm, *Wolley Dod*, 652!

7. S. hastæfolia (Benth. in E. Meyer, Comm. 233); an erect branched shrub, sparingly viscid pulverulent-puberulous on the stem and branches; leaves rather thick, subcoriaceous, ovate-lanceolate, $\frac{1}{2}$–1 in. long, 2–6 lin. broad, acute, cuneate at the base, dentate, often hastate-auriculate at the base, pulverulent-glandular, shortly and stiffly pubescent; petiole 2–5 lin. long; whorls usually 2-flowered, distant, in short branched terminal racemes; calyx broadly campanulate, 7 lin. long, accrescent, membranous after flowering, veined, purplish, villous; upper lip broad, undulate; lower lip shortly 2-lobed; lobes broadly ovate, acute; corolla about 1 in. long, shortly pubescent outside; tube $7\frac{1}{2}$ lin. long, $2\frac{1}{4}$ lin. broad above; upper lip about 6 lin. long, slightly falcate, compressed; lower lip 3 lin. long, $2\frac{1}{4}$ lin. broad; median lobe obovate, $1\frac{1}{4}$ lin. long, $1\frac{1}{2}$ lin. broad, emarginate; lateral lobes smaller, rounded. *DC. Prodr.* xii. 274.

COAST REGION: Clanwilliam Div.; between Clanwilliam and Bosch Kloof, 1000–2000 ft., *Drège*, 7934!

8. S. africana (Linn. Sp. Pl. ed. ii. 38, excl. syn. Plukenet.); a much-branched shrub; branches shortly grey-pubescent; leaves subsessile or petiolate, coriaceous, ovate-lanceolate, oblong, obovate or sometimes lanceolate, 1–1¼ in. long, 2–8 lin. broad, acute, apiculate, obtuse or rounded, often cuneate at the base, entire, crenate-dentate or sometimes coarsely lobed, often rugose, densely grey-tomentose beneath, strigillose to glabrous above; whorls 2- to many-flowered, usually distant, in terminal rarely branched racemes; bracts broadly ovate to suborbicular; calyx broadly campanulate, 4–5 lin. long, accrescent, veined, purplish, villous; lips short and broad; upper lip very shortly and broadly 3-toothed; lower lip broadly 2-lobed; teeth and lobes all apiculate or acute; corolla purple, violet or blue, 8–12 lin. long; tube 4-6 lin. long, 2¾–3½ lin. broad at the apex, annular-pilose inside; upper lip 4–6 lin. long, shortly 2-lobed; lower lip 4–5 lin. long, 5½–6½ lin. broad; median lobe obreniform, 5½–6½ lin. broad; lateral lobes small, rounded. *Thunb. Prodr. 96, and Fl. Cap. ed. Schult.* 449; *Vahl, Enum.* i. 230; *Roem. et Schultes, Syst.* i. 237, *and Mant.* i. 200; *A. Dietr. Sp. Pl.* i. 308; *Benth. Lab.* 216, *in E. Meyer, Comm.* 234, *and in DC. Prodr.* xii. 274. *S. africana cœrulea, Linn. Sp. Pl. ed.* i. 26. *S. africana frutescens, etc., Commel. Hort. Amst.* ii. 181, *t.* 91. *S. rotundifolia, Salisb. Prodr.* 74. *S. colorata, Vahl, Enum.* i. 230. *S. barbata, Lam. Ill.* i. 72. *S. integerrima, Mill. Gard. Dict. ed.* viii. *n.* 12. *S. foliis oblongo-ovatis integerrimis, Mill. Ic.* ii. 150, *t.* 225, *fig.* 2. *S. foliis subrotundis serratis, etc., Mill. Ic.* ii. 150, *t.* 225, *fig.* 1.

SOUTH AFRICA : without locality, *Ecklon*! *Forbes*! *Forster*! *Mund*! *Pappe*! *Thom*, 545! *Thunberg*!

COAST REGION : Malmesbury Div. ; Zwartland, *Zeyher*! Riebecks Castle, *Drège*, 7939c! Tulbagh Div. ; at the foot of Mosterts Berg, near Mitchells Pass, 1000 ft., *MacOwan, Herb. Austr.-Afr.* 1640! Mitchells Pass, 1200 ft., *Schlechter*, 8958! Piquetberg Road, *Tyson*, 2327! Worcester Div. ; near Brand Vlei, *Bolus*, 5223 *partly*! Paarl Div. ; between Paarl and Lady Grey Railway Bridge, *Drège*, 7938! Cape Div. ; various localities on the Cape Peninsula, *Bolus*, 2894! 2895! *Burchell*, 73! 311! 770! *Cooper*, 2892! *Ecklon*, 716! *Drège*! *MacGillivray*, 566! *Milne*, 172! *Wolley Dod*, 627! 628! 736! *Prior*! *Mrs. C. Southey in Herb. Galpin*, 7850! Stellenbosch Div. ; between Stellenbosch and Somerset West, *Drège*, 7939b! Caledon Div. ; Zwart Berg, about 1000 ft., *Galpin*, 4424!

The leaves are very variable in shape, even on the same plant. A form in which they are broadly obovate and rounded at the apex, as in Drège 7938, has been distinguished as var. *obtusa*, Benth. in E. Meyer, Comm. 234.

S. subspathulata, Lehm. in E. Otto, Hamb. Gartenz. vi. 457, from the description appears to be the same as *S. africana*.

9. S. dentata (Ait. Hort. Kew. ed. 1, i. 37); a much-branched shrub; branches minutely grey-pubescent, often becoming glabrous; leaves usually crowded, petiolate or the upper subsessile, spathulate or obovate, sometimes lanceolate or linear, rounded or sometimes subacute, crenate-dentate to pinnatifid, rarely quite entire, often undulate-crisped, shortly grey-tomentose; blade 2–6 lin. long, 1–4 lin. broad; petiole 1–6 lin. long; whorls 2-6-flowered, close together or rather distant, in short terminal rarely branched racemes; bracts

ovate or suborbicular ; calyx broadly funnel-shaped, 5–7 lin. long,
shortly hispid ; upper lip shortly and broadly 3-lobed ; lower lip
broadly 2-lobed ; lobes all rounded ; corolla blue, 9–12 lin. long ;
tube 6½ lin. long, 1¼ lin. broad at the base, 3 lin. broad at the apex,
annular-pilose inside ; upper lip 5–6 lin. long, shortly and broadly
2-lobed ; lower lip 6–7 lin. long ; median lobe transversely oblong,
6½–7 lin. broad, deeply emarginate ; lateral lobes 1½–1¾ lin. broad,
rounded. *Vahl, Enum.* i. 232 ; *Roem. et Schult. Syst.* i. 217, *and*
Mant. i. 181 ; *A. Dietr. Sp. Pl.* i. 264 ; *Benth. Lab.* 217, *and in*
DC. Prodr. xii. 275. *S. angustifolia, Salisb. Prodr.* 73. *S. rigida,*
Thunb. Prodr. 96, *and Fl. Cap. ed. Schult.* 451. *S. crispula, Benth.*
in E. Meyer, Comm. 234, *and in DC. Prodr.* xii. 274.

SOUTH AFRICA : without locality, *Masson* !
CENTRAL REGION : Calvinia Div. ; Bokkeland, *Thunberg* !
WESTERN REGION : Little Namaqualand ; Modder Fontein, *Whitehead* ! Mount
Spektakel, *Morris in Herb. Bolus,* 5777 ! Uitkomst, 2000–3000 ft., *Drège,* 3113a !
between Pedros Kloof and Lily Fontein, 3000–4000 ft., *Drège,* 3113b ! near Klip
Fontein, *MacOwan & Bolus, Herb. Norm. Austr.-Afr.* 681 ! Brackdamm, 2000 ft.,
Schlechter, 11161 ! and without precise locality, *Wyley* ! Vanrhynsdorp Div. ;
Karee Bergen, 2000 ft., *Schlechter,* 8247 !

Drège's specimen, 4742, from the Cederberg Range, Clanwilliam Division, cited
by Bentham in *E. Meyer, Comm.* 235, and in *DC. Prodr.* xii. 275, has longer
leaves than the type, and the teeth, when present, are more acute. It has no
flowers.

10. **S. undulata** (Benth. in DC. Prodr. xii. 275) ; a branched
shrub ; branches shortly villous ; leaves petiolate to subsessile,
obovate-oblong, acute, erose-dentate or lobed at the base, somewhat
rugose, green and almost glabrous above, shortly greyish-pubescent
beneath ; blade ¾–1¼ in. long, 3½–4½ lin. broad ; petiole up to ½ in.
long ; whorls 2–6-flowered, close together in short simple terminal
racemes ; bracts broadly ovate ; calyx campanulate, 5 lin. long,
shortly villous ; lips each 2-lobed ; lobes short and broad, apiculate ;
corolla 11 lin. long ; tube 4 lin. long, 1⅔ lin. broad at the base,
3¼ lin. broad at the apex, annular-pilose inside ; upper lip 7 lin.
long, shortly 2-lobed ; lower lip 6 lin. long, inflated at the base ;
median lobe 3 lin. long, 7 lin. broad, deeply emarginate ; lateral
lobes rounded, 1¼ lin. long, 1¾ lin. broad.

COAST REGION : Clanwilliam Div., *Ecklon* !

11. **S. paniculata** (Linn. Mant. i. 25 and ii. 511) ; an erect much-
branched shrub, 4 ft. high ; branches nearly erect, scabrous,
sometimes sparingly pilose especially at the nodes ; leaves coriaceous,
obovate-cuneate, sometimes obovate-lanceolate, ¼–1¼ in. long, apiculate or cuspidate, entire or toothed, prominently net-veined, usually
green, very shortly adpressed-pubescent or nearly glabrous, slightly
scabrous, densely gland-dotted ; petiole 1–2 lin. long ; whorls
2-flowered, distinct, in long much-branched terminal panicles ; calyx
campanulate, 4 lin. long, setulose, prominently 4-keeled, 5-nerved ;
upper lip minutely 2- or 3-toothed ; lower lip 2-lobed ; lobes deltoid,

1¼–1½ lin. long, shortly acuminate; corolla blue, 9–12 lin. long; tube 4 lin. long, 1 lin. broad at the base, 2½ lin. broad at the apex, annular-pilose inside; upper lip 5½–7 lin. long, shortly 2-lobed; lower lip 5–6 lin. long, crumpled; median lobe transversely oblong or rounded, 3–4½ lin. broad, concave, emarginate; lateral lobes smaller, rounded. *Thunb. Prodr.* 96, *and Fl. Cap. ed. Schult.* 450; *Vahl, Enum.* i. 229; *A. Dietr. Sp. Pl.* i. 308; *Roem. et Schult. Syst.* i. 237, *and Mant.* i. 200; *Benth. Lab.* 217, *in E. Meyer, Comm.* 235, *and in DC. Prodr.* xii. 275; *Hook. f. Bot. Mag. t.* 6790. *S. Chamelæagnea, Berg. Descr. Pl. Cap.* 3. *S. Chamelæagnus, Burm. f. Fl. Cap. Prodr.* 1. *S. minor, etc., Breyne, Exot. Pl. Cent.* i. 169, *t.* 85.

SOUTH AFRICA : without locality, *Miller*! *Thunberg*! *Zeyher*! & *cultivated specimens*!
COAST REGION : Clanwilliam Div. ; between Pakhuis and Biedouw, 2000–3000 ft., *Drège*! Cederberg Range, *Drège*, 3115! Ceres Div. ; near Ceres, *MacOwan & Bolus, Herb. Norm. Austr.-Afr.* 490! Worcester Div. ; near Worcester, *Drège*, 7936! Paarl Div. ; Paarl, *Burchell*, 952! *Prior* ! Dal Josaphat, Drakenstein Mountains, 600 ft., *Tyson*, 2441! Cape Div. ; near Cape Town, *Harvey* ! eastern side of Table Mountain, *Ecklon*, 719 ! kloof between the Lions Head and Table Mountain, *Burchell*, 259 ! Caledon Div. ; Genadendal, 1000 ft., *Schlechter*, 9866 ! Riversdale Div. ; between Little Vet River and Garcias Pass, *Burchell*, 6923 !
CENTRAL REGION : Prince Albert Div. ; Gamka River, *Burke* !
WESTERN REGION : Little Namaqualand, *Wyley*, 90 !

12. **S. albicaulis** (Benth. in E. Meyer, Comm. 234); an erect shrub; stems and branches sharply 4-angled, whitish with very short closely adpressed pubescence, sometimes hispid at the base; leaves coriaceous, rigid, obovate, elliptic, ovate, rarely suborbicular, ¾–1¼ in. long, ½–1 in. broad, rounded or acute, cuneate at the base, crenate-dentate or sometimes lobed at the base, prominently reticulately veined beneath, densely and shortly whitish-pubescent or sometimes hispid beneath, shortly hispid above ; petiole 1–4 lin. long ; whorls 2-flowered, in somewhat lax terminal panicles 6 in. long or more; calyx narrowly campanulate, 5 lin. long, slightly accrescent, densely villous; upper lip trifid ; teeth ovate-triangular, ½–1 lin. long, acuminate, the median tooth smallest ; lower lip bifid ; teeth narrowly ovate-triangular, long-acuminate, 1¼–1½ lin. long; corolla 9–10 lin. long, purple, sparingly villous, with a fringe of hairs round the inside of the tube about the middle; tube 4–5 lin. long, 1½ lin. broad above, shorter than the calyx; upper lip falcate, 5–6 lin. long, shortly and broadly 2-lobed ; lower lip about 5 lin. long, 3 lin. broad ; median lobe broadly obovate, 1¾ lin. long, 2½ lin. broad, emarginate; lateral lobes smaller, rounded. *DC. Prodr.* xii. 274.

VAR. β, **dregeana** (Skan); leaves oblong-lanceolate or sometimes oblong-obovate, 3–8 lin. long, deeply incised-dentate or irregularly pinnatifid. *S. dregeana, Benth. in E. Meyer, Comm.* 234, *and in DC. Prodr.* xii. 274.

SOUTH AFRICA: without locality, *Drège*, 7937 !
COAST REGION : Tulbagh Div. ; near Tulbagh, *Ecklon* ! New Kloof, near Tulbagh, *Burchell*, 1019! Mitchells Pass, *Grey* ! *Wyley* ! 1800 ft., *Schlechter*, 9970 ! Worcester Div. ; near Brand Vlei, 800 ft., *Bolus*, 5223, *partly* ! Var. β,

Clanwilliam Div. ; Sandheights between Pakhuis and Biedouw, 2000–3000 ft., *Drège*, 3114 !
CENTRAL REGION : Ceres Div. ; near Ceres, 1500 ft., *MacOwan & Bolus, Herb. Norm. Austr.-Afr.* 491 !

13. S. rugosa (Thunb. Prodr. 97); stem herbaceous, robust, ascending ; branches glandular-villous ; leaves petiolate or the uppermost sessile, oblong, oblong-lanceolate or sometimes ovate, acute or rounded, cuneate to cordate at the base, irregularly crenate, erose-dentate or shortly lobed, often distinctly rugose, pubescent, up to 8 in. long and 2½ in. broad, often about 2½ in. long and ½–¾ in. broad ; petiole up to 3 in. long; whorls 6–10-flowered, finally distant, in simple or much-branched racemes 6–12 in. long or more ; bracts broadly cordate, abruptly acuminate, shorter than the calyx; calyx campanulate, 4–5½ lin. long, 13-nerved, covered with gland-tipped villous hairs ; upper lip broadly obovate, 3-toothed, recurved; teeth small, often pungent, connivent ; lower lip bifid ; teeth ovate-triangular, 1½–2 lin. long, acuminate-pungent; corolla white or purple and white, 8–10 lin. long; tube 4–5 lin. long, ventricose at the throat; upper lip falcate, 4–6 lin. long, shortly 2-lobed; lower lip 3 lin. long, 3½ lin. broad ; median lobe broadly obovate-cuneate, 2 lin. long, 3½–4 lin. broad ; lateral lobes ovate or oblong, erect, 1¼–2 lin. long, ¾–1¼ lin. broad. *Ait. Hort. Kew. ed.* 1, i. 42 ; *Vahl, Enum.* i. 259 ; *Roem. et Schult. Syst.* i. 241, *and Mant.* i. 204 ; *Thunb. Fl. Cap. ed. Schult.* 451 ; *A. Dietr. Sp. Pl.* i. 316 ; *Benth. Lab.* 235, *in E. Meyer, Comm.* 235, *and in DC. Prodr.* xii. 291.

SOUTH AFRICA : without locality, *Zeyher*, 1332 ! *Ecklon* ! *Thunberg* !
COAST REGION : Swellendam Div., *Zeyher* ! Uitenhage Div. ; hills of Uitenhage Karroo, *Prior* !
CENTRAL REGION : Ceres Div. ; at Ongeluks River, *Burchell*, 1228 ! Prince Albert Div., 4500–5000 ft., *Drège*, 806*b* ! Richmond Div. ; Winterveld, between Nieuwjaars Fontein and Ezels Fontein, 3000–4000 ft., *Drège*, 806*c* ! Middelburg Div.; Culmstock, 3300 ft., *Mrs. Southey in Herb. Galpin*, 5707 ! Colesberg Div., *Shaw* ! *Burke* ! Hopetown Div. ; near Hopetown, *Bolus*, 2032 !
WESTERN REGION : Little Namaqualand ; between Koperberg and Kook Fontein, 2000–3000 ft., *Drège*, 806*a* ! near Kook Fontein, *Bolus*, 9435 !
KALAHARI REGION : Griqualand West ; Griqua Town, *Burchell*, 1861 ! Lower Campbell, *Burchell*, 1801 ! Klip Drift, *Mrs. Barber*, 9 ! Orange River Colony, *Mrs. Barber* ! Transvaal ; north of Silverdale, Vaal River, *Nelson*, 46 !

This species is extremely variable in the size, shape, and texture of the leaves. A narrow-leaved form has been distinguished as var. β, *angustifolia*, Benth. in DC. Prodr. xii. 291, but the specimens cited have not narrower leaves than Thunberg's type.

A specimen from Graaff Reinet in Herb. Bolus (1789 bis) has broadly ovate leaves 4 in. long and 3–3½ in. broad, cordate at the base. The stem is covered with long slender slightly interwoven hairs. Calyx as in *S. rugosa*. Corolla absent. It may be a distinct species.

14. S. Radula (Benth. in DC. Prodr. xii. 291); stem herbaceous, erect, branched, more or less densely covered with white woolly tomentum; leaves petiolate, oblong, oblong-lanceolate or narrowly ovate, 1½–4 in. long, ½–1¾ in. broad, obtuse or rounded at the apex, rounded, truncate or cordate at the base, rather regularly erose-crenate, very rugose above, shortly whitish tomentose beneath ; inflorescence and flowers as in *S. rugosa.*

SOUTH AFRICA : without locality, *Zeyher*, 1333 !

KALAHARI REGION : Transvaal ; Magalies Berg and Mooi River, *Burke!* Malmanie Oog and Buffels Hoek, near Jacobsdahl, *Burtt-Davy*, 86 ! Zeerust, Marico District, about 5000 ft., *Burtt-Davy*, 109 ! 7179 ! Crocodile River, *Burtt-Davy*, 186 ! *Miss Leendertz*, 707 !

Zeyher's and Burtt-Davy's specimens are much less tomentose than the type. A specimen collected in the Matebe Valley, Transvaal, by Holub, has indumentum approaching that of *S. rugosa* to which *S. Radula* is very closely allied.

15. **S. disermas** (Linn. Sp. Pl. ed. ii. 36); stem herbaceous or becoming somewhat woody at the base ; branches ascending, densely glandular-villous ; leaves petiolate, the upper subsessile, oblong-lanceolate or sometimes ovate, rounded to acute, cuneate, rounded or cordate at the base, coarsely erose-dentate, more or less rugose, glandular-pubescent, 1–4¼ in. long, ¾–2 in. broad; petiole up to 1¾ in. long; whorls 6–10-flowered, distant, in simple or branched terminal racemes up to 8 in. long ; bracts broadly ovate, acuminate, the upper shorter than the calyx ; calyx campanulate, 3½ lin. long, 12- or 13-nerved, glandular-pilose ; upper lip broadly ovate, minutely and acutely 3-toothed ; lower lip bifid ; teeth ovate-triangular, 1¼–1½ lin. long, acuminate-pungent ; corolla whitish, 3–6 lin. long ; tube 2–3¾ lin. long, ventricose at the throat ; upper lip 1½–2¾ lin. long, emarginate ; lower lip 1¼–2¾ lin. long ; median lobe transversely oblong, 1 lin. long, 1¼–2 lin. broad, retuse, concave ; lateral lobes ovate-oblong, ½–¾ lin. long, ⅓–½ lin. broad, slightly oblique, rounded to subacute. *Willd. Sp. Pl.* i. 139 ; *Vahl, Enum.* i. 266 ; *Ait. Hort. Kew. ed.* 2, i. 59 ; *Roem. et Schult. Syst.* i. 246 ; *A. Dietr. Sp. Pl.* i. 323 ; *Benth. Lab.* 236, *and in DC. Prodr.* xii. 291. *S. caule fruticoso, etc., Arduini, Animad. Bot. Sp.* i. 9, *t.* 1. *Horminum disermas, Moench, Meth. Suppl.* 140. *H. sylvestre majus, etc., Barrel. Pl. Gall. Obs.* 25, *Ic.* 187.

SOUTH AFRICA : without locality, *Zeyher*, 1334 ! and *cultivated specimens!*
COAST REGION : Uniondale Div. ; Olifants River, *Bolus*, 1789 *bis* !
CENTRAL REGION : Calvinia Div. ; Onder Bokkeveld at Matjes Fontein, 2200 ft., *Schlechter*, 10926 !

This is probably only a small-flowered form of *S. rugosa*. It is doubtful whether the South African material belongs to the true *S. disermas*, stated by Linnæus to be a native of Syria.

16. **S. clandestina** (Linn.), var. **angustifolia** (Benth. in DC. Prodr. xii. 295) ; stem herbaceous, erect, ½–2 ft. high, simple or branched, covered with long and spreading hairs mixed with short and adpressed ones ; leaves petiolate or the upper subsessile, narrowly oblong, rarely ovate-oblong, 1½–4½ (usually about 2) in. long, ¼–1 in. broad, subacute or rounded, irregularly pinnatifid or laciniate, sometimes erose-dentate, puberulous chiefly beneath, more or less pilose on the principal veins, usually rugose ; whorls usually

6-flowered, distant, in branched or sometimes simple racemes 6–12 in. long; rhachis densely pilose; bracts broadly cordate, acuminate, shorter than the calyx; calyx 2¾–3¼ lin. long, pilose outside, densely pilose or villous inside except at the base, 13-nerved; upper lip semi-elliptic, 1½ lin. long, slightly broader, minutely and acutely 3-toothed, often blue; lower lip deeply bifid; teeth lanceolate-triangular, 1½–1¾ lin. long, ½–¾ lin. broad at the base, acuminate-pungent; corolla 2–5 lin. long; tube 1½–3 lin. long, ventricose at the throat; upper lip 1¼–2 lin. long, sparingly pilose outside; lower lip 1¼–1½ lin. long; median lobe transversely oblong, about 1–1¼ lin. long, 1½–1¾ lin. broad, emarginate, concave. *E. Meyer, Comm.* 235 (*as S. clandestina*). *S. cleistogama, De Bary & Paul, Ind. Sem. Hort. Halens.* 1867, 6; *Aschers. in Bot. Zeit.* 1871, 555, *and* 1872, 293. *S. controversa, Ten. Syll. Fl. Neap.* 18. *S. controversa, Benth. Lab.* 241, *and in DC. Prodr.* xii. 295, *as to Burchell,* 1454. *S. Verbenaca, Linn., var. controversa, Briq. Lab. Alp. Marit.* 520. *S. Verbenaca, var. angustifolia, Pugsley in Journ. Bot.* 1908, 144.

SOUTH AFRICA: without locality, *Ecklon*! and a *cultivated specimen*!
COAST REGION: British Kaffraria, *Cooper*, 2983!
CENTRAL REGION: Somerset Div.; at the foot of the Bosch Berg, 2300 ft., *MacOwan*! Graaff Reinet Div.; Graaff Reinet, in cultivated and other places, 2500 ft., *Bolus*, 142! Beaufort West Div.; near Beaufort West, *Zeyher*! Fraserburg Div.; between Klein Quaggas Fontein and Dwaal River, *Burchell*, 1454! Richmond Div.; Uitvlugt, various places near Stylkloof, 4000–5000 ft., *Drège*, 806*d*! Middelburg Div.; Conway Farm, 3600 ft., *Gilfillan in Herb. Galpin*, 5573!
KALAHARI REGION: Griqualand West; Kimberley, "the commonest plant around the town," *Kolbe*, 3160!

The species is a native of South Europe, North Africa, the Orient, the Canaries and Madeira. A plant closely resembling the South African form and regarded by Bentham as the same occurs in Arabia Felix. The flowers of the South African plant are often cleistogamous.

17. **S. Tysonii** (Skan); stems herbaceous, stout, apparently tall and erect, 1½–2¼ lin. in diam. in the upper part, quadrangular, rounded on the angles, broadly furrowed, rather densely often brownish-puberulous; lower leaves not seen; upper shortly petiolate, pinnatifid or runcinate, 2–3 in. long, 1½–2 in. broad, densely covered with short fine matted often brownish hairs beneath and minutely gland-dotted, sparingly covered with short adpressed hairs above; terminal lobe deltoid-ovate, up to 2 in. long and broad, usually acute; lateral lobes few, ovate or oblong, ¼–1 in. long, ¼–½ in. broad, acute, like the terminal lobe erose-dentate or shortly lobed; uppermost leaves sessile or subsessile, ovate or ovate-oblong, sometimes hastate or auriculate at the base; whorls up to 24-flowered or more, crowded or shortly distant, in spike-like terminal racemes or few-branched panicles 4–7 in. long, also in short subcapitate racemes terminating short lateral branches; bracts broadly to narrowly ovate, slightly to much shorter than the calyx, acuminate; pedicels about 1 lin. long, densely hairy; calyx tubular-campanulate, 3⅓ lin. long, softly or somewhat hispidly

pubescent, sparingly gland-dotted ; upper lip $1\frac{3}{4}$–$1\frac{3}{4}$ lin. long, 3-toothed ; teeth subulate or narrowly triangular, $\frac{1}{2}$–1 lin. long ; lower lip $1\frac{1}{2}$–$1\frac{3}{4}$ lin. long, bifid ; teeth linear-triangular, $1\frac{1}{4}$–$1\frac{1}{2}$ lin. long ; corolla intensely red (*Tyson*), blue (*Wood*), $5\frac{1}{2}$ lin. long, sparingly pubescent outside on the upper part ; tube about $3\frac{3}{4}$ lin. long, cylindric in the lower part, inflated at the throat ; upper lip $1\frac{1}{2}$–2 lin. long, $1\frac{1}{4}$ lin. broad, emarginate ; lower lip 2–$2\frac{1}{4}$ lin. long ; median lobe broadly obcordate, $1\frac{1}{2}$–$1\frac{3}{4}$ lin. long, $1\frac{1}{2}$–2 lin. broad ; lateral lobes short, rounded.

EASTERN REGION : Tembuland ; near Emgwali River, *Bolus*, 10249 ! Griqua-land East ; banks of the Umzimkulu River, near Clydesdale, 2500 ft., *Tyson* ! Natal ; Drakensberg, near Charlestown, 5000–6000 ft., *Wood*, 7883 ! Zululand ; Gudena, *Gerrard*, 2031 !

18. **S. scabra** (Linn. f. Suppl. 89) ; a perennial herb with a woody subterranean creeping caudex ; stem branched, erect, more or less villous ; leaves petiolate, oblong-obovate or obovate, lyrate-pinnatifid or -pinnatipartite, rarely only slightly lobed at the base, $\frac{3}{4}$–2 in. long, $\frac{1}{2}$–$1\frac{1}{4}$ in. broad, prominently veined, densely grey-pilose or villous beneath ; terminal lobe largest, semi-orbicular or broadly ovate, $\frac{1}{2}$–$\frac{3}{4}$ in. long, 6–10 lin. broad ; lateral lobes usually 2 each side, with broad sinuses between them, irregularly ovate or oblong, 2–6 lin. long, $1\frac{3}{4}$–4 lin. broad, like the terminal lobe erose-dentate ; whorls usually 4- or 6-flowered, secund, distant, in terminal racemes 2–5 in. long ; calyx tubular-turbinate, $5\frac{1}{2}$–6 lin. long, somewhat hispidly pubescent, often coloured ; tube 3–$3\frac{1}{2}$ lin. long ; upper lip 2–$2\frac{1}{4}$ lin. long, truncate, 3-toothed ; teeth subulate, $\frac{2}{3}$ lin. long ; lower lip $2\frac{1}{4}$–$2\frac{1}{2}$ lin. long, bifid or rarely trifid ; teeth linear-triangular, spinescent, $1\frac{1}{2}$–2 lin. long ; corolla blue, red-purple or purple, 12–14 lin. long, pilose ; tube 10–$12\frac{1}{2}$ lin. long, narrowly cylindric in the lower half, gradually enlarged above, slightly ventricose at the throat ; upper lip obovate, $1\frac{1}{2}$–2 lin. long, emarginate ; lower lip 3 lin. long ; median lobe broadly and deeply cordate, $1\frac{3}{4}$–$2\frac{1}{4}$ lin. long, about 3 lin. broad ; lateral lobes very short and broad, rounded. *Ait. Hort. Kew. ed.* 1, i. 41 ; *Thunb. Prodr.* 97, *and Fl. Cap. ed. Schult.* 452 ; *Vahl, Enum.* i. 259 ; *Roem. et Schult. Syst., Mant.* i. 213 ; *A. Dietr. Sp. Pl.* i. 344. *S. aurita, Benth. Lab.* 305, *in E. Meyer, Comm.* 237, *and in DC. Prodr.* xii. 351, *not of Linn. f. S. graciliflora, Avé-Lall. in Ind. Sem. Hort. Petrop.* x. 57.

SOUTH AFRICA : without locality, *Thunberg* ! and *cultivated specimens* !
COAST REGION : Uitenhage Div. ; Van Stadens River, *Drège*, 7940a ! *Zeyher*, 396 ! Port Elizabeth Div. ; Port Elizabeth, *Drège*, 7940b ! New Brighton, near Port Elizabeth, *Mrs. Southey in Herb. Galpin*, 5881 ! Bathurst Div. ; near Barville Park, *Burchell*, 4086 ! at the mouth of the Great Fish River, *Burchell*, 3750 ! near Port Alfred, *Burchell*, 3827 ! *Bolus*, 10647 !

19. **S. Galpinii** (Skan) ; an erect branched rather slender herb, $1\frac{1}{4}$ ft. high or more, rather densely puberulous on all the green parts ; leaves shortly petiolate or subsessile, obovate or oblong, pinnatipartite or lyrate-pinnatipartite, usually $1\frac{1}{2}$–2 in. long, gland-

dotted; terminal lobe broadly ovate-deltoid or deltoid, $\frac{3}{4}$–1 in. long, $\frac{1}{2}$–1$\frac{1}{4}$ in. broad, rounded or acute, like the lateral crenate or crenate-dentate; lateral lobes usually 2 each side, often retrorse, ovate-oblong, 3–6 lin. long, 2–4 lin. broad, usually rounded; whorls often 4–8-flowered, 4–7 lin. apart, in terminal racemes 2–2$\frac{1}{2}$ in. long; bracts ovate-lanceolate, acuminate, shorter than the calyx; pedicels 1–1$\frac{1}{2}$ lin. long; calyx tubular-campanulate, campanulate when mature, 3$\frac{1}{2}$–3$\frac{3}{4}$ lin. long, pilose, gland-dotted; tube 2 lin. long; upper lip 1$\frac{1}{2}$ lin. long, finally subtruncate, 3-toothed; teeth linear-triangular, $\frac{1}{2}$–1$\frac{1}{4}$ lin. long; lower lip 1$\frac{1}{2}$ lin. long, bifid; teeth lanceolate-triangular, about 1 lin. long; corolla mauve (*Galpin*), about 6 lin. long, pubescent outside; tube cylindric-funnel-shaped, 5 lin. long, $\frac{2}{3}$ lin. broad in the lower part, 1$\frac{1}{4}$ lin. broad at the throat; upper lip obcordate, 1$\frac{1}{4}$ lin. long; lower lip 1$\frac{1}{2}$–2 lin. long; median lobe obcordate, 1$\frac{1}{4}$–1$\frac{1}{2}$ lin. long, 1$\frac{1}{2}$–1$\frac{3}{4}$ lin. broad; lateral lobes scarcely $\frac{1}{2}$ lin. long, rounded.

COAST REGION : Queenstown Div.; mountains near Queenstown, 4000–4500 ft., *Galpin*, 1956 !

Evans, 391, from Southdowns, Natal, is probably *S. Galpinii*. The leaves have a smaller narrower terminal lobe, and the flowers are described as white.
Zeyher, 913, from Albany, referred doubtfully by Bentham to *S. aurita* in DC. Prodr. xii. 352, is perhaps also *S. Galpinii*. The specimen has no flowers.

20. **S. aurita** (Linn. f. Suppl. 88); stem herbaceous, slender, ascending, 2 ft. high or more, simple or sparingly branched, more or less villous; leaves shortly petiolate, broadly ovate, 1$\frac{3}{4}$–2$\frac{1}{4}$ in. long, 1$\frac{1}{4}$–2 in. broad, acute or rounded, often broadly and deeply few-lobed or auriculate at the base, broadly dentate or crenate, shortly villous on both sides; whorls 6- or 8-flowered, lax, distant, secund, in simple racemes 2–6 in. long; bracts ovate, acuminate, very small; calyx 3$\frac{1}{2}$–4 lin. long, pilose outside, puberulous inside; upper lip 1$\frac{3}{4}$ lin. long and broad, somewhat truncate, 3-toothed; teeth $\frac{2}{3}$–$\frac{3}{4}$ lin. long, broad at the base, subulate above; lower lip 2 lin. long, bifid; teeth lanceolate-triangular, 1$\frac{1}{4}$ lin. long, acuminate; corolla pale rose, 7 lin. long, pubescent outside on the upper part; tube narrowly funnel-shaped, 5$\frac{1}{2}$–6 lin. long, scarcely 1$\frac{1}{4}$ lin. broad at the throat; upper lip broadly obcordate, about 1$\frac{1}{4}$ lin. long and broad; lower lip 2 lin. long; median lobe orbicular-obcordate, 1$\frac{1}{4}$ lin. long, 1$\frac{1}{2}$ lin. broad; lateral lobes very small, rounded. *Thunb. Prodr.* 96, *and Fl. Cap. ed. Schult.* 451; *Ait. Hort. Kew. ed.* 2, i. 62; *Roem. et Schult. Syst.* i. 259, *and Mant.* i. 212; *A. Dietr. Sp. Pl.* i. 340. *S. sylvicola, Burch. ex Benth. Lab.* 304, *in E. Meyer, Comm.* 236, *and in DC. Prodr.* xii. 350. *S. sylvatica, E. Meyer in Drège, Zwei Pflanzengeogr. Documente,* 124, 146, 218.

SOUTH AFRICA : without locality. A specimen cultivated in the Upsala Botanic Garden, *Herb. Thunberg* !
COAST REGION : George Div.; in the forest near George, *Burchell*, 6052 ! *Prior* ! Kaimans Gat, *Prior* ! Knysna Div. ; Ruigte Vallei, *Drège*, 7941a ! Albany Div., *Atherstone*, 48 !

KALAHARI REGION: Transvaal; Houtbosch, *Rehmann*, 6153!
EASTERN REGION: Transkei; between Gekau (Geua) River and Bashee River, 1000–2000 ft., *Drège*, 7941*b*! Kentani District, 1200 ft., *Miss Pegler*, 913!

21. **S. pallidifolia** (Skan); a slender erect branched herb $1\frac{1}{4}$ ft. high or more, hispidly puberulous on all the green parts; branches erect-spreading, scarcely $\frac{1}{2}$ lin. in diam.; leaves petiolate, pale green, ovate-triangular, usually with 2 or sometimes 3 small auricles at the base, 10–18 lin. long, 7–12 lin. broad, rounded or rarely subacute, somewhat regularly crenate-dentate, thin, pale green, gland-dotted; petiole 1–4 lin. long; whorls 2–6-flowered, usually 5–8 lin. apart, lax, in slender terminal racemes $1\frac{1}{2}$–$4\frac{1}{2}$ lin. long; bracts ovate-lanceolate, usually about as long as the pedicels, acuminate, occasionally the lowermost larger and foliaceous; pedicels slender, 1–2 lin. long; calyx campanulate, $3\frac{1}{4}$ lin. long, scarcely longer when mature, sparingly pubescent and gland-dotted; tube $1\frac{3}{4}$ lin. long; upper lip $1\frac{1}{3}$ lin. long; teeth lanceolate- or linear-triangular, $\frac{1}{2}$–1 lin. long; lower lip $1\frac{1}{3}$ lin. long; teeth lanceolate- or linear-triangular, 1 lin. long; corolla rose-coloured, 5–6 lin. long, slightly pubescent; tube 4–$4\frac{1}{2}$ lin. long, subcylindric for about 3 lin. from the base where it is $\frac{3}{4}$ lin. broad, $1\frac{1}{2}$ lin. broad at the throat; upper lip obcordate, $1\frac{1}{4}$ lin. long; lower lip $1\frac{3}{4}$–$2\frac{3}{4}$ lin. long; median lobe obcordate, $1\frac{1}{4}$–$1\frac{2}{3}$ lin. long, $1\frac{1}{4}$–2 lin. broad; lateral lobes very small, rounded. *S. scabra, Benth. in DC. Prodr.* xii. 351, *partly, not of Linn. f. S. triangularis, Benth. l.c., partly, not of Thunb.*

COAST REGION: Stockenstrom Div.; Tyumie Berg, *Ecklon*! Katberg, 4000 ft., *Shaw*!
CENTRAL REGION: Somerset Div.; Bosch Berg, near Somerset East, *Burchell*, 3165!

22. **S. triangularis** (Thunb. Prodr. 96); stem herbaceous, erect or ascending, slender, simple or branched, 6–18 in. high, more or less pilose; leaves ovate-deltoid or ovate-rounded, $\frac{1}{2}$–$1\frac{1}{4}$ in. long, $\frac{1}{4}$–1 in. broad, acute or rounded at the apex, truncate, slightly cordate, sometimes cuneate or deeply few-lobed at the base, crenate or crenate-dentate, more or less softly pilose on both sides, sometimes gland-dotted beneath; petiole usually $\frac{1}{2}$–$1\frac{1}{4}$ in. long; whorls loosely 2–6-flowered, distant, in simple racemes 2–8 in. long; bracts ovate-lanceolate, acuminate, much shorter than the calyx; calyx $3\frac{1}{4}$–4 lin. long, pilose and sometimes slightly gland-dotted outside, puberulous inside; upper lip truncate, 3-toothed; teeth broad at the base, subulate above, $\frac{1}{2}$–1 lin. long, the median shorter than the lateral; lower lip bifid; teeth lanceolate-triangular, acuminate, 1–$1\frac{1}{2}$ lin. long; corolla blue, 5–7 lin. long, villous on the upper part outside; tube narrowly funnel-shaped, $3\frac{3}{4}$–6 lin. long, $\frac{2}{3}$–$\frac{3}{4}$ lin. broad at the base, $1\frac{1}{4}$–$1\frac{1}{2}$ lin. broad at the throat; upper lip obovate or elliptic, $1\frac{1}{4}$–$1\frac{1}{2}$ lin. long, entire or emarginate; lower lip $1\frac{3}{4}$–$2\frac{3}{4}$ lin. long; median lobe very broadly obcordate, 1–$1\frac{3}{4}$ lin. long, $1\frac{1}{3}$–2 lin. broad; lateral lobes very small, rounded. *Thunb. Fl. Cap. ed. Schult.* 451;

Roem. et Schult. Syst. i. 228, *and Mant.* i. 187 ; *A. Dietr. Sp. Pl.*
i. 287 ; *Benth. Lab.* 308, *in E.* Meyer, *Comm.* 236, *and in DC. Prodr.*
xii. 351, *partly. S. tenuifolia, Burch. ex Benth. Lab.* 304.

SOUTH AFRICA : without locality, *Ecklon* ! *Thunberg* ! and *cultivated specimens* !
COAST REGION : Uitenhage Div. ; Enon, 1000–2000 ft., *Drège*, 7942 ! hills in
Uitenhage Karroo, *Prior* ! Stockenstrom Div. ; Katberg, 4000 ft., *Shaw* ! British
Kaffraria, *Cooper*, 396 !
CENTRAL REGION : Somerset Div. ; at Blyde River, *Burchell*, 2981 !

23. **S. obtusata** (Thunb. Prodr. 97) ; stem branched and somewhat woody at the base, herbaceous above ; branches ascending, up
to 1 ft. long, terete, glabrous and smooth below, quadrangular
and more or less sparingly covered with very short stiff hairs above ;
leaves elliptic or elliptic-obovate, $\frac{3}{4}$–2 in. long, $\frac{1}{2}$–1$\frac{1}{4}$ in. broad,
usually rounded, often hastate at the base or deeply few-lobed
especially in the lower half, rarely cuneate at the base, mostly
crenate, usually with some rather stiff often long hairs on the
nerves beneath and here and there on the margins ; petiole $\frac{1}{4}$–2 in.
long, usually with long rather stiff spreading hairs, sometimes quite
glabrous ; whorls loosely 2–6-flowered, distant, often secund, in
simple or branched racemes 4–8 in. long ; bracts broadly ovate,
acuminate, much shorter than the calyx ; pedicels 1–4 lin. long ;
calyx 3$\frac{1}{2}$–4$\frac{1}{2}$ lin. long, shortly hispidly pubescent outside and inside,
more or less ciliate on the ribs and margins of the lips ; upper lip
truncate, 3-toothed ; teeth usually broad at the base and subulate
above, $\frac{1}{3}$–1$\frac{1}{4}$ lin. long, lower lip bifid ; teeth lanceolate- or linear-
triangular, 1$\frac{1}{2}$ lin. long ; corolla narrowly funnel-shaped, 6–8 lin.
long, sparingly pilose outside ; tube 4$\frac{3}{4}$–6$\frac{1}{2}$ lin. long, ventricose at
the throat ; upper lip erect, 1$\frac{1}{2}$–1$\frac{3}{4}$ lin. long, 1$\frac{1}{4}$ lin. broad,
emarginate ; lower lip about 2$\frac{1}{2}$ lin. long, 3-lobed ; median lobe
broadly obcordate, 1 lin. long, 1$\frac{1}{2}$ lin. broad ; lateral lobes broad,
very short. *Thunb. Fl. Cap. ed. Schult.* 451 ; *Roem. et Schult. Syst.*
i. 260, *and Mant.* i. 213 ; *A. Dietr. Sp. Pl.* i. 341 ; *Benth. Lab.* 308,
and in DC. Prodr. xii. 351. *S. marginata, Benth. in E. Meyer,
Comm.* 236, *and in DC. Prodr.* xii. 351.

SOUTH AFRICA : without locality, *Thunberg* !
COAST REGION : Uitenhage Div. ; *Ecklon* ! between Coega River and Sunday
River. *Drège*, 7944*a* ! Addo, 400–1000 ft., *Drège*, 7944*b* ! valley and hills of the
Zwartkops River, 50–500 ft., *Zeyher*, 3533 ! Albany Div. ; Grahamstown, *Prior* !
Curries Kloof, Grahamstown, *MacOwan*, 556 !
CENTRAL REGION : Albert Div., *Cooper*, 592 !

The type specimen of *S. obtusata* appears to be somewhat depauperate. It is
more slender than the other specimens cited, the few leaves present are thinner,
mostly smaller, cuneate at the base or sometimes slightly hastate, the whorls are
2-flowered, and the pedicels are longer.

24. **S. lasiostachys** (Benth. in DC. Prodr. xii. 350) ; stem herbaceous, hispid ; hairs tuberculate at the base ; leaves shortly petiolate,
rigid, semi-orbicular or ovate-suborbicular, 1–1$\frac{1}{2}$ in. long or more,
up to 2 in. broad, cordate-truncate at the base, broadly dentate,
hispid and green on both sides ; whorls 6–10-flowered, lax, about

½ in. apart, in a simple raceme ; rhachis densely hispid ; bracts
narrowly ovate, acuminate, very small ; calyx 4½ lin. long, hispid
outside, puberulous inside, otherwise as in *S. aurita* ; corolla about
twice as long as the calyx, villous.

COAST REGION : Uitenhage Div., *Ecklon*, 62 !

25. S. namaensis (Schinz in Verhandl. Bot. Ver. Brandenb.
xxxi. 208) ; a perennial branched herb woody in the lower part, or
almost an undershrub, 1½–3¼ ft. high, erect, the leaves and calyxes
emitting (when rubbed) a camphor-like odour : branches woolly,
leafy ; leaves petiolate, elongate-obovate or oblong, lyrate-pinnatifid,
up to 1¾ in. long, ¼–1 in. broad, narrowed to apex and base, gland-
dotted, pilose, somewhat rugose ; lateral lobes 2–4 each side, rather
smaller than the terminal, obtuse, about 2½ lin. long and 1–1½ lin.
broad, all irregularly crenate ; whorls 2–6-flowered, 4–6 lin. apart,
in lax terminal racemes up to 8 in. long ; bracts ovate-lanceolate to
lanceolate, much shorter than the calyx ; pedicels slender, ½–2½ lin.
long, bluish ; calyx narrowly campanulate, 3½–4 lin. long, sparingly
and shortly pubescent, sparingly gland-dotted, ciliate on the teeth ;
upper lip 1½ lin. long, 3-toothed ; teeth lanceolate-triangular, ½–¾
lin. long, acuminate ; lower lip 1½ lin. long, bifid ; teeth lanceolate-
triangular, about 1 lin. long, acuminate ; corolla whitish, 7½ lin.
long ; tube about 5 lin. long, subcylindric below, rather wider from
the middle to the throat and recurved, annular-pilose 2 lin. from
the base ; upper lip oblong, 3¼–3½ lin. long, 1⅓–1½ lin. broad, 2-
lobed ; lower lip 4¾ lin. long ; median lobe transversely oblong,
narrowed at the base, 2¼ lin. long, 2¾ lin. broad, broadly emarginate ;
lateral lobes rounded-ovate, about 1 lin. long and broad ; stamens
and style long-exserted.

WESTERN REGION : Great Namaqualand ; Tiras, *Schinz*, 30 !

26. S. Burchellii (N. E. Br. in Kew Bulletin, 1901, 130) ; a weak.
perennial herb, somewhat woody at the base, up to about 1 ft. high,
ascending, much branched, rather densely and shortly grey-
tomentose on the branches and underside of the leaves, or in places
almost woolly ; leaves rather crowded, shortly petiolate, subdeltoid
or ovate-oblong, pinnatipartite, ½–1 in. long, bullate-rugose, crisped,
shortly hispid above, gland-dotted ; lobes 2 or 3 on each side,
narrowly oblong, 1–2½ lin. long, ½–1 lin. broad, obtuse, irregularly
crenate or almost lobulate ; whorls usually 2-flowered, rarely up to
6-flowered, 2–4 lin. apart, in terminal racemes 1–2 in. long ; lower
bracts leaf-like ; upper broadly ovate, apiculate or acuminate, much
shorter than the calyx ; pedicels ½–1¼ lin. long ; calyx tubular-
campanulate, 2⅔ lin. long, rather densely covered with short curled
grey hairs, scarcely tomentose, gland-dotted ; tube 1⅗–2 lin. long ;
upper lip ⅔ lin. long, 3-toothed ; teeth ovate-deltoid, ⅓–½ lin. long ;
lower lip bifid ; teeth narrowly deltoid or lanceolate-triangular,
1¼ lin. long, acuminate ; corolla 4–4¼ lin. long, somewhat tomentose

outside; tube subcylindric, 3 lin. long, 1–1¼ lin. broad at the throat, annular-pilose inside 1¾ lin. from the base; upper lip elliptic, 1½ lin. long, emarginate; lower lip 2–2¾ lin. long; median lobe broadly obovate, 1¼ lin. long, 1⅔ lin. broad, slightly emarginate; lateral lobes smaller, rounded. *S. scabra, Benth. Lab.* 305, *partly, not of Linn. f. S. runcinata, var. crispa, Benth. in E. Meyer, Comm.* 237, *and in DC. Prodr.* xii. 352, *partly.*

VAR. β, **hispidula** (Skan); branches and the underside of the leaves covered with short somewhat hispid hairs, not tomentose. *S. scabra, Benth. Lab.* 305, *partly, not of Linn. f. S. runcinata, var. β, crispa, Benth. in DC. Prodr.* xii. 352, *partly.*

SOUTH AFRICA: without locality, var. β, *Thom,* 209! *Ecklon,* 77!
CENTRAL REGION: Victoria West Div.; Winterveld, between Nieuwjaars Fontein and Ezels Fontein, 3000–4000 ft., *Drège,* 803! Richmond Div.; Rhenoster Poort, *Burchell,* 2120!

The locality given by Bentham for Drège's specimen is Nieuwveld.

27. **S. stenophylla** (Burch. ex Benth. Lab. 306); an erect weak perennial much-branched herb, ½–1½ ft. high; branches suberect, giving the plant a pyramidal form, often very slender, almost glabrous or usually with a few short stiff hairs, gland-dotted; leaves shortly petiolate or subsessile, linear to lanceolate-oblong, pinnatifid to pinnatisect, ½–4 (usually 1½–2½) in. long, ¾–10 (usually 3–6) lin. broad, very sparingly shortly and stiffly hairy chiefly on the nerves beneath, densely gland-dotted; lateral lobes usually 6–12 each side, smallest towards the apex, subdeltoid to linear-oblong, ½–6 (usually 1½–4) lin. long, about ¾–1½ lin. broad, subacute to rounded, denticulate, often crisped; whorls usually 6-flowered, distant, in slender racemes 3–7 in. long terminating the numerous branches; bracts ovate, acuminate, shorter than the calyx; pedicels about 1 lin. long; calyx tubular-campanulate, 2¼–2¾ lin. long, sparingly and minutely hispidly hairy, densely gland-dotted; upper lip 1–1¼ lin. long, with 3 short apiculate or acuminate teeth; lower lip 1–1⅓ lin. long, bifid; teeth lanceolate-triangular, ⅔–1 lin. long, acute or acuminate; corolla blue or lilac, 4–5 lin. long; tube 2½–3¼ lin. long, ventricose at the throat, annular-pilose inside about 1 lin. from the base; upper lip suborbicular, 1¼–1½ lin. long and broad, slightly emarginate; lower lip 2–2½ lin. long; median lobe transversely oblong, narrowed at the base, about 1 lin. long, 1⅔ lin. broad, slightly emarginate; lateral lobes short, rounded. *Benth. in E. Meyer, Comm.* 238, *and in DC. Prodr.* xii. 353.

VAR. β, **subintegra** (Skan); leaves (only the upper seen) minutely undulate-denticulate, not deeply toothed to pinnatisect.

SOUTH AFRICA: without locality, *Zeyher,* 1336!
COAST REGION: Uitenhage Div.; hills in Uitenhage Karroo, *Prior*! Fort Beaufort Div.; Kat River Poort, *Drège,* 7946*b*! Ceded Territory (ex Bentham), *Ecklon*! Queenstown Div.; Queenstown, *Cooper,* 2893! plains near Queenstown, 3500 ft., *Galpin,* 1645! Imvane, *Baur,* 82! Shiloh, 3500 ft., *Baur,* 941!
CENTRAL REGION: Graaff Reinet Div.; around Graaff Reinet, *MacOwan,* 995! *Bolus,* 734! Beaufort West Div.; between Beaufort West and Rhenoster Kop,

Salvia.]

LABIATÆ (Skan). 327

2500–3000 ft., *Drège,* 7946a ! Albert Div. ; Stormberg Spruit, *Burke,* Colesberg
Div. ; Colesberg, *Shaw* !
 KALAHARI REGION : Griqualand West ; Griqua Town, *Burchell,* 1881 ! Orange
River Colony ; Witte Bergen, *Bowker,* 658 ! Orange and Caledon Rivers, *Burke* !
Mrs. Hutton ! Harrismith, *Sankey,* 229 ! Bethlehem, *Richardson* ! Bechuanaland ;
Batlapin Territory, *Holub* ! Transvaal ; Potchefstroom, *Fry in Herb. Galpin,* 6167 !
Burtt-Davy, 2707 ! Warm Bath, *Burtt-Davy,* 2607 ! Var. β : Bechuanaland ;
Batlapin Territory, *Holub* !
 EASTERN REGION : Natal ; Weenen County, *Wood,* 3581 ! near Gourton,
4300 ft., *Wood,* 3631 !

Salvia xerobia, Briq. in Bull. Herb. Boiss. 2me sér. iii. 1076, we have not seen.
We suspect that it is the same as *S. stenophylla,* in which the calyx is never quite
glabrous, and its teeth are more or less spinescent-acuminate. The plant is
usually very freely branched.

28. S. runcinata (Linn. f. Suppl. 89) ; a perennial herb with a long
woody subterranean caudex ; stem erect, rather slender, simple or
branched in the upper part, up to 1¾ ft. high or more, densely
covered with slender mostly short hairs ; leaves shortly petiolate or
the upper subsessile, oblong-lanceolate or oblong-obovate, runcinate-
pinnatipartite or coarsely erose-dentate to pinnatifid towards the
apex, ½–3½ in. long, ¼–1½ in. broad, acute, somewhat hispidly
pubescent both sides, scabrous, slightly rugose and crisped ; terminal
lobe large and elliptic in the basal leaves, smaller and triangular in
the upper ; lateral lobes 3–5 each side, more or less distant, oblong
to oblong-lanceolate, 3–7 lin. long, 1½–4 lin. broad, usually acute,
erose-dentate ; whorls usually 2–6-flowered, very distant, in terminal
racemes 3–12 in. long ; bracts ovate, acuminate, shorter than the
calyx ; calyx campanulate, 2½–3½ lin. long, slightly larger in fruit,
somewhat hispidly hairy, densely gland-dotted ; upper lip 1–1¾ lin.
long, truncate, 3-toothed ; teeth subulate from a broad base, ¼–⅓
lin. long, spinescent ; lower lip bifid ; teeth lanceolate-triangular,
¾–1½ lin. long, spinescent ; corolla blue, 2¾–4½ lin. long ; tube
funnel-shaped, 1¾–3 lin. long ; upper lip obovate, 1–1½ lin. long,
emarginate ; lower lip 1¼–2¼ lin. long ; median lobe broadly obovate,
¾–1¼ lin. long, 1–2 lin. broad, emarginate ; lateral lobes smaller,
rounded. *Thunb. Prodr.* 97, *and Fl. Cap. ed. Schult.* 452 ; *Vahl,
Enum.* i. 260 ; *Ait. Hort. Kew. ed.* 2, i. 58 ; *Roem. et Schult. Syst.* i.
260, *and Mant.* i. 213 ; *A. Dietr. Sp. Pl.* i. 343 ; *Benth. Lab.* 305,
and in DC. Prodr. xii. 352 *partly* ; *Jacq. Hort. Schoenbr.* i. 5, *t.* 8. *S.
runcinata, var. major, Benth. in DC. Prodr.* xii. 352. *S. scabra,
Benth. in E. Meyer, Comm.* 236, *and in DC. Prodr.* xii. 351, *partly,
not of Linn. f.*

 VAR. β, **nana** (Skan) ; very dwarf, the entire plant only 6 in. high or less ;
leaves 2–4 in. long, pinnatipartite, often purplish above ; terminal lobe subelliptic,
up to 1½ in. long, 10–11 lin. broad, rounded ; lateral lobes 3 or 4 each side, 3–8
lin. long, 2–3 lin. broad, rounded.

 VAR. γ, **grandiflora** (Skan) ; corolla about 5 lin. long. *S. scabra, var. angustifolia, Benth. in E. Meyer, Comm.* 236, *and in DC. Prodr.* xii. 351.

 SOUTH AFRICA : without locality, *Thunberg* ! and *cultivated specimens* !
 COAST REGION : Caledon Div. ; Caledon, *Prior* ! Riversdale Div. ; near the
Zoetemelks River, *Burchell,* 6624 ! Uitenhage Div. ; Zwartkops River, *Zeyher,*

397 !   Var. γ : Albany Div. ; Sandy Drift, near Grahamstown, *Miss Daly*, 937 !
Trapps Valley, *Miss Daly*, 680 ! and without precise locality, *Bowker* !
CENTRAL REGION : Var. γ : Victoria West Div. ; Nieuwveld, *Drège*, 4750c !
Albert Div. ; New Hantam, *Drège*, 7945 !
KALAHARI REGION : Griqualand West ; left bank of the Vaal River, probably
near Kimberley, *Holub* ! Transvaal ; Waterval Boven, *Burtt-Davy*, 1453 ! Carolina
District, Leeuwpoort, 5000 ft.. *Burtt-Davy*, 7347 ! Fourteen Streams, *Burtt-Davy*,
1556 !   Var. β : Transvaal ; Skinners Court, Pretoria, *Burtt-Davy*, 606 ! road to
Wonderboom, *Miss Leendertz*, 965 !
EASTERN REGION : Transkei ; between the Gekau (Geua) and Bashee Rivers,
Drège, 4750a !

29. **S. sisymbrifolia** (Skan) ; stem herbaceous, 1 ft. high or more,
erect, branched, leafy, somewhat densely covered with white more
or less curled hairs or here and there glabrescent ; leaves shortly
petiolate, lyrate-pinnatisect, $2\frac{1}{2}$–$4\frac{1}{2}$ in. long, $\frac{3}{4}$–$1\frac{3}{4}$ in. broad, very
sparingly pubescent or almost glabrescent above, sparingly or
densely pubescent on the nerves beneath, gland-dotted ; terminal
lobe elliptic or broadly ovate, $\frac{3}{4}$–$1\frac{3}{4}$ in. long, up to $1\frac{1}{4}$ in. broad,
rounded or rarely somewhat acute, like the lateral lobes more or less
coarsely toothed or sometimes lobed ; lateral lobes usually about
5 each side, opposite, subopposite or rarely alternate, often with
naked portions of the midrib between them, very irregular in shape
and size, more or less oblong, 4–13 lin. long, 2–7 lin. broad, usually
rounded ; whorls 6-flowered, 6–9 lin. apart, in slender short racemes
or few-branched panicles terminating the stem and upper branches ;
bracts broadly ovate or ovate-lanceolate, acuminate, ciliate, shorter
than the calyx, or the lower as long as or longer than the calyx ;
pedicels $\frac{1}{2}$–$1\frac{1}{4}$ lin. long ; calyx campanulate, $2\frac{1}{4}$–3 lin. long, pubescent,
gland-dotted ; tube 1–$1\frac{1}{4}$ lin. long ; upper lip 1–$1\frac{1}{3}$ lin. long, 3-toothed,
finally subtruncate ; teeth subulate from a broad base or sometimes
lanceolate, $\frac{1}{2}$–1 lin. long ; lower lip bifid ; teeth lanceolate-triangular,
acuminate, 1 lin. long ; corolla $2\frac{3}{4}$–3 lin. long, lavender, pubescent
outside on the upper part ; tube 2–$2\frac{1}{3}$ lin. long, slightly enlarged
upwards, 1–$1\frac{1}{4}$ lin. broad at the throat, slightly annular-pilose about
$\frac{3}{4}$ lin. from the base ; upper lip ovate or obovate, emarginate, about
1 lin. long and broad ; lower lip $1\frac{1}{2}$ lin. long ; median lobe broadly
obcordate, $\frac{3}{4}$–1 lin. long, $1\frac{1}{4}$ lin. broad ; lateral lobes smaller, rounded ;
stamens and style slightly exserted.

KALAHARI REGION : Transvaal ; Matebe Valley, *Holub*, 1768 ! near Lydenburg,
Wilms, 1109 ! Pretoria, 4500 ft., *Burtt-Davy*, 7079 ! Wonderfontein, Marico
District, about 3590 ft., *Burtt-Davy*, 7233 ! and without precise locality, *Holub* !
EASTERN REGION : Natal ; Colenso, 3300 ft., *Wood*, 4042 ! Zululand ? Ingoma,
Gerrard, 1227 !

30. **S. repens** (Burch. ex Benth. Lab. 306) ; a perennial herb ;
stem creeping and branched at the base underground, the aerial
part ascending, up to 2 ft. high or more, usually simple, weak,
flexuose, very sparingly hispidly puberulous ; leaves somewhat
tufted at the base, and in very distant pairs on the stem, the lower
petiolate, the upper sessile, lanceolate, oblong or sometimes obovate-

oblong, pinnatifid or sometimes only slightly lobed, almost quite glabrous, or very sparingly pubescent chiefly on the nerves beneath; blade 2–4 in. long, $\frac{3}{4}$–$1\frac{1}{3}$ in. broad, usually acute; terminal lobe about 1 in. long and $\frac{1}{2}$ in. broad; lateral lobes 2–5 each side, subdeltoid, acute, 3–5 lin. long, $1\frac{1}{2}$–4 lin. broad, broadly toothed like the terminal lobe; whorls usually 6-flowered, distant, in slender mostly unbranched racemes up to 1 ft. long; bracts ovate, acuminate or lanceolate, shorter than the calyx or the lower sometimes longer; pedicels up to 2 lin. long; calyx narrowly campanulate, $3\frac{1}{4}$–4 lin. long, somewhat hispidly pubescent, gland-dotted; tube $1\frac{1}{4}$–2 lin. long; upper lip $1\frac{2}{3}$–2 lin. long, 3-toothed; teeth subulate, sometimes much broadened at the base, $\frac{1}{3}$–$1\frac{1}{4}$ lin. long; lower lip $1\frac{1}{2}$–2 lin. long, bifid; teeth linear-triangular, $1\frac{1}{4}$–$1\frac{3}{4}$ lin. long; corolla blue or mauve, 6–8 lin. long; tube $3\frac{1}{2}$–6 lin. long, funnel-shaped, somewhat ventricose, annular-pilose inside about $1\frac{1}{4}$ lin. from the base; upper lip oblong-obovate, 2–$2\frac{1}{2}$ lin. long, emarginate or shortly 2-lobed; lower lip $2\frac{1}{3}$–3 lin. long; median lobe broadly obcordate, $1\frac{1}{4}$–2 lin. long, $1\frac{3}{4}$–$2\frac{1}{2}$ lin. broad; lateral lobes short, rounded. *Benth. in DC. Prodr.* xii. 353. *S. incisa, Benth. in DC. Prodr.* xii. 352. *S. subsessilis, Benth. in E. Meyer, Comm.* 237, *and in DC. Prodr.* xii. 352 *partly.*

COAST REGION: Alexandria Div.; Zuurberg Range, 2000–3000 ft., *Drège*, 4761*b partly*! Albany Div.; Blauw Krantz Bridge, near Grahamstown, *Galpin*, 370!

CENTRAL REGION: Somerset Div.; damp places near the Bosch Berg, 3000 ft.; *MacOwan*, 1591! Graaff Reinet Div.; near Graaff Reinet, 2500 ft., *Bolus*, 259! near Wagenpads Berg, on the southern side, *Burchell*, 2830! in valleys of the Sneeuwberg Range, 3800 ft., *Bolus*, 130! Middelburg Div.; Culmstock, about 3800 ft., *Mrs. Southey in Herb. Galpin*, 5882! Wodehouse Div.; Zuur Poort, Stormberg Range, *Ecklon*! Albert Div., *Cooper*, 587! 2886!

Ecklon's specimen, the type of *S. incisa*, Benth., is from the Wodehouse Division, if the number 112 on the sheet is the locality number (see *Linnæa*, xix. 584–598, and xx. 258).

Cooper's 2886 is labelled "Natal," apparently an error for Albert Division, where his other specimen of the plant was collected.

S. natalensis, Briq. & Schinz in Bull. Herb. Boiss. 2me sér. iii. 1078 (Harrismith, Orange River Colony, Wood, 4972, and near the Kei River, Komgha Division, Coast Region, Schlechter, 6232) is, judging from the first mentioned specimen which we have seen through the kindness of Mr. J. Medley Wood, so much like one of the two specimens on the type sheet of *S. subsessilis*, Benth., referred in this work to *S. repens*, that we are unable to find any distinguishing characters for it.

31. **S. monticola** (Benth. in E. Meyer, Comm. 238, partly); a perennial herb; stems arising from a thick woody caudex, 8–10 in. high, simple or sparingly branched, erect or ascending, rather stout, often rigid, puberulous to villous, somewhat scabrous; basal leaves petiolate, oblanceolate, $1\frac{1}{2}$–3 in. long including the petiole, 4–9 lin. broad, rounded, crenate or crenate-dentate above the middle, few-lobed below; cauline leaves subsessile, lanceolate or narrowly oblong, 1–2 in. long, 3–6 lin. broad, pinnatifid to pinnatipartite,

hispidly pubescent chiefly on the nerves beneath, prominently gland-dotted, crisped, rugose ; terminal lobe oblong to elliptic, 3–6 lin. long, 2–4 lin. broad, somewhat acute or rounded, erose-dentate ; lateral lobes usually 4 each side, rather distant, ovate-triangular, 1–3 lin. long, $\frac{3}{4}$–1$\frac{1}{2}$ lin. broad, usually acute, erose-dentate ; whorls 4–6-flowered, 8–10 lin. apart, in terminal racemes 3–4 in. long ; bracts orbicular-ovate, shortly acuminate, much shorter than the calyx ; calyx about 3$\frac{1}{4}$ lin. long, shortly hispid, gland-dotted ; tube 2 lin. long ; upper lip 1$\frac{1}{4}$–1$\frac{1}{2}$ lin. long, finally subtruncate, 3-toothed ; teeth subulate, $\frac{1}{4}$–$\frac{1}{2}$ lin. long ; lower lip 1$\frac{1}{4}$–1$\frac{3}{4}$ lin. long ; teeth lanceolate-triangular, 1–1$\frac{1}{3}$ lin. long ; corolla 5$\frac{1}{2}$–6 lin. long, pubescent ; tube 3$\frac{1}{2}$ lin. long, inflated at the throat, annular-pilose near the base inside ; upper lip 2–2$\frac{3}{4}$ lin. long, 2-lobed ; lower lip 3–4 lin. long ; median lobe transversely oblong or broadly rounded-ovate, 1$\frac{3}{4}$–2$\frac{1}{2}$ lin. long, 3$\frac{1}{2}$ lin. broad ; lateral lobes short, broad, rounded. *DC. Prodr.* xii. 353, *partly.*

VAR. β, **angustiloba** (Skan) ; all the leaves deeply lobed from base to apex ; lobes narrower, more acute, acutely toothed, often less than $\frac{3}{4}$ lin. broad. *S. monticola, Benth. in E. Meyer, Comm.* 238, *and in DC. Prodr.* xii. 353, *partly.*

COAST REGION : Stockenstrom Div. ; Katberg, *Miss Sole,* 374 ! Queenstown Div. ; Winterberg, *Ecklon* ! Cathcart Div. ; Blesbok Flats, near Windvogel Mountain, *Drège,* 7946a !
EASTERN REGION : Transkei ; between the Geua and Bashee Rivers, 1000–2000 ft., *Drège,* 7946c ! Tembuland ? Bramneck, *Baur,* 76 ! Griqualand East ; around Kokstad, *Tyson,* 1893 ! *and in MacOwan & Bolus, Herb. Norm. Austr.-Afr.,* 578 ! Natal, *Cooper,* 2888 ! Var. β : Transkei ; between the Geua and Bashee Rivers, *Drège,* 4751 !

We suspect that *S. Schlechteri,* Briq. in Bull. Herb. Boiss. 2me sér. iii. 1077 (Schlechter 6330, from Umtata, Tembuland, not seen by us) is the same as *S. monticola,* var. *angustiloba.*

32. S. raphanifolia (Benth. in E. Meyer, Comm. 237);

stem herbaceous, erect, about 1$\frac{1}{2}$ ft. high, densely covered with short velvety deflexed hairs ; leaves velvety pubescent, densely gland-dotted ; lower leaves petiolate, lyrate-pinnatifid, up to 3 in. long ; terminal lobe ovate-oblong, $\frac{3}{4}$–1$\frac{1}{4}$ in. long, 4–9 lin. broad, somewhat acute, often hastate at the base, erose-dentate ; lateral lobes ovate or oblong, up to $\frac{1}{2}$ in. long and broad, rounded or somewhat acute ; upper leaves sessile, oblong-ovate, 1$\frac{1}{4}$–1$\frac{3}{4}$ in. long, 6–7 lin. broad, rounded or hastate at the base, erose-dentate, somewhat rugose ; whorls 6–10-flowered, distant, in simple racemes or few-branched panicles 5–8 in. long ; bracts broadly ovate, shortly acuminate, much shorter than the calyx ; pedicels about 1 lin. long, villous ; calyx tubular-campanulate, 2$\frac{1}{2}$ lin. long, pubescent, gland-dotted ; upper lip about 1 lin. long, 3-toothed ; teeth deltoid-ovate, $\frac{1}{2}$–$\frac{2}{3}$ lin. long, acuminate ; lower lip about 1 lin. long, bifid ; teeth broadly lanceolate-triangular, about $\frac{3}{4}$ lin. long ; corolla about 4$\frac{1}{2}$ lin. long, sparingly pubescent outside ; tube 2$\frac{3}{4}$ lin. long, cylindric below, ventricose at the throat, pilose inside chiefly on the lower side 1$\frac{1}{4}$ lin. from the base ; upper lip broadly oblong, 1$\frac{3}{4}$ lin. long,

emarginate; lower lip 2¼ lin. long; median lobe transversely
oblong, narrowed at the base, or obcordate, 1–1¼ lin. long, 1¾–2¼ lin.
broad; lateral lobes short, rounded. *DC. Prodr.* xii. 352.

COAST REGION: Albany Div., *Bowker*! Cathcart Div.; Blesbok Flats, 3000–
4000 ft., *Drège*, 7943! without precise locality, *Prior*!

33. S. Pegleræ (Skan); stem herbaceous, simple or sparingly
branched, up to about 1⅓ ft. high, slender (only about 1 lin. in
diam.), densely covered as well as the rhachis with rather short soft
often recurved greyish hairs which are sometimes minutely gland-
tipped; internodes 1–2½ in. long; cauline leaves petiolate or the
upper sessile or subsessile, elliptic to elliptic-oblong, the larger
1¼–2¼ in. long, 1–1⅓ in. broad, broadly rounded to slightly acute at
the apex, slightly cordate-sagittate or somewhat hastate at the
base, or sometimes with a little irregular lobing at the base, shortly
and irregularly dentate or crenate, softly pubescent on both sides
especially on the nerves beneath, thin; petiole from 11 lin. long in
the lower leaves to 1 lin. long in the upper; whorls usually
6-flowered, ¼–1¼ in. apart, in terminal racemes 3–5 in. long; bracts
ovate, 3¼ lin. long; pedicels 1¼–1¾ lin. long in flower, 2½ lin. long
in fruit; calyx 3–3¾ lin. long in flower up to 4½ lin. long in fruit,
clothed with sessile glands and with rather long soft spreading hairs
which are sometimes minutely gland-tipped; upper lip 1½–2 lin.
long, 3-toothed; teeth broadly deltoid at the base, subulate above,
⅔–1 lin. long, with a broad sinus between them; lower lip bifid,
1¾–2¼ lin. long; teeth narrowly deltoid at the base, subulate above,
1¼–1½ lin. long; corolla lilac or purple, 5½–6 lin. long; tube 3½–4¼
lin. long, about 1¾ lin. broad at the slightly ventricose upper part,
imperfectly annular-pilose inside; upper lip broadly obovate; 1¾ lin.
long and broad, emarginate; lower lip 2¾ lin. long; median lobe
oblate-obovate, 1½ lin. long, 2–2¼ lin. broad, retuse or emarginate;
lateral lobes ⅔ lin. long, ¾ lin. broad, rounded; stamens shorter or
slightly longer than the upper lip.

COAST REGION: East London Div.; Fort Pato, in grassy fields, *Galpin*, 7830!
EASTERN REGION: Transkei; Kentani District, in valleys and along roadsides,
1200 ft., *Miss Pegler*, 196!

Distinguished from *S. raphanifolia*, Benth., by the much broader leaves, the
lower of which are distinctly petiolate. Miss Pegler states that a valuable lotion
for ophthalmia is obtained from it.

34. S. rudis (Benth. in E. Meyer, Comm. 235); stem herbaceous,
simple, rather slender, about 1 ft. high, densely covered as well as
the leaves and inflorescence with somewhat hispid hairs; basal
leaves long-petiolate; cauline shortly petiolate or the upper sessile;
blade elliptic or oblong-obovate, 1–2½ in. long, ½–1¼ in. broad,
rounded to subacute, crenate or erose-dentate, somewhat crisped on
the margins, rugose; whorls usually 6-flowered, distant, in simple
racemes 4–6 in. long; bracts ovate, acuminate, shorter than the
calyx; calyx 4¾ lin. long, somewhat hispidly hairy outside, finely

pubescent inside, 10-nerved; upper lip 2 lin. long, truncate, 3-toothed; teeth subulate, pungent, $\frac{3}{4}$–1 lin. long, the median shorter than the others; lower lip $2\frac{1}{2}$ lin. long, deeply bifid; teeth lanceolate-subulate, pungent, $2\frac{1}{4}$ lin. long, incurved; corolla $7\frac{1}{2}$ lin. long, shortly pilose on the upper part outside; tube funnel-shaped, $4\frac{1}{2}$ lin. long, $2\frac{1}{4}$ lin. broad at the throat, annular-pilose near the base inside; upper lip erect, 3 lin. long, about 2 lin. broad, emarginate; lower lip $2\frac{1}{4}$ lin. long; median lobe transversely oblong, $1\frac{1}{4}$ lin. long, 2 lin. broad; lateral lobes very short, rounded. *DC. Prodr.* xii. 350. *S. subsessilis, Benth. in E. Meyer, Comm.* 237, *and in DC. Prodr.* xii. 352, *partly.*

COAST REGION : Uitenhage Div., *Ecklon ex Bentham.* Alexandria Div.; Zuurberg Range, 2000–3000 ft., *Drège,* 4761*b partly*! Albany Div.; Grahamstown, *Burke*! *Bolton*! *Williamson*!

EASTERN REGION : Natal; between Umzimkulu River and Umkomanzi River, *Drège,* 4748!

35. S. Woodii (Guerke in Engl. Jahrb. xxvi. 76); a perennial
herb with a creeping subterranean woody caudex; stem erect, often fastigiately branched, rarely simple, somewhat densely pilose, $\frac{2}{3}$–$1\frac{3}{4}$. ft. high; lower leaves subsessile or shortly petiolate, up to 4 in. long, $\frac{1}{4}$–$1\frac{1}{4}$ in. broad; upper sessile, gradually smaller, oblong, lanceolate-oblong or oblanceolate-oblong, rounded to acute, crenate or erose-dentate, occasionally few-lobed in the lower half, shortly pilose, usually more or less rugose, gland-dotted; whorls usually 6–10-flowered, distant, in simple or branched racemes 5–12 in. long; bracts broadly ovate or suborbicular, long-acuminate, shorter than the calyx; pedicels 1–2 in. long; calyx campanulate, $3\frac{1}{4}$–$3\frac{3}{4}$ lin. long, slightly longer after flowering, often densely pilose, gland-dotted; upper lip finally subtruncate, 3-toothed; teeth subulate, $\frac{1}{4}$–1 lin. long; lower lip bifid; teeth lanceolate-triangular, $1\frac{1}{4}$–2 lin. long, acuminate; corolla pale blue or lilac, about 5 lin. long, shortly pubescent; tube funnel-shaped, $2\frac{1}{2}$–3 lin. long, annular-pilose inside; upper lip $1\frac{1}{2}$–2 lin. long; lower lip 2–$2\frac{1}{2}$ lin. long.

KALAHARI REGION : Orange River Colony; Bloemfontein, *Rehmann,* 3840. Transvaal; Hoogeveld, Perekopberg, *Rehmann,* 6842. Standerton, *Rehmann,* 6780, 6781! Houtbosch, *Rehmann,* 6165! Aapies Poort, *Rehmann,* 4112! Aapies River, *Miss Leendertz,* 967!

EASTERN REGION : Natal; Weenen County, *Wood.* Bank of the Mooi River, *Wood,* 992, 3621! Mooi River Station, *Kuntze.* Near Newcastle, 3800 ft., *Wood,* 6801!

S. Schenckii, Briq. in Bull. Herb. Boiss. 2me sér. iii. 1079 (Orange River Colony, between Harrismith and the Vaal River, Schenck, 732) we have not seen. We are unable from the description to distinguish it from *S. Woodii.*

36. S. Cooperi (Skan); a perennial herb with a woody subter-
ranean caudex; stem branched, erect, up to $1\frac{1}{2}$ ft. high or more, $\frac{3}{4}$–1 lin. in diam., more or less covered with slender hairs; internodes $1\frac{1}{4}$–3 in. long; lower leaves petiolate, upper sessile, lanceolate-oblong or sometimes elliptic-obovate, the lower about 2 in. long (rarely up to $3\frac{1}{2}$ in. long), 6–11 lin. broad, the upper gradually

becoming smaller, rounded, obtuse or acute at the apex, rounded or slightly narrowed at the base, rather regularly crenate, irregularly erose-dentate or sometimes only slightly undulate, occasionally lobed in the lower part, prominently 5- or 6- nerved each side, sparingly covered with short stiff hairs above, hispid on the nerves beneath or almost glabrous, densely gland-dotted ; petiole 2–10 lin. long ; whorls 2–6-flowered, distant, in a terminal simple or few-branched raceme 3–12 in. long ; bracts ovate, spinescent-acuminate, shorter than the calyx ; pedicels 1–2 lin. long, villous ; calyx campanulate, $3\frac{1}{4}$–$3\frac{3}{4}$ lin. long, about 4 lin. long when mature, rather sparingly pubescent, gland-dotted ; tube $1\frac{3}{4}$ lin. long ; upper lip about 2 lin. long, finally truncate, 3-toothed ; teeth subulate from a broad base, $\frac{1}{2}$–$\frac{3}{4}$ lin. long, spinescent ; lower lip about $1\frac{3}{4}$ lin. long, bifid ; teeth linear- or lanceolate-triangular, $1\frac{1}{2}$ lin. long, spinescent ; corolla pale blue or purple-blue, $6\frac{1}{2}$–7 lin. long, puberulous ; tube broad, $3\frac{3}{4}$–$4\frac{3}{4}$ lin. long, $\frac{3}{4}$–1 lin. broad at the base, broadly ventricose at the throat, slightly recurved ; upper lip obovate-oblong, $2\frac{1}{2}$–3 lin. long, bilobed ; lower lip 3–$4\frac{1}{4}$ lin. long ; median lobe broadly obcordate, $1\frac{3}{4}$–$2\frac{1}{4}$ lin. long, nearly $2\frac{1}{2}$–$3\frac{3}{4}$ lin. broad ; lateral lobes shorter, rounded.

KALAHARI REGION : Orange River Colony ; Besters Vlei, near Witzies Hoek, *Bolus*, 8237 !

EASTERN REGION : Griqualand East ; Vaal Bank, *Haygarth in Herb. Wood*, 4190 ! Natal, *Cooper*, 1279 ! Southdowns, 5000–6000 ft., *Evans*, 389 !

Cooper, 2889, from Natal, a juvenile specimen, is probably the same species. It has deeply lobed basal leaves up to $3\frac{1}{2}$ in. long.

S. Cooperi closely resembles *S. Woodii*, Guerke, and the larger-flowered form of *S. runcinata*, Linn. f., but it has much larger flowers than either. From *S. repens*, Burch., it differs in its dwarfer branched habit and in having usually much smaller less-lobed leaves.

Imperfectly known Species.

37. S. granitica (Hochst. in Flora, 1845, 65) ; an undershrub ; stem simple, 1 ft. high, minutely hairy ; leaves oblong-lanceolate, almost quite entire or serrate-dentate in the upper part, prominently net-veined beneath, shining (glutinous?), somewhat hispid ; whorls few-flowered, in simple or slightly branched racemes ; pedicels short, pubescent ; bracts rather small ; calyx 2-lipped ; teeth cuspidate.

COAST REGION : Caledon Div., at the foot of Babylon's Tower Mountain, in granitic soil, *Krauss*, 1120.

38. S. lanuginosa (Burm. f. Fl. Cap. Prodr. 1) ; lower leaves lunulate ; upper oblong, tomentose.

SOUTH AFRICA : without locality, *Oldenland ex Burmann*.

XVII. CEDRONELLA, Moench.

Calyx tubular or campanulate, 13–15-nerved, subequal or oblique at the mouth, equally 5-toothed. *Corolla-tube* exserted, enlarged at the throat, exannulate inside ; limb 2-lipped ; upper lip erect ;

somewhat flat, 2-fid or emarginate; lower spreading, 3-fid, with the median lobe largest. *Stamens* 4, didynamous, the upper longer than the lower, ascending under the upper lip or exserted; anthers 2-celled; cells parallel, distinct. *Disk* subequal. *Style* shortly 2-fid at the apex; lobes subulate, subequal. *Nutlets* ovoid, smooth.

Herbs or shrubs; leaves toothed or in 1 species 3-foliate, the floral reduced to bracts; whorls loosely few-flowered or densely many-flowered, crowded in a terminal spike or raceme; bracteoles small, setaceous; corolla red, violet, purple or blue, often showy.

DISTRIB. Species 12, chiefly natives of North and Central America; 1 in Japan.

1. **C. triphylla** (Moench, Meth. 412); plant with an odour of balsam; stem tall, woody at the base, herbaceous above, simple or branched, sharply 4-angled, glabrous or nearly so; leaves petiolate, 3-foliolate; leaflets lanceolate, acuminate, 1–3½ in. long, ¼–1¼ in. broad, the median larger than the lateral and stalked, crenate or crenate-serrate, glabrous or pubescent beneath; lateral leaflets subsessile, oblique at the base, sometimes unequally 2-lobed; petiole ¼–1½ (usually ½–1¼) in. long; whorls 6–12-flowered, in dense sometimes slightly interrupted oblong terminal spikes 1–2½ in. long; calyx tubular-campanulate, about 5 lin. long, pubescent, gland-dotted; teeth equal, lanceolate-triangular, subulate at the apex, 1½ lin. long; corolla about 9 lin. long, tawny-purple, marbled; tube narrow at the base, very much dilated towards the throat. *Benth. Lab.* 502, *and in DC. Prodr.* xii. 406. *C. canariensis, Webb et Berth. Phyt. Canar.* iii. 87. *C. canariensis viscosa, etc., Commel. Hort. Amstel.* ii. 81, *t.* 41. *Dracocephalum canariense, Linn. Sp. Pl. ed.* ii. 829.

COAST REGION : Cape Div.; Devils Peak, 900 ft., *Bolus,* 4624 ! thicket near Blockhouse, *Wolley Dod,* 684 ! Caledon Div.; Caledon, *Mrs. Southey in Herb. Galpin,* 7850 !

Introduced. A native of the Canaries and Madeira.

XVIII. ACROTOME, Benth.

Calyx tubular-campanulate, slightly oblique or almost equal at the mouth, 10- or 11-nerved, 5–11-toothed. *Corolla-tube* as long as the calyx or exserted, exannulate inside; limb 2-lipped; upper lip erect, slightly arched, entire or emarginate; lower lip spreading, 3-lobed, the median lobe larger than the lateral. *Stamens* 4, didynamous, the lower pair longer than the upper, included; anthers 1-celled by confluence, of the upper stamens short, ovate, of the lower oblong, twice as long. *Disk* equal. *Style* included, barbate at the apex, almost entire or very shortly 2-fid. *Nutlets* triquetrous, truncate at the apex.

Herb or undershrubs; leaves entire or toothed, the floral similar to the others or gradually reduced in size; whorls few- or densely many-flowered, in the axils of the upper leaves; flowers small, sessile; calyx in the fruiting-stage more or less enlarged.

DISTRIB. Species 6, in South and South Tropical Africa.

Plant 1-1½ ft. high ; whorls usually solitary, densely
very many-flowered, forming a large globose head ... (1) **inflata**.

Plant 3-10 in. high ; whorls usually 2-4, 2-10-flowered :
Leaves oblong or oblong-lanceolate, much longer than
broad ; plant minutely puberulous or very shortly
pubescent :
Calyx 5-toothed   ...   . (2) **pallescens**.

Calyx 8-11-toothed   ...   . (3) **Thorncroftii**.

Leaves usually obovate and often scarcely longer than
broad ; plant densely hispid   ...   . (4) **hispida**.

1. **A. inflata** (Benth. in DC. Prodr. xii. 436) ; an erect villous
branched annual herb 1-1½ ft. high ; leaves usually shortly petiolate,
oblong-lanceolate, ovate-lanceolate or ovate, ¾-2 in. long, ⅓-1 in.
broad or more, broadly crenate-serrate from the middle to the apex,
shortly and densely pilose on both sides ; whorls densely many-
flowered, in usually a solitary globose head ¾-1 in. in diam. in the
axils of the upper leaves ; bracts narrowly linear, pilose, somewhat
shorter than the calyx ; calyx 3-3¼ lin. long in flower, about 7 lin.
long and somewhat inflated in fruit, 5-toothed ; teeth deltoid or
subulate, 1-1½ lin. long, spinescent ; corolla very pale mauve, about
as long as the calyx. *Oliv. in Hook. Ic. Pl. t.* 1467.

COAST REGION : Queenstown Div. ; Shiloh, about 3500 ft., *Baur*, 784 ! plains
at Queenstown and Sterkstroom, 3500-4000 ft., *Galpin*, 1505 !
CENTRAL REGION : Steynsburg Div. ; Zuurberg Range, *Burke* ! Albert Div.,
Cooper, 1382 !
KALAHARI REGION : Bechuanaland ; Eastern Bamanguato Territory, *Holub* !
Transvaal ; Linokana, *Holub* ! near Viljoens Drift, *Gilfillan in Herb. Galpin*,
7148 ! Nylstroom, *Mrs. de Jongh in Herb. Galpin*, 6500 !

Also in Hereroland and Ngamiland.

2. **A. pallescens** (Benth. in DC. Prodr. xii. 436) ; a minutely
puberulous branched undershrub ; branches twiggy, slender, 2½-10
in. long ; leaves very shortly petiolate, distant, oblong or oblong-
lanceolate, 4-8 lin. long, 1-3 lin. broad, obtuse, narrowed at the
base, dentate at the apex ; whorls few, distant, 2-6-flowered ;
bracts minute ; calyx 3-3½ lin. long, 5-toothed, minutely puberulous ;
teeth deltoid at the base, subulate above, ¾-1 lin. long, spinescent ;
corolla about 5 lin. long, densely pilose ; tube 3½ lin. long, slender.
Stachys Steingroeveri, Briq. in Engl. Jahrb. xix. 193, *and in Bull.
Herb. Boiss.* 2ᵐᵉ *sér.* iii. 1096.

SOUTH AFRICA : without locality, *Drège* ! 7951 !
WESTERN REGION : Great Namaqualand, *Steingroever*, 11. Little Namaqualand,
Wyley, 92 !

3. **A. Thorncroftii** (Skan) ; a small perennial herb, slightly woody
at the base ; stems ascending, slender, scarcely 6 in. long, sparingly
branched, rather densely covered with short spreading hairs,
sparingly leafy ; leaves oblong, 4-5½ lin. long, 1⅓-2¾ lin. broad,

entire or 1–3-toothed at the apex obtuse, narrowed at the base, rather thick, shortly hispid, scabrous; petiole up to $\frac{3}{4}$ lin. long; whorls 2–5 at the ends of the stems or branches, 4–7 lin. apart, 6–10-flowered; bracts linear, $1\frac{1}{4}$–$1\frac{3}{4}$ lin. long, ciliate; calyx sub-sessile, tubular, slightly incurved, $3\frac{1}{4}$ lin. long, nearly equal at the mouth, 10–11-nerved, 8–11-toothed, rather densely covered with short stiff spreading hairs; teeth unequal or nearly equal, subulate, often with a deltoid base, $\frac{1}{4}$–$\frac{3}{4}$ lin. long, rather thick; corolla white, $4\frac{1}{2}$–5 lin. long, retrorsely villous outside; tube about 3 lin. long, slightly curved; upper lip broadly ovate, deeply concave, $1\frac{3}{4}$–2 lin. long, shortly ciliate, entire; lower lip 3 lin. long; median lobe broadly ovate, $1\frac{3}{4}$ lin. long, $2\frac{1}{2}$–$2\frac{3}{4}$ lin. broad, crenulate; lateral lobes about $1\frac{1}{4}$ lin. long and 1 lin. broad, rounded.

KALAHARI REGION : Transvaal ; Barberton, *Thorncroft*, 3124 !

4. A. hispida (Benth. in DC. Prodr. xii. 436); an erect or ascending branched herb 3 or 4 in. high, everywhere densely hispid, sometimes rather woody at the base; leaves very shortly petiolate, obovate or sometimes ovate, $\frac{1}{4}$–$\frac{1}{2}$ in. long, $1\frac{1}{2}$–4 lin. broad, obtuse, narrowed at the base, quite entire or few-toothed at the apex; whorls usually 2–4, somewhat distant, 6–10-flowered ; bracts linear, as long as the calyx ; calyx subcampanulate, hispid, 3–$3\frac{1}{2}$ lin. long; teeth 7–10, deltoid or subulate, unequal, $\frac{1}{2}$–$1\frac{1}{4}$ lin. long, acute, somewhat spinescent; corolla 4 lin. long ; tube $2\frac{3}{4}$ lin. long, densely villous on the upper part; upper lip $1\frac{1}{2}$ lin. long, ovate, hirsute, emarginate ; lower lip $1\frac{1}{2}$–2 lin. long; median lobe suborbicular, 1 lin. long, about $1\frac{1}{2}$ lin. broad.

KALAHARI REGION : Transvaal ; near Schoen Spruit, *Burke* ! Vaal River, *Burke* ! Aapies River, *Burke* ! near Carolina, 5000–5600 ft., *Burtt-Davy*, 7397 ! *Bolus*, 12243 ! Krugers Post, in Lydenburg District, *Burtt-Davy*, 7296 ! near Lydenburg, *Wilms*, 1135 ! Witbank, Middelburg District, *Gillfillan in Herb. Galpin*, 7231 !

EASTERN REGION : Natal, *Gerrard*, 1220 !

Bentham in DC. Prodr. xii. 436 distinguished two varieties of *A. hispida*. Var. *elongata*, with lax stems, 6–8 in. long and shorter hairs (Burke, Vaal River); Var. *obliqua*, with shorter hairs, and suboblique calyx with shorter teeth (Burke, Aapies River).

XIX. STACHYS, Linn.

Calyx tubular-campanulate, campanulate or sometimes funnel-shaped, 5- or 10-nerved, equal or oblique at the mouth, rarely more or less distinctly 2-lipped; teeth 5, subequal. *Corolla-tube* cylindric to narrowly funnel-shaped, often enlarged at the mouth, included or exserted, straight or incurved, naked inside or more or less perfectly annular-pilose near the base; limb 2-lipped ; upper lip erect or ascending, usually concave or arched, entire or very shortly emarginate ; lower lip spreading or deflexed, 3-lobed. *Stamens* 4, didynamous, (the lower pair longer), ascending under the upper lip, more or less exserted from the corolla-tube ; anthers 2-celled ; cells distinct, parallel or usually divergent and at length divaricate. *Disk* usually equal. *Style* subequally bifid at the apex. *Nutlets* ovoid or oblong, obtuse or rounded at the apex.

Annual or perennial herbs, undershrubs or sometimes shrubs, with various kinds of indumentum, or sometimes nearly glabrous ; leaves sessile or petiolate, entire or toothed, the upper often reduced to bracts ; whorls 2- to many-flowered (rarely reduced to 1 flower), axillary, or in terminal spikes or racemes ; flowers sessile or very shortly pedicellate, often rather small, sometimes showy, variously coloured.

DISTRIB. Species about 320, most frequent in the temperate regions of both hemispheres.

*Herbs, rarely undershrubs, variously hairy, but not densely covered with a felt-like or wool-like indumentum, sometimes almost glabrous :
　Corolla-tube 7–10 lin. long, often twice or more than twice as long as the calyx :
　　Stem robust, somewhat prickly on the angles ;
　　　calyx 5–7 lin. long ; corolla-tube broad　　... (1) **Thunbergii.**
　　Stem slender, not prickly ; calyx 3½ lin. long ;
　　　corolla-tube slender　...　...　...　... (2) **tubulosa.**
　Corolla-tube less than 6 lin. long, not twice as long as the calyx :
　　†Leaves ovate, large, often 1½ in. and sometimes more than 3 in. long, usually more than 1 in. up to 2½ in. broad :
　　　Calyx-teeth usually long and very narrow, not or scarcely spreading, or if spreading they are more or less curved near the apex ; spike usually unbranched :
　　　　Calyx-teeth usually 1¼–2¼ lin. long ; all the leaves distinctly petiolate :
　　　　　Upper bracts scarcely longer than the calyx ; calyx more or less densely covered with rather short hairs :
　　　　　　Leaves especially beneath densely and softly hairy ; teeth often 20–25 each side　...　...　...　...　... (3) **grandifolia.**
　　　　　　Leaves very sparingly hairy ; teeth usually 12–15 each side　...　...　... (4) **Cooperi.**
　　　　　Upper bracts distinctly longer than the calyx ; calyx densely covered with long hairs　...　...　...　... (5) **Bolusii.**
　　　　Calyx-teeth ¾–1 lin. long ; upper leaves sessile or subsessile　...　...　...　... (7) **Kuntzei.**
　　　　Calyx-teeth rather short and broad, distinctly spreading ; spike branched...　...　... (6) **albiflora.**
　　††Leaves oblong-lanceolate, linear-oblong to linear, or if ovate to ovate-oblong or suborbicular they are rarely more than 1 in. long, often much shorter, and rarely more than ¾ in. broad :
　　　‡Leaves ovate to ovate-oblong, about as long as broad up to twice as long as broad, or if more than twice as long as broad not usually exceeding ½ in. long :
　　　　§Leaves sessile or subsessile (sometimes shortly petiolate in *S. Galpini*) :

Leaves 4–12 lin. long, 3–9 or rarely 10 lin.
broad ; calyx densely covered with
rather long silky hairs or with shorter
velvety hairs :
Leaves so densely covered beneath with
velvety hairs that the surface is
hidden   ...   . (10) **sessilifolia.**
Leaves more or less hairy, but the surface
not hidden by the hairs :
Branches somewhat weak and spreading ; leaves ovate, usually only
slightly longer than broad ; whorls
2-flowered ...   . (11) **Galpini.**
Branches somewhat rigid, erect ; leaves
usually ovate-elliptic, often about
twice as long as broad ; whorls
4-flowered or more   ...   ... (31) **obtusifolia**, var.
                                                    Flanaganii.

Leaves usually 2–6 lin. long, 1½–3 lin.
broad ; calyx glabrous or with few
rather short stiff hairs :
Stems decumbent, with stiff reflexed
hairs ; whorls usually 2-flowered ;
calyx with usually rather short stiff
hairs ...   . (24) **subsessilis.**
Stems erect, glabrous or with a few
spreading hairs; whorls 4–6-flowered;
calyx nearly glabrous   ...   ... (36) **tenella.**
§§Leaves distinctly petiolate (usually sessile or
subsessile in *S. Galpini*) :
Stem erect, usually robust, not or sparingly
branched :
Whorls 2-flowered ; leaves not much
longer than broad :
Leaves very rugose ; calyx about 3 lin.
long ; teeth 1 lin. long ...   ... (8) **Rehmannii.**
Leaves not rugose ; calyx 4–5 lin. long ;
teeth 1¾–2 lin. long :
Leaves about as long as broad ;
whorls of flowers up to 12   ... (12) **transvaalensis.**
Leaves distinctly longer than broad ;
whorls of flowers 1–3 ...   ... (14) **parilis.**
Whorls 4-flowered or more ; leaves usually
about twice as long as broad :
Leaves elliptic, rounded at the apex ... (31) **obtusifolia.**
Leaves ovate-lanceolate, narrowed to
an acute or obtuse apex ..   ... (32) **Tysonii.**
Stem prostrate, decumbent or ascending,
weak, usually much branched :
Plant very densely covered with a greyish
velvety pubescence ...   . (9) **malacophylla.**
Plant variously hairy, usually not densely
covered with a greyish velvety
pubescence,   sometimes   almost
glabrous :
Annual ...   . (27) **arvensis.**
Perennial : calyx - teeth more or less,
acute, but not subulate and
spinescent at the apex...   ... (13) **Rudatisii.**

Perennial : calyx-teeth subulate and more or
less spinescent at the apex :
  Plant dark brown when dry ; stem and
    branches sparingly hairy or glabrous,
    often distinctly scabrous ...   . (16) **scabrida.**
  Plant usually greenish when dry ; stem and
    branches variously hairy or glabrous, not
    distinctly scabrous :
    Leaves scarcely cordate at the base ; whorls
      several, 2-flowered ; calyx purplish ... (22) **flexuosa.**

    Leaves not cordate at the base ; whorls
      solitary, terminating the slender
      branches, 4–6-flowered ; calyx green... (17) **Harveyi.**

    Leaves usually distinctly cordate at the
      base ; whorls several, 2- to many-
      flowered ; calyx green :
      Calyx narrowly obconical at the base,
        somewhat abruptly enlarged above
        the middle ...   . (20) **fruticetorum.**

      Calyx campanulate or tubular-campanu-
        late, not abruptly enlarged above the
        middle :
        Calyx often conspicuously glandular-
          puberulous :
          Whorls usually 4–6-flowered ; calyx-
            teeth very narrow at the long
            spinescent apex...   ...   ... (15) **æthiopica,** var.
              glandulifera.

          Whorls 2 - flowered ; calyx - teeth
            broader, acute, not or only
            slightly spinescent   ...   ... (13) **Rudatisii.**

        Calyx almost glabrous to variously hairy :
          Plant more or less densely short-
            velvety or long-silky hairy :
            Leaves 9–15 lin. long, 6–11 lin.
              broad ; calyx up to 5½ lin. long :
            Leaves deltoid-ovate ; whorls of
              flowers few, distant   ... (14) **parilis.**

            Leaves usually ovate or ovate-
              rounded ; whorls of flowers
              several, usually close   ... (11) **Galpini.**

          Leaves 4–10 lin. long, 4–8 lin.
            broad ; calyx up to 3 lin.
            long :
            Teeth of the leaves small and
              regular, often 10 – 12 or
              more pairs :
              Corolla - tube 2¼ lin. long,
                shorter than the calyx... (18) **serrulata.**

              Corolla-tube 4 lin. long, much
                longer than the calyx ... (9) **malacophylla.**

            Teeth of the leaves relatively
              large and irregular, usually
              not more than 5 pairs   ... (19) **attenuata.**

Plant nearly glabrous or more or less
  stiffly hairy, sometimes densely :
  Calyx with a few long pallid hairs (25) **Priori.**

Calyx with few or many short
    hairs or densely long-hairy :
Leaves shallowly crenate ; bran-
    ches very numerous, slen-
    der, wiry ... ... (23) **cymbalaria.**

Leaves deeply toothed; branches
    fewer, somewhat thicker
    and shorter ... ... (26) **sublobata.**

Leaves usually rather regularly
    crenate ; branches often
    many and slender, but
    scarcely wiry :
    Inflorescence unilateral ... (21) **leptoclada.**
    Inflorescence not unilateral (15) **æthiopica.**

‡‡Leaves oblong, oblong-lanceolate, ovate-lanceolate
    or deltoid-lanceolate to linear, about 3 to
    several times longer than broad, rarely less
    than $\frac{3}{4}$ in. long :
Leaves sessile or subsessile :
    Calyx densely hairy :
    Plant blackish when dry ; leaves usually
        $1\frac{1}{2}$–$2\frac{1}{4}$ in. long, 1–4 lin. broad... ... (28) **nigricans.**
    Plant not blackish when dry ; leaves usually
        $\frac{3}{4}$–2 in. long, 3–6 lin. broad ... ... (29) **sessilis.**
    Calyx glabrous or sparingly and shortly
        hairy :
    Stem glabrous or nearly so ; calyx as long
        as or longer than the corolla-tube ... (34) **humifusa.**
    Stem distinctly hairy ; calyx shorter than
        the corolla-tube... ... ... (35) **rivularis.**

Leaves (except sometimes the upper) distinctly
    petiolate (sometimes very shortly in *S.
    erectiuscula*) :
    Pairs of leaves few, usually approximate near
        the bottom ... ... (30) **simplex.**
    Pairs of leaves many, spaced at nearly equal
        distances from the bottom :
    Leaves oblong, rounded at the apex ; calyx
        rather densely covered with long hairs (31) **obtusifolia,** var.
                         angustifolia.
    Leaves ovate-lanceolate, often subacute ;
        calyx covered (sometimes sparingly)
        with short hairs :
    Calyx with short stiff adpressed hairs ;
        teeth linear-triangular, rigid ... (33) **erectiuscula.**
    Calyx shortly and softly glandular-
        pubescent ; teeth deltoid, not rigid (32) **Tysonii.**

**Shrubs or undershrubs, usually very densely covered
    with whitish or sometimes yellowish felt-like or
    wool-like indumentum (often almost glabrous in
    S. hyssopoides ; sometimes only thinly covered
    with stellate hairs in *S. dregeana* and *S. caffra* ;
    branches glabrescent in *S. Zeyheri*) :
Calyx more or less distinctly 2-lipped ... ... (38) **Burchellii.**

Calyx not 2-lipped :
    †Calyx densely covered with wool-like sometimes
    almost plumose indumentum :

Leaves 2–10 lin. long, 1–4 lin. broad ; calyx-
teeth up to 2¾ lin. long:
Branches angular, scarcely woody :
　Upper whorls of flowers crowded ; calyx-
　　teeth subulate or linear ; corolla only
　　slightly longer than the calyx　　... (47) **integrifolia.**

　Upper whorls of flowers distinct ; calyx-
　　teeth ovate-lanceolate ; corolla half as
　　long again as the calyx　　...　...　(48) **hantamensis.**

　Branches terete, distinctly woody　...　...　(49) **teres.**

Leaves 1–1¾ in. long, ¼–1¼ in. broad ; calyx-
teeth 1–1¼ lin. long ...　...　...　...　(39) **Lamarckii.**

††Calyx densely covered with felt-like indumentum :
　Leaves usually scarcely longer than broad, rarely
　　more than 6 lin. long :
　Branches tomentose ; leaves usually obovate,
　　distinctly cuneate at the base ...　...　(45) **cuneata.**

　Branches glabrescent ; leaves elliptic-ovate,
　　usually rounded at the base　...　...　(46) **Zeyheri.**

Leaves much longer than broad, usually much
　more than 6 lin. long :
　Leaves ovate-elliptic or sometimes obovate ... (40) **multiflora.**

Leaves oblong to linear or spathulate :
　Plant yellowish in the dried state :
　　Indumentum on branches and leaves very
　　　short and felt-like, not floccose :
　　Leaves lanceolate, acute or subobtuse,
　　　narrowed at the base　...　...　(41) **flavescens.**

　　Leaves oblong, obtuse, rounded at
　　　the base　...　...　...　...　(42) **gariepina.**

　　Indumentum longer and somewhat loose,
　　　often floccose, sometimes of scattered
　　　stellate hairs ...　...　...　...　(43) **dregeana.**

　Plant greyish in the dried state (sometimes
　　yellowish in *S. dregeana*) :
　　Indumentum short or very short, not
　　　floccose :
　　Leaves spathulate or linear-spathulate,
　　　rounded and broadest at the apex,
　　　smooth, entire　...　...　...　(44) **spathulata.**

　　Leaves oblong-lanceolate, lanceolate to
　　　linear, usually narrowed to the
　　　apex and often acute, entire or
　　　toothed, often very rugose　... (37) **rugosa.**

　　Indumentum longer and somewhat loose,
　　　often floccose, sometimes of scat-
　　　tered stellate hairs ...　...　... (43) **dregeana.**

†††Calyx very thinly and minutely stellate-tomentose
　or sometimes glabrescent :
　Leaves linear to oblong-lanceolate, usually
　　entire, often glabrescent　...　... (50) **hyssopoides.**

　Leaves lanceolate, usually serrate, often more
　　or less stellate-tomentose especially
　　beneath ...　...　...　...　... (51) **caffra.**

1. **S. Thunbergii** (Benth. Lab. 540); stem herbaceous, erect, stout, usually simple, up to 5 ft. high, somewhat prickly, chiefly on the angles; leaves ovate to ovate-lanceolate, the larger 2–2½ in. long, ¾–1½ in. broad, acute or somewhat obtuse, deeply cordate at the base, crenate, sparingly hispid or almost glabrous, rather thick and leathery, often rugose; petiole 2–7 lin. long; lower bracts similar to the upper leaves, the upper much reduced; whorls usually several, commonly 4- or 6-flowered, distant; pedicels ¾–1½ lin. long; calyx tubular-campanulate, 5–7 lin. long, puberulous; teeth broadly to narrowly lanceolate-triangular, 1¾–3 lin. long, acuminate, somewhat spinescent; corolla red or purple, shortly adpressed-pubescent; tube narrowly funnel-shaped, curved, 8–10 lin. long; upper lip obovate-elliptic, 3½–4 lin. long; lower lip 4½–5 lin. long. *DC. Prodr.* xii. 467; *Bolus & Wolley-Dod in Trans. S. Afr. Phil. Soc.* xiv. 310. *S. hispida, Briq. in Engl. & Prantl, Pflanzenfam.* iv. 3A, 263. *Galeopsis hispida, Thunb. Prodr.* 96, and *Fl. Cap. ed. Schult.* 446.

SOUTH AFRICA : without locality, *Villett*!
COAST REGION : Cape Div. ; Devils Mountain, *Thunberg*! *Carmichael*! *Pappe*! *Prior*! *Wolley-Dod. Bolus in MacOwan & Bolus, Herb. Norm. Austr.-Afr.* 367! near Constantia, *Thunberg*! Caledon Div. ; Genadendal, *Prior*! Mossel Bay Div.; Attaquas Kloof, *Thunberg*! George Div. ; George, *Prior*! Montagu Pass, *Penther*, 1761 (ex *Guerke*). Knysna Div. ; Diep River, *Bolus*, 2435 ! between Cloetes Kraal and Paarde Kraal, *Burchell*, 5153 !

Also a specimen labelled "Beaufort," from Lehmann's Herbarium.

2. **S. tubulosa** (MacOwan in Kew Bulletin, 1893, 13); stem herbaceous, weak, ascending, sparingly branched, 1 ft. high or more, softly pilose; internodes often 2–3 in. long; leaves broadly ovate, 1¼–2 in. long, 1–1¾ in. broad, acute or subacute, deeply cordate at the base, crenate, sparingly pilose, thin ; petiole slender, ½–1¾ in. long, or the upper shorter ; lower bracts leaf-like, the upper much reduced ; whorls few, 4–6-flowered, distant ; calyx almost sessile, campanulate, very thin, 3½ lin. long, sparingly pilose outside ; teeth lanceolate-triangular, about 1½ lin. long, acuminate; corolla pale purple, shortly pubescent; tube narrowly tubular, scarcely enlarged upwards, 7–9 lin. long; upper lip elliptic, 3½ lin. long ; lower lip 2¾ lin. long, subequally 3-lobed, pencilled with dark purple veins ; lobes ovate, scarcely 1 lin. long. *S. dolichodeira, Briq. in Bull. Herb. Boiss.* 2ᵐᵉ *sér.* iii. 1081.

EASTERN REGION : Griqualand East ; in woods on Mount Malowe near the River Umzimvubu, and on the Zuurberg Range, 4000–4500 ft., *Tyson*, 2153 ! 2549 ! and *in MacOwan & Bolus, Herb. Austr.-Afr.* 1297 ! Zuurberg Range, *Wood*, 1895 ! *Schlechter*, 6605 ! Natal, *Mrs. K. Saunders*! *Sanderson*, 608 !

3. **S. grandifolia** (E. Meyer, Comm. 239); stem herbaceous, ascending, 15 in. high or more, ¾–1½ lin. in diam., sparingly branched or simple, pilose ; leaves ovate, the larger 1½–2¼ in. long and 1–1½ in. broad, subacute to rounded, rather deeply and openly

cordate at the base, regularly crenate, shortly greyish-velvety-pilose, especially beneath ; teeth usually 20–25 each side ; petiole 6–17 lin. long ; bracts ovate to ovate-lanceolate, longer than or about as long as the calyx ; whorls 2–8-flowered, many, distant, in a spike up to 6 in. long ; calyx campanulate, $2\frac{3}{4}$–$3\frac{1}{2}$ lin. long, softly pilose ; teeth narrowly lanceolate-triangular, 1–$1\frac{2}{3}$ lin. long, acuminate-spinescent ; corolla white or tinted with lilac, pubescent ; tube rather narrow, $3\frac{1}{2}$–4 lin. long ; upper lip elliptic, $1\frac{3}{4}$–$2\frac{3}{4}$ lin. long ; lower lip $2\frac{1}{2}$–$3\frac{1}{2}$ lin. long. *Benth. in DC. Prodr.* xii. 475.

COAST REGION : Swellendam Div. ; near Grootvaders Bosch, 1000–4000 ft., *Zeyher*, 3536 ! Queenstown Div. ; Chumie (Tyumie) Mountain, *Ecklon* ! and without precise locality, *Cooper*, 343 !

CENTRAL REGION : Somerset Div. ; Bosch Berg, 3000–4000 ft., *MacOwan*, 1494 ! 1499 ! Graaff Reinet Div. ; Sneeuwberg Range, 3800 ft., *Bolus*, 1969 !

KALAHARI REGION : Transvaal ; near Lydenburg, *Wilms*, 1118 !

EASTERN REGION : Transkei ; around Kentani, 1200 ft., *Miss Pegler*, 434 ! Pondoland ; between Umtata River and Umtsikaba River, 1000–2000 ft., *Drège*, 4781*a* ! 4781*b*. Natal ; Polela, *Fourcade*, 4556 ! Durban, *Gerrard*, 2042 !

4. S. Cooperi (Skan in Kew Bulletin, 1909, 420) ; stem herbaceous, decumbent, very weak, branched, 1–2 ft. long, $\frac{2}{3}$–$\frac{3}{4}$ lin. thick, rather softly pubescent or sparingly hispidulous ; branches very slender, up to 1 ft. long or more ; internodes $1\frac{1}{4}$–$2\frac{1}{2}$ in. rarely up to 6 in. long ; leaves ovate, the larger $1\frac{1}{2}$–2 in. long and $1\frac{1}{4}$–$1\frac{2}{3}$ in. broad, acute or obtuse at the apex, openly and deeply cordate at the base, regularly and shallowly crenate, sparingly and softly pubescent or more sparingly hispidulous, sometimes almost glabrous beneath, very thin ; teeth usually 12–15 each side ; petiole very slender, $\frac{1}{2}$–$1\frac{1}{2}$ in. long ; bracts shortly petiolate to subsessile, the lower ovate and longer than the calyx, the upper linear-lanceolate and shorter than the calyx ; whorls few, 6-flowered, distant ; pedicels up to $\frac{1}{2}$ lin. long ; bracteoles minute ; calyx campanulate, $2\frac{3}{4}$–$4\frac{1}{4}$ lin. long, rather densely and softly pubescent or sparingly hispidulous ; teeth linear-triangular, $1\frac{1}{4}$–$2\frac{1}{2}$ lin. long, $\frac{1}{2}$ to nearly 1 lin. broad at the base, often distinctly spreading, sometimes bent at the needle-like apex ; corolla white, pubescent ; tube narrowly funnel-shaped, $3\frac{1}{4}$–$4\frac{1}{4}$ lin. long, curved ; upper lip elliptic, $2\frac{3}{4}$–$3\frac{1}{4}$ lin. long, entire, keeled on the back ; lower lip 4–$4\frac{1}{2}$ lin. long ; median lobe suborbicular to obovate, $1\frac{3}{4}$–3 lin. long, $2\frac{1}{4}$–$4\frac{1}{4}$ lin. broad ; lateral lobes ovate, $\frac{3}{4}$–$1\frac{3}{4}$ lin. long.

COAST REGION : Albany Div., *Cooper*, 15 !

EASTERN REGION : Transkei ; in forests near Kentani, 1200 ft., *Miss Pegler*, 908 !

5. S. Bolusii (Skan) ; a perennial herb ; stem erect or ascending, branched, up to 18 in. high or more, covered as well as the branches, especially on the angles, with rather long slender whitish hairs ; internodes $\frac{3}{4}$–4 in. long ; leaves deltoid-ovate, the larger 1–3 in. long and $\frac{2}{3}$–$1\frac{3}{4}$ in. broad, obtuse or rounded, deeply and openly cordate at the base, regularly sometimes coarsely crenate, somewhat densely covered, especially above, with rather short soft hairs which are adpressed above; petiole $\frac{1}{4}$–1 in. long, hairy as the branches; spike up to 8 in. long; bracts leaf-like, especially the lower, shortly petiolate to subsessile, longer (often much longer) than the calyx; whorls several, usually 6-flowered, up to $2\frac{1}{4}$ in. apart; calyx campanulate-funnel-shaped, $3\frac{1}{4}$–4 lin. long, densely covered with slender whitish hairs which are often about $\frac{3}{4}$ lin. long; teeth narrowly triangular or deltoid, $1\frac{1}{2}$–$2\frac{1}{2}$ lin. long, acuminate-spinescent; corolla-tube 3–$3\frac{1}{4}$ lin. long, dilated upwards; upper lip elliptic or suborbicular, 2–3 lin. long, shortly pubescent outside; lower lip $3\frac{1}{2}$–5 lin. long; median lobe $2\frac{1}{4}$–3 lin. long and broad; lateral lobes very short and broad. *S. æthiopica, Benth. in E. Meyer, Comm.* 239 *partly, not of Linn.*

SOUTH AFRICA : without locality, *Forster*!
COAST REGION: Malmesbury Div. ; near Hopefield and on a hill near Hoetjes Bay, Saldhana Bay, 100–200 ft., *Bolus*, 12809 !
CENTRAL REGION : Richmond Div. ; near Stylkloof, 3000–4000 ft., *Drège*!

Gerrard, 406 (unlocalised), has a different pubescence, and its leaves are not or scarcely cordate at the base. It is probably a distinct species, but the specimen is insufficient for description.

6. S. albiflora (N. E. Br. in Kew Bulletin, 1901, 131);

stem herbaceous, erect, apparently tall, branched above, rather stout, pilose, glandular-puberulous; internodes up to $3\frac{1}{2}$ in. long; leaves broadly ovate, $2\frac{1}{2}$–$3\frac{1}{2}$ in. long, 2–$2\frac{1}{2}$ in. broad, acute, deeply cordate at the base, regularly crenate-dentate, softly pubescent both sides, rather thin; petiole $\frac{3}{4}$–$1\frac{3}{4}$ in. long; bracts ovate, longer than the calyx; whorls 6-flowered, several, rather distant; pedicels $\frac{1}{3}$ lin. long; calyx campanulate, $2\frac{3}{4}$ lin. long, shortly and densely pubescent; teeth spreading, narrowly deltoid, 1–$1\frac{1}{4}$ lin. long, spinescent-acuminate; corolla white, sparingly pubescent outside; tube funnel-shaped, about 3 lin. long; upper lip obovate-oblong, compressed-galeate, $3\frac{1}{4}$ lin. long; lower lip $4\frac{1}{4}$ lin. long.

EASTERN REGION: Natal; Drakensberg Range, 6000–7000 ft., *Evans*, 395 !

7. S. Kuntzei (Guerke in Kuntze, Rev. Gen. Pl. iii. ii. 262);

a herb 8–12 in. high; stem ascending, rather thick, pubescent; branches rather densely pilose; lower leaves petiolate, the upper subsessile, broadly ovate, up to $1\frac{3}{4}$ in. long and $1\frac{1}{4}$ in. broad, obtuse-rounded to subacute, cordate at the base, regularly and densely crenate, everywhere subtomentose, thick, more or less rugose; inflorescence up to 5 in. long; bracts ovate, apiculate, not or scarcely exceeding the calyx; whorls 6-flowered, the lower distant, the upper somewhat crowded; pedicels $\frac{3}{4}$–1 lin. long; calyx tubular, $2\frac{1}{2}$–3 lin. long, densely pilose-pubescent; tube $1\frac{1}{2}$–2 lin. long; teeth lanceolate with a broad base, $\frac{3}{4}$–1 lin. long, rigidly subspinescent-acuminate, enlarged when mature; corolla white, puberulous outside; tube cylindric, exserted, 3–4 lin. long; upper lip oblong, 2 lin. long, emarginate, somewhat tomentose outside; lower lip $2\frac{1}{2}$ lin. long, spreading; median lobe obovate; lateral lobes obliquely rounded, smaller. *S. petrogenes, Briq. in Bull. Herb. Boiss.* 2me *sér.* iii. 1085.

KALAHARI REGION: Orange River Colony, *Cooper*, 835 !
EASTERN REGION: Natal; Van Reenens Pass, *Kuntze*! near Van Reenen, in rocky places, 5800 ft., *Schlechter*, 6969 !

8. **S. Rehmannii** (Skan); a perennial herb with a long underground creeping rootstock; stem erect, shortly branched, moderately thick, 9–12 in. high, densely and shortly villous; internodes 4–9 lin. long; leaves suborbicular-ovate, ½–1 in. long and broad, rounded, cordate at the base, regularly and densely crenate, thick, bullate especially beneath, densely and shortly villous; petiole thick, 1–3 lin. long; inflorescence 1½–2½ in. long; lower bracts similar to the upper leaves, the upper spathulate and shorter than the calyx; whorls often 2-flowered, many, distinct or somewhat crowded; pedicels about ¾ lin. long; calyx tubular-campanulate, about 3¼ lin. long, densely and shortly villous; teeth lanceolate-triangular, 1 lin. long, acuminate; corolla pubescent outside; tube rather narrow, almost cylindric, 3¼ lin. long; upper lip suborbicular, 1¼ lin. long; lower lip 2⅓ lin. long; median lobe suborbicular, about 1 lin. long and ⅓ lin. broad; lateral lobes smaller, rounded.

KALAHARI REGION: Transvaal; Houtbosch, *Rehmann*, 6178 !

9. **S. malacophylla** (Skan in Kew Bulletin, 1909, 421); stem herbaceous, decumbent, simple or sparingly branched, slender, 9–15 in. long or more, ⅓–½ lin. in diam., greyish-velvety-pilose; internodes ½–2¼ in. long; leaves broadly ovate, 5–7 lin. long and broad, rounded, broadly cordate at the base, densely or very densely greyish-velvety-pilose, crenulate; auricles rounded; petiole 2–7 lin. long; lower bracts similar to the leaves but smaller; upper much reduced; whorls 2–6-flowered, several, distant, in a spike 2½–5 in. long; pedicels about ¾ lin. long; calyx campanulate, about 3 lin. long, densely velvety-pilose; teeth deltoid-lanceolate, about 1¼ lin. long, acuminate, somewhat spinescent; corolla mauve, or with a white tube and mauve or mauve-spotted lips, pubescent outside; tube 4 lin. long; upper lip elliptic, 2 lin. long, entire, concave; lower lip 3¼–3¾ lin. long; median lobe suborbicular, 1½ lin. long, 2–2¼ lin. broad; lateral lobes short and broad, rounded.

COAST REGION: Queenstown Div. ; Queenstown, 4000 ft., *Galpin*, 1955 !
Hangklip Mountain, 6300–6500 ft., *Galpin*, 5891 !

10. **S. sessilifolia** (E. Meyer, Comm. 239); a herb everywhere softly white-villous; stem decumbent or ascending, branched, slender, up to 2⅓ ft. long; lower leaves very shortly petiolate, the upper sessile, narrowly ovate or deltoid-ovate, 6–11 (usually 8) lin. long, 4–5 lin. broad, obtuse, rounded or slightly cordate at the base, crenate; bracts similar to the leaves, longer than the calyx; whorls 2–6-flowered, rather crowded at the ends of the branches; pedicels about ½ lin. long; calyx campanulate, 4–5 lin. long, densely villous; tube 1½–1¾ lin. long; teeth narrowly linear-triangular, 2½ lin. long;

corolla 5½–6 lin. long, white, with carmine on the lower lip; tube 2½–3 lin. long, ventricose about 1 lin. from the base to the throat; upper lip obovate-elliptic, 2⅓–3 lin. long, densely pilose outside; lower lip up to nearly 5 lin. long. *Benth. in. DC. Prodr.* xii. 476. *S. Bachmannii, Guerke in Engl. Jahrb.* xxvi. 75.

EASTERN REGION : Pondoland; between St. Johns River and Umtsikaba River, 1000–2000 ft., *Drège*, 4752 ! near Dorkin, *Bachmann*, 1169 !

Tyson, 1331, from Mount Currie, Griqualand East, differs from the type in having somewhat smaller and rather more broadly ovate leaves, most of which are subsessile, and the whorls of flowers are not so close together.

11. **S. Galpini** (Briq. in Bull. Herb. Boiss. 2me sér. iii. 1082) ; a branched herb or undershrub ; stems decumbent ; branches slender, ½–2 ft. long, densely spreading-villous or sometimes hispidulous; leaves subsessile or shortly petiolate ; blade ovate, ovate-rounded or ovate-deltoid, 4–12 lin. long, 3–10 lin. broad, obtuse to acute, rounded or slightly cordate at the base, crenulate, crenate or almost entire, rather thick, everywhere softly and densely rather long- villous ; petiole up to 3 lin. long ; inflorescence moderately dense or very lax, up to 3½ in. long, densely villous ; bracts similar to the leaves or much reduced ; whorls 2-flowered ; calyx campanulate or tubular-campanulate, 3½–4½ lin. long, densely villous ; teeth lanceo- late, sometimes setaceous at the tips, 1½–2¼ lin. long, scarcely spinescent ; corolla white, pilose outside ; tube narrowly funnel- shaped, 2¾–4 lin. long ; upper lip erect, ovate or suborbicular, 1–1½ lin. long ; lower lip 2–3½ lin. long. *S. lupulina, Briq. l.c.*

COAST REGION : Cape Div. ; Claremont Flats, *Schlechter*, 465 !

KALAHARI REGION : Transvaal ; upper slopes of the Saddleback Range, Barber- ton, 4000–5000 ft., *Galpin*, 681 ! High Veld near Carolina, 5600 ft., *Bolus*, 12241 !

EASTERN REGION : Transkei ; around Kentani, 1200 ft., *Miss Pegler*, 187 ! Natal; Durban, *Grant* ! *Peddie* ! Weenen County, 3000–5000 ft., *Sutherland* !

Bolus, 12240, from the High Veld near Carolina, Transvaal, is probably *S. Galpini*, but it differs from the type in having the corolla-tube shorter and broader, sometimes only about 1 lin. long and as much as 2 lin. broad. The few flowers on the Kew material may, however, be abnormal. The anthers appear to be imperfect and the style is much reduced. Schlechter, 465, the type of *S. lupulina*, Briq., we have been able to see through the kindness of Dr. Schinz. It does not appear to be distinct from *S. Galpini*, and has probably been introduced into the Cape Division where Dr. Schlechter found it. In the original description the number is by mistake given as 4651, and the locality as Natal.

12. **S. transvaalensis** (Guerke in Engl. Jahrb. xxviii. 316) ; a perennial herb with a subterranean horizontal stem ; aerial stem erect, about 1 ft. high, few-branched, ⅔–1 lin. broad in the lower part, rather densely whitish-villous ; internodes ¾–2¾ in. long ; branches rather slender, ascending, 5–10 in. long ; leaves deltoid- ovate, the larger ¾–1¼ in. long and broad, obtuse, rounded or sub- acute, broadly and deeply cordate at the base, somewhat coarsely crenate or sometimes crenate-dentate, shortly whitish-villous on both sides ; petiole 2–5 lin. long ; bracts similar to the leaves, gradually smaller, more shortly petiolate, acute ; whorls 2-flowered,

$\frac{1}{2}$–$1\frac{1}{2}$ in. apart, forming a leafy raceme up to about 6 in. long; pedicels $\frac{1}{2}$ lin. long; calyx tubular-campanulate, $4\frac{1}{2}$–5 lin. long, somewhat densely villous; teeth narrowly triangular, $1\frac{3}{4}$–2 lin. long, acuminate; corolla white, $5\frac{1}{2}$–7 lin. long, shortly pubescent; tube narrow, 4–5 lin. long, slightly enlarged upwards; upper lip suborbicular, $1\frac{3}{4}$–2 lin. long and slightly broader, entire; lower lip $3\frac{1}{2}$ lin. long; median lobe semi-orbicular, $1\frac{1}{3}$–$1\frac{3}{4}$ lin. long, 2–3 lin. broad; lateral lobes broadly ovate, $\frac{3}{4}$–1 lin. long.

KALAHARI REGION: Transvaal; Lydenburg District, kloof near Stephan Shoemanns Farm, *Wilms*, 1136! Crocodile Valley near Lydenburg, *Burtt-Davy*, 7662! Crocodile River, *Burke*!

Gerrard, 1223, from Natal, and a specimen collected by Mrs. K. Saunders in Natal have a different indumentum. They possibly belong to another species.

13. **S. Rudatisii** (Skan); a rather weak diffusely branched herb; branches up to 1 ft. long or more, up to about $\frac{3}{4}$ lin. thick, deeply 4-furrowed, hispidulous with somewhat recurved hairs, glandular-puberulous; internodes $1\frac{1}{4}$–$3\frac{1}{4}$ in. long; leaves ovate, $\frac{3}{4}$–$1\frac{1}{2}$ in. long, 5–13 lin. broad, obtuse, cordate at the base, regularly and shallowly crenate, rather densely covered on both sides with sessile or shortly stalked glands mixed with longer eglandular hairs; petiole $\frac{1}{4}$–1 in. long or more; bracts similar to the leaves but smaller, longer than the calyx; whorls few, distant, 2-flowered; pedicels up to $\frac{3}{4}$ lin. long; bracteoles minute; calyx nearly tubular in the flowering stage, afterwards tubular-campanulate, $3\frac{1}{4}$–$3\frac{3}{4}$ lin. long, rather densely covered with short soft spreading hairs, glandular-puberulous; teeth narrowly deltoid or broadly lanceolate, 1–$1\frac{1}{2}$ lin. long, $\frac{1}{2}$–$\frac{2}{3}$ lin. broad at the base, $\frac{1}{3}$–$\frac{1}{2}$ lin. broad at $\frac{1}{2}$ lin. from the apex, acute, but scarcely spinescent; corolla white, puberulous; tube 4 lin. long, narrowly cylindric up to about $1\frac{1}{4}$ lin. from apex, then enlarged, $\frac{3}{4}$ lin. broad below, $1\frac{3}{4}$ lin. broad at apex; upper lip suborbicular, 2 lin. long, deeply concave, entire; lower lip $3\frac{1}{2}$ lin. long; median lobe suborbicular, $1\frac{3}{4}$ lin. across; lateral lobes semi-ovate, $\frac{1}{2}$ lin. long, nearly $\frac{3}{4}$ lin. broad.

EASTERN REGION: Natal; Dumisa, Alexandra County, in shady thickets, at about 2000 ft., *Rudatis*, 405!

Apparently nearest *S. transvaalensis*, Guerke, but it differs in being glandular-puberulous and in having a shorter calyx with shorter, broader teeth.

14. **S. parilis** (N. E. Br. in Kew Bulletin, 1901, 131); stem herbaceous, erect, slender, 10–15 in. high, simple or very sparingly branched, villous-pubescent; leaves deltoid-ovate, 9–15 lin. long, 6–11 lin. broad, obtuse or sometimes apiculate, openly cordate at the base, crenate-dentate or dentate, densely villous both sides; petiole 2–6 lin. long; bracts similar to the leaves but smaller; whorls 2-flowered, few, distant; pedicels 1–$1\frac{1}{2}$ lin. long; calyx campanulate, 4 lin. long, villous; tube 2 lin. long; teeth linear-triangular, 2 lin. long, acuminate-spinescent; corolla white, puberulous; tube about 4 lin. long, rather narrow, slightly enlarged at

the apex ; upper lip suborbicular, 1¾–2 lin. long, suberect, concave, entire ; lower lip about 4½ lin. long.

EASTERN REGION : Natal ; Tiger Cave Valley, on the Drakensberg Range, *Evans,* 387 !

15. S. æthiopica (Linn. Mant. i. 82) ; a perennial herb ; stems decumbent or ascending, up to 1¼ ft. long or more, branched ; branches slender, more or less hairy or rarely nearly quite glabrous ; internodes up to 3 in. long ; leaves very variable in size and shape, ovate or deltoid-ovate, 4–12 lin. long, 3–11 lin. broad, usually 6–9 lin. long and 5–8 lin. broad, obtuse, rounded or acute, usually more or less cordate at the base, regularly crenate or crenate-serrate, sparingly and shortly hispid chiefly on the nerves, sometimes densely pubescent, rarely quite glabrous ; petiole 1–9 lin. long ; lower bracts similar to the leaves ; the upper much reduced ; whorls few to several, 2–6-flowered, distant or approximate ; pedicels ⅓–1 lin. long ; calyx campanulate or narrowly campanulate, 2–4 lin. long, sparingly to densely hispid or pilose ; teeth linear-triangular to lanceolate-triangular or sometimes narrowly deltoid or ovate-lanceo-late, 1–2½ lin. long, usually setaceous or slightly spinescent at the tips ; corolla sparingly pubescent outside, white, purplish or reddish-white ; tube narrowly funnel-shaped, 3¼–4½ lin. long ; upper lip elliptic or obovate-elliptic, 1¼–3½ lin. long, emarginate or entire ; lower lip 3–4 lin. long ; median lobe broadly obovate, transversely oblong or sometimes suborbicular, 1¼–3 lin. long, 2¼–4 lin. broad ; lateral lobes ovate, 1–2 lin. long. *Burm. f. Fl. Cap. Prodr.* 16 ; *Thunb. Prodr.* 96, *and Fl. Cap. ed. Schult.* 447 ; *Willd. Sp. Pl.* iii. 102 ; *Benth. Lab.* 548, *in E. Meyer, Comm.* 239 (*including var. grandiflora, Burch.*), *and in DC. Prodr.* xii. 476 *partly* ; *Baker in Dyer, Fl. Trop. Afr.* v. 467 ; *Guerke in Ann. Naturhist. Hofmus. Wien,* xx. 46 ; *Jacq. Obs. Bot.* iv. 2, *t.* 77. *S. pulchella, Salisb. Prodr.* 83. *Betonica capensis, Burm. f. Fl. Cap. Prodr.* 16. *Sideritis erecta, etc., Pluk. Almagest. Bot.* 345, *t.* 315, *f.* 3. *Prasium hirsutum, Poir. Encycl.* v. 611.

VAR. β, hispidissima (Benth. in E. Meyer, Comm. 239) ; branches and petioles densely covered with pallid slender but stiffish spreading hairs often about 1 lin. long ; leaves covered with similar hairs, especially on the nerves, but on the upper side they are somewhat adpressed.

VAR. γ, parviflora (Skan) ; branches rather densely and somewhat hispidly hairy ; hairs rather short, directed forwards ; leaves (at least the upper) more shortly petiolate ; petiole often only 1 lin. long or less ; corolla only 3–4½ lin. long. *S. æthiopica, Benth. in DC. Prodr.* xii. 476, *partly, not of Linn. See Bolus & Wolley-Dod in Trans. S. Afr. Phil. Soc.* xiv. 310, *in note under S. æthiopica.*

VAR. δ, glandulifera (Skan) ; branches, leaves and especially the calyx, densely glandular-pulverulent or glandular-puberulous, together with few or many scattered slender eglandular hairs.

SOUTH AFRICA : without locality, *Forster* ! *Forbes* ! *Thom,* 438 ! *Harvey,* 541 ! *Cole,* 63 ! *Thunberg* ! Var. γ, *Pappe.*

COAST REGION : Clanwilliam Div. ; Oliphants River, *Penther,* 1709 (ex *Guerke*). Piquetberg Div. ; near Piquetberg, *Penther,* 1744 (ex *Guerke*). Malmesbury Div. ; Hopefield, *Penther,* 1750 (ex *Guerke*). Cape Div. ; various localities,

Wolley-Dod, 625 ! *Burchell*, 346 ! *MacGillivray*, 568 ! *Mrs. Southey in Herb. Galpin*, 7849 ! *Burchell*, 236 ! *Hooker*, 445 ! Caledon Div. ; mountain near Genadendal, *Burchell*, 8633 ! Mossel Bay Div. ; between Little Brak River and Hartenbosch, *Burchell*, 6215 ! Uniondale Div. ; near Haarlem, *Burchell*, 4890 ! Uitenhage Div. ; Klein Winterhoek, *Drège*, 75*d* ! Albany Div. ; Howisons Poort, *Hutton in Herb. Harvey* ! Peddie Div. ; Fredricksburg on the Gualana River, *Gill* ! British Kaffraria, *Cooper*, 398 ! Var. *β* : Worcester Div. ; Hex River Kloof, *Drège*, 75*h* ! Var. *γ* : Cape Div. ; near the Signal Station, *Wolley-Dod*, 3048 ! eastern side of the Lions Rump, *Burchell*, 122 ! Var. *δ* : Uitenhage Div. ; Vanstadens Berg, *MacOwan*, 559 ! between Vanstadens Berg and Bethelsdorp, *Drège*, 75*g* ! Cathcart Div. ; Zwartkei River, *Baur* ! Mount Hope Farm, Upper Zwartkei River, 5100 ft., *Galpin*, 2682 ! East London Div. ; along sea-coast, *Galpin*, 1879 !

CENTRAL REGION : Graaff Reinet Div. ; near Graaf Reinet, 2900 ft., *Bolus*, 248 ! Albert Div., *Cooper*, 1777 ! Colesberg Div., *Shaw* ! Var. *δ* : Graaff Reinet Div. ; Weltevrede, Sneeuwberg Range, 4500 ft., *Bolus*, 2012 ! Middelburg Div. ; between Wolve Kop and Rhenoster Berg, *Burchell*, 2787 !

KALAHARI REGION : Var. *δ*, Orange River Colony ; Bethlehem, *Richardson* ! EASTERN REGION : Var. *δ*, Transkei ; around Kentani, 1200 ft., *Miss Pegler*, 231 ! Natal ; near Greytown, *Wilms*, 2192 ! Inanda, on Mount Edgecumbe, *Wood*, 1126 ! and without precise locality, *Cooper*, 2896 !

Also in Nyassaland.

Rehmann, 2384, from Brand Vley, Worcester Division, is possibly a form of *S. æthiopica.*

O. Kuntze (Rev. Gen. Pl. iii. ii. 262) distinguished a form *rosea*, in which the corolla is rose-coloured. This he collected in Cradock Division.

Bolus, 2012, here referred to var. *glandulifera*, appears to be a very luxuriant specimen, the leaves sometimes being 1¾ in. long and 1½ in. broad. It connects *S. æthiopica* with *S. grandifolia.*

16. **S. scabrida** (Skan) ; a perennial herb drying dark brown ; stems prostrate and creeping or decumbent, much elongated, about ¾ lin. thick or less, sparingly (rarely abundantly) furnished with usually very short recurved hook-like hairs callous at the base, often more or less scabrous, rarely rather densely puberulous or quite glabrous and smooth ; branches ascending, otherwise similar to the stems ; leaves ovate-triangular or sometimes ovate, 5–11 lin. long, 2½–7½ lin. broad near the base, subacute or obtuse, cordate at the base, with a deep open sinus and distant rounded auricles, regularly crenate or crenate-dentate, rather leathery, callous on the margin, quite glabrous or with few (rarely many) stiff hairs above and on the nerves beneath, sometimes with a few callosities or many stiff hairs callous at the base above, occasionally glandular-pulverulent beneath ; midrib and primary lateral nerves rather thick and conspicuous beneath ; petiole 1–4 lin. long, quite glabrous, or more or less pubescent ; lower bracts similar to the leaves, the upper gradually smaller ; whorls 2–6-flowered, lax, few or several, rather distant ; calyx funnel-shaped-campanulate, 3½–4¼ lin. long, some-what sparingly covered with short stiff hairs, sometimes almost glabrous ; teeth lanceolate or lanceolate-triangular, 1½–2½ lin. long, usually quite straight, acuminate-spinescent ; corolla shortly pubes-cent outside ; tube narrowly cylindric below, slightly enlarged above, 3–4 lin. long, annular-pilose inside ; upper lip obovate or elliptic, 2¼–4 lin. long, recurved, emarginate ; lower lip 3–5 lin.

long. *S. æthiopica, Benth. in E. Meyer, Comm.* 239 *partly, and in DC. Prodr.* xii. 476 *partly, not of Linn.*

COAST REGION: Uitenhage Div.; Vanstadens Berg, *Zeyher*, 831! Port Elizabeth Div.; Baakens Valley, *Tyson*, 2240! Albany Div.; near Grahamstown, *MacOwan*, 559 (*in Dublin Herb.*)! *Williamson*! *Bolton*!

CENTRAL REGION: Somerset Div.; Somerset East, 3000 ft., *Bolus*! upper part of the Bruintjes Hoogte, *Burchell*, 3037! 3100!

EASTERN REGION: Transkei; Gekau (Geua) River, *Drège*!

Drège, 75k, collected between Bethelsdorp and Vanstadens River, Uitenhage Div., closely approaches this, but it does not dry so dark brown, and the hairs on the stems are not recurved but turned upwards. Corollas are wanting.

MacOwan, 559, in Herb. Kew. is different from his 559 cited above and has been referred to *S. æthiopica*, var. *glandulifera*.

17. **S. Harveyi** (Skan); a small weak branched herb; branches apparently prostrate or decumbent, up to 6 in. long, only $\frac{1}{4}$–$\frac{1}{3}$ lin. in diam., shortly retrorsely pilose; leaves ovate, 4–10 (usually about 7) lin. long, 3$\frac{1}{2}$–9 lin. broad, obtuse, rounded, subtruncate or sometimes slightly cuneate at the base, rather regularly crenate-serrate, thin, shortly pilose on both sides, slightly glandular-puberulous beneath; teeth moderately large, usually 6–9 each side; petiole very slender, 3–7 lin. long; bracts lanceolate, as long as or shorter than the calyx; flowers about 6, in a single lax whorl terminating the branches; pedicels $\frac{3}{4}$ lin. long; calyx very narrowly campanulate, 3$\frac{1}{2}$ lin. long, covered with short slender spreading hairs and shorter gland-tipped hairs; teeth narrowly lanceolate-triangular, 1$\frac{3}{4}$ lin. long, about $\frac{1}{3}$ lin. broad at the base, somewhat spinescent; corolla shortly pubescent outside, about 5$\frac{1}{2}$ lin. long; tube 4 lin. long, narrowly cylindric up to near the apex where it is curved and slightly enlarged; upper lip reflexed, elliptic, 2$\frac{1}{4}$ lin. long; lower lip 2$\frac{1}{2}$–3 lin. long; median lobe transversely elliptic, about 1 lin. long and 1$\frac{1}{2}$ lin. broad.

COAST REGION: Cape Div.; near Cape Town, *Harvey in Dublin Herb.*!

18. **S. serrulata** (Burch. ex Benth. Lab. 549); an annual herb everywhere rather densely and softly pubescent; stem decumbent, elongated, slender, loosely branched; leaves broadly ovate, 4–10 lin. long, 4–8 lin. broad, somewhat acute, usually slightly and very broadly cordate at the base, regularly crenulate-serrulate; petiole 2–5$\frac{1}{2}$ (usually about 4) lin. long; lower bracts similar to the leaves, upper much reduced; whorls 2–4-flowered, few, distant; pedicels $\frac{1}{4}$–$\frac{1}{2}$ lin. long; calyx tubular-campanulate, about 2$\frac{1}{2}$ lin. long, softly pilose; teeth linear-triangular, setaceous at the tips, 1$\frac{1}{4}$ lin. long; corolla-tube straight, nearly cylindric, 2$\frac{1}{4}$ lin. long; upper lip erect, obovate-oblong, 1$\frac{1}{4}$ lin. long; lower lip 2$\frac{1}{4}$ lin. long. *Benth. in DC. Prodr.* xii. 477 *partly*; *Guerke in Ann. Naturhist. Hofmus. Wien*, xx. 46.

COAST REGION: Clanwilliam Div.; Krantzolei, *Penther*, 1748 (ex *Guerke*). Caledon Div.; near Houwhoek, *Penther*, 1721 (ex *Guerke*). Knysna Div.; near the Keurbooms River, *Burchell*, 5155!

A specimen in the Kew Herbarium from Somerset, collected by Bowker, differs in having larger flowers, and the indumentum consists of somewhat stiffer hairs.

19. S. attenuata (Skan) ; a herb everywhere covered with long slender soft hairs ; stem procumbent, elongated, extremely slender as well as the numerous very long ascending branches ; internodes up to 2¼ in. long ; leaves ovate-deltoid, 4–8 lin. long, 3–6 lin broad, obtuse, usually slightly and broadly cordate at the base, rather irregularly crenate-serrate ; teeth large, 3 or 4 on each side, rarely more, rounded ; petiole very slender, 2–7 lin. long ; bracts shortly petiolate, ovate-lanceolate to narrowly lanceolate, toothed or entire, often very small ; whorls 2-flowered, few, very distant ; calyx campanulate, 2¾–3 lin. long ; teeth linear- or lanceolate-triangular, 1¼–1¾ lin. long, setaceous at the tips ; corolla-tube straight and cylindric below, curved and enlarged at the throat, 2¾–3½ lin. long ; upper lip ovate or oblong-ovate, 2–2¾ lin. long, recurved ; lower lip about 4 lin. long ; median lobe about 2¼ lin. long and 2 lin. broad, emarginate. *S. serrulata, Benth. in E. Meyer, Comm.* 240, *and in DC. Prodr.* xii. 477 *partly, not of Burch.*

COAST REGION : Worcester Div. ; near Bains Kloof, 1000 ft., *Bolus,* 2896 ! Paarl Div. ; Paarl Mountain, 1000–2000 ft., *Drège,* 75b !

20. S. fruticetorum (Briq. in Bull. Herb. Boiss. 2me sér. iii. 1083) ; a branched herb about 8 in. high ; branches divergent-ascending, rather flaccid, retrorsely pilose ; leaves cordate-ovate, up to 1¼ in. long, ¾ in. broad, obtuse-rounded, regularly and rather coarsely crenate-dentate, thin, sparingly pilose ; auricles distant ; petiole up to 7½ lin. long ; inflorescence 1¼–2 in. long ; bracts ovate-elliptic, more or less petiolate, as long as the calyx ; whorls 2–6-flowered ; pedicels ½–1 lin. long ; calyx narrowly obconical at the base, then abruptly and broadly funnel-shaped-campanulate, 3 lin. long, shortly pilose ; tube 2 lin. long ; teeth broadly triangular-lanceolate, equal, about 1½ lin. long, nearly 1 lin. broad at the base, shortly subspinescent-acuminate ; corolla pilose outside ; tube rather broad, 2½–3 lin. long ; upper lip oblong, 2½ lin. long, emarginate ; lower lip 3½ lin. long ; median lobe obcordate ; lateral lobes obliquely rounded, smaller.

SOUTH AFRICA : without locality, *Wright,* 496 ! 503 *partly (in Dublin Herb.)* ! *Prior* !
COAST REGION : Cape Div. ; Simons Bay, *Prior* ! Stellenbosch Div. ; Lowrys Pass, *Schlechter,* 1179 !

21. S. leptoclada (Briq. in Bull. Herb. Boiss. 2me sér. iii. 1084) ; a herb about 1 ft. high ; stem rather stout at the base, spreading-pilose ; flowering branches slender, elongate-incurved, almost glabrous, shining green ; leaves ovate-triangular, up to 1 in. long and 10 lin. broad, obtuse, cordate at the base, strongly and regularly crenate, glabrescent ; auricles with a broad semicircular sinus ; petiole up to 5 lin. long ; infloresceuce up to 6 in. long, dorsiventral, very slender, subscandent ; bracts petiolate, from triangular to elliptic-lanceolate and subulate ; whorls usually 2-flowered, distant ; pedicels very short ; calyx campanulate, or funnel-shaped-campanulate, 2½–3 lin. long, 4 lin. long when mature, very shortly pilose ; tube 1½–2½ lin. long ; teeth lanceolate, equal, 1½ lin. long, ciliolate, subspinescent-acuminate; corolla sparingly pilose outside, about 2½ lin. longer than the calyx ; tube cylindric, exserted ; upper lip ovate, 1 lin. long, emarginate ; lower lip 1½–2 lin. long ; median lobe obovate ; lateral lobes obliquely rounded, smaller.

EASTERN REGION : Natal ; near Bluekranz River, 3700 ft., *Schlechter,* 6865 !

22. **S. flexuosa** (Skan)́; a small weak decumbent branched perennial herb ; branches up to 9 in. long, ¼–⅓ lin. in diam., somewhat wiry at the base, flexuose, sparingly white-pilose and glandular-pulverulent ; hairs rather long, slightly retrorse ; leaves ovate, 4–9 lin. long, 3–7 lin. broad, rounded to subcordate at the base, sometimes shortly tapering to the petiole, crenate, sparingly white-pilose above and on the nerves beneath ; teeth broad, shallow, 5–9 each side ; petiole 1–2½ lin. long ; bracts ovate to oblong, longer to shorter than the calyx ; whorls 2-flowered, several, distinct or distant ; flowers mostly turned to one side ; pedicels scarcely ½ lin. long ; calyx campanulate, 3¾ lin. long, somewhat hispid with rather long white hairs, glandular-puberulous, often purplish, prominently 10-nerved ; teeth lanceolate-triangular, 1¾ lin. long, spinescent ; corolla reddish-purple, shortly pubescent outside ; tube about 3 lin. long, slightly enlarged upwards ; upper lip suborbicular, about 2 lin. long ; lower lip 4½ lin. long ; median lobe semi-orbicular, 2¼ lin. long, 3½ lin. broad.

COAST REGION : Stockenstrom Div. ; mountain side at Old Katberg Pass, 5200 ft., *Galpin,* 2093 !

23. **S. cymbalaria** (Briq. in Bull. Herb. Boiss. 2me sér. iii. 1088) ; a small perennial herb ; stems several, weak, wiry, creeping, 2–6 in. long, sparingly clothed with long slender spreading hairs ; leaves suborbicular-cordate or broadly ovate-cordate, 2½–5 lin. long, 1½–4 lin. broad, obtuse, crenate, sparingly pilose; petiole 2½–4 lin. long; inflorescence up to 2½ in. long ; rhachis almost glabrous ; bracts sessile, elliptic or elliptic-subulate, shorter than the calyx ; whorls 2-flowered, few, distant ; pedicels scarcely ½ lin. long ; calyx narrowly campanulate, 2¼ lin. long, puberulous or subglabrous; tube 1¼–1½ lin. long ; teeth lanceolate-triangular, 1–1¼ lin. long, spinescent-acuminate ; corolla pink ; tube nearly cylindric, curved, about 3 lin. long ; upper lip obovate-orbicular, 1½ lin. long, entire, very slightly puberulous outside ; lower lip 2¾ lin. long ; median lobe obovate or suborbicular, 1¼ lin. long, 1⅔ lin. broad ; lateral lobes obliquely rounded, smaller.

VAR. β, **alba** (Skan) ; hairs on the branches, petioles and leaves much shorter ; petioles 1–2 (rarely up to 4) lin. long ; calyx up to 3½ lin. long ; teeth up to 2 lin. long ; corolla white ; tube 1⅔ lin. long ; upper lip obovate-elliptic, 1⅔ lin. long ; lower lip 3½ lin. long ; median lobe transversely oblong-obovate, 2¼ lin. long, 2½–3 lin. broad.

CENTRAL REGION: Cradock Div. ; on stony mountains, *Cooper*, 516 ! 2894 !
Graaff Reinet Div. ; near Wagenpads Berg, on the southern side, *Burchell*, 2829 !
EASTERN REGION: Var. β : Natal ; Richmond, 3000 ft., *Wood*, 1846 !

Stachys æthiopica, var. *tenella*, O. Kuntze, Rev. Gen. Pl. iii. ii. 262, we have
not seen, but suspect from the very short description given and from the fact that
the plant is from the Cradock Division that it is the same as *S. cymbalaria*, and
that the form *albiflora*, O. Kuntze, is the same as *S. cymbalaria*, var. *alba*.

24. S. subsessilis (Burch. ex Benth. Lab. 548) ; a perennial
herb ; stem decumbent, slender, branched or sometimes almost
simple, up to 15 in. long or more, sparingly clothed with somewhat
stiff hairs or almost glabrous ; leaves shortly petiolate or subsessile,
ovate or ovate-oblong, 2–4 (rarely up to 9) lin. long, 1½–3 (rarely
up to 6) lin. broad, obtuse, usually more or less cordate at the
base, crenate or sometimes subentire, somewhat leathery, sparingly
strigose or sometimes nearly glabrous ; petiole ½–1½ (rarely up to 4)
lin. long ; bracts broadly lanceolate, shorter than the calyx ; whorls
2-flowered, usually close together at the ends of the branches ;
pedicels about ½ lin. long ; calyx campanulate, 2¾ lin. long, hispidly
pubescent ; tube 1¼ lin. long ; teeth lanceolate-triangular, 1½ lin.
long, spinescent-acuminate ; corolla rose, about 5½ lin. long,
puberulous outside ; tube rather broad, curved, 3 lin. long ; upper
lip erect, elliptic, 2 lin. long, entire ; lower lip about 3½ lin. long.
Benth. in E. Meyer, Comm. 240, *and in DC. Prodr.* xii. 476.

SOUTH AFRICA : without locality, *Thunberg* !
COAST REGION : Uitenhage Div., *Zeyher* ! Port Elizabeth Div. ; near Port Eliza-
beth, *Burchell*, 4326 ! *Bolus*, 2432 ! 9139 ! *Drège*, 2302a ! *Tyson*, 2177 ! *Galpin*,
6365 ! Port Elizabeth and near the Boschmans River, *Zeyher*, 746 ! Algoa Bay,
Forbes ! Bathurst Div. ? Glenfilling, 1000 ft., *Drège*, 2302b. Albany Div. ; Elands
Kloof, near Grahamstown, about 2000 ft., *Galpin*, 384 ! King Williamstown Div. ;
Kachu (Yellowwood) River, 1000–2000 ft., *Drège*, 2302c !

There are 3 sheets of specimens in Thunberg's herbarium named *S. æthiopica*
and numbered 1, 2 and 3. Sheet 1 is here referred to *S. subsessilis*.

25. S. Priori (Skan) ; a perennial herb ; stem and branches
apparently decumbent, very slender, much elongated, rather thinly
covered with very slender long white hairs which are often callous
at the base ; internodes often ⅔–1 in. long, sometimes up to 2¼ in.
long ; leaves all distinctly petiolate, deltoid-ovate, the larger 8–10
lin. long and 5–6 lin. broad, obtuse, openly cordate at the base,
regularly crenate, often rather thin, minutely punctate, slightly
callous on the margin, somewhat thinly covered above and on the
principal nerves beneath with very slender long white hairs which
are often callous at the base ; petiole 1–3½ lin. long, hairy like the
branches ; bracts narrowly ovate, the lower slightly longer, the
upper slightly shorter than the calyx, narrowed into a short broad
petiole ; whorls 2 or 3, near together at the ends of the branches,
4- or 5-flowered ; calyx campanulate, about 3 lin. long, thinly
covered chiefly on the nerves and on the margins of the teeth with
rather long slender but stiff hairs ; teeth triangular-lanceolate,

about $1\frac{1}{2}$ lin. long, spinescent-acuminate; corolla-tube $2\frac{3}{4}$ lin. long, annular-pilose inside; upper lip $2\frac{1}{2}$–$2\frac{3}{4}$ lin. long; lower lip about $3\frac{1}{2}$ lin. long and 4 lin. broad.

COAST REGION: Port Elizabeth Div.; Algoa Bay, *Prior*!

26. **S. sublobata** (Skan); a small rather slender perennial herb, prostrate at the base, then ascending, up to 9 in. high; branches somewhat densely covered with retrorse or spreading hispid usually long hairs; internodes 5–12 lin. long; leaves narrowly deltoid, 4–6 lin. long, 2–4 lin. broad at the base, subacute or obtuse, openly cordate at the base, subentire towards the apex, crenate-serrate or shortly lobed in the lower part, thick, somewhat hispid; teeth or lobes usually 4 or 5 each side, rather irregular, up to about $\frac{3}{4}$ lin. long, the lowermost longest and usually unequally bilobed and more or less retrorse, rounded or obtuse; petiole 1–4 lin. long; lower bracts similar to the leaves; upper narrow and shorter than the calyx; whorls 2-flowered, few, distant; pedicels about $\frac{1}{2}$ lin. long; calyx campanulate, $2\frac{1}{3}$–3 lin. long, thinly covered with rather long stiff somewhat adpressed hairs and sessile glands; teeth lanceolate-triangular, 1–$1\frac{1}{4}$ lin. long, acuminate, spinescent; corolla mauve or lilac, pubescent; tube subcylindric, $3\frac{1}{4}$–4 lin. long; upper lip erect, elliptic or suborbicular, $1\frac{3}{4}$–$2\frac{1}{2}$ lin. long; lower lip 3–4 lin. long; median lobe semi-orbicular, $1\frac{1}{2}$–2 lin. long, 2–3 lin. broad, crenate; lateral lobes ovate, about 1 lin. long.

COAST REGION: Swellendam Div.; Barrydale, 1200 ft., *Galpin*, 4425! Oudtshoorn Div.; in thickets near the great Cango cave, about 2100 ft., *Bolus*, 12244!

This is closely allied to *S. subsessilis*, Benth., but it seems to differ in habit, being more freely branched, the branches are much more hairy, and the leaves are thinner and often more deeply toothed or almost lobed.

27. **S. arvensis** (Linn. Sp. Pl. ed. ii. 814); a hispidly pubescent annual herb; stem decumbent at the base, then ascending or erect, sometimes altogether erect, slender, branched usually from the base, a few inches up to 2 ft. long; leaves ovate or ovate-rounded, $\frac{3}{4}$–$1\frac{1}{2}$ in. long, $\frac{1}{3}$–1 in. broad, obtuse, cordate or rounded at the base, crenate or crenate-serrate, more or less hispidly pubescent; petiole up to 14 lin. long, usually 2–6 lin. long; bracts similar to the leaves, the upper narrower, subsessile; whorls 4–6-flowered, several, distant; calyx tubular-campanulate, 3–4 lin. long, slightly oblique, often deflexed, hispidly pubescent; teeth lanceolate, 1–$1\frac{3}{4}$ lin. long, acuminate, subspinescent; corolla scarcely longer than the calyx, pale purplish-rose and white. *Benth. Lab.* 550, *and in DC. Prodr.* xii. 477; *Burm. f. Fl. Cap. Prodr.* 16; *Bolus & Wolley-Dod in Trans. S. Afr. Phil. Soc.* xiv. 310; *Boiss. Fl. Orient.* iv. 747; *Benth. Fl. Austral.* v. 73; *Mart. Fl. Bras.* viii. i. 197; *Curt. Fl. Lond.* ii. *t.* 41; *Engl. Bot. ed.* i. *t.* 1154; *Fl. Dan. t.* 587; *Reichb. Ic. Bot. seu Pl. Crit.* x. 24, *t.* 967.

COAST REGION: Cape Div.; near Cape Town, *Bolus*, 4799! near Mosterts Farm, Mowbray, *Wolley-Dod*, 1743! and without precise locality, *Harvey*! Introduced.

A native of Europe, Temperate Asia, North Africa and some of the Atlantic Islands. Introduced into Australia, New Zealand, Tropical America, and the West Indies.

28. **S. nigricans** (Benth. in E. Meyer, Comm. 238); a herb usually more or less black when dried; stem erect, 1–3 ft. high, simple or sparingly branched in the upper part, 1–1½ lin. in diam. near the base, hispidly pubescent or subglabrescent at the base, often scabrous; leaves sessile or subsessile, erect, narrowly oblong-lanceolate, sometimes linear, usually 1½–2¼ in. long, 1–4 lin. broad, scarcely acute, rounded at the base, crenate or serrate, sometimes almost entire, callous along the margin, strigose, scabrous; inflorescence up to 10 in. long; bracts lanceolate, longer or in the upper part shorter than the calyx; whorls usually 6-flowered; pedicels ¾ lin. long; calyx campanulate, about 3 lin. long, hispidly pubescent; teeth lanceolate, hispidly ciliate, 1¼–1½ lin. long; corolla white, pubescent; tube 2¾ lin. long; upper lip suborbicular, deeply concave, 1¾ lin. long; lower lip 3½ lin. long. *DC. Prodr.* xii. 471; *Wood, Natal Pl.* iii. *t.* 271.

KALAHARI REGION: Transvaal; Umlomati Valley, near Barberton, 4000 ft., *Galpin*, 1129! Berg Plateau, *Mudd*! Swaziland; near Mbabane, 4600 ft., *Bolus*, 12245! *Burtt-Davy*, 2819!
EASTERN REGION: Pondoland; between the Umtata River and St. Johns River, 1000–2000 ft., *Drège*, 4729a! Griqualand East; around Clydesdale, 3000 ft., *Tyson*, 2132! and in *MacOwan & Bolus in Herb. Austr.-Afr.* 1299! Natal; Umvoti River, *Gerrard*, 1224! Durban, *Sanderson*, 143! 235! *Gueinzius*! between the Umzimkulu River and the Umkomanzi River, *Drège*, 4729b; near Newcastle, *Wilms*, 2194! Inanda, 1800 ft., *Wood*, 5! Zululand, *McKenzie*!

29. **S. sessilis** (Guerke in Engl. Jahrb. xxvi. 74); stems herbaceous, erect, simple, 12–16 in. high, hispid-villous; leaves sessile or subsessile, oblong-lanceolate or oblong, ¾–2 in. long, 3–6 lin. broad, obtuse or sometimes subacute, rounded at the base, regularly crenate or crenate-serrate, everywhere more or less adpressed-pilose; inflorescence 3–6 in. long; bracts lanceolate, the lower much longer, the upper shorter than the calyx; whorls many, 6–8-flowered, distant or approximate; pedicels about 1¼ lin. long; calyx campanulate, 2½–3 lin. long, rather densely villous; teeth deltoid-lanceolate, 1¼–1½ lin. long, acuminate; corolla white, very thin, sparingly pubescent outside; tube 2¾ lin. long; upper lip suborbicular, 1½–1¾ lin. long, emarginate or entire; lower lip 3½–4 lin. long; median lobe broadly obovate, 1½–2 lin. long, 2–3 lin. broad; lateral lobes short, broad. *Briq. in Bull. Herb. Boiss.* 2ᵐᵉ ser. iii. 1082; *S. Moore in Journ. Bot.* 1905, 173. *S. pseudo-nigricans, Guerke in Ann. Naturhist. Hofmus. Wien*, xx. 46, *not of Guerke in Engl. Jahrb.* xxviii. 315.

KALAHARI REGION: Basutoland, *Cooper*, 943! Transvaal; Carolina District, one mile north of Robinson's, *Burtt-Davy*, 2974!
EASTERN REGION: Griqualand East; near Newmarket, *Krook*, 1724! Natal; Inchanga, 2000 ft., *Wood*, 4806!

Tyson, 1118, from near Kokstad, Griqualand East, appears to be a depauperate form of *S. sessilis.* The plant is smaller and less hairy, and the spikes shorter, with smaller flowers.

30. **S. simplex** (Schlechter in Journ. Bot. 1897, 221); a perennial herb 8–16 in. high; stems ascending, simple, everywhere densely villous; leaves in pairs rather close together at the base of the stem, with the uppermost pairs smaller and distant, oblong or lanceolate-oblong, $1\frac{1}{4}$–2 in. long, 4–8 lin. broad, obtuse, rounded at the base, regularly serrate-crenate, rather thick, villous on both sides, especially on the underside; petiole $2\frac{1}{2}$–4 lin. long; inflorescence slender, 4–6 in. long; rhachis densely villous; bracts narrowly lanceolate, about as long as the calyx; whorls 4- or 6-flowered, many, the lower distant, the upper closer together; pedicels about $\frac{1}{2}$ lin. long; calyx campanulate, $2\frac{3}{4}$–$3\frac{1}{2}$ lin. long, densely villous; tube $1\frac{3}{4}$–2 lin. long; teeth linear-triangular, 1–$2\frac{1}{2}$ lin. long, equal, subspinescent-acuminate; corolla white, villous on the outside of the lips; tube funnel-shaped, 3–$3\frac{1}{2}$ lin. long; upper lip suborbicular, $1\frac{1}{2}$–$1\frac{3}{4}$ lin. long, emarginate or entire; lower lip $2\frac{1}{2}$–3 lin. long; median lobe obovate or suborbicular, up to $1\frac{1}{2}$ lin. long and 2 lin. broad; lateral lobes obliquely rounded, smaller. *S. chrysotrichos, Guerke in Engl. Jahrb.* xxviii. 316. *S. pascuicola, Briq. in Bull. Herb. Boiss.* 2me sér. iii. 1086.

KALAHARI REGION : Transvaal ; Elandsberg Range, 7600 ft., *Schlechter*, 3844! Lydenburg District, between Middelburg and the Crocodile River, *Wilms*, 1137! Saddleback Mountain, near Barberton, 4000–5000 ft., *Galpin*, 1006! Carolina District, Outspan, one mile north of Robinson's, *Burtt-Davy*, 2974!

31. **S. obtusifolia** (MacOwan in Kew Bulletin, 1893, 13, excl. Tyson, 2561); stem herbaceous, erect, 8–12 in. high, few-branched, scarcely 1 lin. in diam. at the base, pilose-villous, sometimes glandular-pilose ; lower leaves shortly petiolate, the upper subsessile; blade elliptic or sometimes ovate-elliptic, 9–18 (usually 12–15) lin. long and 8–10 lin. broad, broadly rounded, usually slightly cordate at the base, crenate, softly pilose-villous on both sides; petiole 1–4 lin. long; bracts sessile, ovate, the lower longer than, the upper as long as, the calyx ; whorls 3–6-flowered, several, distant, in a spike up to 5 in. long; calyx campanulate, $2\frac{3}{4}$–$3\frac{1}{2}$ lin. long, densely pilose, often glandular-pilose ; teeth ovate-lanceolate or deltoid-lanceolate, 1–$1\frac{1}{4}$ lin. long, acuminate ; corolla yellowish-brown when dry, glandular-pubescent ; tube 3–$3\frac{3}{4}$ lin. long; upper lip obovate-elliptic, $1\frac{1}{4}$–$2\frac{1}{4}$ lin. long, entire ; lower lip $3\frac{1}{4}$–4 lin. long.

VAR. β, **angustifolia** (Skan); stem weaker ; leaves all shortly petiolate; blade oblong or sometimes lanceolate-oblong, 12–18 lin. long, 4–7 lin. broad, narrower at the apex than in the type ; petiole $1\frac{1}{2}$–$3\frac{1}{2}$ lin. long.

VAR. γ, **Flanaganii** (Skan); stem weaker ; leaves all subsessile, not exceeding 9 lin. long and 5 lin. broad ; corolla blackish when dry.

COAST REGION : Var. γ: Stutterheim Div. ; Kabousie River, *Flanagan*, 496! KALAHARI REGION: Var. β: Orange River Colony ; on the mountains at Besters Vlei, near Witzies Hoek, 6200 ft., *Bolus*, 8240!

EASTERN REGION: Tembuland; Bazeia, *Baur,* 75! Griqualand East; around
Clydesdale, near the Umzimvubu River, 2500 ft., *Tyson in MacOwan & Bolus,*
Herb. Norm. Austr.-Afr., 1298!

Tyson, 2561, is evidently different, and is described in this work as a new
species, *S. Tysonii.*

32. S. Tysonii (Skan); a perennial herb; stem erect, about 10
in. high, scarcely ¾ lin. thick near the base, very sparingly branched,
rather densely covered with very slender soft spreading long hairs;
branches very slender, hairy as the stem; leaves all distinctly
petiolate, thin, ovate-lanceolate, the larger 14–17 lin. long and 5–7
lin. broad, subacute or obtuse, distinctly narrowed to the apex,
broadly cordate or subcordate at the base, regularly crenate-
serrate, covered with very slender soft hairs above and on the
principal nerves beneath; petiole 1½–4 lin. long, hairy as the stem;
lower bracts similar to the leaves; upper lanceolate, slightly longer
than the calyx; whorls 4–6-flowered, numerous, rather close
together in a terminal tapering spike; calyx campanulate, 3–3½ lin.
long, shortly and softly glandular-pubescent; teeth deltoid, subulate-
spinescent at the apex, 1¼–1½ lin. long; corolla-tube 3–4 lin. long,
annular-pilose inside; upper lip obovate-oblong, 3–3½ lin. long,
slightly emarginate or entire; lower lip 3½–5 lin. long. *S. obtusifolia, MacOwan in Kew Bulletin,* 1893, 13 *partly.*

EASTERN REGION: Griqualand East; mountain slopes near Clydesdale, 2500 ft.,
Tyson, 2561!

33. S. erectiuscula (Guerke in Engl. Jahrb. xxviii. 315); stem
herbaceous, erect or almost erect, up to 2 ft. high, very sparingly
branched, shortly pilose; internodes ½–3½ in. long; leaves ovate-
lanceolate or deltoid-lanceolate, the larger ¾–2 in. long and 5–7 lin.
broad, obtuse or sometimes apiculate, slightly cordate at the base,
crenate, sparingly pilose on both sides; petiole 1½–6 lin. long;
lower bracts similar to the leaves, but narrower and sessile, the
upper much reduced; whorls 2–10-flowered, many, distinct; pedicels
½–1 lin. long; calyx campanulate, 3–3½ lin. long, sparingly covered
with short stiff adpressed hairs; teeth linear-triangular, 1¼–1½ lin.
long, somewhat spinescent; corolla glandular-pubescent outside; tube
2½–4 lin. long; upper lip nearly erect, obovate-oblong, 3–3½ lin.
long, slightly emarginate or entire; lower lip 3½–5 lin. long; median
lobe obovate-suborbicular, 1¾–2¼ lin. long and broad; lateral lobes
ovate, 1–1¼ lin. long, rounded.

VAR. β, **natalensis** (Skan); blackish when dry, more hispidly hairy or scabrous;
stem up to 16 in. high; leaves somewhat smaller (usually 12–15 lin. long and
3–4 lin. broad), truncate or rounded (not cordate) at the base; whorls 2–6-flowered,
usually less distant.

KALAHARI REGION: Transvaal; near Lydenburg, *Wilms,* 1116! Machadadorp,
Lydenburg District, *Burtt-Davy,* 7660! 7661! Aapies River, near Pretoria, *Miss
Leendertz,* 979!
EASTERN REGION: Var. β: Natal; near Newcastle, 3000–4000 ft., *Wood,* 6349!
6795!

34. S. humifusa (Burch. ex Benth. Lab. 547) ; stem herbaceous, procumbent, up to 1¾ ft. long, about 1½ lin. in diam. in the thickest part, rather succulent, glabrous, very slightly scabrous, shining, acutely 4-angled ; leaves subsessile or sessile, deltoid-lanceolate, usually 9–18 lin. long, 3–6 lin. broad, much narrowed to the apex, cordate at the base, crenate, slightly revolute, glabrous or nearly so ; bracts similar to the leaves but much reduced ; whorls 4–6-flowered, several, distant, in a spike 1½–2½ in. long ; flowers nearly sessile ; calyx campanulate, about 3 lin. long, very shortly and stiffly hairy or almost glabrous ; teeth subulate, 1¼ lin. long, spinescent ; corolla white ; tube 2¾ lin. long, enlarged upwards, slightly curved ; upper lip suborbicular, concave, 1¾ lin. in diam. ; lower lip about 3 lin. long. *Benth. in DC. Prodr.* xii. 476.

SOUTH AFRICA : without locality, *Herb. Prior,* 3650 !
COAST REGION : Bathurst Div. ; near Port Alfred, *Burchell,* 3794 ! Albany Div., *Bowker* !

35. S. rivularis (Wood & Evans in Journ. Bot. 1897, 489) ; a perennial herb ; stems 2 to many, arising from a strong rootstock, erect, rather slender, 6–12 in. high or more, leafy, somewhat hispidly retrorsely pilose ; lower leaves shortly petiolate, the upper subsessile or sessile and smaller, almost erect, deltoid-lanceolate, 5–10 lin. long, 2–4 lin. broad, gradually narrowed to the apex, obtuse or scarcely acute, cordate at the base, crenulate, shortly hispidly pubescent above and chiefly on the nerves beneath, slightly scabrous above ; bracts similar to the leaves but smaller ; whorls 2–6-flowered, several, distant, in a slender spike up to about 3 in. long ; calyx campanulate, 5-nerved, 2¼–2½ lin. long, sparingly hispid-pubescent chiefly on the nerves ; teeth linear-triangular or deltoid at the base and subulate above, spinescent, about 1 lin. long ; corolla white with pink spots on the lower lip ; tube 2¾ lin. long ; upper lip obovate-elliptic, 1¼–1¾ lin. long ; lower lip 3 lin. long. *Wood, Rep. Colonial Herb. Natal,* 1897, 14. *S. Schlechteri, Guerke in Engl. Jahrb.* xxvi. 74.

EASTERN REGION : Tembuland ; Tabase, near Bazeia, 2500 ft., *Baur,* 327 ! Natal ; near the Mooi River, 4000–5000 ft., *Wood,* 4022 ! 6252 ! *Schlechter,* 6837 !

Probably only a hairy variety of *S. humifusa,* Burch.

36. S. tenella (Skan) ; a small perennial herb branched chiefly from the base ; stems several, very slender, erect, a few inches up to 1 ft. high, the thickest up to ½ lin. in diam., very sparingly sprinkled with short stiff hairs or mostly quite glabrous ; lower internodes scarcely longer than the leaves, the upper 2–4 times as long ; leaves subsessile, ovate or sometimes ovate-lanceolate, 3½–6 lin. long, 2–3 lin. broad, obtuse to rounded, usually more or less cordate at the base, crenulate or almost entire, glabrous or with a few hairs on the nerves beneath ; bracts similar to the upper leaves but smaller ; whorls few, 4–6-flowered, distant, in a spike about 3 in. long ; calyx campanulate, about 2¼ lin. long, slightly glandular-puberulous or almost glabrous ; teeth deltoid-lanceolate, $\frac{3}{4}$–1 lin. long, acuminate-spinescent ; corolla about 4 lin. long ; tube $2\frac{1}{2}$–$2\frac{3}{4}$ lin. long ; upper lip suborbicular, $1\frac{1}{4}$ lin. long, entire ; lower lip $2\frac{1}{2}$–3 lin. long.

EASTERN REGION : Griqualand East ; in a marshy place below Kokstad, 4200–5000 ft., *Tyson*, 1790 !

37. **S. rugosa** (Ait. Hort. Kew. ed. 1, ii. 303) ; an erect branched shrub up to 2 ft. high or more, closely and densely white-tomentose on almost all its parts or sometimes glabrescent on the older branches ; leaves sessile, more or less spreading, oblong-lanceolate, lanceolate or sometimes oblanceolate, 1–$3\frac{1}{2}$ in. long, 2–8 lin. broad, usually narrowed, obtuse to acute, rarely rounded at the apex, usually narrowed at the base, quite entire or more rarely serrulate or crenulate, conspicuously rugose ; whorls 2–10-flowered, several, distant ; calyx campanulate or tubular-campanulate, 3–$4\frac{1}{2}$ lin. long ; teeth narrowly deltoid, lanceolate-subulate or lanceolate, $\frac{3}{4}$–$2\frac{1}{4}$ lin. long, acute, rarely spinescent ; corolla yellow, pink or pale violet, or yellow with some pink or white on the upper lip, tomentose outside ; tube $2\frac{1}{3}$–$3\frac{3}{4}$ lin. long ; upper lip elliptic or ovate, $1\frac{1}{2}$–$3\frac{1}{4}$ lin. long ; lower lip 3–$4\frac{1}{2}$ lin. long. *Willd. Sp. Pl.* iii. 104 ; *Benth. Lab.* 559, *in E. Meyer, Comm.* 241, *and in DC. Prodr.* xii. 493 ; *Jacq. Coll. Suppl.* 116, *and Ic. Rar.* iii. 7, *t.* 493. *S. rugosa, var. longiflora, Benth. in E. Meyer, Comm.* 241, *and in DC. Prodr.* xii. 494. *S. jugalis, Burch. ex Benth. Lab.* 562. *S. Deserti, Benth. in DC. Prodr.* xii. 494.

VAR. β, **linearis** (Skan) ; leaves shorter and narrower, $\frac{1}{3}$–$1\frac{1}{4}$ in. long, $\frac{1}{2}$–$1\frac{1}{2}$ lin. (rarely up to 2 lin.) broad, usually more rigid, sometimes recurved ; whorls 2-flowered. *S. linearis, Burch. ex Benth. Lab.* 559, *and in DC. Prodr.* xii. 494. *S. rosmarinifolia, Benth. Lab.* 559, *and in DC. Prodr.* xii. 494. *S. rosmarinifolia, var. Burkei, Benth. in DC. Prodr.* xii. 494. *S. hyssopifolia, Vahl ex Benth. Lab.* 559, *not of Michx. S. recurva, Guerke in Bull. Herb. Boiss.* vi. 549. *Sideritis pallida, Thunb. Prodr.* 95, *and Fl. Cap. ed. Schult.* 445.

VAR. γ, **foliosa** (Skan) ; leaves less rugose or not rugose at all, thinner, 1–$2\frac{1}{4}$ lin. long, mostly 2–3 lin. broad. *S. foliosa, Benth. in E. Meyer, Comm.* 241, *and in DC. Prodr.* xii. 493.

SOUTH AFRICA : without locality, *Ecklon* ! *Scholl* ! *Masson* ! *Hort. Vienna* ! Var. β : *Zeyher*, 1337 !
COAST REGION : Clanwilliam Div. ; Cederberg Range, Packhuis Pass, about 1800 ft., *Schlechter*, 8639 ! Packhuis Berg, 3000 ft., *Bolus*, 9074 ! Var. β : Albany Div. ; Fish River, near Grahamstown, *Burke* ! *Bowker* ! Var. γ : Queenstown Div. ; Tambukiland, *Ecklon* ! Shiloh, *Baur*, 870 !
CENTRAL REGION : Calvinia Div. ; Uien Vallei, Bokkeveld Mountains, 2000–2500 ft., *Drège*, 3096a ! Ceres Div. ; near the Yuk River, *Burchell*, 1233 ! Beaufort West Div. ; Nieuwveld Mountains near Beaufort West, 3000–5000 ft., *Drège*, 3096b ! Var. β : Calvinia Div. ; Bokkeland, *Thunberg* ! Laingsburg Div. ; Babians Krantz near Laingsburg, *Rehmann*, 2883 ! and in *Herb. Bolus*, 5778 ! Somerset Div., *Bowker*, 130 ! 202 ! Cradock Div., *Cooper*, 486 ! *Bowker*, 642 ! Graaff Reinet Div. ; Graaff Reinet, *Sanderson*, 29 ! Murraysburg Div. ; Sneeuwberg Range, *Wyley* ! about 4500 ft., *Bolus*, 2004 ! banks of rivers, 4000 ft., *Tyson*, 384 ! Sutherland Div. ; Roggeveld, *Thunberg* ! *Rehmann*, 3196. Middelburg Div. ; Conway Farm, *Gilfillan in Herb. Galpin*, 2994 ! Colesberg Div. ; near

Colesberg, *Shaw* ! between Riet Fontein and Plettenbergs Beacon, *Burchell*, 2744 !
Philipstown Div. ; Bavers Pan, *Burchell*, 2717 ! Var. γ : Somerset Div. ; Little
Fish River near Somerset East, 3000 ft., *MacOwan*, 1581 ! Graaff Reinet Div. ;
Sneeuwberg Range, 4500–5000 ft., *Bolus*, 1847 ! Richmond Div. ; Zeekoe River,
near its source, *Drège*, 3584b !
WESTERN REGION : Little Namaqualand ; Modder Fontein, 1500–2000 ft.,
Drège, 3111 ! Ookiep, 3200 ft., *Bolus in MacOwan & Bolus, Herb. Norm.
Austr.-Afr.* 683 ! Vanrhynsdorp Div. ; Eenkokerboom, 900 ft., *Schlechter*, 11054 !
on mountains near the Oliphants River, *Thunberg* !

Rehmann, 3854, labelled Bloemfontein, Orange River Colony, is evidently
wrongly localised. It belongs to the variety *linearis*. Rehmann collected other
material of this variety in Worcester and Sutherland Divisions.

Galpin, 2608, from Andriesberg, Queenstown Division, has flowers with a very
short much inflated corolla-tube. It may be a distinct species. In foliage it is
nearest var. *linearis*.

It is doubtful whether there are any constant differences between varieties
β and γ and between them and the type. Baur, 870, MacOwan, 1581, and Bolus,
1847, placed here with var. *foliosa*, are intermediate between the broader-leaved
forms of var. *linearis* and the type specimens of var. *foliosa*.

38. **S. Burchellii** (Benth. Lab. 561) ; an erect branched undershrub 4 ft. high, densely white stellate-tomentose, almost floccose
on the younger branches ; leaves subsessile, oblong-lanceolate or
lanceolate, the lower up to 3 in. long and 6 lin. broad, somewhat
acute, narrowed at the base, serrulate-crenulate, more or less
floccose-tomentose, somewhat green ; upper leaves similar but smaller
and whiter, longer than the flowers ; whorls 6–12-flowered, distinct
but rather close together at the ends of the branches ; calyx tubular-
campanulate, about 3 lin. long, more or less distinctly 2-lipped,
densely floccose-tomentose outside and on the upper part inside ;
tube about 2 lin. long ; upper lip slightly longer than the lower,
3-toothed ; lower lip 2-toothed ; teeth lanceolate, ½–1½ lin. long or
sometimes minute, thick, obtuse or acute, the lower longer than the
upper, spreading ; corolla 4½ lin. long, sulphur-yellow, tomentose
outside ; tube 2½ lin. long ; upper lip elliptic, 1¼ lin. long, entire ;
lower lip 2½ lin. long ; median lobe suborbicular, 1⅓ lin. long, nearly
2 lin. broad ; lateral lobes ¾ lin. long, slightly broader, rounded.
DC. Prodr. xii. 493. *Phlomis micrantha, Burch. Trav.* i. 340.

KALAHARI REGION : Griqualand West ; Asbestos Mountains at Kloof Village,
Burchell, 1672 ! at Klipfontein, *Burchell*, 2624 !

39. **S. Lamarckii** (Benth. Lab. 562) ; a rather robust branched
undershrub, very densely white-woolly-tomentose especially on the
younger parts, glabrescent on the older branches ; stem decumbent
at the base, then erect ; leaves shortly petiolate or subsessile, oblong-
elliptic, sometimes ovate or lanceolate, 1–1¼ in. long, ¼–1¼ in. broad,
obtuse, often minutely apiculate, cuneate-rounded at the base,
crenate or serrate, thick, very rugose ; petiole 1½–4 lin. long ; bracts
similar to the leaves, longer than the calyx ; pedicels 1–1½ lin. long ;
whorls 6- to many-flowered, several, distant or the upper approximate ; calyx tubular-campanulate, 4¼–4½ lin. long, very densely

white-woolly-tomentose ; teeth deltoid, acuminate, 1–1¼ lin. long ;
corolla yellow (*Bolus*), flesh-coloured (*Thunberg*), slightly tomentose
outside ; tube 3–3½ lin. long, slightly curved ; upper lip oblong-
obovate, 1½–2⅓ lin. long ; lower lip 3¾ lin. long. *DC. Prodr.* xii.
492. *S. rugosa, Lam. Ill.* iii. 66, *t.* 509, *fig.* 3, *not of Ait. Sideritis
decumbens, Thunb. Prodr.* 95, *and Fl. Cap. ed. Schult.* 444.

CENTRAL REGION : Calvinia Div. ; Bokkeland, *Thunberg*! Sunderland Div. ;
Roggeveld, *Thunberg*!
WESTERN REGION : Little Namaqualand ; Kaus Mountains, 3000–4000 ft., *Drège*,
3100 ! near Klipfontein, 3000 ft., *Bolus in MacOwan & Bolus, Herb. Norm.
Austr.-Afr.* 682 !

40. **S. multiflora** (Benth. in DC. Prodr. xii. 492) ; a somewhat
robust branched undershrub ; branches (especially the younger)
densely white-tomentose ; leaves ovate-elliptic or sometimes obovate,
7–12 lin. long, usually 4–6 lin. rarely up to 8 lin. broad, rounded,
narrowed at the base, crenulate, densely white-tomentose, thick,
rugose ; petiole 1–2½ lin. long ; bracts similar to the leaves, much
longer than the calyx ; whorls few- to many flowered, rather crowded
at the ends of the branches ; pedicels ¾–1 lin. long ; calyx tubular-
campanulate, 3½–4 lin. long, densely tomentose outside and on the
teeth inside ; teeth narrowly deltoid, about 1 lin. long, acuminate ;
corolla apparently yellow, tomentose outside on the upper part ;
tube about 3 lin. long, slightly curved ; upper lip elliptic, 2¼ lin.
long, entire ; lower lip 3¼ lin. long.

WESTERN REGION : Little Namaqualand, *Wyley*, 88 ! 89 ! *in Dublin Herb.* ;
between Kook Fontein and Hollegat River, 1000–2000 ft. *Drège*, 3099 ! I'us
(Tc Ous ?), 2800 ft., *Schlechter*, 11427 !

Schlechter, 8245, from the Karee Bergen, Vanrhynsdorp Division, should
probably be referred to this species.

S. crenulata, Briq. in Engl. Jahrb. xix. 192, appears to be the same as this
species. We have not seen the·type specimen (Steingroever, 8), which was collected
in South-west Africa (precise locality not recorded), but Schlechter's 11427,
which was received as *S. crenulata*, is undoubtedly *S. multiflora*. Dr. Briquet, in
Bull. Herb. Boiss. 2ᵐᵉ sér. iii. 1087, has identified a specimen collected by Dinter
(1130) at Gubub, in Great Namaqualand, with *S. crenulata*.

41. **S. flavescens** (Benth. in E. Meyer, Comm. 241) ; an erect
branched undershrub, everywhere shortly and densely yellowish-
tomentose ; branches erect, straight, rigid ; leaves sessile, lanceolate,
4–9 lin. long, 1–2 lin. broad, acute or sometimes subobtuse, narrowed
at the base, entire or minutely few-toothed at the apex, thick ;
bracts similar to the leaves, longer or shorter than the calyx ;
whorls 2–6-flowered, rather close together at the ends of the
branches ; calyx tubular-campanulate or campanulate, 3½–4¼ lin.
long ; teeth lanceolate-triangular, 1¼–1½ lin. long, acuminate ;
corolla apparently yellow, tomentose outside on the upper part ;
tube 2½–2¾ lin. long ; upper lip suborbicular, or broadly obovate,
about 1⅔ lin. long ; lower lip 3–3½ lin. long. *DC. Prodr.* xii. 493.

SOUTH AFRICA: without locality, *Ecklon*!
WESTERN REGION: Little Namaqualand; between Pedros Kloof and Lily
Fontein, 3000–4000 ft., *Drège*, 3097! Spektakel Mountain, *Morris in Herb.
Bolus*, 5779!

42. **S. gariepina** (Benth. in DC. Prodr. xii. 493); an undershrub,
everywhere densely yellow-tomentose; leaves oblong, obtuse, rounded
at the base, subdentate, thick, densely tomentose on both sides;
whorls about 6-flowered; calyx campanulate; teeth lanceolate-
linear, not spinescent.

WESTERN REGION: Little Namaqualand; at the mouth of the Orange River,
Ecklon, ex *Bentham*.

43. **S. dregeana** (Benth. in E. Meyer, Comm. 240); an erect
undershrub, 6–16 in. high, densely floccose-tomentose on stem,
branches and leaves; stem simple or branched, rather densely
leafy; leaves sessile, oblong-linear or somewhat spathulate, $\frac{3}{4}$–1$\frac{1}{2}$ in.
long, 1$\frac{1}{2}$–2$\frac{1}{4}$ lin. broad, obtuse or rounded, somewhat narrowed or
broad at the base, more or less crenate or sometimes entire, rather
thick; bracts similar to the leaves, gradually smaller; whorls
many, 2–4-flowered, distant or the upper approximate, in a spike
2–8 in. long; pedicels $\frac{1}{2}$–$\frac{3}{4}$ lin. long; calyx campanulate, 3–4$\frac{1}{2}$ lin.
long, densely tomentose; teeth lanceolate-triangular or broadly
lanceolate, 1$\frac{1}{4}$–1$\frac{1}{2}$ lin. long, acuminate; corolla pink or purple,
densely tomentose outside; tube 2–3 lin. long; upper lip 1$\frac{1}{2}$–2 lin.
long; lower lip 2$\frac{3}{4}$–3$\frac{1}{2}$ lin. long. *DC. Prodr.* xii. 494.

VAR. β, **lasiocalyx** (Skan); much less tomentose on stem, branches and leaves;
leaves oblanceolate, up to 7$\frac{1}{2}$ lin. broad, distinctly narrowed to the base; calyx with
a more woolly indumentum. *S. lasiocalyx, Schlechter in Journ. Bot.* 1898, 317.

VAR. γ, **tenuior** (Skan); similar to var. β, but the plant is more slender, and
the stem, branches and especially the leaves are still less tomentose; leaves
thinner, oblanceolate to lanceolate, up to 5 lin. broad, drying yellowish, usually
only very sparingly covered and mostly on the underside with a stellate tomentum;
calyx less tomentose than in the type.

COAST REGION: Var. γ: Queenstown Div.; summit of Andriesberg, 6600–
6800 ft., *Galpin*, 2031!
CENTRAL REGION: Tarka Div.; Wildchuts Berg, 5000–6000 ft., *Drège*, 7949b!
Cradock Div.; on hills, *Mrs. Barber*, 236! Aliwal North Div.; Witte Bergen,
7000–8000 ft., *Drège*, 7949c. Albert Div., *Cooper*, 643 *partly*! near Gaatje,
5000 ft., *Drège*, 7949a! Var. γ: Barkly East Div.; Ben McDhui, Witte Bergen,
9200 ft., *Galpin*, 6817!
KALAHARI REGION: Var. β: Orange River Colony; grassy slopes of Mont aux
Sources, 8500 ft., *Thode*, 46!

44. **S. spathulata** (Burch. ex Benth. Lab. 559); a dwarf branched
undershrub, erect or ascending, sometimes decumbent at the base,
everywhere densely grey-velvety-tomentose or sometimes glabrescent
at the base of the older branches; branches leafy; leaves sessile or
the lower shortly petiolate, spathulate or linear-spathulate, $\frac{3}{4}$–1$\frac{3}{4}$ in.
long, 1–4 lin. broad near the apex, rounded, entire; bracts lanceo-
late to ovate, longer to shorter than the calyx; whorls 2-flowered,

several, usually close together at the ends of the branches; pedicels
$\frac{1}{4}$-$\frac{1}{2}$ lin. long; flowers fragrant (*Sankey*); calyx campanulate,
$2\frac{1}{2}$-$3\frac{3}{4}$ lin. long, densely and shortly tomentose outside; teeth ovate
or ovate-lanceolate, $\frac{3}{4}$-$1\frac{1}{2}$ lin. long, acute; corolla rose, flesh-coloured
or very pale blue, sparingly tomentose outside; tube 2–2$\frac{1}{4}$ lin. long;
upper lip ovate, 1$\frac{3}{4}$ lin. long; lower lip 2$\frac{1}{2}$–3 lin. long. *Benth. in
DC. Prodr.* xii. 494. *S. minima, Guerke in Bull. Herb. Boiss.* vi. 550.

KALAHARI REGION : Griqualand West ; right bank of the Vaal River, at
Blaauwbosch Drift, *Burchell*, 1738! near Kimberley, 4000 ft., *Bolus*, 5380! and
in *MacOwan & Bolus, Herb. Norm. Austr.-Afr.* 329! Eitalers Fontein, *Rehmann*,
3360. Orange River Colony; Bloemfontein, *Sankey*, 223! Olifants Fontein,
Rehmann, 3532. Bechuanaland ; near Kuruman, *Burchell*, 2435! plains south of
Takun and at the source of the Moshowing River, near Takun, *Burchell*, 2223!
2277! Transvaal ; Magalies Berg, *Burke*! vicinity of Morloas, *Holub*, 813–815!
Potchefstroom Farm, *Burtt-Davy*, 1069!

Drège, 3584a, collected in Cathcart Division, between Kat Berg and Klipplaat
River, 3000–4000 ft., has been referred to this species by Bentham in E. Meyer
Comm. 240. The fragmentary specimen at Kew differs in having shorter
narrowly lanceolate subacute leaves.

45. S. cuneata (Banks ex Benth. Lab. 560) ; an erect branched
undershrub up to 1$\frac{1}{2}$ ft. high or more ; branches elongated, shortly
and densely white-tomentose ; leaves very shortly petiolate or
subsessile, obovate-cuneate or oblong, 3$\frac{1}{2}$–6 (rarely up to 12) lin.
long, 1$\frac{1}{2}$–5 lin. broad, rounded, cuneate at the base, crenate or
crenulate, densely white-tomentose, thick, rugose; petiole up to
$\frac{3}{4}$ lin. long; bracts similar to the leaves, longer than the calyx ;
whorls 2–4-flowered, several, rather close together at the ends of
the branches ; calyx tubular-campanulate, about 3 lin. long, densely
tomentose outside and on the teeth inside ; teeth narrowly deltoid
or lanceolate-triangular, $\frac{3}{4}$–1 lin. long, shortly acuminate ; corolla
red-purple ; tube 1$\frac{3}{4}$–3$\frac{3}{4}$ lin. long; upper lip up to 2$\frac{1}{2}$ lin. long ;
lower lip up to 4$\frac{1}{4}$ lin. long. *Benth. in DC. Prodr.* xii. 493. *S.
denticulata, Burch. ex Benth. Lab.* 560, *and in DC. Prodr.* xii. 493 ;
Vatke in Bot. Zeit. 1875, 462.

SOUTH AFRICA : without locality, *Masson*!
CENTRAL REGION : Calvinia Div. ; Brand Vley, *Johanssen*, 25! Hantam
Mountains, *Meyer*, ex *Vatke*. Beaufort West Div. ; Nieuwveld, near Bokkepoort,
3500–4500 ft., *Drège*, 3096c! Sutherland Div. ; between Kuilen Berg and the
Great Riet River, *Burchell*, 1352! at the Great Riet River, *Burchell*, 1369!
WESTERN REGION : Vanrhynsdorp Div. ; Boschjesmans Karoo, *Drège*, 7950!

Drège, 7950, has thicker leaves with less conspicuous teeth.

46. S. Zeyheri (Skan) ; a small twiggy shrub ; branches except
the white-tomentose youngest ones glabrous or glabrescent ; leaves
elliptic-ovate, broadest in the middle or near the base, 2–2$\frac{1}{2}$ lin.
long, 1$\frac{1}{2}$–3 lin. broad, rounded at apex and base or sometimes
slightly cuneate at the base, thick, crenulate, densely and shortly
white-tomentose ; petiole $\frac{3}{4}$–1 lin. long; whorls often 2-flowered, or
sometimes only 1 flower to the pair of leaves ; calyx tubular-

campanulate, about 3 lin. long, densely and shortly white-tomentose ;
teeth deltoid, acute, $\frac{1}{2}-\frac{2}{3}$ lin. long ; corolla-tube about $1\frac{3}{4}$ lin. long;
upper lip $\frac{3}{4}$ lin. long ; lower lip $1\frac{1}{2}$ lin. long. *S. cuneata, Drège in
Linnæa,* xx. 201, *not of Banks.*

CENTRAL REGION : Calvinia Div. ; between Lospers Plaats and Springbok Kuil
River, *Zeyher,* 1333 !

47. **S. integrifolia** (Vahl ex Benth. Lab. 562 partly) ; an erect
branched shrub or undershrub ; branches quadrangular, elongated,
not so hard as in *S. teres,* very sparingly leafy, densely and shortly
white- or yellowish-tomentose ; leaves sessile or very shortly petiolate, oblong, lanceolate-oblong or sometimes obovate, 5–10 lin. long,
1–4 lin. broad, rounded or obtuse, rarely acute, narrowed or rounded
at the base, entire or few-toothed in the upper part, more or less
covered with stellate hairs, sometimes glabrescent ; bracts usually
about as long as or shorter than the calyx ; whorls 2–8-flowered,
several, the lower distant, the upper close together ; pedicels $\frac{1}{2}-\frac{3}{4}$ lin.
long ; calyx campanulate, $2\frac{3}{4}-5\frac{1}{2}$ lin. long, very densely yellowish-
woolly ; teeth subulate or linear, 1–$2\frac{3}{4}$ lin. long, not spinescent ;
corolla yellow, slightly tomentose outside ; tube $2\frac{1}{4}-3$ lin. long ;
upper lip suborbicular or broadly ovate, $1\frac{1}{4}-1\frac{3}{4}$ lin. long ; lower lip
about 3 lin. long. *Benth. in DC. Prodr.* xii. 492 *partly. S. aurea,
Benth. l.c. S. plumosa, Benth. l.c., not of Griseb. Betonica heraclea,
Linn. Mant.* 83. *Sideritis plumosa, Thunb. Prodr.* 95, *and Fl. Cap.
ed. Schult.* 445.

SOUTH AFRICA : without locality, *Masson* !
COAST REGION : Clanwilliam Div. ; Cederberg Range, *Drège,* 3098 !
CENTRAL REGION : Calvinia Div. ; Bokkeland, *Thunberg* ! Sutherland Div. ;
Roggeveld, *Thunberg* !

48. **S. hantamensis** (Vatke in Bot. Zeit. 1875, 462) ; perennial,
herbaceous ? somewhat tomentose above ; branches angular, ascending,
more than 8 in. long ; leaves sessile, ovate-oblong, acute, quite
entire or few-toothed at the apex, green ; lowermost bracts more
than 5 lin. long ; whorls about 6-flowered, distinct ; calyx sessile,
campanulate, yellowish-villous-woolly ; teeth ovate-lanceolate, acute,
somewhat spinescent ; corolla half as long again as the calyx ;
upper lip short, villous-woolly outside.

CENTRAL REGION : Calvinia Div. ; Hantam Mountains, *Meyer.*

49. **S. teres** (Skan) ; an erect branched shrub 3 ft. high ; branches
terete, more or less spreading, hard, sparingly leafy, densely covered
with a close yellowish-white tomentum ; internodes usually much
longer than the leaves ; leaves very small, subsessile or shortly
petiolate, suborbicular to obovate or oblong-lanceolate, 2–7 lin.
long, 1–3 lin. broad, rounded or obtuse, quite entire or few-toothed
at the apex, cuneate at the base, more or less covered with stellate
hairs on both sides ; bracts conspicuous, usually longer than or as

long as the calyx; whorls 4–6-flowered, several, distant or close together at the ends of the branches; pedicels about 1 lin. long; calyx campanulate, about 4½ lin. long, very densely yellowish-woolly outside; teeth triangular-lanceolate or triangular-linear, 1½–2¼ lin. long, acuminate, not spinescent; corolla yellow, about 5¼ lin. long, slightly tomentose outside; tube about 3½ lin. long; upper lip broadly obovate, 1¾ lin. long; lower lip 3 lin. long. *S. integrifolia, Benth. Lab.* 562, *and in DC. Prodr.* xii. 492 *partly, not of Vahl. Phlomis parvifolia, Burch. Trav.* i. 225.

CENTRAL REGION: Ceres Div.; near the Yuk River, *Burchell,* 1232! 1276!

In Benth. Lab. 562 and in DC. Prodr. xii. 492, Burchell's specimens are referred to *S. integrifolia,* Vahl, the type of which we have not seen. We have, however, seen the types of *Betonica heraclea,* Linn., and *Sideritis plumosa,* Thunb., both of which are reduced by Bentham to *S. integrifolia.* These agree with Drège's specimen on which Bentham based his *S. aurea,* which is therefore in this work referred to *S. integrifolia,* while Burchell's plant, which differs chiefly in having distinctly terete instead of square branches, is described as a new species.

50. **S. hyssopoides** (Burch. ex Benth. Lab. 558); apparently an undershrub; stem ascending or erect, from a few inches up to 2½ ft. high, slender, branched, glabrous above or sometimes very slightly or densely tomentose below; leaves scarcely petiolate, sometimes fascicled, linear, oblanceolate or linear-oblanceolate, usually 1–2¼ in. long, ½–6 lin. broad, rounded, obtuse or obtusely apiculate, rarely acute, much attenuated at the base, quite entire or minutely toothed in the upper part, glabrous, slightly glandular-pulverulent or shortly whitish tomentose; bracts lanceolate or ovate-lanceolate, longer to shorter than the calyx; whorls 2–6-flowered, several, distant, in slender spikes up to 6 in. long or more; pedicels ½–1¼ lin. long; calyx tubular-campanulate, 2¼–3¼ lin. long, very sparingly pubescent, sometimes shortly whitish-tomentose, gland-dotted; teeth narrowly deltoid or lanceolate-triangular, ¾–1 lin. long, shortly acuminate; corolla pink, pale purple or blue, more or less silky-hairy outside; tube funnel-shaped, 2½–3¼ lin. long, curved; upper lip elliptic, 1¾–2¾ lin. long; lower lip 2¾–4 lin. long. *E. Meyer, Comm.* 240; *Benth. in DC. Prodr.* xii. 495. *S. macilenta, E. Meyer, Comm.* 240; *Benth. in. DC. Prodr.* xii. 495. *S. cœrulea, Burch. ex Benth. Lab.* 558; *Benth. in DC. Prodr.* xii. 495.

SOUTH AFRICA: without locality, *Zeyher,* 1340!
COAST REGION: Uitenhage Div., *Zeyher*! Queenstown Div.; Klaas Smits River, near Queenstown, 3500 ft., *Baur,* 985! Cathcart Div.; Klipplaat River, *Ecklon*! between Kat Berg and the Klipplaat River, 3000–4000 ft., *Drège*!
CENTRAL REGION: "Cradock" (Tarka?) Div.; Tafel Berg, *Mrs. Barber,* 641! Graaff Reinet Div.; near Graaff Reinet, 2500 ft., *Bolus,* 76! and in *MacOwan & Bolus, Herb. Norm. Austr.-Afr.* 1343! Murraysburg Div.; near Murraysburg, 4000 ft., *Tyson,* 383! Victoria West Div.; Nieuwveld, between the Brak River and Uitvlugt, 3000–4000 ft., *Drège.* Middelburg Div.; Sneeuwberg Range, 5000–6000 ft., *Drège.* Aliwal North Div.; Leeuwenspruit, between the Kraai River and Witte Bergen, *Drège,* 3588*a*! Albert Div., *Cooper,* 643 *partly*! 645! 784! Hopetown Div.; near the Orange River, *Burchell,* 2653!
KALAHARI REGION: Griqualand West; along the Vaal River, *Burchell,* 1775! Vaal River near Kimberley, *Bolus,* 6829! Orange River Colony, *Hutton*! Mud

River Drift, *Rehmann*, 3600 ! Wolve Kop, *Burke*, 392 ! Bechuanaland ; Barolong
Territory, at Harms Salt-pan, *Holub* ! Transvaal ; Magalies Berg, *Burke* ! Standerton, *Burtt-Davy*, 3163 ! Vereeniging, *Gilfillan in Herb. Galpin*, 6168 !
EASTERN REGION : Transkei, *Mrs. Barber*, 791 ! Natal ; Weenen County, near
the Tugela River, *Wood*, 3553 !

51. S. caffra (E. Meyer ex Benth. in DC. Prodr. xii. 495); a
slender branched erect shrub, 3 to 8 ft. high ; branches elongated,
more or less stellate-pubescent, sometimes glabrescent; leaves
lanceolate, 1–2 (rarely up to 3) in. long, 3½–8 lin. broad, acute or
acuminate, cuneate or somewhat rounded at the base, serrulate or
entire in the lower part, thin, puberulous above, more or less
stellate tomentose beneath or sometimes puberulous beneath ;
petiole 1–2 lin. long ; inflorescence 4–7 in. long ; bracts similar to
the leaves but smaller ; whorls 4–10-flowered, very lax, distant ;
pedicels 1½–2½ lin. long ; calyx campanulate, 1½–2½ lin. long, very
shortly stellate-tomentose outside ; teeth broadly deltoid at the
base, acuminate above, ½–¾ lin. long; corolla white, sparingly
pubescent outside ; tube funnel-shaped, curved, 2¼–2½ lin. long ;
upper lip broadly ovate or suborbicular, 1–1¼ lin. long ; lower lip
2¼–2¾ lin. long. *Guerke in Ann. Naturhist. Hofmus. Wien*, xx. 46.

COAST REGION : King Williamstown Div. ; near Mount Coke, *Galpin*, 7845 ;
Kei Road Station, *Krook*, 1719 (ex *Guerke*). Komgha Div. ; Kei River, *Penther*,
1715 (ex *Guerke*).
KALAHARI REGION : Orange River Colony, *Cooper*, 2902 ! Transvaal ; Lynwood,
Pretoria, 4500 ft., *Burtt-Davy*, 7478 ! fountains near Pretoria, 4200 ft., *Miss
Leendertz*, 617 !
EASTERN REGION : Transkei ; between the Bashee and Kei Rivers, *Drège*, 3585 !
between the Kei and Gekau (Geua) Rivers, 1000–2000 ft., *Drège*, 4759 ! near
Idutywa, 2700 ft., *Schlechter*, 6277 ! near Kentani, 1200 ft., *Miss Pegler*, 422 !
Tembuland ; Bazeia Mountain, 2500 ft., *Baur*, 634 ! Pondoland ; between the
Umtata and St. Johns Rivers, *Drège*, 4750 ! Isnuka, Port St. John, *Galpin*, 2858 !
Griqualand East ; at the foot of Mount Currie, 4500 ft., *Tyson*, 1148 ! Kokstad,
5000 ft., *Tyson in MacOwan & Bolus, Herb. Norm. Austr.-Afr.* 588 ! near
Newmarket, *Krook*, 1717 (ex *Guerke*). Natal ; Emyati, *Gerrard*, 1214 ! Buffalo
River, *Pappe* ! near Byrne, 4000 ft., *Wood*, 1821 ! Olivers Hoek Pass, *Wood*,
3491 ! Van Reenen, 5500 ft., *Wood*, 5195 ! and without precise locality, *Cooper*,
1179 !

Imperfectly known species.

52. S. capensis (Presl, Bot. Bemerk. 100) ; stem shrubby ;
branches puberulous ; leaves petiolate, ovate, obtuse ; bracts oblong-
lanceolate ; whorls 6-flowered, distinct ; calyx puberulous ; teeth
ovate, subpungent ; upper lip of the corolla entire, densely white-
hirsute outside ; lower lip 3-lobed ; lobes rounded, the median
emarginate. *Benth. in DC. Prodr.* xii. 496.

SOUTH AFRICA : without locality, *Krebs.*

53. S. graciliflora (Presl, Bot. Bemerk. 100) ; stem quadrangular,
hispid on the angles and on the petioles with rigid retrorse hairs ;
leaves cordate-ovate, obtuse, crenate, ciliate, otherwise almost
glabrous ; whorls 4-flowered, the upper naked ; flowers ebracteate,

sessile ; calyx somewhat hispid, almost 2-lipped ; teeth spreading, ovate, acuminate, subpungent ; corolla-tube slender, pubescent, twice as long as the calyx ; upper lip rounded, entire, shorter than the lower ; lower lip 3-lobed ; median lobe slightly 4-lobed, twice as large as the lateral. *Benth. in DC. Prodr.* xii. 496.

SOUTH AFRICA : without locality, *Krebs.*

54. S. hispidula (Hochst. in Flora, 1845, 66) ; a dwarf branched undershrub ; branches slender ; leaves shortly petiolate, cordate, obtusely crenate, hispid ; whorls few, about 4-flowered ; calyx elongate-dentate ; teeth linear-lanceolate, as long as the tube, somewhat hispid.

COAST REGION : Humansdorp Div. ; near the Kromme River, *Krauss*, 1125.

55. S. Kraussii (Hochst. in Flora, 1845, 66) ; stems procumbent at the base, nearly glabrous or sparingly retrorsely hairy ; leaves petiolate, ovate or deltoid-lanceolate, ½ in. long or somewhat longer, deeply cordate at the base, obtusely crenate, nearly glabrous ; whorls few, 2- or 3-flowered ; calyx glabrous ; teeth acute ; corolla twice as long as the calyx ; upper lip puberulous.

SOUTH AFRICA : without locality, *Krauss.*

Hochstetter suggested that the affinity of this appears to be with *S. humifusa,* Burch.

56. S. natalensis (Hochst. in Flora, 1845, 65) ; stem ascending, 9 in. high or more, hirsute ; leaves shortly petiolate, broadly ovate, ½ in. long, obtuse, subcordate at the base, crenate-serrate, hirsute-pubescent ; bracts ovate, shorter than the calyx ; calyx hirsute ; teeth subspinose ; corolla puberulous ; tube shorter than the calyx-teeth ; lips exceeding the calyx.

EASTERN REGION : Natal ; summit of Table Mountain, *Krauss,* 1139.

Bentham, in DC. Prodr. xii. 476, refers to this and the two preceding species under *S. æthiopica,* from which, he says, Hochstetter's characters did not enable him to distinguish them.

57. S. nutans (Benth. Lab. 561) ; a pubescent undershrub ; leaves sessile, ovate, obtuse, narrowed at the base, crenate, rugose, villous above, white-woolly beneath ; whorls 2–6-flowered ; calyx pedicellate, very densely white-wholly, reflexed in fruit ; teeth lanceolate, obtuse, not spinescent ; corolla shortly exceeding the calyx, pubescent outside. *DC. Prodr.* xii. 492.

SOUTH AFRICA : without locality, *Dahl in Herb. Vahl.*

Bentham separates this from *S. Lamarckii* by its broader leaves, which are green above, with long rather villous hairs, not white-woolly, and especially by its nodding more deeply toothed calyx.

XX. BALLOTA, Linn.

Calyx funnel-shaped, with a more or less spreading orbicular or oblique limb, 10-nerved, 5–10- or sometimes up to many-toothed. *Corolla-tube* subincluded, annular-pilose inside ; limb 2-lipped ; upper lip erect, subconcave, emarginate or bifid ; lower lip 3-lobed ; median lobe emarginate or crenate. *Stamens* 4, didynamous (lower pair the longer), ascending under the upper lip ; anthers approximate in pairs, 2-celled ; cells finally divaricate, scarcely confluent. *Disk* equal, entire or sinuate-dentate, or produced in front. *Style* 2-fid ; lobes subulate, subequal. *Nutlets* ovoid-oblong, obtuse, smooth.

Perennial herbs or more rarely undershrubs, hirsute, woolly or tomentose ; leaves often rugose, toothed, the upper similar but smaller ; whorls axillary, many- or more rarely few-flowered ; bracteoles subulate or oblong, sometimes spinescent or very small ; upper corolla-lip pubescent or villous.

DISTRIB. Species 33, chiefly in the Mediterranean Region and in the Orient, 1 widely spread in Europe, 1 in Central Asia, 1 in Malaya, and 2 in Tropical Africa.

Marrubium Pseudo-dictamnus, Linn. (*Ballota Pseudo-dictamnus*, Benth.), a native of Crete, is included in Burm. f. Fl. Cap. Prodr. 16.

1. **B. africana** (Benth. Lab. 594) ; an erect branched perennial herb, 1–3 ft. high or more, usually rather densely covered with fine often long very soft sometimes minutely gland-tipped hairs ; stem and branches 4-angled ; leaves orbicular, suborbicular or ovate, the larger $\frac{3}{4}$–2 in. long and 1–2 in. broad, rounded, obtuse or rarely subacute at the apex, cordate, truncate or rounded at the base, rather coarsely and irregularly crenate or crenate-dentate, often somewhat rugose especially when young ; upper leaves similar but smaller ; petiole usually $\frac{1}{2}$–$1\frac{1}{2}$ in. long ; whorls few to many, distant, globose, few- to many-flowered ; bracteoles subulate, about as long as or shorter than the calyx, scarcely spinescent ; calyx sessile, funnel-shaped, $3\frac{1}{2}$–$4\frac{1}{2}$ lin. long, strongly 10-nerved, 10–20-toothed, when mature often with a more or less spreading orbicular limb ; teeth usually irregular, deltoid or narrowly deltoid at the base, often subulate and more or less spinescent at the apex, $\frac{1}{4}$–$1\frac{1}{2}$ lin. long ; corolla rose-coloured or white, 4–$5\frac{1}{2}$ lin. long, pubescent outside ; tube $2\frac{1}{2}$–4 lin. long ; upper lip elliptic or ovate-elliptic, $1\frac{1}{4}$–$1\frac{3}{4}$ lin. long, emarginate or bifid ; lower lip $1\frac{3}{4}$–3 lin. long ; median lobe transversely oblong or broadly obovate, 1–2 lin. long, 2–3 lin. broad ; lateral lobes broadly ovate or rounded, smaller. *DC. Prodr.* xii. 517. *Marrubium africanum, Linn. Sp. Pl. ed.* i. 583 ; *ed.* ii. 816, *and Mant.* ii. 412 ; *Thunb. Prodr.* 96, *and Fl. Cap. ed. Schult.* 447 ; *Willd. Sp. Pl.* iii. 112. *M. crispum, Linn. Sp. Pl. ed.* ii. 1674. *M. Thouini, Schult. ex Weinm. in Syll. Pl. Ratisb.* ii. 23. *Pseudodictamnus africanus foliis subrotundis, etc., Commel. Hort. Amst.* ii. 179, *t.* 90. *P. emarginatus, Moench, Meth. Pl. Suppl.* 139.

SOUTH AFRICA : without locality, *Thunberg* ! *Carmichael* ! *Thom*, 68 ! 194 ! *MacGillivray*, 674 ! *Zeyher*, 1344 ! 3538 ! *Villett* !

COAST REGION : Clanwilliam Div. ; between Clanwilliam and Bosch Kloof, *Drège.* Cape Div.; Greenpoint, *Prior* ! Lion Mountain, *Harvey* ! Lions Head, *Wolley-Dod,* 3137 ! near Simons Bay, *Milne,* 158 ! and without precise locality, *Hooker* ! Riversdale Div. ; near Riversdale, *Schlechter,* 1965 ! Mossel Bay Div. ; between Duyker River and Gauritz River, *Burchell,* 6388 ! Uitenhage Div. ; Olivenhout Kloof, near Enon, *Drège* ! Albany Div. ; Albany, *Miss Bowker* ! Fort Beaufort Div., *Cooper,* 557 ! Victoria East Div. ; near Victoria, *Mr. Barber* ! Queenstown Div. ; Lesseyton Drift, near Queenstown, 3500 ft., *Galpin,* 2592 !

CENTRAL REGION : Ceres Div. ; near Yuk River, *Burchell,* 1255 ! Laingsburg Div. ; Witteberg Range at Matjesfontein, *Rehmann,* 2941 ! Somerset Div. ; between the Zuurberg Range and Klein Bruintjes Hoogte, *Drège* ! foot of Bosch Berg, 2300 ft., *MacOwan,* 1388 ! Cradock Div. ; without precise locality, *Cooper,* 509 ! Graaff Reinet Div. ; at Wagenpads Berg, *Burchell,* 2825 ! Graaff Reinet, *Bolus,* 44 ! Beaufort West Div. ; Nieuwveld, near Bok Poort, *Drège* ! between Beaufort West and Rhenoster Kop, 2500–3000 ft., *Drège* !

WESTERN REGION : Little Namaqualand ; Silver Fontein, near Ookiep, *Drège* ! between Lekkersing and Noagas, *Drège* ! Ara Koop, 2300 ft., *Schlechter,* 11242 !

XXI. LEUCAS, Burm.

Calyx tubular or tubular-campanulate, rarely inflated, 10-nerved, striate, straight or incurved, equal or oblique at the throat, being often produced on the lower side or sometimes on the upper side ; teeth 6–10, equal or unequal. *Corolla-tube* included, annular-pilose or annular-papillose inside or sometimes naked ; limb 2-lipped : upper lip erect, concave, entire or more rarely emarginate ; lower lip spreading, 3-lobed, median lobe largest. *Stamens* 4, didynamous, lower pair longer than the upper, all (or at least the longer) exserted from the corolla-tube and ascending under the upper lip ; anthers approximate in pairs, 2-celled ; cells divaricate, finally confluent. *Disk* equal and entire or sinuate-dentate, or produced in front. *Style* subulate at the apex, with a very short posterior lobe. *Nutlets* ovoid-triquetrous, obtuse or scarcely truncate at the apex.

Annual or perennial herbs or sometimes undershrubs, variously hairy or rarely glabrescent ; leaves entire or toothed ; upper floral leaves similar to the others or reduced ; whorls axillary, few- or many-flowered, often distant ; corolla white, rarely purplish ; upper lip very villous.

DISTRIB. Species about 130, in the tropical regions of the Old World, 1 in Tropical America and the West Indies.

Corolla 6–8½ lin. long :
  Calyx 10-toothed ; bracts much shorter than the
    calyx :
      Leaves ovate-lanceolate to ovate, up to 1 in. broad (1) **glabrata.**
      Leaves linear-lanceolate, up to 5 lin. broad ... (2) **Fleckii.**
  Calyx 6-toothed ; bracts often as long as or longer
    than the calyx ... ... (3) **sexdentata.**
Corolla only about 3 lin. long :
  Leaves ovate to ovate-lanceolate, ½–2 in. broad ;
    calyx curved near the apex, produced on the
    upper side ... ... (4) **martinicensis.**
  Leaves oblanceolate to narrowly obovate, 2½–3½ lin.
    broad ; calyx straight, produced on the lower
    side... ... ... (5) **neuflizeana.**

1. L. glabrata (R. Br. Prodr. 504); a perennial herb, up to 3 ft. high; stem decumbent or ascending, branched, sparingly leafy, usually more or less recurved-pilose on the angles, rarely everywhere somewhat densely pilose, sometimes glabrescent; leaves membranous, ovate-lanceolate or ovate, the larger 1–3 in. long and ½–1 in. broad, attenuated, obtuse or subacute at the apex, rounded or cuneate at the base, coarsely few-toothed or rarely nearly entire, more or less pilose chiefly above and on the nerves beneath; petiole up to 8 lin. long; floral leaves similar to the foliage leaves but smaller; whorls 2–10-flowered, distant or approximate at the ends of the branches; bracteoles setaceous, minute; calyx tubular-campanulate, 3–4½ lin. long, oblique at the mouth, produced on the lower side, often purplish above, shortly pilose chiefly on the nerves, or sometimes glabrous, 10-nerved, 10-toothed; teeth narrowly triangular or the shorter sometimes deltoid, setaceous at the apex, ¼–1½ lin. long, the lower shortest; corolla white, 6–7½ lin. long; tube narrowly funnel-shaped, 3–3¾ lin. long, annular-papillose inside; upper lip oblong, 2½–3¼ lin. long, densely villous especially on the margin at the apex; lower lip 2½–3½ lin. long; median lobe obovate, 1¾–2½ lin. long, emarginate; lateral lobes ovate or elliptic, 1–1½ lin. long; stamens nearly as long as the upper lip of the corolla. *Benth. Lab.* 606, *and in DC. Prodr.* xii. 524; *Jaub. & Spach, Ill. Pl. Orient. t.* 385; *Baker in Dyer, Fl. Trop. Afr.* v. 482; *Schinz in Mém. Herb. Boiss.* x. 60. *L. galeopsidea, Hochst. ex Benth. in DC. Prodr.* xii. 524; *A. Rich. Tent. Fl. Abyss.* ii. 199. *L. natalensis, Sond. in Linnæa,* xxiii. 85. *L. Junodii, Briq. in Ann. Conserv. & Jard. Bot. Genève,* ii. 1898, 109. *Phlomis glabrata, Vahl, Symb.* i. 42.

COAST REGION: Komgha Div.; near Komgha, *Flanagan,* 1200, *in Herb Bolus*! KALAHARI REGION: Bechuanaland; Bakwena Territory, about 3500 ft., *Holub*! EASTERN REGION: Natal; Mooi River, *Gerrard,* 1221! near Weenen, 3000–4000 ft., *Wood,* 4483! junction of the Tugela and Blaaukrantz Rivers, 2000–3000 ft., *Evans,* 672! and without precise locality, *Gueinzius,* 363. Delagoa Bay, *Forbes*! *Monteiro,* 44! *Junod,* 92. Lorenzo Marquez and neighbourhood, *Bolus,* 9747! *Schlechter,* 11572!

Also in Tropical Africa and Arabia.

2. L. Fleckii (Guerke in Engl. Jahrb. xxii. 140); herb 1–1¼ ft. high, of slender habit; stem erect, sparingly branched or simple, pubescent; leaves sessile or very shortly petiolate, linear-lanceolate, up to 2 in. long and 5 lin. broad, somewhat acute, narrowed at the base, irregularly serrate, canescent-pubescent on both sides; whorls few-flowered, ¾–1¼ in. apart; bracts subulate, hirsute, shorter than the calyx; pedicels 1½ lin. long; calyx oblique at the mouth, 3½–4 lin. long in fruit; teeth 10, all subulate, 1–1½ lin. long. *Baker in Dyer, Fl. Trop. Afr.* v. 484.

WESTERN REGION: Great Namaqualand; Tiras, *Schinz,* 43. Aus, *Steingröver,* 39.

Also in Tropical Africa.

3. L. sexdentata (Skan) ; herb ; branches slender, 4-furrowed, obtusely 4-angled, retrorsely pilose ; hairs closely adpressed, greyish ; leaves thin, ovate, 7–13 lin. long, 7–9 lin. broad, rounded at the apex, broadly cuneate at the base, regularly and somewhat coarsely crenate except at the entire base, rather densely adpressed-pilose, minutely gland-dotted ; petiole 3–6 lin. long ; whorls axillary, distinct, many-flowered, up to 8 in. across ; bracts linear to spathulate, 4–6 lin. long ; calyx tubular, 5½ lin. long, slightly inflated about the middle, 10-nerved, oblique at the mouth and somewhat 2-lipped, much longer on the lower side, 6-toothed, densely and shortly pubescent ; uppermost tooth ovate, 1¼ lin. long, shortly acuminate ; 2 lateral teeth narrowly triangular, scarcely 1 lin. long, setaceous at the apex ; 3 lowermost teeth ovate or triangular, ¾–1 lin. long, the median narrower than the others, all setaceous at the apex ; corolla white, 8½ lin. long ; tube narrowly funnel-shaped, about 5 lin. long, annular-papillose inside ; upper lip 3½ lin. long, pilose on back and densely ciliate on the margin in the upper part ; lower lip 5½ lin. long ; median lobe broadly obovate, 2¾ lin. long, deeply emarginate ; lateral lobes broadly ovate, about 1 lin. long, rounded ; stamens nearly equalling the upper lip of the corolla.

KALAHARI REGION : Transvaal ; probably Marico District, *Holub* !

4. L. martinicensis (R. Br. Prodr. 504) ; an erect annual herb 1–4 ft. high ; stem and branches obtusely 4-angled, 4-furrowed, shortly and usually retrorsely pubescent ; leaves ovate or ovate-lanceolate, 1–3 in. long, ½–2 in. broad, obtuse, often cuneate at the base, coarsely crenate-serrate, adpressed-pubescent ; petiole ¼–¾ in. long ; floral leaves narrower ; whorls axillary, distant, globose, densely flowered ; bracts numerous, linear or lanceolate-subulate, 3–5 lin. long, subspinescent at the apex ; calyx tubular, abruptly incurved near the apex, somewhat inflated near the base, oblique at the mouth, about 3½ lin. long, up to 8 lin. long in fruit, 10-nerved, 10-toothed ; teeth subulate, spinescent at the apex, ciliate, unequal, ½–2 lin. long, the uppermost much longer than the others ; corolla white, about 3 lin. long, villous above ; tube nearly straight, 2¼ lin. long, exannulate or imperfectly annular-papillose ; upper lip ¾ lin. long ; lower lip about 1 lin. long ; median lobe obovate, emarginate, scarcely ½ lin. long ; longer stamens slightly exserted from the corolla-tube ; shorter included. *Benth. Lab.* 617, *in E. Meyer, Comm.* 242, *and in DC. Prodr.* xii. 533 ; *Hook. f. Fl. Brit. Ind.* iv. 688 ; *Baker in Dyer, Fl. Trop. Afr.* v. 479. *Phlomis martinicensis, Swartz, Prodr. Veg. Ind. Occ.* 88. *P. caribæa, Jacq. Ic. Pl. Rar.* i. 11, *t.* 110.

COAST REGION : East London Div. ; near the mouth of the Nahoon (Kahoon) River, *Galpin,* 7788 !

KALAHARI REGION : Orange River Colony ; Vaal River, *Burke* ! Transvaal ; near Lydenburg, *Wilms,* 1144 ! Hooge Veld, Standerton, *Rehmann,* 6827 ! Fountains, near Pretoria, 4200 ft., *Miss Leendertz,* 616 ! Waterval Boven, *Burtt-Davy,* 1454 ! Springbok Flats, *Burtt-Davy,* 1741 !

EASTERN REGION : Griqualand East ; in cultivated places around Clydesdale, 2500 ft., *Tyson,* 2061 ! *and in MacOwan & Bolus, Herb. Norm. Austr.-Afr.,* 893 ! Natal ; Bushmans River, *Gerrard,* 361 ! between the Umkomanzi and Umlazi Rivers, *Drège,* 4833 !

Also in Tropical Africa, Arabia, Madagascar, India, Tropical America and the West Indies, often as a weed.

5. L. neuflizeana (Courb. in Ann. Sc. Nat. 4me sér. xviii. 145) ; an erect annual herb, $\frac{1}{2}$–1 ft. high or more ; stem simple or sparingly branched ; branches densely and shortly retrorsely pubescent ; leaves subsessile or sessile, oblanceolate to narrowly obovate, $\frac{1}{2}$–2 in. long, $2\frac{1}{2}$–$3\frac{1}{2}$ lin. broad, obtuse, cuneate at the base, entire in the lower part, toothed above, densely and shortly adpressed-pubescent ; whorls axillary, many, the lower usually distant and the upper close together, sometimes all close together, 5–12-flowered ; bracts setaceous, about $\frac{1}{2}$ lin. long ; calyx $1\frac{1}{2}$–$3\frac{1}{2}$ lin. long, very oblique, very much produced on the lower side, densely and shortly pubescent, 10-nerved, 10-toothed ; teeth narrowly triangular to setaceous, $\frac{1}{4}$–$\frac{1}{2}$ lin. long ; corolla white, 2–3 lin. long, villous ; tube cylindric, $1\frac{1}{4}$–$1\frac{3}{4}$ lin. long, imperfectly annular-papillose ; upper lip $\frac{3}{4}$–$1\frac{1}{4}$ lin. long, densely ciliate ; lower lip 1–$1\frac{1}{4}$ lin. long ; median lobe obovate, $\frac{1}{2}$–$\frac{3}{4}$ lin. long, emarginate ; lateral lobes ovate, about $\frac{1}{3}$ lin. long ; stamens exserted from the corolla-tube ; nutlets subtriquetrous, truncate at the apex, nearly $\frac{2}{3}$ lin. long, minutely tuberculate. *Balf. f., Bot. Socotra,* 242 ; *Baker in Dyer, Fl. Trop. Afr.* v. 480. *L. paucicrenata, Vatke in Linnæa,* xliii. 98.

KALAHARI REGION : Transvaal ; Crocodile River Drift, between Komati River Drift and Barberton, about 500 ft., *Bolus,* 9746 ! Wonderboom Poort, *Miss Leendertz,* 948 !

Also in Tropical Africa, Dessi Island in the Red Sea, and in Socotra.

XXII. LASIOCORYS, Benth.

Calyx tubular-campanulate or campanulate, equal or slightly oblique at the mouth, 10-nerved ; teeth 5, ovate to narrowly deltoid, sometimes with 1 or 2 much smaller additional teeth. *Corolla-tube* included or scarcely exserted, annular-pilose or annular-papillose inside ; limb 2-lipped ; upper lip erect, concave, entire or emarginate, densely villous ; lower lip spreading, 3-lobed ; median lobe broader than the lateral, emarginate. *Stamens* 4, didynamous (the lower pair longer), ascending under the upper lip ; anthers 2-celled ; cells divaricate, finally confluent. *Disk* produced on the lower side into a gland as long as the ovary. *Style* subulate at the apex, with a shorter, sometimes minute posticous lobe. *Nutlets* ovoid-triquetrous, rounded at the apex.

Branched shrubs or undershrubs ; leaves entire or toothed, the upper similar to the others or reduced ; whorls axillary, few- or many-flowered ; bracts subulate, very small ; corolla usually white ; upper lip densely villous.

DISTRIB.   Species 9, 6 in Tropical Africa and Arabia, 2 in Socotra and 1 in South Africa.   Some authors regard the genus as a section of *Leucas* from which it differs in usually having only 5 teeth to the calyx.

1. **L. capensis** (Benth. Lab. 600); a slender divaricately branched shrub, up to 4 ft. high; branches at first obscurely 4-angled, finally terete, canescent-puberulous, finally glabrous, rough with the short persistent bases of the petioles; leaves in distant pairs or somewhat crowded on the short branchlets, petiolate to subsessile, linear-spathulate to obovate-spathulate, sometimes elliptic or lanceolate, 3–13 (often 4–7) lin. long, $\frac{3}{4}$–6 lin. broad, rounded, obtuse or rarely apiculate, narrowed at the base, entire or rarely few-toothed near the apex, shortly adpressed canescent-pubescent or glabrescent, often somewhat coriaceous; petiole up to 3 lin. long; whorls 2–6-flowered, distant or crowded; bracts subulate, $\frac{1}{4}$–1$\frac{1}{4}$ lin. long calyx sessile or shortly pedicellate, tubular-campanulate or campanulate, 3–3$\frac{1}{2}$ lin. long, slightly oblique at the throat, 10-nerved, 5-toothed (sometimes with 1 or 2 smaller additional teeth), coriaceous, adpressed canescent-pubescent; teeth deltoid, ovate or sometimes narrowly triangular, usually $\frac{3}{4}$–1$\frac{1}{4}$ lin. long, shortly acuminate-spinescent, rarely obtuse; corolla white, about 5$\frac{1}{2}$ lin. long; tube 2$\frac{1}{2}$–3$\frac{1}{2}$ lin. long, slightly curved, retrorsely villous above, annular-papillose inside; upper lip obovate, 2$\frac{3}{4}$–3 lin. long, slightly emarginate; lower lip 2$\frac{3}{4}$–3 lin. long; median lobe obovate or semi orbicular. 1$\frac{1}{2}$–1$\frac{3}{4}$ lin. long, 1$\frac{3}{4}$–2 lin. broad, emarginate; lateral lobes about 1 lin. long and broad, rounded. *E. Meyer, Comm.* 241; *DC Prodr.* xii. 534; *Krauss in Flora*, 1845, 66. *Leucas capensis, Engl. Jahrb.* x. 268. *Phlomis capensis, Thunb. Prodr.* 95, *and Fl. Cap. ed. Schult.* 446.

SOUTH AFRICA : without locality, *Ecklon* ! *Zeyher*, 1350 ! *Thunberg* ! *Prior* !
COAST REGION : Uitenhage Div. ; Zwartkops River, *Zeyher*, 736 ! Sunday River and Coega River, *Drège* ! between Sunday River and Addo, *Drège* ! and without precise locality, *Prior* ! Port Elizabeth Div. ; Cradock Place, Port Elizabeth, *Galpin*, 6375 ! Albany Div. ; Grahamstown, *Williamson* ! *MacOwan*, 621 ! Bedford Div. ; Small-deal country, *Burke* ! Victoria East Div., *Cooper*, 401 ! 3108 ! Stockenstrom Div. ; Kat Berg, *Shaw* ! Queenstown Div. ; Kat River, 800–900 ft., *Baur* ! King Williamstown Div. ; near King Williamstown, *Tyson*, 1022 ! 2198 ! banks of the Buffalo River, 1300 ft., *Galpin*, 5940 ! British Kaffraria, *Pappe* !
CENTRAL REGION : Jansenville Div. ; Zwart-Ruggens, 2500–3000 ft., *Drège* ! Somerset Div. ; near Little Fish River and Great Fish River, 2000–3000 ft., *Drège* ! Graaff Reinet and Uitenhage Div., *Bolus*, 1661 !
KALAHARI REGION : Griqualand West ; Albania, *Shaw* ! Asbestos Mountains, at the Kloof Village, *Burchell*, 1660 ! between Griqua Town and Spuigslang Fontein, *Burchell*, 1700 ! Lower Campbell, *Burchell*, 1820 ! Orange River Colony, *Mrs. Hutton* ! Olifants Fontein, *Rehmann*, 3789 ! Bechuanaland ; Batlapin Territory, near the Vaal River, *Holub* ! Transvaal ; near Pietersburg, 4000 ft., *Bolus*, 10859 ! Waterval River, *Wilms*, 1149 !
EASTERN REGION : Natal ; Mooi River Valley, *Gerrard*, 1217 ! 2000–3000 ft., *Sutherland* ! and without precise locality, *Miss Owen* !

XXIII. LEONOTIS, R. Br.

Calyx tubular, 10-nerved, incurved at the apex, more or less
oblique at the mouth, usually 8–10-toothed; teeth often rigid and
spinescent, rarely unarmed or obtuse, more or less unequal, the
uppermost usually much longer and broader than the others, sometimes all but the uppermost minute or obsolete. *Corolla-tube*
usually exserted, cylindric or slightly enlarged above, naked or
imperfectly annular-pilose inside; limb 2-lipped; upper lip erect,
elongated, concave, densely villous; lower lip short, spreading,
3-lobed, marcescent; median lobe scarcely larger than the lateral.
Stamens 4, didynamous, ascending under the upper lip, the lower
pair longer than the upper; anthers approximate in pairs, 2-celled;
cells divaricate, subconfluent. *Disk* equal. *Style* subulate at the
apex, 2-lobed; upper lobe very short. *Nutlets* ovoid-triquetrous,
obtuse or truncate at the apex, glabrous.

Herbs or shrubs; leaves often ovate, sometimes oblong-lanceolate, toothed,
usually petiolate; floral leaves similar but smaller and often sessile; whorls
axillary, densely many-flowered, usually few or solitary near the ends of the
stem or branches; bracteoles numerous, subulate to narrowly lanceolate, often
spinescent; flowers sessile to shortly pedicellate, often deep orange-yellow or red,
sometimes white, usually large.

DISTRIB. Species 32, nearly all in Tropical or South Africa, 1 extending into
Tropical Asia and America.

L. nepetæfolia, R. Br., does not appear to be South African, though Bentham
(Lab. 618, and in DC. Prodr. xii. 535) cites a specimen collected by Forbes at
Delagoa Bay under this name. The specimen, however, is not in the Kew
Herbarium.

Leaves lanceolate, 5 or 6 times as long as broad :
  Leaves pubescent ; calyx-teeth usually very short and
    weak, often obtuse, not or rarely slightly
    spinescent ... ... (1) **Leonurus.**

  Leaves glabrous ; calyx-teeth rigid, very acute,
    spinescent ... ... (2) **Schinzii.**

Leaves ovate-lanceolate to suborbicular-obovate, scarcely
  longer than broad up to 3 (rarely more) times longer
  than broad :
Leaves small, rarely more than 1 in. long and often
  only about ½ in. long :
    Branches and leaves very shortly pubescent or
    merely puberulous ; leaves usually broadly
    ovate and not attenuated at the base ... ... (3) **Leonitis.**

    Branches and leaves rather densely covered with
    somewhat long often spreading hispid hairs ;
    leaves mostly obovate-cuneate, much attenuated
    at the base... ... ... (4) **microphylla.**

Leaves usually more than 1 in. long up to 4 in. long :
  *Leaves covered on both sides or on the underside
    only with a dense velvety often greyish
    pubescence :

†Leaves ovate or suborbicular-ovate, from about as
   long as broad up to 1½ times as long as broad,
   cordate, subcordate or scarcely cuneate at the
   base :

  Leaves suborbicular-ovate ; petiole usually
    distinctly longer than the blade ; all the
    calyx-teeth rather long, rigid and very
    spinescent, the uppermost often much
    longer than the others   ...   .   (5) **mollis.**

  Leaves ovate ; petiole shorter, often much
    shorter than the blade ; all the calyx-teeth
    rather short, not or only slightly spinescent,
    the uppermost not much longer than the
    others :

    Petiole (of the upper leaves) ⅛–½ in. long   ...   (6) **brevipes.**

    Petiole usually 1–1½ in. long :
      Leaves up to 3½ in. long ; stem deeply 4-
      furrowed ; calyx adpressed-pubescent,
      8-toothed   ...   .   (7) **latifolia.**

      Leaves 1½–2¾ in. long ; stem not or scarcely
      furrowed ; calyx spreading-pubescent,
      10-toothed   ...   .   (8) **Galpini.**

††Leaves usually ovate-lanceolate, often 2 or 3 times
   as long as broad, usually distinctly cuneate at
   the base ...   .   (9) **dysophylla.**

**Leaves more or less pubescent to nearly quite
   glabrous, not densely covered with a velvety
   pubescence :
  Calyx usually 10-toothed ; teeth distinct, rather
    regular, slender, spinescent, the uppermost
    only slightly longer and broader than the
    others   ...   .   (13) **Westæ.**

  Calyx 8–10-toothed ; teeth often indistinct, very
    short, irregular, unarmed or slightly spines-
    cent, the uppermost scarcely longer but very
    much broader than the others   ...   ...   (11) **intermedia.**

  Calyx more or less indistinctly 8-toothed, the
    uppermost tooth relatively long compared
    with the other small unarmed or slightly
    spinescent teeth, or often with only 1 tooth
    (the uppermost) :
    Leaves usually less than 3 in. long and 2¼ in.
      broad, scarcely flaccid ; teeth usually 1–2½
      lin. broad at the base, the terminal one
      usually less than 3 lin. long ; petioles 1–2
      rarely up to 2½ in. long   ...   .   (10) **dubia.**

    Leaves usually more than 3 in. long and 2¼
      in. broad, very flaccid ; teeth often 3–4 lin.
      broad at the base, the terminal one 4–9 lin.
      long ; petioles usually more than 2 in. up
      to 3 in. long ...   .   (12) **laxifolia.**

   1. L. Leonurus ([R. Br. in] Ait. Hort. Kew. ed. 2, iii. 410)); shrubby ;
stems up to 7 or 8 ft. high, branched, densely and shortly adpressed-
pubescent ; leaves subsessile to shortly petiolate, oblong-lanceolate
or lanceolate, 2–4 in. long, ¼–1 in. broad, obtuse, much narrowed at
the base, crenate-serrate, shortly pubescent, often rather densely

beneath, prominently veined beneath ; floral leaves similar but
smaller, the uppermost usually sterile ; whorls up to about 7, distant
or approximate, usually densely flowered ; bracteoles narrowly linear,
up to 1 in. long, scarcely spinescent ; pedicels up to 2 lin. long ;
calyx 6½–8 lin. long, slightly oblique at the throat, densely and
shortly pubescent, sometimes with longer spreading hairs ; teeth
8–10, deltoid to narrowly deltoid, ¼–1¼ rarely up to 1¾ lin. long,
subequal or very unequal, acute, usually soft, rarely slightly
spinescent ; corolla orange-scarlet, up to 2¼ in. long or more ; tube
14–18 lin. long ; upper lip 6–9 lin. long ; lower lip 3¼–4 lin. long.
Benth. Lab. 620, *in E. Meyer, Comm.* 243, *and in DC. Prodr.* xii.
536 ; *Krauss in Flora,* 1845, 66 ; *Lynch in Gard. Chron.* 1883, xix.
186, *fig.* 28 ; *Morris in Ann. Bot.* i. 160 ; *Wood & Evans, Natal
Pl.* i. *t.* 53. *Phlomis Leonurus, Linn. Sp. Pl. ed.* ii. 820, *and Mant.* ii.
412 ; *Berg. Descr. Pl. Cap.* 151 ; *Bot. Mag. t.* 478 ; *Thunb. Prodr.* 95,
and Fl. Cap. ed. Schult. 446. *Leonurus africanus, Mill. Dict. ed.* 8,
n. 1. *L. grandiflorus, Moench, Meth. Pl.* 400.

Var. β, albiflora (Benth. in DC. Prodr. xii. 537) ; corolla white or when dry
sulphur-yellow.

South Africa : without locality, *Thunberg* ! *Ecklon* ! *Forbes* ! *Pappe* ! *Harvey*,
562 ! *Wallich* !

Coast Region : Paarl Div. ; Paarl Mountain, 1000–2000 ft., *Drège*, 7955*b* !
Cape Div. ; various localities on the Cape Peninsula and near Cape Town, *Harvey* !
Wright, 505 ! *Prior* ! *MacGillivray*, 569¦! *Milne*, 133 ! *Wolley-Dod*, 662 !
Ecklon, 48 ! *Burchell*, 108 ! *Drège*, 7955*a* ! *Burke* ! *MacOwan & Bolus, Herb.
Norm. Austr.-Afr.*, 591 ! Knysna Div. ; near Stofpad, *Burchell*, 5288 ! Uitenhage
Div. ; Zwartkops River, *Zeyher*, 14 ! and without precise locality, *Cooper*, 1499 !
Alexandria Div. ; Zuurberg Range, 2000–3000 ft., *Drège* ! Albany Div. ; near
Grahamstown, *MacOwan* ! *Williamson* ! *Schlechter*, 2624 ! *Bolton* ! *Miss Daly &
Miss Cherry*, 943 ! and without precise locality, *Atherstone*, 38 ! Stockenstrom Div. ;
Kat Berg, *Hutton* ! 4000 ft., *Shaw*, 1992 ! Stutterheim Div. ; near Dohne Post,
Bowker, 86 ! Komgha Div. ; between Zandplaat and Komgha, 2000–3000 ft.,
Drège. Var. β : Malmesbury Div. ; Mamre, 2000–2500 ft., *Baur*, 160 ! Worcester
Div. ; Hex River, *Drège*, 4829 !

Kalahari Region : Transvaal ; Sable Falls to Pilgrims Rest, common near
Burghers Pass (a form with unusually long calyx-teeth), *Burtt-Davy*, 433 !

Eastern Region : Transkei ; near Kentani, 1200 ft., *Miss Pegler*, 368 !
Pondoland ; St. Johns River, *Drège*, 4828*b* ! Griqualand East ; near Clydesdale,
2500 ft., *Tyson*, 2796 ! Natal ; near Durban, *Plant*, 32 ! between the Umzimkulu
River and the Umkomanzi River, *Drège*, and without precise locality, *Cooper*,
1115 ! *Gerrard*, 19 ! *Sanderson*, 30 ! Var. β : Natal ; Inanda, *Wood*, 164 !
Zululand ; Eshowe, *Mrs. K. Saunders*, 1 !

2. L. Schinzii (Guerke in Engl. Jahrb. xxii. 143) ; plant glabrous
or here and there clothed with very short scattered hairs ; stem
branched ; leaves lanceolate, 1¼–2 in. long, 1½–4 lin. broad, acute,
narrowed at the base, coarsely serrate, glabrous both sides, prominently nerved beneath ; petiole 5–10 lin. long ; bracteoles subulate,
spinescent ; calyx 8½–10 lin. long, glabrous or puberulous above ;
teeth 8, the uppermost larger than the others, all deltoid, rigid, very
acute, spinescent ; corolla orange-yellow. *Baker in Dyer, Fl. Trop.
Afr.* v. 494.

WESTERN REGION : Great Namaqualand ; Homeib, *Schinz*, 40.
Also in Hereroland.

3. **L. Leonitis** ([R. Br. in] Ait. Hort. Kew. ed. 2, iii. 410); shrubby;
stems up to 2 ft. high or more, shortly and densely pubescent,
sparingly branched ; leaves ovate to elliptic, 6–12 lin. long, 4–8 lin.
broad (on the branches often only 2–4 lin. long and 1½–3 lin. broad),
obtuse or rounded at the apex, truncate, rounded or sometimes
cuneate at the base, crenate or crenate-serrate, shortly adpressed-
pubescent above, more or less densely pubescent beneath, rugose ;
petiole 2–5 lin. long ; whorls usually solitary on each stem, densely
flowered ; bracteoles linear-subulate, 1½–5 lin. long, spinescent ;
calyx 7–10 lin. long, densely shortly pubescent, sometimes with a
few longer hairs on the upper part ; teeth usually 8, all more or
less spinescent ; uppermost tooth ovate-deltoid, 1¾–2 lin. long ;
other teeth deltoid to subulate, ¼–1¼ lin. long, or sometimes almost
obsolete ; corolla deep orange-yellow, 1¼–1¾ in. long ; tube 8–10½
lin. long; upper lip 6–10½ lin. long; lower lip 3–4½ lin. long.
L. ovata, Spreng. Syst. ii. 744 ; *Benth. Lab.* 619, *in E. Meyer, Comm.*
242, *and in DC. Prodr.* xii. 535 ; *Krauss in Flora*, 1845, 66 ;
Scott-Elliot in Ann. Bot. iv. 272. *Phlomis Leonotis, Linn. Mant.* i.
83 ; *Thunb. Prodr.* 96, *and Fl. Cap. ed. Schult.* 446. *P. Leonitis,*
Willd. Sp. Pl. iii. 128. *Leonurus minor, etc., Boerh. Ind. Alt. Pl.*
Hort. Lugd.-Bat. i. 180 ; *Mill. Ic.* ii. 108, *t.* 162, *fig.* 1.

VAR. **β, hirtiflora** (Skan) ; calyx densely and shortly pubescent and in the
upper part more or less densely clothed with long spreading hairs. *L. hirtiflora,*
Benth. in DC. Prodr. xii. 536.

SOUTH AFRICA : without locality, *Thunberg* ! *Ecklon* ! Var. *β*, *Thom*, 273 !
COAST REGION : Swellendam Div. ; between Storms Vallei and Attaquas
Kloof, *Drège*, 7952*a* ! George Div. ; near George, *Prior* ! Uniondale Div. ; near
Onzer, *Drège*, 7954*a* ! Humansdorp Div. ; between Humansdorp and Gamtoos
River, *Bolus*, 2431 ! Uitenhage Div. ; Vanstadens Berg, *Zeyher*, 686 ! between
Vanstadens Berg and Bethelsdorp, *Drège*. Addo, 1000–2000 ft., *Zeyher*, 1349
partly ! Port Elizabeth Div. ; Algoa Bay, *Prior* ! Var. *β* : Cape Div. ; Ludwigs-
burg (Ludwigsburg Garden), *Zeyher*, 206 ! Bredasdorp Div. ; Riet Fontein Poort,
Schlechter, 9689 ! near Elim, *Bolus*, 8583 ! Swellendam Div. ; between Storms
Vallei and Attaquas Kloof, *Drège*, 7952*b* ! Albany Div. ; near Grahamstown, *Miss*
Daly & Miss Cherry, 952 !
CENTRAL REGION : Somerset Div. ; Bruintjes Hoogte, lower part, *Burchell*,
3008 !
EASTERN REGION : Tembuland ; Bazeia, 2000–3000 ft., *Baur*, 95 !

Phlomis ocymifolia, Burm. f. Fl. Cap. Prodr. 16, is possibly *L. Leonitis*, R. Br.

4. **L. microphylla** (Skan) ; perennial, rather densely covered on
stem, branches and leaves with short mixed with longer somewhat
hispid hairs ; stem up to 2 ft. high or more, up to 2¼ lin. thick,
freely branched at the base ; branches often very short (1–5 in.
long) and slender, densely leafy ; leaves obovate-cuneate to lanceolate, up to 9 (usually less than 6) lin. long, 1½–6 lin. broad, obtuse
or rounded, narrowed to the base ; crenate-serrate, rather thick ;
petiole 1–8 (rarely up to 12) lin. long ; whorls 1 or 2, distant,

densely-flowered; bracteoles subulate, 3–5 lin. long, spinescent; pedicels $\frac{3}{4}$–4 lin. long; calyx curved, 9–10 lin. long, somewhat densely covered with short hispid hairs, prominently 10-nerved; teeth 8, all spinescent; uppermost tooth deltoid, $1\frac{3}{4}$–$2\frac{1}{2}$ lin. long; other teeth narrowly deltoid to subulate, $\frac{3}{4}$–1 lin. long; corolla deep orange-yellow, 13–20 lin. long; tube 7–11 lin. long; upper lip 6–9 lin. long; lower lip about 4 lin. long.

KALAHARI REGION: Transvaal; Jeppestown Ridges, Johannesburg, 6000 ft., *Gilfillan in Herb. Galpin*, 6169! Meintjies Kop, Pretoria, 4800 ft., *Burtt-Davy*, 3936! Heidelberg, *Miss Leendertz*, 1035!

5. **L. mollis** (Benth. in E. Meyer, Comm. 242); shrubby, at least at the base; stems up to 3 ft. high, very densely covered with short adpressed soft hairs; leaves broadly ovate to suborbicular, usually $\frac{3}{4}$–$1\frac{3}{4}$ in. long and broad, obtuse or rounded at the apex, broadly and shallowly cordate, truncate or slightly cuneate at the base, crenate or crenate-serrate, densely and shortly velvety-pubescent especially beneath, rather thick; petiole $\frac{3}{4}$–2 in. long; whorls solitary or few, densely flowered; bracteoles subulate or linear, up to 7 lin. long, spinescent; calyx $9\frac{1}{2}$–11 lin. long, densely and shortly pubescent and sometimes with a few longer hairs on the upper part; teeth 8, all rigidly spinescent; uppermost tooth deltoid, up to $4\frac{1}{2}$ lin. long including the long spine; other teeth narrowly deltoid, $\frac{1}{2}$–$1\frac{3}{4}$ lin. long; corolla deep orange-yellow, up to about $1\frac{1}{2}$ in. long; tube 8–10 lin. long; upper lip 7–8 lin. long; lower lip $3\frac{3}{4}$–$4\frac{1}{2}$ lin. long. *DC. Prodr.* xii. 536.

VAR. β, **albiflora** (Skan); somewhat less velvety-pubescent; corolla white.

SOUTH AFRICA: without locality, *Mund*!
COAST REGION: Uitenhage Div.; Addo, *Drège*, 7953*b*! *Zeyher*, 1349 partly! Zuurberg Range, *Drège*, 7953*c*! banks of the Coega River, *Prior*! Albany Div.; Howisons Poort, *Hutton*! near Grahamstown, *Bolton*! Queenstown Div.; mountain sides near Queenstown, 4500 ft., *Galpin*, 1825!
CENTRAL REGION: Somerset Div.; Somerset East, *Bowker*, 107! Bruintjes Hoogte, lower part, *Burchell*, 3008! Graaff Reinet Div.; near Graaff Reinet, 2600 ft., *Bolus*, 545! Beaufort West Div.; Nieuwveld Mountains near Beaufort West, 3000–5000 ft., *Drège*, 7953*a*! Philipstown Div.; near Riet Fontein, *Burchell*, 2732! Var. β: Somerset Div.; Bosch Berg, *MacOwan*!
KALAHARI REGION: Orange River Colony; Sand Drift, *Burke*!

6. **L. brevipes** (Skan); stems herbaceous, at least in the upper part, $1\frac{1}{4}$–$1\frac{1}{2}$ lin. in diam., rounded on the angles, 4-furrowed, densely covered with short fine grey recurved hairs; leaves ovate to narrowly ovate, $1\frac{1}{2}$–$2\frac{1}{2}$ in. long, 1–$1\frac{1}{2}$ in. broad, obtusely or subacutely acuminate at the apex, truncate-cuneate to slightly cordate at the base, dentate-serrate, densely greyish velvety-pubescent, especially beneath, moderately thick; terminal tooth narrowly deltoid, 3–5 lin. long, often (as well as some of the lateral teeth) with a minute callous apiculus; petiole of the upper leaves 2–6 lin. long; whorls solitary, globose, compact, many-flowered;

bracteoles subulate to narrowly lanceolate, up to 5 lin. long, often slightly spinescent ; pedicel up to 1 lin. long ; calyx 8–9 lin. long, densely covered with short fine spreading hairs mixed with longer hairs ; teeth 8 ; uppermost tooth broadly deltoid or ovate-deltoid, 1¼ lin. long, spinescent ; other teeth deltoid, ¼–¾ lin. long, usually slightly spinescent ; corolla deep orange-yellow, up to 17 lin. long ; tube up to 9 lin. long ; upper lip up to 8 lin. long ; lower lip 4 lin. long.

KALAHARI REGION : Transvaal ; Medingen Mission Station, Zoutpansberg, in bush, *Burtt-Davy*, 2657 !

Differs from *L. mollis*, Benth., in having longer leaves narrower at the apex, much shorter petioles, and very short calyx-teeth.

7. **L. latifolia** (Guerke in Engl. Jahrb. xxii. 143) ; stem up to 5 ft. high, rather much branched, pubescent as well as the branches ; leaves ovate or suborbicular, 2½–4 in. long, 1¾–3 in. broad, acute, deeply cordate at the base (the upper subcordate), coarsely crenate, pubescent above, canescent or almost velvety beneath, prominently veined beneath ; petiole up to 2½ in. long ; bracteoles lanceolate, acute ; calyx 6½–8 lin. long, pubescent outside, more or less hirsute along the nerves ; teeth 8, the uppermost not much larger than the others, all rigid and shortly spinescent ; corolla brick-red.

EASTERN REGION : Natal ; Biggarsberg Range, *Rehmann*, 7057 ; Van Reenens Pass, 5500–6200 ft., *Kuntze* ; Mooi River, 4900 ft., *Schlechter*, 6839 ; Murchison, Alfred County, *Bachmann*, 1174.

A specimen collected by Drège at the St. Johns River, Pondoland, and Wood, 1303, from Inanda, Natal, should probably be referred to this species, of which we have not seen any of the specimens cited above. In *E. Meyer, Comm.* 242, Drège's specimen is identified with *L. dubia*, E. Meyer, but it appears to be specifically distinct.

8. **L. Galpini** (Skan) ; stem herbaceous, at least in the upper part, 4 ft. high, branched, densely covered with short grey reflexed hairs, somewhat sharply 4-angled, not or scarcely furrowed ; leaves broadly ovate, 1½–3 in. long, 1½–2 in. broad, obtuse or minutely apiculate at the apex, truncate or very slightly cuneate at the base, coarsely dentate-serrate or crenate-dentate, densely greyish velvety-pubescent, especially on the nerves beneath, moderately thick ; petiole 1¼–1¾ in. long ; whorls solitary, rather loose, many-flowered ; bracteoles subulate, up to about 6 lin. long, slightly spinescent ; pedicels up to 1½ lin. long ; calyx about 9 lin. long, densely covered with short fine spreading hairs mixed with much longer hairs, rather thin and somewhat transparent between the ribs ; teeth usually 10 ; uppermost tooth narrowly deltoid, 1¾ lin. long, slightly spinescent ; other teeth narrowly deltoid, ¼–⅓ lin. long, acute but scarcely spinescent, sometimes 1 or more almost obsolete ; corolla deep orange-yellow, 17–18 lin. long ; tube about 9 lin. long ; upper lip about 8 lin. long ; lower lip 4¼ lin. long.

COAST REGION : Queenstown Div. ; mountain sides near Queenstown, 4500 ft., *Galpin,* 1825 !

Differs from *L. mollis,* Benth., in the more coarsely toothed leaves, longer pedicels, longer and thinner calyx-tube, with 10 shorter much less spinescent teeth.

9. **L. dysophylla** (Benth. in E. Meyer, Comm. 242) ; shrubby, at least at the base ; stem robust, upwards of 2 ft. high, densely and usually shortly pubescent ; leaves ovate-lanceolate or sometimes ovate, up to 2¾ in. long or more and 1¼ in. broad, obtuse or subacute at the apex, cuneate at the base, somewhat regularly crenate or crenate-serrate, densely and shortly sometimes yellowish villous, especially beneath, thick ; petiole ½–1½ in. long ; whorls solitary or few, densely flowered ; bracteoles subulate to narrowly lanceolate, up to about 8 lin. long, spinescent ; calyx 9–11½ lin. long, shortly and densely adpressed-pubescent ; teeth 8, all spinescent ; uppermost tooth broadly deltoid-ovate, 1¾–2¼ lin. long ; other teeth narrowly deltoid, ½–¾ lin. long ; corolla deep orange-yellow, 12–18 lin. long ; tube 9–10 lin. long ; upper lip 5–9 lin. long ; lower lip 4–4½ lin. long. *DC. Prodr.* xii. 536. *L. dasyphylla, Drège, Zwei Pfl. Documente,* 198. *L. malacophylla, Guerke in Engl. Jahrb.* xxii. 142 ; *S. Moore in Journ. Bot.* 1903, 406. *L. Leonurus, Rand in Journ. Bot.* 1903, 194, *not of R. Br.*

KALAHARI REGION : Orange River Colony, *Cooper,* 1041 ! Transvaal ; Rooiplaat, *Miss Leendertz,* 770*a* ! Hooge Veld. near Heidelberg, *Wilms,* 1146*a* ! near Lydenburg, *Wilms,* 1146 !

EASTERN REGION : Transkei ; near Kentani, 1200 ft., *Miss Pegler,* 1514 ! Pondoland ; between St. Johns River and Umsikaba River, 1000–2000 ft., *Drège,* 4832*a* ! Griqualand East ; Clydesdale, near the Umzimkulu River, 2500 ft., *Tyson,* 2729 ! *and in MacOwan, Herb. Austr.-Afr.,* 1508 ! Natal ; between the Tugela and Klip Rivers, *Gerrard,* 393 ! Drakensberg, near Ladysmith, *Wilms,* 2111 ! Camperdown, *Rehmann,* 7750 ; Howick, about 3300 ft., *Junod,* 403 ; between the Umzimkulu River and Umkomanzi River, *Drège* ; and without precise locality, *Gerrard,* 596 !

10. **L. dubia** (E. Meyer, Comm. 242 partly) ; stems apparently herbaceous above and woody below, rather slender, branched, rounded on the angles, 4-furrowed, rather densely covered with minute curled hairs ; leaves ovate, usually 1¼–2¼ in. long, rarely up to 3¼ in. long, 1–2¼ in. broad, obtuse to shortly obtusely acuminate at the apex, broadly and shallowly cordate, truncate or slightly cuneate at the base, coarsely crenate-serrate, thinly to densely somewhat velvety-pubescent especially beneath ; lateral teeth usually about ¾–1½ lin. long and broad ; terminal tooth ovate to narrowly deltoid, usually less than 3 lin. long ; petiole slender, ¾–1½ rarely up to 2½ in. long ; whorls often solitary, sometimes 2 or 3, densely flowered ; bracteoles subulate or linear, up to 5 or 6 lin. long, usually slightly spinescent ; pedicels up to 1¾ lin. long ; calyx 7½–8½ lin. long, densely and shortly pubescent ; teeth usually 8 ; uppermost tooth deltoid, 1½–2 lin. long, spinescent ; other teeth very small, obtuse or slightly spinescent, sometimes obsolete ;

corolla deep orange-yellow, up to 1½ in. long; tube up to 10 lin. long; upper lip up to 8 lin. long; lower lip 3½–4 lin. long. *Benth. in DC. Prodr.* xii. 536; *Baker in Dyer, Fl. Trop. Afr.* v. 493. *L. parvifolia, Benth. Lab.* 619.

SOUTH AFRICA: without locality, *Masson, Harvey! Ecklon!*
COAST REGION: Uitenhage Div.; Enon, *Drège,* 4831*a*! and without precise locality, *Zeyher!* Albany Div.; near Grahamstown, 2000 ft., *MacOwan,* 1264! Bedford Div.; Bedford, *Miss Nicol,* 34! King Williamstown Div.; Buffalo River, *Drège!* East London Div.; East London, in river bed, *Galpin,* 5734!

Also in British Central Africa.

Galpin, 2633, from Bailey Poort, Queenstown Division, is probably *L. dubia,* though the leaves are more densely greyish-villous than is usual in this species.

11. **L. intermedia** (Lindl. Bot. Reg. x. t. 850); subshrubby; stem erect, about 4 ft. high, densely and very shortly adpressed-pubescent; leaves ovate to ovate-lanceolate, up to 3 in. long and 2 in. broad, obtusely acuminate, the lower cordate, the upper more or less cuneate at the base, crenate-serrate or crenate, very shortly and sometimes rather densely pubescent, often rugose beneath; petioles of the upper leaves ½–1½ in. long; whorls densely flowered, usually solitary and subterminal, sometimes with a second much smaller one above; bracteoles linear to lanceolate, up to about ½ in. long, usually spinose at the apex; calyx 6–8 lin. long, densely long-villous outside; teeth 8–10, very short, the uppermost only slightly longest, all obtuse or sometimes slightly spinose; corolla orange-yellow, 13–15 lin. long; tube 8–9½ lin. long; upper lip 6 lin. long; lower lip 3½ lin. long. *Benth. in DC. Prodr.* xii. 536 *partly.*

VAR. β, **natalensis** (Skan); calyx puberulous or shortly villous; teeth usually slightly spinose. *L. intermedia, Benth. in DC. Prodr.* xii. 536, *partly.*

COAST REGION: Uitenhage Div.; *Cooper,* 2397! Port Elizabeth Div.; Algoa Bay, *Forbes!*
EASTERN REGION: Var. β: Natal; near Durban, *Peddie! Grant! Williamson!* Bushmans River, *Gerrard,* 362!

12. **L. laxifolia** (MacOwan in Kew Bulletin, 1893, 13); stems apparently herbaceous, slender, more or less covered with short curled hairs; leaves broadly ovate, 2½–4 in. long, 1¼–3½ in. broad, acuminate at the apex, broadly and shallowly cordate, truncate or slightly cuneate at the base, incised-dentate, sparingly sprinkled with minute (rarely rather long) slender hairs, sometimes rather densely puberulous beneath, very thin; lateral teeth often 3 lin. long and 3 lin. broad at the base; terminal tooth usually lanceolate, up to 9 lin. long; petioles very slender, 1½–2¾ in. long; whorls 1–3, loosely many-flowered; bracteoles subulate or linear, up to 7 lin. long, reflexed, often slightly spinescent; pedicels up to 3 lin. long; calyx 8–9½ lin. long, puberulous and sometimes with rather longer hairs chiefly on the nerves near the apex; teeth 8, all spinescent, or sometimes only 1 (the uppermost); uppermost tooth up to 2¾ lin. long; other teeth when present ¼–¾ lin. long; corolla deep orange-

yellow, up to 1⅓ in. long; tube 7–8 lin. long; upper lip 6–8 lin. long; lower lip 3½ lin. long. *L. urticifolia, Briq. in Bull. Herb. Boiss.* 2*me* sér. iii. 1091.

COAST REGION : Albany Div. ; Grahamstown, *Prior*!
EASTERN REGION : Transkei ; near Kentani, 1200 ft., *Miss Pegler*, 355! Griqualand East ; in woods on Mount Malowe, 4500 ft., *Tyson*, 2766, *and in MacOwan & Bolus, Herb. Austr.-Afr.*, 1300! Natal; Ismont, *Wood*, 1837! and without precise locality, *Cooper*, 1182!

Guerke in *Engl. Jahrb.* xxii. 144, distinguishes a form (*pilosa*) which differs from the type in its greater hairiness, especially on the calyx, the ribs of which are furnished with rather long hairs. The specimen cited by him (Rehmann, 7374, from Kar Kloof, Natal) we have not seen. Typical *L. laxifolia* is less hairy than the other specimens referred to above.

13. **L. Westæ** (Skan); stems apparently herbaceous, weak, scarcely furrowed, at least in the upper part, densely covered with very short greyish recurved hairs; leaves ovate-deltoid, 1–2½ in. long, ¾–2 in. broad, obtusely acuminate at the apex, broadly cuneate at the base, coarsely crenate-dentate except at the entire base, somewhat thinly covered above and more densely beneath with short adpressed greyish hairs, rather thin; lateral teeth 1½–3 lin. long, up to 3½ lin. broad; terminal tooth narrowly deltoid, up to 5 lin. long; petiole slender, ¾–2 in. long, densely grey-pilose; whorls solitary, relatively small, up to about 30-flowered in specimens seen; bracteoles subulate or linear, up to about 5 lin. long, spinescent; pedicels up to ½ lin. long; calyx 7 lin. long, densely and very shortly pubescent, with slightly longer hairs on the nerves; teeth usually 10, all subulate or the uppermost sometimes deltoid, spinescent; uppermost tooth 1½ lin. long; other teeth rather regular, ¼–¾ lin. long; corolla deep orange-yellow, up to 1½ in. long; tube up to 9 lin. long; upper lip up to 9 lin. long; lower lip 4 lin. long.

COAST REGION : Port Elizabeth Div. ; Port Elizabeth, *Miss West*, 75!

Near *L. laxifolia*, MacOwan, but the plant is more hairy, leaves smaller and less flaccid, pedicels much shorter, calyx shorter and usually with 10 less unequal teeth.

Imperfectly known species.

14. **L. Bachmannii** (Guerke in Engl. Jahrb. xxii. 143); stem simple, 3–6½ ft. high, pubescent; leaves ovate-lanceolate, 1¾–2 in. long, ⅔–1 in. broad, acute or acuminate, narrowed at the base, crenate-serrate, pubescent both sides; petiole ¾–1¼ in. long; bracteoles subulate, spinescent; pedicels 1½–2 lin. long; calyx 8½–10 lin. long, puberulous or almost glabrous outside; teeth 8, the uppermost only slightly larger than the others, all ending in a short strong rigid spine; corolla orange-yellow.

KALAHARI REGION : Transvaal ; hillsides near Barberton, at about 3000 ft., *Galpin*, 922.
EASTERN REGION : Pondoland ; on hills, *Bachmann*, 1170, 1175.

XXIV. TINNEA, Kotschy et Peyr.

Calyx campanulate, with 2 broad entire or nearly entire lips, ovoid, much enlarged inflated and deeply 2-valved when in fruit. *Corolla-tube* short, broad, enlarged at the throat, scarcely longer than the calyx; limb 2-lipped; upper lip short, broad, erect-spreading, emarginate or 2-lobed; lower lip much larger, spreading, 3-lobed; median lobe much larger than the lateral rounded lobes, emarginate. *Stamens* 4, didynamous (the lower pair longer), ascending under the upper lip, more rarely somewhat exserted; anthers 2-celled; cells short, divergent, finally subconfluent. *Disk* equal. *Ovary* shortly 4-lobed; style shortly bifid; lobes acute, the upper usually shorter. *Nutlets* obovoid-clavate, long-contracted at the base, attached by a lateral areole, furnished on the back with a broad membranous elliptic or orbicular wing. *Seeds* attached laterally.

Shrubs or perennial herbs, pubescent or grey-tomentose; leaves sessile or petiolate, usually entire, the upper similar or gradually reduced to bracts; whorls usually loosely 2-flowered, axillary or in terminal racemes; pedicels 2-bracteolate; flowers fragrant, rather small, usually brownish- or violet-purple.

DISTRIB. Species 20, of which 18 are Tropical African; 1 in Arabia.

1. **T. Galpini** (Briq. in Bull. Herb. Boiss. 2me sér. iii. 1094, *Tinnæa*); an undershrub 6–15 in. high or more; branches slender, terete, densely and shortly pilose-pubescent; leaves sub-sessile or shortly petiolate, elliptic-lanceolate to elliptic-ovate, up to 1 in. long and ½ in. broad, minutely apiculate or obtuse at the apex, slightly narrowed at the base, entire or nearly so, sparingly pilose-pubescent both sides, gland-dotted beneath; petiole up to 1½ lin. long; flowers violet-scented, solitary or sometimes in pairs, axillary or in terminal more or less unilateral racemes up to 6 in. long; whorls remote; bracts ovate-elliptic or ovate, about as long as to much longer than the pedicels; pedicels up to 6 lin. long, densely and shortly pilose-pubescent, minutely bibracteolate; calyx (in flower) broadly campanulate, 3–4 lin. long, densely pilose-pubescent, purplish; lips rounded, 1½–2 lin. long, 3½–4 lin. broad, the upper slightly longer than the lower; calyx (in fruit) ovoid, inflated, membranous, up to 8 lin. long and 5 lin. broad; corolla claret- or prune-colour, about 4 lin. longer than the calyx; tube funnel-shaped, 4–5 lin. long; upper lip about 1–1½ lin. long, about 2¾ lin. broad, emarginate; lower lip 3-lobed; median lobe transversely oblong, 2½ lin. long, 4½–5½ lin. broad, emarginate; lateral lobes broadly rounded, 1½ lin. long, 1¾–2½ lin. broad.

KALAHARI REGION: Transvaal; Barberton and neighbourhood, 3000–4500 ft., *Galpin*, 1212! *Thorncroft*, 39 (*Wood*, 4148)! *Miss Leendertz*, 4114! Klippan, *Rehmann*, 5288! 5289!

XXV. TEUCRIUM, Linn.

Calyx tubular or campanulate, rarely inflated; teeth 5, equal or the uppermost broadest. *Corolla-tube* included or rarely exserted, naked inside; limb as if with only 1 (the lower) lip; lobes 5, the lowermost largest and often concave. *Stamens* 4, didynamous (the lower pair longer), exserted between the uppermost corolla-lobes; anthers 2-celled; cells divergent or more usually divaricate, confluent. *Disk* equal. *Style* 2-fid; lobes subulate, subequal. *Nutlets* obovoid, reticulate-rugose, attached by an oblique or lateral areole which sometimes extends beyond the middle.

Herbs, undershrubs or shrubs, of various habit; leaves entire, toothed or more or less deeply lobed, the upper similar or reduced to bracts; whorls 2- to several-flowered, axillary or forming terminal spikes, racemes or heads.

DISTRIB. Species about 180, widely distributed over the temperate and warmer regions of the world, but chiefly in the northern hemisphere and most frequent in the Mediterranean Region.

T. mauritanum, Linn. [= *T. Pseudo-chamæpitys*, Linn.], and *T. lucidum*, Linn., are included in Burm. f. Fl. Cap. Prodr. 16. They are not South African, but natives of the Mediterranean Region.

Leaves more or less deeply 3-fid, rarely entire :
    Peduncles usually 1-flowered and much shorter than
      the leaves   ...   . (1) **africanum**.
    Peduncles 3–7-flowered ; cymes often as long as or
      longer than the leaves   ...   . (2) **capense**.
Leaves entire or few-toothed   ...   . (3) **riparium**.

1. **T. africanum** (Thunb. Prodr. 95); an erect undershrub, a few inches up to 1¼ ft. high or more, usually much branched; branches very slender, 4-angled, rather densely leafy, grey-puberulous; leaves deeply 3-fid, usually ¾–1 in. long, sometimes up to 2 in. long or more, thinly pubescent above, grey- puberulous or -tomentose beneath; lobes linear or linear-oblong, 2–9 lin. long, usually ½–1 (sometimes up to 1¾) lin. broad, obtuse, revolute at the margin, usually entire, sometimes (especially the median lobe) more or less 3-fid; flowers axillary, solitary or sometimes 2 or more on the same peduncle; peduncle 1½–4 lin. long, bearing below the middle a pair of small bracteoles; calyx campanulate, 2–2½ lin. long, thinly grey-puberulous; teeth linear-triangular or lanceolate, 1¼–1½ lin. long, ⅓–½ lin. broad at the base, acuminate; corolla white, about 2 lin. long; tube ¾–1 lin. long; lobes elliptic or elliptic-oblong, rounded, 1–1¾ lin. long. *Thunb. Fl. Cap. ed. Schult.* 445 ; *Benth. Lab.* 669, *in E. Meyer, Comm.* 243, *and in DC. Prodr.* xii. 577. *T. trifidum, Retz. Obs.* i. 21 ? *T. trifidum, Wendl. Bot. Beobacht.* 50 ? *Ajuga africana, Pers. Syn.* ii. 109.

SOUTH AFRICA : without locality, *Thunberg* ! *Ecklon* !
COAST REGION : George Div. ; Kamanassie Hills, *Prior* ! and without precise locality, *Zeyher* ! *Pappe* ! Uitenhage Div. ; woods of Zwartkops River and Addo,

Zeyher, 63! Enon, *Drège*.   Albany Div.; near Grahamstown, *Bolton*! *William-son*! Queenstown Div.; Engotini, near Shiloh, *Baur*, 31! Queenstown, 3500–4000 ft., *Galpin*, 2012! British Kaffraria, *Cooper*, 276!
CENTRAL REGION: Willowmore Div.; Zwaanepoels Poort, *Drège*, 7948*d*! Somerset Div.; Bruintjes Hoogte, lower part, *Burchell*, 2994! Graaff Reinet Div.; near Graaff Reinet, 2500 ft., *Bolus*, 197! Aberdeen Div.; Camdeboo Mountain, 4000–5000 ft., *Drège*, 7948*b*! Beaufort West Div.; between Beaufort West and Rhenoster Kop, *Drège*. Richmond Div.; between Richmond and Brak Vallei River, *Drège*.

2. **T. capense** (Thunb. Prodr. 95); an erect undershrub, 1–3 ft. high or more; stem usually freely branched, thinly and shortly pubescent; branches twiggy, slender, 4-angled; leaves usually deeply trifid, rarely entire and lanceolate, the larger up to 2 in. long or more, shortly and thinly pubescent above, more pubescent or often canescent beneath; lobes lanceolate to linear, up to about 1 in. long, ½–3¼ lin. broad, acute, entire or 3–5-fid, revolute at the margin; cymes axillary, 3–7-flowered, as long as or longer (rarely shorter) than the leaves; peduncles very slender, 3–15 lin. long; pedicels 1–6 lin. long; bracteoles usually very small and linear, rarely up to 5 lin. long and lanceolate; calyx campanulate, 1¾–2¼ lin. long, thinly covered with short curled or adpressed hairs; teeth narrowly deltoid, lanceolate or linear-triangular, ¾–1¾ lin. long, ⅓–⅔ lin. broad at the base, acute or acuminate; corolla white, 1½–2 lin. long; lobes elliptic or ovate-elliptic, ¾–2 lin. long. *Thunb. Fl. Cap. ed. Schult.* 445; *Benth. Lab.* 667, *in E. Meyer, Comm.* 243, *and in DC. Prodr.* xii. 577. *Ajuga capensis, Pers. Syn.* ii. 109.

SOUTH AFRICA: without locality, *Ecklon*, 30! 53! *Zeyher*, 1351!
COAST REGION: Oudtshoorn Div.; at the foot of the Zwartbergen, *Bolus*, 2437! Humansdorp Div.; Zeekoe River, *Thunberg*! Uitenhage Div.; Zuurberg Range near Bontjes River, 2000 ft., *Drège*, 7948*c*! and without precise locality, *Zeyher*! *Prior*! Albany Div.; on the rocks of Zwartwater Poort, *Burchell*, 3389! Grahamstown and neighbourhood, *Miss Daly & Miss Sole*, 92! *Bolton*! Trapps Valley, *Miss Daly*, 550! Queenstown Div.; plains at Queenstown, 3500 ft., *Galpin*, 1647! British Kaffraria, *Cooper*, 2899!
KALAHARI REGION: Orange River Colony; Vet River, *Burke*! Transvaal; various localities, *Junod*, 1593! *Sanderson*! *McLea in Herb. Bolus*, 5780! *Wilms*, 1084! 1106! *Burtt-Davy*, 1509! 1593! 3929! 7670! *Miss Leenderiz*, 462!
EASTERN REGION: Transkei; Kreilis Country, *Bowker*! Natal; various localities, *Sanderson*, 27! *Gerrard*, 1215! *Wood*, 3566! *Wilms*, 2147!

3. **T. riparium** (Hochst. in Flora, 1845, 66); an erect undershrub, up to 4 ft. high; stem usually simple below and branched above, rather densely pubescent; branches usually short, slender, 4-angled; leaves lanceolate, oblong-lanceolate or linear, the larger 1–2¼ in. long, 2–6 lin. broad, remotely few-toothed near the apex or entire, acute or obtuse at the apex, much narrowed at the base, revolute at the margin, glabrous or minutely hispidulous above, thinly pubescent below; cymes axillary, 2–7-flowered, usually about as long as the leaves, minutely bracteolate; peduncles 5–8 lin. long; pedicels 2–4 lin. long; calyx campanulate, 1¼–1½ lin. long, thinly covered with short adpressed hairs; teeth narrowly deltoid or

lanceolate, about ¾ lin. long, ⅓–⅔ lin. broad at the base, acute; corolla white, about 2 lin. long; tube ¾–1¼ lin. long; lobes elliptic or ovate-elliptic, 1–1⅓ lin. long. *Benth. in DC. Prodr.* xii. 576.

EASTERN REGION: Transkei; near Kentani, 1200 ft., *Miss Pegler*, 332! Tembuland; Bazeia, 2000 ft., *Baur*, 92! Griqualand East; near Clydesdale, 2500 ft., *Tyson*, 2062! *and in MacOwan, Herb. Austr.-Afr.* 1518! Zuurberg Range, 5000 ft., *Tyson*, 1183! Natal; on the Umlaas River, *Krauss*, 153! Durban, *Sanderson*, 91! 96! *Gerrard*, 1216! Inanda, 1800 ft., *Wood*, 82! and without precise locality, *Cooper*, 1138! Zululand, *Mrs. McKenzie*!

XXVI. AJUGA, Linn.

Calyx campanulate, 10-nerved or irregularly many-nerved, subequal, 5-fid or 5-toothed. *Corolla-tube* included or exserted, annular-pilose inside, somewhat enlarged at the throat; limb 2-lipped; upper lip short or very short, subentire, emarginate or 2-fid, sometimes truncate; lower lip elongated, spreading, 3-lobed; median lobe largest, emarginate or 2-fid. *Stamens* 4, didynamous (the lower pair longer), usually exserted from the upper lip; anthers 2-celled; cells divergent or divaricate, finally confluent. *Disk* equal or often produced in front. *Ovary* shortly 4-lobed nearly to the middle; style 2-fid; lobes subulate, subequal. *Nutlets* obovoid, reticulate-rugose, attached by a broad lateral areole which extends beyond the middle.

Annual or more usually perennial herbs, rarely suffruticose at the base, often decumbent or stoloniferous; leaves often coarsely toothed, sometimes incised, rarely quite entire; floral leaves similar or the upper (sometimes all) reduced to bracts; whorls 2- to many-flowered, axillary or in dense or interrupted terminal spikes; flowers usually blue, white or yellow; corolla marcescent.

DISTRIB. Species about 50, chiefly in the extra-tropical regions of the Old World, most numerous in the Orient.

1. **A. Ophrydis** (Burch. ex Benth. Lab. 695); a perennial herb, without stolons; stem rather stout, erect, 4–12 in. high, leafy, more or less pilose or sometimes rather densely white woolly-pilose; leaves sessile or sometimes distinctly petiolate, obovate, obovate-oblong to spathulate, 1¼–3½ in. long, ½–1¾ in. broad, rounded or obtuse, distinctly narrowed at the base, usually coarsely few-toothed, rarely quite entire, glabrous or more or less pilose, subcoriaceous, rigid; floral leaves sessile, ovate to ovate-lanceolate, usually longer than the flowers; whorls few- to many-flowered; lower distant; upper usually close together forming an elongated somewhat crowded spike; bracteoles linear to linear-oblong, scarcely as long as the calyx; calyx sessile or very shortly stalked, campanulate, 3–4½ lin. long, more or less densely somewhat stiffly hairy on the upper part; teeth deltoid to narrowly deltoid, 1–2½ lin. long, ¾–1¼ lin. broad at the base, subacute to acuminate; corolla pale blue or lilac, more rarely white; tube 3½–4½ lin. long, slightly curved, gibbous in front

at the base, enlarged at the throat; upper lip $\frac{3}{4}$–1 lin. long, about 2 lin. broad at the base, emarginate; lower lip $3\frac{1}{2}$–5 lin. long; median lobe broadly obovate, $2\frac{3}{4}$–$3\frac{1}{2}$ lin. long and broad, deeply emarginate; lateral lobes ovate-oblong, $1\frac{1}{4}$–$1\frac{3}{4}$ lin. long. *Benth. in E. Meyer, Comm.* 243, *and in DC. Prodr.* xii. 597.

SOUTH AFRICA : without locality, *Thunberg* ! *Ecklon* !
COAST REGION : Uitenhage Div. ; Zuurberg Range, 2000–3000 ft., *Drège* ! Van Stadens Berg, *Zeyher,* 346 ! Bathurst Div. ; between Blue Krantz and Kaffir Drift Military Post, *Burchell,* 3700 ! Albany Div. ; Slaay Kraal, *Burke* ! Grahamstown and neighbourhood, *MacOwan* ! *Bolton* ! Fort Beaufort Div. ; Kat River, 2000 ft., *Drège* ! Winter Berg, *Mrs. Barber,* 120 ! Stockenstrom Div. ; Kat Berg, *Miss Sole,* 375 ! Cathcart Div. ; Blesbok Flats, near Windvogel Mountain, 3000 ft., *Drège* ! Komgha Div. ; between Zandplaat and Komgha, *Drège* ! British Kaffraria, *Cooper,* 169 ! 334 !
CENTRAL REGION : Somerset Div. ; Somerset East, *Bowker* ; Bosch Berg, 3000–4000 ft., *MacOwan,* 436 !
KALAHARI REGION : Orange River Colony ; Wolve Kop, *Burke* ! Bethlehem, *Richardson* ! Harrismith, *Sankey,* 227 ! Witte Bergen, *Mrs. Barber & Mrs. Bowker,* 763 ! Basutoland ; Drakensberg, 8000 ft., *Mellersh* ! and without precise locality, *Cooper,* 2901 ! Transvaal ; Heidelberg and neighbourhood, *Miss Leendertz,* 1033 ! *Burtt-Davy,* 3125 ! near Lydenburg, 4400 ft., *Burtt-Davy,* 7664 ! Vereeniging, 4700 ft., *Burtt-Davy,* 7029 ! and without precise locality, *McLea in Herb. Bolus,* 5781 !
EASTERN REGION : Transkei ; Kreilis Country, *Bowker* ! Tembuland ; Bazeia, 2000 ft., *Baur,* 272 ! Griqualand East ; around Kokstad, 4800 ft., *Tyson,* 1102 ! *and in MacOwan, Herb. Austr.-Afr.* 1519 ! Natal ; Durban, *Gerrard,* 1218 ! Pietermaritzburg, 2000–3000 ft., *Sutherland* ! *Wilms,* 2200 ! Inanda, *Wood,* 1436 ! Howick, *Mrs. Hutton,* 373 ! and without precise locality, *Sanderson,* 375 ! Swaziland ; mountains above Embabane, 4500 ft., *Burtt-Davy,* 3336 !

ORDER CVI. **PLANTAGINEÆ**.

(By DR. T. COOKE.)

Flowers regular, usually hermaphrodite. *Calyx* inferior, 2-partite ; sepals imbricate, persistent, the anticous free or connate, keeled on the back and with membranous margins. *Corolla* hypogynous, tubular, scarious, marcescent, 4-lobed ; tube ampulliform or cylindric ; lobes 4, imbricate in bud. *Stamens* usually 4, inserted on the corolla-tube ; filaments filiform ; anthers versatile, 2-celled, dehiscing by a long slit. *Ovary* superior, 1–4-celled ; ovules 1 to many in each cell ; style filiform, erect. *Capsule* 1–4-celled, submembranous, 1- or many-seeded, dehiscent or indehiscent. *Seeds* attached to the placenta by the inner face ; testa thin ; albumen fleshy ; embryo cylindric, transverse ; radicle inferior.

Perennial or annual herbs with or without stems ; leaves in stemless plants rosulate, in plants with stems alternate or opposite ; petioles usually dilated at the base ; flowers usually spicate (rarely solitary), each subtended by a persistent bract.

DISTRIB. Genera 3 ; species about 200, cosmopolitan, chiefly in temperate and subtemperate regions.

I. PLANTAGO, Linn.

Flowers hermaphrodite or polygamo-diœcious, each supported by a bract. *Calyx* 4-lobed; segments subequal or the two outer larger. *Corolla-tube* cylindric or ampulliform; lobes 4. *Stamens* 4, inserted on the corolla-tube. *Ovary* usually 2-celled with 1 to several ovules in each cell; style simple. *Capsule* membranous, circumscissilely dehiscing at the middle or near the base. *Seeds* 2 to several; albumen fleshy; embryo straight or curved; radicle inferior.

Annual or perennial herbs, often stemless with the leaves in a radical rosette; leaves various, usually entire; flowers inconspicuous, spicate or capitate, each subtended by a single bract.

DISTRIB. Species nearly 200, cosmopolitan.

Capsules many-seeded :
   Leaves 1–4 in. long; spikes less than 6 in. long   ... (1) **major.**
   Leaves 8–12 in. long; spikes reaching 1 ft. long   ... (2) **dregeana.**
Capsules few-seeded :
   Anticous sepals connate for nearly their entire length ;
     capsules 2-seeded   ...   . (3) **lanceolata.**
Sepals all free :
   Corolla-tube glabrous ; capsules 2-seeded :
     Leaves linear, sessile ; rootstock not woolly :
       Spikes 1½ in. long, cylindric   ...   . (4) **cafra.**
       Spikes 2–8 lin. long, subglobose or ovoid   ... (5) **capillaris.**
     Leaves elliptic or lanceolate, with long petioles ;
      rootstock woolly :
       Flowers closely arranged along the rhachis of
       the spike except near its very base ; leaves
       elliptic-oblong   ...   . (6) **longissima.**
       Flowers in distant fascicles along the rhachis of
       the spike ; leaves lanceolate   ...   ... (7) **remota.**
   Corolla-tube hairy ; capsules 2–4-seeded   ...   ... (8) **carnosa.**

1. **P. major** (Linn. Sp. Pl. ed. i. 112); a perennial stemless herb with an erect stout rootstock; leaves radical, 1–4 in. long, of variable width, ovate or ovate-oblong, acute or subacute, entire or toothed, nearly glabrous, 3–7- (commonly 5-) nerved, tapering into the petiole; petioles usually longer than blades, channelled, sheathing at the base; flowers scattered or crowded, in rather lax spikes 2–4 in. or more long; bracts ¾–1 lin. long, broadly ovate-oblong, obtuse, glabrous, with scabrous margins; sepals 1 lin. long, broadly oblong or rotund-ovate, obtuse, obtusely keeled on the back and with scarious margins; corolla-tube $\frac{9}{10}$ lin. long; lobes $\frac{3}{5}$ lin. long, ovate-lanceolate, acute, reflexed; anthers $\frac{2}{5}$ lin. long; style $\frac{3}{4}$ lin. long; capsules ellipsoid, 1¼ lin. long, the top coming off circumscissilely as a conical lid tipped with the remains of the style; seeds 4–8 in each cell, $\frac{2}{5}$ lin. long, angular dull black. *Decne in DC. Prodr.* xiii.

i. 694 ; *Hook. f. Fl. Brit. Ind.* iv. 705 ; *Baker in Dyer, Fl. Trop. Afr.* v. 503 ; *Barn. Monogr. Plantag.* 10.

SOUTH AFRICA : without locality, *Mund* !
COAST REGION : Bathurst Div. ; damp hollows near the sea coast at Port Alfred, 50 ft., *Galpin*, 2945 !
CENTRAL REGION : Graaff Reinet Div. ; Voor Sneeuw Berg, *Burchell*, 2856 !
Cosmopolitan.

2. **P. dregeana** (Presl, Bot. Bemerk. 105) ; leaves 8–12 in. long, 4–7 in. broad, ovate-oblong with sinuate margins, thin, glabrous, 7-nerved, attenuated into a channelled petiole often much longer than the blade ; spikes reaching 1 ft. long on long terete peduncles ; bracts as long as the calyx, oblong, obtuse, with membranous margins ; sepals 1½ lin. long, broadly ovate or suborbicular, quite glabrous, with membranous margins ; corolla-tube ¾ lin. long ; lobes as long as the tube, deltoid-oblong, subacute, membranous ; filaments 1¼ lin. long, filiform ; anthers ¾ lin. long, apiculate ; style 2 lin. long ; capsules 1¼ lin. long, subglobose, circumscissilely dehiscing about the middle ; seeds numerous, ⅔ lin. long, obscurely angular, rugulose, dark brown. *Decne in DC. Prodr.* xiii. i. 695.

COAST REGION : Albany Div., *Bowker* !
CENTRAL REGION : Graaff Reinet Div. ; near Graaff Reinet, 2500 ft., *Bolus*, 163 !
KALAHARI REGION : Transvaal ; Aapies Poort, near Pretoria, *Rehmann*, 4023 !
EASTERN REGION : Natal ; Botanic Gardens, *Wood*, 3848 !

3. **P. lanceolata** (Linn. Sp. Pl. ed. i. 113) ; perennial, stemless ; rootstock tapering ; leaves 1–12 in. long, ½–1½ in. broad, lanceolate or oblong-lanceolate, acute, entire or toothed, gradually narrowed to a sessile base or to a short petiole, 3–5- (rarely 7-) nerved ; axils woolly ; peduncles longer than the leaves, grooved, angular, erect or ascending, puberulous or glabrous ; spikes ½–2 in. long, ovoid, globose or cylindric ; bracts ovate, acuminate, as long as the calyx, glabrous ; sepals oblong, obtuse, 1½ lin. long, hairy on the nerves at the back near the top, 2 of them connate almost throughout their length, giving rise to a 3-lobed calyx with 1 of the sepals obovate, 2-fid at the apex and 2-nerved ; corolla-tube 1 lin. long ; lobes 1 lin. long, broadly ovate, acuminate, usually with a tubercular thickening at the base of each lobe ; filaments 1¼ lin. long ; anthers 1¼ lin. long ; style 2 lin. long ; capsules 2-seeded. *Decne in DC. Prodr.* xiii. i. 714 ; *A. Rich. Tent. Fl. Abyss.* ii. 206 ; *Hook. f. Fl. Brit. Ind.* iv. 706 ; *Baker in Dyer, Fl. Trop. Afr.* v. 503.

COAST REGION : Cathcart Div. ; Glencairn, 4800 ft., *Galpin*, 2406 ! East London Div. ; West Bank, near East London, 50 ft., *Galpin*, 7350 !
EASTERN REGION : Natal ; Mooi River, *Wood*, 4057 !
Cosmopolitan.

4. **P. cafra** (Decne in DC. Prodr. xiii. i. 719) ; annual ; leaves 6–8 in. long, 1–2 lin. broad, linear, entire or distantly toothed with

subulate teeth, obscurely 3-nerved, sessile, clothed on both sides with lax slender spreading hairs; peduncles terete, 5½ in. long in the only specimen at Kew, clothed with lax spreading hairs; spikes cylindric, 1½ in. long; rhachis with long slender hairs; bracts boat-shaped, rostrate, with broad scarious margins, the lower bracts reaching 2 lin. long, the upper rather shorter; sepals 1–1¼ lin. long, suborbicular, quite glabrous; corolla-lobes scarcely ½ lin. long, ovate, acute; style 1 lin. long; capsules 1¾ lin. long, ovoid, glabrous, coming off circumscissilely from near the base; seeds 1¼ lin. long, ellipsoid, with rounded back and a flat channelled face, black, smooth. *P. Loeflingii, Thunb. Prodr.* 30, *and Fl. Cap. ed. Schult.* 148 (*not of Linn.*). *P. Bellardi, Drège, Zwei Pfl. Documente,* 103.

COAST REGION: Malmesbury Div.; Riebeeks Castle, under 1000 ft., *Drège!*

5. **P. capillaris** (E. Meyer ex Decne in DC. Prodr. xiii. i. 719); a small plant 1 to 6 in. high; leaves narrowly linear or subulate, as long as or often longer than the peduncles, sessile, usually dilated at the base, entire or remotely toothed, or distantly and pinnatifidly lobed with short subulate lobes, clothed with long laxly spreading hairs; peduncles terete, slender, suberect, usually shorter than the leaves, laxly and softly pilose with spreading hairs; spikes dense, sub-globose or ovoid oblong, 2–8 lin. long; rhachis densely clothed with long slender hairs; bracts boat-shaped, rostrate, ovate, acute, with broad scarious margins, 1¾ lin. long, hairy on the back; sepals 1–1¼ lin. long, suborbicular, quite glabrous; corolla-lobes ½ lin. long, suborbicular, quite glabrous; corolla-lobes ½ lin. long, ovate-oblong, subacute, glabrous; style 1 lin. long; capsules a little longer than the sepals, ovoid, rounded at the apex, coming off circumscissilely near the base; seeds oblong-ellipsoid, ½ lin. long, rounded on the back, flattened on the face, black.

COAST REGION: Van Rhynsdorp Div.; Olivants River, *Drège!* Clanwilliam Div.; Vogelfontein, *Schlechter,* 8524! Tulbagh Div.; near Tulbagh, under 1000 ft., *Drège!* Worcester Div.; Hex River Valley, *Wolley-Dod,* 4040!
WESTERN REGION: Little Namaqualand; Klipfontein, 3000 ft., *Bolus,* 684!

6. **P. longissima** (Decne in DC. Prodr. xiii. i. 720); rootstock woolly; leaves 2–6 in. long, 1–3 in. broad, oblong or elliptic-oblong, glabrous, 5–9-nerved, coriaceous; petioles long, exceeding the blades; peduncles longer than the leaves, terete or striate; spikes 6–15 in. long, lax-flowered below; bracts 1 lin. long, ovate, acute, with membranous margins; sepals 1½ lin. long, broadly ovate, acute, apiculate, with broad membranous margins and a strong keel; corolla-tube 1 lin. long; lobes ¾ lin. long, ovate, subobtuse; filaments 1 lin. long, filiform; anthers ¾ lin. long; style hairy, variable in length, very long and conspicuous nearly ¼ in. long, or only 1 lin. long; capsules 1¼ lin. long, subglobose, quite glabrous, 2-seeded; seeds 1 lin. long, elliptic-oblong, with a

rounded back and a flat face, black. *P. Burchellii, Decne in DC. Prodr.* xiii. i. 720. *P. capensis, var.* β, *E. Meyer in Drège, Zwei Pfl. Documente,* 211 ; *var. longissima, Barn. Monogr. Plantag.* 35.

SOUTH AFRICA : without locality, *Zeyher,* 1432 !
KALAHARI REGION : Transvaal ; Pinedene, near Irene, *Burtt-Davy,* 2326 !
Lydenburg, *Wilms,* 1247 ! Mooi River, *Burke* ! Megalies River, near Pretoria, *Burke* ! Lynwood, near Pretoria, *Burtt-Davy,* 7465 !
EASTERN REGION : Tembuland ; Bazeia, 2500 ft., *Baur,* 353 ! Pondoland ; between Umtata River and St. Johns River, *Drège* ! Natal ; Inanda, *Wood,* 1078 ! 1078a ! and without precise locality, *Gerrard,* 1481 !

I have included *P. Burchellii,* Decne, under this, as I cannot find any characters to separate it. It was founded on the specimens collected by *Burke* (not *Burchell*), quoted above. Decaisne misread the name *Burke.*

7. **P. remota** (Lam. Illustr. i. 341) ; rootstock woolly ; leaves 3–6 in. long, ¾–1¼ in. broad, linear-lanceolate, obtuse, entire or remotely denticulate with short callous teeth, coriaceous, attenuated into a long petiole, densely woolly at the very base ; peduncles longer than the leaves, striate ; spikes 2–12 in. long ; rhachis glabrous ; flowers yellow, remote, solitary or 2–3 together ; bracts 1–1¼ lin. long, broadly ovate, obtusely acuminate, fimbriate at the tip and with membranous irregularly toothed margins ; sepals 1½ lin. long, broadly ovate, acute, with membranous toothed or ciliate margins ; keel stout ; corolla-tube 1¼ lin. long ; lobes ¾ lin. long, broadly oblong, irregularly toothed, reflexed ; filaments 1½ lin. long ; anthers 1¼ lin. long, oblong, with an acute triangular tip, yellow ; style 1¾ lin. long ; capsules 2-seeded. *Decne in DC. Prodr.* xiii. i. 721. *P. capensis, Thunb. Prodr.* 29 ; *Fl. Cap. ed. Schult.* 148 ; *Barn. Monogr. Plantag.* 35.

COAST REGION : Cape Div. ; Orange Kloof, *Wolley-Dod,* 2403 ! Devils Mountain, above Rondebosch, 550 ft., *Bolus,* 7025 !
EASTERN REGION : Natal ; in damp ground by the Mooi River, *Wood,* 4049 !

8. **P. carnosa** (Lam. Illustr. i. 341) ; a plant of variable size from 3 in. to more than 1 ft. high ; rootstock woody, descending ; leaves variable, numerous, reaching 5 in. long, ½–5 lin. broad, linear, obtuse, acute or acuminate, entire or remotely dentate, fleshy or coriaceous, rugose, usually coarsely hairy, but sometimes glabrous, sessile ; peduncles erect, adpressedly hairy, conspicuously terete, reaching in well-grown plants 6 in. or more long ; spikes 1–3 in. long, cylindric, dense ; bracts 1 lin. long, broadly ovate, acuminate, with ciliate membranous margins and a thick keel ; sepals 1¼ lin. long, ovate or elliptic, obtuse, with ciliate membranous margins and a strong keel, hairy on the back ; corolla-tube ¾–1 lin. long, hairy outside ; lobes ½–¾ lin. long, ovate, acute, deflexed ; filaments 1 lin. long, filiform ; anthers 1 lin. long (including a long apiculus) ; style 1 lin. long ; capsules 1½ lin. long, ovoid-oblong, obtuse, beaked by the remains of the style, circumscissilely dehiscing below the middle,

2–3- (rarely 4-) seeded, 1 of the seeds when 3 usually infertile ; seeds reaching 1 lin. long, ellipsoid-oblong, obtuse at both ends, rounded on the back, with flattened face, reddish brown. *Barn. Monogr. Plantag.* 22 ; *Decne in DC. Prodr.* xiii. i. 729. *P. hirsuta, Thunb. Fl. Cap.* ed. i. 541 ; *Barn. Monogr. Plantag.* 21.

COAST REGION : Cape Div. ; various localities near Cape Town, *Milne,* 163 ! *Burchell,* 8386 ! *Drège* ! *Wallich* ! *Bolus,* 3066 ! 4766 ! *Wolley-Dod,* 2376 ! Stellenbosch Div. ; near Somerset West, *Bolus,* 2971 ! Caledon Div. ; Zoetemelks Valley, *Burchell,* 7569 ! Bredasdorp Div. ; Zeekoe Vley, 100 ft., *Schlechter,* 10551 ! Humansdorp Div. ; Kromme River Heights, *Bolus,* 2394 ! Uitenhage Div. ; Zwartkops River, *Drège* ! Witte Klip, *MacOwan,* 1940 ! and without precise locality, *Zeyher* ! Port Elizabeth Div. ; Cape Recife, *Burchell,* 4391 ! East London Div. ; seashore on rocks near Bats Cave, *Galpin,* 2809 ! near seashore West Bank, East London, 50 ft., *Galpin,* 5847 ! *Rattray,* 229 !

EASTERN REGION : Tembuland ; Umtata, 2000 ft., *Baur,* 450 !

ORDER CVII. **NYCTAGINEÆ.**

(By DR. T. COOKE.)

Flowers hermaphrodite (rarely unisexual), regular, sometimes dimorphous ; inflorescence various ; bracts often involucrate, free or connate. *Perianth* monophyllous, small, herbaceous or petaloid, persistent, often accrescent ; tube short or long, sometimes circumscissile above the base ; limb 3–5-toothed or lobed, persistent or deciduous. *Stamens* 1–30, hypogynous ; filaments small, usually unequal, free or connate into a cup at the base, involute in bud ; anthers 2-celled, dorsifixed, included or exserted, dehiscing longitudinally. *Ovary* 1-celled ; ovule solitary, erect, campylotropous ; style filiform, involute in bud ; stigma small, simple or multifid. *Fruit* (anthocarp) membranous, indehiscent, enclosed in the persistent base of the perianth-tube, costate, sulcate or winged, sometimes glandular. *Seed* erect ; testa adherent ; albumen soft or floury ; embryo straight or curved ; radicle inferior.

Herbs, shrubs or trees ; leaves usually opposite, entire ; stipules 0 ; flowers in terminal or axillary cymes, panicles or corymbs ; bracts often forming a brightly coloured involucre.

DISTRIB. Species about 150, chiefly American, a few in India, the Mascarene Islands and Pacific Islands.

I. **Mirabilis.**—A herb. *Leaves* opposite. *Flowers* hermaphrodite. *Bracts* large, connate.

II. **Boerhaavia.**—Herbs. *Leaves* opposite. *Flowers* hermaphrodite. *Bracts* small, free.

III. **Pisonia.**—Shrubs. *Leaves* alternate or opposite. *Flowers* polygamo-diœcious. *Bracts* small, free.

IV. **Phæoptilum.**—A spiny shrub. *Leaves* fascicled. *Flowers* polygamo-diœcious. *Bracts* small, free.

I. MIRABILIS, Linn.

Involucre calyx-like, 1- to many-flowered, gamophyllous, 5-lobed ; lobes acuminate. *Perianth* coloured ; tube long, constricted above the ovary ; limb spreading, 5-lobed, plicate, deciduous. *Stamens* 5–6, unequal, exserted ; filaments filiform, incurved, united into a fleshy cup at the base ; anther-cells subglobose. *Ovary* ellipsoid or ovoid ; ovule solitary, erect ; style filiform, exserted ; stigma globose, bearing stalked papillæ. *Fruit* ribbed, enclosed in the hardened base of the perianth and surrounded by the persistent staminal cup. *Seed* filling the pericarp to which the testa adheres ; embryo curved ; cotyledons surrounding the scanty farinaceous albumen.

Di- or tri-chotomously branched glabrous or glandular perennial herbs ; root thickened, tuberous ; leaves opposite, the lower petiolate, the upper sessile ; involucres cymosely arranged ; flowers large, fragrant or not, white, red, yellow or variegated.

DISTRIB. Species about 10 in the hotter parts of America, introduced elsewhere.

1. **M. Jalapa** (Linn. Sp. Pl. ed. i. 177) ; an erect perennial much-branched herb reaching 2 ft. or more high ; root tuberous ; stem glabrous or shortly pubescent ; leaves up to 3 in. long, $1\frac{1}{2}$ in. broad, ovate, acuminate, glabrous or pulverulent above, entire, often with ciliate margins, base rounded, truncate or cordate ; petioles slender, 3–12 lin. long ; flowers inodorous, 3–6 in each cyme ; involucre nearly $\frac{1}{2}$ in. long, glandular when young ; lobes ovate, shortly bristle-tipped ; perianth purple, red, yellow or white, sometimes more or less blotched ; tube $1\frac{1}{2}$ in. long, cylindric below, funnel-shaped at the top ; limb spreading, 1 in. or more in diam. ; stamens exserted ; fruit ovoid, black, 4–5 lin. long, wrinkled-tuberculate, 5-ribbed. *Ait. Hort. Kew. ed.* 1, i. 234 ; *Bot. Mag. t.* 371 ; *Choisy in DC. Prodr.* xiii. ii. 427 ; *Baker & Wright in Dyer, Fl. Trop. Afr.* vi. i. 2. *M. dichotoma, Linn. Syst. ed.* 10, ii. 931, *and Sp. Pl. ed.* ii. 252.

EASTERN REGION : Natal ; near Durban, *Grant* !

Also in Tropical Africa. Known as "*The Marvel of Peru*," of which country it is a native ; established now in many parts of the Old World.

II. BOERHAAVIA, Vaill.

Bracts small, often deciduous, rarely whorled and involucrate. *Perianth-tube* long or short, cylindric, narrowed above the ovary, the lower part persistent and becoming hardened to enclose the fruit, the upper part petaloid and deciduous ; limb funnel-shaped with 5-lobed margin, the lobes plicate. *Stamens* 1–5, more or less exserted ; filaments capillary, unequal, connate below. *Ovary* oblique, stipitate ; ovule erect ; style filiform ; stigma peltate.

Fruit enclosed in the ovoid, turbinate or clavate, obtuse or truncate perianth-tube, round, 5-ribbed or 5-angled, often viscidly glandular. *Seed* filling the pericarp, with testa adhering to it ; embryo hooked ; cotyledons thin, broad, enclosing a soft scanty albumen ; radicle long.

Erect or diffuse, often divaricately branched herbs ; leaves opposite, often in unequal pairs : flowers small, paniculate, umbellate or subcapitate, articulated with the pedicel.

DISTRIB. Species about 20, throughout the tropics and warm temperate regions.

Perianth scarcely 2 lin. long :
   Stems with horizontally spreading hairs ; branches of
      panicles fascicled, 2 together, with a large ciliate
      bract at their origin ...   .   (1) **bracteata.**
   Hairs of stems not horizontally spreading ; branches
      of panicle not fascicled nor furnished with large
      bracts :
      Fruit clavate-oblong, not much tapered towards the
         base ; bracteoles acute   ...   .   (2) **repens,**
                                                                var. **diffusa.**
      Fruit turbinate, much tapered towards the base ;
         bracteoles obtuse ...   .   (3) **adscendens.**
   Perianth 6 lin. or more long   ...   .   (4) **pentandra.**

1. **B. bracteata** (T. Cooke in Kew Bulletin, 1909, 421) ; shrubby, erect ; stem terete, densely villous with horizontally spreading hairs, woody below ; leaves petiolate, 1¼ in. long, ¾ in. broad, to 2¾ in. long and 2½ in. broad, broadly ovate, obtuse, more or less hairy on the nerves and with ciliate margins, rounded or subcordate at the base ; lower petioles long (about ⅓ as long as the blade), villous ; flowers in lax leafy panicles, branches fascicled, two together with a large hairy and ciliate bract 3–6 lin. long at their origin, 1 branch bearing 2 or 3 sessile flowers at its apex, the other again similarly divided ; bracteoles 1 lin. long, ovate-lanceolate, subacute, membranous, with a dark conspicuous midrib ; perianth in bud reaching 1½ lin. long, the ovarian portion ½ lin. long in bud, subglobose, elongating afterwards ; stamens 3 ; fruit 1½ lin. long, oblong, ribbed, glandular.

KALAHARI REGION : Transvaal ; Avoca near Barberton, 1900 ft., *Galpin*, 1240 ! Bechuanaland ; on the rocks at Chue Vley, *Burchell*, 2381 ! EASTERN REGION : Natal ; Tugela, *Gerrard*, 1787 bis !

There is 1 small specimen of this plant in Harvey's Herbarium, Trinity College, Dublin, from Damara land, Tropical Africa, without collector's name.

2. **B. repens,** var. **diffusa** (Hook. f. Fl. Brit. Ind. iv. 709) ; a variable diffuse herb ; root large, fusiform ; stems usually several, prostrate or ascending, reaching 2–3 ft. long, divaricately branched, slender, cylindric, thickened at the nodes, minutely pubescent or nearly glabrous ; leaves in unequal pairs at each node, the larger

1-1¼, the smaller ½-¾ in. long, broadly ovate or suborbicular, rounded at the apex, green and glabrous above, green or white beneath, the margins entire, often pink, more or less undulate, base rounded or subcordate ; petioles nearly as long as the blade, slender ; flowers small, shortly stalked or nearly sessile, 4-10 together in small umbels arranged in slender long-stalked corymbose axillary panicles ; bracteoles small, lanceolate, acute ; perianth 1-1¼ lin. long, the ovarian portion of the tube ¾ lin. long, contracted above the ovary, 5-ribbed, glandular ; limb funnel-shaped, dark pink ; lobes very short, rounded ; stamens 1-3, slightly exserted ; stigma peltate ; fruit 1½ lin. long. clavate-oblong rounded, broadly and bluntly 5-ribbed, glandular. *Baker & Wright in Dyer, Fl. Trop. Afr.* vi. i. 5. *B. diffusa, Linn. Fl. Zeyl.* 4, *and Sp. Pl. ed.* i. 3 ; *Choisy in DC. Prodr.* xiii. ii. 452. *B. procumbens, Roxb. Fl. Ind ed. Carey* i. 148. *Talu-Dama, Rheede, Hort. Malab.* vii. 105, *t.* 56.

KALAHARI REGION : Transvaal ; Shiluvane, *Junod,* 1061 ! Rooiplaat, Pienaars River, *Miss Leendertz,* 774 ! near Pienaars River Mountains, *Schlechter,* 4221 !

EASTERN REGION : Delagoa Bay, *Schlechter,* 11582 ! between Delagoa Bay and Pretoria, *Bolus,* 9749 !

Also in Tropical Africa, Tropical and Subtropical Asia and America.

I have in the "Flora of the Presidency of Bombay" adopted *B. diffusa* as the type, in consequence of its priority ; it having been described by Linnæus in 1747 (*Fl. Zeyl.* 4), the species *repens* having been described in 1753 (*Sp. Pl. ed.* i. 3), but as Sir J. Hooker has adopted *B. repens* as the type, I follow his lead. Heimerl (*Engl. & Prantl, Pflanzenfam.* iii. 1B, 26) is apparently also of opinion that the type species should be *diffusa,* and he places under it no less than 8 forms, many of which are elsewhere regarded as distinct species.

The canescence of the lower surface of the leaves, which is very marked in some, though absent in other specimens, appears to be due to the fact that a very loose and when wetted easily detachable epidermis exists on the lower surface, with a number of raphides in the lower mesophyll. The epidermis becomes more or less detached, and in dry places wrinkled, the corrugations enclosing small portions of air. The raphides are white and in bundles, resembling short stiff white hairs.

3. **B. adscendens** (Willd. Sp. Pl. i. 19) ; prostrate or ascending ; stems and branches terete, glabrous or nearly so ; leaves 1-1½ in. long, ¾-1¼ in. broad, broadly ovate, obtuse, green above, paler and somewhat wrinkled beneath, with waved more or less ciliate margins ; flowers sessile or nearly so in small umbels forming a lax terminal panicle ; bracteoles ½ lin. long, ovate-oblong, obtuse ; lower portion of perianth in fruit 1¼ lin. long, turbinate, much tapered towards the base, obtusely ribbed, glandular. *Vahl, Enum.* i. 285 ; *Baker & Wright in Dyer, Fl. Trop. Afr.* vi. i. 4. *B. ascendens, Choisy in DC. Prodr.* xiii. ii. 451.

EASTERN REGION : Natal ; The Bluff, near Durban, 20-400 ft., *Wood,* 6400 ! 7199 !

Considered by Heimerl (*Engl. & Prantl, Pflanzenfam.* iii. 1B, 26) to be a form of *B. repens,* var. *diffusa* ; indeed the only difference would seem to be in the fruit which in *B. ascendens* is more tapered than in the other plant.

Also in Tropical Africa.

4. B. pentandra (Burch. Trav. S. Afr. i. 432); a procumbent plant; stems long, herbaceous, terete, glabrous or pubescent, often trailing amongst grass and giving off suberect branches from the axils; leaves shortly petiolate, $\frac{3}{4}$–2 in. long, $\frac{3}{4}$–$1\frac{1}{2}$ in. broad, broadly ovate or suborbicular, obtuse or (rarely) subacute, finely serrate, often shortly apiculate, base truncate or shallowly cordate; peduncles much longer than the leaves, axillary and terminal, stout; flowers in distant whorls (rarely reduced to 1 whorl) usually about 6 flowers in each whorl, forming a long panicle above the leaves; pedicels 2–5 lin. long, persistent after the flowers fall; bracteoles linear-subulate, reaching 2 lin. long or more; deciduous portion of the perianth above the ovary reaching 5 lin. long, funnel-shaped, nearly 4 lin. across at the mouth; persistent portion of the perianth $\frac{1}{4}$ in. long or more, not distinctly pentagonal, with a row of globose glands at the top and often on the sides of the tube; stamens 3, much exserted; style $\frac{1}{2}$ in. long, much exserted; stigma disciform-peltate. *Heimerl in Engl. Jahrb.* x. 9; *Baker & Wright in Dyer, Fl. Trop. Afr.* vi. i. 7. *B. Burchellii, Choisy in DC. Prodr.* xiii. ii. 455. *B. grandiflora, A. Rich. Tent. Fl. Abyss.* ii. 209. *B. dichotoma, Hochst. ex A. Rich. l.c.,* not of Vahl.

SOUTH AFRICA : without locality, *Zeyher,* 1433!
COAST REGION: Queenstown Div.; near Queenstown, 3700 ft., *Galpin,* 1801!
CENTRAL REGION: Hopetown Div.; banks of the Orange River near Hopetown, *Bolus,* 1826!
KALAHARI REGION : Bechuanaland; Kosifontein, *Burchell,* 2556! 2571! 2597! Batlapin Territory, *Holub!* Griqualand West; along the Vaal River, *Burchell,* 1765! *MacOwan,* 1215! west of the Vaal River, *Shaw!* between Kimberley and the Vaal River, *Schenck,* 808! Griquatown, *Burchell,* 1897! 1954! Orange River Colony, various localities, *Hutton! Mrs. Barber,* 761! *Sanderson! Burke!* Transvaal; various localities, *Baines! Rogers,* 2383! *Nelson,* 115! *Galpin,* 1241! *Bolus, Herb. Norm. Austr.-Afr.* 1346! *Schlechter,* 4339! *Burtt-Davy,* 1821!
EASTERN REGION : Natal; between Umcomaas River and Umlazi River, *Drège!* The Bluff, *Sanderson,* 75! *Gueinzius!* Mooi River Valley, 2300 ft., *Sutherland! Gerrard,* 1488! near Weenen, *Wood,* 4441! near Clairmont, *Haygarth in Herb. Wood,* 4495! Portuguese East Africa; Ressano Garcia, 1000 ft., *Schlechter,* 11878!

Mrs. Barber remarks concerning this plant in a letter to Dr. Harvey, dated March 16th, 1865, which is preserved in Harvey's Herbarium, Trinity College, Dublin, that the plant is much " valued for its nutritious properties as an herbage plant, and is called *Veld Batatas,* from its resemblance to that plant; when it is plentiful, stock of all kinds fatten rapidly and thrive well; these plants are large and prostrate, and very much branched and jointed, throwing out at every joint an almost upright little stem with an umbel of crimson flowers."

Also in Tropical Africa.

III. PHÆOPTILUM, Radlk.

Perianth funnel-shaped, divided to the middle into 4 (rarely 5) ovate spreading petaloid lobes. *Stamens* 8, shortly exserted; filaments united at the base into a short fleshy cup; anthers versatile, oblong. *Ovary* stipitate, the stipes free from the perianth;

ovule solitary, inserted near the base of the cell; style filiform, exserted; stigma penicillate. *Fruit* enclosed in the indurated longitudinally 4-winged perianth-tube. *Seed* erect, albuminous; embryo hooked.

Spinous shrubs with grey or yellowish bark; leaves short, narrow, linear, coriaceous; flowers polygamo-dioecious, not involucrate, in axillary fascicles, with short pedicels.

DISTRIB. Species 2, in Tropical and South Africa.

1. **P. spinosum** (Radlk. in Abhandl. Naturw. Ver. Bremen, viii. 436); a small shrub with woody stem furnished with numerous sharp spines reaching ½ in. long; branches many, running out into spines; leaves 4–5 lin. long, ¼ lin. broad, linear-cuneate, obtuse, thick, in fascicles along the branches; perianth-tube 4 lin. long, about 5 lin. in diam. at the mouth; lobes 2 lin. long, suborbicular; staminal tube ½ lin. long; filaments about ¼ in. long; fruit a 4-5-winged anthocarp 8 lin. long, 6 lin. broad, enclosed in the persistent perianth-tube. *Baker & Wright in Dyer, Fl. Trop. Afr.* vi. i. 9. *P. Heimerli, Engl. Jahrb.* xix. 133. *Amphoranthus spinosus,* *S. Moore in Journ. Bot.* 1902, 305, *t.* 441, *fig. A* (in p. 408 of the same publication Moore withdraws the genus in favour of Radlkofer's).

SOUTH AFRICA : without locality, *Shaw!*
CENTRAL REGION : Calvinia Div.; near Hantam, *Meyer.*

Also in Tropical Africa.

IV. PISONIA, Linn.

Male inflorescence in paniculate cymes. *Perianth* campanulate; limb 5-lobed or 5-toothed; segments induplicate-valvate, erect or spreading. *Stamens* 5–10, exserted; filaments connate below into a tube or ring; anthers oblong or didymous. *Female inflorescence* in paniculate cymes. *Perianth* tubular, usually enlarged at the base. *Ovary* elongate, ovoid, sessile; ovule solitary; style slender, included or exserted; stigma capitellate, peltate or lacerate. *Fruit* enclosed in the coriaceous or hardened oblong, linear or clavate perianth-base, cylindric, compressed or 5-angled, with 5 viscid ribs or with 5 single or double rows of viscid stipitate glands; utricle elongate, membranous. *Seed* with a hyaline testa adnate to the pericarp; embryo straight; albumen scanty, soft; radicle inferior.

Trees or shrubs unarmed or with axillary spines; leaves opposite or alternate, sessile or petiolate, entire; flowers small, dioecious (rarely monoecious or hermaphrodite), in paniculate, subsessile or pedunculate cymes; bracteoles 2-3, not involucrate.

DISTRIB. Species about 30, cosmopolitan in the tropics, chiefly American, 4 in Mauritius.

1. **P. aculeata** (Linn. Sp. Pl. ed. i. 1026); a large scandent shrub with many curved often nearly opposite stout spines; trunk reaching

6 in. in diam. ; branches numerous, subopposite, terete, finely pubescent or nearly glabrous ; leaves 1–3 in. long, $\frac{3}{4}$–1$\frac{1}{2}$ in. broad, elliptic or elliptic-lanceolate, obtuse, entire, glabrous or nearly so, base tapering ; petioles $\frac{1}{4}$–$\frac{1}{2}$ in. long ; flowers in small dense cymose pubescent clusters forming pedunculate axillary panicles ; bracts and bracteoles scarcely $\frac{3}{4}$ lin. long, ovate-oblong, obtuse, pubescent ; pedicels short, pubescent, fruiting much elongate ; perianth of male flowers campanulate, 1$\frac{1}{4}$ lin. long, pubescent outside, with 5 deep triangular acute teeth ; stamens 6–10, much exserted ; perianth of female flowers tubular, 1$\frac{1}{4}$ lin. long, shortly 5-toothed ; style rather stout ; stigma lacerate ; fruit $\frac{1}{2}$–$\frac{3}{4}$ in. long, oblong or clavate, with long pedicels, 5-ribbed, pubescent between the ribs, each rib muricate with 1 or 2 rows of stalked viscous glands. *Lam. Ill.* t. 861 ; *Wight, Icon. tt.* 1763–64 ; *Choisy in DC. Prodr.* xiii. ii. 440 ; *Hook. f. Fl. Brit. Ind.* iv. 711. *Benth. Fl. Austral.* iv. 279 ; *Baker & Wright in Dyer, Fl. Trop. Afr.* vi. i. 8.

EASTERN REGION : Natal ; Tugela, *Gerrard,* 1597 !

Also in Tropical Africa.

ORDER CVIII. **ILLECEBRACEÆ.**

(By DR. T. COOKE.)

Perianth herbaceous or coriaceous, persistent and often indurated after flowering, 4–5-lobed or 4–5-partite. *Petals* 0. *Stamens* as many as the perianth-lobes and opposite to them (rarely fewer or more), perigynous, often alternating with subulate or petaloid staminodes ; filaments short, sometimes connate at the base ; anthers 2-celled, dehiscing laterally. *Ovary* free, 1-celled ; ovule solitary (rarely ovules 2–4), erect or pendulous from a basal funicle ; style obsolete or produced ; stigmas 2–3. *Fruit* usually a utricle enclosed in the persistent perianth. *Seed* globose, lenticular or reniform ; testa usually smooth ; albumen floury ; embryo straight, curved or annular ; cotyledons oblong ; radicle inferior.

Annual or perennial herbs, rarely shrubs ; leaves usually opposite ; stipules scarious, rarely absent ; flowers minute, commonly green, usually hermaphrodite, cymose, often with scarious bracts.

DISTRIB. Species about 110, chiefly in Europe, the Orient, Africa, North and South America, a few in India, Australia and New Zealand.

[Genera I.–III. were included by Harvey and Sonder (Fl. Cap. i. 132–3) in *Caryophyllaceæ,* to which order they were at that time referred.]

I. **Pollichia.**—An undershrub. *Stipules* 0. *Stigmas* 2. *Ovules* 2. *Embryo* straight or slightly curved.

II. **Herniaria.**—A prostrate herb. *Stipules* small, scarious. *Stigmas* 2. *Embryo* annular.

III. **Corrigiola.**—Annual or perennial herbs. *Stipules* scarious. *Stigmas* 3. *Embryo* annular.

IV. **Scleranthus.**—Annual or perennial herbs. *Stipules* 0. *Stigmas* 2. *Embryo* annular.

I. POLLICHIA, Soland.

Perianth herbaceous, urceolate, the mouth of the tube closed by a thickened lobed disc ; lobes ovate, short, obtuse, erect or spreading. *Stamens* 1–2, inserted on the disc ; filaments very short ; anthers oblong. *Ovary* ovoid, attenuated into a short filiform style ; ovules 2, basal, semi-anatropous, with short funicles ; stigmas 2, minute. *Utricle* globose or ovoid-oblong, membranous, 1–2-seeded. *Seeds* oblong or ovoid ; testa hyaline ; embryo dorsal, straight or slightly curved.

A dichotomously branched undershrub with round stiff branches and weak hairs ; leaves opposite or in false whorls, sessile, lanceolate, acuminate, quite entire, flat ; stipules scarious, free ; flowers minute, in sessile axillary crowded cymes surrounded by white scarious bracts, each flower subtended by a bracteole which is at first scarious, then enlarging and becoming thick and fleshy, oblong, rounded, the bracteoles conniving so as to present the appearance of a succulent berry open at the top.

DISTRIB. Species 1 in Tropical and South Africa.

1. **P. campestris** (Soland. in Ait. Hort. Kew. ed. 1, i. 5) ; a much-branched straggling undershrub ; stem and branches terete, pubescent, or nearly glabrous ; leaves 4–8 lin. long, $\frac{1}{2}$–2 lin. broad, lanceolate, acuminate, arranged in often subsecund pseudo-whorls along the stem and branches, pubescent or glabrous ; stipules scarious, lanceolate, cuspidate ; perianth scarcely 1 lin. long ; lobes $\frac{1}{4}$ lin. long, ovate, subacute ; bracteoles at first scarious, then enlarging and becoming thick and fleshy, each bracteole subtending a flower, oblong, obtuse, all the bracteoles of the head conniving and presenting the appearance of a succulent berry open at the top. *Smith, Spicil.* 1, *t.* 1 ; *DC. Prodr.* iii. 377 ; *A. Rich. Tent. Fl. Abyss.* i. 304 ; *Harv. & Sond. Fl. Cap.* i. 133 ; *Baker & Wright in Dyer, Fl. Trop. Afr.* vi. i. 10. *Neckeria campestris, Gmelin, Syst. Veg.* i. 16.

VAR. β, **marlothiana** (Engl. Jahrb. x. 13) ; leaves wider and more densely ashy-pilose.

COAST REGION : Uitenhage Div. ; by the Zwartkops River, *Zeyher,* 1807 ! near Strandfontein (Sandfontein ?) and Matjesfontein, *Drège* ! Port Elizabeth Div. ; near Port Elizabeth, *Burchell,* 4365 ! Albany Div. ; Howisons Poort, near Grahamstown, *Cooper,* 6 ! 2490 ! Fish River, *Burke* ! Queenstown Div. ; around Queenstown, 3800 ft., *Galpin,* 1789 !

CENTRAL REGION : Albert Div., *Cooper,* 594 !

KALAHARI REGION : Basutoland ; *Cooper,* 2488 bis ! Transvaal ; Vlakfontein, near Amersfoort, *Burtt-Davy,* 4017 ! near Waterval Boven, *Burtt-Davy,* 1416 ! Lydenburg, *Wilms,* 507 ! near Potchefstroom, *Bolus,* 3110 ! Hooge Veld, *Rehmann,* 6675 ! Boschveld, *Rehmann,* 5283 ! Var. β : Bechuanaland ; Kuruman, 3920 ft., *Marloth,* 1115.

EASTERN REGION : Griqualand East ; near Clydesdale, 2500 ft., *Tyson,* 3125 ! Ibisi River, *Wood,* 3001 ! Natal ; Inanda, *Wood,* 517 ! and without precise locality, *Sanderson* ! *Gueinzius* ! *Cooper,* 2489 ! *Krauss,* 2 !

II. HERNIARIA, Linn.

Perianth herbaceous, deeply 4–5-fid ; tube short, turbinate ; segments equal or unequal, obtuse, muticous. *Stamens* 3–5, perigynous, equal or unequal ; filaments setaceous ; anthers short ;

staminodes 4–6, setaceous, minute or 0. *Ovary* ovoid; ovule solitary, basal, erect, with a short funicle; style very short; stigmas 2. *Utricle* included in the perianth, ovoid, membranous. *Seed* erect, lenticular, with a basal funicle; testa shining; embryo annular, surrounding farinaceous albumen; cotyledons linear; radicle elongate, descending. *Herbs* annual or with a perennial base, prostrate, much-branched, glabrous or hirsute; leaves opposite, alternate or fascicled, small, subsessile, quite entire; stipules small, scarious, entire or ciliate; flowers hermaphrodite or unisexual, minute, green, crowded in the axils, subsessile or pedicellate; bracts and bracteoles small.

DISTRIB. Species 8–10 in central and southern Europe, N.W. India, North and South Africa.

1. H. hirsuta (Linn. Sp. Pl. ed. i. 218); a prostrate much-branched herb; branches and branchlets slender; terete, glabrous or pubescent; leaves numerous, 2–3 lin. long, $\frac{1}{2}$–$1\frac{1}{2}$ lin. broad, elliptic, tapering at both ends, sessile or nearly so, hirsute and with entire ciliate margins; stipules scarious, ovate-lanceolate, acute; flowers in axillary, usually few-flowered clusters; perianth 1 lin. long, campanulate, green, hairy, divided more than $\frac{1}{2}$-way down; lobes 5, ovate-oblong, obtuse, strongly ciliate; style short; stigmas 2; seed lenticular, smooth, shining, dark brown. *DC. Prodr.* iii. 367; *A. Rich. Tent. Fl. Abyss.* i. 302; *Harv. & Sond. Fl. Cap.* i. 132; *Hook. f. Fl. Brit. Ind.* iv. 712; *Baker & Wright in Dyer, Fl. Trop. Afr.* vi. i. 12. *H. lenticulata, Thunb. Fl. Cap. ed. Schult.* 245 (*not of Linn.*). *H. incana, var. capensis, Pers. Syn. Pl.* i. 292. *H. capensis, Steud. Nomencl. ed.* i. 401; *Bartl. in Linnæa,* vii. 624. *H. virescens, Saltzm. ex DC. Prodr.* iii. 367.

SOUTH AFRICA : without locality, *Zeyher,* 611.
COAST REGION : Cape Div. ; Simons Bay, *Wright* ! Fish Hoek Station, *Wolley-Dod,* 3593 ! Bredasdorp Div. ; between Cape Agulhas and Pot Berg, *Drège* !
CENTRAL REGION : Aliwal North Div. ; south bank of the Orange River, *Burke* ! Graaff Reinet Div. ; *Bowker,* 12 ! Somerset Div. ; Bosch Berg, 2300 ft., *MacOwan,* 1585 !
KALAHARI REGION : Orange River Colony ; Thaba Uncha, *Burke* ! Transvaal ; Jeppestown Ridges, near Johannesburg, 6000 ft., *Gilfillan in Herb. Galpin,* 6170 ! Griqualand West ; in sandy places, *Bowker,* 10 !
Also in Europe, North Africa, the Orient, India and Tropical Africa.

III. **CORRIGIOLA**, Linn.

Perianth herbaceous, 5-partite, persistent; segments oblong, obtuse, muticous, with membranous margins. *Stamens* 5, perigynous; filaments filiform; anthers oblong; staminodes 5, scale-like, alternating with the stamens. *Ovary* ovoid; ovule solitary, suspended by a basal funicle, amphitropous; style very short; stigmas 3. *Fruit* an indehiscent crustaceous globose- or ovoid-trigonous nut included in the perianth. *Seed* globosely ovoid, pendulous; testa membranous; embryo annular, surrounding copious albumen.

Annual or perennial herbs, rarely fruticose at the base, glabrous, diffusely branched ; leaves opposite and alternate, sessile or shortly petiolate, spathulate or linear, entire, flat, sometimes fleshy or glaucous ; stipules various, scarious ; flowers minute, pedicellate, minutely bracteate and 2-bracteolate, in axillary and terminal cymes or clusters.

DISTRIB.   Species 4–6, cosmopolitan.

Annual ; flowers axillary ...   . (1) **litoralis.**
Perennial ; inflorescence leafless ...   . (2) **telephiifolia.**

1. C. litoralis (Linn. Sp. Pl. ed. i. 271) ; a small prostrate diffuse glabrous annual ; stems numerous, 6–12 in. long, spreading on the ground in every direction, smooth, glabrous ; leaves alternate, $\frac{1}{4}$–$\frac{3}{4}$ in. long, 1 lin. broad, linear-spathulate or oblanceolate, subacute, tapering much at the base, sessile or shortly petiolate, glabrous ; stipules ovate-lanceolate, scarious ; flowers in congested cymes in the axils of the upper leaves ; perianth globose, green, glabrous, $\frac{1}{2}$–$\frac{3}{4}$ lin. long ; stamens shorter than the perianth ; nutlet ovoid-trigonous with prominent angles, verrucose, dark brown. *Baker & Wright in Dyer, Fl. Trop. Afr.* vi. i. 12.   *C. littoralis, DC. Prodr.* iii. 367 ; *A. Rich. Tent. Fl. Abyss.* i. 305 ; *Harv. & Sond. Fl. Cap.* i. 132 ; *Sm. Engl. Bot. ed.* 3. vii. 177, *t.* 670 ; *var. capensis, Fenzl in Drège, Zwei Pfl. Documente,* 175.   *C. capensis, Willd. Sp. Pl.* i. 1507 ; *Thunb. Fl. Cap. ed. Schult.* 272 ; *DC. Prodr.* iii. 367.

COAST REGION : Vanrhynsdorp Div. ; Ebenezer, *Drège* ! Cape Div. ; near Capetown, *Burchell*, 431 ! railway at Retreat Station, *Wolley-Dod*, 1192 ! and without precise locality, *Harvey*, 521 ! Mossel Bay Div. ; in a dry channel of an arm of the Gouritz River, *Burchell*, 6461 ! Uitenhage Div. ; *Zeyher*, 27 ! 1834 ! Port Elizabeth Div. ; around Krakakamma, *Burchell*, 4558/5.

KALAHARI REGION : Transvaal ; Houtbosch, *Rehmann*, 6389 !
EASTERN REGION : Natal ; Umzinyati Valley, *Wood*, 1348 !

Also in Europe, Tropical Africa and temperate South America.

2. C. telephiifolia (Pourr. in Act. Toul. iii. 316) ; strongly resembling *C. litoralis*, but differing by the following characters :— rootstock perennial ; leaves obovate or oblong, thick ; flowers in clusters racemosely arranged at the ends of leafless branches ; perianth and nutlet nearly twice as large. *DC. Prodr.* iii. 367 ; *Bolus & Wolley-Dod in Trans. S. Afr. Phil. Soc.* xiv. iii. 310.

COAST REGION : Cape Div. ; roadsides, near the Jetty, Cape Town, *Ecklon & Zeyher*, 1835 (ex *Sonder*), beyond Simonstown Cemetery, *Wolley-Dod*, 2841 ! Caledon or Swellendam Div. ; by the River Zondereinde, *Zeyher*, 2502 partly (ex *Sonder*).

Also in South Europe and North Africa..

IV. SCLERANTHUS, Linn.

Perianth herbaceous when young, becoming crustaceous in age, usually 5-lobed ; tube campanulate. *Stamens* 1–10 (usually 5), inserted in the throat of the perianth-tube ; filaments subulate ; anthers didymous ; staminodes 0. *Ovary* ovoid ; ovule solitary, pendulous from a basal funicle ; styles 2, distinct, filiform, erect.

Utricle membranous, included in the indurated perianth-tube. *Seed* lenticular ; testa coriaceous ; embryo annular, surrounding farinaceous albumen ; cotyledons linear ; radicle superior.

Low rigid annual or perennial dichotomously branched herbs, glabrous or puberulous ; leaves opposite, connate at the base, subulate, rigid, pungent ; stipules 0 ; flowers small, green, in axillary and terminal sessile or pedunculate clusters, not bracteate.

DISTRIB.   Species about 10, cosmopolitan in the Old World.

1. **S. annuus** (Linn. Sp. Pl. ed. i. 406) ; a small annual much-branched herb, 1–4 in. high ; stems slender, more or less pubescent or nearly glabrous ; leaves $\frac{3}{8}$–$\frac{3}{4}$ in. long, narrowly linear, sometimes fascicled, more or less puberulous ; flowers green, in axillary and terminal clusters ; perianth $1\frac{1}{2}$–2 lin. long, divided to a little below the middle ; tube 10-ribbed ; lobes 5, lanceolate, acute, thick, erect or nearly so, with narrow membranous branches ; stamens 5, short. *DC. Prodr.* iii. 378 ; *A. Rich. Tent. Fl. Abyss.* ii. 304 ; *Sm. Engl. Bot. ed.* 3, vii. 181, *t.* 674 ; *Baker & Wright in Dyer, Fl. Trop. Afr.* vi. i. 13.

COAST REGION : Tulbagh Div. ; Mitchell's Pass, 800 ft., *Bolus*, 5225 ! Cape Div. ; Maitland, *Wolley-Dod*, 620 ! Simons Bay, *Wright* ! and without precise locality, *Harvey*, 431 ! Caledon Div. ; by the Bot River, *Burchell*, 929/² ! Albany Div. ; near Grahamstown, *Schlechter*, 2636 !

CENTRAL REGION : Graaff Reinet Div. ; Sneeuwberg Range, 5000–6000 ft., *Drège* !

Also in Europe, North and Tropical Africa and introduced into North America.

ORDER CIX.   **AMARANTACEÆ.**

(By T. COOKE and C. H. WRIGHT.)

Flowers 2- (rarely 1-) sexual, many of them rudimentary or obsolete, monochlamydeous, chaffy or scarious. *Perianth-segments* usually 5 (less commonly 3 or 4), usually united near the base, much imbricated. *Stamens* hypogynous, 5 (rarely 3 or 4) opposite the perianth-segments ; filaments united at the base into a scarious (sometimes very short) tube, linear to the base, with processes (*staminodes*) on the. tube alternating with them, sometimes the filaments wider at the base and uniting by an acute sinus into a longer cup-like tube without any staminodes ; staminodes resembling filaments or oblong, often fimbriate, or small or nearly obsolete ; anthers attached by the middle of the back, 1–2-celled, oblong, with a longitudinal slit ; pollen minute, globose. *Ovary* superior, 1-celled ; style short or long or 0 ; stigma capitellate, simple, or stigmas 2–3, erect or recurved ; ovules solitary or many, amphitropous, erect or suspended from basal funicles.   *Fruit* a membranous utricle (rarely a berry), irregularly breaking up or circumscissile. *Seed* lenticular, oblong or orbicular-reniform, compressed or rarely turgid ; testa crustaceous, smooth or nearly so; embryo annular, surrounding copious albumen.

Herbs or undershrubs (rarely trees), usually erect, seldom scandent ; leaves simple, entire, opposite or alternate ; flowers small or minute, in spikes or heads or rarely racemose ; bracts and bracteoles usually hyaline.

DISTRIB.   Species about 600, in tropical and warm climates.

Tribe 1.   CELOSIEÆ.—*Anthers* 2-celled.   *Ovules* 2 to many.   *Leaves* alternate.

    I.   **Celosia.**—*Filaments* united at the base.   *Staminodes* none or very short.

    II.   **Hermbstædtia.**—*Filaments* united high up.   *Staminodes* longer than the filaments.

Tribe 2.   AMARANTEÆ.—*Anthers* 2-celled.   *Ovules* solitary.

    * *Ovule erect ; funicle short.*

    III.   **Amaranthus.**—*Leaves* alternate.

    ** *Ovule on a long basal funicle.*

    † Flowers usually several under each bract, some sterile.

    ‡ Staminodes none.

    IV.   **Sericorema.**—*Fruit* glabrous.   *Flowers* fertile and sterile in each cluster.   *Leaves* alternate, narrow.

    XI.   **Pupalia.**—*Fruit* glabrous.   *Flowers* fertile and sterile in each cluster. *Leaves* opposite, broad.

    V.   **Marcellia.**—*Fruit* hairy.   *Flowers* fertile and sterile in each cluster. *Leaves* opposite.

    VI.   **Leucosphæra.**—*Fruit* hairy.   *Flowers* fertile only.   *Leaves* alternate or opposite.

    ‡‡ Staminodes present (except in *Centema subfusca* and *Sericocoma avolans*).

    VII.   **Cyphocarpa.**—*Fruit* horned.   Sterile *flowers* reduced to straight spines.   *Leaves* opposite.

    VIII.   **Sericocoma.**—*Fruit* not horned.   Sterile *flowers* reduced to straight spines, or absent.   *Leaves* alternate or opposite.

    IX.   **Centema.**—*Fruit* not horned.   Sterile *flowers* reduced to straight spines, thickened and united to the base of the fertile flower. *Leaves* opposite.

    X.   **Cyathula.**—*Fruit* not horned.   Sterile *flowers* reduced to hooked spines.   *Leaves* opposite.

    †† Flowers solitary under each bract, bibracteolate, all hermaphrodite.

    XII.   **Psilotrichum.**—*Staminodes* none.   *Leaves* opposite.

    XIII.   **Ærva.**—*Staminodes* present.   *Filaments* connate at the base. *Perianth* woolly.   *Leaves* opposite and alternate.

    XIV.   **Calicorema.**—*Staminodes* present.   *Filaments* connate at the base *Leaves* alternate.

XV. **Achyranthes.**—*Staminodes* present.   *Filaments* connate into a long tube.   *Leaves* opposite.

XVI. **Achyropsis.**—*Staminodes* present.   *Filaments* connate at the base.   *Leaves* opposite or fascicled.

Tribe 3. GOMPHRENEÆ.—*Anthers* 1-celled.   *Ovule* solitary.

XVII. **Telanthera.**—*Fruit* not compressed.   *Stigma* capitate.

XVIII. **Alternanthera.**—*Fruit* much compressed, winged.   *Stigma* capitate.

XIX. **Gomphrena** —*Fruit* compressed.   *Stigmas* 2, linear.

I. CELOSIA, Linn.

Flowers hermaphrodite, bracteate and 2-bracteolate, arranged in dense terminal and axillary spikes, or fasciculate along the floriferous branchlets, sessile or shortly pedicellate, white, silvery or rosy, shining.   *Perianth* scarious, 5-partite; segments (*sepals*) oblong or lanceolate, obtuse or acute, erect in fruit.   *Stamens* 5; filaments subulate or filiform, united at the base into a membranous cup; anthers short or elongate, 2-celled.   *Ovary* ovoid or subglobose; ovules 2 or more on elongated basal funicles; style short or long or 0; stigmas 2–3, subulate.   *Fruit* ovoid or oblong, membranous, circumscissile.   *Seeds* 2 or more, usually erect, lenticular; testa crustaceous, black, smooth, polished; embryo annular, surrounding farinaceous albumen.

Herbs or undershrubs, erect or rambling; leaves alternate, attenuated into the petiole, simple or rarely lobed.

DISTRIB. Species about 40, in the warmer regions of the globe.

1. **C. trigyna** (Linn. Mant. 212); a branched slender somewhat straggling glabrous herb; leaves 1–3 in. long, ovate, hastate-ovate, or lanceolate, with cuneate base, glabrous; petioles of the lower leaves reaching 1 in. long, slender; inflorescence in spike-like, sometimes paniculate racemes 3–12 in. long; cymes interrupted; perianth-segments 1¼ lin. long, oblong-lanceolate, equal, straw-coloured, usually 1-nerved; stamens without interposed teeth; filaments united into a cup at the base, free portion ¼ lin. long, subulate; anthers short, elliptic, 2-celled; style very short, about ½ lin. long; stigmas 3, recurved, 1¼ lin. long; capsule 1 lin. long, subglobose, circumscissile a little below the middle; seeds flattened, subreniform, ½ lin. in diam., smooth and shining, black.   *Moquin in DC. Prodr.* xiii. ii. 240; *A. Rich. Tent. Fl. Abyss.* ii. 211; *Oliv. in Trans. Linn. Soc.* xxix. 140 (*excl. syn.*); *Schinz in Engl. & Prantl, Pflanzenfam.* iii. 1A, 99; *E. G. Baker in Trans. Linn. Soc. ser.* 2, *Bot.* iv. 39; *Baker & C. B. Cl. in Dyer, Fl. Trop. Afr.* vi. i. 19. *C. triloba, Meisn. in Hook. Lond. Journ. Bot.* ii. (1843) 448 (548). *Achyranthes decumbens & A. paniculata, Forsk. Fl. Ægypt-Arab.* 47, 48.   *Lestibudesia trigyna, R. Br. Prodr.* 414.

KALAHARI REGION : Transvaal ; Shiluvane, *Junod,* 570 ! Rimers Creek, near Barberton, *Thorncroft,* 3131 !
EASTERN REGION : Pondoland ; by the St. Johns River, *Drège* ! forest above Tiger Flat, near Port St. John, 300 ft., *Galpin,* 2855 ! Natal ; Durban, *Rehmann,* 8741 ! *Schlechter,* 2790 ! Inanda, *Wood,* 529 ! near the Umgeni River, *Krauss,* 238 ! and without precise locality, *Gerrard,* 25 ! Delagoa Bay ; north-west of Lorenzo Marquez, *Bolus,* 9750 !

Also in Tropical Africa and Madagascar.

II. HERMBSTÆDTIA, Reichb.

Flowers hermaphrodite ; bract 1 ; bracteoles 2, small. *Perianth* 5-fid, scarious ; segments subequal, oblong or oblong-lanceolate. *Fertile stamens* 5, united into a tube ; staminodes developed as processes alternate with and distinct from the fertile filaments, sometimes fused partially or wholly with these ; anthers 2-celled, oblong, sessile or shortly stipitate. *Ovary* ovoid, attenuated into a short or subelongate style ; stigmas 3 (rarely 2, 4 or 5). *Capsule* included in the perianth, ovoid, circumscissile. *Seeds* few or many, erect, lenticular ; testa black, shining, crustaceous ; embryo annular, surrounding farinaceous albumen ; radicle inferior.

Herbs or undershrubs, glabrous or puberulous ; leaves scattered, linear or spathulate-oblong, entire ; flowers arranged on elongate terminal spikes (rarely capitate), white or rosy.

DISTRIB. Species 17, in Tropical and South Africa.

Stigmas 2 ... ... (1) caffra.

Stigmas 3 :
  Inflorescence capitate ... ... (2) glauca.

  Inflorescence spicate :
    Style long ; stigmas exserted... ... ... (3) laxiflora.
    Style short ; stigmas not exserted :
      Leaves sessile or subsessile :
        Leaves linear, recurved : spikes oblong, reach-
        ing 2 in. long ; bracts broadly ovate ½–¾ lin.
        long ... ... (4) odorata.
        Leaves not recurved, the lower spathulate-
        oblong ; spikes reaching 8 in. long ; bracts
        caudate-acuminate, 1½ lin. long ... ... (5) elegans.
      Leaves petiolate :
        Leaves oblanceolate ; spikes ¾–11 in. long,
        conical ; bracts 2 lin. long ... ... (6) transwaaleensis.

Stigmas 4–5 ... ... (7) rubromarginata.

1. H. caffra (Moquin in DC. Prodr. xiii. ii. 246) ; whole plant reaching 2½ ft., suffruticose below ; stems erect, simple below, branched above, terete, striate, glabrous ; lower leaves numerous, 1–1¾ in. long (including a petiole ¼ in. long), 2–3 lin. broad, spathulate-oblong, narrowed towards the base, rather thick ; upper leaves small, distant, linear ; inflorescence of branched spikes, often

in threes, the 2 lower spikes small, the central 2–6 in. long; bracts about 2 lin. long, ovate-lanceolate, hyaline, strongly keeled, very acute, persistent; perianth $1\frac{3}{4}$ lin. long, white; segments oblong-lanceolate, 3-nerved, midrib reaching the apex, lateral nerves shorter; staminodes slightly longer than the stamens, 2-fid at the tip into acute lobes; ripe capsules 2 lin. long, oblong-obovate, brown; style very short, almost 0; stigmas 2, recurved, $\frac{1}{4}$ lin. long; seeds usually 2, lenticular, $\frac{1}{2}$–$\frac{3}{4}$ lin. in diam., black, smooth and polished. *Lestiboudesia caffra, Meisn. in Hook. Lond. Journ. Bot.* ii. (1843) 549. *Pelianthus celosioides, E. Meyer ex Moquin, l.c.*

EASTERN REGION: Natal; margins of woods near the Umlaas River, *Krauss*, 37! near Durban, *Gerrard*, 779! *Drège*! *Mudd*! *Peddie*! Inanda, *Wood*, 77!

The type specimen of this is *Krauss*, 37, which has only 2 stigmas, as is also the case with the other specimens enumerated. Messrs. Baker & Clarke, in *Dyer's Fl. Trop. Africa*, vi. i. 25, have united this with *H. recurva*, C. B. Clarke, but that plant is smaller, and has 3 stigmas.

2. **H. glauca** (Moquin in DC. Prodr. xiii. ii. 247); erect, branched, 12–18 in. high; branches terete, finely striate, pale glaucous-green; leaves $\frac{1}{2}$–1 in. long, $\frac{1}{2}$–$\frac{3}{4}$ lin. broad, sessile, distant, linear or linear-lanceolate, subfleshy, glaucous, glabrous; spikes pedunculate, capituliform, globose, dense-flowered; bracts $1\frac{1}{2}$ lin. long, ovate-lanceolate, hyaline, concave, strongly keeled; perianth reaching 2 lin. long; segments ovate-oblong, mucronulate, 3–5-nerved on the back; staminodes hyaline, broad, 2-fid at the apex into 2 triangular acute lobes, shorter than the stamens; anthers linear-oblong; capsule globosely ovoid, circumscissile near the base; style scarcely $\frac{1}{4}$ lin. long; stigmas 3, nearly $\frac{3}{4}$ lin. long, spreading; seeds 3–5, less than $\frac{3}{4}$ lin. in diam., lenticular, brown-black, shining.

WESTERN REGION: Little Namaqualand; by the Orange River near Verlept-pram, *Drège*! on the hills near Buffels River, 1600 ft., *Schlechter*, 11270! sandy and stony places near Spektakel, 800 ft., *Bolus, Herb. Norm. Austr.-Afr.* 685! Little Bushmanland; Naroep, *Schlechter*!

3. **H. laxiflora** (Lopr. in Malpighia, xiv. 430); suffruticose, rigid; branches elongate, slender, striate, glabrous; leaves $\frac{3}{5}$ in. long, $\frac{1}{8}$ in. broad, petiolate, oblanceolate, oblong-lanceolate or spathulate, narrowed towards the base and decurrent into a short petiole, rounded or obtuse at the apex, upper leaves gradually smaller, sometimes submucronulate; spikes few-flowered, $\frac{3}{4}$–$1\frac{1}{4}$ in. long, 2 lin. wide, obtuse; bracts 1 lin. long, $\frac{1}{2}$ lin. broad, ovate-lanceolate, hyaline, persistent after the flowers have fallen, midrib produced into a mucro; flowers distant in the lower part, close in the upper part of the inflorescence, straw-coloured; perianth 2 lin. long; segments oblong-lanceolate, acute, 3–5-nerved; staminodes shorter than the anthers, 2-fid into lanceolate segments; anthers $\frac{1}{2}$ lin. long, oblong; capsule ovoid, elongate, circumscissile below the middle; style elongate, reaching nearly $\frac{1}{2}$ lin. long; stigmas 3, exserted; seeds 2 or 3, lenticular, $\frac{1}{2}$ lin. in diam., smooth, black. *Lopr. in Engl. Jahrb.* xxx. 105.

EASTERN REGION : Portuguese East Africa ; Ressano Garcia, near Komati Poort, 1000 ft., *Schlechter*, 11876 !

4. H. odorata (T. Cooke) ; a low branched plant less than 1 ft. high ; stem and branches slender, striate, glabrous ; leaves $\frac{1}{4}$–1 in. long, scarcely reaching $\frac{1}{2}$ lin. broad, subsessile, linear-oblong, usually recurved, glabrous or nearly so ; spikes $\frac{1}{2}$–2 in. long, 3–4 lin. broad, simple, continuous ; bracts $\frac{1}{2}$–$\frac{3}{4}$ lin. long, broadly ovate, acute, scarious, persistent ; perianth up to 2 lin. long, rosy (*Burchell*) ; segments oblong, acute at the tip, 3-nerved, central nerve reaching the apex, lateral nerves shorter ; staminodes scarious, quadrate-oblong, as long as or slightly longer than the stamens, 2-fid at the apex into lanceolate lobes ; anthers oblong, 2-celled ; capsule globosely ovoid, circumscissile about the middle ; style scarcely $\frac{1}{8}$ lin. long ; stigmas 3, about $\frac{1}{4}$ lin. long, recurved ; seeds few, about $\frac{1}{2}$ lin. in diam., lenticular, black, shining. *H. elegans, var. recurva, Moquin in DC. Prodr.* xiii. ii. 247. *Celosia odorata, Burch. Trav. S. Afr.* i. 389. *C. recurva, Burch. Trav. S. Afr.* ii. 226.

KALAHARI REGION : Griqualand West ; Griquatown, *Burchell*, 2111 ! between Spuigslang Fontein and the Vaal River, *Burchell*, 1712 !

5. H. elegans (Moquin in DC. Prodr. xiii. ii. 247) ; 10–20 in. high, branched ; branches erect, striate, glabrous or nearly so, rigid, green ; leaves $\frac{3}{4}$–1$\frac{1}{2}$ in. long, 1–2 lin. broad, obtuse or subacute, narrowed towards the base as if petiolate, subcoriaceous, lower leaves spathulate-oblong, the upper narrowly linear, glabrous ; spikes terminal, 2–8 in. long, 4–6 lin. broad, nearly continuous, at first ovate-triangular, at length oblong ; bracts 1$\frac{1}{2}$ lin. long, ovate, caudate-acuminate, hyaline ; perianth 2 lin. long ; segments oblong-lanceolate, acute, scarious, shining, white or rosy, 3–5-(rarely 7-) nerved, lateral nerves much shorter than the central nerve which reaches the apex ; anthers narrowly oblong ; staminodes slightly longer than the stamens, 2-fid at the apex into 2 lanceolate lobes ; capsule circumscissile below the middle ; style very short ; stigmas 3, about $\frac{1}{4}$ lin. long ; seeds $\frac{1}{2}$ lin. in diam., lenticular, smooth and polished, black. *Schinz in Engl. & Prantl, Pflanzenfam.* iii. 1A, 100 ; *Baker & C.B. Cl. in Dyer, Fl. Trop. Afr.* vi. i. 26. *H. recurva, C. B. Cl. in Dyer, Fl. Trop. Afr.* vi. i. 25.

CENTRAL REGION : Hopetown Div. ; near Hopetown, *Bolus* ! KALAHARI REGION : Griqualand West ; Asbestos Mountains, *Burchell*, 2028 ! along the Vaal River, *Burchell*, 1767 ! between Griquatown and Witte Water, *Burchell*, 1963 ! Orange River Colony ; by the Vaal River, *Burke*, 334 ! Thorntree Valley, *Mrs. Barber* ! Bechuanaland ; between Kuruman and the Vaal River, *Cruickshank*, 2548 ! between the Mashowing River and Kuru, *Burchell*, 2409 ! near the source of the Kuruman River, *Burchell*, 2455/² ! Kosifontein, *Burchell*, 2563 ! Barolong Territory, *Holub* ! Bakwena Territory, 3500 ft., *Holub* ! Transvaal ; Pretoria, *Bolus*, 3109 ! Potgieters Rust, *Miss Leendertz*, 1963 ! South African gold fields, *Baines* ! Hermans Kraal, *Schlechter*, 4198 ! Rustenburg, 4000 ft., *Miss Pegler*, 1014 ! EASTERN REGION : Natal ; banks of the Umzinyati River, 3000–4000 ft., *Sutherland* ! Delagoa Bay ; Lorenzo Marquez, *Wilms*, 1257 ! Also in Tropical Africa.

6. **H. transwaaleensis** (Lopr. in Malpighia, xiv. 429); suffruticose, about 20 in. high; branches erect, elongate, glabrous or puberulous, much striate, slender; leaves $\frac{3}{4}$–$1\frac{1}{4}$ in. long, $\frac{1}{8}$–$\frac{1}{3}$ in. broad, petiolate, oblanceolate or oblanceolate-oblong, gradually narrowed towards the base into the petiole, obtuse at the apex, glabrous or nearly so; spikes terminal, $\frac{3}{4}$–$1\frac{1}{4}$ in. long, nearly $\frac{1}{2}$ in. broad at the base, conical, acute at the apex, dense, much thickened; bracts hyaline, ovate-lanceolate, acute, concave, nearly $\frac{1}{8}$ in. long, $\frac{1}{10}$ in. broad, persistent; perianth straw-coloured, $\frac{1}{8}$ in. long; segments 1 lin. broad, ovate-lanceolate, turning black, with hyaline, much dilated margins, 5–7-nerved; staminodes obtusely and irregularly 2-fid; anthers linear-oblong, rounded at the apex, $\frac{3}{4}$ lin. long; capsule elongate-ovoid, circumscissile; style short, terete, rufescent; stigmas 3, revolute, papillose; seeds usually 2, lenticular, $\frac{1}{5}$ lin. in diam., black. *H. transvaalensis, Lopr. in Engl. Jahrb.* xxx. 105.

KALAHARI REGION : Transvaal; near Lydenburg, *Wilms,* 1254.

Plant not seen; the description is taken from Lopriore.

7. **H. rubromarginata** (C. H. Wright); stem woody; branches slightly ribbed, pilose, reddish; leaves oblanceolate, 6 lin. long, 1–$1\frac{1}{2}$ lin. wide, obtuse, tapering downwards, pilose, red along the margins; spikes terminal on the branches, dense, $1\frac{1}{4}$ in. long; bracts and bracteoles about 1 lin. long, ovate, concave, scarious; perianth-segments pink, broadly ovate, about $2\frac{1}{2}$ lin. long, the inner slightly narrower than the outer; filaments subulate; staminodes about as long as the filaments, linear, with 2 acute lobes; ovary ovoid; styles 4–5, short; ovules about 6.

KALAHARI REGION : Transvaal; Warmbath, *Miss Leendertz,* 1326!

This species has the appearance of *H. caffra,* Moquin, but has pilose branches and 4–5 styles.

III. AMARANTHUS, Linn.

Flowers monœcious or polygamous, bracteate and 2-bracteolate. *Perianth-segments* 5 (less commonly 1–3), membranous, equal or subequal, ovate-lanceolate in male, usually oblong or spathulate-oblong in female flowers. *Stamens* usually 5 (rarely 1–3); filaments subulate or filiform, free at the base and without interjected staminodes; anthers oblong or linear-oblong, 2-celled. *Ovary* ellipsoid, compressed; ovule 1, subsessile, erect; style short or 0; stigmas 2–3, subulate or filiform, spreading. *Fruit* usually enclosed in the perianth, orbicular or ovoid, indehiscent or circumscissile, mostly membranous, simple or 2–3-toothed at the apex. *Seed* erect, globose, compressed; testa crustaceous, smooth and shining; embryo annular, surrounding farinaceous albumen; cotyledons linear; radicle inferior.

Annual, erect or decumbent; leaves alternate, mostly simple, usually entire, petiolate; flowers small or minute, arranged in dense axillary heads or in terminal paniculate spikes, white, green, rosy or purplish; bracts and perianth usually persistent.

DISTRIB. Species about 25, common weeds in all the warmer parts of the world.

Sepals 5; stamens 5:
    Leaf-axils furnished with spines ... ... (1) **spinosus.**
    Leaf-axils without spines:
        Sepals 1¼ lin. long, oblong-lanceolate, the mid-rib produced into an acicular point... ... ... (2) **paniculatus.**
        Sepals 1 lin. long, linear-oblong, obtuse, often emarginate ... ... (3) **retroflexus.**

Sepals 3; stamens 3:
    Capsule circumscissile; flowers not pedicellate:
        Bracteoles longer than the perianth; sepals subacute, cuspidate ... ... (4) **Thunbergii.**
        Bracteoles equal to or shorter than the perianth; sepals obtuse, with a short slender mucro ... (5) **Blitum.**
    Capsule indehiscent; flowers pedicellate ... ... (6) **viridis.**

1. **A. spinosus** (Linn. Sp. Pl. ed. i. 991); stem 1–3 ft. or more high, suberect, rather hard, branched near the base, grooved, glabrous; leaves 1¼–3 in. long, ½–1 in. broad, ovate or lanceolate, obtuse, spinous-apiculate, entire, glabrous above, sometimes scurfy beneath, with spines in the lower axils sometimes ⅜ in. long; main nerves numerous, slender, conspicuous below, white; petioles reaching 1½ in. long, slender; flowers very numerous, sessile, in dense axillary clusters and in terminal and axillary dense or interrupted cylindric spikes; bracteoles 1½ lin. long, ovate, cuspidate; sepals 1 lin. long, bristle-pointed, those of the male flowers ovate, acute, those of the female oblong-obovate; stamens 5; capsule ovoid, membranous, circumscissile about the middle; style 0; stigmas 2, reaching ¾ lin. long, divaricate; seeds ½ lin. in diam., lenticular, with obtuse margins, shining, black. *Moquin in DC. Prodr.* xiii. ii. 260; *Wight, Icon. t.* 513; *Hook. f. Fl. Brit. Ind.* iv. 718; *Baker & C. B. Cl. in Dyer, Fl. Trop. Afr.* vi. i. 32.

KALAHARI REGION: Transvaal; Shiluvane, *Junod,* 1062! a common roadside weed around Pretoria, *Burtt-Davy,* 107D! *Miss Leendertz,* 7!

EASTERN REGION: Natal; near Durban, *Wood,* 1762! *Wilms,* 2238! in sandy places near Ixopo, 3800 ft., *Schlechter,* 6679!

Also in Tropical Africa and India.

2. **A. paniculatus** (Linn. Sp. Pl. ed. ii. 1406); a tall handsome plant 3–5 ft. or more high; stem stout, more or less grooved, glabrous or puberulous; leaves 2–3 in. long, 1–1½ in. broad, elliptic-lanceolate, tapering to both ends, obsoletely punctulate, often mucronulate; nerves prominent below; petioles reaching 2½ in. long; flowers numerous; panicles much-branched; spikes erect or spreading, cylindric, gold-coloured or red, the central spikes the longest; bracteoles 2¼ lin. long, exceeding the sepals, narrowly

ovate, hyaline, with a strong midrib produced into a long hard point; perianth 1¼ lin. long; sepals 5, oblong-lanceolate, scarious, midrib produced into a long acicular point; stamens 5; capsules 1–1¼ lin. long, ovoid, narrowed to the tip, circumscissile about the middle; stigmas 3, recurved, 1¼ lin. long; seeds ½–¾ lin. in diam., lenticular, white, red or black.   *Moquin in DC. Prodr.* xiii. ii. 257; *Willd. Amarant.* 32, *t.* 2, *fig.* 4; *Hook. f. Fl. Brit. Ind.* iv. 718. *A. frumentaceus, Buch.-Ham. in Roxb. Fl. Ind.* iii. (1832) 609.

COAST REGION : Queenstown Div. ; Shiloh, *Baur,* 945! plains near Queenstown, 3500 ft., *Galpin,* 2039 !

KALAHARI REGION : Transvaal; near Lydenburg, *Wilms,* 1251 ! Standerton, *Burtt-Davy,* 1805 ! Potchefstroom, *Burtt-Davy,* 1050 !

EASTERN REGION : Natal ; near Mooi River, 3000–4000 ft., *Wood,* 4440!

Cultivated in many warm countries for its grain. In India it supplies the staple food of hill tribes over a large area.

3. **A. retroflexus** (Linn. Sp. Pl. ed. ii. 1407); stem erect, obsoletely striate, more or less pubescent, 2–3 ft. or more high, branched below; younger branches recurved, then ascending; leaves 3–4½ in. long, 1½–2½ in. broad, ovate-lanceolate or rhomboid-ovate, subobtuse, with undulate margins, pale glaucous-green; main nerves numerous, conspicuous below, white; petioles 2–3 in. long, stout; flowers pale green, in dense branched panicles; spikes erect or spreading, the terminal spike rather short, rigid; bracteoles reaching 2 lin. long, recurved, narrowly lanceolate, scarious, midrib strong, produced into a long hard sharp point; sepals linear-oblong, obtuse, hyaline, 1 lin. long, ⅜ lin. broad, rounded, sometimes slightly retuse at the apex, with a green midrib; stamens 5; capsule less than 1 lin. long, ovoid, compressed, membranous, circumscissile about the middle; stigmas 3, recurved, ½ lin. long; seeds ½ lin. in diam., lenticular, dark brown, nearly black.   *Moquin in DC. Prodr.* xiii. ii. 258 ; *Willd. Amarant.* 33, *t.* 11, *fig.* 21.

COAST REGION : Queenstown Div. ; Shiloh, *Baur,* 1139 !

CENTRAL REGION : Somerset Div. ; in fields at the foot of Bosch Berg, 2500 ft., certainly an introduction, *MacOwan,* 1957 !

An introduction from America, rapidly becoming naturalized.

4. **A. Thunbergii** (Moquin in DC. Prodr. xiii. ii. 262) ; stems 1½ ft. high, ascending, striate, glabrous, often tinged with purple, branched ; leaves ½–1 in. long, 2–6 lin. broad, ovate, obovate or spathulate, obtuse, sometimes excised at the apex, cuneate at the base, apiculate and with more or less undulate margins ; nerves elevated beneath ; petioles ⅓–1 in. long ; flowers in small axillary heads ; bracteoles 1¾ lin. long, lanceolate, with a strong midrib produced into a hard sharp point ; perianth 1½ lin. long ; sepals 3, unequal, ovate, subacute, cuspidate ; stamens 3 ; capsule ovoid, 1¼ lin. long, narrowed at the apex, circumscissile about the middle, membranous ; stigmas 3, about ½ lin. long ; seeds ½–¾ lin. in diam., lenticular, with acute margins, dark brown, shining.   *A græcizans, Baker & C. B. Cl. in Dyer, Fl. Trop. Afr.* vi. i. 34 (*not of Linn.*).

COAST REGION: Cape Div. ; Groot Schuur, *Wolley-Dod*, 1142 ! Uitenhage Div.,
Zeyher, 576 ! Queenstown Div. ; near Shiloh, 3500 ft., *Baur*, 971 !
CENTRAL REGION : Graaff Reinet Div. ; near Graaff Reinet, *Bolus*, 357 ! Rich-
mond Div. ; between Richmond and Brak Vallei River, *Drège* ! Albert Div.,
Cooper, 789 !
KALAHARI REGION : Basutoland, *Cooper*, 3049 ! 3500 ; Bechuanaland ; plains
near Takun, *Burchell* ! Transvaal ; Marico District, *Holub* ! Potchefstroom Farm,
Burtt-Davy, 1051 ! 1057 ! Springbok Flats, *Burtt-Davy*, 1740 ! near Standerton,
Burtt-Davy, 1802 !
EASTERN REGION : Natal ; Inanda, *Wood*, 234 !
I do not consider *A. Thunbergii* to be conspecific with *A. græcizans*, Linn.
(*Sp. Pl. ed.* ii. 1405), a Virginian plant, the leaves of which are described by
Linnæus as lanceolate, while those of *A. Thunbergii* are obovate.

5. **A. Blitum** (Linn. Sp. Pl. ed. i. 990) ; a variable weed ; stems
1–2 ft. high, erect or decumbent, grooved, glabrous, often reddish ;
leaves in S. African specimens usually small, rarely reaching 2 in. long
(including the petiole), varying from narrowly oblong to obovate,
attenuated to the base ; petioles variable in length ; inflorescence
in axillary, often ternate clusters ; bracteoles about as long as the
perianth or a little shorter, oblong-lanceolate, with a short recurved
mucro ; sepals 3, ovate-oblong, 1 lin. long, obtuse, with a short very
slender mucro and an inconspicuous keel ; anthers 3, about ½ lin.
long ; capsule ovoid, rather longer than the perianth, circumscissile
about the middle, or sometimes indehiscent ; stigmas 3, short ;
seeds ½ lin. in diam., lenticular, slightly margined, brown-black,
shining. *Moquin in DC. Prodr.* xiii. ii. 263 ; *Hook. f. Fl. Brit. Ind.*
iv. 721 ; *Baker & C. B. Cl. in Dyer, Fl. Trop. Afr.* vi. i. 35.

SOUTH AFRICA : without locality, *Zeyher*, 1438 !
COAST REGION : Queenstown Div. ; bare mountain slopes near Queenstown,
4400 ft., *Galpin*, 2015 !
CENTRAL REGION : Prince Albert Div. ; Boter Kraal, 2100 ft., *Bolus*, 11624 !
Somerset Div. ; Somerset East, *Miss Bowker* !
Also in Tropical Africa and India.

6. **A. viridis** (Linn. Sp. Pl. ed. ii. 1405) ; erect, branched ; stems
grooved, glabrous ; leaf-blades ¼–1½ in. long, ½–¾ in. broad, ovate
or rhomboid-ovate, obtuse, with a short stiff apiculation, narrowed
at the base and running into the petiole, glabrous ; nerves conspicuous beneath ; petioles reaching 1 in. long, slender ; inflorescence
terminal and axillary, the terminal spike-like racemes reaching
2 in. long, the axillary much shorter, often reduced to heads ;
flowers shortly pedicellate ; bracteoles much shorter than the
sepals, lanceolate, acute, membranous ; perianth 1 lin. long ; sepals
narrowly linear-spathulate apiculate with a green midrib ; capsule
1½ lin. long, indehiscent, ovoid-oblong, subobtuse, faintly nerved ;
seeds rather more than ½ lin. long, less than ½ lin. broad, ovoid,
compressed, margined, dark brown, shining. *Baker & C. B. Cl. in
Dyer, Fl. Trop. Afr.* vi. i. 33 ; *Hook. f. Fl. Brit. Ind.* iv. 720.
Euxolus caudatus, Moquin in DC. Prodr. xiii. ii. 274.

COAST REGION : Albany Div. ; Grahamstown, *MacOwan*, 3419 ! perhaps introduced.
Also in Tropical Africa and India.

IV. SERICOREMA, Lopr.

Inflorescence long and laxly spicate, the partial inflorescences consisting of 1–3 perfect flowers, sessile and distant along the rhachis of the spike, supported by 2 sterile flowers represented by a fascicle of recurved spines united into a stalk below. *Bracteoles* each enclosing a tuft of soft woolly hairs which cover the spines and which are enlarged in fruit so as almost to envelop the flower. *Stamens* long, without interposed staminodes. *Ovary* ovoid; style scarcely any, stout; stigma with a tuft of hairs. *Seed* doubled over on itself, somewhat horseshoe-shaped, compressed, red, glabrous; embryo annular.

Erect herbs ; leaves scattered, alternate and fasciculate, linear or oblong, sessile, quite entire.

DISTRIB. Species 2, in Tropical and South Africa.

Sir J. Hooker established *Sericorema* as a section of *Sericocoma* in *Benth. et Hook. f. Gen. Pl.* iii. 30, but suggested that it might be entitled to generic rank.

1. S. remotiflora (Lopr. in Engl. Jahrb. xxvii. 39); stem herbaceous, erect, glabrous, paniculately branched above; branches numerous, ascending, terete, striate, glabrous; leaves sessile, $\frac{1}{4}$–$\frac{3}{4}$ in. long, $\frac{1}{3}$–$\frac{1}{2}$ lin. broad, alternate or fasciculate, linear-subulate, acute, shortly mucronate, thick and subfleshy, glabrous, green, with revolute margins ; inflorescences distant, of 1 fertile flower with 2 sterile ones ; bracteoles 2 lin. long, ovate, acute, membranous, with a strong midrib prolonged into a cuspidate point, each bracteole enclosing a tuft of pale brownish woolly hairs which covers the sterile flower, a stalked fascicle of recurved spines (after flowering the tuft of hairs enlarges and almost conceals the fruit); 2 outer sepals reaching $4\frac{1}{2}$ lin. long, linear-lanceolate, membranous, with a wide midrib running out into a cuspidate mucro; 3 inner sepals 4 lin. long, similar in shape to the outer, all silky villous ; stamens $3\frac{1}{4}$ lin. long ; filaments flat, linear, membranous, about $\frac{1}{4}$ lin. broad; anthers reaching nearly 2 lin. long, linear ; ovary villous; style scarcely any, stout ; stigma with a tuft of hairs; capsule softly villous ; seed doubled on itself, somewhat horseshoe-shaped, appearing 2-lobed, compressed, red. *Trichinium remotiflorum, Hook. Ic. Pl.* t. 596. *Sericocoma remotiflora, Hook. f. in Benth. & Hook. f. Gen. Pl.* iii. 30. *Pupalia remotiflora, Moquin in DC. Prodr.* xiii. ii. 333 ; *Sonder in Linnæa,* xxiii. 97.

CENTRAL REGION : Murrayburg Div. ; mountain near Snyders Kraal, *Tyson,* 422 !
KALAHARI REGION : Griqualand West ; Griquatown, *Burchell,* 1895 ! 1942 ! Albania (Douglas), between the Orange River and Vaal River, 4500 ft., *Bolus,* 1834 ! St. Clair, Douglas, *Orpen,* 224 ! Orange River Colony ; by the Vaal River, *Burke,* 185 ! *Zeyher,* 1434 ! damp places in low valleys, *Mrs. Barber* ! Transvaal ; Warm Bath, *Miss Leendertz,* 1347 ! South African Gold Fields, *Baines* ; Dronkfontein, east of Nylstroom River, *Nelson,* 367 !

Lopriore makes the glabrous ovary a characteristic of this genus. In what may be considered the type of the genus, *Trichinium remotiflorum,* Hook., figured in *Hook. Ic. Pl.* l.c., the ovary is shown to be villous, a condition which I have invariably found to exist.

V. MARCELLIA, Baill.

(LEUCOSPHÆRA, Gilg, partly. SERICOCOMOPSIS, Schinz, partly.)

Fertile flowers 1–3 in each cluster; sterile flowers 1 or more in each cluster, reduced to spines, not hooked. *Perianth-segments* oblong. *Stamens* 5; filaments linear, united below into a cup; staminodes none; anthers 2-celled, oblong. *Ovary* ovoid, densely hairy; ovule solitary, suspended from a basal funicle; style columnar; stigma small.

Undershrubs; leaves opposite, simple, entire; flower-clusters collected into dense spikes near the top of the stems.

DISTRIB. Species about 9, in Tropical Africa.

1. **M. Bainesii** (C. B. Cl. in Dyer, Fl. Trop. Afr. vi. i. 51); a much-branched undershrub, up to 18 in. high; branches terete, clothed with soft white hairs; leaves obovate-elliptic, 10 lin. long, 3 lin. wide, acute, tapering to the base, densely white silky-tomentose on both surfaces; spikes subglobose, about 6 lin. in diam.; clusters consisting of 1–2 fertile and 2–4 sterile flowers; bracts lanceolate, 3 lin. long, acuminate, scarious, silky outside; bracteole ovate, acuminate, much shorter than the bracts; perianth-segments lanceolate, acuminate, 2 lin. long, silky; staminal cup nearly as long as the hairy ovary; utricle densely hairy. *Sericocoma Bainesii,* Hook. f. in Benth. et Hook. f. Gen. Pl. iii. 31; *Schinz in Engl. & Prantl, Pflanzenfam.* iii. 1A, 107, 106, *fig.* 60. *Sericocomopsis Bainesii, Schinz in Engl. Jahrb.* xxi. 185, *and in Bull. Herb. Boiss.* v. *Append.* iii. 65. *Leucosphæra Bainesii, Gilg in Engl. & Prantl, Pflanzenfam. Nachtr.* i. 153; *Lopr. in Engl. Jahrb.* xxvii. 41, xxx. 107.

WESTERN REGION : Great Namaqualand ; Ganab, *Schinz,* 249 ; Nanas, *Fleck,* 176a ; Keetmanshoop, *Fleck,* 172a.

Also in Tropical Africa.

VI. LEUCOSPHÆRA, Gilg, partly.

(SERICOCOMOPSIS, Schinz, partly.)

Fertile flowers in 2-flowered clusters arranged in globose heads; barren flowers none. *Perianth-segments* 5, clothed with white silky hairs. *Stamens* 5; staminodes none. *Ovary* elongate. Otherwise as *Marcellia*, Baill.

DISTRIB. Species 1, endemic. *L. Bainesii*, Gilg, has been transferred to *Marcellia.*

1. **L. Pfeilii** (Gilg in Notizbl. Königl. Bot. Gart. Berlin, i. 328); a small much divaricately branched shrub ; young leafy branches greyish velvety ; leaves alternate or opposite, 5 lin. long, 1½ lin. wide, sessile, lanceolate or oblanceolate, acute, tapering to the base, entire, very densely covered with rather long adpressed grey or whitish hairs on both surfaces ; flowers in globose many-flowered heads 9 lin. in diam. at the ends of the branches ; clusters 2-flowered ; bracts spine-like, a little longer than the flowers, 4 lin. long, plumose ; bracteoles 2, ovate, lateral, short, membranous ; perianth-segments 3½ lin. long, long and densely silky-pilose, three inner slightly narrower than the others ; stamens 5, without a basal membrane or intermediate staminodes.

COAST REGION : Bredasdorp Div. ; Rietfontein-Koes, *Pfeil*, 121.

VII. CYPHOCARPA, Lopr.

(SERICOCOMA, § KYPHOCARPA, Fenzl.)

Inflorescence of 1–4 perfect, and 1–2 sterile flowers reduced to spines, spicately arranged and full of fine hairs. *Perianth-segments* 5, oblong or oblong-lanceolate, subequal or the outer the longer. *Stamens* 5 ; filaments linear, united at the base into a short cup, with interposed staminodes on its edge between the stamens ; anther-cells 2, oblong. *Ovary* ovoid or obovoid, with a distinct horn on one side below its apex ; ovule suspended from a basal funicle ; style slender, about as long as the ovary ; stigma capitate. *Seed* suborbicular, more or less compressed ; embryo annular.

Herbs or shrubs of various habit ; leaves opposite, linear or oblong, simple, entire, sessile or shortly petiolate.

DISTRIB. Species about 8, in Tropical and South Africa.

Leaves ¼–½ in. broad, shortly petiolate ; sepals not
    aristate, subequal   ...   . (1) **trichinioides.**
Leaves not reaching ₁₀ in. broad, sessile ; outer sepals
    aristate, larger than the inner :
    Plant 2 ft. high : spikes 2–3 in. long ...   . (3) **angustifolia.**
    Plant 2 ft. high ; spikes 1½ in. long   ...   . (2) **resedoides.**
    Plant 8 in. high ; spikes less than ¾ in. long ...   ... (4) **Wilmsii.**

1. **C. trichinioides** (Lopr. in Engl. Jahrb. xxvii. 45) ; suffruticose at the base, quite glabrous ; stems simple, erect, leafy below, naked upwards ; leaves opposite or subopposite, shortly petiolate, 1–2 in. long, ¼–½ in. broad, oblong-lanceolate, obtuse or subacute, shortly mucronulate, attenuated at the base, flat, subfleshy ; midrib prominent beneath ; petiole 3–5 lin. long ; spikes 1¼–2¾ in. long, ½–⅝ in. broad, obtuse, usually simple, but sometimes with a short branch at the base ; flowers yellowish ; bracteoles slightly shorter than the sepals,

oblong, slightly mucronulate, woolly; perianth reaching 4 lin. long; sepals subequal, narrowly lanceolate, acute, not mucronate, 3-nerved, clothed with yellowish wool; filaments 1¼ lin. long; staminodes small, ovate; ovary villous, shortly horned; style 2 lin. long, slender; stigma capitate; capsule ovoid, villous; seed doubled on itself, the folded seed rather less than 1 lin. in diam., subglobose, scarcely compressed, smooth, red. *Sericocoma Chrysurus, Meisn. in Hook. Lond. Journ. Bot.* ii. (1843) 547; *Moquin in DC. Prodr.* xiii. ii. 307; *Schinz in Engl. & Prantl, Pflanzenfam.* iii. 1A, 107. *S. trichinioides, Fenzl in Linnæa,* xvii. 324. *Trichinium (misspelt Tichinium) Chrysurus, Meisn. ex Moquin in DC. Prodr.* xiii. ii. 307.

EASTERN REGION: Natal; near Durban, *Krauss,* 294! *Wood in MacOwan & Bolus, Herb. Norm.* 1034! *Grant!* Inanda, 1800 ft., *Wood,* 67!

There is no doubt that the specimens of Krauss (294) in the Kew Herbarium represent Meisner's type of *Sericocoma Chrysurus.* I quite agree with Schinz that *Sericocoma trichinioides,* Fenzl, is conspecific.

2. **C. resedoides** (Lopr. in Engl. Jahrb. xxvii. 44); stem slightly woody, erect, 2 ft. high, glabrous, slightly constricted at the insertion of the leaves when dry; leaves distant, decussate, narrowly linear-lanceolate, mucronate, tapering into the petiole, up to 2 in. long and 1½ lin. wide, glabrous; flowers in clusters of 2–3 fertile and 1–2 sterile collected into terminal spikes 1½ in. long and 4 lin. wide; bracts ovate-triangular, acute, awned, hyaline, ribbed, 2½ lin. long, 1 lin. wide; bracteoles connate at the base, broadly ovate, acute, hyaline, with a thick hairy midrib produced into a long awn; perianth-segments unequal, oblong-ovate, about 2½ lin. long, acute, hyaline, densely long hairy, the strong midrib of the two outer produced into a short awn; filaments subulate; staminodes shortly trapezoid; ovary turbinate, laterally horned, pilose; style subulate, twice as long as the ovary; stigma capitate, *Lopr. in Malpighia,* xiv. 437.

EASTERN REGION: Transvaal; Lydenburg district, by the Watervaal River, *Wilms,* 1260.

3. **C. angustifolia** (Lopr. in Engl. Jahrb. xxvii. 45); suffruticose, glabrate except the inflorescence, often 2 ft. high, branched below; branches terete, rather slender, glabrous; leaves opposite or fasciculate, reaching 2 in. long, scarcely more than 1 lin. broad, sessile, linear, shortly mucronate, glabrous; flowers in dense, usually obtuse spikes 2–3 in. long, ½ lin. broad, the partial inflorescences consisting of 2–4 fertile and 1–2 sterile flowers, each of the latter reduced to a pair of silky-hairy hard sharp terete yellow shortly stalked spines reaching 2½ lin. long; bracteoles broadly ovate, membranous, silky-hairy, the midrib produced into a strong terete prominent spine; perianth 3 lin. long; sepals 5, the 2 outer the largest, oblong-lanceolate, aristate, silky-hairy, the 3 inner oblong-lanceolate, subobtuse, not aristate; stamens 1¼ lin. long; anthers rather more than ¼ lin. long; ovary subglobose, 1 lin.

across, slightly flattened, furnished with a horn at one side, silky-villous; style reaching 1 lin. long, slender; stigma small, capitate; seed doubled on itself, the folded seed 1 lin. in diam., orbicular, compressed, red.   *Cyathula angustifolia, Moquin in DC. Prodr.* xiii. ii. 328 ; *Sonder in Linnæa,* xxiii. 97.   *Sericocoma angustifolia, Hook. f. in Benth. & Hook. f. Gen. Pl.* iii. 30.

CENTRAL REGION : Hopetown Div. ; near Hopetown, 5000 ft., *Muskett in Herb. Bolus,* 2208 !

KALAHARI REGION : Griqualand West ; near Griquatown, *Burchell,* 1945 ! *Holub* ! Asbestos Mountains, *Burchell,* 2065 ! St. Clair, Douglas, *Orpen,* 241 ! Transvaal ; Daspoort, near Pretoria, *Miss Leendertz,* 156 ! Komati Poort, 1000 ft. *Schlechter,* 11819 ! Vaal River, *Burke,* 333 ! *Zeyher,* 1437 ; Boshveld, between Kameel Poort and Elands River, *Rehmann,* 4807 ! Warm Bath, *Miss Leendertz,* 1599 !

4. **C. Wilmsii** (Lopr. in Engl. Jahrb. xxvii. 42) ; plant 8 in. high ; stem erect, herbaceous at first, subwoody later, branched from the base ; leaves decussate, glabrous, 2¼ in. long, ¾ lin. broad, with a thick bristle point ; inflorescence spicate, conical, supported by a pair of leaves immediately below the spike ; spikes 7 lin. long, 6 lin. broad ; partial inflorescences of usually 4 fertile and 4 sterile flowers, 2 of the fertile flowers completely developed and 2 incompletely, each accompanied by 2 lateral sterile flowers which consist of 5 unequal spindle-shaped spines, of which the outer ones are provided with a tuft of hairs and a narrow blade ; bracteoles unequal, obliquely suborbicular with a hyaline blade, densely long-hairy at the base, 2½–3 lin. long, 2–2½ lin. broad, with a strongly developed midrib produced into a spine ; sepals about 2–2½ lin. long, 1–1½ lin. broad, ovate, acute, the 2 outer ending in spines which are a continuation of the midrib, all hyaline, densely pilose at the base ; filaments subulate, connate at the base, 1½ lin. long ; staminodes short, papilla-like ; anthers oblong-elliptic ; ovary turbinate, laterally horned, pilose, especially on the crown, style subulate, 2 or 3 times as long as the ovary ; stigma capitate.

KALAHARI REGION : Transvaal ; Lydenburg district, between Middelburg and the Crocodile River, *Wilms,* 1259, ex *Lopriore.*

I have not seen the specimen ; the description is taken from Lopriore.

VIII. SERICOCOMA, Fenzl.

Flowers spicate or capitellate, hermaphrodite, 1–2 fertile with 1 or more sterile flowers usually reduced to spines, the inflorescences full of fine hairs, bracteate and bibracteolate. *Perianth-segments* 5, thickly coriaceous or chartaceous, connate at the base, ovate- or oblong lanceolate, the inner the narrowest. *Stamens* 5 ; filaments filiform, united at the base into a short cup ; staminodes usually on the rim of the cup between the stamens, rarely (*S. avolans*) 0 ; anthers oblong, 2-celled. *Ovary* ovoid or oblong, not horned, woolly or tomentose; ovule 1, suspended from the apex of an elongate funicle; style about as long as the ovary; stigma capitellate. *Utricle* included in the perianth, membranous, indehiscent. *Seed* inverse, oblong or suborbicular, compressed; embryo annular.

Herbs or undershrubs of various habit; leaves linear or oblong (rarely obovate), sessile, entire, alternate or alternate and opposite on the same plant.

DISTRIB. Species 6, in Tropical and South Africa.

Staminodes present; leaves alternate; flowers in heads
6–8 lin. long ... ... (1) **pungens.**

Staminodes absent; leaves alternate and opposite;
flowers in spikes reaching 2½ in. long ... ... (2) **avolans.**

1. **S. pungens** (Fenzl in Linnæa, xvii. 326); a low much-branched rigid undershrub; branches subterete, pubescent; leaves alternate, subsessile, 3–5 lin. long, 1–1¾ lin. broad, obovate-oblong, thick and fleshy, shortly mucronulate; nerves obscure; heads of flowers terminal, globose or ovate, 6–8 lin. long, 5–6 lin. broad, with much fine pale brown hair; bracteoles unequal, the longer reaching 4 lin. long, spinously aristate; perianth 3¾ lin. long; sepals 5, unequal, the 2 outer longer than the inner, 3-nerved, all spinously aristate and clothed with soft hairs; stamens 5; filaments 1 lin. long, narrowly linear to the base; anthers small, oblong; staminodes small, ovate-oblong; ovary globose, ½ lin. in diam., villous; style up to 1 lin. long, slender; stigma small, capitate; seeds not seen. *Moquin in DC. Prodr.* xiii. ii. 308; *Baker & C. B. Cl. in Dyer, Fl. Trop. Afr.* vi. i. 41. *S. pungens, var. longearistata, Schinz in Engl. Jahrb.* xxi. 181.

WESTERN REGION : Little Namaqualand; hills by the Orange River, near Verleptpram, *Drège*!

Also in Tropical Africa.

2. **S. avolans** (Fenzl in Linnæa, xvii. 328); 1–1¾ ft. high; stems branched; branches glabrous or puberulous; leaves opposite and alternate, sessile or subsessile, 1 in. long, 1½ lin. broad, linear or linear-oblong, acute, shortly mucronate, glabrous, thick, subfleshy; flowers in cylindric obtuse simple terminal spikes 1–2½ in. long, at first pale yellowish, then ash-coloured; partial inflorescences usually of 1 fertile without infertile flowers; bracteoles hyaline, ovate-lanceolate, acute, cuspidately mucronate by the produced midrib; perianth 2 lin. long; sepals subequal or the 2 outer longest, lanceolate, acute, densely woolly; stamens about 1 lin. long; anthers small, ⅓ lin. long, oblong; staminodes 0; ovary 1 lin. long, globosely ovoid, villous; style nearly 1 lin. long; stigma capitate; seed doubled on itself, the folded seed 1 lin. long, ½ lin. broad, oblong, compressed, smooth, red. *Moquin in DC. Prodr.* xiii. ii. 307, *partly. Sericocoma Zeyheri, Engl. in Engl. Jahrb.* x. 6, *partly. S. capensis, Moquin in DC. Prodr.* xiii. ii. 307. *Trichinium Zeyheri, Moquin in DC. Prodr.* xiii. ii. 296. *Cyphocarpa Zeyheri, Lopr. in*

Engl. Jahrb. xxvii. 45.   *Eurotia capensis, E. Meyer ex Moquin* in *DC. Prodr.* xiii. ii. 307.

CENTRAL REGION : Calvinia Div. ; between Lospers Plaats aud Springbok Kuil River, 2000–3000 ft., *Zeyher*, 1439 ! Prince Albert Div. ; between the Dwyka River and Zwartbulletje River, *Drège* ! Graaff Reinet Div. ; near Graaff Reinet, *Bolus* 312 !
KALAHARI REGION : Griqualand West ; St. Clair, Douglas, *Orpen*, 129 !

Much confusion exists regarding *Sericocoma Zeyheri* (*Trichinium Zeyheri*, Moquin) and *Sericocoma avolans*, Fenzl, to both of which species Moquin has referred Zeyher's 1439. In his description of *S. avolans*, Moquin simply copies Fenzl who says that staminodes are present, which is certainly not the case in Zeyher's 1439 at Kew, while in his description of *Trichinium Zeyheri* under the same number, Moquin assigns the absence of staminodes as a reason for placing the plant in the genus *Trichinium* instead of in *Sericocoma*. Engler based his description of *Sericocoma Zeyheri* (Bot. Jahrb. x. 6) upon a plant collected at Kimberley in Griqualand West by Marloth, 785, in which the ovary is horned (and therefore a species of *Cyphocarpa*), and also referred to it *Trichinium Zeyheri*, Moquin, in an authentic specimen of which at Kew (as well as all the others cited above) the ovary is not horned. Lopriore (Engl. Jahrb. xxvii. 45) makes *Cyphocarpa Zeyheri* conspecific with *Trichinium Zeyheri*, Moquin, and therefore accepts the horned ovary described by Engler. I think that *Trichinium Zeyheri* and *Sericocoma avolans* should be regarded as conspecific, under the latter name. This will invalidate one of Lopriore's characters of *Sericocoma* (viz. the presence of staminodes), but according to the Genera Plantarum of Benth. & Hook. f. staminodes in *Sericocoma* are sometimes absent and are said to be so in *S. avolans*. (T. C.)

IX. CENTEMA, Hook. f.

Flowers spicate, 1–2 hermaphrodite, with 1 or more sterile flowers reduced to strong simple spines thickened at the base ; bracts persistent ; bracteoles 2. *Perianth-segments* coriaceous or chartaceous, connate and thickened at the base, ovate-lanceolate, acuminate, 3–5-nerved, the inner the narrower. *Stamens* 5, connected at the base by a membrane and with interposed staminodes, or in one species without staminodes ; anthers 2-celled, linear-oblong. *Ovary* ovoid, attenuated into a slender style ; ovule 1, suspended from an elongate funicle ; stigma obliquely truncate or shortly 2-fid. *Fruit* included in the perianth, membranous, indehiscent, not indurated at the top. *Seed* orbicular or oblong, compressed ; embryo annular.

Erect herbs or undershrubs ; leaves opposite, sessile or shortly petiolate, linear or linear-oblong, entire.

DISTRIB. Species 7, 5 of them in Tropical Africa.

An undershrub 2 ft. high ; staminodes 0   ...   ...   (1) **subfusca**.
A herb 6 in. high ; staminodes short, 2-lobed   ...   (2) **cruciata**.

1. C. subfusca (T. Cooke) ; stem suffruticose, erect. reaching 2 ft. high, glabrous ; branches subterete ; leaves opposite, shortly petiolate, $\frac{7}{8}$–$1\frac{1}{2}$ in. long, $\frac{1}{4}$–$\frac{3}{8}$ in. broad, oblong or subspathulate, subacute, narrowed and subauriculately dilated at the base, entire,

glabrous; petiole very short, $\frac{1}{2}$–1 lin. long; flowers in pedunculate spikes, reaching 3 in. long, $\frac{1}{2}$ in. broad, subobtuse; outer bract 2 lin. long, sessile, broadly ovate, acute, very concave, persistent on the rhachis, keeled; bracteoles 2 lin. long, ovate, acute, strongly keeled, very shortly apiculate, nearly glabrous, each bracteole enclosing 2 hard sharp yellow shining spines about $1\frac{1}{2}$ lin. long, together with much white wool; perianth nearly 4 lin. long; sepals subequal in length, the two outer a little broader than the inner, linear-lanceolate, acute, much indurated at the base, woolly on the back; stamens unequal, exserted; staminodes 0; ovary glabrous, ovoid, $1\frac{1}{4}$ lin. long; style 2 lin. long, slender; stigma shortly 2-fid; seed $1\frac{1}{4}$ lin. long, folded on itself, oblong. *Pupalia subfusca, Moquin in DC. Prodr.* xiii. ii. 332.

EASTERN REGION : Delagoa Bay, *Forbes* ! *Langley*, 5 !

2. **C. cruciata** (Schinz in Engl. Jahrb. xxi. 184); an annual herb about 6 in. high, branched; leaves linear-lanceolate, up to 2 in. long and $1\frac{1}{2}$ lin. wide, acute, aristate, glabrous; flowers in glomerules collected into terminal spikes, each glomerule consisting of 3 fertile and 3 sterile flowers, the latter reduced to spines; bracts lanceolate from an ovate base, concave, aristate, 2 lin. long, glabrous; bracteoles rotundate, $2\frac{1}{2}$ lin. long, about 3 lin. wide, thin; outer perianth-segments ovate-oblong, inner oblong, $2\frac{1}{2}$ lin. long, 1 lin. wide; staminal tube rather large; filaments 2 lin. long; staminodes short, ligulate, 2-lobed; utricle ovoid, pilose; embryo forming a complete ring.

KALAHARI REGION : Transvaal ; Boshveld, *Rehmann*, 5096, between Elands River and Klippan, *Rehmann*, 5100 ! in sandy places near Vaalboschfontein, *Schlechter*, 4227 ! Warmbath, *Miss Leendertz*, 2073 !

X. CYATHULA, Lour.

Flowers 1–2 perfect with others imperfect, in bracteate and bibracteolate spicate or capitate fascicles, the perianth-segments of the imperfect flowers ultimately converted into rigid hooked spines. *Perianth-segments* of perfect flowers 5, scarious, not indurated at the base, subequal or the 3 interior narrower, ovate or oblong-lanceolate, acuminate, or aristate. *Stamens* 5; filaments united at the base by a membrane, with interposed lacerate or linear staminodes; anthers oblong, 2-celled. *Ovary* obovoid; ovule 1, suspended from the apex of an elongate funicle; style filiform; stigma capitellate. *Utricle* closely enclosed in the perianth, areolate at the apex, membranous, indehiscent. *Seed* inverse, oblong; testa thinly coriaceous; embryo peripheric, surrounding farinaceous albumen; cotyledons linear, flat; radicle erect.

Herbs sometimes shrubby at the base, with terete branches ; leaves petiolate, ovate, acuminate, quite entire ; fascicles of flowers small or large, green or white, reflexed after flowering ; bracts ovate, concave, scarious, usually aristate.

DISTRIB. Species about 15, in the warmer parts of Asia, Africa and South America.

Inflorescence elongate ... ... (1) **cylindrica**.

Inflorescence globose :

Leaves ovate, flat ... ... (2) **globulifera**.

Leaves obovate, flat ... ... (3) **natalensis**.

Leaves spathulate, flat :
Perianth-segments oblong ... ... (4) **spathulata**.

Perianth-segments oblong-ovate ... ... (5) **spathulifolia**.

Leaves elliptic, crisped ... ... (6) **crispa**.

1. **C. cylindrica** (Moquin in DC. Prodr. xiii. ii. 328) ; suffruticose ; stems branched, glabrous or nearly so ; leaves ovate, acuminate, shortly mucronulate, 1–2 in. long, $\frac{5}{8}$–$1\frac{1}{4}$ in. broad, rounded or cuneate at the base, more or less hairy on both sides; nerves conspicuous beneath ; petioles $\frac{1}{4}$–$\frac{3}{4}$ in. long, slender ; spikes 1–4 in. long, $\frac{1}{2}$ in. broad, often interrupted near the base, the lower clusters forming opposite subglobose fascicles ; bracteoles of fertile flowers 2 lin. long, 1 lin. broad, ovate, acute, concave, sometimes hooked, scarious, hairy on the upper part of the midrib at the back ; bracteoles of sterile flowers exceeding the perianth, hooked ; perianth of fertile flowers $2\frac{1}{4}$ lin. long ; sepals subequal, the inner 3 very slightly narrower than the 2 outer, oblong-lanceolate; perianth of sterile flowers with 2–4 short hooked rigid spines ; filaments of anthers reaching $\frac{1}{4}$ lin. long ; staminodes shorter than the filaments, subquadrate, laciniate at the apex ; seeds $\frac{3}{4}$ lin. long, rounded at both ends, oblong, light brown. *Lopr. in Engl. Jahrb.* xxvii. 64 *in obs. and* xxx. 28 *in obs.* ; *Schinz in Engl. & Prantl, Pflanzenfam.* iii. 1A, 107 ; *Baker & C. B. Cl. in Dyer, Fl. Trop. Afr.* vi. i. 46. *Pupalia Alopecurus, Fenzl ex Drège, Zwei Pfl. Documente,* 149, 150, 214. *Trichinium latifolium, E. Meyer ex Moquin, l.c.*

SOUTH AFRICA : without locality, *Drège* !
KALAHARI REGION : Orange River Colony, *Cooper,* 3051 ! Transvaal ; hills near Aapies River, *Rehmann,* 4292 !
EASTERN REGION : Pondoland ; between Umtata River and St. Johns River, *Drège* ! Natal ; Isnama, *Sutherland* ! Inanda, *Wood,* 525 ! near Van Reenen, 5000–6000 ft., *Wood,* 5703 ! and without precise locality, *Cooper,* 1143 ! 3052 !

Also in Tropical Africa.

2. **C. globulifera** (Moquin in DC. Prodr. xiii. ii. 329) ; suffruticose, erect ; stem reaching 3 ft. long and as well as the branches obtusely quadrangular, hairy ; leaves petiolate, ovate, acute or subobtuse, $1\frac{1}{2}$–3 in. long, $\frac{3}{4}$–2 in. broad, hairy on both sides, rounded or cuneate at the base ; nerves not prominent ; petioles $\frac{1}{4}$–1 in. long, hairy ; inflorescence in dense globose heads, $\frac{3}{4}$–$1\frac{1}{4}$ in. long, nearly as broad as long ; partial inflorescences usually of 1 narrow fertile

flower with sterile flowers around it, the sepals of the sterile flowers ending in long hooked spines; bracteoles $2\frac{1}{4}$ lin. long, ovate, aristate, sometimes shortly hooked; perianth $2\frac{1}{2}$ lin. long; 2 outer segments lanceolate, aristate, shortly hooked, $2\frac{1}{2}$ lin. long; 3 inner segments 2 lin. long, ovate-lanceolate, apiculate, all pubescent or woolly; filaments $\frac{1}{2}$ lin. long; anthers $\frac{1}{4}$ lin. long, ovoid; staminodes much shorter than the filaments, oblong, laciniate, glabrous; ovary obovoid; style $\frac{3}{4}$ lin. long, slender; seed rather more than $\frac{1}{2}$ lin. long, light brown. *Schinz in Engl. & Prantl, Pflanzenfam.* iii. 1A, 107, *fig.* 61, B ; *Baker & C. B. Cl. in Dyer, Fl. Trop. Afr.* vi. i. 44. *Pupalia holosericea, Fenzl in Drège, Zwei Pfl. Documente,* 51, 56, 152, 158, 214 ; *Moquin in DC. Prodr.* xiii. ii. 329. *Alternanthera lappulacea, Schlechtend. ex Moquin, l.c. Desmochæta uncinulata, Hiern in Cat. Afr. Pl. Welw.* i. 890.

COAST REGION : Albany Div.: near Grahamstown, *Cooper*, 3053 ! Queenstown Div. ; in scrub, Lesseyton Drift, *Galpin*, 2591 ! Shiloh, *Baur*, 793 !
CENTRAL REGION : Graaff Reinet Div. ; at Milk River, Burchell, 2594/2 ! Oude Berg, near Graaff Reinet, 3800 ft., *Drège, Bolus,* 302 ! Aliwal North Div. ; rocky hills near the Orange River, *Burke* ! Kraai River, 4500 ft., *Drège*.
KALAHARI REGION : Orange River Colony ; Bloemfontein, *Rehmann*, 3833 ! Transvaal ; near Lydenburg, *Wilms*, 1261 ! near Pretoria, *Miss Leendertz*, 41 ! *Rehmann*, 4547 !
EASTERN REGION : Pondoland ; between St. Johns River and Umsikaba River, 1000–2000 ft., *Drège.* Griqualand East ; mountains near Matatiele, 5000 ft., *Tyson*, 1619 ! Natal ; Inanda, 1800 ft., *Wood*, 14 ! Umlazi River Heights, under 500 ft., *Drège*, and without precise locality, *Cooper*, 1079 ! *Gerrard*, 1572 !

3. **C. natalensis** (Sond. in Linnæa, xxiii. 97); stem weak, decumbent, several feet long ; branches opposite, sulcate, scabrous ; leaves opposite, obovate or oblong-spathulate, up to $1\frac{1}{2}$ in. long and 10 lin. wide, cuspidate or shortly acuminate, obtuse at the base, adpressed, pilose ; petiole short; heads terminal, solitary or in pairs, globose, 6–8 lin. in diam., woolly ; bracts hyaline, glabrous, shining ; lateral flowers sterile, setaceous, uncinate, golden-yellow, base surrounded by wool ; central flower hermaphrodite ; perianth-segments lanceolate, equal, 3 lin. long, herbaceous with scarious margins, rather woolly outside ; filaments filiform, about half as long as the sepals, hirsute at the base ; anthers oblong, 2-celled ; staminodes much wider than the filaments, truncate, denticulate, hirsute ; style filiform, glabrous, longer than the ovary ; stigma capitate ; utricle glabrous ; seed brownish. *Schinz in Engl. & Prantl, Pflanzenfam.* iii. 1A, 108. *Pupalia natalensis, Sond. l.c.*

EASTERN REGION : Natal ; Durban, *Gueinzius*, 143.

4. **C. spathulata** (Schinz in Bull. Herb. Boiss. iv. 421); an erect plant (probably perennial) clothed with rusty hairs ; leaves spathulate, rotundate or truncate or attenuate at the apex, sessile or with a petiole up to 5 lin. long, $1\frac{1}{2}$ in. long, 9–14 lin. wide, hairy on the nerves and margin ; flower-clusters formed of 1 fertile and several sterile flowers, the latter reduced to hooked spines ; bracts broadly

ovate, villous at the base; perianth-segments oblong, acute, 3 lin. long, about ½ lin. wide, dark green edged with white, hairy at the base, glabrous elsewhere; filaments nearly 2 lin. long; staminodes truncate, fimbriate, about a third as long as the filaments; ovary truncate; style filiform, 1½-2 lin. long.

EASTERN REGION : Delagoa Bay, *Kuntze*, 204, ex *Schinz*.

5. **C. spathulifolia** (Lopr. in Engl. Jahrb. xxvii. 54); stem erect, slender, subwoody, pubescent or tomentose ; leaves spathulate, acute or shortly acuminate, 1-2 in. long, ⅜-1¼ in. broad, running down into a short petiole, hairy on both sides; nerves slightly prominent beneath ; inflorescence capituliform, terminal or lateral, subspherical ; partial inflorescences of 1 solitary fertile and 4 sterile flowers ; bracts hyaline, ovate, concave, acute, aristate ; bracteoles hooked ; perianth 2½ lin. long ; sepals subequal, oblong-ovate, the 2 outer slightly broader than the 3 inner, 3-nerved, all acute or minutely apiculate, softly woolly ; sterile flowers converted into hooked spines with woolly bases ; filaments of anthers subulate ; anthers oblong; staminodes subquadrate, shorter than the filaments; ovary ¾ lin. long, obovoid ; style reaching 1¼ lin. long, very slender ; stigma minute ; ripe utricle 1½-1¾ lin. long, ellipsoid, compressed, areolate at the apex, glabrous ; seed 1¼ lin. long, ellipsoid, rounded at both ends, light brown. *Lopr. in Malpighia*, xiv. 444.

EASTERN REGION : Natal; Umzinyati Falls, *Wood*, 1323 ! Lorenzo Marquez, 150 ft., *Schlechter*, 11640 !

This may be conspecific with *C. spathulata*, Schinz.

6. **C. crispa** (Schinz in Engl. Jahrb. xxi. 188) ; a low shrub, at first hairy, finally more or less glabrous ; leaves elliptic, 5 lin. long, 2 lin. wide, crisped, sessile or shortly petioled ; spikes globose, stalked or sessile, formed of 3-flowered clusters ; sterile flowers none ; bracts and bracteoles lanceolate, with a long awn, hairy on the centre of the back in the lower half ; perianth-segments oblong-lanceolate, acute, glabrous, 4 lin. long, 1 lin. wide ; staminal tube 1 lin. long ; filaments 1½ lin. long ; staminodes dentate or shortly fimbriate ; utricle ovoid, glabrous ; style 2 lin. long ; stigma capitate.

KALAHARI REGION : Transvaal; at Streyd Poort, in the Makapansberg Range, *Rehmann*, 5420 !

This species may on account of the absence of sterile flowers belong to some other genus.

XI. PUPALIA, Juss.

Flowers in fascicles spicately arranged, central flower perfect, the others reduced to stellate bunches of hooked spines. *Perianth* 5-partite, not hardened at the base ; segments subequal, acuminate,

3–5-nerved. *Stamens* 5 ; filaments subulate, very shortly connate at the base, without alternating staminodes ; anthers 2-celled, didymous. *Ovary* ovoid, tapering upwards into a slender style ; stigma capitellate. *Ovule* solitary, pendulous from a long funicle. *Utricle* enclosed in the perianth, ovoid, compressed, membranous, indehiscent. *Seed* lenticular ; testa thinly crustaceous ; aril none ; embryo surrounding the floury albumen ; cotyledons linear, flat.

Herbs or subshrubs, tomentose or nearly glabrous, trichotomously branched ; leaves opposite, petiolate, broad, quite entire ; spikes simple or branched ; flowers green ; bracts scarious.

DISTRIB. Species about 6, in Tropical Africa, the Mascarene Islands and Tropical Asia.

Leaves tomentose on both surfaces ; awns of barren
    flowers yellow ...   . (1) **lappacea.**
Leaves glabrous or sparingly hairy ; awns of barren
    flowers purple ...   . (2) **atropurpurea.**

1. **P. lappacea** (Juss. in Ann. Mus. Par. ii. [1803] 132) ; a large straggling undershrub ; branches terete, tomentose ; leaves 1¼–4 in. long, ½–2 in. wide, elliptic or ovate, acute or acuminate, rounded or shortly cuneate at the base, finely apiculate, tomentose on both surfaces, ciliate ; main nerves conspicuous beneath ; petioles ₁/₁₀–¼ in. long ; flowers in close or distant clusters arranged in terminal spikes 4–10 in. long ; rhachis tomentose ; bracts ovate, acuminate, pungent ; villous, ⅛–⅙ in. long ; bracteoles ovate-oblong, apiculate, concave, ⅛ in. long, ⅛ in. wide ; sterile flowers subtended by a small lanceolate bract, reduced to bunches of unequal stellately spreading hooked awns, woolly below, enlarged and yellow in fruit ; perianth-segments ⅙ in. long, lanceolate, aristate, 3-nerved, densely clothed with white wool ; utricle membranous, very thin, oblong, suddenly and shortly tapering at the apex into the long persistent style ; seed ₁/₁₀ in. long, ₁/₁₂ in. wide, ellipsoid, compressed, smooth, shining, black. *Moquin in DC. Prodr.* xiii. ii. 331 ; *A. Rich. Tent. Fl. Abyss.* ii. 217 ; *Hook. Niger Fl.* 494 ; *Garcke in Peters, Reise Mossamb. Bot.* 504 ; *Aschers. in Schweinf. Beitr. Fl. Aethiop.* 180 ; *Schinz in Engl. & Prantl, Pflanzenfam.* iii. 1A, 93, *figs. E, H, and* 108 ; *in Bull. Herb. Boiss.* iv. *Append.* ii. 164, v. *Append.* iii. 65, *and in Engl. Pfl. Ost-Afr. C.* 173 ; *Engl. Hochgebirgsfl. Trop. Afr.* 207 ; *Durand & De Wild. in Comptes-rendus Soc. Bot. Belg.* xxxvi. 85 ; *Hook. f. Fl. Brit. Ind.* iv. 724 ; *Baker & C. B. Cl. in Dyer, Trop. Afr.* vi. i. 47. *P. styracifolia, Juss. ex Steud. Nomencl. ed.* i. 269, 669. *Achyranthes lappacea, Linn. Sp. Pl. ed.* i. 204. *Desmochæta flavescens, DC. Cat. Hort. Monspel.* 1813, 102. *Pupal lappacea, Hiern in Cat. Afr. Pl. Welw.* i. 891. *Pupal-Valli, Rheede, Hort. Malab.* vii. 81, *t.* 43.

COAST REGION : Uitenhage Div. ; Enon, *Prior* ! Albany Div. ; on the rocks of Zwartwater Poort, *Burchell*, 3398 !
CENTRAL REGION : Graaff Reinet Div. ; near Graaff Reinet, 2600 ft., *Bolus*, 562!

KALAHARI REGION: Griqualand West; on the Asbestos Mountains and on plains at their foot between Witte Water and Griquatown, *Burchell,* 1972! 2053/¹! 2088! Transvaal; between Delagoa Bay and Pretoria, *Bolus,* 9752! Wonderboom Poort, 4550 ft., *Rehmann,* 4546! *Miss Leendertz,* 614! Warmbath, *Miss Leendertz,* 2089! Fourteen Streams, *Burtt-Davy,* 1594! Klippan, *Rehmann,* 5298!
EASTERN REGION: Natal; in "Thorns" near Weenen, 300–4000 ft., *Wood,* 4480!

Also in Tropical Africa and Asia.

2. **P. atropurpurea** (Moquin in DC. Prodr. xiii. ii. 331); a biennial or annual herb, 2–5 ft. high; branches long, straggling, cylindric, glabrous or pubescent, slightly ribbed, often tinged with purple; leaves 1–4 in. long, $\frac{3}{8}$–2 in. wide, ovate or elliptic-lanceolate, acuminate, mucronate, glabrous or with scattered hairs, rounded or cuneate at the base; petiole up to $\frac{1}{2}$ in. long; flowers in sessile clusters in lax pedunculate terminal spikes 1–10 in. long, the imperfect ones reduced to hooked purple awns; bracts broadly ovate, pungent, persistent, $\frac{1}{10}$–$\frac{1}{8}$ in. long; bracteoles similar to the bracts; perianth-segments ovate-lanceolate, aristate, clothed with cottony wool, 3-nerved; utricle thinly membranous; seed oblong-ellipsoid, $\frac{1}{10}$ in. long, $\frac{1}{12}$ in. wide. *A. Rich. Tent. Fl. Abyss.* ii. 218; *Hook. Niger Fl.* 494; *Hook. f. Fl. Brit. Ind.* iv. 723; *Baker & C. B. Cl. in Dyer, Fl. Trop. Afr.* vi. i. 48; *Schinz in Engl. & Prantl, Pflanzenfam.* iii. 1A, 93, *fig. F,* and 108, *and in Engl. Pfl. Ost-Afr. C.* 173. *Achyranthes atropurpurea, Lam. Encycl.* i. 546. *Desmochæta atropurpurea, DC. Cat. Hort. Monsp.* 1813, 102; *Drège, Zwei Pfl. Documente,* 127, 133, 159.

COAST REGION: Uitenhage Div.; in woods at Galgebosch, under 1000 ft., *Drège,* near Strand Fontein and Matjes Fontein, ·under 500 ft., *Drège;* Van Stadens River, under 200 ft., *Drège;* Enon, under 500 ft., *Drège,* and without precise locality, *Zeyher,* 554!
EASTERN REGION: Pondoland; St. Johns River, *Drège!* Natal; near Durban, *Peddie! Rehmann,* 8744! coastland, *Sutherland!* and without precise locality, *Gerrard,* 546!

XII. PSILOTRICHUM, Blume.

Flowers hermaphrodite, bracteate and bibracteolate. *Perianth* chaffy, 5-partite, persistent and becoming hard; segments linear or ovate-oblong, concave, sometimes gibbous at the base, strongly nerved, usually villous outside, glabrous within. *Stamens* 5; filaments unequal, linear or subulate, united at the base into a cup; staminodes 0; anthers short or long, 2-celled. *Ovary* subglobose or oblong; style slender; stigma capitate or bifid; ovule suspended from a long basal funicle. *Utricle* membranous, indehiscent, enclosed in the hardened base of the perianth. *Seed* lenticular; testa coriaceous or crustaceous; aril 0; embryo surrounding the floury albumen; cotyledons rather flat.

Herbs or shrubs, trichotomously branched, glabrous to woolly; branches terete; leaves opposite, ovate or elliptic-lanceolate, quite entire; flowers in spikes or heads, solitary and axillary or paniculately arranged ; flowers white or greenish ; bracts or bracteoles small, hyaline.

DISTRIB. Species about 20 (some imperfectly known), chiefly in Tropical Asia and Africa, a few in the Sandwich Islands.

1. **P. africanum** (Oliv. in Hook. Ic. Pl. t. 1542) ; a branched undershrub ; branches puberulous or glabrous, terete ; leaves opposite, ovate or ovate-elliptic, $\frac{3}{4}$–2 in. long, $\frac{1}{2}$–1 in. wide, acute, usually cuneate at the base, with scattered hairs on both surfaces or glabrous ; petiole 2–3 lin. long, pubescent ; spikes up to 4 lin. long, of few clusters, axillary and terminal on short peduncles ; bracts ovate, acuminate, 1 lin. long, densely silky ; bracteoles 2, broadly ovate, concave, glabrous, midrib stout and excurrent ; perianth-segments about 2 lin. long, the 2 outer slightly larger than the inner, lanceolate, silky outside ; staminal cup $\frac{1}{4}$ lin. long ; filaments subulate, 1 lin. long ; anthers subglobose ; ovary oblong, tapering into a short subulate style ; stigma capitate. *Oliv. in Trans. Linn. Soc. ser. 2, Bot. ii. 348 ; Schinz in Engl. & Prantl, Pflanzenfam. iii. 1A, 111 and 112, fig. 65, and in Engl. Pfl. Ost-Afr. C. 173 ; C. B. Cl. in Dyer, Fl. Trop. Afr. vi. i. 58.*

EASTERN REGION : Natal, *Gerrard*, 594 ! Zululand ; Indulendi, 1000–2000 ft., *Wood*, 3956 ! Portuguese East Africa ; Ressano Garcia, 1000 ft., *Schlechter*, 11880 !

Also in Tropical Africa.

XIII. ÆRVA, Forsk.

Flowers hermaphrodite, polygamous or diœcious, bracteate and 2-bracteolate, small, in dense cylindric terminal and axillary solitary or paniculate spikes. *Perianth* usually of 5 segments, not indurated at the base ; segments equal or the 3 interior narrower, oblong or lanceolate, all or the 3 interior only softly woolly. *Stamens* usually 5 ; filaments usually unequal, linear-subulate, united at the base into a cup with interposed staminodes ; anthers 2-celled. *Ovary* 1-celled ; ovule solitary, suspended from the apex of an elongate funicle ; style short or long ; stigmas 2 or stigma capitellate. *Utricle* enclosed in the perianth, membranous. *Seed* ovoid or reniform, compressed ; testa thinly coriaceous ; embryo peripheric, surrounding farinaceous albumen ; radicle superior.

Woolly herbs or undershrubs ; leaves usually alternate, entire, flat ; flowers minute, white or ferruginous.

DISTRIB. Species 10, in the warmer parts of Asia and Africa.

Spikes running out into terminal leafless panicles ;
  perianth up to 1¼ lin. long ... ... (1) **leucura**.
Spikes not running out into terminal panicles ; perianth
  less than 1 lin. long ... ... (2) **lanata**.

1. **Æ. leucura** (Moquin in DC. Prodr. xiii. ii. 302) ; suffruticose, reaching 2 ft. high ; stem erect, simple or branched, striatulate, pubescent (often in age glabrate) ; leaves alternate, oblong or oblong-obovate, acute or subobtuse, shortly mucronulate, pubescent and pale green above, ashy-pubescent or tomentose beneath, ¾–1½ in. long, ½–¾ in. broad ; nerves not conspicuous ; petiole 2–3 lin. long ; flowers in dense white, softly woolly, cylindric or ovate-conic spikes, varying in length from ½–2 in., ⅛–⅜ in. broad, the uppermost spikes forming a panicle ; bracteoles less than 1 lin. long, ovate, finely aristate ; perianth 1¼ lin. long ; outer 2 sepals lanceolate, shortly and finely aristate ; inner sepals narrower, acute, all densely woolly ; staminodes a little shorter than the filaments, subulate, acute ; style ¼ lin. long ; stigma 2, very short. *Sonder in Linnæa*, xxiii. 96 ; *Baker & C. B. Cl. in Dyer, Fl. Trop. Afr.* vi. i. 39. *Æ. ambigua, Moquin in DC. Prodr.* xiii. ii. 302.

SOUTH AFRICA : without locality, *Zeyher*, 1441 !
KALAHARI REGION : Griqualand West ; Leeuwenkuil Valley, near Griqua Town, *Burchell*, 1892 ! Klipfontein, *Burchell*, 2620 ! Dutoits Pan and Klip Drift, *Tuck*, 9 ! Orange River Colony ; Modder River, *Mrs. Barber*, 15 ! Bechuanaland ; Batlapin Territory, *Holub* ! Transvaal ; near Pretoria, 5200 ft., *Schlechter*, 4146 ! *McLea in Herb. Bolus*, 3132 ! 5782 ! Magalies Berg, *Burke* ! Rimers Creek, near Barberton, *Thorncroft*, 410 ! *Galpin*, 912 ! Klerksdorp, *Nelson*, 222 ! Vaal River, *Burke* ! near Lydenburg, *Wilms*, 1253 ; Springbok Flats, *Burtt-Davy*, 2344 ! Warm Bath, *Miss Leendertz*, 2007 ! Potgieters Rust, *Miss Leendertz*, 1133 ! Houtbosch, *Rehmann*, 5972 !

Also in Tropical Africa.

2. **Æ. lanata** (Juss. in Ann. Mus. Par. ii. [1803] 131) ; 1½–3 ft. high ; stems erect or ascending, suffruticose below, branched, terete, striatulate, pubescent or woolly-tomentose ; leaves alternate, ¾–1¼ in. long, ⅜–⅝ in. wide on the main stem, smaller on the branches ; elliptic or obovate or suborbicular, obtuse or acute, entire, pubescent above, more or less white-woolly (especially when young) beneath, very shortly or not mucronulate ; nerves beneath scarcely conspicuous ; petioles 2–4 lin. long ; flowers white or greenish, in small dense axillary woolly heads or spikes ¼–½ in. long, often crowded and forming globose clusters, the upper spikes not running into leafless terminal panicles ; bracteoles less than ½ lin. long, ovate, concave, apiculate, hyaline ; perianth ¾ lin. long, woolly ; sepals subequal, oblong, the 2 outer very shortly and finely apiculate, all woolly ; stamens united into a tube with interposed staminodes shorter than the filaments ; ovary subglobose ; style about ⅛ lin. long ; stigmas 2, minute ; seeds rather more than ¼ lin. long, slightly broader than long, subreniform, black, smooth, shining. *Moquin in DC. Prodr.* xiii. ii. 303 ; *Sonder in Linnæa*, xxiii. 96 ; *Baker & C. B. Cl. in Dyer, Fl. Trop. Afr.* vi. i. 39 ; *A. Rich. Tent. Fl. Abyss.* ii. 214 ; *Oliver in Trans. Linn. Soc.* xxix. 141 ; *Hook. f. Fl. Brit. Ind.* iv. 728. *Achyranthes lanata, Linn. Sp. Pl. ed.* i. 204.

SOUTH AFRICA : without locality, *Drège* !
KALAHARI REGION : Transvaal ; Scheer Poort, *Miss Leendertz*, 131a !
EASTERN REGION : Natal ; around Durban Bay, *Krauss*, 298 (by error 198)!
near Durban, *Wood*, 200 ! Inanda, *Wood*, 629.
Also in Tropical Africa, India and Malaya.

XIV. CALICOREMA, Hook. f.

Flowers subcapitate or spicate, hermaphrodite, bracteate and
2-bracteolate. *Perianth* coriaceous, 5-partite, not or moderately
indurated at the base ; segments clothed on the back and margins
with straight silky-white hairs not longer than the sepals, the
2 exterior sepals oblong-lanceolate muticous 3-nerved, the 3 interior
narrower. *Stamens* 5 ; filaments filiform, united at the base by a
membranous tube and with interposed short broad erose staminodes ;
anthers 2-celled. *Ovary* ovoid, glabrous, attenuated into an elongated
slender style ; ovule suspended from the apex of an elongate funicle ;
stigma capitellate.

A rigid branched shrub ; branches robust, terete, woody ; leaves scattered,
small, narrow, cylindric, obtuse, glabrous, fleshy, sulcate above, narrower than the
branchlets ; flowers subspicately arranged at the apices of branches, solitary or
fasciculate ; rhachis of the inflorescence robust ; bracts and bracteoles short boat-
shaped hyaline, much shorter than the perianth.

DISTRIB. Species 1 in Tropical and South Africa.

1. **C. capitata** (Hook. f. in Benth. et Hook. f. Gen. Pl. iii. 35) ;
a short much-branched undershrub ; branches alternate, terete,
thick, ashy-grey, the younger tomentose ; leaves all alternate, few,
scattered, cylindric, obtuse, up to ¾ in. long, about ½ lin. broad,
glabrous ; nerves not conspicuous ; heads of flowers reaching ½ in.
long and rather broader than long ; bracteoles 2½ lin. long, broadly
ovate, obtuse, concave, glabrous, hyaline ; perianth reaching 5 lin.
long, silky-hairy ; 2 outer sepals longer and broader than the inner,
all lanceolate, acute, 3-nerved ; filaments rather more than 1 lin.
long ; staminodes very short, ovate, obtuse ; anthers more than ½ lin.
long, oblong ; ovary glabrous ; style 2 lin. long ; stigma capitellate.
Sericocoma capitata, Moquin in DC. Prodr. xiii. ii. 308 ; *Baker &
C. B. Cl. in Dyer, Fl. Trop. Afr.* vi. i. 42.

SOUTH AFRICA : without locality, *Drège*, 2914 !
WESTERN REGION : Little Namaqualand ; banks of the Orange River, *Schlechter*,
11473 ! Little Bushmanland ; Naroep, *Schlechter* !
Also in Tropical Africa.

XV. ACHYRANTHES, Linn.

Flowers hermaphrodite, deflexed when old ; bracts and bracteoles
spinescent. *Perianth-segments* 4-5, aristate, becoming hardened and
ribbed. *Stamens* 2-5 ; filaments subulate, connate at the membra-

nous base ; anthers 2-celled ; staminodes toothed or with a toothed scale on the back. *Ovary oblong, slightly compressed,* 1-celled ; *ovule solitary, pendulous from a long basal funicle ; style filiform ; stigma capitellate.*

Herbs ; leaves opposite, entire, petiolate ; flowers in slender simple or panicled spikes.

DISTRIB. Species about 15, in the warm parts of the Old World.

Leaves pubescent ; bracts 1½ lin. long   ...   . (1) **aspera**
Leaves densely velvety ; bracts 3 lin. long   ...   ... (2) **robusta.**

1. **A. aspera** (Linn. Sp. Pl. ed. i. 204) ; an erect hairy branched herb, 1–4 ft. high ; branches terete or obsoletely quadrangular, striate ; leaves elliptic or obovate, obtuse or subacute, pubescent on both surfaces, 1½–3 in. long, 1–1¾ in. wide ; petiole 3–9 lin. long ; spikes in flower 2–4 in. long, in fruit lengthening to 20 in. ; bracts 1½ in. long, broadly ovate, acuminate, aristate, membranous, persistent ; bracteoles as long as the bracts, broadly ovate, spinescent, becoming hard and falling off with the fruit ; perianth-segments subequal, 2–3 lin. long, ovate-oblong, acute, membranous and white on the margins ; stamens 5 ; staminodes quadrate, fimbriate ; ovary depressed, obovoid, granular at the top ; utricle oblong, truncate, thinly membranous, about 1 lin. long, enclosed in the hardened persistent perianth and bracteoles, smooth, brown ; seed ellipsoid ; embryo curved. *Moquin in DC. Prodr.* xiii. ii. 314 ; *Benth. in Hook. Niger Fl.* 493 ; *Boiss. Fl. Orient.* iv. 993 ; *Schinz in Engl. & Prantl, Pflanzenfam.* iii. 1A, 112, *and* 94, *fig.* 47, D, E, *in Bull. Herb. Boiss.* iv. *Append.* ii. 165, v. *Append.* ii. 66, *and in Engl. Pfl. Ost-Afr. C.* 173 ; *Lopr. in Engl. Jahrb.* xxx. 12, *t.* 1, *figs.* G–J ; *Gilg in Baum, Kunene-Samb. Exped.* 232, 433 ; *Wight, Ic. t.* 1777 ; *Hook. f. Fl. Brit. Ind.* iv. 730 ; *Baker & C. B. Cl. in Dyer, Fl. Trop. Afr.* vi. i. 63.

SOUTH AFRICA : without locality, *Sieber* ! *Wallich* ! *Drège* !
COAST REGION : Cape Div. ; Table Mountain, *Milne,* 212 ! *Prior* ! *Bolus,* 4007 ! *Alexander* ! Groot Schuur, *Wolley-Dod,* 567 ! near Cape Town, 500 ft., *Bolus,* 2913 ! Rondebosch, *Pappe* ! Uitenhage Div. ; near Uitenhage, *Prior* ! near Sand Fontein and Matjes Fontein, *Drège* ! British Kaffraria, *Cooper,* 3050 ! and without precise locality, *Zeyher,* 555 !
CENTRAL REGION : Somerset Div. ; at the foot of the Bosch Berg, *MacOwan,* 1522 ! Albert Div., *Cooper,* 1352 !
KALAHARI REGION : Griqualand West ; in Leeuwenkuil Valley, at Griquatown, *Burchell,* 1894 ! Orange River Colony ; Mudriver Drift, *Rehmann,* 3568 ! Vet River, *Burke* ! Transvaal ; near Lydenburg, *Wilms,* 1263 ! Pretoria, *Miss Leendertz,* 60 ! Rooiplaat, *Miss Leendertz,* 798 ! 799 ! hill sides near Barberton, 2800 ft., *Galpin,* 920 ! Potgieters Rust, *Miss Leendertz,* 1950 !
EASTERN REGION : Natal ; near Durban, *Gerrard,* 492 ! *Wood,* 7203 ! and without precise locality, *Cooper,* 1162 ! Griqualand East ; Clydesdale 2500 ft., *Tyson,* 2681 ! and without precise locality, *Tyson* !
A weed in the hotter parts of the Old World.

2. **A. robusta** (C. H. Wright) ; a robust herb ; branches ribbed, densely pubescent ; nodes swollen ; leaves broadly ovate or almost orbicular, 3 in. long, 2¼ in. wide, densely velvety on both surfaces,

thick ; main nerves prominent on the under surface; petiole rigid, up to 9 lin. long, channelled above, convex beneath ; spikes terminal, up to 10 in. long; rhachis white-woolly ; bracts 3 lin. long, lanceolate, acuminate, scarious, woolly at the base and on .the margins, midrib strong ; bracteoles 1½ lin. long, ovate and scarious at the base, with a very strong long-excurrent midrib, white-woolly ; perianth-segments 2 lin. long, the 2 outer slightly larger than the inner, broadly lanceolate, acuminate, glabrous ; stamens 5; filaments subulate ; anthers oblong; staminodes ½ as long as the filaments, quadrate, ciliate along the top; ovary globose; style filiform, as long as the perianth.

KALAHARI REGION : Transvaal ; Batloaka Kraals, *Nelson,* 408 !
EASTERN REGION : Natal ; near Durban, 100 ft., *Wood,* 7202 ! *Peddie!* Inanda, 1800 ft., *Wood,* 4 ! and without precise locality, *Gerrard,* 544 !

Imperfectly known species.

3. A. acuminata (E. Meyer in Drège, Zwei Pfl. Documente, 159) ; resembling *A. aspera,* L., but differing in the leaves being up to 4 in. long and 3 in. wide, subacuminate, green and subglabrous; the flowers reflexed and the perianth ⅓ longer than the bracts. *Sonder in Linnæa,* xxiii. 96.

EASTERN REGION : Natal ; Durban, *Drège.*

4. A. frumentacea (Burm. f. Fl. Cap. Prodr. 7) ; stems branched ; spikes slender, corn-bearing.

SOUTH AFRICA : without locality.

5. A. hamosa (Burchell, Trav. S. Afr. i. 308, without description).

CENTRAL REGION : Prieska Div. ; at Keikams (Modder Gat) Poort, *Burchell,* 1621/².

The name was given by Burchell to a plant raised in his garden at Fulham from seed he collected "in the pass through the mountains near Modder Gat," of which no specimen appears to have been preserved.

6. A. verticillata (Thunb. in Hoffm. Phytogr. Blaetter, i. 1803, 26) ; stem shrubby, hirsute ; leaves elliptic, rather glabrous, verticillate ; bracts scarious, white. *Moquin in DC. Prodr.* xiii. ii. 318.

SOUTH AFRICA : without locality, *Thunberg.*

This species is not included in Thunberg's *Flora Capensis.*

XVI. ACHYROPSIS, Hook. f.

Flowers hermaphrodite. *Perianth* 4–5-partite, glabrous, not becoming hardened at the base ; segments oblong, subacute, shining. *Stamens* 4–5 ; filaments subulate, connected by a basal membrane ; staminodes quadrate, sometimes fimbriate ; anthers with 2 globose cells. *Ovary* ovoid or oblong, compressed ; ovule solitary, suspended from a long basal funicle. *Utricle* enclosed by the perianth, membranous, indehiscent. *Seed* ovoid or lenticular ; testa brownish, thinly coriaceous ; aril 0 ; embryo surrounding the fleshy albumen ; cotyledons flat.

Erect trichotomously branched undershrubs ; leaves opposite or fascicled, narrow, quite entire ; nerves inconspicuous ; flowers small, white, arranged in axillary spikes or terminal panicles ; bracts membranous ; bracteoles in pairs.

DISTRIB. Species 2, one extending into Tropical Africa.

Flowers 5-merous ; perianth 1½ lin. long ; staminodes
    fringed ... ... (1) **avicularis.**

Flowers usually 4-merous ; perianth 1 lin. long ; staminodes
    not fringed ... ... (2) **leptostachya.**

1. A. avicularis (Hook. f. in Benth. et Hook. f. Gen. Pl. iii. 36, by error *acicularis*) ; an erect undershrub about 1½ ft. high with rather short branches ; stem ribbed, glabrous ; leaves subsessile, narrowly oblong, very shortly mucronate, 6–9 lin. long, about 2 lin. wide, rather thick, green and sparingly pubescent above, paler and slightly villous below ; spikes ½–1½ in. long, conical or oblong, acute ; bracts ovate, entire, obtuse, mucronate, ½ lin. long ; perianth-segments 5, shining, lanceolate, concave, 1½ lin. long ; stamens 5 ; filaments subulate ; staminodes quadrate, fringed along the top ; anthers oblong ; utricle oblong, glabrous, green ; seed ovoid, black, shining. *Achyranthes avicularis, E. Meyer in Drège, Zwei Pfl. Documente*, 159 ; *Moquin in DC. Prodr.* xiii. ii. 311.

EASTERN REGION : Natal ; near Durban, *Drège* !

2. A. leptostachya (Hook. f. in Benth. et Hook. f. Gen. Pl. iii. 36) ; a branched undershrub 1–1½ ft. high ; stem terete, or tetragonous in the upper part, ribbed ; branches pubescent, ascending ; leaves subsessile or shortly petiolate, narrowly oblong, obtuse, very shortly mucronate, up to 1½ lin. long and 5 lin. wide (usually much smaller), sparingly pubescent above, whitish villous beneath ; midrib prominent beneath ; spikes up to 1¼ lin. long, on slender rigid silky peduncles nearly as long as the spikes ; bracts very short, ovate, entire, shortly mucronate, shining ; perianth-segments 4, elliptic, concave, obtuse, glabrous, 1 lin. long ; stamens 4 ; filaments subulate ; staminodes quadrate, not fringed ; anthers ovate ; utricle subglobose, green ; seed nearly lenticular, black. *Baker & C. B. Cl. in Dyer, Fl. Trop. Afr.* vi. i. 66. *A. alba, Hook. f. l.c. Achyranthes alba, Eckl. & Zeyh. ex Moquin in DC. Prodr.* xiii. ii. 311 ; *Sonder in Linnæa*, xxiii. 96. *A. leptostachya, E. Meyer ex Meisn. in Hook. Lond. Journ. Bot.* ii. (1843), 548, *by error* 448. *Psilotrichum densiflorum, Lopr. in Malpighia*, xiv. 453. *Paronychia capensis, Spreng. in Sonder in Linnæa*, xxiii. 96.

COAST REGION : Uitenhage Div. ; by the Zwartkops River, *Zeyher*, 56 ! 3612 ! *Prior* ! Bathurst Div. ; near Port Alfred, *Burchell*, 4031 ! *Galpin*, 2965 ! Komgha Div. ; banks of the Kei River, under 500 ft , *Drège* ! British Kaffraria, *Cooper*, 3049 bis !

KALAHARI REGION : Transvaal ; Linokana, *Holub* ! Streyd Poort, *Rehmann*, 5719 ! Lydenburg, *Wilms*, 1256 !
EASTERN REGION : Pondoland ; among scrub, Isunka, Port St. John 100 ft., *Galpin*, 2865 ! Griqualand East ; near Clydesdale, *Tyson*, 2780 ! and in *Bolus & MacOwan, Herb. Austr.-Afr.*, 1225 ! Natal ; Durban, *Grant* ! coast-land, *Sutherland* ! and without precise locality, *Gerrard*, 604 !

Also in Tropical Africa.

XVII. TELANTHERA, Moquin.

Flowers hermaphrodite, subtended by a bract and 2 bracteoles. *Perianth-segments* 5, equal or unequal, erect, glabrous or villous. *Stamens* 5 ; filaments filiform above, united into a tube below ; staminodes long ligulate, laciniate at the apex ; anthers oblong, 1-celled. *Style* short ; stigma capitate. *Utricle* obovoid, enclosed by the persistent perianth. *Seed* lenticular or oblong ; embryo annular, surrounding the floury albumen.

Erect or decumbent herbs or undershrubs, usually much-branched and hairy ; leaves opposite ; flowers in terminal or axillary clusters.

DISTRIB. Species about 50, chiefly on the shores of South America.

1. **T. maritima,** var. **Sparmanni** (Moquin in DC. Prodr. xiii. ii. 365) ; a somewhat pilose decumbent plant ; stems much-branched, rooting, slightly swollen at the nodes ; leaves rather distant, suborbicular or orbicular, very obtuse, entire ; heads axillary, 4–5 lin. long, consisting of 4–5 flower-clusters ; bracts orbicular-ovate, mucronate, strongly keeled ; bracteoles slightly longer than the bracts, ovate, acuminate ; perianth-segments 2 lin. long, the outer ovate-lanceolate, the inner rather narrower ; staminodes a little longer than the filaments, 3–4-laciniate at the apex, entire at the margins ; anthers oblong ; utricle shorter than the perianth, obovoid, glabrous ; seed lenticular, obtuse at the margin, shining blackish. *T. maritima, Harv. Gen. S. Afr. Pl. ed.* 2, 319.

SOUTH AFRICA : without locality, *Sparmann.*

Also in Senegambia, the type on the shores of Tropical Africa and South America.

XVIII. ALTERNANTHERA, Forsk.

Flowers hermaphrodite, small, white, bracteate and bibracteolate. *Perianth-segments* 5, unequal, the 2 innermost concave. *Stamens* 2–5 ; filaments concave at the base ; anthers 1-celled. *Staminodes* long or short. *Ovary* obovoid or obcordate ; ovule solitary, pendulous from a long basal funicle ; style very short ; stigma capitellate. *Utricle* compressed, sometimes with thickened or winged margins ; cotyledons narrow ; radicle superior.

Herbaceous or slightly woody; leaves opposite; flowers in axillary, often clustered heads.

DISTRIB. Species about 16, chiefly in Australia and Tropical America; 3 in Tropical Africa.

Woody; outer perianth-segments spiny ...   . (1) **Achyrantha.**
Herbaceous; outer perianth-segments not spiny   ... (2) **sessilis.**

1. **A. Achyrantha** (R. Br. Prodr. i. 417); stems slightly woody, procumbent, 1–2 ft. long, terete, pilose, much-branched; leaves ovate or obovate, very obtuse, mucronate, tapering into the petiole, entire, $\frac{1}{2}$–$1\frac{1}{2}$ in. long, 5 lin. wide, glabrous above, softly pilose beneath; heads in axillary and subterminal clusters of 2–3, ovoid, about 5 lin. in diam., spiny; bracts about 3 lin. long, lanceolate, pungent, finely denticulate in the upper part; bracteoles slightly shorter and narrower; perianth-segments slightly unequal, about 2 lin. long, the outer lanceolate-subulate, finely denticulate near the apex, the anticous elliptic, laciniate and shortly mucronate at the apex, the 2 inner smaller, concave, bearded on the back; filaments filiform; staminodes shorter than the filaments, triangular, acute, entire; utricle about $\frac{1}{3}$ as long as the perianth, truncate or slightly bidentate at the apex; seed ovoid, compressed. *Moquin in DC. Prodr.* xiii. ii. 358. *A. echinata, Sm. in Rees, Cyclop. Suppl. n.* 10 (1819); *Moquin in l.c.* 360; *Baker & C. B. Cl. in Dyer, Fl. Trop. Afr.* vi. i. 74; *Seubert in Mart. Fl. Bras.* v. i. 183, *t.* 55. *A. repens, Steud. Nomencl. Bot. ed.* 2, i. 65. *Achyrantha repens, Linn. Sp. Pl. ed.* i. 205, *not of Forsk. Illecebrum Achyrantha, Linn. Sp. Pl. ed.* ii. 299.

KALAHARI REGION: Griqualand West; Barkly West, Patons Farm, "Newlands," *MacOwan,* 3396! Transvaal; Daspoort, *Miss Leendertz,* 585!
EASTERN REGION: Tembuland; Umtata, *Bolus,* 8305!

A native of Tropical South America, but now becoming a troublesome weed in Tropical and South Africa.

2. **A. sessilis** (R. Br. Prodr. i. 417); stem herbaceous, creeping, jointed, slightly pubescent; leaves ovate-lanceolate or obovate-oblong, obtuse, up to $1\frac{1}{4}$ in. long and $\frac{1}{2}$ in. wide, tapering into a short petiole, glabrous or minutely puberulous, entire; heads subsessile in subglobose clusters of 2–4 in the axils of and shorter than the leaves; bracts ovate, mucronate, not spiny; bracteoles a little longer than the bracts; perianth-segments 1 lin. long, ovate, acuminate, minutely denticulate; filaments subulate; staminodes as long as the filaments, subulate, entire; utricle obcordate, slightly longer than the calyx, finely rugose; seed ovoid, much compressed, yellowish. *Drège, Zwei Pfl. Documente,* 158; *Moquin in DC. Prodr.* xiii. ii. 357; *Seubert in Mart. Fl. Bras.* v. i. 184, *t.* 57, *fig.* 2; *Hook. f. Fl. Brit. Ind.* iv. 731. *Alternanthera achyranthoides, Hiern in Cat. Afr. Pl. Welw.* i. 896; *Baker & C. B. Cl. in Dyer, Fl. Trop. Afr.* vi. i. 73. *Illecebrum sessile, Linn. Sp. Pl. ed.* ii. 300.

KALAHARI REGION : Transvaal ; Rooiplaat, *Miss Leendertz,* 765 ! and without precise locality, *Kirk,* 111 !
EASTERN REGION : Natal ; Valley of the Umlazi River, *Drège*! near Durban, *Krauss,* 221 ! Clairmont, *Wood,* 3889 !
Also in Tropical Asia and Australia.

XIX. GOMPHRENA, Linn.

Flowers hermaphrodite, bracteate and bibracteolate. *Perianth* 5-partite or 5-fid, usually woolly at the base ; segments unequal or equal, lanceolate, acuminate, concave, rarely flat and obtuse. *Staminal-tube* long, included or exserted, with 5 emarginate or bifid lobes at the top ; staminodes usually absent ; anthers 1-celled. *Ovary* turbinate or subglobose ; style short or long ; stigmas 2, rarely 3, subulate or filiform ; ovule solitary, suspended from a basal funicle. *Utricle* ovoid or oblong, compressed, sometimes hardened at the base. *Seed* lenticular, smooth ; embryo annular with narrow or obovate cotyledons ; albumen floury.

Erect or prostrate branched herbs, usually thickened at the nodes, hairy ; leaves opposite, sessile or subsessile, quite entire ; flowers in heads (rarely spikes), naked or involucrate, often solitary and sessile at the top of the branches, white or coloured ; bracteoles short or long, concave, keeled or winged on the back or crested.

DISTRIB. Species about 90, in Central and South America.

1. **G. globosa** (Linn. Sp. Pl. ed. i. 224) ; an erect or ascending much-branched annual, 1½–3 ft. high ; stems terete, pilose ; leaves oblong, 1–3 in. long, ½–1 in. wide, mucronate, with numerous long soft hairs on both surfaces ; petiole 2–6 lin. long, slightly amplexicaul at the base ; heads terminal, solitary or 2–3 together, globose, 6 lin. in diam., many-flowered ; rhachis villous ; bracts ovate-triangular, acuminate, mucronate ; bracteoles twice as long as the bracts, oblong, acute, very concave, crested on the back ; perianth-segments 4–5 lin. long, woolly, narrowly lanceolate, keeled, white or purplish ; stamens shorter than the perianth ; ovary oblong ; style rather long, slender ; stigmas 2, linear ; utricle ovoid-oblong, white ; seed compressed, rostrate, yellowish. *Moquin in DC. Prodr.* xiii. ii. 409 ; *Hook. f. Fl. Brit. Ind.* iv. 732 ; *Baker & C. B. Cl. in Dyer, Fl. Trop. Afr.* vi. i. 75.

KALAHARI REGION : Transvaal ; roadsides, Barberton, *Burtt-Davy,* 274 ! Haamans Kraal, Pretoria District, *Burtt - Davy,* 1099 ! Potgieters Rust, *Miss Leendertz,* 1888 !
EASTERN REGION : Natal ; roadsides, near Durban, *Grant* ! *Wood,* 1929 !
A weed throughout the warmer parts of the world.

ORDER CX. CHENOPODIACEÆ.

(By C. H. WRIGHT.)

Flowers hermaphrodite or unisexual, usually regular, naked or bracteate. *Perianth* simple, 3–5-lobed, or absent from the female flowers, accrescent or unchanged after flowering, membranous, herbaceous or chartaceous, naked, tuberculate or winged; segments usually imbricate in bud. *Stamens* as many as the perianth-lobes and opposite them, or fewer, hypogynous or perigynous, usually without staminodes; filaments subulate, filiform or compressed, free, rarely connate at the base; anthers dorsifixed, 2-celled; connective sometimes produced at the apex. *Disk* none, rarely annular. *Ovary* superior, sometimes immersed in the base of the perianth, 1-celled; style short or long and simple with a 2–3-lobed stigma, or styles 2–3, long and papillose at the apex, or stigmas 2–3, sessile, filiform, papillose all over; ovule solitary, amphitropous, erect on a short funicle, or suspended from a long basal funicle. *Fruit* a usually indehiscent utricle enclosed in and falling off with the perianth. *Seed* erect or horizontal, lenticular, subglobose or reniform; testa various, smooth or granular; embryo annular or spiral, surrounding the floury or fleshy albumen, albumen absent in *Salicornia*; cotyledons usually narrow.

Annual or perennial herbs, or shrubs, rarely small trees, glabrous, farinose, lepidote or hairy, sometimes fleshy; stems continuous or jointed, erect or decumbent; leaves alternate, rarely opposite, flat or cylindrical, usually entire, sometimes sinuate, exstipulate; inflorescence various, often of clusters arranged in spikes or panicles, sometimes dichotomously cymose, or flowers solitary and axillary.

DISTRIB. Genera about 80, and species about 520, cosmopolitan; many are weeds of cultivation.

* CYCLOLOBEÆ. *Embryo annular*; *albumen copious, except in* Salicornia.

 I. **Chenopodium.**—Herbs, rarely woody at the base, farinose or glandular-pubescent. *Leaves* alternate. *Flowers* hermaphrodite, without bracts or bracteoles; perianth 3–5-lobed.

 II. **Roubieva.**—A prostrate herb. *Leaves* alternate, pinnately lobed. *Flowers* hermaphrodite or female by abortion, without bracts or bracteoles; perianth very shortly 5-lobed, urceolate and quite enclosing the fruit.

 III. **Exomis.**—A dichotomously branched white scurfy shrub. *Leaves* alternate. *Flowers* unisexual, the female with small bracteoles unchanged in fruit and no perianth.

 IV. **Atriplex.**—Herbs or shrubs, more or less lepidote. *Leaves* usually alternate. *Flowers* unisexual, the female with large accrescent bracteoles and no perianth.

 V. **Chenolea.**—Herbaceous, or woody at the base. *Leaves* alternate. *Flowers* hermaphrodite and female, ebracteate; bracteoles present. *Perianth* in fruit inappendiculate or spiny.

 VI. **Kochia.**—Herbs or small shrubs. *Leaves* usually alternate. *Flowers* hermaphrodite and female, without bracts or bracteoles. *Perianth* in fruit horizontally winged.

 VII. **Salicornia.**—Fleshy herbs or shrubs with articulate branches. *Leaves* opposite, or opposite and alternate. *Flowers* hermaphrodite or polygamous, immersed in the hollows of a fleshy rhachis.

** SPIROLOBEÆ. *Embryo spiral; albumen scanty or none.*

VIII. **Suæda.**—Erect or prostrate herbs or shrubs. *Leaves* ternate, fleshy, more or less terete. *Perianth* in fruit not appendiculate or with horns.

IX. **Salsola.**—Herbs or shrubs. *Leaves* usually alternate, often spiny at the apex. *Perianth* in fruit with a broad horizontal wing above the middle.

I. CHENOPODIUM, Linn.

Flowers hermaphrodite, without bracts or bracteoles. *Perianth* 5- (very rarely 1- 3- or 4-) lobed ; lobes concave, sometimes keeled but not appendaged, unchanged in fruit. *Stamens* 5 or fewer, hypogynous or subperigynous ; filaments sometimes connate at the base ; anther-cells globose or oblong. *Disk* none or annular. *Ovary* usually depressed-globose ; style usually absent ; stigmas 2–5, filiform or subulate, free, rarely connate below ; ovule subsessile. *Utricle* ovoid and erect or globose and depressed, membranous or rather fleshy. *Seed* usually horizontal ; embryo annular or nearly so, surrounding copious floury albumen.

Herbs, rarely woody at the base, annual or perennial, often glandular pubescent, rarely glabrous ; leaves alternate, linear to deltoid, entire to pinnatifid ; flowers minute, in globose clusters, which are solitary and axillary or in terminal spikes or racemes.

DISTRIB. Species about 50, chiefly in temperate regions, rare in the tropics. Except possibly *C. Botrys* and *fœtidum*, the species described have probably been introduced into S. Africa with cultivation.

Flowers in globose axillary clusters :
Perianth 3-merous ; stamen solitary ... ... (1) **Blitum.**
Perianth 5-merous ; stamens 5 ... ... (2) **rubrum.**
Flowers in spicate axillary clusters :
Perianth completely enclosing the fruit ... ... (3) **ambrosioides.**
Perianth not completely enclosing the fruit :
Leaves entire ... ... (4) **polyspermum.**
Leaves toothed ... ... (5) **glaucum.**
Flowers in dense panicles near the apex of the stem :
Leaves longer than the cymes, coarsely serrate ... (6) **murale.**
Upper leaves shorter than the cymes :
Perianth-segments not keeled ... ... (7) **Vulvaria.**
Perianth-segments strongly keeled ... ... (8) **album.**
Flowers in diffuse axillary filiform cymes :
Leaves glandular pubescent ... ... (9) **Botrys.**
Leaves almost glabrous ... ... (10) **fœtidum.**

1. **C. Blitum** (F. Muell. Sel. Pl. Industr. ed. ii. 49) ; a very variable plant ; stem erect, angular, branched ; leaves deltoid or oblong-triangular, acute, deeply sinuate-toothed, thinly membranous, glabrous, 2 in. long, 1½ in. wide ; petiole up to 2½ in. long ; flowers in simple glomerules in the axils of the upper reduced leaves ; perianth reddish ; segments usually 3, ovate, obtuse, not keeled,

becoming succulent, not enclosing the fruit ; stamen solitary ; fruit erect ; seed lenticular, thick at the margin, not shining. *Hook. f. Fl. Brit. Ind.* v. 5. *Blitum virgatum, Linn. Sp. Pl. ed.* i. 4 ; *Bot. Mag. t.* 276 ; *Lam. Ill. t.* 5 ; *Moquin in DC. Prodr.* xiii. ii. 83.

SOUTH AFRICA : Without locality, *Prior* !
CENTRAL REGION : Graaff Reinet Div. ; Sneeuw Berg, 5300 ft., *Bolus*, 1863 !
Also in South Europe, the Orient and India.

2. **C. rubrum** (Linn. Sp. Pl. ed. i. 218) ; plant polymorphic ; stem angular, branched, 1–2 ft. high, glabrous, sometimes reddish ; leaves deltoid or deltoid-ovate, subobtuse, cuneate at the base, sinuate or sinuate-dentate, 2–6 in. long, 8–15 lin. wide (9 by 5 lin. in var. *pseudobotryoides*), rather thick, shining glaucous-green or reddish, nerves prominent beneath ; glomerules simple or slightly compound, the upper subspicate, leafy or not ; flowers dimorphic, $\frac{1}{3}-\frac{1}{2}$ lin. long, sessile, glabrous ; terminal flowers : perianth 5-partite ; stamens 5 ; seed horizontal ; lateral flowers : perianth-segments 2–3, obovate, obtuse, not keeled ; stamens 1–2 ; seed obtuse at the margin, puncticulate, shining ; embryo annular. *Curt. Fl. Lond.* iii. t. 29 ; var. *pseudobotryoides, H. C. Watson in Bot. Exch. Club Rep. for* 1863, 8, 1865, 11, *and* 1868, 13 ; *Bolus & Wolley-Dod in Trans. S. Afr. Phil. Soc.* xiv. iii. 311. *Atriplex rubra, Crantz, Inst.* i. 206. *Blitum rubrum, Reichenb. Fl. Germ. Excurs.* 582 ; *Moquin in DC. Prodr.* xiii. ii. 83. *B. polymorphum, C. A. Meyer in Ledeb. Fl. Alt.* i. 13.

COAST REGION : Cape Div. ; stream-bed near Kenilworth racecourse, *Wolley-Dod*, 2465 !

Also in South Europe, the Orient and the Azores.

The South African plant belongs to the variety *pseudobotryoides*, H. C. Watson, characterized by its prostrate habit and smaller leaves, but its author recorded in the *Botanical Exchange Club Report* for 1868, 13, that under certain conditions it developed into typical *C. rubrum*, Linn.

3. **C. ambrosioides** (Linn. Sp. Pl. ed. i. 219) ; stem herbaceous, erect, branched, 1–2 ft. high, more or less pubescent ; leaves oblong, acute at both ends, sinuate-dentate or subentire, thin, puberulous, up to 4 in. long and $1\frac{1}{4}$ in. wide, shortly petiolate : flowers in clusters spicately arranged amongst the uppermost often linear-lanceolate and entire leaves ; perianth-segments ovate, obtuse ; stamens exserted ; filaments linear ; fruit entirely enclosed by the perianth ; seed rounded at the margins, sometimes vertical. *Willd. Sp. Pl.* i. 1304 ; *Thunb. Prodr.* 48, *Fl. Cap. ed. Schultes*, 246 ; *Drège, Zwei Pfl. Documente*, 61, 106, 111, 133 ; *Wight, Icon. t.* 1786 ; *Moquin in DC. Prodr.* xiii. ii. 72, 460 ; *Hook. f. Flor. Brit. Ind.* v. 4 ; *Baker & C. B. Cl. in Dyer, Fl. Trop. Afr.* vi. i. 79 ; *Bolus & Wolley-Dod in Trans. S. Afr. Phil. Soc.* xiv. iii. 310 ; var. *dentatum, Fenzl ex Drège, Zwei Pfl. Documente*, 106, 172. *Atriplex ambrosioides, Crantz, Inst.* i. 207. *Ambrina ambrosioides, Spach, Hist. Veg. Phan.* v. 297.

SOUTH AFRICA : without locality, *Pappe* ! *Drège*, 8029b ! 8030 !
    COAST REGION : Cape Div. ; flats near Claremont, *Wolley-Dod*, 2464 ! Castle
Ditch (*forma coarctata*), *Wolley-Dod*, 2462 ! Uitenhage Div. ; Uitenhage, *Prior* !
Albany Div. ; Grahamstown, *MacOwan*, 3409a ! Queenstown Div. ; plains near
Queenstown, 3500 ft., *Galpin*, 2042 ! Shiloh, *Baur*, 789 !
    WESTERN REGION : Little Bushmanland ; Henkries, *Schlechter* !
    KALAHARI REGION : Transvaal ; Pretoria, *Miss Leendertz*, 13 ! and without
precise locality, *McLea in Herb. Bolus*, 5783 !
    EASTERN REGION : Natal ; Inanda, *Wood*, 1314 ! and without precise locality,
Gerrard, 245 !
    Also in Tropical Africa and widely spread as a weed in hot countries.

4. C. polyspermum (Linn. Sp. Pl. ed. i. 220) ; a procumbent or
suberect herb, branched from the base ; stem ribbed and more or
less angular ; leaves ovate or ovate-oblong, obtuse, shortly mucronate, quite entire, thin, glabrous, usually about 1½ in. long and
½ in. wide ; petiole slender, 6 lin. long ; flowers very numerous in
(sometimes clustered) spikes in the axils of much reduced leaves,
1 lin. in diam. ; perianth-segments elliptic, obtuse, patent, not completely enclosing the fruit ; stamens exserted ; seed acute at the
margin, shining, obscurely punctate ; embryo annular. *Moquin in
DC. Prodr.* xiii. ii. 62, 460 ; *Flor. Dan.* vii. *t.* 1153 ; *Curt. Fl. Lond.*
ii. *t.* 112. *C. marginatum, Spreng. in Hornem. Hort. Hafn.* i. 256.
Atriplex polysperma, Crantz, Inst. i. 207.

VAR. β, **cymosum** (Cheval. Flor. Env. Paris, ii. 385) ; leaves ovate or lanceolate,
rather obtuse ; flowers in axillary much-branched cymes. *Moquin in DC.
Prodr. l.c.*

COAST REGION : British Kaffraria, *Cooper*, 3062 ! Var. β : British Kaffraria,
Cooper, 3063 !

Also in Europe and N. Asia, introduced in N. America.

5. C. glaucum (Linn. Sp. Pl. ed. i. 220) ; a herb about 1 ft. high ;
stem prostrate or ascending, sulcate, much-branched ; leaves ovate-
oblong, repand or remotely toothed, rarely entire, glabrous above,
farinose and pale green beneath, 1½–2 in. long, ½–⅔ in. wide ; midrib prominent beneath ; flowers in short dense axillary leafless
spikes, glabrous ; perianth-segments obovate-oblong, obtuse, not
keeled, sometimes by abortion 3 or 4, not entirely covering the
fruit ; fruit greenish above ; seed sometimes vertical ; embryo
annular. *Fl. Dan. t.* 1151 ; *Moquin in DC. Prodr.* xiii. ii. 72 ;
Drège, Zwei Pfl. Documente, 49, 58 ; *Hook. f. Fl. Brit. Ind.* v. 4.
Atriplex glauca, Crantz, Inst. i. 207. *Blitum glaucum, Koch, Syn.
Fl. Germ. ed.* ii. 699.

COAST REGION : Queenstown Div. ; on the plains between Table Mountain and
Wildschuts Berg, 4000 ft., *Drège.*
    CENTRAL REGION : Victoria West Div. ; Nieuwveld between Brak River and
Uitvlugt, 3000–4000 ft., *Drège.*

Also in Europe and India.

6. C. murale (Linn. Sp. Pl. ed. i. 219) ; stem ascending up to
1½ ft. high, sulcate, branched ; leaves ovate or deltoid, up to 3 in.

long and 1½ in. wide (usually smaller in South African specimens), coarsely and unequally toothed, acute, cuneate at the base, rather thin, bright pale green on both surfaces, pulverulent above; petiole slender, rather shorter than the blade; cymes axillary, usually shorter than the leaves; flowers sessile, ½ lin. in diam.; perianth-segments elliptic, obtuse, slightly pulverulent outside, faintly keeled, closed over the fruit; stamens exserted; anthers minute, globose; seed lenticular, acute at the margin, dark brown or almost black, ¾ lin. in diam.; embryo annular. *Curt. Fl. Lond.* iii. *t.* 117; *Moquin in DC. Prodr.* xiii. ii. 69, 460; *Thunb. Prodr.* 48, *and Fl. Cap. ed. Schult.* 246; *Drège, Zwei Pfl. Documente,* 94, 104, 113; *Hook. f. Fl. Brit. Ind.* v. 4; *Engl. Pfl. Ost-Afr. C.* 171; *Bolus & Wolley-Dod in Trans. S. Afr. Phil. Soc.* xiv. iii. 311; *Baker & C. B. Cl. in Dyer, Fl. Trop. Afr.* vi. i. 78.

COAST REGION: Cape Div.; Oatland Point, *Wolley-Dod*, 2856! Lion Mountain, under 500 ft., *Drège*! Uitenhage Div.; Uitenhage, *Zeyher*, 471! *Prior*! Albany Div.; near Grahamstown, *MacOwan*, 3414 partly!

KALAHARI REGION: Transvaal; Pretoria, *Miss Leendertz*, 624! *Burtt-Davy*, 834! near Lydenburg, *Wilms*, 1207!

EASTERN REGION: Natal; Inanda, *Wood*, 235! near Pietermaritzburg, *Wilms*, 2241!

A cosmopolitan weed.

7. **C. Vulvaria** (Linn. Sp. Pl. ed. i. 220); stem ascending, branched; leaves rhomboid-ovate, obtuse or acute, entire, ½–1 in. long, ⅓–¾ in. wide, white-pulverulent especially beneath; petiole up to ½ in. long; flowers in subsessile clusters closely placed along the rhachis, 1 lin. in diam.; perianth-segments elliptic-ovate, not keeled, green with white margins, farinose outside; stamens as long as the perianth; seed depressed, rather acute at the margin; embryo annular. *Fl. Dan. t.* 1152; *Moquin in DC. Prodr.* xiii. ii. 64, 460. *C. olidum, Curt. Fl. Lond. fasc.* v. *t.* 20; *Drège, Zwei Pfl. Documente,* 67.

COAST REGION: Albany Div.; near Grahamstown, *MacOwan*, 3414, partly! Kingwilliams Town Div.; banks of the Buffalo River near King Williams Town, *Galpin*, 5939!

WESTERN REGION: Little Namaqualand; on hills at Brakdam, 2000 ft., *Schlechter*, 11159!

CENTRAL REGION: Prince Albert Div.; between Droogeheuvel and Jackhals Fontein, 2500–3000 ft., *Drège.*

KALAHARI REGION: Transvaal; Pretoria, *Miss Leendertz*, 12!

Also in Europe and North Africa.

8. **C. album** (Linn. Sp. Pl. ed. i. 219); an annual herb of variable size; stem erect, sulcate, branched, more or less mealy; branches erect; leaves rhomboid-triangular, obtuse or acute, sinuately toothed throughout or in the upper part only, or quite entire, 2–3 in. long, petiolate, the uppermost oblong or linear-lanceolate and quite entire, pulverulent, green or white; flowers 1 lin. in diam., in glomerules arranged in lateral and terminal spikes;

perianth-segments navicular, strongly keeled and mealy outside, completely enclosing the fruit; stamens about as long as the perianth; filaments complanate; seed depressed, black, shining; embryo annular. *Drège, Zwei Pfl. Documente*, 56, 59, 172 (*var. dentatum, Fenzl, and var. integrifolium, Fenzl*); *Moquin in DC. Prodr.* xiii. ii. 70, 460; *Engl. Pfl. Ost-Afr. C.* 171; *Agric. Gaz. N.S.W.* 1905, 474; *Clark & Fletcher, Farm Weeds of Canada, t.* 40; *Bolus & Wolley-Dod in Trans. S. Afr. Phil. Soc.* xiv. iii. 311; *Baker & C. B. Cl. in Dyer, Fl. Trop. Afr.* vi. i. 77.

COAST REGION: Cape Div.; Sandown Road, Rondebosch, *Wolley-Dod*, 2485!
CENTRAL REGION: Beaufort West Div.; Nieuweveld, between Rhinoster Kop and Ganzefontein, 3000–4500 ft., *Drège*! Richmond Div.; Uitvlugt, near Steelkloof, 4000–5000 ft., *Drège*, 8028! Murraysburg Div.; Murraysburg, 4000 ft., *Tyson*, 57! Graaff Reinet Div.; Graaff Reinet, 2600 ft., *Bolus*, 74!
KALAHARI REGION: Transvaal; Byinsel Farm, Standerton, *Burtt-Davy*, 1801! Warm Bath, Springbok Flats, *Burtt-Davy*, 2339!

A cosmopolitan weed.

9. **C. Botrys** (Linn. Sp. Pl. ed. i. 219); stem herbaceous, erect, ribbed, branched, glandular-pubescent, viscid; leaves oblong, obtuse, deeply pinnately and obtusely (often bipinnately) lobed, glandular-pubescent on both surfaces, glaucous green, up to 2 in. long and 1¼ in. wide, the uppermost often lanceolate and entire; flowers in very numerous divaricate cymes racemosely arranged in the axils of much reduced leaves, ½ lin. in diam.; perianth-segments ovate, obtuse, scarcely enclosing the fruit, not keeled; seed obtuse at the margin, shining; embryo not forming a complete circle. *Moquin in DC. Prodr.* xiii. ii. 75, 460; *Burchell, Trav. S. Afr.* ii. 226; *Sibth. Fl. Græca, t.* 253; *Hook. f. Fl. Brit. Ind.* v. 4; *Engl. Pfl. Ost-Afr. C.* 171; *Britt. & Br. Ill. Fl. N.U.S.A.* i. 574; *Baker & C. B. Cl. in Dyer, Fl. Trop. Afr.* vi. i. 79. *Ambrina Botrys, Moquin, Chenop. Enum.* 37. *Atriplex Botrys, Crantz, Inst.* i. 207. *Botrydium aromaticum, Spach, Hist. Veg. Phan.* v. 299.

COAST REGION: Swellendam Div.; Swellendam, *Bowie*! Albany Div.; Grahamstown, *MacOwan*, 958! Queenstown Div.; near Queenstown, 3500 ft., *Galpin*, 2037! Shiloh, *Baur*, 943!
CENTRAL REGION: Somerset Div.; Somerset East, *Bowker*, 100! Graaff Reinet Div.; Graaff Reinet, 2600 ft., *Bolus*, 388! Albert Div., *Cooper*, 1375!
WESTERN REGION: Little Namaqualand; Orange River near Verleptpram, *Drège.*
KALAHARI REGION: Griqualand West; at Griqua Town, *Burchell*, 1955! Transvaal; Standerton, *Burtt-Davy*, 1780!
EASTERN REGION: Natal; near the Mooi River, *Wood*, 4104! near the Umlaas River, *Wood*, 1832!

Also in South Europe, the Orient, North Africa, Temperate Asia and North America.

10. **C. fœtidum** (Schrad. Mag. Ges. Naturf. Fr. Berl. 1808, 73, not of Lam.); a herb 1–2 ft. high, slightly glandular-pubescent, odour aromatic; stem erect, sulcate, sparingly branched; leaves

oblong, sinuately and obtusely lobed, almost glabrous, pale green
on both sides, the lower 4 in. long, the upper $1\frac{1}{2}$–2 in. long,
6–10 lin. wide; nerves slender, prominent beneath; cymes shorter
than the subtending leaves, much-branched; branches filiform,
rigid; flowers minute, very shortly stalked; perianth-segments
ovate, subacute, glandular, with a finely toothed keel, not entirely
enclosing the fruit; stamens exserted; style short, branches filiform; seed lenticular with obtuse margin, smooth. *Moquin in DC.*
Prodr. xiii. ii. 76; *Engl. Pfl. Ost-Afr. C.* 171. *C. schraderianum,*
Roem. & Schult. Syst. vi. 260; *C. B. Cl. in Dyer, Fl. Trop. Afr.*
vi. i. 80. *Botrydium Schraderi, Spach, Hist. Veg. Phan.* v. 299.
Ambrina fœtida, Moquin, Chenop. Enum. 38.

SOUTH AFRICA : without locality.

Also in Tropical Africa, Central and South America.

Imperfectly known species.

11. **C. mucronatum** (Thunb. Prodr. 48); stem herbaceous, erect,
angular, very thinly pubescent, about 18 in. high; branches few,
near the apex of the stem; leaves triangular, hastate with rounded
angles, very obtuse, mucronate, entire, glabrous, about 1 in. long;
petiole as long as the blade; racemes leafy, near the top of the
plant. *Thunb. Fl. Cap. ed. Schult.* 246; *Willd. Sp. Pl.* i. ii. 1299;
Moquin in DC. Prodr. xiii. ii. 64.

SOUTH AFRICA : without locality, *Thunberg, Verreaux.*

COAST REGION : Uitenhage Div. ; at the foot of the Winterhoek Mountains,
Krauss, 791, *ex Moquin.*

This may be a form of *C. album,* Linn.

II. ROUBIEVA, Moquin.

Flowers minute, hermaphrodite or by abortion female, without
bracts or bracteoles. *Perianth* urceolate, shortly 5-lobed, almost
closed at the mouth in fruit ; lobes rounded, accrescent, coriaceous.
Stamens 5 ; filaments thick ; anthers ovoid. *Disk* 0. *Ovary* globose ;
stigmas 2–5, filiform or subuláte, connate at the base ; ovule subsessile. *Utricle* subglobose or oblong, enclosed in the enlarged
perianth ; pericarp thinly membranous. *Seed* erect, orbicular,
slightly compressed ; testa smooth, crustaceous ; embryo annular,
surrounding the copious albumen.

Branched glandular-puberulous herbs ; leaves small, alternate, subsessile,
sinuate-dentate or subpinnatifid : fruits stalked, often subverticillate.

DISTRIB. Species 2, in tropical and temperate America, one widely spread in
the Old World.

1. **R. multifida** (Moquin in Ann. Sci. Nat. 2me sér. i. 293); a
much-branched aromatic herb ; stems up to 2 ft. long, prostrate,
striate ; leaves pinnatifid with linear or lanceolate lobes, up to

1½ in. long and 9 lin. wide, tapering downwards, shortly petiolate, glandular-puberulous ; nerves prominent beneath ; flowers sub-sessile in axillary clusters ; perianth ½ lin. long, puberulous, in fruit reticulately veined ; lobes ovate, subobtuse ; utricle oblong, whitish with irregular orange spots ; seed shortly beaked, blackish, shining, rugose. *Chenop. Enum.* 43, *and in DC. Prodr.* xiii. ii. 80 ; *Bolus & Wolley-Dod in Trans. Phil. Soc. S. Afr.* xiv. iii. 311. *Atriplex multifida, Crantz, Inst.* i. 207. *Chenopodium Payco, Roem. & Schult. Syst.* vi. 260. *Ambrina pinnatisecta, Spach, Hist. Veg. Phan.* v. 296.

CAPE REGION : Cape Div. ; between Newlands Bridge and village, *Wolley-Dod*, 2449 ! Albany Div. ; Grahamstown, *MacOwan*, 3411 !
KALAHARI REGION : Transvaal ; Standerton, *Burtt-Davy*, 1783 !

Also in the Mediterranean region and South America.

III. EXOMIS, Fenzl.

Male flowers : Sepals 5, ovate or triangular, acute, concave. *Stamens* 5 ; filaments filiform ; anthers ovate. *Hermaphrodite flowers* : *Sepals* 3–5, sometimes none, very minute, slightly united below. *Staminodes* none. *Styles* 2, united below, stigmatic on the inner surface. *Fruit* fleshy, sometimes enclosed in the accrescent bracts ; pericarp adherent to the seed. *Seed* vertical ; testa crustaceous ; albumen copious, floury, surrounded by the annular embryo.

An ashy-grey shrub ; leaves alternate, entire ; flowers in terminal spikes or axillary clusters, the male ebracteate, the hermaphrodite 2-bracteate.

DISTRIB. Species 1, endemic.

1. **E. axyrioides** (Fenzl ex Moquin, Chenop. Enum. 49) ; a shrub 1–2 ft. high ; branches rigid, 1–2 lin. in diam., ribbed when dry ; leaves deltoid-ovate or elliptic, obtuse or subacute, ½–1 in. long, 3–5 lin. wide, quite entire, greyish green, glabrous or the upper-most farinose ; midrib prominent beneath ; petiole 1–3 lin. long ; flowers in axillary clusters, sessile ; bracts narrowed below ; sepals triangular or ovate, ½ lin. long ; stamens about as long as the sepals ; fruit oblong ; pericarp whitish ; seed compressed, obtuse at the margin, smooth, black, shining. *Moquin in DC. Prodr.* xiii. ii. 89 ; *Drège, Zwei Pfl. Documente,* 78 ; *Melliss, Fl. St. Helena,* 314. *Chenopodium pauciflorum, Herb. Vindob. ex Moquin in DC. Prodr. l.c.*

SOUTH AFRICA : without locality, *Bergius* !
COAST REGION : Cape Div. ; kloof between the Lions Head and Table Mountain, *Burchell,* 249 ! Van Kamps Bay, 50 ft., *MacOwan,* 1619 ! and *Herb. Austr.-Afr.,* 1784 ! Castle ditch, *Wolley-Dod,* 2460 ! Tulbagh Div. ; New Kloof, 500–2000 ft.. *Drège,* 8027*b* ! *Schlechter,* 9048 ! Caledon Div. ; Caledon, *Zeyher* ! Hang Klip, *Mund & Maire.* Uitenhage Div. ; Enon, *Baur,* 1004 ! and without precise locality, *Zeyher,* 71 ! Albany Div. ; Grahamstown, *MacOwan* !

CENTRAL REGION: Calvinia Div. ; Oorlogs Kloof, Onder Bokkeveld, 2200 ft., *Schlechter*, 10932 !
KALAHARI REGION : Griqualand West ; St. Clair, Douglas, *Orpen*, 166 ! Orange River Colony ; Bloemfontein, *Potts*, 491 !
This species varies much in the density or otherwise of its habit.

IV. ATRIPLEX, Linn.

Flowers monœcious or diœcious. *Male flowers* without bracts or bracteoles. *Perianth* 3–5-partite ; segments obovate or oblong, obtuse. *Stamens* 3–5, inserted at the base of the perianth ; filaments free or connate at the base ; anthers 2-lobed. *Rudiment of ovary* none or conical. *Female flowers* bibracteolate ; bracteoles accrescent, in fruit dilated at the base and connate into a 2-lipped cup, rarely quite separate. *Perianth* none. *Disk* and *staminodes* rudimentary. *Ovary* ovoid or depressed-globose ; stigmas 2, subulate or filiform, connate at the base ; ovule erect on a short funicle or suspended from a long basal funicle. *Utricle* included in the much enlarged bracteoles ; pericarp membranous. *Seed* erect or inverted, rarely horizontal ; testa membranous, coriaceous or almost crustaceous ; embryo annular, surrounding the floury albumen.

Herbs or shrubs, more or less furfuraceous or covered with lepidote scales ; leaves alternate, rarely opposite, sessile or stalked ; flowers in glomerules either axillary and sessile or collected into spikes or panicles.

DISTRIB. Species about 100, in the temperate and tropical regions of the whole world.

Leaves all opposite and entire ... ... (1) **portulacoides.**
Leaves alternate, rarely the lower opposite :
  Leaves linear to oblong, entire or finely toothed :
    Stem more or less woody :
      Leaves lanceolate to oblong, subacute ... ... (2) **Verreauxii.**
      Leaves oblanceolate, obtuse ... ... (3) **Bolusii.**
    Stem herbaceous :
      Bracteoles 3-lobed, entire or slightly toothed ... (4) **patula.**
      Bracteoles coarsely dentate, spongy ... ... (5) **littoralis.**
  Leaves ovate or subrotund, entire ... ... (6) **glauca.**
  Leaves deltoid to elliptic, entire :
    Bracteoles membranous, more or less toothed ... (7) **Halimus.**
    Bracteoles fleshy, entire ... ... (8) **albicans.**
  Leaves lanceolate to deltoid, coarsely toothed or sinuate :
    Bracteoles in fruit spongy ... ... (9) **halimoides.**
    Bracteoles in fruit not spongy :
      Spikes leafy ... ... (10) **rosea.**
      Spikes naked, except at the base ... ... (11) **laciniata.**

1. **A. portulacoides** (Gmel. Syst. 450) ; an unarmed shrub, procumbent at the base ; branches angular ; leaves opposite, obovate

or oblong, rather obtuse, quite entire, rarely with 1–2 small teeth, rather thick, 1–1½ in. long, 2–4 lin. wide, densely covered with silvery lepidote scales ; petiole about 3 lin. long ; flower-clusters globose, distant on spikes collected into terminal panicles, yellowish ; perianth-segments rotundate, lepidote ; filaments rather longer than the perianth, shortly united below ; bracteoles in fruit obcordate-trapezoid, 3-lobed at the apex, 1–1½ lin. long ; seed dark brown, beaked. *Willd. Sp. Pl.* iv. ii. 957 ; *Fl. Dan. t.* 1889. *Chenopodium portulacoides, Thunb. Fl. Cap. ed. Schult.* 245. *Obione portulacoides, Moquin, Chenop. Enum.* 75, and in *DC. Prodr.* xiii. ii. 112 ; *Reichenb. Ic. Fl. Germ.* xxiv. 145, *t.* 271. *Halimus portulacoides, Dumort. Fl. Belg.* 20.

SOUTH AFRICA : without locality, *Thunberg.*

Also in Western Europe, from Britain southwards and throughout the Mediterranean region.

2. **A. Verreauxii** (Moquin in DC. Prodr. xiii. ii. 98) ; stem herbaceous or slightly woody, erect, very sparingly branched, slightly ribbed, rigid, whitish lepidote ; leaves alternate, lanceolate, subdeltoid or oblong, subacute, cuneate at the base, ½–1 in. long, 1½–3 lin. wide, the lower irregularly denticulate, rather thick, densely lepidote on both surfaces ; midrib prominent beneath ; petiole rather stout, 2–3 lin. long ; flowers collected in dense terminal spikes, male 5-merous ; perianth-segments obovate, densely farinose ; anthers oblong. *A. farinosa, Moquin, Chenop. Enum.* 55, *not of Dumort.*

SOUTH AFRICA : without locality, *Verreaux ex Moquin,* and a specimen without locality or collector's name in Herb. Kew !

The flowers have not been described by Moquin, and those on the Kew specimen (authenticated by him) are very young. The Kew specimen, although labelled South Africa, was probably collected by Curror in Angola.

3. **A. Bolusii** (C. H. Wright) ; stem woody, subterete, densely white furfuraceous ; leaves alternate, oblanceolate, rounded at the apex, 6 lin. long, 2 lin. wide, very densely clothed on both surfaces with whitish scales, quite entire, rather fleshy ; petiole 3 lin. long ; flower clusters in spikes in the upper part of the plant, scaly like the leaves ; bracteoles in fruit rotundate, 10 lin. in diam., sinuate-dentate, free, furfuraceous, membranous, strongly and reticulately veined ; utricle membranous, pellucid, 1½ lin. in diam., compressed ; seed erect ; radicle superior ; style short ; stigmas 2, ¾ lin. long.

WESTERN REGION : Little Namaqualand ; in sandy places near Port Nolloth, 20 ft., *Bolus,* 9457 ! *Pearson,* 509 !

This is allied to *A. leucoclada,* Boiss., from Egypt and Arabia, which differs in having smaller fruiting bracteoles 3-lobed to the middle.

4. **A. patula,** var. **angustifolia** (Syme, Engl. Bot. ed. 3, viii. 29, t. 702) ; an annual much-branched herb, 1–3 ft. high ; branches

divaricate, ribbed, glabrous; lower leaves hastate, $\frac{3}{4}$–$2\frac{1}{2}$ in. long, $\frac{1}{4}$–$1\frac{1}{4}$ in. wide, the upper lanceolate or linear, $\frac{1}{2}$–$1\frac{1}{4}$ in. long, all entire, glabrous; flowers in clusters spicately arranged in the upper part of the branches; bracteoles sessile, about 1 lin. long, free nearly to the base, deltoid, entire or slightly dentate. *Bolus & Wolley-Dod in Trans. S. Afr. Phil. Soc.* xiv. iii. 311. *A. angustifolia, Sm. Fl. Brit.* 1092, *and Engl. Bot. t.* 1774.

COAST REGION : Cape Div. ; ditch near Kenilworth racecourse, *Wolley-Dod*, 2422 ! Caledon Div. ; Vogel Gat, *Schlechter*, 10426 !

Also in Europe and North Africa. A weed of cultivation.

5. **A. littoralis** (Linn. Sp. Pl. ed. i. 1054); stem erect, herbaceous, sulcate, $2\frac{1}{2}$ ft. high, much branched; leaves alternate, linear-lanceolate or linear, 1–3 in. long, $1\frac{1}{2}$–3 lin. wide, entire or slightly sinuate, rather thick; lateral nerves obsolete; petiole short; flower clusters distant on slender spikes paniculately arranged; bracteoles in fruit rhomboid-ovate, coarsely dentate on' the margins and back, free nearly to the base, texture spongy, 2–7 lin. long; styles linear, thrice as long as the ovary. *Moquin in DC. Prodr.* xiii. ii. 96, 460 ; *Reichenb. Ic. Fl. Germ.* xxiv. 136, *t.* 266. *Chenopodium littorale, Thunb. in Act. Upsal.* vii. (1815) 142.

COAST REGION : Uitenhage Div.; Uitenhage, *Zeyher* !

Also in Europe.

6. **A. glauca** (Linn. Sp. Pl. ed. ii. 1493); stem slender, slightly woody, branched, terete, whitish; leaves alternate, sessile, ovate or subrotundate, very obtuse, quite entire, 3–4 lin. long, 3 lin. wide, thick, crisped and somewhat sheathing at the base, silvery pulverulent; nerves not conspicuous beneath; flower clusters spicately arranged; bracts in fruit $1\frac{1}{2}$ lin. long, sessile, rhomboid-deltoid, toothed near the base; pericarp white; seed compressed, fuscous, thick at the margin. *Boiss. Voy. Bot. Espagne,* ii. 542. *Obione glauca, Moquin in DC. Prodr.* xiii. ii. 108. *Chenopodium vestitum, Thunb. Prodr.* 48, *and Fl. Cap. ed. Schult.* 245. *Atriplex maritima hispanica, etc., Dill. Hort. Elth.* 46, *t.* 40, *fig.* 46.

SOUTH AFRICA : without locality, *Thunberg.*

Also in Spain, North Africa and Arabia.

This has the habit of *Exomis axyrioides,* Fenzl.

7. **A. Halimus** (Linn. Sp. Pl. ed. i. 1492); an erect shrub; branches slightly angular, whitish pulverulent; leaves alternate, elliptic or sometimes almost deltoid, very obtuse, entire, about 1 in. long, $\frac{1}{2}$ in. wide, densely silvery pulverulent on both surfaces; petiole 2 lin. long; bracts rhomboid, long acuminate, denticulate near the base, 2–3 lin. long in fruit; flowers in dense globose sessile clusters about $1\frac{1}{2}$ lin. in diam. arranged in terminal panicles; perianth $\frac{3}{4}$ lin. in diam.; segments obovate, concave,

densely farinose outside ; filaments shortly connate at the base ;
bracteoles broadly ovate or rotundate, entire or more or less toothed,
especially below, up to 6 lin. long and broad, reticulately veined ;
stigmas filiform, thrice as long as the ovary. *Moquin in DC. Prodr.*
xiii. ii. 100 ; *Reichenb. Ic. Fl. Germ.* xxiv. 144, *t.* 270. *A. capensis,*
Moquin, Chen. Enum. 63, *and in DC. Prodr.* xiii. ii. 100. *Chenopodium*
Halimus, Thunb. Prodr. 48, *and Fl. Cap. ed. Schult.* 245. *Schizothéca*
Halimus, Fourr. in Ann. Soc. Linn. Lyon, xvii. (1869) 143.

SOUTH AFRICA : without locality, *Thom,* 207 ! 239 ! *Drège !*
COAST REGION : Laingsburg Div. ; M itjesfontein, 3000 ft., *MacOwan,* 3343 !
and *Herb. Austr.-Afr.,* 1943 ! Mossel Bay Div. ; on dry hills by the Gouritz River,
Burchell, 6428 ! Uitenhage Div. ; Uitenhage, *Zeyher !* 616 ! 733 ! Zwanspoels
Kraal near Enon, *Baur,* 1045 ! Bathurst Div. ; by the Kowie River at Port
Alfred, 10 ft., *Galpin,* 2967 ! King Williamstown Div. ; Keiskamma, *Mrs. Hutton !*
CENTRAL REGION : Somerset Div. ; Somerset East, *Bowker,* 211 ! Philipstown
Div. ; by the Orange River, near Petrusville, *Burchell,* 2671 ! Graaff Reinet Div. ;
near Graaff Reinet, *Bolus,* 661 ! Cradock Div. ; plains, Witmoss Station, 2400 ft.,
Galpin, 3078 ! and without precise locality, *Cooper,* 517 ! Hopetown Div. ; Salt
pans near Hopetown, *Shaw !*
KALAHARI REGION : Orange River Colony ; Karroo, Witteberg Range, *Rehmann,*
2877 !
WESTERN REGION : Little Namaqualand ; J'us (TᶜOus ?), 2800 ft., *Schlechter,*
11432 ! Van Rhynsdorp Div. ; Attys, 300 ft., *Schlechter,* 8085 !

Also in Southern Europe and North Africa.

8. **A. albicans** (Ait. Hort. Kew. ed. 1, iii. 430, not of Besser) ;
a shrub 1½–2 ft. high ; branches ribbed, whitish ; leaves rhomboid-
hastate, ½–1½ in. long, ½–1 in. wide, entire, rather thick, greyish-
green, the lower obtuse, the upper acute and farinose ; petiole about
½ lin. long ; flowers in terminal leafless spikes up to 2 in. long ;
bracts sessile, concave, becoming fleshy and enclosing the fruit ;
sepals 3–5, ovate, acute, membranous, sometimes erose ; fruit white
or greenish ; seed compressed, obtuse at the margin, smooth, shining.
Drège, Zwei Pfl. Documente, 78, 102 ; *Burch. Trav. S. Afr.* i. 207,
ii. 21. *A. odorata, Pers. Syn.* i. 293. *Exomis atriplicioides, Moquin,*
Chenop. Enum. 49. *E. albicans, Moquin in DC. Prodr.* xiii. ii. 89.

COAST REGION : Clanwilliam Div. ; Clanwilliam, *MacOwan,* 3308 ! and *Herb.*
Austr.-Afr., 1942 ! Malmesbury Div. ; Laauws Kloof, under 1000 ft., *Drège,*
Tulbagh Div. ; New Kloof, 2000–3000 ft., *Drège.*
CENTRAL REGION : Ceres Div. ; near Yuk River, *Burchell,* 1243 !
WESTERN REGION : Little Namaqualand ; Port Nolloth, *Pearson,* 507 ! near
Ookiep, *Morris in Herb. Bolus,* 5784 ! Van Rhynsdorp Div. ; Bitter Fontein,
Schlechter, 11012 !

9. **A. halimoides** (Lindl. in Mitch. Three Exped. E. Austr. i. 285,
not of Tineo) ; a procumbent herb or undershrub ; stem whitish,
glabrous ; leaves alternate, lanceolate or oblanceolate, acute, coarsely
toothed, 1½ in. long, 1½ in. wide, greyish lepidote on both surfaces,
tapering into a winged petiole ; flowers in axillary clusters ; bracteoles in fruit 6 lin. in diam., spongy and fibrous, depressed so as to

appear turbinate with a horizontal wing and small central opening;
seed brownish, lenticular. *Benth. Fl. Austral.* v. 178. *A. Lindleyi,
Moquin in DC. Prodr.* xiii. ii. 100. *A. inflata, F. Muell. in Trans.
Phil. Inst. Vict.* ii. 75.

COAST REGION : Albany Div.; Grahamstown, *Schönland* !
Introduced from Australia.

10. **A. rosea** (Linn. Sp. Pl. ed. ii. 1493) ; stem much-branched
from the base, herbaceous or slightly woody, obtusely angled,
1–3 ft. high; leaves alternate, spreading, rhomboid to oblong,
sinuate, 1–2 in. long, $\frac{1}{2}$–$1\frac{1}{2}$ in. wide, cuneate at the base, silvery
lepidote on both surfaces; nerves rather prominent beneath;
petiole 3–6 lin. long; flowers monœcious, in terminal leafy spikes
up to 2 in. long, reddish; bracteoles sessile, obcuneate, shortly
cuspidate, 2–3 lin. long in fruit, connate half-way up, strongly
nerved on the back; seed beaked, thick at the margin, fuscous.
Moquin in DC. Prodr. xiii. ii. 92 ; *Reichenb. Ic. Fl. Germ.* xxiv. 138,
t. 267.

CENTRAL REGION : Graaff Reinet Div. ; near Graaff Reinet, 2600 ft., *Bolus,*
656 !
Also in Europe, the Orient and North Africa.

11. **A. laciniata** (Linn. Sp. Pl. ed. i. 1053) ; a herb 2–5 ft. high;
stem simple or sparingly branched above, obscurely angular ; leaves
alternate or the lowest opposite, hastate-deltoid, acute or subobtuse,
sinuately toothed, 1–3 in. long, $\frac{3}{4}$–$\frac{1}{2}$ in. wide, glabrous above, silvery
pulverulent beneath ; petiole 5–10 lin. long ; bracts in fruit $1\frac{1}{2}$–2 lin.
long, rhomboid, acute, sometimes 3-lobed, toothed at the sides,
connate half-way up ; spikes up to 2 in. long, 1 lin. in diam.,
paniculately arranged, with the clusters closely placed, naked or
leafy only at the base. *Moquin in DC. Prodr.* xiii. ii. 93.
A. tatarica, Linn. Sp. Pl. ed. i. 1053 ; *Reichenb. Ic. Fl. Germ.*
xxiv. 142, *t.* 269. *Chenopodium laciniatum, Thunb. Prodr.* 48.
C. sinuatum, Thunb. Fl. Cap. ed. Schult. 245.

SOUTH AFRICA : without locality, *Thunberg* !
Also in Europe, the Orient and North Africa.

Imperfectly known species.

12. **A. microphylla** (Willd. Sp. Pl. iv. ii. 958, not of F. Muell.) ;
stem shrubby, branched from the base, erect, scarcely 1 ft. high,
ashy-grey ; branches terete, virgate ; leaves scattered, ovate, obtuse,
entire, glaucous, 1 lin. long ; flowers in the axils of the leaves.
Moquin in DC. Prodr. xiii. ii. 104 ; *Burch. Trav. S. Afr.* i. 225.
Chenopodium microphyllum, Thunb. Prodr. 48, *and Fl. Cap. ed.
Schult.* 245.

SOUTH AFRICA : without locality, *Thunberg.*

V. CHENOLEA, Thunb.

Flowers hermaphrodite and female, without bracts, bracteolate.
Perianth turbinate, globose or orbicular-depressed, villous or
tomentose, rarely glabrous ; lobes 5, incurved, accrescent, produced
into spines or horns on the back, rarely unarmed. *Stamens* 5,
hypogynous ; filaments compressed, short or long ; anthers oblong.
Disk none. *Ovary* ovoid, attenuate into a long or short style ;
stigmas 2–3, capillary, papillose all over ; ovule subsessile. *Utricle*
enclosed in the crustaceous or coriaceous perianth ; pericarp
membranous or hardened at the apex. *Seed* horizontal, orbicular ;
embryo annular ; albumen scanty.

Herbs or shrubs, erect or decumbent, usually hairy ; leaves alternate, sessile,
linear, lanceolate or terete, quite entire ; flowers axillary, minute, solitary or
clustered.

DISTRIB. Species 3, 2 in North Africa and Arabia and the following.

1. **C. diffusa** (Thunb. Nov. Gen. 10) ; stem decumbent, flexuous,
terete, reddish, canescent when young ; leaves oblong or lanceolate,
6 lin. long, 1½ lin. wide, acute, sessile, fleshy, the upper erect,
imbricate and clothed with silvery silky hairs, the lower spreading
or reflexed and less hairy ; nerves inconspicuous ; flowers axillary,
solitary, minute ; perianth-segments oblong, obtuse, silky ; spinules
minute, obtuse, rather villous. *Echinopsilon diffusus, Moquin in
DC. Prodr.* xiii. ii. 137. *E. sericeus, Moquin, Chenop. Enum.* 89.
Salsola diffusa, Thunb. Prodr. 48, *and Fl. Cap. ed. Schult.* 243. *S.
sericea, Ait. Hort. Kew. ed.* 1, i. 317. *Kochia sericea, Schrad. Neues
Journ.* 1809, 87. *Chenopodium sericeum, Spreng. Syst.* i. 921.

COAST REGION : Cape Div. ; Millers Point, *Wolley-Dod*, 2397 ! about the
ponds and at Salt River, *Burchell*, 666 ! Knysna Div. ; at the mouth of the
Knysna River, *Krauss*, 786, ex *Moquin.* Uitenhage Div., *Zeyher*, 501 ! *Prior* !
East London Div. ; near the mouth of the Nahoon River, *Galpin*, 5674 ! Div. ?
Redhouse, *Mrs. Paterson*, 287 !
CENTRAL REGION : Cradock Div. ; without precise locality, *Cooper*, 3055 ! 3140 !
KALAHARI REGION : Bechuanaland ; bank of the Moshowing River between
Takun and Melito, *Burchell*, 2306 !
EASTERN REGION : Natal ; near Durban, *Wood*, 901 !

VI. KOCHIA, Roth.

Flowers hermaphrodite and female, without bracts or bracteoles.
Perianth subglobose to urceolate, in fruit coriaceous, horizontally
winged on the back ; lobes 5, incurved. *Stamens* 5 ; filaments
short or long, compressed ; anthers large. *Disk* none. *Ovary*
ovoid, attenuate into a slender style ; stigmas 2–3, capillary,
papillose all over ; ovule subsessile. *Utricle* depressed-globose ;
pericarp membranous or coriaceous at the apex. *Seed* horizontal,

orbicular, depressed ; embryo annular, surrounding the scanty albumen.

Herbs or shrubs, hairy, rarely glabrous ; leaves alternate, rarely subopposite, sessile, linear to oblong, flat or terete, sometimes minute, quite entire ; flowers axillary, small, sessile, solitary or clustered.

DISTRIB. Species about 30, in Central Europe, Temperate Asia, North Africa and Australia.

1. **K. pubescens** (Moquin, Chenop. Enum. 92) ; stem suffruticose, procumbent, glabrous, striate, with short lateral branches ; leaves terete, acute, fleshy, up to 4 lin. long and $\frac{1}{2}$ lin. in diam., glabrous, or pubescent when young ; flowers in clusters of 4–6 ; hermaphrodite flower : perianth about 1 lin. long; lobes ovate, acute ; filaments short; anthers oblong ; female flowers : perianth (including wings) $2\frac{1}{2}$ lin. in diam., wings 5, distinct, membranous, lacerate ; ovary globose-depressed ; stigmas capillary, about 4 times as long as the ovary. *DC. Prodr.* xiii. ii. 131. *Salsola fruticosa, Drège, and S. sativa, Zeyh. ex Moquin in DC. Prodr.* xiii. ii. 131, *not of others.*

VAR. β, **cinerascens** (Moquin in DC. Prodr. xiii. ii. 131) ; leaves with ashy-grey hairs.

SOUTH AFRICA : without locality, *Zeyher*, 1442 ! Var. β ; *Zeyher*, 1449.
CENTRAL REGION : Graaff Reinet Div. ; Graaff Reinet, 2800 ft., *Bolus*, 410 !

The Australian plant mentioned by Moquin is *K. villosa*, Lindl.

Imperfectly known species.

2. **K. salsoloides** (Fenzl, Nov. Stirp. Dec. Vindob. 1839, 74) ; a shrub ; stems procumbent, striate, much-branched, almost glabrous, sometimes rusty-tomentose when young ; branches filiform, about 2 ft. long ; leaves filiform, obtuse, flaccid, reflexed, villous or hirsute when young, naked in age; flowers in clusters of 2–7, pubescent ; perianth-wing $1\frac{1}{2}$ lin. wide, with obovate-subrotundate lobes, flabellately nerved, coloured. *Drège, Zwei Pfl. Documente,* 66 ; *Moquin in DC Prodr.* xiii. ii. 131.

CENTRAL REGION: Prince Albert Div. ; Jackhals Fontein and banks of the Gamka River, *Drège*, 8022.

Moquin (*l.c.*) suggests that this may be the same as *K. pubescens*, Moquin.

VII. SALICORNIA, Linn.

Flowers hermaphrodite or polygamous, immersed in clusters of 3–7 in hollows at the articulations of the branches, free or connate. *Perianth* fleshy, 3–4-toothed. *Stamens* 2, rarely 1 ; filament terete ; anther exserted, lobes globose. *Ovary* ovoid, attenuate at the apex; style lacerate at the apex or with two subulate stigmas papillose all over ; ovule erect, subsessile. *Utricle* membranous, surrounded by the persistent spongy perianth, more or less immersed in the rhachis. *Seed* erect, oblong or ellipsoid, compressed, exalbuminous ; testa thinly coriaceous, covered with hooked hairs ; cotyledons conduplicate.

Annual fleshy leafless herbs or shrubs ; stems articulate, erect or decumbent, glabrous ; branches opposite, dilated and sheath-like at the apex of the articulations ; spikes terminal, cylindrical.

DISTRIB.   Species 9, widely spread on sea-coasts.

Perianth of middle flower pointed or narrowly rounded
  in front ; annual   ...   .   (1) **herbacea.**

Perianth of middle flower broad in front :
  Seeds oblong-oval ; perianth-tip flatly arched ; peren-
    nial   ...   .   (2) **fruticosa.**

  Seeds lenticular ; perianth-tip strongly arched   ...   (3) **natalensis.**

1. **S. herbacea** (Linn. Sp. Pl. ed. i. 3) ; an annual herb ; stem erect, 6–12 in. high, glabrous, branched only in the upper part ; branches patent, cylindrical, thick ; articulations compressed, obtusely lobed ; peduncles thickened above ; spikes cylindrical, ½–1 in. long ; perianth of the middle flower pointed or narrowly rounded in front ; fruit oblong, 1 lin. long ; albumen none or scanty. *Moquin in DC. Prodr.* xiii. ii. 144 ; *Ungern-Sternb. Vers. Syst. Salic.* 45 ; *Boiss. Fl. Orient.* iv. 933 ; *Engl. Pfl. Ost-Afr. C,* 171 ; *Volk. in Engl. & Prantl, Pflanzenfam.* iii. 1A, 77, *fig.* 36, *G–L* ; *Baker and C. B. Cl. in Dyer, Fl. Trop. Afr.* vi. i. 86. *S. annua, Smith in Sowerby, Engl. Bot. t.* 415.

COAST REGION : Cape Div. ; Paarden Island, *Drège* !

Also on the coasts of Europe, North and Tropical Africa and America.

2. **S. fruticosa** (Linn. Sp. Pl. ed. ii. 5) ; an erect fleshy shrub up to 3 ft. high, quite glabrous ; branches terete below, tetragonous above, upper articulations about 4 lin. long ; spikes 1 in. long, formed of 3-flowered closely placed clusters ; perianth concave, flat at the top ; seeds oblong-oval. *Thunb. Prodr.* 1 *and Fl. Cap. ed. Schult.* 1 ; *Ungern-Sternb. Vers. Syst. Salic.* 56. *Arthrocnemum fruticosum, Moquin, Chenop. Enum.* 111, *and in DC. Prodr.* xiii. ii. 151 ; *Baker and C.B. Cl. in Dyer, Fl. Trop. Afr.* vi. i. 86.

VAR. β, **capensis** (Ungern-Sternb. Vers. Syst. Salic. 59) ; branches prostrate ; internodes of the youngest branches dilated and laterally compressed ; teeth faintly keeled on the back. *S. indica, Eckl. & Zeyh. ex Ungern-Sternb. l.c.,* not of *Willd.*

VAR. γ, **paardeneilandica** (Ungern-Sternb. l.c.) ; branches long procumbent, rooting ; twigs ascending, usually herbaceous ; teeth very short, rounded ; flowers over 1 lin. long. *S. herbacea, var. procumbens, Drège ex Ungern-Sternb. l.c.*

VAR. δ, **densiflora** (Ungern-Sternb. l.c.) ; branches ascending, woody high up ; teeth shortly acuminate.

COAST REGION : Cape Div. ; rocky shore north of Camps Bay, *Alexander Prior* !
Wolley-Dod, 3056 ! Hout Bay, *Harvey*, 194 ! Knysna Div. ; salt marshes at
Knysna, 10 ft., *Galpin*, 4428 ! Uitenhage Div., *Zeyher*, 5 ! Port Elizabeth
Div. ; Cape Recife, *Burchell*, 4398 ! Div. ? Redhouse, *Mrs. Paterson*, 497 ! Var. β :
Cape Div. ; about the Ponds and at Salt River, *Burchell*, 667 ! beyond
Uitvlugt, *Wolley-Dod*, 1430 ! rocky shore below Paulsberg, *Wolley-Dod*, 3012 !
by Raapenburg Vley, near the Observatory, *Wolley-Dod*, 2690 ! Var. γ : Cape
Div. ; sand flats between Paarden Island and Tygerberg, *Drège* ! Var. δ : Cape
Div. ; Paarden Island, *Drège*.

EASTERN REGION : Transkei; Kentani coast, in patches of mud and sand
within tidal reach, *Miss Pegler*, 648 !

Also on the shores of Europe and North and Tropical Africa.

3. **S. natalensis** (Bunge ex Ungern-Sternb. Vers. Syst. Salic.
62) ; stem 6–8 in. high, creeping below, glabrous ; articulations
funnel-shaped, 2-lobed, about 6 lin. long, and 2 lin. in diam. at the
mouth ; spikes 6–9 lin. long ; flowers in clusters of 3 ; perianth
½ lin. long, becoming much curved in the upper part ; teeth very
short and blunt ; seed lenticular. *S. indica, Drège, Zwei Pfl.
Documente,* 159, *not of Willd. Arthrocnemum indicum, Bunge,
Reliq. Lehmann. Bot.* 459, *not of Moquin.*

COAST REGION : Cape Div. ; Camps Bay, *Burchell*, 844 ! Uitvlugt, *Wolley-
Dod*, 2691 ! Raapenburg Vley, *Wolley-Dod*, 2398 ! Knysna Div. ; Plettensbergs
Bay, sandhills, *Burchell*, 5310 ! East London Div. ; water edge, near the mouth
of the Nahoon River, *Galpin*, 5673.

EASTERN REGION : Natal ; Durban Bay, *Drège* !

VIII. SUÆDA, Forsk.

Flowers hermaphrodite or by abortion unisexual, minute, bracteate
and bibracteolate. *Perianth* globose, turbinate or urceolate, more
or less fleshy ; lobes 5, equal and without appendages, or 1–2 larger
and inflated or horned, rarely all with a small transverse wing on
the back. *Stamens* 5, more or less perigynous ; filaments short ;
anthers rather large. *Disc* elevated or none. *Ovary* sessile by a
broad base or adnate to the tube of the perianth, rounded or
truncate at the apex ; stigmas 2–5, short, subulate, papillose all
over; ovule subsessile. *Utricle* included in the perianth, membranous or almost spongy. *Seed* horizontal to erect, of various
shapes ; testa crustaceous, smooth ; albumen 0 or divided into
2 small masses ; embryo slender, in a flat spiral, usually green.

Herbs or shrubs, erect or prostrate, simple or branched, rarely farinose or
puberulous ; leaves ternate, fleshy, terete or semiterete, rarely almost flat, quite
entire ; flowers axillary, sessile or subsessile, solitary or clustered.

DISTRIB. Species about 40, on sea-shores throughout the world.

Styles 2 ; seed horizontal ...   . (1) **cæspitosa**.

Styles 3 ; seed vertical   ...   . (2) **fruticosa**.

1. **S. cæspitosa** (Wolley-Dod in Journ. Bot. 1901, 401) ; herbaceous ? cæspitose ; stem much-branched, terete, glabrous ; leaves

closely placed, linear, acute, 4–6 lin. long, ½–1 lin. wide, flat above, convex beneath; spikes about 4 in. long, male in the upper part; bracts 3–4 lin. long, linear, acute; flowers about 1 lin. in diam.; perianth-lobes 5, fleshy, concave, obtuse; stamens included; ovary depressed-globose; styles 2, recurved.

COAST REGION: Cape Div.; Paarden Island, *Wolley-Dod*, 3396! Port Elizabeth Div.; New Brighton, 20 ft., *Galpin*, 6468!

The stems in the type specimen appear to be almost woody.

2. S. fruticosa (Forsk. Fl. Ægypt.-Arab. cix. and 70, Ic. 9); a much-branched evergreen shrub, 2–3 ft. high; branches erect-spreading, glabrous; leaves 4–5 lin. long, 1 lin. wide, slightly convex above, much so beneath, acute, glaucous, blackish when dry; flowers in axillary clusters of 3, the central flower alone perfect; perianth-lobes 5, nearly 1 lin. long, oblong, obtuse, membranous at the margins; stamens 5, short; ovary long ovoid; styles 3, spreading; seed erect, slightly beaked. *Moquin in Ann. Sc. Nat.* 1ʳᵉ *sér.* xxiii. 311, *t.* 20, *and in DC. Prodr.* xiii. ii. 156; *Boiss. Fl. Orient.* iv. 939; *Volk. in Engl. & Prantl, Pflanzenfam.* iii. 1A, 80; *Schweinf. in Bull. Herb. Boiss.* iv. *Append.* ii. 157; *Baker and C. B. Cl. in Dyer, Fl. Trop. Afr.* vi. i. 91. *Chenopodium fruticosum, Linn. Sp. Pl. ed.* i. 221. *Salsola fruticosa, Linn. Sp. Pl. ed.* ii. 324; *Sibth. Fl. Gr. t.* 255; *Sowerby, Engl. Bot. t.* 635. *Lerchea obtusifolia, Steud. Nomencl. ed.* i. 187, 474; *Hiern in Cat. Afr. Pl. Welw.* i. 900. *L. maritima,* γ *fruticosa, O. Kuntze, Rev. Gen. Pl.* ii. 549.

SOUTH AFRICA: without locality, *Burchell*! *Schlechter*!
COAST REGION: Uitenhage Div., *Zeyher*!
CENTRAL REGION: Craddock Div.; without precise locality, *Cooper*, 584!
WESTERN REGION: Namaqualand, *Schlechter*, 13a!

Also in Tropical and North Africa, Europe and through the Orient to Western India.

IX. SALSOLA, Linn.

Flowers hermaphrodite, subtended by 2–3 bracteoles. *Perianth* 5-partite; segments concave, thickened on the back and in front furnished with a large horizontal scarious wing, below the wing free or connate into an indurated cup. *Stamens* 5, usually hypogynous; anthers obtuse or with the connective variously produced. *Ovary* globose or ovoid; style long or short; stigmas 2, spreading, subulate; ovule subsessile or pendulous from the top of a long funicle. *Utricle* included in the persistent winged perianth. *Seed* usually horizontal, orbicular; testa membranous; albumen none; embryo spiral.

Annual or perennial herbs or undershrubs of various habit; leaves alternate or rarely opposite, sometimes wide sheathing, short, long or scale-like, sometimes mucronate; flowers small, solitary in the axils of the upper reduced leaves of the branchlets or spicate.

DISTRIB. Species about 40, chiefly in temperate Asia, North and Tropical Africa, 1 in temperate North and South America and 1 in Australia.

Burchell's 2896 collected on the lower part of Bruintjes Hoogte, Somerset Division, probably represents an undescribed species of *Salsola*, but is in young fruit only.

Flowers in spikes ... ... (1) **fœtida.**

Flowers solitary :
  Leaves 2–3 lin. long, 1 lin. wide, tomentose ... ... (2) **Zeyheri.**

  Leaves ½ lin. long and wide, glabrous ... ... (3) **tuberculata.**

  Leaves minute, densely pubescent ... ... (4) **aphylla.**

1. **S. fœtida** (Del. Fl. Égypte, 57) ; a shrub 1–4 ft. high, with the odour of decaying fish ; branchlets many, pallid, ascending, slender, woody ; leaves minute, alternate, orbicular, fleshy with membranous margins ; flowers forming very short dense cylindrical spikes, solitary in the axils of broad ovate imbricate leaves ; bracteoles like the leaves in size and shape ; perianth-segments ovate-triangular ; anthers with a small appendage ; disk membranous, faintly 10-lobed ; stigmas exserted from the perianth, recurved ; wings of fruiting perianth inserted at the middle of the lobes, white, subequal, obcuneate, erose-laciniate. *Boiss. Fl. Orient.* iv. 961 ; *Hook. f. Fl. Brit. Ind.* v. 18 ; *Barbey, Levant, t.* 8, *fig.* 11 ; *Franch. in Journ. de Bot.* i. 134 ; *Baker & C. H. Wright in Dyer, Fl. Trop. Afr.* vi. i. 87. *Caroxylon fœtidum, Moquin in DC. Prodr.* xiii. ii. 178 ; *Aschers. in Schweinf. Beitr. Fl. Aethiop.* 183. *Chenopodium baryosmon, Roem. & Schult. Syst. Veg.* vi. 269.

CENTRAL REGION : Graaff Reinet Div. ; Graaff Reinet, *Bolus,* 596 !

Also in North and Tropical Africa, Western Asia and India.

Zeyher's 1443, which bears old flowers only, somewhat resembles *S. fœtida,* Del., but differs in having longer, more acute bracts. Its flowers have a cup-shaped disk from which 5 complanate filaments spring.

2. **S. Zeyheri** (Schinz in Bull. Herb. Boiss. v. App. iii. 62) ; an erect much-branched shrub ; branches terete, not articulate, pubescent, ashy grey ; leaves alternate, 2–3 lin. long, about 1 lin. wide, thick and fleshy, concave above, broadly keeled beneath, tomentose ; flowers subsolitary ; bracteoles triangular-ovate or ovate-orbicular, obtuse, thick, very concave, rather tomentose ; perianth-segments lanceolate, rather obtuse, villous on the back ; filaments dilated below ; anthers oblong-hastate, with a minute terminal yellow appendage ; disk fleshy, with 5 entire very obtuse lobes ; style elongate ; stigmas lanceolate-subulate, compressed ; wings of the fruiting perianth inserted near the base, membranous, unequal, 1½–2¼ lin. long, 3 obovate-reniform and 3–4 lin. wide, 2 lanceolate and ½–1 lin. long. *Baker & C. H. Wright in Dyer, Fl. Trop. Afr.* vi. i. 89. *S. aphylla, Hiern in Cat. Afr. Pl. Welw.* i. 900. *Caroxylon Zeyheri, Moquin in DC. Prodr.* xiii. ii. 176.

SOUTH AFRICA : without precise locality, *Zeyher*, 1447 !
WESTERN REGION : Little Namaqualand ; Port Nolloth, *Pearson*, 571 !
Also in Tropical Africa.

3. **S. tuberculata** (Fenzl ex Moquin in DC. Prodr. xiii. ii. 178) ;
a low much-branched shrub ; branches alternate, divaricate, ashy-
grey ; leaves alternate, densely imbricate and congested into globose
nodules, triangular-ovate, subacute, ½ lin. long and wide, glabrous,
those near the flowers suborbicular and rather thick ; bracts orbicular, obtuse, keeled, membranous at the margins ; flowers solitary ;
perianth-lobes ovate-lanceolate, subobtuse, pubescent ; anthers oblong-hastate, dorsifixed, produced above into an ovate obtuse
appendage ; disk obtusely 5-lobed ; stigmas obtuse, compressed ;
wings of the fruiting perianth inserted just below the apex of the
lobes, unequal, less than 1 lin. long, thinly membranous, rosy.
Schinz in Bull. Herb. Boiss. v. *App.* iii. 62 ; *Baker & C. H. Wright
in Dyer, Fl. Trop. Afr.* vi. 90. *Caroxylon tuberculatum, Moquin in
DC. Prodr.* xiii. ii. 178.

SOUTH AFRICA: without locality, *Drège*, 3000a ! *Zeyher*, 1446 !
CENTRAL REGION : Beaufort West Div. ; on the Karoo near Beaufort West,
Henderson, 9 !
Also in Tropical Africa.

4. **S. aphylla** (Linn. f. Suppl. 173) ; a much-branched shrub,
reaching a height of 6 ft. ; branches slender, terete, not jointed,
pallid, pubescent ; leaves alternate, ovate, minute, amplexicaul,
densely pubescent ; flowers solitary ; bracteoles suborbicular,
keeled, pubescent ; perianth-segments ovate-lanceolate, obtuse,
slightly pubescent ; filaments dilated below ; anthers oblong, with
a minute terminal appendage ; disk fleshy, obtusely 5-lobed ; style
elongate ; stigmas subulate ; wings of the fruiting perianth inserted
below the middle of the segments, obovate, membranous, yellowish
or dull purple, the two inner narrower than the others. *Drège,
Zwei Pfl. Documente*, 67 ; *Schinz in Bull. Herb. Boiss.* v. *Append.*
iii. 61 ; *Baker & C. H. Wright in Dyer, Fl. Trop. Afr.* vi. i. 89.
Caroxylon Salsola, Thunb. Diss. Med. Afr. (1785), 8 ; *Moquin in
DC. Prodr.* xiii. ii. 176.

SOUTH AFRICA : without locality, *Forsyth* ! *Zeyher*, 1445 !
CENTRAL REGION : Laingsburg Div.: Witteberg Range, near Màtjesfontein,
Rehmann, 2947 ! Beaufort West Div. ; Karoo at Beaufort West, *Henderson*, 5 ! 6 !
Albert Div. ; without precise locality, *Cooper*, 1383 !
WESTERN REGION : Little Namaqualand ; between Buffels River and Pedros
Kloof, *Drège* ! Van Rhynsdorp Div. ; near Rhynsdorp, *Schlechter*, 8091 !
KALAHARI REGION : Griqualand West ; between the Kloof Village in the
Asbestos Mountains and English Drift, *Burchell*, 2106 ! Orange River Colony ;
Mud River Drift, *Rehmann*, 3612 ! Transvaal ; near Bloemhof, *Burtt-Davy*, 1496 !

Vernacular name "*Brak Ganna*" (Henderson).
Also in Tropical Africa.

Imperfectly known species.

5. **S. bullata** (Fenzl ex Drège, Zwei Pfl. Documente, 94, name
only ; Schinz in Bull. Herb. Boiss. v. App. iii. 62).

6. S. Calluna (Drège, Zwei Pfl. Documente, 50, 56, name only) ; a very dwarf shrub ; stem and branches glabrous ; leaves orbicular, 1 lin. in diam., collected into alternate nodules ; flowers unknown. *Drège in Linnæa*, xx. 204 ; *Moquin in DC. Prodr.* xiii. ii. 191 ; Schinz in Bull. Herb. Boiss. v. App. iii. 62.

CENTRAL REGION : Richmond Div. ; vicinity of Stylkloof about 10 miles west of Richmond, 4000–5000 ft., *Drège* ; Aliwal North Div. ; at the union of the Stormberg Spruit and Orange River, 4200 ft., *Drège*.

Zeyher's 1448 is quoted by Moquin both under this species and under *S. fœtida*, Del. ; from the latter it is obviously distinct. *Henderson's* 10 (known as *Rooi Ganna*) from the Karoo at Beaufort West agrees with Zeyher's plant.

7. S. candida (Fenzl ex Drege, Zwei Pfl. Documente, 96, name only ; Drège in Linnæa, xx. 204 ; Moquin in DC. Prodr. xiii. ii. 191 ; Schinz in Bull. Herb. Boiss. v. App. iii. 62).

COAST REGION : Vanrhynsdorp Div. ; near Mieren Kasteel, 1000–2000 ft., *Drège*.

8. S. geminiflora (Fenzl ex Drège, Zwei Pfl. Documente, 108, name only) ; dwarf, pubescent ; leaves opposite.

COAST REGION : Vanrhynsdorp Div. ; Ebenezer, under 500 ft., *Drège*, 8024a !

ORDER CXI. **PHYTOLACCACEÆ.**

(By A. W. HILL.)

Flowers hermaphrodite or unisexual. *Perianth* inferior, herbaceous or coriaceous, rarely coloured, 4–5-partite, regular or nearly so ; lobes equal or unequal, imbricate. *Stamens* 4–5 or many, usually inserted on a hypogynous disk ; filaments subulate, sometimes connate at the base ; anther-cells parallel, dehiscing longitudinally. *Ovary* superior ; carpels one or many, concrete or distinct ; style none or short ; stigmas as many as the carpels, linear or capitate ; ovules solitary, basal, with a short funicle. *Fruit* of one or many carpels, fleshy or dry. *Seed* erect, compressed ; testa membranous or crustaceous ; embryo peripheric, enclosing the albumen ; cotyledons foliaceous or subcylindrical ; radicle long.

Shrubs or herbs, rarely trees ; leaves alternate, entire ; stipules none or small ; flowers usually racemose, green or whitish, small.

DISTRIB. Species about 60 concentrated in Tropical America.

I. **Microtea.**—*Carpels* 2 or more, united into a 1-celled ovary ; styles 2–5.

II. **Phytolacca** —*Carpels* several, free, or united into a several-celled ovary ; styles free.

The genus *Adenogramma*, Reichb., placed as a suborder of *Caryophylleæ* in Harvey & Sonder's Fl. Capensis, i. 149, has been placed by some authors in

Phytolaccaceæ, a view which we cannot adopt. In Bentham & Hooker's Genera Plantarum, i. 858, the genus is classified under *Ficoideæ*, and Walter in Engler's Das Pflanzenreich, iv. 83, p. 25, upholds this view.

I. MICROTEA, Sw.

Flowers hermaphrodite. *Perianth* 5-partite; lobes equal or nearly so, usually erect in fruit. *Stamens* 3–8, hypogynous; filaments free, filiform; anthers subglobose, 2-celled. *Ovary* 1-celled; styles 2–5, free or united at the base, papillose above. *Fruit* obovoid, more or less warted or spiny, seated on the persistent perianth. *Seed* vertical; embryo peripheric, surrounding the central farinaceous albumen.

Annual, usually branched, herbs; leaves alternate, quite entire; flowers small, white, in long slender spikes or racemes; bracts membranous, persistent.

DISTRIB. Species about 10, chiefly in tropical South America; 4 in South Africa, 1 of which extends into the tropics.

Fruits smooth　…　…　…　…　…　…	(1) **polystachya.**
Fruits muricate　…　…　…　…　…　…	(2) **tenuissima.**
Fruits ribbed :	
Ribs of fruit prominent; spikes elongate, rigid　…	(3) **Burchellii.**
Ribs of fruit inconspicuous; spikes slender　…　…	(4) **gracilis.**

1. **M. polystachya** (N. E. Br. in Kew Bulletin, 1909, 135); undershrub, branched, about 1 ft. high; leaves spathulate-linear, subcoriaceous, ½–¾ in. long, slightly mucronate, apex incurved; inflorescence spicate, 6–9 in. long, rigid; bracts and bracteoles orbicular-lanceolate, acute or acuminate, irregularly toothed or subentire, half as long as perianth; perianth-lobes 5, equal, elliptic or ovate-elliptic, obtuse; stamens 4; filaments longer than the perianth-lobes; ovary ovoid; stigmas 4, filiform, spreading; fruit globose, smooth, ¾ lin. in diam. *Lophiocarpus polystachyus, Turcz. in Bull. Soc. Nat. Mosc.* xvi. 56. *Wallinia polystachya, Moquin in DC. Prodr.* xiii. ii. 143.

WESTERN REGION : Little Namaqualand; by the Orange River, near Verleptpram, *Drège*, 2940! hills at I'us, 2800 ft., *Schlechter*, 11417!

2. **M. tenuissima** (N. E. Br. in Kew Bulletin, 1909, 134); annual, 4–11 in. high, slender; stems simple or branched from the base; leaves few, filiform, ½–1 in. long, mucronate; inflorescence spicate, 1½–6 in. long; bracteoles 2–3 under each flower, about as long as the perianth; perianth-lobes 5, 1 lin. in diam., subregular, obovate, finely 1-nerved; stamens 3–4; filaments longer than the perianth; anthers subglobose, dorsifixed; ovary ovoid, warted, ½ lin. long; styles 2–4; fruit obovoid, 1 lin. long, muricate. *Lophiocarpus tenuissimus, Hook. f. in Hook. Ic. Pl.* xv. 50, *t.* 1463, *figs.* 10–11; *Baker & C. H. Wright in Dyer, Fl. Trop. Afr.* vi. i. 96.

KALAHARI REGION : Transvaal ; near Pretoria, *Rehmann*, 4018 ! *Kirk*, 47 !
sandy places near Batsabelo,˙5000 ft., *Schlechter*, 4058 ! in thickets near Potgieters
Rust, 3800 ft., *Bolus*, 11010 !
Also in Tropical Africa.

3. **M. Burchellii** (N. E. Br. in Kew Bulletin, 1909, 135) ; undershrub, 8–10 in. high ; stems much-branched, ascending ; leaves
sparse, fasciculate, $\frac{1}{2}$–$\frac{3}{4}$ in. long, sessile, terete, linear, subacute,
mucronate ; inflorescence spicate, 3–7 in. long ; flowers $\frac{1}{12}$ in. in
diam., fasciculate, sessile ; bracts ovate, persistent, shorter than the
flowers ; bracteoles obtusely 3-lobed ; perianth-lobes 5, external
smaller, suberect, incurved ; stamens 4–5, more or less alternating
with the perianth-lobes, one (apparently always) opposite the
external lobe ; filaments filiform, longer than the perianth ; anthers
small, extrorse ; ovary ovoid, substipitate ; stigmas divaricating in
pairs ; fruit subglobose, slightly compressed, longer than the perianth,
with 4 main prominent and several shorter well-defined intermediate
ribs. *Lophiocarpus Burchellii, Hook. f. in Benth. & Hook. f. Gen.
Pl.* iii. 50 ; *Ic. Pl.* xv. 49, *t.* 1463, *figs.* 1–9.

KALAHARI REGION: Griqualand West ; at Griqua Town, *Burchell*, 1934 !
Bechuanaland ; Batlapin Territory, *Holub* !

4. **M. gracilis** (A. W. Hill in Kew Bulletin, 1910, 56) ; an
annual, $3\frac{1}{2}$–9 in. high ; stems usually simple below and branching
above ; leaves elliptic-linear, acute, mucronulate, subherbaceous,
$\frac{3}{4}$–$1\frac{1}{2}$ in. long, fasciculate, dense ; inflorescence slender, spicate,
somewhat lax, $2\frac{1}{2}$–6 in. long, the lower portion being barren ; bracts
nearly equal in length to the perianth, triangular-ovate, acute, subentire or slightly 3-toothed ; perianth-lobes 5, $\frac{1}{2}$ lin. long, unequal,
broadly or narrowly elliptic, obtuse ; stamens 4 ; filaments longer
than the perianth ; ovary ovoid ; stigmas 2–4, erect, short, more or
less united ; fruit globose, about $\frac{1}{2}$ lin. in diam., with 4 main ribs
slightly marked and a few indefinite and interrupted intermediate ribs.

KALAHARI REGION : Transvaal ; Komati Poort, 1000 ft., *Schlechter*, 11806 !

II. PHYTOLACCA, Linn.

Flowers hermaphrodite or diœcious. *Perianth* green or slightly
coloured, 5-partite ; lobes equal, oblong, obtuse, spreading or reflexing
at a late stage. *Stamens* 5–25, inserted at the base of the perianth,
rudimentary in the female flowers ; filaments subulate, sometimes
connate at the base ; anthers oblong, incumbent. *Ovary* globose ;
carpels 6–12, free or more or less connate ; styles as many as the
carpels ; ovules solitary, basal, campylotropous. *Fruit* depressed-
globose, fleshy ; carpels free or connate. *Seeds* reniform, compressed,
beaked or obtuse at the base ; testa black, crustaceous, shining ;

embryo annular, enclosing the endosperm ; cotyledons semiterete ; radicle long.

Herbs or shrubs, rarely trees, erect or scandent ; leaves alternate, entire, petiolate ; flowers in dense racemes ; pedicels bracteate and bibracteolate.

DISTRIB. Species 26, tropical or subtropical, mainly natives of America, a few in Africa, Eastern Asia, Himalaya and Asia Minor.

Leaves elliptic-lanceolate, emarginate   ...   .   (1) **heptandra.**

Leaves ovate- or elliptic-lanceolate, acute, mucronate or
    slightly mucronulate :
    Stamens longer than the perianth-lobes   ...   ...   (2) **dodecandra.**

Stamens shorter than the perianth-lobes :
    Pedicels 3-4 lin. long ; bracts broadly subulate,
        situated midway on the pedicel ...   .   (3) **americana.**

    Pedicel very short ; bracts elongate subulate, close
        to the perianth ...   .   (4) **octandra.**

1. **P. heptandra** (Retz. Obs. vi. 29) ; a herb 1-3 ft. high ; stems numerous, erect, angular, smooth ; leaves alternate, elliptic-lanceolate, emarginate, mucronulate, 2-3½ in. long, ⅓-½ or 1 in. broad, decurrent into the petiole, margin entire, slightly inrolled, sometimes waved ; petiole ¼-½ in. long, slightly decurrent into the stem-angles ; racemes glabrous, few- or many-flowered, somewhat drooping, 2-6 in. long ; peduncle as long as the leaves ; bracts subulate ; pedicels as long as or longer than the flowers ; bracteoles 2, linear-subulate ; flowers hermaphrodite ; perianth greenish or reddish-green, urceolate, 1½-2 lin. long ; lobes obovate or elliptic, obtuse ; stamens 7 or 8, about equal in length to the perianth-lobes ; carpels 7 ; fruit orange or yellow. *Walter in Engl. Das Pflanzenr.* iv. 83, 39, *fig.* 14. *P. stricta, Hoffm. in Comm. Goett.* xii. (1796) 27, *t.* 3 ; *Heimerl in Engl. & Prantl, Pflanzenfam.* iii. 1B. 11. *P. resediformis & P. resedifolia, Hort. Berol. ex Moquin in DC. Prodr.* xiii. ii. 30. *Pircunia stricta, Moquin, l.c.* 30 (*incl. vars. resediformis & latifolia*).

SOUTH AFRICA : without locality, *Zeyher,* 1856 !
    COAST REGION : Fort Beaufort Div. ; Kat River Poort, *Drège* ! Kunap River, *Baur,* 1048 ! Queenstown Div. ; Shiloh, *Baur,* 771 ! Hangklip Mountain, near Queenstown, *Galpin,*1804 ! British Kaffraria, *Cooper,* 366 !
    CENTRAL REGION : Graaff Reinet Div. ; Wagenpads Berg, *Burchell,* 2821 ! 2966 ! near Graaff Reinet, *MacOwan,* 1483 ! Albert Div., *Cooper,* 1358 ! Aliwal North Div. ; bank of the Orange River, *Burke* !
    KALAHARI REGION : Orange River Colony, near Harrismith, *Sankey,* 239 ! and without precise locality, *Cooper,* 829 ! Transvaal ; hills above Aapies River, *Rehmann,* 4290 ! near Lydenburg, *Wilms,* 1264 ! and without precise locality, *Sanderson.*
    EASTERN REGION : Transkei ; valleys near Kentani, *Miss Pegler,* 744 ! Griqualand East ; around Kokstad, *Tyson,* 1985 ! Natal ; Itafamasi, *Wood,* 1178 ! hills near Byrne, *Cooper,* 1817 ! near Pietermaritzburg, *Wilms,* 2240 ! and without precise locality, *Gerrard,* 1985 ! *Gueinzius,* 76 !

2. **P. dodecandra** (L'Hérit. Stirp. 143, t. 69) ; a woody climber ; stems slender, sometimes 15-20 ft. long ; leaves alternate, ovate or oblong, acute, 3-4 in. long, cuneate or rounded at the base, distinctly petioled ; racemes dense, many-flowered, at first 3-4 in., finally

½–1 ft. long; rhachis pubescent; pedicels erect-patent, as long as
or longer than the flowers; bracts minute, lanceolate, greenish;
flowers hermaphrodite; perianth greenish, campanulate, 1–1½ lin.
long; segments ovate, finally reflexed; stamens 10–20, much
longer than the perianth; carpels 5–8; fruit bright red; pulp
staining the fingers yellow; carpels about 5, not connate. *Hiern in
Cat. Afr. Pl. Welw.* i. 901; *Baker & C. H. Wright in Dyer, Fl. Trop.
Afr.* vi. i. 97; *Walter in Engl. Pflanzenr.* iv. 83, 42, *fig.* 15.
P. abyssinica, Hoffm. in Comm. Goett. xii. 25, *t.* 2; *T. Thoms. in
Speke, Nile, Append.* 646; *Oliv. in Trans. Linn. Soc.* xxix. 140,
and ser. 2, *Bot.* ii. 348; *Heimerl in Engl. & Prantl, Pflanzenfam.*
iii. 1B, 11; *Engl. Hochgebirgsfl. Trop. Afr.* 209, *and Pfl. Ost-Afr. C.*
175; *Wood, Natal Pl. t.* 263; *Schweinf. in Bull. Herb. Boiss.* iv.
Append. ii. 164, *not of Hook. & Arn. P. elongata, Salisb. Prodr.*
345. *P. lutea, Marsigl. ex Steud. Nom. ed.* 1, 618. *P. scandens,
Hilsenb. et Boj. ex Moquin in DC. Prodr.* xiii. ii. 30. *Pircunia
abyssinica, Moquin in DC. Prodr.* xiii. ii. 30; *Schweinf. Beitr. Fl.
Aethiop.* 58; *var. latifolia, A. Rich. Tent. Fl. Abyss.* ii. 222.
P. saponacea, Welw. Apont. 558.

EASTERN REGION: Natal; between Umzimkulu River and Umkomanzi River,
Drège! Inanda, *Wood,* 949! and without precise locality, *Gerrard,* 130!

Also in Tropical Africa.

3. **P. americana** (Linn. Sp. Pl. ed. i. 441); a herb 6–12 ft. high;
stems branched, grooved, smooth; leaves elliptic-ovate or ovate-
lanceolate, acute, slightly mucronulate, cuneate at the base, 3½–5
in. long, 1–1¾ in. broad, distinctly petioled; racemes on long
peduncles, glabrous, many-flowered, erect or slightly drooping;
flowers on pedicels ¼–⅓ in. long; bracteoles placed midway, shortly
subulate; perianth-lobes orbicular-ovate, obtuse, apex concave,
greenish-white or purplish, 1–1½ lin. long; stamens 10–12, equal
in length to the perianth-lobes; fruit umbilicate, syncarpous;
carpels about 10; styles persistent. *P. decandra, Linn. Sp. Pl.
ed.* ii. 631; *Desf. Fl. Atl.* i. 369; *Bot. Mag. t.* 931; *Moquin in
DC. Prodr.* xiii. ii. 32; *Heimerl in Engl. & Prantl, Pflanzenfam.*
iii. 1B, 10, *fig.* 2 L, M, N *& fig.* 3; *Walter in Engl. Jahrb.* xxxvii.
Beibl. 85, 4, *figs.* 12–14, *and in Engl. Pflanzenr.* iv. 83, 52.
P. decandra, var. acinosa, Moquin, l.c. 33. *P. vulgaris, Crantz,
Instit.* ii. 484.

COAST REGION: Cape Div.; foot of Table Mountain, 400 ft., *Bolus,* 4800!
George Div.; near George, 650 ft., *Schlechter,* 2333.

4. **P. octandra** (Linn. Sp. Pl. ed. ii. 631); a perennial herb;
stems stout, erect, smooth, branched, 2–3 ft. long, grooved; leaves
ovate-lanceolate, acute, mucronate, cuneate at the base, 3–4 in. long,
distinctly petioled; racemes erect, dense, usually pubescent; pedicels
very short; bracteoles elongate-subulate, acute, just below and
almost equal in length to the perianth-lobes; flowers hermaphrodite;
perianth green; lobes orbicular-ovate, obtuse, 1 lin. long; stamens

about 8 ; fruit globose, umbilicate, syncarpous, purplish-black ; carpels about 8 ; styles persistent. *Moquin in DC. Prodr.* xiii. ii. 32 ; *Griseb. Fl. Brit. West Ind.* 58 ; *Benth. Fl. Austral.* v. 143 ; *Baker & C. H. Wright in Dyer, Fl. Trop. Afr.* vi. i. 98 ; *Walter in Engl. Pflanzenr.* iv. 83, 58. *P. octandra, var. grandiflora, Moquin, l.c.* 32. *P. americana, var. mexicana, Linn. Sp. Pl. ed.* i. 441.

EASTERN REGION : Natal ; Fields Hill, 1000 ft., *Wood,* 1934 !

Also in Tropical Africa and America.

Mr. Wood states that the plant was unknown in Natal until the railway cuttings were made on Fields and Bothas Hills between Durban and Maritzburg. Before the railway was opened for traffic the plant appeared in profusion on the banks among the excavated soil. It was said in 1885 that it had not been found more than 200–300 yards from the railway line.

ORDER CXII. **POLYGONACEÆ**.

(By C. H. WRIGHT.)

Flowers regular, hermaphrodite or polygamo-diœcious. *Perianth* inferior, coloured or greenish ; tube short ; lobes 4–6, imbricate. *Stamens* usually 6–9, inserted at the base of the perianth ; filaments free or connate at the base ; anthers 3-celled, dehiscing longitudinally. *Disk* annular. *Ovary* superior, sessile, trigonous or lenticular ; styles 2–3, distinct ; stigmas dilated or capitate ; ovule solitary, orthotropous, basal, sessile or stipitate. *Fruit* an indehiscent trigonous or lenticular nut. *Seeds* similar in shape to the nut ; testa membranous ; albumen abundant ; embryo usually more or less excentric ; cotyledons flat, narrow or broad ; radicle long.

Herbs or shrubs ; leaves alternate, with the base of the petiole dilated into a membranous sheath ; flowers small, racemose or axillary, usually fascicled in the axils of persistent membranous bracts.

DISTRIB. Species about 600, cosmopolitan.

I. **Oxygonum.**—*Flowers* polygamous. *Perianth* accrescent and hardened at the base ; limb marcescent ; segments 5. Herbs.

II. **Polygonum.**—*Flowers* usually hermaphrodite. *Perianth* usually persistent, but not accrescent ; segments 5. Herbs.

III. **Rumex.**—*Flowers* hermaphrodite or unisexual. *Perianth-segments* 6, rarely 4, the outer unchanged in fruit, the inner enlarged and membranous. Herbs, rarely shrubs.

IV. **Emex** —*Flowers* monœcious. *Perianth-tube* accrescent and hardened in fruit ; 3 outer segments accrescent and ending in spreading spines, 3 inner smaller and erect. Rigid herbs.

I. **OXYGONUM**, Burch.

Flowers polygamous. *Perianth-tube* in the hermaphrodite flowers constricted above the ovary, in the male flowers almost obsolete ; limb coloured, 5-lobed, marcescent. *Stamens* 8, inserted on the

perianth ; filaments filiform ; anthers oblong. *Ovary* included in
the perianth-tube ; styles 3, filiform, connate at the base ; stigmas
capitate. *Perianth-tube* accrescent in the fruiting stage and
hardened, often with wings or spines on the three angles, enclosing
the nut. *Seed* erect, turbinate, 3-angled ; embryo straight, subcentral in the albumen ; cotyledons flat, oblong ; radicle short,
superior.

Annual or perennial herbs ; leaves alternate, petioled, entire or pinnatifid ;
ochreæ membranous, truncate ; flowers red or whitish, fascicled in the axils of
the bracts, forming long lax racemes.

Distrib. Species about 11 ; also in Arabia and Tropical Africa.

Fruit broadly winged ... ... (1) **alatum.**
Fruit without wings or spines :
  Fruit obovoid, ribbed ... ... (2) **Zeyheri.**
  Fruit ovoid, acuminate, smooth ... ... (3) **dregeanum.**
Fruit toothed at the base ... ... (4) **delagoense.**
Fruit with 3 spines at the base :
  Leaves entire or with 1–2 short teeth ... ... (5) **calcaratum.**
  Leaves pinnately or bipinnately lobed ; lobes narrow (6) **canescens.**
  Leaves pinnately lobed ; lobes broad ... ... (7) **atriplicifolium,**
    var. **sinuatum.**

1. **O. alatum** (Burch. Trav. i. 548) ; an annual herb, much-
branched from the crown of the root ; stems erect, finely pubescent,
up to 1 ft. long ; leaves lanceolate or rhomboid, entire or pinnatisect,
1–1½ in. long, acute, narrowed gradually at the base into a short
petiole ; ochreæ greenish-white, funnel-shaped, ciliate-dentate or
nearly entire ; racemes very lax, 4–8 in. long ; flowers 2–3 to a
cluster ; pedicels much longer than the ovate membranous bract ;
perianth-limb pinkish or milk-white, 2 lin. long ; perianth-tube in
fruit ovoid, 4 lin. long, broadly winged at the 3 angles. *Meisn. in
DC. Prodr.* xiv. 38 ; *Hook. Ic. Pl.* xiv. 14, *t.* 1321 ; *Hiern in Cat.
Afr. Pl. Welw.* i. 902 ; *Schinz in Bull. Herb. Boiss.* iv. *App.* iii. 57.

Var. β : **Marlothii** (Engl. Jahrb. x. 6) ; fruit rounded, not triangular.

South Africa : without locality, *Zeyher,* 1451 !
Western Region : Great Namaqualand ; Amhub, *Schinz,* 501 !
Kalahari Region : Griqualand West ; plain at the foot of the Asbestos
Mountains, between Kloof village and Wittewater, *Burchell,* 2074 ! Orange River
Colony ; Sand River, *Burke* ! along the Orange River, *Mrs. Barber* ! var. β :
Bechuanaland ; Kuruman, 3900 ft., *Marloth,* 1016 !
Eastern Region : Natal ; near Pietermaritzburg, *Bolus,* 10884 !

2. **O. Zeyheri** (Sond. in Linnæa, xxiii. 100) ; herbaceous, 1–1½ ft.
high, much-branched from the ground ; branches erect, angular,
slender, glabrous or finely pubescent ; leaves 1 in. long, 3-lobed or
rarely pinnatifid ; lobes linear, mucronate, entire or toothed, the
terminal twice as long as the lateral ; ochreæ 3–4 lin. long, with
several subulate teeth ; spikes terminal, interrupted, 2–4 in. long ;

perianth yellow, 2½ lin. long; filaments subulate, with a tuft of brown hairs on the upper side near the base; fruit 3½ lin. long, obovoid, ribbed. *Meisn. in DC. Prodr.* xiv. 38.

COAST REGION: Stellenbosch Div.; Stellenbosch, *Mrs. de Jongh in Herb. Galpin.*, 4429!
KALAHARI REGION: Transvaal; Rustenberg, 4000 ft., *Miss Pegler*, 991! The Willows, near Pretoria, *Burtt-Davy*, 2528! Marabastadt, *Nelson*, 118! Magaliesberg, *Burke! Zeyher*, 1451b. Swaziland; High Veld between Carolina and Mbabane, 5400 ft., *Bolus*, 12260!

3. O. dregeanum (Meisn. in Linnæa, xiv. 487); stem erect or ascending, sparingly branched, terete, faintly sulcate; leaves membranous, 1–1½ in. long, 3–7 lin. wide, lanceolate-oblong, rarely subspathulate, entire or few-toothed, 1-nerved, subobtuse, with or without a mucro, tapering into a 1–4 lin.-long petiole, glabrous; ochreæ greenish, membranous, 4–6 lin. long, truncate, with setæ up to 2 lin. long; racemes 6–10 in. long, straight; bracts membranous, 2–3 lin. long, oblique and subulate at the mouth; flowers hermaphrodite, about 4 in a fascicle, rarely solitary; pedicels 2–3 lin. long, straight or finally decurved, articulated at the apex; perianth whitish or pale-yellow tinged with purple; tube ovoid, 1 lin. long; lobes lanceolate, 2 lin. long; filaments white; anthers dark purple; fruit oblong, trigonous, slightly ribbed between the angles. *Meisn. in Hook. Lond. Journ. Bot.* ii. 551. *O. Dregei, Meisn. in DC. Prodr.* xiv. 38. *O. alatum, var. dregeanum, Sond. in Linnæa,* xxiii. 98.

KALAHARI REGION: Transvaal; near Lydenburg, *Atherstone*! Barberton, *Thorncroft*, 2781! Saddleback Range, 4000–4500 ft., *Galpin*, 615! Jeppes Town Ridges, Johannesburg, 6000 ft., *Gilfillan in Herb. Galpin.*, 6171! Bosch Veld, Elands River, *Rehmann*, 4996; Belfast, *Jenkins*, 6799! Orange River Colony, *Cooper*, 833!
EASTERN REGION: Griqualand East; Enshlenzi, 2500 ft., *Tyson in MacOwan and Bolus, Herb. Austr.-Afr.* 1235! Pondoland; Fort William, 2500 ft., *Tyson*, 2718! Natal; Clairmont, *Wood*, 1321! Durban, *Krauss*, 283! *Peddie! Cooper*, 3059! Coast, *Wood*, 313! Inanda, *Wood*, 628! Verulam, *Wood*, 757! Berg Plateau, *Mudd*! and without precise locality, *Cooper*, 3060! *Gerrard*, 356!

4. O. delagoense (O. Kuntze, Rev. Gen. Pl. iii. ii. 268); herbaceous; stem erect or ascending, up to 3 ft. high; branches slender, more or less distinctly triquetrous, papillose; leaves from linear to lanceolate, 1–2 in. long, 1–3 lin. wide, acuminate, sometimes trifid, pubescent; ochreæ 3 lin. long, tubular, pubescent, setæ about 10, pallid, shorter than the tubular part; racemes terminal, lax, slender, up to 6 in. long; bracts oblique, subulate, not setose, pubescent; pedicels 2–4 in the axil of each bract, shortly exserted; perianth nearly 3 lin. long, cylindrical and densely pubescent below, campanulate and less densely so above; lobes twice as long as the tube, oblong, acute; stamens half as long as the perianth; anthercells diverging at the base; fruit triquetrous, acuminate, not winged nor spiny, slightly toothed at the base, smooth between the angles.

VAR. β, **robustum** (O. Kuntze, l.c. 269) ; stem shorter and less branched than in the type, 1 lin. in diam. ; leaves three times wider.

EASTERN REGION : Delagoa Bay ; in maize stubble, *Scott! Monteiro*, 16 ! *Schlechter*, 12013 ! Var. β : Delagoa Bay, *Kuntze*. Lorenzo Marques, *Wilms*, 1280 !

5. **O. calcaratum** (Burch. ex Meisn. in DC. Prodr. xiv. 38) ; stems much-branched, virgate, glabrous, ribbed ; branches virgate ; leaves entire or with 1–2 short teeth below the middle, about ½–1 in. long, 1 lin. wide, glabrous ; ochreæ about 3 lin. long, brown, entire above, more or less toothed below ; flowers about 3 together ; pedicel slender, 1½ lin. long ; perianth campanulate, 3 lin. long, pubescent outside below ; lobes longer than the tube, oblong ; fruit 3 lin. long, conical, ribbed, pubescent, with 3 horns ½ lin. long below.

KALAHARI REGION : Bechuanaland ; near the sources of the Kuruman River, *Burchell*, 2459 ! 2487/1 !

6. **O. canescens** (Sond. in Linnæa, xxiii. 100) ; stem about 6 in. high, much-branched, woody ; branches angular, clothed with downward pointing white hairs ; leaves pinnately or bipinnately lobed, hairy like the branches ; lobes linear, hair-pointed ; ochreæ with setaceous teeth from a triangular base ; racemes 2–3 in. long, interrupted ; flowers solitary ; pedicels eglandular ; perianth yellow, 2½ lin. long ; segments oblong, acute ; filaments more slender than in *O. Zeyheri*, with a tuft of hairs near the base ; styles 3, as long as the stamens ; ovary pubescent, with 3 basal horns. *Meisn. in DC. Prodr.* xiv. 38.

KALAHARI REGION : Transvaal ; Marabastadt, *Nelson*, 118 ! Rustenburg, *Collins*, 46 ! Aapies River, *Zeyher*, 1451a, *Burke* !

There are also specimens at Kew marked "Grahamstown, Atherstone," but there is doubt as to the correctness of this labelling.

7. **O. atriplicifolium**, var. **sinuatum** (Baker in Dyer, Fl. Trop. Afr. vi. i. 101) ; stems diffuse, ascending, finely pubescent along one side ; leaves deltoid to oblong, deeply pinnately lobed, acute, abruptly cuneate at the base, puberulous, 1½ in. long, 9 lin. wide ; petiole 3 lin. long ; ochreæ 3 lin. long, pubescent, setæ as long as the tube ; racemes lax, usually about 6 lin. long, slender ; bracts similar to the ochreæ, 2–3-flowered ; perianth 1½ lin. long, pale pink ; lobes ovate ; fruiting-perianth 3 lin. long, conical at the base and apex, with a spreading spine from each of the three angles below the middle. *O. sinuatum, Dammer in Engl. & Prantl, Pflanzenfam.* iii. 1A, 30. *O. cordofanum, Dammer, l.c. O. canescens, var. subglabra, Schinz in Bull. Herb. Boiss.* iv. *App.* iii. 57. *Ceratogonum sinuatum, Hochst. & Steud. ex A. Rich. Tent. Fl. Abyss.* ii. 231 ; *Meisn. in DC. Prodr.* xiv. 40. *C. cordofanum, Meisn. l.c.* 39. *C. atriplicifolium, A. Rich. Tent. Fl. Abyss.* ii. 231 ; *Oliv. in Trans. Linn. Soc.* xxix. 141. *Diplopyramis æthiopica, Welw. Apont.* 591 : *J. Britt. in Journ. Bot.* 1895, 75.

KALAHARI REGION : Transvaal ; Boschveld, Elands River, *Rehmann* 4994 ! near Hammanskraal, 4900 ft., *Schlechter*, 4194 ! Pietersburg, 4000 ft., *Bolus*, 10884 ! Pretoria, *Miss Leendertz*, 589 ! Potgieters Rust, *Miss Leendertz*, 1877 ! Portuguese East Africa ; woods at Masuku, 100 ft., *Schlechter*, 12112 !

II. POLYGONUM, Linn.

Flowers hermaphrodite, very rarely polygamous. Perianth coloured, deeply 5- (rarely 4-) lobed or partite, persistent ; segments equal or the outer three rather larger, but little accrescent. Stamens usually 8, inserted near the base of the perianth ; filaments filiform ; anthers oblong, the two cells united only by a short connective. Ovary trigonous with 3 styles or lenticular with 2 styles ; stigmas capitate, usually entire ; ovule solitary, basal. *Fruit* a trigonous or lenticular nut enclosed in the persistent perianth. *Seed* similar in shape to the nut ; embryo eccentric or lateral in the albumen ; cotyledons usually narrow, longer or shorter than the incumbent or accumbent radicle.

Herbs or shrubs, erect, prostrate or scandent ; leaves alternate ; ochreæ membranous, clasping the stem, often fringed with bristles ; flowers usually fascicled in the axils of membranous bracts ; fascicles often arranged in terminal racemes, spikes or panicles ; pedicels articulated.

DISTRIB. Species about 150, cosmopolitan.

Perianth 4-merous :
    Leaves oval and oblong, 2–6 lin. long ... ... (1) **atraphaxoides.**
    Leaves lanceolate, 4–5 in. long ... ... (11) **glutinosum.**
Perianth 5-merous :
    Leaves not more than 1 in. long, linear to oblong :
        Nutlet shorter than the perianth ... ... (2) **aviculare.**
        Nutlet longer than the perianth ... ... (3) **maritimum.**
    Leaves not more than 1½ in. long, ovate or deltoid-
        ovate :
        Nutlet lenticular ; styles 2 ... ... (4) **alatum.**
        Nutlet triquetrous ; styles 3 ... ... (5) **Convolvulus.**
    Leaves more than 2 in. long, oblong ... ... (6) **amphibium.**
    Leaves more than 2 in. long, more or less lanceolate :
        Styles 3 :
            Peduncles glabrous ... ... (7) **serrulatum.**
            Peduncles bairy ... ... (8) **barbatum.**
            Peduncles glandular... ... ... (9) **meisnerianum.**

        Styles 2 :
            Leaves densely hairy beneath :
                Bracts not ciliate ... ... (10) **lanigerum.**
                Bracts ciliate ... ... (12) **tomentosum.**
            Leaves hairy on the nerves and margins only :
                Ochreæ not ciliate at the mouth :
                    Peduncle not glandular ... ... (13) **senegalense.**
                Peduncle glandular :
                    Perianth glandular ... ... (14) **lapathifolium.**

Perianth not glandular   ...   . (15) **strigosum.**
Ochreæ ciliate at the mouth:
Ochreæ naked:
Leaves oblong-lanceolate   ...   . (16) **pedunculare.**
Leaves narrowly lanceolate...   ...   ... (17) **hystriculum.**
Ochreæ densely hairy ...   . (18) **acuminatum.**

1. **P. atraphaxoides** (Thunb. Prodr. 77); an undershrub up to 2 ft. high; stem erect, glabrous, much-branched, terete; leaves oval or oblong, 2–6 lin. long, 1–3 lin. wide, acute (rarely obtuse), subsessile, glabrous, minutely pitted below; ochreæ membranous, nearly as long as the short internodes, whitish or pale brown, long laciniate; flowers in a dense leafy spike at the apex of the branches; perianth 1½ lin. long; lobes 4, oval, the 2 outer concave, keeled, 2 inner flat; stamens 6, much shorter than the perianth; filaments broad, flat, shortly united at the base; ovary compressed, orbicular; styles 2, very short; nut lenticular; cotyledons incumbent. *Meisn. in DC. Prodr.* xiv. 84; *Engl. & Prantl, Pflanzenfam.* iii. 1A, 27, *fig.* 13c; *Bolus & Wolley-Dod in Trans. S. Afr. Phil. Soc.* xiv. 311. *P. Atraphaxis, Thunb. Fl. Cap. ed. Schult.* 385. *P. undulatum, Berg. Descr. Pl. Cap.* 135. *Atraphaxis undulata, Linn. Syst. Veg. ed.* xiv. 345; *Drège, Zwei Pfl. Documente,* 87, 128; *Meisn. in Linnæa,* xiv. 489.

SOUTH AFRICA: without locality, *Villette*! *Smith*! *Oldenburg,* 841!
COAST REGION: Cape Div.; between Cape Town and Table Mountain, on the plain, *Burchell,* 33! Table Mountain, *Ecklon,* 121! *Galpin,* 4433! near Camps Bay, 300 ft., *Bolus,* 2438! Paarl Div.; Paarl Mountains, 1000–2000 ft., *Drège*! French Hoek, 2000 ft., *Schlechter,* 10276! Bredasdorp Div.; Elands Kloof, 600 ft., *Schlechter,* 9759! Caledon Div., *Thom,* 972! Uniondale Div.; near Ongelegen in Longkloof, *Bolus,* 2438! between Van Stadens Berg and Bethelsdorp, under 1000 ft., *Drège*! Van Stadens Hoogte, *MacOwan,* 2075! Sand Fontein, *Burke*! at or near the Lead Mine, *Burchell,* 4492! Albany Div.; Assegai Bosch, *Baur,* 1097! and without precise locality, *Harvey,* 29! *Bowker*!

2. **P. aviculare** (Linn. Sp. Pl. ed. i. 362); a polymorphic species; stem procumbent, branched from the base, upper part ascending, more rarely erect, ribbed, glabrous or minutely scabrous; leaves oblong, lanceolate or linear, up to 1 in. long and 3 lin. wide, but usually smaller, glabrous or scabrous on the margin, entire; ochreæ silvery, membranous, long laciniate; flowers in 3–5-flowered axillary fascicles; pedicels short, jointed at the apex; perianth 1 lin. long, tapering downwards; stamens 8; styles 3; nut ovoid-trigonous, not longer than the perianth, minutely rugose, not shining. *Meisn. in Linnæa,* xiv. 486, *and in DC. Prodr.* xiv. 97; *Engl. Bot. ed.* 3, t. 1229; *Hook. f. Fl. Brit. Ind.* v. 26; *Gage in Rec. Bot. Surv. Ind.* ii. 379, 420; *Bolus & Wolley-Dod in Trans. S. Afr. Phil. Soc.* xiv. 311; *Baker & C. H. Wright in Dyer, Fl. Trop. Afr.* vi. i. 105; *var. dregeanum, Meisn. in Linnæa,* xiv. 486. *P. dregeanum, Meisn. in Linnæa,* xiv. 487; *Drège, Zwei Pfl. Documente,* 58, 131. *P. Dregei, Meisn. in DC. Prodr.* xiv. 98. *P. herniarioides, Drège, l.c.* 93, not

Polygonum.] POLYGONACEÆ (Wright). 465

of Del. ; *var. prostratum, Meisn. Syn. Polyg.* 62, *in Linnæa,* xiv. 486, *and in DC. Prodr.* xiv. 93. *P. Roxburghii, var. longifolium, Meisn. in DC. Prodr.* xiv. 93.

SOUTH AFRICA : without locality, *Oldenburg,* 2168 ! *Bowker* !
COAST REGION : Cape Div. ; Claremont Flats, *Wolley-Dod,* 621 ! Riversdale Div. ; between Zoetemelks River and Little Vet River, *Burchell,* 6821 ! Queenstown Div. ; Engotini, near Shiloh, *Baur,* 966 !
CENTRAL REGION : Victoria West Div. ; Nieuwe Veld, *Drège* ! Albert Div. ; without precise locality, *Cooper,* 1377 !
WESTERN REGION : Little Namaqualand ; near Verleptpram, *Drège* !
KALAHARI REGION : Griqualand West ; along the Vaal River, *Burchell,* 1759/1 ! Basutoland ; without precise locality, *Cooper,* 3057 ! Transvaal ; Pretoria, 4400 ft., *Miss Leendertz,* 411 ! Townlands, Zeerust, *Evans,* 9 ! Groot Vlei, Heidelburg Distr., *Gilfillan in Herb. Galpin.,* 7866 ! Potchefstroom, *Burtt-Davy.* 1771 ! Mathibis Kom, between Lorenzo Marques and Komati River, *Bolus,* 9755 ! Farm Ludlow, Springbok Flats, *Burtt-Davy,* 2494 !
EASTERN REGION : Griqualand East ; near Kokstad, *Haygarth in Herb. Wood.,* 4273 ! Natal ; Van Reenans Pass, 5000 ft., *Wood,* 4564 !

A native of the North Temperate Zone of the Old World ; now widely dispersed. This species much resembles *P. plebeium,* R. Br. (*Prodr.* 420), with which it is easily confused in the flowering state, but can be distinguished when in fruit by the nut of the latter being smooth and shining.

3. **P. maritimum** (Linn. Sp. Pl. ed. i. 361) ; stem prostrate, much-branched and woody at the base ; leaves linear- or oval-oblong, about 7 lin. long and 2 lin. wide, rather fleshy ; ochreæ scarious, 4 lin. long, at first ovate and entire, finally laciniate ; flowers 1–3 in axillary clusters ; perianth 1½ lin. long, scarcely enlarged in fruit, pink or almost white ; lobes 5, obovate, obtuse, 1-nerved ; stamens 8, about half as long as the perianth ; filaments shortly subulate from a broadly ovate base ; ovary trigonous ; styles 3 ; nutlet brown, trigonous, longer than the perianth. *Thunb. Fl. Cap. ed. Schult.* 385 ; *Meisn. in DC. Prodr.* xiv. 88 ; *Engl. Bot. ed.* 3, *t.* 733 ; *Bolus & Wolley-Dod in Trans. S. Afr. Phil. Soc.* xiv. 311.

COAST REGION : Cape Div. ; Camps Bay, *Alexander Prior* ! *Wolley-Dod,* 2884 !

Also in Western and South Europe, North Africa, the Atlantic Islands and North America.

4. **P. alatum** (Buch.-Ham. ex D. Don, Prodr. Fl. Nep. 72) ; an erect or procumbent annual ; stem slender, glabrous, internodes long ; leaves ovate or deltoid-ovate, up to 1 in. long and 9 lin. wide, entire, acute, minutely verrucose ; petiole 3–6 lin. long, broadly winged, wing often cordate at the base ; ochreæ tubular, oblique, membranous, entire ; flowers in capitate few-flowered cymes subtended by a reduced leaf at the apex of slender branches, which are glandular or slightly hairy at their tips ; perianth 1½ lin. long ; lobes about as long as the tube, oblong, obtuse ; stamens 6–8, much shorter than the perianth ; filaments narrowly lanceolate ; anthers dark brown, cells discrete ; ovary compressed ; styles 2, nearly as long as the ovary, united nearly half-way ; nut lenticular, chestnut-brown, shining, very minutely verrucose. *Spreng. Syst. Veg. Cur. Post.* 154 ;

2 H

Hook. f. Fl. Brit. Ind. v. 41 ; *Gage in Rec. Bot. Surv. Ind.* ii. 404, 427 ; *Baker & C. H. Wright in Dyer, Fl. Trop. Afr.* vi. i. 104. *P. punctatum, var. alatum, Meisn. Monogr. Polyg.* 85.

KALAHARI REGION : Transvaal ; Houtbosch (Woodbush) Mountains, 6300 ft., *Schlechter*, 4715 ; Belfast, *Burtt-Davy*, 1394 !
EASTERN REGION : Griqualand East ; Malowe, 4500 ft., *Tyson*, 3093 Natal ; near Byrne, *Wood*, 3432 !

Also in Tropical Africa, Madagascar and Tropical Asia, probably introduced into South Africa.

5. **P. Convolvulus** (Linn. Sp. Pl. ed. i. 364) ; an annual ; stem twining, slender ; leaves cordate-ovate, acuminate, 1½ in. long, 1 in. wide, glabrous, or minutely puberulous on the upper side of the nerves ; petiole 6–9 lin. long, slender ; ochreæ short, obtuse, glabrous ; flowers in axillary clusters collected into long terminal slender racemes ; pedicels 1 lin. long, articulated near the apex ; perianth 1½ lin. long ; 3 outer segments navicular, herbaceous, 2 inner flat, obovate, obtuse ; stamens 8, much shorter than the perianth ; anthers pale buff ; ovary triquetrous ; stigmas 3, capitate, subsessile ; nut nearly 2 lin. long, enclosed in the accrescent perianth, triquetrous, dark brown, shining. *Flor. Dan. t.* 744 ; *Meisn. Monogr. Polyg.* 63, *t. 4, fig. P., and in DC. Prodr.* xiv. 135 ; *Engl. Bot. ed.* 3, *t.* 1227 ; *Hook. f. Fl. Brit. Ind.* v. 53 ; *Gaye in Rec. Bot. Surv. Ind.* ii. 417 ; *Bolus & Wolley-Dod in Trans. S. Afr. Phil. Soc.* xiv. 311.

COAST REGION : Cape Div. ; about Rondebosch Camp, fide *Bolus & Wolley-Dod.* Cathcart Div. ; Glencairn, 4500 ft., *Galpin*, 2408 !
KALAHARI REGION : Transvaal ; Beginsel Farm, near Standerton, *Burtt-Davy*, 1785 !

Probably an introduced weed.   Also in North Africa, Europe and Temperate Asia.

6. **P. amphibium** (Linn. Sp. Pl. ed. i. 361) ; perennial ; stem trailing at the base and rooting from the nodes ; leaves floating, oblong, obtuse, broadly rounded at the base, 4 in. long, 1½ in. wide, smooth at the margins, long petioled ; ochreæ membranous, truncate, not ciliate ; perianth bright red, 2 lin. long ; stamens 5 ; styles 2, united half-way up ; nut lenticular, shining, much shorter than the perianth. *Meisn. in Linnæa,* xiv. 484, *and in DC. Prodr.* xiv. 115 ; *A. Rich. Tent. Fl. Abyss.* ii. 224 ; *Engl. Hochgebirgsfl. Trop. Afr.* 202 ; *Hook. f. Fl. Brit. Ind.* v. 34 ; *Engl. Bot. ed.* 3, *tt.* 1241–2 ; *Gage in Rec. Bot. Surv. Ind.* ii. 423 ; *Baker and C. H. Wright in Dyer, Fl. Trop. Afr.* vi. i. 106.

CENTRAL REGION : Somerset Div. ; Little Fish River, *MacOwan*, 2135 !
KALAHARI REGION : Orange River Colony ; Sand River, *Burke*, 489 ! Vredefort, *Barrett-Hamilton* ! Bechuanaland ; Moshowing River, between Takun and Molito, *Burchell*, 2282 ! Transvaal ; Standerton, Beginsel Farm, common in deep water of vleis, *Burtt-Davy*, 1808 ! Hooge Veld, between Trigardsfontein and Standerton, *Rehmann*, 6746 ! Klerksdorp, Schoon Spruit, *Nelson*, 226 !

7. **P. serrulatum** (Lag. Gen. et Sp. Nov. 14); annual; stem slender, glabrous, 2–3 ft. long, erect or decumbent at the base; leaves lanceolate, acuminate, rounded at the base, subsessile, glabrous or slightly hairy, the lower 3–4 in. long; ochreæ ciliate with long bristles; spikes slender, cylindrical, often several to a stem, 1½–2 in. long; peduncle very slender, glabrous; bracts rigidly ciliate; perianth pink, eglandular, 1 lin. long; stamens 6–8; styles usually 3; nut usually trigonous, polished, shorter than the perianth. *Meisn. in DC. Prodr.* xiv. 110; *Engl. Hochgebirgsfl. Trop. Afr.* 202; *Gage in Rec. Bot. Surv. Ind.* ii. 425; *Bolus & Wolley-Dod in Trans. S. Afr. Phil. Soc.* xiv. 311; *Baker & C. H. Wright in Dyer, Fl. Trop. Afr.* vi. i. 107. *P. salicifolium, Del. Fl. Ægypt. Illustr.* 12; *Meisn. l.c. P. strictum, Meisn. in Linnæa,* xiv. 485, *as to the South African plant. P. abyssinicum, A. Rich. Tent. Fl. Abyss.* ii. 225. *P. scabrum, Hiern in Cat. Afr. Pl. Welw.* i. 903, *an Poir*?

SOUTH AFRICA: without locality, *Banks & Solander! Oldenburg! Pappe! Cunningham!*

COAST REGION: Cape Div.; Cape Flats, near Rondebosch, *Burchell,* 175! Camps Bay, *Burchell,* 383! Black River, *Wolley-Dod,* 622! about the Ponds and at Salt River, *Burchell,* 671! Caledon Div.; Palmiet River, near Grabouw, 700 ft., *Bolus,* 4187! Uitenhage Div.; at or near Uitenhage, *Burchell,* 4234! on marshy ground near the Zwartkops River, *Zeyher,* 227! Div.? Five Islands District, *Bowie!*

CENTRAL REGION: Somerset Div.; near Somerset East, 2800 ft., *MacOwan,* 1055!

KALAHARI REGION: Orange River Colony, *Mrs. Barber!* Parys, 4000 ft., *Rogers,* 2385! Transvaal; Barberton, *Miss Thorncroft,* 28! 4978! MacMac, *Mudd!* Lydenburg, *Wilms,* 1285! Komati Poort, 600 ft., *Rogers,* 84! Pretoria, *Miss Leendertz,* 164!

EASTERN REGION: Transkei; Gekau (Gcua) River, below 1000 ft., *Drège!* Kentani, 1200 ft., *Miss Pegler,* 402! Natal; Durban, 0–1000 ft., *Grant!* *Sutherland! Peddie! Gerrard,* 653! near York, *Wood,* 4321! Polela, *Fourcadi in Herb. Wood.,* 4238! Drakensberg, *Rehmann,* 7021! Alexandra District, Dumisa, *Rudatis,* 287! Swaziland; Hlalikulu, *Miss Stewart,* 23!

Also in Tropical Africa.

8. **P. barbatum** (Linn. Sp. Pl. ed. i. 362); perennial; stem slender, erect, hairy upwards or throughout; leaves lanceolate, acuminate, narrowed to the base, nearly sessile, usually hairy on both surfaces, the lower 4–6 in. long; racemes cylindrical, slender, 2–3 in. long; peduncles hairy; bracts conspicuously ciliate with rigid bristles; perianth pink, eglandular, 1 lin. long; stamens 6–8; ovary shortly conical; styles 3; nutlet trigonous, small, smooth. *Thunb. Fl. Cap. ed. Schult.* 385; *Wight, Ic. t.* 1798; *Meisn. in DC. Prodr.* xiv. 104; *A. Rich. Tent. Fl. Abyss.* ii. 226; *Aschers. in Schweinf. Beitr. Fl. Aethiop.* 170; *Hook. f. Fl. Brit. Ind.* v. 37; *Engl. Hochgebirgsfl. Trop. Afr.* 201; *Schweinf. in Bull. Herb. Boiss.* iv. *App.* ii. 155; *De Wild. Études Fl. Bas et Moyen Congo,* i. 238; *Gage in Rec. Bot. Surv. Ind.* ii. 425; *Baker & C. H. Wright in Dyer, Fl. Trop. Afr.* vi. i. 109.

COAST REGION: Cape Div.; Muizenberg, *Wallich,* 394!

Also in Tropical Africa and Tropical Asia.

9. **P. meisnerianum** (Cham. & Schlecht. in Linnæa, iii. 40);
herbaceous; stem erect, branched, terete, furnished (especially at
the nodes) with retrorse hairs; ochreæ up to 9 lin. long, truncate
and shortly ciliate at the mouth, sparsely hairy; leaves shortly
petioled, narrowly lanceolate, more or less cordate or hastate at the
base, about 3 in. long and 6 lin. wide, sparsely hairy; clusters of
flowers arranged in terminal lax pseudodichotomous panicles; rhachis
slender, covered with stalked glands; bracts small, glandular;
perianth rosy, 2 lin. long; lobes 5, elliptic, obtuse; stamens 5–8,
much shorter than the perianth; ovary triquetrous; style 3-partite;
stigmas capitate; nutlet triquetrous, smooth, shining. *Meisn. in
Mart. Fl. Bras.* v. i. 19, *t.* 1, *fig.* 2, *and .in DC. Prodr.* xiv. 132.
P. chamissœanum, Wedd. in Ann. Sci. Nat. 3ᵐᵉ *sér.* xiii. 254. *P.
refractum, Mart. ex Meisn. in DC. Prodr.* xiv. 132.

KALAHARI REGION : Transvaal ; near Lydenburg, *Wilms,* 1283 ! Hooge Veld,
Rehmann, 6585 ! Belfast, pools at Lakenvlei, *Burtt-Davy,* 1319 !

Also in Brazil.

10. **P. lanigerum** (R. Br. Prodr. 419); perennial; stem stout,
erect, 4–5 ft. high, white-tomentose upwards; leaves shortly
petioled, lanceolate, acute, clothed densely beneath and thinly
above with persistent white tomentum, the lower 6–9 in. long,
1½ in. broad at the middle; ochreæ long, membranous, ciliate;
racemes dense or moderately dense, oblong or oblong-cylindrical,
1–3 in. long; bracts orbicular, shortly ciliate; peduncles pubescent,
eglandular; perianth eglandular, 1½–2 lin. long; stamens usually 6;
styles 2; nut lenticular, orbicular, black, shining. *Meisn. in Linnæa,*
xiv. 485, *and in DC. Prodr.* xiv. 117 ; *Hook. f. Fl. Brit. Ind.* v. 35 ;
Benth. Fl. Austral. v. 271 ; *Gage in Rec. Bot. Surv. Ind.* ii. 424.

KALAHARI REGION : Bechuanaland ; banks of the Moshowing River between
Takun and Molito, *Burchell,* 2281 !
EASTERN REGION : Natal ; Pondoland ; between Umtata River and St. Johns
River, *Drège* ! Umgazi River, *Miss Pegler,* 1562 ! Umhlanga, *Wood,* 1234 ! Durban
Flats, *Mudd* ! *Wood in MacOwan & Bolus, Herb. Norm. Austr.-Afr.* 1347 !
Rehmann, 8748 !

11. **P. glutinosum, var. capense** (Meisn. in DC. Prodr. xiv. 120);
stem ascending, branched, glabrescent; leaves lanceolate, 4–5 in.
long, 1 in. wide, the lower whitish hairy on both surfaces, the upper
glabrescent on both surfaces, densely fuscous glandular-punctate
beneath, nerves densely adpressed hirsute; ochreæ shortly ciliate;
racemes usually in pairs, narrowly cylindrical, dense-flowered;
perianth-segments 4, glandular-punctate outside; stamens 6;
nutlet lenticular. *P. glutinosum, Meisn. in Linnæa,* xiv. 484.

COAST REGION : Komgha Div. ; banks of the Kei River, 500 ft., *Drège.*

The type in India.

12. **P. tomentosum** (Willd. Sp. Pl. ii. 447); perennial; stems
stout, hairy, 3–4 ft. long; leaves shortly petioled, oblong-lanceolate,

Polygonum.] POLYGONACEÆ (Wright). 469

acute, narrowed gradually to the base, the lower 6–8 in. long,
1½–2 in. wide at the middle, persistently hairy on both surfaces;
ochreæ large, membranous and clasping the stem to the tip, ciliate
with long rigid bristles; spikes dense, cylindrical, 2–3 in. long;
peduncles clothed with adpressed hairs; bracts strongly ciliate;
perianth pink, not glandular, 1½ lin. long; stamens usually 7; ovary
globose; styles 2; nut lenticular, orbicular, black, shining, shorter
than the perianth. *Meisn. in Linnæa,* xiv. 483, *and in DC. Prodr.*
xiv. 124 (*incl. vars. sericeo-velutinum, denudatum and strigillosum*);
Oliv. in Trans. Linn. Soc. xxix. 142; *Gartenfl.* 1874, 291, *t.* 810;
Engl. Hochgebirgsfl. Trop. Afr. 202; *Hiern in Cat. Afr. Pl. Welw.*
i. 905; *Dammer in Engl. Pfl. Ost-Afr. C.* 170; *De Wild. Études
Fl. Bas et Moyen Congo,* i. 238; *Schuster in Bull. Herb. Boiss.* 2ᵐᵉ
sér. viii. 706; *Baker & C. H. Wright in Dyer, Fl. Trop. Afr.* vi. i.
110; *var. limogenes, Hiern in Cat. Afr. Pl. Welw.* i. 905. *P.
setulosum, A. Rich. Tent. Fl. Abyss.* ii. 227. *P. limogenes, Vatke ex
Engl. Hochgebirgsfl. Trop. Afr.* 202.

SOUTH AFRICA : without locality, *Banks & Solander*! *Harvey,* 537!
COAST REGION: Cape Div.; near Klein Constantia, *Wolley-Dod,* 433! 1934!
Paarl Div.; Paarl, *Alexander Prior*! Caledon Div.; Zoetemelks Valley, *Burchell,*
7584! George Div.; about the sources of the Keurbooms River, *Burchell,* 5072!
Knysna Div.; Knysna, *Pappe*! Uitenhage Div.; Uitenhage, *Cooper,* 1492!
Alexander Prior! in moist spots in the channel of tha Zwartkops River, *Drège*!
Zeyher, 114! Algoa Bay, *Cooper,* 3061! Port Elizabeth Div.; Port El.zabeth, on
sand hills and rocky shores, *Drège*!
KALAHARI REGION : Orange River Colony; Parys, *Rogers,* 2386! 2387! and
without precise locality, *Cooper,* 3058! Transvaal; Hooge Veld, between Porter
and Trigardsfontein, *Rehmann,* 6642! Lydenburg, *Wilms,* 1282! Pretoria, hills
above Apies River, *Rehmann,* 4291! Bereaparle, *Miss Leendertz,* 605! Houtbosch
(Woodbush) Mountains, *Nelson,* 499! Komati Poort, *Rogers,* 5135!
EASTERN REGION: Transkei Div.; near Bazeia, 2000 ft., *Baur,* 71! Natal:
ranges 30–60 miles from the sea, 2000–3500 ft., *Sutherland*! Durban, *Grant*!
Gerrard, 598! Inanda, *Wood,* 60! 324! river bank, Ipolweni, *Wood,* 4302!
Umzimkulu River, *Drège*! Alexandra Distr., Dumisa, *Rudatis,* 330!

Also in Tropical Africa, North Africa and Tropical Asia.

13. **P. senegalense** (Meisn. Monogr. Polyg. 54); perennial; stems
robust, erect, glabrous, 4–5 ft. high; leaves distinctly petioled,
oblong-lanceolate, glabrous except on the midrib and margins,
acute, narrowed very gradually to the base, the lower 6–8 in. long,
1½–2 in. broad at the middle; ochreæ large, truncate, not ciliate
with bristles; racemes few or several, cylindrical, moderately
dense, 2–3 in. long; bracts broadly ovate, not fringed; pedicels
2–3-nate, finally about as long as the bract; perianth pale pink,
1½ lin. long; stamens usually 7; styles 2; nut orbicular, with
flattened faces, shining, nearly black. *Meisn. in DC. Prodr.* xiv.
123; *Aschers. in Schweinf. Beitr. Fl. Aethiop.* 171; *Engl. Hochgebirgsfl. Trop. Afr.* 202; *Durand & Schinz, Études Fl. Congo,* i. 236;
Schweinf. in Bull. Herb. Boiss. iv. *App.* ii. 156; *Hiern in Cat. Afr.
Pl. Welw.* i. 904; *Baker & C. H. Wright in Dyer, Fl. Trop. Afr.*
vi. i. 111. *P. macrochæton, Fresen. in Flora,* 1838, 601; *A. Rich.
Tent. Fl. Abyss.* ii. 225. *P. nodosum, Pers. Syn.* i. 440; *Garcke in*

Peters, Reise Mossamb. Bot. 503; *Engl. Hochgebirgsfl. Trop. Afr.*
202; *Hemsl. in Journ. Linn. Soc.* xxvi. 343; *Gage in Rec. Bot.
Surv. Ind.* ii. 396. *P. quadrifidum, Meisn. in Linnæa,* xiv. 485, *not
of Buch.-Ham. P. lapathifolium, var. nodosum, Hook. f. Fl. Brit.
Ind.* v. 35.

COAST REGION : Cape Div. ; in sandy plains around Diep River, near Constantia,
below 100 ft , *Drège.* Districts of George and Albany, *Bowie* ! Div. ? Stony Vale,
Gill !
CENTRAL REGION: Prince Albert Div. ; by the River Gamka between Blaauwe-
krans and Wilgebosch Fontein, 2000 ft., *Drège.*
EASTERN REGION : Pondoland ; near St. Johns River, 500–1000 ft., *Drège.*
Also in Egypt, Tropical Africa, Madagascar and Tropical Asia.

14. P. lapathifolium, sub-sp. **maculatum** (Dyer & Trim. in Journ.
Bot. 1871, 36) ; annual; stem glabrous, erect, 2–3 ft. long ; leaves
lanceolate, green, glabrous except for some short bristles on the
underside of the midrib ; ochreæ short, membranous, not ciliate;
petiole short, with short adpressed bristles ; racemes cylindrical,
not very dense, $1\frac{1}{2}$–2 in. long, several to a stem ; peduncles slender,
rough with glands ; bracts orbicular, not ciliate ; perianth pale
pink, glandular, 1 lin. long; styles 2 ; nut lenticular, orbicular,
black, shining, nearly as long as the perianth. *Baker & C. H.
Wright in Dyer, Fl. Trop. Afr.* vi. i. 108. *P. lapathifolium, Bolus
& Wolley-Dod in Trans. S. Afr. Phil. Soc.* xiv. 311. *Persicaria
maculata, S. F. Gray, Nat. Arr. Brit. Pl.* ii. 270.

COAST REGION : Cape Div. ; Vley between Diep River and Retreat, *Wolley-Dod,*
1220 ! near Muizenberg Vley, *Wolley-Dod,* 2633 ! Uitenhage Div. ; Uitenhage,
Zeyher, 771 ! *Alexander Prior* ! Div. ? Five Islands District, *Bowie,* 22 !
CENTRAL REGION : Graaff Reinet Div. ; Graaff Reinet, 2500 ft., *Bolus,* 662 !
Albert Div. ; *Cooper,* 604 !
KALAHARI REGION : Griqualand West ; at Griquatown, *Burchell,* 1914 ! Orange
River Colony ; Harrismith, *Sankey,* 252 ! Transvaal ; Standerton, *Burtt-Dary,*
905 ! Lydenburg, *Wilms,* 1284 ! Zuikerbosch Kop, *Nelson,* 391 ! Pyramid, 4500 ft.,
Rogers, 1032 ! Pretoria, *Miss Leendertz,* 10 ! Rock River, *Burke* ! along Apies
River, *Burtt-Davy,* 835 ! Potchefstroom, 3000 ft., *Bolus,* 3108 ! and without
precise locality, *Holub* ! Springbok Flats, *Burtt-Dary,* 2499 !
EASTERN REGION : Transkei ; Kentani, 1200 ft., *Miss Pegler,* 258 ! Natal ;
Mount Edgcumbe, *Wood,* 1124 ! Durban, *Grant* !
Also in Tropical Africa and the North Temperate Zone.

15. P. strigosum (R. Br. Prodr. 420) ; annual; stem slender,
branched, furnished with recurved bristles ; leaves lanceolate,
truncate at the base, hairy, membranous, distinctly petioled, the
lower 5–6 in. long ; ochreæ not ciliate ; racemes subglobose, 3 lin.
long, at the ends of the slender glandular-hispid branches of a lax
panicle ; bracts orbicular, not ciliate ; perianth pink, not glandular,
1 lin. long ; stamens 5 ; styles 2 ; nut lenticular, subglobose,
shorter than the perianth. *Meisn. in DC. Prodr.* xiv. 134 ; *Benth.
Fl. Austral.* iv. 268 ; *Hook. f. Fl. Brit. Ind.* v. 47 ; *Dammer in
Engl. Pfl. Ost-Afr. C.* 170 ; *Baker & C. H. Wright in Dyer, Fl. Trop.
Afr.* vi. i. 106.

EASTERN REGION : Natal, *ex Dammer.*

Also in Tropical Africa, Tropical Asia and Australia.

16. **P. pedunculare** (Wall. Cat. no. 1718); stem erect, glabrous
or slightly puberulous ; leaves oblong-lanceolate, acuminate, cuneate
at the base, minutely ciliate on the margins, otherwise glabrous,
thinly membranous, up to 6 in. long and 1 in. wide ; petiole slender,
¾ in. long ; ochreæ obtusely truncate, very shortly ciliate at the
mouth only, membranous, strongly nerved ; panicle terminal,
pseudo-dichotomous, axis finely glandular ; bracteoles ovate, very
short, ciliate, brown, membranous ; pedicels about as long as or
shorter than the bracteoles ; perianth-lobes oblong, obtuse ;
stamens 5 ; styles 2, united about half-way ; nutlet orbicular,
biconvex, straw-coloured. *Wight, Ic. t.* 1802 ; *Meisn. in DC. Prodr.*
xiv. 133, *partly* ; *Hook. f. Fl. Brit. Ind.* v. 48 ; *Gage in Rec. Bot.
Surv. Ind.* ii. 426 ; *Schuster in Bull. Herb. Boiss.* 2ᵐᵉ *sér.* viii. 710 ;
Baker & C. H. Wright in Dyer, Fl. Trop. Afr. vi. i. 107.

KALAHARI REGION : Transvaal ; Hooge Veld, Bronkers Spruit, *Rehmann,* 6577,
Vilgeris to Porter, *Rehmann,* 6585.

Also in Tropical Africa, Tropical Asia and Australia.

17. **P. hystriculum** (Schuster in Bull. Herb. Boiss. 2ᵐᵉ sér. viii.
705) ; stem erect, 12–20 in. high, reddish-fuscous, glabrous,
thickened at the nodes ; internodes 4–9 lin. long ; ochreæ naked,
ciliate at the mouth with very rigid setæ nearly 1 lin. long ; leaves
8–24 in. long, 2–3½ lin. wide, narrowly lanceolate, tapering into
the petiole, when young scabrous with adpressed bristles, when
adult scabrous only on the margin and midrib, containing crystals of
calcium oxalate, central part black ; spikes dense, 9–18 lin. long, erect
or inclined ; bracts turbinate, contiguous, naked at the mouth, rarely
with a few setæ at the mouth ; pedicels naked ; perianth rose-coloured,
veins straight or subarcuate ; stamens 6 ; style bipartite above, arms
subarcuate, ⅓ longer than the perianth ; nutlet orbicular, shining,
1 lin. long ; cotyledons accumbent.

WESTERN REGION : Great Namaqualand ; Hinaab, *Fleck,* 323a.

18. **P. acuminatum**, var. **capense** (Meisn. in DC. Prodr. xiv. 114) ;
stem erect, branched, terete, with adpressed hairs when young,
afterwards glabrous ; leaves lanceolate or linear-lanceolate, 5 in.
long, 9 lin. wide, acuminate, shortly cuneate at the base, at first
sparsely short-hairy above, finally scabrous only on the principal
nerves and margin, glandular beneath ; ochreæ tubular, 1 in. long,
truncate, densely clothed with adpressed hairs, ciliate at the mouth
with bristles 4 lin. long ; racemes 2–3 at or near the apex of the
stem, on pubescent peduncles 1½–2 in. long ; flowers few in a cluster ;
pedicels slightly longer than the bracts ; bracts 1 lin. long, rounded,
ciliate ; perianth 2 lin. long, pink ; segments oblong, obtuse ;

stamens about 6, nearly as long as the perianth ; ovary compressed, orbicular ; styles 2, 1 lin. long ; stigmas capitate ; nut plano-convex, 1½ lin. long, 1 lin. wide, almost black.

SOUTH AFRICA : without locality, *Bowie* !
COAST REGION : Riversdale Div. ; Corente River Farm, *Muir in Herb. Galpin*, 5295 ! Uitenhage Div. ; without precise locality, *Zeyher*, 14.
EASTERN REGION : Albany Div. ; by river sides, *Miss Bowker* ! Delagoa Bay, 100 ft., *Schlechter*, 12003 !

The type occurs in Tropical Africa and Tropical America.

III. RUMEX, Linn.

Flowers hermaphrodite or unisexual. *Perianth-segments* usually 6, three inner (*valves*) much enlarged in fruit, three outer unchanged. *Stamens* 6, inserted at the base of the perianth ; filaments short ; anthers linear-oblong. *Ovary* trigonous ; styles 3, spreading ; stigmas fimbriate ; ovule solitary, erect. *Nutlet* trigonous, included in the persistent inner perianth. *Seed* similar in shape to the nutlet ; embryo on one side of the albumen, straight or curved ; cotyledons linear or oblong.

Herbs, more rarely shrubs ; leaves alternate, often cordate or hastate ; ochreæ membranous ; flowers in whorls arranged in leafy or leafless terminal panicles.

DISTRIB. Species about 100. Cosmopolitan, but most numerous in the temperate regions. Some of the following have probably been introduced into South Africa.

Inner perianth-segments toothed or fimbriate :
    Teeth of perianth-segments hooked ... ... (1) **nepalensis.**
    Teeth of perianth-segments not hooked :
        Teeth of perianth-segments stout ... ... (2) **pulcher.**
    Teeth of perianth-segments setaceous :
        Clusters many-flowered ... ... (3) **garipensis.**
        Clusters few-flowered ... ... (4) **fimbriatus.**
Inner perianth-segments entire :
    Leaves usually sagittate :
        Inner perianth-segments not warted... ... ... (5) **sagittatus.**
        Inner perianth-segments warted ... ... (6) **Acetosa.**
    Leaves cordate or hastate (see also 14. *Woodii*) :
        Pedicels jointed at the apex ... ... (7) **Acetosella.**
    Pedicels jointed at or below the middle :
        Leaves hastate ... ... (8) **lativalvis.**
        Leaves cordate-ovate ... ... (9) **cordatus.**
    Pedicels not jointed ... ... (10) **aquaticus.**
    Leaves not cordate, hastate or sagittate :
    Leaves much tapering to the base :
        One perianth-segment with a small wart ... ... (11) **Meyeri.**
    Two or three perianth-segments with large warts (12) **ecklonianus.**
    Perianth-segments not warted ... ... (14) **Woodii.**

Leaves not usually much tapering to the base :
Inner perianth-segments not warted :
  Pedicels jointed at the middle ... ... (13) **dregeanus.**
  Pedicels jointed below the middle ... ... (14) **Woodii.**
Inner perianth-segments (or one of them) warted :
  Inner perianth-segments suborbicular ... ... (15) **crispus.**
  Inner perianth-segments deltoid-ovate :
    Panicle leafy below ... ... (16) **Hydrolapathum.**
    Panicle leafless ... ... (17) **linearis.**
  Inner perianth-segments oblong, obtuse ... (18) **sanguineus.**
  Inner perianth-segments linear-oblong, subacute (19) **conglomeratus.**

1. **R. nepalensis** (Spreng. Syst. ii. 159) ; stem erect, 2–4 ft. high ; radical leaves up to 14 in. long and 5 in. wide, ovate-oblong or triangular-ovate, acute or obtuse, more or less cordate at the base, upper narrowed at the base and often sessile ; panicle terminal ; whorls distant ; pedicels slender ; inner perianth-segments orbicular-ovate, one or all with a large wart on the back, edged with hooked laciniæ, strongly reticulate. *Meisn. Syn. Polyg.* 64, *in Linnæa,* xiv. 492, *and in DC. Prodr.* xiv. 55 ; *Drège, Zwei Pfl. Documente,* 51 ; *Wight, Ic. Pl. t.* 1810 ; *Hook. f. Fl. Brit. Ind.* v. 60 ; *Baker and C. H. Wright in Dyer, Fl. Trop. Afr.* vi. i. 117. *R. obtusifolius, T. Thoms. in Speke, Nile, Append.* 645, *not of Linn.* ; *var., Oliv. in Trans. Linn. Soc.* xxix. 141 ; *var. Steudelii, Hook. f. in Journ. Linn. Soc.* vii. 214. *R. Steudelii, Hochst. ex A. Br. in Flora,* 1841, i. 278 ; *Engl. Hochgebirgsfl. Trop. Afr.* 204 ; *var. cordifolius, A. Rich. Tent. Fl. Abyss.* ii. 229. *R. steudelianus, Meisn. in DC. Prodr.* xiv. 56 ; *var. cordifolius, Meisn. l.c.· R. ramulosus, E. Meyer ex Meisn. in DC. Prodr.* xiv. 55. *R. hamulosus, E. Meyer ex Meisn. in DC. Prodr.* xiv. 693.

SOUTH AFRICA : without locality, *Oldenburg,* 527 !
COAST REGION : British Kaffraria ; Shiloh, 3500 ft., *Baur,* 1134 !
CENTRAL REGION : Aliwal North Div. ; Kraai River, 4500 ft., *Drège* ! Graaff Reinet Div. ; Sneeuw-Berg Range, near Quaggas Drift, 3800 ft., *Bolus,* 2593 !
KALAHARI REGION : Orange River Colony ; Harrismith, *Sankey,* 253 ! Basutoland ; without precise locality, *Cooper,* 2986 ! Transvaal ; Fountain Grove, *Miss Leendertz,* 654 !
EASTERN REGION : Transkei ; Kentani, 1200 ft., *Miss Pegler,* 1407 ! Natal ; in swamps on the Drakensberg, *Evans,* 383 !

Also in Tropical Africa and from Asia Minor and Java.

A specimen collected by Wood (4005) near the mouth of the Umgeni, Natal, consists of a plant with old fruit and two detached leaves. The fruit agrees with that of *R. obtusifolius,* Linn., but the leaves, which are oblong, 8 in. long, 2 in. wide, acute at both ends and on petioles 5 in. long, are more like those of *R. crispus,* Linn. It comes nearest to *R. nepalensis,* Spreng., of the South African species, but may be a hybrid.

2. **R. pulcher** (Linn. Sp. Pl. ed. i. 336) ; a biennial or perennial ; stem erect, up to 2 feet high, ribbed ; leaves oblong or panduriform, sometimes slightly cordate, lower obtuse, upper acute, minutely and irregularly crenulate, petiolate ; panicle terminal with spreading

branches ; whorls distant ; pedicels moderately stout, jointed below
the middle ; inner perianth-segments oblong in fruit, conspicuously
toothed, veins reticulate, with an oblong (often rough) wart on the
back. *Meisn. in DC. Prodr.* xiv. 58 ; *Engl. Bot. ed. 3, t.* 1214 ;
Bolus & Wolley-Dod in Trans. S. Afr. Phil. Soc. xiv. 312 ; *Dammer
in Engl. & Prantl, Pflanzenfam.* iii. 1A. 18 ; *Reichb. Ic. Fl. Germ.*
xxiv. *t.* 183, *figs.* 1–6.

COAST REGION : Cape Div...; roadside near Wynberg, *Wolley-Dod,* 2020 !
Simons Bay, *MacGillivray,* 598 !

Also in Europe, North Africa and Western Asia.

3. **R. garipensis** (Meisn. in Linnæa, xiv. 491) ; an erect glabrous
herb 1–2 ft. high, branched from the base ; lowest leaves lanceolate-
oblong, obliquely cordate at the base, 4 in. long, central lanceolate,
uppermost linear, all entire and acute ; petioles from 1½ in. long in
the lower part to about ½ in. long in the uppermost ; whorls many-
flowered, distant below, approximate above ; pedicels erect or
recurved, 1–3 lin. long, slender ; inner perianth-segments ovate or
ovate-oblong, 1 lin. long, acuminate, with 3–4 unequal teeth on
each side, one bearing a tubercle, the others naked. *Meisn. in DC.
Prodr.* xiv. 60 (*incl. var. elatus*) ; *Drège, Zwei Pfl. Documente,* 93.
R. maritimus, E. Meyer ex Meisn. in DC. Prodr. xiv. 60, *not of
Linn.*

VAR. β, **humilis** (Meisn. in Linnæa, xiv. 491) ; much smaller in all its parts ;
whorls approximate or almost confluent into a leafy raceme. *Drège, Zwei Pfl.
Documente,* 94.

WESTERN REGION : Little Namaqualand ; between Verleptpram and the mouth
of the Orange River, *Drège* ! Var. β, at the mouth of the Orange River, *Drège* !

4. **R. fimbriatus** (Poir. Encycl. v. 65, not of R. Br.) ; root
tuberous ; stem herbaceous, nodose, branched, decumbent ; leaves
subcordate, subrepand, fleshy, petiolate ; racemes terminal ; clusters
3–4-flowered ; pedicels short, slender, jointed at the middle ; inner
perianth-segments (especially towards the apex) finely laciniate-
fimbriate, almost plumose. *Campderá, Monogr. Rum.* 138 ; *Schult.
f. Syst. Veg.* vii. 1471 ; *Meisn. in Linnæa,* xiv. 501, *and in DC.
Prodr.* xiv. 62.

SOUTH AFRICA, ex *Poiret.*

5. **R. sagittatus** (Thunb. Prodr. 67) ; stem erect, flexuous ; leaves
sagittate with obtuse or acute basal lobes, entire, up to 3 in. long
and 1½ in. wide, glabrous, membranous ; petiole up to 1½ in. long ;
panicle terminal, leafless ; whorls few-flowered ; pedicels slender,
jointed near the base ; flowers usually diœcious ; inner perianth-
segments in fruit orbicular-cordate, quite entire, strongly reticulate,
not warted at the base. *Thunb. Fl. Cap. ed. Schult.* 341 ; *Meisn. in
Linnæa,* xiv. 498, *and in DC. Prodr.* xiv. 68 (*incl. var. latilobus*) ;

Drège, Zwei Pfl. Documente, 90, 129, 151, 158 ; *Bolus & Wolley-Dod in Trans. S. Afr. Phil. Soc.* xiv. 312. *R. luxurians, Linn. f. Suppl.* 212, *not Mant.* 64. *R. Burchellii, Campderá, Monogr. Rum.* 135.

VAR. β, **angustilobus** (Meisn. in Linnæa, xiv. 499); leaves about as long as the petioles, hastate- or sagittate-triangular, lobes lanceolate or triangular-oblong, acuminate, lateral shorter than the terminal. *Drège, Zwei Pfl. Documente,* 125, 135.

VAR. γ, **megalotys** (Meisn. in Linnæa, xiv. 499); leaves shorter than the petiole, hastate-triangular, lobes lanceolate or oblong-triangular, lateral divaricate, usually longer and narrower than the terminal. *Drège, Zwei Pfl. Documente,* 90.

VAR. δ, **cordifolius** (Meisn. in Linnæa, xiv. 499); leaves shorter than the petiole, lower cordate, ovate or suborbicular, obtuse or subacute, auricles obtuse or rotundate.

SOUTH AFRICA : without locality, *Harvey,* 540 !

COAST REGION : George Div. ; Longkloof, mountains near the source of the Keurbooms River, *Burchell,* 5086 ! Uitenhage Div. ; Uitenhage, *Alexander Prior* ! Port Elizabeth Div. ; near Port Elizabeth, under 100 ft., *Drège* ! Albany Div. ; Grahamstown, 2000 ft., *MacOwan,* 198 ! and *Herb. Austr.-Afr.* 1944 ! Var. β : Knysna Div. ; Bosch River, under 100 ft., *Drège* ! Alexandria Div. ; between Hoffmanns Kloof and Drie Fontein, 1000–2000 ft., *Drège* !

CENTRAL REGION : Somerset Div. ; foot of the Bosch Berg, *MacOwan,* 198 ! Graaff Reinet Div. ; along the Sundays River, near Monkey Ford, *Burchell,* 2890 ! near Graaff Reinet, 2500 ft., *Bolus,* 667 !

WESTERN REGION : Var. γ : Little Namaqualand ; Silver Fontein, near Ookiep, 2000–3000 ft., *Drège* ! Var. δ : Little Namaqualand ; Silver Fontein, 2500 ft., *Drège ex Meisner.*

KALAHARI REGION : Orange River Colony ; Harrismith, *Sankey,* 251 ! and without precise locality, *Cooper,* 1073 ! Transvaal, near Lydenburg, *Wilms,* 1279 ! Vlakfontein, near Amersfoort, *Burtt-Davy,* 4041 ! Sabie, *Burtt-Davy,* 1546 ! Pretoria, *Miss Leendertz,* 604 ! Shilouvane, *Junod,* 646 ! and without precise locality, *Nelson,* 5 !

EASTERN REGION : Pondoland ; between St. Johns River and Umsikaba River, 1000-2000 ft., *Drège* ! Natal ; Durban, *Peddie* ! Inanda, *Wood,* 96 ! Umlazi River Heights, under 500 ft., *Drège,* and without precise locality, *Gerrard,* 300 ! Swaziland ; on grassy slopes at Hlatikulu, *Miss Stewart,* 97 !

6. **R. Acetosa** (Linn. Sp. Pl. ed. i. 337); perennial, glabrous ; stem 1–2 ft. high, simple ; radical leaves sagittate, 3–6 in. long, on long petioles, glaucous beneath, upper sessile ; panicle leafless, with erect branches ; flowers diœcious ; whorls dense ; pedicels jointed below the middle ; inner perianth-segments in fruit orbicular, quite entire, warted at the base. *Drège, Zwei Pfl. Documente,* 44 ; *Meisn. in Linnæa,* xiv. 496, *and in DC. Prodr.* xiv. 64 ; *Engl. Bot. ed.* 3, *t.* 1223.

SOUTH AFRICA : without locality, *Oldenburg,* 372 !
COAST REGION : Stockenstrom Div. ; Katberg, 3000-4000 ft., *Drège* !
Also in the North Temperate and Arctic regions.

7. **R. Acetosella** (Linn. Sp. Pl. ed. i. 338) ; stem herbaceous, erect, simple or sparingly branched; 3 in. to 1 ft. high ; leaves hastate, more rarely oblong, up to 1¼ in. long but usually much smaller, terminal lobe much larger than the lateral and up to 8 lin. wide, quite entire, petioled ; panicle terminal, leafless, with erect branches ;

pedicels short, jointed at the apex ; inner perianth-segments ovate, entire. *Thunb. Prodr.* 67, *and Fl. Cap. ed.* Schult. 341 ; *Meisn. in Linnæa*, xiv. 496, *and in DC. Prodr.* xiv. 63 ; *Drège, Zwei Pfl. Documente,* 107 ; *Engl. Bot. ed.* 3, *t.* 1224 ; *Bolus & Wolley-Dod in Trans. S. Afr. Phil. Soc.* xiv. 311.

SOUTH AFRICA : without locality, *Burke.*
COAST REGION : Cape Div. ; Rondebosch, between the Steen Berg and Constantia, below 1000 ft., *Drège!* Wynberg Flats, *Alexander Prior!* Westenford, *Wolley-Dod,* 1896 ! Tokay, *Ecklon,* 711 ! Humansdorp Div. ; Storms River, 600 ft., *Galpin,* 4430 !
KALAHARI REGION : Transvaal ; Bezuidenhout Valley, Johannesberg, *Gilfillan, in Herb. Galpin,* 7235 ! Pretoria Kopjes, *Miss Leendertz,* 956 !
EASTERN REGION : Tembuland ; Cala, 4000 ft., *Miss Pegler,* 1685 ! Natal ; Mooi River, 4000 ft., *Wood,* 4032 !

Almost cosmopolitan.

8. R. lativalvis, var. acetosoides (Meisn. in Linnæa, xiv. 497) ; root tuberous ; stem ½–2 ft. or more high, erect, glabrous, often flexuous ; leaves hastate, terminal lobe oblong-lanceolate, 1–2 in. long, 4–6 lin. wide, 3–4 times as large as the ovate or oblong lateral ; petiole of lowest leaves 3–4 in. long ; uppermost leaves lanceolate or linear ; panicle monœcious, leafless ; pedicels capillary, 1–3 lin. long, patent, jointed below the middle ; inner perianth-segments broadly triangular-ovate becoming almost reniform, 2 lin. long, 2–3 lin. wide, entire, naked, nerves very slender. *Meisn. in DC. Prodr.* xiv. 67 ; *Drège, Zwei Pfl. Documente,* 90, 106.

VAR. β, **decipiens** (Meisn. in Linnæa, xiv. 497) ; lobes of the leaves linear or narrowly lanceolate, lateral half as long as the terminal but sometimes wider and often curved upwards. *Drège, Zwei Pfl. Documente,* 100, 112.

SOUTH AFRICA : without locality, *Oldenburg!*
COAST REGION : Clanwilliam Div. ; Clanwilliam 300 ft., *Schlechter,* 8408 ! Cape Div. ; Rondebosch to Hout Bay, under 1000 ft., *Drège* ; var. β : Cape Div. ; between Paarden Island, Tygerberg and Blueberg, under 500 ft., *Drège* ; Paarl Div. ; Klein Drakenstein Mountains, under 1000 ft., *Drège!* Uitenhage, *Harvey!*
WESTERN REGION : Little Namaqualand ; Silver Fontein, near Ookiep, *Drège!*

9. R. cordatus (Desf. Cat. Hort. Par. ed. ii. 40) ; root pyriform ; stem annual, subsimple, erect, ribbed ; lower leaves cordate-ovate, obtuse, 1½ in. long, 1¼ in. wide, on a petiole 1½ in. long, entire, glabrous, cauline hastate, smaller ; panicle terminal, leafless ; branches erect ; flowers polygamo-diœcious in lax whorls ; pedicels slender, jointed at the middle ; inner perianth-segments broadly ovate, entire, strongly reticulate. *Poir. Encycl. Suppl.* iv. 324 ; *Drège, Zwei Pfl. Documente,* 67, 112, 115 ; *Meisn. in Linnæa,* xiv. 500, *and in DC. Prodr.* xiv. 68 ; *Bolus & Wolley-Dod in Trans. S. Afr. Phil. Soc.* xiv. 312. *R. tuberosus, Thunb. Prodr.* 67, *and Fl. Cap. ed. Schult.* 341, *not of Linn. R. sarcorhizus, Link, Enum.* i. 351 ; *Meisn. in Linnæa,* xiv. 501.

SOUTH AFRICA: without locality, *Pappe*! *Sieber*, 134! *Burke*! *Harvey*, 540! *Alexander Prior*!
COAST REGION: Malmesbury Div. ; Groene Kloof (Mamre), *Bolus*, 4321! near Hopefield, *Bolus*, 12811! Cape Div. ; Cape Flats between Cape Town and Simons Bay, *Burchell*, 8547! sand flats beteen Paarden Island, Tygerberg and Blueberg, *Drège*; Lion Mountain, *Ecklon*, 710! near Cape Town, *Bolus*, 2911! Camps Bay, *Alexander Prior*! Caledon Div. ; Bavians Kloof near Genadendal, *Drège*. Knysna Div. ; near Groene Vallei, *Burchell*, 5634!
WESTERN REGION: Little Namaqualand ; Kamies Berg, Kaspars Kloof, Elleboog Fontein and Geelbeks Kraal, *Drège*!
EASTERN REGION: Transkei ; Kentani, near Columba Mission, 1000 ft., *Miss Pegler*, 725!

10. **R. aquaticus** (Linn. Sp. Pl. ed. i. 336) ; perennial ; stem 1–3 ft. high, erect, striate ; radical leaves cordate-ovate or triangular, flat or undulate, glabrous, up to 10 in. long and 6 in. wide, central leaves oblong, acuminate, cordate or rounded at the base, uppermost lanceolate, rounded at the base, all petioled ; panicle terminal, leafy at the base only ; branches erect ; pedicels long, slender, not jointed ; flowers polygamo-diœcious ; inner perianth-segments in fruit ovate-triangular, truncate at the base, slightly warted, membranous, entire, rarely sparingly denticulate, reticulate. *Thunb. Fl. Cap. ed. Schult.* 340 ; *Meisn. in DC. Prodr.* xiv. 42 ; *Reichenb. Ic. Fl. Germ.* xxiv. 20, *t.* 160.

SOUTH AFRICA: ex *Thunberg*.
Also in Western Europe, Temperate Asia and North America.

11. **R. Meyeri** (Meisn. in Linnæa, xiv. 494) ; a herb 1–3 ft. high ; stem more or less branched, erect, glabrous, slightly sulcate, often purple ; lower leaves oblong-lanceolate, tapering downwards, 6 in. long, 1–1½ in. wide, upper lanceolate or subspathulate, all petioled, slightly crisped ; ochreæ ½–1 in. long ; panicle terminal, leafless ; branches few, spreading ; whorls densely many-flowered, the lower separate, the upper confluent ; pedicels ½–1 lin. long, spreading or recurved, jointed near the apex ; perianth tapering downwards ; inner segments ovate or ovate-oblong, entire, subobtuse, one with a small wart at the base, nerves indistinct. *Drège, Zwei Pfl. Documente*, 51, 55, 62. *R. meyerianus, Meisn. in DC. Prodr.* xiv. 50.

CENTRAL REGION: Aliwal North Div. ; by the Kraai River, 4500 ft., *Drège*! Beaufort West ; Rhenoster Kop, *Drège* ; Middleberg Div. ; Sneeuw Berg Range, *Drège*.

12. **R. ecklonianus** (Meisn. in Linnæa, xiv. 493) ; stem erect, almost simple, glabrous ; leaves lanceolate, acuminate at both ends, flat or obscurely undulate, petiolate, up to 7 in. long and ⅔ in. wide, petiolate ; flowers in a terminal leafless panicle with ascending branches ; whorls many-flowered, approximate, sometimes almost contiguous ; pedicels short ; inner perianth-segments ovate, subacute, entire, all or only 2 bearing large warts ; fruit pendulous. *Drège, Zwei Pfl. Documente*, 94, 129. *R. Ecklonii, Meisn. in DC. Prodr.* xiv. 50 ; *Bolus & Wolley-Dod in Trans. S. Afr. Phil. Soc.* xiv. 312.

Coast Region: Cape Div. ; North Hoek, *Milne,* 164 ! Riversdale Div.;
between Zoetmelks River and Little Vet River, *Burchell,* 6817 ! Riversdale,
Schlechter, 1997 ! Uitenhage Div. ; Uitenhage, *Alexander Prior* ! Zwartkops River,
Ecklon ! *Zeyher,* 106 ! Albany Div. ; near Grahamstown, 2000 ft., *MacOwan,*
1410 ! Howisons Poort, *Cooper,* 3056 ! and without precise locality, *Bowker* !
Alexander Prior ! British Kaffraria, *Cooper,* 362 !
Central Region: Graff Reinet Div. ; Graaff Reinet, 2500 ft., *Bolus,* 431 !
Thornton, 200 ! Colesberg, *Shaw* !
Western Region: Little Namaqualand ; mouth of the Orange River, *Drège* !
Kalahari Region: Griqualand West ; Griquatown, *Burchell,* 1926 ! Transvaal ;
various localities, *Wilms,* 1278 ! *Burtt-Davy,* 903 ! 1250 ! 1526 ! 1618 ! 2503 !
Miss Leendertz, 1542 !
Eastern Region: Tembuland ; Bazeia, *Baur,* 282 ! Griqualand East ; Vaal
Bank, *Haygarth in Herb. Wood,* 4191 ! Natal ; Upper Umlaas River, *Wood,* 3161 !

13. R. dregeanus (Meisn. in Linnæa, xiv. 496) ; a glabrous erect
herb ; upper leaves oblong-lanceolate, sometimes the uppermost
hastate, 1 in. long, 5–6 lin. wide, subacute ; panicle terminal, leafless,
about 1 ft. long ; branches ascending ; whorls 6–12-flowered ;
flowers diœcious ; pedicels 3–5 lin. long, slender, arching, jointed at
the middle ; inner perianth-segments oval-orbicular, deeply cordate
at the base, naked, reticulately veined, not warted. *Drège, Zwei
Pfl. Documente,* 159. *R. Dregei, Meisn. in DC. Prodr.* xiv. 68.

Eastern Region: Natal ; near Durban, below 200 ft., *Drège.*
This may be conspecific with *R. Woodii,* N. E. Br.

14. R. Woodii (N. E. Br. in Kew Bulletin, 1909, 187) ; a herb
1–2½ ft. high ; radical leaves linear-lanceolate, lanceolate, elliptic-
lanceolate, more rarely hastate-lanceolate or ovate, acute or obtuse,
3–6 in. long (including the petiole), ¾–1¾ in. wide ; cauline leaves
becoming gradually smaller upwards ; ochreæ entire, truncate,
membranous, 4–7 lin. long ; panicle leafless, glabrous, 2¾–8 in.
long, with suberect branches ; flowers diœcious ; pedicels filiform,
jointed below the middle ; outer perianth-segments in fruit reflexed, linear-lanceolate, acute, concave ; inner elliptic- or orbicular-
cordate, obtuse, 5–8 lin. long, 4–7 lin. wide, entire, thinly scarious,
reticulate, not warted ; nutlet lanceolate, subacuminate, triquetrous
or almost winged.

Central Region: Somerset Div. ; Boschberg, 4000 ft., *MacOwan,* 1857 !
Kalahari Region: Orange River Colony ; low-lying Veld at Bethlehem,
Richardson ! Transvaal ; north and south of Carolina in sandy soil, 5800 ft.,
Burtt-Davy, 2714 ! Ermelo Experimental Farm, 5575 ft., *Burtt-Davy,* 3919 !
Wemmers Hoek, Lydenburg, 5400 ft., *Burtt-Davy,* 7625 !
Eastern Region: Natal ; Itafamasi, *Wood,* 644 ! near Lambonjwa River,
4000 ft., *Wood,* 3583 !

15. R. crispus (Linn. Sp. Pl. ed. i. 335) ; an erect robust herb
2–4 ft. high ; radical leaves 5–9 in. long, acute, acute to cordate at
the base, undulate ; cauline leaves lanceolate, the upper almost
linear, all petioled ; panicles terminal, leafless or with 1–2 leaves in
the lower part ; whorls approximate, many-flowered ; pedicels
slender, jointed near the base ; inner perianth-segments sub-

orbicular, entire, about 2 lin. in diam., sometimes wider than long, strongly reticulate, all with a large wart on the back. *Meisn. in DC. Prodr.* xiv. 44; *Engl. Bot. ed.* 3, *t.* 1218; *Bolus & Wolley-Dod in Trans. S. Afr. Phil. Soc.* xiv. 312.

SOUTH AFRICA : without locality, *Wallich* !
COAST REGION : Cape Div.; Vygeskraal River, *Wolley-Dod*, 3592 !
KALAHARI REGION : Transvaal ; Pretoria, along the Aapies River, *Burtt-Davy*, 833 !

Also in Europe, North Africa and Temperate Asia ; introduced into North America.

16. **R. Hydrolapathum** (Huds. Fl. Angl. ed. ii. 154); an erect robust herb ; radical leaves 1–3 ft. long, 2–7 in. wide, lanceolate-oblong, acute, petioled ; cauline leaves gradually becoming smaller upwards, passing from lanceolate to linear ; panicle terminal, usually leafy below; whorls separate, many-flowered ; pedicels short, jointed near the base ; inner perianth-segments $2\frac{1}{2}$ lin. long in fruit, triangular-ovate, truncate or rounded at the base, entire or faintly toothed, prominently veined, warted. *Meisn. in DC. Prodr.* xiv. 47; *Engl. Bot. ed.* 3, *t.* 1220; *Reichb. Ic. Fl. Germ.* xxiv. 24, *t.* 165.

KALAHARI REGION : Transvaal ; Pretoria, Bereaparle, *Miss Leendertz*, 394 !
Also in Europe.

17. **R. linearis** (Campderá, Monogr. Rum. 90); stem branched ; leaves shortly petioled, oblong-linear, acuminate, serrate, 3–4 in. long, 5–6 lin. wide ; ochreæ long, acute ; racemes terminal and axillary, the upper leafless ; whorls distant ; pedicels short, jointed at the base ; inner perianth-segments deltoid-ovate, rather obtuse, warted. *Roem. & Schult. Syst.* vii. 1407 ; *Meisn. in DC. Prodr.* xiv. 50.

VAR. β, **affinis** (Meisn. in DC. Prodr. xiv. 50); dwarf ; stem ascending, branched ; leaves narrowly lanceolate, tapering towards both ends, shortly petioled ; panicle short, leafless, few-branched ; whorls confluent, dense ; pedicels shorter than the fruiting perianth, jointed at the apex ; inner perianth-segments in fruit deltoid-ovate, subobtuse, reticulate, all or two warted. *R. linearis, Meisn. in Linnæa,* xiv. 495 ; *Drège, Zwei Pfl. Documente,* 58.

SOUTH AFRICA : without locality, *Sonnerat ex Campderá*.
CENTRAL REGION : Var. β : Victoria West Div.; Nieuwveld, between Brak River and Uitvlugt, 3000–4000 ft., *Drège*.

18. **R. sanguineus** (Linn. Sp. Pl. ed. i. 334) ; a perennial ; stem up to 4 ft. high, slender, red ; leaves up to 6 in. long, oblong-lanceolate contracted above the usually cordate base, slightly undulate, entire or minutely crenulate, petiolate, midrib red ; panicle lax, usually leafless ; whorls distant, many flowered ; pedicels slender, jointed near the base ; inner perianth-segments oblong, obtuse, rounded at the base, entire, one or all with a sub-globose tubercle. *Meisn. in DC. Prodr.* xiv. 49 ; *Engl. Bot. ed.* 3, *t.* 1211.

KALAHARI REGION : Transvaal ; near Lydenberg, *Wilms,* 1276 !

Also in Europe, Western Asia ; introduced into North America.

19. **R. conglomeratus** (Murr. Prodr. Stirp. Gœtt. 52) ; a perennial ; stem up to 4 ft. high, simple or slightly branched, glabrous ; leaves oblong-lanceolate, subacute, rounded or cordate at the base, irregularly crenate, 8 in. long, 2 in. wide, petioled ; panicle terminal, more or less leafy, branches spreading ; whorls many-flowered, distinct ; pedicels slender, jointed below the middle ; inner perianth-segments linear - oblong in fruit, subacute, quite entire, with oblong warts on the back. *Meisn. in Linnæa,* xiv. 492, *and in DC. Prodr.* xiv. 49. *R. Nemolapathum, Ehrh. Beitr.* i. 181 ; *Meisn. in Linnæa, l.c., not of Wallr.*

SOUTH AFRICA : without locality, *Drège.*
COAST REGION : Cape Div. ; Kenilworth, near Cape Town, 100 ft., *Bolus,* 7039 ! Kloof below Constantia Nek, *Wolley-Dod,* 2425 ! Paarl, ex *Meisner.*

Also in Europe, North Africa and Western Asia ; introduced into North America.

Imperfectly known species.

20. **R. lanceolatus** (Thunb. Prodr. 67) ; stem erect, 2 ft. high, branched, rather flexuous, sulcate, glabrous ; leaves lanceolate, acute, rounded or shortly attenuate at the base, about 6 in. long, margins reflexed, very slightly crisped, glabrous, petiolate ; racemes terminal and axillary, solitary, undivided, slender, interrupted at the base ; whorls 3–8-flowered ; pedicels 2–3 times as long as the perianth, jointed above the middle ; inner perianth-segments ovate-oblong, suddenly narrowed at the middle into a ligulate point, entire ; fruit unknown. *Thunb. Fl. Cap. ed. Schult.* 340 ; *Campderá, Monogr. Rum.* 148 ; *Meisn. in Linnæa,* xiv. 493, *and in DC. Prodr.* xiv. 50.

SOUTH AFRICA : without locality, *Thunberg, Drège.*

An imperfectly known species. A plant collected in South Africa by Wallich agrees in many respects with it, but its perianth-segments are not suddenly contracted at the middle. Also one collected in Natal, near Pietermaritzburg by Wilms (2243c). A similar plant was collected at Genadendal, Caledon Div., by Alexander Prior.

21. **R. spathulatus** (Thunb. Prodr. 67) ; stem erect, striate, glabrous, purple, about 1 ft. high ; branches few ; leaves obovate, obtuse, the upper lanceolate, erecto-patent, 1 in. long ; petiole a little shorter than the blade ; flowers verticillate, erect ; inner perianth-segments warted. *Thunb. Fl. Cap. ed. Schult.* 340 ; *Meisn. in DC. Prodr.* xiv. 50.

SOUTH AFRICA : without locality, ex *Thunberg.*

A very imperfectly known species which Meisner suggests may be the same as *R. Meyeri,* Meisn.

IV. **EMEX**, Neck.

Flowers monœcious. *Male flowers :* Perianth-segments 5–6, equal, patent. Stamens 4–6 ; filaments filiform ; anthers ovate. *Female flowers :* Perianth urceolate ; tube ovoid ; lobes 6 in 2 series. Ovary included in the tube, trigonous ; styles 3, short, patent ; stigmas dilated, fimbriate. Perianth in fruit enlarged and hardened ; tube 3- or 6-angled, more or less transversely ribbed between the angles ; outer lobes patent, spiny, inner erect, obtuse or aristate. *Nutlet* free within the perianth-tube, triquetrous ; pericarp membranous or scarcely crustaceous. *Seed* subterete ; embryo much curved, lateral or almost peripheral ; cotyledons narrow, longer than the radicle.

Rigid herbs ; leaves alternate, petioled ; ochreæ more or less membranous, quickly splitting up or falling off ; flowers in axillary fascicles or the upper by the abortion of the leaves appearing racemose ; male flowers pedicelled, female sessile.

DISTRIB. A second species in the Mediterranean region.

1. **E. australis** (Steinh. in Ann. Sci. Nat. 2me sér. ix. 195, t. 7, fig. 16) ; a herb ; stem branched, 3 lin. in diam., ribbed when dry, glabrous ; leaves ovate or oval, with the outer basal angles rounded, obtuse, cuneate into the 2 in.-long petiole, 3 in. long, 2 in. wide, entire or obscurely crenate, glabrous ; outer fruiting perianth 6 lin. long, less indented between the angles than in *E. spinosa,* Linn., spines spreading ; inner segments broadly ovate, mucronate, strongly reticulately nerved. *Benth. Fl. Austr.* v. 262 ; *Wood, Natal Pl. t.* 360. *E. Centropodium, Meisn. in Linnæa,* xiv. 490, *and in DC. Prodr.* xiv. 40 ; *Bolus & Wolley-Dod in Trans. S. Afr. Phil. Soc.* xiv. 312 ; *Burtt-Davy in Trans. Agric. Journ.* vii. (1909) 654. *E. Podocentrum, Meisn. ex Drège, Zwei Pfl. Documente,* 93, 99, 112, 128. *E. spinosa, var. capensis, Campderá, Monogr. Rum.* 59. *Rumex spinosus, Thunb. Prodr.* 67, *and Fl. Cap. ed. Schult.* 341, *not of Linn. Podocentrum, Burch. ex Meisn. in Linnæa,* xiv. 489. *Vibo australis, Greene, Man. Bot. San Francisc. Bay,* 44.

SOUTH AFRICA : without locality, *Oldenburg* ! *Forster* ! *Zeyher* ! *MacOwan,* 2222 !

COAST REGION : Cape Div. ; Sand Flats between Paarden Island, Tygerberg and Blueberg, *Drège* ; near Cape Town, *Wilms,* 3552 ! Camp Ground, *Wolley-Dod,* 623 ! Wynberg Flats, *Alexander Prior* ! Paarl Div. ; by the Berg River near Paarl, *Drège* ; Riversdale Div. ; near Riversdale, *Schlechter,* 1865 ! between Zoetemelks River and Little Vet River, *Burchell,* 6818 ! Uitenhage Div. ; Winterhoek Mountains, *Krauss,* 784 ; Port Elizabeth Div. ; near Port Elizabeth, *Drège* ! Albany Div. ! Grahamstown, *MacOwan,* Fish River, 400–500 ft., *Baur,* 1056 !

CENTRAL REGION : Graaff Reinet Div. ; along the Sundays River near Monkey Ford, *Burchell,* 2877 !

WESTERN REGION : Little Namaqualand ; between Verleptpram and the mouth of the Orange River, under 1000 ft., *Drège* !

KALAHARI REGION : Griqualand West ; Asbestos Mountains at the Kloof Village, *Burchell*, 1687 !

EASTERN REGION : Natal ; without precise locality, *Cooper*, 12í6 !

Also in Australia.

This species differs from *E. spinosa*, Campd., in the larger, less rugose outer fruiting perianth with longer spinescent segments, and the inner erect segments being broader and more rounded.

ORDER CXIII. **PODOSTEMACEÆ.**

(By A. W. HILL.)

Flowers regular, hermaphrodite or diœcious. *Perianth* inferior, membranous, 3–5-partite or represented by minute scales or entirely absent. *Stamens* hypogynous, 1 to many ; filaments free or united ; anthers oblong, 2-celled, dehiscing longitudinally ; pollen 1–2-celled. *Ovary* sessile or stipitate, 1–3-celled ; styles 2–3 ; stigmas capitate or decurrent ; placentas axile or parietal ; ovules many, anatropous. *Capsule* 1–3-celled, dehiscing by valves, often septifragal. *Seeds* many, minute, sessile, exalbuminous ; cotyledons flat ; radicle very short.

Submerged herbs of various habit, often resembling mosses, foliaceous and frondose hepaticæ and algæ ; flowers minute, variously arranged.

DISTRIB. Species about 130, chiefly in the tropics and subtropics of the Southern hemisphere.

I. **Tristicha.**—*Flowers* hermaphrodite. *Perianth* equally 3-partite.

II. **Sphærothylax.**—*Flowers* hermaphrodite. *Perianth* of 2 small linear segments.

III. **Hydrostachys.**—*Flowers* diœcious. *Perianth* none.

I. **TRISTICHA,** Thouars.

Flowers hermaphrodite. *Perianth* membranous, 3-partite ; segments oblong, obtuse, slightly imbricate. *Stamen* 1, hypogynous ; filament long ; anthers oblong. *Ovary* oblong, 3-celled ; styles 3, short, linear ; stigmas decurrent on the inner side of the styles ; placentas axile ; ovules numerous. *Capsule* oblong, crustaceous, septicidally and septifragally 3-valved.

Herbs with a moss-like habit, growing under water and attached to rocks by a flat-lobed thallus ; stem slender, much-branched. Leaves small, sessile, entire, pellucid, 1-nerved. Flowers terminal or axillary ; pedicels stiffly erect, each with 3 bracts at the base.

DISTRIB. Species 3, variable and difficult of limitation.

Also in Tropical Africa, the Mascarene Islands and Tropical America.

1. **T. hypnoides** (Spreng. Syst. iv. Cur. Post. 10) ; stems cæspitose, often very short, branched, more or less compressed ; leaves broadly

ovate to elliptic, usually trifarious, about ½ lin. long, the upper
sometimes longer and narrower; flowers solitary, terminal on short
branches near the apex of the stem; peduncle up to 9 lin. long in
fruit, terete, rigid, erect; perianth-segments 3, equal, oblong-elliptic,
cymbiform, pellucid; filament compressed; anthers ovoid, obtuse;
ovary shortly stalked, ellipsoid, trigonous; capsule ellipsoid, 1 lin.
long; valves 3-nerved. *Weddell in DC. Prodr.* xvii. 44; *Tulasne
in Ann. Sci. Nat.* 3ᵐᵉ *sér.* xi. 112, *in Arch. Mus. Hist. Nat. Par.* vi.
(1852), 186, *and in Mart. Fl. Bras.* iv. i. 272; *Engl. Hochgebirgsfl.
Trop. Afr.* 228; *Warming in Kgl. Danske Selsk. Skrifter, ser.* 6, ix.
ii. 107, *figs.* 1–6, *and* xi. i. 28, 61, *figs.* 20–23, *excl syn. T. trifaria;
Baker & C. H. Wright in Dyer, Fl. Trop. Afr.* vi. i. 121. *T. bryoides,
Gardn. in Calc. Journ. Nat. Hist.* vii. (1847), 178, *and Wight,
Ic. t.* 1920. *T. Philocrena, Steud. Nomencl. Bot. ed.* 2, ii. 715. *T.
alternifolia, var. pulchella, Warm. l.c.* xi. i. 36. *Dufourea hypnoides,
A. St. Hil. in Mém. Mus. Paris,* x. (1823), 472. *Philocrena pusilla,
Bong. in Mém. Acad. Petersb.* 6ᵐᵉ *sér.* i. (1832), 80, *t.* 6.

COAST REGION : Albany Div.; Gwacwaba River, near King-Williams Town,
submerged or nearly so, *Sim,* 100 !
CENTRAL REGION : Calvinia Div.; Doorn River, *Drège,* 2991 !

Also in Tropical Africa.

II. SPHÆROTHYLAX, Bischoff.

(ANASTROPHEA, Wedd.)

Flowers hermaphrodite, subtended by 2 minute scales. *Stamens* 2;
filaments united to the apex; anthers 2, sessile at the top of the
column, 2-celled, or solitary and 4-celled according to *Weddell,*
dehiscence longitudinal. *Staminodia* 2, small, linear. *Ovary* ovoid,
8-ribbed; valves 2, unequal, the larger persistent, the smaller
deciduous. *Seeds* ellipsoid, compressed.

Stems slender and branched, arising from a Marchantia-like base. Leaves
elongate, laciniate. Spathellæ produced both on the thallus and on the elongated
stems.

DISTRIB. Species 4, three in Tropical Africa.

1. S. algiformis (Bisch. ex Krauss in Flora, 1844, 426, t. 1);
stem ½–¾ in. long, about ½ lin. broad, flattened, somewhat thickened
in every part closely applied to substratum, more or less repeatedly
subdichotomously branched; branches for the most part spreading;
leaves linear-spathulate, about ¼ lin. long; flowers arising from
axils of branches; gemmæ (*Tulasne*) very minute, composed of very
few scale-like leaves and occasionally scarcely visible; pedicels
bearing fruits ascending, ½–¾ lin. long, surrounded at the base with
remnants of withered spathellæ; capsule scarcely ½ lin. long, apex
obtuse, bare or crowned with short subulate stigmas; valves, the
smaller with 5 nerves breaking away, the larger with 7 nerves
persisting on pedicel for some time; seeds fairly large, blackish-

green.   *Drège in Linnæa*, xx. 244 ; *Tulasne in Ann. Sci. Nat.* 3me *sér.*
xi. 105 ; *Monogr.* 161 ; *Wedd. in DC. Prodr.* xvii. 78..

EASTERN REGION : Tembuland ; Umtata River, under 1000 ft., *Drège* !

III. HYDROSTACHYS, Thouars.

Flowers dioecious.   *Perianth* none.   *Male* : Stamen solitary ;
filament very short ; anther oblong, 2-celled.   *Female* : Ovary
oblong, 1-celled ; placentas 2, parietal.   *Styles* 2, long, filiform,
connate in the lower half.   *Capsule* oblong, seated in the hollow
of a concave accrescent bract, the apex of which is often reflexed,
2-valved.   *Seeds* many to each placenta.

Stem short, tuberous ; leaves large, pinnately branched, usually bearing protuberances on the petiole and rhachis ; spikes simple, tufted, peduncled, much
resembling the fruiting spikes of *Plantago major*, Linn., each flower seated in the
axil of a persistent bract.

DISTRIB.   Species about 15, chiefly in the Mascarene Islands, three in Tropical
Africa.

1. **H. natalensis** (Wedd. in DC. Prodr. xvii. 88) ; leaves 4–14
in. long. 3–4-pinnatisect ; rhachis compressed above the base and
flattened, tapering towards the apex, about 1 lin. broad, more or
less densely covered with imbricate subulate or ovate acute protuberances ½–1 lin. long ; pinnæ subopposite, numerous, spreading,
lanceolate, brush-like, ½–1¼ in. long, densely covered with subulate
acute protuberances ; pinnules 1–1½ lin. long, finely divided into
numerous irregular linear lacinulæ ; female spikes 5–12 in. long,
simple, crowded above with flowers, flowering portion 2½–6 in. long ;
peduncles stout, 2–2½ in. in diam., naked below ; bracts boat-
shaped, obovate, 1½ lin. long, rounded above, margins membranous ;
nerves 3–5, the median one decurrent into the axis ; capsule ovate-
oblong, about 1 lin. long, naked at the base ; male spikes 5–10 in.
long, flowering portion cone-like, ¾–2½ in. long ; peduncles simple or
branching below, rugose or covered with minute protuberances ;
bracts crowded, imbricate, orbicular-ovate, obtuse or subacute,
about 1 lin. in diam., transversely rugose on the back, with thickened
margin ; anther-cells separated.

EASTERN REGION: Natal ; Umgeni Falls, *Sanderson* ! *Mudd* ! *Schlechter*, 3314 !
in the river at Ulwimbu, 2500 ft., *Sutherland* ! on stones in Umvoti River, *Wood*,
292 ! and in *MacOwan & Bolus, Herb. Norm. Austr.-Afr.* 1015 ! *Reynolds*, 1643 !
rapids in mountain streams, *McKen*, 13 !

ORDER CXIV. **CYTINACEÆ.**

(By A. W. HILL.)

Flowers regular, hermaphrodite or unisexual. *Perianth* simple, sometimes fleshy; tube adnate to the ovary and often produced above it, solid in the male flowers; lobes 3–10, imbricate or valvate, 1–2-seriate. *Stamens* 8 to many, free or united; anthers surrounding a central column and dehiscing by apical slits, or forming a lobed ring inside the perianth-tube with long sinuous cells like those of some *Cucurbitaceæ*, dehiscing longitudinally but folded so that the slits are near together and across the ring. *Ovary* inferior, 1-celled; placentas parietal or numerous and pendulous from the top of the cell, entirely covered with ovules; stigma sessile, flat or cushion-like, lobed. *Fruit* a berry, globose or.turbinate. *Seeds* minute, albuminous; embryo small.

Fleshy root or branch parasites, leafless or with the leaves reduced to scales.

DISTRIB. Species about 40, in the warmer regions of both hemispheres.

I. **Cytinus.**—*Flowers* unisexual. *Scale-leaves* present.
II. **Hydnora.**—*Flowers* hermaphrodite. *Scale-leaves* absent.

I. **CYTINUS, Linn.**

Flowers monœcious or diœcious, bracteate and often 2-bracteolate. *Male flowers*: Perianth tubular-campanulate or infundibuliform; lobes 4–9. *Anthers* connate in an exserted head, 2-celled, extrorse. *Ovary* rudiment none. *Female flowers*: Perianth almost absent; style columnar; stigma globose, grooved, apex obtuse or very shortly lobed radially; ovary with 8–14 scarcely exserted parietal placentæ.

Herbs coloured, somewhat fleshy, parasitic on roots; stems short, thick, simple, bearing alternate coloured scales; flowers arranged in a simple terminal spike, single, sessile or shortly pedicellate in axil of bract, with 2 opposite bracteoles or ebracteolate.

DISTRIB. Species 4, Mediterranean region, South Africa, Mexico.

1. **C. dioicus** (Juss. in Ann. Mus. Par. xii. (1808), 443); stems 4–6 in. high; erect, fleshy, simple or branched, 1–3-flowered; scales lax, imbricate, oblong, denticulate; flowers 2 in. long, shortly pedicellate; bracts oblong, obtuse; bracteoles oblong-spathulate, concave; male flowers: perianth infundibuliform, with 6 grooves, papillose towards the base; septa 6, alternating with perianth-lobes, extending to the staminal column; lobes elongate-oblong, obtuse, suberect, papillose outside, margins membranous, more or

less fimbriate, with a linear-lanceolate lacinule at the base on one
side; staminal column exserted, with 6 grooves, smooth; anthers
7–8, connectives produced above into broad processes; female
flowers: perianth shorter than in the male; tube much ribbed;
lobes broader and shorter than in male flower; ovary compressed,
with 6 ribs or angles; placentæ 12–14; style columnar; stigma
large globose, lamellæ 12–14, cuneate-subulate, densely papillose;
ovules minute. *Hook. Ic. Pl. t.* 336; *R. Br. in Trans. Linn. Soc.*
xix. 246; *Griffith in Trans. Linn. Soc.* xix. 323. *Phelypæa sanguinea,*
Thunb. Prodr. 1, *and Fl. Cap. ed. Schult.* 2. *Hypolepis sanguinea,*
Pers. Syn. ii. 598; *Harvey, Gen. S. Afr. Plant. ed.* 1, 300. *Aphyteia*
multiceps, Burchell, Trav. S. Afr. i. 213, *in note.*

COAST REGION: Clanwilliam Div.; Zuur Fontein, *Schlechter,* 8559! Tulbagh
Div.; hills near Saron, *Schlechter,* 4847! Cape Div.; near Wynberg and between
Cape Town and Tyger Berg, *Drège! Bolus,* 9287! by the Kuils River, *Pappe!*
near Cape Town, *Griffith!* Camps Bay, *Harvey!* Stellenbosch Div.; by the
Eerste River, *Scott!* Caledon Div.; near Hemel en Aarde, *Zeyher,* 1510!
CENTRAL REGION: Middelburg Div.; Conway Farm, *Gilfillan in Herb. Galpin,*
5571!

Parasitic upon the roots of *Eriocephalus racemosus,* Linn., *Trichogyne radicans,*
DC. (*T. reflexa,* Less), and according to Dr. Bolus also grows upon the roots of
Agathosma ciliata, Link. Mr. Gilfillan's plant differs somewhat from the other
specimens in having the perianth-lobes more or less glabrous.

II. HYDNORA, Thunb.

Flowers hermaphrodite. *Perianth* superior; tube short or long;
lobes 3–5, valvate, fleshy, triquetrous above, channelled or concave
in the lower part inside. *Anthers* forming a flexuous 3–4-lobed ring
in the perianth-tube, transversely divided into numerous cells.
Ovary inferior, 1-celled, with many placentas pendulous from the
top of the cell and ovuliferous all over; stigmas cushion-shaped,
3–5-lobed and bearing numerous radiating lamellæ; ovules with a
single integument. *Fruit* subglobose, filled with gelatinous pulp.
Seeds globose, free in the pulp; endosperm copious, horny;
perisperm formed of a single layer of cells; embryo with a long
suspensor dilated at the apex.

Parasitic fungus-like plants with a fleshy warted subterranean rhizome; flowers
large, arising singly from the rhizome, with an unpleasant odour.
DISTRIB. Species 9, 6 in Tropical Africa, 2 in S. Africa, 1 in America.

Perianth-lobes broadly induplicate, margins with ramenta (1) **africana.**
Perianth-lobes dilated above, margins naked ... ... (2) **triceps.**

1. **H. africana** (Thunb. in Vet. Akad. Handl. Stockh. 1775, 69,
t. 2, figs. 1–3, and 1777, 144, t. 4, figs. 1–2); rhizome horizontal
angular, tuberculate, $\frac{1}{4}$–$\frac{3}{4}$ in. in diam.; tubercles obtuse; roots
absent; flower erect, sessile or shortly pedunculate; perianth fleshy,
tubular, flesh-coloured, externally rugose, smooth within, 5–6 in.
long; segments 3 or very rarely 4, 2–3 in. long, oblong-ovate,

subacute, apices inflexed, connivent, margins ½–¾ in. thick, furnished
with ciliate ramenta, a large snow-white spongy body is present on
the inner side of each segment ; staminal column inserted below the
middle of the perianth-tube forming a very short subcylindric ring ;
anthers numerous 2–celled, with the cells opening extrorsely and
introrsely ; ovary slightly broader than the perianth-tube ; stigma
short, cushion-like ; fruit 3–4 in. in diam., thick, fleshy, subtrilobed.
Thunb. Nov. Gen. Pl. 24, *and Fl. Cap. ed. Schultes,* 499 ; *F. Meyer
in Nov. Act. Acad. Nat. Cur.* xvi. 2, 775, *t.* 58 ; *R. Br. in Trans.
Linn. Soc.* xix. 234, 245, *tt.* 27–30 ; *Wedd. in Ann. Sc. Nat.* 3ᵐᵉ sér.
xiv. 173, *t.* 8, *figs.* 5–10 ; *Chatin, Anat. t.* 92 *bis* ; *Griff. in Trans. Linn.
Soc.* xix. 319 ; *Hook. f. in DC. Prodr.* xvii. 109 ; *Marloth in Trans.
S. Afr. Phil. Soc.* xvi. 465, *with fig. H. Acharii, Hook. f. l.c.
Aphyteia Hydnora, Acharius, Diss. de Planta Aphyteia,* 1776, 10,
with plate, and in Linn. Amœn. Acad. viii. 315 ; *Syst. Veg. ed.* xiv.
609 ; *Gærtn. De Fruct.* ii. 262, *t.* 137, *fig.* 3 ; *Lam. Illustr. t.* 568 ;
Tratt. Archiv. ii. 145, *t.* 190.

SOUTH AFRICA : without locality, *Ludwig, Zeyher,* 1511 !
COAST REGION : Robertson Div. ; Karoo near Kokmans Kloof, *Mund* ! Karoo,
on roots of *Euphorbia* and *Cotyledon, Thunberg.*
CENTRAL REGION : Jansenville Div. ; Zwartruggens, parasitic on roots of
Euphorbia, MacOwan, Herb. Austr.-Afr., 1724 !
"*Planta Aphyteia,*" mentioned in *DC. Prodr.* xvii. 109, is the title of the
dissertation by Acharius, and is not used by him as a plant name.

Marloth (*l.c.*) draws attention to a white spongy body on the inner face of
each perianth-segment, which is eaten by a beetle (*Dermestes vulpinus*), the
insect responsible for the pollination of the flower. The white body on decaying
emits an offensive odour of putrefaction.

2. **H. triceps** (Drège & E. Meyer in Nov. Act. Acad. Nat. Cur.
xvi. 2, 779, *t.* 59) ; rhizome stout, 4-angled, angles crested with
obtuse tubercles ; flower shortly pedunculate ; perianth fleshy,
3-angled above ; tube 3–4 in. long, 1–1½ in. across, broadening
above across the segments to 3–4 in. in diam., broadly clavate with
impressed apex ; segments 3, dilated, broadly oblong, connate above
leaving apertures only near the base, margins naked ; staminal
column inserted near apex of the tube ; anthers numerous, opening
extrorsely only ; stigma obscurely trilobed ; ovary and fruit as in
H. africana. R. Br. in Trans. Linn. Soc. xix. 245 ; *Harv. Thes.
Cap.* ii. *tt.* 187–188.

WESTERN REGION : Little Namaqualand ; between Silverfontein, Koperberg
and Kaus, 2000 ft., *Drège* ; Modderfontein, Whitehead ; near Ookiep, on roots of
Euphorbia, Hofmeyer !

ORDER CXV. **PIPERACEÆ.**

(By C. H. WRIGHT.)

Flowers hermaphrodite or unisexual. *Perianth* absent in the
Tropical African genera. *Stamens* 2–6, rarely more, hypogynous ;
filaments usually free ; anthers erect, often articulated on the

filament, cells 2, distinct or confluent, dehiscing longitudinally. *Ovary* sessile, 1-celled and 1-ovuled in the South African genera ; stigmas 1–5 ; ovule orthotropous. *Fruit* in the South African genera indehiscent, baccate. *Seed* globose, ovoid or oblong ; testa usually membranous or rather fleshy ; endosperm small ; perisperm copious, farinaceous ; embryo minute.

Herbs or shrubs, erect or climbing ; leaves usually alternate and entire, rarely opposite or verticillate ; stipules none or adnate to the petiole ; flowers minute, usually forming dense spikes, each subtended by a peltate bract.

DISTRIB. Species about 1000, spread through the warmer regions of both hemispheres ; most numerous in Tropical America.

    I. **Piper.**—Shrubs. *Stamens* 2–6 ; anther-cells usually distinct. *Stigmas* 2–3.

    II. **Peperomia.**—Herbs. *Stamens* 2 ; anther-cells usually confluent. *Stigma* 1.

I. PIPER, Linn.

Flowers hermaphrodite or unisexual, usually forming dense cylindrical spikes, rarely racemes. *Perianth* none. *Stamens* 2–4, rarely more ; filaments short ; anther-cells usually distinct. *Ovary* sessile, 1-celled, 1-ovuled, obtuse or rostrate ; stigmas 2–4, distinct, erect or recurved. *Berry* small, usually globose, often immersed in the succulent rhachis, more rarely stalked. *Seed* similar in shape to the berry ; testa thin ; endosperm usually hard.

Erect or scandent shrubs ; branches jointed at the nodes ; leaves alternate, entire, equal or unequal at the base, penninerved ; stipules adnate to the petiole or connate into a leaf-opposed sheath ; flowers usually sessile ; spikes terminal or leaf-opposed.

DISTRIB. Species about 600, spread through the warmer regions of both hemispheres.

Leaves 5–7-nerved from the base ; berry sessile ;
    stigmas 2 ... ... (1) **capense.**
Leaves penninerved ; berry stalked ; stigmas 3 ... ... (3) **borbonense.**

1. **P. capense** (Linn. f. Suppl. 90) ; a shrub, erect or more or less climbing ; branches terete, swollen at the nodes, glabrous ; leaves ovate or more rarely elliptic, equilateral or nearly so, shortly acuminate, rounded or shortly cordate at the base, 5–7-nerved from the base, membranous, pellucid-dotted, about 4 in. long, $2\frac{1}{2}$–3 in. wide, glabrous above, usually villous on the under-surface of the nerves especially towards their base ; petiole channelled above, about 1 in. long, glabrous ; stipules lanceolate, membranous, deciduous ; catkins terminating short lateral branches which appear to spring from the middle of the petiole, $1\frac{1}{2}$–2 in. long in flower, cylindrical ; peduncles about 9 lin. long ; bracts peltate, glabrous except at the base of the very short stalk ; stamens 3–2 ; anther-cells separated by a wide connective ; ovary ovoid ; stigmas 2, recurved ; fruit obtuse, compressed, sessile. *Thunb. Fl. Cap. ed. Schultes,* 443 ; *Drège, Zwei Pfl. Documente,* 124, 125, *and in Linnæa,* xx. 215 ; *C. DC. in DC. Prodr.* xvi. i. 339, *and in Engl.*

Jahrb. xix. 224 ; *Engl. Pfl. Ost-Afr. C.* 159. *P. Volkensii, C. DC. in Engl. Jahrb.* xix. 225. *Coccobryon capense, Miq. Syst. Piper.* 343, *and in Nov. Act. Acad. Nat. Cur.* xxi. *Suppl.* 59, *t.* 61. *Cubeba capensis, Miq. Syst. Piper.* 303. *Peperomia capensis, Loudon, Hort. Brit. ed.* i. 13 ; *Steud. Nomencl. Bot. ed.* 2, ii. 301.

SOUTH AFRICA : without locality, *Forster* ! *Mund* !
COAST REGION : Swellendam Div. ; near Swellendam, *Zeyher*, 3871 ! Knysna Div. ; Kynsna, in the forest by the quarry, *Burchell*, 5397 ! Karratera River, under 1000 ft., *Drège* ! Bosch River, in wood, under 500 ft., *Drège* !
KALAHARI REGION : Transvaal : Barberton, *Kirk*, 91 ! *Thorncroft*, 2965 ! Makwongwa forest, 3300 ft., *Galpin*, 907 ! Houtbosch (Woodbush) *Rehmann*, 5970 !
EASTERN REGION : Pondoland ; Egossa, near St. Andrews, 1000 ft., *Tyson*, 3137 ! Natal ; Inanda, *Wood*, 1061 ! and without precise locality, *Gerrard*, 1456.

Also in Tropical Africa.

2. **P. borbonense** (C. DC. in DC. Prodr. xvi. i. 339) ; a parasite on trees ; branches woody, glabrous ; leaves oblong-ovate, obtusely acuminate, unequally cordate or cuneate at the base, 2½ in. long, 1¼ in. wide, penninerved, glabrous ; petiole 4 lin. long, slightly hairy beneath ; peduncle longer than the petiole, glabrous ; bracts subrotundate, pubescent beneath and on the margin ; stamens 3 ; exserted ; female spike dense-flowered ; rhachis slightly hairy ; stigmas 3 ; berry ovoid, shorter than its pedicel. *Engl. Jahrb.* xix. 224. *Cubeba costulata, Miq. Syst. Piper.* 299. *C. borbonense, Miq. l.c.* 301.

SOUTH AFRICA : without locality, *Verreaux, Gueinzius.*

Also in the Mascarene Islands.

II. PEPEROMIA, Ruiz & Pav.

Flowers hermaphrodite, spicate. *Perianth* none. *Stamens* 2 ; filaments short ; anthers transversely oblong or subglobose ; cells 2 ; usually confluent. *Ovary* sessile or subsessile, obtuse or acute ; ovule solitary, erect ; stigma undivided, often penicillate. *Fruit* minute ; pericarp thin. *Seed* similar in shape to the fruit ; testa membranous or coriaceous ; endosperm farinaceous.

Herbs, annual or perennial, sometimes climbing ; leaves exstipulate, alternate, opposite or verticillate, entire, penniveined or triplinerved ; spikes terminal or leaf-opposed, solitary or several together ; bracts sessile, usually peltate.

DISTRIB. Species about 400, spread through the warmer regions of both hemispheres, concentrated in America.

Leaves whorled, glabrous above, pilose beneath ; stem
  glabrous ...   . (1) **reflexa.**
Leaves opposite, pubescent on both surfaces ; stem
  densely pubescent   ...   . (2) **arabica.**
Leaves alternate or the upper opposite :
  Spikes terminal :
    Leaves glabrous on both surfaces   ...   . (3) **retusa.**
    Leaves glabrous above, puberulous beneath, ciliate (4) **Bachmannii.**
  Spikes leaf-opposed ; leaves glabrous on both surfaces (5) **nana.**

1. **P. reflexa** (A. Dietr. Sp. Pl. i. 180); stems creeping, tufted, 3-6 in. long, once or twice dichotomously forked, tetragonal, deeply sulcate when dried ; leaves usually 4 in a whorl, obovate, obtuse, 3-4 lin. long, bright green, glabrous above, pilose beneath, rigidly coriaceous, obscurely triplinerved, subsessile ; spikes terminal, solitary, 6-18 in. long, distinctly peduncled, rhachis pilose, deeply pitted ; bracts round, subsessile ; ovary ovoid, stigma terminal, penicillate ; fruit oblong, narrowed to a point, deeply immersed in the pits of the rhachis. *Miq. Syst. Piper.* 169 ; *and in Hook. Lond. Journ. Bot.* iv. (1845), 426 ; *Wight, Ic. t.* 1923, *fig.* 1 ; *Drège, Zwei Pfl. Documente,* 124, 136, 141, 146, *and in Linnæa,* xx. 215 ; *Baker, Fl. Maurit.* 298 ; *C. DC. in DC. Prodr.* xvi. i. 451 ; *Hook. f. Fl. Brit. Ind.* v. 99 ; *Engl. Pfl. Ost-Afr. C.* 159 ; *Dawe in Rep. Bot. Uganda Protect.* 1906, 55 ; *Baker & C. H. Wright in Dyer, Fl. Trop. Afr.* vi. i. 155. *P. Rehmanni, C. DC. in Engl. Jahrb.* xix. 227. *Piper reflexum, Linn. f. Suppl.* 91 ; *Thunb. Fl. Cap. ed. Schult.* 443. *Micropiper pusillum, Miq. Comm. Phyt.* 62, *t.* 5, *fig. B.*

VAR. β, **capense** (C. DC. in DC. Prodr. xvi. i. 451) ; leaves petioled, glabrous on both surfaces, subpellucid, 3-nerved ; peduncle glabrous, twice as long as the leaves. *Dahlstedt in Kgl. Sv. Vet. Akad. Handl.* xxxiii. ii. 179, *t.* 3, *fig.* 28. *P. reflexa, forma capensis, Miq. Syst. Piper.* 169.

SOUTH AFRICA : without locality, var. β *Sparrmann, Wahlberg, Mund & Maire.*

COAST REGION : Knysna Div. ; between Keurbooms River and Bitou River, *Burchell,* 5275 ! Karratera River, under 1000 ft., *Drège* ! and without precise locality, *Rehmann,* 489. Uitenhage Div. ; Zuurberg Range, 2000-3000 ft., *Drège* ; Port Elizabeth Div. ; around Krakkakamma, *Burchell,* 4557 ! *Zeyher.* Albany Div. ; wooded kloof west of Grahamstown, *Burchell,* 3596 ! Bathurst Div. ; Glenfilling, in woods and thickets, under 1000 ft., *Drège.* Stutterheim Div. ; near Fort Cunynghame, *Galpin,* 2465 ! Queenstown Div. ; Rockwood, Bongolo, *Galpin,* 2508 ! Var. β : Uitenhage Div. ; Van Standens Berg, *Zeyher,* 3873.

CENTRAL REGION : Somerset Div. ; on Bosch Berg, *Burchell,* 3209 !

KALAHARI REGION : Transvaal ; Houtbosch, *Rehmann,* 5969, Reitfontein, *Miss Leendertz,* 873 !

EASTERN REGION : Transkei ; between Gekau (Gcua) and Bashee Rivers, 1000-2000 ft., *Drège.* Tembuland : Bazeia, *Baur,* 18 ! Natal : Van Reenen, *Schlechter,* 6995 ! and without precise locality, *Cooper,* 1276 ! *Gerrard,* 1667 ! Zululand ; Eshowe, *Mrs. K. Saunders,* 3 ! Var. β : Natal ; without precise locality, *Gueinzius.*

Throughout the warmer regions of both hemispheres.

2. **P. arabica** (Decsne ex Miq. Syst. Piper. 121); suffruticose, 6-12 in. high ; stem erect from a decumbent base, terete, densely pubescent ; leaves opposite, elliptic or oblong-rhomboid, obtuse or acute, up to 1½ in. long and 1 in. wide, the lower smaller, more or less cuneate at the base, succulent when fresh, membranous when dry, pubescent on both surfaces, obscurely pellucid-punctate, 5-nerved ; petiole 2-6 lin. long, channelled above, densely pubescent ; spikes axillary and terminal, 2½ in. long, filiform when dry ; peduncle 6 lin. long, pubescent ; flowers distant, in pseudo-whorls ; bracts peltate, very shortly stalked ; filaments short ; anthers globose, pallid ; ovary ovoid, glabrous ; berry globose, black ; seed globose, black, shining, areolate. *Miq. Ill. Piper.* 18, *t.* 12 ; *C. DC.*

in DC. Prodr. xvi. i. 442.   *P. caffra, E. Meyer in Drège, Zwei Pfl. Documente,* 151.

COAST REGION : East London Div. ; on rockeries, First Creek, 200 ft., *Galpin,* 3158 !
EASTERN REGION : Transkei ; near the north of the Kei River, *Flanagan,* 2588 ; Pondoland ; St. Johns River, *Drège* ! Natal : Umzinyati Falls, *Wood,* 1222 ! and without precise locality, *Gueinzius.*
Also in Madagascar, Socotra and Arabia ; a variety in German East Africa.

3. **P. retusa** (A. Dietr. Sp. Pl. i. 155) ; stem creeping, ascending, about 4 in. high, simple or more or less forked, glabrous ; leaves alternate or the uppermost opposite, obovate, elliptic, or almost orbicular, about 6 lin. long but varying much on the same branch, obtuse or emarginate, glabrous on both surfaces, fleshy, 1- or 3-nerved, with a marginal nerve all the way round ; petiole about 3 lin. long, glabrous ; spikes terminal, filiform, about 9 lin. long ; peduncle 5 lin. long ; flowers scattered ; bracts subsessile, peltate ; ovary ovoid ; stigma minute, puberulous ; berry subglobose, black, slightly sunk in the rhachis. *Miq. Syst. Piper.* 132; *Drège, Zwei Pfl. Documente,* 124 ; *C. DC. in DC. Prodr.* xvi. i. 446 *(incl. var. alternifolia)* ; *Bolus & Wolley-Dod in Trans. S. Afr. Phil. Soc.* xiv. 312. *Piper retusum, Linn. f. Suppl.* 91 ; *Thunb. Fl. Cap. ed. Schult.* 443 ; *Vahl, Enum.* i. 346 ; *Willd. Sp. Pl.* i. 165.

VAR. *β,* **ciliolata** (C. DC. in DC. Prodr. xvi. i. 447) ; leaves ciliate at the apex, sometimes emarginate.

COAST REGION : Cape Div. ; Cape Town, *Alexander Prior* ! Table Mountain, Skeleton Gorge, 1500 ft., *Galpin,* 4435 ! Devils Mountain, among stones in a shady valley, 1200 ft., *Bolus,* 3307 ! and by the Waterfall, *Wolley-Dod,* 858 ! Kerstenbosch, *Bergius.* Knysna Div. ; Karratera River, under 1000 ft., *Drège* ! near the Keurbooms River, *Burchell,* 5171 ! Kraatjes Kraal, in the forest and by the rivulet, near Yzer Nek, *Burchell,* 5222 ! Var. *β* : Cape Div. ; Devils Mountain, *Fischer.* George Div. ; in the forest near George, *Burchell,* 6053 !
CENTRAL REGION : Somerset Div. ; on the Bosch Berg, *Burchell,* 3205 !
KALAHARI REGION : Transvaal ; Spitz Kop, *Wilms,* 1354 !
EASTERN REGION : Tembuland ; Bazeia, *Baur* ! Natal ; without precise locality, *Gerrard,* 1518 !

4. **P. Bachmannii** (C. DC. in Engl. Jahrb. xix. 227) ; a small herb growing on old trees or stones ; branches glabrous, about 2 in. long and ¼ lin. in diam. ; leaves alternate, obovate, rounded at the base and apex, 6 lin. long, 4½ lin. wide, glabrous above, puberulous beneath, ciliate on the margin, thinly membranous, pellucid, 3-nerved and with a slender marginal nerve continued to the apex ; petiole 1½ lin. long ; spikes terminal, 9 lin. long ; peduncle 4 lin. long.

EASTERN REGION : Pondoland, without precise locality, *Rehmann,* 419.

5. **P. nana** (C. DC. in Journ. Bot. 1866, 135) ; a dwarf herb ; leaves alternate, orbicular or subreniform, obtuse at the apex, rounded or truncate at the base, glabrous on both surfaces, very

thin, pellucid, 3-nerved, without veinlets, midribs continued to apex, lateral nerves more slender ; petiole glabrous, pellucid ; spikes leaf-opposed, lax-flowered ; stigma simple ; berry sessile, ovoid, acute, slightly sunk in the rhachis. *C. DC. in DC. Prodr.* xvi. i. 404, *and in Engl. Jahrb.* xix. 228.

EASTERN REGION : Pondoland ; in woods, 650–1600 ft., *Beyrich*, 99.

Also in the Mascarene Islands.

ORDER CXVa. **MONIMIACEÆ.**

(By C. H. WRIGHT.)

Flowers unisexual in the South African genus, in some others hermaphrodite.   *Perianth* inferior, regular or irregular ; tube globose ; limb 4- or more lobed or oblique.   *Disk* adnate to the perianth-tube.   *Stamens* indefinite, in two or more rows ; filaments short, usually flat, often with a gland on each side of the base ; anthers erect ; cells 2, distinct or confluent at the apex, dehiscing longitudinally or by valves.   *Carpels* distinct, usually many, more or less immersed in the disk ; style long or short ; stigma terminal ; ovule solitary, erect or pendulous, usually anatropous.   *Fruit* indehiscent, included in the accrescent perianth-tube.   *Seed* solitary ; testa membranous ; albumen fleshy or oily.

Trees or shrubs ; leaves opposite or alternate, entire or toothed, penninerved, exstipulate ; flowers axillary, racemose, cymose or fascicled ; bracts small or absent.

DISTRIB.   Species about 150, widely dispersed in the warmer regions of both hemispheres.

I. **XYMALOS**, Baill.

Flowers diœcious.   *Male flower* : Perianth 4–6-partite ; lobes ovate or lanceolate.   *Stamens* 10–15 ; anthers subsessile, 2-celled, dehiscence longitudinal.   Rudiment of *ovary* none.   *Female flower* : Perianth 3–5-lobed.   *Staminodes* absent, or represented by a ring of hairs around the base of the ovary.   *Ovary* obovoid or turbinate, glabrous, 1-celled ; ovule solitary, pendulous, anatropous ; stigma sessile, discoid or subhemispherical, wider than the top of the ovary. *Fruit* fleshy, smooth, crowned by the persistent stigma.   *Seed* compressed-ellipsoid, albuminous ;   embryo small ;   cotyledons roundish, flat.

Shrubs or small trees, glabrous except the inflorescence ; leaves nearly opposite, coriaceous, shortly petioled, minutely pellucid-punctate ; racemes solitary or geminate in the axils of the leaves ; bracts ovate or oblong.

DISTRIB.   Species 2, in Tropical Africa.

1. X. monospora (Baill. in Bull. Soc. Linn. Paris, i. 650); a shrub or small tree; branches glabrous; leaves alternate or almost opposite, up to 6 in. long and 2½ in. wide, varying from obovate to elliptic, cuneate at the base, usually irregularly serrate, more rarely entire, glabrous; main lateral nerves forming loops about 3 lin. within the margin; petiole about 6 lin. long; inflorescence diœcious; male flowers subtended by an ovate pubescent bract shorter than the perianth; perianth nearly 1 lin. long, deeply 4-lobed, pubescent; stamens about 10; anthers shortly elliptic, subsessile; female flowers: perianth 3-5-lobed, less than 1 lin. long, pubescent outside, and with a ring of hairs inside near the base; ovary oblong, slightly longer than the perianth. *Warb. in Engl. & Prantl, Pflanzenfam.* iii. 6A, 53, *fig.* 21, *A. B.*; *Oliv. in Hook. Ic. Pl. t.* 2444; *Perkins in Engl. Pflanzenr. Monimiaceæ*, 23; *Sim, For. Fl. Cape Col.* 288, *t.* 121; *Burtt-Davy in Transv. Agric. Journ.* v. 416, 426, *t.* 171; *Baker & C. H. Wright in Dyer, Fl. Trop. Afr.* vi. i. 169. *X. usambarensis, Engl. Jahrb.* xxx. 310. *Xylosma monospora, Harv. Thes. Cap.* ii. 52, *t.* 181. *Toxicodendron acutifolium, Benth. in Journ. Linn. Soc.* xvii. 214. *Paxiodendron usambarense, Engl. Pfl. Ost-Afr. C.* 182, *incl. var. serratifolia.*

COAST REGION: Victoria East Div.; in woods at the source of the Chumie River, *Tyson in Herb. MacOwan*, 2962! Cathcart Div.; Amatola Range, *MacOwan*, 2962! British Kaffraria; without precise locality, *Mrs. Barber*, 10!
KALAHARI REGION: Orange River Colony; without precise locality, *Cooper*, 1204! Transvaal; Houtbosch mountains, *Nelson*, 428!
EASTERN REGION: Transkei; Kentani, *Miss Pegler*, 835! Natal; Inanda, *Wood*, 986! 1315! and without precise locality, *Cooper*, 1251! *Gerrard*, 1921! Swaziland; Forbes Reef, 5100 ft., *Burtt-Davy*, 2732!

Also in Tropical Africa. Known in South Africa as "wild lemon," "lemonwood" and "limoen hout."

ORDER CXVI. **LAURINEÆ.**

(By O. STAPF.)

Flowers hermaphrodite, polygamous or diœcious, regular. *Perianth* inferior, very rarely superior; tube (receptacle) ovoid, turbinate, campanulate or rarely oblong, sometimes growing out and persisting after flowering; lobes usually 6, equal or more or less unequal, in 2 whorls, or the perianth divided almost to the very base into 6 equal or subequal segments; æstivation imbricate. *Stamens* typically in 4 whorls at the base of and opposite to the perianth-lobes or in the upper part of the receptacle, often one or the other (usually the fourth) reduced to staminodes or entirely suppressed; filaments usually present, more or less flattened, varying from very short to several times the length of the anther, those of the third whorl mostly with a pair of large globose glands at the sides or the base or behind them (i.e., between the second

and third whorl), very rarely the glands fused with the receptacle
into a disk, or also the outer stamens with glands at the base;
anthers continuous with the filaments, 2- or 4-valved, valves
superposed or more or less collateral, dehiscing from the base
upwards, introrse, or those of the third whorl more often extrorse.
Ovary superior, often more or less surrounded by the receptacle or, ultimately quite enclosed in it, very rarely inferior,
1-celled; style terminal, short or long, simple; stigma small, obtuse
or unilaterally widened or discoid. *Ovule* solitary, anatropous,
pendulous from near the apex of the ovary. *Fruit* baccate, fleshy
or more or less drupaceous, indehiscent, more or less surrounded
by or entirely free or enclosed in the persistent and accrescent
perianth or its receptacular portion, often borne on an enlarged
pedicel. *Seed* pendulous, exalbuminous; testa membranous, often
adnate to the pericarp and indistinct in the mature state. *Embryo*
straight; cotyledons thick, fleshy, sometimes very tightly adpressed
to each other and not separable; radicle superior; plumule
distinct.

Trees or shrubs, very rarely (*Cassytha*) twining parasitic herbs, all parts with
aromatic oil glands; leaves alternate, rarely opposite or subopposite, coriaceous
and evergreen, rarely membranous and annual, penniveined or 3–5-nerved, usually
with a distinct network of veins, very rarely (*Cassytha*) reduced to small scales;
stipules 0; leaf-buds often scaly; flowers small, greenish or yellowish in axillary
or subterminal, rarely terminal, cymose or racemose inflorescences, rarely solitary;
bracts caducous or subpersistent, sometimes forming involucres below the partial
inflorescences; bracteoles 0, except in *Cassytha*.

DISTRIB. Species about 1000, in the tropics and subtropical regions; few in
Africa.

I. **Cryptocarya.**—*Anthers* 2-valved. *Fruit* completely enclosed in the persistent and accrescent receptacle. Trees or shrubs.

II. **Ocotea.**—*Anthers* 4-valved; valves superposed. *Fruit* baccate, seated on
or in the enlarged cupular receptacle. Trees (the South African
species a tall tree) or shrubs.

III. **Cassytha.**—Twining parasitic herbs destitute of chlorophyll, with the
leaves reduced to small scales.

I. CRYPTOCARYA, R. Br.

Flowers hermaphrodite. *Perianth* herbaceous; receptacle ovoid
to turbinate or subcylindric, after flowering constricted above, persistent; lobes 6, in 2 whorls, subequal, deciduous. *Stamens* in
4 whorls, the outer 2 whorls fertile and inserted at the base of the
perianth-lobes, the third fertile, and like the fourth, which is
staminodial, inserted in the upper part of the receptacle; anthers
2-valved, of the two outer whorls introrse, of the third extrorse;
filaments short, those of the third whorl with a pair of sessile or
stipitate glands at the base or in front of it; staminodes ovoid and
shortly stipitate or attenuated at the base. *Ovary* sessile, enclosed
in the receptacle; style shortly exserted. *Fruit* globose or oblong,
enclosed in the enlarged indurated or somewhat fleshy receptacle,

smooth or longitudinally ribbed ; pericarp membranous or indurated, more or less free from the receptacle. *Testa* membranous, not or imperfectly separable from the pericarp.

Trees or shrubs ; leaves alternate, rarely subopposite, penninerved or 3-nerved, coriaceous ; flowers small, in subterminal or axillary panicles, rarely solitary.

DISTRIB. Over 40 species in the tropics (mostly in the Indo-Malayan region) and 6 in extra-tropical South Africa.

Leaves linear to linear-lanceolate, about 8 times as long
as broad ...   .   (1) **angustifolia.**

Leaves elliptic or ovate, about twice (or less) as long as
broad :
   Leaves more or less cuneate and acute at the base ;
   flowers in axillary panicles or few- to 1-flowered
   cymes :
     Leaves tri- or tripli-nerved at the base, usually very
     obtuse   ...   .   (2) **latifolia.**
     Leaves not tri- or tripli-nerved at the base, acute or
     acuminate :
       Leaves broad-elliptic, obtusely acuminate ; panicles
       lax, the upper often reduced to few-flowered
       cymes or solitary flowers   ...   ...   (3) **Woodii.**
       Leaves elliptic-lanceolate, acute at both ends ;
       panicles dense and short   ...   ...   (4) **myrtifolia.**
   Leaves rounded at the base, almost permanently
   hairy below ; flowers in 3–1-flowered axillary or
   terminal cymes, or solitary on short leafy
   branchlets :
     Reticulation raised on both sides of the leaves,
     these therefore punctate or foveolate   ...   (5) **Sutherlandii.**
     Reticulation impressed above, raised below ;
     upper side of leaf therefore finely rugose
     under the lens   ...   .   (6) **Wyliei.**

1. **C. angustifolia** (E. Meyer in Drège, Zwei Pfl. Documente, 97, 99, 176 ; name only) ; a tall shrub, 8–12 ft. high ; branches glabrous except at the growing tips which are fulvo-pubescent ; leaves linear-lanceolate, acute at both ends, with a callous acute point at the tips, 2–3½ in. long, 3–5 lin. broad, coriaceous, glossy above, glaucous underneath, fulvo-pubescent in bud, very soon glabrous, lateral nerves numerous, spreading and like the very close reticulation faintly raised on both sides ; petioles 1–2½ lin. long ; panicles axillary, up to 3 in. long (including the peduncle), few- to over 12-flowered, the lowest and longest often leafy, the uppermost much reduced and short, very finely pubescent ; bracts lanceolate to subulate, deciduous, small ; bracteoles minute, subulate ; pedicels hardly any or up to 1 lin. long, slender ; perianth subcampanulate, up to over 1½ lin. long, greyish to fulvously pubescent within and without excepting the obovoid receptacle which is glabrous within ; segments subequal, elliptic-oblong, up to 1 lin. long ; filaments very minutely pubescent ; anthers as long as and broader than the filaments, ovate, obtuse, minutely pubescent on the back and at the tips, about ⅖ lin. long ; staminal glands distinctly stipitate, capitate ; staminodes ¼ lin.

long, obtuse ; ovary and style glabrous, the latter almost 1 lin. long ; fruit oblong in outline, 8–9 lin. long, 4–5 lin. in diam. ; receptacle thin, fleshy ; pericarp crustaceous. *Meisn. in DC. Prodr.* xv. i. 74.

COAST REGION : Vanrhynsdorp Division ; near the Olifants River, between Ebenezer and Gift Berg, *Drège* ! Paarl Div. ; between Paarl and Lady Grey Railway Bridge and on Great Draakenstein Mountains, *Drège* ! Tulbagh Div. ; Mosterts Hoek, in moist places, *MacOwan, Herb. Norm. Austr.-Afr.*, 244 ! Worcester Div. ; near the Hex River, *Burke* ! Swellendam Div. ; dry hills near the Breede River, *Burchell*, 7481 ! by the Buffeljagts River, *Burchell*, 7287 ! *Zeyher*, 1128 !

2. **C. latifolia** (Sonder in Linnæa, xxiii. 101) ; a large tree ; branches fulvo-pubescent ; leaves elliptic to elliptic-oblong, very obtuse, rarely subacute, subcuneate at the base, 2–3 in. long, 1–1¼ in. broad, thinly coriaceous, cinnamon-colour below, finely pubescent and at length glabrous above, finely fulvo-tomentellous (particularly along the nerves) when young, slowly becoming more or less glabrous ; lateral nerves about 4 on each side, the lowest pair from the base or almost so, remote from the next pair, very oblique and long produced beyond the middle, transverse nerves loose, slightly raised below and above, reticulation faint and close or obscure ; petioles 3–4 lin. long ; panicles axillary, up to 1½ in. long (including the peduncle), few- to over 12-flowered, the larger sometimes with 1 or 2 small foliage leaves in the upper part densely fulvo-pubescent or almost velvety, the uppermost often reduced to 3 or 2 on very short peduncles ; bracts minute, very early deciduous ; flowers subsessile in small clusters at the ends of the branchlets ; perianth campanulate, up to over 1½ lin. long, fulvo-tomentellous without and within ; receptacle oblong, constricted at the insertion of the segments ; segments ovate-oblong, subequal, about 1 lin. long ; filaments pubescent ; anthers equalling the filaments, ovate-oblong, subacute, outer 6 about ⅜ lin. long, inner smaller ; staminal glands sessile or subsessile ; staminodes subsessile, ovate-lanceolate, acuminate ; fruit unknown. *Meisn. in DC. Prodr.* xv. i. 74.

EASTERN REGION : Natal ; near Durban, *Gueinzius* ; Berea, *McKen*, 13 ! and without precise locality, *Gerrard*, 1656 !

This is probably the unnamed *Cryptocarya* mentioned by Sim, *For. Flor. Cap.* 289, which he describes as a large tree, up to 60 ft. high with a trunk 3 ft. in diameter, and producing fruits ¾ in. across, with a green skin which blackens after falling. It is known in Pondoland and lower Natal as "Umtungwa."

3. **C. Woodii** (Engl. in Bot. Jahrb. xxvi. 391) ; a small tree 10–15 ft. high ; branches slender, glabrous excepting the minutely pubescent growing tip ; leaves broad-elliptic, shortly cuneate at the base, obtusely, and sometimes long, acuminate, 1¼–1¾ in. long, ¾–1¼ in. broad (rarely up to 3 by 1½ in.), thinly coriaceous, quite glabrous excepting when in bud and then fulvo-pubescent, lateral nerves about 4 on each side like the very close and fine reticulation slightly raised on each side, petioles 1½–2½ lin. long ; panicles axillary, often reduced to few-flowered cymes or solitary flowers, up to

$\frac{1}{2}$ in. long (including the very slender peduncles), more or less finely and adpressedly pubescent; bracts ovate, fulvo-pubescent, very early deciduous; bracteoles minute, lanceolate, with dark glandular tips, somewhat below the receptacle; pedicels of the cymes rarely over 1 lin. long and often much shorter, of solitary flowers up to 3 lin. long, filiform; perianth campanulate, up to almost $1\frac{3}{4}$ lin. long; very minutely pubescent without and within, excepting the subcylindric receptacle which is glabrous within; segments subequal, elliptic-oblong, obtuse, $\frac{3}{4}$–$\frac{4}{5}$ lin. long; filaments pubescent; 6 outer anthers elliptic in outline, obtuse, $\frac{3}{8}$ lin. long, inner 3 smaller and narrower, and acute, all glabrous or pubescent at the base of the connective, rather longer than the filaments; staminal glands stipitate; staminodes lanceolate, acutely acuminate, $\frac{1}{4}$ lin. long, ovary and style glabrous; fruit globose or subglobose, dark brown, 6–8 lin. long, 5–6 lin. (according to *Sim*, 6–9 lin.); receptacle fleshy; pericarp thick crustaceous. *C. acuminata*, *Schinz ex Sim*, *For. Fl. Cap.* 289, *t.* 158, *fig.* 1.

COAST REGION: King Williamstown Div.; Perie Forest, *Galpin*, 3279! by the riverside near King Williamstown, *Sim*, 1157! Komgha Div.; near Komgha, *Schlechter*, 6153! East London Div.; wooded ravine near the mouth of Kwenqura River, *Galpin*, 5808!

EASTERN REGION: Transkei; Tsomo Forest, *Barber*, 12! by streams near Kentani, *Miss Pegler*, 712! Natal; Inanda, *Wood*, 766! 4684, and without precise locality, *Gerrard*, 72! 1658! *Cooper*, 1173! *Gerrard & McKen*, 702!

The fruits of Wood's specimen from the Inanda forest are produced at the top into a beak, a condition of which there is hardly an indication in the other specimen; otherwise, however, the specimens agree perfectly with the remainder. Native name *Tunga* according to Sim.

4. **C. myrtifolia** (Stapf); branches slender, growing tips of branchlets fulvo-pubescent, soon glabrescent; leaves broadly elliptic-lanceolate to elliptic, acute at both ends, or tips acutely acuminate, 1–$1\frac{1}{2}$ in. long, $\frac{1}{2}$–$\frac{3}{4}$ in. broad, coriaceous, finely pubescent when quite young, at length quite glabrous; lateral nerves about 6 on each side, very faint, exceedingly closely and finely reticulate, pallid below; petioles slender, about $1\frac{1}{2}$ lin. long; panicles axillary, $\frac{1}{2}$–$1\frac{1}{2}$ in. long (including the slender peduncles), the lower often leafy in the upper part, the uppermost much reduced and short, all somewhat dense, minutely fulvo-pubescent; bracts ovate, small, very early deciduous; bracteoles obsolete; pedicels hardly any or up to over $\frac{1}{2}$ lin. long; perianth campanulate, over $1\frac{1}{2}$ lin. long, fulvo-pubescent without and excepting the oblong receptacle within; segments subequal, elliptic-oblong, obtuse, almost 1 lin. long; filaments pubescent; anthers ovate, apiculate, $\frac{3}{8}$ lin. long, very minutely pubescent on the back along the connective; staminal glands stipitate; staminodes ovate, mucronulate-acute; ovary and style glabrous, the latter $\frac{3}{4}$ lin. long; fruit globose, 4 lin. in diam.; receptacle very thin, fleshy; pericarp crustaceous.

EASTERN REGION: Natal; Inanda, *Wood*, 1402! and without precise locality, *Gerrard*, 1657!

5. C. Sutherlandii (Stapf); a shrub; branches very slender, densely fulvo-pubescent to almost tomentose in the younger parts; leaves ovate, rounded at the base, subacute or subacuminate, often with a minute mucro, $\frac{1}{2}$–1 in. long, $\frac{1}{3}$–$\frac{1}{2}$ in. broad, thinly coriaceous, finely pubescent above when quite young, soon glabrous, fulvo-tomentose below, never becoming quite glabrous, lateral nerves 4–6 on each side, very faint or quite obscure, reticulation very close, usually faintly raised below, more or less concealed by the hairs, slightly impressed above; petioles 1–1$\frac{1}{2}$ lin. long; flowers in 3–1-flowered axillary or terminal cymes on short leafy branchlets; peduncle filiform, finely villous, up to 3 lin. long; bracts lanceolate with an acute fleshy mucro, up to almost 1 lin. long; bracteoles similar but smaller; pedicels almost as long as the bracts; perianth campanulate, not quite 1 lin. long, glabrous within and without or nearly so; segments subequal, ovate to elliptic, almost $\frac{1}{2}$ lin. long; filaments glabrous or sparingly and minutely pubescent; anthers ovate, obtuse, glabrous; staminal glands stipitate; staminodes ovate, subacute; fruits globose, 3 lin. in diam.; receptacle very thin, fleshy; pericarp crustaceous.

EASTERN REGION : Natal; in stony places near Murchison, *Wood,* 3083! Umlaasi location, near Bevaan River, *Wood,* 3388! Coast-land, *Sutherland*!

6. C. Wyliei (Stapf); branches very slender, densely fulvo-pubescent to tomentose in the younger parts; leaves ovate, rounded at the base, subacuminate, sometimes minutely mucronate, $\frac{3}{4}$–1$\frac{1}{2}$ lin. long, $\frac{1}{2}$–1 in. broad, thinly coriaceous, dull subglaucous, finely pubescent above, soon glabrous with the exception of the midrib, fulvo or rusty-tomentose underneath when young, then loosely hairy with the exception of the permanently tomentose nerves, lateral nerves 4–5 on each side, slightly raised below, impressed above, reticulation close, faintly raised on both sides; petioles 1–1$\frac{1}{2}$ lin. long, pubescent; flowers solitary, axillary or terminal on short leafy branchlets, borne on slender rusty tomentose pedicels 1–5 lin. long; perianth finely pubescent without and excepting the receptacle within; segments subequal, rotundate-ovate to broad-elliptic, subacute or obtuse, slightly over $\frac{1}{2}$ lin. long; filaments pubescent; anthers ovate, obtuse, $\frac{1}{4}$ lin. long, glabrous; staminal glands sessile; staminodes very small triangular-ovate, acute; fruit globose, black, almost $\frac{1}{2}$ in. in diam.; receptacle thin, fleshy; pericarp crustaceous.

EASTERN REGION ; Zululand; Ngoye, 2000–3000 ft., *Wylie in Herb. Wood,* 10391!

II. OCOTEA, Aubl.

Flowers usually diœcious or hermaphrodite. *Perianth* herbaceous, with or without a receptacle or tube, 6- or 8-lobed or partite; lobes or segments equal, usually deciduous. *Hermaphrodite* : stamens in 3

or 4 whorls, the outer 3 fertile, the fourth (if present) staminodial;
anthers 4-valved; valves in superposed pairs, of the 2 outer whorls
introrse, of the third extrorse or subextrorse, very rarely introrse;
filaments very short or 0, or longer than the anthers, of the third
whorl with a sessile, very rarely stipitate, gland at each side of the
base; staminodes, if present, slender; ovary ovoid, ellipsoid or subglobose, usually glabrous, longer or shorter than the style. *Male*:
as in the hermaphrodite flowers, but ovary sterile, stalk-like or quite
suppressed. *Female*: as in the hermaphrodite flowers, but stamens
rudimentary, barren. *Fruit* baccate, ellipsoid or globose, seated on
or in an enlarged cupular receptacle, which is either truncate or
6-toothed or 6-lobed from the persistent perianth-lobes.

Trees or shrubs; leaves alternate, membranous or coriaceous, glabrous or
hairy; flowers small, in cymes, arranged in axillary or subterminal panicles.

DISTRIB. Species about 200, mostly in Tropical America, 1 in Tropical
Africa, 1 in South Africa, and a few in the Mascarene Islands.

1. O. bullata (E. Meyer in Drège, Zwei Pfl. Documente, 205,
name only); a tree, 60–80 ft. high, with a straight clean trunk,
3–5 ft. in diam.; bark dark brown, rugged and scaly in old trees;
young branches very minutely greyish pubescent at their tips, soon
glabrous, drying dark brown or blackish; leaves alternate, elliptic
to oblong, shortly acuminate, obtuse or subacute at the base, 2–4 in.
long, 1–2 in. broad, coriaceous, glabrous; lateral nerves about 6 on
each side of the midrib, closely and prominently reticulated on both
sides; usually large pits (*acarodomitia*) with ciliolate orifices on the
underside in the axils of the lowest 1–2 pairs of nerves, the pits
corresponding to large hollow tubercles on the upper side; petiole
½–1 in. long, channelled above; panicles from the axils of some of
the uppermost leaves, lax, including the peduncles 2–3 in. long, about
1 in. wide, very finely and scantily pubescent at least in the upper
part; peduncles ½ to over 1 in. long; bracts ovate, concave, very
early deciduous, greyish-silky-pubescent; pedicels 1 (rarely 2) lin.
long; flowers polygamous, perianth yellowish-white, finely pubescent
without, 2½ lin. across when quite open; receptacle hemispheric,
¾ lin. high, glabrous within; segments spreading, subequal, ovate-
elliptic, obtuse, ciliolate, glabrous within; stamens of the male and
hermaphrodite flowers with linear glabrous filaments as long as the
anthers; glands sessile, subglobose, on each side of the base of the
stamens of the third whorl; staminodes narrow, acute, about ⅜ lin.
long; stamens and staminodes of the female very much reduced;
ovary immersed, but free, in the receptacle, like the slender style
glabrous; stigma discoid; fruit oblong, ¾ in. long, ⅓ in. in diam
seated in the cup-shaped enlarged receptacle which equals about ⅓ of
the fruit. *Sim, For. Fl. Cap.* 289, *t.* 122; *Thonn. Blütenpfl. Afr.*
t. 52; *Burtt-Davy in Transv. Agr. Journ.* v. 467, *t.* 172. *Oreodaphne*
bullata, Nees, Syst. Laur. 449; *Meisn. in DC. Prodr.* xv. i. 118.
Laurus bullata, Burch. Trav. i. 72.

COAST REGION : Swellendam Div. ; Grootvaders Bosch, *Zeyher*, 3629 ! Knysna Div. ; Bosch River, *Drège* ; in the forest at Knysna, *Burchell*, 5409 ! 5432 ! forest near Yzer Nek, *Burchell*, 5236 ! Humansdorp Div. ; by the Kromme River, *Drège* !

This is the "*Stinkwood*," one of the most valuable timber trees of Cape Colony. According to Sim, l.c. 290, this tree ranges "from Cape Town to the Transvaal eastern forests," and is at its best in the Knysna and Natal forests. He also mentions it from the Transkeian Mountains and Pondoland. Apart from the specimens enumerated above, all of which are typical *Ocotea bullata*, there are two sheets of Gerrard's at Kew which are possibly referable to *Ocotea*. The specimens consist of barren shoots and seem to represent a species distinct from *O. bullata*. The leaves are broad-elliptic, very obtuse, with only 4 lateral nerves on each side, and they, like the branchlets, are much more pubescent than those of *G. bullata*. There is also no trace of acarodomitia so commonly found in the latter species. Under the circumstances it is not improbable that the "Stinkwood" of Natal is a species as yet undescribed.

III. CASSYTHA, Linn.

Flowers hermaphrodite, sometimes dimorphic (dioecious ?). *Perianth* with a turbinate or ovoid receptacle (very small during flowering) and a 6-partite limb, after flowering tightly constricted at the junction of limb and tube ; segments unequal, outer 3 much smaller. *Stamens* in 4 whorls of 3, of the 3 outer fertile, of the fourth staminodial, rarely also those of the second whorl reduced to staminodes ; anthers 2-celled, of the 2 outer whorls introrse, of the third extrorse ; filaments of the third whorl with subsessile glands at each side of the base ; staminodes subsessile or stipitate. *Ovary* during flowering hardly immersed in the receptacle, which afterwards grows out and envelops it. *Fruit* completely enclosed in the succulent receptacle, often crowned by the persistent limb. *Seed* with a coriaceous testa. *Cotyledons* tightly adpressed to each other, distinct only when young.

Twining, parasitic herbs, destitute of chlorophyll, adhering to their hosts by means of uniseriate haustoria ; leaves reduced to minute scales ; flowers small, racemose, spicate or capitate, sessile or pedicelled from the axils of scale-like bracts, supported by a pair of bracteoles.

DISTRIB. Species about 16, mostly Australian.

Inflorescences loosely spicate, 1–2 in. long   ...   ... (1) **filiformis.**

Inflorescences capitate, subsessile or peduncled ; peduncles rarely over 4 lin. long :

 Whole plant glabrous excepting the sometimes very minutely hispidulous growing tips and inflorescences and the sometimes ciliolate bracts and outer perianth-segments ...   . (2) **ciliolata.**

 Whole plant hairy, inflorescences and growing tips rusty tomentose ...   . (3) **pondoensis.**

1. **C. filiformis** (Linn. Sp. Pl. ed. i. 35) ; stems filiform, bright yellow, glabrous or more or less pubescent to tomentose ; leaf-scales ovate to lanceolate-subulate, acute ; inflorescences loosely spicate, usually solitary, spreading, 1–2 in. long, few- to 10-flowered ; flowers hermaphrodite, white, up to about 1 lin. long ; bracteoles ciliolate ;

outer perianth-segments very similar to the bracteoles, inner broadly ovate-oblong, obtuse, 3 times longer ; filaments of the first whorl broader than the anther, thinly membranous, of the second narrower than the anther; staminodes triangular, fleshy ; fruit crowned with the persistent limb, 2–3 lin. in diam. *Meisn. in DC. Prodr.* xv. i. 255 ; *Engl. Pfl. Ost-Afr. C.* 182 ; *Hiern in Cat. Afr. Pl. Welw.* i. 915 ; *Engl. ex Gilg in Baum, Kunene-Samb. Exped.* 238 ; *De Wild. in Études Fl. Bas et Moyen Congo,* i. 244 ; *Stapf in Dyer, Fl. Trop. Afr.* vi. i. 188. *C. guineensis, Meisn. l.c. (incl. var. Livingstonii). C. americana, Meisn. l.c.* 256. *Cassyta guineensis, Schum. & Thonn. Beskr. Guin. Pl.* 199 ; *Benth. in Hook. Niger Fl.* 497. *C. americana, Nees, Syst. Laur.* 644.

EASTERN REGION : Pondoland ; between Umtentu River and Umzimkula River, *Drège* ! Delagoa Bay : Lourenço Marques, *Schlechter,* 11547 !

Widely spread throughout the tropics.

2. C. ciliolata (Nees, Syst. Laur. 646); stems filiform, yellow, glabrous or sometimes the growing tips rufous-hispidulous ; leaf-scales ovate, subacute, slightly produced downwards at the obtuse and often very minutely hispidulous base, up to 1 lin. long ; inflorescences peduncled, few- (usually 3-) flowered, capitate or shortly spicate ; peduncles simple or more rarely divided, glabrous or sometimes sparingly and very minutely hispidulous 1–6 (usually about 3) lin. long ; with two lateral scale-leaves at the base resembling the supporting scale-leaves, but usually narrower and more acute ; bracts and bracteoles broad-ovate to rotundate, subacute or obtuse, $\frac{1}{2}$–$\frac{3}{4}$ lin. long, minutely and sometimes sparingly ciliolate or eciliolate ; perianth at the time of flowering 1$\frac{1}{2}$–2 lin. long, quite glabrous apart from the outer segments which resemble the bracteoles and are like these often more or less ciliolate ; segments somewhat fleshy ; inner segments ovate-oblong, obtuse, 1$\frac{1}{4}$–1$\frac{1}{2}$ lin. long ; stamens of the first 3 whorls fertile ; anthers ovate ; filaments glabrous, slightly narrowed towards the base, those of the second whorl very short, of the fourth whorl reduced to thick triangular sessile staminodes ; fruit yellow, globose, ellipsoid, 2$\frac{1}{2}$–3 lin. long, crowned by the persistent perianth. *Meisn. in DC. Prodr.* xv. i. 254. *C. triflora E. Meyer in Drège, Zwei Pfl. Documente,* 97, 171 (*name only*). *C. capensis, Meisn. l.c. (incl. var. spicata*).

COAST REGION : Paarl Div. ; Paarl Mountain, *Drège* ! Cape Div. ; Table Mountain, *Ecklon,* 160 ! *Zeyher,* 3630b ! Mountains near Cape Town, *Burke* ! *Bolus,* 2445 ! *Harvey,* 560 ! 483 ! *Burchell,* 331 ! 920 ! 486 ! 8434 ! Simons Bay, *Wright* ! Knysna Div. ; hills at Plettenbergs Bay, *Burchell,* 5332 ! 5342 ; Uitenhage Div. ; Uitenhage, *Zeyher,* 731 ! *Zeyher,* 3630a.

C. glabella, E. Meyer (not of R. Br.) in *Drège, Zwei Pfl. Documente,* 88, from Table Mountain belongs very probably here, whilst *Cassyta* sp. 8037, E. Meyer, l.c., is *Cuscuta cassytoides,* Nees, as is also Burchell, 3178, quoted by Meisner under *C. capensis.*

3. C. pondoensis (Engl. in Engl. Jahrb. xxvi. 392) ; stems filiform, hairy all over, the young parts rusty tomentose, hairs at length

discoloured and often more or less curled ; leaf-scales ovate, sub-obtuse, up to 1½ lin. long, very slightly produced downwards at the obtuse base, loosely hairy ; inflorescences peduncled, capitate, 3–5-flowered, rarely reduced to a single flower, rusty-tomentose ; peduncles solitary or in fascicles of 2–4, up to 3 lin. long, with 2 lateral scale-leaves at the very base, resembling the subtending scale-leaf, but usually more elliptic and more obtuse ; bracts and bracteoles broad-ovate to rotundate, ½–1 lin. long, subhyaline, rusty-pubescent and ciliate ; perianth greenish, at the time of flowering 1½ lin. long ; receptacle rusty-pubescent ; segments sub-hyaline, outer rotundate or rotundate-ovate, ½–¾ lin. long, ciliate; otherwise glabrous, inner ovate-oblong, when flattened out, subobtuse or, owing to the more or less inflexed sides, lanceolate-triangular and almost acute, 1¼ lin. long, glabrous without, minutely pubescent within, 3-nerved ; stamens of the first 3 whorls fertile ; anthers ovate ; filaments broad-linear, those of the first whorl ciliate, of the second very short, of the fourth reduced to triangular thick staminodes. Fruit up to 3 lin. in diam. *C. rubiginosa*, E. *Meyer in Drège, Zwei Pfl. Documente*, 154, 171 ; *name only*. *C. pubescens*, E. *Meyer*, *l.c.* 154, *not of R. Br.*

EASTERN REGION: Pondoland ; on a rocky hill by the Umtentu River and between Umtentu River and Umzimkulu River, *Drège*! various localities, *Bachmann*, 515 ! 516 ! 517 !

ORDER CXVII. **PROTEACEÆ.**

(By J. HUTCHINSON, E. P. PHILLIPS and O. STAPF.)

Flowers hermaphrodite or unisexual and diœcious, rarely polygamous. *Perianth* corolline, simple, inferior, tetramerous, actinomorphic or more or less zygomorphic, valvate, usually tubular in bud, with a more or less differentiated widened limb, variously divided when opening. *Stamens* 4, opposite the perianth-segments ; filaments usually more or less adnate to the perianth-segments, rarely free ; anthers free, erect, with introrse parallel thecæ, very rarely laterally cohering. *Ovary* sessile or stipitate, with or without hypogynous scales or an annular or cupular disc at the base, 1-celled, sometimes oblique or excentric ; style terminal, short or more often long ; stigma small. *Ovules* numerous and biseriate or few or only 1, pendulous and orthotropous or laterally attached and amphitropous or anatropous ; micropyle always inferior. *Fruit* a nut, drupe, follicle or capsule. *Seeds* several or 2 or 1, often compressed and winged ; testa thin or coriaceous ; albumen 0 ; embryo with 2 equal or unequal, compressed or thick, and fleshy cotyledons, and a short radicle, rarely with more than 2 cotyledons.

Trees or shrubs, rarely perennial herbs. Leaves spirally arranged, rarely verticillate or opposite, entire or variously and sometimes decompoundly divided, mostly coriaceous, exstipulate. Flowers solitary or in pairs in the axils of persistent or deciduous bracts, arranged in racemes, umbels, spikes or small or large heads, the latter often involucrate.

DISTRIB. Over 50 genera with nearly 1000 species, of which 14 genera inhabit South Africa, a few of these extending into Tropical Africa, the remainder mostly in Australia, few in the Indo-Malayan region, the Pacific Islands and South America.

Tribe 1. PERSONIEÆ.—*Flowers* solitary in the axils of leaves or in bracteate racemes or spikes. *Perianth* actinomorphic, with the segments separating as the flower opens, or soon afterwards. *Filaments* free or adnate to the perianth-segments at the base only or up to below the limb. *Ovules* 2 or 1, rarely more, pendulous and orthotropous, rarely laterally attached and amphitropous. *Fruit* indehiscent. *Cotyledons* thick, unequal, 2 or more.

I. **Brabeium.**—*Flowers* hermaphrodite or polygamous in dense axillary spike-like racemes. *Filaments* attached to the base of the perianth-segments. *Ovary* sessile, woolly ; ovules 2, pendulous. *Fruit* a drupe with a corky exocarp and a woody endocarp. *Leaves* verticillate, serrate.

Tribe 2.—PROTEEÆ.—*Flowers* solitary in the axils of bracts, usually in heads, more rarely in spikes or racemes. *Perianth* actinomorphic or more or less zygomorphic, with all the segments more or less deeply separating or with only one detaching itself from the remainder of the perianth. *Anthers* sessile or subsessile at the base of the limb. *Ovule* 1, laterally attached or ascending, amphitropous. *Fruit* a dry nut, sometimes winged.

* *Flowers diœcious, actinomorphic. Leaves entire.*

II. **Aulax.**—*Male flowers* in spike-like racemes ; female flowers in involucrate heads ; involucre made up of whorls of flattened dorsally foliate and laterally bracteate branchlets, resembling pectinate bracts ; flowers solitary on the inner side of the branchlets and in spirals on a central axis or only on the latter. *Nuts* laid bare.

III. **Leucadendron.**—*Male* and *female flowers* in bracteate heads ; bracts persistent, the outer often forming an involucre, accrescent and indurated in the female heads which on maturity become strobiliform. *Nuts* hidden within the bracts, more or less compressed, sometimes winged.

** *Flowers hermaphrodite, more or less zygomorphic, at least on opening. Inflorescence capitate or spicate. Leaves entire.*

† Flowers capitate.

IV. **Protea.**—Anticous (abaxial) *perianth-segment* entirely separating from the others, which remain fused into a sheath widened at the base and a 3-toothed or 3-awned lip.

V. **Leucospermum.**—Anticous (abaxial) *perianth-segment* separating more or less between the base and the limb, never quite free ; limb recurved in the open flower.

†† Flowers spicate.

VI. **Faurea.**—*Perianth* splitting anticously (abaxially) to the base or almost so.

*** *Flowers hermaphrodite, actinomorphic or slightly zygomorphic* (Spatalla) ; *perianth-segments in the open flower cohering at the base only, recurved Inflorescence capitate. Leaves entire or variously divided.*

† Flowers in typically many-flowered, or sometimes by reduction few-flowered, terminal or axillary heads.

‡ Leaves undivided.

VII. **Mimetes.**—*Heads* medium-sized, aggregated and partly hidden in the axils of the upper leaves.

VIII. **Orothamnus.**—*Heads* large, much exserted from the axils of the upper leaves, surrounded by large spathulate-oblong, obtuse, coloured involu:ral bracts.

IX. **Diastella.**—*Heads* small, terminal and solitary, surrounded by small inconspicuous bracts.

‡‡ Leaves pinnately or bipinnately divided, very rarely entire and then usually terete or narrowly linear.

X. **Serruria.**—*Involucral bracts* rarely conspicuous and coloured, or if so then glabrous.

†† Flowers in 1-flowered or typically few- (2-4-, rarely up to 9-) flowered heads congested in terminal spikes, racemes or head-like glomerules, never in the axils of the upper leaves.

‡ Leaves undivided.

§ Involucre calycoid, more or less bilabiate, toothed.　Inflorescence cylindric.

XI. **Spatalla.**—*Corolla* slightly zygomorphic, the posticous (adaxial) lobe being larger and more densely villous or bearded than the others. *Ovary* somewhat oblique ; stigma obliquely discoid.

XII. **Spatallopsis.**—*Corolla* actinomorphic.　*Ovary* not oblique ; stigma obliquely capitate.

§§ Involucre of as many free bracts as flowers.　Inflorescence globose.

XIII. **Sorocephalus.**—*Corolla* actinomorphic.　*Ovary* not oblique ; style constricted above the ovary ; stigma oblong.

‡‡ Leaves all pinnately divided or the upper undivided.　Involucral bracts imbricate, the two lateral exterior.

XIV. **Nivenia.**—*Inflorescence* spicate or globose ; partial heads with as many involucral bracts as there are flowers.　*Stigma* clavate to capitate.

I. BRABEIUM, Linn.

Flowers hermaphrodite, or polygamous by abortion, actinomorphic. *Perianth* cylindric in bud with a globose limb ; segments 4, on flowering separating to the base, spathulate-linear, revolute. *Stamens* slightly shorter than the perianth-segments ; filaments linear, from the base of the perianth-segments ; anthers linear-oblong ; connective produced into a small apical gland. *Hypogynous disc* annular. *Ovary* sessile, long-hairy ; style terete, clavate above, glabrous, subpersistent ; stigma small, terminal. *Ovules* 2, pendulous, orthotropous. *Fruit* a drupe, densely velvety ; exocarp corky, traversed by fibres ; endocarp woody. *Seed* solitary ; embryo with hard thick equal cotyledons.

A small tree or shrub ; leaves in whorls of 6, shortly petioled, undivided, serrate, coriaceous ; flowers pedicelled, 2-nate in the axils of early deciduous bracts, arranged in dense spike-like axillary racemes.

DISTRIB. Species 1, confined to the Western portion of Cape Colony.

1. B. stellatifolium (Linn. Sp. Pl. ed. i. 121) ; a small tree or shrub, 8–10 ft. high ; young tops purplish ; branches fulvous or rufous-tomentose when young, at length glabrescent or quite glabrous; leaves lanceolate, acute or subobtuse and apiculate, distantly serrate, 3–6 in. long, ½–1 in. broad, coriaceous, fulvously silky-tomentose in bud, usually very soon glabrescent, prominently reticulated, midrib very prominent below; petiole short; racemes cylindric, about 3 in. (the female sometimes up to 6 in.) long, solitary or sometimes ternate on a short common peduncle; axis of raceme fulvous-villous; bracts closely imbricated, deciduous before the opening of the flowers, obovate to elliptic, obtuse, 2–2¼ lin. long, 1½ lin. broad, membranous, fulvously to rufously villous; flowers white, sweet-scented; pedicel 2–2½ lin. long, finely greyish-villous; segments free to the base, spathulate-linear, obtuse, 2¾ lin. long, revolute, sparingly hairy; filaments 1 lin. long; anthers scarcely ½ lin. long; apical gland minute, ovate; style 1¾ lin. long, glabrous, long persistent, covered with long fulvous or rufous hairs; mature fruit borne on pedicels 2–2½ lin. long, obovoid, constricted at one or both ends, 1–1½ in. long, densely rufously velvety. *Linn. Sp. Pl. ed.* ii. 177; *Linn. Mant.* 332; *R. Br. in Trans. Linn. Soc.* x. 165; *Meisn. in DC. Prodr.* xiv. 344; *Pappe, Silva Cap. ed.* i. 29; *Sim, For. Fl. Cape,* 299, *t.* 132. *B. stelluli-folium, Murray (Linn.) Syst. Veg. ed.* xiii. 764; *Houtt. Handl.* vi. 424, *t.* 37; *Lam. Ill. t.* 847; *Willd. Sp. Pl.* iv. 972; *Roem. & Schult. Syst. Veg.* iii. 399. *B. stellatum, Thunb. Prodr.* 31; *Fl. Cap. ed. Schult.* 156. *B. stellare, Knight, Prot.* 98. *Brabyla capensis, Linn. Mant.* 137. *Amygdalus æthiopica, etc., Breyne, Cent.* i. *t.* 1. *Arbor æthiopica hexaphylla, Pluk. Almag.* 47, *t.* 265, *fig.* 3.

SOUTH AFRICA : without locality, *Thunberg* ! *Grey* ! *Gueinzius* !
COAST REGION : Tulbagh Div.; Mitchells Pass, *Bolus,* 4639 ! Cape Div.; near Paradise, east side of Table Mountain, *Pappe* ! near Cape Town, *Burchell,* 426 ! Phillips, Paarl Div.; near Paarl, *Thunberg.* Paarl Mountain, *Drège* ! Stellenbosch Div. ; Hottentots Holland, *MacOwan, Herb. Austr.-Afr.* 1522 ! Caledon Div. ; by the Palmiet River, ex *Meisner.* Swellendam Div. ; near the Buffeljagts and Zondereinde Rivers, *Thunberg.* Riversdale Div. ; between Garcias Pass and Krombecks River, *Burchell,* 7175 !

The red, reticulated wood is used for ornamental joiners' and turners' work. The seed may be eaten after prolonged soaking in water, but is considered unwholesome when fresh.

II. AULAX, Berg.

Flowers diœcious, actinomorphic. *Male flowers:*—*Perianth* very indistinctly differentiated into tube and limb, tubular, straight in bud, on opening separating into 4 linear segments; the channelled limbs about twice as long as the claws. *Anthers* linear on very short filaments, inserted at the base of the perianth-limbs; connective not produced. *Hypogynous scales* 0. *Ovary* rudimentary, small, with a subulate style thickened upwards. *Female flowers:*—*Perianth* tubular in bud, widened towards the base, cylindric above, in the open flower consisting of a subangular elongate-conical tube and 4 more or less spreading or recurved lobes, as long as or shorter than the tube; lobes filiform from a widened base, the filiform portion corresponding to the limb of the male perianth, channelled. *Staminodes* inserted at the base of the limb, very short, filiform. *Hypogynous scales* 0. *Ovary* oblong-ovoid, densely covered with long hairs ; style filiform, subclavate upwards ; stigma lateral at the end of the style, oblong, deeply and longitudinally grooved, papillose. *Ovule* 1, ascending from near the base, anatropous. *Nut* somewhat compressed, angular, hairy. *Cotyledons* thick, equal.

Glabrous, densely foliate shrubs ; leaves scattered, undivided, entire, narrow, coriaceous ; male flowers in spike-like, bracteate racemes, terminal on leafy, short, equal or unequal branchlets, which are usually gathered into fascicles or pseudo-whorls at the end of longer shoots ; bracts subulate, persistent with the pedicels ; female flowers in involucrate heads arranged more or less like the male racemes, but usually fewer and surrounded by the crowded uppermost leaves, which often exceed them considerably ; heads made up of whorls of 10–15 highly modified bract-like branchlets, forming a persistent involucre, and of a short conical central axis ; involucral branchlets dorsally flattened, with a few more or less reduced leaves on the back (representing an outer involucre) and with the margins pectinate upwards, the marginal segments being formed by subulate persistent bracts bearing modified subulate axes (arrested ♂ flowers) in their axils, whilst the inner faces of the branchlets are either smooth and naked or some distance above the base bear a solitary ♀ flower ; involucre at length woody and sometimes spreading out star-like when dry ; central axis short, bearing the bracteate flowers at the base and upwards to a varying height, sometimes ending in a tuft of barren bracts ; flowers yellow or whitish-yellow.

DISTRIB. Three species in the South-west corner of Cape Colony.

Female flowers confined to the central axis of the head ;
　　nut with the lateral ribs more prominent than the
　　others and long bearded, the ventral and the dorsal
　　shortly hairy; the faces between the angles with
　　oblique or transverse raised veins:
　　Leaves linear-oblanceolate to oblanceolate-cuneate, very
　　　　obtuse... 　　...　　...　　...　　..　　...　　... (1) **cneorifolia.**

　　Leaves long needle-shaped 　　...　　...　　...　　... (2) **pinifolia.**

Female flowers on the central axis and on the involucral
　　branches ; nut more rounded in cross section, long
　　villous all over, with secondary ribs between the flat
　　broad primary ones 　　...　　...　　...　　...　　... (3) **pallasia.**

1. **A. cneorifolia** (Knight, Prot. 15) ; shrub, 2 to 6 ft. high ; leaves linear-oblanceolate to oblanceolate-cuneate, obtuse to very obtuse with a minute callous apiculus, flat, $1\frac{1}{2}$–3 in. long, $1\frac{1}{2}$–5 lin. broad, venation raised, distinct in the wider leaves ; male racemes 1–2 in. long, rather stiff ; bracts $1\frac{1}{2}$–2 lin. long, equalling or exceeding the pedicels ; perianth whitish or pale yellow, 3–$3\frac{1}{2}$ lin. long including the 2–$2\frac{1}{4}$ lin.-long limb ; anthers about 1 lin. long, usually at length recurved ; style $2\frac{1}{2}$–3 lin. long, very slender ; female heads $\frac{2}{3}$–1 in. high and 1–$1\frac{1}{2}$ in. across ; involucral branches not bearing flowers, more or less fused below into a shallow receptacle, quite glabrous, at length hard and woody ; dorsal leaves linear-lanceolate and acute to spathulate and obtuse, 6–12 lin. long ; central axis stout, ending in a sharp point, 4–6 lin. long ; bracts linear-lanceolate, 2–1 lin. long ; perianth 4 lin. long including the

slightly more than 1 lin.-long spreading segments ; nuts quadrangular,
obovoid, dorsally more convex than ventrally, about $2\frac{1}{2}$ lin.
long, angles prominent, lateral fringed all round with a fulvous silky
spreading beard up to 2 lin. long, dorsal and ventral angles and
the faces between them shortly tomentose, faces obliquely and
prominently veined. *A. umbellata, R. Br. in Trans. Linn. Soc. x.*
50 ; *Lindley in Bot. Reg. t.* 1015 (δ); *Meisn. in DC. Prodr.* xiv.
212. *Protea aulacea, Thunb. Diss. Prot.* 31, *t.* 2 (δ); *Prodr.* 26 ;
Fl. Cap. ed. Schult. 131 ; *Lam. Ill.* i. 237 ; *Willd. Spec. Pl.* i. 520 ;
Poir. Encycl. v. 651. *P. umbellata, Thunb. Diss. Prot.* 32 ; *Prodr.*
26 ; *Fl. Cap. ed. Schult.* 131 ; *Linn. f. Suppl.* 118 ; *Lam. Ill.* i. 237 ;
Willd. Spec. Pl. i. 520 ; *Andr. Bot. Repos. t.* 248 ($\female$); *Poir. Encycl.*
v. 650. *Sim, For. Fl. Cape,* 293, *t.* 123, *fig.* 1. *P. cneorifolia,*
Salisb. Prodr. 49. *P. lanceolata, DC. ex Meisn. l.c.*

SOUTH AFRICA : without locality, *Thunberg,* δ ! *Thom,* 95, δ ! 784, δ ! 936, δ !
COAST REGION : Cape Div. ; Table Mountain, *Thunberg,* $\female$! Caledon Div. ; near
Palmiet River and Houw Hoek, *Drège* ; Houw Hoek Mountains, *Zeyher,* 3633,
δ, $\female$! *Pappe,* δ, $\female$! *MacOwan & Bolus, Herb. Norm. Austr.-Afr.* 772, δ, $\female$!
Schlechter, 7450, δ ! 7449, $\female$! Hermanus, *Galpin,* 4437, δ ! Bredasdorp Div. ;
near Elim, *Schlechter,* 7622, $\female$! 7623, δ ! *Bolus,* 7862, δ ! Riversdale Div. ;
Platte Kloof, *Thunberg,* δ !

2. **A. pinifolia** (Berg. Descr. Pl. Cap. 33) ; a shrub, up to 6 ft.
high ; leaves more or less needle-shaped and semiterete, more rarely
distinctly widened upwards and there up to 1 lin. wide, acute or
subacute, $1\frac{1}{2}$–6 in. long, usually much curved ; male racemes $\frac{3}{4}$–1 in.
long ; bracts up to 2 lin. long, exceeding or in old racemes equalling
the ultimately lengthened pedicels ; perianth yellow, 3–$3\frac{1}{2}$ lin. long
including the 2–$2\frac{1}{4}$ lin.-long limb ; anthers about 1 lin. long,
usually at length recurved ; style $2\frac{1}{2}$–3 lin. long, very slender ;
female heads $\frac{1}{2}$–1 in. long and 1–$1\frac{1}{2}$ in. across when mature ;
involucral branches not bearing flowers, more or less fused below
into a shallow receptacle, quite glabrous, at length hard and woody ;
dorsal leaves lanceolate-linear to subulate, acute, $\frac{3}{4}$–$1\frac{1}{2}$ (or more) in.
long ; central axis ending in a compact tuft of barren (at length
hardened) subulate bracts, including them $\frac{1}{2}$ in. long ; bracts
subulate 3–1 lin. long ; perianth 4 lin. long, including the slightly
more than 1 lin.-long spreading or recurved segments ; style $2\frac{1}{2}$ lin.
long, long persistent ; nut obovoid, dorsally more convex than
ventrally, $2\frac{1}{2}$–3 lin. long, ribbed, lateral ribs fringed all round with
a fulvous silky beard up to 2 lin. long, dorsal and ventral ribs and
faces shortly tomentose or the ventral rib with longer hairs, faces
transversely and more or less prominently veined. *Knight, Prot.*
15 ; *R. Br. in Trans. Linn. Soc.* x. 49 ; *Meisn. in DC. Prodr.* xiv.
212. *Aulax umbellata, E. Meyer in Drège, Zwei Pfl. Documente,* 97
(*and* 167 *partly*), not of *R. Br. Leucadendron pinifolium, Linn.*
Mant. 36. *L. cancellatum, Linn. Sp. Pl. ed.* i. 91. *Protea pinifolia,*
Linn. Mant. alt. 187 ; *Thunb. Diss. Prot.* 25 ; *Prodr.* 26 ; *Fl. Cap.*
ed. Schult. 127 ; *Willd. Spec. Pl.* i. 515 ; *Lam. Ill.* i. 237 ; *Andr.*
Bot. Repos. t. 76 ($\female$); *Poir. Encycl.* v. 651. *Protea bracteata,*

Thunb. Diss. Prot. 27, *t.* 1 (♂) ; *Prodr.* 26 ; *Fl. Cap. ed. Schult.* 128 ;
Linn. fil. Suppl. 118 ; *Willd. Spec. Pl.* i. 517 ; *Lam. Ill.* i. 238 ;
Poir. Encycl. v. 652. *Lepidocarpendron, etc., Boerh. Ind. Pl. Hort.
Lugd. Bot.* ii. 193, *t.* 193 (♀)? *Pini foliis planta, etc., Burm. Rar.
Afr. Pl.* 193, *t.* 70, *fig.* 3 (♂). *Conophorus Capensis pinifolius, Petiv.
Op.* i, 3, *no.* 458, *t.* 25, *fig.* 7 (♀).

SOUTH AFRICA : without locality, *Ecklon*, ♂ ! *Thunberg*, ♂ ! *Sparmann*, ♂ !
COAST REGION : Paarl Div. ; Paarl Mountain, *Drège*, ♂, ♀ ! Cape Div. ;
Muizenberg, *Ecklon*, 122*b*, ♀ ! Table Mountain, *Burchell*, 527, ♂, ♀ ! near Cape Town,
Schmieterlich, 183, ♂ ! Stellenbosch Div. ; Hottentots Holland, *Thunberg*, ♂ !
Caledon Div. ; Knoflooks Kraal and Little Houw Hoek, *Zeyher*, 3634, ♂, ♀ !
Nieuwe Kloof, *Burchell*, 8148, ♀ ! lower part of the Lange Bergen, near Garcias
Pass, *Burchell*, 6951, ♂, ♀ ! George Div. ; Lange Kloof, *Drège*, ♀ ! Knysna Div. ;
between Plettenbergs Bay and Knysna, *Burchell*, 5352, ♂, ♀ !

Protea bracteata was described from female specimens, and the specimen
named so in Thunberg's own herbarium at Upsala is no doubt identical with
Berg's *A. pinifolia* ; but two specimens in Montini's and Alstroemer's Herbarium
in the Stockholm collections, received from Thunberg, and also named *P. bracteata*,
are *A. pallasia*.

3. **A. pallasia** (Stapf) ; a shrub, up to 4 ft. high ; leaves intermediate in shape between those of the two preceding species from
linear-filiform to linear with a long attenuated base, acute to obtuse,
2–5 in. long, $\frac{1}{2}$–2 (rarely 3) lin. wide, rather straight, flat or the
narrowest semiterete and channelled above ; male racemes 1–1$\frac{1}{2}$ in.
long ; bracts up to 2 lin. long, exceeding or the upper equalling the
pedicels ; perianth 3–4 lin. long, including the 2$\frac{1}{2}$ lin.-long limb ;
anthers over 1 lin. long, usually at length recurved ; style 3 lin.
long, very slender ; female heads $\frac{1}{2}$–$\frac{3}{4}$ in. long and $\frac{3}{4}$–1 in. across ;
involucral branches bearing 1 flower at or above the lower $\frac{1}{3}$,
narrow, not fused at the base into a receptacle at length hard and
woody, usually sparingly and minutely puberulous, particularly
below, ofteu deeply divided, the lateral divisions without flowers
and pectinate-bracteate, pectination throughout looser than in the
preceding species ; dorsal leaves few, mostly needle-shaped, acute,
incurved, rarely flat, linear to oblanceolate-linear and obtuse,
$\frac{1}{2}$–1 in. long, forming with the uppermost leaves a dense outer
involucre, central axis rather slender, including the terminal barren
bracts over $\frac{1}{2}$ in. long, usually bearing flowers to more than $\frac{1}{2}$ way ;
bracts subulate, 2–1 lin. long ; perianth 3$\frac{1}{2}$ lin. long, including the
slightly finely filiform more than 1 lin.-long spreading or recurved
segments ; style 2$\frac{1}{2}$ lin. long ; nut ellipsoid, shortly contracted at
the base, elliptic in cross-section, with 4 rather broad primary
and 4 slender secondary ribs, more or less fulvous-villous all over,
the hairs on the ribs, and particularly those of the lateral, longer
than the rest. *A. umbellata, E. Meyer in Drège, Zwei Pfl. Documente,* 85, 97 (*and* 167 *partly*), *not of R. Br.*

SOUTH AFRICA : without locality, *Thunberg*, ♀ ! *Thom*, 331, ♀ ! 469, ♂ !
Forster, ♂ ! *Gueinzius*, ♂ !
COAST REGION : Tulbagh Div. ; Witsen Berg, *Burchell*, 8695, ♀ ! 8731, ♀ !
Winterhoeks Berg, *Bolus*, 5226, ♀ ! Worcester Div. ; Dutoits Kloof, *Drège*, ♀ !

hills near Worcester, *Rehmann*, 2510, ♀ ! Paarl Div. ; Paarl Mountain, *Drège*, ♀ !
Ecklon & Zeyher, ♂, ♀ ! Stellenbosch Div. ; Lowrys Pass, *Drège*, ♂ ! *Burchell*,
8239, ♂, ♀ ! *Schlechter*, 7254, ♂, ♀ ! Caledon Div. ; Houw Hoek Mountains,
Burchell, 8020, ♀ ! Knoflooks Kraal and Little Houw Hoek, *Zeyher*, 3634, ♂, ♀ !
Bavians Kloof, near Genadendaal, *Burchell*, 7784, ♂, ♀ ! Bot River, *Burchell*,
930, ♀ ! Swellendam Div. ; Mountain peak near Swellendam, *Burchell*, 7319, ♂ !
Riversdale Div. ; Platte Kloof (?) *Thunberg*, ♂.

Male specimens of *Aulax* are only distinguishable by their foliage, and as the
leaves of *A. pallasia* approach sometimes to one or the other of the two
remaining species the determination of some of the males referred here to
A. pallasia is perhaps open to doubt. These doubtful males are nearly all
specimens without definite localisation. Burchell 8239 and 930 approach more
than any others to *A. cneorifolia* so far as the leaves (including those of the
involucres) are concerned, but the structure of the involucral branches is clearly
that of *A. pallasia*. One or the other of the lower marginal bracts of the
involucral branches of *A. pallasia* may have a more or less perfect male flower in
its axil.

III. LEUCADENDRON, R. Br.

Flowers dioecious, regular. *Male flower* : *Perianth* linear, straight
or slightly incurved, with a somewhat thickened limb ; segments at
length separated to the middle or nearly to the base, differentiated
into limb and claw, spreading and recurved ; limb linear, oblong or
elliptic. *Anthers* sessile at the base of the limb, oblong or linear ;
connective sometimes shortly produced beyond the cells. *Hypogynous
scales* 4, free, linear or filiform, or absent. *Pistil* rudimentary,
consisting of a short slender style and a clavate entire or slightly
bifid stigma. *Female flower* : *Perianth* more or less as in the
male, but usually more deeply divided and the segments not so
differentiated into limb and claw. *Staminodes* usually linear.
Ovary ovoid, trigonous or compressed ; style usually slender and
gradually widened towards the apex, usually persistent ; stigma
terminal, oblique or lateral, entire or bifid, rarely 2-lobed. *Ovule*
solitary, attached laterally or ascending from near the base. *Fruit*
ovoid, transversely ellipsoid, trigonous or flattened and winged,
sometimes emarginate.

Trees, shrubs, or decumbent under-shrubs ; leaves acicular, linear, lanceolate to
obovate, entire, hardened at the apex into a blunt (rarely acute) callus, coriaceous,
glabrous, pilose or clothed with adpressed silky silvery indumentum ; male
flowers usually numerous, arranged in conical, globular or cylindric heads, each
flower subtended by a bract ; heads terminal, sessile or rarely pedunculate, sometimes surrounded by an involucre of imbricate bracts within the upper leaves ;
female flowers solitary, subtended by woody bracts aggregated in cone-like
heads ; bracts erect or spreading, free or rarely partially united, usually truncate,
rounded or retuse at the apex, rarely subacute.

DISTRIB. Species about 62, endemic, many very imperfectly known, and
described from specimens representing one sex only ; the following key must
therefore be regarded as provisional and artificial.

*Leaves pubescent or tomentose with adpressed silky,
　　silvery indumentum :
　　Lower leaves 2½ in. long or more, distinctly nerved ;
　　　　male heads large, 1½–2½ in. in diam.　　...　　...　(1) **argenteum**.

Lower leaves less than 2½ in. long, rarely with distinct
  nervation ; male heads much smaller, usually less
  than 1 in. in diam. :
  Leaves elliptic or ovate-elliptic, ½–1 in. broad,
    strongly nerved ...   .   (2) **nervosum.**

  Leaves oblanceolate, linear-oblanceolate or linear,
    ⅓ in. broad or less, without distinct nervation :
  Leaves acute or acutely mucronate, the inner of
    those surrounding the male heads more or
    less broadened at the base and sub-bracteate,
    but without a definite involucre of bracts
    closely adpressed to the flowers within :
    Leaves acute or with a short apiculus, those
      surrounding the male inflorescence not
      exceeding it by more than twice ; male
      perianth-tube hairy ...   .   (3) **uliginosum.**

    Leaves with a long apiculus, those surrounding
      the male inflorescence exceeding it by
      more than twice ; male perianth - tube
      glabrous   ...   .   (4) **salignum.**

  Leaves obtuse or subobtuse, sometimes with a
    blunt callose apiculus, those surrounding the
    male heads narrowed to the base, often with
    a definite involucre of closely imbricate bracts
    within :
    Male perianth-limb glabrous ; male heads ovoid,
      oblong-cylindric or obconic, usually with
      a distinct involucre of several series of
      bracts :
      Male heads oblong-cylindric or obconic ;
        female bracts ovate-lanceolate, the inner
        long villous   ...   .   (5) **plumosum.**

      Male heads ovoid ; female bracts broader than
        long, truncate or emarginate, glabrous ...   (42) **minus.**

    Male perianth-limb hairy ; male heads (where
      known) depressed-globose or subglobose,
      with a more or less indistinct involucre of
      few series of bracts (conspicuous in 9,
      schinzianum) :
      Lower leaves narrowly linear or subacicular,
        not widened in the upper part ...   ...   (6) **aurantiacum.**

      Lower leaves oblanceolate or linear-oblan-
        ceolate, widened in the upper part :
        Female heads large, 2–2½ in. in diam.   ...   (7) **proteoides.**

        Female heads (where known) less than
        1½ in. in diam. :
        Leaves more or less densely tomentose ;
          female heads depressed-globose or
          subobconic ; female bracts as long as
          or longer than broad, the outer ones
          usually pointed :
        Leaves surrounding the inflorescence
          usually much larger than those
          below ; outer involucral bracts of
          the female heads numerous and
          acuminate, inner obovate and
          subacute :

Bracts of the male inflorescence
shortly pubescent or tomentellous, those of the female
densely villous-tomentose ; male
and female heads dissimilar ... (8) **Schlechteri.**

Bracts of the male and female lax
villous with long weak hairs ;
male and female heads very
similar ... ... (9) **schinzianum.**

Leaves surrounding the inflorescence
about the same length as those
below ; outer involucral bracts of
the female few, scarcely acuminate, inner broadly ovate :
Indumentum of the male perianth
fulvous ; fruits narrowly ob-
ovoid, ivory white and smooth (10) **nitidum.**

Indumentum of the male perianth
whitish or dull ; fruits transversely oblong-ellipsoid, rugose,
dull and nearly black when dry (11) **elatum.**
Leaves more or less adpressed-pilose or
pubescent, rarely somewhat tomentose ; female heads more or less
oblong-cylindric or ellipsoid (not
known in 14, *sericeum*) ; bracts much
broader than long :
Leaves ¾–2 in. long, 1½–5 lin. broad :
Leaves oblanceolate, 2½–4½ lin. broad,
those surrounding the female
inflorescence about 1¾ in. long
and 4–5 lin. broad ; male
perianth-limb thinly pilose,
becoming glabrous ... ... (12) **sericocephalum.**

Leaves narrowly - oblanceolate or
linear - oblanceolate, 1½–2 lin.
broad, those surrounding the
female inflorescence about 1 in.
long and 1–2 lin. broad ;
male perianth-limb permanently
villous ... ... (13) **cinereum.**

Leaves very small, 2½–3 lin. long,
¾–¾ lin. broad ... ... (14) **sericeum.**

**Leaves glabrous or rarely thinly pilose with weak spread-
ing hairs or hairy margins (sometimes subtomentose
in 54, *daphnoides*) :
Male heads pedunculate :
Male flowers and female bracts (where known)
hairy :
Male heads many-flowered, solitary or subsolitary
at the ends of the shoots ... ... (15) **tortum.**

Male heads about 6-flowered, crowded at the ends
of the shoots ... ... (16) **ericifolium.**

Male flowers and female bracts glabrous ... ... (17) **abietinum.**

Male heads (where known) sessile :
 †Leaves ¾ in. long or less :
  ‡Leaves linear-acicular or acicular, not widened
   upwards, about ½ lin. broad or less, rarely
   ¾ lin. broad :
  Male heads oblong-cylindric or ellipsoid ; male
   perianth glabrous ; female bracts retuse,
   glabrous ; leaves of the female thinly pilose
   or if glabrous then reflexed :
  Male heads exserted or only slightly clasped
   by the upper leaves ; leaves of the female
   at length strongly reflexed ; female head
   with 15–20 series of bracts ; fruits emarginate ... ...   (17) **abietinum.**

  Male heads more or less clasped and hidden
   by the upper leaves ; leaves of the female
   not reflexed ; female head with about 10
   series of bracts ; fruits mucronate   ...   (18) **scabrum.**

 Male heads subglobose or depressed-globose ;
  male perianth (where known) hairy ; leaves
  of the female glabrous or rarely pilose when
  young, not reflexed :
  Female bracts ligulate ; male perianth-lip
   glabrous or nearly so   ...   .   (19) **corymbosum.**

  Female bracts not ligulate ; male perianth-lip
   (where known) very hairy :
  Female bracts villous - tomentose outside ;
   leaves flat or nearly so :
  Male heads with numerous very hairy
   involucral bracts ; fruiting head
   about 1 in. long   ...   .   (27) **truncatum.**

  Male heads with few inconspicuous involucral bracts ; fruiting head about
   ½ in. long   ...   .   (20) **fusciflora.**

  Female bracts glabrous except at the ciliate
   margin ; leaves terete ...   .   (21) **sorocephalodes.**

 ‡‡Leaves lanceolate, oblanceolate, spathulate or
  elliptic, mostly widened above, more than
  1 lin. broad.
 Flower-heads surrounded by numerous spread-
  ing, densely adpressed-tomentose leaves ...   (22) **radiatum.**

 Flower-heads surrounded by few erect glabrescent (rarely spreadingly pilose) leaves or by
  erect coloured glabrous or hairy bracts :
  Male heads with numerous highly coloured
   glabrescent rounded bracts ; young leaves
   adpressed-pubescent   ...   .   (23) **dubium.**

  Male heads with very hairy, often scarcely
   coloured, acute or acuminate bracts or
   hardly bracteate ; young leaves glabrous
   or rarely loosely pilose :
  Leaves lanceolate or lanceolate-elliptic, not
   widened above, often closely imbricate   (24) **imbricatum.**

  Leaves obovate, oblanceolate or spathulate :
   Leaves about 2–3 lin. broad, obovate,
    oblanceolate or narrowly oblong-
    oblanceolate :

Male heads with a very distinct involucre of bracts, the outer ones
acuminate ; leaves oblanceolate,
subacute, about ¾ in. long, pube-
scent at the base ...   . (25) **pubescens.**

Male heads with few bracts within the
upper leaves ; leaves narrowly oblanceolate, ¾ in. long or more,
often pilose, especially when young (43) **lanigerum.**

Male heads with few indistinct bracts ;
leaves obovate, rounded at the
apex, 4–7 lin. long, quite glabrous (26) **coriaceum.**

Leaves less than 1½ lin. broad, spathulate
or spathulate-linear :
Leaves ½–1 in. long ; female heads 1 in.
in diam.   ...   . (27) **truncatum.**

Leaves 3–5 lin. long ; female heads less
than ¾ in. in diam. ...   . (28) **levisanum.**

††Leaves over ¾ in. long, usually more than 1 in. :
Leaves of two kinds on the same shoot, the lower
acicular, the upper narrowly oblanceolate or
all acicular :
Female inflorescence oblong-cylindric ; branches
and perianth-limb glabrous ; fruits compressed, winged :
Male and female inflorescences with an involucre of imbricate, densely villous or
ciliate bracts ; fruits transversely oblong-
elliptic   ...   . (29) **platyspermum.**

Male and female inflorescences naked or
nearly so at the base ; fruits broadly and
longitudinally elliptic, emarginate   ... (30) **æmulum.**

Female inflorescence ovoid-globose ; branches
hairy ; perianth densely villous ; fruits not
compressed or winged   ...   . (31) **Dregei.**

Leaves of one kind on the same shoot, flat, never
acicular :
Leaves linear, not or scarcely broadened in the
upper part, glabrous or hairy, 2½ lin. broad
or less :
Male perianth very hairy :
Leaves obtuse ; outer male bracts tomentose, not coloured, shorter than the
flowers ...   . (32) **Galpinii.**

Leaves acute ; outer male bracts glabrous
or slightly pubescent, broad at the
base, coloured, as long as or longer
than the flowers   ...   . (33) **ramosissimum.**
Male perianth or at least the limb quite
glabrous :
Leaves produced at the apex into long acute
subulate points :
Inner bracteate leaves glabrous ; female
bracts glabrous except on the ciliate
margins   ...   . (34) **strictum.**

Inner bracteate leaves villous ; female
bracts villous-tomentose outside   ... (4) **salignum.**

Leaves obtuse or subacute, not produced
into long points :
  Leaves with villous margins ; female
    bracts glabrous   ...   . (42) **minus.**
  Leaves with glabrous margins ; female
    bracts hairy all over or only at the
    base :
    Leaves surrounding the male inflor-
      escence very broad, pubescent and
      coloured at the base ; lower leaves
      $2-2\frac{1}{2}$ in. long ; branches pilose
      with rather long weak hairs   ... (35) **eucalyptifolium.**
    Leaves surrounding male inflorescence
      glabrous and not or scarcely
      broadened at base ; lower leaves
      $1-1\frac{1}{2}$ in. long ; branches glabrous
      or rarely slightly adpressed-
      pubescent :
      Male heads without an involucre of
        imbricate bracts within upper
        leaves, rarely a few of the latter
        subbracteate ; female bracts to-
        mentose all over or pubescent in
        lower half ; ovary winged :
        Female bracts whitish-tomentose
          all over   ...   . (36) **adscendens.**

        Female bracts pubescent only in
          the lower half, glabrous in the
          upper part   ...   . (37) **Phillipsii.**

      Male heads with a distinct involucre
        of bracts ; female bracts glabrous
        except on the ciliate margins ;
        ovary not winged   ...   ... (38) **meyerianum.**
Leaves oblanceolate, oblong-lanceolate or spathulate, glabrous or hairy or with hairy margins or tips, if less than 3 lin. broad then
  distinctly widened in the upper part :
  Leaves rather narrow, 2–3 lin. broad (rarely
    4 lin.), glabrous or pilose, rarely with dis-
    tinct nerves :
    Branches glabrous, rarely with a few weak
      hairs :
      Leaves mostly with hairy tips, usually
        glaucous ; male inflorescence ellipsoid
        or ellipsoid-globose ; female inflores-
        cence nearly completely hidden by
        the broadened coloured leaves ; bracts
        velvety-tomentose   ...   . (39) **concinnum.**
      Leaves with glabrous tips, not glaucous ;
        male inflorescence depressed-globose ;
        female inflorescence surrounded by
        leaves with more or less narrow
        bases ; bracts glabrous or slightly
        pubescent :
        Leaves less than $1\frac{1}{2}$ in. long ; female
          head broadly ovoid   ...   ... (40) **decurrens.**
        Leaves $2-2\frac{1}{2}$ in. long ; female head
          cylindric ...   . (41) **glabrum.**

Branches villous or shortly pubescent :
  Leaves with densely pubescent margins ;
    male heads with a distinct involucre
    of bracts ; female bracts glabrous ... (42) **minus.**

Leaves thinly pilose or nearly glabrous,
  but not densely hairy on the margin ;
  male heads with an indistinct involucre of bracts ; female bracts
  tomentose ... ... (43) **lanigerum.**

Leaves broader than in the preceding, 4–7 lin.
  broad, glabrous or rarely subtomentose,
  mostly with distinct ascending nerves :
Branches glabrous or very rarely minutely
  puberulous :
  Leaves 1½–1¾ in. broad ... ... (44) **crassifolium.**

Leaves ½–¾ in. (rarely 1 in.) broad :
  Leaves spathulate and much attenu-
    ated to a narrow base or obovate-
    oblanceolate, those surrounding
    the female inflorescence narrow at
    the base and not hiding it ; female
    bracts densely rusty-villous :
    Leaves spathulate, falcate, without
      distinct nerves ... ... (45) **spathulatum.**

    Leaves obovate-oblanceolate, shortly
      narrowed to the base, 3-nerved,
      not falcate ... ... (46) **pseudo-**
                    spathulatum.

  Leaves more or less oblanceolate, only
    slightly attenuated to a fairly
    broad base ; female heads mostly
    nearly hidden by the broadened
    surrounding leaves ; female bracts
    shortly velvety - tomentose or
    glabrous :
    Branches glabrous ; leaves glaucous,
      the lower 1–1¾ in. long ; female
      bracts shortly velvety - tomen-
      tose all over the outside ... (39) **concinnum.**

    Branches glabrous ; leaves not
      glaucous, 1–1½ in. long, 5–7 lin.
      broad ; female bracts shortly
      velvety-tomentose only in the
      upper half, glabrous in the
      lower part ... ... (47) **discolor.**

    Branches minutely puberulous or
      nearly glabrous ; leaves 1½–3 in.
      long ; female bracts glabrous in
      the upper part, slightly pubes-
      cent below ... ... (48) **squarrosum.**

Branches densely and softly pubescent or
  villous with long weak spreading hairs :
  Upper leaves retuse ... ... (49) **retusum.**

  Upper leaves not retuse :
    Leaves 1–1¼ in. broad, obovate or
      obovate-elliptic ... ... (50) **ovale.**

Leaves ¾ in. broad or less :
Male heads less than 1 in. in diam. ;
    female head elongated, cylin-
    dric ; leaves tomentose on the
    margin :
    Male heads elongated ; leaves
       oblong-lanceolate ... ... (51) **decorum**
    Male heads globose ; leaves ob-
      ovate - oblanceolate or ob-
      lanceolate ... ... (52) **concolor.**
Male heads 1–2 in. in diam. ; female
    heads broadly obovoid or de-
    pressed-globose :
    Male heads about 1 in. in diam. ;
      inner leaves glabrous outside ;
      bracts in 2–3 series, glabrous
      or nearly so ; female bracts
      with a very dense villous tuft
      of rust-coloured hairs below
      the middle, glabrous above ... (53) **venosum.**
    Male heads about 1¾–2 in. in diam. ;
      inner leaves silky-villous out-
      side ; bracts in about 2 series ;
      female bracts long-villous in
      the lower half and along the
      margin ... ... (54) **daphnoides.**
    Male heads 1½–2 in. in diam. ;
      inner leaves slightly villous or
      nearly glabrous ; bracts 5–6-
      seriate ; female bracts ad-
      pressed - pilose in the lower
      part ... ... (55) **grandiflorum.**

1. **L. argenteum** (R. Br. in Trans. Linn. Soc. x. 52) ; a tree
20–30 ft. high ; branchlets stout, terete, about ⅓ in. in diam.,
densely pubescent with closely adpressed hairs and often long pilose ;
leaves sessile, 2½–5½ in. long, ½–1¼ in. broad, gradually increasing
in length towards the inflorescence, lanceolate or elongate-lanceolate,
slightly narrowed to a broad insertion at the base, narrowed to a
subacute hardened glabrous apex, entire, rather thinly coriaceous,
densely pubescent on both surfaces with adpressed silvery silky hairs ;
male inflorescence terminal, solitary, more or less globose, 1½–2
in. in diam. ; floral bracts 3½ lin. long, oblong, obtuse, coriaceous,
ciliate, villous ; perianth-tube 3 lin. long, cylindric, glabrous ;
segments 3 lin. long ; limb 1½ lin. long, oblong-linear, shortly and
bluntly acuminate, shortly villous outside ; anthers ¼ lin. long, linear :
style 5¼ lin. long, pubescent in the lower half ; stigma 1 lin. long,
clavate-obtuse, furrowed ; hypogynous scales 1–2½ lin. long, linear-
filiform ; female inflorescence similar to the male in position and
shape ; bracts very broadly ovate or semiorbicular, rigidly coriaceous,
velvety-tomentellous outside, glabrous and slightly shining within ;
perianth-tube inflated, constricted at the mouth, membranous,
glabrous ; segments 8 lin. long, villous with long simple hairs ; limb
1 lin. long, oblong, villous with short hairs ; staminodes ⅚ lin. long,

linear, minutely glandular at the apex; style 6½ lin. long, filiform, glabrous; stigma 1 lin. long, clavate, bifid at the apex; fruit 5 lin. long, 4½ lin. broad, oblong-obovoid, sparingly pubescent in the lower part. *Bot. Reg. t.* 979; *Meisn. in DC. Prodr.* xiv. 213. *Protea argentea, Linn. Sp. Pl. ed.* i. 94; *Thunb. Diss. Prot.* 55; *Linn. Syst. Veg. ed.* xiv. 141; *Gærtn. Fruct.* i. 239, *t.* 51; *Lam. Ill.* i. 237, *t.* 53, *fig.* 1; *Willd. Sp. Pl.* i. 529; *Thunb. Fl. Cap. ed. Schult.* 136. *Leucadendros africana, arbor tota argentea, etc., Pluk. Phytogr. t.* 200, *fig.* 1, *and Almag.* 212. *Argyrodendros africana foliis sericis et argenteis, Comm. Hort.* ii. 51, *t.* 26. *Conocarpodendros foliis argenteis, etc., Boerh. Ind. Pl. Hort. Lugd. Bat.* ii. 195, *t.* 195. *Scolymocephalos africanus folio crasso, etc., Weinm. Phyt.* iv. 293, *t.* 900.

SOUTH AFRICA: without locality, *Sieber,* 5! *Bergius!* *Thunberg!*
COAST REGION: Cape Div.; Table Mountain, *Ecklon,* 45! *Pappe!* Devils Peak, *Burchell,* 8475! Rondebosch, *Drège!* Lions Head, *MacOwan,* 2926! & *in Herb. Norm. Austr.-Afr.* 905! *Phillips!*

2. **L. nervosum** (Phillips & Hutchinson); a shrub about 6 ft. high; branchlets longitudinally sulcate, pilose with long weak hairs; leaves 1–2 in. long, ½–1 in. broad, elliptic or ovate-elliptic, mucronate, rigidly coriaceous, very shaggy-pubescent when young, becoming closely adpressed-pubescent when older with a cartilaginous margin, prominently nerved on both surfaces; male inflorescence solitary, terminal, about ¾ in. long and ½ in. in diam., oblong-ellipsoid, the surrounding leaves long-villous in the lower part; floral bracts ¾ lin. long, oblong, obtuse, somewhat concave, pilose; perianth-tube ½ lin. long, glabrous; segments 3¼ lin. long, spathulate-linear; limb ¾ lin. long, oblanceolate, incurved above, adpressed-pubescent outside; anthers ¾ lin. long, linear; style 2¾–3 lin. long, terete, pilose at the base; stigma ½ lin. long, clavate, subacute; hypogynous scales ½ lin. long, linear; female flowers not known.

COAST REGION: Caledon Div.; mountains near Genadendal, *Burchell,* 7862!

3. **L. uliginosum** (R. Br. in Trans. Linn. Soc. x. 63); a straight erect shrub 6–7 ft. high; branchlets adpressed silky-pubescent; leaves ½–1½ in. long, 1½–3 lin. broad, oblong-linear or lanceolate-linear, acute or sharply mucronate, rigidly but rather thinly coriaceous, densely adpressed silky-pubescent or tomentose; male inflorescence conic or ellipsoid, ⅓–¾ in. long, scarcely ½ in. in diam., surrounded by several leaves about ¾ in. long, the latter often rusty-pubescent; floral bracts ¼ lin. long, oblong, subacute, pilose; perianth-tube ¾ lin. long, somewhat compressed, pilose; segments 1 lin. long, spathulate-linear, pilose; limb ¼ lin. long, elliptic, subobtuse, pilose; anthers sessile, ¼ lin. long, oblong; style 1½ lin. long, filiform, glabrous; stigma ¼ lin. long, ellipsoid; hypogynous scales ½ lin. long, linear; young female head ½ in. long, oblong-cylindric; bracts 1 lin. long, 2 lin. broad, transversely oblong, tomentose; perianth-tube 1 lin. long, compressed, pilose; segments

$\frac{1}{2}$ lin. long, linear, pilose; limb $\frac{1}{8}$ lin. long, oblong, pilose; staminodes $\frac{1}{8}$ lin. long, linear; style $\frac{3}{4}$ lin. long, linear, broadening above, glabrous; stigma truncate; ovary $\frac{1}{2}$ lin. long, compressed, oblong, pilose; hypogynous scales $\frac{2}{3}$ lin. long, linear; mature head ellipsoid, 1–1$\frac{1}{4}$ in. long, scarcely 1 in. in diam.; bracts tomentose outside; fruits compressed, wrinkled, elliptic, glabrous, about 2 lin. long. *Meisn. in DC. Prodr.* xiv. 223. *L. floridum, R. Br. l.c.*; *Meisn. l.c. L. cuspidatum, Klotzsch ex Meisn. l.c.?. L. salignum, var. lineari-folium, Meisn. l.c.?. L.? coniferum, Meisn. l.c.* 227, *partly. Protea saligna, Andr. Bot. Rep. t.* 572. *P. conifera, Thunb. Diss. Prot.* 53 (*excl. specimens* α, β, γ *and* δ *of Herb.*); *Thunb. Fl. Cap. ed. Schult.* 135 (*excl. spec.* α, β, γ *and* δ *of Herb.*). *P. saligna, Thunb. Diss. Prot.* 39; *Linn. Syst. Veg. ed.* xiv. 140; *Thunb. Fl. Cap. ed. Schult.* 136. *P. concinna, Salisb. Prodr.* 50. *Frutex æthiopicus conifer, etc., Breyn. Cent.* 21, *t.* 9.

SOUTH AFRICA: without locality, *Krauss*, 1043, 1045, *Thom*, 181! *Sieber*, 7! *Zeyher*! *Ecklon*! *Mund*! *Bergius*! *Labillardière*! *Grey*! *Thunberg*! *Brown*!

COAST REGION: Cape Div.; Claremont Flats, *Dümmer*, 503! Kenilworth Racecourse, *Dümmer*, 1776! Cape Flats near Rondebosch, *Burchell*, 210! between Hout Bay and Wynberg, *Drège*! near Wynberg, *Oldenburgh*! *Roxburgh*! *Wallich*! *Brown*! *Bolus*, 3846! 3847! Stellenbosch Div.; between Stellenbosch and Cape Flats, *Burchell*, 8365! Riversdale Div.; near Milkwoodfontein, *Galpin*, 4443! 4444! Mossel Bay Div.; Robinson Pass, *Bolus*, 12263! 2447! Oudtshoorn Div.; near the Oliphants River, *Gill*! George Div.; near the Touw River, *Burchell*, 5740! 5752! on mountains, *Bowie*! Post Berg, near George, *Bowie*!

4. L. salignum (R. Br. in Trans. Linn. Soc. x. 62); branches terete, glabrous; branchlets stiff, shortly pubescent; leaves subequal in the sexes, those surrounding the inflorescences longer than those on the remainder of the shoot, 1$\frac{1}{4}$–2$\frac{1}{2}$ in. long, 1$\frac{1}{2}$–3$\frac{1}{2}$ lin. broad, oblong-linear or lanceolate-linear, produced into a long subulate acute apex, thinly but rigidly coriaceous, silky-pubescent with long closely adpressed hairs on both surfaces, at length becoming glabrous and longitudinally wrinkled or striate; male inflorescences solitary and terminal, surrounded by several coloured leaves, $\frac{1}{2}$ in. long, about 5 lin. in diam., ellipsoid; bracts 1 lin. long, obovate, obtuse, densely villous; perianth-tube $\frac{3}{4}$ lin. long, glabrous; segments 1$\frac{1}{2}$ lin. long, spathulate-linear, glabrous; limb $\frac{3}{4}$ lin. long, oblong-elliptic, obtuse, glabrous; anthers $\frac{2}{3}$ lin. long, oblong; style 1$\frac{1}{2}$ lin. long, cylindric, thickened at the base; stigma $\frac{1}{3}$ lin. long, clavate, obtuse; hypogynous scales $\frac{5}{6}$ lin. long, filiform; female inflorescence surrounded by several leaves, $\frac{3}{4}$ in. long, $\frac{1}{2}$ in. in diam., ellipsoid; bracts 1$\frac{1}{2}$ lin. long, 4$\frac{1}{2}$ lin. broad, transversely linear-oblong, rounded above, tomentose; perianth-segments 2 lin. long, cohering, glabrous; limb $\frac{1}{4}$ lin. long, suborbicular, glabrous; style 1 lin. long, linear, widened above, doubly bent just above the base, obliquely inserted; stigma obovate, oblique; hypogynous scales $\frac{1}{2}$ lin. long, filiform; ovary $\frac{3}{4}$ lin. long, winged, the wings produced into two arms above, glabrous; fruits compressed, narrowly winged, about 2 lin. long, elliptic, black, glabrous. *Meisn. in DC. Prodr.* xiv. 223 (*excl. syn. P. sericea, Thunb., et spec. Zeyher*, 3646). *Protea argentea* β, *Linn. Sp. Pl. ed.* i. 94 (*excl. syn. Brayn. et Tournef.*). *P. conifera* a, *Linn. Sp. Pl. ed.* ii. 138 (*excl. syn.*). *P. saligna, Linn. Mant. alt.* 194, ♂ (*excl. syn. Berg. et Breyn.*); *Lam. Ill.* i. 236; *Poir. Encycl.* v. 648, ♀. *P. diversifolia, Willd. Enum. Hort. Berol.* 139. *Conocarpodendron folio tenui angusto, etc., Boerh. Ind. Pl. Hort. Lugd. Bat.* ii. 204.

SOUTH AFRICA: without locality, *Forster*! *Bergius*! *Mund*! *Ecklon*! COAST REGION: Cape Div.; Devils Mountain, *Burchell*, 8485! *MacOwan, Herb. Norm. Austr.-Afr.* 787! Table Mountain, *Burchell*, 531! 532! *Milne,* 30! *MacGillivray,* 638! *Ecklon,* 468! Slopes of Millers Point, *Wolley-Dod,* 2924! Smitwinkel Bay, *Wolley-Dod,* 2684! Cape Flats near Wynberg, *MacOwan, Herb. Norm. Austr.-Afr.* 786! Simons Bay, *Wright*! near Cape Town, *Pappe*! Stellenbosch Div.; mountains of Lowrys Pass, *Burchell,* 8192! Caledon Div.; mountains of Klein River Kloof, *Zeyher,* 3647! near Palmiet River, *Bolus,* 4202! Zwart Berg, *Ludwig*! Humansdorp Div.; Witte Els Bosch, Zitzikamma, *Galpin,* 4446! Uitenhage Div.; Van Stadens Drift, *Drège,* 102!

5. **L. plumosum** (R. Br. in Trans. Linn. Soc. x. 53); an erect bush 5–7 ft. high; branches pubescent, rarely glabrous; leaves ½–2½ in. long, 1–3 lin. broad, linear or lanceolate-linear, acute to obtuse at the apex, narrowed to the base, adpressed-pubescent; male inflorescences 6 lin. long, terminal or axillary, clustered at the end of the branches; bracts 3 lin. long, spathulate-linear, glabrous below, villous and bearded above; perianth-tube 3¼ lin. long, cylindric, slightly widening above, pubescent; lobes 2¼ lin. long, spathulate-linear, glabrous; limb 1¾ lin. long, linear, sub-acuminate, acute; anthers 1½ lin. long, linear, with a minute linear gland at the apex; style 4 lin. long, filiform; stigma ½ lin. long, narrow-cylindric; female inflorescence sessile, 1–1¾ in. long, terminal, usually solitary, rarely 3-nate; involucral bracts ovate or ovate-oblong, obtuse or subobtuse, pubescent or glabrous when young, becoming tomentose outside with age, ciliate; perianth-tube 1¾ lin. long, villous with very long hairs; segments 10 lin. long, linear, channelled within, pilose with very long hairs outside; limb 1¾ lin. long, linear, subobtuse or subacute, glabrous; staminodes ¾ lin. long, linear, with a minute linear gland at the apex; style 1 in. long, cylindric, furrowed above, becoming filiform below, glabrous; stigma lateral, 2 lin. long, linear, curved; hypogynous scales ¼ lin. long, linear; ovary ⅓ lin. long, elliptic, covered with long hairs. *Meisn. in DC. Prodr.* xiv. 213; *Spreng. Syst.* i. 456. *Protea parviflora, Linn. Mant. alt.* 195; *Thunb. Diss. Prot.* 35, *t.* 4; *Linn. Syst. Veg. ed.* xiv. 140; *Lam. Ill.* i. 235; *Thunb. Prodr.* 27; *Willd. Sp. Pl.* i. 524; *Poir. Encycl.* v. 643; *Thunb. Fl. Cap. ed. Schult.* 133. *P. obliqua, Thunb. Diss. Prot.* 35; *Linn. f. Suppl.* 117; *Thunb. Prodr.* 27; *Thunb. Fl. Cap. ed. Schult.* 133. *P. arcuata, Lam. Ill.* i. 234, *excl. var.* β. *P. plumosa, Ait. Hort. Kew. ed.* 1, i. 127. *Gissonia collina, Knight, Prot.* 33.

SOUTH AFRICA: without locality, *Labillardière*! *Thom,* 265! *Thunberg*! *Ludwig*! *Gueinzius*! COAST REGION: Clanwilliam Div.; Packhuis Berg, *Schlechter,* 8622! 8623! Piquetberg Div.; Piquetberg Range, *Schlechter,* 5208! Tulbagh Div.; Mitchells

Pass, *Bolus*, 3167 ! New Kloof, *Burchell*, 1021 ! near Tulbagh Waterfall, *Phillips*, 528 ! Witsenberg Range near Tulbagh, *Burchell*, 8703 ! *Schlechter*, 1392 ! Worcester Div. ; near Worcester, *Cooper*, 1609 ! Hex River Kloof, *Drège* ! Paarl Div. ; near Paarl, *Wilms*, 3571 ! 3578 ! *Pappe* ! *Drège* ! Cape Div. ; near Wynberg, *Bolus*, 3845 ! Hottento's Kloof, *Pearson*, 4933 ! Stellenbosch Div. ; Hottentots Holland, *Zeyher*, 1453 ! Caledon Div. ; Baviaans Kloof near Genadendal, *Burchell*, 7810 ! Mossel Bay Div. ; Attaquas Kloof, *Gill* !

6. **L. aurantiacum** (Buek in Drège, Zwei Pfl. Documente, 117, 198, partly, as to spec. b) ; a shrub 2–3 ft. high, with the appearance of a small tree ; branches rather thick, marked by the scars of fallen leaves, slightly pubescent or glabrous ; young branchlets rather slender, silky-villous ; leaves 1–2¾ in. long, ½–1¾ lin. broad, the lower ones linear, subterete, those surrounding the inflorescence flat and a little wider towards the top, contracted into a hardened glabrous mucro at the apex, villous with adpressed silky hairs ; male inflorescences solitary, terminating short branchlets, surrounded and overtopped by the uppermost leaves, depressed-globose, scarcely more than ½ in. in diam. ; bracts lanceolate, acute, up to 2 lin. long, densely villous outside ; perianth-tube ¾ lin. long, thinly pilose ; segments 1½ lin. long, villous ; limb ½ lin. long, elliptic-oblong, obtuse, densely villous outside ; anthers ½ lin. long, linear ; style 1¼ lin. long, filiform, pilose at the base ; stigma ⅓ lin. long, spindle-shaped, subobtuse ; female inflorescences 1¼–1½ in. long, 1½–2 in. in diam., broadly-ovoid or ovoid-globose ; bracts about ½ in. long and broad, ovate-triangular to narrowly lanceolate, rather abruptly acuminate to a glabrous apex, becoming obtuse in fruit, rigidly coriaceous, glabrous, shining and striate within, densely villous outside ; perianth-tube 3–4 lin. long, inflated, membranous, glabrous ; segments linear-filiform, 5½ lin. long, long-villous ; limb ¾ lin. long, linear, suboutuse, long-villous ; staminodes ½ lin. long, linear ; style 5½ lin. long, glabrous ; stigma minutely bifid ; fruit 3 lin. long, 1⅓ lin. in diam., oblong-obovoid, subacute at the base, white villous with weak ascending hairs. *Meisn. in DC. Prodr.* xiv. 217. *L. proteoides, E. Meyer in Pl. Drège, as to spec. b, ex Meisn. l.c. L. cinereum, Meisn. l.c.* 216, *partly, as to syn. Protea alba, Thunb. Diss. Prot.* 31 ; *Linn. Syst. Veg. ed.* xiv. 139 ; *Thunb. Fl. Cap. ed. Schult.* 130 ; *Lam. Ill.* i. 236 ; *Poir. Encycl.* v. 647. *P. cinerea, Ait. Hort. Kew. ed.* 1, i. 127, *not of Willd.*

SOUTH AFRICA : without locality, *Thunberg* ! *Sparrman* !
COAST REGION : Swellendam Div. ; summit of a mountain peak near Swellendam, *Burchell*, 7335 ! George Div. ; on the Cradock Berg, near George, *Burchell*, 5892 ! Uniondale Div. ; between Avontuur and Klip River, in Lange Kloof, *Drège* !
CENTRAL REGION : Prince Albert Div. ; Zwartberg Pass, *Bolus*, 11625 !

7. **L. proteoides** (E. Meyer MSS.) ; branchlets stout, slightly wrinkled, about 4 lin. in diam., glabrous ; leaves crowded, 1¼–2½ in. long, 1½–4½ lin. broad, linear-oblanceolate or spathulate-oblanceolate, rounded or obtuse at the apex, thick and coriaceous, pubescent on both surfaces with small adpressed silky hairs ;

male inflorescence not known; female inflorescence terminal or terminating a lateral branch, solitary, globose, 2–2½ in. in diam.; bracts oblong or lanceolate-oblong, obtusely pointed, densely adpressed silky-pubescent outside, glabrous within; perianth-tube 4 lin. long, swollen in the middle, narrowing to both ends, membranous, glabrous; segments 4 lin. long, linear, villous outside with very long hairs; limb 1 lin. long, linear, subacute, villous; staminodes $\frac{5}{6}$ lin. long, linear; style 5 lin. long, filiform, glabrous; stigma $\frac{1}{3}$ lin. long, truncate, minutely bifid; fruit 2½ lin. long, narrowly ellipsoid-obovoid, villous. *L. aurantiacum, Buek in Drège, Zwei Pfl. Documente* 64, 198, *partly, as to spec.* ♀ *a.*

CENTRAL REGION: Prince Albert Div.; Great Zwartberg Range, near Vrolykheid, *Drège*, ♀, a!

8. L. **Schlechteri** (Phillips & Hutchinson); branches terete, densely adpressed-pubescent; young branchlets adpressed-tomentose with whitish hairs; leaves linear-oblanceolate or oblanceolate, those on the shoot below the inflorescence ½–¾ in. long, 1¼–2 lin. broad, those surrounding the inflorescence larger, 1¼–2 in. long, 2–2½ lin. broad, all contracted at the apex into a hardened glabrous mucro, rigidly coriaceous, densely tomentose with silky adpressed hairs on both surfaces; male inflorescences solitary at the apex of each branchlet, depressed-globose, scarcely ½ in. in diam.; bracts 1 lin. long, lanceolate, acute, villous; perianth-tube 1¾ lin. long, rather densely pilose; segments 1¼ lin. long, shortly villous; limb ⅔ lin. long, lanceolate-elliptic, subacute, villous outside; anthers ⅓ lin. long, linear; style filiform; stigma ⅓ lin. long, clavate, subacute; female inflorescence terminal, depressed-globose, ¾–1¼ in. in diam.; bracts 4–5 lin. long, 2–2½ lin. broad, obovate, the outer ones abruptly narrowed into an acute apex, densely villous with silky hairs outside, glabrous and longitudinally striate within; perianth-segments 4½ lin. long, linear, villous; limb ⅔ lin. long, ovate, subacute, villous; staminodes ¼ lin. long, oblong, minutely apiculate; hypogynous scales ½ lin. long, ovate, acuminate; style 3½ lin. long, obliquely inserted, filiform, glabrous; stigma 2-lobed; ovary 1½ lin. long, glabrous.

COAST REGION: Clanwilliam Div.; Cedarberg Range, at Ezelsbank, *Schlechter*, 8829! 8830!

9. L. **schinzianum** (Schlechter in Engl. Jahrb. xxvii. 113); an erect shrub, branched from the base, 3–4 ft. high; branchlets slender, straight, subterete, slightly villous or puberulous, at length glabrescent; leaves erect, 4–9 lin. long, 1–2 lin. broad, oblanceolate or spathulate-oblanceolate, narrowed to the base, hardened into a subacute glabrous mucro at the apex, rigidly coriaceous, shortly adpressed-pubescent on both surfaces, subglabrous on the margin; male and female inflorescences very similar, solitary, terminating the main branches or short lateral branchlets, subglobose, ½–¾ in.

in diam.; involucral bracts of the male about 4-seriate, ovate to ovate-lanceolate, acute or shortly acuminate, scaly, up to 3 lin. long and 2 lin. broad, loosely long villous on the outside; perianth-tube cylindric, $3\frac{1}{2}$ lin. long, adpressed-villous; segments about $1\frac{1}{2}$ lin. long, pubescent in the lower half; limb oblong-elliptic, acute, 1 lin. long, shortly villous outside; anthers subacute, $\frac{3}{4}$ lin. long; style $\frac{1}{2}$ lin. long, glabrous; stigma subacute, somewhat club-shaped, $\frac{2}{3}$ lin. long; female bracts very similar to the male; perianth-tube $\frac{1}{2}$–$\frac{3}{4}$ lin. long, glabrous at the base, adpressed-pubescent above; segments 4–$4\frac{1}{2}$ lin. long, villous with ascending hairs; limb $\frac{3}{4}$–1 lin. long, narrowly lanceolate, subacute, villous outside; staminodes $\frac{1}{2}$ lin. long; style $3\frac{1}{2}$ lin. long, filiform, glabrous; stigma clavate, bifid or bilobed; ovary $\frac{3}{4}$ lin. long, oblong, compressed, winged, minutely puberulous; fruits not seen.

CENTRAL REGION: Ceres Div.; Cold Bokkeveld, *Schlechter*, 8871! 8872!

10. **L. nitidum** (Buek in Drège, Zwei Pfl. Documente, 74, 198); branches subterete, adpressed-pubescent or glabrous; young lateral flowering branchlets very short, silky-pubescent; leaves 3–7 lin. long, 1–2 lin. broad, erect, oblanceolate, narrowed to the base, contracted at the apex into a small glabrous obtuse mucro, coriaceous, adpressed silky-tomentose on both surfaces; male inflorescences solitary at the apex of each short branchlet, about $\frac{1}{2}$ in. in diam., subglobose; bracts 2–$2\frac{1}{4}$ lin. long, ovate or ovate-lanceolate, subacute, finely pubescent, minutely ciliate, the outer with narrowly membranous margins; perianth-tube 2 lin. long, narrowly cylindric, widening upwards; segments $1\frac{1}{4}$ lin. long, spathulate-linear, densely pilose; limb $\frac{3}{4}$ lin. long, oblong-elliptic, subacute; anthers $\frac{1}{2}$ lin. long, linear; style $2\frac{1}{2}$ lin. long, filiform, slightly widening at the base; stigma $\frac{1}{2}$ lin. long, clavate, obtuse; female inflorescence solitary and terminal, $\frac{1}{2}$–$\frac{3}{4}$ in. in diam.; bracts 3 lin. long, about 3–4 lin. broad, suborbicular or transversely oblong, very shortly and bluntly pointed, villous, ciliate; perianth-tube $1\frac{1}{2}$ lin. long, glabrous below, villous above; segments 3 lin. long, villous; limb 1 lin. long, lanceolate-linear, subacute, villous; staminodes $\frac{1}{3}$ lin. long, linear; filaments much swollen; style $2\frac{1}{4}$ lin. long, narrowed to the base; stigma truncate; ovary $\frac{3}{4}$ lin. long, ellipsoid, acute, pubescent, keeled on one side; fruit narrowly obovoid, about $3\frac{1}{2}$ lin. long, ivory white, smooth and glossy. *Meisn. in DC. Prodr.* xiv. 217. *L. cinereum, E. Meyer ex Meisn. l.c., not of R. Br.*

COAST REGION: Clanwilliam Div.; Sneeuwkop Mountain, near Wupperthal, *Bodkin in Herb. Bolus*, 9075! Cedarberg Range, at Ezelsbank, *Drège*, a!

11. **L. elatum** (Buek in Drège, Zwei Pfl. Documente, 74, 198); branches terete, glabrous or adpressed-pubescent; young branchlets slender, shortly adpressed-pubescent or puberulous; leaves $\frac{1}{2}$–$1\frac{1}{4}$ in. long, $1\frac{1}{2}$–$3\frac{1}{2}$ lin. broad, those of the male plant smaller than those

of the female, narrowly oblanceolate, attenuate to the base,
contracted at the apex into a small blunt hardened mucro, shortly
adpressed-pubescent on both surfaces, at length becoming nearly
glabrous; male inflorescence terminating short branchlets, subglobose, 4–5 lin. in diam.; bracts 1½--2 lin. long, ovate or lanceolate-
oblong, acute, pubescent or pilose, ciliate; perianth-tube 1¾ lin.
long, widening above, pilose; segments 1¼ lin. long, pubescent;
limb ¾ lin. long, oblong, obtuse or subacute, pubescent; anthers
¾ lin. long, linear; style 2 lin. long, cylindric, hairy in the lower
half; stigma clavate; hypogynous scales 1 lin. long, filiform-linear;
young female inflorescences not seen; mature heads about 1 in. in
diam.; bracts about 5 lin. long and ½ in. broad, thick and hard,
very broadly ovate, silky-tomentose outside, glabrous, shining and
longitudinally sulcate within; fruits 3½ lin. long, 4½ lin. broad,
about 2½ lin. thick, transversely oblong-ellipsoid, with an obtuse
rib around the edge, dull and almost black, rugose. *Meisn. in DC.
Prodr.* xiv. 217.

COAST REGION: Clanwilliam Div.; Cedarberg Range, Ezelsbank, *Drège*, b!
Schlechter, 8823! 8824! *Bolus*, 5797!

12. **L. sericocephalum** (Schlechter in Engl. Jahrb. xxvii. 114);
an erect shrub 2–4 ft. high; branches terete, villous with long
weak hairs; leaves of the male much smaller than those of the
female, the former ½–1 in. long, 1½–3 lin. broad, the latter 1½–2
in. long, 3–5 lin. broad, all oblanceolate, contracted at the apex
into an obtuse mucro, more or less adpressed-pubescent on both
surfaces, those surrounding the female inflorescence silky-villous
towards the base; male inflorescences depressed-globose, about
½ in. in diam., surrounded but not hidden by a few leaves;
involucral bracts 2 lin. long, lanceolate, acute, pubescent, ciliate;
floral bracts 1¼ lin. long, oblanceolate, acute, densely pilose above,
ciliate; perianth-tube 2 lin. long, pilose; segments 1½ lin. long,
spathulate - linear, pilose; limb ½ lin. long, linear, subobtuse,
glabrous; anthers ½ lin. long, linear; style 2½ lin. long, cylindric-
filiform, pilose with long hairs; stigma ⅓ lin. long, clavate-cylindric;
hypogynous scales 1¼ lin. long, linear; female inflorescence nearly
hidden by the surrounding leaves, about 1 in. long, ovoid; bracts
4½ lin. long, 6 lin. broad, ovate, subacute, densely tomentose;
perianth-segments 6 lin. long, linear, imbricate below, long and
densely villous; limb ½ lin. long, subacute, incurved above,
glabrous; staminodes ¼ lin. long, linear; hypogynous scales 1¼ lin.
long, broadly linear; style obliquely inserted, 5 lin. long, terete,
widened near the apex, glabrous; stigma oblique; ovary 1 lin. long,
ellipsoid, villous.

COAST REGION: Clanwilliam Div.; Zeekoe Vley, *Schlechter*, 8486! 8487!

13. **L. cinereum** (R. Br. in Trans. Linn. Soc. x. 57); a shrub;
branchlets rather slender, subterete, adpressed-pubescent, at length

glabrous ; leaves overlapping, ½–1 in. long, 1½–2 lin. broad, oblanceolate or linear-oblanceolate, narrowed to the base, contracted at the apex into a subacute hardened glabrous mucro, rigidly coriaceous, densely adpressed-pubescent with silky hairs on both surfaces, glabrescent towards the margin ; male inflorescence terminal, solitary, somewhat depressed-globose, ¾ in. in diam. ; bracts 1¼ lin. long, linear, acute, densely villous ; perianth-tube 1½ lin. long, pubescent above, glabrous below ; segments 2 lin. long, villous ; limb ½ lin. long, oblong - linear, subacute, villous ; anthers ½ lin. long, linear ; style 3 lin. long, cylindric above, linear below ; stigma ¼ lin. long, clavate ; female inflorescences terminal, surrounded by a few upper leaves, ellipsoid-globose, about ½ in. in diam. ; bracts 1¼ lin. long, 3 lin. broad, transversely oblong, villous, ciliate, very shortly and bluntly pointed ; perianth-segments villous ; style about 1½ lin. long, linear ; stigma oblong, truncate ; fruit ovoid-ellipsoid, 2½ lin. long, 1¾ lin. broad, thinly villous.   *Meisn. in DC. Prodr.* xiv. 216 (*excl. syns.*).   *L. ? verticillatum, Meisn. l.c.* 228. *Protea verticillata, Thunb. Phytogr. Blaett.* 12 ; *Thunb. Fl. Cap. ed. Schult.* 136.   *Sorocephalus verticillatus, Roem. & Schult. Syst. Veg.* iii. 391 ; *Steud. Nomencl. ed.* 2, ii. 401.

SOUTH AFRICA : without locality, *Drège,* 8046 ! *Thunberg*!
COAST REGION : Clanwilliam Div. ; Zwartbosch Kraal, *Schlechter,* 5161 ! Paarl Div. ; between Mosselbanks River and Berg River, *Burchell,* 977 ! Cape Div. ; Koeberg, *Pappe*! between Koeberg and Drooge Valley, *Zeyher,* 4954 !

14. **L. sericeum** (R. Br. in Trans. Linn. Soc. x. 65) ; branchlets erect, slender, shortly pubescent ; leaves 2½–3 lin. long, ½–¾ lin. broad, narrowly oblanceolate, obtuse or subacute, coriaceous, adpressed silky pilose ; male heads solitary or congested at the apices of the shoots, clasped but not overtopped by the upper leaves, surrounded by a distinct involucre of imbricate bracts, obovoid or obconic, about 3½ lin. long and in diam. ; involucral bracts 3–4-seriate, ovate, caudate-acuminate, about 1½ lin. long, scaly, minutely puberulous, very shortly ciliate ; floral bracts ¾ lin. long, linear-oblanceolate, acute, shortly pubescent in the upper part on the outside, glabrous within ; perianth-tube 2 lin. long, cylindric, shortly villous ; segments spathulate, 1¼ lin. long, pubescent outside ; limb ⅔ lin. long, lanceolate-elliptic, acute, pubescent outside ; anthers linear-oblong, ⅓ lin. long ; style filiform, glabrous ; stigma small ; female flowers not known.   *Meisn. in DC. Prodr.* xiv. 225.   *Protea sericea, Thunb. Diss. Prot.* 39 ; *Linn. f. Syst. ed.* xiv. 140 ; *Suppl.* 118 ; *Thunb. Fl. Cap. ed. Schult.* 136.

SOUTH AFRICA : without locality, *Thunberg*! and a specimen without collector's name, from *Herb. Forsyth,* at Kew !

Meisner, l.c. 223, referred *Protea sericea,* Thunb., doubtfully to *L. salignum,* R. Br. ; they are, however, quite distinct species.

15. **L. tortum** (R. Br. in Trans. Linn. Soc. x. 56) ; branchlets rather slender, subterete, thinly pubescent or glabrous ; leaves of

the male smaller than those of the female, the former 3–5 lin. long, narrowly linear-oblanceolate, the latter $\frac{1}{2}$–1 in. long, linear, all obtuse at the apex, flattened, coriaceous, glabrous; male inflorescence terminal, small, shortly pedunculate, about 4 lin. in diam., globose; bracts 1 lin. long, oblong-linear, subacuminate, acute, villous; perianth-tube 1 lin. long, villous; segments $1\frac{1}{4}$ lin. long, spathulate-linear, villous; limb $\frac{1}{2}$ lin. long, oblong, obtuse, villous; anthers $\frac{1}{3}$ lin. long, linear; style $1\frac{1}{2}$ lin. long, filiform, glabrous; stigma $\frac{1}{3}$ lin. long, clavate, acute; female inflorescence broadly ovoid or subglobose, about 1 in. in diam.; bracts 4 lin. long and broad, broadly obovate, white-villous outside; perianth-segments $3\frac{3}{4}$ lin. long, dilated and keeled below, villous above; limb $\frac{1}{4}$ lin. long, oblong, villous; staminodes $\frac{1}{4}$ lin. long, linear; ovary $\frac{3}{4}$ lin. long, ovoid, covered with long hairs; style 2 lin. long, cylindric, glabrous; stigma flat, oblique; fruit $2\frac{1}{3}$ lin. long, obovoid, subacuminate, villous. *Bot. Reg. t.* 826; *Meisn. in DC. Prodr.* xiv. 216 (*excl. syn.*, *Protea torta, Thunb.*, *et spec. Zeyher*, 3651). *L. pedunculatum, Meisn. l.c. L. pruinosum, Mund fide Meisn. l.c.? L. inflexum, Klotzsch* (*partly*), *fide Meisn. l.c.* 215. *Protea torta, Jacq. Hort. Schoenbr.* iv. 1, *t.* 401, *not of Thunb. P. cinerea, Willd. Sp. Pl.* i. 521, *partly. P. densa, Willd. Enum. Hort. Berol. Suppl.* 7, *acc. to Link, Enum. Pl. Hort. Berol.* i. 114. *P. hirta and P. passerina, Hort. ex Meisn. l.c.*

SOUTH AFRICA: without locality, *Thom*, 520! *Gueinzius*!

COAST REGION: Bredasdorp Div.; Kars River Valley, *Ludwig*! *Mund*, 3! Caledon Div.; Bot River, *Schlechter*, 9452! 9453! Swellendam Div.; Swellendam, *Pappe*!

16. **L. ericifolium** (R. Br. in Trans. Linn. Soc. x. 66); branches terete, glabrous or shortly pubescent; leaves ericoid, 3–5 lin. long, subterete, subacute, glabrous; male inflorescences small, about 6-flowered, axillary, pedunculate, collected at the ends of the branchlets; peduncle bracteate, about $\frac{1}{4}$ in. long, shortly rusty-pubescent; bracts $\frac{3}{4}$ lin. long, oblong, obtuse, pilose; perianth-tube $\frac{3}{4}$ lin. long, cylindric, shortly villous; segments $1\frac{1}{2}$ lin. long, pilose; limb $1\frac{1}{4}$ lin. long, linear-lanceolate, subobtuse, pubescent with glabrescent tips; anthers $\frac{3}{4}$ lin. long, linear; style 1 lin. long, cylindric, glabrous; stigma $\frac{1}{2}$ lin. long, clavate; female inflorescences not known. *Meisn. in DC. Prodr.* xiv. 225. *L. comosum*, ♂, *Buek in Drège, Zwei Pfl. Documente*, 116, 198. *L. scabrum, Steud. ex Meisn. l.c., not of R. Br. Protea ericifolia, Poir. Encycl. Suppl.* iv. 557.

SOUTH AFRICA: without locality, *Roxburgh*!

COAST REGION: Swellendam Div.; between Sparrbosch and Tradouw, *Drège*!

17. **L. abietinum** (R. Br. in Trans. Linn. Soc. x. 64); a shrub about 3 ft. high, the female plant much resembling a small *Abies*; branches terete, pubescent; young branchlets hirsute, in the female plant arising in clusters from below the old persistent female heads;

leaves of the male plant much smaller than those of the female, the former 3–4 lin. long, the latter $\frac{3}{4}$–1 in. long, all acicular, subacute, glabrous, the older ones of the female plant becoming reflexed and arcuate; male inflorescences much more numerous than the females, very small, about $\frac{1}{4}$ in. long, subcylindric or ellipsoid, terminating crowded branchlets; bracts $\frac{1}{3}$ lin. long, ovate, subobtuse, ciliate; perianth-tube $\frac{1}{3}$ lin. long, cylindric, glabrous; segments 1 lin. long, spathulate-glabrous; limb $\frac{3}{4}$ lin. long, oblong-elliptic, subobtuse; anthers $\frac{1}{2}$ lin. long, oblong; style $\frac{3}{4}$ lin. long, filiform, glabrous; stigma $\frac{1}{3}$ lin. long, cylindric-ellipsoid, furrowed; hypogynous scales $\frac{1}{4}$ lin. long, filiform; female inflorescences solitary, terminating the young branchlets which arise from below the persistent head of the previous season, small and nearly hidden by the upper leaves; bracts $\frac{3}{4}$–$1\frac{1}{3}$ lin. long, transversely oblong, very shortly pointed; perianth-segments $1\frac{1}{4}$ lin. long, linear, pubescent below; style $\frac{3}{4}$ lin. long, linear, narrowed to the base; stigma oblique, flat, elliptic; ovary $\frac{1}{4}$ lin. long, orbicular; hypogynous scales $\frac{1}{8}$ lin. long, linear; old persistent female heads 1–$1\frac{1}{2}$ in. long, 1 in. in diam., oblong-cylindric; bracts 15–20-seriate, spreading, transversely oblong, widely retuse, woody, glabrous; fruits compressed, winged, oblong-obovate, emarginate, $2\frac{1}{2}$ lin. long, $2\frac{1}{4}$ lin. broad, glabrous. *Spreng. Syst.* i. 458; *Meisn. in DC. Prodr.* xiv. 225. *Protea teretifolia, Andr. Bot. Rep. t.* 461 (*see note below*). *P.* (*Leucadendron abietinum*), *Poir. Encycl. Suppl.* iv. 559. *P. abietina, Poir. ex Ind. Kew.* ii. 631; *Chasme teretifolia, Knight, Prot.* 16.

SOUTH AFRICA: without locality, *Mund*!

COAST REGION: Caledon Div.; Klein River Mountains, *Zeyher*, 3654! near the mouth of the Bot River, *MacOwan, Herb. Norm. Austr.-Afr.* 908! Riversdale Div.; between Great Valsch River and Zoetemelks River, *Burchell*, 6560!

A cultivated male specimen in the Kew Herbarium raised from seeds obtained from Burchell's no. 6560 differs somewhat from the wild male plant; the leaves are longer, falcate and sometimes reflexed and pilose just as in the female.

The plant figured by Andrews (t. 461) shows the male and female flowers on the same individual. This, however, never occurs in the genus and the figure was evidently made up from plants representing both sexes.

18. **L. scabrum** (R. Br. in Trans. Linn. Soc. x. 65); young branchlets arising in clusters from below the previous season's inflorescence, straight, slender, pubescent or glabrous, almost hidden by the closely overlapping leaves; leaves erect, 2–$4\frac{1}{2}$ lin. long, ericoid, subacute, slightly pubescent when young, at length quite glabrous; male inflorescences terminal, 3–4 lin. long, ellipsoid; bracts obsolete; perianth-tube $\frac{1}{4}$ lin. long; lobes 1 lin. long, forming 2 lips, the posterior and 2 lateral lobes connate, 3-toothed, membranous, glabrous; anthers $\frac{1}{2}$ lin. long, oblong; style $\frac{1}{2}$ lin. long, filiform; stigma $\frac{1}{4}$ lin. long, linear; young female inflorescences not seen; old persistent head oblong-globose, about $\frac{3}{4}$ in. in diam.; bracts rigidly coriaceous, glabrous; fruit compressed, very broadly obovate, mucronate, narrowly winged, 3 lin. long, $3\frac{1}{2}$ lin. broad, mottled, glabrous. *Meisn. in DC. Prodr.* xiv. 225. *Protea* (*Leuca-*

dendron scabrum), *Poir. Encycl. Suppl.* iv. 559. *P. scabra, Poir. ex Ind. Kew.* ii. 633. *P. thyoides, Smith ex Meisn. l.c.*

SOUTH AFRICA : without locality, *Mund* !
COAST REGION : Caledon Div. ; Zoetemelks River, *Bowie* ! Uniondale Div. ; Lange Kloof, *Bowie* !

19. L. corymbosum (Berg. in Vet. Akad. Handl. Stockh. 1766, 325) ; branches straight, terete, rather sparingly pubescent ; young branchlets clustered, pubescent ; leaves ericoid or acicular, those of the male smaller than those of the female, the former 3–5 lin. long, the latter ½–¾ in. long, all acute or subacute and glabrous except those at the apices of barren branchlets which are pilose with long weak spreading hairs ; male inflorescences terminal on short lateral branchlets, often subcorymbosely arranged, subglobose, 4–5 lin. in diam. ; bracts ½ lin. long, ovate, subacute, long-ciliate ; perianth-tube ¾ lin. long, cylindric, pilose above ; segments 1 lin. long, spathulate-linear, pilose in the lower part ; limb ½ lin. long, oblong-elliptic, obtuse, glabrous ; anthers ⅓ lin. long, oblong ; style 1½ lin. long, filiform, glabrous ; stigma ¼ lin. long, clavate ; hypogynous scales ½ lin. long, filiform ; female inflorescences arranged similarly to the males, surrounded by a few leaves, 5–7 lin. in diam. ; bracts 3½ lin. long, obovate, ligulate in the upper half, obtuse, densely pilose outside, glabrescent towards the tips ; perianth-tube 1¾ lin. long, split and glabrous below, villous above ; segments 1 lin. long, linear, pilose or villous ; limb ½ lin. long, oblong, obtuse or subacute, glabrous ; staminodes ¼ lin. long, linear ; style 1¾ lin. long, filiform, bent below and obliquely inserted on the ovary ; stigma ¼ lin. long ; ovary 2 lin. long, oblong, hirsute ; hypogynous scales ¾ lin. long, linear ; fruiting heads turbinate, about ¾ in. long and ⅓–½ in. in diam. ; fruits obovoid, shortly pointed, 2¼ lin. long, 1½ lin. broad, not winged, pilose near the base. *Berg. Descr. Pl. Cap.* 21 ; *R. Br. in Trans. Linn. Soc.* x. 57 ; *Bot. Reg. t.* 402 ; *Meisn. in DC. Prodr.* xiv. 217. *L. linifolium, R. Br. in Trans. Linn. Soc.* x. 216 ; *Meisn. l.c.* 228. *Protea corymbosa, Thunb. Diss. Prot.* 29 ; *Willd. Sp. Pl.* i. 518 ; *Thunb. Fl. Cap. ed. Schult.* 129 ; *Andr. Bot. Rep. t.* 495. *P. bruniades, Linn. f. Suppl.* 117. *P. linifolia, Jacq. Hort. Schoenb.* i. *t.* 26. *P. bruniæfolia, Knight, Prot.* 32. *P. ericæfolia, Willd. ex Meisn. l.c.* 218.

SOUTH AFRICA: without locality, *Thom* ! *Thunberg* ! *Ludwig*, 15 ! *Masson* ! *Roxburgh* ! *Niven, Ecklon, Gueinzius.*
COAST REGION: Tulbagh Div. ; Tulbagh, *Pappe* ! Ceres Road, *Schlechter*, 9076 ! 9077 ! Mitchells Pass, *Bolus*, 7450 ! Worcester Div. ; near Darling Bridge, *Bolus*, 2787 ! Paarl Div. ; between Paarl and French Hoek, *Drège* ! near Paarl, *Burchell*, 961 ! Klapmuts, *Rehmann*, 2268 ! Bredasdorp Div. ; near Elim, *Bolus*, 7856 ! *Schlechter*, 7617 ! Swellendam Div.; near Breede River, *Thunberg* !

20. L. fusciflora (R. Br. in Trans. Linn. Soc. x. 216); a shrub 4–5 ft. high, many-stemmed and much-branched ; branches erect ; young branchlets rather slender, minutely pubescent or puberulous ;

leaves of both sexes subequal, $\frac{1}{3}-\frac{3}{4}$ in. long, acicular, obtuse or
subacute, thick and coriaceous, glabrous; male inflorescences
numerous, often crowded towards the apices of the shoots, $\frac{1}{4}-\frac{1}{3}$ in.
in diam., subglobose, with a few outer bracts; floral bracts $\frac{1}{2}-1$ lin.
long, oblong or lanceolate-oblong, shortly acuminate, obtuse or
subobtuse, densely pilose or villous; perianth-tube 1 lin. long,
cylindric above, gradually narrowed to the base, shortly villous;
segments 2 lin. long, spathulate-linear, pubescent; limb $\frac{1}{2}$ lin. long,
oblong or lanceolate-oblong, subobtuse, shortly villous; anthers
$\frac{1}{3}$ lin. long, linear; style $2\frac{1}{4}$ lin. long, very densely hairy in the
lower part; stigma $\frac{1}{4}$ lin. long, clavate, obtuse; hypogynous scales
$1\frac{1}{4}$ lin. long, linear, acute; female inflorescence solitary at the apex
of each branchlet, about $\frac{1}{2}$ in. long, ovoid; bracts 2 lin. long, $4\frac{1}{2}$ lin.
broad, tomentose outside; perianth-tube $1\frac{1}{2}$ lin. long, glabrous;
segments $2\frac{1}{2}$ lin. long, linear, villous; limb $\frac{1}{4}$ lin. long, oblong,
subobtuse, pubescent; staminodes $\frac{1}{4}$ lin. long, linear-oblong;
hypogynous scales 2 lin. long, linear; ovary compressed, $1\frac{1}{2}$ lin.
long, ovate, covered with long hairs; style cylindric; stigma $\frac{1}{3}$ lin.
long, ellipsoid, obtuse; fruit $2\frac{1}{2}$ lin. long, ellipsoid, bluntly beaked,
villous. *Meisn. in DC. Prodr.* xiv. 228. *L. stellare, Steud. Nomencl.
ed.* i. 475. *L. stellatum, Sweet, Hort. Brit. ed.* i. 345. *L. canalicu
latum, E. Meyer in Drège, Zwei Pfl. Documente,* 198, *partly.
L. brunioides, Meisn. l.c.* 215, *partly, excl. syn. L. meyerianum,
Buek. L. inflexum, Klotzsch, partly, fide Meisn. l.c.* 216. *L. tortum,
var. inflexum, Meisn. l.c. L.? tenuifolium, Meisn. l.c.* 227. *L.
imbricatum, var.? canaliculatum, Meisn. l.c.* 215, *partly. Protea torta,
Thunb. Diss. Prot.* 31 ; *Fl. Cap. ed. Schult.* 130. *P. tenuifolia,
Thunb. Fl. Cap. ed. Schult.* 135. *P. Thunbergii, Steud. Nomencl.
ed.* 2, ii. 401. *P. fusciflora, Jacq. Hort. Schoenb.* i. *t.* 27. *P.
stellaris, Sims, Bot. Mag. t.* 881. *P. globulariæfolia, Knight, Prot.*
30, *partly. P. squarrosa, Knight, Prot.* 127, *partly.*

SOUTH AFRICA : without locality, *Thom* ! *Thunberg* !
COAST REGION : Clanwilliam Div. ; between Lange Valley and Oliphants River,
Drège ! Worcester Div. ; near Worcester, *Rehmann,* 2511 ! Caledon Div. ; by the
Zondereinde River, near Appels Kraal, *Zeyher,* 3652 ! Swellendam Div. ; between
Breede River and Zondereinde River, *Burchell,* 7489 !

21. **L. sorocephalodes** (Phillips & Hutchinson) ; branches terete,
pubescent with weak whitish hairs; leaves acicular, 4 – 6 lin.
long, scarcely $\frac{1}{2}$ lin. in diam., terete, obtuse, glabrous; male
inflorescence not seen; female inflorescence terminal, solitary,
ovoid-globose, a little over $\frac{1}{2}$ in. in diam.; bracts about 6-seriate,
$3\frac{1}{2}$ lin. long, ovate, obtuse or subacute, glabrous except on the ciliate
margin; perianth-segments free to the base, 6 lin. long, dilated
below, linear above, villous; limb $\frac{2}{3}$ lin. long, linear, subacute, long
villous with rusty hairs; staminodes $\frac{1}{4}$ lin. long, linear; style $2\frac{1}{4}$ lin.
long, narrowed to the base, glabrous; stigma oblique; ovary $1\frac{1}{3}$ lin.
long, narrowly oblong, densely villous; fruits 3 lin. long, oblong-
cylindric, slightly narrowed to the base, villous with rather weak

hairs. *Sorocephalus Dregei, Buek in Drège, Zwei Pfl. Documente,* 117, 222; *Meisn. in DC. Prodr.* xiv. 305. *Leucadendron?* *scoparium, E. Meyer ex Meisn. l.c.*

COAST REGION: Uniondale Div. ; between Avontuur and Klip River, in Lange Kloof, *Drège,* c!

22. **L. radiatum** (Phillips & Hutchinson); a robust undershrub about 2 ft. high ; branches spreading ; branchlets erect, straight, fairly stout, slightly sulcate, pilose with weak hairs ; leaves $\frac{1}{2}$–$\frac{3}{4}$ in. long, $1\frac{1}{2}$–$2\frac{1}{2}$ lin. broad, those surrounding the inflorescence a little larger, all oblanceolate, contracted at the apex into an obtuse short mucro, rigidly coriaceous, thinly pubescent on both surfaces, especially towards the margin; male inflorescences terminal, surrounded by several silky adpressed-tomentose leaves, subglobose, about $\frac{1}{2}$ in. in diam. ; floral bracts $\frac{3}{4}$ lin. long, ovate-lanceolate, obtuse, villous; perianth-tube 1 lin. long, villous ; segments $1\frac{3}{4}$ lin. long, spathulate-linear, coiled, pubescent; limb $\frac{1}{2}$ lin. long, oblong, obtuse, pubescent ; anthers $\frac{1}{3}$ lin. long, linear ; style 2 lin. long, terete, glabrous ; stigma $\frac{1}{4}$ lin. long, clavate, subacute ; hypogynous scales $\frac{1}{2}$ lin. long, filiform-linear ; female inflorescence solitary, terminal, ovoid, about $\frac{1}{2}$ in. long and in diam. ; bracts $1\frac{1}{4}$ lin. long, $2\frac{1}{4}$ lin. broad, transversely oblong, tomentose ; perianth-tube $1\frac{1}{4}$ lin. long, split below, pilose ; segments $\frac{3}{4}$ lin. long, linear, villous ; limb $\frac{1}{3}$ lin. long, oblong, obtuse, hirsute ; hypogynous scales $\frac{1}{2}$ lin. long, linear ; style 1 lin. long, filiform, glabrous, obliquely inserted ; stigma small, truncate, minutely bifid ; ovary $\frac{1}{2}$ lin. long, ellipsoid, glabrous ; fruiting head ellipsoid, about 1 in. long, $\frac{3}{4}$ in. in diam. ; bracts about 15-seriate, transversely oblong, truncate, about $2\frac{1}{2}$ lin. long and $\frac{1}{2}$ in. broad, adpressed-pubescent outside, becoming nearly glabrous at the tips ; fruits not seen.

COAST REGION : Riversdale Div. ; on the summit of Kampsche Berg, *Burchell,* 7110!

23. **L. dubium** (Buek in Drège, Zwei Pfl. Documente, 73, 198) ; branchlets very shortly pubescent ; leaves closely overlapping, those of the male smaller than those of the female, the former 2–3 lin. long, 1–$1\frac{3}{4}$ lin. broad, the latter 3–7 lin. long, $1\frac{3}{4}$–$2\frac{1}{4}$ lin. broad, all obovate or those around the female inflorescence oblanceolate, obtuse or subacute, coriaceous, softly adpressed-pubescent on both surfaces, at length becoming glabrous ; male inflorescences subcorymbose, solitary and terminal, scarcely $\frac{1}{2}$ in. in diam., surrounded by several broad purple bracts ; floral bracts $2\frac{1}{2}$ lin. long, oblanceolate-spathulate, pubescent, ciliate ; perianth-tube $1\frac{1}{4}$ lin. long, cylindric, pilose ; segments $1\frac{1}{2}$ lin. long, pilose ; limb $\frac{3}{4}$ lin. long, oblong, subacute, glabrous ; anthers $\frac{2}{3}$ lin. long, linear ; style 2 lin. long, cylindric, glabrous ; stigma $\frac{1}{2}$ lin. long, oblong-cylindric ; hypogynous scales? ; female inflorescences hidden by the coloured bract-like leaves ; floral bracts 4–5 lin. long, oblong, shortly and

obtusely acuminate, pubescent, ciliate ; perianth-tube $1\frac{1}{2}$ lin. long, ventricose below, covered with very long hairs ; segments $4\frac{1}{2}$ lin. long, linear, pilose with long hairs ; limb $\frac{2}{3}$ lin. long, oblong-linear, acute, glabrous ; anthers $\frac{1}{2}$ lin. long, linear ; style $3\frac{1}{2}$ lin. long, filiform-cylindric, narrowed to the base, glabrous ; stigma not seen ; ovary $\frac{1}{2}$ lin. long, ellipsoid, glabrous ; mature heads subglobose, villous ; fruits transversely ellipsoid, 3 lin. long, $3\frac{1}{2}$ lin. broad, not winged. *L. buxifolium, car. dubium, Meisn. in DC. Prodr.* xiv. 215.

COAST REGION : Clanwilliam Div. ; near Honig Valley and Koude Berg, *Drège!* near Ezelsbank, *Drège! Schlechter,* 8802! 8803! Sneeuwkop, *Bodkin in Herb. Bolus,* 9077 !

24. **L. imbricatum** (R. Br. in Trans. Linn. Soc. x. 55) ; branchlets terete, glabrous or finely puberulous ; leaves of the male smaller than those of the female, the former 3–4 lin. long, 1–$1\frac{1}{4}$ lin. broad, narrowly lanceolate, the latter $\frac{1}{2}$–$\frac{3}{4}$ lin. long, $1\frac{1}{4}$–2 lin. broad, narrowly oblong, all obtuse or subacute, very thick and coriaceous, glabrous, wrinkled when dry ; male inflorescences crowded, nearly $\frac{1}{2}$ in. in diam., transversely ellipsoid or obovoid ; bracts $1\frac{1}{2}$ lin. long, lanceolate, subobtuse, villous, ciliate ; perianth-tube 1 lin. long, minutely pubescent ; segments $2\frac{1}{2}$ lin. long, pubescent ; limb $\frac{2}{3}$ lin. long, oblong, subacute ; anthers $\frac{1}{2}$ lin. long, linear ; style $3\frac{1}{2}$ lin. long, very densely hairy on the lowermost third, filiform above ; stigma $\frac{1}{3}$ lin. long, clavate-cylindric, obtuse ; hypogynous scales 1 lin. long, linear ; female inflorescences about 5 lin. in diam., subglobose ; bracts 2 lin. long, 4 lin. broad, tomentose, very minutely ciliate ; perianth-tube $1\frac{1}{2}$ lin. long, compressed, curved, pubescent above, glabrous below ; segments $1\frac{1}{4}$ lin. long, pubescent ; limb $\frac{1}{3}$ lin. long, oblong, subacute, pubescent ; staminodes $\frac{1}{4}$ lin. long, linear ; hypogynous scales $\frac{1}{2}$ lin. long, linear, acuminate, acute ; style $2\frac{1}{2}$ lin. long, cylindric, furrowed above, glabrous ; stigma $\frac{1}{4}$ lin. long, subclavate ; ovary $1\frac{1}{3}$ lin. long, ovoid, covered with long white hairs. *Meisn. in DC. Prodr.* xiv. 214, *excl. spec. Ludwig, var. dregeanum and part of var. canaliculatum. L. buxifolium, R. Br. in Trans. Linn. Soc.* x. 55 ; *Meisn. l.c.* 215, *excl. vars. L. canaliculatum, E. Meyer in Drege, Zwei Pfl. Documente,* 198, *partly, as to Drege ♂ et ♀ a. L. angustatum, E. Meyer, l.c.* 113, 198, *not of R. Br. L.? læve, Meisn. l.c.* 227. *Protea lævis, Thunb. Fl. Cap. ed. Schult.* 133. *P. (Leucadendron imbricatum), Poir. Encycl. Suppl.* iv. 557. *P. imbricata, Poir. ex Ind. Kew.* ii. 632. *P. Wendlandi, Poir. Encycl. Suppl.* iv. 556. *P. imbricata, Wendl. Hort. Herrenhaus. t.* 14, *excl. syn., fide Meisn. l.c.* 215. *P. polygaloides, Willd. ex Meisn. l.c. P. cinerea, Willd. Herb. ex Meisn. l.c. (not Willd. Sp. Pl.). P. levisana, Linn. Herb. ex Meisn. l.c., not Linn. Syst.*

SOUTH AFRICA: without locality, *Thunberg! Masson! Niven in Herb. Stockholm!*
COAST REGION : Clanwilliam Div. ; Vogelfontein, *Schlechter,* 8519 ! 8520 !

Piquetberg Div. ; near Piquetberg Road. *Tyson*, 2292 ! Malmesbury Div. ; Zwartland, *Zeyher* ! *Pappe* ! between Groene Kloof and Saldanaha Bay, *Drège* ! Tulbagh Div. ; Saron, *Schlechter*, 10635 ! 10636 ! Vogelvalley, *Pappe* ! Worcester Div. ; by the Doorn River, near Mordkuil, Boschjesveld Range, *Drège* ! Cape Div. ; between Paarde Berg and Tiger Berg, *Ludwig* ! Swellendam Div. ; *Pappe* !

The specimen in Herb. Stuttgart collected by Ludwig at Kars River and quoted by Meisner under this species is *L. tortum*, R. Br.

25. **L. pubescens** (R. Br. in Trans. Linn. Soc. x. 66) ; branchlets slender, terete, shortly pubescent or tomentellous ; leaves $\frac{2}{3}$–1 in. long, $1\frac{1}{2}$–3 lin. broad, oblanceolate or narrowly oblanceolate, contracted at the apex into a very short hardened subacute mucro, rather rigidly coriaceous, papillose and sometimes very sparingly pubescent on both surfaces, with distinct ascending nerves ; male inflorescence terminal, solitary, surrounded by a distinct involucre of bracts, about 5 lin. in diam., subglobose ; involucral bracts about $2\frac{1}{2}$ lin. long, oblong, the outer acuminate, obtuse, pubescent, long ciliate ; floral bracts $1\frac{1}{2}$ lin. long, linear, subobtuse, densely ciliate at the apex only ; perianth-tube $1\frac{1}{3}$ lin. long, pilose ; segments $1\frac{1}{2}$ lin. long, spathulate-linear, pilose ; limb $\frac{2}{3}$ lin. long, linear, obtuse, glabrous ; anthers $\frac{3}{4}$ lin. long, linear ; style 2 lin. long, filiform, long-pilose ; stigma $\frac{1}{2}$ lin. long, cylindric, obtuse. *Meisn. in DC. Prodr.* xiv. 226. *L. globularia, R. Br. l.c.* 65 ; *Meisn. l.c. L.? acutum, Meisn. l.c.* 228. *Protea globularia, Lam. Ill.* i. 236, t. 53, fig. 2. *P. virgata, Thunb. Fl. Cap. ed. Schult.* 133. *P. (Leucadendron pubescens), Poir. Encycl. Suppl.* iv. 559.

SOUTH AFRICA : without locality, *Drège* ! *Thunberg* ! *Roxburgh* !

COAST REGION : Piquetberg Div. ; Piqueniers Kloof, *Schlechter*, 7942 ! Tulbagh Div. ; between New Kloof and Elands Kloof, *Drège* !

26. **L. coriaceum** (Phillips & Hutchinson) ; branches terete, shortly pubescent or glabrous ; leaves subequal in both sexes, 4–7 lin. long, 2–$3\frac{1}{4}$ lin. broad, oblanceolate or spathulate-obovate, rounded into a very obtuse mucro at the apex, rigidly or thickly coriaceous, glabrous ; male inflorescences solitary and terminal, scarcely $\frac{3}{4}$ in. in diam., subdepressed-globose ; bracts 1 lin. long, lanceolate, subobtuse ; perianth-tube $1\frac{1}{4}$ lin. long, tubular, pilose above, glabrous below ; segments 2 lin. long, spathulate-linear, adpressed-pilose ; limb $\frac{2}{3}$ lin. long, oblong, subacute, pubescent ; anthers $\frac{1}{2}$ lin. long, linear ; style $2\frac{1}{2}$ lin. long, cylindric, with very long hairs at the base ; stigma $\frac{1}{3}$ lin. long, ellipsoid, subacute ; hypogynous scales $1\frac{1}{2}$ lin. long, filiform ; female inflorescences similar to the male but nearly hidden by the upper leaves ; bracts 2 lin. long, ovate, acuminate, villous, ciliate ; perianth-segments $3\frac{1}{4}$ lin. long, spathulate-linear, villous above, glabrous below ; limb $\frac{1}{2}$ lin. long, oblong, obtuse or subobtuse, villous ; staminodes $\frac{1}{3}$ lin. long, linear ; hypogynous scales $1\frac{1}{4}$ lin. long, linear, long-acuminate, acute ; style $2\frac{1}{3}$ lin. long, linear, glabrous, inserted laterally ; stigma oblique ; ovary $\frac{1}{2}$ lin. long, compressed, winged, produced above into 2 flat membranous horns.

2 M 2

27. **L. truncatum** (Meisn. in DC. Prodr. xiv. 228); branches terete, purplish, glabrous or slightly pubescent ; young branchlets slender, those of the female much longer than those of the male ; leaves of the male shorter than those of the female, the former 3–5 lin. long, ½–¾ lin. broad, the latter 1–1⅓ in. long, 1¼–1½ lin. broad, all linear or linear-oblanceolate, obtuse or subacute, glabrous ; male inflorescences surrounded by a conspicuous involucre of numerous lanceolate or ovate-lanceolate tomentose bracts, depressed-globose, about ⅓ in. in diam. ; floral bracts 2–2¼ lin. long, linear, acute, villous ; perianth-tube 1¼–2 lin. long, at length split nearly half-way down ; limb ½–¾ lin. long, lanceolate, nearly glabrous ; anthers ½ lin. long ; style 1½–3 lin. long, slender, glabrous ; stigma ⅓ lin. long, clavate ; young female inflorescence not seen ; fruiting head about 1 in. in diam., subglobose ; bracts woody, transversely oblong or broadly obovate and truncate, about 3 lin. long and 4½ lin. broad, rusty-villous in the lower part, whitish tomentose above. *L. globularia, E. Meyer in Drege, Zwei Pfl. Documente*, 114, 198, *not of R. Br. L. tortum, Meisn. in DC. Prodr.* xiv. 216, *as to Zeyher*, 3651. *Protea truncata, Thunb. Fl. Cap. ed. Schult.* 134.

28. **L. levisanum** (Berg. in Vet. Akad. Handl. Stockh. 1766, 324) ; an erect shrub 2–4 ft. high ; young branchlets numerous, pilose ; leaves of the male slightly smaller than those of the female, the former 2½–4 lin. long, ¾–1½ lin. broad, the latter 3–5 lin. long, ¾–2 lin. broad, all oblanceolate or spathulate-oblanceolate, obtuse or subacute, thick and coriaceous, those towards the tips of the shoots long-pilose, the remainder glabrous ; male inflorescences rather crowded, about ½ in. in diam., depressed-globose ; involucral bracts 1–2-seriate, subulate-lanceolate, about as long as the flowers, densely villous ; floral bracts ¾–⅚ lin. long, lanceolate, subacute, pilose ; perianth-tube ¾ lin. long, pilose ; segments 2¾ lin. long, pubescent below, glabrous above ; limb ½ lin. long, oblong, obtuse, glabrous ; anthers ⅓ lin. long, linear ; style 1¾ lin. long, filiform above, glabrous ; stigma ⅓ lin. long, clavate ; female inflorescences terminal, solitary, about ½ in. in diam., ovoid ; bracts 2 lin. long, 2½ lin. broad, obovate, with a very short blunt point, villous ; perianth-segments nearly free, dilated and keeled at the base, the anterior and posterior inside the lateral, 3 lin. long, pilose above, glabrous below ; limb ⅓ lin. long, oblong, obtuse, pilose ; staminodes ⅙ lin. long, linear ; style 2⅓ lin. long, cylindric ; stigma obovoid,

bifid; ovary $\frac{2}{3}$ lin. long, elliptic, 4-angled, covered with long white hairs; fruits obovoid, slightly 3-keeled, $1\frac{3}{4}$ lin. long, $1\frac{1}{3}$ lin. in diam., villous between the angles; bracts of the fruiting head transversely oblong, truncate or slightly emarginate, about $\frac{1}{2}$ in. broad, villous outside. *R. Br. in Trans. Linn. Soc.* x. 55; *Meisn. in DC. Prodr.* xiv. 216. *L. hirsutum, Hoffmansegg. Verz. Pfl.* 72; *Meisn. l.c.* 227. *Protea fusca, Linn. Sp. Pl. ed.* i. 95. *P. levisana, Thunb. Diss. Prot.* 37 (*excl. var. γ of Herb.*); *Fl. Cap. ed. Schult.* 135. *P. tenuifolia, Salisb. Prodr.* 49. *P. hirsuta, Thunb. in Hoffm. Phytog. Blaetter,* i. 12; *Roem. et Schultes, Syst.* iii. 355; *Thunb. Fl. Cap. ed. Schult.* 131. *P. spatulæfolia, Knight, Prot.* 31. *P. hirsuta, Willd. ex Meisn. l.c.* 216. *Brunia levisanus, Linn. Sp. Pl. ed.* ii. 289. *Conocarpodendron, foliis subrotundis, etc., Boerh. Ind. Pl. Hort. Lugd. Bat.* ii. 202, *t.* 202. *Brunia foliis oblongis, etc., Burm. Pl. Afric.* 267, *t.* 100, *fig.* 2. *Scolymocephalus seu Conocarpodendron foliis brevissimis, Weinm. Phyt.* iv. 296, *t.* 904, *f.g. a.*

SOUTH AFRICA: without locality, *Zeyher! Grey! Mund! Bergius! Sieber, 7! Thunberg! Ecklon,* 333! 352! *Wallich & Hartman! Andersson!*
COAST REGION: Cape Div.; Kommetjes, *Galpin,* 4448! Vygeskraal Farm, *Wolley-Dod,* 1866! Wynberg, *Zeyher! Schlechter,* 1695! Cape Flats, *Bowie! Pappe! Zeyher,* 4683! *Burchell,* 216! 696! *Schmieterlich,* 192! *MacOwan,* 2841! *Zeyher,* 3947! *Ludwig!* Zwart River near Rondebosch, *Zeyher,* 102! Stellenbosch Div.; between Stellenbosch and Cape Flats, *Burchell,* 8349! near Eerste River, *Bolus,* 4201! Caledon Div.; Grabouw near Palmiet River, *Bolus,* 4196!

29. **L. platyspermum** (R. Br. in Trans. Linn. Soc. x. 63); an erect shrub about 6 ft. high; branchlets yellow, more or less angular, glabrous; lower leaves $1\frac{1}{2}$–$2\frac{1}{2}$ in. long, linear, upper leaves $1\frac{1}{2}$–$2\frac{1}{2}$ lin. long, $1\frac{1}{2}$–$3\frac{1}{2}$ lin. broad, linear-oblanceolate, obtuse, all coriaceous and glabrous; male inflorescence terminal, small, about 4 lin. long and $2\frac{1}{2}$ lin. in diam., surrounded by a few larger leaves and several overlapping very hairy outer bracts; bracts 3 lin. long, spathulate-linear, densely villous, ciliate; perianth-tube $\frac{3}{4}$ lin. long, glabrous; segments $1\frac{1}{2}$ lin. long, linear, pilose below, glabrous above; limb $\frac{5}{8}$ lin. long, linear, glabrous; anthers $\frac{3}{4}$ lin. long, linear; style 1 lin. long, filiform, glabrous; stigma $\frac{2}{3}$ lin. long, cylindric, acute; hypogynous scales $\frac{1}{4}$ lin. long, filiform; female inflorescences $1\frac{1}{4}$ in. long, $\frac{1}{2}$ in. in diam., narrowly ellipsoid or conical, surrounded by numerous outer linear-lanceolate coriaceous very hairy bracts; floral bracts $1\frac{3}{4}$ lin. long, $4\frac{1}{4}$ lin. broad, semi-circular, glabrous; perianth-tube $1\frac{1}{2}$ lin. long, compressed, dilated at the base, villous; segments $\frac{2}{3}$ lin. long, linear, glabrous; limb $\frac{1}{3}$ lin. long, ovate, obtuse, glabrous; staminodes $\frac{1}{8}$ lin. long; style 1 lin. long, linear, widened above; stigma shortly 2-lobed; ovary $\frac{3}{4}$ lin. long, $1\frac{1}{4}$ lin. broad, flattened, glabrous; hypogynous scales $1\frac{1}{4}$ lin. long, lanceolate or ovate; old female heads persistent, $1\frac{1}{2}$–$2\frac{1}{2}$ in. long, $1\frac{1}{4}$–$1\frac{1}{2}$ in. in diam.; scales very rigid and thick, spreading; fruits transversely oblong-elliptic, flattened, nucleus 4 lin. long, about 7 lin. broad, surrounded by a membranous wing about $1\frac{1}{2}$ lin. broad, shining, glabrous. *Meisn. in DC. Prodr.* xiv. 224. *L.*

comosum, R. Br. l.c. 64. *L. ericifolium, Drège, partly, fide Meisn.
l.c. Protea polysperma, Poir.* Encycl. Suppl. iv. 556. *P. comosa,
Thunb.* Diss. Prot. 28, *partly (specimen a of Herb.);* Willd. Sp. Pl. i.
517, *partly;* Lam. Ill. i. 238, *partly;* Poir. Encycl. v. 655, *partly.*

SOUTH AFRICA: without locality, *Thunberg! Gueinzius!*
COAST REGION: Caledon Div.; Knoflooks Kraal and Little Houwhoek, *Zeyher,*
3653! Donker Hoek Mountain, *Burchell,* 8003! Bredasdorp Div.; near Elim,
Bolus, 7868! *Schlechter,* 9635! Swellendam Div.; between Sparrbosch and
Tradouw, *Drège,* ♀.

30. **L. æmulum** (R. Br. in Trans. Linn. Soc. x. 64); an erect
shrub, about 5 ft. high; branches robust, erect, glabrous; leaves
$\frac{3}{4}$–$2\frac{1}{4}$ in. long, $\frac{1}{2}$–3 lin. broad, acicular or linear-oblanceolate, acute
or subacute, glabrous; male inflorescences small and subglobose,
terminal, surrounded by numerous enlarged leaves; bracts 1 lin.
long, ovate, subacuminate, subacute, glabrous, ciliate; perianth-
tube $\frac{3}{4}$ lin. long, glabrous; segments $1\frac{1}{2}$ lin. long, linear, glabrous;
limb $\frac{3}{4}$ lin. long, oblong, obtuse or subobtuse, glabrous; anthers
$\frac{1}{2}$ lin. long, linear; style $1\frac{2}{3}$ lin. long, filiform; stigma $\frac{1}{4}$ lin. long;
hypogynous scales $\frac{1}{2}$ lin. long, linear; female inflorescences terminal,
about $1\frac{1}{2}$ in. long and $\frac{1}{2}$ in. in diam., conical; bracts $1\frac{1}{2}$ lin. long,
$2\frac{1}{4}$ lin. broad, transversely oblong, shortly pointed, glabrous;
perianth-tube $1\frac{1}{4}$ lin. long, glabrous at the base, hirsute above;
lobes $\frac{3}{4}$ lin. long, broadly linear, hirsute; limb $\frac{1}{2}$ lin. long,
oblong, subacute, sometimes concave, glabrous; staminodes $\frac{1}{3}$ lin.
long, linear; style $1\frac{1}{4}$ lin. long, linear, narrowed to the base;
stigma cyathiform; ovary $\frac{1}{3}$ lin. long, suborbicular; hypogynous
scales $\frac{1}{4}$ lin. long, ovate, shortly pointed; mature female heads
persistent, $2\frac{1}{4}$ in. long, $1\frac{1}{2}$ in. in diam.; bracts spreading, rigid,
glabrous or nearly so; fruits compressed, broadly elliptic, winged,
emarginate, $4\frac{1}{2}$ lin. long, 4 lin. broad, glabrous. *E. Meyer in Drège,
Zwei Pfl. Documente,* 117, 121, 124, 198; *Meisn. in DC. Prodr.*
xiv. 224. *Protea comosa, Thunb. Diss. Prot.* 28, *partly (specimen
β of Herb.);* Linn. Syst. ed. xiv. 138, *partly;* Thunb. Fl. Cap. ed.
Schult. 129, *partly;* Willd. Sp. Pl. i. 517, *partly;* Lam. Ill. i. 238,
partly; Poir. Encycl. v. 655, *partly.* P. (*Leucadendron æmulum*),
Poir. Encycl. Suppl. iv. 558, *partly.* P. *æmula, Poir. ex Ind. Kew.*
ii. 631, *partly.*

VAR. β, **homœophyllum** (Meisn. l.c.); leaves all of one kind on each shoot,
acicular. *L. abietinum, E. Meyer in Drège, Zwei Pfl. Documente,* 79, 81, 198, not
of *R. Br. Protea incurva, Andr. Bot. Rep. t.* 429. *Chasme ramentacea, Knight,*
Prot. 17.

SOUTH AFRICA: without locality, *Thunberg! Gueinzius!* Var. β, *Ludwig!*
COAST REGION: Swellendam Div.; summit of a peak near Swellendam,
Burchell, 7352! George Div.; Hooge Kraal River, *Drège!* Uniondale Div.;
between Avontuur and Klip River, in Lange Kloof, *Drège!* Var. β: Worcester
Div.; Dutoits Kloof, *Drège!*

According to Knight (l.c.) the variety β grows also in the mountains of Lange
Kloof in Uniondale Div., where it was gathered by Niven and Masson.

31. **L. Dregei** (E. Meyer in Drège, Zwei Pfl. Documente, 64, 198); branches stout, shortly adpressed-pubescent; leaves acicular, 1–$1\frac{1}{4}$ in. long, acute or subacute, glabrous when mature, pilose with weak spreading hairs when young; male flowers not known; female head terminal, solitary, about $1\frac{1}{4}$ in. in diam., ellipsoid-globose; bracts rigidly coriaceous, ovate, shortly adpressed-pubescent in the lower, glabrous in the upper half; perianth-tube 2 lin. long, ventricose at the base, membranous, glabrous; segments 4 lin. long, linear, long-villous; limb $\frac{2}{3}$ lin. long, linear, villous; staminodes $\frac{1}{4}$ lin. long, linear; style 5 lin. long, filiform, glabrous; stigma minutely bifid; fruit $2\frac{1}{3}$ lin. long, obovoid, not winged, pilose. *Meisn. in DC. Prodr.* xiv. 217.

CENTRAL REGION: Prince Albert Div.; Zwartberg Range, near Vrolykheid, *Drège*!

32. **L. Galpinii** (Phillips & Hutchinson); branchlets of the male rather slender, of the female stouter, finely pubescent, rather sharply angular when young, at length becoming terete; leaves of the male slightly smaller than those of the female, the former $\frac{1}{2}$–1 in. long, 1–$1\frac{1}{2}$ lin. broad, the latter 1–$1\frac{3}{4}$ in. long, 1–$1\frac{3}{4}$ lin. broad, all linear-oblanceolate, obtuse, flat, longitudinally striate or wrinkled, glabrous on both surfaces; male inflorescences solitary and terminating short lateral branchlets, $\frac{1}{3}$–$\frac{1}{2}$ in. in diam., globose; bracts $\frac{3}{4}$ lin. long, lanceolate, densely villous; perianth-tube 1 lin. long, cylindric, pubescent; lobes $1\frac{1}{4}$ lin. long, pubescent; limb $\frac{1}{2}$ lin. long, lanceolate, subacute, pubescent; anthers $\frac{1}{3}$ lin. long, linear; style $1\frac{2}{3}$ lin. long, filiform, hairy below; stigma $\frac{1}{4}$ lin. long, ellipsoid or clavate; female inflorescence $\frac{3}{4}$–$1\frac{1}{4}$ in. long, $\frac{3}{4}$–1 in. in diam., oblong-ellipsoid; bracts 3 lin. long, 5 lin. broad, villous at the base, otherwise pubescent, ciliate; perianth-segments $4\frac{3}{4}$ lin. long, widened, membranous and glabrous below, villous above; limb $\frac{1}{4}$ lin. long, linear; style $2\frac{3}{4}$ lin. long, cylindric-filiform, with a few long hairs at the base; stigma truncate or minutely bifid at the apex; ovary $1\frac{1}{4}$ lin. long, elliptic, long hairy.

SOUTH AFRICA: without locality, *Thom,* 69! 71! 538! *Bowie*!
COAST REGION: Riversdale Div.; Milkwoodfontein, *Galpin,* 4439!

33. **L. ramosissimum** (Buek ex Meisn. in DC. Prodr. xiv. 221); branches slightly sulcate, puberulous; young branchlets more or less rusty adpressed-pilose; leaves $1\frac{1}{4}$–$2\frac{1}{2}$ in. long, $1\frac{1}{2}$–$2\frac{1}{2}$ lin. broad, oblanceolate-linear, contracted at the apex into a rather long sharp mucro, chartaceous, glabrous or slightly pubescent; male heads surrounded by a few leaves the inner of which gradually become shorter and broadened at the base, ovoid-triangular or subglobose, about $\frac{1}{2}$ in. in diam.; floral bracts $\frac{3}{4}$ lin. long, oblong, subacute, villous and ciliate on the upper half; perianth-tube 1 lin. long, subcompressed, pilose; segments $1\frac{1}{4}$ lin. long, spathulate-linear, pilose; limb $\frac{1}{2}$ lin. long, oblong, obtuse, pilose; anthers $\frac{1}{3}$ lin. long,

oblong ; style 1½ lin. long, terete, narrowing above, somewhat
flattened below, pubescent ; stigma ⅓ lin. long, clavate, obtuse ;
hypogynous scales ½ lin. long, linear ; female flowers not seen.
L. coniferum, Sieber, ex Meisn. l.c., 222. *L. virgatum, Meisn. l.c., as
to Sieber,* 191, ♂.

SOUTH AFRICA : without locality, *Drège,* 8045 ! *Sieber,* 191 !

34. **L. strictum** (R. Br. in Trans. Linn. Soc. x. 60) ; a shrub
5–7 ft. high ; lower branches yellow (*Burchell*) ; young branchlets
terete, pubescent or puberulous ; leaves in the male smaller than
those of the female, the former 1–2 in. long, the latter about 2½ in.
long, all 1¼–2 lin. broad, linear, contracted at the apex into a long
acute mucro about 1 lin. long, thinly and rigidly coriaceous or
chartaceous, glabrous or thinly long-pilose towards the base, rather
closely longitudinally striate ; male inflorescences numerous, solitary
at the apices of short lateral branchlets, about ¼ in. in diam., subglobose ; bracts ½ lin. long, ovate, subacute, glabrous except for a
small tuft of hairs at the apex ; perianth-tube ⅔ lin: long, glabrous ;
limb ⅔ lin. long, oblong, subobtuse, incurved above ; anthers ½ lin.
long, linear ; style ¾ lin. long, filiform, glabrous ; stigma ¼ lin. long,
cylindric ; hypogynous scales ¼ lin. long, filiform ; female inflorescences solitary at the apex of each branchlet, surrounded by several
imbricate bract-like leaves, about ½ in. long and 4½ lin. in diam.,
ovoid ; bracts 1½ lin. long, 3 lin. broad, transversely oblong, rounded
above, glabrous, scantily ciliate ; perianth-segments compressed,
1¾ lin. long, slightly imbricated, pilose ; limb ⅛ lin. long, oblong ;
staminodes obsolete ; style 1½ lin. long, filiform, glabrous ; stigma
ovoid, lateral ; ovary ⅓ lin. long, obovoid, nearly glabrous ; style
obliquely inserted ; fruits 2½ lin. long, 2¼ lin. broad, slightly compressed, oblong, a little emarginate at the apex, acute at the sides
but scarcely winged, very sparingly setulose ; old persistent female
heads 1¼–1½ in. long, 1¼ in. in diam., ellipsoid ; lower bracts
recurved, the remainder spreading. *Meisn. in DC. Prodr.* xiv. 221.
L. ? coniferum, Meisn. l.c. 227, *partly. Protea conifera, Thunb. Diss.
Prot.* 53 (*as to specimen* γ) ; *Fl. Cap. ed. Schult.* 135 (*as to specimen* γ).
P. conifera, Andr. Bot. Rep. t. 541. *P. conica, Lam. Illustr.*
i. 237 ?, *fide Meisn. l.c. Euryspermum salicifolium, Salisb. Parad.
t.* 75.

SOUTH AFRICA : without locality, *Thunberg* !
COAST REGION : Tulbagh Div. ; at the waterfall, *Pappe* ! Paarl Div. ; near
French Hoek, *Bolus,* 6996 ! Caledon Div. ; Donker Hook Mountain, *Burchell,*
8004 ! Nieuw Kloof, *Burchell,* 8158 ! Houw Hoek Mountains, *Zeyher,* 3648 !
Baviaans Kloof near Genadendal, *Drège* ! Riversdale Div. ; Gysmans Hoek, *Muir,*
397, and in *Herb. Galpin,* 5300 ! George Div. ; Cradock Berg, *Burchell,* 5980 !
Uitenhage Div. ; Van Stadensberg, *Zeyher,* 3650 !

35. **L. eucalyptifolium** (Buek in Drège, Zwei Pfl. Documente,
123, 198) ; a shrub or tree 10–15 ft. high ; branches slightly
flexuous, subterete, pilose with weak hairs and minutely puberulous

with small adpressed hairs, becoming glabrous when older; branchlets pubescent or silky tomentose; leaves subequal in both sexes,
those around the inflorescence longer than those below, gradually
broadening at the base and passing into the bracts, $1\frac{1}{4}$–$3\frac{1}{2}$ in. long,
2–3 lin. broad, linear, conspicuously and acutely apiculate, gradually
narrowed to the base, coriaceous; longitudinally striate, glabrous or
pilose towards the base; male inflorescences solitary at the apex of
rather long lateral branchlets, $\frac{1}{2}$–$\frac{3}{4}$ in. long, 4–5 lin. in diam.,
nearly hidden by the broadened coloured bases of the surrounding
leaves; bracts $\frac{3}{4}$ lin. long, obovate, concave, glabrous, ciliate above;
perianth-tube $\frac{1}{4}$ lin. long, glabrous; segments $1\frac{1}{4}$ lin. long, spathulate-linear, glabrous; limb $\frac{1}{2}$ lin. long, elliptic or oblong, obtuse,
glabrous; anthers $\frac{1}{3}$ lin. long, oblong; style 1 lin. long, cylindric-
filiform, glabrous; stigma $\frac{1}{3}$ lin. long, clavate, obtuse; hypogynous
scales $\frac{1}{4}$ lin. long, filiform; female inflorescence terminal, conical,
$\frac{3}{4}$ in. long, $\frac{1}{2}$ lin. in diam.; bracts $2\frac{1}{3}$ lin. long, 4 lin. broad, transversely oblong, narrowed to the base, densely tomentose in the
upper half, ciliate; perianth-segments $1\frac{3}{4}$ lin. long, linear, glabrous;
limb $\frac{1}{6}$ lin. long, obovate, glabrous; style 1 lin. long, linear, filiform
below, glabrous; stigma terminal, truncate; ovary $1\frac{1}{4}$ lin. long,
$1\frac{1}{2}$ lin. broad, flattened, suborbicular, glabrous; old female heads
cylindric-ellipsoid, $1\frac{1}{2}$ in. long, 1 in. in diam.   *Meisn. in DC. Prodr.*
xiv. 221.   *L. salignum, var. longifolium, Meisn. l.c.* 223.

SOUTH AFRICA : without locality, *Hohenacker*!
COAST REGION : Tulbagh Div. ; Witzen Berg, *Burchell*, 8648 ! George Div.;
near George, *Drège*, 8044 ! forest near Tuuw River, *Burchell*, 5727 ! Swellendam
Div. ; Puspas Valley, Voormansbosch, *Zeyher*, 3646 ! Knysna Div. ; on the Paarde
Berg, *Burchell*, 5193! Millwood Goldfields, *Tyson*! Uniondale Div. ; near
Avontuur, *Bolus*, 2450 ! Humansdorp Div. ; Witte Els Bosch, *Galpin*, 4445 !
Uitenhage Div. ; Van Staadens, near Port Elizabeth, *Paterson*, 891! between
Maitland and Van Stadens River, *Burchell*, 4646 !

36. **L. adscendens** (R. Br. in Trans. Linn. Soc. x. 61); a low-
growing shrub 9–12 in. high ; branches slightly angular, glabrous
or rarely adpressed-pubescent ; leaves on the branches much smaller
and narrower at the base than those surrounding the inflorescence,
the former $\frac{1}{2}$–1 in. long, narrowly oblanceolate, the latter $1\frac{1}{4}$–$2\frac{1}{2}$ in.
long, in the male narrowly linear-oblanceolate, in the female lanceo-
late or linear-lanceolate, all hardened but scarcely apiculate at the
apex, coriaceous, glabrous and longitudinally striate on both
surfaces; male inflorescences solitary, terminal, conical or sub-
globose, $\frac{1}{2}$–$\frac{3}{4}$ in. long, $\frac{1}{3}$–$\frac{2}{3}$ in. in diam. ; bracts $\frac{2}{3}$ lin. long, linear-
oblong, obtuse, villous ; perianth-tube $\frac{1}{2}$ lin. long, glabrous ; segments
$1\frac{1}{2}$ lin. long, spathulate-linear, glabrous; limb $\frac{3}{4}$ lin. long, oblong,
obtuse, glabrous; anthers $\frac{1}{2}$ lin. long, oblong; style 1 lin. long,
filiform-cylindric, glabrous ; stigma $\frac{1}{2}$ lin. long, clavate ; hypogynous
scales $\frac{1}{3}$ lin. long, filiform ; female inflorescences terminal, solitary,
ellipsoid or subglobose, about 1 in. long and $\frac{3}{4}$ in. in diam. ; bracts
$1\frac{3}{4}$ lin. long, ovate, obtuse, pubescent, ciliate ; perianth-segments
imbricate, the two lateral exterior, $2\frac{1}{3}$ lin. long, linear, narrowed

below, villous in the middle ; limb $\frac{1}{2}$ lin. long, lanceolate, subacute, glabrous ; staminodes $\frac{1}{8}$ lin. long ; hypogynous scales $\frac{3}{4}$ lin. long, linear ; style $1\frac{3}{4}$ lin. long, linear, narrowed to the base, glabrous ; stigma $\frac{1}{8}$ lin. long, flat, oblique ; ovary $\frac{1}{3}$ lin. long, oblong, winged ; fruits winged, emarginate.   *Meisn. in DC. Prodr.* xiv. 222. *L. salignum, Berg. in Vet. Akad. Handl. Stockh.* 1766, 323, not of *R. Br. L. virgatum, R. Br. l.c.* 60. *L. glabrum, var. angustifolium, Meisn. l.c.* 221. *L. coniferum, Meisn. l.c.* 227, *partly. L. virgatum, Drège a,* ♀ *, ex Meisn. l.c.* 221, *partly. Protea pallens and P. conifera, Linn. Mant. alt.* 193. *P. pallens* (α *et* β *of Herb.), Thunb. Diss. Prot.* 53 ; *Fl. Cap. ed. Schult.* 134. *P. conifera, Thunb. Diss. Prot.* 53 (*as to spec.* α, β *et* δ *of Herb.). P. argentea, Linn., var.* β ? *Sp. Pl. ed.* i. 94. *P. pallida, Salisb. Prodr.* 49. *P. virgata, Poir. Encycl. Suppl.* iv. 556. *P. stricta, Don ex Steud. Nomencl. ed.* ii. 401, *fide Meisn. l.c.* 252. *P. obliqua and P. involucrata, Willd. ex Meisn. l.c.* 222. *Thymelæa capitata, angusto, etc., Pluk. Mant.* 181, *t.* 229, *fig.* 6. *Conocarpodendron folio angusto rigido, etc., Boerh. Ind. Pl. Hort. Lugd. Bat.* ii. *t.* 200 (*and t.* 203 ?). *Scolymocephalus minor, Weinm. Phyt.* iv. 295, *t.* 903 *a. Frutex æthiopicus conifer foliis, etc., Breyn. Exot. Pl. Cent.* 21, *t.* 9.

VAR. β, **pallens** (Phillips & Hutchinson) ; male inflorescence completely hidden by the upper vegetative leaves.   *Protea pallens* (γ *of Herb. partly), Thunb. Diss. Prot.* 53, *excl. syn.*

SOUTH AFRICA : without locality, *Harvey,* 379 ! 383 ! *Wahlberg* ! *Hooker* ! *Thom,* 269 ! *Mund* ! *Sieber,* 8 ! 19 ! 383 ! *Thunberg* ! *Ludwig* ! *Ecklon* ! *Gueinzius* ! *Hohenacker,* 5 !   Var. β, *Thunberg* !

COAST REGION : Clanwilliam Div. ; between Lange Valley and Oliphants River, *Drège* ! near Wupperthal, *Leipoldt,* 488 ! Tulbagh Div. ; *Pappe* ! Saron, *Schlechter,* 7878 ! 7879 ! Worcester Div. ; Worcester, *Cooper,* 1593 ! Cape Div. ; Mountains and Flats around Cape Town, *Burchell,* 41 ! 903 ! 8483 ! *Bolus,* 3702 ! *Phillips,* 271 ! *Dümmer,* 1239 ! *Wolley-Dod,* 2738 ! *Zeyher,* 3644 ! 4684 ! 4685, partly ! *Ecklon,* 469 ! *Pearson,* 4922 ! 5046 ! *Andersson* ! *Drège* ! *Wilms,* 3573 ! 3574 ! *Galpin,* 4451 ! Stellenbosch Div. ; Stellenbosch, *Andersson* ! *Caledon* Div. ; Mountains of Baviaans Kloof near Genadendal, *Burchell,* 7853 ! Zoetemelks Valley, *Burchell,* 7592 ! between Zwart Berg and the Zondereinde River, *Zeyher,* 3645 ! Robertson Div. ; Sand Berg, near Robertson, *Pearson* ! Oudtshoorn Div. ; 20 miles from Oudtshoorn, *Britten,* 96 ! Uniondale Div. ; between Avontuur and the Keurbooms River, *Burchell,* 5045 ! 5064 ! Uitenhage Div. ; Van Staadens Berg, *Paterson,* 888 ! 889 ! 890 ! Bethelsdorp, *Paterson,* 699 ! Albany Div. ; hill near Stones Hill Road, near Grahamstown, *Misses Daly & Cherry,* 1030 ! Var. β : Clanwilliam Div. ; Zwartbosch Kraal, *Schlechter,* 5179 !

CENTRAL REGION : Calvinia Div. ; Oorlogs Kloof, *Schlechter,* 10954 ! 10955 !

37. **L. Phillipsii** (Hutchinson) ; a shrub up to 9 ft. high ; branchlets adpressed-puberulous or pubescent ; leaves $1\frac{1}{4}$–$1\frac{3}{4}$ in. long, 1–$1\frac{1}{2}$ lin. broad, linear, with a subacute callous apex, straight or more often somewhat falcate, rigidly coriaceous, glabrous ; male heads subglobose, about 5 lin. in diam., surrounded by a few narrow-based leaves which are a little longer than those below, rarely with a few of the inner ones becoming smaller and bract-like ; floral bracts oblong, obtuse, $\frac{1}{2}$ lin. long, slightly keeled on the back, thinly pubescent towards the base ; perianth-tube $\frac{1}{2}$ lin. long,

cylindric, glabrous; segments spathulate-linear, $1\frac{1}{4}$ lin. long, glabrous; limb $\frac{2}{3}$ lin. long, narrowly oblong, obtuse, glabrous; anthers $\frac{1}{2}$ lin. long; style $1\frac{3}{4}$ lin. long, slender, glabrous; stigma clavate, $\frac{1}{3}$ lin. long; female heads oblong-ellipsoid, about $\frac{1}{2}$ in. long, surrounded by a small involucre of short acuminate ciliate bracts; floral bracts transversely oblong-elliptic, about 3 lin. long and 4 lin. broad, coriaceous, rather densely adpressed-pubescent across the middle, glabrous around the margin and near the base; perianth-tube glabrous; segments linear, glabrous; staminodes very small; ovary compressed, suborbicular, slightly emarginate, winged, about $1\frac{1}{4}$ lin. in diam., glabrous; style $1\frac{1}{2}$ lin. long, glabrous; stigma clavate; fruiting head about 1 in. long, surrounded by a few more or less coloured leaves; fruit compressed, winged, suborbicular, slightly emarginate, $2\frac{1}{2}$ lin. in diam., glabrous.

SOUTH AFRICA: without locality, *Drège*! *Harvey*! *Bowie*!

COAST REGION: Caledon Div.; Houw Hoek Mountains, *Zeyher*, 3648 partly! Knysna Div.; Paarde Berg, *Burchell*, 5192! Uniondale Div.; near Avontuur, *Bolus*, 2448!

38. **L. meyerianum** (Buek in Drège, Zwei Pfl. Documente, 70, 198); branches terete, glabrous; young branchlets rather slender, glabrous, purplish; leaves subequal in both sexes, 1–2 in. long, $\frac{1}{2}$–$1\frac{1}{4}$ lin. broad, linear, hardened at the apex into an obtuse mucro, rigidly coriaceous, glabrous and slightly glaucous; male inflorescences solitary or rarely 2 or 3 crowded together on short branchlets, broadly obovoid, 3–4 lin. long, 5–7 lin. in diam.; bracts $2\frac{1}{3}$ lin. long, ovate, obtuse, glabrous; perianth-tube 2 lin. long, cylindric, glabrous; lobes 2 lin. long, linear, glabrous; limb $\frac{3}{4}$ lin. long, oblong-linear, obtuse, glabrous; anthers 1 lin. long, linear; style 4 lin. long, narrowed to the base, coarsely pilose on the middle third; stigma flat, oblique; hypogynous scales 2 lin. long, filiform; female inflorescence solitary, terminal, surrounded by several leaves, ovoid, 3–4 lin. long, 3 lin. in diam.; bracts 2 lin. long, $2\frac{1}{4}$ lin. broad, ovate, subacuminate, subobtuse, concave, glabrous except for the long villous base; perianth-tube 2 lin. long, compressed, villous; segments 1 lin. long, linear, glabrous; limb $\frac{1}{3}$ lin. long, oblong, subacute, incurved above, glabrous; staminodes $\frac{1}{4}$ lin. long; style $2\frac{1}{2}$ lin. long, cylindric, glabrous; stigma 2-lobed; ovary $\frac{1}{2}$ lin. long, ellipsoid, villous; hypogynous scales 1 lin. long, linear; fruiting head broadly ovoid, about $\frac{1}{2}$ in. high and $\frac{3}{4}$ in. in diam.; bracts rusty-villous near the base outside; fruit transversely oblong-ellipsoid, $2\frac{1}{2}$ lin. long, 4 lin. broad, slightly keeled, glabrous, black and wrinkled when dry.

CENTRAL REGION: Calvinia Div.; Oorlogs Kloof, *Schlechter*, 10956! 10957! near Groen River and Waterval River, *Drège*!

Meisner, in *DC. Prodr.* xiv. 215, wrongly reduced this species to his *L. brunioides* = *L. fusciflorum*, R. Br.

39. **L. concinnum** (R. Br. in Trans. Linn. Soc. x. 61); branches terete, purple when dry, glabrous; young branchlets of the male

rather slender, of the female a little stouter; lower leaves $1\frac{1}{4}$–$1\frac{3}{4}$ in. long, 3–7 lin. broad, oblanceolate or oblanceolate-elliptic, obtuse or subacute, coriaceous, glaucous, with hairy tips and margins especially when young, otherwise glabrous; leaves surrounding the female inflorescence about $2\frac{1}{2}$ in. long; male inflorescences scarcely $\frac{1}{2}$ in. in diam., subglobose, surrounded by about 2 series of bracts, the latter 2–3 lin. long, ovate, very long-acuminate, acute, glabrous; floral bracts $1\frac{1}{4}$ lin. long, lanceolate, acute, villous; perianth-tube $\frac{1}{2}$ lin. long, glabrous; segments $2\frac{1}{2}$ lin. long, glabrous; limb $\frac{3}{4}$ lin. long, elliptic, subobtuse, glabrous; anthers $\frac{1}{2}$ lin. long, linear; style $2\frac{3}{3}$ lin. long, terete, very slightly narrowing above, glabrous; stigma $\frac{2}{3}$ lin. long, clavate; hypogynous scales $\frac{3}{4}$ lin. long, linear; young female head 6 lin. long, subglobose; bracts $1\frac{1}{4}$–$1\frac{1}{2}$ lin. long, 2 lin. broad, ovate or semicircular, obtuse or rounded at the apex, velvety-tomentose; perianth-segments $1\frac{3}{4}$ lin. long, slightly imbricate, spathulate-linear, glabrous; limb $\frac{1}{3}$ lin. long, oblong, subacute, incurved at the apex: staminodes $\frac{1}{6}$ lin. long, oblong; hypogynous scales $\frac{1}{2}$ lin. long, linear; style $1\frac{1}{2}$ lin. long, swollen and semiterete above, becoming compressed and linear below, glabrous; stigma oblique; ovary $\frac{1}{4}$ lin. long, ellipsoid, glabrous.   *Meisn. in DC. Prodr.* xiv. 222.   *L. bueki-anum, Meisn. l.c.* 214, *partly, as to spec. Drège, b.   L. discolor, Buek in Drège, Zwei Pfl. Documente,* 198, *partly, as to spec. Drège, b. L. glabrum, var. obtusatum, Meisn. l.c.* 221, *partly.   Protea concinna, Poir. Encycl. Suppl.* iv. 556.

VAR. β, **latifolium** (Meisn. l.c. 223); glabrous; leaves elliptic-oblong, with an acuminate callus.   *L. acuminatum, Buek in Drège, Zwei Pfl. Documente,* 198.

COAST REGION: Clanwilliam Div.; Jackalls Vley, *Niven*! Pakhuis Pass, *Bolus*, 9076! *Leipoldt*, 266! Rondegat, *Schlechter*, 10787! Olifants River, *Drège*, 2408c, ♀! *Schlechter*, 7990! Paarl Div.; Paarl Mountain, *Drège*, 2408a. Bosch Kloof, *Drège*, 2408b! Var. β: Vanrhynsdorp Div.; Gift Berg, *Drège*.

We have not seen the variety, which was founded on a fragmentary sterile specimen in Sonder's Herbarium.

40. **L. decurrens** (R. Br. in Trans. Linn. Soc. x., 59); a decumbent undershrub much branched from the base; branches subterete, glabrous (rarely thinly pilose, *Schlechter*, 5227); leaves erect or suberect, those surrounding the inflorescence longer than those on the shoot, the latter 1–$1\frac{1}{4}$ in. long, 2–3 lin. broad, the former $1\frac{1}{4}$–$1\frac{1}{2}$ in. long, 2–3 lin. broad, all oblanceolate or linear-oblanceolate, contracted at the apex into an obtuse hardened mucro, gradually narrowed to the base, rigidly coriaceous, glabrous; male inflorescences solitary at the ends of the crowded branchlets, sub-globose, $\frac{1}{2}$–1 in. in diam.; bracts 2 lin. long, ovate, subacute, glabrous, the inner 3 lin. long, ligulate; perianth-tube 4 lin. long, cylindric, glabrous; segments 3 lin. long, linear, glabrous; limb $1\frac{1}{2}$ lin. long, oblong-linear, obtuse, glabrous; anthers $1\frac{1}{4}$ lin. long, linear; style $5\frac{1}{2}$ lin. long, filiform, villous below; stigma $\frac{1}{2}$ lin. long, cylindric; hypogynous scales 3 lin. long, filiform; female inflo-

rescence solitary at the apex of each main shoot, with the surrounding leaves much attenuated to the base, broadly ovoid, or ovoid-cylindric, about $\frac{3}{4}$ in. long and in diam.; bracts $2\frac{1}{3}$–$3\frac{1}{2}$ lin. long, transversely oblong or very broadly ovate, obtuse, pubescent below, ciliate; perianth-tube 3 lin. long, flattened, pilose; segments $1\frac{1}{2}$ lin. long, spathulate-linear, glabrous; limb $\frac{3}{4}$ lin. long, lanceolate, obtuse, glabrous; staminodes $\frac{1}{3}$ lin. long, linear; hypogynous scales $\frac{3}{4}$ lin. long, linear; style $3\frac{1}{2}$ lin. long, linear, slightly sinuate below, glabrous; stigma flat, oblique; ovary 1 lin. long, ellipsoid, villous. *Meisn. in DC. Prodr.* xiv. 220. *L. pyramidale, Steud. Nomencl. ed.* ii. 35; *Meisn. l.c.* 228. *L. spathulatum, Meisn. l.c.* 213, *as to spec. Ludwig. L. glaberrimum, Schlechter in Engl. Jahrb.* xxvii. 111. *Protea pallens, Thunb. Diss. Prot.* 53, *partly. P. pyramidalis, Thunb. Fl. Cap. ed. Schult.* 135. *P. chamælea, Lam. Ill.* i. 237? (*teste Meisn.*).

South Africa: without locality, *Brown*! *Ludwig*! *Thunberg*!

Coast Region: Clanwilliam Div.; near Kromme River, 2900 ft., *Bolus*, 5785! 8393! Sneeuw Kop, *Bodkin in Herb. Bolus*, 9078! Piquetberg Div.; Piquet Berg, *Schlechter*, 5227! *Bodkin in Herb. Bolus*, 7560! Tulbagh Div.; Ceres Road, *Schlechter*, 8990! 8991! Tulbagh, *Pappe*! Witzen Berg, *Pappe*! Worcester Div.; between Slangenheuvel, Frenchhoek and Donkerhoek, *Drège*! Hex River Valley, *Wolley-Dod*, 4042! Caledon Div.; Zwart Berg, *Bowie*! George Div.; on mountains, *Bowie*!

Central Region: Ceres Div.; Cold Bokkeveld, *Schlechter*, 8916!

41. **L. glabrum** (R. Br. in Trans. Linn. Soc. x. 60); a shrub; branches quite glabrous and purplish; leaves 2–$2\frac{1}{2}$ in. long, 3–4 lin. broad, oblanceolate, obtuse or subacute, rigid, glabrous, with indistinct nerves and purplish margins; male heads ovoid-globose, about $\frac{1}{2}$ in. in diam., surrounded by a few leaves and fewer glabrous ciliate bracts; perianth-tube 1 lin. long, glabrous; segments $\frac{3}{4}$ lin. long, spathulate, glabrous; limb $\frac{1}{2}$ lin. long, oblong-elliptic, obtuse, glabrous; anthers $\frac{1}{3}$ lin. long; style glabrous; stigma $\frac{1}{2}$ lin. long, clavate; female inflorescence about $\frac{3}{4}$ in. long and scarcely $\frac{1}{2}$ in. in diam., bracteate at the base; involucral bracts ovate, ciliate, otherwise glabrous; floral bracts transversely oblong, $1\frac{1}{4}$ lin. long, about 3 lin. broad, glabrous except on the shortly ciliate margin; perianth-segments free to the base, $1\frac{1}{2}$ lin. long, spathulate-linear, glabrous; limb $\frac{3}{4}$ lin. long, obtuse; staminodes $\frac{1}{3}$ lin. long, elliptic; style $1\frac{1}{2}$ lin. long; stigma lateral, clavate, glabrous; ovary oblong, produced at each side into a short point, glabrous; fruiting head about 2 in. long and $1\frac{1}{4}$ in. in diam.; bracts spreading, truncate, glabrous; fruit suborbicular, compressed, winged, emarginate, 3 lin. broad, glabrous and shining. *Meisn. in DC. Prodr.* xiv. 220, *excl. vars.*

Coast Region: Caledon Div.; Klein River Mountains, *Zeyher*, 3642!

42. **L. minus** (Phillips & Hutchinson); a shrub about 4 ft. high; branches ascending; young branchlets terete, densely villous,

almost hidden by the closely overlapping leaves; leaves subequal, those surrounding the inflorescence becoming bract-like and broadening at the base, 1–1¾ in. long, 1½–3 lin. broad, linear-oblong or slightly linear-oblanceolate, acute or subacute at the apex, coriaceous, longitudinally wrinkled or striate, densely ciliate, otherwise thinly pubescent; male inflorescence solitary and terminal, quite hidden by the surrounding bract-like leaves, about 4 lin. in diam., subglobose; floral bracts 1½ lin. long, oblong, rounded at the apex, glabrous, ciliate; perianth-tube ¾ lin. long, subcompressed, glabrous; segments 1 lin. long, spathulate-linear, glabrous; limb ½ lin. long, linear, obtuse or subobtuse; anthers ⅓ lin. long, linear; style 1½ lin. long, filiform, glabrous; stigma ⅓ lin. long, clavate; hypogynous scales ½ lin. long, filiform; female inflorescence nearly ¾ in. long, 5 lin. in diam., oblong-ellipsoid; bracts 1½ lin. long, 4 lin. broad, transversely linear-oblong, rounded above, glabrous; perianth-segments 2 lin. long, linear, glabrous; limb ¼ lin. long, ovate, subobtuse, concave; staminodes ⅙ lin. long, linear; style 1¼ lin. long, linear; stigma truncate, terminal; ovary ½ lin. long, oblong, much compressed. *L. decorum, var. minus, Buek in Drège, Zwei Pfl. Documente,* 116, 198. *L. pubescens, Meisn. in DC. Prodr.* xiv. 226, *as to prec. syn., not of R. Br.*

VAR. β, glabrescens (Phillips & Hutchinson); branchlets shortly adpressed-pubescent; leaves glabrous or nearly so, except for the shortly pubescent margin.

COAST REGION : Caledon Div. ; tops of the mountains of Baviaans Kloof near Genadendal, *Burchell,* 7676 ! Genadendal, *Drège* ! Var. β : Caledon Div. ; Donker Hoek Mountain, *Burchell,* 8006 !

43. L. lanigerum (Buek in Drège, Zwei Pfl. Documente, 198); branchlets terete, rather densely pubescent with weak hairs; leaves subequal in the two sexes, ¾–1¼ in. long, 1¼–2 lin. broad, narrowly lanceolate or oblong-linear, obtuse at the apex, rigidly coriaceous, thinly pilose, at length nearly glabrous; male inflorescences solitary and terminal, surrounded by several leaves, ½–¾ in. in diam., subglobose; bracts ¾ lin. long, elliptic, obtuse, villous; perianth-tube 1 lin. long, cylindric, glabrous; segments 1½ lin. long, spathulate-linear, glabrous; limb ¾ lin. long, oblong, subacute, glabrous; anthers ⅔ lin. long, linear; style 2¼ lin. long, cylindric, glabrous; stigma ½ lin. long, clavate, subacute; hypogynous scales ¾ lin. long, linear; female inflorescences solitary at the apices of branchlets which arise from below the old persistent heads, nearly hidden by the surrounding leaves, about the same size and shape as the male; bracts 1½ lin. long, 2¼ lin. broad, subquadrangular, tomentose; perianth-segments 1¾ lin. long, linear, glabrous; limb ⅓ lin. long, oblong-ovate, obtuse; staminodes ⅛ lin. long, linear; hypogynous scales ½ lin. long, linear; style 1½ lin. long, linear, widened above; stigma oblique; ovary ½ lin. long, subglobose, compressed, with a broad marginal wing. *Meisn. in DC. Prodr.* xiv. 222 (*by error lanigenum*). *L. floridum, Drège ex Meisn. l.c., not of R. Br.*

VAR. β, **lævigatum** (Meisn. l.c.) ; branches shortly pubescent ; leaves glabrous or nearly so, except when quite young. *L. heterophyllum, E. Meyer in Drège, Zwei Pfl. Documente,* 114, 198. *L. rubricallosum, Buek in Drège, Zwei Pfl. Documente,* 198. *L. æmulum, Schlechter in Engl. Jahrb.* xxvii. 113, *not of R. Br.*

COAST REGION : Paarl Div. ; between Mosselbanks River and Berg River, *Burchell,* 978 ! Paarl Mountain, *Drège*! Stellenbosch Div. ; between Lowrys Pass and Jonkers Hoek, *Burchell,* 8310 ! Var. β : Tulbagh Div. ; Mosterts Hoek, *Pappe*! Mitchells Pass, *Bolus,* 5227 ! Ceres Road, *Schlechter,* 8971 ! 8972 ! Worcester Div. ; Dutoits Kloof, *Drège,* d !
CENTRAL REGION : Var. β : Ceres Div. ; between Hex River Mountains and the Warm Bokkeveld, *Drège* !

44. L. crassifolium (R. Br. in Trans. Linn. Soc. x. 66) ; branches glabrous, glaucous ; leaves 2½–3½ in. long, 1¼ in. broad at the widest part, obovate, rounded at the apex, narrowed to the base, coriaceous, glabrous ; female head (in damaged condition) almost hidden by the upper leaves. *Meisn. in DC. Prodr.* xiv. 226. *Protea crassifolia, Poir. Encycl. Suppl.* iv. 557.

SOUTH AFRICA : without locality, *Masson* !

45. L. spathulatum (R. Br. in Trans. Linn. Soc. x. 54) ; branches glabrous, purplish ; leaves 1½–2¼ in. long, 4–7 lin. broad, falcate, spathulate-oblanceolate, obtuse or subacute, attenuated to the base, with a narrow cartilaginous margin, glabrous, with indistinct nerves ; male heads about 8 lin. long, globose, surrounded at the base by a few imbricate ovate subacuminate slightly acute glabrous ciliate bracts ; perianth-limb 1½ lin. long, obtuse, glabrous ; anthers 1 lin. long ; stigma clavate ; female heads (in fruit) subglobose, 1¼–1½ in. in diam. ; involucral bracts lanceolate, subacute, up to ½ in. long, glabrous except on the ciliate margin ; floral bracts very broadly ovate, densely rusty-villous in the lower part, with glabrous tips and margin ; perianth-tube 2¼ lin. long, membranous, glabrous ; segments 6½ lin. long, villous in the lower half ; limb 1 lin. long, elliptic, obtuse, glabrous ; staminodes ½ lin. long, oblong ; style 4½ lin. long, bifid at the apex, glabrous : fruit transversely ellipsoid, not winged, 2¾ lin. long, 3 lin. broad, slightly pilose. *E. Meyer in Drège, Zwei Pfl. Documente,* 77, 199, *partly, as to spec.* Drège, b ; *Meisn. in DC. Prodr.* xiv. 213, *partly, excl. spec. Ludwig. Protea mutica, Poir. Encycl. Suppl.* iv. 555.

SOUTH AFRICA : without locality, *Nelson* ! *Thunberg* ! *Roxburgh* !
COAST REGION : Clanwilliam Div. ; Jackals Vley, *Niven* ! Tulbagh Div. ; Witzen Berg, *Zeyher,* 1454 ! *Pappe*! between New Kloof and Elands Kloof, *Drège,* b !

46. L. pseudospathulatum (Phillips & Hutchinson) ; branchlets slightly sulcate, glabrous ; leaves 1–1½ in. long, 5–7 lin. broad, obovate or oblanceolate, rounded to a subacute apex, narrowed to and trinerved at the base, rigidly coriaceous, glabrous on both surfaces, with a thin cartilaginous margin ; male heads depressed-globose, about 1 in. in diam., surrounded by an involucre of about 2 series of bracts; involucral bracts coloured, ovate, subacuminate, up to 4½ lin. long and 3 lin. broad, coriaceous, glabrous except on the ciliate margin; floral bracts 4½ lin. long, oblong-lanceolate or linear-lanceolate, subacute, ciliate; perianth-tube 4 lin. long, villous above; segments 3 lin. long, glabrous; limb 2 lin. long, linear-oblong, glabrous; anthers 1¼ lin. long, linear; style 5½ lin. long; stigma 1 lin. long; female head subglobose, scarcely 1 in. in diam. in fruit; bracts in about 5 series, rusty-villous in the lower half, glabrous above; fruits transversely ellipsoid, 2 lin. long, 5 lin. broad, glabrous.

COAST REGION: Clanwilliam Div.; near Honig Valley and on Koude Berg, *Drège* (*L. concolor*), aa! Worcester Div.; Hex River Valley, *Wolley-Dod*, 4042!

47. **L. discolor** (Buek in Drège, Zwei Pfl. Documente, 198, partly); branches purple, longitudinally sulcate, glabrous; leaves 1–1½ in. long, 5–7 lin. broad, shortly oblanceolate, narrowed to the base, subobtuse at the apex, rigidly coriaceous, margin somewhat cartilaginous and shortly tomentose especially when young, those surrounding the female heads very broad and overlapping, broadly lanceolate or elliptic, obtuse, about 2 in. long and ¾ in. broad, with distinct nerves; male flowers not known; young female inflorescence not seen; mature head subglobose or ellipsoid-globose, about 1¼ in. in diam.; bracts about 10-seriate, oblong, rounded at the apex, 4½ lin. long, 2 lin. broad, the lower half of each spreading and glabrous, the upper part ascending and hirsute-tomentose outside, glabrous within; seeds flattened and slightly 3-sided, winged, broadly obovate, 3 lin. long, 2½ lin. broad, black, glabrous. *L. buekianum, Meisn. in DC. Prodr.* xiv. 214, *as to spec. Drège, c.*

COAST REGION: Piquetberg Div.; Piquet Berg, *Drège*, 8038c!

48. **L. squarrosum** (R. Br. in Trans. Linn. Soc. x. 58); branches longitudinally wrinkled or sulcate, glabrous or minutely puberulous; leaves 1½–3 in. long, 4–7 lin. broad, oblanceolate, contracted into a hardened obtuse apex, rigidly coriaceous, very distinctly nerved, glabrous; male heads subglobose, about ¾ in. in diam., at first hidden by overlapping glabrous bracts; floral bracts ovate-lanceolate, subacute, ¾ lin. long, ½ lin. broad, coriaceous, glabrous; perianth-tube ½ lin. long, glabrous; segments 2 lin. long, spathulate, glabrous; limb 1 lin. long, obtuse; anthers linear-oblong, ¾ lin. long; style 1½ lin. long; stigma ⅔ lin. long, clavate; young female inflorescence ovoid, about ¾ in. long; involucral bracts glabrous; floral bracts 2 lin. long, 2¾ lin. broad, ovate, obtuse, coriaceous, glabrous; perianth-tube 1 lin. long, compressed, glabrous; segments 1¼ lin. long, linear, slightly widened at the apex, glabrous; limb ¼ lin. long, elliptic, glabrous; staminodes ⅓ lin. long; style 2 lin. long, gradually narrowed to the base, glabrous; stigma some-

what flattened, minutely bifid at the apex; ovary ¼ lin. long, suborbicular; fruiting head oblong-cylindric, about 2 in. long and 1¼ in. in diam.; bracts glabrous or nearly so; fruit compressed, obovate, emarginate, about 3 lin. long and 3½ lin. broad, winged, glabrous. *Meisn. in DC. Prodr.* xiv. 219. *L. decorum, var. zeyherianum, Meisn. l.c.* 218, *partly, as to part of Zeyher,* 3635. *Protea arcuata, Lam. Ill.* i. 234, *excl. var. β? P. obliqua a, Poir. Encycl.* v. 642 ? *excl. syn. Thunb., Linn. & Boerh. P. strobilina, Linn. Mant. alt.* 192, *ex Brown, l.c., not of Thunb.*

SOUTH AFRICA: without locality, *Niven*! *Ludwig*! *Brown*! *Forster*! *Hooker*! *Roxburgh*!
COAST REGION: Worcester Div.; Dutoits Kloof, *Drège*! Caledon Div.; Donker Hook Mountain, *Burchell,* 7987! mountains near Grietjes Gat, *Zeyher,* 3635!

49. **L. retusum** (R. Br. in Trans. Linn. Soc. x. 53); branchlets terete, shortly tomentellous; leaves 1–1½ in. long, 4–5½ lin. broad, oblanceolate, rounded and retuse at the apex, thinly and rigidly coriaceous, prominently veined, slightly glaucous, glabrous; male flowers not known; female fruiting head a little over 1 in. in diam., broadly ovoid, surrounded by several leaves narrowed to the base; bracts densely rusty-villous in the lower, tomentellous in the upper half; perianth-tube 3 lin. long, globose, membranous, glabrous; segments 3¼ lin. long, linear, with a single furrow, villous in the lower part, becoming glabrous above; limb ½ lin. long, oblong, obtuse, glabrous; staminodes ¼ lin. long, linear; style 2 lin. long, filiform, glabrous; stigma ¾ lin. long, linear, unequally bifid at the apex; fruit 3¼–4 lin. long, broadly obovoid, glabrous. *E. Meyer in Drège, Zwei Pfl. Documente,* 97, 198; *Meisn. in DC. Prodr.* xiv. 213. *Protea retusa, Poir. Encycl. Suppl.* iv. 555.

COAST REGION: Paarl Div.; Paarl Mountain, *Drège,* a! Piquetberg Div.; Twenty-four Rivers, *Roxburgh,* 3!

50. **L. ovale** (R. Br. in Trans. Linn. Soc. x. 59); a shrub; branches terete, shortly tomentose; leaves 1½–2 in. long, 1–1¼ in. broad, those around the inflorescence a little longer, obovate or obovate-elliptic, rounded to an obtuse callous apex, rigidly coriaceous, very minutely papillose all over, otherwise glabrous, with about 5 distinct ascending nerves; male heads not known; female heads nearly hidden by the surrounding leaves, ellipsoid-globose, 1½ in. long, 1¼ in. in diam.; floral bracts 5 lin. long, 4 lin. broad, ovate, subacute, pubescent on the outside; perianth-tube 4 lin. long, ventricose at the base, pubescent; lobes 1½ lin. long, glabrous; limb ½ lin. long, ovate, obtuse, glabrous; staminodes ¼ lin. long, linear; style 3½ lin. long; stigma slightly bifid at the apex; ovary about 1 lin. in diam., orbicular; fruits winged. *Meisn. in DC. Prodr.* xiv. 219, *excl. var. L. rubrum, Burm. f. Fl. Cap. Prodr.* 4, *name only. Protea strobilina, Thunb. Diss. Prot.* 54; *Fl. Cap. ed. Schult.* 136, *not of Linn. Conocarpodendron; acaulon; folio rigido, etc., Boerh. Ind. Alt. Pl. Hort. Lugd. Bat.* ii. *t.* 201.

South Africa: without locality, *Thunberg*!

Coast Region: Caledon Div.; in the valley of the Palmiet River, near Grabouw, 700 ft., *Bolus*, 5528!

On Thunberg's sheet of *P. strobilina* there are two species; the right-hand specimen is the true plant and evidently the one described; the left-hand one is *L. spathulatum*, R. Br.

Protea lenta, Salisbury, Prodr. 50, is according to the author possibly identical with *Protea strobilina*, Thunb.; the description, however, is too incomplete to decide the point.

51. L. decorum (R. Br. in Trans. Linn. Soc. x. 58); branchlets terete, softly tomentose or pubescent; leaves subequal in the two sexes, $\frac{3}{4}$–$2\frac{1}{4}$ in. long, 2–9 lin. broad, oblong-lanceolate or very slightly oblanceolate, contracted into a hardened subacute apex, rigidly but rather thinly coriaceous, ciliate, with a distinct midrib and several ascending lateral nerves; male inflorescence solitary, terminal, surrounded by several leaves, up to $\frac{3}{4}$ in. long, 6–7 lin. in diam., ellipsoid-cylindric, longer than broad; bracts $1\frac{3}{4}$–2 lin. long, oblong-lanceolate, obtuse, densely villous; perianth-tube $1\frac{1}{2}$–2 lin. long, glabrous; segments $1\frac{3}{4}$ lin. long, spathulate-linear, glabrous; limb $\frac{2}{3}$–1 lin. long, oblong, obtuse, glabrous; anthers $\frac{1}{2}$ lin. long, linear; style $2\frac{3}{4}$ lin. long, terete above, flattened below, glabrous; stigma $\frac{3}{4}$ lin. long, clavate, minutely bifid; hypogynous scales 1 lin. long, filiform; female inflorescence similar to the male; bracts 3 lin. long, $2\frac{1}{4}$ lin. broad, ovate, obtuse or subacute, hairy in the lower part, ciliate; perianth-segments $2\frac{3}{4}$ lin. long, deeply channelled, glabrous or villous on the middle part of the keel; limb $\frac{1}{2}$ lin. long, ovate, concave, obtuse, glabrous; staminodes $\frac{1}{6}$ lin. long, oblong; style $1\frac{3}{4}$ lin. long, linear, widened above, glabrous; stigma oblique; ovary $\frac{1}{4}$ lin. long, oblong, glabrous; hypogynous scales $\frac{1}{2}$ lin. long, linear; fruiting head ellipsoid-cylindric, about 2 in. long and $1\frac{1}{4}$ in. in diam.; bracts about 20-seriate, broadly-obovate, shortly and somewhat rusty-pubescent on the outside; fruits flattened, elliptic, winged, about $2\frac{1}{2}$ lin. long and $1\frac{3}{4}$ lin. broad, glabrous. *Meisn. in DC. Prodr.* xiv. 218, *including vars. Protea laureola, Lam. Ill.* i. 234; *Poir. Encycl.* v. 641. *P. venosa, Thunb. Fl. Cap. ed. Schult.* 134, *not of Lam. P. venulosa, Steud. Nomencl. ed.* 2, ii. 401; *Meisn. l.c.* 247. *P. marginata, Willd. ex Meisn. l.c.* 218. *Euryspermum grandiflorum, Salisb. Parad.* 105?

South Africa: without locality, *Hooker*! *Armstrong*! *Pappe*! *Thunberg*! *Andersson*! *Ludwig*!

Coast Region: Cape Div.; Mountains and Flats near Cape Town; *Masson*! *Brown*! *Burchell*, 290! 768! 8586! *Ecklon*, 44! *Drège*! *Schmieterlich*, 190! *Wolley-Dod*, 1789! *Bolus*, 3737! *Tyson*, 2982! Caledon Div.; Grietjes Pass, *Zeyher*, 3635 partly! Grabouw, *Bolus*, 4203! Baviaans Kloof, near Genadendal, *Burchell*, 7846!

52. L. concolor (R. Br. in Trans. Linn. Soc. x. 58); branches lax-villous; leaves $1\frac{1}{2}$–$2\frac{1}{2}$ in. long, $\frac{1}{2}$–$\frac{3}{4}$ in. broad, oblanceolate or obovate-oblanceolate, obtuse at the apex, narrowed to the base, coriaceous, densely villous along the margins, prominently nerved ; male heads subglobose, $\frac{1}{2}$–1 in. in diam., surrounded by a rosette of leaves and a few ciliate but otherwise glabrous bracts ; floral bracts 2 lin. long, oblong-lanceolate, obtuse, villous ; perianth-tube 2 lin. long, cylindric, glabrous ; segments 1$\frac{3}{4}$ lin. long, spathulate-linear, glabrous ; limb 1 lin. long, oblong, obtuse ; anthers $\frac{5}{6}$ lin. long, linear-oblong ; style 3 lin. long, filiform ; stigma $\frac{3}{4}$ lin. long, clavate, minutely bifid ; young female inflorescence not seen ; old female head subglobose, about 2 in. in diam. ; bracts reflexed, leathery, tomentose outside ; fruit obovoid, slightly flattened, truncate at the top, 3$\frac{1}{2}$ lin. long and broad, black, smooth, glabrous except for a ring of rust-coloured hairs around the base. *Meisn. in DC. Prodr.* xiv. 219, *excl. vars. and Zeyher* 3636. *Protea arcuata* β, *Lam. Ill.* i. 234. *P. globosa, Andr. Bot. Rep. t.* 307 ; *Bot. Mag. t.* 878. *P. strobilina, Don ex Steud. Nomencl. ed.* 2, ii. 401, *fide Meisn. l.c.*

SOUTH AFRICA : without locality, *Masson* !
COAST REGION : Cape Div. ; Kasteels Berg, *Wolley-Dod*, 1786 ! Table Mountain, *Bolus*, 2910 ; Kommetjes, *Galpin*, 4447 ! Uitenhage Div. ; Van Stadens Berg, *Zeyher*, 3638 !

53. **L. venosum** (R. Br. in Trans. Linn. Soc. x. 59) ; a shrub about 4 ft. high ; branchlets terete, softly villous-tomentose ; leaves oblanceolate, with a subacute callous apex, narrowed to the base, 1$\frac{1}{2}$–3$\frac{1}{4}$ in. long, 4–7 lin. broad, thinly coriaceous, often a little glaucous, glabrous, distinctly trinerved from above the base ; male heads with a distinct involucre of 2–3-seriate bracts, depressed-globose, about 1 in. in diam. ; involucral bracts broadly ovate, the outer shortly cuspidate, all coloured and glabrous, up to $\frac{1}{2}$ in. long ; perianth-tube glabrous, 4 lin. long ; segments linear, 2$\frac{1}{2}$ lin. long, $\frac{1}{4}$ lin. broad, glabrous ; anthers 1$\frac{1}{2}$ lin. long, linear ; style linear, glabrous ; female heads when young very similar to the male ; bracts 3 lin. long, 6 lin. broad, very broadly triangular-ovate, glabrous in the upper part, with a dense fringe of rust-coloured hairs in the lower part ; perianth-tube 3 lin. long, villous, glabrous at the base ; lobes 1 lin. long, linear, glabrous ; staminodes $\frac{1}{3}$ lin. long, linear ; style 3 lin. long, narrowed to the base ; stigma bifid ; ovary 1$\frac{1}{4}$ lin. long, obovoid, glabrous ; hypogynous scales 2 lin. long, linear ; fruiting head ovoid-globose, about 1$\frac{1}{2}$ in. long and 2 in. in diam. ; bracts $\frac{1}{2}$ in. long, nearly $\frac{3}{4}$ in. broad at the base, densely rusty-villous in the lower half ; fruit boat-shaped, 2$\frac{1}{2}$ lin. long, 4 lin. broad, about 2$\frac{1}{2}$ lin. thick across the top, glabrous except for a ring of yellow hairs around the base. *Meisn. in DC. Prodr.* xiv. 220.

VAR. β, **oblongifolium** (Meisn. l.c.) ; leaves 1$\frac{1}{2}$–2 in. long, about $\frac{3}{4}$ in. broad, obovate-oblong, with a small subacute mucro at the apex ; male heads globose ; bracts glabrous, scarcely ciliate ; perianth-tube sparsely pilose ; segments glabrous. *Protea coriacea, Willd. ex Meisn. l.c.*

SOUTH AFRICA : without locality, *Roxburgh* ;
COAST REGION : Tulbagh Div. ; Witzen Berg, *Pappe* ! near Tulbagh, *Rehmann*,

2255! Tulbagh Waterfall, *Schlechter*, 9010! 9011! *Phillips*, 526! New Kloof, *Burchell*, 997! Stellenbosch Div.; Lowrys Pass, *Burchell*, 8206! *Bolus*, 5554! Hottentots Kloof, *Pearson*, 4905! 4923! Caledon Div.; Zwart Berg, *MacOwan*, *Herb. Austr.-Afr.*, 1521! Houw Hoek, *Ludwig*! Var. β: Caledon Div.; Little Houw Hoek, *Zeyher*, 3637!

54. L. daphnoides (Meisn. in DC. Prodr. xiv. 226); branches terete, densely and softly pubescent; leaves $1\frac{3}{4}$–$2\frac{1}{4}$ in. long, 4–7 lin. broad, lanceolate, contracted at the apex into an obtuse mucro, coriaceous, densely and softly pubescent on both surfaces or nearly glabrous, with 2–3 distinct lateral nerves on each side of the midrib; male inflorescence solitary, terminal, depressed-globose, about $1\frac{1}{2}$ in. in diam., surrounded by leaves broadened at the base, and about 2 series of broad mucronate imbricate bracts; floral bracts about $\frac{1}{2}$ in. long, $1\frac{1}{6}$ lin. broad, spathulate-oblanceolate, acute, glabrous within, tawny-villous in the lower half on the outside, long ciliate in the lower three-fourths; perianth-tube 3 lin. long, narrowly obconic, sparingly pilose; segments 3 lin. long, linear; limb $2\frac{1}{2}$ lin. long, linear, obtuse, glabrous; anthers 2 lin. long, linear; style about 5 lin. long, terete, sparingly pubescent below; stigma not seen; female inflorescence solitary, terminal, surrounded by numerous coloured overlapping leaves, a little over 1 in. in diam., subglobose; bracts 7 lin. long, ovate, acute, densely pilose with long hairs below, glabrous above; perianth-tube 5 lin. long, compressed, very densely pilose; lobes $2\frac{1}{2}$ lin. long, spathulate-linear, villous below; limb 1 lin. long, linear, obtuse, glabrous; staminodes $\frac{1}{4}$ lin. long, linear; style 7 lin. long, terete, gradually thickening above, glabrous; stigma deeply 2-lobed. *L. concolor*, vars. *insigne* and *lanceolatum*, *Meisn. in DC. Prodr.* xiv. 219. *L. grandiflorum, Buek in Drège, Zwei Pfl. Documente*, 198, *not of R. Br. L. retusum, Drège, fide Meisn. l.c.*, *not of R. Br.   Protea daphnoides, Thunb. Fl. Cap. ed. Schult.* 134.

SOUTH AFRICA: without locality, *Thunberg*! *Ludwig*!
COAST REGION: Paarl Div.; Paarl Mountain, *Drège*, 586b! French Hoek, *MacOwan, Herb. Norm. Austr.-Afr.* 906! 907!

55. L. grandiflorum (R. Br. in Trans. Linn. Soc. x. 59); a shrub about 4 ft. high, erect; branchlets softly tomentose or pubescent; leaves 1–$2\frac{1}{4}$ in. long, 5–10 lin. broad, those surrounding the inflorescence attaining 3 in. long, all oblanceolate, constricted into an obtuse apex, distinctly 3–5-nerved from near the base, coriaceous, glabrous; male inflorescence depressed-globose, $1\frac{1}{2}$–2 in. in diam., surrounded by an involucre of 5–6-seriate closely imbricate coloured bracts; involucral bracts broadly ovate, glabrous; floral bracts $4\frac{1}{2}$–6 lin. long, linear, obtuse, glabrous; perianth-tube 6 lin. long, cylindric, glabrous; lobes $3\frac{3}{4}$ lin. long, linear, glabrous; limb 3 lin. long, oblong-linear, obtuse; anthers $2\frac{1}{2}$ lin. long, linear; style 5 lin. long, glabrous; stigma bifid; female inflorescence nearly hidden by the surrounding leaves, about $1\frac{1}{2}$ in. long and $1\frac{1}{4}$ in. in diam., ovoid or ellipsoid; bracts 5 lin. long, 4 lin. broad, ovate, subacute, pubescent outside; perianth-tube 4 lin. long, cylindric above, ventricose at the base, thinly rusty-pubescent; lobes 1½ lin. long, glabrous; limb ½ lin. long, oblong-ovate, obtuse, glabrous; staminodes ¼ lin. long, linear; style 3½ lin. long; stigma bifid; ovary about 1 lin. long and ⅔ lin. broad, flattened, elliptic, thinly pilose; fruits not seen. *Meisn. in DC. Prodr.* xiv. 219. *L. ciliatum,* E. *Meyer in Drège, Zwei Pfl. Documente,* 116, 198. *L. concolor, Meisn. l.c., partly, as to Zeyher,* 3636. *L. concolor, var.? ciliatum, Meisn. l.c.* *L. rugosum, Meisn. l.c.* 227. *Protea rugosa, Thunb. Fl. Cap. ed. Schult.* 135. *P. decora, Salisb. Prodr.* 50 (*teste Meisn.*). *P. ciliata, Desf. Tabl. ed.* ii. 45 (*name only*).

South Africa : without locality, *Thunberg* ! *Masson* ! *Brown* ! *Zeyher* ! *Drège* !
Coast Region : Tulbagh Div. ; New Kloof, *Burchell*, 997 ! Caledon Div. ! Donker Hook Mountain, *Burchell*, 7984 ! Houw Hoek Mountains, *Zeyher*, 3636 ! Bot River, *Burchell*, 937 ! Genadendal, *Drège* ! Baviaans Kloof, near Genadendal, *Burchell*, 7869 ! Uitenhage Div. ; between Maitland River and Van Stadens River, *Burchell*, 4647 !

Imperfectly known species.

56. **L. angustatum** (R. Br. in Trans. Linn. Soc. x. 54) ; a shrub ; branches straight, glabrous ; leaves few, erect, 8–9 lin. long, 1½ lin. broad, linear-spathulate, obtuse, glabrous ; inflorescence subglobose ; scales ovate, the outer broader ; perianth plumose ; fruit the size of a vetch seed, smooth, compressed, clothed with short indumentum. *Meisn. in DC. Prodr.* xiv. 214, *incl. var. latifolium. L. lineare, Steud. Nomencl. ed.* 2, ii. 34 ? *Protea linearis, Houtt. Handl.* iv. 116, *t.* 19, *fig.* 2 ? ; *R. Br. l.c.* 217.

South Africa : without locality, *Masson.*

Known to us only from Brown's description ; the type specimen does not appear to be in existence, but the figure of *Protea fusciflora,* which Brown thought was a variety of *L. angustatum,* is referred by us to *L. fusciflorum,* R. Br.

57. **L. caudatum** (Link, Enum. Hort. Berol. i. 115) ; branches purple, hairy ; leaves 2 in. long, 6 lin. broad, lanceolate, with a callus at the apex, softly hirsute. *Meisn. in DC. Prodr.* xiv. 698.

South Africa : described by Link from a cultivated plant.

58. **L. empetrifolium** (Gandog. in Bull. Soc. Bot. Fr. xlviii. p. xcviii.) ; erect ; branches puberulous ; leaves imbricate, 3 lin. long, "ovate-linear," dilated at the base, shortly narrowed to and mucronulate at the apex, glabrous ; heads 3–3½ lin. in diam., sessile, globose ; floral bracts ovate-cuspidate, tomentellous outside ; corolla-tube shortly grey-pubescent ; stigma ovate, acute.

South Africa : without locality, *Drège.*

59. **L. humifusum** (E. Meyer in Drège, Zwei Pfl. Documente, 64, 118, 198) ; branches purplish, glabrous ; leaves 1¼–2¼ in. long,

6–9 lin. broad, oblong-oblanceolate, obtuse at the apex, slightly narrowed to the base, rather thinly and rigidly coriaceous, somewhat glaucous, glabrous, with 5 more or less distinct ascending nerves; male heads with an involucre of about 8 series of purple-coloured bracts; involucral bracts 5–6 lin. long, 4½–5 lin. broad, oblong-ovate, glabrous and shining on both surfaces; floral bracts oblanceolate, about 3¼ lin. long, glabrous; perianth-tube 3½ lin. long, cylindric, glabrous; segments 3 lin. long, linear, subobtuse, scarcely differentiated into a limb, glabrous; anthers 2 lin. long; style and stigma not seen; female flowers not present. *L. ovale, var. humifusum, Meisn. in DC. Prodr.* xiv. 220.

Locality uncertain, but either :—
COAST REGION : Caledon Div. ; Donker Hoek and Ezelsjagt Mountains, *Drège* ! or
CENTRAL REGION : Prince Albert Div. ; Great Zwart Bergen, *Drège* !
The specimens at Kew are very fragmentary.

60. L.? involucratum (Meisn. in DC. Prodr. xiv. 228); leaves linear-lanceolate, acute and somewhat oblique at the apex, narrowed to the base, glabrous; heads sessile, small, tomentose; bracts ovate-oblong, acuminate, concave, glabrous, somewhat coloured at the base; bracts of the fruiting head villous. *Protea involucrata, Lichtenst. ex Spreng. Syst.* i. 457. *Leucospermum? involucratum, Roem. & Schult. Syst.* iii. 363.

COAST REGION : Cape Div. ; Steen Berg, *Lichtenstein.*

61. L. marginatum (Link, Enum. Hort. Berol. i. 115); leaves lanceolate, acute, narrowed to the base, glabrous, with silky-pilose margins; heads surrounded by lanceolate whitish-yellow leaves; bracts pilose; corolla glabrous. *Meisn. in DC. Prodr.* xiv. 228. *Protea marginata, Willd. Enum. Hort. Berol. Suppl.* 7. *P. ciliaris, Hort. ex Roem. & Schult. Syst.* iii. 356, *and P. ciliata, Breit. Hort. Breit.* 380, *fide Meisn. l.c. Leucospermum marginatum, Spreng. Syst.* i. 464.

SOUTH AFRICA : described by Link from a cultivated plant.
From the description this is very probably identical with *L. decorum,* R. Br.

62. L. sessile (R. Br. in Trans. Linn. Soc. x. 54); branches pubescent; leaves 1¾–2¼ in. long, 5–7 lin. broad, lanceolate, subobtuse, subdistinctly veined, glabrous; old female head 2 in. long; bracts deeply concave, villous. *Meisn. in DC. Prodr.* xiv. 214. *Protea (Leucadendron sessile), Poir. Encycl. Suppl.* iv. 557.

SOUTH AFRICA : without locality, *Masson* !
Brown's type is very imperfect and consists of an old female head and a few leaves.

63. L. callosum (Hoffmgg. Verz. Pfl. 72). *Protea callosa, Wendl. ex Steud. *Nomencl. ed.* 2, ii. 399, *name only.*
SOUTH AFRICA : formerly cultivated.

64. L. flavescens (Link, Enum. Hort. Berol. i. 115). *Protea flavescens, Willd. Enum. Hort. Berol. Suppl. 7, name only.*

SOUTH AFRICA : formerly cultivated at Berlin.

65. L. cuneiforme (Burm. f. Fl. Cap. Prodr. 4); the leaves are described as tricuspidate, so the plant evidently does not belong to this genus ; it may be a *Leucospermum.*

SOUTH AFRICA : without locality, *Burmann.*

66. L. filamentosum (Burm. f. Fl. Cap. Prodr. 4) ; leaves ovate-lanceolate, imbricate, glabrous ; heads terminal ; stamens very long, persistent.

SOUTH AFRICA : without locality, *Burmann.*

67. L. glomiflorum (Knight, Prot. 59) ; a low decumbent shrub with slender branches ; leaves 1–1½ in. long, 1½–2 lin. broad, linear-lanceolate, often falcate, quite entire, slightly pubescent when old ; heads of flowers large, upon rather long peduncles.

COAST REGION : Caledon Div. ; Great Houw Hoek, *Niven.*

This is very probably a species of *Leucospermum.*

68. L. ? glutinosum (Hutchinson) ; a stout shrub ; leaves 1½–2 in. long, about 3 lin. broad in the males and 5 lin. in the females, spathulate-elliptic, obtuse, not quite smooth when old, especially the lower ones ; bracts glutinous, bearded externally towards the base ; female flowers only a little exserted. *Protea glutinosa, Knight, Prot. 27.*

COAST REGION : Piquetberg Div. ; Twenty Four Rivers, *Niven.*

Known only from Knight's description ; the plant referred to by him is evidently a species of *Leucadendron* and it may be identical with *L. spathulatum,* R. Br., which is found on the neighbouring Witzenberg Range.

69. L. gnaphaliifolium (Knight, Prot. 60) ; a tall shrub 7–8 ft. high ; leaves 8–10 lin. long, 2–3 lin. broad, elliptic-lanceolate, generally quite entire, with a narrow callous apex, exceedingly pubescent, slightly nerved ; stigma broadly conical.

SOUTH AFRICA : without locality, *Niven* ?

70. L. gracile (Knight, Prot. 59) ; stem decumbent ; leaves 4–7 lin. long, 1–1½ lin. broad, distant from one another, linear, entire, pubescent ; bracts short ; style narrow.

COAST REGION : Caledon Div. ; Klein River, *Niven.*

71. L. inflexum (Link, Enum. Hort. Berol. i. 115). *Protea inflexa, Willd. Enum. Hort. Berol. Suppl. 7, name only.*

SOUTH AFRICA : formerly cultivated at Berlin.

72. L. polifolium (Burm. f. Fl. Cap. Prodr. 4) ; leaves ovaté, obtuse, pubescent.

SOUTH AFRICA : without locality, *Burmann.*

73. L. polygaloides (Link, Enum. Hort. Berol. i. 115).   *Protea polygaloides, Willd. Enum. Suppl. 7, name only.*

SOUTH AFRICA : formerly cultivated at Berlin.

74. L. splendens (Burm. f. Fl. Cap. Prodr. 4); leaves lanceolate, acuminate, glabrous ; head foliaceous.

SOUTH AFRICA : without locality, *Burmann.*

75. L. undulatum (Link, Enum. Hort. Berol. i. 115, name only).

SOUTH AFRICA : formerly cultivated at Berlin.

IV. PROTEA, Linn.

Flowers hermaphrodite, zygomorphic.   *Perianth* tetramerous, tubular in bud, slender, more or less widened towards the base, early divided into an anticous (abaxial) segment with a long very slender claw and a narrow oblong or linear limb and a posticous (adaxial) portion, consisting of the fused posticous and lateral segments, their claws forming a long anteriorily open sheath and their limbs a linear to oblong, concave or subtubular, equally or unequally 3-lobed lip, the lobes being either short, tooth-like or produced into slender often filiform awn-like and very hairy processes, the whole perianth variously hairy or glabrous.   *Anthers* sessile or subsessile, inserted low down on and shorter than the perianth-limbs, linear, rarely oblong ; tip of connective produced into a small fleshy gland.   *Hypogynous scales* 4, free, variously shaped, rarely absent.   *Ovary* covered with long hairs ; style rigid, straight or curved, terete or laterally compressed, sometimes bulbously thickened at the base or with a glandular (?) depression on the inner (adaxial) side, glabrous or hairy ; stigma slender, mostly finely grooved, gradually passing into the style or suddenly bent or kneed at the junction with the style.   *Ovule* 1, sublaterally attached, subpendulous, anatropous.   *Nut* densely bearded, crowned by the persistent style.

Small trees, shrubs or acaulescent perennial plants with glabrous or hairy stems ; leaves alternate, coriaceous, entire, hairy or glabrous ; flowers in many-flowered, sessile or subsessile, terminal or lateral, usually solitary heads, enclosed in an involucre of numerous imbricate, coriaceous to scarious, glabrous or hairy, sometimes bearded, variously coloured bracts ; receptacle flat, convex or conical, bearing numerous short, persistent, free or coalescent paleæ.

DISTRIB.   Mostly in the south-western parts of Cape Colony, but also extending northwards into tropical Africa, few north of the equator.   Species about 100.

The working out of the genus *Protea* met with particular difficulties owing to the great number of imperfectly described species and the scanty material in the herbaria.   A considerable number of Proteas were introduced into cultivation in Europe at the end of the 18th and the beginning of the 19th centuries. Not a few of them were figured when they came into flower, the plates generally being accompanied by the most meagre and vague descriptions ; others were only briefly described from the cultivated specimens, whilst usually no specimens were preserved or only fragments, and even these have been lost in many cases.   Interest in those plants soon waned, and with it the plants disappeared from the gardens.   Nor were many of them collected again, and for all we know, they may have become extinct. At the same time there is some chance of finding them again, and they have therefore been put on record and described as fully as possible from the plates. Another source of difficulty is the apparent variability of the vegetative parts of the plants. This is borne out in a few cases by the material in the herbaria, and in the others by the collectors' notes. In such instances it has been possible to connect extreme forms, differing in their habit or in the shape and size of their leaves ; but in others similarly divergent forms were treated as specifically distinct, as it was considered inexpedient to reduce them on purely hypothetical grounds. The species group themselves quite naturally round certain forms as centres, or they stand rather isolated, as e.g. *P. mellifera.* These relationships have found expression in the creation of a number of sections. Some of them will possibly be found to overlap or to run into each other when more material is available and a closer examination of the structure of the flower and fruit is possible. The sections are therefore in a sense provisional, at least in their limitation.

SYNOPSIS OF SECTIONS.

A. Small trees, shrubs or undershrubs with a distinct overground stem (see 62, *P. cynaroides* and 44, *P. tenax*) ; heads always terminal.

Heads large 4–8 (rarely 3) in. long ; inner involucral bracts exceeding or equalling the flowers ; perianth-lip 3-awned, awns 3–15 lin. long, villous or woolly:
Leaves sessile :
Inner involucral bracts with usually broad rounded tips fringed with a long villous whitish, fulvous or deep-purple to black beard ...   I. SPECIOSÆ.

Inner involucral bracts with a long claw and an oblong or oblanceolate, obtusely pointed, shortly villous or glabrous limb ... ...   II. LIGULATÆ.

Inner involucral bracts not clearly differentiated into limb and claw, glabrous ... ...   III. MELLIFERÆ.

Leaves long petioled ; involucral bracts acute, tomentellous all over ... ...   VIII. CYNAROIDEÆ.

Heads medium-sized to small, rarely up to 4 in. long ; inner involucral bracts equalling, or shorter than, the styles, or if exceeding them, then the heads small ; perianth-lip 3-toothed, teeth rarely exceeding 1 lin. :
Heads medium-sized, 2½–4 in. long ; involucral bracts silky-tomentose or finely silky-pubescent, rarely nearly glabrous ; flowers (or styles) 1½ to over 3 in. long, exserted from the involucre or equalling it ; perianth-lip 5–9 lin. long, glabrous or hairy :
Perianth-sheath very slender, soon spirally coiled up and withdrawn from the long exserted styles ; lip more or less glabrous apart from the villously tufted teeth ... ...   IV. EXSERTÆ.

Perianth-sheath firmer, not spirally coiled up ; lip villous or pubescent all over ... ...   V. LASIO-CEPHALÆ.

Heads medium-sized to small, not over 2 in. long, or if so, then the base contracted into a scaly stipes ; involucral bracts glabrous or nearly so, rarely tomentose (41, *P. caffra*) ; flowers under 1¾ in. long, exserted from or enclosed in the involucre ; perianth-lip 2–5 (rarely up to 6 or 7)

lin. long, glabrous or sparingly hairy, sometimes
with an apical tuft, particularly when young,
rarely permanently pubescent (39, *P. convexa*) :
  Leaves linear to obovate, ($\frac{1}{2}$ rarely $\frac{1}{3}$) to 3 in.
  broad ; perianth-lip 3–5 (rarely up to 6
  or 7) lin. long ... ...   VI. LEIO-
                                     CEPHALÆ.

  Leaves narrowly linear to filiform or acicular ;
  perianth-lip 2–4 lin. long ... ... VII. PINIFOLIÆ.

B. Main stem underground ; heads on the ground, terminal or lateral at the base
of barren shoots (see also 62, *P. cynaroides*, and 44, *P. tenax*).

  Heads terminal, solitary :
    Heads surrounded by an outer involucre of foliaceous
    bracts different in shape from the normal foliage
    leaves ... ... X. OBVALLATÆ.

    Heads surrounded by normal foliage leaves :
    Heads 6–2 in. long ; flowers from over 2 to 1 in.
    long ; styles gently curved or almost straight,
    long-subulate from a linear-lanceolate base,
    gradually passing into the narrow ovary ... IX. PARACYNA-
                                       ROIDEÆ.

    Heads 2–1 in. long ; flowers 1$\frac{1}{2}$–1$\frac{3}{4}$ in. long ; styles
    distinctly curved to sickle-shaped, subulate
    from a bulbously thickened base ... ... XI. MICRO-
                                       GEANTHEÆ.

  Heads lateral, crowded at the base of barren shoots ... XII. HYPOCEPHALÆ.

SYNOPSIS OF SPECIES.

§ 1. SPECIOSÆ. Heads always terminal, large, 3–6 in. long ; inner
involucral bracts elongated, distinctly exceeding the flowers, with usually
broad rounded (sometimes cuspidate or acute in 3, *P. barbigera*) tips,
fringed with a long dense villous whitish, fulvous, deep-purple or black
beard ; flowers 2$\frac{1}{2}$–3$\frac{1}{2}$ in. long, rarely less (11, *P. patens*) ; perianth-lip
hairy all over or at least at both ends, 3-awned, awns 3–12 lin. long,
mostly more or less villous ; style very gently curved to almost straight,
laterally compressed and linear below, terete upwards, with a more or
less distinct glandular(?) depression at the base on the inner (adaxial)
side ; separated from the subulate stigma by a sudden bend or knee.
Shrubs or undershrubs with a distinct overground stem.

Leaves more than 1 in. wide :
  Leaves obovate to elliptic with a broad base, 1$\frac{1}{2}$–3 in.
  broad ... ... (1) **grandiceps.**

  Leaves obovate-oblong to oblong-lanceolate, 1–1$\frac{1}{2}$ in.
  broad :
    Leaves obovate-oblong, obtuse, up to 3 times as long
    as broad ; beard of involucral bracts reddish ... (2) **speciosa.**

    Leaves oblong-lanceolate, more than 3 times as long as
    broad :
      Outer involucral bracts permanently villous-edged,
      beards of the inner whitish ; flowers 3$\frac{1}{2}$ in. long,
      long-villous all along ; lip 1$\frac{1}{2}$ in. long including
      the flexuous awns which are 1 in. long ... (3) **barbigera.**

      Outer involucral bracts at length glabrescent ; beards
      of the inner deep-purple to black, at least
      partly ; flowers 3 in. long, long-villous only on
      the tips ; lip 1 in. long including the straight
      awns which are less than $\frac{1}{2}$ in. long ... ... (4) **marginata.**

Leaves less than 1 in. wide :
Heads 4–5 in. long ; outer involucral bracts glabrous or
glabrescent :
Stem and leaf-bases (particularly those near the head)
hirsute with long soft hairs ; beards pallid　　...　(5) **incompta.**

Stem and leaves glabrous, or the former tomentellous ;
beards black or purple :
Beard of inner involucral bracts forming a round
apical tuft of long hairs ...　...　...　...　(6) **comigera.**

Beard of inner involucral bracts forming a fringe
along the edges of the tips :
Inner involucral bracts tomentose or villous on
the back inside the long beards :
Inner involucral bracts with black tips and a
black or white beard　　...　...　...　(7) **Lepidocarpodendron.**

Inner involucral bracts with whitish or pinkish
tips and a black beard　　...　...　...　(8) **neriifolia.**

Inner involucral bracts glabrous or finely pubescent
on the back inside the short beards ...　...　(9) **pulchella.**

Heads 3½–3 in. long ; outer involucral bracts pubescent,
hirsute or villous :
Leaves glabrous with the exception of the tomentose
base and the ciliate margin ; outer involucral
bracts pubescent or hirsute　　...　...　...　(10) **fulva.**

Leaves loosely covered with long hairs all over ; outer
involucral bracts villous on the back and bearded
along the margin ...　...　...　...　...　(11) **patens.**

§ 2. LIGULATÆ. Heads always terminal, large, 4–6 in. long ; inner
involucral bracts exceeding, rarely only equalling, the flowers, with a
long slender claw and an oblong or oblanceolate, obtusely pointed, shortly
villous or glabrous limb ; flowers 2½–3½ in. long ; perianth-lip tomentose
to villous all over, 3-awned, awns up to 15 lin. long, tomentose or villous ;
style as in section 1, *Speciosæ*. Shrubs or undershrubs with a distinct
overground stem.

Leaves (excepting sometimes those close to the head)
elliptic to oblong-elliptic, rounded or subcordate at the
base, 1½–2½ in. broad, 2 to 3 times as long :
Leaves elliptic-cordate ; lateral awns of perianth-lip
10–12 lin. long ...　...　...　...　...　...　(12) **latifolia.**

Leaves ovate to ovate-oblong, uppermost sometimes
oblong-lanceolate ; lateral awns of perianth-lip 5–6
lin. long　　...　...　...　...　...　...　(13) **compacta.**

Leaves oblanceolate to linear, narrowed towards the base,
½–1½ in. broad (rarely slightly more), more than
3 times as long :
Leaves 8–21 lin. broad ; awns of perianth-lip 3½–9 lin.
long :
Perianth-tube adpressedly silky-pubescent or shortly
tomentose or finely villous :
Flowers 3¼–3½ in. long ; awns 9 lin. long :
Larger leaves broad towards the base to subcordate ;
involucral scales more or less acute ...　...　(14) **magnifica.**

All the leaves narrowed at the base ; involucral
bracts obtuse to rounded at the lips ...　...　(15) **macrophylla.**

Flowers 2½–3½ in. long ; awns 3–5 lin. long :
Leaves very obtuse to round and emarginate at
the tips, long attenuated at the base, almost
petioled, 1–1½ in. wide   ...   .   **(16) obtusifolia.**

Leaves subobtuse to subacute or, if obtuse, not
emarginate, ½–1 in. wide, rarely wider :
Leaves linear-oblong, very shortly attenuated at
the base   ...   .   **(17) triandra.**

Leaves oblanceolate, long attenuated at the base :
Outer involucral scales ovate, very obtuse ;
awns of perianth-lip 2½ lin. long, with
purplish villous tips   ...   .   **(18) Susannæ.**

Outer involucral scales ovate-lanceolate, acuminate ; awns of perianth-lip 3½–4 lin.
long, often spreading   ...   .   **(19) calocephala.**

Perianth-tube spreadingly villous   ...   .   **(20) Rouppelliæ.**

Leaves 3–5 lin. broad :
Awns of the long and spreadingly villous perianth-lip
up to 15 lin. long   ...   .   **(21) longifolia.**

Awns of the perianth-lip up to 9 lin. long :
Centre of flower-head umbonate :
Outer involucral bracts with black tips   ...   ...   **(22) ignota.**

Outer involucral bracts without black tips   ...   **(23) umbonalis.**

Centre of flower-head slightly convex   ...   ...   **(24) ligulæfolia.**

§ 3. . MELLIFERÆ. Heads always terminal, 5 in. long ; inner involucral
bracts linear-oblong to broad-linear, not clearly differentiated into limb
and claw, glabrous ; flowers about 3 in. long ; perianth-lip glabrous,
excepting the bearded awns which are 3 lin. long ; style gradually
attenuated from the widened base, slender, subulate, gently curved or
flexuous, without a glandular depression at the base, gradually passing
into the slender-spindleshaped acute stigma.   A tall shrub.

Only species   ...   .   **(25) mellifera.**

§ 4. EXSERTÆ. Heads terminal, medium-sized to rather large, 2½–4 in.
long ; inner involucral bracts exceeded by the stigmata, not clawed,
with rounded villous tips, bordered with a short dense silky fringe ;
perianth-sheath very slender, spirally coiled up after flowering and thus
withdrawn from the long exserted styles, lip 3-toothed, more or less
glabrous excepting an apical coma ; styles straight or nearly so, subulate
from a very slightly broader base, somewhat suddenly attenuated into
the filiform obtuse or capitate stigma.   Shrubs or undershrubs.

Flowers over 3 in. long ; perianth-lip 8–9 lin. long ;
teeth up to 2 lin. long ; stigma very slender,
6 lin. long   ...   .   **(26) longiflora.**

Flowers up to 2½ in. long ; perianth-lip 5–7 lin. long ;
teeth up to 1 lin. long ; stigma 3 lin. long :
Leaves densely woolly all over when young ; flowers
scarcely 2 in. long ; style 1¾ in. long, including
the stigma ...   .   **(27) subvestita.**

Leaves never woolly all over ; flowers over 2 in. long ;
style 2½ in. long, including the stigma :
Style slender ; stigma not capitate   ...   ...   **(28) lacticolor.**

Style stout ; stigma distinctly capitate   ...   ...   **(29) Mundii.**

§ 5. LASIOCEPHALÆ. Heads terminal, medium-sized, 2½–3 in. long; involucral bracts gradually increasing in size inwards, firm, silky-tomentose or finely pubescent, rarely nearly glabrous, inner equalling the flowers or shorter, densely ciliate or nearly glabrous on the margins; flowers 1½–2½ in. long; lip 6–9 lin. long, 3-dentate, villous or pubescent all over; styles very slender, subulate from a very slightly widened base, slightly curved, almost imperceptibly passing into the stigma or separated from it by a minute bend. Shrubs or undershrubs.

Flowers 2–2½ in. long:
Leaves falcate, linear-oblong, long attenuated towards
the base ... ... (30) **curvata.**

Leaves oblong to obovate-oblong or lanceolate, shortly
attenuated or broad at the base, not falcate:
Involucral bracts silky-pubescent only when quite
young; perianth-lip adpressedly pubescent ... (31) **grandiflora.**

Involucral bracts more or less permanently silky-pubes-
cent to villous; perianth-lip shaggy-pubescent:
Stem glabrous ... ... (32) **trigona.**
Stem pilose, at least when young ... ... (33) **abyssinica.**

Flowers under 2 in. long:
Branches and leaves pilose, the latter distinctly veined;
perianth-sheath villous down to the widened base ... (34) **hirta.**

Branches and leaves glabrous, the latter indistinctly
veined:
Perianth-sheath densely hairy:
Innermost involucral bracts slightly produced beyond
the preceding series, much shorter than the
flowers ... ... (35) **Dykei.**

Innermost involucral bracts much produced beyond
the preceding series, equalling the flowers ... (36) **rupicola.**

Perianth-sheath glabrous, excepting for a few hairs
towards the lip ... ... (37) **glabra.**

§ 6. LEIOCEPHALÆ. Heads terminal, medium-sized to small, not over 2 in. long, or, if so, then the base contracted into a scaly stipes; involucral bracts gradually increasing in size inwards, firm, glabrous or nearly so (see 41, *P. caffra*); inner bracts equalling the flowers or shorter, their tips rounded with glabrous, rarely pubescent, margins; flowers under 1¾ in. long; lip 3–7 lin. long, glabrous or with scanty rigid hairs along the sides, and more often at the tips, forming a frequently fugacious tuft, rarely permanently and scantily pubescent on the sides and tips (39, *P. convexa*); styles more or less curved, sometimes strongly so in the upper part, or sickle-shaped, subulate from a bulbously thickened or obliquely widened base, with a sudden small bend or gradually passing into the very slender stigma. Shrubs or undershrubs.

Leaves very glaucous, obovate to obovate-oblong, 3–8 in.
by 1½ (rarely 1)–3 in., quite glabrous and smooth:
Perianth-lip with deciduous brown rigid hairs, glabrescent; perianth-sheath glabrous ... ... (38) **recondita.**

Perianth-lip permanently whitish-pubescent on the sides
and tips; perianth-sheath ciliate ... ... (39) **convexa.**

Leaves not glaucous, or, if slightly so, then narrower than
in the preceding species:
Leaves oblong, woolly when young, 2–2½ by ¾–1 in.;
perianth-lip glabrous or with a few hairs on the
teeth ... ... (40) **punctata.**

Leaves much narrower in proportion :
   Leaves over 3 in. long, lanceolate to linear-lanceolate :
      Heads more or less stipitate ; stipes scaly :
         Leaves about 1 in. broad :
            Involucral bracts tomentose, at length more or
               less glabrous ...   .   **(41) caffra.**

            Involucral bracts glabrous or finely pubescent ...   **(42) rhodantha.**

      Leaves less than 1 in. broad   ...   .   **(43) multibracteata.**

   Heads rounded at the base :
      Upper parts of branches and leaf-bases or the whole
      leaves more or less softly hirsute ; leaves very
      variable, from obovate to linear-lanceolate,
      4–6 in. by ⅓–2 in.   ...   .   **(44) tenax.**
      Branches and leaves glabrous :
         Leaves oblong to oblanceolate, 3–4 by 1¼ in.,
         with a distinct thin translucent cartilaginous
         margin   ...   .   **(45) transvaalensis.**
         Leaves oblanceolate to linear-lanceolate, 2–5 in.
         by ½ (rarely ⅓)–⅔ in., margin not carti-
         laginous, or, if so, then usually indistinct
         and not translucent :
            Heads 2½ in. in diam. ; flowers 1¼ in. long ...   **(46) Flanagani.**
            Heads 2 in. or less in diam. ; flowers 1 in. long   **(47) simplex.**

   Leaves under 3 in. long, more or less oblanceolate to
   linear-oblanceolate :
      Perianth-lip 7 lin. long ; young heads turbinate-
      oblong ; leaves oblanceolate, obtuse   ..   ...   **(48) lanceolata.**
      Perianth-lip less than 5 lin. long :
         Leaves over 2 in. long :
            Receptacle slightly convex ; leaves subobtuse,
            obscurely nerved ; perianth-lip 4½ lin. long   **(49) Doddii.**
            Receptacle conical, ¾ in. high ; leaves acute,
            distinctly veined ; perianth-lip 3 lin. long   **(50) Marlothii.**
         Leaves 2 to less than 1 in. long :
            Leaves narrowly oblanceolate, acute, 4 to 6
            times as long as broad :
               Heads erect, about 2 in. long   ...   ...   **(51) effusa.**
               Heads pendulous, about 1½ lin. long   ...   **(52) pendula.**
            Leaves broadly oblanceolate, obtuse, 2–3 times
            as long as broad ; heads 2½ in. long,
            pendulous   ...   .   **(53) sulphurea.**

   § 7. PINIFOLIÆ. Heads terminal, small, 1–2 in. long ; involucral bracts
gradually increasing in size inwards, glabrous or nearly so ; inner bracts
exceeding the flowers, their tips rounded, glabrous ; flowers 1 in. long or
less ; perianth-lip 2–4 lin. long, 3-toothed, usually glabrous or with
scanty deciduous hairs (see also 55, *P. cedromontana*). Styles more or
less falcate, compressed-subacute from an obliquely widened base,
gradually or with a minute bend passing into the short slender stigma.
Small shrubs or undershrubs. Leaves narrowly linear to filiform or
needle-shaped.

Leaves linear, 1–2 lin. broad :
   Leaves 3–7 in. long :
      Leaves green when dry, slightly rough ; perianth-lip
      4 lin. long, glabrous, excepting at the tips, sheath
      slender   ...   .   **(54) canaliculata.**

Leaves somewhat glaucous, smooth ; perianth-lip up
　　to 2 lin. long, hairy along the margins, sheath
　　wide ...　　...　　...　　...　　...　　...　　...　(55) **cedromontana.**
Leaves under 2½ in. long:
　　Heads obovate-oblong when young ; inner involucral
　　　bracts narrowly lanceolate, acutely acuminate ;
　　　leaves long and acutely acuminate...　　...　　...　(56) **odorata.**
　　Heads ovoid-globose when young ; inner involucral
　　　bracts rounded at the tips ; leaves not acuminate :
　　　Leaves acute, mucronate ; branches glabrous ;
　　　perianth-sheath wide　　...　　...　　...　　...　(57) **scolymo-
　　　　　　　　　　　　　　　　　　　　　　　　　　cephala.**

　　　Leaves subobtuse to subacute ; branches whitish-
　　　　tomentellous above ; perianth - sheath very
　　　　slender　　...　　...　　...　　...　　...　　...　(58) **Harmeri.**
Leaves filiform to acicular :
　　Lowest involucral bracts produced into foliaceous
　　　appendages :
　　　Branches glabrous ; leaves 2–2½ in. long　　...　　...　(59) **pityphylla.**
　　　Branches villous ; leaves up to 1¼ in. long　　...　　...　(60) **witzen-
　　　　　　　　　　　　　　　　　　　　　　　　　　bergiana.**
　　Lowest involucral bracts not appendaged ; leaves acicular,
　　　pungent, up to ¾ in. long　　...　　...　　...　　...　(61) **rosacea.**

　　§ 8. CYNAROIDEÆ. Heads terminal, 5–8 in. long ; involucral bracts very
　　　numerous, gradually increasing inwards, acute, whitish-tomentellous all
　　　over ; flowers 3 in. long or longer ; perianth-lip up to 1 in. long,
　　　3-toothed, tomentose ; style curved in the upper part, compressed and
　　　linear up to the middle, then subulate, passing with a sudden bend into
　　　the subulate stigma. Leaves subrotundate to oblong, long-petioled.
　　　Stem up to 6 ft. high, rarely almost suppressed.

Only species　　...　　...　　...　　...　　...　　...　　...　(62) **cynaroides.**

　　§ 9. PARACYNAROIDEÆ. Heads large to medium-sized, 6–2 in. long,
　　　sessile, rarely stipitate, surrounded by normal foliage leaves ; flowers 1
　　　to over 2 in. long ; styles gently curved or almost straight, long-subulate
　　　from a linear-lanceolate base, passing gradually or rarely with a sudden
　　　bend into the subulate stigma. Main stem underground with leaf-tufts
　　　and flower-heads close to the ground.

Involucral bracts lanceolate, acutely acuminate, flat, whitish
　　woolly-tomentose, at length more or less glabrescent
　　and dark from the base upwards ; leaves attenuated
　　into long petioles :
　　Leaves obovate-lanceolate, including the petiole ¾–1¼ ft.
　　　long :
　　　Head 6–7 in. long ; flowers over 2 in. long　　...　　...　(63) **cryophila.**
　　　Head 2–3 in. long ; flowers 1–1½ in. long　　...　　...　(64) **Scolopendrium.**
　　Leaves linear-oblanceolate with wavy margins, including
　　　the petiole about ½ ft. long ; heads up to 2 in. long ;
　　　flowers 1 in. long　　...　　...　　...　　...　　...　(65) **scabriuscula.**
Involucral bracts from ovate to linear-oblong, the innermost
　　much elongated, acuminate or obtuse, convex on the
　　back, finely silky-pubescent when young, soon more or
　　less glabrescent and reddish :
　　Leaves linear up to 3 lin. broad, long attenuated at the
　　　base ; heads 3 in. long　　...　　...　　...　　...　(66) **aspera.**
　　Leaves stoutly filiform :
　　　Heads 4–5 in. long ; leaves glabrous :
　　　　Leaves up to 1 ft. long, smooth ; outer involucral
　　　　bracts acuminate　　...　　...　　...　　...　　...　(67) **lorea.**

Leaves up to ⅔ ft. long, often more or less rough;
    outer involucral bracts obtuse   ...   .   **(68) repens.**
Heads 2½–3 in. long; leaves with long soft hairs   ...   **(69) echinulata.**

§ 10. OBVALLATÆ. Heads medium-sized, 2–2½ in. long, sessile, surrounded by an outer involucre of subsessile ovate to obovate or oblanceolate
leaves, different from the petioled oblanceolate long foliage leaves; ovary
cylindric passing imperceptibly into the terete subulate and almost
straight style. Stigma subulate, hardly differentiated. Main stem underground with leaf-tufts and flower-heads close to the ground.

Only species ...   .   **(70) turbiniflora.**

§ 11. MICROGEANTHEÆ. Heads medium-sized to small, 2–1½ in. long,
sessile, rounded to turbinate at the base, surrounded by normal foliage
leaves; flowers ¾–1½ in. long; styles distinctly curved to sickle-shaped,
subulate from a (at least in old flowers) bulbously thickened base.
Stigma subulate, hardly differentiated. Main stem underground with
very short ascending or decumbent leaf- and flower-bearing branches.

Outer involucral bracts not produced into foliaceous
    appendages:
Leaves rough with small tubercles; inner involucral
    bracts pubescent :
Leaves lorate, long attenuated towards the base, 3–7
    lin. broad   ...   .   **(71) scabra.**
Leaves narrowly linear, 1–2 lin. broad ...   .   **(72) tenuifolia.**
Leaves smooth; involucral bracts glabrous, excepting on
    the sometimes ciliate margins :
Leaves obovate to oblanceolate, 1–3 in. broad; heads
    1–1½ in. long   ...   .   **(73) acaulis.**
Leaves narrowly oblanceolate to linear, long attenuated
    at the base :
Leaves oblanceolate, ¾–1 in. broad, distinctly veined;
    heads 2 in. long:
Stem glabrous; leaves dull; flowers 9 lin. long;
    perianth-lip 3 lin. long   ...   .   **(74) glaucophylla.**
Stem hairy; leaves more or less glossy; flowers
    1½ in. long; perianth-lip 6 lin. long   ...   **(75) Burchellii.**
Leaves narrowly oblanceolate-linear to linear with a
    long attenuated base or stoutly filiform-linear,
    usually obscurely or not at all veined:
Leaves narrowly oblanceolate-linear to linear, 6–1½
    lin. broad:
Leaves obscurely (the broadest sometimes more
    or less distinctly) veined, slightly glossy
    with cartilaginous margins; heads 1½ in.
    long; inner involucral bracts somewhat
    elongated   ...   .   **(76) angustata.**
Leaves not veined, dull, glaucous, without car-
    tilaginous margins; heads 1–1½ in. long;
    inner involucral bracts broad   ...   ...   **(77) lævis.**
Leaves stoutly filiform-linear, canaliculate, terete
    below, 1 lin. broad   ...   .   **(78) revoluta.**
Outer involucral bracts produced into foliaceous appendages
    resembling the foliage leaves; these linear to narrowly
    oblanceolate, 1¾–3 in. long, ¾–1¾ lin. broad ...   ...   **(79) montana.**

§ 12. HYPOCEPHALÆ. Heads crowded at the base of the shoots or
(80, *P. humiflora*) scattered along the branches, not terminal, 1–2 in.

long; involucral bracts gradually increasing upwards, shortly pubescent,
the inner equalling or exceeding the flowers, more or less spathulate, tips
ciliate or villous; flowers less than 1 in. long; perianth-lip 1½–2½ lin.
long, glabrous excepting at the tips; style as in the preceding section.
Small shrubs with erect or decumbent branches.

Leaves linear or acicular:
　Leaves linear, 1½–2 lin. broad, flat ...　　...　　... (80) **humiflora.**
　Leaves acicular, ⅓–⅔ lin. broad:
　　Leaves 1¼–2¼ in. long, decurrent ...　　...　　... (81) **decurrens.**
　　Leaves ½–1 in. long, not decurrent　　...　　... (82) **acerosa.**
　Leaves broad and cordate at the base:
　　Leaves very few, not crowded, broadly ovate or
　　　orbicular, 2–4½ by 1½–4½ in.　　...　　...　　... (83) **cordata.**
　　Leaves very numerous and crowded, ovate or ovate-
　　　lanceolate, 1–2¼ in. by ½–1½ in.　　...　　... (84) **amplexicaulis.**

1. **P. grandiceps** (Tratt. Thesaur. ed. i. (1805), 5 (?), t. 12); a
shrub, 4–5 ft. high; branches glabrous, rarely with some long hairs
when young; leaves elliptic, obtuse, slightly narrowed to and often
subcordate at the base, 3½–5 in. long, 1½–2 in. broad, coriaceous,
glaucous, with often red cartilaginous margins, prominently veined
above and beneath, glabrous or ciliate with long loose soft hairs
when young; heads sessile, 4½ in. long, 4–6 in. in diam.;
involucral bracts 8-seriate, more or less silky-pubescent, outer ovate-
oblong, obtuse, with glabrous or ciliate margins, inner oblong,
widened at the apex, more or less red or rose-purple with a
white beard 4–5 lin. long, exceeding the flowers; perianth-sheath
very slender, membranous, 2 in. long, gradually expanded and
faintly 7-nerved and 3-keeled below, rufously hirsute with the
exception of the glabrous base; lip 9 lin. long, upwards more or less
hirsute and ciliate with long pale hairs, 3-awned, awns glabrous
or nearly so, the lateral 3 lin. long, linear, obtuse, the median
1½ lin. long; stamens all fertile; anthers linear, 3 lin. long;
filaments ½ lin. long, flattened; apical glands oblong-lanceolate,
½ lin. long; ovary obovoid, covered with long reddish-brown hairs;
style tapering upwards, 2 in. long, laterally flattened, jointed on
the ovary, but not disarticulating, glabrous; stigma 3 lin. long,
linear, passing abruptly into the style. *P. speciosa, Gawl. Recens.* 8;
Sims, Bot. Mag. t. 1183; *Tratt. Thesaur. ed.* ii. 1819, 5, *t.* 12, *not of
Linn.; var. latifolia, Andr. Bot. Rep. t.* 110.　*P. coccinea, R. Br. in
Trans. Linn. Soc.* x. 77; *Roem. & Schult. Syst. Veg.* iii. 343; *Meisn.
in DC. Prodr.* xiv. 230.　*P. obtusa, Knight ex Sweet, Hort. Brit. ed.*
i. 346; *Loud. Encycl. Pl. ed.* i. 80; *Hort. Brit. ed.* i. 37.
P. rangiferina, Hort. teste Roem. & Schult. l.c. Mant. 263.　*P. villi-
fera, Lindl. Bot. Reg. t.* 1023.　*Erodendrum obtusum, Knight, Prot.* 38.

SOUTH AFRICA: without locality, *Boos! Forster!*
COAST REGION: Worcester Div.; Dutoits Kloof, *Drège!* Swellendam Div.;
Swellendam, *Zeyher!* Cape Div.; Devils Peak, *Niven! Ecklon,* 651! Table
Mountain, *Ecklon,* 651! Camps Bay, *Zeyher,* 4682!

The involucral bracts are described as scarlet in Salisbury's herbarium and
shown as rose-coloured or purple with white beards in the plates quoted; but

according to Knight the beard varies to reddish-purple. The *Botanical Magazine* plate (t. 1183) shows, moreover, a black line below the beard.

No copy of the first (incomplete) issue of Trattinick's Thesaurus, containing the name, description and figures of *P. grandiceps*, was accessible to us. But from the text to t. xii. and the preface of the second edition, it is perfectly clear that plate xii. was the same in both issues, whilst the name and the description were cancelled in the second owing to the identification of the plant with Andrews' t. 110, which had erroneously been referred to *P. speciosa*.

2. **P. speciosa** (Linn. Mant. alt. 191); a bush 3–4 ft. high; branches finely tomentellous when young, then glabrous, with reddish bark; leaves 4–4½ in. long, 1½–1¾ in. broad, obovate to oblong-obovate, obtuse, sometimes subapiculate, stoutly coriaceous, ciliate and villous at the base, usually at length quite glabrous; heads sessile, 5 in. long, 3–5 in. in diam.; involucral bracts 9–11-seriate; outer ovate-oblong, obtuse, silky pubescent, generally bearded at the apex, inner much elongated, oblong-spathulate, obtuse, whitish, silky pubescent or upwards tomentose with a dense reddish-brown beard, 4–5 lin. long; perianth-sheath very slender, membranous, 2¼ in. long, gradually dilated and faintly 7-nerved and 3-keeled below, densely pubescent above the dilated part; lip 1 in. long, densely pubescent, villosulous above, 3-awned, lateral awns 5 lin. long, linear, acuminate, tomentose with penicillate tips; median awn 1¼ lin. long; stamens all fertile; anthers linear, 3 lin. long; filaments ¾ lin. long, flattened concave; apical glands ¾ lin. long, linear-oblong; ovary 2 lin. long, oblong, covered with numerous long golden hairs; style 2½ in. long, gently curved, laterally flattened, somewhat swollen above the ovary, pubescent in the lower third; stigma 3 lin. long, linear, suddenly bent at the junction with the style. *Thunb. Diss. Prot.* 42, 56, *partly*; *Fl. Cap. ed. Schult.* 139; *R. Br. in. Trans. Linn. Soc.* x. 78 (*excl. the two last synonyms*); *Roem. & Schult. Syst. Veg.* iii. 343, *partly*; *Meisn. in DC. Prodr.* xiv. 231, *partly*; *var. obovata, Meisn. l.c.*; *Roupell, Cape Flow. t.* 6. *P. barbata, Lam. Ill.* i. 236. *P. Lepidocarpodendron, β, Linn. Syst. ed.* xiii. 118. *Leucadendron speciosum, Linn. Mant.* i. 36 (*excl. cit. Clus.*). *Erodendrum speciosum, Knight, Prot.* 39 *with plate. Lepicarpodendron folio oblongo, etc., Boerh. Ind. Pl. Hort. Lugd. Bat.* ii. 185, *t.* 185. *Scolymocephalus foliis longis, etc., Weinm. Phyt.* iv. 288, *t.* 893.

VAR. β, **angustata** (Meisn. in DC. Prodr. xiv. 231); leaves 3½–5¼ in. long, 5–12 lin. broad, gradually widening from the base upwards.

SOUTH AFRICA: without locality, *Niven*! *Oldenburg*, 612! *Nelson*! Var. β: *Zeyher*, 3666a partly!

COAST REGION: Clanwilliam Div.; Olifants River, near Brackfontein, *Zeyher*! Cape Div.; Simons Town, *Wright*, 639! Table Mountain, *Thunberg*! *Brown*! *Bolus*, 4485! Stellenbosch Div.; Hottentots Holland, *Roxburgh*, 38! Caledon Div.; Mountains near Hemel en Aarde, *Zeyher*, 3666a partly! Houw Hoek, *MacOwan in Herb. Austr.-Afr.* 1764 partly! Klein River, *Krauss*, 1042. Swellendam Div.; Between Sparrbosch and Tradouw, *Drège*! Var. β: Bredasdorp Div.; Koude River, *Schlechter*, 9610! Caledon Div.; Houw Hoek, *MacOwan in Herb. Austr.-Afr.* 1764, partly! Appels Kraal and River Zondereinde, *Zeyher*, 3666β! near Onrust River, *Zeyher*, 3666! Zwarteberg Kloof, *Ludwig*, 10!

3. **P. barbigera** (.Meisn. in DC. Prodr. xiv. 233) ; a large bush, 8–9 ft. high ; branches tomentose towards the heads ; leaves sessile, $3\frac{1}{4}$–$7\frac{1}{2}$ in. long, $\frac{3}{4}$–$1\frac{3}{4}$ in. broad, oblong-lanceolate, acute or subobtuse, with thickened margins and a prominent midrib, loosely and softly hairy, particularly towards the base, at length glabrous ; heads sessile, 5–6 in. long, about 6 in. wide ; involucral bracts 9–10-seriate, silky pubescent ; outer ovate-oblong, subacute to obtuse, bearded ; inner oblanceolate-oblong or the innermost narrowed downwards into a long narrow claw, subacuminate or cuspidate, with a whitish or (dry) fulvous beard, 2 lin. long, equalling the flowers ; perianth-sheath 2 in. long, very gradually dilated and 7-nerved and 3-keeled below, densely pubescent to villous excepting at the glabrous base ; lip $1\frac{1}{2}$ in. long, pubescent or villous on the sides, less so or glabrescent on the back, long-villous upwards, produced into 2 densely ciliate undulating awns, 1 in. long, cilia secund, up to 4 lin. long, passing into more or less purple woolly tufts upwards ; fertile stamens 3, $2\frac{1}{4}$ lin. long ; anthers linear ; apical glands $\frac{3}{4}$ lin. long, ovate-lanceolate, acuminate ; barren anther $2\frac{1}{2}$ lin. long, oblong, acuminate, acute, eglandular, with a filiform filament $\frac{1}{2}$ lin. long ; ovary 1 lin. long, obovoid, covered with long reddish hairs ; style $2\frac{1}{4}$ in. long, gently curved, laterally compressed below, more or less terete above, with a short ventral groove, ending in a projecting point above the ovary, pubescent ; stigma 3 lin. long, linear, acute, kneed at the junction with the style. *P. speciosa, Drège, partly, ex Meisn. l.c.* *P. macrophylla, Buek ex Meisn. l.c. Scolymocephalus africanus foliis angustis villosis, Weinm. Phyt. iv. t. 894 (here ?).*

SOUTH AFRICA : without locality, *Gueinzius* !
COAST REGION : Clanwilliam Div. ; Blue Berg, *Drège.* Worcester Div. ; Dutoits Kloof, *Drège* ; Matroos Berg, *Lamb in Herb. Bolus*, 9370 ! Tulbagh Div. ; Witzenberg, *Zeyher* ! Tulbagh Waterfall, *Phillips*, 527 ! Stellenbosch Div. ; Hottentots Holland Mountains near Lowrys Pass, *MacOwan, Herb. Austr.-Afr.* 1763 ! Swellendam Div. ; between Sparrbosch and Tradouw, *Drège* !

4. **P. marginata** (Thunb. in Hoffm. Phytogr. Blaett. i. 15) ; branches tomentellous, at length glabrescent ; leaves shortly petiolate, 3–6 in. long, $\frac{3}{4}$–$1\frac{1}{2}$ in. broad, lanceolate, acute or subacute, woolly towards the base, particularly on the midrib, soon more or less glabrous, distinctly veined, with thickened margins ; petiole tomentose or glabrous ; heads sessile, 5 in. long, about 4 in. wide ; involucral bracts 10–11-seriate ; outer ovate, subacuminate, obtuse, recurved, silky pubescent, at length partly glabrescent ; inner oblong or linear-oblong, obtuse to subacute or innermost long-produced, spathulate and clawed, silky-pubescent, long-bearded, beard blackish-purple, often mixed with white ; perianth-sheath 2 in. long, very gradually dilated and 7-nerved and 3-keeled at the glabrous base, pubescent to villous above ; lip 1 in. long, pubescent, more or less glabrescent on the back, 3-awned, lateral awns 5 lin. long, linear, acuminate, ciliate below the woolly tips ; cilia secund, up to $2\frac{1}{2}$ lin. long, purple ; median awn $4\frac{3}{4}$ lin. long, filiform ; fertile stamens 3 ; filaments $\frac{1}{2}$ lin. long, dilated concave ; anthers

2 o 2

linear, 3 lin. long; apical glands $\frac{1}{2}$ lin. long, ovate, acuminate, acute; barren stamen stalked, acute, eglandular; ovary 2 lin. long, oblong in outline, covered with long reddish-brown hairs; style 2 in. long, tapering, very gently curved, laterally compressed and slightly dilated above the ovary, pubescent; stigma $3\frac{1}{2}$ lin. long, linear, acuminate, obtuse, kneed and slightly bent at the junction with the style.

SOUTH AFRICA: without locality, *Thunberg! Banks! Drège! Labillardiere! Gueinzius,* 188!
COAST REGION : Clanwilliam Div. ; Olifants River, *Marloth,* 3204! Blue Berg, *Schlechter,* 8475! Tulbagh Div. ; Tulbagh, *Pappe! MacOwan!* Saron, *Schlechter.* 7881! Paarl Div. ; Drakenstein Mountains, near Bains Kloof. *Bolus,* 4066. Swellendam Div. ; Tradouw, *Drège,* 3360! Port Elizabeth Div. ; Algoa Bay (wrong locality?), *Cooper,* 3069!

This species differs from *P. neriifolia,* R. Br., by the broader leaves and shorter awns.

5. **P. incompta** (R. Br. in Trans. Linn. Soc. x. 83); an erect shrub, 3–4 ft. high; branches densely pilose with long soft hairs, at length glabrescent; leaves 2–4 in. long, $\frac{1}{2}$–$1\frac{1}{4}$ in. broad, lanceolate or ovate-lanceolate, acute, rounded at the base, reticulately veined, with a strong midrib, younger softly pilose to hirsute, particularly those near the head, at length glabrescent; heads sessile, $2\frac{1}{2}$–$3\frac{1}{2}$ in. long, about $3\frac{1}{4}$ in. wide, surrounded by leaves, the innermost of which are narrower to linear, gradually shorter and very hirsute, often widened and coriaceous at the base, gradually passing into appendaged bracts; involucral bracts 5–6-seriate, outer oblong, obtuse, green, glabrous; inner oblong, widened above, incurved, shortly white-bearded, slightly exceeding the flowers; perianth-sheath $1\frac{3}{4}$ in. long, loosely hairy above, glabrous below, faintly 5-nerved, gradually dilated and faintly 3-keeled at the base; lip villous on the sides, 3-awned, lateral awns 5 lin. long, linear. flexuous, loosely long-villous with white hairs; stamens all fertile anthers linear, 3 lin. long; apical glands $\frac{1}{2}$ lin. long, oblong, obtuse. filaments $\frac{3}{4}$ lin. long, swollen; ovary $1\frac{1}{4}$ lin. long, ellipsoid, densely covered with long reddish-brown hairs; style $2\frac{1}{4}$ in. long, tapering and almost straight above, laterally compressed below, ventrally furrowed, glabrous; stigma 3 lin. long, subulate, gradually passing into the style. *Roem. & Schult. Syst. Veg.* iii. 347 ; *Krauss in Flora,* 1845, 75 (*by error incorrupta*) ; *Krauss in Beitr. Fl. Cap- und Natall.* 139 ; *Meisn. in DC. Prodr.* xiv. 234. *P. macrocephala, Thunb. in Hoffm. Phytogr. Blaett.* i. 13 (?). *Erodendrum incomptum, Knight. Prot.* 37. *Scolymocephalus seu Lepidocarpodendron,* etc., *Weinm. Phyt.* iv. *t.* 898.

VAR. β, **Susannæ** (E. P. Phillips) ; leaves with a silky silvery tomentum branches densely silky above.

SOUTH AFRICA: without locality, *Drège! Sieber* 1! *Boos! Scholl! Thunberg Oldenburg,* 61A! *Bergius! Ludwig! Ecklon,* 331!
COAST REGION: Cape Div. ; Devils Peak, *Ecklon,* 652! *Bolus,* 4575! *Wilms.* 3566! *MacOwan, Herb. Norm. Austr.-Afr.* 785! Lion Mountain, *Drège!*

Pappe, 14 ! Table Mountain, *Krauss* ! Between Rondebosch and Wynberg, *Burchell*, 772 ! Wynberg, *Roxburgh* ! near Cape Town, *Fry in Herb. Galpin*, 5019 ! Caledon Div. ; Little Houw Hoek, *Zeyher*, 3658 ! Var. β : Riversdale Div. ; Mountains above Platte Kloof, *Muir*, 390 ; George Div. ; Barbiers Kloof, near George, *Bowie* !

If *Protea macrocephala*, Thunb., the original of which we have not seen, should really prove to be identical with R. Brown's *P. incompta*, as is very probable, the former name would have to stand.

6. P. comigera (Stapf) ; an undershrub, about 3 ft. high, of bushy growth ; branches glabrous, purplish ; leaves lanceolate or oblanceolate, narrowed towards the base, more or less undulate, acute or subacute, up to 4 in. long and $\frac{2}{3}$–$\frac{4}{5}$ in. broad, coriaceous with thickened pinkish margins, glabrous, with a prominent midrib and obscure venation ; heads sessile, oblong-turbinate, 4 in. long, 3 in. across ; involucral bracts 7–8-seriate, outermost green with black tips, the following pinkish upwards with black tips, the inner and innermost yellowish on the back with pinkish sides, ciliate and the innermost crowned with a large round tuft of long black hairs, all more or less lanceolate and acute, the inner much elongated ; perianth-sheath 1½–1¾ in. long, purplish, pubescent and villous upwards ; lip 4–5 lin. long, awned, awns about ½ in. long, purplish-villous ; anthers 3 lin. long with a red apical gland ; style 1¾ in. long, purple upwards, glabrous ; stigma 4 lin. long, geniculate at the junction with the style. *P. grandiflora, var. foliis undulatis, Andr. Bot. Rep. t.* 301.

SOUTH AFRICA : without locality, *Niven.*

Only known from the figure and description quoted, which were made from a plant introduced into Hibbert's garden by Niven in 1800.

7. P. Lepidocarpodendron (Linn. Mant. alt. 190) ; a large bush, 6–8 ft. high ; branches glabrous (in the type) ; leaves sessile, 3¼–4 in. long, 4–8 lin. broad, linear-lanceolate, subobtuse, glabrous, with a prominent midrib and thickened pink margins ; heads sessile, 3½–4 in. long, about 2½ in. wide ; involucral bracts 8-seriate ; outer ovate-oblong, brown, silky-pubescent, with a white fringe of cilia and sometimes a white apical beard, inner oblong to linear, pubescent, green, innermost spathulate, clawed, with a black densely tomentose limb surrounded by a white or black beard, exceeding the flowers ; perianth-sheath 2 in. long, slightly and very gradually dilated and 7-nerved and 3-keeled below ; fulvously or rufously villous excepting at the very base which is glabrous ; lip 7 lin. long, 3-awned, lateral awns 5 lin. long, filiform or capillary, long ciliate, cilia secund, purplish or fulvous ; median awn 2 lin. long, filiform ; fertile stamens 3 ; anthers 3 lin. long, linear, apical glands ½ lin. long, ovate, acuminate ; filaments ¼ lin. long, flattened ; barren stamen 2 lin. long, acute, eglandular, with a filiform filament, 1 lin. long ; ovary densely covered with long reddish-brown hairs ; style 2 in. long, almost straight, tapering upwards, laterally compressed, pubescent, with a short ventral

groove (gland?) at the base, ending below a small knob; stigma
3 lin. long, linear, acute, kneed at the junction with the style.
P. speciosa, Thunb. Diss. Prot. 42, 56, *partly*; *Willd. Sp. Pl.* i. 531,
partly; *Thunb. Fl. Cap. ed Schult.* 139, *partly.* *P. cristata, Lam. Ill.*
i. 235; *Poir. Encycl.* v. 644. *P. scabrida, Thunb. in Hoffm. Phytogr.*
Blaett. i. (1803), 14. *P. lepidocarpon, R. Br. in Trans. Linn.*
Soc. x. 80; *Roem. & Schult. Syst. Veg.* iii. 346; *Meisn. in DC.*
Prodr. xiv. 232. *Leucadendron Lepidocarpodendron, Linn. Sp. Pl.*
ed. i. 91. *Lepidocarpodendron foliis angustis, etc., Boerh. Ind. Pl.*
Hort. Ludg. Bat. ii. 188. *Scolymocephalus africanus, etc., Weinm.*
Phyt. iv. 289, *t.* 895.

VAR. β, villosa (E. P. Phillips); branches pilose; leaves oblong-linear, acute,
ciliate, younger pilose, at length glabrous. *P. melaleuca, R. Br. in Trans. Linn.*
Soc. x. 79; *Tratt. Thes. t.* 11; *Meisn. in DC. Prodr.* xiv. 232. *P. speciosa, var.*
nigra, Andr. Bot. Rep. t. 103. *P. lepidocarpon, Sims, Bot. Mag. t.* 674.
P. villosa, Hort. Bollw. ex Meisn. l.c. 233. *P. glauca, d, Drège ex Meisn. l.c.*
P. nigrita, DC. Herb. ex Meisn. l.c. *Erodendrum neriifolium, Knight, Prot.* 40.
Scolymocephalus melaleucus, O. Kuntze, Rev. Gen. Pl. iii. 280.

SOUTH AFRICA: without locality, *Ludwig*! *Sparmann*! *Forokier*! *Thunberg*!
Banks! *Boos*! *Sieber*! *Oldenburg*, 374! *Bergius*! *Scholl*! Var. β, *Martin*!
Brown.

COAST REGION: Clanwilliam Div.; Blue Berg, *Schlechter*, 8475! Cape Div.;
mountains near Cape Town and on the Cape Peninsula, *Burchell*, 402! *Scholl*!
Osbeck! *Zeyher*, 3657! *Ecklon*, 653! *Krauss*! *Drège*; *Jameson*! *Bolus*, 4026!
Wilms, 3567! *MacOwan, Herb. Norm. Austr.-Afr.* 784! *Phillips*! Var. β: Cape
Div.; Lion Mountain, *Bolus*, 2901! Swellendam Div.; between Sparrbosch and
Tradouw, *Drège*. Port Elizabeth Div.; Van Stadensberg, *Drège*. Caledon Div.;
Genadendal, *Drège*!

We have seen Martin's specimen, preserved in the Herb. Mus. Palat. Vindob.,
and it undoubtedly agrees with Brown's description of *P. melaleuca* and also
with the post-linnean figures which he cites. We cannot regard this plant as
specifically distinct from *P. Lepidocarpodendron,* Linn., but the hairy branches,
and the hairy and acute leaves warrant its being kept separate as a variety.

8. **P. neriifolia** (R. Br. in Trans. Linn. Soc. x. 81); an erect
shrub, 3–5 ft. high; branches tomentellous; leaves 3–5¾ in. long,
4–11 lin. broad, linear to linear-oblong, subobtuse, penninerved
with a conspicuous midrib, glabrous or woolly at the base; head
sessile, 4½–5 in. long, about 3 in. in diam.; involucral bracts
11-seriate; outer squarrose, recurved, densely silky-pubescent or
glabrescent; inner oblong or linear-oblong, densely silky-pubescent
or tomentose above, innermost with an oblanceolate limb and a
slender claw bearing a purplish-black beard up to 2½ lin. long,
exceeding the flowers; perianth-sheath 2½ in. long, base dilated,
glabrous, faintly 5–7-nerved and 3-keeled, otherwise loosely pubescent to densely villous; lip 1¼ in. long, 3-awned, villous along the
sides, glabrous on the back, long and densely ciliate above; lateral
awns 9 lin. long, densely ciliate, cilia secund and up to 3 lin. long,
purple or fulvous; median awn 3½ lin. long, filiform; fertile
stamens 3; anthers linear, 3 lin. long, apical glands ½ lin. long,
ovate, acuminate, acute, keeled on the inner face; filaments 2 lin.
long, dilated, concave; barren anther 2 lin. long, linear, acute,

eglandular, with a filament 2 lin. long ; ovary oblong, 3 lin. long,
covered with long reddish-brown hairs ; style 2½ in. long, tapering
upwards, compressed up to the middle, then terete, shortly villous ;
stigma 3 lin. long, linear, obtuse, distinctly kneed and bent
at the junction with the style. *Bot. Reg. t.* 208; *Meisn. in
DC. Prodr.* xiv. 233 (*incl. var. glauca*). *P. pulchella, Bot. Reg.
t.* 20 ; *Reichenb. Fl. Exot.* iv. *t.* 218; *Geel, Sert. Bot., not of
Andr. P. glauca, Buek ex Meisn. l.c. Cardui generis elegantissimi,
etc., Clus. Exot.* 38, *fig.* 15 *on p.* 39.

SOUTH AFRICA : without locality, *Pillans in Herb. Bolus,* 12535 !
COAST REGION : Tulbagh Div. ; Roode Zand, *Bowie!* De Liefde, *Drège!*
Worcester Div. ; Dutoits Kloof and Hex River, *Drège!* Bains Kloof, *Bolus,*
4066 ! Paarl Div. ; Paarl Mountain, *Drège* ! Cape Div. ; Table Mountain, *Brown* !
Zeyher, 3657 ! Constantia, *Bowie* ! near Cape Town, *Ludwig,* 10 ! Lions Head,
MacOwan in Herb. Norm. Austr.-Afr. 784 partly ! Below Platteklip, *Phillips* !
Caledon Div. ; Genadendal, *Drège* ! *MacOwan* ! Swellendam Div. ; Swellendam,
Bowie ! *Fry in Herb. Galpin* 4984 ! Riversdale Div. ; Garcias Pass, *Phillips,* 504 !
510 ! 514 ! Knysna Div. ; Between Plettenbergs Bay and Knysna, *Burchell,*
5350 ! Uitenhage Div. ; Vanstadens Berg, *Zeyher,* 3659 !

The involucral bracts are white or pink and the inner capped with a dense
beard of purple-black hairs, sometimes interspersed with white. The awns are
of a light tawny colour. This plant was distributed by MacOwan in the Herb.
Norm. Austr.-Afr. as *P. lepidocarpon,* R. Br., but can be at once distinguished
from this species by the inner involucral bracts being black-bearded at the apex
and not black-tomentose on the back as in *P. Lepidocarpodendron,* Linn.
(*P. lepidocarpon,* R. Br.).

9. **P. pulchella** (Andr. Bot. Rep. t. 270) ; a stout shrub, 2–5 ft.
high ; branches tomentellous, at length glabrous ; leaves 1½–6½ in.
long, 3–7 lin. broad, linear to linear-lanceolate, acute, narrowed at
the base, with a prominent midrib, pubescent when young, sometimes ciliate, at length glabrous ; head sessile, 3½–4 in. long, about
2–4 in. in diam., surrounded by the upper foliage leaves ; involucral
bracts 10–12-seriate ; the lowermost tomentose or pubescent ; outer
ovate, subacuminate, obtuse or subobtuse, green with reddish tips,
glabrous or ciliate, inner oblong to linear-oblanceolate or broad-
linear, glabrous or pubescent below the rose- or carmine-purple to
black beard, exceeding the flowers ; perianth-sheath 2 in. long,
glabrous, whitish pubescent to villous above the gradually dilated
5-nerved and 3-keeled base ; lip 6 lin. long, pubescent on the sides,
more or less glabrous on the back, 3-awned, lateral awns 3 lin. long,
filiform, villous, intermixed with long purple to black hairs ; median
awn 1 lin. long ; fertile stamens 3 ; anthers linear, 2 lin. long ;
apical glands ½ lin. long, lanceolate, subacute ; filaments ¼ lin. long,
dilated, concave ; barren anther acute, eglandular ; ovary 1½ lin.
long, oblong, covered with long reddish-brown hairs ; style 2 in.
long, very gently curved, compressed below, terete and ventrally
grooved above, pubescent ; stigma 3 lin. long, linear, acute,
slightly kneed at the junction with the style. *R. Br. in Trans.
Linn. Soc.* x. 81; *Meisn. in DC. Prodr.* xiv. 233 ; *var. speciosa,
Andr. Bot. Rep. t.* 442. *P. speciosa, var. foliis glabris, Andr. l.c.
t.* 277 ; *var. rosea, Wendl. Coll. t.* 73.

VAR. β, **undulata** (E. P. Phillips); leaves with undulating margins. *Eroden-drum pulchellum, Knight, Prot.* 36.

SOUTH AFRICA: without locality, *Drèje! Ludwig! Martin! Boos!*
COAST REGION: Paarl Div.; Paarl Mountain, *Drège,* 1459! 2399! Tulbagh
Div.; Saron, *Schlechter,* 7869! Cape Div.; Kraaifontein, *Dümmer,* 1531! Lions
Head, *MacOwan,* 2908! *MacOwan, Herb. Austr.-Afr.* 1520! *Marloth,* 3328!
Table Mountain and Devils Peak, *Drège!* Stellenbosch Div.; Stellenbosch, *Rox-burgh! Marloth,* 3228! Between Lowrys Pass and Jonkers Hoek, *Burchell,* 8330!
Var. β :, Stellenbosch Div.; Stellenbosch, *Niven!*

This species has been confounded in Herbaria with *P. calocephala,* Meisn., and
MacOwan, in Herb. Austr. Afr. 1520, distributed specimens of it as such. We
have seen Meisner's type of *P. calocephala,* which is at once distinguished from
P. pulchella, Andr., in having broader leaves, and the inner involucral bracts not
bearded at their apices.

10. **P. fulva** (Tausch in Flora, 1842, i. 285); branches tomentose;
leaves 3½–4 in. long, 5–8 lin. broad, oblong-linear, subacute, narrowed at the base into a very short petiole, distinctly veined, with
a prominent midrib, densely tomentose at the base, otherwise
glabrous, ciliate when young; head sessile, 3½ in. long, about 4 in.
in diam.; involucral bracts 13-seriate; outer ovate, subacuminate,
obtuse, pubescent or hirsute; inner from oblong, acuminate, obtuse
to linear or linear-oblong, densely villous, rufo-ciliate at the apex,
equalling the flowers ; perianth-sheath 1¾ in. long, dilated, 3-keeled
and 3-nerved below, densely pilose, becoming glabrous at the base ;
lip 8 lin. long, 3-awned, pilose; lateral awns 4 lin. long, ciliate
with long brown hairs; median awn 2 lin. long, filiform; fertile
stamens 3; anthers linear, 2½ lin. long, subsessile; apical glands
¾ lin. long, ovate, acuminate, acute; barren stamen acute,
eglandular; ovary 2½ lin. long, oblong, covered with long reddish-
brown hairs; style 2 in. long, tapering above, kneed at the base,
pubescent, becoming villous below; stigma 2¾ lin. long, furrowed
subacute.

SOUTH AFRICA: Raised from seeds sent by Ecklon from the Cape and
flowered in the gardens of *Prince Salm-Dyck* in 1838!

11. **P. patens** (R. Br. in Trans. Linn. Soc. x. 82); a procumbent
shrub; branches densely tomentose; leaves sessile, oblong to
oblong-linear, subacute, with undulating thickened pink margins,
3–4½ in. long, 6–9 lin. broad, prominently ribbed, loosely villous;
heads sessile, 3 in. long, about 2½ in. in diam.; involucral bracts
12-seriate, outer ovate, subacute, silky-tomentose, inner ovate to
oblong, silky-pubescent, equalling the flowers, all bracts fringed
with a short dense beard, which is fulvous in the lower, dark purple
in the upper bracts; perianth-sheath 1¼ in. long, gradually dilated
and faintly 5–7-nerved and 3-keeled below, densely villous excepting
at the glabrous base; lip 9 lin. long, 3-awned, pubescent on the
sides, glabrous on the back; lateral awns 3–7 lin. long, linear,
very densely villous, the apical hairs black; median awn ½ lin.
long, filiform with an apical tuft of black hairs; stamens all fertile,
3 lin. long; anthers linear, 3½ lin. long, apical glands ½ lin. long,

ovate-lanceolate, with a swelling on the inner face; filaments almost
1 lin. long, widened upwards; ovary small, densely covered with
long reddish-brown hairs; style 1¼ in. long, curved, tapering
upwards, laterally compressed, somewhat swollen above the ovary,
terete upwards, red, pubescent; stigma 3 lin. long, linear, obtuse,
conspicuously kneed at the junction with the style.   *Roem. &
Schult. Syst. Veg.* iii. 347; *Meisn. in DC. Prodr.* xiv. 232.   *P.
speciosa, var. patens, Andr. Bot. Rep. t.* 543.   *Erodendrum holoseri-
ceum, Knight, Prot.* 40.

COAST REGION: Div.? Wilde River, *Niven*!

12. P. latifolia (R. Br. in Trans. Linn. Soc. x. 75); a bush
5–8 ft. high; branches more or less tomentose, at length glabrous;
leaves 2¼–4 in. long, 1½–2¼ in. broad, very crowded, oblong-elliptic
to elliptic, obtuse, cordate at the base, subglaucous, woolly-ciliate
when young, at length glabrous, prominently veined; head sessile,
4¾ in. long, about 2¾ in. in diam.; involucral bracts 9–12-seriate;
outer ovate, obtuse, adpressedly pubescent to tomentose, ciliate;
inner flesh-colour to carmine, with an oblong limb and a linear
claw, densely pubescent and fringed above with white cilia, exceeding the flowers; perianth-sheath 2¼ in. long, dilated, 5–7-nerved
and 3-keeled below, densely pubescent excepting at the glabrous
base; lip 1 in. long, villous, 3-awned; lateral awns 10 lin. long,
terete, tomentose, indumentum intermixed with long more or less
dark-purple hairs, tips penicillate; median awn 5 lin. long, filiform;
fertile stamens 3; filaments ½ lin. long, dilated; anthers linear,
2½ lin. long; apical glands ¾ lin. long, lanceolate, acuminate;
barren stamen eglandular; ovary covered with numerous long
reddish hairs; style 2⅓ in. long, flattened, finely pubescent almost
all along; stigma 3 lin. long, obovoid, covered with numerous
long reddish hairs. *Bot. Mag. t.* 1717; *Roem. & Schult. Syst. Veg.*
iii. 342; *Meisn. in DC. Prodr.* xiv. 230. *P. radiata, Andr. Bot.
Rep. t.* 646; *Bonpl. Malm.* 144, *t.* 59; *Geel, Sert. Bot.*; *Reichenb.
Fl. Exot.* iv. 35, *t.* 273. *Erodendrum eximium, Knight, Prot.* 41.
P. spectabilis, Lichtenst. ex Spreng. Syst. Veg.. i. 461; *Meisn. l.c.*
231. *Mimetes spectabilis, Roem. & Schult.* iii. *Mant.* 268; *Steud.
Nomencl. ed.* ii. 147.

VAR. β, auriculata (Phillips); leaves 3–3½ in. long, 1–1¾ in. broad, oblong-
elliptic, obtuse, distinctly veined; style strongly compressed and very much
dilated below. *P. auriculata, Tausch in Flora,* 1842, i. 285.

SOUTH AFRICA: without locality, *Roxburgh*!
COAST REGION: Caledon Div.; Houw Hoek, *Bowie*! Zwart Berg, *Niven*!
Riversdale Div.; Garcias Pass, *Galpin*, 4453! *Bolus,* 11367! *Phillips,* 513!
Humansdorp Div.; Kromme River, *Drège.* Uitenhage Div.; Van Stadens Berg,
Drège, 3033! *Zeyher,* 384! 3656! *MacOwan*! Var. β: specimen cultivated in
the gardens of *Prince Salm Dyck* in 1838!

The involucral bracts are a pale pink and the awns of the lip covered with
purple hairs. Andrews, however, also figures a perfectly green variety. This
species approaches *P. compacta,* R. Br., but can be readily distinguished from it
by the pubescent style.

The texture of the leaves of var. *auriculata* differs considerably from that of typical *P. latifolia* which moreover is easily distinguished from the type by the considerable dilatation of the style base.   *Erodendrum coronarium*, Knight, Prot. 41, collected by Masson near Great Hout Hoek, and introduced into Kew in 1790, must come very near this species.

13. **P. compacta** (R. Br. in Trans. Linn. Soc. x. 76); branches finely tomentellous, at length glabrous; leaves $3\frac{1}{4}$–$4\frac{1}{4}$ in. long, $\frac{3}{4}$–$1\frac{1}{2}$ in. broad, strongly imbricate, ovate to ovate-lanceolate or elliptic-oblong, obtuse with a callous point, subcordate or rounded or slightly narrowed at the base, coriaceous, prominently veined, glabrous with the margins shortly villous or at length glabrous; head sessile, 4 in. long, about $2\frac{1}{2}$ in. in diam.; involucral bracts 8-seriate; outer ovate, obtuse, villous-pubescent or more or less glabrescent, with a dense fringe of woolly cilia; inner more or less flesh-colour to carmine with an oblong limb and linear claw, finely villous-tomentose, tips densely ciliate, exceeding the flowers; perianth-sheath 2 in. long, dilated, 5-nerved and 3-keeled below, finely tomentose, glabrous at the base; lip over 1 in. long, 3-awned; lateral awns 3–$4\frac{1}{2}$ lin. long, filiform, flexuous, tawny to purplish-tomentose; median awn 1 lin. long, filiform; fertile stamens 3, subsessile; filaments $\frac{3}{4}$ lin. long, flattened; anthers linear, $4\frac{1}{2}$ lin. long; apical gland almost 1 lin. long, lanceolate-oblong; barren stamen $4\frac{1}{2}$ lin. long, linear, eglandular; ovary 1 lin. long, oblong, densely covered with long light-golden hairs; style 2 in. long, finely grooved on the convex side, glabrous; stigma $2\frac{1}{2}$ lin. long, linear, obtuse, strongly kneed and bent at the junction with the style.   *Meisn. in DC. Prodr.* xiv. 238.   *P. formosa, R. Br. in Trans. Linn. Soc.* x. 79.   *P. coronata, Andr. Rep. t.* 469, *not of Lam. P. spectabilis, Willd. ex Meisn. l.c.   Erodendrum formosum, Salisb. Parad. t.* 76.   *Leucospermum* (*errore pro Erodendro*) *formosum, Knight ex Loud. Encycl.* i. 82.

COAST REGION : Cape Div. ; Kraaifontein, *Dümmer,* 1531 ! Stellenbosch Div. ; Hottentots Holland, *Niven* ! Caledon Div. ; Houw Hoek, *Masson* ! *Zeyher,* 3660 ! Klein River, *Krauss,* 1073 ! Hermanus, *Galpin,* 4454 ! Bredasdorp Div. ; Elands Kloof, *Ludwig* ! near Elim, *Zeyher,* 1820 ! *Bolus,* 8592 ! *Schlechter,* 7673 ! Koude River, *Schlechter,* 10457 ! Uitenhage Div. ; Van Stadens Berg, *Zeyher,* 3656 !

14. **P. magnifica** (Link, Enum. alt. i. 113); a shrub with stout villous stems; leaves broad- to narrow-oblong or lanceolate, shortly acute, somewhat narrowed at the base or subcordate, undulate, the largest up to 4 in. long and 2 in. wide, coriaceous, subglaucous, hairy along the thickened margins, distinctly veined; heads sessile, turbinate-obovoid, 6 in. long, 5 in. wide; involucral bracts 7–8-seriate, creamy-silky-pubescent to tomentose, the lower ovate to ovate-lanceolate, densely white ciliate, the inner much elongated, linear-lanceolate, acute, with a dense fringe of white cilia, equalling or slightly exceeding the creamy flowers; perianth-sheath $2\frac{1}{4}$ in. long, hairy; lip 7–8 lin. long, hairy, awned, lateral awns 9 lin. long, long-hairy; stamens all fertile; anthers yellow, 4 lin. long; apical gland red, 1 lin. long; ovary villous; style rather stout, whitish;

stigma suddenly geniculate at the base. *P. speciosa, Andr. Bot. Rep. t.* 438, *not of Linn. Erodendrum magnificum, Knight, Prot.* 37.

SOUTH AFRICA : Figured from a specimen in Hibbert's collection.

Andrews' statement that this is identical with a plant collected by Roxburgh in the Hottentot Mountains and named *P. speciosa* in the Lambert Herbarium is incorrect. Roxburgh's plant is typical *P. speciosa,* Linn.

15. **P. macrophylla** (R. Br. in Trans. Linn. Soc. x. 78) ; a small tree 5–10 ft. high ; branches minutely greyish-tomentellous above, at length glabrous ; leaves 3¾–9 in. long, ¾–1¾ in. broad, linear-oblong or oblong, obtuse, narrowed at the base, finely villosulous or woolly, particularly towards the base, at length glabrous, reticulately veined, with thickened margins ; head sessile, 4–5¼ in. long, about 3½ in. in diam. ; involucral bracts 7–10-seriate, whitish or tawny tomentose to tomentellous ; outer ovate, obtuse ; inner with an oblong limb passing into a long linear claw, minutely ciliate round the apex, mostly exceeding or at least equalling the flowers ; perianth-sheath 2¼–2½ in. long, dilated, 5–7-nerved and 3-keeled below, densely pubescent excepting at the glabrous base ; lip 1⅛ in. long, 3-awned, flexuous, villous or tomentose ; lateral awns 8–9 lin. long, densely villous ; median awn 4 lin. long, filiform ; fertile stamens 3 ; anthers linear, 2–2½ lin. long : apical glands ½ lin. long, ovate, acuminate, acute ; filaments ½ lin. long, flattened concave ; barren stamen linear, acute, eglandular ; ovary 1 lin. long, obovate, densely covered with long brown hairs ; style 2 in. long, laterally compressed, with a short ventral groove (gland ?) above the ovary, scantily pubescent below the middle ; stigma 2½–3 lin. long, slender, linear, obtuse, slightly kneed and bent at the junction with the style. *Roem. & Schult. Syst. Veg.* iii. 344 ; *Meisn. in DC. Prodr.* xiv. 234, *incl. vars. dregeana and zeyheriana. Erodendrum lorifolium, Knight, Prot.* 41.

COAST REGION : Clanwilliam Div. ; Olifants River, near Brackfontein, *Zeyher* ! Worcester Div. ; Dutoits Kloof, *Drège* ! Ladismith Div. ; Zwartberg Range, near Ladismith, *Marloth,* 3203 ! Mossel Bay Div. ; Attaquas Kloof, *Niven,* 37 ! Uitenhage Div. ; " De Hoek," *Geard in Herb. Galpin,* 2981 ! Albany Div. ; Grahamstown, *Misses Daly & Sole,* 455 !

CENTRAL REGION : Somerset Div. ; Bosch Berg, *MacOwan,* 848 !

16. **P. obtusifolia** (Buek in Drège, Zwei Pfl. Documente, 123, 213, name only) ; branches tomentose above ; leaves 3½–4¾ in. long, 1–1½ in. broad, oblong-elliptic, obtuse, emarginate, attenuate at the base into a petiole, distinctly veined, glabrous and vernicose, ciliate below ; head sessile, 3¾ in. long, about 2½ in. in diam. ; involucral bracts 13-seriate ; outer ovate, acuminate, obtuse, pubescent, at length glabrous, ciliate ; inner oblong or spathulate-linear, glabrous, ciliate, slightly exceeding the flowers ; perianth-sheath 2 in. long, dilated, 3-keeled and 7-nerved below, shortly tomentose or villosulous ; lip 9–10 lin. long, 3-awned, villosulous ; lateral awns 3–4½ lin. long, terete, villous, tomentose with longer hairs intermixed ; median

awn 1½ lin. long, filiform; filaments ½ lin. long, swollen; anthers linear, 2 lin. long; apical glands ¾ lin. long, oblong, obtuse; ovary covered with long reddish-brown hairs; style 2 in. long, flattened, strongly keeled above the ovary, pubescent below and along the keel up to the middle, glabrous upwards; stigma 2½ lin. long, furrowed, obtuse, tapering into the style.   *Meisn. in DC. Prodr.* xiv. 235.

COAST REGION : Caledon Div. ; Onrust River, *Zeyher*, 3661 ! Bredasdorp Div. ; between Cape Agulhas and Pot Berg, *Drège* !

Very closely allied to *P. calocephala*, Meisn., but differing in having distinctly veined leaves, the inner involucral bracts only just exceeding the flowers, besides small differences in the floral characters.

17. **P. triandra** (Schlechter in Engl. Jahrb. xxvii. 110); a shrub 1½–2 ft. high ; branches tomentellous ; leaves 2¾–4½ in. long, ½–1¼ in. broad, oblong-lanceolate, obtuse, callously mucronate, pinnately veined, with a prominent midrib, glabrous excepting on the thickened woolly-ciliate margins ; head sessile, 4–4½ in. long, 3–3¼ in. in diam. ; receptacle slightly convex ; paleæ ovate, acute ; involucral bracts 12–14-seriate ; outer ovate, obtuse, glabrous or pubescent below, ciliate, with a small beard of cilia at the apex ; inner oblong-linear, widening above, tomentose, ciliate, rose-coloured, exceeding the flowers ; perianth-sheath 1¾ in. long, dilated, 3-keeled and 5-nerved below, shortly tomentose, becoming glabrous at the base ; lip 12 lin. long, 3-awned, shortly tomentose ; lateral awns 4–5 lin. long, terete ; median awn 2½ lin. long ; fertile stamens 3 ; filaments ¾ lin. long, swollen, channelled ; anthers linear, 4 lin. long ; apical glands ⅚ lin. long, lanceolate, obtuse ; ovary 1 lin. long, pubescent, covered with long golden hairs ; style 2 in. long, furrowed above, compressed below, pubescent at the very base, otherwise glabrous ; stigma 5 lin. long, trigonous, with a blunt ovate swelling at the apex, kneed and strongly curved at the junction with the style.

COAST REGION : Bredasdorp Div. ; Koude River, *Schlechter*, 9621 !

18. **P. Susannæ** (Phillips in Kew Bulletin, 1910, 229) ; branches tomentellous ; leaves 1½–5 in. long, 5–10 lin. broad, lanceolate or oblong-lanceolate, obtuse, narrowed at the base, with a distinct mid-rib and more prominent venation on the under surface, pubescent or tomentose when young, becoming glabrous with age ; head subsessile, 3¾ in. long, about 3 in. in diam. ; involucral bracts 9-seriate ; outer ovate, obtuse, pubescent to tomentose, at length glabrescent ; inner oblong, concave, tomentose below, subequalling the flowers ; perianth-sheath 2–2¼ in. long, dilated, 3-keeled and 7-nerved below, densely tomentose above the glabrous base, 9 lin. long, 3-awned above, tomentose ; lateral awns 2½–3 lin. long, terete, purple-villous ; median awn 2 lin. long, filiform ; fertile stamens 3 ; filaments ½ lin. long, dilated, concave ; anthers linear, 2½ lin. long ; apical glands ¾ lin. long, lanceolate, obtuse ; barren stamen linear,

acute, eglandular; filament filiform; ovary ½ lin. long, obovate-oblong in outline, covered with long light yellow hairs with golden-coloured tips; style 2½ in. long, tapering above, furrowed, gradually thickened downwards with a pubescent ventral keel and a pubescent base; stigma 2½–3 lin. long, oblong-linear, trigonous with acute angles, subacute, swollen and recurved at the apex, conspicuously sinuate at the junction with the style.

COAST REGION : Bredasdorp Div.; Elim, *Schlechter*, 7718 ! Riversdale Div. ; Rivers lale, *Muir in Herb. Galpin*, 5305 !

19. **P. calocephala** (Meisn. in DC. Prodr. xiv. 239); branches tomentose above; leaves 2½–4½ in. long, 8–10 lin. broad, oblanceo-late, acute or obtuse, long attenuated at the base, indistinctly veined, the young leaves densely villous, becoming glabrous with age; head sessile, 4 in. long, about 4 in. in diam.; involucral scales 13-seriate ; outer ovate, acuminate, obtuse, glabrous or the lowermost pubescent, ciliate ; inner with a lanceolate to oblong-lanceolate limb, narrowed into a slender linear claw, ciliate, the innermost pubescent, greatly exceeding the flowers ; perianth-sheath 1¾ in. long, dilated, 3-keeled and 7-nerved below, shortly tomentose, glabrous at the base ; lip 8 lin. long, 3-awned, shortly tomentose ; lateral awns 3½–4 lin. long, terete, shortly tomentose with long hairs intermixed and an apical tuft ; median awn 1¼ lin. long, filiform ; anthers linear, 2 lin. long, apical glands ½ lin. long, lanceolate, subacute, keeled on the inner face ; filaments ½ lin. long, swollen ; ovary covered with long reddish-brown hairs ; style 1¾ in. long, subterete above, compressed below, dilated with a wide groove or cavity (gland ?) above the ovary, pubescent up to ⅓, then glabrous ; stigma 2 lin. long, linear, obtuse and slightly recurved at the apex, furrowed, slightly bent at the junction with the style.

SOUTH AFRICA : without locality, *Ludwig* !
COAST REGION : Riversdale Div. ; Riversdale, *Muir* !

20. **P. Rouppelliæ** (Meisn. in DC. Prodr. xiv. 237) ; a small tree 8–15 ft. high ; branches villous or tomentose above, at length glabrescent ; leaves 2–5¾ in. long, ¾–1½ in. broad at the widest part, 2 lin. broad at the base, oblong-lanceolate or obovate-spathulate, acute, the younger densely villous or tomentose, at length glabrous, narrowed at the base, reticulately veined ; head shortly peduncled, 3¼–4½ lin. long, 2–4 in. in diam. ; involucral bracts 10-seriate, silky-tomentose, deep pink to pinky white ; outer ovate, obtuse, recurved to revolute, ciliate ; inner with an obovate to obovate-oblong limb, gradually passing into the claw, shortly ciliate above, exceeding the flowers ; perianth-sheath 1¼ in. long, dilated and 3-keeled and 7-nerved below, loosely villous above the dilated portion ; lip 1¼ in. long, 3-awned, spreadingly villous ; lateral awns 7 lin. long, linear, acuminate, purple tomentose to villous ; median awn 4 lin. long ; fertile stamens 3 ; filaments ½ lin. long, flattened ; anthers linear,

1½ lin. long; apical glands ¼ lin. long, oblong, acute; barren
stamen acute, eglandular; ovary 2 lin. long, obovate in outline,
densely covered with numerous long golden hairs; style 2 in.
long, curved, somewhat flattened, keeled below on the convex side, usually
more or less villosulous; stigma 2 lin. long, curved and kneed
at the junction with the style. *P. lanuginosa, K. Schum. in Just,
Jahresb.* xxvi. i. 364. *Scolymocephalus lanuginosus, O. Kuntze, Rev.
Gen. Pl.* iii. 279 (*from descript.*).

SOUTH AFRICA : without locality, *Sim,* 2457 !
KALAHARI REGION : Orange River Colony ; Nelson's Kop, *Cooper,* 952 ! Witzies
Hoek, *Bolus,* 8242 ! *Flanagan,* 1849 ! Transvaal ; Observatory Ridge, near
Johannesburg, *Burtt-Davy,* 4004 ! Eersteling (Zoutpansberg), *Miss Leendertz,*
891 ! Houtbosch, *Bolus,* 10951 ! Barberton, *Galpin,* 974 ! Magalisberg, *Zeyher.*
1457 ! *Burke,* 31(♀?) ! Lydenburg, *Wilms,* 1273 ! Shilovane, *Junod,* 5523 !
Roodepoort, *Rand,* 1315 ! Swaziland ; Embabaan, *Burtt-Davy,* 2789 ! and with-
out precise locality, *Burtt-Davy,* 353 !
EASTERN REGION : Pondoland, Egossa, *Sim,* 2547 ! Port St. John, *Galpin,*
3198 ! Griqualand East ; Pot River Berg, *Galpin,* 6822 ! Natal ; Groenberg,
Wood, 7918 ! near Van Reenens Pass, *Krook,* 1586 ! *Wood,* 5632 ! Slopes of the
Drakensberg, near the Tugela Falls, *Wood,* 3514 ! and without precise locality,
Gerrard, 1887 !
The hairiness of the style is variable ; in Galpin 974 it is slightly pubescent,
in Zeyher 1457 densely villous, and gradations between these two extremes
are to be met with.

21. **P. longifolia** (Andr. Bot. Rep. t. 132) ; a bush 5–7 ft. high ;
branches minutely tomentellous or glabrescent ; leaves 2½–7 in.
long, 3–5 lin. broad, narrow strap-shaped, obtuse, long attenuated
at the base, distinctly pinnately veined above, glabrous ; head
sessile, 4–6 in. long, about 4 in. in diam. ; involucral bracts 9–10-
seriate, glabrous ; outer ovate to oblong-ovate, obtuse, green, often
with black tips ; inner with a lanceolate limb, gradually passing
into the claw, equalling the flowers or shorter ; perianth-sheath
2 in. long, dilated and 3-keeled below, pubescent, becoming glabrous
at the base ; lip 2¼ in. long, villous at the base and on the sides,
glabrous on the back, 3-awned ; lateral awns 1¼ in. long, linear,
long-villous, lower hairs whitish, upper dark purple to black ;
median awn 4 lin. long, filiform, black-ciliate at the apex ; fertile
stamens 3 ; filaments ½ lin. long, dilated ; anthers linear, 3½ lin.
long ; apical glands 2 lin. long, ovate-lanceolate, acuminate, acute ;
barren anther linear, acute, eglandular, with a filiform filament ;
ovary 2½ lin. long, obovate in outline, covered with long reddish-
brown hairs ; style 2 in. long, very slightly curved, terete, or
somewhat compressed below, with a swelling (gland ?) at the base
on the ventral side, faintly grooved on the convex side, pubescent ;
stigma 2½–3 lin. long, obtuse, conspicuously kneed and curved at
the junction with the style. *R. Br. in Trans. Linn. Soc.* x. 83,
partly ; *Roem. & Schult. Syst. Veg.* iii. 347, *partly* ; *Meisn. in DC.
Prodr.* xiv. 238, *partly. P. coronata, Lam. Ill.* i. 236, *partly.
P. dodonæifolia, Buek ex Meisn. l.c.* 239 (?). *P. vidua, Gawl. Recens.*
39. *Erodendrum longipenne, Knight, Prot.* 35. *Lepidocarpodendron
foliis angustis, etc., Boerh. Ind. Pl. Hort. Ludg. Bat.* ii. 186, *t.* 186.

VAR. β, **minor** (E. P. Phillips) ; leaves 3–4½ in. long, 1½–3½ lin. wide ; head sessile 3¼ in. long, about 2¼ in. in diameter.

COAST REGION : Stellenbosch Div. ; Lowrys Pass, *Schlechter*, 7796 ! Hottentots Holland Kloof, *Ludwig* ! Caledon Div. ; Houw Hoek Mountains, *Zeyher*,. 3662 ! *Drège* ! *MacOwan*, 2637 ! and in *Herb. Norm. Austr.-Afr.* 902 ! *Niven*, 5 ! Zwart Berg, *Pappe*, 15 ! Donker Hoek Mountain, *Burchell*, 7988 ! Var. β : Bredasdorp Div. ; near Elim, *Bolus*, 7860 !

Andrews described this as "*P. longifolia nigra*," his idea being that it was one of three varieties which had to be separated from *P. speciosa* and made to represent a distinct species *P. longifolia.*

In the Stockholm Herbarium there are fragments of the type of *P. dodonæifolia*, Buek, which may not belong here. The only leaf present is 8 lin. wide and has not dried black as is usual in *P. longifolia.* The specimen was collected by Drège at Hooge Kraal near the Zoetemelks river, Riversdale Division.

22. **P. ignota** (Phillips in Kew Bulletin, 1910, 229) ; branches glabrous ; leaves 6 in. or more long, 3–4 lin. wide, narrow strap-shaped, subacute, long attenuated at the base ; head sessile, tur-binate-obovoid, 4 in. long ; involucral bracts 7-seriate ; outer ovate, obtuse, with a red band and black tips ; inner linear or linear-oblong, greenish with rose tips, equalling the flowers ; perianth-sheath loosely villosulous ; lip 3-awned ; awns densely villous with long purple hairs ; style 2 in. long, narrowing above, compressed, white-villous ; stigma subulate, acute, kneed and curved at the junction with the style.   *P. longifolia, Ker-Gawl. in Bot. Reg. t.* 47 ; *Meisn. in DC. Prodr.* xiv. 238, *partly.*

SOUTH AFRICA : Known only from the figure in the Botanical Register ; allied to *P. ligulæfolia*, Sweet. It was introduced by Masson.

23. **P. umbonalis** (Sweet, Hort. Brit. ed. i. 346) ; stem 6–10 ft. high ; branches glabrous ; leaves 5–7 in. long, narrow strap-shaped, subacute, attenuated at the base, the younger pubescent ; head sessile, turbinate, over 4 in. long ; involucral bracts 10-seriate, glabrous ; outer ovate, acute ; inner linear-oblong, exceeding the flowers ; perianth-sheath . 2 in. long, hairy ; lip 1¼ in. long, with 2 long and 1 short awn ; awns densely ciliate ; style 2¼ in. long, almost straight ; stigma kneed and curved at the junction with the style.   *P. longifolia, var. cono turbinato, Andr. Bot. Rep. t.* 144. *P. longifolia, R. Br. in Trans. Linn. Soc.* x. 83, *partly ; Meisn. in DC. Prodr.* xiv. 238, *partly.   Erodendrum umbonale, Knight, Prot.* 35 ; *Salisb. ex Steud. Nomencl. ed.* 2, i. 589.

SOUTH AFRICA : Known only from the figure in Andrews' Botanists' Repository, which was prepared from a specimen, received from Schoenbrum.

It is somewhat doubtful whether Knight's *Erodendron umbonale* is actually the plant figured in Andrews' Repository, t. 144. He describes it as having pale green leaves with red nerves and margins, whereas Andrews' plant is represented as having uniformly dark green leaves. Knight's plant came from Hottentots Holland.

24. **P. ligulæfolia** (Sweet, Hort. Brit. ed. i. 346) ; leaves 5–7 in. long, narrow strap-shaped, subacute, long-attenuated at the base ;

head sessile, almost 4 in. by 3 in., obovoid; involucral bracts about
7-seriate; outer broad lanceolate; inner all acute, glabrous, green
with reddish tips and edges, linear, exceeding the flowers; perianth-
sheath 2 in. long, hairy; lip 1½ in. long, produced into 2 long and
1 short awn; awns densely purple-ciliate; style narrowing above,
2¼ in. long, almost straight; stigma kneed and curved at the
junction with the style. *P. longifolia*, var. *ferruginoso-purpurea*,
Andr. Bot. Rep. t. 133. *P. longifolia, R. Br. in Trans. Linn. Soc.*
x. 83, *partly*; *Meisn. in DC. Prodr.* xiv. 238, *partly.* *Erodendrum
ligulæfolium, Knight, Prot.* 36.

SOUTH AFRICA: Known only from the figure in Andrews' Botanists' Repository,
which was prepared from a specimen in Hibbert's Collection received from the
Imperial Garden at Schoenbrunn and probably introduced there by Scholl.

25. **P. mellifera** (Thunb. Diss. Prot. 34, 52); a large bush, 6–8 ft.
high; branches glabrous; leaves 2¼–3½ in. long, 2–4 lin. broad,
linear-oblanceolate or oblong, acute, attenuated at the base, with a
distinct midrib, glabrous; head shortly stipitate, 5 in. long, about
2½ in. in diam., stipes scaly; receptacle convex; paleæ ovate, acute;
involucral bracts 14–18-seriate, very viscid, those of the stipes
silky pubescent, the others glabrous, dark red to whitish-green
with pinkish tips and margins, but usually rosy pink; outer ovate,
subacuminate, subacute, ciliate; inner linear-oblong to broad-linear,
subacuminate or acute, exceeding the whitish flowers; perianth-
sheath 1⅔ in. long, membranous, dilated, keeled and 7-nerved
below, glabrous, excepting at the ciliate base; lip 1⅓ in. long,
3-awned, glabrous; lateral awns 3¼ lin. long, filiform, with a tuft
of white hairs at their apices; median awn similar, 2¼ lin. long;
stamens all fertile [sometimes the lower half of the anticous stamen
is devoid of pollen]; filaments 1 lin. long, dilated; anthers linear,
1 in. long; apical glands 1 lin. long, linear; ovary 2 lin. long,
obovate-oblong in outline, covered with long golden hairs; style
2–3½ in. long, tapering above, furrowed, compressed, slightly swollen
and deeply furrowed above the ovary, glabrous; stigma 1 in. long,
furrowed. *Lam. Ill.* i. 236; *Bot. Mag. t.* 346; *Poir. Encycl.* v. 646;
Willd. Sp. Pl. i. 522; *R. Br. in Trans. Linn. Soc.* x. 84; *Roem. &
Schult. Syst. Veg.* iii. 348; *Wendl. Hort. Herrenhaus.* iii. 3, *t.* 13;
Roupell, Cap. Flcw. t. 7; *Meisn. in DC. Prodr.* xiv. 239. *P. repens,
Linn. Mant. alt.* 189, *not of Thunb. Leucadendron repens, a, Linn.
Spec. Pl. ed.* i. 91; *Syst. Nat. ed.* xii. ii. 110; *Berg. in Vet. Acad.
Handl. Stockh.* 1766, 323. *Erodendrum mellifluum, Knight, Prot.* 34.
Scolymocephalus mellifer, O. Kuntze, Rev. Gen. Pl. iii. 280. *Lepi-
docarpodendron, foliis angustis, brevioribus, etc., Boerh. Ind. Pl. Hort.
Ludg. Bat.* ii. 187, *t.* 187. *Scolymocephalus seu Lepidocarpodendron
folii saligno β, Weinm. Phyt.* iv. 289, *t.* 896. *Conifera africana, foliis
angustis, etc., Sloane in Philos. Trans.* xvii. 66, *with plate, ex R. Br. l.c.*

VAR. β, **albiflora** (Andr. Bot. Rep. t. 582); as in the type, but the involucral
bracts are pure white.

SOUTH AFRICA: without locality, *Drège! Sieber, 2! Forster! Ludwig! Ecklon,*
316! *Pappe! Boos! Scholl! Wawra,* 107!

COAST REGION : Worcester Div. ; *Cooper,* 1592 ! Cape Div. ; various localities around Cape Town, *Thunberg* ! *Andersson* ! *Wright* ! *Wolley-Dod,* 1232 ! *Burchell,* 810 ! *Bolus,* 3738 ! 3956 ! *Scholl,* 165 ! 790 ! *MacOwan, Herb. Norm. Austr.-Afr.* 783 ! *Ecklon,* 654 ! *Drège* ! *Krauss* ! ; *Zeyher,* 3662 ! *Phillips,* 269 ! *Pappe,* 16 ! ; Stellenbosch Div. ; Between Stellenbosch and Cape Flats, *Burchell,* 8375 ! Caledon Div. ; Houw Hoek Mountains, *Zeyher,* 3662 ! Albany Div. ; Grahamstown, *Zeyher,* 3663 ! Coldstream, *Misses Daly & Sole,* 249 ! Bothas Berg, *MacOwan,* 771 ! Var. *β* : Paarl Div. ; mountains between Wellington and Bains Kloof, ex *Phillips.*

We have seen no Herbarium specimens of the variety, but Mr. Phillips saw living plants growing on the mountain-sides along the road leading from Wellington to Bains Kloof.

P. mellifera, Thunb., is named in the Linnean Herbarium *P. repens* and was published as such by Linnæus in his Mantissa, p. 189. Eighteen years previously he had published a description of this plant under the name of *Leucadendron repens* (Linn. Sp. Pl. ed. i. 91) ; his var. *β* being *P. repens,* Thunb. The specific name *repens* is so inapplicable to this plant, while Thunberg's name is so suitable and has been in such general use that we have retained it.

26. **P. longiflora** (Lam. Ill. i. 234) ; a bush 8–10 ft. high ; branches tomentose, at length glabrous ; leaves $1\frac{1}{4}$–$3\frac{3}{4}$ in. long, 4–16 lin. broad, elliptic, more or less oblong, rather narrow, obtuse or subacute, rounded or subcordate at the base, reticulately veined, the younger pilose, at length glabrous, ciliate ; head sessile, 4 in. long, 2–$2\frac{1}{2}$ in. in diam. ; involucral bracts 11-seriate ; outer ovate, subacuminate, obtuse or subacute, minutely whitish silky-pubescent, ciliate ; inner elongated-oblong, yellowish-white silky-pubescent, with a fringe of long cilia equalling the flowers ; perianth-sheath $2\frac{1}{2}$ in. long, slender and thin above the middle, dilated and 3-keeled below, shortly villous, the upper slender half spirally coiled in old flowers ; lip 8–10 lin. long, 3-toothed, sparingly hairy on the sides, glabrous on the back or nearly all over ; lateral teeth up to 2 lin. long, hirsute or villous ; stamens all fertile, subsessile ; anthers 6 lin. long ; apical glands $\frac{1}{4}$ lin. long, lanceolate, swollen on the inner face ; ovary $1\frac{1}{2}$ lin. long, ovoid, densely covered with long reddish-brown hairs ; style 3 in. long, slender, straight, subterete, grooved on one side, finely glabrous ; stigma 6–8 lin. long, filiform, with slightly thickened tips, gradually tapering into the style. *Poir. Encycl.* v. 640 ; *R. Br. in Trans. Linn. Soc.* x. 76 ; *Roem. & Schult. Syst. Veg.* iii. 343 ; *Bot. Mag. t.* 2720 ; *Meisn. in DC. Prodr.* xiv. 234 ; var. *latifolia, Klotzsch in Krauss, Beitr. Fl. Cap-und Natal.* 140 ; *Klotzsch in Flora,* 1845, 76. *P. ovata, Thunb. in Mém. Acad. Pétersb.* 1813–14, 548, *t.* 17, *and in Fl. Cap. ed. Schult.* 139 ; *Meisn. l.c.* 247. *P. obliqua, Hort. Bollwill,* 1835, *ex Meisn. l.c.* ; *Kerner, Hort. Sempervir.* xi. *t.* 489. *Leucospermum ? ovatum, Roem. & Schult. Syst. Veg.* iii. *Mant.* 266. *Leucadendron aureum, Burm. f. Fl. Cap. Prodr.* 4. *Erodendrum æmulum, Knight, Prot.* 38. *Conocarpodendron folis subrotundo, etc., Boerh. Ind. Pl. Hort. Lugd. Bat.* ii. 199, *t.* 199.

VAR. *β,* **ovalis** (E. P. Phillips) ; leaves $2\frac{1}{4}$ in. long, $1\frac{1}{4}$ in. broad, elliptic-cordate, younger ciliate and pubescent at the base.

SOUTH AFRICA: without locality, *Drège*, 3967 ! *Sparrman* ! *Scholl*, 723 !
COAST REGION : Swellendam Div. ; foot of the Langeberg Range, near Swellendam, *Burchell*, 7432 ! *MacOwan & Bolus, Herb. Norm. Austr.-Afr.* 782 ! *MacOwan*, 2827 ! *Fry in Herb. Galpin*, 4987 ! Riversdale Div. ; summit of Kampsche Berg, *Burchell*, 7109 ! George Div. ; Cradock Pass, *Pappe* ! Montagu Pass, *Bolus*, 8688 ! Postberg and Zwart River, near George, *Bowie* ! near George, *Krauss*, 1075 ! Mossel Bay Div. ; Attaquas Kloof, *Niven*, 35 ! *Drège*, 2007 ! Uitenhage Div. ; Winterhoek Mountains, *Krauss* ! Var. *β* : cultivated specimen from Glasgow Botanic Garden !

P. calycina, Schnevogt, Ic. Pl. Rar. t. 48, representing a plant, grown in the Utrecht Gardens in 1794, is very probably referable to this species although the leaves are shown as shortly petioled and acute.

27. **P. subvestita** (N. E. Br. in Kew Bulletin, 1901, 132); a shrub 8 ft. high ; branches tomentose, at length glabrous ; leaves sessile, 1½–3½ in. long, 6–12 lin. broad, oblong, obtuse or subacute, narrowed at the base, distinctly veined beneath, younger densely woolly, becoming glabrous with age ; head sessile, 2 in. long, about 1¾ in. in diam. ; involucral bracts 6–8-seriate ; outer ovate, acute, silky-pubescent, ciliate ; inner oblong, silky-pubescent, densely fringed with hairs round the upper margin, exceeding the perianths, but not the styles ; perianth-sheath 1½ in. long, very slender and thin above, gradually dilated, 5-nerved and 3-keeled below, the slender upper half at length spirally coiled up, rufo-hirsute except at the base ; lip about 5 lin. long, glabrous, 3-toothed ; lateral teeth ¾ lin. long, white woolly ; stamens all fertile, subsessile ; anthers linear, 2¾ lin. long ; apical glands, ⅙ lin. long, ovate, obtuse, swollen on the inner face ; ovary 1¾ lin. long, oblong, covered with long reddish hairs ; style 1¾ in. long, straight, tapering above, compressed below, subterete above, grooved on one side, glabrous ; stigma 2½ in. long, very slender, obtuse, rather abruptly passing into the style.

EASTERN REGION: Pondoland ; Fakus Territory, *Sutherland* ! Griqualand East ; Tent Kop, *Galpin*, 6824 ! Insizwa Range, *Krook*, 1575 ! Natal ; hill near Van Reenen, *Wood*, 5631 ! Div. ? summit of Omaqua Mountain, *Thode*, 47 !

Near *B. hirta*, Klotzsch, but differing in the shape of the outer involucral bracts.

28. **P. lacticolor** (Salisb. Parad. Lond. t. 27); a small tree 10 ft. high ; branches tomentellous, at length glabrous ; leaves 3–3½ in. long, 7–12 in. broad, lanceolate to lanceolate-oblong, subacute, obtuse at the base, minutely punctate, distinctly veined, tomentose at the base and midrib or glabrous ; head sessile, 2¼–2¾ in. long, about 2–2½ in. in diam. ; involucral bracts 6-seriate, silky-pubescent, yellowish-white ; outer ovate, ciliate ; inner oblong or spathulate-oblong, with a long silky fringe at the apex, shorter than the styles ; perianth-sheath very slender and thin, 1¾ in. long, dilated, 3-keeled and faintly 5-nerved below, spirally twisted in old flowers, rufo-pilose ; lip 7 lin. long, 3-toothed, otherwise glabrous ; sparingly hairy along the edges ; lateral teeth hirsute, ¾–1 lin. long ; median tooth ½ lin. long ; stamens all fertile ; filaments ¼ lin. long, swollen, channelled ; anthers linear, 4½ lin. long, apical glands ⅓ lin.

long, ovate-lanceolate, subacute, keeled on the inner face; style $2\frac{1}{2}$ in. long, slightly curved, tapering above, faintly keeled on one side, glabrous; stigma $4\frac{1}{2}$ lin. long, filiform, slightly thickened at the apex, rather abruptly passing into the much stouter style. *P. ochroleuca, Smith, Exot. Bot.* ii. 43, *t.* 81. *P. penicillata, E. Meyer in Drège, Zwei Pfl. Documente,* 82, 124; 213, *partly* ; *Meisn. in DC. Prodr.* xiv. 235, *partly. P. latericolor, Meisn. l.c.* 248, *in error. Erodendrum bombycinum, Knight, Prot.* 38; *Salisb. ex Roem. & Schult. Syst. Veg.* iii. *Mant.* 263.

VAR. β, **angusta** (E. P. Phillips) ; leaves narrowing at the base ; stigmas not so distinctly clavate as in the type ; teeth of lip subequal, $\frac{1}{4}$ lin. long.

VAR. γ, **orientalis** (E. P. Phillips) ; differing from the type in that the protruding styles in old heads form a brush-like tuft. *P. orientalis, Sim, For. Fl. Cap.* 296, *t.* 128.

COAST REGION : Worcester Div. ; Dutoits Kloof, *Drège* ! Caledon Div. ; near the River Zondereinde, *Niven* ! George Div. ; Kaymans Gat, *Drège* ! Outeniqua woods, *Krauss* ! Knysna Div. ; Knysna, *Pappe* ! *Mrs. Newdigate* ! Uitenhage Div. ; Winterhoek Mountains, *Krauss* ! Stockenstrom Div. ; Kat Berg, *Baur,* 1072 ! Cathcart Div. ; Toise River Railway Station, *Flanagan,* 1703 ! Stutterheim Div. ; summit of Dohne Peak, *Galpin,* 2426 ! Var. γ: King Williamstown Div. ; Perie, *Sim,* 1478 !

CENTRAL REGION : Var. β : Somerset Div. ; Bosch Berg, *MacOwan,* 1481 !

29. P. Mundii (Klotzsch in Otto & Dietr. Gartenzeit., 1838, 113)

; branches tomentellous to tomentose above; leaves $1\frac{1}{2}$–$4\frac{1}{4}$ in. long, 6–16 lin. broad, lanceolate or lanceolate-elliptic, subobtuse, narrowing at the base, distinctly veined, glabrous or the youngest leaves sometimes loosely pilose; head sessile, $2\frac{3}{4}$–3 in. long, about 2 in. in diam. ; involucral bracts 11–12-seriate ; outer ovate, obtuse, silky on the back, green, ciliate ; inner oblong or spathulate-oblong, whitish pubescent to tomentose, fringed with white cilia, shorter than the styles; perianth-sheath $1\frac{1}{2}$ in. long, slender and thin above the middle, gradually dilated and 5-nerved below, not keeled, the upper half at length coiled up, loosely hairy; lip 7 lin. long, tridentate, glabrous, with a dense tuft of hairs at the apex; lateral teeth 1 lin. long; median tooth $\frac{3}{4}$ lin. long; stamens all fertile; filaments $\frac{1}{2}$ lin. long, channelled down the middle ; anthers linear, 3 lin. long ; apical glands $\frac{1}{3}$ lin. long, ovate, subacuminate, subacute, keeled on the inner face ; ovary covered with a tuft of long brown hairs ; style 2 in. long, almost straight, keeled on one side, compressed above the ovary, then more or less terete, glabrous ; stigma 3 lin. long, furrowed, subcapitate at the apex, abruptly and obliquely passing into the much stouter style. *P. penicillata, E. Meyer in Drège, Zwei Pfl. Documente,* 118, 126 ; 213, *partly* ; *Meisn. in DC. Prodr.* xiv. 235, *partly. P. longiflora, Lam., var. Mundii, Link, Klotzsch & Otto, Ic. Pl. Rar.* i. 55, *t.* 22 ; *Krauss, Beitr. Fl. Cap-und Natall.* 140 *and in Flora,* 1845, 76. *P. ovalis, Buek, ex. Meisn. l.c.*

COAST REGION : Worcester Div. ; ·Dutoits Kloof, *Drège* ! Kynsna Div. ; between Plettenbergs Bay and Knysna, *Burchell,* 5351 ! Plettenbergs Bay, *Pappe* ! Paarde Berg, *Burchell,* 5194 ! Uniondale Div. ; Outeniqua Mountains, near Avontuur, *Bolus,* 2452 ! Humansdorp Div. ; near the Kromme River, *Drège* !

Uitenhage Div. ; Van Stadens Berg, *Drège. Burchell,* 4687 ! Stutterheim Div. ;
Mountain slopes near Stutterheim, *Flanagan,* 1703 !
EASTERN REGION : Transkei ; mountains near Barzeia, *Baur,* 624 !

30. **P. curvata** (N. E. Br. in Kew Bulletin, 1901, 131) ; a
tree 15 ft. or more high ; branches thick, rough, glabrous or
minutely pubescent at the apex ; leaves 4–7 in. long, $4\frac{1}{2}$–7 lin.
broad, falcate, linear-oblanceolate, obtuse, long attenuated at the
base, indistinctly veined, with a distinct midrib, glabrous ; head
sessile, $2\frac{1}{2}$ in. long, about 2 in. in diam. ; young heads ovoid ;
involucral bracts 8–9-seriate, tomentose on the lower half, at
length glabrous, minutely ciliate ; outer ovate, subacuminate,
obtuse, inner oblong, not equalling the flowers ; perianth-sheath
pubescent excepting the lower part of the $1\frac{1}{2}$ in.-long dilated
3-keeled and 7-nerved base ; lip 11 lin. long, deeply 3-toothed,
keeled, tomentose , lateral teeth $1\frac{1}{2}$ lin. long, tomentose excepting
at the glabrous tips ; median tooth 1 lin. long, filiform ; stamens
all fertile ; anthers linear, 5 lin. long ; apical glands 2 lin. long,
lanceolate, acute, swollen on the inner face ; ovary 1 lin. long,
covered with long yellow-brown hairs ; style 2 in. long, flattened
below, trigonous above, furrowed, glabrous ; stigma 5 lin. long,
obtuse, faintly bent at the junction with the style.

KALAHARI REGION : Transvaal ; hill-sides near Barberton, *Galpin,* 973 !

31. **P. grandiflora** (Thunb. Diss. Prot. 56) ; branches glabrous ;
leaves very variable in shape and size, linear-oblong to obovate,
3–7 in. long, $\frac{1}{2}$–2 in. broad, obtuse, glaucous, distinctly veined,
glabrous ; head subsessile, $2\frac{3}{4}$ in. long, about 3 in. in diam., globose
when young, usually contracted into a short scaly stipes ; involucral
bracts 13-seriate, finely silky-canescent, at length glabrescent, outer
ovate, obtuse, inner oblong or spathulate-oblong, convex, not equalling the corolla ; perianth-sheath 1 in. long, pubescent towards the
lip, otherwise glabrous, excepting on the ciliate margins of the
dilated 7-nerved and 3-keeled base ; lip 8 lin. long, 3-toothed,
whitish-tomentose excepting on the glabrescent back ; lateral teeth
oblong, 1–$1\frac{1}{2}$ lin. long ; median tooth up to $\frac{3}{4}$ lin. long ; stamens all
fertile, subsessile ; anthers linear, 5 lin. long ; apical gland $\frac{1}{4}$ lin.
long, lanceolate, swollen on the inner face ; ovary covered with
long fulvous or reddish hairs ; style 2 in. long, curved, narrowed
upwards from a widened and compressed base, glabrous ; stigma
$4\frac{1}{2}$ lin. long, hardly differentiated from the style, very slender,
slightly thickened at the apex. *Lam. Ill.* i. 234 ; *Thunb. Prodr.* i.
27 *and Fl. Cap. ed. Schult.* 137 ; *Willd. Sp. Pl.* i. 530 ; *Poir. Encycl.
Suppl.* v. 640 ; *R. Br. in Trans. Linn. Soc.* x. 85 *incl. var. β* ; *Roem.
& Schult. Syst. Veg.* iii. 348 ; *Mant.* 265 ; *Meisn. in DC. Prodr.* xiv.
236 ; *var. angustifolia,* Ker in *Bot. Reg. t.* 569 ; *var. latifolia, Bot.
Mag. t.* 2447. *P. marginata, Lam. Ill.* i. 235 (*according to Poiret,
l.c.*). *P. laurifolia, Buek ex Meisn. l.c. Leucadendron cinaroides β,
Linn. Spec. Pl. ed.* 1, i. 92. *Erodendrum grandiflorum, Knight, Prot.*

42. *Lepidocarpodendron folio saligno, etc., Boerh. Ind. Pl. Hort.*
Lugd. Bat. ii. 183, *t.* 183. *Scolymocephalus foliis oblongis, etc.*
Weinm. Phyt. iv. *t.* 891.

SOUTH AFRICA : without locality, *Thunberg* ! *Labillardiere* ! *Niven* ! *Mund* !
COAST REGION : Paarl Div. ; Paarl Mountain, *Drège* ! Cape Div. ; Camps Bay,
Zeyher ! Table Mountain, *Pappe* ! *Bolus,* 4801 ! *Phillips* ! *Krauss* ! Simons Town,
Wright ! Caledon Div. ; Baviaans Kloof, near Genadendal, *Burchell,* 7902 ! Vogel
Gat, *Schlechter,* 10420 ! Swellendam Div. ; near Swellendam, *Fry in Herb.*
Galpin, 4982 ! Humansdorp Div. ; by the Kromme River, *Burchell,* 4885 !
CENTRAL REGION : Ceres Div. ; Leeuwen Fontein, 2500 ft., *Pearson,* 3684 !

32. P. trigona (Phillips in Kew Bulletin, 1910, 230) ; branches
hairy at the insertion of the leaves, otherwise glabrous ; leaves
$2\frac{1}{4}$–$4\frac{1}{2}$ in. long, 9–12 lin. broad, oblong-lanceolate to oblong, obtuse,
slightly narrowed at the base, distinctly veined, glabrous or sub-
glaucous ; head constricted into a very short scaly stipes, $2\frac{1}{2}$ in.
long, about 3 in. in diam. ; receptacle slightly concave ; involucral
bracts 9–11-seriate ; outer ovate, subacuminate, obtuse, densely
silky-pubescent to tomentose below, glabrescent above ; inner
oblong, concave, silky-pubescent on the back, glabrescent on the
sides and tips, shorter than the flowers ; perianth-sheath $1\frac{1}{4}$ in.
long, glabrous at the dilated, 3-keeled and 7-nerved base, increasingly hairy above ; lip 8 lin. long, 3-toothed, densely villous on
the sides, glabrescent on the back ; teeth subequal, 1 lin. long,
ovate, acuminate, densely whitish-tomentose ; stamens all fertile,
sessile ; anthers linear, $4\frac{1}{2}$ lin. long ; apical glands $\frac{1}{4}$ lin. long,
ovate or suborbicular, obtuse, swollen on the inner face ; ovary
$1\frac{1}{2}$ lin. long, elliptic, covered with long reddish-yellow hairs ;
hypogynous scales $\frac{3}{4}$ lin. long, ovate, subacuminate, obtuse ; style
$1\frac{1}{2}$ in. long, more or less trigonous and grooved, glabrous ; stigma
$4\frac{1}{2}$ lin. long, filiform, obtuse, faintly swollen and bent at the junction
with the style.

KALAHARI REGION : Transvaal ! Derde Poort, near Pretoria, *Miss Leendertz,* 679 !
Very near to and probably not specifically distinct from *P. abyssinica,* Willd.

33. P. abyssinica (Willd. Sp. Pl. i. 522) ; a tree 12–15 ft. high ;
branches pilose especially at the insertion of the leaves, or, in the S.
African specimens, usually glabrous ; leaves $2\frac{3}{4}$–6 in. long, 4–10 lin.
broad, oblong-lanceolate or lanceolate, subacute or obtuse, narrowing
to the base, coriaceous, prominently veined, pilose above and beneath,
or more often glabrous ; head often contracted into a scaly stipes,
$2\frac{1}{2}$ in. long, about $2\frac{1}{2}$ in. in diam. ; involucral bracts 11-seriate,
densely silky-tomentose ; outer broad-ovate, obtuse ; inner oblong,
or cuneate-oblong, convex, shorter than the flowers ; perianth-sheath
$1\frac{1}{4}$ in. long, dilated and 3-keeled below, fulvously villous, glabrous
at the base ; lip 9 lin. long, 3-toothed, villous to the tips of the
teeth, excepting on the glabrous or glabrescent back ; lateral teeth
$1\frac{1}{4}$ lin. long ; median tooth $\frac{3}{4}$ lin. long, lanceolate, acuminate ;
stamens all fertile ; filaments $\frac{1}{2}$ lin. long, expanded, concave ;

anthers linear, 5 lin. long ; apical glands $\frac{1}{4}$ lin. long, elliptic ; ovary
1$\frac{3}{4}$ lin. long, obovate, covered by a dense tuft of long reddish-brown
hairs ; hypogynous scales $\frac{1}{2}$ lin. long, ovate, obtuse ; style 1$\frac{3}{4}$ in.
long, more or less curved and faintly grooved, glabrous ; stigma
4–5 lin. long, filiform, obtuse, slightly bent at the junction with the
style. *R. Br. in Trans. Linn. Soc.* x. 85 ; *Meisn. in DC. Prodr.*
xiv. 237 ; *Rich. Tent. Fl. Abyss.* ii. 232 ; *Baker & C. H. Wright in
Dyer, Flor. Trop. Afr.* vi. i. 199. *P. Gaguedi, Gmel. Syst.* 225.
Guaguedi (native name), Bruce, Abyss. v. 52, *with a plate.*

KALAHARI REGION : Transvaal ; various localities, *Mrs. Saunders*, 81 ! *Burtt-
Davy*, 144 ! 354 ! ; 5647 ! *Miss Pegler*, 941 ! *Pole Evans*, 2963 !

Also in Tropical Africa.

34. **P. hirta** (Klotzsch in Flora, 1845, 76, and in Beitr. Fl.
Cap- und Natall. 140) ; branches hirsute ; leaves 1–4 in. long,
$\frac{1}{2}$–1$\frac{1}{2}$ in. broad, lanceolate or oblong-lanceolate, attenuate at the
base, with subprominent venation, loosely villous to glabrescent ;
head 2$\frac{1}{2}$ in. long, about 2$\frac{1}{2}$ in. in diam., terminal, sometimes
lateral, turbinate at the base and sometimes shortly constricted
into a scaly stipe ; involucral bracts 8-seriate, silky-tomentose,
outer ovate, obtuse or subacute ; inner oblong, exceeding
the flowers ; perianth-sheath over 1 lin. long, slightly expanded
below, densely whitish-villous excepting at the glabrous base ; lip
7 lin. long, whitish-villous, 3-toothed ; lateral teeth 1 lin. long,
intermediate much shorter ; stamens all fertile, subsessile ; filaments
$\frac{1}{2}$ lin. long, flattened ; anthers linear, 2–3 lin. long ; apical glands
$\frac{1}{4}$ lin. long, subsessile, linear, swollen on the inner face ; ovary
1 lin. long, globose, covered with a tuft of long whitish hairs ; style
1$\frac{1}{4}$ in. long more or less curved, slender, glabrous ; stigma 3 lin.
long, filiform, obtuse, very slightly bent at the junction with the style.

KALAHARI REGION : Transvaal ; Magaliesberg Range, *Burke*, 318 ! *Zeyher*, 1455 !
around Pretoria, *Jause*, 80 ! *Crawley*, 5069 ! *Reck*, 3796 ! 4311 ! *Burtt-Davy*,
9101 ! Ridges near Johannesburg, *Mrs. De Jongh in Herb. Galpin*, 1477 ! *Burtt-
Davy*, 4002 ! by the Koster River, Rustenburg, *Burtt-Davy*, 156 !
EASTERN REGION : Natal ; Inanda, *Wood*, 577 ! near Durban, *Wood*, 8044 ! near
the Umlaas River, *Krauss*, 202, and without precise locality, *Gerrard*, 1516 ! *Mrs.
K. Saunders.*

35. **P. Dykei** (Phillips) ; branches glabrous, with almost black
bark ; leaves narrowly obovate-oblanceolate, obtuse, 1–1$\frac{3}{4}$ in. long,
5–6 lin. broad, glabrous, dull, coriaceous, faintly veined ; heads
sessile, somewhat over 2 in. long and wide ; receptacle conical ;
involucral bracts about 8-seriate, white-silky-pubescent and ciliate ;
outer ovate, obtuse, inner narrowly oblong, obtuse, much shorter
than the flowers ; perianth-sheath 1$\frac{1}{4}$–1$\frac{1}{2}$ in. long, slender, 7-nerved,
widened and 3-keeled at the base for about $\frac{1}{2}$ in., white-pubescent
outside and along the margins within all along to about 3 lin. from
the base ; lip 4–4$\frac{1}{2}$ lin. long, white-pubescent on the back, shortly
villous on the sides, 3-awned, lateral awns stout, about 3 lin. long,

white villous with interspersed reddish hairs, middle awn somewhat
shorter ; anticous segment with a finely filiform 1- (rarely 3-)nerved
claw and a very narrow limb, otherwise in size and tomentum as
the fused segments ; fertile stamens 3, the anticous more or less
imperfect ; filament $\frac{1}{2}$–$\frac{3}{4}$ lin. long; anther $1\frac{1}{2}$ lin. long ; apical gland
oblong, $\frac{1}{2}$ lin. long ; ovary obconical, $7\frac{1}{2}$ lin. long ; style laterally
much compressed, at the base wider than the ovary, gradually
tapering, not quite 2 in. long, incurved, glabrous ; stigma $1\frac{1}{2}$ lin.
long, capitate ; hypogynous scale obliquely ovate, $\frac{3}{4}$ lin. long.

COAST REGION : Uitenhage Div. ; Coxcombe Mountain, *Dyke*, 2676 ! *& in Herb.*
Marloth, 4977 !

36. **P. rupicola** (Mund ex Meisn. in DC. Prodr. xiv. 236);
branches glabrous ; leaves oblanceolate, obtuse with a minute
callous blunt point, cuneately attenuated at the base, $1\frac{1}{4}$–2 in. long,
$\frac{1}{3}$ to almost $\frac{1}{2}$ in. broad, thickly coriaceous, subglaucous, drying
reddish-brown, glabrous, obscurely veined ; head subsessile, about
2 in. long and wide ; involucral bracts about 6–7-seriate, those of
the first 5 or 6 series gradually increasing in size, those of the
last exceeding the preceding by almost 1 in. ; outer ovate-oblong to
elliptic-oblong, obtuse, finely silky-pubescent, soon glabrescent ;
inner oblong-linear to spathulate-linear, obtuse, fulvously villosulous,
more or less glabrescent at length in the upper part, equalling the
dull red flowers ; perianth-sheath somewhat over 1 in. long, slender,
fulvously and spreadingly villous down to the middle of the widened
5-nerved and 3-keeled base, then glabrous ; lip 4 lin. long, 3-toothed,
membranous, fulvously villous along the sides and on the teeth,
glabrous on the back ; lateral teeth $1\frac{1}{4}$ lin. long, intermediate $\frac{3}{4}$ lin.
long ; stamens all fertile ; filaments filiform, slightly wider above,
1–$1\frac{1}{4}$ lin. long ; anthers oblong-linear, $1\frac{1}{2}$–$1\frac{3}{4}$ lin. long ; apical glands
$\frac{1}{3}$ lin. long, ovate, subacute ; ovary covered with reddish-fulvous
hairs ; style rather strongly curved, $1\frac{1}{2}$ in. long, laterally compressed,
$\frac{3}{4}$ lin. broad below, gradually tapering upwards ; stigma $1\frac{1}{2}$ lin. long,
obtuse, grooved, set off from the style by a slight and sudden bend.

COAST REGION : Tulbagh Div. ; top of Great Winterhoek Mountain, *Mund* ;
5000 ft., *Bolus*, 4194 !

37. **P. glabra** (Thunb. Diss. Prot. 42) ; branches glabrous ; leaves
$1\frac{1}{4}$–$2\frac{1}{2}$ in. long, 6–10 lin. broad, lanceolate or obovate-lanceolate,
obtuse or subacute, attenuated at the base, indistinctly nerved,
glabrous ; head sessile, $1\frac{1}{2}$–2 in. long, about $1\frac{3}{4}$ in. in diam. ;
involucral bracts 6-seriate, silky-pubescent, at length becoming
glabrous, ciliate ; outer ovate, obtuse ; inner oblong, becoming
convex, shorter than the flowers ; perianth-sheath over 1 in. long,
glabrous, with a few hairs towards the lip ; lip 6 lin. long, 3-toothed
above, tomentose to villous excepting at the glabrous back, ciliate
at the apex ; lateral teeth $\frac{3}{4}$ lin. long, villous ; median tooth $\frac{1}{2}$ lin.
long ; stamens all fertile ; filaments $\frac{1}{4}$ lin. long, flattened ; anthers
linear, $2\frac{1}{2}$ lin. long ; apical glands $\frac{1}{4}$ lin. long, ovate, obtuse, con-

spicuously swollen on the inner face; ovary $\frac{3}{4}$ lin. long, oblong,
covered with a tuft of light fulvous hairs; style $1\frac{1}{4}$ in. long, more
or less grooved, glabrous; stigma $2\frac{1}{2}$ lin. long, linear, truncate,
slightly curved at the junction with the style.   *Thunb. Fl. Cap. ed.
Schult.* 138 ; *Meisn.. in DC. Prodr.* xiv. 247.   *P. buekiana, Meisn.
in DC. Prodr.* xiv. 236.   *P. pyrifolia, Buek ex Meisn. l.c.   P.
grandiflora, β angustifolia, Drège ex Meisn. l.c.   P. Banksii, Klotzsch
ex Meisn. l.c.   Leucadendron Thunbergii, Endl. Gen. Suppl.* iv. ii. 75.

SOUTH AFRICA : without locality, *Thunberg* !
COAST REGION : Clanwilliam Div. ; Lange Kloof, *Schlechter*, 8391 ! between
Clanwilliam and Bosch Kloof, *Drège* !
CENTRAL REGION : Calvinia Div. ; between Grasberg River and Watervals
River, *Drège* !

38. **P. recondita** (Buek ex Meisn. in DC. Prodr. xiv. 237) ;
branches glabrous, glaucous; leaves 3–4$\frac{1}{2}$ in. long, 1–2 in. broad at
the widest part, 2 lin. wide at the base, obovate, cuneate, obtuse,
distinctly veined, glabrous, glaucous; head sessile, 1$\frac{1}{2}$–2$\frac{1}{2}$ in. long,
about 2 in. in diam., globose ; involucral bracts 7–8-seriate, glabrous ;
outer ovate, subacuminate, subobtuse ; inner oblong, slightly convex,
equalling the flowers ; perianth-sheath 12 lin. long, glabrous excepting
for a few reddish setulæ near the lip, dilated, 5-nerved and 3-keeled
below ; lip 3 lin. long, glabrous or sparingly setulose ; stamens all
fertile ; filaments $\frac{1}{6}$ lin. long and broad, thin ; anthers linear-elliptic,
2 lin. long ; apical glands $\frac{1}{6}$ lin. long, ovate, subobtuse ; ovary 4 lin.
long, oblanceolate in outline, swollen above, covered with long
reddish-brown hairs ; style over 1 in. long, falcate from a short,
almost straight base, 1 lin. wide, tapering upwards, compressed,
glabrous ; stigma 1$\frac{1}{2}$–2 lin. long, linear, obtuse, grooved, almost
imperceptibly passing into the style.

COAST REGION : Clanwilliam Div. ; Cedarberg Range, at Ezelsbank, *Drège*,
2404 !

39. **P. convexa** (Phillips in Kew Bulletin, 1910, 235) ; a shrub
up to 10 ft. high ; leaves 5–9 in. long, 2–3$\frac{1}{2}$ in. broad at the widest
part, obovate-oblong, obtuse, bluntly-mucronate, prominently veined,
glaucous, glabrous ; head sessile, 2 in. long, about 3 in. in diam. ;
receptacle 9 lin. high, hemispherical ; paleæ ovate, subacute ;
involucral bracts 10–12-seriate ; outer elliptic-ovate, obtuse, glabrous,
ciliolate ; inner spathulate-oblong or spathulate-linear, glabrous or
minutely pubescent on the back ; perianth-sheath 10 lin. long,
dilated, 3-keeled and usually 9- (sometimes 7-)nerved below, whitish-
pubescent above, glabrous below excepting on the ciliate margins ;
lip 3 lin. long, 3-toothed, sides pubescent, back glabrescent, top
villous, the villi concealing the subequal teeth, which are $\frac{1}{2}$ lin. long ;
stamens all fertile ; anthers subsessile, 2 lin. long ; apical glands
$\frac{1}{3}$ lin. long, ovate, subobtuse ; ovary 5 lin. long, linear-oblong,
covered with long golden-yellow hairs ; style 9 lin. long, falcate,

flattened and up to ¾ lin. broad below, much attenuated above, glabrous; stigma 1½ lin. long, subulate, grooved, obtuse, almost imperceptibly passing into the style.

COAST REGION : Laingsburg Div. ; on the Witteberg Range near Matjesfontein, *Marloth*, 3209 !

40. **P. punctata** (Meisn. in DC. Prodr. xiv. 238); branches tomentose ; leaves 1½–2 in. long, 6–14 lin. broad, elliptic-oblong or oblong-oblanceolate, obtuse or subacute, rounded or attenuated at the base, with the venation more prominent beneath than above, the younger leaves villous-tomentose, soon woolly glabrous ; head sessile, 1½–2 in. long, about 1⅓ in. in diam. ; receptacle flat ; involucral scales 5-seriate, silky-tomentose, at length becoming glabrous ; outer ovate, obtuse ; inner oblong, ciliate round the apex, equalling the flowers ; perianth-sheath tube 8 lin. long, membranous, rufous-hirsute excepting at the 3-keeled glabrous base ; lip 7 lin. long, tridentate, glabrous ; teeth subequal, ½ lin. long, obtuse ; stamens all fertile ; filaments ¼ lin. long, dilated and concave ; anthers linear, 5¾ lin. long ; apical glands ¼ lin. long, lanceolate, with a swelling on the inner face ; style 1½ in. long, slightly curved or flexuous, compressed and dilated below, glabrous ; stigma 5¾ lin. long, filiform, gradually passing into the style. *P. carlescens, E. Meyer ex Meisn. l.c.   P. coriacea, Buek ex Meisn. l.c.*

COAST REGION : Clanwilliam Div. ; Cedarberg Range, near Ezelsbank, *Drège* !

CENTRAL REGION : Prince Albert Div. ; Great Zwartberg Range, *Drège*, 2009 ! *Hallack in Herb. Galpin*, 3075 !

41. **P. caffra** (Meisn. in DC. Prodr. xiv. 237) ; branches glabrous ; leaves 3¾–4¾ in. long, 8–11 lin. broad, lanceolate or oblong-lanceolate, obtuse or retuse, narrowing at the base, indistinctly veined, glabrous ; head contracted at the base into a scaly stipes, 2½ in. long, about 2½ in. in diam., globose ; involucral bracts 16-seriate, densely tomentose, at length glabrous ; outer ovate, obtuse ; inner oblong-elliptic, incurved, not equalling the flowers ; perianth-sheath glabrous, base dilated, 3-keeled with distinct intracarinal nerves ; lip 6 lin. long, 3-toothed, 3-keeled, glabrous ; stamens all fertile ; filaments ¼ lin. long, flattened, concave ; anthers linear, 4½ lin. long ; apical glands ¼ lin. long, elliptic, swollen on the inner face ; ovary 3 lin. long, narrowly oblong, covered with a tuft of long reddish-brown hairs ; style 1¼ in. long, curved in the lower half, tapering from the base upwards, slender above ; stigma up to 6½ lin. long, finely filiform, grooved, obtuse, almost imperceptibly passing into the style.

KALAHARI REGION : Basutoland, Leribe, *Mrs. Dieterlen* ! Transvaal ; Magaliesberg Range, *Zeyher*, 1458 ! Heidelberg, *Burtt-Davy*, 5646 ! Zeerust, *Burtt-Davy*, 107 ! Rustenburg, *Collins*, 34 !

42. P. rhodantha (Hook. f. in Bot. Mag. t. 7331); branches glabrous; leaves up to 5½ in. long and 14 lin. broad, lanceolate, subacute, slightly narrowed at the base, distinctly veined, with the midrib prominent in the lower half, glabrous; head contracted at the base into a scaly stipes, 2–2½ in. long, and as wide, receptacle shortly conical; involucral bracts about 10-seriate; outer ovate, subacuminate, green or more or less pink, evanescently silky-pubescent below, ciliolate; inner oblong, rose-colour, glabrous, shorter than the flowers; perianth-sheath 12 lin. long, dilated, 3-keeled and 5-nerved below, membranous, glabrous, excepting at the sides which are rufo-pubescent within, middle portion spirally coiled in old flowers; lip 5 lin. long, orange, 3-toothed, 3-keeled, glabrous excepting at the minutely hirsute tips; teeth subequal, ½ lin. long; stamens all fertile, subsessile; anthers linear, 3 lin. long; apical glands ⅛ lin. long, elliptic, swollen on the inner face; ovary up to 2 lin. long, oblong in outline, covered with long reddish-yellow hairs; style 1¼ lin. long, distinctly swollen above the ovary, almost equally wide up to ¾, then strongly curved and tapering, glabrous; stigma 3–3½ lin. long, filiform, obtuse, slightly wavy at the junction with the style. *P. Bolusii, Phillips in Kew Bulletin,* 1910, 231.

KALAHARI REGION : Transvaal; Pilgrims Rest (a plant raised from seeds at Kew), *Horn*! Swaziland, between Dalriach and Forbes Reef, *Bolus,* 12265!

43. P. multibracteata (Phillips in Kew Bulletin, 1910, 230); branches glabrous; leaves 3–5 in. long, 4–8 lin. broad, linear-lanceolate to linear, subacute or subobtuse, narrowing at the base, distinctly or indistinctly veined, glabrous; heads peduncled, contracted at the base into a scaly stipes 4–10 lin. long, 2½–3 in. long, about 2¼ in. in diam.; involucral bracts 14–20-seriate; outer ovate, subacuminate, obtuse, finely silky-pubescent on the back; inner oblong or spathulate-oblong, equalling or slightly shorter than the flowers; perianth-sheath 1¼–1½ in. long, dilated, 3-keeled and 7-nerved below, membranous, densely pilose within; lip 5–6 lin. long, 3-toothed, usually glabrous or very sparingly pilose; lateral teeth ¾ lin. long; median tooth ½ lin. long; stamens all fertile, subsessile : anthers linear, 4 lin. long; apical glands ¼–⅓ lin. long, ovate, obtuse; ovary 1½ lin. long, obovoid, covered with long reddish-brown hairs; style about 1½ in. long, slightly curved or flexuous, tapering above, somewhat compressed, glabrous; stigma 3½–4½ lin. long, filiform, obtuse, passing almost imperceptibly into the style. *P. Pegleri, Phillips in Kew Bulletin,* 1910, 230. *P. natalensis, Phillips, l.c.* 231. *P. Baurii, Phillips, l.c.* 232.

COAST REGION: King Williamstown Div.; Perie Mountains, *Perke in Herb. Galpin,* 5828!·British Kaffraria, *Cooper,* 86!
EASTERN REGION : Transkei; near Kentani, *Miss Pegler,* 274! Tembuland; Bazeia, *Baur,* 721! Natal; between Umlazi River and Durban, *Krauss,* 176! on the Drakensberg, *Cooper,* 951!

44. P. tenax (R. Br. in Trans. Linn. Soc. x. 88); stem decumbent; branches softly hirsute, then glabrescent; leaves 4–6 in. long, 4–6 lin. broad above, 1½ lin. broad below, oblanceolate to linear-oblanceolate, subacute, long attenuate at the base, more or less vernicose, indistinctly veined, with a distinct midrib, glabrous or scantily hirsute with long soft hairs, particularly towards the base and along the margin; head sessile, 2 in. long, about 2½ in. in diam.; receptacle conical; paleæ ovate, acute; involucral bracts 6-seriate; outer ovate, obtuse, silky-pubescent, at length glabrous, inner spathulate-oblong, concave, silky-pubescent, ciliate, equalling the flowers; perianth-sheath 9 lin. long, dilated, 3-keeled and 5-nerved below, the membranous upper part at length spirally coiling up, glabrous; lip 4 lin. long, 3-keeled, glabrous, dentate; lateral teeth ¾ lin. long, linear-lanceolate, villous; median tooth ½ lin. long; stamens all fertile; filaments ¼ lin. long, dilated, concave; anthers linear, 2 lin. long; apical glands ¼ lin. long, ovate, swollen on the inner face; hypogynous scales ovate, obtuse; ovary 2 lin. long, oblong in outline, covered with numerous long dark-brown hairs; style 10 lin. long, slightly curved, tapering above, compressed, swollen above the ovary, glabrous; stigma 2 lin. long, grooved, obtuse, almost imperceptibly passing into the style.   *Roem. & Schult. Syst. Veg.* iii. 350; *Meisn. in DC. Prodr.* xiv. 244.   *P. caulescens, E. Meyer ex Meisn. l.c., partly. P. undulata, Phillips in Kew Bulletin,* 1910, 233.   *Erodendrum tenax, Salisb. Parad. t.* 70.   *E. fœtidum, Knight, Prot.* 46.

Var. β, **latifolia** (Meisn. in DC. Prodr. xiv. 244); shrub 1½–2½ ft. high; leaves 3–5½ in. long, 6–16 lin. broad, lanceolate or obovate-cuneate.   *P. caulescens, E. Meyer ex Meisn. l.c., partly.   P. magnoliæfolia, Buek ex Meisn. l.c.*

COAST REGION: Uniondale Div.; plains in Lange Kloof, *Niven!* Uitenhage Div.; Galgebosch and Van Stadens Berg, *Drège,* 3361! Alexandria Div.; Zuurberg Range, *Drège,* 2009! Var. β: Uitenhage Div.; Van Stadens Berg, *Pappe! Burchell,* 4755! Albany Div.; Assegaibosch, *Zeyher,* 3667! between Riebeck East and Grahamstown, *Burchell,* 3508! Zwartwater Poort, *Burchell,* 3436! near Grahamstown, *MacOwan,* 1203! *Pym,* 1174! *Misses Daly & Sole,* 473! Howisons Poort, *Galpin,* 26!

45. P. transvaalensis (Phillips in Kew Bulletin, 1911, 84); branches glabrous, bark almost black lower down; leaves 3–4 in. long (rarely less than 2½ in. long), ¾–1¼ in. broad, oblong to oblanceolate, obtuse, narrowed at the base, distinctly pinnately veined, glabrous, with thin transparent cartilaginous margins; head (immature) sessile, 1¾–2 in. long, about 1 in. in diam., oblong; receptacle slightly convex; involucral bracts 10-seriate, more or less minutely fulvously ciliate; outer ovate, subobtuse, very minutely pubescent below or almost glabrous; inner oblong, obtuse, glabrous, or sometimes minutely pubescent near the apex; perianth-sheath over 1 in. long, dilated and 7-nerved below, glabrous outside, fulvously pubescent within from the widened base upwards; lip 7 lin. long, 3-toothed, glabrous for the greater part, but sparsely hairy towards the teeth and along the sides (but the limb of the anticous perianth-segment fulvously pilose); lateral teeth $\frac{3}{4}$ lin. long, oblong, fulvously villous; median tooth $\frac{1}{2}$ lin. long, narrower than the 2 lateral teeth; stamens subsessile; filaments concave; anthers linear, $4\frac{1}{2}$ lin. long; apical glands $\frac{1}{8}$ lin. long, ovate, swollen on the inner face; hypogynous scales $\frac{1}{3}$ lin. long, $\frac{1}{6}$ lin. broad, oblong, obtuse; ovary 1 lin. long, covered with long hairs; style over 1 in. long, slender, glabrous; stigma $4\frac{1}{2}$ lin. long, filiform, obtuse, furrowed, almost imperceptibly passing into the style.

KALAHARI REGION : Transvaal ; Goedgeluk, Zoutpans Berg, *Burtt-Davy*, 5179 !

46. P. Flanaganii (Phillips in Kew Bulletin, 1910, 232); branches glabrous; leaves $3\frac{1}{2}$–5 in. long, occasionally $1\frac{1}{2}$ in. long, 3–7 lin. broad above, $\frac{3}{4}$–$1\frac{1}{4}$ lin. broad at the base, linear-oblanceolate to strap-shaped, obtuse or subacute, narrowed or attenuated at the base, glabrous; head sessile, $2\frac{1}{4}$ in. long, about $2\frac{3}{4}$ in. in diam.; involucral bracts 8–10-seriate ; outer ovate, obtuse, minutely silky-pubescent on the lower half or glabrous; inner oblong or spathulate-oblong, exceeding the flowers; perianth-sheath 12–14 lin. long, dilated, 3-keeled and 7-nerved below, fulvous-pubescent or villosulous within in the upper part; glabrous outside, ciliate, lip $5\frac{1}{2}$ lin. long, 3-toothed, glabrous or with scattered stiff hairs; lateral teeth $\frac{3}{4}$ lin. long, broadly oblong, ciliate ; median tooth $\frac{1}{2}$ lin. long, ovate, acuminate ; stamens all fertile, sessile; anthers linear, $3\frac{1}{2}$–$4\frac{1}{2}$ lin. long ; apical glands $\frac{1}{3}$ lin. long, elliptic, obtuse, swollen on the inner face ; ovary 2 lin. long, obovate-oblong, covered with yellowish-red hairs ; style $1\frac{1}{4}$ in. long, somewhat flexuous, glabrous, bulbously thickened above the ovary, then constricted, then widening and compressed to the middle, whence gradually tapering ; stigma $3\frac{1}{2}$–$4\frac{1}{2}$ lin. long, very slender, obtuse, passing gradually or with a small bend into the style.

COAST REGION: Komgha Div. ? ; Gwenkala River, *Flanagan*, 804 !
EASTERN REGION : Transkei ; near Kentani, *Miss Pegler*, 274 !

The material of Pegler, 274, consists of a barren shoot and a detached head. The smaller leaf measurements all refer to the barren shoot.

47. P. simplex (Phillips in Kew Bulletin, 1910, 232); stem simple, $\frac{1}{2}$–$1\frac{1}{2}$ ft. high, glabrous; leaves 2–$3\frac{1}{2}$ in. long, 3–8 lin. broad, oblanceolate to linear-lanceolate, acute or subobtuse, long attenuated at the base, indistinctly veined, with a distinct midrib, glabrous ; head sessile, $1\frac{1}{2}$ in. long, about $1\frac{1}{2}$ in. in diam. ; involucral bracts 6-seriate, glabrous or the outermost sometimes minutely silky-pubescent ; outer ovate, subacuminate, obtuse ; inner oblong, equalling the flowers ; perianth-sheath 9 lin. long, dilated, 3-keeled and 7-nerved below, fulvous-villosulous within in the upper part; lip $5\frac{1}{2}$ lin. long, 3-toothed, glabrous; lateral teeth 3 lin. long, oblong; median tooth $\frac{1}{2}$ lin. long, ovate, acuminate, setulose ; stamens all fertile, subsessile; anthers linear, $2\frac{3}{4}$ lin. long ; apical glands ovate, obtuse, swollen on the inner face ; hypogynous scales 1 lin. long, elliptic, acuminate, obtuse ; ovary $2\frac{1}{2}$ lin. long, oblong in outline, covered with reddish-brown hairs; style 10½ lin. long, slightly flexuous, bulbously thickened above the ovary, then constricted and subterete; stigma 2¾ lin. long, very slender, obtuse, slightly wavy at the junction with the style.

KALAHARI REGION : Swaziland ; near Embabaan, *Burtt-Davy,* 2767 ! 2896 !
EASTERN REGION: Tembuland ; Bazeia Mountains, *Baur,* 608 ! Griqualand East ; Pot River Berg, *Galpin,* 6823 ! Natàl ; *Gerrard,* 721 ! *Sutherland* !

48. P. lanceolata (E. Meyer in Drège, Zwei Pfl. Documente, 213, name only) ; a bush 6 ft. high ; branches glabrous ; leaves 1–2¾ in. long, 3–5 lin. broad, oblanceolate, obtuse, attenuated at the base, indistinctly veined, glabrous ; head sessile, 2 in. long, about 1¼ in. in diam. ; involucral bracts 9–11-seriate, glabrous ; outer ovate, subacuminate, obtuse, with dark margins and tips ; inner oblong, shorter than the flowers ; perianth-sheath much widened, 7-nerved and 3-keeled and glabrous up to 5–6 lin. from the base, then much attenuated, sparingly setulose outside and rufo-pubescent within, 1 in. long ; lip 7–8 lin. long, glabrous or with a few setulæ at the apex, 3-toothed, 3-keeled ; teeth subequal, ½ lin. long ; stamens all fertile, sessile ; anthers linear, 6 lin. long ; apical glands ⅓ lin. long, oblong, obtuse ; style over 1¼ in. long, somewhat flexuous upwards, glabrous, bulbously thickened above the ovary, very slender above ; stigma finely filiform, 4–5 lin. long, obtuse, imperceptibly passing into the style ; fruit 3–4 lin. long, cylindric-obovoid, crowned by the persistent bulbous base of the style. *Meisn. in DC. Prodr.* xiv. 240.

COAST REGION : Riversdale Div. ; Zandhoogte, *Muir in Herb. Galpin,* 5306 ! Hooge Kraal, near Zoetemelks River, *Drège* ! Mossel Bay Div. ; Honig Klip, *Drège* ! ; Between Mossel Bay and Cape St. Blaize, *Burchell,* 6258 !

49. P. Doddii (Phillips in Kew Bulletin, 1911, 82) ; a small shrub 12–14 in. high ; branches glabrous ; leaves 1¾–2¼ in. long, 2¼–2½ lin. broad, linear to linear-oblanceolate, obtuse, narrowed at the base, indistinctly veined, with a somewhat sunken midrib, glabrous ; head sessile, 2 in. long, about 1½ in. in diam. ; receptacle slightly convex ; paleæ ovate, acute ; involucral bracts 12-seriate ; outer ovate, acute or subacute, glabrous or the very lowest silky-pubescent and ciliate ; inner oblong, obtuse, concave, glabrous, equalling the flowers ; perianth-sheath 10–11 lin. long, dilated, 3-keeled and 7-nerved below, glabrous except on the puberulous sides within ; lip 4½ lin. long, 3-toothed, 3-keeled, glabrous with little tufts of setulæ on the tips of the teeth ; teeth equal, ⅔ lin. long ; stamens sessile ; anthers linear, 3 lin. long ; apical glands ⅛ lin. long, ovate, obtuse ; ovary 1 lin. long, obovate-elliptic in outline, covered with long brownish-yellow hairs ; style 1¼ in. long, bulbously thickened at the base, then slender, flexuous, glabrous ; stigma 3 lin. long, filiform, obtuse, passing with a slight bend into the style.

COAST REGION : East London Div. ; between Gonubie and Quinera Rivers, *Dodd in Herb. Galpin,* 7936 !

50. P. Marlothii (Phillips in Kew Bulletin, 1910, 233); branches glabrous; leaves $2\frac{1}{4}$–3 in. long, 3–7 lin. broad, oblanceolate, acute, mucronate, narrowing at the base, distinctly veined, glabrous; heads sessile, 3 in. long, about 3 in. in diam.; receptacle 9 lin. high, conical; involucral bracts 13-seriate, glabrous; outer orbicular-ovate, obtuse, oblong, shorter than the flowers; perianth-sheath 9 lin. long, expanded and 5-nerved below, setose outside in the upper part, otherwise glabrous; lip 3 lin. long, 3-toothed, setose; teeth subequal, $\frac{1}{4}$ lin. long; stamens all fertile; anthers subsessile, linear, $1\frac{3}{4}$ lin. long : apical glands $\frac{1}{4}$ lin. long, ovate, obtuse; hypogynous scales $\frac{3}{4}$ lin. long, $\frac{1}{3}$ lin. broad, elliptic, obtuse; ovary $1\frac{1}{2}$ lin. long, obovoid, covered with long yellow-brown hairs; style 13 lin. long, curved, very much compressed below, glabrous; stigma $1\frac{3}{4}$ lin. long, obtuse, passing into the style.

COAST REGION: Worcester Div.; Matroos Berg, *Marloth*!

51. P. effusa (E. Meyer in Drège, Zwei Pfl. Documente, 82, name only); branches glabrous; leaves 1–2 in. long, 3–4 lin. broad, lanceolate, acute, mucronate, narrowing at the base, distinctly veined; head sessile, 2 in. long, about 3 in. in diam.; involucral bracts 7-seriate, glabrous; outer ovate, subacuminate, obtuse, with membranous margins; inner oblong, convex, incurved above, exceeding the flowers; perianth-sheath 9 lin. long, dilated and 5-nerved below, rufous-setose outside in the upper part, otherwise glabrous; lip 3 lin. long, 3-toothed, rufous-setose; teeth subequal, $\frac{1}{3}$ lin. long; stamens all fertile; filaments $\frac{1}{6}$ lin. long, flat; anthers linear, $1\frac{3}{4}$ lin. long ; apical glands $\frac{1}{6}$ lin. long, ovate, subacute, swollen on the inner face; ovary small, covered with long golden hairs; style 10–12 lin. long, falcate, compressed, suddenly and obliquely widened above the ovary (to $\frac{3}{4}$ lin.), then gradually tapering, glabrous; stigma $1\frac{3}{4}$–2 lin. long, filiform obtuse, grooved, almost imperceptibly passing into the style. *Meisn. in DC. Prodr.* xiv. 240.

COAST REGION: Worcester Div.; Dutoits Kloof, *Drège*!

52. P. pendula (R. Br. in Trans. Linn. Soc. x. 87); branches softly hirsute above, at length glabrous; leaves $\frac{3}{4}$–$1\frac{3}{4}$ in. long, 2–$2\frac{1}{2}$ lin. broad, narrowly oblanceolate-acute, with a recurved mucro, distinctly veined beneath, very minutely and loosely pubescent, at length glabrous; head sessile, $1\frac{1}{2}$ in. long, about $1\frac{1}{2}$ in. in diam., pendulous; involucral bracts 7–8-seriate; outer ovate, subacuminate, obtuse, silky-pubescent or tomentose on the lower half, with membranous margins, ciliate; inner oblong, incurved above, slightly concave, minutely pubescent outside, exceeding the flowers; perianth-sheath 9 lin. long, dilated, 3-keeled and 3-nerved below, rufous-pilose outside in the uppermost part; lip $2\frac{1}{2}$ lin. long, 3-toothed, setose below with a few stiff rufous hairs; teeth subequal, glabrous, $\frac{1}{4}$ lin. long : stamens all fertile ·

anthers linear, 1¼ lin. long; apical glands ⅛ lin. long, ovate; ovary oblong, covered with long reddish-brown hairs; style 10 lin. long, curved to falcate, compressed, obliquely dilated above the ovary, then gradually tapering, glabrous; stigma 1½ lin. long, filiform, obtuse, almost imperceptibly passing into the style. *Roem. & Schult. Syst. Veg.* iii. 350; *Meisn. in DC. Prodr.* xiv. 241.

SOUTH AFRICA: without locality, *Masson*! also in *Herb. Forsyth at Kew*!

COAST REGION: Tulbagh Div.; Witzenberg Range, *Zeyher*, 3687 ex *Meisner*.

53. **P. sulphurea** (Phillips in Kew Bulletin, 1910, 234); a depressed shrub, 6 ft. high; branches glabrous; leaves 10–14 lin. long, 3½–6 lin. broad, narrowly obovate-cuneate or oblanceolate, subacute, mucronate, indistinctly veined, minutely rugulose, glaucous or (in the dry state) yellowish, glabrous; heads subsessile, 2½ in. long, about 3 in. in diam., pendulous; receptacle 12 lin. high, conical; involucral bracts 9–10-seriate, glabrous; outer ovate, subacute, minutely ciliate; inner oblong or spathulate-oblong, exceeding the flowers; perianth-sheath 11 lin. long, dilated, 3-keeled and 6–7-nerved below, sparingly setose outside to the upper part, otherwise glabrous; lip up to 4 lin. long, 3-toothed and slightly recurved, with a few scattered setose hairs; lateral teeth ⅔ lin. long; median tooth ½ lin. long; stamens all fertile; filaments ¼ lin. long, channelled; anthers linear, 2¼ lin. long; apical glands ⅙ lin. long, ovate, acute; hypogynous scales ⅔ lin. long, oblong, obtuse; ovary covered with long spreading yellow-brown hairs; style 1¼ in. long, falcate, compressed, tapering towards both ends, glabrous; stigma 2½ lin. long, obtuse, grooved, imperceptibly passing into the style.

CENTRAL REGION: Laingsburg Div.; Witteberg Range, near Matjesfontein, *Marloth*, 3208, *Pearson*!

54. **P. canaliculata** (Haw. in Andr. Bot. Rep. t. 437); a small decumbent plant; branches glabrous; leaves 5–7 in. long, ¾–1½ lin. broad, linear, acute or subacute, narrowed at the base, indistinctly veined, glabrous; head sessile, 1¾ in. long, about 1½ in. in diam.; involucral bracts 9-seriate; outer ovate, obtuse, silky-pubescent, ciliate; inner oblong, concave, pubescent, at length glabrous, ciliate, equalling the flowers; perianth-sheath 9 lin. long, glabrous, upper half narrow, white and almost hyaline between reddish nerves, with wavy margins, lower half dilated, 5-nerved and faintly 3-keeled; lip over 4 lin. long, tridentate, glabrous; teeth subequal, over ½ lin. long, ovate, with a cylindric apiculus, long villously penicillate; stamens all fertile; filaments ½ lin. long, dilated, concave; anthers linear, 2½ lin. long; apical glands ⅓ lin. long, ovate, obtuse; ovary 1½ lin. long, covered with long reddish-brown hairs; style 10 lin. long, curved, compressed and as wide as the ovary below, then gradually tapering; stigma 2 lin. long,

grooved, subacute, almost imperceptibly passing into the style.
R. Br. in Trans. Linn. Soc. x. 88; *Roem. & Schult. Syst. Veg.* iii.
351; *Meisn. in DC. Prodr.* xiv. 241. *Erodendrum pæoniflorum,*
Knight, Prot. 46.

SOUTH AFRICA: without locality, *Roxburgh* !
COAST REGION: Uniondale Div. ; Lange Kloof, *Niven* !

55. **P. cedromontana** (Schlechter in Engl. Jahrb. xxvii. 109);
branches glabrous; leaves 3–4 in. long, 1–1½ lin. broad, linear,
acute or subacute, long attenuated at the base, indistinctly veined,
glabrous; head sessile, 1¼ in. long, 1 in. in diam., oblong; involucral bracts 6-seriate, glabrous, with membranous margins;
outer ovate, subacuminate, obtuse; inner erect, oblong, equalling
the flowers; perianth-sheath 5 lin. long, rather wide and gradually
passing into the dilated, faintly 3-keeled glabrous base, membranous, glabrous or nearly so except near the lip; lip 2 lin. long,
3-toothed, with rufous rigid hairs at the base and more or less so
along the lateral keels; teeth oblong, subobtuse, equal, ½ lin. long,
glabrous; stamens all fertile; filament flattened, ¼ lin. long;
anthers linear, 1½ lin. long; apical glands ovate, ⅙ lin. long; ovary
up to 2 lin. long, oblong, covered with reddish hairs; style 9 lin.
long, falcate, compressed, obliquely widened at the base, then
gradually tapering, glabrous; stigma 1½ lin. long, filiform,
obtuse.

COAST REGION: Clanwilliam Div.; Honig Valley and Ezelsbank, *Drège* !
Schlechter, 8808 !

56. **P. odorata** (Thunb. Prodr. 187; Fl. Cap. ed. Schult.
130); branches glabrous; leaves 1–2 in. long, ½–1 lin. broad,
linear-subulate, ending in a fine pungent mucro, glabrous, midrib
prominent; head sessile, obovoid, 1 in. long, ¾ in. wide; involucral
bracts 5–6-seriate, glabrous, white, adpressed; outer ovate to ovate-
lanceolate, subacuminate; inner lanceolate, acutely acuminate,
almost pungent, exceeding the flowers; perianth-sheath 4 lin. long,
3½ lin. long, glabrous, gradually dilated, 3-keeled and 5-nerved
below; lip 3 lin. long, 3-toothed, glabrous, crimson; teeth subequal,
½ lin. long, with a tuft of stiff white or fulvous hairs; stamens
all fertile; filaments ¼ lin. long, hardly flattened; anthers linear,
1½ lin. long; apical glands ¼ lin. long, oblong, obtuse; ovary ¾ lin.
long, obovate in outline, covered with long white hairs; style
5 lin. long, almost straight, slender, widened above the ovary,
then gradually tapering; stigma 1¼ lin. long, filiform, obtuse,
almost imperceptibly passing into the style. *P. mucronifolia,*
Salisb. in Parad. Lond. t. 24; *Bot. Mag. t.* 933; *Andr. Bot.*
Rep. t. 500; *R. Br. in Trans. Linn. Soc.* x. 86; *var. Brownii,*
Meisn. in DC. Prodr. xiv. 241. *Leucadendron* (?) *odoratum,*
Steud. Nomencl. ed. 2, ii. 35; *Meisn. l.c.* 228. *Erodendrum mucroni-*

folium, Knight, Prot. 48.   *P. odoratissima, Masson, in Herb. Ait. ex Meisn. l.c.* 240.   *P. mucronata, Hort. ? ex Steud. Nomencl. ed.* 2, ii. 400.

VAR. *β,* **Gueinzii** (Stapf); differs from the type in having narrower channelled leaves with recurved margins. *P. mucronifolia, var. Gueinzii, Meisn. in DC. Prodr.* xiv. 241.

SOUTH AFRICA : Var. *β :* without locality, *Roxburgh,* 47 ! *Gueinzius* !
COAST REGION : Clanwilliam Div. ; sandy flats near Berg Vley, *Niven* !

57. **P. scolymocephala** (Reichard, Syst. Pl. i. 271) ; a small bush 2–3 ft. high ; branches glabrous ; leaves ¾–2½ in. long, ¾–2 lin. broad, narrowly linear-oblanceolate, acute, mucronate, long attenuated at the base, indistinctly veined, glabrous ; head sessile, ¾–1 in. long, about 1 in. in diam. at the base, somewhat flattened ; receptacle conical ; paleæ acute ; involucral bracts 6–7-seriate, glabrous, pale green, with membranous margins, ciliolate ; outer ovate, obtuse ; inner spreading, oblong, obtuse ; perianth-sheath 5–6 lin. long, much curved, glabrous, gradually widened from the middle downwards, 5-nerved and faintly 3-keeled ; anticous limb hairy on the back ; lip 1½ lin. long, 3-toothed, pinkish, with an apical tuft of white hairs ; lateral teeth ¼ lin. long ; median tooth smaller ; stamens all fertile, subsessile ; anthers linear-oblong, ¾ lin. long ; ovary up to 2 lin. long, oblong, covered with long brown hairs ; style 5–6 lin. long, compressed and more or less obliquely widened above the ovary, then gradually tapering and falcate, glabrous ; stigma ¾ lin. long, finely filiform, almost imperceptibly passing into the style. *P. Scolymus, Thunb. Diss. Prot.* 33 ; *Thunb. Prodr.* 26 ; *Willd. Sp. Pl.* i. 522 ; *Wendl. Sert. Hannov.* i. iv. 4, *t.* 20 ; *Andr. Bot. Rep. t.* 409 ; *Bot. Mag. t.* 698 ; *Poir. Encycl.* v. 647 ; *R. Br. in Trans. Linn. Soc.* x. 86 ; *Roem. & Schultes, Syst. Veg.* iii. 349 ; *Thunb. Fl. Cap. ed.* 1, 483 ; *Meisn. in DC. Prodr.* xiv. 239. *P. angustifolia, Salisb. Prodr.* 49. *Erodendrum scolymiflorum, Knight, Prot.* 48. *Leucadendron scolymocephalum, Linn. Sp. Pl. ed.* i. 92 ; *Berg. in Vet. Acad. Handl. Stockh.* 1766, 323.—*Lepidocarpodendron acaulon, ramis numerosis, etc., Boerh. Ind. Pl. Hort. Ludg. Bat.* ii. 192, *with plate. Scolymocephalus foliis angustis, etc., Weinm. Phyt.* iv. 288, *t.* 893.

SOUTH AFRICA : without locality, *Ludwig* ! *Thunberg*! *Sieber* ! *Grey* ! *Ecklon,* 324 ! *Bergius.*
COAST REGION : Piquetberg Div. ; between twenty-four Rivers and Pikeniers Kloof, *Drège* ! Paarl Div. ; between Mosselbanks River and Berg River, *Burchell,* 975 ! Cape Div. ; Devils Mountain, *Ecklon,* 656 ! near Wynberg, *Burchell,* 783 ! *Drège* ! *Bolus,* 2907 ! 3845 ! Red Hill, *Jameson* ! Table Mountain, *MacOwan, Herb. Austr.-Afr.* 1950 ! Hout Bay Valley, *Phillips,* 518 ! Muizen Berg, *Wilms,* 3572 ! Simons Bay, *Wright* ! Stellenbosch Div. ; between Stellenbosch and Cape Flats, *Burchell,* 8345 !

58. **P. Harmeri** (Phillips in Kew Bulletin, 1911, 83) ; a bush about 3 ft. high ; branches greyish tomentellous above, becoming glabrous ; leaves 1¾–2½ in. long, 1¼–1¾ lin. broad, linear, obtuse to subacute with a callous point attenuated at the base, margins recurved ; youngest leaves finely villous at the base ; head sessile,

1 in. long, about 1 in. in diam., globose; receptacle convex; involucral bracts 10–11-seriate; outer ovate, obtuse, glabrous or the
lowest very finely pubescent, with membranous ciliate margins;
inner oblong-spathulate, obtuse, brick-red, recurved above, glabrous
or minutely pubescent, not equalling the styles; perianth-sheath
7 lin. long, $\frac{1}{3}$ lin. broad above, dilated, 3-keeled and 3-nerved below,
glabrous or hirsute at the apex; lip 2 lin. long, 3-toothed, rufously
setulose, glabrescent on the back; teeth subequal, $\frac{1}{6}$ lin. long, the
middle one smaller; stamens all fertile, subsessile; anthers linear,
$1\frac{1}{2}$ lin. long; apical glands $\frac{1}{6}$ lin. long, ovate, obtuse, swollen on
the inner face; ovary 1 lin. long, oblong-obovate in outline, covered
with long brown hairs; hypogynous scales $\frac{1}{2}$ lin. long, $\frac{1}{8}$–$\frac{1}{4}$ lin.
broad, oblong, obtuse; style 10 lin. long, falcate, arching over the
centre of the head, terete above, flattened and hollow below,
glabrous; stigma $1\frac{1}{4}$ lin. long, linear, obtuse, grooved, passing into
the style.

CENTRAL REGION: Laingsburg Div.; hill near Matjesfontein, *Harmer*!

59. **P. pityphylla** (Phillips in Kew Bulletin, 1910, 234); branches
glabrous; leaves $2\frac{1}{2}$–3 in. long, about $\frac{1}{8}$ lin. wide, needle-shaped,
acute, pungent, channelled and prominently costate on the upper
face, glabrous; head sessile, $1\frac{1}{2}$–$1\frac{3}{4}$ in. long, about $2\frac{1}{2}$ in. in diam.,
cernuous; involucral bracts 7-seriate, glabrous; outer ovate,
acuminate, obtuse or acute, the lowest produced into long foliaceous
appendages resembling the leaves; inner oblong, slightly concave,
exceeding the flowers; perianth-sheath 8 lin. long, dilated, 3-keeled
and 7-nerved below, scarious, rufously setulose within in the upper
part, otherwise glabrous; lip $2\frac{2}{3}$ lin. long, 3-toothed, 3-keeled, setose
below; teeth subequal, $\frac{1}{4}$ lin. long; stamens all fertile; filaments
$\frac{1}{4}$ lin. long, dilated, concave; anthers oblong-linear, $1\frac{2}{3}$ lin. long;
apical glands $\frac{1}{6}$ lin. long, ovate, subacute, somewhat swollen on the
inner face; ovary 1 lin. long, obovate-oblong, covered with long
reddish-yellow hairs; hypogynous scales $\frac{1}{2}$ lin. long, oval-oblong;
style up to 11 lin. long, widened and much compressed from the
base upwards for 3 lin., then much constricted and strongly bent
and subulate, the slender portion obliquely arching inwards,
glabrous; stigma $1\frac{2}{3}$ lin. long, obtuse.

COAST REGION: Ceres Div.; Mitchells Pass, *MacOwan*, 2907! *MacOwan, Herb.
Austr.-Afr.* 913! *Bodkin in Herb. Bolus*, 6089!

60. **P. witzenbergiana** (Phillips in Kew Bulletin, 1910, 234);
decumbent; branches villous; leaves 10–14 lin. long, $\frac{1}{2}$–$\frac{3}{4}$ lin.
broad, needle-shaped, acute, mucronate, channelled on the upper
face, convex below, minutely punctate, glabrous or scantily pilose;
heads sessile, subglobose and obtuse in bud, when expanded 2 in.
long, about 2 in. in diam.; involucral bracts 9–10-seriate, glabrous;
outer ovate, produced into long foliaceous appendages, resembling
leaves, ciliolate; inner oblong, eciliolate, slightly convex on the

back, exceeding the flowers; perianth-sheath 7 lin. long, dilated, setulose-ciliate along the margin and densely rufously hairy within, otherwise glabrous, dilated, 3-keeled and 7-nerved below; lip 3 lin. long, glabrous, or sometimes with a few scattered bristles; stamens all fertile; filaments $\frac{1}{6}$ lin. long, swollen, channelled; anthers linear, $1\frac{1}{2}$ lin. long; apical glands $\frac{1}{8}$ lin. long, ovate, acute, swollen on the inner face; ovary obovate-oblong, covered with a tuft of long fulvous hairs; style 8–9 lin. long, obliquely widened and compressed above the ovary, then subulate, falcate, glabrous; stigma $1\frac{1}{2}$ lin. long, grooved, obtuse, passing with an obscure bend into the style.

COAST REGION : Tulbagh and Ceres Div. ; Witzenberg Range, *Zeyher*, 3687 ! *Burchell*, 8676 !

61. P. rosacea (Linn. Mant. alt. 189); small shrub with a simple stem 4–6 in. long ; branches numerous, gracefully curved, glabrous ; leaves 6–10 lin. long, $\frac{1}{4}$–$\frac{1}{3}$ lin. broad, linear, acicular, pungent, with a shallow groove along each side of the midrib, glabrous ; head sessile, $1\frac{1}{2}$ in. long, about $1\frac{1}{2}$ in. in diam., pendulous ; involucral bracts 8-seriate, glabrous, bright rose to crimson ; outer ovate to ovate-oblong, more or less obtuse, ciliate ; inner oblong, obtuse, but often apparently acuminate owing to the involute upper margin, exceeding the flowers ; perianth-sheath 5 lin. long, almost hyaline above, dilated, 3-keeled and 5-nerved below, glabrous, ciliate ; lip $1\frac{1}{2}$ lin. long, 3-toothed, oblong, rufously setulose above, ciliate ; teeth subequal, 1 lin. long ; stamens all fertile ; filaments $\frac{1}{4}$ lin. long, flattened ; anthers linear, 1 lin. long ; apical glands $\frac{1}{6}$ lin. long, ovate, acute, keeled on the inner face ; ovary covered with long light brown hairs ; style 7 lin. long, narrowed from the base upwards, curved inwards or at length almost erect, flattened below, glabrous ; stigma 1 lin. long, cylindric, obtuse, hardly swollen at the junction with the style. *Lam. Ill.* i. 238 ; *Poir. Encycl.* v. 653 ; *Smith, Exot. Bot.* i. 85, *t.* 44. *P. nana, Thunb. Diss. Prot.* 51 ; *Murr. Syst. Veg. ed.* xiv. 139 ; *Willd. Sp. Pl.* i. 519 ; *R. Br. in Trans. Linn. Soc.* x. 87 ; *Thunb. Fl. Cap. ed.* 1, 475 ; *Meisn. in DC. Prodr.* xiv. 241. *P. acuifolia, Salisb. Parad. t.* 2. *Leucadendron nanum, Berg. in Vet. Acad. Handl. Stockh.* 1766, 325. *Erodendrum acuifolium, Knight, Prot.* 49.

COAST REGION : Tulbagh Div. ; mountain near Tulbagh Waterfall, *Bergius, Bolus*, 5229 ! *Phillips* ! near Tulbagh, *Pappe* ! Witzenberg Range, near Tulbagh, *Zeyher*, 1459 ! *Burchell*, 8670 ! Worcester Div. ; Dutoits Kloof, *Drège* ! CENTRAL REGION : Ceres Div. ; near Ceres, *Bolus, Herb. Norm. Austr.-Afr.*, 1090 !

62. P. cynaroides (Linn. Mant. alt. 190) ; a bush, up to 6 ft. high or sometimes acaulescent ; branches glabrous ; leaves petioled ; blade $2\frac{1}{4}$–$5\frac{1}{4}$ in. long, 2–$3\frac{1}{4}$ in. broad, varying from subrotundate and obtuse to elliptic and acute, cuneate at the base, prominently and reticulately veined on both sides, punctate, glabrous ; petiole up to

2 Q 2

4½ in. long, terete; head sessile, 5–8 in. long, about 5–8 in. in diam.; involucral bracts 12–13-seriate; outer ovate to ovate-lanceolate, acute, like the inner at first more or less densely whitish- or greyish-tomentose, often at length glabrescent or the lowest quite glabrous; inner lanceolate-oblong, acuminate, acute, mostly permanently tomentose, exceeding the flowers; perianth-sheath over 2 in. long, pubescent outside on the upper part, and pubescent to villous within from the widened base upwards, particularly along the sides, dilated, 7-nerved, faintly 3-keeled and glabrous below; lip 1 lin. long, tomentose, produced into 3 tomentose or villous awns; lateral awns 2 lin. long; median awn 2 lin. long; stamens all fertile; filaments ¼ in. long, flattened; anthers linear, 4 lin. long; apical glands ¾ lin. long, oblong, obtuse; ovary 2 lin. long, oblong, covered with long whitish hairs; style 2½ in. long, laterally much flattened, slightly curved inwards, pubescent, at least below; stigma 4 lin. long, filiform, obtuse, kneed and bent at the junction with the style. *Thunb. Diss. Prot.* 58; *Willd. Sp. Pl.* i. 534; *Thunb. Prodr.* 28; *Lam. Ill.* i. 234; *Poir. Encycl.* v. 639; *Bot. Mag. t.* 770; *Andr. Bot. Rep. t.* 288; *R. Br. in Trans. Linn. Soc.* x. 75; *Thunb. Fl. Cap. ed.* i. 514; *Roem. & Schultes, Syst. Veg.* iii. 342; *Roupell, Cap. Flow. t.* 8; *Meisn. in DC. Prodr.* xiv. 245 (*incl. vars. obtusifolia, elliptica and glabrata*). *P. cynaroides, var. elliptica, Klotzsch in Flora,* 1845, 75. *P. petiolata, Buek, ex Meisn. l.c. P. Woodwardii, Endl. Gen. Suppl.* iv. ii. 77. *Erodendrum cynaræflorum, Knight, Prot.* 43. *Leucadendron cinaroides,* α, *Linn. Sp. Pl. ed.* i. 92.—*Lepidocarpodendron folio subrotundo, Boerh. Ind. Pl. Hort. Lugd. Bat.* ii. 184, *t.* 184. *Scolymocephalus africanus folio, etc., Weinm. Phyt.* iv. 287, *t.* 892.

Coast Region: Tulbagh Div.; mountains above Tulbagh Waterfall, *Brodie!* Worcester Div.; Dutoits Kloof, *Drège!* Cape Div.; Table Mountain, *Thunberg!* *Burchell,* 660! *Drège! Phillips!* Swellendam Div.; near Swellendam, *Burchell,* 7381! *Fry in Herb. Galpin,* 4983! George Div.; Kaymans Gat, *Drège.* Knysna Div.; Outemqua Mountains, *Drège.* Humansdorp Div.; Zitzikamma, *Krauss!* Port Elizabeth Div.? *Cooper,* 3068! Uitenage Div.; Van Stadens Berg, *Ecklon.* Albany Div.; near Grahamstown, *Cooper,* 53! *MacOwan,* 1202!

Meisner's var. *glabrata* is based on Krauss's specimen from Zitzikamma, which is at the same time the type of Klotzsch's var. *elliptica,* a variety kept up by Meisner himself. It has less hairy involucral scales than the usual form, but they are by no means glabrous. As to Andrews, t. 288, also referred by Meisner to his var. *glabrata,* there is nothing in the plate to show that the scales of the plant figured were glabrous.

63. **P. cryophila** (Bolus in Trans. Royal Soc. S. Africa, i. 163); stem short; leaves 12–13½ in. long, 1½–2 in. broad, oblanceolate, subacute, attenuated into a long petiole, distinctly veined, coriaceous, glabrous; petiole flat above, convex on the back; head sessile, erect, 6–7 in. long, about 5 in. in diam.; involucral bracts 10–12-seriate; outer lanceolate, long-acuminate, acute, the lowest glabrous or glabrescent, the following increasingly white-tomentose to felted, inner ovate-lanceolate, long-acuminate, recurved at the apex, very densely white-felted, shorter than the flowers;

perianth-sheath 2½ in. long, dilated, 3-keeled and 7-nerved below for ¾ in., then attenuated, membranous and very densely white-ciliate on the inner margin, except for a short space above; lip 8 lin. long, 3-toothed, densely whitish-villous on the sides, glabrous on the back; lateral teeth 1½ lin. long; median tooth 1 lin. long; stamens all fertile; filaments ⅔ lin. long, oblong-linear; anthers linear, 3½ lin. long; apical glands ⅔ lin.· long, oblong-linear, obtuse; ovary 6-7 lin. long, cylindric, constricted into a short beak and ending with an annular thickening at the junction of the style, covered with long whitish hairs; style 2½ in. long, gently curved, compressed and widened above the ovary, then subulate, terete, glabrous; stigma 3½ lin. long, subacute, faintly grooved, very slender, imperceptibly passing into the style. *P. chionantha, Bolus in Trans. S. Afric. Philos. Soc.* xvi. 399, *not of Engler & Gilg, in Warb. Kunene-Samb.-Exped.* 225.

COAST REGION: Clanwilliam Div.; summit of Sneeuwkop, *Bodkin in Herb. Bolus,* 8676!

64. **P. Scolopendrium** (R. Br. in Trans. Linn. Soc. x. 94); an acaulescent plant; leaves petioled; blade 7-8 in. long, 2-2½ in. broad, broadly oblanceolate, obtuse, prominently pinnately veined, with a prominent midrib, rugulose, glabrous; petiole 4½ in. long, flat above, convex dorsally; head subsessile, 2¾ in. long, about 3½ in. in diam.; involucral bracts 6-seriate, outermost lanceolate, glabrous, chestnut-brown, following lanceolate-ovate, acutely acumi-nate, more or less ciliate and pubescent on the back, at length glabrous; inner narrowly lanceolate to linear-lanceolate with tomentose lips shorter than the flowers; perianth-sheath 12-14 lin. long, dilated, 3-keeled and 7-nerved below, up to 8 lin. from the base, then rather suddenly attenuated, densely white ciliate and villous along the margins and higher up all over within, glabrous outside or nearly so; lip 5-6 lin. long, 3-toothed, densely villous on the sides, more or less glabrous on the back; lateral teeth 1 lin. long, oblong; median tooth ¾ lin. long; stamens all fertile; filaments ½ lin. long, thick, oblong-elliptic; anthers linear, 3 lin. long; apical glands ½ lin. long, lanceolate or oblong, hardly swollen on the inner face; ovary 1½ lin. long, oblong, covered with long whitish hairs; style 1¼ in. long, obliquely widened from the base upwards and much compressed, thin, slender, subulate, terete and curved, glabrous; stigma 3 lin. long, finely grooved, obtuse, imper-ceptibly passing into the style. *Roem. & Schultes, Syst. Veg.* iii. 354; *Meisn. in DC. Prodr.* xiv. 243; *St. Lager in Ann. Soc. Bot. Lyon,* vii. 132. *P. scolopendrina, Ind. Kew.* ii. 633. *Erodendrum scolo-pendriifolium, Knight, Prot.* 43.

COAST REGION: Tulbagh Div.; Great Winterhoek Mountains, 5000 ft., *Bolus,* 5232! *Niven!*

65. **P. scabriuscula** (Phillips in Kew Bulletin, 1910, 236, excl. syn.); an acaulescent plant; leaves 6-10 in. long, 3-6 lin. broad

at the widest part, linear-oblanceolate, acute, long attenuated at the base into a petiole, rough, glabrous or those surrounding the flower-heads somewhat hirsute below, with very wavy margins; head sub-sessile, 2½ in. long, about 1¾ in. in diam.; involucral bracts 6-seriate, lanceolate, acuminate, acute, densely fulvously woolly-tomentose, lower at length becoming glabrous and dark chestnut-brown; inner not equalling the flowers; perianth-sheath 12 lin. long, gradually dilated, faintly 3-keeled and 5–7-nerved below, membranous, densely villous on the outside excepting at the glabrous base, hairs pallid; lip 3 lin. long, 3-toothed, hairy along the sides and below, glabrescent on the back; tips with a pallid tuft of hairs, at length glabrous; lateral teeth ½ lin. long, linear, obtuse, median tooth slightly shorter; stamens all fertile; anthers subsessile, linear, 2½ lin. long; apical gland linear-oblong, ¾ lin. long, red; ovary 1½ lin. long, oblong, covered with long white hairs; style 14 lin. long, slightly curved, subulate from an obliquely lanceolate com-pressed base, up to over 1 lin. wide in the lower quarter, pubescent below, otherwise glabrous; stigma 1½–2 lin. long, obtuse, almost imperceptibly passing into the style.

CENTRAL REGION : Ceres Div. ; Gydouw, _Bolus,_ 7557 ! ; _Schlechter,_ 10000 !

66. **P. aspera** (Phillips in Kew Bulletin, 1910, 236); an acaulescent plant ; leaves 4½–6 in. long, 1½–3 lin. broad, straight or falcate, linear, obtuse, mucronate, long attenuated at the base, rough with minute tubercles, glabrous or the innermost with a few scattered long hairs; head shortly stipitate, 3–3¼ in. long, about 2 in. in diam.; involucral bracts 9-seriate, finely pubescent outside, ciliate ; outer ovate, obtuse ; inner oblong to linear-oblong, and densely pubescent near the tips, shorter than the flowers; perianth-sheath 2 in. long, dilated, 3-keeled, and faintly 7-nerved below, densely villous-pilose within and without, except at the lower part of the widened base; lip 6 lin. long, 3-awned, tomentose, ending in a woolly tuft ; awns ovate, acuminate, white-woolly with a few dark cilia; lateral awns 1½ lin. long; median awn 1 lin. long; stamens all fertile ; filaments ¼ lin. long, swollen, expanded, deeply furrowed ; anthers linear, 3¼ lin. long; apical glands ⅓ lin. long, ovate and acute or ovate-oblong and subacute; ovary 2 lin. long, oblong-obovate in outline, covered with long reddish-brown hairs; style 2 in. long, narrowing from the base upwards, trigonous below, then more or less flattened, sparingly pubescent in the lowest quarter ; stigma 3½ lin. long, furrowed, subobtuse.

COAST REGION : Bredasdorp Div. ; near Elim, 200 ft., _Bolus,_ 7861 !

67. **P. lorea** (R. Br. in Trans. Linn. Soc. x. 93) ; an acaulescent plant ; leaves up to 12 in. long, ½ lin. broad, cylindric, acute, mucronate, glabrous, longitudinally grooved ; head long-stipitate, 4 in. long, about 2½ in. in diam.; stipes up to over 1 in. long; involucral bracts many-seriate, silky-pubescent, ciliate; lowest

ovate, acute, following ovate-lanceolate more or less acuminate or
subacute; inner elongate-oblong to linear, convex, more or less
acuminate, subacute, exceeding the flowers; perianth-sheath $2\frac{1}{2}$ in.
long, dilated, 3-keeled and 7-nerved below, then gradually narrowed
and slender, glabrous, densely white-ciliate for more than 1 in.;
lips 8 lin. long, 3-awned, very densely woolly along the margins,
more or less glabrous on the back; lateral awns 3 lin. long, linear-
oblong, hidden in dense whitish wool; median awn 2 lin. long,
ovate, acuminate; stamens all fertile; filaments $\frac{1}{2}$ lin. long, swollen,
deeply furrowed; anthers linear, 3 lin. long; apical glands $\frac{3}{4}$ lin.
long, oblong or lanceolate; ovary 1 lin. long, oblong in outline,
covered with long white hairs; style $2\frac{1}{2}$ in. long, narrowing from
the widened hollow base upwards, glabrous; stigma $3\frac{3}{4}$ lin. long,
grooved, obtuse, kneed and slightly bent at the junction with the
style. *Roem. & Schultes, Syst. Veg.* iii. 353; *Meisn. in DC. Prodr.*
xiv. 242. *P. aulax, Hibbert in Herb. Smith. ex Meisn. l.c. P.
coronata, Curt. ex Steud. Nomencl. ed.* i. 658. *Erodendrum pinifolium, Knight, Prot.* 45.

SOUTH AFRICA : without locality, *Masson*.
COAST REGION: Paarl Div. ; French Hoek, *Niven* ! Stellenbosch Div. ;
Hottentots Holland, *MacOwan, Herb. Norm. Austr.-Afr.* 781 ! Riversdale Div. ;
Garcias Pass, *Burchell*, 6958 !

68. **P. repens** (Thunb. Diss. Prot. 34); an acaulescent plant;
leaves $2\frac{1}{2}$–8 in. long, $\frac{1}{2}$–1 lin. broad, filiform to linear, acute,
coriaceous, glabrous, smooth or more or less tubercled and rough,
the uppermost sometimes long softly pilose at the base, upwards
with strongly recurved margins, hence dorsally channelled; heads
shortly stipitate, 4 in. long, about 3 in. in diam. ; involucral bracts
10-seriate, silky-pubescent without, ciliolate; outer ovate, obtuse;
inner much elongated, oblong; innermost narrowed downwards,
equalling the flowers; perianth-sheath 2 in. long, dilated, 7-nerved
at the base, densely and long pilose to villous outside except at the
glabrous base; lip 9 lin. long, densely pubescent to villous, 3-awned;
lateral awns $2\frac{3}{4}$ lin. long, shaggy with yellowish hairs; median awn
$1\frac{1}{4}$ lin. long; stamens all fertile; filaments $\frac{1}{4}$ lin. long, swollen,
expanded, deeply furrowed; anthers linear, $4\frac{1}{2}$ lin. long; ovary
1 lin. long, obovate in outline, covered with long reddish-brown
hairs; style 2 in. long, slightly curved or almost straight below,
somewhat widened and flattened below, then terete and tapering
above, glabrous or very sparingly puberulous; stigma finely grooved,
3–4 lin. long, subacute, passing into the style. *Lam. Ill.* i. 236;
Thunb. Prodr. 26; *Fl. Cap. ed.* i. 486; *Willd. Sp. Pl.* i. 523; *Poir.
Encycl.* v. 646; *R. Br. in Trans. Linn. Soc.* x. 92; *Roem. & Schultes,
Syst. Veg.* iii. 353; *Meisn. in DC. Prodr.* xiv. 242, *with vars. P.
canaliculata, Herb. Linn Soc. ex Meisn. l.c. P. Strobus, Meisn. l.c. not
of Andr. Leucadendron repens, β, Linn. Sp. Pl. ed.* i. 92. —*Lepidocarpodendron fol. longiss. etc., Boerh. Ind. Pl. Hort. Ludg. Bat.* ii.
190, *t.* 190; *Weinm. Phyt.* iv. 290, *t.* 897.

SOUTH AFRICA: without locality, *Thunberg*! *Gueinzius*!
COAST REGION: Worcester Div.; Dutoits Kloof, *Drège*! Paarl Div.; by the
Breede River, near Darling Bridge, *Bolus*, 5230! French Hoek, *Miss Treleaven*!
Niven! Cape Div.; Constantia ex *Boerhaave*. Port Elizabeth Div.; Algoa Bay,
Cooper, 3067!

69. **P. echinulata** (Meisn. in DC. Prodr. xiv. 242); an acaulescent
plant; leaves 5–7½ in. long, ½ lin. broad, filiform, flexuous, acute,
mucronate, channelled below in the upper part, coriaceous, tubercled, scabrous, with long soft hairs, at length glabrous; head
sessile or subsessile, 2½ in. long, about 2 in. in diam.; involucral
bracts 10-seriate; outer ovate to oblong-ovate, obtuse, glabrous or
ciliate; inner oblong; innermost elongated, pubescent or more or
less tomentellous or glabrescent in part, the margins fringed with
dense fulvous cilia equalling the flowers; perianth-sheath 1¾ in.
long, very slender, suddenly expanded and faintly 7-nerved below,
fulvous-villous except at the very base; lip 6 lin. long, 3-toothed
or nerved, fulvously villous, ciliate; lateral awns 2½ lin. long,
long villous, acuminate; median tooth over 1 lin. long; stamens
all fertile; filaments ½ lin. long, obovate, thickened; anthers linear,
4½ lin. long; apical glands ½ lin. long, lanceolate, swollen on the
inner face; ovary 2 lin. long, oblong, covered with long reddish-
brown hairs; style 1¾ in. long, curved and twisted, narrowing from
the base upwards, more or less triquetrous below and flattened
upwards, at least when dry, glabrous; stigma 5 lin. long, finely
channelled, obtuse, imperceptibly passing into the style. *Erodendrum restionifolium, Knight, Prot.* 45.

VAR. β, **minor** (E. P. Phillips); head 1¼ in. long, about 1¼ in. in diam.
COAST REGION: Worcester Div.; Brand Vley, *Niven,* 32! Caledon Div.;
between Genadendal and Donkers Hoek, *Burchell,* 7913! between Houw Hoek
and Bot River, *Bowie*! Klein River Mountains, *Zeyher,* 3670! *Ludwig*!
CENTRAL REGION: Var. β: Ceres Div.; Klein Vley, Cold Bokkeveld, *Schlechter*,
10215!

70. **P. turbiniflora** (R. Br. in Trans. Linn. Soc. x. 93); an
acaulescent plant; leaves 5–11½ in. long, 1½–1¾ in. broad at the
widest part, broadly oblanceolate to elliptic-oblanceolate, acute,
attenuated at the base into a long-winged petiole, prominently
pinnately veined, minutely rugulose, softly pilose when young, at
length glabrous, with undulating margins; head sessile, 2–2½ in.
long, about 1¾ in. in diam., erect, surrounded by subsessile short
obovate or obovate-lanceolate acute glabrous leaves, 2 in. long;
involucral bracts about 5-seriate, densely pubescent to tomentose,
ciliate; outer ovate-oblong, obtuse; inner oblong to linear-oblong,
densely ciliate or shortly bearded at the apex, equalling the
flowers; perianth-sheath 1¼ in. long, dilated, 3-keeled and obscurely
7-nerved at the glabrous base, densely hairy above; lip 6 lin. long,
pilose, produced into 3 long finely filiform wavy densely villous
awns, 3 lin. long; stamens all fertile; filaments ¼ lin. long, spathulate; anthers linear, 2½ lin. long; apical glands ¼ lin. long, ovate,
obtuse, swollen on the inner face; ovary 2 lin. long, cylindric,

covered with long brown hairs; style almost 1 in. long, terete, passing into the cylindric ovary, tapering upwards, slightly curved, glabrous; stigma 2½ lin. long, filiform, subacute, finely grooved, with a minute bend at the junction with the style. *Roem. & Schultes, Syst. Veg.* iii. 353; *Meisn. in DC. Prodr.* xiv. 244. *P. cæspitosa, Andr. Bot. Rep. t.* 526. *Erodendrum turbiniflorum, Salisb. Parad. t.* 108. *E. cæspitosum, Knight, Prot.* 43.

COAST REGION: Paarl Div.; Bushmans Kloof (Boshiesmans Gat), *Niven!* Caledon Div.; tops of mountains near Genadendal, *Burchell*, 7751!

71. **P. scabra** (R. Br. in Trans. Linn. Soc. x. 91); stems and branches subterranean with the flower-heads and surrounding leaves close to the ground; leaves 4½–10 in. long, 3–7 lin. (of young shoots up to 1 in.) broad, linear, linear-lanceolate or oblanceolate, flat or with revolute margins and then almost cylindric, acute to sub-obtuse or even obtuse, attenuated at the base, the broader ones distinctly veined, the lateral nerves joining into a more or less conspicuous marginal nerve, rough with small tubercles, when young loosely pilose with long flexuous white hairs, soon glabrous; head sessile, 1½–2 in. long, about 2 in. in diam.; involucral bracts 8-seriate, finely rufo-tomentose or pubescent, at length becoming glabrous; outer ovate, subacute, ciliate; inner oblong or linear-oblong, ciliate, exceeding the flowers; perianth-sheath 12 lin. long, gradually dilated, 3-keeled and 7-nerved below, loosely villous with the exception of the glabrous base; lip 5 lin. long, 3-toothed, villous; teeth subequal, ¾–1 lin. long, villous; stamens all fertile, subsessile; anthers linear, 3 lin. long; apical glands ½ lin. long, lanceolate or oblong, subacute; ovary 2 lin. long, obovate; densely covered with long reddish-brown hairs; style 1¼–1½ in. long, strongly curved, bulbously thickened at the base, terete above, glabrous; stigma 3½ lin. long, finely subulate, subobtuse. *Roem. & Schultes, Syst. Veg.* iii. 352; *Meisn. in DC. Prodr.* xiv. 243, *incl. vars.*

SOUTH AFRICA: without locality, *Roxburgh, Drège! Grey! Gueinzius!* COAST REGION: Caledon Div.; Houw Hoek, *Burchell*, 8075! *Zeyher*, 3673! *MacOwan*, 2911! & *in Herb. Norm. Austr.-Afr.* 903! Onrust River, and Haartebeest River, *Zeyher*, 3672! Ganzekraal, *Burchell*, 7549! Zoetemelks Valley, *Burchell*, 7593! Swellendam Div.; near Swellendam, *Zeyher*, 3671! Uniondale Div.; Long Kloof, *Mund.*

72. **P. tenuifolia** (R. Br. in Trans. Linn. Soc. x. 90); stems and branches subterranean with the flower-heads and surrounding leaves close to the ground; leaves 6–9½ in. long, 1–2 lin. broad, linear, with revolute margins, long attenuated at the base, acute, rough with small tubercles, when young loosely pilose with long flexuous white hairs, soon quite glabrous; heads sessile, 1½ in. long, about 1 in. in diam.; involucral bracts 7-seriate, finely rufo- or fulvo-pubescent, at length glabrescent, ciliate; outer ovate, obtuse; inner oblong or linear-oblong, not equalling the flowers; perianth-

sheath 8–9 lin. long, rather abruptly dilated, 3-keeled and 7-nerved
below, villous on the outside, hairs rufous or fulvous, glabrous at
the base; lip $3\frac{1}{2}$–4 lin. long, 3-toothed, tomentose to villous on the
sides, glabrescent on the back; teeth subequal, $\frac{1}{2}$ lin. long, villous;
stamens all fertile; filaments $\frac{1}{4}$ lin. long, ovate; anthers linear,
$2\frac{1}{3}$ lin. long; apical glands $\frac{1}{3}$ lin. long, lanceolate-ovate, subacute,
swollen on the inner face; ovary 3 lin. long, obovoid, covered with
long dark-brown hairs; style 12 lin. long, curved, subulate, tapering
from the bulbously much thickened base, glabrous; stigma 2 lin.
long, obtuse, imperceptibly passing into the style. *Roem. & Schultes,
Syst. Veg.* iii. 352 ; *Meisn. in DC. Prodr.* xiv. 242, *partly.* *P. revoluta,
Buek, ex Meisn. l.c.* *P. lorea, Drège ex Meisn. l.c., partly.* *P. scabra
var. stenophylla, Meisn. l.c.* 243.

SOUTH AFRICA : without locality, *Drège*!
COAST REGION : Swellendam Div. ; near Swellendam, *Niven*! *Fry in Herb.
Galpin*, 4985 ! Sparrbosch, *Drège.* Caledon Div. ; Great Houw Hoek and River
Zondereinde, *Pappe*!

73. P. acaulis (Thunb. Diss. Prot. 56); stems subterranean or
prostrate on the ground and up to 1 ft. long, glabrous; leaves
extremely variable, 3–$8\frac{1}{2}$ in. long, $\frac{3}{4}$–3 in. broad at the widest
part, rotundate-obovate, obovate-lanceolate or oblanceolate, usually
obtuse, rarely acute, often apiculate, long attenuated into a petiole,
distinctly veined, with a callous margin, glabrous, often glaucous;
head sessile, sometimes contracted into a short stipes, $1\frac{1}{4}$–$1\frac{1}{2}$ lin.
long, about $1\frac{1}{2}$ in. in diam.; involucral bracts 8-seriate, minutely
ciliate, otherwise glabrous, outer ovate, obtuse; inner oblong,
equalling the flowers; perianth-sheath 10–11 lin. long, rather wide,
gradually dilated, 3-keeled and 7-nerved at the base, densely ciliate
in the middle part, otherwise glabrous; lip 3 lin. long, 3-toothed,
pubescent; teeth subequal, $1\frac{1}{2}$ lin. long; stamens all fertile;
filaments $\frac{1}{6}$ lin. long, flattened; anthers linear, $1\frac{1}{2}$ lin. long; apical
glands $\frac{1}{4}$ lin. long, lanceolate, obtuse; ovary 1 lin. long, elliptic-
ovoid, covered with long reddish-brown hairs; style 11 lin. long,
sickle-shaped, subulate from a narrow lanceolate more or less
swollen base, terete above, glabrous; stigma $1\frac{1}{4}$ lin. long, finely
subulate, obtuse, gradually passing into the style. *Thunb. Prodr.*
27 ; *Willd. Sp. Pl.* i. 529 ; *Murr. Syst. Veg. ed.* xiv. 141 ; *R. Br. in
Trans. Linn. Soc.* x. 89 ; *Thunb. Fl. Cap. ed.* i. 503 ; *Roem. &
Schultes, Syst. Veg.* iii. 351 ; *Bot. Mag. t.* 2065 ; *Roem. & Schultes,
Syst. Veg.* iii. *Mant.* 265 ; *Meisn. in DC. Prodr.* xiv. 244 ; *var.
arenaria, Meisn. l.c.* 245. *P. nana, Lam. Ill.* i. 233, *not of Thunb.* ;
Poir. Encycl. v. 639 ; *Kerner, Hort. Sempervir.* ix. *t.* 399, *ex Roem. &
Schultes, Syst. Veg.* iii. 351. *P. arenaria, Buek in Drège, Zwei Pfl.
Documente,* 213. *P. acaulis, a, Drège ex Meisn. l.c., partly.* *Ero-
dendrum limoniifolium, β, Knight, Prot.* 47. *Leucadendron acaulon,
Linn. Sp. Pl. ed.* 1, i. 92.—*Lepidocarpodendron acaulon, etc., Boerh.
Ind. Pl. Hort. Ludg. Bat.* ii. 191, *t.* 191, *fig. b. Scolymocephalus seu
Lepidocarpodendron acaulon, etc., Weinm. Phyt.* iv. 291, *t.* 897.

SOUTH AFRICA : without locality, *Bergius* ! *Roxburgh, Labillardière* ! *Forster* ! *Ecklon*, 6 ! *Gueinzius* !
COAST REGION : Clanwilliam Div. ; near Ezelsbank, *Drège.* Piquetberg Div. ; near Piquetberg, *Drège* ! Tulbagh Div. ; Witzenberg Range, *Pappe* ! Paarl Div. ; Paarl Mountain, *Drège* ! Cape Div. ; hills and flats around Cape Town, *Thunberg* ! *Masson* ! *Burchell*, 1 ! 8575 ! *Ecklon*, 11 ! *Jameson* ! *Krauss*, 1071, *Bolus*, 3761 ! *MacOwan, Herb. Norm. Austr.-Afr.* 780 ! *Phillips* ! *Pappe* ! Caledon Div. ; various localities, *Burchell*, 7935 ! 8008 ! 8105 ! 8636 ! *Zeyher*, 3668 ! Bredasdorp Div. ; Kars River, *Ludwig.* Uitenhage Div. ; Winterberg Range, *Krauss*, 1071.

74. P. glaucophylla (Salisb. Parad. Lond. t. 11) ; main stem subterranean ; flowering and leaf-bearing branches 1–3½ in. long, prostrate ; leaves 5¼–8 in. long, 8–12 lin. broad, elongate-lanceolate, subacute, sometimes acutely apiculate, long attenuated at the base, distinctly veined, glabrous, glaucous ; head sessile, or shortly contracted into a scaly stipes, 2 in. long, about 2 in. in diam. ; involucral bracts 12-seriate, glabrous ; outer ovate, obtuse, minutely ciliate ; inner oblong, equalling the flowers ; perianth-sheath 9 lin. long, rather slender, somewhat abruptly dilated and 3-keeled below, excepting on the ciliate margins of the middle part, glabrous ; lip 3 lin. long, 3-toothed, pubescent or hirsute, glabrous on the back ; lateral teeth ½ lin. long, oblong, subacute, villous, at least when young, median somewhat shorter and more acute ; stamens all fertile, subsessile ; anthers linear, 1¾ lin. long ; apical glands ¼ lin. long, ovate, acuminate, slightly swollen on the inner face ; ovary 2 lin. long, oblong, covered with long reddish-brown hairs ; style 10 lin. long, sickle-shaped, subulate from the narrow-linear compressed lower half, somewhat smaller at the base, glabrous ; stigma 2 lin. long, subulate, obtuse, gradually passing into the style. *P. elongata, R. Br. in Trans. Linn. Soc.* x. 90 ; *Meisn. in DC. Prodr.* xiv. 245. *P. angustata, Drège ex Meisn. l.c. Erodendrum glaucophyllum, Knight, Prot.* 47, *incl. var.*

COAST REGION : Tulbagh Div. ; near Tulbagh, *Niven, Drège* ! *Pappe* ! Riversdale Div. ; between the Little Vet River and Garcias Pass, *Burchell*, 6859 !

75. P. Burchellii (Stapf) ; stems subterranean, branched ; branches rising slightly above the ground, their upper portions finely hairy, at least when young ; leaves oblanceolate-linear to linear, attenuated below, acute, with a callous point, 6–9 in. long, ¾–1 in. broad, glossy, prominently veined on both sides, lateral nerves running into the narrow more or less thickened margin, glabrous or finely hairy near the base ; head subglobose, sessile, rounded at the base, 2½ in. long, over 1½ in. in diam. ; outer involucral bracts ovate, obtuse to subobtuse, finely greyish pubescent and ciliate when quite young, soon glabrous, dark chestnut-brown, inner elongated, oblong, obtuse, not quite equalling the flowers ; perianth-sheath 1¼ in. long, slender in the upper ⅔, then widened, finely 3-keeled and 5-nerved, densely fulvo-pubescent down to the widened ciliolate (but otherwise glabrous) base ; lip 6 lin. long, villous, excepting on the glabrescent back ; lateral teeth filiform, acute, 1½ lin. long, the middle one much shorter and finer ; stamens all fertile ;

filaments ¼ lin. long, widened upwards ; anthers linear, 2 lin. long ; apical gland oblong, obtuse, ¼ lin. long ; ovary subobovate-oblong, densely covered with rufous hairs ; style subulate and terete from a compressed narrowly and obliquely lanceolate base, constricted at the junction with the ovary, strongly curved from below the middle, 1¾ in. long, pubescent up to the middle ; stigma subulate, obtuse, almost imperceptibly passing into the style.

CoAST REGION : Stellenbosch Div. ; between Lowrys Pass and Jonkers Hoek, *Burchell,* 8332 !

76. P. angustata (R. Br. in Trans. Linn. Soc. x. 90) ;

stems under-ground or the flowering branches raised 1–2 in. above ground, glabrous, rather slender ; leaves linear-oblanceolate to narrowly oblanceolate, long attenuated below, subobtuse to acute with a callous point, up to 6 in. long, ⅛–½ in. wide, slightly glossy or dull, obscurely or the broadest prominently veined on both sides with the lateral nerves running into the thickened margin, glabrous ; heads globose, sessile, rounded at the base or contracted into a scaly stipes, 1–1¼ in. in diam. ; outer involucral bracts ovate, obtuse, inner elongated, oblong, all quite glabrous excepting on the ciliolate margins, reddish to dark brown, not equalling the flowers ; perianth-sheath 8 lin. long, gradually widened from the upper ⅓ downwards, finely 3-keeled and 5-nerved, densely ciliate from the glabrous lower ⅓ upwards and more or less villous on the back above the middle ; lip 3 lin. long, pubescent to villous ; lateral teeth linear-oblong, subobtuse, ½ lin. long, middle one narrower and shorter, shortly villous ; stamens all fertile ; filaments linear, ⅓ lin. long ; anthers linear, 1 lin. long ; apical gland oblong, acute ; ovary oblong, 3½ lin. long, densely covered with rufous hairs ; style subulate from an obliquely linear-lanceolate compressed base, strongly curved from the middle, about 10 lin. long, glabrous or slightly pubescent at the base ; stigma subulate, obtuse, 1½ lin. long, imperceptibly passing into the style.

CoAST REGION : Caledon Div. ; Vogelgat, *Schlechter,* 9539 ! Tulbagh Div. ; New Kloof, *Drège,* 1450 ! Caledon Div. ; near the mouth of the Klein River, *Zeyher,* 3669 ! Great Houw Hoek, *Niven* ! mountains near the Zondereinde River, *Zeyher,* 3668 !

This is possibly an extreme, narrow-leaved state of *P. acaulis.*

77. P. lævis (R. Br. in Trans. Linn. Soc. x. 91) ;

main stem subterranean, producing prostrate or ascending, glabrous branches up to over 3 in. long ; leaves 3½–6½ in. long, 1½–3 lin. broad, linear to oblanceolate-linear, acute, narrowed at the base, glabrous, subglaucous, veinless, with acute margins ; head sometimes con-tracted into a short scaly stipes about 1½ in. long, about 1 in. in diam. ; involucral bracts 12–14-seriate, glabrous ; outer elliptic-ovate, obtuse or subobtuse, ciliate ; inner oblong, ciliate, equalling the flowers ; perianth-sheath 8 lin. long, gradually dilated from the middle downwards, distinctly 3-nerved, faintly 5-nerved below,

glabrous, excepting on the very densely ciliate margins; lip 2
lin. long, 3-toothed, pubescent or shortly tomentose excepting on
the more or less glabrous back; teeth subequal, ¼ lin. long, villous;
stamens all fertile; filaments ⅛ lin. long, obovate; anthers linear,
1½ lin. long; apical glands ⅓ lin. long, ovate, subacuminate, acute;
ovary obovate-oblong, covered with long reddish-yellow hairs;
style 9 lin. long, sickle-shaped, subulate from an obliquely lanceolate
compressed base, glabrous; stigma 1 lin. long, finely subulate,
obtuse. *Poir. Encycl. Suppl.* iv. 562; *Roem. & Schultes, Syst. Veg.*
iii. 352; *Mant.* 265; *Bot. Mag. t.* 2439; *Meisn. in DC. Prodr.* xiv.
241. *P. longifolia, Salisb. Parad. Lond. t.* 37. *P. Zeyheri, Phillips
in Kew Bulletin,* 1910, 235. *Erodendrum longifolium, Knight, Prot.* 46.

SOUTH AFRICA : without locality, *Masson* !
COAST REGION : Clanwilliam Div. ; Ezelsbank, *Drège* ! *Schlechter,* 8804 ! Honig
Valley and Koude Berg, *Drège* ! Piquetberg Div. ; Piquet Berg, *Drège.* Paarde
Kloof in the Witzenberg Range, *Zeyher,* 1460 ! Caledon Div. ; mountains near
Houw Hoek, *MacOwan,* 2912 ! Knysna Div. ; Outeniqua Mountains, *Drège.*
Uniondale Div. ; Long Kloof, *Masson, Niven* !

This also approaches *P. acaulis* very closely in the structure of the heads, and
differs mainly in the veinless leaves and thin acute cartilaginous margin.

78. **P. revoluta** (R. Br. in Trans. Linn. Soc. x. 90); stem
subterranean; leaf-bearing branches prostrate or ascending, 2–3 in.
long, glabrous; leaves 6–8 in. long, ¾–1½ lin. broad, stoutly linear-
filiform, semi-terete, subacute, channelled upwards, glabrous; head
about 1¼ in. long, about 1 in. in diam.; involucral bracts 10–15-
seriate, glabrous, ciliolate when young; outer ovate, subobtuse to
obtuse; inner oblong, equalling the flowers; perianth-sheath 7 lin.
long, dilated from the middle downwards, 3-keeled and 7-nerved
below, glabrous, with the exception of the ciliate margins; lip 1½–1¾
lin. long, 3-toothed, 3-keeled, pubescent at first, then glabrous; teeth
subequal, ⅛–¼ lin. long; stamens all fertile, subsessile; anthers
linear, ¾ lin. long; apical glands ⅛ lin. long, ovate, lanceolate,
acute; ovary 2 lin. long, covered with long reddish-brown hairs;
style 10–11 lin. long, sickle-shaped, subulate from the obliquely
and narrowly lanceolate lower half, swollen at the base, then compressed, glabrous; stigma 1 lin. long, imperceptibly passing into
the style, obtuse.

SOUTH AFRICA : without locality, *Niven, Roxburgh.*
COAST REGION : Caledon Div. ; Great Houw Hoek, *Zeyher,* 3669 β !

79. **P. montana** (E. Meyer in Drège, Zwei Pfl. Documente, 213, name
only); main stem subterranean; leaf-bearing and flowering branches
up to 4 in. long, prostrate; leaves 1¾–3 in. long, ¾–1¾ lin. broad, linear
to narrowly oblanceolate, acute, mucronate, long attenuated at the
base, indistinctly veined, adpressedly hairy, soon glabrous; head
sessile, 2 in. long, about 1½ in. in diam. surrounded by the upper
leaves; involucral bracts 6–7-seriate; outer ovate, produced into
long foliaceous appendages, silky-pubescent; inner oblong or

spathulate-oblong, silky-pubescent on the back, ciliate, equalling the flowers; perianth-sheath 11 lin. long, dilated, 3-keeled and 5–7-nerved below, membranous, rufo-pubescent to the middle, glabrous at the very base; lip 5 lin. long, 3-awned, glabrous below, excepting on the ciliate margin, then increasingly pubescent and ending in a woolly tuft; lateral awns 2 lin. long, linear, woolly; median awn 1 lin. long, linear, woolly; stamens all fertile; filaments ½ lin. long, swollen; anthers linear, 2¼ lin. long; apical glands ¼ lin. long, ovate, subacute; ovary 2 lin. long, oblong-elliptic, covered with long reddish-brown hairs; style 11 lin. long, falcate, narrowing from the base upwards, flattened above, glabrous; stigma 2½ lin. long, obtuse, finely channelled, almost imperceptibly passing into the style. *Meisn. in DC. Prodr.* xiv. 240.

COAST REGION : Caledon Div. ; on the Great Zwartberg Range; near Vrolykheid, *Drège* !

80. **P. humiflora** (Andr. Bot. Rep. t. 532); a shrub with loosely and diffusely procumbent glabrous branches to over 1½ ft. long ; leaves spreading, 2–3¼ in. long, 1½–2 lin. broad, linear, acute, not or very slightly attenuated at the base, often shortly decurrent, indistinctly veined, but with a distinct midrib, green, glabrous ; heads solitary at the bases of branches, often contracted into a short stipes, up to over 1½ in. long, and 2 in. in diam.; receptacle 9 lin. high, conical; paleæ ovate, acute ; involucral bracts about 7-seriate, densely hirsute; outer ovate, obtuse, finely and adpressedly pubescent to glabrous; inner oblong or spathulate-oblong, rufo-pubescent to silky-tomentose, equalling the flowers or slightly shorter ; perianth-sheath up to 7 lin. long, much dilated, from above the middle downwards, 3-keeled and 7-nerved at the base, thinly membranous from below, glabrous ; lip 2 lin. long, 3-toothed, cylindric, glabrous, excepting at the fugaciously hairy tips, ciliate ; teeth subequal, ½ lin. long ; stamens subsessile ; anthers linear, 1 lin. long ; apical glands ⅛ lin. long, ovate, obtuse, swollen on the inner face ; style 9 lin. long, strongly curved to sickle-shaped, compressed in the lower half, bulbously thickened at the base, glabrous ; stigma 1¼ lin. long, obtuse, slightly bent at the junction with the style ; young fruit 4 lin. long, cylindric, covered with long brown hairs. *P. humiflorens, Willd. Enum. Hort. Berol. Suppl.* 7 (name only) ? *ex Meisn. P. humilis, R. Br. in Trans. Linn. Soc.* x. 95 ; *Spreng. Syst. Veg.* i. 463 ; *Meisn. in DC. Prodr.* xiv. 246, *partly. P. humifusa, Hort. ex Meisn. l.c. Pleuranthe glastiflora, Knight, Prot.* 50.

SOUTH AFRICA : without locality, *Masson* ! *Drummond* !
COAST REGION : Stellenbosch Div.; Hottentots Holland, *Niven*. Riversdale Div.; between Great Valsch River and Zoetemelks River and hills near Zoetemelks River, *Burchell*, 6561 ! 6759 ! George Div. ; Montague Pass, *Marloth in Herb. MacOwan*, 3405 !

In Andrews' figure, which represents an "entire plant, just as we found it growing in the conservatory of G. Hibbert," 4 or 5 flower-heads are shown, each springing apparently from the base of a leafy branch and all clustered together close to the ground. In all the dried specimens the flower-heads are scattered, but young buds may sometimes be seen close to them. In all cases, however, the heads are a long distance from the base of the branches, and this seems to be the natural condition. Andrews represents the involucral bracts, the perianth-lips and the styles as purple.

81. **P. decurrens** (Phillips in Kew Bulletin, 1910, 236); main stem underground, throwing out a number of leaf- and flower-bearing prostrate or ascending mostly divided glabrous branches; leaves $1\frac{1}{4}$–$2\frac{1}{4}$ in. long, $\frac{2}{3}$ lin. broad, linear to acicular, acute, shortly decurrent, glabrous; heads lateral, solitary or in often large clusters, but then usually only one of each cluster in flower, the others forming small buds, $1\frac{1}{2}$ in. long, about $\frac{1}{2}$ in. in diam., obovoid and obtuse when young, contracted at the base into a scaly stipes or peduncle up to 5 lin. long; involucral bracts 9–10-seriate, very densely whitish silky-pubescent to tomentose and ciliate; outer ovate, obtuse; inner oblong or spathulate-oblong, villous at the apex, exceeding the flowers; perianth-sheath 9 lin. long, dilated from the middle downwards, 3-keeled and 7-nerved in the lower half, thinly membranous above, glabrous; lip 2 lin. long, 3-toothed above, oblong, glabrous, excepting at the fugaciously villous tips; teeth subequal, $\frac{1}{4}$ lin. long; stamens all fertile; filaments $\frac{1}{6}$ lin. long, swollen; anthers linear, $1\frac{1}{2}$ lin. long; apical glands $\frac{1}{8}$ lin. long, ovate; ovary 2 lin. long, covered with long reddish-brown hairs; style 10 lin. long, strongly curved to sickle-shaped, slender, subulate upwards, from the widened and bulbously thickened base keeled; stigma $1\frac{1}{2}$ lin. long, subulate, obtuse, obscurely bent at the junction with the style. *P. humilis, Meisn. in DC. Prodr.* xiv. 246, *partly, not of R. Br.*

SOUTH AFRICA: without locality, *Drummond*!
COAST REGION: Swellendam Div.; mountains near Swellendam, *Zeyher*, 3676! *Bolus, Herb. Norm. Austr.-Afr.* 1348!

82. **P. acerosa** (R. Br. in Trans. Linn. Soc. x. 95); main stem subterranean, producing erect branched shoots up to $1\frac{1}{2}$ ft. high, with clusters of flower-heads at their base; branches glabrous; leaves 7–12 lin. long, $\frac{1}{3}$ lin. broad, linear, acute, mucronate; heads lateral, often very numerous, but only a few in flower at a time, contracted into a slender scaly stipes or peduncle up to 4 lin. long, $1\frac{3}{4}$ in. long, about $1\frac{1}{2}$ in. in diam.; young heads obovoid, obtuse; involucral bracts 12–14-seriate, finely but densely silky-pubescent outside; outer ovate, obtuse; inner oblong to spathulate-oblong, with thinner (often wavy) wide margins, exceeding the flowers; perianth-sheath 7 lin. long, dilated from above the middle downwards, 3-keeled and 7-nerved in the lower half, thinly membranous above, glabrous; lip $1\frac{2}{3}$ lin. long, 3-toothed, glabrous, excepting for a few minute hairs at the tips, obtuse, ellipsoid; teeth subequal, $\frac{1}{4}$ lin. long; stamens all fertile; filaments $\frac{1}{8}$ lin. long, obovate, swollen, deeply furrowed; anthers oblong, 1 lin. long; apical glands $\frac{1}{6}$ lin. long, ovate, obtuse; ovary 1 lin. long, covered with

long reddish-brown hairs; style 9 lin. long, strongly curved from
the middle upwards, subulate upwards, distinctly compressed and
slightly dilated in the lower half, glabrous, somewhat bulbously
thickened at the base; stigma 1 lin. long, oblong-linear, obtuse,
furrowed, slightly swollen at the junction with the style. *Roem. &
Schultes, Syst. Veg.* iii. 355; *Bot. Reg. t.* 351; *Meisn. in DC.
Prodr.* xiv. 246. *Protea abietina, Buek, ex Meisn. l.c.*

VAR. β, **virgata** (Meisn. in DC. Prodr. xiv. 246); leaves 1½–2½ in. long,
filiform. *Pleuranthe subulæfolia, Knight, Prot.* 50. *Protea virgata, Andr. Bot.
Rep. t.* 577.

SOUTH AFRICA: without locality, *Masson,* ex *Brown.* Var. β, *Kolbing,* 8!
COAST REGION: Caledon Div.; tops of mountains near Genadendal, *Burchell,*
7703! near Palmiet River and Houw Hoek, *Drège! Zeyher,* 3677! *Mund!*
Var. β: Caledon Div.; mountains near Genadendal and the Zondereinde River,
Niven! Burchell, 7607!

83. **P. cordata** (Thunb. Diss. Prot. 59, t. v. fig. 1); stem
underground, throwing up annual leaf-bearing erect simple shoots
resembling pinnate leaves, with clusters of flower-heads at their
base, shoots glabrous; leaves 3–5 on each shoot, 2–4½ in. long,
1½–4½ in. broad at the widest part, ovate-orbicular, obtuse, cordate
at the base, palmately 8–10-veined, glabrous, with smooth or wavy
margins; heads contracted into a slender scaly stipes or peduncle,
2 in. long, about 2 in. in diam.; involucral bracts 10-seriate; outer
ovate to ovate-lanceolate, obtuse, finely pubescent and ciliate to
glabrous; inner oblong to obovate-oblong, glabrous with densely
ciliolate margins to finely but densely and silky rufo-pubescent
without, exceeding the flowers; perianth-sheath 7 lin. long, dilated
from above the middle downwards, strongly 3-keeled and sub-
5-nerved, glabrous; lip 1½ lin. long, 3-toothed, oblong, obtuse,
glabrous; teeth subequal, ⅙ lin. long, incurved; filaments ¼ lin.
long, obovate, swollen; anthers elliptic-linear, curved, ¾ lin.
long; apical glands ⅛ lin. long, ovate; ovary 3 lin. long,
obovate, elliptic in outline, covered with long brown hairs;
style 7–9 lin. long, strongly curved above the middle, bulbously
thickened at the base, then gradually tapering and subulate
upwards; stigma ¾ lin. long, cylindric, obtuse, somewhat swollen
at the junction with the style. *Lam. Ill.* i. 233; *Thunb. Prodr.* 28;
Murr. Syst. Veg. ed. xiv. 142; *Willd. Sp. Pl.* i. 534; *Andr. Bot.
Rep. t.* 289; *Poir. Encycl.* v. 639; *Thunb. Fl. Cap. ed.* i. 515;
R. Br. in Trans. Linn. Soc. x. 94; *Roem. & Schultes, Syst. Veg.*
iii. 354; *Kerner, Hort. Sempervir.* viii. t. 377, ex *Roem. & Schultes, l.c.*;
Meisn. in DC. Prodr. xiv. 245. *P. cordifolia, Sims, Bot. Mag. t.* 649;
Reichenb. Fl. Exot. iv. 35, t. 274.

COAST REGION: Stellenbosch Div.; mountains of Lowrys Pass, *Burchell,* 8190!
Hottentots Holland Mountains, *Thunberg, Zeyher,* 3674! Caledon Div.; Houw
Hoek Mountains, *Burchell,* 8061! hills near the Zondereinde River, *Thunberg!*

84. **P. amplexicaulis** (R. Br. in Trans. Linn. Soc. x. 95); main
stem underground, throwing up densely leafy simple or branched
decumbent glabrous shoots up to 2 ft. long with usually clustered or

approximate flower-heads at their bases, rarely higher up; leaves spreading horizontally, 1–2½ in. long, ½–1½ in. broad at the widest part, ovate or lanceolate-ovate, acute, more or less amplexicaul at the base, distinctly palmately 5–8-veined, glabrous, with undulating often red margins; heads lateral, suddenly contracted into a short scaly stipes or peduncle, 2 in. long, about 2 in. in diam.; young heads spherical, becoming ovoid with age; involucral bracts 10–12-seriate, more or less finely pubescent to (the inner) densely tomentose, indumentum rufous; outer ovate, obtuse; inner oblong to spathulate or obovate-oblong, spreading in the adult head; perianth-sheath 7 lin. long, broad and much dilated, 3-keeled and 5–7-nerved below, glabrous; lip 2½ lin. long, 3-toothed, glabrous; teeth subequal ⅓ lin. long; filaments ¼ lin. long, swollen; anthers oblong-linear, 1¾ lin. long; apical glands ¼ lin. long, ovate, sub-acute; ovary 1¾ lin. long, oblong, covered with long reddish-yellow hairs; style 9 lin. long, more or less falcate, somewhat thickened and widened at the base, then gradually tapering and subulate upwards, curved, glabrous; stigma 2 lin. long, subobtuse, gradually passing into the style; fruit bottle-shaped, 5 lin. long. *Meisn. in DC. Prodr.* xiv. 246. *P. repens, Andr. Bot. Rep. t.* 453 (*not of Thunb. or Linn.*). *Erodendrum amplexicaule, Salisb. Parad. t.* 67. *Pleuranthe amplexicaulis, Knight, Prot.* 51.

SOUTH AFRICA: without locality, *Labillardiere* ! *Thom* ! *Ludwig* ! *Bowie* !
COAST REGION : Worcester Div. ; Dutoits Kloof, *Drège* ! Caledon Div. ; Zwart Berg, near Caledon, *Niven*, 14 ! *Zeyher*, 3675 ! Baviaans Kloof, *Burchell*, 7705 !

Imperfectly known species.

85. P. acuminata (Sims, Bot. Mag. t. 1694) ; an erect plant ; branches twiggy ; leaves linear-lanceolate or linear, acute, flat, veiny above ; head terminal ; inner involucral bracts obtuse, concave above, with the margins black-pubescent ; styles curved. *Poir. Encycl. Suppl.* iv. 562 ; *Roem. & Schult. Syst. Veg.* iii. 351 ; *Mant.* 265.

SOUTH AFRICA : Known only from the figure in the Botanical Magazine and probably allied to *P. canaliculata.*

P. arborea, Link, Enum. Pl. Hort. Berol. i. 113.

P. arborescens, Hort. ex Steud. Nomencl. ed. 2, ii. 399.

P. asplenifolia, Link, l.c.

P. brassicæfolia, Hort. ex Steud. l.c.

P. carinata, Hort. Breit. ex Schult. Mant. iii. 266.

P. carnosa, Hort. ex Roem. & Schult. Syst. iii. 356.

P. cerifera, Hort. ex Roem. & Schult. l.c.

P. ciliaris, Hort. ex Roem. & Schult. l.c.

P. ciliata, Breit. Hort. Breit. 380 ; Schult. Mant. iii. 266.

P. concolor, Hort. ex Steud. Nomencl. ed. 2, ii. 399.

P. declinata, Hort. ex Meisn. in DC. Prodr. xiv. 247.

P. **ericoides,** Schult. Mant. iii. 266.

P. **glauca,** Brouss. ex Roem. & Schult. Syst. iii. 356.

P. **graminea,** Hort. ex Link, Enum. Hort. Berol. i. 113.

P. **grandis,** Hort. ex Meisn. in DC. Prodr. xiv. 247.

P. **hirtella,** Hort. ex Meisn. l.c. 248.

P. **incana,** Hort. ex Meisn. l.c.

P. **lanceolata,** Hort. ex Meisn. l.c.

P. **leucantha,** Hort. ex Meisn. l.c.

P. **linguiformis,** Hort. ex Meisn. l.c.

P. **micrantha,** Hort. ex Meisn. l.c.

P. **mucronata,** Nois. ex Schult. Mant. iii. 266.

P. **multifida,** Nois. ex Schult. l.c.

P. **nigra,** Regel, Cat. Pl. Hort. Aksakov. 116.

P. **passerina,** Desf. ex Schult. l.c.

P. **Pinaster,** Hort. ex Steud. Nomencl. ed. 2, ii. 400.

P. **pinastrifolia,** Hort. Loud. ex Dum. Cours. Bot. Cult. ed. 2, i. 591.

P. **Radula,** Hort. ex Link, Enum. Hort. Berol. i. 113.

P. **radulifolia,** Donn, Cat. Hort. Cantab. ed. iv. 25.

P. **reflexa,** Hort. ex Meisn. l.c. 248.

P. **retroflexa,** F. G. Dietr. Vollst. Lexik. Gaertn. vii. 556.

P. **setacea,** Hort. ex Meisn. l.c. 248.

P. **spiralis,** Hort. ex Meisn. l.c.

P. **staticæfolia,** Hort. ex Meisn. l.c.

P. **subacaulis,** Hort. ex Meisn. l.c.

P. **tenella,** Hort. ex Meisn. l.c.

P. **tenera,** Hort. ex Steud. Nomencl. ed. 2, ii. 401.

P. **uliginosa,** Hort. ex Meisn. l.c.

P. **viscosa,** Hort. ex Meisn. l.c.

V. LEUCOSPERMUM, R. Br.

Flowers hermaphrodite, subzygomorphic, rarely quite actinomorphic. *Perianth* cylindric in bud with an ovoid or ellipsoid limb, 2- or 4-partite to or beyond the middle, always with a distinct tube ; segments differentiated into a slender claw and a spoon-shaped limb ; the posticous (adaxial) and the lateral claws usually permanently united, or separating at the tips or all along, with a fleshy, often finely 2-keeled, more or less decurrent swelling at the upper end and there recurved or revolute in the open flower, anticous claw like the others, but free or adhering to them by the tips only ;

limbs soon free, deflexed or inflexed on the revolute tips of the claws. *Stamens* 4; anthers oblong to ovate, sessile or subsessile, inserted at the base of the limb; connective always very marked on the back of the anther, sometimes produced into a small point. *Hypogynous scales* 4, linear or subulate. *Ovary* sessile, pubescent, surrounded by a basal ring of hairs; style filiform or subulate, straight or nearly so, subpersistent; stigma conical, ovoid, obliquely turbinate or cylindrical, large or small, epapillose. *Ovule* 1, laterally attached. *Fruit* a smooth, whitish, often shining nut with a crustaceous pericarp. *Seed* solitary.

Erect or sometimes procumbent shrubs; leaves usually very crowded, entire or toothed at the apex, coriaceous, like the stems usually with a fine dense whitish or greyish covering of minute curled hairs and sometimes an additional indumentum of copious or scanty long spreading hairs, or sooner or later glabrous; flowers in terminal, or by overtopping pseudolateral, heads; heads solitary or in clusters of 2–3, rarely more, shortly peduncled or sessile, bracteate, all the bracts of the head fertile or usually the outer barren, forming a more or less distinct involucre, the fertile bracts distributed over a cylindric, conical or flat receptacle, like the barren ones persistent and often at length indurated, nearly always very densely tomentose on the back; perianth more or less hairy, usually yellow (rarely red) or reddish upwards; styles yellow; stigmas often red.

DISTRIB. About 32 species in extratropical South Africa, mostly in the coast region, one extending into Rhodesia.

A. Flower-heads large, including the styles 2½ (rarely 2)–3 in. long; receptacle cylindrical, much longer than wide (excepting 12, *præmorsum*).

Section 1. CONOCARPODENDRON. Heads large, including the styles 2½ (rarely 2)–3 in. long, with or without a distinct involucre of barren bracts; receptacle cylindrical, much longer than wide (excepting 12, *præmorsum*); fertile bracts persistent, more or less indurated; the adaxial and lateral perianth-claws permanently united into a sheath, or flattened-out and revolute in the upper part; abaxial claw adhering long to the top of the sheath, otherwise free; limbs reflexed or inflexed; styles 1½–3 in. long, tapering and markedly quadrangular (or sometimes sub–8-angular) upward; stigma various. Small, mostly erect shrubs, rarely attaining to 6–10 ft.

*Stigma obliquely turbinate (hoof-shaped) with a large oblique discoid face:
  Leaves glabrous or nearly so when adult :
    Leaves more or less ovate, cordate at the base, entire
      or very broadly and shortly 3-toothed, 1½–2½ by
      1 in. ... ... (1) **nutans.**
    Leaves oblong, obtuse at the base, coarsely 3–6-toothed,
      the larger 3–3½ by 1 in. ... ... (2) **mixtum.**
    Leaves all permanently greyish-tomentose, attenuated
      at the base ... ... (3) **Bolusii.**
**Stigma conic-ovoid, very much wider at the base than
  the style :
  †Leaves oblong to elliptic, or obovate or if oblong-linear,
    then at most 6 times as long as broad, 1¼–1½ in.
    broad :
    Basal barren bracts of flower-head not forming a distinct involucre :

2 R 2

Basal barren bracts of flower-head densely and long-
hirsute, hidden by the hairs ; leaves distinctly
veined, cordate or obtuse at the base, 2–4 by
½–1½ in. :

Branches stout, densely hirsute besides being
finely greyish-tomentose : leaves very much
crowded, more or less permanently hirsute
near the base ...   .   (4) **conocarpum.**

Branches not stout, nor hirsute besides being
finely greyish-tomentose ; leaves moderately
crowded, glabrous except at the shortly
villous or puberulous base   ...   .   (5) **glabrum**

Basal barren bracts of flower-head shortly and
densely tomentose, small ; leaves attenuated at
the base :

Leaves very stoutly coriaceous, dull, pale to
glaucous-green, veinless or the veins very fine
and impressed ...   .   (6) **attenuatum.**

Leaves moderately coriaceous, green, slightly dull
to almost glossy ; veins slender, raised   ...   (7) **Gerrardii.**

Basal bracts of flower-head forming a distinct short
involucre, ovate, gradually acuminate, sparingly
hairy to glabrous upwards and sidewards :

Leaves 1½–3 in. by ½–1 in., uppermost coming up
close to the head :

Leaves deeply 6–7-toothed   ...   .   (8) **incisum.**

Leaves entire or very shortly 3- (rarely 5-)
toothed   ...   .   (9) **ellipticum.**

Leaves 1–1½ by ¼ to almost ⅓ in. ; flower-heads
more or less exserted on a loosely bracteate
peduncle from the uppermost leaves...   ...   (10) **tottum.**

††Leaves narrowly linear, often with recurved margins,
1½–3 by ¹⁄₁₆–⅓ in.   ...   .   (11) **lineare.**

***Stigma cylindric or subsubulate :

Flowers erect :

Leaves cuneate to oblanceolate-cuneate, more or less
truncate and about 7-toothed, very obscurely
veined ; basal bracts of flower-head long linear,
acuminate and long-pilose ...   .   (12) **præmorsum.**

Leaves elliptic-oblong to lanceolate, shortly attenuated
at the base, entire or 3-toothed, prominently
veined ; basal bracts broad-ovate to ovate-lanceo-
late, loosely hairy or glabrous   ...   .   (13) **grandiflorum.**

Flowers deflexed   ...   ., ...   (14) **reflexum.**

B. Flower-heads medium-sized to small, including the styles from less than
1–1½ lin. long ; receptacle flat or slightly convex, rarely broad-conic (15,
hypophyllum) or oblong (30, *obtusatum*).

Section 2. HYPOPHYLLOIDEA. Heads medium-sized, including the styles
1–1½ in. long, with or without (17, *Muirii*) a distinct involucre of barren
bracts ; receptacle broad-conic (15, *hypophyllum*), slightly convex or flat ;
fertile bracts persistent, more or less indurated, at least below ; adaxial
and lateral perianth-claws united into a sheath recurved at the top, often
twisted, abaxial claw free ; limbs deflexed, the abaxial early free ; styles
filiform, ¼–⅘ in. long, slightly quadrangular upwards ; stigma more or
less cylindric. Shrubs up to 6 ft. high or procumbent.

Receptacle conical, as high as wide at the base ; flower-
  heads distinctly peduncled, peduncle up to 1 in. long,
  loosely bracteate ; bracts like those of the involucre
  very broadly ovate   ...   . (15) **hypophyllum**.
Receptacle low-convex or flat ; flower-heads shortly peduncled
  and with imbricate bracts, or sessile among the crowded
  uppermost leaves ; bracts acuminate :
  Involucral bracts densely and shortly tomentose all over
    or absent :
    Leaves linear, 1½–2½ in. by 1–3 lin., greyish-tomentose
      or glabrous, margins often recurved ; bracts very
      densely imbricate passing from the short peduncle
      into the involucre   ...   . (16) **tomentosum**.
    Leaves oblong to oblanceolate or linear-oblong, 1¼–2 in.
      by 2–6 lin. :
      Leaves glabrous or sometimes finely tomentose at
        the base, pale green ; flower-heads shortly
        peduncled without a distinct involucre...   ... (17) **Muirii**.
      Leaves permanently whitish-tomentellous ; flower-
        heads sessile, with a densely imbricated in-
        volucre   ...   . (18) **candicans**.
  Involucral bracts of the sessile flower-heads long-acu-
    minate, scarious upwards, brown, more or less
    glabrous and ciliate   ...   . (19) **parile**.

Section 3.   CRINITÆ.   Heads rather small with a flat top, including the style,
  1–1¼ in. long, with a distinct though sometimes very short involucre of
  barren bracts ; receptacle flat ; floral bracts persistent, rarely indurated ;
  adaxial and lateral perianth-claws united up to the limb (or free upwards)
  into a narrow and sometimes flattened-out more or less straight sheath ;
  abaxial claw free ; limbs deflexed, the abaxial early free ; styles filiform,
  somewhat quadrangular about the middle, tapering and almost capillary
  upwards, ¾ to over 1 in. long ; stigma minute and slender, cylindric.

Involucral bracts much exceeding the flowers, up to 6 lin.
  long, penicillate ...   . (20) **crinitum**.
Involucral bracts shorter than the flowers :
  Involucral bracts ovate to ovate-lanceolate, finely acumi-
    nate, reaching to or exceeding the middle of the
    flower-head :
    Leaves oblong to linear-oblong, 3–6 lin. broad   ... (21) **oleæfolium**.
    Leaves linear, about 2 lin. broad   ...   . (22) **diffusum**.
  Involucral bracts usually very short, not reaching up to
    the middle of the flower-head :
    Leaves linear, about 2 lin. broad   ...   . (23) **stenanthum**.
    Leaves cuneate-obovate, up to over 1 in. wide   ... (24) **Mundii**.

Section 4.   DIASTELLOIDEÆ.   Heads small, usually under 1 in. long including
  the styles, with a distinct (though sometimes scanty) involucre of barren
  bracts ; receptacle flat ; floral bracts persistent, narrow, not indurated ;
  adaxial and lateral perianth-claws usually more or less separating at
  length, or already when the flower opens and then quite like the abaxial
  claw which is always free ; limbs deflexed ; style filiform, obscurely quadrangular, 5–11 lin. long ; stigma minute.   Small shrubs, erect (up to 5 ft.
  high) or procumbent.

  Stigma conic-ovoid from a broad base :
    Leaves very closely imbricate, quite concealing the
      stem, suborbicular to oblong-elliptic   ...   ... (25) **buxifolium**.

Leaves imbricate, but usually not so close as to
conceal the stem, sometimes loosely scattered,
oblanceolate to oblong-linear ...   . (26) **puberum.**

Stigma clavate or cylindric :
Leaves more or less distinctly veined, more or less
glabrescent, but rarely quite glabrous when
adult ; flower-heads with a distinct involucre
of barren lanceolate bracts :
Heads rather lax, turbinate ; fruit not beaked ... (27) **royenifolium.**

Heads very compact and many-flowered, hemi-
spherical ; fruit with a distinct conical beak (28) **prostratum.**

Leaves veinless, glabrous when adult ; flower-heads
with a scanty involucre of few barren bracts :
Leaves obovate to spathulate, about 12 by 3–4 lin. (29) **cartilagineum.**

Leaves spathulate-linear to linear :
Leaves about 6 by 1–1¼ lin. ; floral bracts
tomentose ; perianth 4½–5 lin. long   ... (30) **obtusatum.**

Leaves 9–15 by 1 lin. ; floral bracts glabrous
excepting at the ciliate margins ; perianth
not much over 3 lin. long ...   . (31) **zwartbergense.**

1. **L. nutans** (R. Br. in Trans. Linn. Soc. x. 98) ; shrub, 4 ft.
high ; branches minutely woolly tomentose ; leaves ovate, oblong
or subelliptic, obtuse, rounded or truncate at the apex, entire or
callously 2–4-toothed, cordate at the base, 1–2¾ in. long, ½–1¼ in.
broad, dull green, almost glaucous, distinctly veined, glabrous or
minutely crisped-pubescent when young, usually decreasing in
length upwards ; heads solitary, without a definite involucre of
barren bracts, peduncled, 1½–2 in. in diam. excluding the styles ;
peduncle 6–9 lin. long, tomentose, with numerous imbricate, often
recurved, ovate, suddenly or gradually acuminate, tomentose bracts ;
receptacle cylindric, about 9 lin. high, 3 lin. in diam. ; floral bracts
broad ovate or obovate to obovate-cuneate, suddenly contracted
into a subulate acumen, 4–6 lin. long, very densely and long-
tomentose up to the truncate and minutely tomentose top, acumen
finely tomentose and ciliate ; adult flower-bud almost 1 in. long ;
perianth-tube 3–4 lin. long, glabrous ; adaxial and lateral claws
permanently connate into a concave or flattened-out sheath 6–7 lin.
long, and flattened out over 1 lin. wide, shortly villous or pubescent
along the sides only, abaxial claw free except towards the apex,
hairy all over ; limbs oblong-ovate, subacute, 1½ lin. long ; stamens
subsessile ; anthers oblong-ovate, 1 lin. long ; hypogynous scales
linear, obtuse, ⅚ lin. long ; ovary oblong, ¾ lin. long, very finely
pubescent, surrounded by whitish hairs, 1½–2 lin. long ; style
nearly 2 in. long, rather stout, acutely quadrangular above,
glabrous ; stigma obliquely turbinate (hoof-shaped), face with a
prominent longitudinal ridge ; fruit subglobose, 3¼ lin. long, very
sparsely puberulous. *Meisn. in DC. Prodr.* xiv. 253. *L. Meisneri*
and *L. nutans, var. integrum, Gandog. in Bull. Soc. Bot. France,* xlviii.
p. xciv. *Leucadendrum cordifolium, Knight, Prot.* 54. *L. nutans,*

O. Kuntze, Rev. Gen. Pl. ii. 579.   *Protea nutans, Poir. ex Steud.*
Nomencl. ed. 2, ii. 400.

SOUTH AFRICA : without locality, *Masson* ! *Ludwig* ! *Gueinzius* !
COAST REGION : Caledon Div. ; Houw Hoek and Little Houw Hoek, *Niven* !
Mund, 16 ! *Bolus*, 5237 ! *Schlechter*, 7332 ! Zwart Berg, near Caledon, *Zeyher*,
3678 ! *Pappe* ! Zandfontein, *Galpin*, 4459 ! Bredasdorp Div. ; between Caledon
and Elim, *Bolus*, 7858 ! Koude River, *Schlechter*, 9595 ! Swellendam Div. ;
mountains near Swellendam, *MacOwan*, 2825 ; also from gardens at Cape Town,
Burchell, 744 ! and between Cape Town and George, *Bowie* !

2. **L. mixtum** (Phillips in Kew Bulletin, 1910, 332) ; branches
tomentose ; leaves oblong, coarsely 3–6-toothed or the lowest and
some of the uppermost entire, subobtuse or obtuse at the base,
3–4¼ in. long (or those near the base and the top of the branch
1¼–2 in. long), ½–1¼ in. broad, glabrous or pubescent to minutely
villous below and on the margins ; heads solitary, without a
definite involucre of barren bracts, peduncled, about 2 in. in diam.
excluding styles ; peduncle about 1 in. long, with numerous, imbricate, often recurved, ovate, suddenly or gradually acuminate,
tomentose bracts ; receptacle cylindric ; floral bracts obovate to
obovate-cuneate, abruptly contracted into a subulate acumen,
6–7 lin. long, very densely tomentose up to the truncate and
minutely tomentose top, then finely and densely pubescent ; adult
flower-bud 1¼–1½ in. long ; perianth-tube 3½–4 lin. long, glabrous ;
adaxial and lateral claws permanently connate into a concave or
flattened-out sheath about 8 lin. long, and flattened out up to
2 lin. wide, pubescent along the margins, abaxial claw free except
towards the apex, hairy all over ; limbs ovate-oblong, acute, 1½–1¾
lin. long ; anthers subsessile, oblong-ovate, 1 lin. long ; hypogynous
scales, linear, obtuse, ¾ lin. long ; ovary oblong, 1¼ lin. long, densely
and very finely whitish pubescent up to or beyond the middle,
surrounded by whitish hairs 1½–2 lin. long ; style 2 in. long,
rather stout, acutely quadrangular above, glabrous ; stigma obliquely turbinate (hoof-shaped) ; face with a prominent longitudinal
ridge.

SOUTH AFRICA : without locality or collector's name ʽKew Herbarium).

This is possibly only an extreme variation of *L. nutans*, characterised by
longer, relatively narrow and more toothed leaves with obtuse or subobtuse—
not cordate—bases. The differences in the flowers are mainly connected with size.

3. **L. Bolusii** (Phillips in Kew Bulletin, 1910, 330 ; not of
Gandoger) ; branches greyish-villous ; leaves oblanceolate-oblong,
obtuse and bluntly 3–5-toothed at the apex, narrowed at the base,
2¼–3½ in. long, ½ to over ¾ in. broad, minutely and densely greyish-
tomentose, distinctly veined, not densely crowded below the flower-
heads ; heads few at the end of the branches or solitary (?), without
a definite involucre of barren bracts, peduncled, 2 in. long, about
1½ in. in diam. excluding the styles ; peduncle 1–1¼ in. long,
tomentose, loosely bracteate ; bracts ovate, acuminate, tomentose ;
floral bracts broadly obovate, suddenly contracted into a subulate

acumen, including the latter 6–7 lin. long and up to 3½ lin. broad, densely tomentose up to a straight line across and just below the truncate top, this and the acumen shortly tomentose ; adult flower-bud rather under 1 in. long ; perianth-tube 4 lin. long, glabrous ; adaxial and lateral claws permanently connate into a concave or more or less flattened-out sheath, about 6 lin. long, pubescent along the margins, abaxial claw free except towards the apex, hairy all over ; limbs ovate-oblong, subacute, up to 2 lin. long, minutely hirsute ; anthers subsessile, ovate-oblong, 1 lin. long ; hypogynous scales linear, acute, ¾ lin. long ; ovary oblong up to 2 lin. long, densely and very finely whitish-pubescent, surrounded by whitish hairs, 2 lin. long ; style about 1¾ lin. long, rather stout, acutely quadrangular above, glabrous ; stigma obliquely turbinate (hoof-shaped), face with a prominent longitudinal ridge ; fruit unknown.

COAST REGION : Bredasdorp Div. ; near Elim, *Bolus*, 8586 !

4. L. conocarpum (R. Br. in Trans. Linn. Soc. x. 99) ; a shrub or small tree, 6–8 ft. high ; branches stout, minutely woolly tomentose and at the same time softly hirsute with long hairs ; leaves crowded up to the flower-heads and surrounding them, usually concealing the stem upwards, obovate to oblanceolate-oblong, obtuse or acute, 4–9-toothed at the apex, very rarely entire, each tooth with a blunt callous point, 1¾–3½ in. long, ½–1½ in. broad, at first more or less hirsute with long soft hairs, then glabrescent with the exception of the villous base and margins, prominently veined ; heads golden-yellow, usually solitary, rarely geminate, shortly peduncled, 2–2¾ in. long, excluding the styles about 1½–2 in. in diam., with an obscure involucre of few barren bracts ; peduncle very stout, up to ½ in. long, with ovate sub-acuminate very hairy bracts ; receptacle conic-cylindric, about 1 in. long ; involucral bracts caudate-acuminate from an ovate base, 8–10·lin. long, 2–3 lin. broad, fulvously villous or softly hirsute ; floral bracts ovate or obovate, and abruptly contracted into a linear or subulate acumen, densely hirsute to felted ; adult flower-bud about 1½ in. long ; perianth-tube 4–5 lin. long, more or less finely pubescent above, otherwise glabrous ; adaxial and lateral claws permanently connate into a concave or more or less flattened-out flexuous sheath, 9–12 lin. long, like the free abaxial claw loosely and softly hirsute ; limbs oblong, subacute, 3 lin. long, densely hirsute ; anthers subsessile, oblong, almost 2 lin. long ; hypogynous scales linear, 1¼ lin. long ; ovary oblong, 1½–2 lin. long, greyish-pubescent, surrounded by whitish or yellowish hairs of the same length ; style over 1¾ in. long, acutely quadrangular above, with secondary less prominent angles between the primary, glabrous ; stigma oblong-conical, subobtuse, 2 lin. long, 8-grooved. *Meisn. in DC. Prodr.* xiv. 254 ; *Roupell, Cape Flow. t.* 6, *fig.* 1 ; *Pappe, Silva Cap.* 28 ; *Sim, For. Fl. Cape,* 297, *t.* 129, *fig.* ii. *L. MacOwanii, Gandog. in Bull. Bot. Soc. France,* xlviii. *p.* xciv.

Leucadendron conocarpodendron, Linn. Sp. Pl. ed. i. 93 ; *Berg. in Vet. Acad. Handl. Stockh.* 1766, 321 ; *Roem. & Schult. Syst. Veg.* iii. 358. *L. crassicaule, Knight, Prot.* 55. *Protea conocarpa, Thunb. Diss. Prot.* 22 ; *Fl. Cap. ed. Schult.* 126 ; *Lam. Ill.* i. 239, *t.* 53, *fig.* 3 ; *Poir. Encycl.* v. 656. *P. tortuosa, Salisb. Prodr.* 48.—*Leucadendro similis Africana arbor, etc., Pluk. Phyt. t.* 200, *fig.* 2 (*leaf only*). *Conocarpodendron, folio crasso, nervoso, etc., Boerh. Ind. Pl. Hort. Lugd. Bat.* ii. 196, *t.* 196. *Scolymocephalus africanus folio crasso, etc., Weinm. Phyt.* iv. 292, *t.* 899, *fig. b.*

SOUTH AFRICA : without locality, *Thunberg* ! *Roxburgh* ! *Sieber,* 6 ! *Ludwig* ! *Thom* ! *Forster* ! *Gueinzius* !
COAST REGION : Cape Div. ; Mountains and Flats around Cape Town, *Burchell,* 400 ! 809 ! 8517 ! *Brown* ! *Bowie* ! *Drège* ! *Ecklon,* 320 ! 470 ! *Zeyher* ! *Milne* ! *MacGillivray,* 640 ! *Wright* ! *Harvey,* 702 ! *Bolus,* 2909 ! *MacOwan, Herb. Norm. Austr.-Afr.,* 774 ! *Wilms,* 3554 ! *Wolley-Dod,* 570 ! *Rogers,* 3001 ! *Phillips* ! Caledon Div. ; between Houw Hoek and Palmiet River, *Burchell,* 8170 ! Zandfontein, *Galpin,* 4462 !

This is the "*Kreupelboom*" or "*Poudboom*" of the Dutch. The reddish, tough but soft wood is used for waggon felloes, and for making charcoal. The bark yields good material for tanning.

5. **L. glabrum** (Phillips in Kew Bulletin, 1910, 331) ; an erect shrub, up to 6 ft. high ; branches very minutely tomentose, sometimes with a few long spreading hairs ; leaves scattered or moderately crowded, more or less exposing the stem, the uppermost loosely surrounding the flower-heads, obovate-oblong to oblanceolate, coarsely 8–13-toothed at the apex, rarely less than 8-toothed or entire, with the teeth subacute with a callous point, narrowed at the base, 2–4 in. long, $\frac{1}{2}$–1$\frac{1}{2}$ in. broad, glabrous or more or less puberulous to shortly villous at the base, distinctly veined ; heads golden-yellow, solitary, subsessile or shortly peduncled, about 2$\frac{1}{2}$ in. long excluding the styles and about as wide, with an obscure involucre of few barren bracts ; receptacle conic-cylindric, over 1 in. long ; peduncle stout, rarely up to $\frac{1}{2}$ in. long, with ovate subulate-acuminate bracts, tomentose below, hirsute above, up to 6 lin. long ; involucral bracts similar to those of the peduncle, but more caudate-acuminate and very densely hirsute-tomentose all over, passing into the broad-obovate, abruptly caudate floral bracts, up to 7 lin. long and 3$\frac{1}{2}$ lin. wide ; adult flower-bud up to over 1$\frac{3}{4}$ lin. long ; perianth-tube 5 lin. long, glabrous ; adaxial and lateral claws permanently connate into a concave or more or less flattened-out slightly flexuous sheath with the exception of the upper part which is at length generally free for 1–2 lin. and variously bent or recurved, with long scattered hairs along the margins, otherwise glabrous below and finely fulvous-tomentose upwards ; abaxial claw coherent at the upper end with the sheath or at length quite free, hirsute and finely fulvous-tomentose all over except at the more or less glabrous base ; limbs oblong, acute, 3 lin. long, red within, densely hirsute ; anthers subsessile, oblong, 2$\frac{1}{4}$ lin. long ; hypogynous scales lanceolate-subulate to subulate, $\frac{3}{4}$ lin.

long; ovary oblong, 1 lin. long, densely pubescent, surrounded by yellowish hairs up to 2 lin. long; style 2 in. long, acutely quadrangular above with less prominent secondary ridges between the primary; stigma conical, subobtuse, $2\frac{1}{2}$ lin. long, 8-grooved, carmine with yellowish or greenish tips.

SOUTH AFRICA : without locality, *Mund*!

COAST REGION : George Div. ; near Touw River, *Burchell*, 5726 ! 5754! Springfield and edge of the Forest near the Poort, *Bowie*! between Cape Town and George, *Rogers*! Knysna Div. ; Plettensberg Bay, *Bowie*!

6. **L. attenuatum** (R. Br. in Trans. Linn. Soc. x. 96); a bush, 3–5 ft. high; branches spreading, minutely tomentose, often with scattered long hairs or villous-hirsute, at length glabrescent; leaves scattered, uppermost loosely surrounding the flower-heads, not much closer than the lower, oblong-oblanceolate, oblanceolate, rarely sublinear, truncate or subobtuse, 3–7-toothed at the apex, with usually short and broad teeth ending in callous tips, rarely entire, long cuneate at the base, $1\frac{3}{4}$–$3\frac{1}{2}$ in. long, $\frac{1}{4}$ to over $\frac{3}{4}$ in. broad, very stoutly coriaceous, indistinctly nerved, more or less rusty-pubescent to tomentose when young, soon glabrescent, excepting sometimes the base and lower margins; heads solitary or geminate, $1\frac{1}{2}$–2 in. long excluding the styles, $1\frac{1}{4}$–$1\frac{1}{2}$ in. wide, golden-yellow or tinged with red or almost red, peduncled, without a definite involucre of barren bracts; peduncle $\frac{1}{2}$–1 in. long, finely tomentose, bearing loosely imbricate, more or less spreading, ovate, acuminate, shortly tomentose bracts; receptacle cylindrical, 1–$1\frac{1}{4}$ in. long; floral bracts ovate to obovate, acuminate or subacuminate, 3–$3\frac{1}{2}$ lin. long, $1\frac{1}{2}$–$2\frac{3}{4}$ lin. broad, densely tomentose; adult flower-bud about $1\frac{1}{6}$ to over $1\frac{1}{2}$ in. long; perianth-tube about 3–$3\frac{1}{2}$ lin. long, glabrous below, densely and finely tomentose above; adaxial and lateral claws permanently connate into a concave or partly flattened-out sheath about 10 lin. long, excepting the upper part which is generally free for 1–$1\frac{1}{2}$ lin. and variously bent or recurved, with (very rarely without) long scattered hairs along the finely fulvous-tomentose margins, glabrous on the back at the base, finely fulvous-tomentose above; abaxial claw coherent at the upper end with the sheath or at length free, finely fulvous-tomentose and hirsute all over; limbs lanceolate, acute, 2 lin. long, finely tomentose and hirsute; anthers subsessile, oblong, 1–$1\frac{1}{2}$ lin. long; hypogynous scales linear or cylindric, $\frac{3}{4}$–1 lin. long; ovary oblong, $\frac{1}{2}$ lin. long, densely pubescent below; style $1\frac{1}{3}$–$1\frac{1}{2}$ in. long, stout, quadrangular above with faint secondary ridges between the primary, glabrous; stigma conical or almost oblong, up to 1 lin. long, 8-grooved; fruit ellipsoid-globose, whitish, 2 lin. long, sparingly and minutely puberulous. *Roem. & Schult. Syst. Veg.* iii. 357; *Meisn. in DC. Prodr.* xiv. 256 (*incl. vars. Dregei* and (?) *ambiguum*); *Sim, For. Fl. Cape*, 297, t. 131, fig. iii. *L. Zeyheri, Meisn. l.c.* 255 (*incl. var. truncatum*). *L. truncatum, Buek ex Meisn. l.c.* 256. *L. truncatum, var. septemdentatum, Gandog. in Bull. Soc. Bot. France*,

xl. *p.* xcv. *L. saxosum, S. Moore in Trans. Linn. Soc.* xl. 185.
Protea conocarpa, Thunb. Diss. Prot. 22, *partly*; *Fl. Cap. ed. Schult.*
126, *partly.* P. *elliptica, Thunb. Diss. Prot.* 22; *Fl. Cap. ed. Schult.*
126. *P. attenuata, Poir. Encycl. Suppl.* iv. 566. *P. tridentata, Hort.
ex Meisn. l.c.* 248. *Leucadendrum phyllanthifolium, Knight, Prot.* 55.
L. cervinum, Knight, l.c. 55 (?). *L. formosum and L. truncatum, O.
Kuntze, Rev. Gen. Pl.* ii. 578. *L. attenuatum, O. Kuntze, l.c.* 579.

SOUTH AFRICA : without locality, *Thunberg* (*P. conocarpum*, β of Herb. Thunberg and *P. elliptica* in Stockholm Herbarium)! *Mund* ! *Roxburgh* ! *Ludwig* !

COAST REGION : Bredasdorp Div. ; Between Cape Aghullas and Pot Berg,
Drège! Rietfontein Poort, *Schlechter*, 9696 ! Swellendam Div. ; hills by the
Buffeljagts River, *Zeyher*, 3682 ! near Swellendam, *Niven*, 49 ! *Burchell*, 7376 !
Zeyher, 3681 ! Riversdale Div. ; Riversdale, *Pappe* ! *Muir* ! at and near Garcias
Pass, *Galpin*, 4460 ! 4461 ! Mossel Bay Div. ; Attaquas Kloof, *Gill* ! George
Div. ; on mountains, *Bowie* ! Knysna Div. ; Knysna, *Pappe* ! between Knysna
and Goukamma River, *Burchell*, 5563 ! Uniondale Div. ; near Avontuur, *Bolus*,
1583 ! Uitenhage Div. ; Strandfontein and Matjesfontein, *Drège* ! Uitenhage,
Zeyher, 380 ! Van Stadens Berg, *Drège* ! Galgebosch, *Pappe* ! between Uitenhage
and Algoa Bay, *Burchell*, 4275 ! Port Elizabeth Div. ; around Krakakamana,
Burchell, 4539 ! Port Elizabeth, *Zeyher*, 3680 ! *Laidley.* Algoa Bay, *Cooper*, 1583 !
Albany Div. ; Howisons Poort, *MacOwan*, 62 ! *and in Herb. Norm. Austr.-Afr.* 778 !
Schönland, 349 ! *Cooper*, 1545 ! Coldstream, *Misses Daly & Sole*, 264 ! Bathurst
Div. ; mouth of Great Fish River, *Burchell*, 3759 ! Komgha Div. ; near Keimouth,
Flanagan, 488 !

KALAHARI REGION : Swaziland, *Saltmarsh in Herb. Galpin*, 1045 !
EASTERN REGION : Transkei Div. ; Transkei, *Miss Pegler*, 714.

Also in Gazaland.

The flowers of specimens collected by Galpin at Garcias Pass (4461) are
destitute of the long spreading hairs generally found intermixed with the short
dense tomentum of the perianth, otherwise they agree with his 4460 from the
same locality, which represents the common state. The size of the heads and
flowers and the shape and size of the stigma are somewhat variable in this
species, but it has not been possible to establish any correlation between these
and other characters or refer those forms to definite areas.

7. **L. Gerrardii** (Stapf) ; a small undershrub, ½ ft. high ; branches
slender, minutely tomentose upwards with intermixed spreading
hairs, glabrescent below ; leaves scattered, uppermost loosely
surrounding the flower-heads, oblong oblanceolate, rounded at the
apex with 3–2 small callously pointed teeth or the smaller quite
entire and acute, long-cuneate towards the base, often narrowed
into a slender short petiole, 2–2¼ in. long, ⅓–⅔ in. broad, thinly
coriaceous, prominently nerved, in the adult state almost glabrous
with traces of a fine tomentum and some very fine long hairs ;
heads solitary, shortly peduncled, without a definite involucre of
barren bracts, about 1¼ in. long and wide ; peduncle ¼ in. long,
finely tomentose, bearing loosely imbricate, more or less spreading,
caudate-acuminate bracts, shortly tomentose below and hirsute on
the tails ; receptacle cylindrical, ½–⅔ in. long ; floral bracts ovate,
caudate or acuminate up to 6 lin. long (including the sometimes
4 lin.-long tails), densely tomentose, hirsute on the tails ; adult
flower-bud about 1 in. long ; perianth, stamens and pistil as in
L. attenuatum, but smaller in all parts ; anthers subsessile, under

1 lin. long; style up to over $1\frac{1}{2}$ in. long; stigma oblong, slightly wider at the base $\frac{3}{4}$–1 lin. long.

EASTERN REGION : Natal or Zululand, *Gerrard*, 1664 !

8. **L. incisum** (Phillips in Kew Bulletin, 1910, 331); branches minutely tomentose and long hirsute; leaves very much crowded, concealing the stem, the uppermost close to the flower-head, much reduced, oblong, coarsely and deeply 6–7- (rarely 5-)toothed towards the apex, with acute callously pointed teeth, obtuse at the base, $2\frac{1}{2}$ in. long, $\frac{1}{2}$–$\frac{2}{3}$ in. broad, distinctly veined, glabrous except at the scantily pilose base and lower margins; head subsessile, $2\frac{1}{2}$ in. long, over 2 in. in diam., with few barren bracts at the base; receptacle cylindric, $1\frac{1}{2}$ in. long; barren and fertile bracts similar, ovate to obovate, gradually acuminate to subcaudate, including the acumen up to 8 lin. long, 3 lin. broad, very minutely tomentose below, glabrescent upwards, long-ciliate all round; adult flower-bud almost $1\frac{1}{4}$ in. long; perianth-tube 3 lin. long, glabrous; adaxial and lateral claws permanently united into a concave or partly flattened-out purple sheath revolute above, about 10 lin. long, hirsute along the margins, minutely and sparsely pubescent on the back; abaxial claw adhering to the sheath or at length free, hirsute and pubescent all over; limbs reflexed on the revolute top of the claw, oblong, acute, 2 lin. long, finely tomentose and hirsute, hairs partly red; anthers subsessile, ovate-oblong, 1 lin. long, connective wide; hypogynous scales linear, obtuse, $\frac{1}{2}$ lin. long; ovary oblong, densely and minutely pubescent; style 2 in. long, acutely quadrangular above with or without secondary ridges between the angles, glabrous; stigma obliquely conic-ovoid, almost $1\frac{1}{2}$ lin. long, unequally grooved.

COAST REGION: Worcester Div. ; Breede River Valley, near Darling Bridge, *Bolus*, 5235 !

9. **L. ellipticum** (R. Br. in Trans. Linn. Soc. x. 98); a sometimes very large dense bush, up to 12 ft. high; branches more or less hirsute and minutely tomentose; leaves much crowded up to the flower-heads, mostly erect and concealing the stem, sometimes spreading, oblong to elliptic or ovate-oblong, entire or 3- (rarely 5-) toothed at the apex with callous-pointed teeth, rounded or cordate at the base, $1\frac{1}{2}$–3 in. long, $\frac{1}{2}$–1 in. broad, more or less distinctly veined, finely tomentose when young or glabrous or villous near the base; heads greenish-yellow, tinged with red, solitary, peduncled, 2–$2\frac{1}{2}$ in. long, excluding the styles up to $2\frac{1}{4}$ in. in diam., with a definite short involucre of imbricate barren bracts; peduncle $\frac{1}{2}$ in. long, covered with closely imbricate, ovate or ovate-lanceolate, acuminate, very finely silky-pubescent, then glabrous, ciliate bracts; receptacle cylindric over 1 in. long, 2 lin. in diam. ; barren and floral bracts similar, ovate, acuminate, 6–7 lin. long, very finely silky-pubescent below, then glabrous, ciliate ; adult flower-bud up to $1\frac{1}{4}$ in. long ;

perianth-tube 2–4 lin. long, glabrous; adaxial and lateral claws permanently united into a sheath concave or flattened-out and revolute above, 9–10 lin. long, hirsute or glabrous on the back below; abaxial claw long-adhering to the top of the sheath, hirsute all over; limbs ovate-oblong, acute, up to 2 lin. long, inflexed on the revolute top of the claw, finely tomentose and hirsute, some of the hairs red; anthers over ½ lin. long, on broad filaments, broadly ovate-oblong, 1 lin. long, connective wide; hypogynous scales 1 lin. long, obtuse; ovary oblong, over 1 lin. long, densely and minutely whitish pubescent; style almost 2 lin. long, robust, acutely quadrangular above, glabrous; stigma orange-red, obliquely conic-ovoid, 1¼ lin. long, unequally grooved; fruit globose-ellipsoid, whitish, 3¼ lin. long, very sparingly and minutely puberulous. *Roem. & Schult. Syst. Veg.* iii. 358; *Meisn. in DC. Prodr.* xiv. 255. *L. medium, R. Br. l.c.* 97; *Meisn. l.c. Protea conocarpa, Thunb. Diss. Prot.* 22, *partly; Fl. Cap. ed. Schult.* 126, *partly. P. media, Poir. Encycl. Suppl.* iv. 566. *P. formosa, Andr. Bot. Rep. t.* 17; *Tratt. Thesaur. t.* 10, *not of R. Br. P. tomentosa, Salisb. l.c. (teste Knight) not of Thunb. Leucadendrum ellipticum, Knight, Prot.* 53; *O. Kuntze, Rev. Gen. Pl.* ii. 579. *L. formosum, Knight, l.c.* 54.

SOUTH AFRICA: without locality, *Oldenburg!* *Thunberg* (sheet *a* in Herb. Thunberg and in Herb. Stockholm)! *Roxburgh!* *Brown!* *Mund!*

COAST REGION: Clanwilliam Div.; Jakhals Vley Mountains, *Niven,* 53! Blue Berg, *Drège,* 8056! Tulbagh Div.; near Tulbagh, *Zeyher,* 1461! *Pappe!* *Bolus,* 5236! Tulbagh Waterfall, *Niven,* 47! *MacOwan, Herb. Norm. Austr.-Afr.* 904! *Phillips,* 524! Tulbagh Road, *Schlechter,* 8997!

10. **L. tottum** (R. Br. in Trans. Linn. Soc. x. 97); a slender shrub; branches horizontally spreading, from more or less finely greyish-hirsute and minutely tomentose to glabrous; leaves rather scattered, more or less spreading and exposing the stem, oblong, obtuse or subacute, with a callous point, entire, rarely 3-toothed, narrow or broad at the base, 1–1½ (rarely 2) in. long, ¼ to almost ½ in. broad, distinctly veined, glabrous or more or less hairy, particularly when young; heads solitary, peduncled, exserted from the uppermost leaves, 1½–2 in. long excluding the styles, and as wide, with a definite involucre of imbricate reddish sometimes spreading bracts; peduncle up to 1 in. long, covered with imbricate ovate acute to acuminate glabrous ciliate bracts; receptacle cylindric, about 1 in. long; barren and fertile bracts similar, broad-ovate to (the inner) obovate, acuminate to caudate- or subulate-acuminate, 5–7 lin. long, 3½–4 lin. broad, outermost more or less glabrous on the back, inner increasingly adpressedly-hirsute; adult flower-bud 1¾–2 in. long; perianth-tube 3–4 lin. long, pubescent above, glabrous below; adaxial and lateral claws permanently united into a concave or upwards flattened-out and revolute sheath, about 14–16 lin. long, hirsute all over like the apically adhering or free abaxial claw; limbs ovate-oblong, acute, almost 2 lin. long, inflexed

on the revolute top of the claw, hirsute, some of the hairs red ;
anthers sessile, ovate, over ¾ lin. long ; hypogynous scales linear,
acute, up to ¾ lin. long ; ovary oblong to ellipsoid 1 lin. long,
minutely whitish-pubescent surrounded by whitish hairs of the same
length ; style 1¾–2 in. long, acutely quadrangular above, glabrous ;
stigma obliquely ovoid, obtuse, not or obscurely grooved.   *Roem. &
Schult. Syst. Veg.* iii. 357 ; *Meisn. in DC. Prodr.* xiv. 257.   *Protea
totta, Linn. Mant. alt.* 191 ; *Thunb. Diss. Prot.* 42 *and Fl. Cap. ed.
Schult.* 139 ; *Lam. Ill.* i. 235 ; *Willd. Sp. Pl.* i. 532 ; *Poir. Encycl.*
v. 644.   *Leucadendron horizontale, Knight, Prot.* 53.   *L. Totta,
O. Kuntze, Rev. Gen. Pl.* ii. 579.

SOUTH AFRICA : without locality, *Oldenburg* ! *Thunberg* ! *Drège* ! *Brown* ! *Zeyher*,
1462 ! *Cooper*, 3070 !
COAST REGION : Worcester Div. ; Dutoits Kloof, *Drège* ! *MacOwan in Herb.
Norm. Austr.-Afr.* 776 ! Worcester, *Cooper*, 1597 ! Tulbagh Waterfall, *Niven*, 47 !
Pappe ! *MacOwan*, 2505 ! *Zeyher* !
CENTRAL REGION : Ceres Div. ; Klein Vley, Cold Bokkeveld, *Schlechter*, 10206 !

11. **L. lineare** (R. Br. in Trans. Linn. Soc. x. 96) ; a slender low-
growing subdecumbent or erect bush ; branches glabrous ; leaves
scattered, exposing the stem, on decumbent branches more or less
turned skywards, linear, subobtuse with a callous point, narrowed
towards the base, entire, flat or with recurved margins, indistinctly
veined, glabrous ; heads usually solitary, rarely 2-nate, exserted from
(rarely overtopped by) the uppermost leaves, peduncled, 1½–1¾ in.
long excluding the styles, and about as wide, with a definite involucre of barren closely imbricate bracts ; peduncle ½ to almost 1 in.
long, covered with spreading ovate acute pubescent or finely
tomentose bracts, 2–3 lin. long ; receptacle cylindric, about 1 in.
long, 1½ lin. in diam. ; barren and fertile bracts similar, ovate or
(the inner) elliptic to obovate, acute to (the inner) acutely acuminate,
up to 6 lin. long, 2½–3½ lin. broad, tomentose and hirsute-ciliate ;
adult flower-bud 1¼ in. long ; perianth-tube 3½–4 lin. long, rather
wide from a narrow base, glabrous below, pubescent upwards ;
adaxial and lateral claws permanently united into an upwardly
flattened-out and revolute sheath, about ¾ in. long, like the apically
adherent or free abaxial claw softly hirsute all over ; limbs ovate-
oblong, acute, 1½ lin. long, hirsute, at length inflexed on the revolute
top of the claw ; anthers subsessile, ovate-oblong, ¾ lin. long ; hypogynous scales linear, subacute, 1 lin. long ; ovary oblong, under 1
lin. long, minutely greyish-pubescent, surrounded by whitish hairs,
1 lin. long ; style 1¼ in. long, slender and acutely quadrangular
upwards, glabrous ; stigma shortly conical from a broad base, somewhat oblique, subobtuse, up to 1 lin. long.   *Roem. & Schult. Syst.
Veg.* iii. 556 ; *Meisn. in DC. Prodr.* xiv. 256 (*incl. vars.*).   *L. lineare,
var. calocephalum, Gandog. in Bull. Soc. Bot. France,* xlviii. *p.* xciv.
Protea linearis, Thunb. Diss. Prot. 33, *t.* 4, *fig.* 2, *and in Fl. Cap. ed.
Schult.* 131 ; *Lam. Ill.* i. 237 ; *Willd. Sp. Pl.* i. 521.   *Leucadendrum
fallax, Knight, Prot.* 52.   *L. lineare, O. Kuntze, Rev. Gen. Pl.* ii. 579.

SOUTH AFRICA : without locality, *Oldenburg* ! *Masson* ! *Brown* ! *Ludwig* !
COAST REGION : Worcester Div. ; Dutoits Kloof, *Drège* ! Bayers Kloof, *Schlechter*,
9202 ! Paarl Div. ; plains near Paarl, *Niven*, 48 ! Paarl Mountains, *Thunberg* !
Drège ! *Zeyher* ! *MacOwan*, 2839 ! *and in Herb. Norm. Austr.-Afr.* 777 ! *Alexander* ! French Hoek Pass, *Bolus*, 5234 !

In Schlechter's 9202, Drège and Brown's specimens the leaves are revolute ; in
Masson's specimen they are 3–4 lin. broad.

12. **L. præmorsum** (Buek ex Meisn. in DC. Prodr. xiv. 257) ;
branches minutely whitish-woolly-tomentose and loosely hirsute ;
leaves very much crowded up to the flower-heads and overtopping
them, concealing the stem, cuneate to oblanceolate-cuneate, usually
truncate and very bluntly 5–7-toothed at the upper end, 2–3½ in.
long, 7–10 (the uppermost 5–6) lin. broad, very indistinctly veined,
young minutely but densely tomentose, at length glabrous ; heads
solitary, subsessile among the top leaves, 1¼ in. long excluding the
styles and as much in diam., with a scanty involucre of barren bracts ;
peduncle very short with a few imbricate bracts, stout, subulate
from a widened base, densely and minutely tomentose and long-
hairy, up to ½ in. long ; receptacle very low to flat (?) ; barren and
fertile bracts similar, broadly or (the inner) narrowly obovate-
cuneate, suddenly contracted into a tail- or awn-like acumen and
including it up to 7 lin. long, body very densely hirsute-tomentose,
acumen hirsute ; adult flower-bud 1¼ in. long ; perianth-tube
gradually tapering downwards, 4–5 lin. long, finely tomentose,
excepting at the glabrous base ; adaxial and lateral claws permanently united into a slender concave sheath almost up to the top,
there free for a short distance and recurved or revolute or twisted,
finely tomentose and hirsute, about 8 lin. long ; abaxial claw
adherent to the sheath for about 2 lin. from the middle upwards,
finely tomentose and hirsute like the sheath ; limbs more or less
deflexed or recurved on the claw, lanceolate, acute, 2½ lin. long,
finely tomentose and hirsute ; anthers sessile, linear, oblong, apiculate, 1½ lin. long ; hypogynous scales subulate, 1½ lin. long ; ovary
oblanceolate in outline, constricted at the junction with the style,
2 lin. long, finely pubescent ; style up to almost 2 in. long, slender
upwards and acutely tetragonous, glabrous ; stigma cylindric, subobtuse, suboblique, slightly grooved, 1½ lin. long. *L. attenuatum*,
var. *præmorsum, Meisn. in DC. Prodr.* xiv. 257.

COAST REGION : Clanwilliam Div. ; Honig Valley and Koude Berg, *Drège* !

13. **L. grandiflorum** (R. Br. in Trans. Linn. Soc. x. 100) ; a stout
shrub ; branches greyish-tomentose and hirsute ; leaves very much
crowded up to the heads, concealing the stem, broad elliptic-oblong
to lanceolate, rounded, rarely truncate, entire or 3-toothed at the
apex, somewhat narrowed at the base, 1½–3 in. long, 7–4 lin. broad,
distinctly veined with the midrib prominent beneath, minutely
but densely greyish-tomentose, at length more or less glabrous ;
heads sessile among the top leaves, yellow, excepting the red stigmas,

1½–2½ in. long, and almost as wide, with a definite involucre of
imbricate barren bracts; receptacle cylindric, ¾–1 in. long, 2–3 lin.
wide; barren and floral bracts similar, ovate to ovate-lanceolate, more
or less, sometimes long, acuminate, up to ¾ in. long, finely tomentose
to subglabrous, or (the inner) very densely hirsute-tomentose on the
back, usually with long straight hairs besides, margins long hirsute-
ciliate; adult flower-bud 1½–2 in. long; perianth-tube 4½–5 lin.
long, contracted into a narrow base, finely pubescent above, glabrous
beneath; adaxial and lateral claws permanently united into a concave or flattened-out flexuous or twisted sheath, recurved or revolute
at the top, softly and loosely villous, over 1 in. long, abaxial claw
adhering to the top of the sheath, very slender, villous like the
sheath; limbs lanceolate, acute, 3¼ lin. long, long conniving and
deflexed, at length spreading out like the prongs of a fork and
inflexed on the revolute end of the claws, villous or softly hirsute;
anthers sessile, linear-lanceolate, long-apiculate, 3 lin. long; hypogynous scales subulate, acute, over 1 lin. long; ovary oblong,
minutely whitish-pubescent, 1 lin. long; style 2 to over 2½ in. long,
robust, more or less quadrangular above, glabrous; stigma subcylindric, slightly oblique at the base, obtuse, finely grooved, 2½–3½
lin. long. *Meisn. in DC. Prodr.* xiv. 254; *Sim, For. Fl. Cape*, 297,
t. 129, *fig.* 1. *L. Gueinzii, Meisn. l.c.* 100. *Protea conocarpa, Thunb.
Diss. Prot.* 22, *partly, and Fl. Cap. ed. Schult.* 126, *partly. P.
villosa, Poir. Encycl. Suppl.* iv. 566. *P. villosiuscula, Herb. Banks
ex R. Br. l.c. P. erosa, Lichtenst. ex Meisn. l.c. Leucadendrum grandiflorum, Salisb. Parad. t.* 116; *Knight, Prot.* 54.   *L. Gueinzii, O.
Kuntze, Rev. Gen. Pl.* ii. 579.

SOUTH AFRICA : without locality, *Oldenburg* ! *Thunberg* (a of Herbarium left-
hand sp.) ! *Mund* ! *Alexander* ! *Gueinzius* !
COAST REGION : Paarl Div. ; mountains near Paarl, *Niven* ! *Roxburgh* ! *Drège* !
Zeyher ! *Bolus*, 5570 ! *Schlechter*, 9211 ! Wellington, *Pappe* ! French Hoek,
Ludwig ! *MacOwan*, 2905 ! Caledon Div. ; Houw Hoek, *Zeyher* ! Riversdale Div. ;
Garcias Pass, *Galpin*, 4458 ! George Div. ; Devils Kop, *Bowie* ! Paschalsdorp,
Zeyher ! Knysna Div. ; near Knysna, *Pappe* !

14. **L. reflexum** (Buek ex Meisn. in DC. Prodr. xiv. 254);
branches minutely greyish-woolly-tomentose; leaves rather crowded
at the base of the long flowering branches, more distant and exposing
the stem above, oblong or oblanceolate-oblong, obtuse, 3- (rarely 2-)
toothed or entire, ¾–1½ in. long, 2½–4½ lin. wide, minutely but
densely greyish-tomentose, obscurely veined; heads solitary or if
paired only one fully developed at a time, peduncled, far exserted
from the uppermost leaves, 1¾–2½ in. long and about as wide,
without a definite involucre of barren bracts; peduncle 1–2 in.
long, stout, densely greyish-villous with ovate acuminate greyish-
villous or tomentose bracts; receptacle oblong-cylindric, up to
1 in. long, 2 lin. wide; floral bracts ovate, or (the inner) obovate-
cuneate, gradually and long or (the inner) abruptly acuminate, up
to 5 lin. long, tomentose above, densely villous towards the base

or the inner very densely hirsute-tomentose up to the acumen ;
adult flower-bud up to almost 2 in. long; flowers reflexed ; perianth-
tube curved, gibbous, 4–5 lin. long, finely villous above, glabrous
at the contracted base, adaxial and lateral claws permanently united
into a flattened-out sheath revolute in the upper part, softly and
spreadingly villous, over 1 in. long ; abaxial claw very slender,
long adhering by the upper end to the sheath, softly and spreadingly
villous ; limbs lanceolate-oblong, acute, up to 3 lin. long, usually
inflexed on the revolute end of the claw-sheath ; anthers sessile,
linear-lanceolate, long-apiculate, over 2 lin. long; hypogynous
scales subulate, 1 lin. long ; ovary oblong, 1 lin. long, finely
greyish-pubescent ; style up to more than 3 in. long, deflexed or
curved upwards, slightly tapering and quadrangular upwards ;
stigma subulate, not wider than the style and set off from it by a
minute bend, 3 lin. long. *Meisn. in DC. Prodr.* xiv. 254. *Leucadendron reflexum, O. Kuntze, Rev. Gen. Pl.* ii. 579.

COAST REGION : Clanwilliam Div. ; Honig Valley, Koudeberg and Wupperthal,
Drège, 2415 ! Pakhuis Pass, *Bolus,* 9079 ! *MacOwan in Herb. Austr.-Afr.* 1946 !
Koude Berg, *Schlechter,* 8755 ! Cederberg Range, *Mader !*

15. L. hypophyllum (R. Br. in Trans. Linn. Soc. x. 102) ; a
small procumbent shrub ; branches 1–2 ft. long, trailing on the
ground, very rarely erect and then short, finely crisped-pubescent
to tomentose with few or many long spreading hairs or almost
glabrous ; leaves loosely scattered, those of procumbent branches
all turning skywards, very variable, linear-cuneate to lanceolate,
obtuse, truncate or rarely subacute, entire or 3- (rarely 4–7-)toothed
at the apex, flat or with recurved margins, 2–4½ in. long, 1–6 lin.
broad, obscurely veined or veinless, when quite young usually finely
pubescent to tomentose or almost villous, mostly very soon quite
glabrous, rarely permanently tomentose ; heads solitary or 2–3-
nate, peduncled, with a definite involucre of numerous densely
imbricate barren bracts, ⅔ to over 1 in. long excluding the styles,
up to 1½ in. in diam. ; peduncles of decumbent branches bent
skywards, ½ to over 1 in. long, very stout upwards, tomentose,
bearing scattered ovate obtuse to subacute finely tomentose bracts ;
receptacle conical, 4–6 in. long, at the base 3–4 lin. broad ; involucral bracts broad-ovate, shortly and abruptly acuminate or acute,
finely tomentose or glabrous upwards, the larger 3–4 lin. long ;
floral bracts broad-obovate, very shortly acute or acuminate,
3–5 lin. long, densely tomentose excepting at the glabrescent tips
or all over ; adult flower-bud ⅔ to over ¾ in. long, perianth-tube
3–5 lin. long, cylindric, glabrous below, pubescent above ; adaxial
and lateral claws permanently united into a sheath more or less
flattened out and revolute in the upper part, 2–4 lin. long, finely
tomentose ; abaxial claw soon free, often twisted ; limbs oblong,
acute, 1¼–1½ lin. long, tomentose ; anthers sessile, ovate-oblong,
1 lin. long ; hypogynous scales subulate, under ½ lin. long ; ovary
oblong, 1 lin. long, very finely pubescent ; style up to about

10 lin. long, straight, thickened and terete below, quadrangular upwards, gradually passing into the clavate obtuse stigma, which is $\frac{1}{2}$–$\frac{2}{3}$ lin. long ; fruit globose-ellipsoid, up to 3 lin. long, whitish, almost glabrous. *Roem. & Schult. Syst. Veg.* iii. 361 ; *Meisn. in DC. Prodr.* xiv. 257, *incl. vars. vulgare, canaliculatum and stenophyllum. L. canaliculatum, Buek in Drège, Zwei Pfl. Documente,* 199. *L. diffusum, Sieber ex Meisn. l.c. Leucadendron Hypophyllocarpodendron, Linn. Sp. Pl. ed.* i. 93 ; *Berg. in Vet. Acad. Handl. Stockh.* 1766, 321 ; *Descr. Pl. Cap.* 16. *Protea Hypophyllocarpodendron, Linn. Mant. alt.* 191. *P. hypophylla, Thunb. Diss. Prot.* 23, *and Fl. Cap. ed. Schult.* 126 ; *Murr. Syst. Veg. ed.* xiv. 137 ; *Lam. Ill.* i. 239 ; *Willd. Sp. Pl.* i. 513 ; *Poir. Encycl.* v. 655. *P. hypophylla, vars. angustifolia and latifolia, Klotzsch in Flora,* 1845, 76. *P. heterophylla, Thunb. Diss. Prot.* 24, *and Fl. Cap. ed. Schult.* 127, *partly from his specimens. Leucadendrum hypophyllum, Knight, Prot.* 56.—*Thymelæa capitata rapunculoides, etc., Pluk. Mant. t.* 440, *fig.* 3. *Conocarpodendron folio rigido angusto, etc., Boerh. Ind. Pl. Hort. Lugd.-Bat.* ii. 198, *t.* 198. *Scolymocephalus seu Conocarpodendron folio angusto, etc., Weinm. Phyt.* iv. 294, *t.* 902, *fig. a.*

SOUTH AFRICA : without locality, *Bergius* ! *Oldenburg* ! *Thunberg* (sheets α and γ of his Herbarium) ! *Masson* ! *Sparrman* ! *Niven* ! *Brown* ! *Ludwig* ! *Ecklon* ! *Zeyher,* 1464 ! *Wahlberg* ! *Sieber,* 3 !

COAST REGION : Malmesbury Div. ; Mamre (Groene Kloof), *Bolus,* 4323 ! Paarl Div. ; by the Berg River, near Paarl, *Drège* ! Cape Div. ; mountains and flats in the vicinity of Cape Town, *Burchell,* 215 ! 710 ! 8561 ! *Bowie*! *Brown* ! *Ecklon,* 472 ! *Wallich* ! *Zeyher,* 1465 ! *Thom* ! *Hooker,* 403 ! *Drège* ! *Milne,* 34 ! *Bolus,* 2908 ! *MacOwan,* 2504 ! *and in Herb. Norm. Austr.-Afr.* 775 ! *Schlechter,* 59 ! *Wolley-Dod,* 618 ! *Galpin,* 4467 ! *Dümmer,* 4491 ! near Simons Town, *Wright* ! *MacGillivray,* 637 ! near Cape Point, *Phillips,* 521 ! Stellenbosch Div. ; between Jonkers Hoek and Lowrys Pass, *Burchell,* 8302 ! Caledon Div. ; near Caledon, *Templeman in MacOwan, Herb. Austr.-Afr.* 1642 ! Bredasdorp Div. ; near Elim, *Bolus,* 8584 ! *Schlechter,* 9684 !

16. **L. tomentosum** (R. Br. in Trans. Linn. Soc. x. 101, excl. vars. β and γ) ; an erect or procumbent shrub, 4–5 ft. high ; branches finely crisped-pubescent to tomentose with or without long spreading hairs ; leaves scattered, those of procumbent branches all turning skywards, linear, obtuse, entire or callously 3-toothed at the apex, flat or with recurved margins, 1$\frac{1}{4}$–2$\frac{1}{4}$ in. long, 1–2$\frac{1}{2}$ lin. broad, veinless, when quite young usually very finely tomentose, at length glabrescent or quite glabrous ; heads usually 2–3 at the end of the branches or solitary, very shortly peduncled, with a definite involucre of numerous densely imbricate barren bracts, about $\frac{3}{4}$ in. long excluding the styles, up to 1 in. in diam. ; peduncles rarely over 3 lin. long, obconical, very stout upwards, densely covered with ovate or ovate-lanceolate obtusely subacuminate shortly tomentose and upwards ciliate bracts, increasing upwards in size and passing into the involucre ; receptacle convex, up to 2 lin. high and 4–5 lin. wide at the base ; involucral bracts like the upper bracts of the peduncle, but more acuminate, up to 3 lin. long ; floral bracts oblanceolate to obovate-cuneate, shortly and

abruptly acuminate or apiculate, 3 lin. long, densely tomentose excepting at the glabrescent tips; adult flower-bud about 8 lin. long; perianth-tube 2½–3 lin. long, slender, cylindric, glabrous; adaxial and lateral claws united into a sheath more or less flattened out and revolute in the upper part, about 3½ lin. long, or separating from below upwards, sheath glabrous excepting on the finely pubescent margins, abaxial claw soon free, pubescent all over; limbs lanceolate or lanceolate-oblong, acute, 1½ lin. long, sparingly and shortly pubescent to hirsute, soon glabrescent; anthers sessile, linear-oblong, 1 lin. long; hypogynous scales subulate, hyaline, ½ lin. long; ovary oblong, over ½ lin. long, whitish-pubescent, gradually passing into the style, surrounded by whitish hairs 2–2½ lin. long; style about 8 lin. long, filiform, quadrangular upwards, gradually passing into the subclavate-oblong grooved obtuse stigma which is ⅔–¾ lin. long. *Roem. & Schult. Syst. Veg.* iii. 360, *partly*; *Meisn. in DC. Prodr.* xiv. 258, *excl. vars.* β *and* δ. *Protea tomentosa, Thunb. Diss. Prot.* 24, *and Fl. Cap. ed. Schult.* 127, *partly from his specimens*; *Linn. f. Suppl.* 118; *Lam. Ill.* i. 239; *Willd. Sp. Pl.* i. 514; *Poir. Encycl.* v. 656. *L. Ecklonii, Buek ex Meisn. l.c. Leucadendrum tomentosum, Knight, Prot.* 57; *O. Kuntze, Rev. Gen. Pl.* ii. 579.

SOUTH AFRICA: without locality, *Thunberg* (sheet β of his herbarium)! *Niven*! COAST REGION: Malmesbury Div.; around Mamre and between Mamre and Saldanha Bay, *Drège*!

This, in the Herbarium of the British Museum, is written up in R. Brown's handwriting as *L. tomentosum a.*

17. **L. Muirii** (Phillips in Kew Bulletin, 1910, 332); a bush, 6 ft. high; branches minutely greyish woolly-tomentose, at length glabrescent; leaves loosely scattered, oblanceolate to cuneate-linear, obtuse, bluntly 3–7-toothed at the apex, 1¼–2 in. long, 2–4½ lin. broad, glabrous excepting at the crisped-tomentose base; heads 2-nate, shortly peduncled, up to ¾ in. long excluding the styles, ¾ to almost 1 in. in diam., without a definite involucre of barren bracts; peduncle up to 4 lin. long, tomentose, bearing stout spreading tomentose bracts up to 2 lin. long and semicylindric from a broader base; receptacle convex, 2 lin. high, 3–3½ lin. wide at the base; floral bracts ovate to obovate, abruptly acuminate, 2½ lin. long, densely tomentose to the base of the acumen; adult flower-bud about 6–7 lin. long; perianth-tube 1–1½ lin. long, glabrous; adaxial and lateral claws slender, flexuous, at length more or less free below, but united at the upper usually recurved end or the tips free or more frequently the four claws united in 2 pairs, those of each pair holding together near the upper end only, 3–3½ lin. long, finely pubescent; limbs lanceolate-oblong, acute, 1¼–1½ lin. long, hirsute; anthers sessile, ovate-lanceolate, 1 lin. long; hypogynous scales lanceolate-linear, 7 lin. long, hyaline; ovary oblong, ¾ lin. long, whitish-pubescent; style 8 lin. long, filiform, very slightly stouter below, quadrangular above,

glabrous; stigma clavate-cylindric, obtuse, grooved, ¾ lin. long, set off from the style by an oblique slight thickening.

COAST REGION : Riversdale Div. ; Milkwoodfontein, *Galpin*, 4457 ! Zandhoogte, *Muir in Herb. Galpin*, 5309 !

18. L. candicans (Loud. Hort. Brit. ed. i. 38) ; an erect shrub, 4–5 ft. high ; branches minutely, but densely, greyish-woolly-tomentose with or without long spreading hairs ; leaves moderately crowded up to the flower-heads, spreading or obliquely erect, oblong to oblong-lanceolate, entire or 2–5-toothed at the apex, somewhat narrowed at the base, 1¼–2¼ in. long, 3–6 lin. broad, veinless or obscurely veined, greyish- or whitish-tomentose ; heads solitary or in clusters of 2–4 at the end of the branches, very shortly peduncled among the top-leaves, about ¾ to almost 1 in. long excluding the styles, 1 in. or more in diam., with a definite involucre of barren bracts ; peduncle turbinate, up to 4 lin. long and wide (at the upper end), densely covered with ovate acuminate densely tomentose upward-increasing bracts ; receptacle flat, over 4 lin. across ; involucral bracts like the upper bracts of the peduncle, up to 4 lin. long ; floral bracts ovate to lanceolate, clawed downwards, acuminate, densely tomentose excepting on the acumen, up to 6 lin. long ; adult flower-bud 7–9 lin. long ; perianth-tube 2–3 lin. long, slender, glabrous ; abaxial and lateral claws united into a sheath flattened out and recurved or revolute in the upper part, with the tips at length free, 4–5 lin. long, glabrous apart from the pubescent margins ; abaxial claw soon free, pubescent all over ; limbs oblong, acute, 1–1½ lin. long, finely tomentose, deflexed or recurved on the claws ; anthers sessile, linear-oblong, ¾ lin. long, apiculate ; hypogynous scales subulate, hyaline, ⅔ lin. long ; ovary oblong, whitish-pubescent, contracted into a short glabrous beak, 1½ lin. long ; style straight or almost so, 6–8 lin. long, filiform, tapering from a somewhat stouter base, quadrangular upwards, glabrous ; stigma oblong-cylindric, obtuse, grooved ; fruit ellipsoid, 4 lin. long, almost glabrous, shining. *L. tomentosum, var. candicans, Meisn. in DC. Prodr.* xiv. 258. *L. tomentosum, var. β, R. Br. in Trans. Linn. Soc.* x. 102. *L. spathulatum, var. dentatum, Meisn. l.c. Protea tomentosa, Thunb. Diss. Prot.* 24, *and Fl. Cap. ed. Schult.* 127, *partly from his specimens. P. candicans, Andr. Bot. Rep. t.* 294. *Leucadendrum rodolentum, Knight, Prot.* 58.

SOUTH AFRICA : without locality, *Niven* ! *Brown* ! *Auge* ! *Thunberg* (sheet α of his Herbarium) !
COAST REGION : Clanwilliam Div. ; Olifants River, *Drège* ! *Bolus*, 5786 ! Zeekoe Vley, *Schlechter*, 8579 ! Piquetberg Div. ; .Pikeniers Kloof, *Zeyher*, 1466 ! Malmesbury Div. ; Zwartland, *Zeyher* ! Hopefield, *Bolus*, 12816 ! *Schlechter*, 5313 ! Worcester Div. ; Brand Vley, *Schlechter*, 9921 ! Caledon Div. ; near Caledon, *Templeman in MacOwan, Herb. Austr.-Afr.* 1643 !

19. L. parile (Phillips) ; an erect shrub, 4–5 ft. high ; branches densely greyish-woolly-tomentose, usually with long spreading

hairs; leaves rather crowded up to the flower-heads, but not
concealing the stem, linear or oblong-linear, obtuse, entire or
minutely 3-toothed at the apex (teeth with a blunt callous point),
slightly narrowed at the base, $1\frac{3}{4}$–$2\frac{1}{2}$ in. long, $1\frac{1}{2}$–3 lin. broad,
minutely and densely greyish-tomentose; heads subsessile, in
clusters of 2–4 at the ends of the branches, about $\frac{3}{4}$ in. long
excluding the styles, and $\frac{3}{4}$–1 in. in diam., with a definite involucre
of closely imbricate reddish-brown barren bracts; peduncle very
short, obconical, densely covered with barren bracts passing into
those of the involucre; receptacle flat; barren bracts of peduncle
and involucre very numerous, increasing upwards, ovate to ovate-
lanceolate, acuminate, finely pubescent below, glabrous above,
ciliate, the uppermost up to $\frac{1}{2}$ in. long; floral bracts unguiculate-
lanceolate or oblanceolate, gradually and often long-acuminate,
about 5 lin. long, $1\frac{1}{2}$–$2\frac{1}{2}$ lin. broad, densely hirsute-tomentose on
the back of the claw, glabrous and ciliate above; adult flower-bud
about 7 lin. long; perianth-tube up to 2 lin. long, slender, glabrous;
adaxial and lateral claws permanently united into a very narrow
sheath flattened out and reflexed in the upper part, minutely
ciliate, otherwise glabrous; abaxial claw minutely pubescent, soon
free; limbs oblong-lanceolate, acute, $1\frac{1}{2}$ lin. long, glabrous or
sparingly hairy; hypogynous scales linear-subulate, hyaline, 1 lin.
long; ovary linear-oblong, shortly villous, contracted above into
a filiform glabrous beak, over $1\frac{1}{2}$ lin. long, surrounded by whitish
hairs up to 2 lin. long; style about 7 lin. long, tapering from a
thickened base, slender and quadrangular upwards; stigma cylindric, obtuse, grooved, $\frac{3}{4}$ lin. long, distinct from the style. *L.
tomentosum, var.* γ, *R. Br. in Trans. Linn. Soc.* x. 102; *Roem. &
Schult. Syst. Veg.* iii. 360, *var.* γ *only*; *var. Dregei, Meisn. in DC.
Prodr.* xiv. 258. *Leucadendrum parile, Knight, Prot.* 57.

COAST REGION: Malmesbury Div.; vicinity of Mamre and between there and
Saldanha Bay, *Drège*! *Bolus*, 4324! *Zeyher*, 1467! Paarde Berg, *Niven*!

20. **L. crinitum** (R. Br. in Trans. Linn. Soc. x. 103, descr. and
syn.); a bush 3–4 ft. high; branches more or less spreading,
minutely woolly-tomentose and hirsute; leaves usually very much
crowded, particularly upwards and concealing the stem, more or
less oblong, obtuse with a callous point, entire or rarely bluntly
3-toothed, obtuse at the base, 1–2 in. long, 3–9 lin. broad,
distinctly veined, very minutely rugulose, villous or tomentose
when young, then glabrescent to glabrous; heads subsessile among
the crowded top leaves, terminal or overtopped by young branches
and thus apparently lateral, solitary or in clusters of 2–3, 1–$1\frac{1}{2}$ in.
long, 1–2 in. in diam., with a definite involucre of imbricate barren
bracts exceeding the perianths; peduncle up to $2\frac{1}{2}$ lin. long, stout,
densely bracteate; receptacle slightly convex, 4–6 lin. in diam.;
barren bracts of peduncle and involucre gradually increasing
upwards, the innermost much produced, lanceolate to narrow-
lanceolate, acute to (the inner) long- and subulate-acuminate and

up to more than 1 in. long, finely tomentose or pubescent to glabrescent, ciliate and the long acumina of the inner bracts more
or less penicillate; floral bracts narrowly lanceolate, very long-
acuminate, 6–7 lin. long, densely and long-tomentose up to 2–3 lin.,
then finely pubescent, ciliate and subpenicillate; adult flower-bud
about ½ in. long; perianth-tube 2–3 lin. long, slightly widened at
the upper end, glabrous; adaxial and lateral claws united into a
slender straight sheath, 3–4 lin. long, flattened out and recurved
at the upper end, at length sometimes splitting downwards, like
the free abaxial claw loosely villous; limbs deflexed, lanceolate,
acute, 1 lin. long, hirsute; hypogynous scales linear, ¾ lin. long;
ovary oblong, whitish-pubescent, ¾ lin. long, surrounded by white
hairs ¾ lin. long; style straight or nearly so, finely filiform from a
stouter base, capillary upwards, 10–12 lin. long, glabrous; stigma
cylindric, very slender, slightly thickened at the base, ⅔ lin. long;
fruit ellipsoid, 3½ lin. long, sparingly and minutely puberulous.
Roem. & Schult. Syst. Veg. iii. 361; *Meisn. in DC. Prodr.* xiv. 260,
excl. vars. β and *γ. L. oleæfolium, R. Br. l.c.* 104 (*var. prima*). *L.
oleæfolium, var. Brownii, Meisn. l.c.* 261. *L. penicillatum, Buek in
Drège, Zwei Pfl. Documente,* 24, 199. *L. penicillatum, β, trichanthum,
Gandog. Bull. Soc. Bot. France,* xlviii. *p.* xcv. *Protea crinita, Thunb.
Diss. Prot.* 21; *Fl. Cap. ed. Schult.* 125; *Willd. Sp. Pl.* i. 511; *Poir.
Encycl.* v. 657. *P. criniflora, Linn. f. Suppl.* 117. *P. erosa, Willd.
ex Meisn. l.c. P. molle, Klotzsch ex Meisn. l.c., not of R. Br. Leuca-
dendron crinitum, Steud. Nomencl. ed.* 2, ii. 399; *O. Kuntze, Rev.
Gen. Pl.* ii. 579. *L. penicillatum, O. Kuntze, Rev. Gen. Pl.* ii. 579.

SOUTH AFRICA: without locality, *Auge! Oldenburg! Thunberg!*
COAST REGION: Paarl Mountains, *Drège!* Stellenbosch Div.; Hottentots
Holland, *Niven,* 46! Lowrys Pass, *MacOwan!* Caledon Div.; Houw Hoek
Mountains, *Burchell,* 8098! 8133! *Bowie! Zeyher,* 3683! *MacOwan,* 2910! *and
in Herb. Norm. Austr.-Afr.* 910! *Pillans! Schlechter,* 9386!

21. **L. oleæfolium** (R. Br. in Trans. Linn. Soc. x. 104, var. *altera*);
a bush 3–4 ft. high; branches villous or tomentose, rather slender;
leaves densely crowded, particularly upwards and there concealing
the stem, linear-oblong, obtuse, rarely truncate, entire with a
callous point or bluntly 3–5-toothed at the apex, slightly narrowed
towards the base, 1–1¾ in. long, 2–5 lin. broad, more or less
distinctly veined, sometimes very minutely rugulose, tomentose
or villous when young, then more or less glabrescent to glabrous;
heads subsessile among and usually exceeded by the crowded top
leaves, terminal, sometimes overtopped by young branches and
then apparently lateral, ½–¾ in. long excluding the styles, up to
1 in. wide, with a definite involucre of imbricate barren bracts
shorter than the perianths; peduncle up to 3 lin. long, moderately
stout, bearing lanceolate to linear-subulate, pubescent to villous
bracts; receptacle flat, up to 5 lin. wide; barren bracts of peduncle
and involucre gradually increasing upwards and inwards, rather
numerous, those of the involucre ovate-lanceolate to narrow-lanceolate, acute to acuminate, finely pubescent or villous all over, at

length sometimes glabrescent, ciliate, up to 5 lin. long; floral bracts
linear-lanceolate to subulate, acutely acuminate, up to 5 lin. long,
long-tomentose all along; adult flower-bud about 8 lin. long;
perianth-tube 2½–3 lin. long, very slender, glabrous; adaxial and
lateral claws united into a slender straight sheath, 3½–5 lin. long,
flattened out and recurved at the upper end, like the free very
slender abaxial claw loosely hairy; limbs deflexed, narrowly
lanceolate-oblong, acute, 1 lin. long, hirsute; hypogynous scales
linear, ⅔ lin. long; ovary oblong, whitish-pubescent, up to 1 lin.
long, surrounded by hairs up to 1 lin. long; style straight or nearly
so, finely filiform from a stouter base, capillary upwards, 12 lin.
long, glabrous; stigma cylindric, very slender, slightly thickened
at the base, ½–⅓ lin. long; fruit ellipsoid, 3½ lin. long, sparingly
and minutely puberulous. *Meisn. in DC. Prodr.* xiv. 261, *excl.
var. β, Brownii. L. molle, R. Br. l.c.* 103; *Meisn. l.c. (partly?).
L. cryptanthum, Buek in Drège, Zwei Pfl. Documente,* 82, 116, 199.
L. medium (c), *Drège ex Meisn. l.c.* 260. *L. medium, f. Zeyheri,
Gandog. in Bull. Soc. Bot. France,* xlviii. *p.* xcv. *Leucadendron oleæfolium, Berg. in Vet. Acad. Handl. Stockh.* 1766, 320, *and Descr. Fl.
Cap.* 15. *L. molle, O. Kuntze, Rev. Gen. Pl.* ii. 579. *Leucadendrum
criniflorum, Knight, Prot.* 58. *Protea mollis, Poir. Encycl. Suppl.* iv.
567, *not* 577.

SOUTH AFRICA: without locality, *Roxburgh*! *Ludwig*! *Drège*! *Thom*, 682!
COAST REGION: Clanwilliam Div.; Blue Berg, *Drège*! Worcester Div.;
Dutoits Kloof, *Drège*! Bains Kloof, *Schlechter*, 9095! Stellenbosch Div.;
Hottentots Holland, *Niven*! Jonkers Hoek, *Niven*, 39! Caledon Div.; mountains
near Genadendaal, *Burchell*, 7691! *Schlechter*, 9839! *Galpin*, 4464! *Drège*!
Zwart Berg, *Bowie*! *Zeyher*, 3683! 3684!

22. **L. diffusum** (R. Br. in Trans. Linn. Soc. x. 104, excl.
synonyms); a procumbent bush 2½ ft. high; branches more or less
minutely woolly-tomentose and finely hirsute; leaves loosely
scattered on decumbent branches turned skywards, linear, subacute,
entire or minutely and acutely 2–3-toothed at the apex, slightly
attenuated downwards, about 1½ in. long, up to 2 lin. broad,
faintly veined, minutely crisped-pubescent to tomentose, sometimes
with some long hairs; heads solitary, terminal or by overtopping
apparently lateral, peduncled, rounded at the base, ¾ in. long,
1 in. in diam., with a definite involucre of imbricate barren
bracts; peduncle up to ½ in. long, slender, with ovate-lanceolate
acuminate finely tomentose and ciliate recurved bracts; receptacle
flat; involucral bracts ovate to ovate-oblong and finely acuminate,
5 lin. long, tomentose on the back with longer hairs on the sides
and upwards; floral bracts lanceolate to linear, acuminate, 2½–5
lin. long, densely tomentose; adult flower-bud 8–9 lin. long;
perianth-tube 3 lin. long, glabrous below, pubescent above,
slightly widened upwards; adaxial and lateral claws united into
a straight sheath up to 5 lin. long, flattened out and recurved in
the upper part, yellow-tomentose; limbs deflexed, lanceolate-
oblong, acute, 1 lin. long, tomentose; anthers sessile, ovate-

oblong, minutely apiculate, $\frac{2}{3}$ lin. long; hypogynous scales linear, acute, $\frac{2}{3}$–1 lin. long; ovary oblong, $\frac{1}{2}$ lin. long, whitish-pubescent, surrounded by whitish hairs, $\frac{3}{4}$ lin. long; style 10–12 lin. long, finely filiform, capillary upwards, glabrous; stigma slender, cylindric, obtuse, finely grooved, $\frac{1}{3}$ lin. long. *Roem. & Schult. Syst. Veg.* iii. 362; *Meisn. in DC. Prodr.* xiv. 259, *partly. L. pedunculatum, Klotzsch ex Krauss in Flora,* 1845, 76. *Leucadendron pedunculatum, O. Kuntze, Rev. Gen. Pl.* ii. 579.

SOUTH AFRICA: without locality, *Masson*! *Roxburgh*!
COAST REGION: Bredasdorp Div.; Zoetendals Valley, *Krauss,* 1069. Riversdale Div.; Muiskraal near Garcias Pass, *Galpin,* 4456!

Leucadendron saxatile (Knight, Prot. 58) probably belongs here. It was collected by Niven on dry rocks near Groote River in Ladismith Division.

23. **L. stenanthum** (Schlechter in Engl. Jahrb. xxvii. 112); a decumbent shrub; branches slender, finely greyish woollytomentose and greyish-hirsute, lower down at length glabrescent with a reddish bark; leaves scattered, or on short branches crowded and concealing the stem, on decumbent branches all turned skywards, linear, somewhat attenuated downwards, obtuse or subobtuse, entire, $\frac{3}{4}$–1$\frac{1}{4}$ in. long, 1$\frac{1}{2}$–2$\frac{1}{2}$ lin. broad, minutely greyish-tomentose with or without long soft hairs, obscurely veined; heads solitary, sessile among the top leaves or overtopped by young branches and apparently lateral, rounded at the base, shortly peduncled, 8–10 lin. long excluding the styles, 7–9 lin. wide, with a definite scanty short involucre of barren bracts; peduncle up to 3 lin. long, slender, bearing linear tomentose spreading bracts; receptacle flat, about 3 lin. wide; involucral bracts ovate-lanceolate to lanceolate, acuminate to caudate-acuminate, finely tomentose, ciliate upwards, up to 3 lin. long; floral bracts oblanceolate-cuneate to narrowly obovate-cuneate, acute or shortly and abruptly subulate-acuminate, up to 2 lin. long, densely tomentose up to the minutely pubescent tips; adult flower-bud 8 to about 9 lin. long; perianth-tube 2$\frac{1}{2}$ lin. long, glabrous; adaxial and lateral claws united into a straight slender sheath, 4 lin. long, $\frac{1}{2}$ lin. wide, flattened out and recurved in the upper part, at length splitting from the top, minutely hirsute along the margins; abaxial claw slender, free, minutely hirsute or pubescent all over with red glands; limbs deflexed, oblong-lanceolate, $\frac{3}{4}$ lin. long, subacute, minutely hirsute; anthers sessile, oblong-lanceolate, $\frac{1}{2}$ lin. long; hypogynous scales linear, somewhat fleshy and clavate upwards, 1 lin. long; ovary lanceolate in outline, 1 lin. long, minutely pubescent, gradually passing into the filiform style which is capillary above and 1 in. long; stigma slender, cylindric, $\frac{1}{4}$–$\frac{1}{3}$ lin. long.

SOUTH AFRICA: without locality, *Masson*!
COAST REGION: Caledon Div.; Bot River, 2000 ft., *Schlechter,* 9446! Shaws Mountain, near Caledon, *Bodkin in Herb. Bolus*!

24. **L. Mundii** (Meisn. in DC. Prodr. xiv. 261); a bush 3 ft. high; branches densely minutely greyish crisped-tomentose, sometimes with fine spreading hairs; leaves densely crowded, usually up

to the flower-heads, concealing the stem, cuneate-obovate, obtuse or subobtuse, 2–12-toothed towards the apex, $1\frac{1}{2}$–$2\frac{1}{2}$ in. long, 10–15 lin. broad, densely greyish crisped-tomentose, sometimes at length almost glabrous; heads in clusters of 2–3, shortly peduncled among the top leaves, sometimes overtopped by young branches and then apparently lateral, rounded at the base, about $\frac{3}{4}$ in. long and wide, with a very short involucre of small ovate to ovate-oblong, closely imbricate, tomentose barren bracts recurved at the tips; peduncles up to 4 lin. long, slender, with scattered small ovate subacute tomentose bracts; receptacle flat; floral bracts obovate to spathulate, abruptly cuspidate or acuminate, 2–3 lin. long, densely hirsute-tomentose up to the minutely hirsute or pubescent tips; adult flower-bud 8–9 lin. long; perianth-tube purplish and widened upwards from a very slender base, up to 4 lin. long, glabrous below, pubescent upwards; adaxial and lateral claws at first united into a straight sheath, 3–$3\frac{1}{2}$ lin. long, red and recurved above, finely whitish-pubescent, soon splitting downwards to the middle or base; abaxial claw always free, pubescent; limbs deflexed, lanceolate-oblong, subacute, 1 lin. long, minutely hirsute; anthers sessile, oblong, apiculate, $\frac{2}{3}$ lin. long; hypogynous scales linear, obtuse, $\frac{1}{2}$–$\frac{3}{4}$ lin. long; ovary oblong, $\frac{3}{4}$ lin. long, finely pubescent, surrounded by white hairs $1\frac{1}{2}$ lin. long; style 12–13 lin. long, filiform from a slightly stouter base, capillary upwards, glabrous; stigma slender, cylindric, obtuse, grooved, $\frac{1}{2}$ lin. long. *L. purpureum, Mund ex Meisn. l.c. L. crinitum, Klotzsch ex Meisn. l.c., not of R. Br. Leucadendron Mundii, O. Kuntze, Rev. Gen. Pl.* ii. 579.

COAST REGION: Swellendam Div.; Tradouw Mountains, *Mund*! Riversdale Div.; Garcias Pass, *Galpin*, 4478! *Bolus*, 11366! *Phillips*, 517!

25. **L. buxifolium** (R. Br. in Trans. Linn. Soc. x. 100); a shrub 2–5 ft. high; branches greyish crisped-tomentose to villous; leaves densely imbricate up to the flower-heads, concealing the stem, obovate or broad-elliptic to oblong-elliptic, very obtuse to subobtuse, with callous tips, slightly narrowed at the base, $3\frac{1}{2}$–10 lin. long, $1\frac{1}{2}$–4 lin. broad, greyish-tomentose; heads sessile among the leaves, usually in clusters of 2–6, sometimes solitary, terminal or owing to overtopping by branches apparently lateral, turbinate at the base, up to $\frac{1}{2}$ in. long and wide, rarely larger, sometimes 2 or 3 heads almost fused into one, with a definite involucre of imbricate barren bracts; receptacle almost flat, about 2 lin. wide; involucral bracts numerous, ovate, shortly acuminate or acute or obtuse, 2–$2\frac{1}{2}$ lin. long, sparingly hairy on the back, long-ciliate, with a terminal tuft of whitish hairs; floral bracts lanceolate, gradually and finely acuminate, about 3 lin. long, densely tomentose; adult flower-bud 5–6 lin. long; perianth-tube 2 lin. long, much attenuated downwards, glabrous below, finely pubescent above; abaxial and lateral claws more or less united at first, at length usually more or less free and like the abaxial straight with recurved tips, 2–$2\frac{1}{2}$ lin. long, villous; limbs deflexed, oblong, acute, 1–$\frac{3}{4}$ lin. long,

hirsute; anthers sessile, oblong, $\frac{2}{3}$ lin. long; subhypogynous scales linear, $\frac{3}{4}$ lin. long; ovary oblong, finely pubescent, $\frac{2}{3}$ lin. long; style $5\frac{1}{2}$ lin. long, finely filiform, glabrous, 5–6 lin. long; stigma ovoid, $\frac{1}{2}$ lin. long. *Meisn. in DC. Prodr.* xiv. 259. *L. buxifolium, f. epacridea, Gandog. in Bull. Soc. Bot. France,* xlviii. *p.* xciv. *Protea pubera, Thunb. Diss. Prot.* 43 ; *and Fl. Cap. ed. Schult.* 140. *P. buxifolia, Poir. Encycl. Suppl.* iv. 566. *P. villosa, Willd. in Spreng. Syst. Veg.* i. 464. *Leucadendrum puberum, Knight, Prot.* 61. *L. truncatulum, Knight, l.c. Leucadendron buxifolium, O. Kuntze, Rev. Gen. Pl.* ii. 579.

SOUTH AFRICA : without locality, *Oldenburg* ! *Ludwig* ! *Brown* ! *Thom,* 711 ! 940 ! COAST REGION : Clanwilliam Div. ; Alexanders Hoek, *Schlechter,* 5128 ! Stellenbosch Div. ; Hottentots Holland, *Thunberg* ! Caledon Div. ; Houw Hoek and Little Houw Hoek Mountains, *Niven,* 200 ! *Burchell,* 8042 ! *Bowie* ! *Zeyher* ! *Schlechter,* 5495 ! *Pillans* ! Bot River, *Burchell,* 928 ! Hermanus, *Galpin,* 4466 ! Klein River Mountains, *Niven* ! *Zeyher,* 3685 ! near Caledon, *Templeman,* Hartebeest River, *Zeyher* ! Zwart Berg, *MacOwan* ! Bredasdorp Div. ; near Elim, *Bolus,* 7857 ! 8588 ! Koude River, *Schlechter,* 9603 !

26. L. puberum (R. Br. Trans. Linn. Soc. x. 100) ; a shrub $1\frac{1}{2}$–5 ft. high ; branches virgate, divaricate, decumbent or ascending, often $\frac{3}{4}$–1 ft. long, minutely crisped-tomentose, usually with soft straight or flexuous hairs ; leaves usually crowded to densely imbricate and concealing the stem, sometimes scattered, oblong, oblanceolate or oblong-linear, acute to obtuse with a callous point, mostly entire, sometimes minutely 3-toothed at the apex, obtuse or narrowed at the base, $\frac{2}{3}$–$1\frac{1}{4}$ in. long, 2–4 lin. broad, more or less distinctly veined, greyish, minutely crisped-tomentose with or without longer spreading hairs, often villous when young ; heads solitary or in clusters of 2–4, terminal or overtopped by young branches and then apparently lateral, 7–8 lin. long excluding the style, $\frac{3}{4}$–1 in. wide, peduncled, with a definite involucre of imbricate barren bracts, rounded at the base ; peduncles slender, from very short up to more than $\frac{1}{2}$ in. long, shortly tomentose, bearing ovate-lanceolate to lanceolate tomentose bracts ; receptacle convex, low, small ; involucral bracts lanceolate, long-acuminate, up to 4 lin. long, villous ; floral bracts obovate-cuneate, suddenly and long acuminate, $2\frac{1}{2}$–4 lin. long, 1–$1\frac{3}{4}$ lin. broad, densely tomentose excepting on the pubescent acumen ; adult flower-bud 6–8 in. long ; perianth-tube $2\frac{1}{2}$–$3\frac{1}{2}$ lin. long, subcylindric, pubescent from the middle upwards ; adaxial and lateral claws about 3 lin. long, united into a straight villous to hirsute sheath with free recurved or revolute tips red inside, soon more or less splitting to the base ; abaxial claw very slender, free villous ; limbs deflexed, ovate-oblong, acute, $\frac{1}{2}$–$\frac{3}{4}$ lin. long, hirsute ; hypogynous scales linear, $\frac{2}{3}$ lin. long ; ovary oblong, $\frac{1}{2}$ lin. long, minutely pubescent ; style 8–$10\frac{1}{2}$ lin. long, finely filiform, tapering upwards, glabrous, straight or flexuous ; stigma conic-ovoid, subobtuse, $\frac{1}{4}$–$\frac{1}{3}$ lin. long ; fruit ellipsoid, obtuse, almost glabrous, $3\frac{1}{2}$ lin. long. *Roem. & Schult. Syst. Veg.* iii. 359 ; *Meisn. in DC. Prodr.* xiv. 258 *incl. vars. a,* β *and* δ*, but excl. var.* γ*. L. Bolusii, Gandog. in Bull. Soc. Bot. France,*

xlviii. *p.* xcv. *L. lemmerzianum, Schlechter in Engl. Jahrb.* xxvii. 111.
Protea pubera, Linn. Mant. alt. 192 ; *Thunb. Diss. Prot.* 43, *and
Fl. Cap. ed. Schult.* 140 ; *Lam. Ill.* i. 234 ; *Willd. Sp. Pl.* i. 533,
partly ; *Poir. Encycl.* v. 642. *P. heterophylla, Thunb. Diss. Prot.*
24, *and Fl. Cap. ed. Schult.* 127, *partly from his specimens* ; *Murr.
Syst. Veg. ed.* xiv. 138. *Leucadendrum calligerum, Knight, Prot.* 60.
Leucadendron pubigerum, Linn. ex Meisn. l.c. 259. *L. puberum,
O. Kuntze, Rev. Gen. Pl.* ii. 579.

Var. *β*, **patulum** (Meisn. in DC. Prodr. xiv. 259) ; a small shrub with mostly
short and slender, often crowded branches ; leaves at length usually almost
glabrous, 1½–2½ (rarely to 3) lin. broad ; heads rather smaller than in the type,
about 6 (rarely to 7) lin. long excluding the styles. *L. patulum, R. Br. in Trans.
Linn. Soc.* x. 100 ; *Roem. & Schult. Syst. Veg.* iii. 360. *Protea heterophylla,
Thunb. Diss. Prot.* 24, *and Fl. Cap. ed. Schult.* 127, *partly from his specimens* ;
Willd. Sp. Pl. i. 515. *Leucadendron heterophyllum, O. Kuntze, Rev. Gen. Pl.* ii. 579.

SOUTH AFRICA: without locality, *Nelson* ! *Thunberg* ! *Masson* ! *Thom*, 657 ! 691 !
Var. *β* : *Nelson* ! *Masson* !
COAST REGION : Clanwilliam Div. ; between Pakhuis and Bidouw, *Drège* ! Blue
Berg, *Schlechter*, 8463 ! Alexanders Hoek, *Schlechter*, 5128 ! Piquetberg Div. ;
Twenty-four Rivers, *Niven*, 41 ! Tulbagh Div. ; Witzenberg Range, *Burchell*,
8726 ! New Kloof, near Tulbagh, *Ludwig* ! *Zeyher*, 1463 ! *Pappe* ! *Bolus*, 5233 !
MacOwan, 2601 ! *and Herb. Norm. Austr.-Afr.* 779 ! 1945 ! Worcester Div. ;
Brand Vley, *Niven* ! Paarl Div. ; Elands Kloof, *Ludwig* ! Stellenbosch Div. ;
Gordons Bay, *Bolus*, 8077 ! Caledon Div. ; various localities, *Burchell*, 7822 !
7939 ! *Bowie* ! *Ludwig* ! *Pappe* ! *Schlechter*, 5558 ! Riversdale Div. ; Garcias Pass,
Galpin, 4463 ! *Phillips*, 516 ! *Muir* ! Var. *β* : Bredasdorp Div. ; hills near Elim,
Schlechter, 9663 ! *Bolus*, 7873 ! 8585 !
CENTRAL REGION : Calvinia Div. ; Uien Valley, Bokkeveld Mountains, *Drège* !

27. **L. royenifolium** (Stapf) ; a small shrub with procumbent
branches and numerous ascending branchlets ; branchlets mostly
rather slender, minutely greyish crisped-pubescent or tomentose and
more or less hirsute ; leaves rather crowded, obliquely spreading,
not concealing the stem, linear- to narrowly elliptic-oblong or
oblanceolate, acute with a callous point, rarely minutely 2–3-toothed
at the apex, slightly narrowed towards the base, 5–10 lin. long,
2–3 lin. broad, faintly but distinctly veined, sparingly crisped-
pubescent and adpressedly hirsute when young, soon glabrescent
and usually at length quite glabrous ; heads solitary or in clusters
of 2–4, terminal or overtopped by young branches and then
apparently lateral, subsessile, turbinate, rather lax when fully out,
5–6 lin. long, 6–8 lin. broad, with a definite involucre of imbricate
barren bracts ; peduncle very short, covered with ovate acute
tomentose bracts passing into those of the involucre ; receptacle
flat ; involucral bracts ovate, acuminate, up to 2 lin. long,
tomentose ; floral bracts obovate, abruptly and shortly acuminate
or cuspidate, 2–3 lin. long, densely tomentose ; adult flower-bud
about 5½ lin. long ; perianth-tube cylindric, 2½–3 lin. long, pubescent
upwards ; adaxial and lateral claws united into a straight villous
or shortly hirsute sheath, about 1½–2 lin. long, with free recurved
tips, soon separating downwards ; abaxial claw free, otherwise like
the others ; limbs ovate-oblong, acute, not quite 1 lin. long, hirsute ;

anthers sessile, linear-oblong, ½ lin. long; hypogynous scales subulate, hyaline, ½ lin. long; ovary oblong, under 1 lin. long, finely pubescent below, gradually passing into the style; style 6–7 lin. long, filiform, tapering upwards; stigma oblong-cylindric, obtuse, ¼ lin. long; fruit ellipsoid, whitish, 3½ lin. long, subacute, not beaked. *L. puberum,* ? *var. dubium, Meisn. in DC. Prodr.* xiv. 259. *Leucadendrum royenæfolium, Knight, Prot.* 59.

COAST REGION : George Div.; Devils Kop, *Niven,* 43 ! Uniondale Div.; on a rocky hill near Haarlem, *Burchell,* 4993 ! between Avontuur and Klip River, *Drège* ! mountains near Uniondale, *Bolus,* 2453 !

28. L. **prostratum** (Stapf); an apparently trailing low shrub; branches slender, villous or hirsute, at length glabrescent; leaves loosely scattered on decumbent shoots, all turned skywards, linear, acute or subacute, rarely obtuse, hardly narrowed at the base, ¾–1¾ in. long, ½–2 lin. broad, more or less distinctly veined in the broader leaves, villous when young, then loosely pubescent; heads solitary or 2-nate, terminal, subterminal or overtopped and apparently lateral, peduncled, hemispherical, very many-flowered and compact, 5–7 lin. long, about 9 lin. across, rounded at the base, with a definite involucre of imbricate barren bracts; peduncle slender, 5–15 lin. long, tomentose, bearing scattered lanceolate, often reflexed, tomentose bracts; receptacle convex, very low; involucral bracts lanceolate, acute or acuminate, up to 3 lin. long, finely tomentose; floral bracts obovate-cuneate, abruptly acuminate or apiculate, 2½–3 lin. long, densely tomentose; adult flower-bud 6–7 lin. long; perianth-tube tapering downwards, 3½ lin. long, pubescent from the middle upwards; claws more or less equal and free with recurved tips, 2 lin. long, villous or tomentose; limbs deflexed, oblong, subacute, 1 lin. long, tomentose; anthers sessile, oblong, apiculate, ¾ lin. long; hypogynous scales filiform, ½ lin. long; ovary oblong, ⅓ lin. long, pubescent, surrounded by hairs, ¾ lin. long; style 6–7 lin. long, finely filiform, glabrous; stigma subclavate or cylindric, subobtuse, ½ lin. long, grooved; fruit ellipsoid, grey, including the short whitish conical beak 3½ lin. long, loosely pubescent. *L. diffusum, Meisn. in DC. Prodr.* xiv. 259, *partly, not of R. Br. Protea prostrata, Thunb. Fl. Cap. ed. Schult.* 133 ; *Roem. & Schult. Syst. Veg.* iii. 355 ; *R. Br. in Trans. Linn. Soc.* x. 221 ; *Meisn. l.c.* 252. *P. cinerea, Ait. Hort. Kew. ed.* 1, i. 127 ; *Poir. Encycl.* v. 651. *Leucadendron* ? *prostratum, Meisn. l.c.* 221.

SOUTH AFRICA : without locality, *Thunberg* ! *Thom,* 933 ! *Ludwig* !
COAST REGION : Caledon Div.; Diep Gat, *Galpin,* 4465 ! Zwart Berg, near Caledon, *Bowie* ! *Templeman in MacOwan, Herb. Austr.-Afr.* 1641 ! Bot River, *Schlechter,* 9446, partly ! Houw Hoek, *Schlechter,* 5513 ! *Bowie* ! mountains near Hemel and Aarde, *Zeyher,* 3686 ! Bredasdorp Div.; mountains near Elim, *Bolus,* 7872 ! 8587 ! *Schlechter,* 7640 !

29. L. **cartilagineum** (Phillips); a small shrub, up to more than 1 ft. high; branches very minutely crisped-pubescent, at length glabrescent, with a bright reddish bark; leaves scattered, spathu-

late-obovate, rounded at the apex, attenuated into a linear base, 10–15 lin. long, 3–5 lin. broad, thickly coriaceous, pallid, veinless or with faint impressed veins, glabrous; heads terminal, sessile or subsessile among the top leaves, solitary or in clusters of 2–3, 6–7 lin. long excluding the styles, 8–9 lin. wide, without a distinct involucre of barren bracts; peduncle if present with a few whitish-tomentose oblong or lanceolate acuminate bracts; receptacle small, flat; floral bracts ovate or ovate-lanceolate, subacute to acuminate, the outermost sometimes up to 2½ lin. long and forming an involucre, densely villous; adult flower-bud up to 6 lin. long; perianth-tube 2½ lin. long, pubescent above; claws all alike, free, straight, with recurved or revolute tips, finely tomentose; limbs lanceolate-oblong, acute, 1¼ lin. long, velvety-tomentose; anthers sessile, oblong, apical, up to almost 1 lin. long; hypogynous scales filiform, flat, ¾ lin. long; ovary cylindric, ⅔ lin. long, densely and minutely pubescent; style 5–6 lin. long, filiform, glabrous; stigma subclavate, ½–⅔ lin. long, passing into the style. *Leucadendron cartilagineum, R. Br. in Trans. Linn. Soc.* x. 67; *Meisn. in DC. Prodr.* xiv. 226.

SOUTH AFRICA: without locality or collector's name in the British Museum! Also *Roxburgh in Herb. Banks, and Niven in Herb. Martins* ex *Meisner.*
WESTERN REGION: Little Namaqualand; Roode Berg and Ezels Kop, *Drège*!

30. **L. obtusatum** (Phillips); a very small, densely branched shrub; branches finely crisped-tomentose above, glabrescent below, with reddish bark; leaves very much crowded, erect, more or less concealing the stem, linear-spathulate, obtuse, long narrowed into a linear base, flat or concave above, 4–7 lin. long, 1–1½ lin. broad, very thick, very minutely crisped-tomentose, at length quite glabrous, glaucous, veinless; heads solitary, terminal, sessile, 4–5 lin. long excluding the styles, 6–8 lin. wide, without a definite involucre of barren bracts; receptacle convex, 1½ lin. wide; floral bracts obovate to oblanceolate, acute to long-acuminate, 1½–3 lin. long, densely tomentose; adult flower-bud up to 4 lin. long; perianth-tube ¾ to almost 1 lin. long, very slender, glabrous below, claws all alike, free, very slender, with recurved tips up to 2½ lin. long, whitish-tomentose; limbs oblong, subacute, 1 lin. long, whitish-tomentose or villous; anthers sessile, linear-oblong, ¾ lin. long; hypogynous scales subulate, ¾ lin. long; ovary pubescent, ⅔ lin. long, surrounded by whitish hairs 1½ lin. long; style filiform, straight, 4½–5½ lin. long, loosely pubescent; stigma clavate, ½–⅔ lin. long, subobtuse. *Protea obtusata, Thunb. in Hoffm. Phytog. Blaett.* i. (1803), 15; *Thunb. Fl. Cap. ed. Schult.* 133. *Leucadendron obtusatum, Meisn. in DC. Prodr.* xiv. 227.

SOUTH AFRICA: without locality, *Thunberg*!
COAST REGION: Worcester Div.; Matroos Berg, *Davidson*!

31. **L. zwartbergense** (Bolus in Trans. S. Afr. Phil. Soc. xviii. 399); a small decumbent shrub, about 9 in. high; branches

minutely greyish crisped-tomentose above, at length glabrous below,
with a reddish bark ; leaves rather crowded, slightly curved, linear
subobtuse, long narrowed towards the base, 9–15 lin. long, $\frac{3}{4}$–1$\frac{1}{4}$ lin.
broad, flat or with recurved margins and then channelled on the
back, finely pubescent when young, soon quite glabrous, pale green
or glaucous ; heads sessile, solitary or in pairs at the end of very
short densely leafy branches, subglobose, 5–5$\frac{1}{2}$ lin. long and wide,
with a few barren basal bracts not forming a definite involucre,
linear-oblong, 2–2$\frac{1}{2}$ lin. long, glabrous apart from the dense cilia,
brown ; receptacle oblong, 2$\frac{1}{2}$ lin. long, 1 lin. in diam. ; floral
bracts subulate, 1 lin. long, glabrous, ciliate at the apex, hidden
among the white villi of the receptacle ; adult flower-buds up to
3 lin. long ; perianth-tube $\frac{3}{4}$ lin. long, very slender, glabrous ;
claws all alike, free, filiform, flexuous with recurved tips, 1$\frac{1}{2}$ lin.
long, whitish-villous ; limbs deflexed, oblong, subacute, $\frac{3}{4}$ lin. long,
villous ; anthers sessile, linear, $\frac{1}{2}$ lin. long ; hypogynous scales
filiform, over $\frac{3}{4}$ lin. long ; ovary slender, cylindric, $\frac{1}{2}$–$\frac{2}{3}$ lin. long,
spreadingly whitish-pubescent ; style finely-filiform, glabrous, 2 lin.
long ; stigma subclavate, obtuse, not quite $\frac{1}{2}$ lin. long.

COAST REGION : Oudtshoorn Div. ; Zwartzberg Pass, *Bolus,* 12267 !

Imperfectly known species.

32. **L. spathulatum** (R. Br. in Trans. Linn. Soc. x. 101) ; a
small shrub ; branches slender, loosely villous or shortly hirsute ;
leaves rather crowded, spathulate or spathulate-oblanceolate, subacute to obtuse, narrowly attenuated at the base, entire or shortly
2–3-dentate, about 1 in. long, 2$\frac{1}{2}$–4 lin. broad, faintly but distinctly
nerved, finely pubescent or glabrescent ; heads solitary, subsessile
among the top leaves, 1$\frac{1}{4}$ in. in diam. ; peduncle very short,
tomentose, with ovate acute to acuminate tomentose bracts ;
involucral bracts few, ovate-lanceolate, acuminate, 2$\frac{1}{2}$–3 lin. long,
tomentose ; floral bracts ? ; adult flower-bud 8–9 lin. long, densely
brown velvety-tomentose ; limbs lanceolate, dark red inside ;
anthers sessile, 1$\frac{1}{2}$ lin. long, with orange-coloured pollen ; style
9 lin. long, filiform ; stigma oblong-cylindric, subobtuse. *Roem. &
Schult. Syst. Veg.* iii. 360. *L. spathulatum, var. Nivenii, Meisn. in
DC. Prodr.* xiv. 258. *Protea spathulata, Poir. Encycl. Suppl.* iv.
567. *Leucadendrum bellidifolium, Knight, Prot.* 56, *from the
description. Leucadendron spathulatum, O. Kuntze, Rev. Gen. Pl.*
ii. 579.

SOUTH AFRICA : without locality, *Niven* !

A perfectly distinct species of uncertain affinity. The flowers of the only
specimen in the Banksian herbarium at the British Museum are in the bud
state and the head is so dried that it is impossible to examine it closely without
destroying part of it.

33. **Leucadendrum gnaphaliifolium** (Knight, Prot. 60) ; a tall
shrub 7–8 ft. high ; leaves crowded, elliptic-lanceolate, generally

quite entire with a slender callous point, 8–10 lin. long, 2–3 lin. broad, exceedingly pubescent, slightly nerved; stigma broadly conical.

SOUTH AFRICA : without locality or collector's name, but probably *Niven.* Said to grow only in low, dry situations.

34. Leucadendrum xeranthemifolium (Knight, Prot. 60); a tall shrub, 6–7 ft. high; stem pubescent; leaves rather crowded, lanceolate-cuneate, entire, obtuse, 6–8 lin. long, 1½–2 lin. broad, slightly pubescent when old; stigma broad conical.

COAST REGION : Clanwilliam Div. ; Jakhals Vley Mountains, *Niven.*

This may be only a small-leaved state of *L. puberum.*

VI. FAUREA, Harv.

Flowers hermaphrodite, zygomorphic. *Perianth* tubular in bud with an oblong or obovoid to clavate limb, on opening split anticously by the emerging style down to or nearly to the base so that the anticous (abaxial) segment becomes detached with the exception of the base which remains usually more or less united with the perianth, the other segments permanently united into a flattened-out, spreading or recurved sheath bearing the more or less separating or cohering spoon-shaped limbs. *Stamens* 4, inserted at the base of the limb, all fertile; filaments very short; anthers linear or linear-oblong; connective usually produced into a small apical gland. *Hypogynous scales* 4, free, lanceolate, subulate or triangular, indurated and persistent on the disc-shaped torus. *Ovary* ovoid or ellipsoid, covered with long hairs; style more or less curved, filiform, rigid, glabrous, long persistent; stigma terete or quadrangular, gradually or with a small bend passing into the style. *Ovule* 1, lateral. *Fruit* a long-villous nut.

Trees or shrubs ; leaves alternate, petioled, entire, coriaceous, usually more or less glabrous, and shining above at least when mature and more or less prominently veined ; flowers in terminal solitary spikes or racemes, each flower subtended by a small bract ; perianth at length deciduous.

DISTRIB. Species about 14 ; mostly in tropical Africa, 5 in South Africa and 1 in Madagascar.

Adult flower-buds 4½–5 lin. long, with a limb up to
  1½ lin. long and correspondingly small anthers and
  stigmas :
    Flowers pedicelled   ...   .   (1) **Galpinii.**
    Flowers sessile   ...   .   (2) **saligna.**
Adult flower-buds 7–10 lin. long, with a limb 2½ to
  over 3 lin. long and correspondingly large anthers
  and stigmas :
  Indumentum of branchlets and spikes very fine, reddish.
    Leaves glabrous :
      Adult flower-buds 7 lin. long, with a rather stout
      tube and a limb not over 2½ lin. long ...   ...   (3) **natalensis.**

Adult flower-buds 9–10 lin. long, slender, with a
limb to over 3 lin. long ...   .   (4) **Macnaughtonii.**

Indumentum of branchlets and spikes densely greyish-
tomentose, that of the branchlets coarse.   Leaves
tomentose, subglabrous only when quite old   ...   (5) **speciosa.**

1. **F. Galpinii** (Phillips) ; a tree or shrub, 8 ft. high ; branchlets
glabrous or minutely pubescent when young, with a dark brown or
blackish bark ; leaves shortly petioled, lanceolate, acute at both
ends, 2–4 in. long, $\frac{3}{4}$–1 in. broad, coriaceous, drying blackish or
dark brown on the upper and reddish-brown on the under surface,
glabrous or minutely pubescent when quite young ; lateral nerves
numerous, very oblique, not joining into a submarginal nerve, like
the veins raised, particularly above ; petiole rarely over 2 lin. long,
glabrous or obscurely pubescent ; inflorescence racemose, shortly
peduncled, cylindrical, usually 2 (rarely up to $3\frac{1}{2}$) in. long ; rhachis
greyish- or whitish-tomentose ; bracts broad-ovate, $\frac{1}{2}$ lin. long,
tomentose ; pedicels up to $1\frac{1}{2}$ lin. long ; adult flower-bud slightly
curved, with a slender tube gradually tapering upwards and an
oblong subacute or obtuse limb, minutely and adpressedly pubescent,
$4\frac{1}{2}$–$5\frac{1}{2}$ lin. long ; perianth-sheath flattened out, spreading, recurved
in the upper part ; limbs oblong, subacuminate, $1\frac{1}{2}$ lin. long, those
of the sheath separating only at the tips ; anthers subsessile, linear-
oblong, 1–$1\frac{1}{4}$ lin. long ; apical glands ovoid, subacute ; hypogynous
scales subulate from a triangular base ; ovary ovoid, $\frac{1}{2}$–$\frac{3}{4}$ lin. long,
covered with whitish or (when dry) fulvous hairs up to 3 lin. long ;
style 4 lin. long, curved upwards, particularly in the mature state ;
stigma cylindric, slightly wider than the style and gradually or
with a slight bend passing into it, 1 lin. long ; fruit subglobose, up
to $1\frac{1}{2}$ lin. in diam.

KALAHARI REGION : Transvaal ; Saddleback Mountain, near Barberton, *Galpin*,
944 ! Zoutpansberg Range, at Potato Bosch, *Eastwood*, 2435 ! Shilovane, *Junod*,
5539 ! near Pilgrims Rest, *Burtt-Davy*, 5650 ! Macamac Falls, *Burtt-Davy*, 5651 !

2. **F. saligna** (Harv. in Hook. Lond. Journ. Bot. vi. 373, t. 15) ;
a shrub or a tree, 8–20 ft. high ; branchlets glabrous or pubescent
when young, with a greyish or pale brown bark ; leaves petioled,
lanceolate, sometimes subfalcate, long and gradually tapering and
acute at both ends, 3–6 in. long, $\frac{1}{2}$–1 (rarely to $1\frac{1}{2}$) in. broad,
thinly coriaceous, concolorous, densely pubescent in bud, very soon
becoming glabrous and shining above ; lateral nerves numerous,
joining into a more or less distinct submarginal nerve, like the
veins slightly raised ; petiole up to $\frac{1}{2}$ in. long, glabrous or pubescent ;
inflorescence spicate, shortly peduncled, cylindric, 3–6 (rarely 2)
in. long ; rhachis greyish-pubescent ; bracts concave, broad-ovate,
acute or obtuse, $\frac{1}{2}$ lin. long ; adult flower-bud slightly curved, with
a tube gradually tapering upwards and an oblong or oblong-obovoid
obtuse limb, minutely and adpressedly pubescent, $4\frac{1}{2}$–$5\frac{1}{2}$ lin. long ;
flower pale yellowish ; perianth splitting to or almost to the base,

sheath flattened out, spreading, recurved in the upper part; limbs oblong, subobtuse, $1\frac{1}{2}$ lin. long, those of the sheath usually at length separating; anthers subsessile, linear-oblong, $1-1\frac{1}{4}$ lin. long, apical gland minute, ovoid-globose; hypogynous scales triangular, usually acute or acuminate, to over $\frac{1}{2}$ lin. long; ovary ovoid, $\frac{1}{2}-\frac{3}{4}$ lin. long, covered with whitish or (when dry) yellow hairs up to 3 lin. long; style $3\frac{1}{2}-4\frac{1}{2}$ lin. long, gently curved upwards; stigma subclavate-cylindric, obtuse, up to 1 lin. long, gradually or with a faint bend passing into the style; fruit globose-ovoid, $2-2\frac{1}{2}$ lin. long, long-villous. *Meisn. in DC. Prodr.* xiv. 344; *Welw. in Trans. Linn. Soc.* xxvii. 65; *Engl. Hochgebirgsfl. Trop. Afr.* 195, *and Pfl. Ost-Afr. C.* 164; *Hiern, in Cat. Afr. Pl. Welw.* i. 921; *Engl. & Gilg in Baum, Kunene-Samb. Exped.* 226; *C. H. Wright in Dyer, Fl. Trop. Afr.* vi. 209; *Sim, For. Fl. Cape Col.* 297.

KALAHARI REGION : Transvaal; Magaliesberg Range, *Burke!* *Zeyher,* 1480! 1481! *Wahlberg, Sanderson!* Boshveld, *Rehmann!* Barberton, *Galpin,* 868! near Nylstroom, *Nelson,* 112! *Burtt-Davy,* 2059! 2592! Rustenberg, *Collins,* 137! *Miss Pegler,* 1009! Warm Bath, *Bolus,* 12268! *Burtt-Davy,* 5648! Potgeiters Rust, *Rogers,* 329! *and in Herb. Miss Leendertz,* 1266! Tweefontein, *Schlechter,* 4259! Marico district, *Burtt-Davy,* 7574! Zoutpansberg Range, *Eastwood,* 2433! *Legat,* 158!

EASTERN REGION : Natal; Inanda, *Wood,* 6! 1189!

Also in Tropical Africa, although it is doubtful whether all the specimens referred to it represent typical *F. saligna.*

Rogers in Herb. Leendertz 1266 has very long bracts subtending the flowers, longer than in any other specimen we have seen. It may be a distinct species, but the material is too young to decide definitely.

3. **F. natalensis** (Phillips); branchlets glabrous or finely and sparingly pubescent upwards, with a blackish bark; leaves petioled, oblong to elliptic-lanceolate, acute at both ends, 2–3 in. long, $1-1\frac{1}{4}$ in. broad, coriaceous, drying olive-green or brown on both sides, glabrous, lateral nerves numerous, very oblique, joining in short faint loops near the margin, like the veins raised on both sides; petiole $\frac{1}{3}-\frac{1}{2}$ in. long, glabrous; inflorescence spicate, shortly peduncled, cylindrical, stout, very dense, $3\frac{1}{2}-5$ in. long; rhachis reddish, minutely pubescent, stout; bracts very broad, acute, $\frac{1}{2}-\frac{3}{4}$ lin. long, glabrescent; adult flower-buds more or less curved upwards or the subterminal almost straight, with a somewhat stout tube and a wide base and a clavate subobtuse or subacute limb, up to 7 lin. long, very finely reddish-tomentellous; perianth-sheath recurved, spreading and flattened out from below the middle; limbs lanceolate-oblong, subacute, $2\frac{1}{2}$ lin. long, those of the sheath at length separating, but more or less tightly conniving; anthers subsessile, linear-oblong, $1\frac{3}{4}-2$ lin. long; apical gland ovoid, minute; hypogynous scales subulate from a triangular base to over 1 lin. long; ovary ovoid, 1 lin. long, covered when dry with fulvous hairs up to almost 5 lin. long; style 6 lin. long, curved upwards from below the middle, glabrous; stigma linear in outline, quad-

rangular, obtuse, up to 2 lin. long, passing with a sudden bend
into the style.

EASTERN REGION: Natal, *Gerrard*, 1505!

4. **F. Macnaughtonii** (Phillips); a tall forest tree, up to 60 ft.
high, with a trunk 30 in. in diam. ; branchlets glabrous with a
greyish-brown bark; leaves petioled, lanceolate to elliptic-lanceolate, acute at both ends, 3–6 in. long, ½–1 in. wide, coriaceous,
drying olive-green, glossy above, glabrous ; lateral nerves numerous,
very oblique, joining in short faint loops near the margin, like the
veins raised on both sides ; petiole up to ⅓–1 in. long, glabrous :
inflorescence spicate, shortly peduncled, very dense, cylindrical,
stout, 4–6 in. long; rhachis very minutely reddish-pubescent ;
bracts very broadly ovate, acute, ½–¾ lin. long, reddish-pubescent ;
adult flower-bud gently curved upwards, with a somewhat stout
tube and a clavate subobtuse limb, not much wider than the tube,
9–10 lin. long, very finely reddish-tomentellous ; perianth-sheath
abruptly spreading and flattened out from below the middle ;.
limbs linear-oblong, subacute, over 3 lin. long, those of the sheath
permanently united, except at the tips, and conniving ; anthers
subsessile, linear, 2¾ lin. long; apical gland ovoid, subacute :
hypogynous scales subulate-lanceolate, 1 lin. long; ovary ovoid,
covered with whitish hairs, up to 4 lin. long; style 8 lin. long,
slightly curved, glabrous ; stigma linear in outline, quadrangular,
to over 2 lin. long, passing with an obscure bend into the style.
F. saligna, MacOwan in Agric. Journ. Cape of Good Hope, xii. 714, *not
of Harv. F. arborea, Sim, For. Fl. Cape Col.* 297, *t.* 130, *not of Engl.*

COAST REGION: Knysna Div. ; Gouna Forest at Klipkop near Knysna,
McNaughton in MacOwan, Herb. Austr.-Afr., 1948 ! *and in Herb. MacOwan*, 3312 !

According to MacOwan this tree is very rare in the locality cited above and
flowers very rarely. Sim also records it on the authority of Mr. McNaughton
from Blaauwkrantz and Zitzikamma, adding that there are only about 60 trees
known in all, apart from some seedlings. He remarks on its absence from the
Kaffrarian forests and the Transkei, but says that it is not very rare in the
Egossa Forests and has been seen in the St. John's and Pondoland forests.
There are no specimens at hand from any of those forests, and it may be that
the Eastern *Faurea* referred to by him under his *F. arborea* is really *F. natalensis*,
Phillips, which resembles the former very much.

5. **F. speciosa** (Welw. in Trans. Linn. Soc. xxvii. 63, t. 20) ;
a bush or tree, 10–20 ft. high ; branchlets greyish-tomentose or
pubescent, after the peeling of the bark reddish ; leaves shortly
petioled, broad-lanceolate to oblong, acute at both ends, 5–6 in.
long, 1½–3 in. broad, coriaceous, concolorous, densely greyish-
tomentose when quite young, then more or less glabrescent and at
length almost entirely glabrous, very coriaceous ; lateral nerves
numerous, joining into a conspicuous submarginal nerve, like the
veins much raised ; petiole stout, 2–7 lin. long ; inflorescence
spicate, very shortly peduncled, very dense, cylindrical, very
stout, 5–7 in. long; rhachis stout, tomentose ; flowers spirally

arranged, almost contiguous with their bases or slightly distant, sometimes almost verticillate, with the whorls close, bracts very broadly-ovate, acute, $\frac{3}{4}$ lin. long, tomentose; adult flower-bud curved upwards, with a cylindric tube, widened at the base, and an obovoid-oblong subobtuse limb, much wider than the tube, 9–10 lin. long, greyish-tomentose; perianth at length splitting down to the base with the abaxial segment long or permanently adhering to the lower third; sheath spreading and flattened out from below the middle; limbs linear-oblong, subacute, to over 3 lin. long, those of the sheath at length more or less separating and opening out, or permanently conniving; anthers subsessile, linear, 2–2½ lin. long; apical gland ovoid, subobtuse; hypogynous scales triangular, acute, 1 lin. long; ovary ovoid, covered when dry with fulvous hairs, at length up to 5 lin. long; style 9–11 lin. long, more or less curved, glabrous; stigma linear in outline, obtuse, quadrangular, 1½–2 lin. long, very slightly wider than the style and set off from it by a minute bend; fruit subglobose, 3 lin. long, long-villous, with a very thick shell. *Engl. Hochgebirgsfl. Trop. Afr.* 195; *Glied. Veg. Usambara,* 60, 63; *Pfl. Ost-Afr. C,* 164; *Jahrb.* xxx. 301; *Hiern in Cat. Afr. Pl. Welw.* i. 922; *Engl. & Gilg in Baum, Kunene-Samb. Exped.* 227; *C. H. Wright in Dyer, Fl. Trop. Afr.* vi. 211. *Trichostachys speciosa, Welw. Syn. Explic.* 19.

KALAHARI REGION : Transvaal ; Woodbush Mountains, *Mrs. Barber,* 2 ! *Hutchins* ! Barberton, *Galpin,* 402 ! *Bolus,* 9756 ! Elands Hoek, *Rogers,* 391 ! and without precise locality, *Burtt-Davy,* 333 ! 356 !

Also in Tropical Africa.

VII. MIMETES, Salisb.

Flowers hermaphrodite, actinomorphic. *Perianth-tube* very short or wanting ; segments. 4, filiform or linear-filiform, often villous ; limb linear or oblong-linear, usually villous, rarely glabrescent. *Stamens* 4 ; filament short and fleshy, often fused with the perianth ; anthers linear; connective produced into an acute or rounded apical appendage. *Hypogynous scales* 4, free, usually filiform or linear, rarely ovate-lanceolate. *Ovary* sessile, pubescent ; style exserted, erect, terete, mostly glabrous, usually terminal, rarely oblique ; stigma smooth, terete or linear, rarely subquadrangular, acute or obtuse, sometimes swollen at the apex, more or less kneed or sinuate at the junction with the style ; ovule solitary. *Fruits* ovoid, glabrescent, with a slightly hardened pericarp.

Erect or subdecumbent undershrubs with simple tomentose or villous stems : leaves mostly oblong-elliptic or ovate, with a callous entire or 3-dentate apex, mostly densely adpressed silky-tomentose or villous ; flower-heads sessile, solitary in the upper leaf-axils, 3–12-flowered ; involucral bracts usually shorter than the flowers, membranous or coriaceous, often villous or tomentose, frequently coloured ; receptacle densely setose.

DISTRIB. Species about 9, confined to the South-western part of Cape Colony.

Stigma with a swollen ovoid apex clearly differentiated
from the lower cylindric portion :
    Callus at the apex of the leaves entire ; perianth-limb
        long-villous ... ... (1) **capitulata.**

    Callus at the apex of the leaves 3-toothed, rarely sub-
        entire ; perianth-limb nearly glabrous .. ... (2) **saxatilis.**

Stigma linear, cylindric, not swollen at the apex :
    Callus at the apex of the leaves 3–5-fid or 3–5-toothed :
        Stigma very acute ... ... , ... (3) **lyrigera.**

        Stigma obtuse or rounded ... ... (4) **splendida.**

    Callus at the apex of the leaves quite entire :
        Leaves 1½–2½ in. long ; stigma 3 lin. long :
            Leaves silvery-tomentose, nearly 3 times as long
                as broad ; involucral bracts thinly pubescent ;
                flowers exserted about ½ their length ... (5) **integra.**

            Leaves yellowish-tomentose, not twice as long as
                broad ; involucral bracts densely velvety-
                tomentose ; flowers exserted about ¾ their
                length or more ... ... (6) **argentea.**

        Leaves usually less than 1½ in. long (rarely 1¾ in.) ;
            stigma ¾–1½ lin. long :
                Perianth-limb glabrous or nearly so ... ... (7) **pauciflora.**

                Perianth-limb villous with long weak hairs :
                    Heads 7–12-flowered ... ... (8) **hirta.**

                    Heads about 3-flowered ... . (9) **palustris**

1. **M. capitulata** (R. Br. in Trans. Linn. Soc. x. 106) ; branches
villous ; leaves 6–14 lin. long, 3–7 lin. broad, lanceolate, lanceolate-
ovate or ovate, gradually narrowed to an obtuse callous apex, a
little narrowed to the base, entire, coriaceous, indistinctly 3-nerved,
densely adpressed-villous with silky hairs ; heads sessile, 1–1¼ in.
long, 10–12-flowered, in the axils of the leaves at the ends of the
branches ; involucral bracts 5–6-seriate, varying from linear-lanceolate to ovate-lanceolate, more or less narrowed to the base, membranous, pubescent outside, long-ciliate ; receptacle long-setose ;
perianth-segments free or nearly so, about 12 lin. long, linear-
filiform, slightly widened for about 2½ lin. at the base, plumose ;
limb 1¾ lin. long, linear-lanceolate, subacute, long-villous on the
back ; stamens with filaments ⅓ lin. long ; anthers ¾ lin. long,
linear ; apical gland ⅓ lin. long, lanceolate, subacute, concave ;
hypogynous scales ½ lin. long, linear, obtuse, brown ; ovary shortly
stalked, 1 lin. long, oblong, pubescent ; style nearly 2 in. long,
subcylindric, glabrous ; stigma 1½ lin. long, with an ovoid acuminate
subacute apex and a distinct kink at the junction with the style.
Roem. & Schultes, Syst. Veg. iii. 380 ; *Meisn. in DC. Prodr.* xiv. 262.
Protea capitulata, Poir. Encyl. Suppl. iv. 568.

SOUTH AFRICA : without locality, *Gueinzius ! Brown !*
COAST REGION : Stellenbosch Div. ; Hottentots Holland Mountains, *Zeyher*,
3690 ! Caledon Div. ; near Grietjes Gat, ex *Zeyher* ! near the Bot River, *Pillans
in Herb. Bolus*, 9361 !

2. M. saxatilis (Phillips in Kew Bulletin, 1911, 84); branches grey-tomentose or villous; leaves closely imbricate, 1¼–1¾ in. long, 8–15 lin. broad, obovate-elliptic or obovate, with a blunt glabrous 3-toothed callous apex, rounded at the base, coriaceous, indistinctly 5–6-nerved from near the base, densely adpressed-pubescent with whitish hairs; heads sessile, about 1 in. long, 10–12-flowered, solitary in the axils of the leaves at the ends of the branches; involucral bracts 5–6-seriate, linear-lanceolate to ovate, all acutely acuminate, about half as long as the flowers, coriaceous, slightly rugose, densely ciliate, otherwise nearly glabrous; receptacle long-setose; perianth-tube very short, glabrous; segments 9–12 lin. long, linear-filiform, densely pilose up to the limb; limb 1½ lin. long, linear, obtuse or subacute, glabrous or with a few scattered hairs; anthers sessile, 1½ lin. long; filaments oblong, swollen, channelled; apical gland ¼ lin. long, lanceolate, acute; hypogynous scales 1 lin. long, linear; ovary ½ lin. long, pubescent; style 1⅓–1½ in. long, cylindric; stigma 1½ lin. long, with an ovoid obtuse apex ⅓ lin. long, and with a distinct kink at the junction with the style.

SOUTH AFRICA: without locality, *Thunberg*!
COAST REGION: Bredasdorp Div.; Elim, *Schlechter*, 7716! Mier Kraal, *Schlechter*, 10521!

Mimetes? nitens, Roem. & Schultes, Syst. Veg. iii. 384; Meisn. in DC. Prodr. xiv. 266 (*Protea nitens*, Thunb. Fl. Cap. ed. i. 514; ed. Schultes, 140) may belong here, but the original specimen consists of two leaves only.

3. M. lyrigera (Knight, Prot. 65); stems 4–5 ft. high, sometimes decumbent; branches densely pubescent or softly tomentose; leaves 1¼–3 in. long, 2–12 lin. broad, mostly oblong or oblong-lanceolate, the lower often broader in the lower half, 3–5-fid at the apex or rarely a few entire on each shoot, obtuse at the base, coriaceous, indistinctly 3-nerved, or if broadened in the lower half then with 6–9 fairly distinct nerves, pubescent when young, at length becoming glabrous; heads sessile, 2 in. long including the styles, 4–10-flowered, axillary towards the ends of the branches; involucral bracts 3–4-seriate, linear or lanceolate, acuminate, acute, minutely pubescent or almost glabrous; receptacle long-setose; perianth-tube 2 lin. long, pubescent inside; segments 1½ in. long, filiform, long-pilose; limb 2½–3 lin. long, linear, acute, villous; anthers 1¼ lin. long, linear; apical gland ½ lin. long, linear, acute; hypogynous scales ¾ lin. long, ovate-lanceolate, acute, white; ovary ¼ lin. long, pubescent; style exserted, 2 in. long, more or less flattened, glabrous; stigma 2¼ lin. long, subquadrangular, furrowed, very acute. *M. cucullata, R. Br. in Trans. Linn. Soc. x. 107; Roem. & Schultes, Syst. Veg. iii. 380; Meisn. in DC. Prodr. xiv. 263, incl. vars.; O. Kuntze, Rev. Gen. Pl. iii. 278. M. Ludwigii, Steud. ex Meisn. l.c. M. mixta, Gandog. in Bull. Soc. Bot. France,* xlviii. *p.* xciii. *M. cucullata, vars. Dregei and laxa, Gandog. l.c. Protea cucullata, Linn. Mant. 189; Thunb. Diss. Prot. 23; Murr. Syst. Veg.*

ed. xiv. 137 ; *Lam. Encycl.* v. 656 ; *Thunb. Fl. Cap. ed. Schultes,* 126.
Leucadendron cucullatum,. Linn. Sp. Pl. ed. i. 93 ; *ed.* ii. 136 ; *Berg.
in Kongl. Vet. Acad. Handl. Stockh.* 1766, 320 ; *Berg. Descr. Pl.
Cap.* 14.—*Protea foliis lanceolatis obtusis flores, etc., Royen, Fl.
Seyd. Prodr.* 184. *Leucadendros africana, s. Scolymocephalos, etc.,
Pluk. Almag.* 212, *t.* 304, *fig.* 6. *Hypophyllocarpodendron foliis
inferioribus, etc., Boerh. Ind. Alt. Pl. Hort. Ludg. Bat.* ii. 206.
*Scolymocephalus seu Hypophyllocarpodendron foliis tribus, etc., Weinm.
Phyt.* iv. 297, *t.* 905, *fig. b.*

VAR. β, **Hartogii** (E. P. Phillips) ; mature leaves conspicuously hirsute on the
margin with white hairs. *Mimetes fimbriæfolius, Knight, Prot.* 65. *M. Hartogii,
R. Br. in Trans. Linn. Soc.* x. 108 ; *Roem. & Schultes, Syst. Veg.* iii. 381 ; *Meisn.
in DC. Prodr.* xiv. 263. *Protea cucullata,* β, *Lam. Ill.* i. 239, *excl. plate. Hypophyllocarpodendron foliis lanuginosis, etc., Boerh. Ind. Alt. Pl. Hort. Lugd. Bat.* ii.
205, *t.* ·205. *Scolymocephalus seu Hypophyllocarpodendron foliis lanuginosis, etc.,
Weinm. Phyt.* iv. *t.* 906, *a.*

SOUTH AFRICA : without locality, *Oldenburg* ! *Thunberg* ! *Nelson* ! *Niven* !
Ludwig ! *Harvey* ! *Drège* ! *Thom,* 424 ! 935 ! *Ecklon & Zeyher* ! *Brentel,* 40 !
Andersson ! *Gueinzius* ! Var. β, *Bergius,* 273 ! *Niven* ! *Mund* ! *Krauss,* 1037.
COAST REGION : Tulbagh Div. ; Witzenberg Range, *Burchell,* 8692 ! *Pappe* !
near Tulbagh Waterfall, *MacOwan* ! *Phillips,* 525 ! Worcester Div. ; Dutoits
Kloof, *Drège* ! Cape Div. ; Table Mountain, *Burchell,* 661 ! *Milne,* 28 ! Camps
Bay, *Burchell,* 351 ! near Simons Town, *Wright* ! *Bolus,* 4198 ! *Wolley-Dod,* 291 !
Caledon Div. ; Lowrys Pass, *Kuntze* ! between the Palmiet River and Lowrys
Pass, *Burchell,* 8186 ! near Grabouw, *Bolus,* 4198 ! Baviaans Kloof, *Burchell,*
7789 ! *Schlechter,* 9844 ! Houw Hoek, *Zeyher,* 1478 ! *MacOwan, Herb. Norm.
Austr.-Afr.,* 773 ! *Schlechter,* 7399 ! Zwart Berg, *Zeyher* ! Swellendam Div. ;
mountains near the Zondereinde River, *Zeyher,* 1478 β ! Voormans Bosch, *Zeyher,*
3693 ! mountains near Swellendam, *Galpin,* 4480 ! Riversdale Div. ; between
Little Vet River and Garcias Pass, *Burchell,* 6889 ! Garcias Pass, *Galpin,* 4479 !
Mossel Bay Div. ; Attaquas Kloof, *Drège* ! Oudtshoorn Div. ; near Oudtshoorn,
Britten ! Var. β : Cape Div. ; Stinkwater, near Camps Bay, *Zeyher,* 4681 ! Table
Mountain, near Oude Kraal, *Pappe* ! mountains near Simonstown, *Wright* !
Jameson ! *Pappe* !

4. **M. splendida** (Knight, Prot. 66) ; shrub 5–6 ft. high ; branches
velvety-tomentose ; leaves 1–2¼ in. long, 5–9 lin. broad, oblong or
elliptic-lanceolate, obtuse or subacute, 3-toothed or rarely entire at
the glabrescent apex, slightly narrowed or rounded at the base,
indistinctly 5–7-nerved, silky tomentose with adpressed hairs ; heads
sessile, including the styles 2 in. long, 8–11-flowered, axillary,
aggregated at the end of the branches ; involucral bracts 3–4-seriate,
narrowly oblong, ovate or elliptic, coriaceous, somewhat rugose,
ciliate, glabrous ; receptacle finely setose ; perianth-tube 1 lin.
long, pilose ; segments 13–16 lin. long, linear-filiform, long rusty-
pilose ; limb 3–4 lin. long, linear, subacute, long-pilose on the
back ; filaments oblong, fused with the perianth ; anthers 2½–3 lin.
long, linear ; apical glands ⅓ lin. long, ellipsoid, obtuse ; hypogynous scales 1 lin. long, linear ; ovary ½ lin. long, pubescent ; style
1½–1¾ in. long, cylindric, glabrous ; stigma 3 lin. long, linear,
obtuse or rounded at the apex, sinuate at the junction with the
style ; fruit 3¼ lin. long, ellipsoid ; testa hard. *M. Hibbertii,
R. Br. in Trans. Linn. Soc.* x. 108 ; *Roem. & Schultes, Syst. Veg.*

iii. 381 ; *Meisn. in DC. Prodr.* xiv. 264. *Protea Hibbertii, Poir. Encycl. Suppl.* iv. 568.

SOUTH AFRICA : without locality, *Roxburgh* ! *Thom,* 275 ! *Ludwig* ! COAST REGION : Caledon Div. ; mountains near Houw Hoek, *Pappe* ! Swellendam Div. ; near Swellendam, *Kennedy* ! Grootvaders Bosch, *Bowie* ! *Mund* ! *Pappe* ! Tradouw mountains, *Bowie* ! George Div. ; mountains near George, *Drège* ! lower part of the Cradock Berg, *Burchell,* 6028 ! Barbiers Kraal, *Niven,* 75 !

5. **M. integra** (Hutchinson) ; branches densely villous ; leaves 2–2¼ in. long, about ¾ in. broad, lanceolate or elliptic-lanceolate, with a subacute entire callous apex, slightly narrowed to the base, with several distinct ascending nerves, densely adpressed silvery tomentose ; heads much shorter than the leaves, including the styles 1½–1¾ in. long, 7–8-flowered, crowded towards the ends of the branches ; outer involucral bracts 3–4 lin. long, 2½ lin. broad, ovate, obtuse, intermediate about ½ in. long and 3 lin. broad, oblong, rounded at the apex, thinly adpressed-pubescent outside, glabrous and striate within, the innermost (about 8) 6–7 lin. long, linear, densely villous outside, glabrous within ; receptacle finely setose ; perianth-tube very short ; segments about 1 in. long, filiform, thinly subadpressed-pubescent ; limb 3½–4 lin. long, linear, acute, shortly pubescent and with a few long ascending hairs outside ; filaments thick, about 1 lin. long ; anthers 1⅓ lin. long, linear-filiform ; apical gland very small ; hypogynous scales 1 lin. long, subulate-filiform ; ovary ½ lin. long, narrowly ovoid, shortly pubescent ; style about 1⅓ in. long, filiform, glabrous ; stigma 3 lin. long, linear, obtuse, only slightly differentiated from the style ; fruits not seen. *M. Massoni, Meisn. in DC. Prodr.* xiv. 264, *partly, as to Zeyher, not of R. Br.*

COAST REGION : Caledon Div. ; banks of the Zondereinde River, near Appels Kraal or neighbouring mountains, *Zeyher,* 3688 !

6. **M. argentea** (Knight, Prot. 67) ; shrub 4 ft. high ; branches velvety-tomentose ; leaves 1½–2½ in. long, 1–1½ in. broad, elliptic or oblong-elliptic, with a subobtuse callus at the apex, entire, slightly narrowed to the base, indistinctly 9-nerved, very densely tomentose with adpressed silky hairs ; heads subsessile, 1½ in. long including the styles, 7–9-flowered, axillary ; involucral bracts about 3-seriate, coriaceous, the outer ovate-oblong, silky tomentose, the inner linear, long-villous ; receptacle densely setose with long weak hairs ; perianth-tube very short, rusty-villous ; segments nearly 1 in. long, linear, rusty-villous ; limb 4½ lin. long, villous ; stamens 3 lin. long ; filaments swollen, fused with the perianth ; anthers 2¾ lin. long, linear ; apical gland ⅓ lin. long, ovoid, acute ; hypogynous scales 1¼ lin. long, linear, subacute, white ; ovary ¾ lin. long, oblong, pubescent ; style 1½ in. long, filiform, glabrous ; stigma 3 lin. long, linear, obtuse, furrowed, kneed at the junction with the style. *Mimetes Massoni, R. Br. in Trans. Linn. Soc.* x.

648   PROTEACEÆ (Phillips & Hutchinson).   [*Mimetes.*

109 : *Roem. & Schultes, Syst. Veg.* iii. 381 ; *Meisn. in DC. Prodr.* xiv.
264 *partly, excl. Zeyher*, 3688.   *Protea Massonii, Poir. Encycl. Suppl.*
iv. 568.

SOUTH AFRICA : without locality, *Roxburgh* !
COAST REGION : Paarl Div. ; mountains near French Hoek, *Masson* ! Caledon
Div. ; near the Zondereinde River, *Niven*, 74 !

7. M. pauciflora (R. Br. in Trans. Linn. Soc. x. 106); shrub
2–3 ft. high ; stem subsimple ; branches villous ; leaves decreasing
in size from the base upwards, ½–1½ in. long, 3–9 lin. broad, mostly
obovate or obovate-elliptic, rarely oblong, very obtuse at the base,
entire, coriaceous, indistinctly 3–5-nerved, villous with long weak
subadpressed hairs ; heads sessile, 1½–2 in. long, subcylindric,
mostly 3–4-flowered, axillary, aggregated towards the ends of the
branches ; involucral bracts 4-seriate, ovate, oblong or lanceolate,
acute, membranous, coloured, many-nerved, varying from densely
shaggy villous to almost glabrous ; receptacle densely setose ;
perianth-tube 1½ lin. long, 4-keeled, glabrous ; segments 1¼ in.
long, linear-filiform, long-villous ; limb 3 lin. long, linear-lanceolate,
subacute, glabrous or somewhat pilose towards the base ; filament
½ lin. long, oblong, swollen, furrowed ; anthers 1¼ lin. long, linear ;
apical gland ⅛ lin. long, lanceolate ; hypogynous scales ¾ lin. long,
linear, obtuse, white ; ovary ¾ lin. long, subglobose, pubescent ;
style 1¾ in. long, linear, glabrous, only slightly exserted ; stigma
1½ lin. long, cylindric, subacute, obtuse, slightly kneed at the
junction with the style. *Roem. & Schultes, Syst. Veg.* iii. 380 ;
Meisn. in DC. Prodr. xiv. 263.   *Protea pauciflora, Poir. Encycl.
Suppl.* iv. 568.

SOUTH AFRICA : without locality, *Roxburgh* ! *Thom*, 274 !
COAST REGION : Swellendam Div. ; lower part of the Tradouw mountains,
Bowie ! George Div. ; between Cape Town and George, *Rogers* ! lower part of
the Cradock Berg, *Burchell*, 6014 ! near George, *Pappe* ! *Alexander* ! Uniondale
Div. ; on the mountains dividing Long Kloof from the coast, *Bowie* ! Knysna
Div. ; Millwood, *Tyson*, 3030 ! *and in MacOwan, Herb. Austr.-Afr.* 1523 !

8. M. hirta (Knight, Prot. 66) ; branches densely villous ; leaves
¾–1¾ in. long, 3–8 lin. broad, ovate-elliptic to oblanceolate, with a
blunt callus at the apex, or very slightly pointed, narrowed to the
base, entire, coriaceous, indistinctly 3–5-nerved, more or less
adpressed-villous ; heads sessile, 2½ in. long, including the styles,
7–12-flowered, axillary, aggregated towards the ends of the branches ;
involucral bracts 5–7-seriate, linear-lanceolate, lanceolate or ovate,
gradually and acutely acuminate, membranous, many-nerved,
glabrous except for the long-ciliate margin, the outermost pilose
towards the apex ; receptacle setose ; perianth-tube 2–3 lin.
long, ventricose, glabrous ; segments 11–15 lin. long, linear, villous
in the upper part, glabrescent below ; limb 1–1⅓ lin. long, linear,
subobtuse, long-pilose on the back, with a tuft of long hairs at the
apex ; anthers linear, with a small apical gland ; hypogynous
scales ⅔ lin. long, linear, obtuse, brown ; ovary ½–¾ lin. long,

oblong, pubescent ; style 2–2¼ in. long, much exserted, trigonous above, subobliquely inserted on the ovary, glabrous ; stigma ¾–1 lin. long, furrowed, subacute, distinctly sinuate at the junction with the style. *R. Br. in Trans. Linn. Soc.* x. 105 ; *Roem. & Schultes, Syst. Veg.* iii. 379 ; *Reichenb. Ic. Exot.* 62, *t.* ·92 ; *Meisn. in DC. Prodr.* xiv. 262. *M. capitulata, Sieber ex Meisn. l.c.   M. decapitata, Meisn. l.c.   Protea hirta, Linn. Mant.* 188 ; *Thunb. Diss. Prot.* 57 ; *Lam. Ill.* i. 234 ; *Thunb. Prodr.* 27 ; *Willd. Sp. Pl.* i. 532 ; *Poir. Encycl.* v. 641 ; *Thunb. Fl. Cap. ed. Schultes,* 139.   *Leucadendron hirtum, Linn. Sp. Pl. ed.* ii. 136.—*Lepidocarpodendron foliis sericeis, etc., Boerh. Ind. Alt. Pl. Hort. Lugd. Bat.* ii. 194, *t.* 194. *Scolymocephalus africanus argenteus, etc., Weinm. Phyt.* iv. 292, *t.* 899, *a.*

SOUTH AFRICA : without locality, *Bowie* ! *Ludwig* ! *Thom* ! *Ecklon,* 334 ! *Sieber,* 4 !

COAST REGION : Cape Div. ; Table Mountain, *Thunberg* ! *Bodkin in Herb Bolus,* 4929 β ! Cape Flats, *Zeyher,* 527 ! Rondebosch, *Jameson* ! hills near Simons Bay, *Brown* ! *Schlechter,* 1201 ! *Pappe* ! plain above Smitwinkel Bay, *Bolus,* 4929 ! Cape Point, *MacOwan, Herb. Norm. Austr.-Afr.* 911 ! *Phillips* ! Stellenbosch Div. ; Hottentots Holland Mountains, *Zeyher,* 3690 ! *Mrs. de Jongh in Herb. Galpin,* 3508 ! Bredasdorp Div. ; near Elim, *Bolus,* 8591 ! *Schlechter,* 9647 ! Knysna Div. ; near Plettenbergs Bay, *Bowie* !

9. **M. palustris** (Knight, Prot. 66, excl. syn. Boerh.) ; branches decumbent, rusty-villous ; leaves ½–1 in. long, 3–5 lin. broad, imbricate, ovate-lanceolate or ovate, shortly pointed and entire at the apex, a little narrowed to the base, with no visible nerves, coriaceous, adpressed villous, with more or less rust-coloured hairs on both surfaces ; heads longer than the leaves, including the styles about 1⅓ in. long, 3-flowered, crowded at the ends of the branches ; outer involucral bracts about 4 lin. long, ovate-lanceolate, acute, thinly villous especially towards the margin ; intermediate about ¾ in. long and ¼ in. broad, oblong-oblanceolate, acute, thinly adpressed-villous outside, glabrous and striate within, the innermost few, almost linear and more densely villous ; receptacle finely and long-setose ; perianth-tube about 1 lin. long ; segments ¾ in. long, linear-filiform, rather densely setose with long ascending hairs ; limb 1½ lin. long, narrowly lanceolate, subacute, densely setose outside ; filaments thick, ⅓ lin. long ; anthers ⅔ lin. long ; apical gland suborbicular, about ⅕ lin. in diam. ; hypogynous scales and ovary not seen ; style exserted, 1⅓ in. long, sulcate, glabrous ; stigma scarcely 1 lin. long, rounded at the apex.

SOUTH AFRICA : without locality, in wet marshes, *Niven* !

Imperfectly known species.

10. **M. floccosa** (Knight, Prot. 65) ; stem 4–5 ft. high, with long branches ; leaves 7–8 lin. long, 4–6 lin. broad, elliptic, entire, very pubescent ; perianth-segments thinly bearded towards the apex.

COAST REGION : Stellenbosch Div.; Hottentots Holland, *Masson.* Known to us only from the description.   Masson's specimen appears to be lost.

VIII. OROTHAMNUS, Pappe.

Flowers hermaphrodite, actinomorphic.   *Perianth-tube* short, pubescent ; segments 4, linear.   *Stamens* 4 ; filaments fused with the perianth-segments ; anthers linear ; connective produced into a distinct apical gland.   *Hypogynous scales* 4, free, linear.   *Ovary* sessile ; style erect, grooved, glabrous ; stigma linear, grooved, obtuse ; ovule solitary.

An erect shrub with long-pilose branches ; leaves ciliate ; flower-heads sessile, crowded at the extremities of the branches, many-flowered ; involucral bracts large, coloured, the outermost densely shaggy.

DISTRIB.   Species 1, confined to the South-Western part of Cape Colony.

1. O. Zeyheri (Pappe in Bot. Mag. t. 4357) ; an erect shrub 6–8 ft. high ; branches long-pilose ; leaves 1–2¼ in. long, ¾–1¼ in. broad, slightly imbricate, obovate or oblanceolate-spathulate, with a very obtuse blackish apex, slightly narrowing at the base or rarely the upper leaves attenuated, distinctly 5–6-nerved, rigidly subcoriaceous, densely ciliate when young, otherwise glabrous or rarely scantily pilose ; heads sessile, 2–2½ in. long, many-flowered, 1–3 or rarely more at the extremity of a branch, drooping ; involucral-bracts rose-red and petaloid, 4–5-seriate, 1¾–2 in. long, 4–12 lin. broad, spathulate-oblong, rounded at the apex, many-nerved, membranous, pilose, ciliate, the outermost densely shaggy-pilose ; perianth-tube 3 lin. long, cylindric, pubescent ; segments lemon-yellow, 1¼ in. long, linear, pilose ; limb 4 lin. long, linear, scantily pilose ; filaments swollen, fused with the perianth ; anthers 3½ lin. long, linear ; apical gland ¼ lin. long, ovate, obtuse ; hypogynous scales ⅗ lin. long, linear, obtuse, brown ; style 1¾ in. long, grooved, glabrous ; stigma 3½ lin. long, sulcate, obtuse ; ovary 1 lin. long, globose ; fruit 3 lin. long, oblong, smooth and shining. *Flor. des Ser.* iv. *t.* 338.   *Mimetes Zeyheri, Meisn. in DC. Prodr.* xiv. 264.

COAST REGION : Stellenbosch Div. ; Hottentots Holland Mountains, *Zeyher,* 3689 !

IX. DIASTELLA, Knight.

Flowers hermaphrodite, actinomorphic.   *Perianth* cylindric in bud with an ellipsoid limb ; segments free to or slightly connate at the base, villous, clearly differentiated into a slender claw and a short broader limb ; limb oblong or lanceolate, subacute or obtuse, villous or rarely glabrous.   *Stamens* 4 ; anthers oblong, sessile, shorter than the limb and inserted at its base ; connective produced into a small apical appendage.   *Hypogynous scales* absent.   *Ovary* sessile, pubescent ; style filiform, straight, subpersistent, glabrous or pilose in the lower part ; stigma cylindric, short, obtuse, epapillose, gradually tapering into the style. *Ovule* 1, pendulous. *Fruit* a yellowish-white, ellipsoid, beaked nut with a thin reticulated pericarp. *Seed* solitary ; testa membranous, hyaline ; embryo straight ; cotyledons large, flat and thin, broad ; radicle very small.

Erect or prostrate undershrubs or shrubs 1–8 ft. high ; branches pilose or villous ; leaves crowded or more often lax, elliptic, spathulate or suborbicular, rarely ericoid, entire or sometimes minutely toothed at the apex, usually hairy ; flowers in terminal solitary sessile subglobose or obconic bracteate heads ; involucral bracts never or only slightly exceeding the flowers, ciliate, mostly coloured ; floral bracts linear or filiform, densely villous ; receptacle flat or slightly concave long-setose with fine rust-coloured hairs.

DISTRIB. Species 5, confined to the South-Western portion of Cape Colony.

Leaves flat, linear-oblong to orbicular, 1½ lin. broad or
   more ; perianth-limb villous :
 Heads rather small, about ½ in. in diam. or less (rarely
   more), mostly few- (rarely about 45-)flowered ;
   involucral bracts obtuse or gradually acuminate,
   never cuspidate :
  Leaves crowded and more or less closely imbricate ;
    heads 6–8 lin. in diam. ; perianth-limb long
    villous   ...   .   (1) **bryiflora.**
  Leaves laxly arranged and mostly spreading ; heads
    about 5 lin. in diam. ; perianth-limb rather
    shortly or long-villous :
   Leaves rarely over ¼ in. long, with a minute
     callous entire apex ; a low spreading shrub ..   (2) **serpyllifolia.**
   Leaves about ¾ in. long, with a conspicuous callous
     apex, the latter sometimes bifid ; an erect
     shrub 7–8 ft. high   ...   .   (3) **myrtifolia.**
 Heads rather large, about ¾ in. in diam., many-flowered
   (about 50) ; involucral bracts mostly cuspidate-
   acuminate ; leaf-apex often toothed   ...   ...   (4) **parilis.**
Leaves subacicular, less than ½ lin. broad ; perianth-
  limb glabrous   ...   .   (5) **ericæfolia.**

1. **D. bryiflora** (Knight, Prot. 62) ; a small erect shrub about 4 ft. high ; branches softly tomentose, at length becoming pubescent ; leaves 3–6 lin. long, 1½–3 lin. broad, elliptic, obtuse or rounded at the apex, rounded or slightly narrowed at the base, rigidly coriaceous, densely hirsute, becoming glabrous with age ; heads sessile, terminal, solitary, 4–5½ lin. long, 6–9 lin. in diam. ; involucral bracts 3–4 lin. long, lanceolate to oblong, obtuse, pilose, ciliate ; perianth-segments joined at the base, at length becoming free, 4 lin. long, linear, villous with long hairs ; limb ¾ lin. long, oblong, sub-acute, villous ; anthers sessile, ½ lin. long, linear ; apical gland ⅛ lin. long, ovate, obtuse ; ovary ⅓ lin. long, oblong in outline, pubescent ; style 4¾ lin. long, filiform, very shortly pubescent on the lower half ; stigma ½ lin. long, cylindric, obtuse. *Mimetes thymelæoides, R. Br. in Trans. Linn. Soc.* x. 109 ; *Roem. & Schultes, Syst. Veg.* iii. 382 ; *Meisn. in DC. Prodr.* xiv. 265. *Leucadendron thymelæoides, Berg. in Vet. Acad. Handl. Stockh.* 1766, 324 ; *Berg. Descr. Fl. Cap.* 19. *Protea thymelæoides, Poir. Encycl. Suppl.* iv. 568. *P. pubera, Thunb.*

Diss. Prot. 43, *partly* ;   *Fl. Cap. ed. Schultes,* 140.   *P. divaricata, Willd. ex Meisn. l.c.*   *P. villosa, Jacq. ex Meisn. l.c.*

SOUTH AFRICA : without locality, *Oldenburg* ! *Thunberg* (var. δ of his Herbarium) ! *Groendahl* ! *Sparrman* ! *Robertson* ! *Herb. Forsyth* ! *& Herb. Salisbury,* at Kew !
COAST REGION : Caledon Div. ; Houw Hoek Mountains, *Burchell,* 8125 ! between Houw Hoek and Palmiet River, *Burchell,* 8169 !

2. **D. serpyllifolia** (Knight, Prot. 62) ; a decumbent shrub, 3–4 ft. high ; branches pilose ; leaves 2–9 lin. long, $1\frac{1}{2}$–3 lin. broad, suborbicular, elliptic, oblong or lanceolate, obtuse, rounded or narrowed to the base, flat, pilose, ciliate ; heads sessile, 3–5 lin. long, many-flowered, terminal, solitary or very rarely 3-nate at the ends of the branches ; involucral bracts 2–3-seriate, $2\frac{1}{2}$–5 lin. long, oblong, linear to ovate, obtuse or acuminate, pubescent, ciliate ; perianth-segments $2\frac{3}{4}$–$3\frac{1}{2}$ lin. long, linear, villous ; limb $\frac{1}{2}$–$\frac{3}{4}$ lin. long, oblong-linear, subobtuse, shortly villous ; anthers subsessile, $\frac{1}{2}$ lin. long ; filament fused with the perianth ; apical gland lanceolate, $\frac{1}{10}$–$\frac{1}{8}$ lin. long ; ovary $\frac{1}{4}$–$\frac{1}{3}$ lin. long, pubescent ; style $2\frac{3}{4}$–$3\frac{1}{2}$ lin. long, filiform, pubescent in the lower half or near the base ; stigma $\frac{1}{6}$–$\frac{1}{4}$ lin. long, cylindric, obtuse. *D. vacciniifolia, Knight, Prot.* 63. *Leucadendron divaricatum, Berg. in Vet. Acad. Handl. Stockh.* 1766, 324 ; *Berg. Descr. Pl. Cap.* 19. *Protea divaricata, Linn. Mant.* 194 ; *Thunb. Diss. Prot.* 58 ; *Murr. Syst. Veg. ed.* xiv. 142 ; *Lam. Ill.* i. 235 ; *Willd. Sp. Pl.* i. 533 ; *Poir. Encycl.* v. 643 ; *Thunb. Fl. Cap. ed. Schultes,* 140. *Mimetes divaricata, R. Br. in Trans. Linn. Soc.* x. 111 ; *Roem. & Schultes, Syst. Veg.* iii. 383 ; *Meisn. in DC. Prodr.* xiv. 265. *M. parviflora, Klotzsch in Krauss, Beitr. Fl. Cap- und Natal.* 141 ; *Klotzsch in Flora,* 1845, 77. *M. intermedia, Buek ex Meisn. in DC. Prodr.* xiv. 265.

SOUTH AFRICA : without locality, *Nelson* ! *Masson* ! *Thunberg* ! *Brown* ! *Mund* ! *Gueinzius* !
COAST REGION : Cape Div. ; mountains near Kalk Bay, *Bolus,* 2906 ! near Cape Town, *Niven,* 45 ! *Bolus* ! Klaver Vley, *Wolley-Dod,* 300 ! near Simons Town, *Wright* ! *Jameson* ! *Pappe* ! *Milne,* 137 ! *MacGillivray,* 635 ! *Bolus in Herb. Norm. Austr. Afr.* 303 ! *Schlechter,* 312 ! Red Hill, *Wolley-Dod,* 1833 ! *Jameson* ! Smitswinkel Bay, *Bolus in Herb. Norm. Austr.-Afr.* 302 ! *Wolley-Dod,* 2737 ! *Phillips,* 519 ! False Bay, *Robertson* ! Caledon Div. ; Houw Hoek, *Pappe* ! *Zeyher,* 3694 ! *Galpin,* 4477 !

This appears to be an extremely variable species. In Zeyher, 3694 (*Mimetes intermedia,* Buek), the leaves are oblong and attenuated at the base and the involucral bracts very acuminate, whilst in a specimen of typical *Mimetes divaricata,* R. Br. (*Wolley-Dod,* 1833), the leaves are suborbicular, rounded at the base and the involucral bracts oblong and very obtuse. Between these two forms, however, almost every intermediate stage occurs.—E. P. P.

3. **D. myrtifolia** (Knight, Prot. 63) ; a bushy shrub, 7–8 ft. high ; branches shortly and thinly villous ; leaves $\frac{1}{2}$–1 in. long, 1–2 lin. broad, linear or linear-lanceolate, with a broad very obtuse callus at the apex, narrowed to the base, entire or rarely bifid, rigidly coriaceous, pilose ; heads sessile, 4–6 lin. long, about 5 lin. in diam. ;

involucral bracts up to 4 lin. long, $\frac{3}{4}$–$1\frac{1}{4}$ lin. broad, the outer oblong, obtuse, pubescent, ciliate ; the intermediate ones ovate-lanceolate, the innermost linear, villous ; perianth-segments connate at the base, soon becoming quite free, 3 lin. long, linear, attenuated to the base, villous with long hairs ; limb $\frac{3}{4}$ lin. long, lanceolate, acute, villous ; anthers sessile, $\frac{1}{2}$ lin. long, oblong ; apical glands $\frac{1}{8}$ lin. long, ovoid, obtuse ; ovary $\frac{1}{2}$ lin. long, very minutely pubescent ; style $3\frac{1}{2}$ lin. long, terete, filiform, gradually tapering to the apex, very minutely pubescent at the base, otherwise glabrous ; stigma $\frac{1}{6}$ lin. long, cylindric, obtuse. *Protea myrtifolia, Thunb. Diss. Prot.* 41 ; *Murr. Syst. Veg. ed.* xiv. 141 ; *Willd. Sp. Pl.* i. 530 ; *Poir. Encycl.* v. 641 ; *Thunb. Fl. Cap. ed. Schultes,* 137. *Mimetes myrtifolia, var. β, R. Br. in Trans. Linn. Soc.* x. 110 ; *Roem. & Schultes, Syst. Veg.* iii. 382 ; *Meisn. in DC. Prodr.* xiv. 265.

SOUTH AFRICA : without locality, *Oldenburg* ! *Auge* ! *Thunberg* ! *Masson* ! *Roxburgh* ! *Brown* !
COAST REGION : Tulbagh Div. ; by the river near Tulbagh (Roode Zand), *Niven,* 36 ! Ceres Road, *Schlechter,* 9085 !

4. **D. parilis** (Knight, Prot. 62) ; a small shrub 1–3 ft. high ; stem simple ; branches pubescent or villous ; leaves $\frac{1}{2}$–1 in. long, $1\frac{1}{2}$–$2\frac{1}{2}$ lin. broad, oblong or oblong-lanceolate, obtuse with an entire or trifid callous apex, a little narrowed to the base, coriaceous, some-times subdistinctly 3–5-nerved, more or less thinly villous ; heads sessile, 4–6 lin. long, $\frac{1}{2}$–$\frac{3}{4}$ in. in diam. ; involucral bracts about $\frac{1}{2}$ in. long, ovate or lanceolate, acutely acuminate, the outer adpressed-pilose outside, the intermediate more glabrescent and densely ciliate ; innermost bracts linear, densely villous ; perianth-segments nearly free, 4 lin. long, linear-filiform, villous with long hairs ; limb $\frac{3}{4}$ lin. long, narrowly lanceolate, acute, villous ; anthers sessile, $\frac{1}{2}$ lin. long, elliptic ; apical gland $\frac{1}{8}$ lin. long, ovate, subobtuse ; ovary $\frac{2}{3}$ lin. long, oblong, pubescent ; style $4\frac{3}{4}$ lin. long, filiform, terete, gradually narrowed to the apex, pubescent at the base, persistent ; stigma $\frac{1}{4}$ lin. long, cylindric, subobtuse, furrowed ; fruit $3\frac{1}{4}$ lin. long, ellipsoid, glabrous, reticulate. *Mimetes myrtifolia, var. a, R. Br. in Trans. Linn. Soc.* x. 110 ; *Roem. & Schultes, Syst. Veg.* iii. 382 ; *Meisn. in DC. Prodr.* xiv. 265. *Leucospermum parile, Knight ex Loud. Encycl. Pl.* 82.

SOUTH AFRICA : Without locality, *Brown* !
COAST REGION : Tulbagh Div. ; between New Kloof and Elands Kloof, *Drège* ! Ceres Road, *Schlechter,* 9085 ! Great Winter Hoek, *Niven.*

R. Brown made two varieties of his *Mimetes myrtifolia,* both of which I have seen at the British Museum. His var. *a* is the same as Drège's specimen, a small shrub 14 in. high, his var. β equals *Diastella myrtifolia,* Knight (*Niven,* 36), a bushy shrub 7–8 ft. high. These two plants are undoubtedly distinct species, and Knight recognised them as such.—E. P. P.

5. **D. ericæfolia** (Knight, Prot. 64) ; a shrub with decumbent pilose branches ; leaves 3–7 lin. long, ericoid, with a blunt callous

apex or very shortly mucronate, channelled or concave on the upper
surface, glabrous or sometimes scantily pilose; heads sessile, $2\frac{1}{2}$–4
lin. long, solitary, terminal; involucral bracts 3–4-seriate, the
outer small, ovate, obtuse or acuminate, shortly pubescent, the
inner 2–$3\frac{1}{2}$ lin. long, lanceolate, acuminate, subacute, all ciliate;
perianth-segments free to the base, but often cohering for some time
at the apex (limb), $2\frac{1}{2}$–3 lin. long, linear, pilose; limb $\frac{1}{2}$ lin. long,
oblong-lanceolate, obtuse, glabrous; anthers sessile, $\frac{1}{3}$ in. long,
linear; ovary $\frac{1}{4}$ lin. long, oblong, pubescent; style 3 lin. long,
filiform, terete, gradually tapering to the apex, glabrous; stigma
$\frac{1}{6}$ lin. long, cylindric, subobtuse.   *Leucadendron proteoides, Linn. Sp.
Pl. ed.* i. 91; *Berg. in Vet. Acad. Handl. Stockh.* 1766, 326; *Descr.
Pl. Cap.* 24.   *Protea purpurea, Linn. Mant.* 195; *Thunb. Diss. Prot.*
28; *Prodr.* 26; *Willd. Sp. Pl.* i. 518; *Poir. Encycl.* v. 654; *Thunb.
Fl. Cap. ed. Schultes,* 129.   *P. salsaloides, Thunb. ex Meisn. in DC.
Prodr.* xiv. 266.   *Mimetes purpurea, R. Br. in Trans. Linn. Soc.* x.
111; *Roem. & Schultes, Syst. Veg.* iii. 383; *Meisn. l.c., incl. vars.; O.
Kuntze, Rev. Gen. Pl.* iii. 278.   *M. homomalla, Reichenb. f. ex Meisn.
l.c.   M. Buekii, Gandog. in Bull. Soc. Bot. France,* xlviii. *p.* xciii.

SOUTH AFRICA: without locality, *Nelson*! *Thunberg*! *Forster*! *Sparrman*!
Ecklon & Zeyher, 78! *Thom*! *Sieber*, 95! 189! *Wahlberg*!
COAST REGION: Malmesbury Div.; Paarde Berg, *Masson*! Paarl Div.;
between Mosselbanks River and Berg River, *Burchell*, 979! French Hoek,
Schlechter, 9229! Cape Div.; Cape Flats, *Burchell*, 217! *Niven*! *Bowie*! *Drège*!
Ecklon, 529! *Zeyher*, 1477! *Pappe*! *Bolus*, 2905! *Schlechter*, 182! *Wolley-Dod*,
617! *Wahlberg*! *Schmieterlich*, 185! Stellenbosch Div.; near Eerste River,
Bolus, 2905! between Stellenbosch and Cape Flats, *Burchell*, 8346! between
Tiger Berg and Simons Berg, *Drège*!

Imperfectly known species.

6. **D. humifusa** (Knight, Prot. 63); stems prostrate; leaves $\frac{1}{2}$ lin.
broad, $\frac{1}{2}$ in. long, smooth, flower-heads narrow.

COAST REGION: Stellenbosch Div.; Hottentots Holland, *Roxburgh*.
This is probably a mere form of *D. ericæfolia*, Knight.

X. SERRURIA, Salisb.

Flowers hermaphrodite, actinomorphic.   *Perianth-segments* free or
slightly connate at the base, usually villous or hirsute, rarely
glabrous, differentiated into a slender claw and a short subacute
more or less oblong limb, the posticous limb often glabrous or more
shortly pubescent than the three others.   *Stamens* 4; anthers oblong,
sessile, shorter than the limb and inserted at its base; connective
rarely slightly produced.   *Hypogynous scales* present or absent,
filiform.   *Ovary* sessile or very shortly stipitate, villous or pubescent;
style slender, straight, glabrous or rarely pubescent in the lower
part; stigma subclavate or cylindric, epapillose, mostly slightly

longitudinally grooved, gradually tapering into the style.   *Ovule* 1, laterally attached in the middle or slightly above.   *Nut* ovoid or subglobose, sometimes beaked by the persistent style-base.

Leafy shrubs ; branches erect or prostrate ; leaves usually crowded, often much dissected into cylindric acute segments, rarely entire ; flower-heads in terminal corymbs or panicles on a common peduncle or solitary on simple axillary and terminal peduncles ; heads often with a small involucre of barren bracts ; floral bracts mostly villous, usually shorter than the flowers ; receptacle conical or subglobose, mostly hairy.

DISTRIB.   About 50 species, confined to the South-Western region.

Section 1.   PLEIOCEPHALÆ.   Flower-heads racemose, paniculate or corymbose on a common peduncle.

Perianth-limb glabrous ; stigma ovoid-globose ...   ...   (1) **meisneriana.**

Perianth-limb hairy (nearly glabrous in 14, *Burmannii*) ;
  stigma cylindric or subclavate, rarely ellipsoid :
  Heads supported by a long common peduncle, usually
    much exserted from the leaves :
    Stems procumbent ; leaves evenly distributed and
      given off more or less at right angles ; perianth
      shortly pubescent or long-villous :
      Perianth very shortly pubescent ..   ..   (2) **hyemalis.**
      Perianth densely villous with rather coarse long
        hairs ...   .   (3) **flagellaris.**
    Stems erect ; leaves evenly arranged or in pseudo-
      whorls, always given off at an acute angle ;
      perianth shortly villous with rather closely ad-
      pressed hairs :
      Leaves clustered in pseudo-whorls around the base
        of the peduncle ; floral bracts glabrous   ...   (4) **elongata.**
      Leaves more or less evenly distributed along the
        stems ; floral bracts pubescent or almost
        villous ...   .   (5) **Leipoldtii.**
  Heads supported by a very short common peduncle,
    always more or less enclosed or clasped by the
    leaves :
    *Ultimate peduncles with a solitary bract at the
      base or one or two towards the apex, otherwise
      ebracteate :
      Leaves large and stout, ultimate segments nearly
        ¾ lin. thick ; petiole nearly 1 lin. broad ;
        primary branches of the inflorescence rather
        elongated   ...   .   (6) **anethifolia.**
      Leaves usually small, segments about ¼˙ lin.
        (rarely ½) thick ; petiole almost filiform ;
        primary branches of the inflorescence mostly
        rather short :
        Mature leaves glabrous or rarely thinly and
          spreadingly pilose :
          Bracts and ultimate peduncles glabrous, the
            former strongly ribbed. ...   .   (7) **Bolusii.**
        Bracts and ultimate peduncles more or less
          hairy, the former not or only slightly
          ribbed :
          †Perianth-limb hirsute or villous :
            Heads 5–8-flowered :

Perianth-limb shortly adpressed and
   more or less silvery-hirsute :
   Ultimate peduncles very short and
      thick   ...   .   (8) **adscendens.**

Ultimate peduncles long and slender   (9) **Knightii.**

Perianth-limb long rusty-villous   ...   (10) **pauciflora.**

Heads 15–25-flowered :
   Perianth-limb shortly hirsute with ad-
      pressed hairs   ...   .   (11) **subsericea.**

Perianth-limb long-villous with more
   or less spreading hairs :
   Branches villous or long-pilose ;
      bracts at the base of the
      ultimate peduncles villous in
      the lower part ...   .   (12) **biglandulosa.**

   Branches puberulous or shortly pu-
      bescent ; bracts at the base of
      the ultimate peduncles nearly
      glabrous...   ...   .   (13) **Kraussii.**

   ††Perianth-limb glabrous or nearly so   ...   (14) **Burmannii.**

Mature leaves permanently covered with a dense
   silvery adpressed indumentum   ...   ...   (15) **candicans.**

**Ultimate peduncles covered with several imbricate
   pustulate purplish bracts...   ...   .   (16) **glomerata.**

Section 2. MONOCEPHALÆ. Flower-heads solitary on a simple axillary or
terminal peduncle.

*Stems usually procumbent, rarely erect ; leaves of the
   annual flowering shoots simple or trifurcate, only
   those towards the base rarely pinnate or bipinnate :
   Perianth-limb glabrous ; stigma much thicker than the
      style ...   .   (17) **flagellifolia.**

Perianth-limb villous or tomentose ; stigma usually
   not much thicker than the style :
   Claw and lower part of limb of the perianth-segments
      glabrous   ...   .   (18) **trilopha.**

   Claw and limb of the perianth-segments hairy
      all over :
      Bracts lanceolate or subulate-lanceolate, very
         villous ; stems subsimple, procumbent, giving
         off the leaves more or less at right angles :
         Branches and petioles pilose with long weak
            hairs ; leaf-segments thick ...   .   (19) **pinnata.**

         Branches and petioles glabrous ; leaf-segments
            slender   ...   .   (20) **gracilis.**

      Bracts acuminate from a rounded or ovate
         membranous base, villous or glabrescent ;
         stems subsimple, prostrate or ascending :
         Outer bracts villous, 2 lin. broad or less :
            Branches diffuse ; leaves slender, rarely
               simple   ...   .   (21) **diffusa.**

            Branches erect ; leaves stouter, mostly
               simple   ...   .   (22) **simplicifolia.**

Outer bracts glabrous outside, 3–4 lin. broad ...   (40) æmula,
                                          var. **heterophylla.**

Bracts conspicuously ribbed and glabrous or subulate and long-ciliate, but otherwise glabrous ;
  stems ascending, much-branched :
  Bracts very narrow and long-ciliate, not ribbed   (23) **ciliata.**

  Bracts ovate-lanceolate, glabrous, with 3–4
    prominent ribs along the back   ...   ...   (24) **nervosa.**

**Stems erect or ascending, very rarely procumbent ; leaves
  of the annual flowering shoots pinnate or bipinnate,
  never simple and only rarely a few trifurcate :
  Leaves permanently silky-silvery adpressed-tomentose :
    Heads solitary and subsessile ; style hairy in the
      lower half ...   .   (25) **Dodii.**

  Heads clustered and pedunculate ; style glabrous :
    Peduncle ½ in. long ...   .   (26) **argentifolia.**

    Peduncle 1–1½ in. long   ...   .   (27) **Aitoni.**

Leaves glabrous or thinly pilose with long weak spreading hairs :
  Style hairy in the lower ½ or ⅔ or in the middle ⅓ :
    Heads solitary, terminal, sessile ; style hairy only
      in the middle third   ...   .   (28) **brevifolia.**

    Heads several together, or if solitary then long-
      pedunculate ; style hairy in the lower ½ or ⅔ :
      Indumentum of the perianth short and
        adpressed :
        Leaves ½–1¼ in. long, thick, divided almost
          from the base   ...   .   (29) **flava.**

        Leaves usually larger, more slender and with
          longer ultimate segments, divided in the
          upper half or two-thirds ...   .   (30) **acrocarpa.**

      Indumentum of the perianth long-villous :
        Leaves divided from or very near the base,
          narrowly ovate in outline when spread
          out ...   .   (31) **longipes.**

        Leaves divided from considerably above the
          base, more or less fan-shaped in outline   (32) **artemisiæfolia.**

  Style glabrous :
    Heads solitary and sessile or subsessile at the apex
      of each branch or leafy lateral branchlet, or if
      shortly pedunculate then more or less clasped
      at the base by the leaves :
      Limb of the posticous perianth-segment glabrous
        or very minutely pubescent, or villous some-
        times in 35, *millefolia* :
        Leaves divided more or less in the upper
          half :
          Leaves 1–1½ in. long ; heads large, many-
            and densely flowered ...   .   (33) **hirsuta.**

          Leaves ¾–1 in. long ; heads small, few- and
            lax flowered   ...   .   (34) **ventricosa.**

        Leaves divided almost from the base, very
          small   ...   .   (35) **millefolia.**

Limb of the posticous perianth-segment as hairy
or almost as hairy as the others (see also 35,
millefolia) ; leaves divided in the upper
half or two-thirds :
 Heads sessile or subsessile :
  Heads with no distinct involucre of barren
  outer bracts, usually hidden by the
  leaves ; indumentum of perianth-claw
  much shorter than that of limb   ... (36) **vallaris.**

  Heads with a distinct involucre of barren
  outer bracts ; indumentum of claw
  the same as that of the limb ...   ... (37) **rostellaris.**

 Heads distinctly pedunculate   ...   ... (38) **cyanoides.**

Heads several and more or less corymbose towards
the apex of each shoot, or if solitary then fairly
long-pedunculate :
 Outer bracts large and broad, (white ?), considerably longer than the flowers, glabrous
 and not ciliate   ...   . (39) **florida.**

 Outer bracts usually small and narrow, if broad
 (see 40, *æmula* and 41, *scariosa*) then ciliate
 or villous :
  Leaves usually well over 1 in. long :
   Stems erect or ascending; outer bracts
   about ¼ in. broad, with a broad membranous margin :
    Indumentum of perianth long-villous ... (40) **æmula.**

    Indumentum of perianth very short   ... (41) **scariosa.**

   Stems erect ; bracts narrow, with narrow
   membranous margins or not membranous :
    Peduncles glabrous or nearly so, covered
    with numerous narrow bracts :
     Limb clothed with short adpressed
     hairs ; bracts long-ciliate or glab-
     rous and pustulate outside :
      Bracts long-ciliate, linear or subulate (23) **ciliata,**
                                             var. **congesta.**

      Bracts glabrous or nearly so, ovate or
      lanceolate   ...   . (42) **fœniculacea.**

     Limb clothed with very long spreading
     hairs ; bracts very shortly or
     scarcely ciliate   ...   . (43) **barbigera.**

    Peduncles tomentose, scarcely bracteate (44) **fucifolia.**

  Stems prostrate :
   Bracts on the peduncle scattered ; heads
   distributed along the branches   ... (45) **cygnea.**

   Bracts on the peduncle imbricate ; head
   solitary at the apex of each shoot ... (46) **incrassata.**

Leaves ¾ in. long or less, rarely 1 in. :
 Leaves filiform ; heads crowded, subsessile (47) **Roxburghii.**

 Leaves stout ; heads solitary, pedunculate (48) **callosa.**

1. S. meisneriana (Schlechter in Engl. Jahrb. xxvii. 108); stems erect; branches terete, glabrous or minutely pubescent; leaves $1\frac{1}{2}$–$3\frac{1}{2}$ in. long, bipinnately divided in the upper $\frac{3}{4}$ or $\frac{2}{3}$; ultimate segments $\frac{3}{4}$–$1\frac{1}{4}$ in. long, cylindric, acutely or subacutely mucronate, narrowly channelled on the upper surface, glabrous; heads numerous, corymbose on a long common peduncle, exserted from the leaves; peduncle $1\frac{1}{2}$–5 in. long, glabrous, bearing a few subulate or lanceolate acuminate acute bracts about $2\frac{1}{2}$ lin. long; primary branches up to $\frac{3}{4}$ in. long; ultimate peduncles 3–5 lin. long, with a solitary bract at the base of each; floral bracts $3\frac{1}{2}$–$4\frac{1}{2}$ lin. long, broadly ovate or suborbicular, with a broad membranous margin and a broad thick keel which passes into an abrupt short acumen; flowers straight in bud, with an ellipsoid limb; perianth-tube 2 lin. long, glabrous; segments 3–4 lin. long, linear-spathulate, very sparingly and shortly setulose; limb 1 lin. long, $\frac{1}{2}$ lin. broad, ovate-elliptic, obtuse or subobtuse, glabrous; anthers sessile, ellipsoid, subacute; ovary $\frac{1}{2}$–$\frac{3}{4}$ lin. long, ellipsoid, densely villous; style $2\frac{3}{4}$–4 lin. long, subquadrangular, glabrous; stigma $\frac{1}{3}$ lin. long, ellipsoid-globose, subobtuse, grooved. *S. glaberrima, var. pinnata, Meisn. in DC. Prodr.* xiv. 284.

COAST REGION: Cape Div.; mountains near Constantia, *Schlechter,* 542. Caledon Div.; Babylons Tower Mountains, *Zeyher,* 3700! *Ecklon & Zeyher,* 14! mountains between Houw Hoek and Bot River, *Pappe!* mountains near Bot River, *Schlechter,* 9442!

On Zeyher's sheet of 3700 in the Stockholm Herbarium there are two species, the middle and larger specimen being the true plant, whilst the remaining smaller examples on each side of it are *S. flagellifolia,* Knight; they probably belong to his 3701 which is that species.

2. S. hyemalis (Knight, Prot. 84); stems prostrate, glabrous; leaves 2–5 in. long, entire or 2–3-furcate; segments cylindric, acutely mucronate, with a narrow groove on the upper surface, glabrous; heads few, crowded at the apex of a common terminal peduncle, a little exserted from the leaves; peduncle $\frac{3}{4}$–2 in. long, slender, glabrous, bearing lanceolate-linear acute bracts $1\frac{1}{2}$–2 lin. long; ultimate peduncles 1–$1\frac{1}{2}$ lin. long, stout, shortly pubescent; floral bracts up to $4\frac{1}{2}$ lin. long, broadly ovate, acuminate, glabrous or very shortly and scantily pubescent; flowers straight in bud; perianth-tube $2\frac{1}{2}$ lin. long, slightly inflated, ribbed, minutely pubescent; segments 6 lin. long, linear, very shortly adpressed-pubescent; limb $1\frac{1}{2}$ lin. long, oblong-lanceolate, subobtuse, sparingly adpressed-pubescent outside; anthers $1\frac{1}{4}$ lin. long, linear; ovary $\frac{3}{4}$ lin. long, densely villous; style $6\frac{1}{2}$ lin. long, linear, 4-angled, glabrous; stigma purple, $1\frac{1}{4}$ lin. long, linear, suboctuse, furrowed, very slightly bent at the junction with the style. *S. decumbens. R. Br. in Trans. Linn. Soc.* x. 126; *Roem. & Schultes, Syst. Veg.* iii. 372; *Meisn. in DC. Prodr.* xiv. 294. *Protea decumbens, Thunb. Diss. Prot.* 14, t. 1; *Murr. Syst. Veg. ed.* xiv. 136; *Lam. Ill.* i. 239;

2 u 2

Thunb. Prodr. 25 ; *Willd. Sp. Pl.* i. 506 ; *Poir. Encycl.* v. 657 ;
Thunb. Fl. Cap. ed. Schultes, 121. *P. procumbens, Linn. fil. Suppl.* 116.

South Africa : without locality, *Thunberg* !
Coast Region : Cape Div. ; mountains near Simons Town, *Wright*, 624 !
Pappe, 27 ! *Fair in Herb. Bolus*, 7948 ! False Bay, *Roberts* ! South-west of Slankop,
Wolley-Dod, 1801 ! Stellenbosch Div. ; Hottentots Holland Mountains, *Thunberg* !

3. S. flagellaris (R. Br. in Trans. Linn. Soc. x. 127) ; stems
decumbent, glabrous or sometimes pilose with a few scattered hairs ;
leaves 2–4 in. long, pinnately or bipinnately divided in the upper
half, sometimes slightly sheathing at the base, glabrous or rarely
thinly pilose ; segments terete and bluntly or subacutely mucronate ;
heads very numerous, paniculate or corymbose on a long and rather
stout common terminal peduncle, usually much exserted from the
leaves ; peduncle up to 2 in. long, bearing lanceolate-linear acute
bracts 3–5 lin. long ; primary branches up to ¾ in. long ; ultimate
peduncles 1–2 lin. long, pubescent, with a solitary bract at the base
of each ; floral bracts 3½–5½ lin. long, ovate or suborbicular with a
long acute acumen, pubescent ; flowers erect in bud ; perianth-tube
1¾–2¼ lin. long, nearly glabrous, soon splitting to the base ;
segments 4¾–5½ lin. long, hirsute or villous with adpressed hairs ;
limb 1¼ lin. long, narrowly oblong, subacute, villous outside ;
anthers 1 lin. long, oblong-linear ; hypogynous scales ¾ lin. long,
linear or filiform ; ovary ½–⅔ lin. long, globose, shortly villous with
club-shaped hairs in the lower and slender ones in the upper part ;
style 4½–5½ lin. long, grooved, glabrous ; stigma 1 lin. long,
cylindric, subacute, gradually passing into the style or sometimes
with a faint kink at the junction. *Roem. & Schultes, Syst. Veg.*
iii. 373 ; *Meisn. in DC. Prodr.* xiv. 294. *Protea flagellaris, Poir.*
Encycl. Suppl. iv. 573.

South Africa : without locality, *Thom* ! *Grey* ! *Niven* ! *Bergius*, 269, *Mund*,
Ludwig.
Coast Region : Cape Div. ; Muizenberg, *Zeyher* ! Table Bay, *Brown* ! Simons
Bay, *Roxburgh* ! *Wright*, 625 ! *MacGillivray*, 673 ! *Milne*, 32 ! *Bolus*, 4686 ! Fish
Hoek, *Fair in Herb. Bolus*, 7287 ! Elsje Peak, *Wolley-Dod*, 2865 !

4. S. elongata (R. Br. in Trans. Linn. Soc. x. 132) ; stems erect
or ascending, glabrous ; leaves arranged in pseudowhorls at the base
of the peduncle, 2–5 in. long, bipinnately or more divided in the
upper ½ or ⅓, glabrous or the young ones sometimes villous ; ultimate
segments cylindric, obtusely mucronate, about ½ lin. thick ; heads
numerous, paniculate or corymbose on a long common peduncle, far
exserted from the leaves ; peduncle 6–12 in. long, glabrous ;
primary branches up to 2¼ in. long, mostly several-headed, with a
lanceolate acuminate acute bract 2–4 lin. long at the base of each ;
ultimate peduncles 4–6 lin. long, glabrous, not or scarcely bracteate ;
floral bracts purplish, about 3 lin. long and 1½ lin. broad, with a
suborbicular basal part and a thick midrib ending in a mucro,
glabrous ; flowers straight in bud ; perianth-tube 1½ lin. long,

glabrous, soon splitting to the base; segments 3¼–4 lin. long,
spathulate-linear, shortly adpressed-villous; limb 1 lin. long,
narrowly oblong, subacute, villous; anthers ¾ lin. long, linear;
ovary ½ lin. long, villous; style 3¼ lin. long, cylindric, glabrous;
stigma ⅔ lin. long, oblong or subclavate, obtuse, slightly swollen at
the junction with the style; fruit 1 lin. long, more or less ellipsoid,
shortly beaked and stipitate, rusty-setose. *Roem. & Schultes, Syst.
Veg.* iii. 377; *Meisn. in DC. Prodr.* xiv. 297. *S. crithmifolia,
Knight, Prot.* 83; *R. Br. in Trans. Linn. Soc.* x. 132. *Leucadendron
elongatum, Berg. in Vet. Acad. Handl. Stockh.* 1766, 327; *Berg. Descr.
Pl. Cap.* 27. *Protea glomerata, Thunb. Diss. Prot.* 18; *Willd. Sp.
Pl.* i. 509. *P. thyrsoides, Lam. Ill.* i. 240?; *Poir. Encycl.* v. 660?
P. helvola, Willd. ex Meisn. l.c. name only.

SOUTH AFRICA: without locality, *Thunberg*! *Niven*! *Ludwig*!
COAST REGION: Clanwilliam Div.; Alexanders Hoek, *Schlechter*, 5130; Paarl
Div.; mountains around French Hoek, *MacOwan, Herb. Norm. Austr.-Afr.* 912!
Caledon Div.; mountains of Klein River Kloof, *Zeyher*, 3702! Houw Hoek,
Schlechter, 5493! 5510! Baviaans Kloof near Genadendaal, *Burchell*, 7709! 7831.
Genadendaal, *Pappe*! mountains near Hermanus, *Galpin*, 4473! Zwart Berg,
Pappe! Bredasdorp Div.; near Elim, *Schlechter*, 9648! *Bolus*, 8590!

5. **S. Leipoldtii** (Phillips & Hutchinson); a small shrub less than
1 ft. high; branches erect, glabrous; leaves 2½–4 in. long, simple or
pinnately or bipinnately divided in the upper half, glabrous;
segments very acutely mucronate, slightly furrowed on the upper
surface, about ½ lin. thick; heads few in a corymbose raceme on a
long slender terminal common peduncle, exserted from the leaves;
peduncle 2½–3 in. long, glabrous; ultimate peduncles up to 1¼ in. long,
adpressed-pubescent towards the apex, with a discoid swelling and
an articulation where it falls off at the base; bracts at the base of
each peduncle about 2 lin. long, glabrous; floral bracts 2½–3 in.
long, ovate-lanceolate, acuminate, subacute, with a flat glandular
surface at the apex, recurved, pubescent or tomentose or the outer
nearly glabrous; perianth-tube 1¼ lin. long, soon splitting to the
base, glabrous; segments 5 lin. long, spathulate-linear, adpressed-
tomentose; limb 1 lin. long, elliptic, subacute, adpressed-tomentose;
anthers ¾ lin. long; ovary 1 lin. long, ellipsoid, villous; style 3¾ lin.
long, glabrous; stigma ½ lin. long, subcylindric, somewhat acute,
strongly sinuate at the junction with the style. *S. elongata?* Drège,
Zwei Pfl. Documente, 74, 221, not of *R. Br.*; *Meisn. in DC. Prodr.*
xiv. 297. *S. triternata, var., Drège ex Meisn. l.c.*

SOUTH AFRICA: without locality, *Drège*, 8073, partly!
COAST REGION: Clanwilliam Div.; along the banks of the Tratra River, near
Wupperthal, *Leipoldt*, 644! and in *Herb. Bolus*, 9384! Ezels Bank, *Drège*!

6. **S. anethifolia** (Knight, Prot. 84); an erect shrub 2–3 ft. high;
branches glabrous; leaves 2½–5½ in. long, pinnately or bipinnately
divided in the upper ½ or ⅔, glabrous when mature; segments
terete, with an acute oblique callous apex, about ½ lin. thick,

furrowed on the upper surface; heads many-flowered, numerous, in a rather dense corymb up to 4 in. broad on a common peduncle, scarcely exserted from the leaves; branches and ultimate peduncles pubescent or tomentose, the latter ½–1 in. long, with a solitary bract and a discoid swelling where it articulates at the base; floral bracts 1¼ lin. long, ovate or ovate-lanceolate, acute and shortly pointed, pubescent; perianth-tube 1–1¼ lin. long, glabrous, soon splitting to the base; segments 3–4 lin. long, spathulate-linear, acute, villous with adpressed white hairs; limb ½ lin. long, villous; anthers ½ lin. long; ovary ⅔–1 lin. long, subglobose, densely villous; style 1¼–2½ lin. long, glabrous; stigma ⅕ lin. long, ellipsoid or subovoid, obtuse. *S. triternata, R. Br. in Trans. Linn. Soc.* x. 131; *Roem. & Schultes, Syst. Veg.* iii. 376; *Meisn. in DC. Prodr.* xiv. 297. *S. tridentata, D. Dietr. Syn.* i. 520 (*by error*). *S. argentiflora, Buek in Drège, Zwei Pfl. Documente*, 77, 221. *Protea triternata, Thunb. Diss. Prot.* 18; *Murr. Syst. Veg. ed.* xiv. 136; *Thunb. Prodr.* 25; *Willd. Sp. Pl.* i. 509; *Poir. Encycl.* v. 660; *Thunb. Fl. Cap. ed. Schultes*, 123. *P. argentiflora, Andr. Bot. Rep. t.* 447. *P. glomerata, Willd. ex Meisn. l.c., partly.*

SOUTH AFRICA: without locality, *Auge*! *Oldenburg*! *Thunberg*! *Masson.*
COAST REGION: Tulbagh Div.; near Tulbagh Waterfall, *Roxburgh*! *Ecklon*! *Bolus, Herb. Norm. Austr.-Afr.*, 382! *Pappe*! *Schlechter*, 9002! *Phillips*, 523! between New Kloof and Elands Kloof, *Drège*! *Zeyher*; Roode Zand, *Niven*! Winterhoek, near Tulbagh, *Bolus*, 4808!

7. **S. Bolusii** (Phillips & Hutchinson); stems erect; branches glabrous; leaves ¾–1¾ in. long, pinnately or bipinnately divided in the upper half, glabrous; ultimate segments 4–6 lin. long, narrowly cylindric, acutely mucronate; heads 4–12, about 4 lin. long and in diam., few-flowered, corymbose on a common peduncle; peduncle about ¾ in. long, bearing linear-lanceolate glabrous bracts about 2½ lin. long which soon become strongly reflexed; ultimate peduncles 2–3 lin. long, glabrous, with a solitary bract at the base of each; floral bracts 1½–3 lin. long, ovate, long-acuminate, acute, ribbed, glabrous or very minutely pubescent; flowers curved in bud; perianth-tube ¾ lin. long, pubescent above, glabrous below; segments 2½–3 lin. long, shortly adpressed-hirsute; limb ⅚–1 lin. long, oblong, subacute, adpressed-hirsute; anthers ¾ lin. long; hypogynous scales ½–¾ lin. long, filiform; ovary ¼–½ lin. long, ellipsoid, covered with long white hairs; style 3–3½ lin. long, swollen and articulated at the base, glabrous; stigma ⅔–¾ lin. long, cylindric, obtuse, furrowed; fruit 2½ lin. long, oblong-ellipsoid, beaked, villous.

SOUTH AFRICA: without locality, *Thom*, 787!
COAST REGION: Bredasdorp Div.; near Elim, *Schlechter*, 9651, partly! *Bolus*, 8589!

8. **S. adscendens** (R. Br. in Trans. Linn. Soc. x. 127); branches prostrate or ascending, purplish, glabrous or pubescent; leaves 1–2½ in. long, bipinnately divided in the upper half or third, glabrous; ultimate segments 2–5 lin. long, narrowly cylindric,

subobtusely mucronate; heads numerous in a dense raceme or
corymbose panicle on a common peduncle, enclosed by or only
slightly exserted from the leaves, 6–8-flowered; peduncle up to
$\frac{3}{4}$ in. long, pubescent; ultimate peduncles very short and thick, up
to $1\frac{1}{2}$ lin. long, with a solitary ovate-lanceolate purple glabrescent
bract about 2 lin. long at the base of each; floral bracts $2\frac{1}{4}$–3 lin.
long, ovate, acutely acuminate, at length recurved, pubescent;
flowers curved in bud; perianth-tube $\frac{1}{2}$–$1\frac{3}{4}$ lin. long, glabrous;
segments $3\frac{1}{4}$–4 lin. long, adpressed-tomentose; limb $\frac{2}{3}$–$1\frac{1}{4}$ lin. long,
oblong-elliptic, subacute, tomentose; anthers $\frac{1}{2}$–1 lin. long; hypogynous scales $\frac{1}{2}$ lin. long, filiform; ovary $\frac{1}{2}$ lin. long, villous; style
about $3\frac{1}{2}$ lin. long, glabrous; stigma $\frac{1}{2}$ lin. long, subclavate, obtuse.
Roem. & Schultes, Syst. Veg. iii. 373; *Meisn. in DC. Prodr.* xiv. 294.
S. rubricaulis, R. Br. l.c. 128, *excl. syn. Thunb.*; *Roem. & Schultes,
l.c.*; *Meisn. l.c. S. compar, R. Br. l.c.* 129; *Roem. & Schultes, l.c.* 375;
Meisn. l.c. 295. *Protea adscendens, Lam. Ill.* i. 239; *Poir. Encycl.*
v. 658; *Steud. Nomencl. ed.* 2, ii. 399. *P. compar, Poir. Encycl.
Suppl.* iv. 574.

VAR. β, **decipiens** (Hutchinson); branches roughly villous; heads crowded
and very shortly pedunculate or subsessile; bracts roughly hairy. *S. decipiens,
R. Br. l.c.* 129; *Roem. & Schultes, Syst. Veg.* iii. 374; *Meisn. in DC. Prodr.* xiv.
295, *incl. vars. S. glomerata, Meisn. l.c.* 294, *as to syn. Thunb. Protea patula,
Thunb. Diss. Prot.* 16; *Fl. Cap. ed. Schultes,* 122; *Murr. Syst. Veg. ed.* xiv. 136.

SOUTH AFRICA: without locality, *Thom,* 421! *Ludwig*! *Gueinzius*! Var. β:
Sieber!
COAST REGION: Clanwilliam Div.; between Kromme River and Berg Valley,
Drège! Piquetberg Div.; Piquet Berg, *Drège,* 8073 partly! *Schlechter,* 5184!
Caledon Div.; mountains of Klein River Kloof, *Zeyher,* 3708! Donker Hoek
mountain, *Burchell,* 7942! near Hermanus, *Bolus,* 9896! *Galpin,* 4475! Bot River,
Schlechter, 9436! Var. β: Malmesbury Div.; near Groene Kloof (Mamre),
Bolus, 4327! between Groene Kloof and Dassenberg, *Drège,* 8074a! Zwartland,
Zeyher, 3711! Caledon Div.; Bot River, *Schlechter,* 9443!

9. **S. Knightii** (Hutchinson); a much-branched shrub about 3 ft.
high; branches diffuse, purplish, thinly pilose or puberulous,
becoming at length nearly glabrous; leaves slender, up to $2\frac{1}{2}$ in.
long, bipinnately divided in the upper $\frac{1}{2}$ or $\frac{2}{3}$, thinly pilose or
glabrous; ultimate segments up to $\frac{3}{4}$ in. long, subulate or
almost filiform, very acute; heads usually very numerous, in
a broad corymb on a short common peduncle, enclosed by the
leaves, individual heads about 4 lin. long, 5–7-flowered; ulti-
mate peduncles $\frac{1}{4}$–$\frac{3}{4}$ in. long, slender, densely pubescent, with a
solitary subulate pubescent bract about 2 lin. long at the base of
each; floral bracts more or less ovate, with a gland-tipped
glabrescent acumen, villous, about $1\frac{1}{2}$ lin. long; flowers slightly
curved in bud; perianth-tube about $\frac{3}{4}$ lin. long, pilose; segments
$2\frac{1}{2}$ lin. long, linear-filiform, shortly adpressed-tomentose; limb $\frac{3}{4}$ lin.
long, elliptic, subacute, tomentose; anthers $\frac{2}{3}$ lin. long; ovary
pubescent; style $2\frac{1}{2}$ lin. long, glabrous; stigma $\frac{1}{3}$ lin. long, subclavate, somewhat acute; fruit about 2 lin. long, obovoid, shortly

villous in the upper half. *S. fasciflora, Knight, Prot.* 85? *excl. all syns. except part of Thunb. Protea Serraria, var.* 1, *Thunb. Diss. Prot.* 18, *not of Linn.* ; *Fl. Cap. ed. Schultes,* 123.

SOUTH AFRICA : without locality, *Thom,* 597 ! *Thunberg* !
COAST REGION: Tulbagh Div. ; near Mitchells Pass, *Bolus,* 5256 ! 5257 ! near Tulbagh, *Schlechter,* 7474 ! *Pappe* ! Worcester Div., *Cooper,* 1616 ! Brand Vley, *Schlechter,* 9925 ! Stellenbosch Div. ; mountains of Lowrys Pass, *Burchell,* 8286 ! Caledon Div. ; Zwart Berg, *Schlechter,* 9774 ! between Caledon and Elim, *Bolus,* 7865 ! Bredasdorp Div. ; Elim, *Schlechter,* 9644 ! Swellendam Div. ; between Zuurbraak and Buffeljagts River Drift, *Burchell,* 7270/1 ! between Swellendam and Buffeljagts River, *Zeyher,* 3710 ! Riversdale Div. ; between Garcias Pass and Krombeks River, *Burchell,* 7176 ! Div. ? Ruyterbosch, *Britten,* 138 !

10. **S. pauciflora** (Phillips & Hutchinson); branches pilose with long hairs or sometimes only shortly pubescent; leaves $1\frac{1}{4}$–3 in. long, bipinnately divided in the upper half, glabrous; ultimate segments up to 1 in. long, terete, very acute ; heads corymbose on a short common peduncle, more or less enclosed by the leaves, about 3 lin. long and in diam., 5–7-flowered ; ultimate peduncles 2–4 lin. long, tomentose, with a solitary bract at the base of each ; floral bracts $1\frac{3}{4}$–2 lin. long, ovate, sharply acuminate, densely pilose or villous below the acumen ; flowers erect in bud ; perianth-tube $\frac{3}{4}$ lin. long, pilose, glabrous near the base ; segments 3–4 lin. long, spreadingly villous ; limb $\frac{3}{4}$ lin. long, elliptic-lanceolate, subacute, villous ; anthers $\frac{1}{2}$–$\frac{2}{3}$ lin. long ; hypogynous scales $\frac{1}{2}$ lin. long, filiform ; ovary villous; style about 3 lin. long, glabrous; stigma $\frac{1}{3}$–$\frac{1}{2}$ lin. long, oblong, obtuse. *S. compar, Meisn. in DC. Prodr.* xiv. 295, *as to Zeyher,* 3712, *not of R. Br.*

SOUTH AFRICA: without locality, *Thom,* 575 ! *Niven* (in Stockholm Herbarium) !
COAST REGION : Caledon Div. ; mountains of Klein River Kloof, *Zeyher,* 3712 ! *Pappe* !

11. **S. subsericea** (Hutchinson); branches purplish, pilose or subvillous, at length becoming nearly glabrous ; leaves $\frac{1}{2}$–$1\frac{1}{2}$ in. long, bipinnately divided in the upper $\frac{1}{2}$ or $\frac{2}{3}$, thinly pilose or glabrous ; ultimate segments $\frac{1}{4}$–$\frac{1}{2}$ in. long, acutely mucronate ; heads corymbose on a common peduncle, 2–4 lin. long, 15–25-flowered ; peduncle $\frac{1}{4}$–$\frac{1}{2}$ in. long, tomentose ; ultimate peduncles with a solitary linear or subulate pubescent gland-tipped bract about 1 lin. long at the base of each ; floral bracts $\frac{3}{4}$–$1\frac{1}{4}$ lin. long, ovate, acuminate, acute, pilose or hirsute ; flower-buds slender and slightly incurved at the apex ; perianth-tube $\frac{3}{4}$ lin. long, glabrous ; segments $2\frac{1}{2}$ lin. long, almost filiform, shortly adpressed-hirsute ; limb $\frac{3}{4}$ lin. long, linear-oblanceolate, hirsute ; anthers $\frac{1}{2}$ lin. long ; ovary obovoid, pubescent ; style $\frac{1}{4}$ in. long, filiform, glabrous ; stigma $\frac{1}{3}$ lin. long, narrowly clavate ; fruits obovoid, shortly villous. *Serruria Burmanni, var. β, R. Br. in Journ. Linn. Soc.* x. 131. *S. Burmanni, var. subsericea, Meisn. in DC. Prodr.* xiv. 296, *excl. Drège,* 8070a. *S. candicans, Drège ex Meisn. l.c. S. Burmanni, b & d, E. Meyer in Drège, Zwei Pfl. Documente,* 83, 119, 221.

Serraria Serraria, var. subsericea, O. Kuntze, Rev. Gen. Pl. iii. 2,
280. *Leucadendron Serraria, var. β, Linn. Sp. Pl. ed.* i. 93, *excl.
syn. Burm.*

SOUTH AFRICA : without locality, *Thunberg*! *Ludwig, Gueinzius*! *Harvey*, 696!
Drège, 8072! *Armstrong*!
COAST REGION : Malmesbury Div.; between Klipfontein and Predikstoel,
Zeyher, 3699! near Groene Kloof, *Bolus*, 4326! Worcester Div.; Goudini,
Drège, b! Paarl Div.; Drakenstein Mountains, *Drège*, d! Cape Div.; near
Raapenberg Vley, *Wolley-Dod*, 2111! Cape Flats, *Ecklon*, 754! *Bolus*, 2903!
Pappe! *Gamble*, 22162! Wynberg, *Gamble*, 22177!

12. **S. biglandulosa** (Schlechter in Engl. Jahrb. xxiv. 451);
branches villous, at length becoming pilose ; leaves $1\frac{1}{4}$–$2\frac{3}{4}$ in. long,
bipinnately divided in the upper half, adpressed-pilose when young,
soon becoming glabrous ; segments terete, acutely mucronate ; heads
few, corymbose on a common terminal peduncle, enclosed by the leaves,
many-flowered, about $\frac{1}{2}$ in. long and in diam. ; peduncle $\frac{1}{2}$–$\frac{3}{4}$ in. long,
white-villous ; ultimate peduncles $\frac{1}{2}$ in. long, with a solitary linear-
spathulate glabrescent bract at the base, white-villous ; floral bracts
2 lin. long, ovate-lanceolate, acutely acuminate, villous ; perianth-
tube 1 lin. long, glabrous, soon splitting to the base ; segments
2–$2\frac{1}{2}$ lin. long, spathulate-linear, villous with spreading hairs ; limb
$\frac{2}{3}$ lin. long, ovate-lanceolate, acute, villous ; anthers $\frac{1}{2}$ lin. long ;
ovary villous ; style $2\frac{1}{2}$–$2\frac{3}{4}$ lin. long, glabrous ; stigma $\frac{1}{4}$ lin. long,
cylindric, subobtuse ; fruit $1\frac{1}{2}$ lin. long, ellipsoid-globose, shortly
stipitate, sparingly pilose.

COAST REGION : Stellenbosch Div. ; Lowrys Pass, *Schlechter*, 7258!

13. **S. Kraussii** (Meisn. in DC. Prodr. xiv. 296); branches
puberulous ; leaves 2–3 in. long, bipinnately divided in the upper $\frac{2}{3}$,
glabrous ; ultimate segments 3–6 lin. long, with an acute callous
apex ; heads several, subcorymbose on a very short common
peduncle, clasped by the leaves, about 4 lin. long and in diam.,
20–30-flowered ; peduncle $\frac{1}{2}$–$\frac{3}{4}$ in. long, tomentose, bearing one or
two subulate very acute bracts 2–$2\frac{1}{4}$ lin. long ; ultimate peduncles
up to 4 lin. long, tomentose, with a solitary bract at the base ; flowers
straight in bud ; floral bracts ovate, villous ; perianth-tube 1 lin.
long, glabrous, soon splitting to the base ; segments $2\frac{1}{4}$ lin. long,
spathulate-linear, villous with spreading hairs ; limb $\frac{2}{3}$ lin. long,
elliptic, acute, villous ; anthers $\frac{1}{2}$ lin. long, oblong ; hypogynous
scales $\frac{1}{2}$ lin. long, filiform ; ovary $\frac{1}{4}$ lin. long, ellipsoid, villous with
long erect hairs ; style $2\frac{1}{2}$ lin. long, glabrous ; stigma $\frac{1}{4}$ lin. long,
oblong, obtuse.

SOUTH AFRICA : without locality or collector in Herb. Kew !
COAST REGION : Stellenbosch Div. ; Hottentots Holland, *Ludwig*! *Gueinzius*.

14. **S. Burmanni** (R. Br. in Trans. Linn. Soc. x. 130, excl. var. β);
an erect shrub about 2 ft. high ; stem simple at the base, branched

above; branches slender, pubescent; leaves very slender, 1–1½ in.
long, bipinnately divided in the upper ⅔, glabrous or thinly pilose;
ultimate segments up to 7 lin. long, almost filiform, very acute;
heads mostly numerous, corymbose on a common peduncle, enclosed
by the leaves, about 15-flowered, nearly ½ in. long; ultimate
peduncles rather slender, up to ¼ in. long, tomentose, with a solitary
subulate very acute glabrescent bract at the base of each; floral
bracts up to 2 lin. long, with a long linear acumen from an ovate
base or the inner subulate, glabrous or nearly so; perianth-tube
glabrous, soon splitting to the base; segments 3½ lin. long, filiform;
claw very shortly setulose-pubescent; limb ¾ lin. long, lanceolate,
subacute, glabrous or nearly so; anthers ½ lin. long; ovary ellipsoid,
glabrous; style slender, 3 lin. long, glabrous; stigma slightly
clavate. *Roem. & Schultes, Syst. Veg.* iii. 375, *excl. var. β*; *Meisn. in
DC. Prodr.* xiv. 296, *excl. var. β and part of a. S. fasciflora, Knight,
Prot.* 85, *partly.* S. *Burmanni, var. vulgaris, Meisn. l.c.* S. *fœniculacea,
Sieber, partly, ex Meisn. l.c.* 296, *not of R. Br.   Leucadendron
Serraria, vars. a and γ, Linn. Sp. Pl. ed.* i. 93; *ed.* ii. 137.
Protea Serraria, Linn. Mant. alt. 188; *partly of his Herb.*; *Willd.
Sp. Pl.* i. 508, *partly*; *Lam. Ill.* i. 240, *partly*; *Poir. Encycl.* v. 660,
partly.—Abrotanoides arboreum, etc., Pluk. Mant. 1, *t.* 329, *fig.* 1.
Abrotanum africanum foliis tenuissimis, etc., Seba, Thesaur. ii. 64,
t. 63, *fig.* 6.   *Serraria foliis tenuissime divisis, etc., Burm. Rar.
Afr. Pl.* 264, *t.* 99, *fig.* i.

SOUTH AFRICA: without locality, *Thunberg*! *Thom*! *Pappe*! *Sieber*, 10!
Drège, 8070a!
COAST REGION: Paarl Div.; between Mosselbanks River and Berg River,
Burchell, 976! Cape Div.; Cape Flats between Cape Town and Simons Town,
Burchell, 8542! Simons Bay, *Wright*! Table Mountain, *Bolus*, 4729! between
Wynberg and Constantia, *Burchell*, 786! behind Wynberg Butts, *Wolley-Dod*,
609! Orange Kloof, *Wolley-Dod*, 826! Stellenbosch Div.; between Stellenbosch
and Cape Flats, *Burchell*, 8366! near the River Eerste, *Bolus*, 4199! between Lowrys
Pass and Jonkers Hoek, *Burchell*, 8333! Caledon Div.; Donker Hoek mountain,
Burchell, 7941! near Caledon, *Bolus*, 9916! Swellendam Div.; hills of
Swellendam, *Bowie*! Riversdale Div.; Garcias Pass, *Galpin*, 4474! Mossel Bay
Div.; Attaquas Kloof, *Gill*! George Div.; Montagu Pass, *Schlechter*, 5839!

15. S. candicans (R. Br. in Trans. Linn. Soc. x. 130); stems
erect; branches rusty-villous when young, at length becoming
whitish-tomentose or pubescent; leaves 1–1½ in. long, bipinnately
divided in the upper ¾, very densely whitish adpressed-tomentose,
rusty-tomentose when young; ultimate segments terete, acutely
mucronate; heads several and corymbose on a short common
peduncle, enclosed by the leaves; peduncle rusty-tomentose; ultimate
peduncles about 2 lin. long, tomentose, with a solitary bract about
2 lin. long at the base of each; floral bracts 2½ lin. long, ovate,
acuminate, with a flat glandular apex, villous outside; flowers erect
in bud; perianth-tube 1 lin. long, glabrous; segments 3½ lin. long,
pilose with spreading hairs; limb ¾ lin. long, oblong, subacute;
anthers ⅔ lin. long, oblong; style 3 lin. long, glabrous; stigma ½

lin. long, subclavate ; fruit 2 lin. long, oblong, villous. *Roem. &
Schultes, Syst. Veg.* iii. 375 ; *Meisn. in DC. Prodr.* xiv. 296.

SOUTH AFRICA : without locality or collector in the British Museum Herbarium !
COAST REGION : Cape Div. ; Paarde Berg, near Salt River, *Zeyher*! *Pappe*,
42 ! Caledon Div. ; mountains of Baviaans Kloof near Genadendaal, *Burchell*,
7867 !

16. **S. glomerata** (R. Br. in Trans. Linn. Soc. x. 128, excl. syn.
Thunb.) ; branches ascending, glabrous or sparingly pilose ; leaves
1–2$\frac{1}{2}$ in. long, pinnately or bipinnately divided in the upper $\frac{1}{2}$ or $\frac{2}{3}$,
glabrous ; ultimate segments 2–5 lin. long, narrowly cylindric,
acutely mucronate, furrowed on the upper surface ; heads densely
crowded and corymbose on a very short common terminal peduncle,
enclosed by the upper leaves ; ultimate peduncles up to 4 lin. long,
covered by ovate-lanceolate acute long-acuminate glabrous slightly
ciliate bracts 1$\frac{1}{3}$–2$\frac{1}{4}$ lin. long ; floral bracts 2–2$\frac{1}{4}$ lin. long, about
1$\frac{1}{2}$ lin. broad, suborbicular or obovate, acuminate, pubescent, slightly
warted outside ; perianth-tube pubescent ; segments 3$\frac{1}{2}$ lin. long,
linear, adpressed-villous ; limb 1 lin. long, lanceolate-elliptic, subacute, adpressed-villous ; anthers $\frac{3}{4}$ lin. long, oblong ; ovary $\frac{1}{2}$ lin.
long, globose, surrounded by numerous long hairs ; style 3$\frac{3}{4}$–4 lin.
long, glabrous, swollen at the base ; stigma $\frac{3}{4}$ lin. long, subclavate,
obtuse. *Roem. & Schultes, Syst. Veg.* iii. 374 ; *Meisn. in DC. Prodr.*
xiv. 293, *excl. part of syn. S. rubricaulis, R. Br. l.c. as to syn.
Thunb. S. fœniculacea, Sieber, partly, ex Meisn. l.c.* 294, *not of R. Br.
Leucadendron glomeratum, Linn. Sp. Pl. ed.* ii. 137. *Protea glomerata,
Linn. Mant. alt.* 187 ; *Syst. Veg. ed.* xiv. 136. *P. sphærocephala,
Thunb. Diss. Prot.* 16 ; *Fl. Cap. ed. Schultes,* 122.—*Serraria foliis
tenuissime divisis, etc., Burm. Rar. Afr. Pl.* 265, *t.* 99, *fig.* 2.

SOUTH AFRICA : without locality, *Thunberg*! *Ludwig*! *Oldenburg*, 582 !
Sieber, 188 ! *Bergius, Zeyher* ! *Rutherford* !
COAST REGION : Cape Div. ; Constantia, *Jameson*! Cape Flats, *Zeyher*, 1476 !
Pappe! *Burchell*, 213 ! 8578 ! *Bolus*, 4807 ! *Gamble*, 22412 ! Wynberg, *Wallich*!
Ludwig! *Ecklon* ; Kommetjes, *Galpin*, 4476 ! Simons Bay, *Wright*, 627 !

17. **S. flagellifolia** (Knight, Prot. 84) ; stems prostrate, slender,
subterete, glabrous ; internodes 1–1$\frac{1}{2}$ in. long ; leaves usually all
growing at right angles to the stem, 1–4 in. long, simple or 2–3-
furcate, rarely pinnately divided ; segments terete except for a
narrow channel on the upper surface, about $\frac{1}{2}$ lin. thick, acutely
mucronate, glabrous ; heads solitary and axillary towards the end of
the shoots, pedunculate, 5–7 lin. long, 6–12-flowered ; peduncle
$\frac{3}{4}$–1$\frac{1}{4}$ in. long, glabrous, bearing a few ovate-lanceolate acuminate
acute glabrous bracts with membranous margins ; floral bracts 2$\frac{1}{2}$–3$\frac{1}{4}$
lin. long, broadly ovate or suborbicular, obtuse or slightly acuminate,
glabrous ; flowers straight in bud ; perianth-tube 1–1$\frac{1}{4}$ lin. long,
glabrous ; segments 3 lin. long, glabrous or sparingly setulose in the
lower part ; limb 1 lin. long, $\frac{2}{3}$ lin. broad, elliptic, subacute, glabrous ;

anthers ¾ lin. long ; ovary 1 lin. long, ellipsoid, long-villous ; style
2½–3 lin. long, glabrous ; stigma ½ lin. long, ellipsoid, obtuse.
S. glaberrima, R. Br. in Trans. Linn. Soc. x. 112 ; *Roem. & Schultes,
Syst. Veg.* iii. 363 ; *Meisn. in DC. Prodr.* xiv. 284, *excl. var. Protea
glaberrima, Poir. Encycl. Suppl.* iv. 569. *P. decumbens, Willd. ex
Meisn. l.c.,* name only, not of *Thunb.*

South Africa : without locality, *Gueinzius* !
Coast Region : Stellenbosch Div. ; Hottentots Holland Mountains, *Masson* !
Caledon Div. ; at the foot of mountains near Bot River, *Bolus, Herb. Norm.
Austr.-Afr.* 1349 ! Houw Hoek Mountains, *Zeyher,* 3701 ! *Schlechter,* 5482 !
Klein Houw Hoek, *Roxburgh* ! rocks on the Zwart Berg, near Caledon, *Bowie* !

18. **S. trilopha** (Knight, Prot. 90) ; a low decumbent plant ;
branches thinly pilose or nearly glabrous ; leaves ¼–1 in. long,
divided into 3 filiform acute furrowed segments, rarely entire,
glabrous, thinly pilose when young ; heads terminal, solitary,
shortly pedunculate, many-flowered, ¾–1 in. long, about 1 in. in
diam., subglobose ; peduncle ¼–¾ in. long, villous, bearing linear-
lanceolate acutely acuminate glabrous ciliate bracts about 2 lin.
long ; floral bracts 2¼–4 lin. long, ovate-oblong to lanceolate,
acutely acuminate, with a thick glabrescent midrib, villous towards
the margin ; flowers erect in bud ; perianth-tube 1–1½ lin. long,
glabrous ; segments spathulate-linear, with a glabrous claw ; limb
1–1¼ lin. long, elliptic, subacute, glabrous in the lower, bearded in
the upper part ; anthers ¾ lin. long ; ovary ½ lin. long, villous ;
style 3–4 lin. long, glabrous ; stigma 1–1½ lin. long, clavate, subobtuse. *S. arenaria, R. Br. in Trans. Linn. Soc.* x. 117 ; *Roem. &
Schultes, Syst. Veg.* iii. 366 ; *Meisn. in DC. Prodr.* xiv. 291. *Protea
arenaria, Poir. Encycl. Suppl.* iv. 571. *P. phylicoides, Willd. Sp.
Pl.* i. 510, *partly.*

South Africa : without locality, *Thom* !
Coast Region : Cape Div. ; Cape Flats, *Ecklon,* 3b ! *Zeyher,* 1471 ! *Pappe,* 9 !
Harvey, 695 ! *Bolus,* 4594 ! east of Plumstead, *Wolley-Dod,* 1613 ! Wynberg,
Niven.

19. **S. pinnata** (R. Br. in Trans. Linn. Soc. x. 116, partly, excl. syn.
Andr.) ; a shrub with long prostrate stems, the latter long-pilose
with weak hairs, rather densely so when young ; internodes about
¼ in. long ; leaves 1–1¼ in. long, usually trifurcate in the upper
third, the petiole or lower part pilose and often stouter than the
segments, the latter terete, acutely mucronate ; heads terminal,
solitary or two at the end of a shoot, but each on its own peduncle,
shortly pedunculate, 1–1½ in. in diam., about ¾ in. long ; peduncle
densely pubescent, bearing a few bracts similar to those of *S. gracilis ;*
floral bracts up to ½ in. long, narrowly lanceolate, gradually and
very acutely acuminate, shortly and sparingly pubescent outside,
glabrous within, with purple tips ; flowers erect in bud ; perianth-
tube about 2 lin. long, glabrous at the base ; segments 5–6 lin. long,
linear, adpressed-pilose outside ; limb 1¼ lin. long, oblong, subobtuse,

rather densely adpressed-pilose outside ; anthers sessile, 1 lin. long ; style ½ in. long, slender, glabrous ; stigma as in *S. gracilis. Roem. & Schultes, Syst. Veg.* iii. 365, *partly, excl. syn. Andr. ; Poir. Encycl. Suppl.* iv. 571, *partly ; Meisn. in DC. Prodr.* xiv. 290, *excl. var.*

COAST REGION : Paarl Div. ; Little and Great Drakenstein Mountains, *Drège* ! Paarl Mountain, *Drège* ! Stellenbosch Div. ; between Tyger Berg and Simons Berg, *Drège* !

20. **S. gracilis** (Knight, Prot. 81) ; a small shrub with decumbent stems ; stems terete, purplish, glabrous, with very short internodes ; leaves ¾–1½ in. long, usually trifurcate in the upper half, rarely a few entire, glabrous ; segments terete, slender, acutely mucronate ; heads solitary, terminal, shortly pedunculate, ¾–1 in. in diam., a little over ½ in. long, many-flowered ; peduncle ½–1 in. long, broadening upwards, more or less tomentose, bearing a few bracts, the latter lanceolate or subulate-lanceolate, acuminate, acute, 1½–3 lin. long, glabrous, purplish ; floral bracts 4½–6 lin. long, narrowly lanceolate, long and gradually acuminate, adpressed-pubescent outside, glabrous within, tips purple ; flowers erect in bud, bright red ; perianth-tube 2 lin. long, pubescent in the upper part, glabrous at the base ; segments 5 lin. long, linear, adpressed-pilose ; limb 1¼ lin. long, oblong-lanceolate, subobtuse, long adpressed-pilose outside ; anthers sessile, 1 lin. loug ; ovary densely villous ; style 5 lin. long, slender, glabrous ; stigma about ¾ lin. long, gradually passing into the style, obtuse or almost truncate. *S. pinnata, R. Br. in Trans. Linn. Soc.* x. 116, *partly, as to syn. Andr. ; Roem. & Schultes, Syst. Veg.* iii. 365, *partly ; Poir. Encycl. Suppl.* iv. 571, *partly. S. pinnata, var. longifolia, Meisn. in DC. Prodr.* xiv. 291. *Protea pinnata, Andr. Bot. Rep. t.* 512.

COAST REGION : Paarl Div. ; near French Hoek, *Bolus, Herb. Norm. Austr.-Afr.* 1350 ! *MacOwan,* 2904 ! Tulbagh Div. ; Tulbagh Waterfall, *Niven.*

21. **S. diffusa** (R. Br. in Trans. Linn. Soc. x. 115) ; branches prostrate, glabrous or slightly pubescent towards the tips ; leaves 1–1¼ in. long, ternately divided in the upper third or simple and then very short, glabrous, often slightly pubescent when young ; segments 3½–7 lin. long, terete, acutely mucronate ; heads solitary, terminal, shortly pedunculate, about ¾ in. long and in diam., many-flowered ; peduncle ¾–1 in. long, pubescent, bearing a few linear-lanceolate subacute glabrous often reflexed bracts ; floral bracts 4–6 lin. long, ovate to lanceolate, acutely acuminate, villous ; flowers straight in bud ; perianth-tube about 1¼ lin. long, glabrous ; segments 3¾–4½ lin. long, linear, long-villous with spreading yellow or whitish hairs ; limb 1½ lin. long, elliptic or oblong-elliptic, subacute, villous ; anthers ¾–1 lin. long ; ovary villous ; style 4 lin. long, swollen at the base ; stigma ¾ lin. long, obtuse. *Roem. & Schultes, Syst. Veg.* iii. 365 ; *E. Meyer in Drège, Zwei Pfl. Documente,* 78, 221 ; *Meisn. in DC. Prodr.* xiv. 286. *S. furcellata, R. Br. l.c.* 118 ;

Meisn. l.c. 285. *S. scariosa, Drège ex Meisn. l.c.* 285, *not of R. Br. Protea Brownii, Poir. Encycl. Suppl.* iv. 570. *P. cyanoides, Thunb. Diss. Prot.* 15; *Fl. Cap. ed. Schultes*, 122. *Leucadendron glomeratum, Linn., partly, ex Meisn. l.c.* 285.

SOUTH AFRICA: without locality, *Thunberg*!
COAST REGION: Tulbagh Div.; New Kloof, *Drège*! *Schlechter*, 7498! near Roode Zand, *Roxburgh*! near Wilde River, *Niven*; Paarl Div.; mountains near Paarl, *Pappe*, 8!

22. **S. simplicifolia** (R. Br. in Trans. Linn. Soc. x. 115); branches simple, finely pubescent or glabrous; leaves $1\frac{1}{4}$–$2\frac{1}{4}$ in. long, entire or rarely 2–3-furcate or pinnately divided, linear, subterete, furrowed above, glabrous; heads solitary, terminal and axillary, pedunculate, about $\frac{3}{4}$ in. long and in diam., many-flowered; peduncle 1–$1\frac{1}{4}$ in. long, tomentose, bearing lanceolate-ovate acuminate acute glabrous bracts 2–3 lin. long; floral bracts 2–4 lin. long, ovate, acuminate, acute, densely villous; flowers erect in bud; perianth-tube $\frac{3}{4}$ lin. long, glabrous; segments $3\frac{1}{2}$ lin. long, spreadingly villous; limb 1 lin. long, oblong, obtuse, villous; anthers 1 lin. long; hypogynous scales $\frac{2}{3}$ lin. long, filiform; ovary $\frac{1}{2}$ lin. long, villous; style 3 lin. long, glabrous, swollen at the base; stigma $\frac{3}{4}$ lin. long, subclavate, obtuse. *Roem. & Schultes, Syst. Veg.* iii. 365; *Spreng. Syst. Veg.* i. 466; *Schultes, Mant.* 267; *Drège, Zwei Pfl. Documente*, 114, 221; *Meisn. in DC. Prodr.* xiv. 284. *Protea plumigera, Thunb. in Mém. Acad. Petersb.* 1818, 14, *t.* 14; *Thunb. Fl. Cap. ed. Schult.* 121; *Schultes, Mant.* 267. *P. simplicifolia, Poir. Encycl. Suppl.* 570.

SOUTH AFRICA: without locality, *Thunberg*!
COAST REGION: Malmesbury Div.; Groene Kloof (Mamre) and neighbourhood, *Zeyher*, 1469! *Bolus*, 4325! Zwartland, *Pappe*! between Groene Kloof and Dassenberg, *Drège*! Tulbagh Div.; Tulbagh Waterfall, *Roxburgh*!

Serruria linearis, Knight, Prot. 82, is very probably this species; the type, which was gathered at Groene Kloof by *Niven*, is apparently not now in existence.

23. **S. ciliata** (R. Br. in Trans. Linn. Soc. x. 123); branches glabrous or rarely pilose with a few weak hairs when young; leaves slender, $\frac{1}{2}$–$\frac{3}{4}$ in. long, trifurcate in the upper $\frac{1}{2}$ or $\frac{1}{3}$, very rarely a few bipinnately divided, glabrous or pilose with a few scattered weak hairs; ultimate segments 2–4 lin. long, cylindric, acutely mucronate; heads solitary, terminal, sessile or subsessile, 5–7 lin. long, $\frac{1}{2}$–$\frac{3}{4}$ in. in diam., many-flowered; outer bracts 3–4 lin. long, subulate, acute, long-ciliate; floral bracts $2\frac{3}{4}$–$4\frac{1}{2}$ lin. long, acutely long-acuminate from an ovate base, sometimes pustulate, glabrous or slightly pilose, very long-ciliate; perianth-tube 1 lin. long, glabrous; segments $2\frac{3}{4}$–$3\frac{3}{4}$ lin. long, spathulate-linear, villous with spreading hairs; limb $\frac{3}{4}$ lin. long, elliptic or oblong-elliptic, subacute, villous; anthers $\frac{2}{3}$ lin. long; hypogynous scales $\frac{1}{4}$–$\frac{2}{3}$ lin. long, filiform; ovary $\frac{1}{4}$–$\frac{1}{2}$ lin. long, ellipsoid, villous; style 3–$3\frac{1}{2}$ lin. long, cylindric, glabrous, not swollen at the base; stigma $\frac{1}{2}$–$\frac{2}{3}$ lin. long,

clavate-cylindric, obtuse, furrowed.   *Roem. & Schultes, Syst. Veg.* iii.
370 ; *Meisn. in DC. Prodr.* xiv. 293.   *Protea ciliata, Poir. Encycl.
Suppl.* iv. 572.   *P. glomerata, Willd. ex Meisn. l.c.*

VAR. β, **congesta** (Hutchinson) ; leaves nearly always bipinnate, rarely a few
trifurcate ; heads mostly crowded at the apex of each shoot. *S. arenaria,
Knight, Prot.* 87 ? *S. emarginata, Sweet, Hort. Brit. ed.* i. 348 ? *S. congesta, R. Br.
in Trans. Linn. Soc.* x. 123 ; *Roem. & Schultes, Syst. Veg.* iii. 370 ; *Meisn. in
DC. Prodr.* xiv. 293, *incl. vars., excl. syn. Andr. Protea abrotanifolia minor,
Andr. Bot. Rep. t.* 536 ? *P. congesta, Poir. Encycl. Suppl.* iv. 573.   *P. phylicoides,
Willd. ex Meisn. l.c.*

SOUTH AFRICA : var.   β : without locality, *Thom* ! *Drège,* 8069 ! *Forster* !
Ludwig ! *Bowie* !

COAST REGION : Stellenbosch Div. ; between Stellenbosch and Cape Flats,
Burchell, 8360 ! between Tyger Berg and Simons Berg, *Drège* ! Physsers Hoek,
Roxburgh !   Var.   β : Cape Div. ; near Tyger Berg, *Ecklon,* 45 ! *Pappe* ! *Bolus,*
5238 ! Vygeskraal Farm, *Wolley-Dod,* 1838 ! Cape Flats, *Pappe,* 36 ! Kuils
River, *Pappe,* 20 ! near Cape Town, *Bolus,* 2904 ! near Durban Road Station,
Wolley-Dod, 1853 ! sand-dunes near Riet Valley, *Zeyher,* 1474 ! Swellendam
Div. ; on secondary hills, *Bowie* !

24. **S. nervosa** (Meisn. in DC. Prodr. xiv. 290) ; branches
purplish, glabrous or very scantily pilose with weak hairs ; leaves
$\frac{3}{4}$–1 in. long, pinnately divided in the upper half, glabrous or
scantily pilose ; ultimate segments 3–4 lin. long, slender, narrowly
cylindric, acutely mucronate ; heads terminal, solitary, subsessile,
$\frac{1}{2}$ in. long and in diam., many-flowered ; peduncle 1–2 lin. long,
glabrous, bearing subulate acuminate acute glabrous purple bracts
about 2 lin. long ; floral bracts $3\frac{1}{2}$–$4\frac{1}{2}$ lin. long, ovate, long-
acuminate, subacute, ribbed, minutely pubescent towards the
margin ; flowers curved in bud ; perianth-tube $1\frac{1}{4}$ lin. long,
glabrous below ; segments $3\frac{1}{2}$ lin. long, spathulate-linear, hirsute
with adpressed hairs ; limb 1 lin. long, oblong, subobtuse, shortly
hirsute ; anthers sessile, $\frac{3}{4}$ lin. long, oblong-linear ; hypogynous
scales 1 lin. long, filiform ; ovary $\frac{1}{2}$ lin. long, ellipsoid, long-villous ;
style 4 lin. long, with an ellipsoid swelling at the base, glabrous ;
stigma $\frac{3}{4}$ lin. long, slightly clavate, obtuse.

SOUTH AFRICA : without locality, *Ludwig* !
COAST REGION : Bredasdorp Div. ; near Elim, *Bolus,* 7864 !

25. **S. Dodii** (Phillips & Hutchinson) ; branches pilose ; leaves
$1\frac{1}{4}$–2 in. long, bipinnately divided in the upper $\frac{2}{3}$, adpressed-
tomentose or almost villous with silvery hairs ; ultimate segments
2–4 lin. long, subobtuse ; heads terminal, solitary, shortly pedunculate, 8 lin. long, about $\frac{3}{4}$ in. in diam., enclosed within the upper
leaves, 12–15-flowered ; peduncle 3 lin. long, densely puberulous,
bearing ovate acutely acuminate puberulous shortly ciliate brácts
1 lin. long ; floral bracts 2 lin. long, broadly ovate or almost transversely oblong, shortly acuminate, obtuse, shortly tomentose outside ;
flowers curved in bud ; perianth-tube $1\frac{1}{2}$ lin. long, glabrous below,
pubescent above ; segments $\frac{1}{2}$ in. long, adpressed-hirsute ; limb $1\frac{1}{2}$
lin. long, elliptic, acute, adpressed-hirsute, that of the posticous

segment more shortly pubescent; anthers 1 lin. long; hypogynous
scales 1 lin. long, filiform; ovary villous; style ½ in.
long, rather densely pubescent in the lower half; stigma almost 1 lin. long,
subclavate, obtuse.

COAST REGION : Worcester Div. ; Els Kloof, Hex River, *Wolley-Dod*, 4050 !

26. **S. argentifolia** (Phillips & Hutchinson); branches more or
less thinly tomentose, becoming pubescent when older; leaves
¾–1¼ in. long, about 1¼ in. broad, bipinnately divided in the upper
half, rather densely adpressed-tomentose; segments nearly ½ lin.
thick, terete, obtuse; heads axillary and terminal, crowded at the
ends of the branches, solitary on each peduncle, about 7 lin. long
and ¾ in. in diam., many-flowered; peduncle 5–6 lin. long, tomentose, bearing one or two subulate-lanceolate shortly pubescent
bracts; floral bracts 2 lin. long, broadly obovate or suborbicular,
cuspidate, subacute, densely pubescent; perianth-tube ½ lin. long,
glabrous; segments 4½ lin. long, spreadingly villous; limb 1 lin.
long, oblong, acute, villous; anthers ¾ lin. long; ovary villous;
style 3½ lin. long, glabrous; stigma ¾ lin. long, cylindric, obtuse.

COAST REGION : Ceres Div. ; Wagenbooms River, *Schlechter*, 10155 !

27. **S. Aitoni** (R. Br. in Trans. Linn. Soc. x. 114); branches
tomentose or pilose; leaves ¾–1¾ in. long, bipinnately divided in
the upper ⅔, densely adpressed-tomentose; segments terete, obtuse;
heads terminal and axillary, corymbose at the end of the shoots,
about ¾ in. in diam., many-flowered; peduncle simple, 1–1½ in.
long, tomentose, bearing very few lanceolate-linear acutely acuminate villous bracts 1½–3 lin. long; floral bracts 2–3 lin. long, ovate
to obovate or suborbicular, acuminate, villous; perianth-tube 1½
lin. long, glabrous; segments ½ in. long, spreadingly villous; limb
1 lin. long, oblong or oblanceolate, villous; anthers nearly 1 lin.
long; ovary villous; style 3½–4 lin. long, glabrous; stigma ⅔ lin.
long, obtuse; fruit 1 lin. long, oblong, shortly beaked, villous.
Roem. & Schultes, Syst. Veg. iii. 364; *Meisn. in DC. Prodr.* xiv. 288.
S. subumbellata, E. Meyer in Drège, Zwei Pfl. Documente, 75, 221.
*S. tomentosa, Meisn. l.c. Nivenia Zahlbruckneri, Ostermeyer in Ann.
Nat. Hofmus. Wien,* xxiv. 297, *t.* vi. *Protea Aitoni, Poir. Encycl.
Suppl.* iv. 570.

VAR. β, **multifida** (Meisn. l.c.); leaves shortly and spreadingly pilose; outer bracts
silky-tomentose. *S. multifida, E. Meyer in Drège, Zwei Pfl. Documente*, 74, 221.

SOUTH AFRICA : without locality, *Masson* !
COAST REGION: Piquetberg Div. ; mountains around Piquetberg, *Drège!*
Schlechter, 5192 ! *Penther*, 1590. Var. β : Clanwilliam Div. ; Ezelsbank, *Drège*.

28. **S. brevifolia** (Phillips & Hutchinson); a branched, erect
shrub about 5 ft. high ; branches erect, slightly rufous-tomentellous
and pilose with long weak hairs when young, becoming at length
puberulous and slightly glaucous ; leaves 3–4 lin. long, pinnately or
bipinnately divided from near the base, shortly tomentose when
quite young, at length glabrous ; segments terete, obtuse ; heads

solitary, terminal, sessile, $\frac{1}{2}$–1 in. long, about $\frac{3}{4}$ in. in diam.; involucral bracts about 5 lin. long, lanceolate, gradually acuminate, acute, nearly glabrous or tomentellous outside, long-ciliate; floral bracts 4 lin. long, ovate-lanceolate, acute, densely villous towards the apex; perianth-tube $1\frac{1}{4}$ lin. long, glabrous; segments $2\frac{3}{4}$ lin. long, spreadingly villous; limb $\frac{3}{4}$ lin. long, lanceolate, obtuse, that of the posticous segment much more shortly villous than the others; anthers $\frac{1}{2}$ lin. long; ovary 2 lin. long, pilose; style 3 lin. long, tapering upwards from a swollen base, pubescent in the middle $\frac{1}{3}$ of its length; stigma $\frac{1}{6}$ lin. long, ovoid or ellipsoid, obtuse.

COAST REGION : Caledon Div. ; mountains of Baviaans Kloof near Genadendaal, *Burchell*, 7864 !

29. **S. flava** (E. Meyer in Drège, Zwei Pfl. Documente, 74, 221); branches tomentose or densely pilose; leaves $\frac{1}{2}$–$1\frac{1}{4}$ in. long, bipinnately divided in the upper $\frac{2}{3}$, pilose with weak whitish hairs; ultimate segments terete, with an obtuse callous apex; heads terminal, solitary, pedunculate, 1 in. long, about 6–8 lin. in diam.; peduncle $\frac{3}{4}$–$1\frac{3}{4}$ in. long, softly tomentose, bearing lanceolate subacute tomentose bracts 2–3 lin. long; floral bracts $2\frac{1}{2}$–$3\frac{1}{2}$ lin. long, ovate to lanceolate, shortly and subacutely acuminate, adpressed-villous; flowers slightly curved in bud; perianth-tube 2–$2\frac{1}{4}$ lin. long, glabrous at the base; segments $\frac{1}{2}$ in. long, spathulate-linear, adpressed-hirsute; limb $1\frac{1}{2}$ lin. long, elliptic, subacute or subobtuse, hirsute, that of the posticous segment more shortly pubescent than the others; anthers 1–$1\frac{1}{4}$ lin. long; ovary 2 lin. long, densely villous; style $6\frac{1}{2}$–7 lin. long, cylindric, pubescent on the lower half or third; stigma $\frac{3}{4}$–$\frac{5}{6}$ lin. long, oblong, obtuse, furrowed, slightly curved, constricted at the junction with the style. *Meisn. in DC. Prodr.* xiv. 287.

COAST REGION : Clanwilliam Div.; Storm Vley, near Wupperthal, *Leipoldt*, 484 ! Cederberg Range, between Wupperthal and Ezelsbank, *Drège*! near Ezelsbank and Kers Kop, *Bodkin in Herb. Bolus*, 9080 ! *Schlechter*, 8795 !

30. **S. acrocarpa** (R. Br. in Trans. Linn. Soc. x. 113); branches pilose with long hairs and puberulous with very short ones; leaves $\frac{3}{4}$–2 in. long, bipinnately divided in the upper half, pilose or glabrous; segments terete, bluntly mucronate, furrowed on the upper surface; heads shortly pedunculate, solitary, 5–7 lin. long, about $\frac{3}{4}$ in. in diam., 10–15-flowered; peduncles $\frac{3}{4}$–$1\frac{1}{4}$ in. long, densely pubescent, bearing a few ovate-lanceolate acutely acuminate glabrescent ciliate bracts $1\frac{3}{4}$–2 lin. long; involucral bracts $2\frac{1}{2}$ lin. long, ovate, acuminate, densely pubescent or hirsute, ciliate; flowers curved in bud; perianth-tube 2 lin. long, glabrous towards the base; segments 4–6 lin. long, spathulate-linear, adpressed yellow-villous; limb 1 lin. long, elliptic, subacute, adpressed yellow-villous; anthers sessile, $\frac{3}{4}$ lin. long; hypogynous scales $\frac{3}{4}$ lin. long,

filiform; ovary $\frac{1}{3}$ lin. long, obovoid, very densely villous; style 6–7 lin. long, cylindric, curved and glabrous above, pubescent on the lower half; stigma $\frac{1}{2}$ lin. long, ovoid, subobtuse; fruit very minutely stalked, 3 lin. long, oblong-ellipsoid, beaked, villous. *Roem. & Schultes, Syst. Veg.* iii. 364; *Drège, Zwei Pfl. Documente*, 115, 221; *Meisn. in DC. Prodr.* xiv. 287, *with vars. S. saxicola, Buek ex E. Meyer in Drège, Zwei Pfl. Documente*, 221. *S. adscendens, E. Meyer in Drège, l.c.* 76. *Protea acrocarpa, Poir. Encycl. Suppl.* iv. 570.

COAST REGION : Piquetberg Div. ; Piquet Berg, *Drège*! Tulbagh Div. ; on the Witzenberg Range, near Tulbagh, *Burchell*, 8658! Worcester Div. ; Brand Vley, *Roxburgh*! Hex River Mountains, *Rehmann*, 2714! Caledon Div. ; mountains near Genadendal, *Ludwig*! *Burchell*, 7852! 8623! *Drège*! *Pappe*! near the Zondereinde River, *Schlechter*, 5644! Donkerhoek Mountain, *Burchell*, 7943! *Pappe*! *Bolus*, 5242! Swellendam Div. ; near Swellendam, *Pappe*, 3! 11! mountains of Houw Hoek, *Zeyher*, 3707! *Galpin*, 4472! *Schlechter*, 7330! between Zuurbraak and Buffeljagts River Drift, *Burchell*, 7271! hills by the Buffeljagts River, *Zeyher*, 3706 β! near Breede River, *Burchell*, 7466! Zuurbraak, *Galpin*, 4471! Hessaques Kloof, *Zeyher*, 3706!

31. **S. longipes** (Phillips & Hutchinson); branches pilose with long weak hairs from a puberulous surface; leaves $\frac{3}{4}$–$1\frac{1}{4}$ in. long, narrowly ovate in outline when spread out, bipinnately divided almost from the base, glabrous or very sparingly pilose with long weak hairs; ultimate segments 2–4 lin. long, with a subobtuse callous apex; heads terminal and axillary, up to 5 collected at the end of the branches, many-flowered; peduncle $1\frac{1}{4}$–$2\frac{3}{4}$ in. long, finely puberulous, bearing 5–8 remote ovate-lanceolate acutely acuminate glabrous or slightly puberulous bracts; floral bracts $\frac{1}{4}$ in. long, about 2 lin. broad, very abruptly and shortly acuminate, villous; perianth-tube $1\frac{1}{4}$ lin. long, glabrous; segments nearly $\frac{1}{2}$ in. long, villous; limb 1 lin. long, elliptic, subobtuse, that of the posticous segment very shortly pubescent, the others long-villous; anthers $\frac{3}{4}$ lin. long; hypogynous scales $\frac{1}{3}$ lin. long, linear; ovary villous; style 5 lin. long, pubescent on the lower $\frac{2}{3}$; stigma $\frac{3}{4}$ lin. long, furrowed, obtuse.

COAST REGION : Tulbagh Div. ; Mitchells Pass, *Bolus, Herb. Norm. Austr.-Afr.*, 383!

This may be only a variety of *S. artemisiæfolia*, Knight ; the leaves are divided almost from the base and when spread out they are narrowly ovate in outline.

32. **S. artemisiæfolia** (Knight, Prot. 80); branches erect, pilose or villous; leaves 1–2 in. long, bipinnately divided in the upper $\frac{1}{2}$ or $\frac{3}{4}$, pilose or nearly glabrous; ultimate segments $\frac{1}{4}$–$\frac{1}{2}$ in. long, narrow, subacutely mucronate; heads solitary, terminal, long-pedunculate, $\frac{3}{4}$–1 in. long, about 1 in. in diam., globose, many-flowered; peduncle 1–3 in. long, pubescent, bearing ovate or lanceolate acutely acuminate glabrous bracts $\frac{3}{4}$–2 lin. long; floral bracts $2\frac{1}{2}$–$3\frac{1}{2}$ lin. long, ovate to spathulate-obovate, with a short recurved acumen, tomentose; flowers more or less straight in bud; perianth-tube 2 lin. long, glabrous; segments $4\frac{1}{2}$–5 lin. long,

spreadingly villous; limb 1–1½ lin. long, elliptic, subacute, villous; anthers oblong; hypogynous scales ¾ lin. long, linear, acute; ovary ½ lin. long, villous; style 4 lin. long, pubescent on the lower ⅔; stigma ¾ lin. long, oblong, obtuse, furrowed. *S. peduncularis, Knight, Prot.* 81? *S. pedunculata, R. Br. in Trans. Linn. Soc.* x. 119; *Roem. & Schultes, Syst. Veg.* iii. 367; *Meisn. in DC. Prodr.* xiv. 288. *S. sphærocephala, Steud. Nomencl. ed.* 2, ii. 571. *Protea sphærocephala, Houtt. Handl.* iv. 99, *t.* 19, *fig.* 1. *P. pedunculata, Lam. Ill.* i. 240. *P. glomerata, Andr. Bot. Rep. t.* 264.

COAST REGION: Tulbagh Div.; New Kloof, near Tulbagh, *Ludwig*! *Drège*! Roode Zand, *Niven*, 18! *Pappe*! *MacOwan*, 2837! *& in Herb. Norm. Austr.-Afr.*, 769! *Schlechter*, 9021! *Bolus*, 5241! *Burchell*, 994! Witzenberg Range, *Pappe*! Worcester Div.; Pinaars Kloof, *Burke*! *Zeyher*, 1470! Bainskloof, *Pappe*!

33. **S. hirsuta** (R. Br. in Trans. Linn. Soc. x. 120); branches villous when young, becoming long-pilose when older; leaves 1–1½ in. long, bipinnately divided in the upper ½ or ⅔, long-pilose when young, soon becoming quite glabrous; ultimate segments 2½–6 lin. long, cylindric, acutely mucronate; heads terminal, solitary, subsessile, 1–1¼ in. long, ¾–1 in. in diam.; peduncle up to 3 lin. long, covered with long lanceolate-linear very acute ciliate bracts; floral bracts ¼–½ in. long, lanceolate or ovate-lanceolate, acutely acuminate, long-villous or densely pilose; perianth-tube 1–2 lin. long; segments 4½–5 lin. long, spreadingly villous; limb 1–1⅓ lin. long, oblong-linear, subacute, villous; anthers ⅔–1 lin. long; hypogynous scales ½ lin. long, subulate; ovary villous; style 4–5 lin. long, swollen at the base, glabrous; stigma ¾–1 lin. long, subclavate, obtuse. *Roem. & Schultes, Syst. Veg.* iii. 368; *Meisn. in DC. Prodr.* xiv. 291. *Protea phylicoides, Thunb. Diss. Prot.* 19, *partly, as to spec.* α *and* β *of Herb.*; *Fl. Cap. ed. Schultes*, 124, *partly, as preceding*; *Murr. Syst. Veg. ed.* xiv. 137 (*excl. syn. Berg.*); *Willd. Sp. Pl.* i. 510, *partly. P. hirsuta, Poir. Encycl. Suppl.* iv. 572. *P. serraroides, Soland. ex Meisn. l.c., name only.*

SOUTH AFRICA: without locality, *Grey*!

COAST REGION: Cape Div.; Table Mountain, *Gamble*, 22187! Tyger Berg, *Pappe*! near Simonstown, *Pappe*! *Schlechter*, 1104! *Wolley-Dod*, 2925! *Jameson*! *Wright*, 626! Caledon Div.; Bot River, *Mund*, 43!

Steudel (Nomencl. ed. 2, ii. 571) reduced *Protea erecta*, Thunb. (Fl. Cap. i. 454; *S. erecta*, Meisn. in DC. Prodr. xiv. 298) to this species. On examining Thunberg's type specimen we find the plant was collected in New Holland. It is identical with *Isopogon anethifolius*, Knight.

34. **S. ventricosa** (Phillips & Hutchinson); stems ascending, much-branched; branches purplish, pubescent; leaves ¾–1 in. long, pinnately or rarely bipinnately divided in the upper half, glabrous, the younger long-pilose with very weak hairs; ultimate segments 3½–5 lin. long, filiform, furrowed on the upper surface, acutely mucronate; heads solitary, terminal, subsessile or very shortly

pedunculate, $\frac{1}{2}$–$\frac{3}{4}$ in. long, about $\frac{3}{4}$ in. in diam., many-flowered; peduncle 2–4 lin. long, pilose, bearing linear-lanceolate acute glabrous purple bracts 2–3 lin. long; floral bracts 3 lin. long, ovate-lanceolate, subacutely acuminate, concave, glabrous or very scantily pubescent, 5-nerved, ciliate; perianth-tube $\frac{3}{4}$ lin. long, subglobose, glabrous; segments 4$\frac{3}{4}$ lin. long, spathulate-linear, villous with spreading white hairs; limb 1 lin. long, lanceolate or lanceolate-elliptic, acute, villous; anthers $\frac{3}{4}$ lin. long; hypogynous scales $\frac{3}{4}$ lin. long, filiform; ovary $\frac{1}{2}$ lin. long, globose, villous; style 4 lin. long, cylindric, glabrous, globose at the base; stigma $\frac{3}{4}$ lin. long, cylindric, furrowed, obtuse.

COAST REGION: Bredasdorp Div.; Mountains near Koude River, *Schlechter*, 9604!

35. S. millefolia (Knight, Prot. 79); branches erect or ascending, pilose or villous; leaves $\frac{1}{2}$–$\frac{3}{4}$ in. long, bipinnately divided almost from the base, long-pilose when young, at length becoming nearly glabrous; ultimate segments 2–4 lin. long, narrowly cylindric, obtusely mucronate; heads very shortly pedunculate or subsessile, solitary at the apex of each shoot or lateral branchlet, $\frac{3}{4}$–1 in. long, about the same in diam., many-flowered; bracts 3 lin. long, ovate or obovate, acutely acuminate, villous; perianth-tube 1$\frac{1}{4}$ lin. long, glabrous, soon splitting to the base; segments 4$\frac{1}{4}$ lin. long, spreadingly villous; limb 1 lin. long, elliptic, subacute, long-villous, the posticous sometimes glabrous; anthers $\frac{3}{4}$ lin. long; hypogynous scales $\frac{1}{2}$ lin. long, filiform; ovary villous; style $\frac{1}{4}$ in. long, glabrous; stigma $\frac{2}{3}$ lin. long, subclavate, obtuse: fruit 2$\frac{2}{3}$ lin. long, oblong, villous when young. *S. abrotanifolia, Knight, Prot.* 79. *S. Stilbe, R. Br. in Trans. Linn. Soc.* x. 120; *Roem. & Schultes, Syst. Veg.* iii. 368; *Meisn. in DC. Prodr.* xiv. 289. *S.? triplicato-ternata, Roem. & Schultes, Syst. Veg.* iii. 378. *S.? pilosa, Roem. & Schultes, l.c.; Meisn. l.c.* 298. *S. Thunbergii, Endl. Gen. Suppl.* iv. ii. 79. *S. commutata, Endl. l.c. S. Andrewsii, Endl. l.c., partly. S. hirsuta, Berg. ex Meisn. l.c.* 290, *not of R. Br. S. Brownii, Meisn. l.c.* 290. *Protea villosa, Thunb. Prodr.* 187; *Fl. Cap. ed. Schultes,* 125: *R. Br. in Trans. Linn. Soc.* x. 220, *mentioned.* *P. triternata, Kenn. in Andr. Bot. Rep. t.* 337, *not of Thunb. P. Stilbe, Poir. Encycl. Suppl.* iv. 571; *Meisn. l.c. P. abrotanifolia, hirta, Andr. l.c. t.* 522.

SOUTH AFRICA: without locality, *Masson*! *Roxburgh*! *Mund*! *Ludwig*!
COAST REGION: Van Rhynsdorp Div.; top of Windhoek Mountains (Giftberg Range), *Niven*, 25! Clanwilliam Div.; Brakfontein, *Niven*, 24! Jakhals Vley, *Niven*, 26! between Berg Valley and Lange Valley, *Drège*! Nieuwoudtville, *Leipoldt in Herb. Bolus*, 9378! Pakhuis Pass, *Bolus*, 9081! Boontjes River, *Schlechter*, 8668! Piquetberg Div.; Pikeniers Kloof, *MacOwan*, 3275! & *in Herb. Austr.-Afr.*, 1949! Paarl Div.; between Mosselbanks River and Berg River, *Burchell*, 974! Cape Div.; Tyger Berg, *Niven*, 19! *Pappe*!

36. S. vallaris (Knight, Prot. 78); branches tomentose or pubescent, sometimes becoming glabrous below; leaves $\frac{3}{4}$–1$\frac{1}{2}$ in.

long, bipinnately divided in the upper $\frac{1}{2}$ or $\frac{2}{3}$, pilose when young, usually becoming glabrous with age; ultimate segments 2–6 lin. long, cylindric, acutely mucronate; heads terminal, solitary, sessile, 7–12 lin. long, about $\frac{3}{4}$–1 in. in diam., usually surrounded by the upper leaves; floral bracts $2\frac{1}{2}$–4 lin. long, ovate or lanceolate-ovate, sharply long- (rarely shortly) acuminate, villous; perianth-tube $1\frac{1}{4}$–$1\frac{3}{4}$ lin. long, subglobose, glabrous below; segments 3–$4\frac{3}{4}$ lin. long, spathulate-linear, spreadingly villous; limb 1–$1\frac{1}{4}$ lin. long, linear-oblong or lanceolate-linear, subacute, villous, that of the posticous segment glabrous; anthers $\frac{3}{4}$–1 lin. long, linear; hypogynous scales $\frac{2}{3}$–1 lin. long, filiform-linear, acutely acuminate; ovary $\frac{1}{2}$–$\frac{3}{4}$ lin. long, elliptic-ovate, covered with long hairs; style $2\frac{3}{4}$–$4\frac{1}{2}$ lin. long, cylindric, distinctly swollen at the base, glabrous; stigma $\frac{3}{4}$–1 lin. long, narrowly cylindric, subacute, furrowed. *S. villosa, R. Br. Trans. Linn. Soc.* x. 122; *Roem. & Schultes, Syst. Veg.* iii. 369; *Meisn. in DC. Prodr.* xiv. 292. *S. Niveni, Meisn. l.c.* 291, *as to Zeyher,* 3705, *not of R. Br. Protea villosa, Lam. Ill.* i. 240. *P. phylicoides, Thunb. Diss. Prot.* 19 (*partly*), *var.* γ *of Herb.; Fl. Cap. ed. Schultes,* 124, *partly; Poir. Encycl.* v. 659, *partly.*

SOUTH AFRICA : without locality, *Nelson* ! *Ludwig* ! *Grey* !
COAST REGION : Cape Div. ; near Constantia, *Brown* ! *Niven* ! *Ludwig* ! mountains near Simons Bay or False Bay, *Thunberg* ! *Niven* ! *Kirk* ! *Bolus*, 4687 ! Muizen Berg, *Zeyher*, 3705 ! *Pappe* ! *Bolus*, 4802 ! ridge beyond Smitwinkel Vley, *Wolley-Dod*, 2707 ! Kalk Bay hills, *Wolley-Dod*, 1010 !

37. **S. rostellaris** (Knight, Prot. 88); a small diffuse shrub; branches ascending, pilose when young, becoming glabrous; leaves 1–$1\frac{1}{4}$ in. long, bipinnately divided in the upper half, glabrous; ultimate segments $\frac{1}{4}$–$\frac{1}{2}$ in. long, terete, acutely mucronate; heads sessile, terminal, solitary, $\frac{1}{2}$–$\frac{3}{4}$ in. long, about $\frac{1}{2}$ in. in diam., many-flowered; involucral bracts about $\frac{1}{2}$ in. long, lanceolate, very acute, glabrous or nearly so, purplish; floral bracts 5 lin. long, lanceolate or ovate-lanceolate, acutely acuminate, villous; perianth-tube $1\frac{1}{2}$ lin. long, glabrous; segments $4\frac{1}{2}$ lin. long, villous with white spreading hairs; limb $1\frac{1}{3}$ lin. long, oblong-elliptic, subacute, the posticous one very minutely pubescent, the other three spreadingly long-villous; anthers 1 lin. long; hypogynous scales $1\frac{1}{4}$ lin. long, linear; ovary villous; style 5 lin. long, glabrous; stigma $\frac{3}{4}$ lin. long, cylindric. *S. Niveni, R. Br. in Trans. Linn. Soc.* x. 121; *Roem. & Schultes, Syst. Veg.* iii. 369; *Meisn. in DC. Prodr.* xiv. 291, *excl. Zeyher,* 3705. *S. plumosa, Meisn. l.c. Protea decumbens, Andr. Bot. Rep. t.* 349. *P. Niveni, Poir. Encycl. Suppl.* iv. 572.

SOUTH AFRICA : without locality, *Ludwig* !
COAST REGION : Caledon Div. ; on the Zwart Berg, *Niven* ! *Pappe* ! *Zeyher*, 3704 ! mountains near Caledon, *Bolus*, 9889 !

38. **S. cyanoides** (R. Br. in Trans. Linn. Soc. x. 117); branches tomentose or pubescent towards the apex, soon becoming glabrous; leaves 1–$2\frac{1}{2}$ in. long, pinnately or bipinnately divided in the upper

½ or ⅔, glabrous or the younger sometimes pubescent; ultimate segments ¼–½ in. long, terete, acutely mucronate; heads solitary, terminal, shortly pedunculate, about ¾ in. in diam., many-flowered; peduncle ¼–1 in. long, pilose or villous, bearing lanceolate acutely acuminate minutely pustulate glabrous purplish bracts 2–3 lin. long; involucral bracts 3–4½ lin. long, ovate or ovate-lanceolate, acutely acuminate, pilose or villous; floral bracts similar, keeled; flowers straight in bud; perianth-tube 1½ lin. long, glabrous; segments 4–5 lin. long, spreadingly villous; limb 1½–2 lin. long, oblong, villous, that of the posticous segment more shortly pubescent; hypogynous scales 1 lin. long, filiform; ovary villous; style 3–4 lin. long, glabrous; stigma 1 lin. long, subclavate, obtuse, furrowed. *Roem. & Schultes, Syst. Veg.* iii. 366 ; *Meisn. in DC. Prodr.* xiv. 290. *Leucadendron cyanoides, Linn. Sp. Pl. ed.* i. 93 ; *Berg. in Vet. Acad. Handl. Stockh.* 1766, 326 ; *Berg. Descr. Pl. Cap.* 27. *Protea cyanoides, Linn. Mant. alt.* 188 ; *Lam. Ill.* i. 239 ; *Poir. Encycl.* v. 658.

SOUTH AFRICA : without locality, *Pappe* !
COAST REGION : Tulbagh Div. ; Great Winterhoek, *Zeyher* ! between Bains Kloof and Mitchells Pass, *Bolus*, 5239 ! Worcester Div.; Dutoits Kloof, *Drège* ! Cape Div. ; Simons Bay, *Wright*, 630 ! 631 ! Cape Flats, *Pappe*, 10 ! *Bolus*, 4983 ! Orange Kloof, *Wolley-Dod*, 2708 ! mountains near Constantia, *Schlechter*, 1225 ! Stellenbosch Div. ; between Lowrys Pass and Jonkers Hoek, *Burchell*, 8319 ! between Tyger Berg and Simons Berg, *Drège*, 8066 ! Caledon Div. ; Zwart Berg, *Templeman in Herb. MacOwan* !

39. **S. florida** (Knight, Prot. 92) ; a shrub ; branches erect or ascending, purplish, glabrous ; leaves 1¾–2½ in. long, pinnately or bipinnately divided, rather broad at the base, glabrous ; ultimate segments very acute ; heads axillary and terminal, few together, pedunculate, 1¼–1½ in. long, about 1½ in. in diam. ; peduncles 1–3 in. long, glabrous, bearing large lanceolate to ovate-lanceolate very acutely acuminate glabrous (white?) bracts ; involucral bracts 1–1½ in. long, 4–7 lin. broad, lanceolate to obovate, acutely acuminate, membranous, white (?), becoming pinkish-yellow when dry ; floral bracts about ¾ in. long, linear or linear-lanceolate, tapered to a fine subulate apex, very long-ciliate ; perianth-tube 1½ lin. long, ellipsoid, glabrous ; segments 4½ lin. long, glabrous except on the limb ; limb 1½ lin. long, oblong-linear, thinly villous with very long hairs, that of the posticous segment glabrous or nearly so ; anthers 1¼ lin. long ; ovary pubescent, surrounded by long hairs ; style 3½ lin. long, with a globose swelling at the base, glabrous ; stigma 1 lin. long ; fruit oblong-ellipsoid, shortly beaked, rusty pilose. *R. Br. in Trans. Linn. Soc.* x. 126 ; *Roem. & Schultes, Syst. Veg.* iii. 372 ; *Meisn. in DC. Prodr.* xiv. 285. *Protea florida, Thunb. Diss. Prot.* 15, *t.* 1, *fig.* 1 ; *Murr. Syst. Veg. ed.* xiv. 136 ; *Lam. Ill.* i. 240 ; *Willd. Sp. Pl.* i. 506 ; *Poir. Encycl.* v. 662 ; *Thunb. Fl. Cap. ed. Schultes*, 121.

SOUTH AFRICA : without locality, *Wedgwood* !
COAST REGION : Paarl Div. ; French Hoek, *Thunberg* ! *Masson* ! *MacOwan*, *Herb. Norm. Austr.-Afr.*, 1524 ! *Kriel in Herb. Bolus*, 6335 !

40. S. æmula (R. Br. in Trans. Linn. Soc. x. 125); a shrub 2 ft. high; branches erect or ascending, pubescent or puberulous, rarely nearly glabrous; leaves 1½–2 in. long, pinnately or bipinnately divided in the upper ⅔, glabrous; ultimate segments subacutely mucronate; heads solitary on simple peduncles, usually several in a corymb at the apex of each shoot, ½–¾ in. long, ¾–1¼ in. in diam.; peduncles ¾–1¼ in. long, pubescent or glabrous, bearing numerous lanceolate to subulate glabrous shortly ciliate bracts 2–4 lin. long; involucral bracts ovate or ovate-lanceolate, acute, with a broad midrib and broad membranous margins, up to 7 lin. long, 2–4 lin. broad; floral bracts 4–5 lin. long, lanceolate, pilose or densely long-villous; flowers straight in bud; perianth-tube 1⅓ lin. long, glabrous; segments 3–3½ lin. long, spreadingly pilose; limb 1½–2 lin. long, narrowly elliptic, acute, densely villous, that of the posticous segment nearly glabrous; anthers 1–1½ lin. long; ovary beaked, densely villous; style 3–4 lin. long, glabrous; stigma 1½–2 lin. long, cylindric, obtuse, furrowed.   *Roem. & Schultes, Syst. Veg.* iii. 372; *Meisn. in DC. Prodr.* xiv. 292.   *S. furcellata, E. Meyer in Drège, Zwei Pfl. Documente,* 118, 221, *not of R. Br.*   *S. florida, var., E. Meyer, l.c.* 79, 81, 82, 221, *not of R. Br.*   *S. subcorymbosa, Meisn. l.c.* 285.   *Protea æmula, Poir. Encycl. Suppl.* iv. 573.

VAR. *β*, **heterophylla** (Hutchinson); leaves on the flowering shoots mostly simple or trifurcate.   *S. heterophylla, Meisn. l.c.* 284.

SOUTH AFRICA : without locality, *Brehm* ! Var. *β* : *Ludwig* !
COAST REGION : Worcester Div.; Dutoits Kloof, *Drège* ! Paarl Div.; Drakenstein Mountains, *Drège* ! French Hoek, *Roxburgh* ! *Schlechter*, 9232 ! Caledon Div.; mountains of Baviaans Kloof near Genadendaal, *Burchell*, 7674 ! 7824 ! 7859/2 ! near Grietjes Gat, *Bolus*, 4200 ! Shaws Mountain, *Galpin*, 4470 ! Onrust River, *Schlechter*, 9500 ! between Bot River and Zwart Berg, *Zeyher*, 3696 ! Donker Hoek, *Drège* ! *Grey* ! Var. *β* : Caledon Div.: Hermanus, *Galpin*, 4468 ! *Bolus*, 9828 ! mountains near Bot River, *Pappe* ! *Zeyher*, 3695 ! *Ludwig*, *Ecklon*. Swellendam Div.; near Swellendam, *Pappe* !

41. S. scariosa (R. Br. in Trans. Linn. Soc. x. 118); branches glabrous or pubescent; leaves 1–1½ in. long, bipinnately divided in the upper ½ or ⅓, glabrous; leaf-segments spreading, 3–5 lin. long, linear, acute; heads pedunculate, 8–9 lin. long, about 7 lin. in diam., 2–3-nate at the end of the branches; peduncle 10–12 lin. long, glabrous, bearing lanceolate-ovate subacutely acuminate glabrous bracts; floral bracts 6 lin. long, lanceolate, acutely acuminate, membranous, pilose, glandular; perianth-tube ¾–1 lin. long, ventricose, glabrous below; segments 4–4¼ lin. long, spathulate-linear, shortly hirsute; limb 1¼ lin. long, linear, acute, pubescent; anthers 1 lin. long; apical gland ⅛ lin. long, ovate, subacute; ovary ¾ lin. long, beaked, villous; style 3 lin. long, cylindric-filiform, glabrous; stigma ¾ lin. long, cylindric, obtuse, slightly swollen at the junction with the style.   *Roem. & Schultes, Syst. Veg.* iii. 367; *Meisn. in DC. Prodr.* xiv. 286.   *P. sphærocephala, Poir. Encycl.* v. 658.

SOUTH AFRICA : without locality, *Roxburgh* !
Protea coarctata, Thunb. Fl. Cap. ed. Schultes, 122, referred doubtfully to this species by Meisner, is an Australian plant and identical with *Petrophila pulchella*, R. Br.

42. **S. fœniculacea** (R. Br. in Trans. Linn. Soc. x. 122); branches erect, pilose with weak hairs; leaves 1–1½ in. long, bipinnately divided in the upper half, glabrous or rarely pilose with weak hairs, the ultimate segments 3–5 lin. long, narrow-cylindric, acutely mucronate; heads pedunculate, ½–¾ in. long, ½–¾ in. in diam., terminal and axillary, clustered at the ends of the branches, many-flowered; peduncles 3–4 lin. long, thinly pubescent, bearing lanceolate acutely acuminate glabrous usually ciliate bracts; floral bracts 2–3 lin. long, 1–1½ lin. broad, suborbicular, acutely and abruptly acuminate, glabrous, ciliate; flowers curved in bud; perianth-tube ¾ lin. long, pubescent; segments 3¾ lin. long, linear, hirsute with short adpressed hairs; limb 1 lin. long, elliptic, subacute, shortly villous; anthers ¾ lin. long; hypogynous scales ⅔ lin. long, filiform; ovary ¼ lin. long, villous; style 3½ lin. long, glabrous; stigma ⅔ lin. long, subclavate, obtuse. *Roem. & Schultes, Syst. Veg.* iii. 370; *Meisn. in DC. Prodr.* xiv. 290. *S. odorata, Sweet, Hort. Brit.* ed. i. 348? *P. abrotanifolia, odorata, Andr. Bot. Rep. t.* 545? *Protea fœniculacea, Poir. Encycl. Suppl.* iv. 572.

Coast Region : Cape Div. ; near Zeekoe Vley, *MacOwan, Herb. Norm. Austr.-Afr.*, 803 ! near Simons Town, *Pappe* !

43. **S. barbigera** (Knight, Prot. 90); branches erect, minutely puberulous or nearly glabrous; leaves 1½–2 in. long, bipinnately divided in the upper ½ or ⅔, glabrous; ultimate segments subterete, ½–¾ in. long, about ⅓ lin. thick, acute or subacute; heads 3–4 at the apex of each shoot, corymbose on separate peduncles, nearly 1 in. in diam., many-flowered; peduncles up to 1 in. long, glabrous or nearly so, furnished with numerous subulate glabrous bracts 3–4 lin. long; involucral bracts very numerous, about ½ in. long, linear, acute, the outer not ciliate and quite glabrous except for a tuft of hairs at the apex, a few of the inner very sparingly villous; floral bracts 3–4½ lin. long, narrowly lanceolate, very acute, long villous outside; flowers straight in bud; perianth-tube very short; segments about 5 lin. long, with an almost glabrous claw; limb 1½ lin. long, long-villous, the posticous one glabrous except at the tip; anthers 1¼ lin. long, linear; ovary villous; style 4½ lin. long, filiform, with a very narrow stigma. *S. parilis, Knight, Prot.* 91. *S. phylicoides, R. Br. in Trans. Linn. Soc.* x. 125 ; *Roem. & Schultes, Syst. Veg.* iii. 371 ; *Meisn. in DC. Prodr.* xiv. 292. *S. nitida and S. squarrosa, R. Br. l.c.* 124 ; *Roem. & Schultes, l.c.* 370, 371 ; *Meisn. l.c.* 292, 293. *S. eriocephala, Steud. Nomencl.* ed. 2, ii. 571. *Leucadendron phylicoides, Berg. in Vet. Acad. Handl. Stockh.* 1766, 328; *Berg. Descr. Pl. Cap.* 29. *L. Serraria, Burm. ex Meisn. l.c.* 292. *Protea abrotanifolia, Andr. Bot. Rep. t.* 507. *P. eriocephala, Roem. & Schultes, Syst. Veg.* iii. 379. *P. squarrosa, Poir. Encycl. Suppl* iv. 573. *P. glomerata, Thib. ex Meisn. l.c.,* not of *R. Br.*

South Africa : without locality, *Stanger* ! *Webb* ! *Ludwig* !
Coast Region : Stellenbosch Div. ; Lowrys Pass, *Schlechter*, 1136 ! Caledon Div. ; Houw Hoek Mountains, *Zeyher*, 3697 ! *Schlechter*, 5478 ! *MacOwan, Herb. Norm. Austr.-Afr.*, 771 ! *Galpin*, 4469. Zwart Berg and Houw Hoek, *Bowie* ! *Pappe* ! near Caledon, *Bolus* !

44. S. fucifolia (Knight, *Prot.* 81); branches more or less tomentose or pubescent ; leaves $\frac{1}{2}$–2$\frac{1}{4}$ in. long, bipinnately divided in the upper $\frac{1}{2}$–$\frac{3}{4}$, pilose, or young leaves villous, rarely glabrous ; leaf-segments filiform-cylindric, obtusely-mucronate, furrowed on the upper surface ; heads axillary and terminal, solitary, $\frac{1}{2}$–1 in. in diam., many-flowered ; peduncles $\frac{3}{4}$–1$\frac{1}{2}$ in. long, pubescent, bearing 2–5 linear bracts ; floral bracts 2$\frac{1}{2}$–3 lin. long, ovate or obovate, pointed, tomentose, sometimes pubescent and ciliate ; flowers curved in bud ; perianth-tube 1–1$\frac{1}{2}$ lin. long, glabrous ; segments 3$\frac{1}{2}$–4$\frac{1}{4}$ lin. long, spathulate-linear, villous with adpressed or spreading hairs ; limb $\frac{3}{4}$ lin. long, elliptic, acute, villous, shortly bearded ; anthers $\frac{1}{2}$ lin. long, oblong ; ovary $\frac{3}{4}$ lin. long, densely villous ; style 3–3$\frac{1}{2}$ lin. long, cylindric, sometimes swollen in the middle third, glabrous ; stigma $\frac{1}{4}$–$\frac{1}{3}$ lin. long, oblong, obtuse, furrowed ; fruits beaked, villous. *S. elevata, R. Br. in Trans. Linn. Soc. x.* 114 ; *Roem. & Schultes, Syst. Veg.* iii. 364 ; *Meisn. in DC. Prodr.* xiv. 287, *with vars. S. hirsuta, Buek in Drège, Zwei Pfl. Documente,* 221. *S. subumbellata, Buek in Drège, l.c. S. vestita, Buek in Drège, l.c.* 221 ; *Meisn. in DC. Prodr.* xiv. 288. *Protea elevata, Poir. Encycl. Suppl.* iv. 570.

SOUTH AFRICA : without locality, *Drège,* 8070a.
COAST REGION : Van Rhynsdorp Div. ; Gift Berg, *Drège* ! Clanwilliam Div. ; near Clanwilliam, *Mader,* 176 ! Kardouw Mountains, *Zeyher,* 1472 ! Blauw Berg, *Mund,* 34 ! *Schlechter,* 8461 ! Pakhuis Mountains, *Schlechter,* 10804 ! Lange Kloof, *Schlechter,* 8400 ! Storm Vley, near Wupperthal, *Leipoldt,* 485 ! Piquetberg Div. ; Piquet Berg, *Niven,* 17 ! *Drège,* 1476a !

45. S. cygnea (R. Br. in Trans. Linn. Soc. x. 113) ; branches prostrate, glabrous ; leaves 1–3 in. long, bipinnately divided in the upper $\frac{1}{2}$ or $\frac{2}{3}$, glabrous, sometimes scantily pilose when young ; segments 2$\frac{1}{2}$–8 lin. long, cylindric, obtusely mucronate, furrowed on the upper surface ; heads solitary, axillary, pedunculate, $\frac{1}{2}$–$\frac{3}{4}$ in. long, about $\frac{3}{4}$ in. in diam., many-flowered ; peduncle simple, 1–2 in. long, glabrous, bearing a few scattered ovate or ovate-lanceolate acute glabrous bracts ; floral bracts 1$\frac{3}{4}$–3 lin. long, ovate, caudate-acuminate, acute, glabrous or shortly ciliate ; flowers curved in bud ; perianth-tube very short, glabrous ; segments 5$\frac{1}{2}$ lin. long, shortly adpressed-pubescent outside ; limb $\frac{3}{4}$–1 lin. long, elliptic, subacute, adpressed-pubescent, that of the posticous segment nearly glabrous ; anthers sessile, $\frac{1}{2}$–$\frac{3}{4}$ lin. long, oblong or elliptic ; hypogynous scales $\frac{1}{2}$ lin. long, linear ; ovary $\frac{1}{2}$–1$\frac{1}{2}$ lin. long, ovoid, very densely villous ; style 4$\frac{1}{2}$–5$\frac{1}{2}$ lin. long, cylindric, curved above, glabrous ; stigma $\frac{1}{3}$–$\frac{1}{2}$ lin. long, oblong or oblong-ellipsoid, slightly furrowed. *Roem. & Schultes, Syst. Veg.* iii. 363 ; *Meisn. in DC. Prodr.* xiv. 286. *S. cyanea, E. Meyer in Drège, Zwei Pfl. Documente,* 78, 221. *S. colorata, Zeyh. ex Meisn. l.c. S. helvola, Steud. Nomencl. ed.* 2, ii. 400. *Protea cygnea, Poir. Encycl. Suppl.* iv. 569. *P. helvola, Lichtenst. ex Roem. & Schultes, Syst. Veg.* iii. 379, *fide Meisn. l.c.*

SOUTH AFRICA : without locality, *Hooker* !
COAST REGION : Tulbagh Div. ; Witzenberg Range, *Zeyher*, 1473 ! *Pappe*!
Great Winterhoek, *Pappe* ! *Bolus*, 5243 ! Mitchells Pass, *Bolus*, 5335 ! and in
Herb. Norm. Austr.-Afr. 381 ! *Schlechter*, 8952 ! New Kloof, *Drège*! Worcester
Div. ; Breede River Valley, *Bolus*, 2902 !

46. S. incrassata (Buek in Drège, Zwei Pfl. Documente, 98, 221); branches pilose with long weak hairs ; leaves 1–2 in. long, bipinnately divided in the upper half, the upper glabrous, the lower often pilose; segments 1½–5 lin. long, terete, with an obtuse callous apex, furrowed on the upper surface ; heads pedunculate, ¾–1 in. long, about 1 in. in diam., terminal, solitary, many - flowered ; peduncle 1 in. long, glabrous or scantily pubescent, bearing lanceolate acutely acuminate ciliate but otherwise glabrous bracts 1½–2½ lin. long ; floral bracts 2½–3 lin. long, ovate, acutely acuminate, glabrous or scantily pubescent ; perianth-tube 1½–2 lin. long, glabrous below ; segments 6–6½ lin. long, linear, hirsute or villous with adpressed hairs outside ; limb 1–1½ lin. long, oblong-lanceolate, acute, villous ; anthers ⅝–1 lin. long, oblong ; hypogynous scales 1–1¼ lin. long, filiform or linear ; ovary 1¼ lin. long, ovoid, villous ; style 5½ lin. long, cylindric, glabrous ; stigma ¾ lin. long, subcylindric, obtuse, furrowed, slightly bent at the junction with the style. *Meisn. in DC. Prodr.* xiv. 286.

COAST REGION : Paarl Div. ; near Paarl, *Drège*, 8065 !

47. S. Roxburghii (R. Br. in Trans. Linn. Soc. x. 130) ; branches erect, glabrous or pubescent ; leaves ½–¾ in. long, bipinnately divided in the upper half, pilose when young, at length quite glabrous ; ultimate segments filiform, acutely mucronate ; heads axillary and terminal, several on separate peduncles at the ends of the branches, about ⅓ in. in diam., many-flowered ; peduncles very short, clothed with long subulate ciliate glabrous bracts about 4 lin. long ; floral bracts 3–4 lin. long, ovate, long and acutely acuminate, villous in the lower part ; perianth-tube ¾ lin. long, glabrous ; segments 3¼ lin. long, subadpressed-villous ; limb 1 lin. long, oblong, subacute, villous ; anthers ⅔ lin. long ; ovary ½ lin. long, villous ; style 2½ lin. long, glabrous ; stigma ¾ lin. long, cylindric. *Roem. & Schultes, Syst. Veg.* iii. 375 ; *Meisn. in DC. Prodr.* xiv. 295. *Protea Roxburgii, Poir. Encycl. Suppl.* iv. 574. *P. triternata, Thib. ex Meisn. l.c.*

COAST REGION : Piquetberg Div.; Twenty-four Rivers, *Drège*! Malmesbury Div. ; near Paarde Berg, *Roxburgh* ! Zwartland, Riebeek's Kasteel and Paarde Berg, *Zeyher*, 1475 ! Worcester Div. ; near Wellington, *Pappe*! Paarl Div. ; near Berg River, *Pappe*!

48. S. callosa (Knight, Prot. 80) ; a small more or less decumbent shrub ; branches pilose or subvillous, at length becoming nearly glabrous ; leaves ½–1 in. long, bipinnately divided in the upper half, long-pilose when young, becoming glabrous or nearly so ;

ultimate segments 1–2½ lin. long, cylindric, obtuse; heads terminal and solitary, 1–1¼ lin. long, about 1 in. in diam., many-flowered; peduncles 6–8 lin. long, pilose; floral bracts 3 lin. long, ovate, acutely acuminate, villous; flowers straight in bud; perianth-tube 2 lin. long, pubescent or glabrous; segments 5 lin. long, spreadingly villous; limb 1 lin. long, oblong, villous; anthers ¾–1 lin. long; ovary villous; style 5 lin. long, glabrous; stigma ⅔ lin. long, subclavate, obtuse; fruit 1½–2 lin. long, ellipsoid, beaked, villous. *S. scoparia, R. Br. in Trans. Linn. Soc.* x. 119; *Roem. & Schultes, Syst. Veg.* iii. 368; *Meisn. in DC. Prodr.* xiv. 289. *S. Dregei, Meisn. l.c. Protea scoparia, Poir. Encycl. Suppl.* iv. 571.

COAST REGION : Malmesbury Div.; near Moorreesburg, *Bolus,* 9979! Tulbagh Div.; Winterhoek Mountain, *Niven,* 8! Klein Berg River, *Niven*! Paarl Div.; near the Berg River, *Zeyher,* 3698!

Imperfectly known species.

49. **S. Zeyheri** (Meisn. in DC. Prodr. xiv. 297); branches glabrous; leaves erect, bipinnatifid, glabrous; segments spreading, slender, acutely mucronate; corymb exceeding the leaves; branches hairy, simple or 2–3-branched at the apex; heads globose, many-flowered; bracts oblong-lanceolate, glabrous, spreading; flowers adpressed - pubescent; limb glabrous; style exserted; stigma subcapitate.

COAST REGION : Caledon Div.; by the Zondereinde River near Appels Kraal, *Zeyher.*

We have not seen a specimen of this species, but it is evidently closely allied to *S. meisneriana,* Schlechter.

50. **S. Bergii** (R. Br. in Trans. Linn. Soc. x. 220); stem woody; branches terete, glabrous; branches simple, erect, glabrous; leaves much divided; segments subulate, acute, grooved above; heads subglobose, about 1 in. long, terminal, slightly pedunculate; bracts wedge-shaped, acuminate from a truncate base, villous, the lower one glabrous; flowers silky-villous, with the posticous limb glabrous; style filiform; stigma obtuse. *Meisn. in DC. Prodr.* xiv. 299. *Leucadendron sphærocephalum, Berg. Descr. Pl. Cap.* 26.

SOUTH AFRICA : without locality, *Bergius.*

51. **S. colorata** (Buek ex Drège, Zwei Pfl. Documente, 109, 221, name only).

COAST REGION : Clanwilliam Div.; between Berg Valley and Lange Valley, *Drège.*

52. **S. anemonefolia** (Knight, Prot. 83); a decumbent plant; leaves 1–3 in. long, 3-pinnatifid from very near their middle, pubescent; peduncles 1–3, a little longer than the head; style bowed from the base.

COAST REGION: Paarl Div. ; Drakenstein Mountains, *Niven.*

This and the following species described by Knight are known to us only from his descriptions. The type specimens were either not preserved or they have been destroyed ; we have not been able to trace them.

53. S. chlamydiflora (Knight, Prot. 91) ; stem 3–4 ft. high, finely cottony ; leaves 2–2½ in. long, 2-pinnatifid from below their middle ; heads 1–3, scarcely shorter than the peduncle ; bracts externally pubescent.

COAST REGION : Paarl Div. ; French Hoek, *Niven.*

54. S. collina (Knight, Prot. 86) ; a weak decumbent shrub ; leaves 1½–2 in. long, incurved erect, 2-pinnatifid from their middle, glossy ; panicle short, broadly conical ; bracts long, wedge-shaped, thinly cottony at the base ; style a little curved.

COAST REGION : Cape Div. ; Table Mountain, *Niven.*

55. S. concinna (Knight, Prot. 88) ; a slender shrub ; branches weak ; leaves 7–11 lin. long, 3-fid, pinnatifid and 2-pinnatifid from below their middle, almost glossy ; heads nearly sessile, higher than the leaves ; bracts linear-attenuated, all thinly fringed.

COAST REGION : Tulbagh Div. ; Tulbagh Waterfall, *Niven.*

56. S. delphiniifolia (Knight, Prot. 82) ; stem prostrate ; leaves 1–1½ in. long, 3–5-fid from below their middle, pubescent ; heads 3–7, as long as the peduncles ; stigma very broad at the top.

COAST REGION : Paarl Div. ; mountains near Paarl, *Niven.*

57. S. elumbis (Knight, Prot. 83) ; leaves 1½–3 in. long, 3-pinnatifid from below their middle, acute, smooth ; heads panicled ; bracts glossy, their margins hardly scarious ; stigma cylindrical.

COAST REGION : Stellenbosch Div. ; Hottentots Holland Mountains, *Niven.*

58. S. fallax (Knight, Prot. 90) ; stem 2 ft. high ; leaves 5–7 lin. long, closely 2-pinnatifid from below their middle, glossy ; head nearly sessile ; bracts fringed ; petals bearded below the limb, besprinkled with a great many glands.

COAST REGION : Paarl Div. ; near Paarl, *Niven.*

59. S. foliosa (Knight, Prot. 89) ; a subdecumbent shrub with rigid branches, seldom more than 1 ft. high ; leaves 4–6 lin. long, close, 3-fid from below their middle, glossy ; head nearly sessile ; bracts thickly cottony ; petals smooth below the limb.

COAST REGION : Clanwilliam Div. ; Blaauw Berg, *Niven.*

This is apparently closely allied to *S. trilopha,* Knight.

60. S. frondosa (Knight, Prot. 85) ; stem 1–2 ft. high ; leaves at the bottom of the branches small and imperfect, so that they appear clustered near the flowers, 1–1½ in. long, 2-pinnatifid from below their middle, thinly silky ; head generally solitary, peduncled, as high as the leaves ; bracts recurved, narrowly wedge-shaped, hairy ; style straight.

SOUTH AFRICA : without locality, *Niven.*

61. S. gremiiflora (Knight, Prot. 88) ; a small bush ; leaves 1–1¼ in. long, 2-pinnatifid from below their middle, smooth ; heads 1–3, clustered, lower than the leaves ; bracts cottony, especially the upper ones.

COAST REGION : Malmesbury Div. ; Zwartland, *Niven.*

62. S. montana (Knight, Prot. 80) ; stem 5–6 ft. high ; leaves 1–1½ in. long, closely 3-pinnatifid from their base, pubescent ; peduncles 1–3, cottony ; bracts suddenly smooth above their base externally.

COAST REGION : Paarl Div. ; mountains near the Brede River, *Niven.*

63. S. pulchella (Knight, Prot. 89) ; leaves ½–¾ in. long, trifid from the middle, rarely pinnatifid, glabrous ; heads subsessile ; bracts long, linear, attenuated, hirsute ; anthers obtuse.

COAST REGION : Cape Div. ; Fish Hoek, *Niven ?*
Knight quotes *Protea cyanoides,* Thunb. (*Serruria cyanoides,* R. Br.), as a synonym of this species, but his description does not agree with Thunberg's plant.

64. S. quinquemestris (Knight, Prot. 87) ; stem 3–4 ft. high, very branching ; leaves 6–8 lin. long, closely 2-pinnatifid from below their middle, slightly pubescent ; heads 3–5, clustered ; bracts all over very closely cottony ; stigma large

COAST REGION : Malmesbury Div. ; Paarde Berg, *Niven.*

65. S. rangiferina (Knight, Prot. 86) ; a low shrub, about 2 ft. high ; leaves 1–1½ in. long, recurved, thinly 2-pinnatifid from below their middle, somewhat pubescent ; panicle short, broadly pyramidal ; bracts long, wedge-shaped, silky-cottony at their base ; style exceedingly bowed.

COAST REGION : Paarl Div. ; near the Brede River, *Niven.*

66. S. zanthophylla (Knight, Prot. 86) ; leaves 1½–2 in. long, 2-pinnatifid almost from their middle, glandular, thinly silky ; heads clustered, nearly sessile ; bracts broad, a little incurved, rhomb-wedge-shaped, almost smooth ; style reclined.

COAST REGION : Tulbagh Div. ; Tulbagh Kloof, *Niven.*

XI. SPATALLA, Salisb.

Flowers hermaphrodite, slightly zygomorphic. *Perianth* cylindric in bud with an ellipsoid recurved limb, 4-partite to below the middle, rarely only to the middle, always with a distinct tube ; segments differentiated into claw and limb, the posticous (adaxial) one larger and thicker than the 3 anticous, all more or less equally hairy ; limbs often slightly recurved or rarely the claws spirally twisted, obtuse or subacute, rarely long-acuminate. *Stamens* 4 ; anthers broadly ovoid or rounded, sessile or subsessile, inserted at the base of the limb ; connective always slightly produced into a small globular point. *Hypogynous scales* 4, free, usually linear or subulate. *Ovary* sessile or very shortly stipitate, hairy ; style slender, terete, often slightly bent towards the apex, subpersistent, glabrous ; stigma very small, situated in the middle of a flat or concave oblique usually obovate disk. *Ovule* 1, laterally attached. *Fruit* a pubescent brown shortly stipitate ovoid or ellipsoid nut with a thin pericarp.

Usually small erect or spreading shrubs ; leaves often crowded and numerous, ericoid or acicular, entire, often mucronate, mostly terete or rarely flat on the upper side, straight or falcate, incurved or at length recurved, mostly thinly pilose when young, at length becoming glabrous or nearly so ; flowers in 1-flowered or few-flowered (3–4) involucres arranged in terminal sessile or pedunculate spikes or racemes ; involucres calycoid, small, rarely membranous, when 1-flowered bilabiate, posticous lip entire, anticous bifid or bilobed, trifid or trilobed or tripartite to near the base, each tooth opposite a flower, when 3–4-flowered then subregularly tripartite and often with a second smaller involucre containing a rudimentary flower ; bracts subtending the secondary peduncles or the involucres solitary, usually linear or lanceolate ; perianth villous or tomentose.

DISTRIB. Species 21, confined to the south-western coast region of Cape Colony, excepting 1 species which occurs in the Prince Albert Division.

*Involucres 1-flowered, not containing a second rudi-
        mentary involucre :
    Involucres more or less membranous, nearly glabrous
        outside, midrib of the anticous lip with a line of
        dense hairs up the inside ; three anticous perianth-
        segments at length spirally coiled and reflexed
        below the limb, the latter long-acuminate   ... (1) squamata.

    Involucres not or scarcely membranous, usually very
        hairy outside ; midrib of anticous lip glabrous
        inside (except in 2, *ericoides*) ; three anticous
        perianth-segments not or rarely coiled, the limb
        obtuse and not acuminate.
    †Anticous lip of involucre 2–3-fid or 2–3-lobed to about
        the middle or rarely entire :
    Involucres sessile or on stalks less than ½ lin. long :
        Leaves about ¼ in. long, concave and pilose on the
            upper surface, glabrous and convex below ;
            midrib of anticous lip of involucre hairy
            inside   ...   . (2) ericoides.

Leaves $\frac{1}{2}$–$\frac{3}{4}$ in. long, subterete with a very narrow
   groove on the upper side, silky-pilose except
   when quite old ; midrib of anticous lip of
   involucre glabrous inside   ...   . (3) **sericea.**

Involucres on stalks 1 lin. long or more, but rarely
   $\frac{3}{4}$ lin. long :
Anticous lip of involucre entire or bifid :
   Leaves permanently long-pilose ; racemes sessile (4) **mollis.**

   Leaves soon glabrous ; racemes pedunculate ... (7) **Galpinii.**

Anticous lip of involucre trifid or trilobed :
   Leaves more or less strongly incurved ; racemes
      pedunculate :
      Leaves 1$\frac{1}{4}$–1$\frac{1}{2}$ in. long; subtending bracts
        3–4 lin. long ; anticous lip of involucre
        3-lobed nearly to the middle ... ... (5) **longifolia.**

      Leaves $\frac{1}{2}$–1$\frac{1}{4}$ in. long; subtending bracts
        about 1$\frac{1}{2}$ lin. long ; anticous lip of
        involucre shortly 3-toothed :
        Leaves thicker in the upper part, nearly
          $\frac{1}{2}$ lin. in diam., strongly curved ... (6) **curvifolia.**

        Leaves slender, not or only slightly thicker
          in the upper part, about $\frac{1}{4}$ lin. in diam.,
          slightly curved ... ... (7) **Galpinii.**

   Leaves straight or only slightly incurved ;
      racemes sessile or if shortly pedunculate
      then lax-flowered :
      Racemes lax-flowered ; subtending bracts
        shorter than the secondary peduncles ... (8) **gracilis.**

      Racemes rather dense-flowered ; subtending
        bracts longer than the secondary
        peduncles :
        Leaves $\frac{1}{2}$ in. long, rather lax ; bracts less
          than 2 lin. long... ... ... (9) **brachyloba.**

        Leaves 1–1$\frac{1}{2}$ in. long, rather crowded ;
          bracts about 4 lin. long ... ... (10) **cylindrica.**

††Anticous lip of involucre 3-partite to the base or
   nearly to the base :
   Branches quite glabrous ; perianth-tube as long as
      the claws ... ... (11) **colorata.**

   Branches silky-pubescent or villous ; perianth-tube
      usually much shorter than the claws :
   Involucres on stalks 1 lin. long or more :
      Leaves spreading almost at right angles,
        arcuate, at length partially recurved ;
        racemes clustered and more or less
        pedunculate at the apex of each shoot ... (12) **bombycina.**

      Leaves more or less erect or ascending, not or
        rarely slightly recurved ; racemes solitary
        and sessile at the apex of each shoot :
        Leaves $\frac{1}{2}$–$\frac{3}{4}$ in. long ; bracts about 4 lin.
          long, usually exceeding the involucre (13) **parilis.**

        Leaves $\frac{1}{3}$–$\frac{1}{2}$ in. long ; bracts 2–3 lin. long,
          reaching to the top of the involucre ... (14) **Bolusii.**

Involucres sessile or subsessile :
Perianth-tube nearly as long as the claws of
the segments, claws glabrous in the
lower half except on the margin...   ... (15) **prolifera.**

Perianth-tube much shorter than the claws,
the latter tomentose outside :
Branches of previous years' growth leafless ;
leaves on the young shoots crowded,
slender, scarcely ¼ lin. in diam. ; hairs
on the perianth-limb slightly longer
than those on the claw   ...   ... (16) **Burchellii.**

Branches of previous years' growth leafy ;
leaves lax, stouter, nearly ½ lin. in
diam. ; hairs on the perianth-limb
much longer than those on the claw ... (17) **barbigera.**

**Involucres 3–4-flowered, usually containing a second
smaller subsessile involucre with a rudimentary
flower :
Leaves 3–4 lin. long, about ⅓ lin. thick, stiff ; racemes
few-flowered ...   . (18) **Wallichii.**

Leaves 6–12 lin. long, about ¼ lin. thick, often slender ;
racemes usually many-flowered :
Leaves nearly straight, abruptly mucronate...   ... (19) **mucronifolia.**

Leaves strongly falcate, gradually mucronate :
Involucres on peduncles ¾–1¾ lin. long ; perianth-
segments straight in the open flower ; fruits
stipitate ...   . (20) **procera.**

Involucres subsessile ; three perianth-segments
spirally twisted in the open flower ; fruits not
stipitate ...   . (21) **thyrsiflora.**

1. **S. squamata** (Meisn. in DC. Prodr. xiv. 310) ; branches thinly
pilose or nearly glabrous , leaves ¼–⅓ in. long, scarcely ⅓ lin. broad,
linear, acute, glabrous or scantily pilose ; involucres 1-flowered,
arranged in sessile solitary or 2–3 terminal spikes ; spikes ½–1¼ in.
long ; subtending bracts 2½ lin. long, linear or lanceolate, subacutely
acuminate, membranous or chaffy, glabrous or nearly so outside,
ciliate ; involucre 2½–3 lin. long, coloured, bilabiate, more or less
membranous or chaffy, thinly pilose outside or nearly glabrous ;
upper lip ovate, entire, lower 3-toothed, 3-nerved, with a dense line
of hairs on the middle nerve within, teeth about ½ lin. long, ovate,
subacute ; perianth-tube ½ lin. long, glabrous ; segments 3¼ lin. long,
the 3 anticous twisting spirally and reflexed below the limb, the
posticous one straight ; claws shortly pubescent with crisped hairs ;
limb ¾ lin. long, ovate with a long acuminate fleshy appendage at
the apex, villous outside ; anthers ⅓ lin. long ; hypogynous scales 3,
⅓ lin. long, ovate-lanceolate, fleshy, glabrous ; ovary ½ lin. long ;
style 3¼ lin. long, subterete ; stigmatic disk ½ lin. long, obovate,
with a small central conical stigma.

SOUTH AFRICA : without locality, *Thom,* 164 ! *Ludwig* !

COAST REGION: Bredasdorp Div.: Koude River, *Schlechter*, 9612! near Elim, *Bolus*, 7666! 7867!

Meisner places this species amongst the set with 3–4-flowered involucres. His type however is very imperfect and almost devoid of flowers, but the specimens quoted above, which seem to be the same, all have 1-flowered involucres.

2. **S. ericoides** (Phillips in Kew Bulletin, 1910, 334); branches thinly adpressed-pilose ; leaves 3–4 lin. long, erect, ericoid, subacute, thinly pilose and concave on the upper surface, glabrous and convex below ; involucres 1-flowered, arranged in solitary sessile terminal spikes ; spikes cylindric, about 1 in. long ; subtending bracts 2 lin. long, ½ lin. broad, lanceolate, acute, membranous, long adpressed-villous outside ; involucre about 2 lin. long, bilabiate, silky-villous outside, with a ring of long hairs at the base within and a line of hairs extending from the base up the inside of the midrib of the anticous lip; posticous lip ovate, entire, anticous 3-toothed, teeth ovate, subacute; perianth-tube very short, glabrous; segments 3–3½ lin. long, linear-spathulate ; claw shortly tomentellous ; limb ovate, obtuse, villous outside ; anthers sessile ; hypogynous scales ½ lin. long, linear ; ovary villous ; style 2¼ lin. long, subterete, glabrous ; stigmatic disk ⅓ lin. long, obovate, flat on the face.

SOUTH AFRICA : without locality or collector's name in the Capetown and Kew Herbaria !

3. **S. sericea** (R. Br. in Trans. Linn. Soc. x. 147); branches adpressed-pilose ; leaves ½–¾ in. long, terete except for a single very narrow groove on the upper side, silky-pilose except when old ; involucres 1-flowered, arranged in sessile solitary terminal spikes ; spikes ½–1 in. long, ovate-lanceolate ; subtending bracts 2½ lin. long, linear, acute, pilose, equalling or slightly exceeding the involucre ; involucre about 1¾ lin. long, bilabiate, pilose outside, glabrous within except for a ring of hairs at the base ; upper lip lanceolate, entire, lower trilobed to near the middle, lobes lanceolate, acute ; perianth-tube ¾ lin. long, glabrous ; segments 2½ lin. long, linear-spathulate ; claw tomentose ; limb ½ lin. long, ovate, obtuse, villous outside ; anthers sessile ; hypogynous scales ¾ lin. long, linear ; ovary villous ; style 2¼ lin. long, subterete, glabrous ; stigmatic disk ¼ lin. long, obovate, with a minute central conical stigma. *Roem. & Schultes, Syst. Veg.* iii. 394; *Meisn. in DC. Prodr.* xiv. 308, *excl. syn. E. Meyer.* S. *ramulosa, Sieber ex Meisn. l.c.,* name only. S. *polystachya, Zeyher ex Meisn. l.c.,* name only, not of *R. Br. Protea sericifolia, Poir. Encycl. Suppl.* iv. 578.

SOUTH AFRICA : without locality, *Roxburgh*! *Sieber*, 471!
COAST REGION : George Div. ; north of Cradock Berg, *Thom*, 505 ! Uniondale Div. ; Lange Kloof, *Mund* !

4. **S. mollis** (R. Br. in Trans. Linn. Soc. x. 144); branches long-pilose ; leaves ½–¾ in. long, terete, subacutely apiculate, minutely punctate, long and rather thinly pilose, at length becoming nearly

glabrous; heads 1-flowered, in sessile terminal solitary racemes; racemes $\frac{3}{4}$–1 in. long; subtending bracts $1\frac{1}{2}$ lin. long, linear, acute, ciliate; secondary peduncles 1 lin. long, pilose; involucre $1\frac{1}{4}$ lin. long, bilabiate, densely pilose outside, long-setose within at the base but otherwise glabrous; upper lip entire, ovate, lower entire or bifid, teeth ovate, acute; perianth-tube $\frac{3}{4}$ lin. long, glabrous; segments $1\frac{1}{2}$ lin. long, linear-spathulate, densely tomentose; limb $\frac{1}{3}$ lin. long, ovate, subobtuse, densely villous outside; anthers sessile, $\frac{1}{4}$ lin. long, oblong; hypogynous scales $\frac{1}{4}$ lin. long, linear; ovary $\frac{1}{4}$ lin. long, shortly pilose; style $1\frac{3}{4}$ lin. long, cylindric, slightly pubescent for a short distance at the base; stigmatic disk $\frac{1}{4}$ lin. long, lateral, obovate, with a small conical swelling in the middle. *Roem. & Schultes, Syst. Veg.* iii. 392; *Meisn. in DC. Prodr.* xiv. 306. *S. pilosa, Phillips in Kew Bulletin,* 1910, 335. *Protea mollis, Poir. Encycl. Suppl.* iv. 577 *not of* 567.

SOUTH AFRICA: without locality, *Roxburgh*!
COAST REGION: Stellenbosch Div.; Hottentots Holland Mountains, *Zeyher*, 3720 partly!

There are several specimens on the type sheet at the British Museum, and the lower lip of the involucre is sometimes entire and bifid on the same inflorescence..

5. **S. longifolia** (Knight, Prot. 77); a shrub 4–7 ft. high; branches permanently adpressed-pubescent with whitish hairs; leaves $1\frac{1}{4}$–$1\frac{1}{2}$ in. long, $\frac{1}{3}$ lin. thick, incurved, terete, acutely apiculate, slightly narrowed to the base, narrowly channelled on the upper side, pubescent when young, at length becoming glabrous; heads 1-flowered, in pedunculate solitary or 2 terminal racemes; racemes cylindric, $1\frac{1}{2}$–2 in. long, over $\frac{1}{2}$ in. in diam.; peduncle about 1 in. long, silky-pubescent, bearing a few subulate pubescent bracts about $1\frac{1}{4}$ lin. long; subtending bracts 3–4 lin. long, foliaceous, subacutely apiculate, silky-pilose; secondary peduncles about half as long as the bracts, adpressed-pubescent; involucre $1\frac{1}{2}$ lin. long, bilabiate, silky-pilose outside; upper lip ovate-lanceolate, acute, lower trilobed to about the middle; perianth-tube $\frac{3}{4}$ lin. long, glabrous towards the base; segments 2–$2\frac{1}{2}$ lin. long, linear-spathulate, tomentose; limb $\frac{1}{2}$ lin. long, ovate, subacute, villous; anthers sessile, $\frac{1}{4}$ lin. long; ovary $\frac{3}{4}$ lin. long, subglobose, densely villous; style 2 lin. long, cylindrical; stigmatic disk $\frac{1}{2}$ lin. long, obovate. *S. nivea, R. Br. in Trans. Linn. Soc.* x. 145; *Roem. & Schultes, Syst. Veg.* iii. 392; *Meisn. in DC. Prodr.* xiv. 307. *S. bracteata, R. Br. l.c.* 146; *Roem. & Schultes, l.c.* 394; *Meisn. l.c.* 308. *Protea nivea, Poir. Encycl. Suppl.* iv. 578. *P. bracteolaris, Poir. l.c., excl. syn. Linn.*

SOUTH AFRICA: without locality, *Niven*! *Masson*!
COAST REGION: Paarl Div.; French Hoek, *Niven*, 37!

6. **S. curvifolia** (Knight, Prot. 77); a small shrub up to 3 ft. high; branches silky-pubescent when young, soon becoming glabrous; leaves $\frac{1}{2}$–$1\frac{1}{4}$ in. long, $\frac{1}{3}$–$\frac{1}{2}$ lin. thick, strongly incurved,

acicular, subobtusely apiculate, narrowed to the base, pilose or villous when young, soon becoming glabrous; heads 1-flowered, in pedunculate solitary or 2–3 terminal racemes; racemes cylindric, 1–2½ in. long, nearly ½ in. in diam.; peduncle ¾–1¾ in. long, adpressed-pubescent, with a few subulate subacute puberulous bracts about 1 lin. long; subtending bracts about 1½ lin. long, linear, thinly pubescent; secondary peduncles slightly shorter than the bracts, adpressed-pubescent; involucre about 1½ lin. long, bilabiate, densely adpressed-pubescent; upper lip ovate, entire, lower shortly 3-toothed; perianth-tube ⅔ lin. long, glabrous; segments 2–2⅓ lin. long, linear-spathulate, densely villous; limb about ⅓ lin. long, ovate, subobtuse, villous; anthers subsessile, ⅓ lin. long; hypogynous scales ⅔ lin. long, linear; ovary globose, densely villous; style 2 lin. long, glabrous; stigmatic disk ⅓ lin. long, obovate, with a minute conical swelling in the middle; fruit 2 lin. long, cylindric, obtuse, tomentose. *S. pedunculata, R. Br. in Trans. Linn. Soc.* x. 144; *Roem. & Schultes, Syst. Veg.* iii. 392; *Meisn. in DC. Prodr.* xiv. 306. *S. ramulosa, R. Br. l.c.* 145, *partly*; *Roem. & Schultes, l.c.* 393, *partly*; *Meisn. l.c.* 307, *partly*. *S. abietina, Roem. & Schultes, l.c.* 397, *fide Meisn. l.c.* 307. *Protea racemosa, Linn. Herb. ex Meisn. l.c.* 307, *not of Linn. Mant.* *P. abietina, Licht. ex Roem. & Schultes, l.c.* 397; *Willd. ex Meisn. l.c.* 307. *P. pedunculata, Poir. Encycl. Suppl.* iv. 578.

SOUTH AFRICA: without locality, *Bowie*!

COAST REGION: Cape Div.: Table Mountain, *Schlechter*, 180a! Stellenbosch Div.; Hottentots Holland, *Niven*, 36! *Zeyher*, 3721, partly! *Ludwig*! Caledon Div.; between Palmiet River and Lowrys Pass, *Burchell*, 8179! Little Houw Hoek, *Roxburgh*! Baviaans Kloof, *Lichtenstein*, 112; Zwart Berg, near Caledon, *Pappe*!

7. **S. Galpinii** (Phillips in Kew Bulletin, 1910, 334); branches very shortly pubescent, at length becoming glabrous, reddish-purple; leaves slender, slightly incurved, ½–¾ in. long, about ¼ lin. in diam., subterete, acutely mucronate, narrowly channelled on the upper side, adpressed-pilose when young, at length becoming quite glabrous; involucres 1-flowered, arranged in shortly pedunculate solitary or paired terminal racemes, the latter dense-flowered, 1–1½ in. long, conical; peduncle about ½ in. long, shortly and silky-pubescent; subtending bracts about 2½ lin. long, linear, subacute, ciliate towards the base, exceeding or subequalling the secondary peduncles, the latter 2 lin. long, silky-pubescent; involucre 1¼ lin. long, bilabiate, adpressed-pilose outside; upper lip entire, ovate-lanceolate, lower 3-toothed (2-toothed in *Galpin*, 4485), teeth ovate, acute; perianth-tube ⅔ lin. long, glabrous; segments 2½ lin. long, claw more shortly villous than the limb, the latter ½ lin. long, ovate-lanceolate, subacute; anthers sessile, rounded; hypogynous scales ⅔ lin. long, linear; ovary densely villous; style 1¾ lin. long, subterete; stigmatic disk ½ lin. long, obovate.

SOUTH AFRICA: without locality, *Thom*, 939!

COAST REGION : Stellenbosch Div. ; Hottentots Holland Mountains, *Pappe* !
Caledon Div. ; Klein River, *Schlechter*, 7608 ! Hermanus, *Galpin*, 4485 ! Bredasdorp Div. ; near Elim, *Bolus*, 7869 ! 8594 !

8. S. gracilis (Knight, Prot. 77) ; branches very shortly
adpressed-pubescent or glabrous ; leaves $\frac{3}{4}$–$1\frac{1}{2}$ in.
long, acicular,
terete, acute or subacute, thinly adpressed-pubescent when young,
soon becoming quite glabrous ; heads 1-flowered, in lax shortly
pedunculate terminal racemes ; racemes $1\frac{1}{4}$–2 in. long ; bracts about
1 lin. long, subulate, adpressed-pubescent ; secondary peduncles
about twice as long as the bracts, shortly pubescent ; involucre
1 lin. long, bilabiate, silky-pubescent outside ; upper lip entire, ovate,
lower lip shortly 3-toothed, teeth more or less triangular ; perianth-
tube $\frac{3}{4}$ lin. long, glabrous in the lower part ; segments $1\frac{1}{2}$ lin. long,
linear-spathulate, densely whitish-tomentose ; limb $\frac{2}{3}$ lin. long, ovate,
densely tomentose ; anthers sessile, $\frac{1}{4}$ lin. long ; hypogynous scales
filiform ; ovary $\frac{3}{4}$ lin. long, globose, densely villous ; style $1\frac{1}{4}$ lin.
long, glabrous ; stigmatic disk $\frac{1}{3}$ lin. long, obovate ; fruit 2 lin. long,
cylindric, densely pubescent.   *S. ramulosa, R. Br. in Trans. Linn.
Soc.* x. 145, *partly* ; *Roem. & Schultes, Syst. Veg.* iii. 393 ; *Spreng.
Syst.* i. 471 ; *Meisn. in DC. Prodr.* xiv. 307, *partly.*   *S. laxa, R. Br.
l.c.* 146 ; *Roem. & Schultes, l.c.* ; *Meisn. l.c.   Leucadendron racemosum, Linn. Sp. Pl. ed.* i. 91 ; *Berg. in Vet. Akad. Handl. Stockh.* 1766,
325 ; *Berg. Descr. Pl. Cap.* 23.   *Protea racemosa, Linn. Mant. alt.*
187 ; *Thunb. Diss. Prot.* 25 ; *Thunb. Fl. Cap. ed. Schultes,* 127.
P. laxa, Poir. Encycl. Suppl. iv. 578.—*Protea foliis setaceis, etc.,
Linn. Hort. Cliff.* 496.

SOUTH AFRICA : without locality, *Thunberg* ! *Thom,* 699 !
COAST REGION : Tulbagh Div. ; New Kloof, *Roxburgh* ! *Lichtenstein,* 110,
Ludwig ! Caledon Div. ; Houw Hoek Mountains, *Niven,* 16 ! *Mund,* 52 !
Roxburgh ! *Zeyher,* 77, 3721 partly ! *Pappe* ! *Ludwig, Gueinzius* ! *MacOwan,* 2726 !
Bolus, Herb. Norm. Austr.-Afr., 360 ! *Schlechter,* 5465 ! 7384 ! *Galpin,* 4487 !
Pillans ! Hermanus, *Galpin,* 4486 ! Swellendam Div. ; on mountains, *Bowie* !

9. S. brachyloba (Phillips in Kew Bulletin, 1910, 333) ; a shrub
about 1 ft. high ; branches terete, pilose or thinly villous when
young, at length becoming glabrous ; leaves $\frac{1}{2}$ in. long, terete,
subacutely mucronate, the youngest very scantily pilose, soon
becoming glabrous ; involucres 1-flowered, arranged in sessile
solitary or paired terminal racemes ; racemes $\frac{3}{4}$–1 in. long ; subtending bracts $1\frac{1}{4}$ lin. long, linear, subacute, adpressed-pilose,
slightly exceeding the secondary peduncles, the latter 1 lin. long,
pilose ; involucre 1 lin. long, bilabiate, adpressed-pilose outside ;
upper lip ovate, entire, lower lip 3-toothed, teeth ovate, obtuse,
glabrous within ; perianth-tube $\frac{3}{4}$ lin. long, glabrous ; segments
about 2 lin. long, linear-spathulate ; claw villous ; limb elliptic,
obtuse, densely villous ; anthers subsessile ; hypogynous scales
lanceolate, acuminate, acute ; ovary ovoid, villous ; style 2 lin. long,
subterete, slightly hairy at the base ; stigmatic disk lateral, $\frac{1}{4}$ lin.
long, obovate, with a minute conical central projection.

COAST REGION: Stellenbosch Div.; Hottentots Holland Mountains, near Lowrys Pass, *MacOwan, Herb. Austr.-Afr.*, 1762! and without number in the *Cape Herbarium.*!

10. **S. cylindrica** (Phillips in Kew Bulletin, 1910, 334); branches adpressed-pilose when young, at length glabrous where devoid of leaves; leaves 1–1½ in. long, subterete, obtuse, adpressed-villous, at length becoming glabrous; involucres 1-flowered, arranged in sessile solitary terminal racemes; racemes 1¾–2 in. long, more or less cylindric; subtending bracts 4¼ lin. long, subterete, with a slightly swollen callous obtuse apex, ciliate towards the base, exceeding the secondary peduncles, the latter 1½–2 lin. long, shortly pubescent; involucre 1 lin. long, bilabiate, adpressed-pilose outside; upper lip ovate-lanceolate, entire, lower 3-lobed to about the middle, lobes ovate, subacute; perianth-tube ½ lin. long, glabrous; segments about 2½ lin. long, linear-spathulate; claw whitish-tomentose; limb ovate, subacute, densely villous outside; anthers ¼ lin. long, rounded; hypogynous scales ⅔ lin. long, linear; ovary subglobose, villous; style 2 lin. long, subterete; stigmatic disk ½ lin. long, obovate.

COAST REGION: Stellenbosch Div.; mountains of Lowry's Pass, *Burchell*, 8212! *Schlechter*, 7230!

11. **S. colorata** (Meisn. in DC. Prodr. xiv. 308); a shrub up to 1½ ft. high; branches glabrous, bark reddish-purple; leaves ½–¾ in. long, terete, acutely mucronate, glabrous; involucres 1-flowered, arranged in sessile solitary terminal racemes; racemes ½–¾ in. long; subtending bracts 2 lin. long, linear, acute, long-ciliate; secondary peduncles nearly 1 lin. long, slightly pubescent; involucre 1½ lin. long, bilabiate, pilose outside; upper lip subulate-lanceolate, entire, lower tripartite almost to the base, segments subulate-lanceolate, acute, long-ciliate; perianth-tube about 1 lin. long, as long as the claw, glabrous; segments about 2 lin. long, linear-spathulate, villous outside; limb elliptic, obtuse, densely villous outside; anthers sessile; hypogynous scales ½ lin. long, narrowly linear; ovary pubescent; style 2½ lin. long, slender, subterete, bent near the apex; stigmatic disk oblique, ⅙ lin. long, obovate.

SOUTH AFRICA: without locality, *Ecklon*, 18!
COAST REGION: Swellendam Div.; by the River Zondereinde, near Appels Kraal, *Zeyher*, 3718! Knofflooks Kraal, *Pappe*!

12. **S. bombycina** (Knight, Prot. 76); an erect shrub 4 ft. high; branches leafy, silky-villous; leaves ¾–1 in. long, arcuate, spreading almost at right angles, at length partially reflexed, subterete, long and acutely mucronate, silky-pilose, becoming nearly glabrous when old; involucres 1-flowered, arranged in pedunculate racemes in terminal clusters of 3–4; racemes 1–1¾ in. long; peduncles about 4 lin. long, pilose; subtending bracts 3 lin. long, linear-lanceolate, acutely acuminate, pubescent; secondary peduncles 1 lin. long,

pubescent; involucre 2 lin. long, bilabiate, pubescent outside; upper lip linear-lanceolate, entire, lower tripartite almost to the base, segments linear-lanceolate, acutely acuminate; perianth-tube short, nearly glabrous; segments 3 lin. long, linear-spathulate, villous outside; limb ovate, subacute, villous outside; anthers sessile; ovary pilose; style 2½ lin. long, subterete, glabrous; stigmatic disk ¼ lin. long, obovate; fruit 1¼ lin. long, brown, ellipsoid, pilose. *S. polystachya, R. Br. in Trans. Linn. Soc.* x. 148; *Roem. & Schultes, Syst. Veg.* iii. 395; *Meisn. in DC. Prodr.* xiv. 308. *Protea polystachya, Poir. Encycl. Suppl.* iv. 579.

SOUTH AFRICA : without locality, *Roxburgh* !
COAST REGION : Caledon Div. ; Zoetemelks Valley, *Niven* ! mountains of Baviaans Kloof, near Genadendal, *Burchell*, 7796 !

13. **S. parilis** (Knight, *Prot.* 75); a shrub about 5 ft. high; branches pilose; leaves erect or suberect, never recurved, ½–¾ in. long, terete, acutely mucronate, silky-pilose when young, at length becoming glabrous; involucres 1-flowered, arranged in sessile solitary terminal racemes, the latter more or less conical, 1–1½ in. long; subtending bracts 3–4 lin. long, linear, acutely mucronate, silky-pubescent, exceeding the secondary peduncles, the latter 1–1½ lin. long, pubescent; involucre 2½ lin. long, bilabiate, pilose outside; upper lip entire, lanceolate, acute; lower lip tripartite to near the base, segments 1½ lin. long, subulate, very acute; perianth-tube about ½ lin. long, glabrous; segments 3–3½ lin. long, linear-spathulate, densely villous outside; limb ovate, subacute; anthers rounded; hypogynous scales small; ovary ellipsoid, villous; style 2½ lin. long, subterete, bent above; stigmatic disk obovate, with a minute conical projection in the middle. *S. pyramidalis, R. Br. in Trans. Linn. Soc.* x. 148; *Roem. & Schultes, Syst. Veg.* iii. 395; *Meisn. in DC. Prodr.* xiv. 308. *Protea pyramidalis, Poir. Encycl. Suppl.* iv. 578.

SOUTH AFRICA : without locality, *Thunberg* !
COAST REGION : Swellendam Div. ; mountains near Swellendam, *Roxburgh* ! *Mund*, 53 ! *Bolus*, 8094 ! and *Herb. Norm. Austr.-Afr.* 686 ! Riversdale Div. ; near the waterfall at Garcias Pass, *Burchell*, 6988 ; lower part of the Langeberg Range at Garcias Pass, *Burchell*, 6955 ! *Schlechter*, 1772 ! *Galpin*, 4488 ! *Phillips*, 511 !

14. **S. Bolusii** (Phillips in Kew Bulletin, 1910, 333); branches whitish-villous; leaves ⅓–½ in. long, suberect, slightly incurved, terete, acutely mucronate, thinly and spreadingly villous; involucres 1-flowered, arranged in sessile solitary terminal racemes; racemes ovoid or ellipsoid, ½–1 in. long; bracts 2–3 lin. long, linear, acute, pilose, reaching the top of the involucre; secondary peduncles ¾–1 lin. long, pilose; involucre 1¾ lin. long, bilabiate, adpressed-pilose outside; upper lip subulate-lanceolate, entire, lower tripartite to near the base; segments subulate-lanceolate, acute; perianth-tube about ¾ lin. long, glabrous; segments 2–2½ lin. long, linear-

spathulate, densely pubescent; limb ovate, subobtuse, densely villous outside; anthers ovate; hypogynous scales small; ovary whitish-villous; style $2\frac{3}{4}$ lin. long, subterete, bent below the stigma; stigmatic disk $\frac{1}{4}$ lin. long, obovate.

COAST REGION : Riversdale Div. ; Garcias Pass, *Bolus*, 11361 !

15. S. prolifera (Knight, Prot. 75); an erect shrub about 2 ft. high; branches erect, leafy, slender, pubescent; leaves erect, 3–4 lin. long, about $\frac{1}{4}$ lin. in diam., flat on the upper side, convex below, acutely mucronate, thinly pilose when young, soon becoming glabrous; involucres 1-flowered, arranged in sessile solitary terminal spikes; spikes few-flowered, about $\frac{1}{2}$ in. long, more or less ellipsoid; subtending bracts $\frac{1}{4}$ in. long, linear-subulate, ciliate; involucre $2\frac{1}{2}$ lin. long, bilabiate, pubescent outside, upper lip lanceolate, entire, lower tripartite to near the base, segments subulate, subacute; perianth-tube as long as the claws of the segments, glabrous; segments 2–$2\frac{1}{2}$ lin. long, linear-spathulate; claw nearly glabrous; limb obtuse, villous; anthers sessile, rounded; ovary globose, villous; style $2\frac{1}{4}$ lin. long, subterete; stigmatic disk $\frac{1}{4}$ lin. long, obovate. *R. Br. in Trans. Linn. Soc.* x. 147; *Roem. & Schultes, Syst. Veg.* iii. 394; *Meisn. in DC. Prodr.* xiv. 308, *excl. Ecklon*, 18. *Protea prolifera, Thunb. Diss. Prot.* 29, *t.* 4; *Linn. f. Suppl.* 118; *Lam. Ill.* i. 238; *Thunb. Prodr.* 26; *Willd. Sp. Pl.* i. 518; *Poir. Encycl.* v. 654; *Thunb. Fl. Cap. ed. Schultes*, 129.

SOUTH AFRICA : without locality, *Masson* ! *Roxburgh* !
COAST REGION : Stellenbosch Div. ; Hottentots Holland Mountains, *Thunberg* ! *Zeyher*, 3719, *Pappe*, 10 !

16. S. Burchellii (Phillips in Kew Bulletin, 1910, 333); a shrub $1\frac{1}{2}$ ft. high; branches of the previous years' growth devoid of leaves, adpressed-pilose, becoming glabrous with age; leaves $\frac{1}{3}$–$\frac{1}{2}$ in. long, scarcely $\frac{1}{4}$ lin. in diam., crowded, slender, terete, acutely mucronate, glabrous or sometimes with a few scattered hairs when young; involucres 1-flowered, arranged in sessile solitary terminal spikes; spikes few-flowered, 3–5 lin. long; subtending bracts 2 lin. long, linear-subulate, subacutely acuminate, ciliate, equalling the involucre; involucre $1\frac{1}{2}$ lin. long, bilabiate, pubescent outside; upper lip entire, lower deeply tripartite; segments ovate-lanceolate, subacutely acuminate; perianth-tube pubescent; segments 2–$2\frac{1}{2}$ lin. long, linear-spathulate; claw tomentose; limb ovate, obtuse, densely villous; anthers rounded; hypogynous scales $\frac{3}{4}$ lin. long, subulate, acute; ovary $\frac{1}{2}$ lin. long, globose, densely villous; style $1\frac{3}{4}$ lin. long, terete; stigmatic disk $\frac{1}{4}$ lin. long, obovate.

COAST REGION : George Div. ; Cradock Berg, near George, *Burchell*, 5899 !

17. S. barbigera (Knight, Prot. 76); branches leafy, pilose; leaves lax, $\frac{1}{4}$–$\frac{1}{2}$ in. long, nearly $\frac{1}{2}$ lin. in diam., straight or slightly curved, linear, acutely mucronate, channelled on the upper surface, more or less adpressed-pilose; spikes 2–3 terminating each shoot, conical when young, 1–1¼ in. long; bracts 2½ lin. long, linear-subulate, acute, pubescent on the back, ciliate, exceeding the involucres in the young inflorescence, at length somewhat shorter; involucre 1-flowered, 2 lin. long, bilabiate, pubescent outside; upper lip entire, lower deeply tripartite, segments ovate-lanceolate, acute, ciliate; perianth-tube shorter than the claws of the segments, glabrous; segments 1¾ lin. long, linear-spathulate, claw tomentose; limb ½ lin. long, ovate, subobtuse, long-villous outside; anthers sessile; ovary whitish-villous; style 2¾ lin. long, terete, glabrous; stigmatic disk ¼ lin. long, rectangular, with a minute central conical production; fruits shortly stipitate, 1½ lin. long, ellipsoid, pubescent. *S. sericea, Meisn. in DC. Prodr.* xiv. 308, *as to following syn., not of R. Br.   Phylica abietina, E. Meyer in Drège, Zwei Pfl. Documente,* 65, 210, *not of Eckl. & Zeyh.*

COAST REGION : Riversdale Div. ; Platte Kloof, *Niven* ! George Div. ; Montagu Pass, *Schlechter,* 5831 !
CENTRAL REGION : Prince Albert Div. ; Great Zwartberg Range, *Drège* !

18. **S. Wallichii** (Phillips in Kew Bulletin, 1910, 336) ; a much-branched shrub ; branches verticillate, tomentose or woolly-pubescent, at length becoming glabrous ; leaves 3–4 lin. long, subterete, very acutely mucronate, somewhat incurved towards the apex, pilose, at length glabrous ; involucres 3–4-flowered, arranged in shortly pedunculate solitary or 2–3 terminal racemes, the latter ¾–1¼ in. long, few-flowered ; secondary peduncles 1 lin. long, villous ; subtending bracts 2 lin. long, lanceolate or linear-lanceolate, acute, 3–5-nerved, pilose outside ; involucre 2 lin. long, sub-regularly tripartite to near the base ; segments ovate ; perianth-tube 1 lin. long, glabrous towards the base ; segments 2½ lin. long, linear-spathulate, villous outside ; limb ⅔ lin. long, ovate, subacute ; anthers rounded ; hypogynous scales small ; ovary obovoid, densely pilose ; style 2½ lin. long, terete, curved above ; stigmatic disk ¼ lin. long, obovate, with a minute central conical projection ; fruit shortly stipitate, 2¼ lin. long, ellipsoid, pilose, tipped by the persistent style.

COAST REGION : Clanwilliam Div. ; Cederberg Range, *Wallich* !

19. **S. mucronifolia** (Phillips in Kew Bulletin, 1910, 335) ; branches pubescent or almost tomentose ; leaves 5–8½ lin. long, suberect, subterete, very acutely and rather abruptly mucronate, thinly pilose ; involucre 3-flowered, arranged in sessile or shortly pedunculate solitary or 2–5 terminal racemes ; racemes 1–1½ in. long ; subtending bracts 2 lin. long, linear or linear-lanceolate, acute, pilose, equalling or slightly exceeding the secondary peduncles, the latter 1½ lin. long, pubescent ; involucre 2½ lin. long, sub-regularly tripartite to the base ; segments 2 lin. long, ovate, shortly acutely acuminate, ciliate ; perianth-tube ½–¾ lin. long, pubescent.

in the upper part; segments $3\frac{3}{4}$–4 lin. long, linear-spathulate; claws tomentose; limb ovate-elliptic, obtuse, tomentose or shortly villous; anthers rounded; hypogynous scales 1 lin. long, linear; ovary $\frac{3}{4}$ lin. long, oblong-ellipsoid, villous; style $2\frac{3}{4}$ lin. long, terete, obliquely inserted; stigmatic disk $\frac{1}{4}$ lin. long, elliptic, with a small conical projection on the face. *S. Thunbergii, var. Dregei, Meisn. in DC. Prodr.* xiv. 310.

COAST REGION: Clanwilliam Div. ; Cederberg Mountains, at Pakhuis Pass, *Bolus,* 9083 ! Pakhuisberg, *Schlechter,* 8611 ! 10814 ! Ezelsbank ? *Drège* !

20. **S. procera** (Knight, Prot. 76); a tall slender shrub up to 8 ft. high; branches hirsute or pubescent; leaves $\frac{3}{4}$–1 in. long, falcate, incurved, terete, rather slender, very acutely mucronate, subadpressed-pilose, at length glabrous; involucres 3-flowered, and containing another smaller involucre with a rudimentary flower, arranged in sessile or subsessile solitary or 2–6 terminal racemes, the latter $1\frac{1}{4}$–2 in. long; subtending bracts $1\frac{1}{2}$–$2\frac{1}{4}$ lin. long, linear or linear-lanceolate to ovate, acute, pubescent outside, shorter than or slightly exceeding the secondary peduncles, the latter $\frac{3}{4}$–$1\frac{3}{4}$ lin. long, hirsute; involucre $1\frac{1}{2}$–2 lin. long, subregularly tripartite to the base; segments ovate, acutely acuminate, pubescent outside; perianth-tube $\frac{1}{2}$–$\frac{3}{4}$ lin. long, glabrous; segments $2\frac{1}{2}$ lin. long, linear-spathulate; claw villous; limb elliptic, subacute, densely villous outside; anthers rounded; hypogynous scales $\frac{1}{2}$ lin. long, linear; ovary oblong-ellipsoid, rusty-pubescent; style $2\frac{1}{4}$ lin. long, sub-terete; stigmatic disk $\frac{1}{4}$ lin. long, almost circular, with a small central conical projection; fruit $1\frac{1}{2}$ lin. long, stipitate, ellipsoid, crowned by the persistent style. *S. nana, Knight, l.c. S. incurva, vars. a* and *β, R. Br. in Trans. Linn. Soc.* x. 149 ; *Roem. & Schultes, Syst. Veg.* iii. 395 ; *Meisn. in DC. Prodr.* xiv. 309, *incl. vars. laxior, densior* and *Zeyheri, Meisn. Protea incurva, Thunb. Diss. Prot.* 26, *t.* 3 ; *Murr. Syst. Veg. ed.* xiv. 138 ; *Willd. Sp. Pl.* i. 516 ; *Thunb. Fl. Cap. ed. Schultes,* 128.

SOUTH AFRICA : without locality, *Roxburgh* ! *Gueinzius* ! *Hornstedt* ! COAST REGION : Tulbagh Div. ; Tulbagh Waterfall, *Niven,* 34 ! *Roxburgh* ! Witsenberg Range and Tulbagh, *Zeyher,* 1480 ! Worcester Div. ; Dutoits Kloof, *Drège* ! Paarl Div. ; French Hoek, *Niven,* 35 ! Stellenbosch Div. ; Hottentots Holland Mountains, *Pappe* ! *Zeyher,* 3720 partly !

The bracts subtending the secondary peduncles are very variable in length and shape even on the same specimen.

21. **S. thyrsiflora** (Knight, Prot. 74); a decumbent shrub; branches tomentellous; leaves $\frac{3}{4}$–1 in. long, falcate, subterete, subacutely and gradually mucronate, very shortly adpressed-pilose when young, at length glabrous and slightly shining; involucres 3-flowered, and containing another smaller involucre with a rudi-mentary flower, arranged in dense sessile solitary terminal spikes, the latter about $\frac{3}{4}$ in. long; subtending bracts about 2 lin. long, ovate, slightly acuminate, pilose, ciliate; involucre 2–3 lin. long,

tripartite; segments lanceolate, pubescent; perianth-tube 1 lin.
long, sparingly pubescent; segments about 4 lin. long; claws
shortly villous, the adaxial one straight, the other three spirally
coiled in the open flower; limb ovate, villous; anthers oblong;
hypogynous scales linear-lanceolate; ovary pubescent; style about
5 lin. long, curved, glabrous; stigma ¼ lin. long, subovoid, obtuse;
fruits sessile, 2 lin. long, 1 lin. broad, oblong-ellipsoid, subacute,
pubescent.   *Sorocephalus spatalloides, R. Br. in Trans. Linn. Soc.*
x. 141; *Roem. & Schultes, Syst. Veg.* iii. 390; *Meisn. in DC. Prodr.*
xiv. 304.

COAST REGION : Caledon Div. ; Zwart Berg, Viven, 29 ! Mund !

XII. SPATALLOPSIS, Phillips.

Flowers hermaphrodite, actinomorphic. *Perianth* straight and
cylindric in bud with an ellipsoid limb, 4-partite to below the
middle; tube glabrous or nearly so; segments differentiated into
a slender hairy claw and an elliptic or ovate obtuse villous limb,
the latter often becoming recurved. *Stamens* 4 ; anthers ellipsoid,
sessile, inserted at the base of the limb; connective slightly produced into a small globular point. *Hypogynous scales* 4, free,
linear. *Ovary* sessile or subsessile, hairy; style slender, terete,
often slightly bent towards the apex, subpersistent, glabrous;
stigma conical or subclavate, terminal. *Ovule* 1, laterally attached.
Fruits as in *Spatalla.*

Habit of *Spatalla* ; flowers in 3–4-flowered involucres arranged in terminal spikes
or racemes ; involucres calycoid, small, more or less regularly 4-partite to near the
base ; bract subtending the secondary peduncles or the involucre solitary, linear to
ovate ; perianth tomentose or villous.

DISTRIB. Species 5, confined to the South-Western Region of Cape Colony.

Involucres and subtending bracts glabrous or nearly so
   outside, the latter very shortly ciliolate or glabrous
   on the margin ; leaves quite glabrous when mature :
   Leaves 3½–5 lin. long, subacute ; subtending bracts
      1 lin. long, reaching to the base of the involucre   (1) confusa.
   Leaves 6–7 lin. long, acutely mucronate ; subtending
      bracts 2½ lin. long, reaching almost to the top of
      the involucre...   ...   . (2) caudata.
Involucres and subtending bracts densely pubescent or
   villous, the latter ciliate, or when both only slightly
   pubescent then the leaves permanently pilose with
   long weak hairs :
   Leaves 2½–4½ lin. long :
      Leaves 2½–3 (rarely 4) lin. long, obtuse or subobtuse ;
         spikes mostly several together ; bracts and
         involucres shortly whitish-pubescent ...   ... (3) ericæfolia.
      Leaves 3–4½ lin. long, sharply and conspicuously
         mucronate ; spikes solitary ; bracts and in-
         volucres rusty-villous   ...   . (4) caudæflora.
   Leaves 8–10 lin. long   ...   . (5) propinqua.

1. S. confusa (Phillips in Kew Bulletin, 1910, 289, partly);
branches sparingly pubescent or nearly glabrous; leaves 3½–5 lin.
long, ⅓–½ lin. thick, subacute, glabrous except when quite young;
involucres 2–3-flowered, arranged in subsessile terminal solitary or
subsolitary racemes, the latter ½–1 in. long; subtending bracts as
long as the secondary peduncles, glabrous or nearly so, minutely
ciliolate; secondary peduncles 1 lin. long, whitish-pubescent;
involucre rather unequally 4-partite, glabrous outside; perianth-
tube 1½ lin. long, glabrous; segments 2 lin. long, with a linear-
filiform minutely puberulous or nearly glabrous claw, and an
ovate-elliptic obtuse whitish-villous limb; anthers sessile; hypogynous scales ½ lin. long; ovary villous; style 2¾ lin. long, filiform,
glabrous; stigma ⅛ lin. long, clavate. *Spatalla brevifolia, E. Meyer
in Drège, Zwei Pfl. Documente, 74, 222, not of R. Br.*

COAST REGION: Clanwilliam Div.; Ezelsbank, *Drège,* 8079! *Schlechter,* 8838!
CENTRAL REGION: Prince Albert Div.; Zwartberg Pass, *Bolus,* 11627!

2. S. caudata (Phillips in Kew Bulletin, 1910, 290); branches
sparingly pilose when young, soon becoming glabrous or nearly so;
leaves somewhat spreading and arcuate, 6–7 lin. long, about ½ lin.
broad, acutely mucronate, widely channelled on the upper side,
convex below, thinly pubescent or pilose when young, soon becoming
quite glabrous; involucres 3–4-flowered, arranged in sessile solitary
or 3–4 terminal racemes, the latter 1–2¼ in. long; subtending bracts
2½ lin. long, 1–1¼ lin. broad, ovate, gradually and subacutely
acuminate, membranous, glabrous on both sides, very sparingly
ciliate; secondary peduncles ½ lin. long, pubescent; involucre sub-
regularly 4-partite to near the base; segments 2 lin. long, 1–1¼ lin.
broad, ovate, acute, glabrous or thinly pubescent outside, sparingly
ciliate; perianth-tube 1 lin. long, glabrous; segments 2 lin. long,
with a linear tomentose claw and an ovate obtuse villous limb;
anthers ⅓ lin. long; hypogynous scales small and filiform; ovary
obovoid, villous; style 2½ lin. long, filiform, glabrous; stigma
obconic; fruit shortly stipitate, 2 lin. long, ellipsoid, pubescent.
*Spatalla caudæflora, Knight, Prot. 75, as to syn. Protea caudata,
Thunb., partly. S. caudata, R. Br. in Trans. Linn. Soc. x. 150;
Roem. & Schultes, Syst. Veg. iii. 396; Meisn. in DC. Prodr. xiv. 310,
Protea caudata, Thunb. Diss. Prot. t. 2, not description, ex R. Br. l.c.*

SOUTH AFRICA: without locality, *Thunberg!* Herb. *Forsyth in Herb. Kew.*!
COAST REGION: Caledon Div.; Palmiet River, *Masson!*

The specimens named *Protea caudata* in Thunberg's own Herbarium are all
identical with the above and evidently the plant figured by him. Another,
however, collected by him and in Baron Alströmer's Herbarium at Stockholm is
evidently the one he described, which is *S. caudæflora.*

3. S. ericæfolia (Phillips in Kew Bulletin, 1910, 288); branches
purplish, pilose; leaves erect, 2½–3 lin. long, obtuse or subobtuse,
very slightly incurved towards the apex, broadly channelled on the
upper side, convex below, about ⅓ lin. broad, slightly shining, thinly

pilose ; involucres 3–4-flowered, arranged in solitary or more often
2–4 terminal spikes, the latter ½–1 in. long, more or less cylindric ;
subtending bracts about 2 lin. long, linear or linear-subulate, subacute,
ciliate, shortly pubescent ; involucre rather unequally 4-partite, the
abaxial lobe a little narrower than the others, which are about 1½ lin.
long, ovate-lanceolate, subacute, thinly pubescent outside ; perianth-
tube 1¼ lin. long, sparingly pubescent ; segments 2 lin. long, with a
filiform pubescent claw and an ovate obtuse more densely and
longer-pubescent limb ; anthers ellipsoid, sessile ; hypogynous scales
¾ lin. long, linear, acute ; ovary densely pilose ; style ¼ in. long ;
stigma more or less ovoid, about ⅕ lin. long ; fruit obovoid, nearly
2 lin. long, tipped by the persistent style, pilose. *Spatalla ericæfolia,
Knight, Prot. 74. S. brevifolia, R. Br. in Trans. Linn. Soc.* x. 151.
S. confusa, Phillips in Kew Bulletin, 1910, 289, *as to Schlechter,*
10180, 10225. *Sorocephalus tulbaghensis, Phillips, l.c.* 1911, 86.

SOUTH AFRICA : without locality, *Masson* ! and without locality, *Herb. Forsyth
in Herb. Kew.* !
COAST REGION : Tulbagh Div. ; near Tulbagh, *Pappe* !
CENTRAL REGION : Ceres Div. ; Verkeerde Vley, *Niven* ! Schoongezigt, *Schlechter,*
10180 ! Gydow Berg, *Schlechter,* 10225 !

4. **S. caudæflora** (Phillips in Kew Bulletin, 1910, 289) ; branches
pilose, at length becoming glabrous and purplish ; leaves 3–4½ lin.
long, linear, widely grooved on the upper side, convex below,
sharply and conspicuously mucronate, thinly pilose with long weak
hairs ; involucres 3–4-flowered, arranged in terminal solitary spikes
1–1½ in. long ; subtending bracts 2½ lin. long scarcely 1 lin. broad,
ovate-lanceolate, villous outside ; involucre unequally 4-partite ;
segments about 2 lin. long, more or less ovate, acute, villous outside ;
perianth-tube 1 lin. long, glabrous towards the base ; segments 1¾
lin. long, with a linear pubescent claw and an elliptic obtuse villous
limb, the latter ½ lin. long ; anthers ellipsoid ; hypogynous scales
½ lin. long, linear ; ovary shortly stipitate, densely pubescent ; style
2¼ lin. long, filiform, glabrous ; stigma broadly clavate ; fruit 1½ lin.
long, shortly stipitate, oblong-cylindric, hirsute. *Spatalla caudæflora,
Knight, Prot.* 75, *excl. Protea caudata, Thunb., partly. S. Thunbergii,
R. Br. in Trans. Linn. Soc.* x. 150 ; *Roem. & Schultes, Syst. Veg.* iii.
396 ; *Meisn. in DC. Prodr.* xiv. 310, *excl. var. Dregei. Protea
caudata, Thunb. Diss. Prot.* 26, *as to description, ex R. Br. l.c.*

SOUTH AFRICA : without locality, *Thunberg* !
COAST REGION : Caledon Div. ; Zwart Berg, *Niven* !

5. **S. propinqua** (Phillips in Kew Bulletin, 1910, 290) ; branches
sparingly pilose ; leaves 8–10 lin. long, acutely mucronate, channelled
on the upper surface, pilose ; involucres 3–4-flowered, arranged in
crowded terminal racemes, the latter 2–3 in. long ; subtending
bracts 2½ lin. long, lanceolate, subacute, puberulous outside :
secondary peduncles ½ lin. long, pubescent ; involucre subregularly
4-partite to near the base ; segments broadly ovate, acute, ciliate ;

perianth-tube 1¼ lin. long, glabrous ; segments 1¾ lin. long, linear, villous ; limb ovate, obtuse, densely villous ; hypogynous scales small ; ovary ¾ lin. long, villous ; style 2¾ lin. long, cylindric, glabrous ; stigma ⅛ lin. long, subconic ; fruits shortly stipitate, 1 lin. long, obovoid, hirsute. *Spatalla propinqua, R. Br. in Trans. Linn. Soc.* x. 150 ; *Meisn. in DC. Prodr.* xiv. 309. *Sorocephalus setaceus, R. Br. l.c.* 140 ? *Roem. & Schultes, Syst. Veg.* iii. 389 ; *Meisn. l.c.* 303. *Protea australis, Poir. Encycl. Suppl.* iv. 579. *Soranthe setacea, O. Kuntze, Rev. Gen. Pl.* ii. 582.

SOUTH AFRICA : without locality, *Auge* ! *Roxburgh* !

XIII. SOROCEPHALUS, R. Br.

Flowers hermaphrodite, actinomorphic. *Perianth* cylindric and straight in bud with an ellipsoid or ovoid limb, 4-partite to near the base ; segments differentiated into claw and limb ; claws straight or spirally twisted in the open flower, hairy ; limb more or less elliptic or ovate, obtuse or subobtuse, usually villous, rarely pubescent or entirely glabrous. *Stamens* 4 ; anthers all perfect, ovoid or ellipsoid, sessile or subsessile, inserted at the base of the limb ; connective mostly slightly produced at the apex. *Hypogynous scales* 4, free, linear. *Ovary* sessile or shortly stipitate, hairy ; style slender, terete, straight, often constricted at the base, glabrous ; stigma terminal, conical, ellipsoid or subglobose, obtuse. *Ovule* 1, laterally attached. *Fruit* a usually glabrescent shining ellipsoid or cylindric sessile or shortly stipitate nut with a more or less hardened pericarp.

Small erect shrubs ; leaves numerous, usually erect, terete or subterete, rarely quite flat, acutely or obtusely mucronate, smooth or slightly scabrous, glabrous or thinly pilose ; flower-heads 2–6-flowered, arranged in dense short spikes or racemes, each head subtended by a more or less membranous glabrous or hairy bract ; floral bracts free from one another, similar to the preceding.

DISTRIB. Species 13, confined to the South-Western portion of Cape Colony.

The name *Sorocephalus*, R. Br. (1810), is here retained in favour of *Soranthe*, Knight (1809), on account of its inclusion in the list of *nomina conservanda* adopted by the members of the Vienna Congress in 1905.

Perianth-limb glabrous :
 Branches rough with the scars of fallen leaves ; branch-
  lets short ; leaves ¾–1 in. long ... ... (1) **imberbis.**
 Branches leafy, smooth ; branchlets elongated ; leaves
  1–2¼ in. long :
  Flower-clusters ovoid ; bracts densely villous outside (2) **longifolius.**
  Flower-clusters depressed-globose ; bracts glabrous
   or nearly so outside ... ... (3) **scabridus.**
Perianth-limb hairy, usually villously bearded :
 Leaves 1½–2 lin. broad, lanceolate, flat, with distinct
  lateral nerves or ribs, scabrous ; bracts glandular-
  hairy ... ... (4) **imbricatus.**
 Leaves usually less than 1 lin. broad, linear, terete or
  flat only on the upper surface, not nerved, rarely
  scabrous ; bracts never glandular :

Leaves flat or shallowly and broadly concave on
   the upper surface :
 Leaves keeled or grooved below, 4–9 lin. long :
  Leaves mostly pilose, somewhat scabrous when
     older ; bracts subtending the flower-heads
     more or less glabrous outside   ...   ... (5) phylicoides.

  Leaves usually glabrous, smooth ; subtending
     bracts densely villous outside   ...   ... (6) lanatus.

 Leaves smooth and convex below, not keeled or
     grooved, 3–4 lin. long ...   . (7) Schlechteri.

Leaves terete except for a very narrow groove on the
   upper surface :
 Leaves straight or if falcate then the perianth-
     limb bearded, less than ¾ in. long, usually
     less than ½ lin. thick, mostly pilose when
     young :
  Perianth-claws straight or slightly recurved in
     the open flower ; fruits (where known) less
     than 2 lin. long :
   Leaves tuberculate or scabrous, especially
      when young   ...   . (8) clavigerus.

   Leaves smooth :
    Leaves filiform, scarcely ¼ lin. thick ; floral
       bracts glabrous or nearly so outside ... (9) tenuifolius.

    Leaves stout, nearly ½ lin. thick ; floral
       bracts glabrous outside   ...   ... (10) salsoloides.

    Leaves stout, nearly ½ lin. thick ; floral
       bracts rather densely pubescent outside (11) rupestris.

  Perianth-claws strongly spirally twisted in the
     open flower ; fruits 2¼ lin. long   ...   ... (12) teretifolius.

 Leaves strongly falcate, ¾–1 in. long, ¾–1 lin.
     thick, glabrous when young ; perianth-limb
     shortly pubescent   ...   . (13) crassifolius.

1. S. imberbis (R. Br. in Trans. Linn. Soc. x. 140) ; a small
shrub about 2 ft. high ; branches rough with the scars left by the
fallen leaves, rather densely pubescent when young ; leaves ¾–1 in.
long, slender, acicular, acutely apiculate, with a very narrow groove on
the upper surface, otherwise terete, glabrous ; heads about 6-flowered,
crowded in dense racemes partially clasped by the leaves ; bracts
subtending the secondary peduncles about ¼ in. long, narrowly
lanceolate, acuminate, pubescent ; secondary peduncles about 1 lin.
long, tomentose ; floral bracts similar to those subtending the
peduncles but smaller ; perianth-tube ¾ lin. long, glabrous ; segments 2¼ lin. long ; claws pubescent, at length spirally coiled ; limb
¾ lin. long, elliptic-lanceolate, subacute, quite glabrous outside ;
anthers ellipsoid ; hypogynous scales linear ; ovary ellipsoid,
adpressed-pubescent ; style nearly 4 lin. long, glabrous ; stigma
⅙ lin. long, ellipsoid, subacute ; fruits 2 lin. long, 1½ lin. in diam.,
oblong-ellipsoid, slightly longitudinally wrinkled, brown, shining,
glabrous. *Roem. & Schultes, Syst. Veg.* iii. 389 ; *Meisn. in DC.
Prodr.* xiv. 303, excl. var. *longifolius. Protea imberbis, Poir. Encycl.*

Suppl. iv. 576. *Soranthe pinifolia, Knight, Prot.* 72. *S. imberbis,*
O. Kuntze, Rev. Gen. Pl. ii. 582.

COAST REGION: Swellendam Div. ; Tyger Hoek, *Niven*!

2. **S. longifolius** (Phillips in Kew Bulletin, 1911, 85) ; branches
densely pilose when young, soon becoming glabrous, with a purplish
smooth bark ; leaves erect, 1–2¼ in. long, almost filiform, long and
acutely apiculate, terete except for a narrow groove on the upper
surface, glabrous ; heads 7–9-flowered, crowded in dense terminal
racemes ¾–1 in. long ; bracts subtending the secondary peduncles
½ in. long, lanceolate, acutely acuminate, densely pubescent or
villous outside ; secondary peduncles about 1 lin. long, pubescent or
tomentose ; floral bracts similar to the others but narrower ;
perianth-tube about 1 lin. long, cylindric, glabrous towards the
base ; segments 3¾ lin. long ; claws villous, spirally coiled ; limb
¾ lin. long, ovate, obtuse, quite glabrous ; anthers ½ lin. long ;
hypogynous scales ⅓ lin. long, linear ; ovary ellipsoid, pubescent or
nearly glabrous ; style 4 lin. long, terete, glabrous ; stigma ⅓ lin.
long, ovoid, obtuse ; fruits 3 lin. long, ellipsoid-cylindric, obtuse,
slightly wrinkled, shining, glabrous. *Sorocephalus imberbis, var.*
longifolius, Meisn. in DC. Prodr. xiv. 303.

COAST REGION: Swellendam Div. ; banks of the Zondereinde River, near
Appels Kraal, *Zeyher,* 3718 !

3. **S. scabridus** (Meisn. in DC. Prodr. xiv. 303) ; branches slender,
very thinly pilose when young ; leaves erect, 1–1¼ in. long, almost
filiform, subacutely apiculate, terete except for a narrow groove on
the upper surface, very thinly pilose when young, soon becoming
scabrous ; heads sessile, 2–4-flowered, crowded in subglobose
terminal clusters about ¾ in. in diam. ; bracts subtending the
individual heads up to 3 lin. long, lanceolate, subacute, glabrous or
nearly so outside, long-ciliate ; floral bracts similar to the preceding
but narrow ; perianth-tube very short,· nearly glabrous ; segments
3½–4 lin. long ; claws villous, remaining apparently united owing to
the interlacing hairs ; limb 1¼ lin. long, ovate, obtuse, quite
glabrous ; anthers sessile ; hypogynous scales very small ; ovary
pubescent ; style slender, 4½ lin. long, glabrous ; stigma obovoid,
⅓ lin. long ; fruits not seen. *S. nivalis, Mund, and S. imberbis,*
Klotzsch ex Meisn. l.c. 304. *Soranthe scabrida, O. Kuntze, Rev. Gen.*
Pl. ii. 582.

COAST REGION : Tulbagh Div. ; Winterhoek Mountain, *Zeyher,* 3718b !

4. **S. imbricatus** (R. Br. in Trans. Linn. Soc. x. 142) ; branches
pilose, at length becoming glabrous ; leaves 3½–5½ lin. long, erect
and densely imbricate, lanceolate, subacutely apiculate, flat,
distinctly nerved and scabrous on both surfaces, thinly ciliate when
young, otherwise glabrous, those surrounding the heads coloured ;
heads 4-flowered, crowded in dense solitary or geminate more or less

globose terminal racemes about $\frac{3}{4}$ in. in diam. ; bracts subtending
the secondary peduncles 5 lin. long, $1\frac{1}{2}$ lin. broad, ovate-lanceolate,
obtuse, rough with short glandular hairs, ciliate; secondary
peduncles about $1\frac{1}{4}$ lin. long, nearly glabrous; floral bracts 3 lin.
long, about 1 lin. broad, ovate-lanceolate, subacute, glandular-
puberulous, ciliate; perianth-tube about $\frac{1}{2}$ lin. long, ellipsoid,
glabrous; segments mauve, $3\frac{1}{2}$–4 lin. long ; claws filiform, glabrous
or minutely puberulous, not spirally twisted ; limb $\frac{2}{3}$ lin. long,
ovate-elliptic, subobtuse, densely and villously bearded in the upper
half, glabrous or slightly pubescent in the lower part ; anthers
nearly as long as the limb ; hypogynous scales $\frac{2}{3}$ lin. long, linear ;
ovary ovoid, pilose ; style $4\frac{1}{2}$ lin. long, slender, glabrous ; stigma
$\frac{1}{4}$ lin. long, subovoid, mostly acute.   *Roem. & Schultes, Syst. Veg.* iii.
391 ; *E. Meyer in Drège, Zwei Pfl. Documente,* 77, 222 ; *Meisn. in
DC. Prodr.* xiv. 305 ; *Spreng. Syst.* i. 470.   *Protea imbricata,
Thunb. Diss. Prot.* 38, *t.* 5, *fig.* 2 ; *Linn. f. Suppl.* 116 ; *Lam. Ill.* i.
235 ; *Thunb. Prodr.* 27 ; *Willd. Sp. Pl.* i. 527 ; *Poir. Encycl.* v. 643 ;
Andr. Bot. Rep. t. 517 ; *Thunb. Fl. Cap. ed. Schultes,* 136.   *Soranthe
glanduligera, Knight, Prot.* 71.   *S. imbricata, O. Kuntze, Rev. Gen.
Pl.* ii. 582.

SOUTH AFRICA : without locality, *Thunberg* ! *Brown, Verreaux.*
COAST REGION : Tulbagh Div. ; mountains near Tulbagh Waterfall, *Niven* !
Bolus, 379 ! between New Kloof and Elands Kloof, *Drège* ! Winterhoek Mountain,
near Tulbagh, *Bolus,* 5258 !

5. **S. phylicoides** (Meisn. in DC. Prodr. xiv. 304) ; an erect shrub
3 ft. high ; branches thinly pilose ; leaves 5–9 lin. long, linear,
subobtusely mucronate, flat on the upper surface, subconvex or
slightly keeled below, sometimes thinly pilose especially when young,
becoming glabrous and slightly scabrous when older ; heads about
4-flowered, numerous in subglobose terminal clusters nearly $\frac{3}{4}$ in. in
diam. ; bracts subtending the secondary peduncles $2\frac{1}{2}$ lin. long,
$1\frac{1}{4}$ lin. broad, ovate, subacute, long-ciliate ; secondary peduncles
1 lin. long, glabrous or nearly so ; bracts subtending the flowers
lanceolate, submembranous, long-ciliate ; perianth-tube $1\frac{1}{2}$ lin. long,
narrowly cylindric, glabrous in the lower, pubescent in the upper
part ; segments $3\frac{1}{2}$ lin. long ; claws slender, pubescent, slightly
twisted ; limb $\frac{2}{3}$ lin. long, ovate-elliptic, subobtuse, villous outside ;
anthers $\frac{1}{2}$ lin. long ; ovary ovoid, pilose or nearly glabrous ; style
about 4 lin. long, filiform, constricted at the base, glabrous ; stigma
$\frac{1}{6}$ lin. long, subellipsoid, somewhat obtuse ; fruits $2\frac{1}{2}$ lin. long, 1 lin.
thick, cylindric, glabrous, black and shining.   *Soranthe phyllicodes,
O. Kuntze, Rev. Gen. Pl.* ii. 582.

SOUTH AFRICA : without locality, *Masson* ! *Mund* !
COAST REGION : Tulbagh Div. ; near Tulbagh, *Pappe* !
CENTRAL REGION : Ceres Div. ; Witzen Berg and Skurfde Berg, *Zeyher,* 1468 !
Gydow, *Schlechter,* 9990 !

6. **S. lanatus** (R. Br. in Trans. Linn. Soc. x. 142) ; branches
pilose when young, at length glabrous, purplish ; leaves erect,

$\frac{1}{3}$–$\frac{1}{2}$ in. long, linear, subobtusely mucronate, keeled below, flat or slightly concave above, very sparingly pilose when young, soon becoming glabrous; heads about 6-flowered, crowded in dense terminal ovoid or subglobose spikes $\frac{3}{4}$–1 in. long and about $\frac{1}{2}$–$\frac{3}{4}$ in. in diam.; axis of the spike tomentose; bract subtending each individual head up to $2\frac{1}{2}$ lin. long, $1\frac{1}{4}$ lin. broad, ovate to ovate-lanceolate, acute, long-villous towards the margin; floral bracts $2\frac{1}{2}$ lin. long, about $\frac{1}{2}$ lin. broad, lanceolate, acute, densely ciliate; perianth-tube 2 lin. long, very narrowly cylindric, pubescent; segments about 2 lin. long; claws slender, sometimes slightly recurved but not spirally twisted, shortly villous; limb $\frac{1}{2}$ lin. long, elliptic, obtuse, villous outside; anthers $\frac{1}{3}$ lin. long; hypogynous scales $\frac{1}{2}$ lin. long, linear; ovary ovoid, pubescent; style $3\frac{1}{2}$–4 lin. long, filiform, glabrous; stigma $\frac{1}{6}$ lin. long, subellipsoid, obtuse; fruits 3 lin. long, 1 lin. in diam., cylindric-oblong, shining, black, glabrous. *Roem. & Schultes, Syst. Veg.* iii. 390; *Meisn. in DC. Prodr.* xiv. 304, *excl. var. teretifolius. S. spatalloides, Sieber ex Meisn. l.c.* 305, *name only. Protea lanata, Thunb. Diss. Prot.* 51, *t.* 3; *Murr. Syst. Veg. ed.* xiv. 139; *Lam. Ill.* i. 238; *Thunb. Prodr.* 26; *Willd. Sp. Pl.* i. 519; *Poir. Encycl.* v. 653; *Thunb. Fl. Cap. ed. Schultes,* 130. *Soranthe ciliciiflora, Knight, Prot.* 72; *Salisb. ex Spreng. Syst.* i. 470. *S. lanata, O. Kuntze, Rev. Gen. Pl.* ii. 582.

SOUTH AFRICA: without locality, *Thunberg*! without collector in *Herb. Forsyth* (at Kew)! and in *Herb. Lindley*!
COAST REGION: Malmesbury Div.; Zwartland, *Niven*!

7. **S. Schlechteri** (Phillips in Kew Bulletin, 1911, 85); branches glabrous, light-purplish when young; leaves 3–4 lin. long, linear, subobtusely apiculate, convex on the lower side, slightly concave above, glabrous; heads 4-flowered, in terminal racemes 4–5 lin. long; bracts about $2\frac{1}{2}$ lin. long, 1–$1\frac{1}{4}$ lin. broad, ovate or ovate-lanceolate, subacute, glabrous within, pilose outside towards and densely long-ciliate on the margin; secondary peduncles 1 lin. long; involucral bracts 2–$2\frac{1}{4}$ lin. long, ovate-lanceolate, subacute, pilose outside, densely ciliate; perianth-tube 1–$1\frac{1}{4}$ lin. long, cylindric, glabrous; segments about 2 lin. long, with a very slender nearly glabrous claw, and an oblong-elliptic obtuse densely villous limb $\frac{1}{2}$ lin. long; anthers nearly $\frac{1}{2}$ lin. long; ovary ovoid, villous; style $2\frac{3}{4}$ lin. long, filiform, glabrous; stigma $\frac{1}{6}$ lin. long, narrowly ellipsoid, subacute.

CENTRAL REGION: Ceres Div.; Gydouw Berg, 6200 ft., *Schlechter,* 10230!

8. **S. clavigerus** (Hutchinson); an erect shrub about 3 ft. high; branches thinly villous or pilose; leaves $\frac{3}{4}$ in. long, scarcely $\frac{1}{2}$ lin. thick, rather sharply mucronate, narrowly grooved on the upper side, otherwise terete, rather densely pilose when young, soon becoming quite glabrous; heads terminal, subsessile, about $\frac{3}{4}$ in. in diam.; bracts $3\frac{1}{2}$ lin. long, subulate-linear, acute, villous with long weak hairs on the outside; perianth-tube $\frac{5}{6}$ lin. long, glabrous;

segments 4½ lin. long; claws almost filiform, shortly pubescent, not coiled; limb ¾ lin. long, ovate-elliptic, subacute, villous outside; anthers ½ lin. long; ovary obovoid, pubescent; style 5 lin. long, filiform, constricted at the base, glabrous; stigma ¼ lin. long, ellipsoid, subacute; fruits nearly 2 lin. long, smooth, beaked. *Soranthe clavigera, Knight, Prot.* 73.

COAST REGION: Stellenbosch Div.; French Hoek Kloof, *Niven*, 40 !

9. **S. tenuifolius** (R. Br. in Trans. Linn. Soc. x. 141); branches pilose when young, at length becoming glabrous, purplish; leaves erect, 4–6 lin. long, subulate-linear, acutely or subacutely apiculate, subterete except for a narrow groove above, pilose when young, soon becoming glabrous; heads 7–9-flowered, crowded in terminal solitary or geminate clusters about 5 lin. in diam.; bract subtending each secondary peduncle 2½ lin. long, ovate or lanceolate, obtuse or subacute, densely ciliate towards the base; secondary peduncles 1–2 lin. long, villous; floral bracts 2½ lin. long, ovate to lanceolate, acute or subacute, villous, ciliate; perianth-tube ¾ lin. long, glabrous; segments 2½ lin. long, spathulate-linear; claws slender, villous, not twisting spirally; limb ½ lin. long, ovate-lanceolate, obtuse, villous outside; anthers oblong; hypogynous scales linear; ovary ellipsoid, pubescent; style 2¾ lin. long, slender, glabrous; stigma ⅙ lin. long, ovoid, obtuse. *Roem. & Schultes, Syst. Veg.* iii. 390; *Krauss in Flora,* 1845, 77, *and in Beitr. zur Fl. Cap- und Natal.* 141; *Meisn. in DC. Prodr.* xiv. 304. *Protea tenuifolia, Poir. Encycl. Suppl.* iv. 577. *Soranthe tenuifolia, Knight, Prot.* 72.

COAST REGION: Worcester Div.; Breede River, *Niven*, 20 ! Swellendam Div.; Swellendam, *Krauss*, 1061 ?(ex Meisn. l.c.).

10. **S. salsoloides** (R. Br. in Trans. Linn. Soc. x. 140); branches erect, glabrous or minutely pubescent; leaves falcately incurved or some nearly straight, 5–8 lin. long, about ½ lin. broad, terete except for a very narrow groove on the upper surface, subacute, glabrous except when in young bud; heads 1-flowered, crowded in subglobose terminal spikes about ½ in. in diam.; subtending bracts up to 1¾ lin. long, acute, narrowly lanceolate, glabrous outside, very shortly and sparingly ciliolate; floral bracts similar to the preceding; perianth-tube ¾ lin. long, nearly glabrous; segments about 3 lin. long; claws straight or slightly curved in the open flower, not spirally twisted, pubescent; limb 1 lin. long, ovate-lanceolate, sub-obtuse, villously bearded outside; anthers sessile; hypogynous scales small, linear; ovary pubescent; style about 4½ lin. long, slender, straight, glabrous; stigma ⅓ lin. long, ellipsoid or subclavate; fruits slightly stipitate, about 1 lin. long, broadly ovoid, brown, slightly pubescent. *Roem. & Schultes, Syst. Veg.* iii. 389; *Meisn. in DC. Prodr.* xiv. 303; *Krauss, Beitr. Fl. Cap- und Natal.* 141. *Protea salsoloides, Poir. Encycl. Suppl.* iv. 576. *Soranthe salsolodes, O. Kuntze, Rev. Gen. Pl.* ii. 582.

SOUTH AFRICA : without locality, *Roxburgh* ! *Viven, Ludwig, Ecklon*, 17.
COAST REGION : Caledon Div. ; summit of mountains of Baviaans Kloof, near
Genadendal, *Krauss*, 1058.

11. **S. rupestris** (Phillips in Kew Bulletin, 1911, 86) ; an alpine
snrublet ; branches blackish-villous, at length becoming nearly
glabrous and closely tubercled with the projections left by the fallen
leaves ; leaves $\frac{1}{2}$ in. long, scarcely $\frac{1}{2}$ lin. thick, subacutely mucronate,
nearly terete, thinly pilose when young, soon becoming quite
glabrous ; heads 1-flowered, arranged in a solitary dense subglobose
spike nearly $\frac{3}{4}$ in. in diam. ; subtending bract about $1\frac{3}{4}$ lin. long,
lanceolate, rusty-villous outside ; floral bracts 2–2$\frac{1}{2}$ lin. long, lanceolate, subacute, keeled and pubescent on the back, ciliate , perianth-
tube 1 lin. long, cylindric, glabrous towards the base, pubescent
above ; segments nearly 4 lin. long ; claws slender, not spirally
twisted, pubescent ; limb $\frac{1}{2}$ lin. long, ovate-lanceolate, obtuse,
tomentose outside ; anthers scarcely $\frac{1}{2}$ lin. long ; ovary nearly
glabrous ; style 3–4 lin. long, slender, glabrous ; stigma $\frac{1}{6}$ lin. long,
ellipsoid, obtuse ; fruits obovoid, 1$\frac{1}{2}$ lin. long, 1 lin. in diam.,
shining, nearly glabrous. *Soranthe rupestris, Knight, Prot.* 72. *S.
montana, Knight, l.c.* 73.

COAST REGION : Stellenbosch Div. ; tops of Stellenbosch Mountains, *Niven*, 28 !·
Caledon Div. ; mouth of the Klein River, *Niven* !

12. **S. teretifolius** (Phillips in Kew Bulletin, 1911, 85) ; branches
minutely pubescent ; leaves $\frac{1}{2}$ in. long, about $\frac{1}{2}$ lin. thick, narrowly
grooved on the upper side, otherwise terete, thinly and weakly
pilose when quite young, soon becoming glabrous ; heads terminal,
in clusters of 2–3, shortly pedunculate, several-flowered, subglobose,
about $\frac{1}{2}$ in. in diam. ; peduncles up to $\frac{1}{4}$ in. long, tomentose ; bracts
oblong, coriaceous, nearly glabrous ; bracts subtending the flowers
3–4 lin. long, linear-oblong, subacute, coriaceous, convex and thinly
villous outside, more densely villous towards the apex ; perianth-
tube 1–1$\frac{1}{4}$ lin. long, narrowly cylindric, glabrous towards the base,
pubescent above ; segments about 3 lin. long ; claws spirally
twisted and pubescent ; limb $\frac{2}{3}$ lin. long, ovate-oblong, subobtuse,
densely villous outside ; anthers $\frac{1}{4}$ lin. long ; ovary ellipsoid, acute
at both ends, nearly glabrous ; style 3$\frac{1}{4}$ lin. long, slender, constricted
at the base, glabrous ; stigma $\frac{1}{4}$ lin. long, oblong-ellipsoid, obtuse ;
fruits 2$\frac{1}{2}$ lin. long, 1$\frac{1}{4}$ lin. in diam., oblong, ellipsoid, smooth, dull
brown, glabrous. *S. lanatus, Buek in Dregè, Zwei Pfl. Documente,*
82, 222, *not of R. Br. S. lanatus, var. teretifolius, Meisn. in DC.
Prodr.* xiv. 305. *Leucadendron ? scoparium, E. Meyer, ex Meisn. l.c.*
(*as to specimen a in Herb. Drège*).

COAST REGION : Worcester Div. ; Dutoits Kloof, *Drège* !

13. **S. crassifolius** (Hutchinson) ; branches densely and shortly
tomentellous ; leaves $\frac{3}{4}$–1 in. long, $\frac{2}{3}$ lin. thick, terete except a
narrow groove on the upper side, subacutely mucronate, glabrous ;

heads 2–3 together at the apex of a common peduncle, about ¼ in. in diam., 4–6-flowered; peduncle up to 5 lin. long, tomentellous; bracts subtending the flowers about 2 lin. long, nearly 1 lin. broad at the base, long-acuminate, slightly pubescent in the lower, glabrous in the upper part; perianth-tube ¾ lin. long, glabrous below, puberulous above; segments soon becoming reflexed and spirally coiled, about 3 lin. long; claw tomentellous; limb scarcely 1 lin. long, subacute, coriaceous, thinly pubescent outside; anthers ⅔ lin. long, narrow; hypogynous scales very small; ovary ellipsoid, pubescent; style slightly constricted above the ovary, 3½ lin. long, slender, glabrous; stigma ⅓ lin. long, narrowly ellipsoid or subclavate, obtuse; fruits 1½ iin. long, ¾ lin. in diam., narrowly obovoid, brown, slightly pubescent.

COAST REGION : Caledon Div.; near Genadendal, *Schlechter*, 9832! tops of the mountains of Baviaans Kloof, near Genadendal, *Burchell*, 7723!

Imperfectly known species.

14. S. spatalloides (Buek in Drège, Zwei Pfl. Documente, 222, name only, not of R. Br.).

REGION? either from Dutoits Kloof in Worcester Div., or from the Great Zwartberg Range near Vrolykheid, in Prince Albert Div., *Drège*!

The original specimen in the Stockholm Herbarium is very imperfect; it is probably a species of *Leucadendron* and may be a form of *L. sorocephalodes*, Phillips & Hutchinson.

XIV. NIVENIA, R. Br.

Flowers hermaphrodite, actinomorphic. *Perianth* cylindric in bud with an ellipsoid limb, 4-partite to near the base; tube short, usually glabrous, rarely villous; segments differentiated into a slender hairy claw and an elliptic or oblong obtuse or rarely apiculate villous limb. *Stamens* 4; anthers sessile or subsessile, inserted at the base of the limb, oblong or elliptic; connective usually very slightly produced at the apex. *Hypogynous scales* linear or subulate. *Ovary* sessile, pubescent, surrounded by a basal ring of hairs; style mostly slender, straight, subpersistent, glabrous, or hairy on part of its length; stigma narrowly clavate or subellipsoid, gradually passing into the style, rarely broadly ovoid and capitate, epapillose. *Ovule* 1, laterally inserted. *Fruit* a smooth ivory-white shining nut with a basal ring of hairs, often beaked and tipped by the persistent basal portion of the style. *Seed* solitary.

Small erect shrubs; leaves usually crowded, dimorphic with the lower variously pinnately divided and the upper broadly spathulate or flabellate and tapered to a petiolar basal portion, rarely linear or oblanceolate, or all alike and much dissected with linear obtuse segments and more or less fan-shaped in outline, glabrous when mature or rarely permanently silky-tomentose; flowers in partial 4-flowered heads, the latter arranged in terminal usually solitary sessile or pedunculate cylindric or rarely subglobose spikes; partial heads each subtended by a solitary coriaceous entire bract; floral bracts 4, imbricate, the two lateral exterior, coriaceous, hairy outside ; perianth equally hairy all over or rarely the claws with shorter indumentum than that on the limb, white or carmine, so far as known.

DISTRIB. Species about 13, confined to the South-Western Region of Cape Colony.

To preserve uniformity in nomenclature (see note under *Sorocephalus*) it has been considered advisable to adopt the name *Nivenia*, R. Br. (1810) in preference to that of *Paranomus*, Salisb. (1807), although unlike *Sorocephalus*, it is not included in the list of *nomina conservanda* adopted by the Vienna Congress.

Stigma broadly ovoid, capitate ; style hairy on the
   middle third of its length ... ... (1) **parvifolia.**

Stigma clavate or more or less cylindric, gradually passing
   into the style :
  Leaves dimorphic (rarely all alike and then undivided),
    the upper entire and broadly spathulate or
    flabellate or rarely spathulate-oblanceolate, the
    lower variously divided ; style glabrous :
    Partial flower-heads strongly reflexed at the time of
     flowering ... ... (2) **reflexa.**
    Partial flower-heads never reflexed :
     Upper leaves broadly spathulate, ½ in. broad or
      more :
      Branches glabrous ; ultimate segments of the
       divided leaves quite terete ; perianth-limb
       not apiculate ... ... (3) **spathulata.**
      Branches glabrous ; ultimate leaf-segments
       broadly concave above ; perianth-limb
       apiculate ... ... (4) **Muirii.**
      Branches usually pubescent ; ultimate leaf-
       segments broadly concave or flat with
       incurved margins ; perianth-limb not apicu-
       late ... ... (5) **Sceptrum.**
     Upper leaves narrowly spathulate-oblanceolate or
      oblong, 1½–2 lin. broad ... ... (6) **diversifolia.**
  Leaves all alike or if slightly dimorphic then the upper
    linear and similar to the segments of the lower ;
    style hairy (except in 7, *Dregei*) :
    Style glabrous ... ... (7) **Dregei.**
    Style hairy :
     Inflorescence small and subglobose, about ⅓ in.
      long ; leaves ½–1 in. long ... ... (8) **capitata.**
     Inflorescence elongated, more or less cylindric,
      usually more than 1 in. long ; leaves over
      1 in. long :
      Mature leaves glabrous or rarely slightly hairy
       at the time of flowering :
       Inflorescence usually dense ; bracts densely
        villous :
        Bract subtending the partial flower-heads
         ovate, caudate-acuminate, densely
         villous with interlacing hairs ; floral
         bracts obtuse, rigidly coriaceous ... (9) **crithmifolia.**
        Bract subtending the partial flower-heads
         subulate-linear, villous with straight
         hairs ; floral bracts long-acuminate,
         membranous ... ... (10) **Lagopus.**

Inflorescence usually rather lax : bracts
shortly tomentose or tomentellous, rarely
with a few longer hairs towards the
apex   ...   . (11) **spicata.**

Mature leaves densely and softly silky-tomentose
at the time of flowering, rarely becoming
nearly glabrous in the fruiting stage :
Leaf-segments ¼ in. long ; bracts rounded at
the apex ; style 6 lin. long   ...   ... (12) **tomentosa.**

Leaf-segments ¾–1 in. long ; bracts shortly
acuminate ; style 4 lin. long   ...   ... (13) **mollissima.**

1. **N. parvifolia** (R. Br. in Trans. Linn. Soc. x. 135) ; branches
subterete, subtomentellous when young, becoming puberulous ;
leaves dimorphic ; the lower up to 3½ in. long, 3–4 times dichotomously divided, with the ultimate segments linear or rarely
subspathulate, subterete, obtuse or subacute, glabrous ; the upper
½–1 in. long, entire, broadly spathulate or flabellate, with an obtuse
triangular apex, much narrowed and slender towards the base,
erect, distinctly nerved, coriaceous, glabrous except when quite
young and then tomentellous ; heads 4-flowered, spicate ; spikes
terminal, solitary or up to 3 together, sessile or shortly pedunculate,
up to 3 in. long, more or less cylindric and about ¾ in. in diam. ;
flowering axis tomentose ; bearing linear-subulate acute shortly
ciliate bracts about 4 lin. long ; floral bracts 4, oblong-oblanceolate,
rounded at the apex, about 4 lin. long, 2–2½ lin. broad, coriaceous,
silky adpressed-villous outside, glabrous within ; flowers with
numerous long white hairs round the base ; perianth-tube obconic,
1 lin. long, glabrous ; segments 4–5 lin. long ; claws straight, very
densely white-villous outside ; limb ¾ lin. long, elliptic, obtuse,
densely villous outside ; anthers sessile, ⅔ lin. long, with the
connective slightly produced at the apex ; hypogynous scales 1 lin.
long, terete ; ovary oblique, puberulous ; style ½ in. long, rather
stout, hairy on the middle third of its length, otherwise glabrous ;
stigma capitate, broadly ovoid, nearly ½ lin. long and broad ; fruits
ivory-white, smooth and shining, about 4 lin. long, oblong-ellipsoid,
acutely beaked, glabrous except for a few long white hairs at the
base. *Roem. & Schultes, Syst. Veg.* iii. 386 ; *Meisn. in DC. Prodr.*
xiv. 300. *N. spathulata, Drège ex Meisn. l.c., name only. N. Sceptrum, forma dissecta, Gandog. in Bull. Soc. Bot. France,* xlviii. p. xcvi.
*Protea spathulata, Thunb. Diss. Prot. t. 5, excl. his herbarium specimens
and description. Paranomus adiantifolius, Knight, Prot.* 70. *P. parvifolius, O. Kuntze, Rev. Gen. Pl.* ii. 580.

COAST REGION : Caledon Div. ; River Zondereinde, *Niven,* 201 ! Knoflooks
Kraal and Little Houw Hoek, *Zeyher,* 3717 ! Houw Hoek, *Schlechter,* 9423 !
Pillans !

2. **N. reflexa** (Phillips & Hutchinson) ; branches puberulous or
tomentellous, rarely nearly glabrous ; leaves dimorphic ; the lower
up to 3½ in. long, bipinnately divided in the upper half, with the
ultimate segments linear or rarely oblanceolate, obtuse, concave on

the upper surface, glabrous; the upper erect, entire, 1–2 in. long,
½–1 in. broad, broadly obovate in the upper, gradually narrowed to
the petiolar base in the lower half, coriaceous, glabrous on both
surfaces, margins not or only very slightly cartilaginous; heads
4-flowered, spicate; spikes terminal, solitary, sessile or on a
peduncle up to 1½ in. long, 1½–2½ in. long, 1¼–1½ in. in diam.,
subcylindric; flowering axis rusty-tomentose, bearing subulate-linear
very acute glabrescent bracts about 4 lin. long; floral bracts 4, 1¾
lin. long, 1⅓ lin. broad, ovate, obtuse, densely rusty-villous outside,
glabrous within; flowers surrounded at the base by numerous long
hairs; perianth-tube about 1 lin. long, nearly glabrous; segments
1¼ in. long, very slender; claws shortly pubescent or tomentellous;
limb nearly 2 lin. long, linear, apiculate, shortly tomentose outside;
anthers 1–1½ lin. long, linear; hypogynous scales ⅔ lin. long, linear,
acute; ovary with a ring of long hairs at the base, otherwise only
very shortly pubescent; style very slender, 1½ in. long, glabrous,
gradually tapered into a very narrow cylindric obtuse stigma ¾ lin.
long and about ⅛ lin. thick; fruits 3½ lin. long, ellipsoid, shortly
beaked, glabrous. *N. Sceptrum, Meisn. in DC. Prodr.* xiv. 299, *as
to Zeyher,* 3713 β, *and the following synonym. Leucospermum spathulatum, Drège ex Meisn. l.c., not of R. Br.*

Coast Region: Uitenhage Div; Van Stadens Berg, *Drège! Zeyher,* 3713b!
Van Stadens Berg, near Galgebosch, *Burchell,* 4686!

3. **N. spathulata** (R. Br. in Trans. Linn. Soc. x. 135); branches
terete, glabrous; leaves dimorphic; the lower 2 in. long, bipinnately
divided in the upper half, with the ultimate segments quite terete,
about ⅓ lin. in diam., obtuse, glabrous; the upper ½–¾ in. long,
erect, entire, broadly flabellate or suborbicular, obtusely and very
shortly mucronate, abruptly narrowed into a petiolar base, rigidly
coriaceous, glabrous and dull on both surfaces, with narrow cartilaginous purplish margins; heads 4-flowered, spicate; spikes terminal,
solitary, about 1¼ in. long, ellipsoid or subcylindric; flowering axis
very densely villous-tomentose, bearing ovate-lanceolate caudate-
acuminate densely villous bracts about 4 lin. long with a glabrous
acumen; floral bracts 4, about 3 lin. long and 1 lin. broad, lanceolate,
acute or subobtuse, coriaceous, glabrous and shining within, silky-
villous outside; flowers surrounded by numerous long white hairs;
perianth-tube about 1 lin. long, nearly glabrous; segments ½ in.
long; claws rather thick, straight in open flower, villous outside;
limb 1⅓ lin. long, obtuse, not acuminate, villous outside; anthers
linear; hypogynous scales ¾ lin. long, linear; ovary with a few long
white hairs towards the base, otherwise shortly pubescent; style
½ in. long, slender, broader and flattened towards the base, glabrous,
gradually passing into the stigma which is 1 lin. long, clavate,
obtuse. *Roem. & Schultes, Syst. Veg.* iii. 385; *Meisn. in DC. Prodr.*
xiv. 300. *N. marginata, R. Br. l.c.* 134; *Roem. & Schultes, l.c.;
Meisn. l.c. N. parvifolia, E. Meyer in Drège, Zwei Pfl. Documente,*

116, 204, *not of R. Br. N. marginata, Drège, and N. parvifolia, Buek ex Meisn. l.c.* 300, *names only. Protea spathulata, Thunb. Diss. Prot.* 44, *excl. fig.; Prodr.* 28 ; *Lam. Illustr.* i. 235 ; *Willd. Sp. Pl.* i. 533 ; *Poir. Encycl.* v. 642 ; *Thunb. Fl. Cap. ed. Schultes,* 125. *Paranomus flabellifer, Knight, Prot.* 70. *P. marginatus, O. Kuntze, Rev. Gen. Pl.* ii. 580.

SOUTH AFRICA : without locality, *Lindley* ! *Roxburgh* !
COAST REGION : Swellendam Div. ; between Spaarbosch and Tradouw, *Drège* !
Riversdale Div. ; Platte Kloof, *Masson* ! between Garcias Pass and Muis Kraal, *Bolus,* 11363 !

4. **N. Muirii** (Phillips & Hutchinson) ; branches terete, quite glabrous, bark purplish ; leaves dimorphic ; the lower 2½–3 lin. long, flabellately divided in the upper half, with the segments linear or linear-oblanceolate, convex below, concave above, obtuse, glabrous ; the upper 1–1½ in. long, ¾–1¼ in. broad, suborbicular or flabellate, obtusely mucronate, abruptly narrowed to the base, erect, entire, rigidly coriaceous, distinctly nerved, dull and glabrous on both surfaces, with purplish subtranslucent margins ; heads 4-flowered, spicate ; spikes terminal, solitary, sessile, about 4 in. long, ¾ in. in diam., cylindric ; flowering axis densely tomentose or villous, bearing lanceolate obtuse densely villous bracts 4–5 lin. long ; floral bracts 4, 4–6 lin. long, 2–3½ lin. broad, oblong or oblong-ovate, obtuse, glabrous and shining within, densely white silky-tomentose outside ; flowers surrounded at the base by numerous long white hairs ; perianth-tube ¾ lin. long, glabrous towards the base ; segments 6½ lin. long ; claws straight in the open flower, white-villous outside ; limb 1½ lin. long, acuminate, densely white-villous outside ; anthers linear, 1 lin. long ; hypogynous scales 1 lin. long, linear, acute ; ovary subcylindric, long-villous with white hairs in the lower half, otherwise minutely puberulous ; style ½ in. long, slender, gradually passing into the stigma which is 1 lin. long, clavate, subobtuse ; fruits 2½ lin. long, cylindric, subobtuse, shortly pubescent and with a ring of long white hairs round the base.

COAST REGION : Riversdale Div. ; mountains at Garcias Pass, *Galpin,* 4481 ! Kampsche Berg, *Muir,* 276 !

5. **N. Sceptrum** (R. Br. in Trans. Linn. Soc. x. 134); a shrub 4–5 ft. high, erect, branched ; branches ascending, terete, shortly pubescent or glabrous ; leaves dimorphic ; the lower up to 4 in. long, bipinnately divided in the upper ⅔, with the ultimate segments ¾–1¾ in. long, 2–3½ lin. broad, rounded at the apex, coriaceous, glabrous, margins incurved ; the upper erect, entire, 1½–2½ in. long, cuneate-obovate or oblanceolate, with an obtuse callous apex, gradually narrowed to the petiolar base, rigidly coriaceous, glabrous on both surfaces, often distinctly nerved, margins not or scarcely cartilaginous ; heads 4-flowered, spicate ; spikes terminal, solitary, subsessile or shortly pedunculate, 1½–2½ in. long, about 1 in. in diam., subcylindric ; flowering axis tomentose ; bract subtending

each partial head about 3 lin. long, ovate-lanceolate, acuminate, subacute, tomentose outside; floral bracts 4, 2–2½ lin. long, about 1½ lin. broad, more or less oblong, rounded at the apex, coriaceous, glabrous within, tomentose outside; flowers surrounded at the base by numerous long hairs; perianth-tube 1–1½ lin. long, glabrous towards the base; segments ½ in. long; claws shortly tomentose outside; limb 1½ lin. long, linear, subobtuse, tomentose outside; anthers 1 lin. long, linear; hypogynous scales 1 lin. long, linear, acute; ovary pubescent; style rather slender, 6–7½ lin. long, glabrous, gradually narrowed into the stigma which is 1 lin. long, subcylindric, obtuse; fruits 3 lin. long, ovoid, shortly beaked, ivory-white and shining, minutely pubescent and with a dense ring of long white hairs around the base. *Roem. & Schultes, Syst. Veg.* iii. 385; *Meisn. in DC. Prodr.* xiv. 299, *incl. var. splendens, Meisn.? N. alopecuroides, Lam. ex St. Lag. in Ann. Soc. Bot. Lyon,* vii. 130. *Protea Sceptrum gustavianum, Sparrm. in Vet. Acad. Handl. Stockh.* 1777, 53, *t.* i.; *Linn. f. Suppl.* 116. *P. Sceptrum, Thunb. Diss. Prot.* 21; *Willd. Sp. Pl.* i. 511; *Thunb. Fl. Cap. ed. Schultes,* 125; *Poir. Encycl.* v. 662. *P. alopecuroides, Lam. Illustr.* i. 240. *Paranomus sceptriformis, Knight, Prot.* 69.

SOUTH AFRICA: without locality, *Sparrman, Thom,* 617! *Roxburgh! Masson!*
COAST REGION: Stellenbosch Div.; Lowrys Pass, *Burchell,* 8224! 8268! *Bolus,* 5832! *Niven!* Hottentots Holland and Zwart Berg, *Bowie!* Caledon Div.; mountains of Baviaans Kloof, near Genadendal, *Burchell,* 7713! Zwart Berg, *Zeyher,* 3713! *Pappe!* Zandfontein, *Galpin,* 4482! Bredasdorp Div.; Elim, *Schlechter,* 7665!

6. **N. diversifolia** (Phillips & Hutchinson); branches glabrous; leaves dimorphic; the lower bipinnately divided in the upper third, 1½–3 in. long, the ultimate segments 1½–4 lin. long, terete, obtuse; upper 5–10½ lin. long, entire, spathulate-oblanceolate or oblong, subacuminate, glabrous; heads 4-flowered, spicate; spikes sessile, 2 in. long, cylindric, about 1¼ in. in diam., surrounded by the upper leaves which gradually become bract-like with a green acumen and purplish thinner margins; bract subtending each partial head nearly 5 lin. long, lanceolate, long-acuminate, densely rusty-villous outside; floral bracts 5–7 lin. long, lanceolate-elliptic, long-acuminate, densely villous with long hairs; perianth-tube 1½ lin. long, glabrous; segments 7½ lin. long; claws very slender, rusty-villous; limb 1¼ lin. long, linear, subacute, densely villous; anthers linear; hypogynous scales 1 lin. long, linear, acuminate; ovary ellipsoid; style 7½ lin. long, filiform, glabrous; stigma 1 lin. long, cylindric, obtuse. *Sorocephalus diversifolius, R. Br. in Trans. Linn. Soc.* x. 143; *Roem. & Schultes, Syst. Veg.* iii. 391; *Meisn. in DC. Prodr.* xiv. 306. *Protea diversifolia, Poir. Encycl. Suppl.* iv. 577. *Soranthe diversifolia, O. Kuntze, Rev. Gen. Pl.* ii. 582.

SOUTH AFRICA: without locality, *Roxburgh!*
COAST REGION: Tulbagh Div.; Roode Zand (near Tulbagh), *Niven!*

7. **N. Dregei** (Buek in Drège, Zwei Pfl. Documente, 64, 204); branches puberulous, at length becoming glabrous or nearly so;

leaves all of one kind and divided, or if a few of the upper entire
then they are linear, up to 3 in. long, pinnately divided in the
upper half or in some specimens only 2–3-lobed, erect, coriaceous,
glabrous except when young and then tomentose, the segments in
the more divided leaves linear, obtuse, those of the less divided a
little broader, all concave on the upper surface ; heads 4-flowered,
spicate ; spikes sessile or subsessile, terminal, solitary or geminate,
about $1\frac{1}{4}$ in. long, nearly 1 in. in diam., subcylindric ; flowering axis
tomentose, bearing ovate acuminate rigidly coriaceous shortly
pubescent bracts about 3 lin. long and 2 lin. broad ; floral bracts 4,
imbricate, the outer 2 larger than the inner, up to 5 lin. long and
$2\frac{3}{4}$ lin. broad, oblong, rounded at the apex, rigidly coriaceous, shortly
woolly-tomentose outside, glabrous within ; flowers surrounded at the
base by numerous long hairs ; perianth-tube $1\frac{1}{2}$ lin. long, nearly
glabrous ; segments about 7 lin. long ; claws slightly recurved in
the open flower, shortly tomentose ; limb $1\frac{1}{2}$ lin. long, narrowly
oblong-lanceolate, subacute, shortly and densely villous ; hypogynous
scales $1–1\frac{1}{2}$ lin. long, linear-filiform ; ovary surrounded by a ring of
long white hairs, puberulous ; style 7 lin. long, rather slender,
glabrous, gradually passing into the stigma which is $1\frac{1}{4}$ lin. long,
cylindric, grooved ; fruits $2\frac{3}{4}$ lin. long, ovoid-ellipsoid, tipped by a
persistent portion of the style, brightly shining, surrounded by a
ring of white hairs. *Meisn. in DC. Prodr.* xiv. 300. *Sorocephalus
diversifolius, Drège ex Meisn. l.c.* 301, *name only, not of* B. *Br.
Paranomus Dregei, O. Kuntze, Rev. Gen. Pl.* ii. 580. *Soranthe
Dregei, O. Kuntze, l.c.* 582.

CENTRAL REGION: Prince Albert Div. ; Great Zwartberg Range, *Drège* !
Zwartberg Pass, *Bolus*, 11629 !

8. N. capitata (R. Br. in Trans. Linn. Soc. x. 138) ; branches
tomentellous ; leaves all alike, $\frac{1}{2}$–1 in. long, pinnately or bipinnately
divided in the upper $\frac{3}{4}$, very shortly pubescent when young, glabrous
when mature, the ultimate segments 2–6 lin. long, subterete except
for a narrow groove on the upper side, almost filiform, subobtuse ;
heads 4-flowered, crowded in a terminal solitary sessile subglobose
cluster about $\frac{1}{2}$ in. in diam. ; bract subtending each partial head
$2\frac{1}{2}$ lin. long, elliptic, obtusely acuminate, at length recurved, slightly
pilose ; floral bracts 2 lin. long, ovate, subacutely acuminate, rigidly
coriaceous, pubescent ; perianth-tube $\frac{3}{4}$ lin. long, sparingly pilose ;
segments $2\frac{3}{4}$ lin. long ; claws villous ; limb $\frac{2}{3}$ lin. long, lanceolate,
subobtuse, villous ; anthers $\frac{1}{2}$ lin. long ; linear ; hypogynous scales
$\frac{2}{3}$ lin. long, linear, acute ; ovary subcylindric, surrounded by a ring
of long hairs at the base, otherwise pubescent ; style 3 lin. long,
pilose on the middle third of its length ; stigma $\frac{1}{6}$ lin. long,
subclavate, rather obtuse. *Roem. & Schultes, Syst. Veg.* iii. 388 ;
E. Meyer in Drège, Zwei Pfl. Documente, 79, 204 ; *Meisn. in DC.
Prodr.* xiv. 302. *Protea capitata, Poir. Encycl. Suppl.* iv. 575.
Paranomus capitatus, O. Kuntze, Rev. Gen. Pl. ii. 580.

COAST REGION : Worcester Div. ; Dutoits Kloof, *Drège* ! Brand Vley, *Roxburgh* !

9. N. crithmifolia (R. Br. in Trans. Linn. Soc. x. 136) ; branches tomentellous except when quite young and then more or less tomentose ; leaves all alike, 2–2½ lin. long, bi- or tripinnately divided in the upper two-thirds, glabrous except when quite young, coriaceous, ultimate segments terete, rarely flat, with an obtuse callous apex ; heads 4-flowered, spicate ; spikes terminal, solitary or several together, up to 3½ in. long, cylindric, shortly or rarely rather long-pedunculate ; peduncle and flowering axis stout, densely rusty-tomentose ; bract subtending each partial head about ¼ in. long, nearly 2 lin. broad, long and caudate-acuminate from an ovate base, coriaceous, glabrous within, very densely villous with interlacing hairs outside ; floral bracts ovate-oblong, obtuse or subobtuse, up to 4 lin. long and 1½ lin. broad in the flowering stage, becoming harder, longer and shortly pointed in the fruiting stage ; perianth-tube densely long-villous, glabrous at the base ; segments 5 lin. long ; claws shortly tomentellous ; limb 1 lin. long, narrowly elliptic, obtuse, villously bearded besides the shorter tomentellous indumentum ; anthers ⅔ lin. long, oblong ; hypogynous scales 1 lin. long, linear, acute ; ovary surrounded by a ring of dense long hairs at the base ; style 5 lin. long, villous or pubescent from near the base to within 1 lin. of the stigma ; stigma ½ lin. long, ellipsoid, obtuse ; fruits about 3½ lin. long, 2 lin. thick, ellipsoid, acutely beaked, smooth and shining, ivory-white, surrounded by a ring of long hairs at the base. *Roem. & Schultes, Syst. Veg.* iii. 387 ; *Meisn. in DC. Prodr.* xiv. 301. *Protea Lagopus, Andr. Bot. Rep. t.* 243. *N. Bolusii, Gandog. in Bull. Soc. Bot. France,* xlviii. *p.* xcvi. *Paranomus crithmifolius, Knight, Prot.* 69, *excl. syn. Thunb., Linn. and Berg. ; O. Kuntze, Rev. Gen. Pl.* ii. 580.

SOUTH AFRICA : without locality, *Lindley* ! *Thunberg* (named *Protea Lagopus* β in his herbarium) !
COAST REGION : Stellenbosch Div. ; Hottentots Holland Mountains, *Mund,* 50 ! *Niven* ! *Bolus,* 4197 ! Caledon Div. ; Genadendal, *Drège* ! mountains of Baviaans Kloof, near Genadendal, *Burchell,* 7814 ! near Bot River, *Bolus,* 1352 ! Oudtshoorn Div. ; Oliphants River, *Gill* !

10. N. Lagopus (R. Br. in Trans. Linn. Soc. x. 137) ; branches pilose ; leaves all alike, 1½–2 in. long, bipinnately divided in the upper half, pubescent or puberulous when young, soon becoming glabrous, ultimate segments slender, terete, obtuse, up to ½ in. long ; heads 4-flowered, spicate ; spikes terminal and solitary or sometimes clustered and pedunculate, up to 2½ in. long, about ½ in. in diam., very densely flowered ; flowering axis tomentose ; bract subtending each partial head about 4 lin. long, subulate-linear, very acute, densely villous outside ; floral bracts about 2½ lin. long and 1¼ lin. broad, ovate, long and gradually acutely acuminate, almost membranous, very densely long-villous outside ; flowers at the time of opening 4 lin. long or less ; perianth-tube shortly pubescent, or nearly glabrous ; segments 3½ lin. long ; claws tomentellous or

shortly pubescent, becoming spirally coiled in the open flower; limb
1 lin. long, lanceolate, subobtuse, villous outside; anthers sessile;
hypogynous scales linear; ovary surrounded by a ring of hairs;
style 3½–4 lin. long, thinly pubescent in the lower half; stigma
ellipsoid, subobtuse, gradually passing into the style, about ⅓ lin.
long.   *Roem. & Schultes, Syst. Veg.* iii. 387 ; *E. Meyer in Drège,
Zwei Pfl..Documente,* 78, 119, 204 ; *Meisn. in DC. Prodr.* xiv. 302,
*incl. var. sericea, Meisn. l.c.?   N. micrantha, Schlechter in Engl.
Jahrb.* xxvii. 107.   *Protea Lagopus, Thunb. Diss. Prot.* 19 (*excl.
specimen β of his herbarium*) ; *Willd. Sp.* i. 510.   *Paranomus abrotani-
folius and P. cumuliflorus, Knight, Prot.* 68.

SOUTH AFRICA : without locality, *Thunberg,* sheet *a* ! *Roxburgh* !
COAST REGION : Tulbagh Div. ; near Tulbagh Waterfall, *Bolus,* 1351 ! New
Kloof, *Drège, a* ! Worcester Div. ; near Brede River Station, *Drège, b* ! Bredas-
dorp Div. ; Elim, *Schlechter,* 9640 ! Swellendam Div. ; on mountains, *Niven* !

11. N. spicata (R. Br. in Trans. Linn. Soc. x. 136, excl. syn.
Thunb.) ; branches tomentellous or puberulous ; leaves all alike,
about 2 in. long, bipinnately divided in the upper half, puberulous
when young, soon becoming quite glabrous, ultimate segments up to
¾ in. long, terete except for a narrow groove on the upper surface,
obtuse ; heads 4-flowered, spicate ; spikes terminal, solitary or subsolitary, up to 3½ in. long, lax, pedunculate ; peduncle mostly about
1 in. long, densely yellow-tomentose, bearing a few subulate very
acute shortly pubescent bracts 4–5 lin. long ; bract subtending
each partial head about 2 lin. long, subulate-lanceolate, acute,
shortly tomentose outside ; floral bracts about 2 lin. long and broad,
broadly ovate, caudate-acuminate, coriaceous, shortly and softly
tomentose or tomentellous outside, glabrous within ; perianth-tube
about 1¼ lin. long, glabrous at the base, shortly pubescent above ;
segments about ½ in. long : claws shortly and softly tomentose,
recurved or coiled in the open flower ; limb 1 lin. long, lanceolate-
elliptic, subobtuse, villous outside ; anthers ¾ lin. long, linear-oblong ;
hypogynous scales ½ lin. long, linear ; ovary surrounded by a ring of
long hairs ; style 7 lin. long, sparingly pilose in the lower half ;
stigma gradually passing into the style, about ¾ lin. long, clavate,
subobtuse.   *Roem. & Schultes, Syst. Veg.* iii. 386 ; *Meisn. in DC.
Prodr.* xiv. 301.   *N. media, R. Br. l.c.* 137 ; *E. Meyer in Drège, Zwei
Pfl. Documente,* 72, 73, 204 ; *forma Zeyheri, Gandog. in Bull. Soc.
Bot. France,* xlviii. *p.* xcvi.   *N. intermedia, Steud. Nomencl. ed.* 2,
ii. 196, *name only.   N. laxa, Schlechter in Journ. Bot.* 1897,
282 ?   *Leucadendron spicatum, Berg. in Vet. Acad. Handl. Stockh.*
1766, 327 ; *Berg. Descr. Pl. Cap.* 25.   *Protea spicata, Linn. Mant.
alt.* 187 ; *Willd. Sp. Pl.* i. 511, *excl. syn. Thunb.* ; *Andr. Bot. Rep.
t.* 234.   *Paranomus bracteolaris, Knight, Prot.* 68.   *P. medius and
P. spicatus, O. Kuntze, Rev. Gen. Pl.* ii. 580.

COAST REGION : Clanwilliam Div. ; Blau Berg, Honig Valley and Koude Berg,
Drège, a ! Pakhuis Pass, *Bolus,* 9082 ! and without precise locality, *Mader,* 95 !
Piquetberg Div. ; Pikeniers Kloof, *MacOwan,* 1947 ! *Schlechter,* 10770 ! Tulbagh
Div. ; New Kloof, *Drège* ! Worcester Div. ; Matroosberg, *Marloth,* 2248.

Caledon Div. ; Babylons Tower Mountains, *Zeyher*, 3716 ! Zwart Berg, *MacOwan*, 770 ! *Galpin*, 4484 ! near Caledon, *Bolus*, 9917 ! 9927 ! Bredasdorp Div. ; near Elim, *Bolus*, 7863 ! Swellendam Div. ; on mountains, *Niven* ! Riversdale Div. ; near Garcias Pass, *Galpin*, 4483 !
CENTRAL REGION : Ceres Div. ; Witzen Berg and Scurfde Berg, *Zeyher*, 1479 !

12. N. tomentosa (Phillips & Hutchinson); branches terete densely and softly tomentose or subvillous with whitish hairs ; leaves all alike, erect, $1\frac{1}{4}$–$1\frac{1}{2}$ in. long, bipinnately divided in the upper two-thirds, more or less flabellate in outline, permanently silky-tomentose with silvery-white hairs, ultimate segments 1–3 lin. long, terete, obtuse ; heads 4-flowered, spicate ; spikes terminal, solitary, subsessile, about $1\frac{1}{2}$ in. long, subcylindric ; flowering axis densely tomentose ; bract subtending each flower-head 4 lin. long, about 3 lin. broad, broadly ovate and somewhat cupular, shortly acuminate, coriaceous, adpressed-pilose outside, glabrous and shining within ; floral bracts 4, about 4 lin. long, 2–$2\frac{1}{2}$ lin. broad, ovate-oblong, rounded at the apex, thinly coriaceous, densely adpressed-pilose outside, glabrous within ; flowers surrounded by numerous long white hairs ; perianth-tube 2 lin. long, densely adpressed-pilose ; segments about $\frac{1}{2}$ in. long ; claws straight, densely villous ; limb $1\frac{1}{2}$ lin. long, lanceolate, subobtuse, densely villous with long hairs outside ; anthers scarcely 1 lin. long, narrowly elliptic ; hypogynous scales 1 lin. long, linear, subacute ; ovary subovoid, pubescent ; style $\frac{1}{2}$ in. long, pubescent nearly to the apex, gradually passing into the stigma which is $\frac{2}{3}$ lin. long, subcylindric, subacute, glabrous ; fruit about 4 lin. long, obliquely oblong, beaked, obtusely 4-angled, minutely puberulous, with a dense ring of long hairs at the base. *Nivenia mollissima, E. Meyer in Drège, Zwei Pfl. Documente*, 74, 204, not of *R. Br.*

COAST REGION : Clanwilliam Div. ; Ezelsbank, *Drège* ! Koude Berg, *Schlechter*, 8770 !

13. N. mollissima (R. Br. in Trans. Linn. Soc. x. 138) ; branches tomentellous with short crisped hairs ; leaves all alike, bipinnately divided in the upper two-thirds, tomentose, rarely becoming nearly glabrous in the fruiting stage; ultimate segments about 1 in. long, sub-obtuse or subacute, apiculate ; heads 4-flowered, spicate ; spikes terminal, solitary or clustered, 2–$3\frac{1}{4}$ in. long, about $\frac{1}{2}$ in. in diam. ; peduncle $\frac{3}{4}$–$1\frac{1}{2}$ in. long, densely tomentose, bearing a few scattered bracts ; bract subtending each partial head $1\frac{1}{2}$–$2\frac{1}{2}$ lin. long, linear, linear-lanceolate or ovate, acuminate, acute or subacute, villous ; floral bracts $1\frac{1}{2}$–$3\frac{1}{2}$ lin. long, ovate, shortly acuminate, subacute or sub-obtuse, tomentose ; perianth-tube 1–$1\frac{1}{4}$ lin. long, pubescent or villous ; segments 4–$4\frac{1}{4}$ lin. long, spathulate-linear ; claws tomentose ; limb $\frac{3}{4}$–$\frac{5}{6}$ lin. long, elliptic or ovate-elliptic, suboobtuse, villously bearded ; anthers $\frac{1}{2}$–$\frac{2}{3}$ lin. long, linear ; hypogynous scales linear ; ovary minutely pubescent, surrounded by a ring of long hairs at the base ; style $4\frac{1}{2}$ lin. long, pilose or villous on the middle third of its length ; stigma ¼ lin. long, subcylindric, subacute ; fruit 2 lin. long, obliquely ovoid, long-beaked and tipped by the persistent style. *Roem. & Schultes, Syst. Veg.* iii. 388 ; *Meisn. in DC. Prodr.* xiv. 302.   *N. spicata, R. Br. in Trans. Linn. Soc.* x. 136 *as to syn. Thunb. N. candicans, R. Br. in l.c.* 221, *in obs.* ; *Roem. & Schultes, l.c.* 389 ; *Meisn. l.c. Protea spicata, Thunb. Diss. Prot.* 20 ; *Murr. Syst. Veg. ed.* xiv. 37 ; *Thunb. Fl. Cap. ed. Schultes,* 124.   *P. candicans, Thunb. Prodr. Append.* 186 ; *Fl. Cap. ed. Schultes,* 123.   *P. mollissima, Poir. Encycl. Suppl.* iv. 575.   *Paranomus argenteus, Knight, Prot.* 68 ?   *P. candicans and P. mollissimus, O. Kuntze, Rev. Gen. Pl.* ii. 580.   *Serruria ? albicans, Roem. & Schultes, Syst. Veg.* iii. 378.

SOUTH AFRICA : 'without locality, *Sparrman* ! *Thunberg* ! *Masson* ! *Roxburgh* !
COAST REGION : Stellenbosch Div. ; Hottentots Holland Mountains, *Thunberg* ! Uniondale Div. ; Lange Kloof, *Masson.*

Imperfectly known species.

14. N. ? concava (R. Br. in Trans. Linn. Soc. x. 221, in obs.) ; branches very villous ; leaves subsessile, imbricate, ovate, concave, somewhat wrinkled, glabrous, with a subcallous apex ; heads crowded, globose ; bracts short, ovate, acute, like the perianth pubescent.   *Roem. & Schultes, Syst. Veg.* iii. 388 ; *Meisn. in DC. Prodr.* xiv. 302.   *Protea concava, Lam. Illustr.* i. 234 ; *Poir. Encycl.* v. 642.   *Paranomus concavus, O. Kuntze, Rev. Gen. Pl.* ii. 580.

Said to be from the Cape.

15. Paranomus longicaulis (Knight, Prot. 70) ; stem 5–6 ft. high ; branches long ; lower leaves bipinnatifid, upper spathulate ; spike short, very close ; bracts very hairy.

COAST REGION : Riversdale Div. ; Gouritz River, *Niven.*

Known to us only from Knight's description ; it may be identical with one of the species with dimorphic leaves which occurs in the same region.

ADDENDA AND CORRIGENDA.

VERBENACEÆ.

1. **Vitex mooiensis,** var. **Rudolphi** (H. H. W. Pearson in Hook. Ic. Pl. sub t. 2705); young parts clothed with tawny pubescence; leaves whorled or opposite, pubescent; petioles pubescent; calyx glandular-pubescent, 5-toothed; tube 1-1½ lin. long; teeth about ½ lin. long.

EASTERN REGION: Delagoa Bay; Ressano Garcia, in stony places at 1000 ft., *Schlechter*, 11935!

LABIATÆ.

Page 310, line 8 from the bottom, for **(22) Pegleræ,** read **(33) Pegleræ.**

PROTEACEÆ.

1. **Brabeium stellatifolium** (Linn.). Add to localities on p. 505 :—

Clanwilliam Div.; near Clanwilliam, *Leipoldt*, 72! Stellenbosch Div.; Lowrys Pass, *MacOwan*! Swellendam Div.; Duivels Bosch, near Swellendam, *Pappe*! Riversdale Div.; near Riversdale, *Schlechter*, 1930!

1. **Aulax cneorifolia** (Knight). Add to localities on p. 507 :—

SOUTH AFRICA: without locality, *Ludwig*, ♂, ♀ !

3. **Aulax pallasia** (Stapf). Add to localities on p. 508 :—

SOUTH AFRICA: without locality, *Ludwig*, ♂, ♀ ! Caledon Div.: Houw Hoek Mountains, *MacOwan*, 2978!

Page 512, line 34, for *fusciflora,* read *fusciflorum.*

6. **Leucadendron aurantiacum** (Buek). On p. 520, lines 12 and 11 from below, delete *P. cinerea, Ait. Hort. Kew. ed.* 1, i. 127, *not of Willd.*

13. **Leucadendron cinereum** (R. Br.). Add as synonym on p. 524, line 20, *Protea cinerea, Ait. Hort. Kew. ed.* 1, i. 127, *not of Willd.* Add to localities on p. 524 :—

SOUTH AFRICA: without locality, *Masson*! Malmesbury Div.; near Hopefield, *Bolus*, 12812!

17. Leucadendron abietinum (R. Br.).   Add to localities on p. 526 :—

Riversdale Div. ; Zoetmelksfontein, *Muir*, 450 !

20. Leucadendron fusciflorum (R. Br.).   Add to the synonymy on p. 528 :—*L. Thunbergii, Endl. Gen. Suppl.* 4, ii. 75.

24. Leucadendron imbricatum (R. Br.).   On p. 530, lines 7 and 6 from below, read *P. polygaloides, Willd. Herb., ex Meisn. l.c.*

29. Leucadendron platyspermum (R. Br.).   Add to localities on page 534 :—

Bredasdorp Div. ; near Elim, *Schlechter*, 9636 !

31. Leucadendron Dregei (E. Meyer).   Add to localities on p. 535 :—

Oudtshoorn Div. ; Zwartberg Pass, *Bolus*, 12262 !

34. Leucadendron strictum (R. Br.).   Add to localities on p. 536 :—

Riversdale Div. ; Platte Kloof, *Muir*, 391 !

36. Leucadendron adscendens (R. Br.).   On p. 538, lines 12 and 13, delete *Protea argentea, Linn., var.* β? *Sp. Pl. ed.* i. 94.

44. Leucadendron crassifolium (R. Br.).   Add to the synonymy on p. 543 :—*L. spathulatum, Buek ex Meisn. in DC. Prodr.* xiv. 226, *not of R. Br.*

48. Leucadendron squarrosum (R. Br.).   On p. 545, line 7, delete *Protea arcuata, Lam. Ill.* i. 234, *excl. var.* β?

53. Leucadendron venosum (R. Br.).   Add as synonyms on p. 547 : —*L. conchiforme, K. Schum. in Just, Jahresb.* xxvi. i. 364 ? *Protea conchiformis, O. Kuntze, Rev. Gen. Pl.* iii. 278 ?

54. Leucadendron daphnoides (Meisn.).   Add to the synonymy on p. 548 :—*L. insigne, Buek ex Meisn. in DC. Prodr.* xiv. 219.

61. Leucadendron marginatum (Link).   On p. 550, lines 17, 16 and 15 from below, for *P. ciliaris, Hort. ex Roem. & Schult. Syst. Veg.* iii. 356, *and P. ciliata, Breit. Hort. Breit.* 380, *fide Meisn. l.c.*, read *P. ciliaris, Wendl. ex Meisn. l.c., and P. ciliata, Hort. Angl. ex Meisn. l.c.*

76. Leucadendron ? lineare (Burm. f. Fl. Cap. Prodr. 4).

SOUTH AFRICA : without locality, *Burmann*.

77. Leucadendron xanthoconus (K. Schum. in Just, Jahresb. xxvi. i. 364); a shrub 2 ft. high; branches erect, like the young leaves white silky-pubescent; leaves about 2 in. long, 2 lin. broad, linear-oblanceolate, long-attenuated to the base, with a short spiny apex, those surrounding the head exceeding it and more or less ovate; bracts close, ascending, tomentellous on the lower half outside, glabrous on the upper part, about ½ in. long and ¼ in. broad; flowers not known; fruits compressed, cordate at the base, winged, black, glabrous, punctate, 2 lin. long, 3 lin. broad. *Protea xanthoconus, O. Kuntze, Rev. Gen. Pl.* iii. 278.

COAST REGION : Caledon Div.; near Caledon, *Kuntze.*
This is probably either *L. uliginosum* or *L. salignum,* R. Br.

3. Protea barbigera (Meisn.). Add to localities on p. 563 :—

CENTRAL REGION : Ceres Div.; Klein Vley, *Schlechter,* 10054 !

7. Protea Lepidocarpodendron (Linn.). Add to localities on p. 566 :—

Bredasdorp Div.; above Bredasdorp, Collector ? *in Herb. Albany Museum,* 202 !

8. Protea neriifolia (R. Br.). Add to localities on p. 567 :—

Port Elizabeth Div.; Walmer, *Mrs. Paterson,* 677 !

9. Protea pulchella (Andr.). Add to localities on p. 568 :—

Clanwilliam Div.; Packhuis Berg, *Schlechter,* 10819 !

15. Protea macrophylla (R. Br.). Add to localities on p. 571 :—

Albany Div.; Bothas Berg, *MacOwan* !

25. Protea mellifera (Thunb.). Add to localities on p. 577 : —

Port Elizabeth Div.; Port Elizabeth, *Kemsley,* 326 !

28. Protea lacticolor (Salisb.). On p. 578, lines 4 and 3 from below read :—lip 7 lin. long, 3-toothed, sparingly hairy along the edges, otherwise glabrous.
Add to localities on p. 579 :—

EASTERN REGION: Tembuland ; Cala, *Kolbe,* 60 !
The specimens collected by Pappe and Mrs. Newdigate at Knysna should be referred to *P. Mundii,* while those from Stutterheim (*Flanagan,* 1703) and Bazeia (*Baur,* 624) should be transferred to this species.

29. Protea Mundii (Klotzsch). Add to localities on p. 580 :—

Uitenhage Div.; Van Staadens Berg, *Zeyher,* 385 ! Van Staadens River, *Drège* !

31. Protea grandiflora (Thunb.).   Add to localities on p. 581 :—

George Div. ; Robinson Pass, *Taylor*, 318 !

33. Protea abyssinica (Willd.).   Add to localities on p. 582 :—

Transvaal ; Elandshoek, *Rogers*, 390 !

34. Protea hirta (Klotzsch).   Add to localities on p. 582 :—

Transvaal ; Pretoria Kopjes, *Miss Leendertz*, 694 ! Witwatersrand, *Hutton*, 892 !
Elandspruit, *Schlechter*, 3864 !

39. Protea convexa (Phillips).   On p. 585 add note :—

From a communication recently received from Mr. Phillips it would appear that
the description of the plant as a shrub up to 10 ft. high is wrong, the original
label should have read 0·3 m.   He describes it as possessing a subterranean stem.
The specimens at Kew are not in a condition to decide that question.   At the
same time they may very well represent the ascending portions of decumbent
branches springing from an underground stem.   If this is the case, *P. convexa*
would be better placed near *P. glaucophylla* in the section *Microgeantheæ*.

43. Protea multibracteata (Phillips).   Delete from localities on
p. 586 :—

Transkei ; near Kentani, *Miss Pegler*, 274 (see under 46, *P. Flanaganii*).

Add to localities :—

East London Div. ; East London, *Rattray*, 124 !

44. Protea tenax (R. Br.).   Add to localities on p. 587 :—

Port Elizabeth Div. ; Bethelsdorp, *Mrs. Paterson*, 142 !

47. Protea simplex (Phillips).   Add to localities on p. 589 :—

Pondoland ; Insizwa Mountains, 6800 ft., *Schlechter*, 6504 ! Natal ; Howick,
Hutton, 193 !

57. Protea scolymocephala (Richard).   Add to localities on
p. 593 :—

Piquetberg Div. ; Piquet Berg, *Schlechter*, 5185 !

62. Protea cynaroides (Linn.).   Add to localities on p. 596 :—

Caledon Div. ; Zwart Berg, *Schlechter*, 10346 ! Oudtshoorn Div. ; Oudtshoorn,
Taylor ! Port Elizabeth Div. ; Port Elizabeth, *Kemsley*, 306 !

63. Protea cryophila (Bolus).   Add to localities on p. 597 :—

Clanwilliam Div. ; Wupperthal, *MacOwan in Herb. Albany Museum* !

72. Protea tenuifolia (R. Br.). Add to localities on p. 602 :—

Swellendam Div. ; Voormans Bosch, *Zeyher*, 3671 *in Herb. Albany Museum* !

The specimen (*Zeyher*, 3671) in the Albany Museum is certainly *P. tenuifolia*, but it approaches so closely to *Zeyher*, 3673, which was referred to *P. scabra*, R. Br., that *P. tenuifolia* may actually represent only a narrow-leaved state of that species. Meisner indeed quotes both under *P. scabra*, δ *stenophylla*.

78. Protea revoluta (R. Br.). Add at end of description :— *Roem. & Schultes, Syst. Veg.* iii. 352 ; *Meisn. in DC. Prodr.* xiv. 243.

80. Protea humiflora (Andr.). Add to localities on p. 606 :—

Bredasdorp Div. ; Pot River, *Pappe* !

4. Leucospermum conocarpum (R. Br.). On p. 617, line 4 from above, read :—*Diss. Prot.* 22, *partly* ; *Fl. Cap. ed. Schult.* 126, *partly.*

6. Leucospermum attenuatum (R. Br.). On p. 618, line 4 from below, delete query, and on p. 619 add to localities :—

SOUTH AFRICA : without locality, *Ecklon & Zeyher*, 7 !

9. Leucospermum ellipticum (R. Br.). Add to localities on p. 620 :—

Worcester Div. ; Breede River Valley, between Worcester and Tulbagh, *Ludwig* !

20. Leucospermum crinitum (R. Br.). P. 630, line 18, "*prima*" was not intended by R. Brown for a varietal name to *L. oleæfolium*.

21. Leucospermum oleæfolium (R. Br.). P. 630, line 32, "*altera*" was not intended by R. Brown for a varietal name. Add to localities on p. 631 :—

Swellendam Div. ; mountains between Zondereinde River and Breede River, *Zeyher*, 3684b ! and read in last line, *Zeyher*, 3684a, instead of 3683.

25. Leucospermum buxifolium (R. Br.). On p. 634, line 9 from above, read :—*Protea pubera, Thunb. Diss. Prot.* 43, *partly* ; *Fl. Cap. ed. Schult.* 140, *partly.*

27. Leucospermum royenifolium (Stapf). Add to localities on p. 636 :—

SOUTH AFRICA : without locality, *Thunberg in Herb. Swartz* !

28. Leucospermum prostratum (Stapf). Add to localities on p. 636 :—

SOUTH AFRICA : without locality, *Ecklon & Zeyher*, 6 !

Page 636, line 10 from the bottom, for 221 read 227.

INDEX.

[Synonyms are printed in *italics*.]

LONDON: PRINTED BY WILLIAM CLOWES AND SONS, LIMITED,
DUKE STREET, STAMFORD STREET, S.E., AND GREAT WINDMILL STREET, W.